The Microbiology Place

(www.microbiologyplace.com).

This rich website includes chapter quizzes, practice tests, Microbe Review questions, web links, flashcards, a glossary with pronunciations, and videos, plus 30 multi-step, tutorial-style microbiology animations, a gradebook, an e-book, and a link to Research Navigator™, which is a powerful online research tool with access to three exclusive databases of reliable source material, including the *Journal of Applied Microbiology*, the *Annual Review of Microbiology*, *Science*, *The New York Times Search by Subject Archive*, and the "Best of the Web" Link Library. The Microbiology Place is also provided on a CD-ROM included with each new copy of the text.

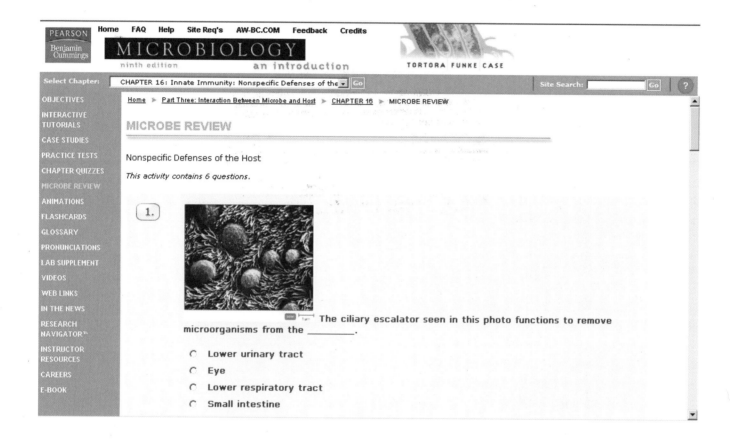

The Latin rules of grammar pertain to singular and plural forms of scientific names.

	Gender		
	Feminine	**Masculine**	**Neuter**
Singular	-a	-us	-um
Plural	-ae	-i	-a
Examples	alga, algae	fungus, fungi	bacterium, bacteria

a-, an- absence, lack. Examples: abiotic, in the absence of life; anaerobic, in the absence of air.

-able able to, capable of. Example: viable, having the ability to live or exist.

actino- ray. Example: actinomycetes, bacteria that form star-shaped (with rays) colonies.

aer- air. Examples: aerobic, in the presence of air; aerate, to add air.

albo- white. Example: *Streptomyces albus* produces white colonies.

ameb- change. Example: ameboid, movement involving changing shapes.

amphi- around. Example: amphitrichous, tufts of flagella at both ends of a cell.

amyl- starch. Example: amylase, an enzyme that degrades starch.

ana- up. Example: anabolism, building up.

ant-, anti- opposed to, preventing. Example: antimicrobial, a substance that prevents microbial growth.

archae- ancient. Example: archaeobacteria, "ancient" bacteria, thought to be like the first form of life.

asco- bag. Example: ascus, a baglike structure holding spores.

aur- gold. Example: *Staphylococcus aureus*, gold-pigmented colonies.

aut-, auto- self. Example: autotroph, self-feeder.

bacillo- a little stick. Example: bacillus, rod-shaped.

basid- base, pedestal. Example: basidium, a cell that bears spores.

bdell- leech. Example: *Bdellovibrio*, a predatory bacterium.

bio- life. Example: biology, the study of life and living organisms.

blast- bud. Example: blastospore, spores formed by budding.

bovi- cattle. Example: *Mycobacterium bovis*, a bacterium found in cattle.

brevi- short. Example: *Lactobacillus brevis*, a bacterium with short cells.

butyr- butter. Example: butyric acid, formed in butter, responsible for rancid odor.

campylo- curved. Example: *Campylobacter*, curved rod.

carcin- cancer. Example: carcinogen, a cancer-causing agent.

caseo- cheese. Example: caseous, cheeselike.

caul- a stalk. Example: *Caulobacter*, appendaged or stalked bacteria.

cerato- horn. Example: keratin, the horny substance making up skin and nails.

chlamydo- covering. Example: chlamydoconidia, conidia formed inside hypha.

chloro- green. Example: chlorophyll, green-pigmented molecule.

chrom- color. Examples: chromosome, readily stained structure; metachromatic, intracellular colored granules.

chryso- golden. Example: *Streptomyces chryseus*, golden colonies.

-cide killing. Example: bactericide, an agent that kills bacteria.

cili- eyelash. Example: cilia, a hairlike organelle.

cleisto- closed. Example: cleistothecium, completely closed ascus.

co-, con- together. Example: concentric, having a common center, together in the center.

cocci- a berry. Example: coccus, a spherical cell.

coeno- shared. Example: coenocyte, a cell with many nuclei not separated by septa.

col-, colo- colon. Examples: colon, large intestine; *Escherichia coli*, a bacterium found in the large intestine.

conidio- dust. Example, conidia, spores developed at the end of aerial hypha, never enclosed.

coryne- club. Example: *Corynebacterium*, club-shaped cells.

-cul small form. Example: particle, a small part.

-cut the skin. Example: Firmicutes, bacteria with a firm cell wall, gram-positive.

cyano- blue. Example: cyanobacteria, blue-green pigmented organisms.

cyst- bladder. Example: cystitis, inflammation of the urinary bladder.

cyt- cell. Example: cytology, the study of cells.

de- undoing, reversal, loss, removal. Example: deactivation, becoming inactive.

di-, diplo- twice, double. Example: diphlococci, pairs of cocci.

dia- through, between. Example: diaphragm, the wall through or between two areas.

dys- difficult, faulty, painful. Example: dysfunction, disturbed function.

ec-, ex-, ecto out, outside, away from. Example: excrete, to remove materials from the body.

en-, em- in, inside. Example: encysted, enclosed in a cyst.

entero- intestine. Example: *Enterobacter*, a bacterium found in the intestine.

ninth edition

MICROBIOLOGY

an introduction

GERARD J. TORTORA
BERGEN COMMUNITY COLLEGE

BERDELL R. FUNKE
NORTH DAKOTA STATE UNIVERSITY

CHRISTINE L. CASE
SKYLINE COLLEGE

PEARSON
Benjamin Cummings

San Francisco Boston New York
Cape Town Hong Kong London Madrid Mexico City
Montreal Munich Paris Singapore Sydney Tokyo Toronto

Executive Editor: Leslie Berriman
Associate Editor: Tamara Keller
Assistant Editor: Blythe Robbins
Editorial Assistant: Jon Duke
Executive Marketing Manager: Lauren Harp
Managing Editor: Wendy Earl
Production Editor: Janet Vail
Art and Photo Coordinator: Linda Jupiter

Photo Editor: Maureen Spuhler
Text and Cover Design: Yvo Riezebos
Copy Editor: Carla Breidenbach
Proofreader: Martha Ghent
Indexer: Karen Hollister
Illustrations: Precision Graphics
Compositor: Techbooks/GTS
Senior Manufacturing Buyer: Stacey Weinberger

Cover photograph: Colorized transmission electron micrograph (TEM) of *Legionella pneumophila* bacteria. Dr. Linda Stannard, UCT/Photo Researchers, Inc.

Text art and photo credits appear following the Glossary.

Library of Congress Cataloging-in-Publication Data

Tortora, Gerard J.
   Microbiology: an introduction / Gerard J. Tortora, Berdell R. Funke, Christine L. Case—9th ed.
     p. cm.
   Includes index.
   ISBN 0-8053-4791-7
   1. Microbiology. I. Funke, Berdell R. II. Case, Christine L., 1948– III. Title.
   QR41. 2.T67 2006
   579—dc22

                                    2006041674

ISBN 0-8053-4791-7
1 2 3 4 5 6 7 8 9 10—CRK—10 09 08 07 06

Gerard J. Tortora Jerry Tortora is a professor of biology and teaches microbiology, human anatomy, and physiology at Bergen Community College in Paramus, New Jersey. He received his M.A. in Biology from Montclair State College in 1965. He belongs to numerous biology/microbiology organizations, such as the American Society of Microbiology (ASM), Human Anatomy and Physiology Society (HAPS), American Association for the Advancement of Science (AAAS), National Education Association (NEA), New Jersey Educational Association (NJEA), and the Metropolitan Association of College and University Biologists (MACUB). Jerry is the author of a number of biological science textbooks. In 1995, he was selected as one of the finest faculty scholars at Bergen Community College and was named Distinguished Faculty Scholar. In 1996, Jerry received a National Institute for Staff and Organizational Development (NISOD) excellence award from the University of Texas and was selected to represent Bergen Community College in a campaign to increase awareness of the contributions of community colleges to higher education.

Berdell R. Funke Bert Funke received his Ph.D., M.S., and B.S. in microbiology from Kansas State University. He has spent his professional years as a professor of microbiology at North Dakota State University. He taught introductory microbiology, including laboratory sections, general microbiology, food microbiology, soil microbiology, clinical parasitology, and pathogenic microbiology. As a research scientist in the Experiment Station at North Dakota State, he has published numerous papers in soil microbiology and food microbiology.

Christine L. Case Chris Case is a registered microbiologist and a professor of microbiology at Skyline College in San Bruno, California, where she has taught for the past 35 years. She received her Ed.D. in curriculum and instruction from Nova Southeastern University and her M.A. in microbiology from San Francisco State University. She was Director for the Society for Industrial Microbiology (SIM) and is an active member of the ASM and Northern California SIM. She received the ASM and California Hayward outstanding educator awards. In addition to teaching, Chris contributes regularly to the professional literature, develops innovative educational methodologies, and maintains a personal and professional commitment to conservation and the importance of science in society. Chris is also an avid photographer, and many of her photographs appear in this book.

PREFACE

The ninth edition of *Microbiology: An Introduction* is the leading textbook in the non-majors microbiology market. In the 24 years since the publication of the first edition, more than one million students have used this book at more than 1,000 colleges and universities, making it the best-selling introductory microbiology text around the world. The ninth edition continues to be a comprehensive, beginning text, assuming no previous study of biology or chemistry. The text is appropriate for students in a wide variety of programs, including the allied health sciences, biological science, environmental sciences, animal science, forestry, agriculture, home economics, and the liberal arts. We have been gratified to hear from instructors and students alike that the book has become a favorite among their textbooks—a learning tool that is both effective and enjoyable.

quality

HALLMARKS OF MICROBIOLOGY: AN INTRODUCTION

We have retained in this new edition features that made the previous editions so popular. These include:

- **An appropriate balance between microbiological fundamentals and applications, and between medical applications and other applied areas of microbiology.** As in previous editions, basic microbiological principles are given greater emphasis than applications, and health-related applications are emphasized. Applications are integrated throughout the text, and considerable attention is devoted to microorganisms in habitats outside the human body. We hope students will gain an appreciation for the fascinating diversity of microbial life, the central roles of microorganisms in nature, and the importance of microorganisms to our daily lives.

- **Straightforward presentation of complex topics.** Each section of the text has been revised with the student in mind, to maintain the clarity of explanation for which our book has become known. Step-by-step diagrams closely coordinated with the narrative descriptions further aid student comprehension of concepts.

- **Integrated learning objectives and end-of-chapter** questions help students check their understanding of key chapter concepts and learn critical problem-solving skills needed in clinical and industrial situations.

- **Applications and discovery-oriented boxes** focus on modern, practical uses of microbiology and biotechnology and emphasize the process of scientific discovery. The application boxes show real people working in science to provide students with examples of career opportunities in microbiology.

ORGANIZATION

We have organized the book in what we think is a useful fashion, while recognizing that the material might be effectively presented in a number of other sequences. For those who wish to use a different order, we have made each chapter as independent as possible and have included numerous cross-references. Fundamentals of microbiology are included in Chapters 1 through 9, and other chapters can be used to illustrate and emphasize topics for specific courses. The various diseases are organized into chapters according to the host organ system most affected. The Instructor's Guide, written by Christine L. Case, provides detailed guidelines for organizing the material in several other ways. A taxonomic guide to diseases covered in the text is included in Appendix F.

NEW TO THE NINTH EDITION

Cutting edge techniques in biotechnology and clinical identification, RNAi and FISH, are explained and illustrated. A section on the new field of Forensic Microbiology is included. Diseases In Focus boxes allow students to compare diseases that present similar symptoms. Taxonomy and nomenclature as well as disease incidence data are current through June 2005. Chapter 17, that introduces the important topic of adaptive immunology, has been completely revised for clarity, accuracy, and readability. In text references to 30 animations, found on The Microbiology Place website and CD-ROM, that explore key concepts in microbiology are indicated with the animation icon: ❋ A visual introduction to the ninth edition follows the Preface.

CHAPTER-BY-CHAPTER REVISIONS

Every chapter in this edition has been thoroughly revised, and data in the text, tables, and figures have been updated through June 2005 where possible. The main changes for each chapter are summarized below.

PART ONE
FUNDAMENTALS OF MICROBIOLOGY

Chapter 1, The Microbial World and You

- Applications of modern molecular biology to taxonomy have resulted in new or changed names and taxa for many microorganisms. These new names and taxa, approved by the appropriate international nomenclature committee, are used throughout this edition
- The emerging infectious disease discussion includes West Nile Virus, bovine spongiform encephalopathy, Avian influenza, Ebola hemorrhagic fever, severe acute respiratory syndrome (SARS), and cryptosporidiosis

Chapter 2, Chemical Principles

- New study questions use microbiological examples

Chapter 3, Observing Microorganisms Through a Microscope

- Quorum sensing and bacterial interactions are discussed
- New section on scanning acoustic microscopy (SAM)

Chapter 4, Functional Anatomy of Prokaryotic and Eukaryotic Cells

- Revised discussion of gram stain, plasma membrane
- New section on acid fast cell walls
- Gram-positive and gram-negative flagella are compared

Chapter 5, Microbial Metabolism

- A discussion of anoxygenic photosynthesis

Chapter 6, Microbial Growth

- The discussion of the toxic forms of oxygen has been rewritten to better relate to destruction of microbes by phagocytosis

Chapter 7, The Control of Microbial Growth

- New products and newly-approved uses are included

Chapter 8, Microbial Genetics

- Important enzymes in DNA replication, expression, and repair are listed in a new table
- New Morbidity & Mortality Weekly Report box describes the use of genomics to track a disease (West Nile encephalitis)

Chapter 9, Biotechnology and Recombinant DNA

- RNA interference (RNAi) is illustrated in a new figure
- Forensic microbiology is included

PART TWO
A SURVEY OF THE MICROBIAL WORLD

Chapter 10, Classification of Microorganisms

- New discussion of several bacteria that have been recently discovered by means of RNA analysis
- Ribotyping and FISH are described

Chapter 11, The Prokaryotes: Domains Bacteria and Archaea

- Phylogenic classification of prokaryotes is in accordance with *Bergey's Manual of Systematic Bacteriology*, second edition (2005)

Chapter 12, The Eukaryotes: Fungi, Algae, Protozoa, and Helminths

- New Clinical Problem Solving box describes leishmaniasis in the Middle East
- Fungal names have been updated

Chapter 13, Viruses, Viroids, and Prions

- New Morbidity & Mortality Weekly Report box describing crossing the species barrier using Avian influenza as the example
- The one-step bacteriophage curve has been replaced with one-step virus growth curves showing virus growth and patterns of infection
- New photographs compare viral entry by fusion and endocytosis

PART THREE
INTERACTION BETWEEN MICROBE AND HOST

Chapter 14, Principles of Disease and Epidemiology

- New section on factors that determine the distribution and composition of normal microbiota with an expanded Table 14.1
- New discussion of probiotics
- New Clinical Problem Solving box describes a nosocomial outbreak
- Data for figures (AIDS, Lyme disease, and typhoid fever) and tables are current through June 2005

Chapter 15, Microbial Mechanisms of Pathogenicity

- New section on the conjuctiva
- New section on membrane ruffling
- Revised Table 15.2 on diseases caused by exotoxins

Chapter 16: Innate Immunity: Nonspecific Defenses of the Host

- New section on toll-like receptors (TLRs)
- New section on dendritic cells
- New sections on transferrins and antimicrobial peptides

- New summary Table 16.3 covers innate immunity
- New Applications of Microbiology box on serum collection

Chapter 17, Adaptive Immunity: Specific Defenses of the Host

- Completely rewritten chapter includes an historical context to introduce concepts and terminology in an easy-to-follow format
- Art has been redrawn for accuracy and ease of understanding

Chapter 18: Practical Applications of Immunology

- The discussion of monoclonal antibodies has now been moved to this chapter
- Definitions for terms such as "sensitivity" and "specificity" are now included

Chapter 19: Disorders Associated with the Immune System

- The discussion of certain allergies, especially to peanuts and latex, has been improved and expanded
- The relationship between the immune system and cancer and the discussion of Acquired Immunodeficiency Syndrome (AIDS) have been rewritten and expanded

Chapter 20: Antimicrobial Drugs

- Several new antimicrobial drugs have been introduced, such as the antiviral *adefovir dipivoxil* (*Hepsera*), the antiprotozoan agent *nitazoxanide*, and a new antifungal *tinidazole*
- There is now a discussion of RNAi to selectively block protein synthesis in important pathogens

PART FOUR
MICROORGANISMS AND HUMAN DISEASE

Chapter 21: Microbial Diseases of the Skin and Eyes

- New discussion of the causes of toxic shock syndrome (TSS) makes a clear distinction between TSS caused by staphylococci and streptococci
- New Clinical Problem Solving box on infections in the gym

Chapter 22: Microbial Diseases of the Nervous System

- New discussion of lyssavirus-related encephalitis
- Expanded the discussion of amebic encephalitis includes primary amebic meningoencephalitis and granulomatous amebic encephalitis
- New Clinical Problem Solving box on a recent case of human rabies

Chapter 23: Microbial Diseases of the Cardiovascular and Lymphatic Systems

- Rat bite fever now has a full discussion
- New discovery that the animal reservoir for the Ebola virus is fruit bats
- New Clinical Problem Solving box on tularemia from pet rodents

Chapter 24: Microbial Diseases of the Respiratory System

- An important disease endemic in Asia and Australia, melioidosis, has been introduced; cases of this disease also occurred among European visitors in the tsunami-affected areas surrounding the Indian Ocean in 2004
- New discussion of the influenza-like epidemic of the SARS-associated corona-virus
- Updated discussion of the 1918 Spanish flu pandemic and a description of the 2005 resurrection of the 1918 virus
- New Clinical Problem Solving box on a legionellosis outbreak

Chapter 25: Microbial Diseases of the Digestive System

- New discussion of the important disease condition, *Clostridium-difficile-associated-diarrhea*
- New discussion of waterborne *Vibrio* bacterial infections related to Hurricane Katrina
- New Clinical Problem Solving box on a foodborne salmonellosis outbreak; includes relative risk calculation

Chapter 26: Microbial Diseases of the Urinary and Reproductive Systems

- New discussion of vaccines that promise to block development of cervical cancer caused by certain strains of human papilloma viruses
- New Clinical Problem Solving box with updated information on antibiotic resistant gonococci

PART FIVE
ENVIRONMENTAL AND APPLIED MICROBIOLOGY

Chapter 27: Environmental Microbiology

- New discussion of the Blue Flag campaign by an international body that identifies beaches in the world that meet sanitation standards
- New Microbiology In the News box on illnesses after Hurricane Katrina resulting from the breakdown of sewage disposal and water purification systems

Chapter 28: Applied and Industrial Microbiology

- The discussion of radiation and high pressure for food preservation has been expanded

ACKNOWLEDGEMENTS

In preparation for this textbook, we have benefited from the guidance and advice of a large number of microbiology instructors across the country. The reviewers listed on the next page provided constructive criticism and valuable suggestions at various stages of the revision. We gratefully acknowledge our debt to these individuals.

We also thank the staff at Benjamin Cummins for their dedication to excellence. Leslie Berriman, our executive editor, successfully kept us all focused on where we wanted this revision to go. Carla Breidenbach's careful attention to continuity and detail in her copyedit of both text and art served to keep concepts and information clear throughout. Janet Vail and Wendy Earl expertly guided the text through the production process. Linda Jupiter effectively managed the large art program. The photo researcher, Maureen Spuhler, made sure we had clear and striking images throughout the book. Yvo Reizebos created the interior design and did a wonderful job with the cover. Techbooks/GTS did their usual outstanding job moving this book quickly and beautifully through composition; the skilled team was led by Sandie Sigrist. Stacey Weinberger guided the book through the manufacturing process. Leslie Austin expertly guided the production of the supplements, with support from David Novak.

We would all like to acknowledge our spouses and families, who have provided invaluable support throughout the writing process.

Finally, we have an enduring appreciation for our students, whose comments and suggestions provide insight and remind us of their needs. This text is for them.

Gerard J. Tortora
Berdell R. Funke
Christine L. Case

NINTH EDITION REVIEWERS

Joel Adams-Stryker, *Evergreen Valley College*
Joan Baird, *Rose State College*
Tobie Bogart, *Jefferson State Community College*
Kari Cargill, *Montana State University*
Lee Couch, *University of New Mexico*
Katherine Foreman, *Moraine Valley Community College*
Joe Francis, *The Master's College*
Stephen Greenwald, *Gordon College*
Keith R. Hench, *Kirkwood Community College, Iowa City*
Dawn Janich, *Community College of Philadelphia*
Richard D. Karp, *University of Cincinnati*
V. Karunakaran, *St. Francis Xavier University*
Judy Kaufman, *Monroe Community College*
Michael A. Lawson, *Missouri Southern State University*
Jean Lu, *University of Tennessee, Martin*
Anne LaGrange Loving, *Passaic County Community College*
Brian K. Martin, *University of Iowa*
Richard L. Myers, *Southwest Missouri State*
Klaus R. L. Nüsslein, *University of Massachusetts, Amherst*
Judy L. Penn, *Shoreline Community College*
Laraine Powers, *East Tennessee State University*
Thomas F. Reed, *Brevard Community College*
Stephen Y. K. Seah, *University of Guelph*
Julie Shaffer, *University of Nebraska, Kearney*
Teri Shors, *University of Washington, Oshkosh*
Ralph E. Smith, *Colorado State University*
David S. Treves, *Indiana University Southeast*
Jane A. Weston, *Genesee Community College*
Ruth A. Wrightsman, *Saddleback College*

Morbidity & Mortality Weekly Report boxes review the latest epidemiological cases from the Centers for Disease Control and Prevention, such as Avian Flu, West Nile Virus, and AIDS.

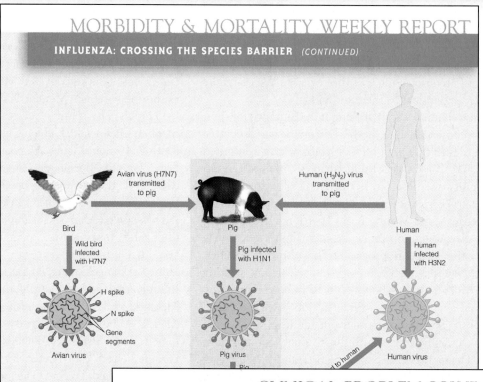

MORBIDITY & MORTALITY WEEKLY REPORT

INFLUENZA: CROSSING THE SPECIES BARRIER *(CONTINUED)*

Avian virus (H7N7) transmitted to pig

Human (H₃N₂) virus transmitted to pig

Bird

Wild bird infected with H7N7

Pig

Pig infected with H1N1

Human

Human infected with H3N2

H spike

N spike

Gene segments

Avian virus

Pig virus

Human virus

Model for antigenic shift i...
virus at the same time, the viru...
a hemagglutinin and/or neura...
humans and spread from pers...
previously seen in influenza vi...

Clinical Problem Solving boxes use case histories adapted from recent issues of *Morbidity & Mortality Weekly Report* to encourage critical thinking in the examination of a clinical problem. Topics include a parasitic disease afflicting soldiers in Afghanistan, a hospital-acquired infection following liposuction, and an antibiotic-resistant sexually transmitted disease.

CLINICAL PROBLEM SOLVING

OUTBREAK

You will see questions as you read through this problem. The questions are those that epidemiologists ask themselves and each other as they solve a clinical problem. As you read through the problem, try to answer each question as though you were an epidemiologist.

1. A 64-year-old man saw his primary care physician complaining of fever, malaise, and a cough. His vaccinations were up-to-date, including DTaP. His condition worsened over several days; he had difficulty breathing and his temperature rose to 40.4°C (104.7°F). He was hospitalized, and his lungs showed signs of mild inflammation with thin, watery secretion. A Gram stain of bacteria isolated from the patient is shown in the figure. *What diseases are possible?*

2. The same day, a 37-year-old man went to the emergency department because he had shortness of breath, fatigue, and cough. The day before he had had fever and chills, with a maximum body temperature of 38.6°C (101.4°F). *What additional tests would you do on both patients?*

Hotel guests with legionellosis	
Age	37–70 yrs (average: 60)
Gender	6 male; 2 female
Number of nights at hotel	1–4 (average: 3)
Diabetes mellitus	4
Immunocompromised	1
Smoker	5
Showered in hotel	8
Used whirlpool spa at hotel	1
Used hotel swimming pool	6

3. Both patients had an antibody titer > 1024 against *Legionella pneumophila* serogroup 1. The local health department (LHD) was contacted because two patients were hospitalized with legionellosis. *What do you need to know now?*

4. One week before hospitalization, both men stayed in the same hotel within one day of each other. The LHD identified six additional cases of legionellosis at other hospitals. The LHD gave a follow-up questionnaire to all eight patients to ascertain travel that preceded the illness, including location, accommodations, dates, and information about exposures to common sources for infection (see the table). *What are likely sources of infection?*

5. Epidemic legionellosis usually results from exposure of susceptible individuals to an aerosol generated by an environmental source of water contaminated with *Legionella*. *Why is it important to identify the source?*

6. Retrospective identification of cases allows further investigation and control and remediation efforts. *L. pneumophila* of the same monoclonal antibody type was recovered from the hot water storage tanks, cooling tower, and showers and faucets in rooms occupied by patients and well guests. *Why didn't other hotel guests get sick?*

7. During outbreaks, attack rates tend to be highest in specific high-risk groups, including the elderly, smokers, and immunocompromised persons. *What are your recommendations for remediation?*

Shower necks and faucets were disinfected with bleach. The spa filter was cleaned, and the potable water system was hyperchlorinated.

Hotels have been common locations for legionellosis outbreaks since the disease was first recognized among hotel guests in Philadelphia in 1976. Active surveillance led to more rapid identification of other cases.

Surveillance data submitted to CDC indicate that approximately 21% of legionellosis cases each year are travel associated. However, identification of travel-associated clusters is hindered because the incubation period is long enough for people to disperse from the point source of infection. Prompt recognition and investigation of clusters can implicate a point source for infection and guide remediation and control efforts. Recognizing the benefits of enhanced surveillance, CDC plans to work with state health departments on new strategies to improve surveillance for travel-associated legionellosis at the national, state, and local levels.

SOURCE: Adapted from reports in *MMWR* 53(52): 1202, 1/7/05, and *MMWR* 54(7): 170, 2/25/05.

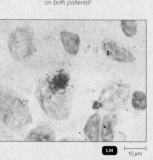

LM
10 µm

Gram stain shows bacteria within a tissue sample.

APPLICATIONS OF MICROBIOLOGY

ARE BACTERIA MULTICELLULAR?

The idea that bacteria are unicellular has changed recently. Bacterial cells do not act as unicellular organisms when they are growing in a colony. Instead, the cells interact and exhibit multicellular organization. Researchers cite a number of examples that suggest cells in a colony are not identical and may have different structures and functions. Multicellular organization is indicated where bacteria influence the behavior of neighboring cells (Figure A).

Myxobacteria

Myxobacteria are found in decaying organic material and fresh water throughout the world. Although they are bacteria, many myxobacteria never exist as individual cells. *Myxococcus xanthus* appears to hunt in packs. In its natural aqueous ho[...] spherical c[...] bacteria, w[...] tive enzym[...]

On solid substrates, other myxobacterial cells glide over a solid surface, leaving slime trails that are followed by other cells. When food is scarce, the cells aggregate to form a mass. Cells within the mass differentiate into a fruiting body that consists of a slime stalk and clusters of spores as shown in Figure B.

Vibrio

Vibrio fischeri is a bioluminescent bacterium that lives as a symbiont in the light-producing organ of squid and certain fish. When free-living, the bacteria are at low concentration and do not give off light. However, when they grow in their host, they are highly concentrated, and each cell is induced to produce the enzyme luciferase, which is used in the chemical pathways of bioluminescence.

sensing. *Quorum sensing* is the ability of bacteria to communicate and coordinate behavior. Bacteria that use quorum sensing produce and secrete a signaling chemical called an *inducer*. As the inducer diffuses into the surrounding medium, other bacterial cells move toward the source and begin producing inducer. The concentration of inducer increases with increasing cell numbers. This, in turn, attracts more cells and initiates synthesis of more inducer.

In law, a quorum is the minimum number of members necessary to conduct the business. Apparently, that's true among bacteria too. *Pseudomonas fluorescens* bacteria that form a pellicle on the surface of a liquid culture medium can use *both* O$_2$ from the air and nutri[...]

Applications of Microbiology boxes focus on modern, practical uses of microbiology and biotechnology. Topics include bioremediation, serum collection, and ensuring a safe blood supply.

DISEASES IN FOCUS

TYPES OF ARBOVIRAL ENCEPHALITIS

Arboviral encephalitis is usually characterized by fever, headache, and altered mental status ranging from confusion to coma. Vector control to decrease contacts between humans and mosquitoes is the best prevention. Mosquito control includes removing standing water and using insect repellent while outdoors.

Disease	Pathogen	Mosquito vector	Reservoir	U.S. distribution	Epidemiology	Mortality
Western equine encephalitis	WEE virus	*Culex*	Birds, horses		Severe disease; frequent neurological damage, especially in infants	5%
Eastern equine encephalitis	EEE virus	*Aedes, Culiseta*	Birds, horses		More severe than WEE; affects mostly young children and younger adults; relatively uncommon in humans	>30%
St. Louis encephalitis	SLE virus	*Culex*	Birds		Mostly urban outbreaks; affects mainly adults over 40	20%

NEW! Diseases In Focus boxes bring together several diseases of the same organ system, helping students to differentiate the diseases and learn important facts, such as symptoms and diagnosis, method of transmission, and treatment. Each box begins with an introductory paragraph to focus the discussion, followed by a table comparing the diseases. Featured topics include macular rashes, types of arboviral encephalitis, viral hemorrhagic fevers, bacterial pneumonia, and characteristics of viral hepatitis.

MICROBIOLOGY IN THE NEWS

ILLNESSES AFTER HURRICANE KATRINA

After natural disasters that destroy sewage disposal and water purification systems, the risk for illness related to infectious diseases is a public health concern. When the number of infectious diseases increases after a natural disaster, they usually are caused by infectious agents normally present in the community or local environment.

By the end of 2005, 18 wound-associated *Vibrio* cases had been reported as part of the human aftermath of Hurricane Katrina; five of these patients died. Three deaths were associated with *V. vulnificus* and two with *V. parahaemolyticus*; this is an increase over the normal reported incidence of *Vibrio* wound infections in the Gulf Coast states. Four people

O139. He was hospitalized for two days in Mississippi. No deaths were associated with these nonwound cases.

Among hurricane evacuees from the New Orleans area, a cluster of infections with methicillin-resistant *Staphylococcus aureus* (MRSA) was reported in approximately 30 pediatric and adult patients at an evacuee facility in Dallas, Texas.

Over 1100 Hurricane evacuees residing in three facilities reported symptoms of acute gastroenteritis. Three-fourths of the patients with acute gastroenteritis symptoms were adults (over age 18). Noroviruses are the most common cause of outbreaks of acute gastroenteritis in the United States, and their presence was confirmed in patient stool samples by

areas and bedding. Initial isolation procedures also were difficult to maintain over time because family members already traumatized by displacement and personal loss were separated from each other because of illness.

Excess moisture and standing water contribute to the growth of mold in homes and other buildings. Residents returning to a home that has been flooded need to be aware that mold may be present and may be a health risk. People who are sensitive to mold may experience stuffy nose, irritated eyes, wheezing, or skin irritation. People allergic to mold may have difficulty in breathing and shortness of breath. People with weakened immune systems and with chronic lung diseases may develop

Microbiology In the News boxes make sense of stories in today's headlines. Topics include environmental and health-related effects of Hurricane Katrina, antibiotics in animal feed, and biological weapons.

Step-by-step diagrams break complex processes into smaller, more manageable pieces. The step numbers are color-coded to correspond to step numbers in the text.

Clear, simple human figures set complex processes in context.

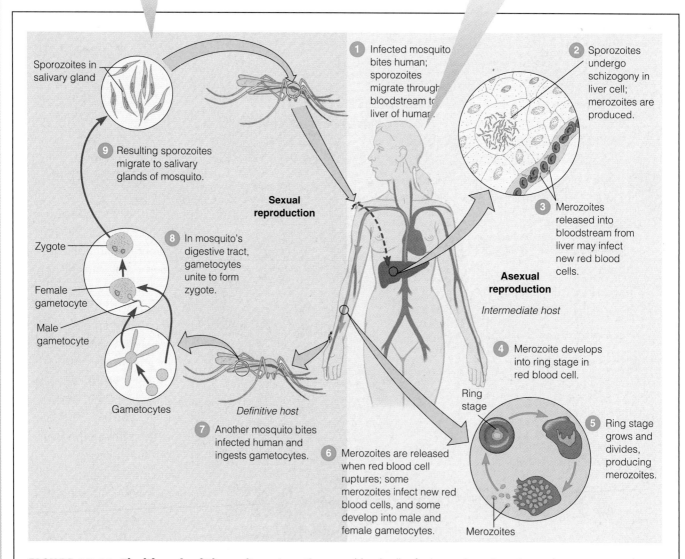

Sporozoites in salivary gland

9 Resulting sporozoites migrate to salivary glands of mosquito.

Sexual reproduction

Zygote

Female gametocyte

Male gametocyte

8 In mosquito's digestive tract, gametocytes unite to form zygote.

Gametocytes

Definitive host

7 Another mosquito bites infected human and ingests gametocytes.

1 Infected mosquito bites human; sporozoites migrate through bloodstream to liver of human.

2 Sporozoites undergo schizogony in liver cell; merozoites are produced.

3 Merozoites released into bloodstream from liver may infect new red blood cells.

Asexual reproduction

Intermediate host

4 Merozoite develops into ring stage in red blood cell.

Ring stage

5 Ring stage grows and divides, producing merozoites.

6 Merozoites are released when red blood cell ruptures; some merozoites infect new red blood cells, and some develop into male and female gametocytes.

Merozoites

FIGURE 12.19 **The life cycle of *Plasmodium vivax*, the apicomplexan that causes malaria.** Asexual reproduction (schizogony) of the parasite takes place in the liver and in the red blood cells of a human host. Sexual reproduction occurs in the intestine of an *Anopheles* mosquito after the mosquito has ingested gametocytes.

Q   **What is the definitive host for *Plasmodium*?**

PHOTOS, TABLES, AND GRAPHS

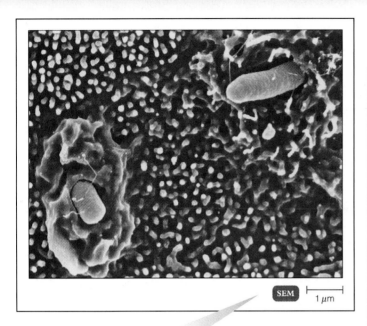

SEM | 1 μm

TEM/SEM/LM icons indicate whether the source is a transmission electron micrograph, scanning electron micrograph, or light micrograph. A red icon indicates that a micrograph has been colorized.

Tables and graphs have been revised to improve readability and aid student comprehension.

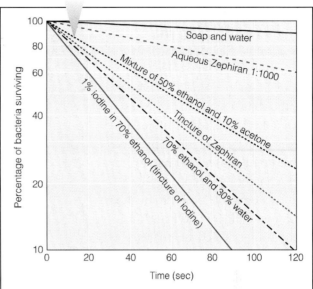

FIGURE 7.10 A comparison of the effectiveness of various antiseptics. The steeper the downward slope of the killing curve of the antiseptic, the more effective it is. A 1% iodine in 70% ethanol solution is the most effective; soap and water are the least effective. Notice that a tincture of Zephiran is more effective than an aqueous solution of the same antiseptic.

Q **Why is the tincture of Zephiran more effective than the aqueous solution?**

Orientation figures help students keep "the big picture" in mind.

Figure-legend questions require students to apply concepts presented in the text to answer questions about tables, graphs, and art.

Consistent use of symbols and colors enables students to progress from familiar parts of illustrated processes to unfamiliar ones with confidence. Molecules such as phosphate groups and ATP are the same color and shape throughout the book.

FOR THE INSTRUCTOR

Media Manager Instructor CD-ROMs. This valuable teaching resource includes JPEG and PowerPoint® files of every illustration, photograph, and table from the text—more than 1,000 images in all. All images include the customizable "Label Edit" feature. The Media Manager also offers prepared PowerPoint lecture outlines for each chapter; "layered," step-by-step, customizable PowerPoint slides for complex figures; 30 Flash® animations on selected key topics in microbiology; 22 QuickTime® videos of microorganisms; Active Lecture Questions (for use with personal response "clicker" systems); and an editable electronic Word file of the Test Bank. 0-8053-7805-7

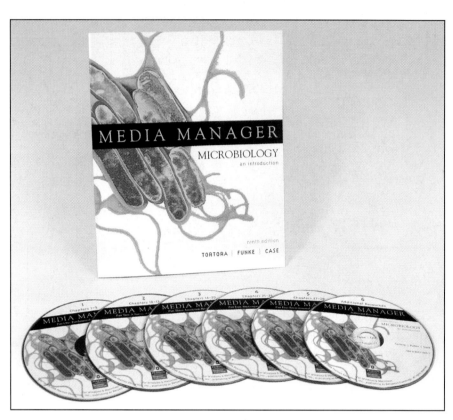

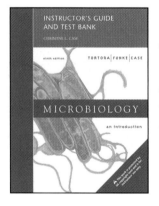

Instructor's Guide and Test Bank by Christine L. Case. The Instructor's Guide contains teaching tips, alternative course outlines, ideas for using special features, and answers to text questions. The Test Bank features more than 1,200 multiple choice questions with answers as well as three to five essay questions per chapter. 0-8053-7803-0

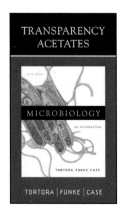

Full-Color Transparency Acetates. More than 400 images from the text have been modified with enlarged labels for effective classroom presentation. 0-8053-7797-2

Computerized Test Bank. This cross-platform, easy-to-use testing program allows instructors to view and edit electronic questions from the Test Bank, create multiple tests, and print in a variety of formats. 0-8053-7811-1

Instructor's Visual Guide. Printed thumbnail views of all the assets of the Media Manager CD-ROMs allow the option of preparing for class away from the computer. 0-8053-7804-9

CourseCompass™. This nationally hosted online course management system offers pre-loaded, book-specific content including testing and assessment, web links, illustrations, and photos. Visit www.coursecompass.com for a demonstration and more information about CourseCompass.

Blackboard and WebCT. Pre-loaded, book-specific content and test item files accompanying the text are available in open-access Blackboard and open-access WebCT formats. Contact your local Benjamin Cummings sales representative for more information. To locate your rep, visit www.aw-bc.com and use the "Find Your Rep" search feature.

AND STUDENTS

FOR THE STUDENT

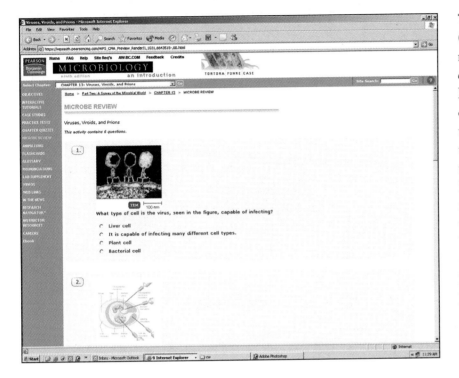

The Microbiology Place website
(www.microbiologyplace.com). This rich website includes chapter quizzes, practice tests, Microbe Review questions, web links, flashcards, a glossary with pronunciations, and videos, plus 30 multi-step, tutorial-style microbiology animations, a gradebook, an e-book, and a link to Research Navigator™, which is a powerful online research tool with access to three exclusive databases of reliable source material, including the *Journal of Applied Microbiology*, the *Annual Review of Microbiology, Science, The New York Times Search by Subject Archive*, and the "Best of the Web" Link Library. The Microbiology Place is also provided on a CD-ROM included with each new copy of the text.

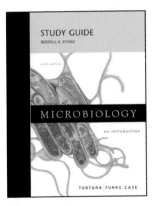

Study Guide by Berdell R. Funke. Students can master key concepts and earn a better grade with the help of the clear writing and creative, thought-provoking exercises found in this Study Guide. The Study Guide includes concise explanations of key concepts, definitions of important terms, art labeling exercises, critical thinking problems, and a variety of self-test questions with answers. 0-8053-7809-X

Study Card for Microbiology: An Introduction. This six-paneled, full-color study card provides students with a quick reference to the three most challenging topics in microbiology: metabolism, genetics, and immunology. 0-8053-8323-9

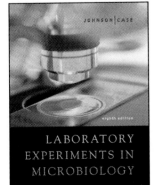

Laboratory Experiments in Microbiology by Ted R. Johnson and Christine L. Case. This popular lab manual includes 57 experiments covering a broad spectrum of topics in microbiology. Experiments encourage students to develop critical thinking and technical skills and show them the importance of microbes, both in nature and in our daily lives. Material with direct application to clinical and commercial labs is included, and emphasis is placed on lab safety. 0-8053-8292-5

The Microbe Files: Cases in Microbiology for the Undergraduate by Marjorie K. Cowan. *The Microbe Files* provides allied health and non-major microbiology students with a fascinating series of short cases that help them apply what they have learned in the course. 0-8053-4928-6

BRIEF CONTENTS

CONTENTS

Part One Fundamentals of Microbiology

Chapter 3 continues

Chapter 8 continues

Part Three   Interaction Between Microbe and Host

14 Principles of Disease and Epidemiology 420

15 Microbial Mechanisms of Pathogenicity 451

16 Innate Immunity: Nonspecific Defenses of the Host 474

17 Adaptive Immunity: Specific Defenses of the Host 502

18 Practical Applications of Immunology 527

Chapter 18 continues

1 The Microbial World and You

The overall theme of this textbook is the relationship between microbes (very small organisms that usually require a microscope to be seen) and our lives. This relationship involves not only the familiar harmful effects of certain microorganisms, such as disease and food spoilage, but also their many beneficial effects. In this chapter we will introduce you to some of the many ways microbes affect our lives. They have been fruitful subjects of study for a number of years, as you will see in the short history of microbiology that opens the chapter. We then discuss the incredible diversity of microorganisms and their ecological importance in maintaining balance in the environment by recycling chemical elements such as carbon and nitrogen between the soil and the atmosphere. We will also examine how microbes are used in commercial and industrial applications to produce foods, chemicals, and drugs (such as penicillin) and to treat sewage, control pests, and clean up pollutants. Finally, we will discuss microbes as the cause of such diseases as influenza A, SARS (severe acute respiratory syndrome), West Nile encephalitis, mad cow disease, diarrhea, hemorrhagic fever, and AIDS.

UNDER THE MICROSCOPE

Normal Microbiota on the Surface of the Human Tongue. Normal microbiota are microbes in and on the human body that normally do not cause disease and can actually provide protection against harmful microorganisms.

MICROBES IN OUR LIVES

LEARNING OBJECTIVE
- List several ways in which microbes affect our lives.

For many people, the words *germ* and *microbe* bring to mind a group of tiny creatures that do not quite fit into any of the categories in that old question, "Is it animal, vegetable, or mineral?" **Microbes,** also called **microorganisms,** are minute living things that individually are usually too small to be seen with the unaided eye. The group includes bacteria (Chapter 11), fungi (yeasts and molds), protozoa, and microscopic algae (Chapter 12). It also includes viruses, those noncellular entities sometimes regarded as being at the border between life and nonlife (Chapter 13). You will be introduced to each of these groups of microbes shortly.

We tend to associate these small organisms only with major diseases such as AIDS, uncomfortable infections, or such common inconveniences as spoiled food. However, the majority of microorganisms make crucial contributions to the welfare of the world's inhabitants by helping to maintain the balance of living organisms and chemicals in our environment. Marine and freshwater microorganisms form the basis of the food chain in oceans, lakes, and rivers. Soil microbes help break down wastes and incorporate nitrogen gas from the air into organic compounds, thereby recycling chemical elements in the soil, water, and air. Certain microbes play important roles in *photosynthesis*, a food- and oxygen-generating process that is critical to life on Earth. Humans and many other animals depend on the microbes in their intestines for digestion and the synthesis of some vitamins that their bodies require, including some B vitamins for metabolism and vitamin K for blood clotting.

Microorganisms also have many commercial applications. They are used in the synthesis of such chemical products as acetone, organic acids, enzymes, alcohols, and many drugs. The process by which microbes produce acetone and butanol was discovered in 1914 by Chaim Weizmann, a Russian-born chemist working in England. When World War I broke out in August of that year, the production of acetone was very important for making cordite (a smokeless form of gunpowder used in munitions). Weizmann's discovery played a significant role in determining the outcome of the war.

The food industry also uses microbes in producing vinegar, sauerkraut, pickles, alcoholic beverages, green olives, soy sauce, buttermilk, cheese, yogurt, and bread (see the box on the facing page). In addition, enzymes from microbes can now be manipulated such that the microbes produce substances they normally do not synthe-size. These substances include cellulose, digestive aids, and drain cleaner, plus important therapeutic substances such as insulin.

Though only a minority of microorganisms are **pathogenic** (disease-producing), practical knowledge of microbes is necessary for medicine and the related health sciences. For example, hospital workers must be able to protect patients from common microbes that are normally harmless but pose a threat to the sick and injured.

Today we understand that microorganisms are found almost everywhere. Yet not long ago, before the invention of the microscope, microbes were unknown to scientists. Thousands of people died in devastating epidemics, the causes of which were not understood. Entire families died because vaccinations and antibiotics were not available to fight infections.

We can get an idea of how our current concepts of microbiology developed by looking at a few of the historic milestones in microbiology that have changed our lives. Before doing this, however, we will first take a look at the major groups of microbes and how they are named and classified.

NAMING AND CLASSIFYING MICROORGANISMS

NOMENCLATURE

LEARNING OBJECTIVE
- Recognize the system of scientific nomenclature that uses two names: a genus and a specific epithet.

The system of nomenclature (naming) for organisms in use today was established in 1735 by Carolus Linnaeus. Scientific names are latinized because Latin was the language traditionally used by scholars. Scientific nomenclature assigns each organism two names—the **genus** (plural: *genera*) is the first name and is always capitalized; the **specific epithet (species** name) follows and is not capitalized. The organism is referred to by both the genus and the specific epithet, and both names are underlined or italicized. By custom, after a scientific name has been mentioned once, it can be abbreviated with the initial of the genus followed by the specific epithet.

Scientific names can, among other things, describe an organism, honor a researcher, or identify the habitat of a species. For example, consider *Staphylococcus aureus* (staf-i-lō-kok′kus ô′rē-us), a bacterium commonly found on human skin. *Staphylo-* describes the clustered arrangement of the cells; *coccus* indicates that they are shaped like spheres. The specific epithet, *aureus,* is Latin for golden,

APPLICATIONS OF MICROBIOLOGY

WHAT MAKES SOURDOUGH BREAD DIFFERENT?

Imagine being a miner during the California gold rush. You've just made bread dough from your last supplies of flour and salt when someone yells, "Gold!" Temporarily forgetting your hunger, you run off to the gold fields. Many hours later you return. The dough has been rising longer than usual, but you are too cold, tired, and hungry to care. Later you find that your bread tastes different from previous batches; it is slightly sour. During the gold rush, miners baked so many sour loaves that they were nicknamed "sourdoughs."

Conventional bread is made from flour, water, sugar, salt, shortening, and a living microbe, yeast. The yeast belongs to the Kingdom Fungi and is named *Saccharomyces cerevisiae*. When flour is mixed with water, an enzyme in the flour breaks its starch into two sugars, maltose and glucose. After the ingredients for the bread are mixed, the yeast metabolizes the sugars and produces alcohol (ethanol) and carbon dioxide as waste products. This metabolic process is called *fermentation*. The dough rises as carbon dioxide bubbles get trapped in the sticky matrix. The alcohol, which evaporates during baking, and the carbon dioxide gas form spaces that remain in the bread.

Originally, breads were leavened by wild yeast from the air, which had been trapped in the dough. Later, bakers kept a starter culture of yeast—dough from the last batch of bread—to leaven each new batch of dough. Sourdough bread is made with a special sourdough starter culture that is added to flour, water, and salt. Perhaps the most famous sourdough bread made today comes from San Francisco, where a handful of bakeries have continuously cultivated their starters for more than 150 years and have meticulously

maintained the starters to keep out unwanted microbes that can produce different and undesirable flavors. After bakeries in other areas made several unsuccessful attempts to match the unique flavor of San Francisco sourdough, rumors attributed the taste to a unique local climate or contamination from bakery walls. Ted F. Sugihara and Leo Kline from the United States Department of Agriculture (USDA) set out to debunk these rumors and to determine the microbiological basis for the bread's different taste so that it could be made in other areas.

The USDA workers found that sourdough is eight to ten times more acidic than conventional bread because of the presence of lactic and acetic acids. These acids account for the sour flavor of the bread. The workers isolated and identified

the yeast in the starter as *Saccharomyces exiguus*, a unique yeast that does not ferment maltose and thrives in the acidic environment of this dough. The sourdough question was not answered, however, because the yeast did not produce the acids and did not use maltose. Sugihara and Kline searched the starter for a second agent capable of fermenting maltose and producing the acids (see the figure). The bacterium they isolated, so carefully guarded all those years, was classified into the genus *Lactobacillus*. Many members of this genus are used in dairy fermentations and are found naturally on humans and other mammals. Analyses of cell structure and genetic composition showed that the sourdough bacterium is genetically different from other previously characterized lactobacilli. It has been given the name *Lactobacillus sanfranciscensis*.

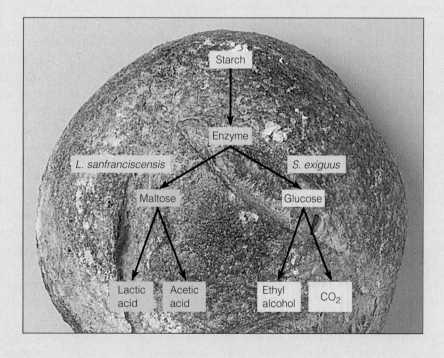

the color of many colonies of this bacterium. The genus of the bacterium *Escherichia coli* (esh-ë-rik′-ë-ä kō′lī or kō′lē) is named for a scientist, Theodor Escherich, whereas its specific epithet, *coli*, reminds us that *E. coli* live in the colon, or large intestine.

TYPES OF MICROORGANISMS

> **LEARNING OBJECTIVE**
> • Differentiate among the major characteristics of each group of microorganisms.

BACTERIA

Bacteria (singular: *bacterium*) are relatively simple, single-celled (unicellular) organisms. Because their genetic material is not enclosed in a special nuclear membrane, bacterial cells are called **prokaryotes** (prō-kar′e-ōts), from Greek words meaning prenucleus. Prokaryotes include both the bacteria and the archaea.

Bacterial cells generally appear in one of several shapes. *Bacillus* (bä-sil′lus) (rodlike), illustrated in Figure 1.1a, *coccus* (spherical or ovoid), and *spiral* (corkscrew or curved) are among the most common shapes, but some bacteria are star-shaped or square (see Figures 4.1–4.5, pages 79–80). Individual bacteria may form pairs, chains, clusters, or other groupings; such formations are usually characteristic of a particular genus or species of bacteria.

Bacteria are enclosed in cell walls that are largely composed of a carbohydrate and protein complex called *peptidoglycan*. (By contrast, cellulose is the main substance of plant and algal cell walls.) Bacteria generally reproduce by dividing into two equal cells; this process is called *binary fission*. For nutrition, most bacteria use organic chemicals, which in nature can be derived from either dead or living organisms. Some bacteria can manufacture their own food by photosynthesis, and some can derive nutrition from inorganic substances. Many bacteria can "swim" by using moving appendages called *flagella*. (For a complete discussion of bacteria, see Chapter 11.)

ARCHAEA

Like bacteria, **archaea** (är′kē-ä) consist of prokaryotic cells, but if they have cell walls, the walls lack peptidoglycan. Archaea, often found in extreme environments, are divided into three main groups. The *methanogens* produce methane as a waste product from respiration. The *extreme halophiles* (*halo* = salt; *philic* = loving) live in extremely salty environments such as the Great Salt Lake and the Dead Sea. The *extreme thermophiles* (*therm* = heat) live in hot sulfurous water such as hot springs at Yellowstone National Park. Archaea are not known to cause disease in humans.

FUNGI

Fungi (singular: **fungus**) are **eukaryotes** (yū-kar′ē-ōts), organisms whose cells have a distinct nucleus containing the cell's genetic material (DNA), surrounded by a special envelope called the nuclear membrane. Organisms in the Kingdom Fungi may be unicellular or multicellular (see Chapter 12, page 345). Large multicellular fungi, such as mushrooms, may look somewhat like plants, but they cannot carry out photosynthesis, as most plants can. True fungi have cell walls composed primarily of a substance called *chitin*. The unicellular forms of fungi, *yeasts*, are oval microorganisms that are larger than bacteria. The most typical fungi are *molds* (Figure 1.1b). Molds form visible masses called *mycelia*, which are composed of long filaments (*hyphae*) that branch and intertwine. The cottony growths sometimes found on bread and fruit are mold mycelia. Fungi can reproduce sexually or asexually. They obtain nourishment by absorbing solutions of organic material from their environment—whether soil, seawater, fresh water, or an animal or plant host. Organisms called *slime molds* have characteristics of both fungi and amoebas. They are discussed in detail in Chapter 12.

PROTOZOA

Protozoa (singular: **protozoan**) are unicellular, eukaryotic microbes (see Chapter 12, page 361). Protozoa move by pseudopods, flagella, or cilia. Amoebas (Figure 1.1c) move by using extensions of their cytoplasm called *pseudopods* (false feet). Other protozoa have long *flagella* or numerous shorter appendages for locomotion called *cilia*. Protozoa have a variety of shapes and live either as free entities or as *parasites* (organisms that derive nutrients from living hosts) that absorb or ingest organic compounds from their environment. Protozoa can reproduce sexually or asexually.

ALGAE

Algae (singular: **alga**) are photosynthetic eukaryotes with a wide variety of shapes and both sexual and asexual reproductive forms (Figure 1.1d). The algae of interest to microbiologists are usually unicellular (see Chapter 12, page 357). The cell walls of many algae, like those of plants, are composed of a carbohydrate called *cellulose*. Algae are abundant in fresh and salt water, in soil, and in association with plants. As photosynthesizers, algae need light, water, and carbon dioxide for food production and growth, but they do not generally require organic compounds from the environment. As a result of photosynthesis, algae produce oxygen and carbohydrates that are then utilized by other organisms, including animals. Thus, they play an important role in the balance of nature.

VIRUSES

Viruses (Figure 1.1e) are very different from the other microbial groups mentioned here. They are so small that

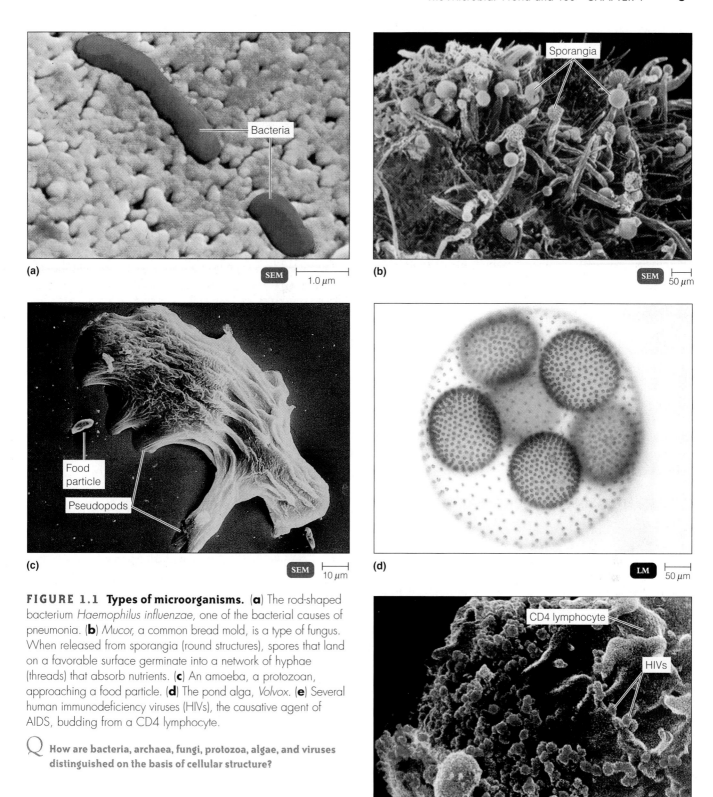

FIGURE 1.1 Types of microorganisms. (**a**) The rod-shaped bacterium *Haemophilus influenzae*, one of the bacterial causes of pneumonia. (**b**) *Mucor*, a common bread mold, is a type of fungus. When released from sporangia (round structures), spores that land on a favorable surface germinate into a network of hyphae (threads) that absorb nutrients. (**c**) An amoeba, a protozoan, approaching a food particle. (**d**) The pond alga, *Volvox*. (**e**) Several human immunodeficiency viruses (HIVs), the causative agent of AIDS, budding from a CD4 lymphocyte.

Q **How are bacteria, archaea, fungi, protozoa, algae, and viruses distinguished on the basis of cellular structure?**

most can be seen only with an electron microscope, and they are acellular (not cellular). Structurally very simple, a virus particle contains a core made of only one type of nucleic acid, either DNA or RNA. This core is surrounded by a protein coat. Sometimes the coat is encased by an additional layer, a lipid membrane called an envelope. All living cells have RNA *and* DNA, can carry out chemical reactions, and can reproduce as self-sufficient units. Viruses can reproduce only by using the cellular machinery of other organisms. Thus, on the one hand viruses are considered to be living when they multiply within host cells they infect. In this sense, viruses are parasites of other forms of life. On the other hand, viruses are not considered to be living because outside of living hosts they are inert. (Viruses will be discussed in detail in Chapter 13.)

MULTICELLULAR ANIMAL PARASITES

Although multicellular animal parasites are not strictly microorganisms, they are of medical importance and therefore will be discussed in this text. The two major groups of parasitic worms are the flatworms and the roundworms, collectively called **helminths** (see Chapter 12, page 370). During some stages of their life cycle, helminths are microscopic in size. Laboratory identification of these organisms includes many of the same techniques used for the identification of microbes.

CLASSIFICATION OF MICROORGANISMS

> **LEARNING OBJECTIVE**
> * List the three domains.

Before the existence of microbes was known, all organisms were grouped into either the animal kingdom or the plant kingdom. When microscopic organisms with characteristics of animals or plants were discovered late in the seventeenth century, a new system of classification was needed. Still, biologists could not agree on the criteria for classifying the new organisms they were seeing until the late 1970s.

In 1978, Carl Woese devised a system of classification based on the cellular organization of organisms. It groups all organisms in three domains as follows:

1. Bacteria (cell walls contain a protein-carbohydrate complex called peptidoglycan)
2. Archaea (cell walls, if present, lack peptidoglycan)
3. Eukarya, which includes the following:
   * Protists (slime molds, protozoa, and algae)
   * Fungi (unicellular yeasts, multicellular molds, and mushrooms)
   * Plants (includes mosses, ferns, conifers, and flowering plants)
   * Animals (includes sponges, worms, insects, and vertebrates)

Classification will be discussed in more detail in Chapters 10–12.

A BRIEF HISTORY OF MICROBIOLOGY

The science of microbiology dates back only two hundred years, yet the recent discovery of *Mycobacterium tuberculosis* (mī-kō-bak-ti′rē-um tü-bėr-ku-lō′sis) DNA in 3000-year-old Egyptian mummies reminds us that microorganisms have been around for much longer. In fact, bacterial ancestors were the first living cells to appear on Earth. While we know relatively little about what earlier people thought about the causes, transmission, and treatment of disease, the history of the past few hundred years is better known. Let's look now at some key developments in microbiology that have helped the field progress to its current high-technology state.

THE FIRST OBSERVATIONS

> **LEARNING OBJECTIVE**
> * Explain the importance of observations made by Hooke and van Leeuwenhoek.

One of the most important discoveries in the history of biology occurred in 1665 with the help of a relatively crude microscope. After observing a thin slice of cork, an Englishman, Robert Hooke, reported to the world that life's smallest structural units were "little boxes," or "cells," as he called them. Using his improved version of a compound microscope (one that uses two sets of lenses), Hooke was able to see individual cells. Hooke's discovery marked the beginning of the **cell theory**—the theory that *all living things are composed of cells*. Subsequent investigations into the structure and functions of cells were based on this theory.

Though Hooke's microscope was capable of showing cells, he lacked the staining techniques that would have allowed him to see microbes clearly. The Dutch merchant and amateur scientist Anton van Leeuwenhoek was probably the first to actually observe live microorganisms through the magnifying lenses of more than 400 microscopes he constructed. Between 1673 and 1723, he wrote a series of letters to the Royal Society of London describing the "animalcules" he saw through his simple, single-lens microscope. Van Leeuwenhoek made detailed drawings of "animalcules" in rainwater, in his own feces, and in material scraped from his teeth. These drawings have since been identified as representations of bacteria and protozoa (Figure 1.2).

(a) Van Leeuwenhoek using his microscope

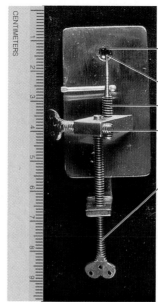

Lens

Location of specimen on pin

Specimen-positioning screw

Focusing control

Stage-positioning screw

(b) Microscope replica

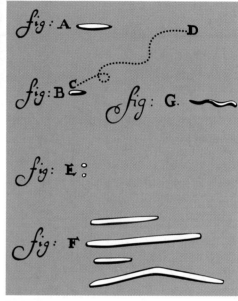

(c) Drawings of bacteria

FIGURE 1.2 Anton van Leeuwenhoek's microscopic observations.
(a) By holding his brass microscope toward a source of light, van Leeuwenhoek
was able to observe living organisms too small to be seen with the unaided eye.
(b) The specimen was placed on the tip of the adjustable point and viewed from
the other side through the tiny, nearly spherical lens. The highest magnification
possible with his microscopes was about 300× (times). **(c)** Some of van
Leeuwenhoek's drawings of bacteria, made in 1683. The letters represent
various shapes of bacteria. C–D represents a path of motion he observed.

Q **What was van Leeuwenhoek's major contribution to microbiology?**

THE DEBATE OVER SPONTANEOUS GENERATION

> **LEARNING OBJECTIVES**
> * Compare spontaneous generation and biogenesis.
> * Identify the contributions to microbiology made by
>   Needham, Spallanzani, Virchow, and Pasteur.

After van Leeuwenhoek discovered the previously "invisible" world of microorganisms, the scientific community of
the time became interested in the origins of these tiny living things. Until the second half of the nineteenth century, many scientists and philosophers believed that some
forms of life could arise spontaneously from nonliving matter; they called this hypothetical process **spontaneous
generation.** Not much more than 100 years ago, people
commonly believed that toads, snakes, and mice could be
born of moist soil; that flies could emerge from manure;
and that maggots, the larvae of flies, could arise from decaying corpses.

EVIDENCE PRO AND CON

A strong opponent of spontaneous generation, the Italian
physician Francesco Redi, set out in 1668 (even before
van Leeuwenhoek's discovery of microscopic life) to
demonstrate that maggots did not arise spontaneously
from decaying meat. Redi filled two jars with decaying
meat. The first was left unsealed; the flies laid their eggs
on the meat, and the eggs developed into larvae. The second jar was sealed and, because the flies could not lay
their eggs on the meat, no maggots appeared. Still, Redi's
antagonists were not convinced; they claimed that fresh
air was needed for spontaneous generation. So Redi set up
a second experiment, in which a jar was covered with a
fine net instead of being sealed. No larvae appeared in the
gauze-covered jar, even though air was present. Maggots
appeared only when flies were allowed to leave their eggs
on the meat.

Redi's results were a serious blow to the long-held belief that large forms of life could arise from nonlife. However, many scientists still believed that small organisms,

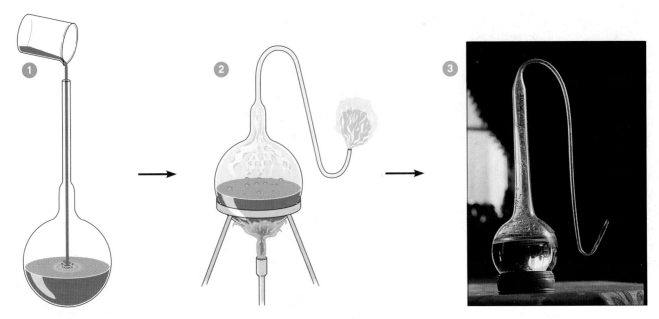

FIGURE 1.3 Pasteur's experiment disproving the theory of spontaneous generation. ① Pasteur first poured beef broth into a long-necked flask. ② Next he heated the neck of the flask and bent it into an S-shaped curve; then he boiled the broth for several minutes. ③ Microorganisms did not appear in the cooled solution, even after long periods, as you can see in this recent photograph of an actual flask used by Pasteur in a similar experiment.

Q **What are aseptic techniques, and how did Pasteur contribute to their development?**

such as van Leeuwenhoek's "animalcules," were simple enough to be generated from nonliving materials.

The case for spontaneous generation of microorganisms seemed to be strengthened in 1745 when John Needham, an Englishman, found that even after he heated nutrient fluids (chicken broth and corn broth) before pouring them into covered flasks, the cooled solutions were soon teeming with microorganisms. Needham claimed that microbes developed spontaneously from the fluids. Twenty years later, Lazzaro Spallanzani, an Italian scientist, suggested that microorganisms from the air probably had entered Needham's solutions after they were boiled. Spallanzani showed that nutrient fluids heated *after* being sealed in a flask did not develop microbial growth. Needham responded by claiming the "vital force" necessary for spontaneous generation had been destroyed by the heat and was kept out of the flasks by the seals.

This intangible "vital force" was given all the more credence shortly after Spallanzani's experiment, when Anton Laurent Lavoisier showed the importance of oxygen to life. Spallanzani's observations were criticized on the grounds that there was not enough oxygen in the sealed flasks to support microbial life.

THE THEORY OF BIOGENESIS

The issue was still unresolved in 1858, when the German scientist Rudolf Virchow challenged spontaneous generation with the concept of **biogenesis,** the claim that living cells can arise only from preexisting living cells. Arguments about spontaneous generation continued until 1861, when the issue was resolved by the French scientist Louis Pasteur.

With a series of ingenious and persuasive experiments, Pasteur demonstrated that microorganisms are present in the air and can contaminate sterile solutions, but air itself does not create microbes. He filled several short-necked flasks with beef broth and then boiled their contents. Some were then left open and allowed to cool. In a few days, these flasks were found to be contaminated with microbes. The other flasks, sealed after boiling, were free of microorganisms. From these results, Pasteur reasoned that microbes in the air were the agents responsible for contaminating nonliving matter such as the broths in Needham's flasks.

Pasteur next placed broth in open-ended long-necked flasks and bent the necks into S-shaped curves (Figure 1.3). The contents of these flasks were then boiled and cooled. The broth in the flasks did not decay and showed no signs of life, even after months. Pasteur's unique design allowed air to pass into the flask, but the curved neck trapped any airborne microorganisms that might contaminate the broth. (Some of these original vessels are still on display at the Pasteur Institute in Paris. They have been sealed but, like the flask shown in Figure 1.3, show no sign of contamination more than 100 years later.)

Pasteur showed that microorganisms can be present in nonliving matter—on solids, in liquids, and in the air. Furthermore, he demonstrated conclusively that microbial life can be destroyed by heat and that methods can be devised to block the access of airborne microorganisms to nutrient environments. These discoveries form the basis of **aseptic techniques,** techniques that prevent contamination by unwanted microorganisms, which are now the standard practice in laboratory and many medical procedures. Modern aseptic techniques are among the first and most important things that a beginning microbiologist learns.

Pasteur's work provided evidence that microorganisms cannot originate from mystical forces present in nonliving materials. Rather, any appearance of "spontaneous" life in nonliving solutions can be attributed to microorganisms that were already present in the air or in the fluids themselves. Scientists now believe that a form of spontaneous generation probably did occur on the primitive Earth when life first began, but they agree that this does not happen under today's environmental conditions.

THE GOLDEN AGE OF MICROBIOLOGY

> ### LEARNING OBJECTIVES
> - Identify the importance of Koch's postulates.
> - Explain how Pasteur's work influenced Lister and Koch.
> - Identify the importance of Jenner's work.

For about 60 years, beginning with the work of Pasteur, there was an explosion of discoveries in microbiology. The period from 1857 to 1914 has been appropriately named the Golden Age of Microbiology. During this period, rapid advances, spearheaded mainly by Pasteur and Robert Koch, led to the establishment of microbiology as a science. Discoveries during these years included both the agents of many diseases and the role of immunity in the prevention and cure of disease. During this productive period, microbiologists studied the chemical activities of microorganisms, improved the techniques for performing microscopy and culturing microorganisms, and developed vaccines and surgical techniques. Some of the major events that occurred during the Golden Age of Microbiology are listed in Figure 1.4.

FERMENTATION AND PASTEURIZATION

One of the key steps that established the relationship between microorganisms and disease occurred when a group of French merchants asked Pasteur to find out why wine and beer soured. They hoped to develop a method that would prevent spoilage when those beverages were shipped long distances. At the time, many scientists believed that air converted the sugars in these fluids into alcohol. Pasteur found instead that microorganisms called yeasts convert the sugars to alcohol in the absence of air. This process, called **fermentation** (see Chapter 5, page 134), is used to make wine and beer. Souring and spoilage are caused by different microorganisms called bacteria. In the presence of air, bacteria change the alcohol in the beverage into vinegar (acetic acid).

Pasteur's solution to the spoilage problem was to heat the beer and wine just enough to kill most of the bacteria that caused the spoilage; the process, called **pasteurization,** is now commonly used to reduce spoilage and kill potentially harmful bacteria in milk as well as in some alcoholic drinks. Showing the connection between spoilage of food and microorganisms was a major step toward establishing the relationship between disease and microbes.

THE GERM THEORY OF DISEASE

As we have seen, the fact that many kinds of diseases are related to microorganisms was unknown until relatively recently. Before the time of Pasteur, effective treatments for many diseases were discovered by trial and error, but the causes of the diseases were unknown.

The realization that yeasts play a crucial role in fermentation was the first link between the activity of a microorganism and physical and chemical changes in organic materials. This discovery alerted scientists to the possibility that microorganisms might have similar relationships with plants and animals—specifically, that microorganisms might cause disease. This idea was known as the **germ theory of disease.**

The germ theory was a difficult concept for many people to accept at that time because for centuries disease was believed to be punishment for an individual's crimes or misdeeds. When the inhabitants of an entire village became ill, people often blamed the disease on demons appearing as foul odors from sewage or on poisonous vapors from swamps. Most people born in Pasteur's time found it inconceivable that "invisible" microbes could travel through the air to infect plants and animals or remain on clothing and bedding to be transmitted from one person to another. But gradually scientists accumulated the information needed to support the new germ theory.

1665	Hooke—First observation of cells
1673	van Leeuwenhoek—First observation of live microorganisms
1735	Linnaeus—Nomenclature for organisms
1798	Jenner—First vaccine
1835	Bassi—Silkworm fungus
1840	Semmelweis—Childbirth fever
1853	DeBary—Fungal plant disease

Louis Pasteur (1822–1895)

1857	Pasteur—Fermentation
1861	Pasteur—Disproved spontaneous generation
1864	Pasteur—Pasteurization
1867	Lister—Aseptic surgery
1876	*Koch—Germ theory of disease
1879	Neisser—*Neisseria gonorrhoeae*
1881	*Koch—Pure cultures
	Finley—Yellow fever
1882	*Koch—*Mycobacterium tuberculosis*
	Hess—Agar (solid) media
1883	*Koch—*Vibrio cholerae*
1884	*Metchnikoff—Phagocytosis

GOLDEN AGE OF MICROBIOLOGY

	Gram—Gram-staining procedure
	Escherich—*Escherichia coli*
1887	Petri—Petri dish
1889	Kitasato—*Clostridium tetani*
1890	*von Bering—Diphtheria antitoxin
	*Ehrlich—Theory of immunity
1892	Winogradsky—Sulfur cycle
1898	Shiga—*Shigella dysenteriae*
1908	*Ehrlich—Syphilis
1910	Chagas—*Trypanosoma cruzi*
1911	*Rous—Tumor-causing virus (1966 Nobel Prize)

Robert Koch (1843–1910)

1928	*Fleming, Chain, Florey—Penicillin
	Griffith—Transformation in bacteria
1934	Lancefield—Streptococcal antigens
1935	*Stanley, Northrup, Sumner—Crystallized virus
1941	Beadle and Tatum—Relationship between genes and enzymes
1943	*Delbrück and Luria—Viral infection of bacteria
1944	Avery, MacLeod, McCarty—Genetic material is DNA
1946	Lederberg and Tatum—Bacterial conjugation
1953	*Watson and Crick—DNA structure
1957	*Jacob and Monod—Protein synthesis regulation
1959	Stewart—Viral cause of human cancer
1962	*Edelman and Porter—Antibodies
1964	Epstein, Achong, Barr—Epstein-Barr virus as cause of human cancer
1973	Berg, Boyer, Cohen—Genetic engineering
1975	Dulbecco, Temin, Baltimore—Reverse transcriptase
1978	Woese—Archaea
	*Nathans, Smith, Arber—Restriction enzymes (used for recombinant DNA technology)
	*Mitchell—Chemiosmotic mechanism
1981	Margulis—Origin of eukaryotic cells
1982	*Klug—Structure of tobacco mosaic virus
1983	*McClintock—Transposons

Rebecca C. Lancefield (1895–1981)

1988	*Deisenhofer, Huber, Michel—Bacterial photosynthesis pigments
1994	Cano—Reported to have cultured 40-million-year-old bacteria
1997	*Prusiner—Prions

FIGURE 1.4 Milestones in microbiology, highlighting those that occurred during the Golden Age of Microbiology. An asterisk (*) indicates a Nobel laureate.

 Why was the Golden Age of Microbiology so named?

In 1865, Pasteur was called upon to help fight silkworm disease, which was ruining the silk industry throughout Europe. Years earlier, in 1835, Agostino Bassi, an amateur microscopist, had proved that another silkworm disease was caused by a fungus. Using data provided by Bassi, Pasteur found that the more recent infection was caused by a protozoan, and he developed a method for recognizing afflicted silkworm moths.

In the 1860s, Joseph Lister, an English surgeon, applied the germ theory to medical procedures. Lister was aware that in the 1840s, the Hungarian physician Ignaz Semmelweis had demonstrated that physicians, who at the time did not disinfect their hands, routinely transmitted infections (puerperal, or childbirth, fever) from one obstetrical patient to another. Lister had also heard of Pasteur's work connecting microbes to animal diseases. Disinfectants were not used at the time, but Lister knew that phenol (carbolic acid) kills bacteria, so he began treating surgical wounds with a phenol solution. The practice so reduced the incidence of infections and deaths that other surgeons quickly adopted it. Lister's technique was one of the earliest medical attempts to control infections caused by microorganisms. In fact, his findings proved that microorganisms cause surgical wound infections.

The first proof that bacteria actually cause disease came from Robert Koch in 1876. Koch, a German physician, was Pasteur's young rival in the race to discover the cause of anthrax, a disease that was destroying cattle and sheep in Europe. Koch discovered rod-shaped bacteria now known as *Bacillus anthracis* (bä-sil′lus an-thrā′sis) in the blood of cattle that had died of anthrax. He cultured the bacteria on nutrients and then injected samples of the culture into healthy animals. When these animals became sick and died, Koch isolated the bacteria in their blood and compared them with the bacteria originally isolated. He found that the two sets of blood cultures contained the same bacteria.

Koch thus established a sequence of experimental steps for directly relating a specific microbe to a specific disease. These steps are known today as **Koch's postulates** (see Figure 14.3, page 426). During the past 100 years, these same criteria have been invaluable in investigations proving that specific microorganisms cause many diseases. Koch's postulates, their limitations, and their application to disease will be discussed in greater detail in Chapter 14.

VACCINATION

Often a treatment or preventive procedure is developed before scientists know why it works. The smallpox vaccine is an example of this. On May 4, 1796, almost 70 years before Koch established that a specific microorganism causes

anthrax, Edward Jenner, a young British physician, embarked on an experiment to find a way to protect people from smallpox.

Smallpox epidemics were greatly feared. The disease periodically swept through Europe, killing thousands, and it wiped out 90% of the American Indians on the East Coast when European settlers first brought the infection to the New World.

When a young milkmaid informed Jenner that she couldn't get smallpox because she already had been sick from cowpox—a much milder disease—he decided to put the girl's story to the test. First Jenner collected scrapings from cowpox blisters. Then he inoculated a healthy 8-year-old volunteer with the cowpox material by scratching the person's arm with a pox-contaminated needle. The scratch turned into a raised bump. In a few days, the volunteer became mildly sick but recovered and never again contracted either cowpox or smallpox. The process was called *vaccination*, from the Latin word *vacca*, meaning cow. Pasteur gave it this name in honor of Jenner's work. The protection from disease provided by vaccination (or by recovery from the disease itself) is called **immunity.** We will discuss the mechanisms of immunity in Chapter 17.

Years after Jenner's experiment, in about 1880, Pasteur discovered why vaccinations work. He found that the bacterium that causes fowl cholera lost its ability to cause disease (lost its *virulence*, or became *avirulent*) after it was grown in the laboratory for long periods. However, it—and other microorganisms with decreased virulence—was able to induce immunity against subsequent infections by its virulent counterparts. The discovery of this phenomenon provided a clue to Jenner's successful experiment with cowpox. Both cowpox and smallpox are caused by viruses. Even though cowpox virus is not a laboratory-produced derivative of smallpox virus, it is so closely related to the smallpox virus that it can induce immunity to both viruses. Pasteur used the term *vaccine* for cultures of avirulent microorganisms used for preventive inoculation.

Jenner's experiment marked the first time in a Western culture that a living viral agent—the cowpox virus—was used to produce immunity. Physicians in China had immunized patients by removing scales from drying pustules of a person suffering from a mild case of smallpox, grinding the scales to a fine powder, and inserting the powder into the nose of the person to be protected.

Some vaccines are still produced from avirulent microbial strains that stimulate immunity to the related virulent strain. Other vaccines are made from killed virulent microbes, from isolated components of virulent microorganisms, or by genetic engineering techniques.

THE BIRTH OF MODERN CHEMOTHERAPY: DREAMS OF A "MAGIC BULLET"

LEARNING OBJECTIVE

- Identify the contributions to microbiology made by Ehrlich and Fleming.

After the relationship between microorganisms and disease was established, medical microbiologists next focused on the search for substances that could destroy pathogenic microorganisms without damaging the infected animal or human. Treatment of disease by using chemical substances is called **chemotherapy.** (The term also commonly refers to chemical treatment of noninfectious diseases, such as cancer.) Chemotherapeutic agents prepared from chemicals in the laboratory are called **synthetic drugs.** Chemicals produced naturally by bacteria and fungi to act against other microorganisms are called **antibiotics.** The success of chemotherapy is based on the fact that some chemicals are more poisonous to microorganisms than to the hosts infected by the microbes. Antimicrobial therapy will be discussed in further detail in Chapter 20.

THE FIRST SYNTHETIC DRUGS

Paul Ehrlich, a German physician, was the imaginative thinker who fired the first shot in the chemotherapy revolution. As a medical student, Ehrlich speculated about a "magic bullet" that could hunt down and destroy a pathogen without harming the infected host. Ehrlich launched a search for such a bullet. In 1910, after testing hundreds of substances, he found a chemotherapeutic agent called *salvarsan,* an arsenic derivative effective against syphilis. The agent was named salvarsan because it was considered to offer salvation from syphilis and it contained arsenic. Before this discovery, the only known chemical in Europe's medical arsenal was an extract from the bark of a South American tree, *quinine,* which had been used by Spanish conquistadors to treat malaria.

By the late 1930s, researchers had developed several other synthetic drugs that could destroy microorganisms. Most of these drugs were derivatives of dyes. This came about because the dyes synthesized and manufactured for fabrics were routinely tested for antimicrobial qualities by microbiologists looking for a "magic bullet." In addition, *sulfonamides* (sulfa drugs) were synthesized at about the same time.

A FORTUNATE ACCIDENT—ANTIBIOTICS

In contrast to the sulfa drugs, which were deliberately developed from a series of industrial chemicals, the first antibiotic was discovered by accident. Alexander Fleming, a Scottish physician and bacteriologist, almost tossed out some culture plates that had been contaminated by mold. Fortunately, he took a second look at the curious pattern of growth on the

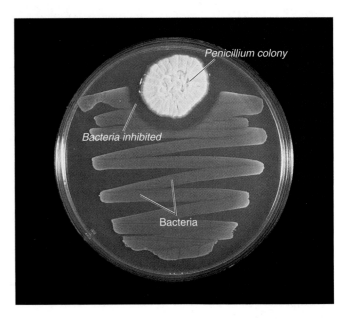

FIGURE 1.5 Discovering antibiotics. The antibiotic secreted by the *Penicillium* fungus during its own growth inhibited the growth of the bacteria.

Q **What are some of the problems associated with antibiotics?**

contaminated plates. Around the mold was a clear area where bacterial growth had been inhibited (Figure 1.5). The actual plate is shown in Figure 20.1. Fleming was looking at a mold that could inhibit the growth of a bacterium. The mold was later identified as *Penicillium chrysogenum* (pen-i-sil′lē-um krī-so′jen-um), and in 1928 Fleming named the mold's active inhibitor *penicillin.* Thus, penicillin is an antibiotic produced by a fungus. The enormous usefulness of penicillin was not apparent until the 1940s, when it was finally tested clinically and mass-produced.

Since the early discoveries of antibiotics, thousands of others have been discovered. Unfortunately, antibiotics and other chemotherapeutic drugs are not without problems. Many antimicrobial chemicals are too toxic to humans for practical use; they kill the pathogenic microbes, but they also damage the infected host. For reasons we will discuss later, toxicity to humans is a particular problem in the development of drugs for treating viral diseases. Viral growth is dependent on life processes of normal host cells. Thus, there are very few successful antiviral drugs because a drug that would interfere with viral reproduction would also likely affect uninfected cells of the body.

Another major problem associated with antimicrobial drugs is the emergence and spread of new varieties of microorganisms that are resistant to antibiotics. Over the years, more and more microbes have developed resistance to antibiotics that at one time were very effective against them. Drug resistance results from genetic changes in microbes that enables them to tolerate a certain amount of an antibiotic that would normally inhibit them (see the box in Chapter 24). These changes might include the production by microbes of chemicals (enzymes) that in-activate antibiotics, changes in the surface of a microbe that prevent an antibiotic from attaching to it, and pre-vention of an antibiotic from entering the microbe.

The recent appearance of vancomycin-resistant *Staphylococcus aureus* and *Enterococcus faecalis* has alarmed health care professionals because it indicates that some previously treatable bacterial infections may soon be impossible to treat with antibiotics.

MODERN DEVELOPMENTS IN MICROBIOLOGY

LEARNING OBJECTIVES

- Define *bacteriology, mycology, parasitology, immunology,* and *virology.*
- Explain the importance of recombinant DNA technology.

The quest to solve drug resistance, identify viruses, and de-velop vaccines requires sophisticated research techniques and correlated studies that were never dreamed of in the days of Koch and Pasteur.

The groundwork laid during the Golden Age of Micro-biology provided the basis for several monumental achievements during the twentieth century (Table 1.1). New branches of microbiology were developed, including immunology and virology. Most recently, the development of a set of new methods called recombinant DNA technol-ogy has revolutionized research and practical applications in all areas of microbiology.

BACTERIOLOGY, MYCOLOGY, AND PARASITOLOGY

Bacteriology, the study of bacteria, began with van Leeuwenhoek's first examination of tooth scrapings. New pathogenic bacteria are still discovered regularly. Many bacteriologists, like their predecessor Pasteur, look at the roles of bacteria in food and the environment. One intriguing discovery came in 1997, when Heide Schulz discovered a bacterium large enough to be seen with the unaided eye (0.2 mm wide). This bacterium, which she named *Thiomargarita namibiensis* (thī′o-mä-gär-e-tä na′mib-ē-ėn-sis), lives in the mud on the African coast. *Thiomargarita* is unusual because of its size and its ecological niche. The bacterium consumes hy-drogen sulfide, which would be toxic to mud-dwelling animals (Figure 11.26, page 339).

Mycology, the study of fungi, includes medical, agri-cultural, and ecological branches. Recall that Bassi's work leading up to the germ theory of disease was on a fungal pathogen. Fungal infection rates have been rising during the past decade, accounting for 10% of hospital-acquired infections. Climatic and environmental changes (severe drought) are thought to account for the tenfold increase in *Coccidioides immitis* (kok-sid-ē-oi′dēz im′mi-tis) infections in California. New techniques for diagnosing and treating fungal infections are currently being investigated.

Parasitology is the study of protozoa and parasitic worms. Because many parasitic worms are large enough to be seen with the unaided eye, they have been known by people for thousands of years. One hypothesis is that the medical symbol, the caduceus, represents the removal of parasitic guinea worms (Figure 1.6).

Text continues on page 16.

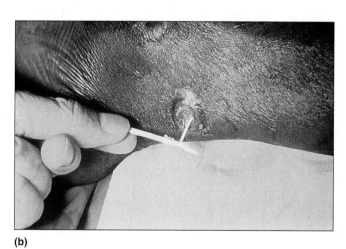

(a)   (b)

FIGURE 1.6 Parasitology: the study of protozoa and parasitic worms.
(**a**) The caduceus, the symbol of the medical profession, may have been designed after the procedure for removing parasitic guinea worms. (**b**) A doctor removes a guinea worm *(Dracunculus medinensis)* from the subcutaneous tissue of a patient by winding it onto a stick.

 How do bacteriology, mycology, and parasitology differ?

TABLE 1.1	Selected Nobel Prizes Awarded for Research in Microbiology		
Nobel Laureates	Year of Presentation	Country of Birth	Contribution
Emil A. von Behring	1901	Germany	Developed a diphtheria antitoxin.
Ronald Ross	1902	England	Discovered how malaria is transmitted.
Robert Koch	1905	Germany	Cultured tuberculosis bacteria.
Paul Ehrlich	1908	Germany	Developed theories on immunity.
Elie Metchnikoff	1908	Russia	Described phagocytosis, the intake of solid materials by cells.
Alexander Fleming, Ernst Chain, and Howard Florey	1945	Scotland England England	Discovered penicillin.
Selman A. Waksman	1952	Ukraine	Discovered streptomycin.
Hans A. Krebs	1953	Germany	Discovered chemical steps of the Krebs cycle in carbohydrate metabolism.
John F. Enders, Thomas H. Weller, and Frederick C. Robbins	1954	United States	Cultured poliovirus in cell cultures.
Joshua Lederberg, George Beadle, and Edward Tatum	1958	United States	Described genetic control of biochemical reactions.
James D. Watson, Frances H. C. Crick, and Maurice A. F. Wilkins	1962	United States England New Zealand	Identified the physical structure of DNA.
François Jacob, Jacques Monod, and André Lwoff	1965	France	Described how protein synthesis is regulated in bacteria.
Peyton Rous	1966	United States	Discovered cancer-causing viruses.
Robert Holley, Har Gobind Khorana, and Marshall W. Nirenberg	1968	United States India United States	Discovered the genetic code for amino acids.
Max Delbrück, Alfred D. Hershey, and Salvador E. Luria	1969	Germany United States Italy	Described the mechanism of viral infection of bacterial cells.
Gerald M. Edelman and Rodney R. Porter	1972	United States England	Described the nature and structure of antibodies.
Renato Dulbecco, Howard Temin, and David Baltimore	1975	United States	Discovered reverse transcriptase and described how RNA viruses could cause cancer.
Daniel Nathans, Hamilton Smith, and Werner Arber	1978	United States United States Switzerland	Described the action of restriction enzymes (now used in recombinant DNA technology).

TABLE 1.1   *(continued)*

Nobel Laureates	Year of Presentation	Country of Birth	Contribution
Peter Mitchell	1978	England	Described the chemiosmotic mechanism for ATP synthesis.
Paul Berg	1980	United States	Performed experiments in gene splicing (recombinant DNA technology).
Aaron Klug	1982	South Africa	Described the structure of tobacco mosaic virus (TMV).
Barbara McClintock	1983	United States	Discovered transposons (small segments of DNA that can move from one region of a DNA molecule to another).
César Milstein, Georges J.F. Köhler, and Niels Kai Jerne	1984	Argentina Germany Denmark	Developed a technique for producing monoclonal antibodies (single pure antibodies).
Susumu Tonegawa	1987	Japan	Described the genetics of antibody production.
Johann Deisenhofer, Robert Huber, and Hartmut Michel	1988	Germany	Described the structure of bacterial photosynthetic pigments.
J. Michael Bishop and Harold E. Varmus	1989	United States	Discovered cancer-causing genes called oncogenes.
Joseph E. Murray and E. Donnall Thomas	1990	United States	Performed the first successful organ transplants by using immunosuppressive agents.
Edmond H. Fisher and Edwin G. Krebs	1992	United States	Discovered protein kinases, enzymes that regulate cell growth.
Richard J. Roberts and Phillip A. Sharp	1993	Great Britain United States	Discovered that a gene can be separated onto different segments of DNA.
Kary B. Mullis	1993	United States	Discovered the polymerase chain reaction to amplify (make multiple copies of) DNA.
Michael Smith	1993	Canada	Discovered a procedure to modify DNA to make new proteins.
Peter C. Doherty and Rolf M. Zinkernagel	1996	Australia Switzerland	Discovered how cytotoxic T cells recognize virus-infected cells prior to destroying them.
Stanley B. Prusiner	1997	United States	Discovered and named proteinaceous infectious particles (prions) and demonstrated a relationship between prions and deadly neurological diseases in humans and animals.
Peter Agre and Roderick MadKinnon	2003	United States	Discovered water and ion channels in plasma membranes.
Aaron Ciechanover, Avram Hershko, and Irwin Rose	2004	Israel Israel United States	Discovered how cells dispose of unwanted proteins in proteasomes.
Barry Marshall and J. Robin Warren	2005	Australia	Discovered that *Heliobacter pylori* causes peptic ulcers.

New parasitic diseases in humans are being discovered as laborers become exposed while clearing rain forests. Previously unknown parasitic diseases are also being found in patients whose immune systems have been suppressed by organ transplants, cancer chemotherapy, and AIDS.

Bacteriology, mycology, and parasitology are currently going through a "golden age of classification." Recent advances in **genomics,** the study of all of an organism's genes, have allowed scientists to classify bacteria and fungi according to their genetic relationships with other bacteria, fungi, and protozoa. Previously these microorganisms were classified according to a limited number of visible characteristics.

IMMUNOLOGY

Immunology, the study of immunity, actually dates back in Western culture to Jenner's first vaccine in 1796. Since then, knowledge about the immune system has accumulated steadily and expanded rapidly during the twentieth century. Vaccines are now available for numerous diseases, including measles, rubella (German measles), mumps, chickenpox, pneumococcal pneumonia, tetanus, tuberculosis, influenza, whooping cough, polio, and hepatitis B. The smallpox vaccine was so effective that the disease has been eliminated. Public health officials estimate that polio will be eradicated within a few years because of the polio vaccine. In 1960, interferons, substances generated by the body's own immune system, were discovered. Interferons inhibit replication of viruses and have triggered considerable research related to the treatment of viral diseases and cancer. One of today's biggest challenges for immunologists is learning how the immune system might be stimulated to ward off the virus responsible for AIDS, a disease that destroys the immune system.

A major advance in immunology occurred in 1933, when Rebecca Lancefield proposed that streptococci be classified according to serotypes (variants within a species) based on certain components in the cell walls of the bacteria. Streptococci are responsible for a variety of diseases, such as sore throat (strep throat), streptococcal toxic shock, and septicemia (blood poisoning). Her research permits the rapid identification of specific pathogenic streptococci based on immunological techniques.

VIROLOGY

The study of viruses, **virology,** actually originated during the Golden Age of Microbiology. In 1892, Dmitri Iwanowski reported that the organism that caused mosaic disease of tobacco was so small that it passed through filters fine enough to stop all known bacteria. At the time, Iwanowski was not aware that the organism in question was a virus in the sense that we now understand the term.

In 1935, Wendell Stanley demonstrated that the organism, called tobacco mosaic virus (TMV), was fundamentally different from other microbes and so simple and homogeneous that it could be crystallized like a chemical compound. Stanley's work facilitated the study of viral structure and chemistry. Since the development of the electron microscope in the 1940s, microbiologists have been able to observe the structure of viruses in detail, and today much is known about their structure and activity.

RECOMBINANT DNA TECHNOLOGY

Microorganisms can now be genetically engineered to manufacture large amounts of human hormones and other urgently needed medical substances. In the late 1960s, Paul Berg showed that fragments of human or animal DNA (genes) that code for important proteins can be attached to bacterial DNA. The resulting hybrid was the first example of **recombinant DNA.** When recombinant DNA is inserted into bacteria (and other microbes), it can be used to make large quantities of the desired protein. The technology that developed from this technique is called **recombinant DNA technology,** or **genetic engineering,** and it had its origins in two related fields. The first, **microbial genetics,** studies the mechanisms by which microorganisms inherit traits. The second, **molecular biology,** specifically studies how genetic information is carried in molecules of DNA and how DNA directs the synthesis of proteins.

Although molecular biology encompasses all organisms, much of our knowledge of how genes determine specific traits has been revealed through experiments with bacteria. Until the 1930s, all genetic research was based on the study of plant and animal cells. But in the 1940s, scientists turned to unicellular organisms, primarily bacteria, which have several advantages for genetic and biochemical research. For one thing, bacteria are less complex than plants and animals. For another, the life cycles of many bacteria require less than an hour, so scientists can cultivate very large numbers of individuals for study in a relatively short time.

Once science turned to the study of unicellular life, progress in genetics began to occur rapidly. In 1941, George W. Beadle and Edward L. Tatum demonstrated the relationship between genes and enzymes. DNA was established as the hereditary material in 1944 by Oswald Avery, Colin MacLeod, and Maclyn McCarty. In 1946, Joshua Lederberg and Edward L. Tatum discovered that genetic material could be transferred from one bacterium to another by a process called conjugation. Then, in 1953, James Watson and Francis Crick proposed a model for the structure and replication of DNA. The early 1960s witnessed a further explosion of discoveries relating to the way DNA controls protein synthesis. In 1961, François

Jacob and Jacques Monod discovered messenger RNA (ribonucleic acid), a chemical involved in protein synthesis, and later they made the first major discoveries about the regulation of gene function in bacteria. During the same period, scientists were able to break the genetic code and thus understand how the information for protein synthesis in messenger RNA is translated into the amino acid sequence for making proteins.

MICROBES AND HUMAN WELFARE

LEARNING OBJECTIVE
- List at least four beneficial activities of microorganisms.

As mentioned earlier, only a minority of all microorganisms are pathogenic. Microbes that cause food spoilage, such as soft spots on fruits and vegetables, decomposition of meats, and rancidity of fats and oils, are also a minority. The vast majority of microbes benefit humans, other animals, and plants in many ways. The following sections outline some of these beneficial activities. In later chapters, we will discuss these activities in greater detail.

RECYCLING VITAL ELEMENTS

Discoveries made by two microbiologists in the 1880s have formed the basis for today's understanding of the biochemical cycles that support life on Earth. Martinus Beijerinck and Sergei Winogradsky were the first to show how bacteria help recycle vital elements between the soil and the atmosphere. **Microbial ecology,** the study of the relationship between microorganisms and their environment, originated with the work of Beijerinck and Winogradsky. Today, microbial ecology has branched out and includes the study of how microbial populations interact with plants and animals in various environments. Among the concerns of microbial ecologists are water pollution and toxic chemicals in the environment.

The chemical elements carbon, nitrogen, oxygen, sulfur, and phosphorus are essential for life and abundant, but not necessarily in forms that organisms can use. Microorganisms are primarily responsible for converting these elements into forms that can be used by plants and animals. Microorganisms, primarily bacteria and fungi, play a key role in returning carbon dioxide to the atmosphere when they decompose organic wastes and dead plants and animals. Algae, cyanobacteria, and higher plants use the carbon dioxide during photosynthesis to produce carbohydrates for animals, fungi, and bacteria. Nitrogen is abundant in the atmosphere but must be made into a usable form by bacteria so that it be available for plants and animals. Only bacteria can make this conversion naturally.

SEWAGE TREATMENT: USING MICROBES TO RECYCLE WATER

With our society's growing awareness of the need to preserve the environment, people have become more conscious of the responsibility to recycle precious water and prevent the pollution of rivers and oceans. One major pollutant is sewage, which consists of human excrement, waste water, industrial wastes, and surface runoff. Sewage is about 99.9% water, with a few hundredths of 1% suspended solids. The remainder is a variety of dissolved materials.

Sewage treatment plants remove the undesirable materials and harmful microorganisms. Treatments combine various physical and chemical processes with the action of beneficial microbes. Large solids such as paper, wood, glass, gravel, and plastic are removed from sewage; left behind are liquid and organic materials that bacteria convert into such by-products as carbon dioxide, nitrates, phosphates, sulfates, ammonia, hydrogen sulfide, and methane. (We will discuss sewage treatment in detail in Chapter 27.)

BIOREMEDIATION: USING MICROBES TO CLEAN UP POLLUTANTS

In 1988, scientists began using microbes to clean up pollutants and toxic wastes produced by various industrial processes. For example, some bacteria can actually use pollutants as energy sources; others produce enzymes that break down toxins into less harmful substances. By using bacteria in these ways—a process known as **bioremediation**—toxins can be removed from underground wells, chemical spills, toxic waste sites, and oil spills, such as the *Exxon Valdez* disaster of 1989 (see the box in Chapter 2, page 33). In addition, bacterial enzymes are used in drain cleaners to remove clogs without adding harmful chemicals to the environment. In some cases, microorganisms indigenous to the environment are used; in others, genetically modified microbes are used. Among the most commonly used bioremedial microbes are certain species of bacteria of the genera *Pseudomonas* (sū-dō-mō′nas) and *Bacillus*. *Bacillus* enzymes are also used in household detergents to remove spots from clothing.

INSECT PEST CONTROL BY MICROORGANISMS

Besides spreading diseases, insects can cause devastating crop damage. Insect pest control is therefore important for both agriculture and the prevention of human disease.

The bacterium *Bacillus thuringiensis* (bä-sil′lus thür-in-jē-en′sis) has been used extensively in the United States to control such pests as alfalfa caterpillars, bollworms, corn borers, cabbageworms, tobacco budworms, and fruit tree leaf rollers. It is incorporated into a dusting powder that is applied to the crops these insects eat. The bacteria produce protein crystals that are toxic to the digestive systems of the insects. The toxin gene has been inserted into some plants to make them insect resistant.

By using microbial rather than chemical insect control, farmers can avoid harming the environment. Many chemical insecticides, such as DDT, remain in the soil as toxic pollutants and are eventually incorporated into the food chain.

MODERN BIOTECHNOLOGY AND RECOMBINANT DNA TECHNOLOGY

LEARNING OBJECTIVE

- List two examples of biotechnology that use recombinant DNA technology and two examples that do not.

Earlier, we touched on the commercial use of microorganisms to produce some common foods and chemicals. Such practical applications of microbiology are called **biotechnology.** Although biotechnology has been used in some form for centuries, techniques have become much more sophisticated in the past few decades. In the last several years, biotechnology has undergone a revolution through the advent of recombinant DNA technology to expand the potential of bacteria, viruses, and yeast cells and other fungi as miniature biochemical factories. Cultured plant and animal cells, as well as intact plants and animals, are also used as recombinant cells and organisms.

The applications of recombinant DNA technology are increasing with each passing year. Recombinant DNA techniques have been used thus far to produce a number of natural proteins, vaccines, and enzymes. Such substances have great potential for medical use; some of them are described in Table 9.1 on page 268.

A very exciting and important outcome of recombinant DNA techniques is **gene therapy**—inserting a missing gene or replacing a defective one in human cells. This technique uses a harmless virus to carry the missing or new gene into certain host cells, where the gene is picked up and inserted into the appropriate chromosome. Since 1990, gene therapy has been used to treat patients with adenosine deaminase (ADA) deficiency, a cause of severe combined immunodeficiency disease (SCID), in which cells of the immune system are inactive or missing; Duchenne's muscular dystrophy, a muscle-destroying disease; cystic fibrosis, a disease of the secreting portions of the respiratory passages, pancreas,

salivary glands, and sweat glands; and LDL-receptor deficiency, a condition in which low-density lipoprotein (LDL) receptors are defective and LDL cannot enter cells. The LDL remains in the blood in high concentrations and increases the risk of atherosclerosis and coronary artery disease because it leads to fatty plaque formation in blood vessels. Results are still being evaluated. Certain genetic diseases may also be treatable by gene therapy in the future, including hemophilia, an inability of the blood to clot normally; diabetes, elevated blood sugar levels; sickle cell disease, an abnormal kind of hemoglobin; and one type of hypercholesterolemia, high blood cholesterol.

Beyond medical applications, recombinant DNA techniques have also been applied to agriculture. For example, genetically altered strains of bacteria have been developed to protect fruit against frost damage, and bacteria are being modified to control insects that damage crops. Bacteria have also been used to improve the appearance, flavor, and shelf life of fruits and vegetables. Potential agricultural uses of recombinant DNA include drought resistance, resistance to insects and microbial diseases, and increased temperature tolerance in crops.

MICROBES AND HUMAN DISEASE

LEARNING OBJECTIVES

- Define *normal microbiota* and *resistance.*
- Define and describe several infectious diseases.
- Describe emerging infectious diseases.

NORMAL MICROBIOTA

We all live from birth until death in a world filled with microbes, and we all have a variety of microorganisms on and inside our bodies. These microorganisms make up our **normal microbiota,** or *flora** (Figure 1.7). The normal microbiota not only do us no harm, but also in some cases can actually benefit us. For example, some normal microbiota protect us against disease by preventing the overgrowth of harmful microbes, while others produce useful substances such as vitamin K and some B vitamins. Unfortunately, under some circumstances normal microbiota can make us sick or infect people we contact. For instance, when some normal microbiota leave their habitat, they can cause disease.

---

*At one time, bacteria and fungi were thought to be plants and thus the term *flora* was used.

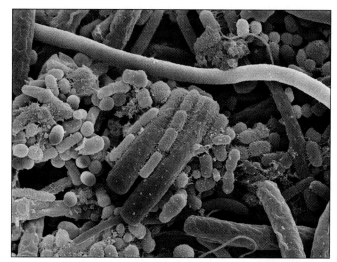

SEM ⊢————⊣
2 μm

FIGURE 1.7 Several types of bacteria found as part of the normal microbiota on the surface of the human tongue.

Q **How are normal microbiota beneficial?**

When is a microbe a welcome part of a healthy human, and when is it a harbinger of disease? The distinction between health and disease is in large part a balance between the natural defenses of the body and the disease-producing properties of microorganisms. Whether our bodies overcome the offensive tactics of a particular microbe depends on our **resistance**—the ability to ward off diseases. Important resistance is provided by the barrier of the skin, mucous membranes, cilia, stomach acid, and antimicrobial chemicals such as interferons. Microbes can be destroyed by white blood cells, the inflammatory response, fever, and by specific responses of our immune system. Sometimes, when our natural defenses are not strong enough to overcome an invader, they have to be supplemented by antibiotics or other drugs.

INFECTIOUS DISEASES

An **infectious disease** is one in which pathogens invade a susceptible host, such as a human or an animal. In the process, the pathogen carries out at least part of its life cycle inside the host, and disease frequently results. By the end of World War II, many people believed that infectious diseases were under control. They thought malaria would be eradicated through the use of the insecticide DDT to kill mosquitoes, that a vaccine would prevent diphtheria, and that improved sanitation measures would help prevent cholera transmission. Malaria is far from eliminated. Since 1986, local outbreaks have been identified in New Jersey, California, Florida, New York, and Texas, and the disease infects 300 million people world-

wide. In 1994, diphtheria appeared in the United States, brought by travelers from the newly independent states of the former Soviet Union, which were experiencing a massive diphtheria epidemic. The epidemic was brought under control in 1998. Cholera outbreaks still occur in less-developed parts of the world.

EMERGING INFECTIOUS DISEASES

These recent outbreaks point to the fact that infectious diseases not only are not disappearing, but seem to be reemerging and increasing. In addition, a number of new diseases—**emerging infectious diseases (EIDs)**—have cropped up in recent years. These are diseases that are new or changing and are increasing or have the potential to increase in incidence in the near future. Some of the factors that have contributed to the emergence of EIDs are evolutionary changes in existing organisms; the spread of known diseases to new geographic regions or populations by modern transportation; and increased human exposure to new, unusual infectious agents in areas that are undergoing ecologic changes such as deforestation and construction. An increasing number of incidents in recent years highlights the extent of the problem.

Avian influenza A (H5N1) caught the attention of the public in 2003 when it killed millions of poultry and 24 people in eight countries in southeast Asia. Avian influenza viruses occur in birds worldwide. Certain wild birds, particularly waterfowl, do not get sick but carry the virus in their intestines and shed it in saliva, nasal secretions, and feces. Most often, the wild birds spread influenza to domesticated birds, in which the virus causes death.

Influenza A viruses are found in many different animals, including ducks, chickens, pigs, whales, horses, and seals. Normally, each subtype of influenza A virus is specific to certain species. However, influenza A viruses normally seen in one species sometimes can cross over and cause illness in another species, and all subtypes of influenza A virus can infect birds. While it is unusual for people to get influenza infections directly from animals, sporadic human infections and outbreaks caused by certain avian influenza A viruses and pig influenza viruses have been reported.

Human infections with avian influenza viruses detected since 1997 have not resulted in sustained human-to-human transmission. However, because influenza viruses have the potential to change and gain the ability to spread easily between people, monitoring for human infection and person-to-person transmission is important (see the box in Chapter 13 on page 406).

Severe acute respiratory syndrome (SARS) first appeared in southern China in late 2002 and has subsequently spread worldwide in 2003. It is a respiratory illness caused by a new variety of coronavirus. (Coronaviruses are associated with the common cold and other upper respiratory tract infections.) Symptoms of SARS include fever, malaise, muscle aches, nonproductive (dry) cough, difficulty in breathing, chills, headache, and diarrhea. The disease is primarily spread through person-to-person contact. There is no effective treatment for SARS, and the death rate is 5–10%, usually among the elderly and in persons with other medical problems. The most recent cases were laboratory acquired in April, 2004.

West Nile encephalitis (WNE) is a disease caused by West Nile virus, which can produce encephalitis (inflammation of the brain). WNE was first diagnosed in the West Nile region of Uganda in 1937. In 1999 the virus made its first North American appearance in humans in New York City. In 2004, West Nile virus infected over 2000 people in 47 states. West Nile virus is now established in nonmigratory birds in 47 states. The virus, which is carried by birds, is transmitted between birds, and to horses and humans, by mosquitoes. West Nile virus may have arrived in the United States in an infected traveler or in migratory birds.

In 1996, countries worldwide were refusing to import beef from the United Kingdom, where hundreds of thousands of cattle born after 1988 had to be killed because of an epidemic of **bovine spongiform encephalopathy** (en-sef-a-lop′a-thē), also called **BSE** or **mad cow disease.** BSE first came to the attention of microbiologists in 1986 as one of a handful of diseases caused by an infectious protein called a *prion.* Studies suggest that the source of disease was cattle feed prepared from sheep infected with their own version of the disease. Cattle are herbivores (plant-eaters), but their growth and health are improved by adding protein to their feed. **Creutzfeldt-Jakob disease** (kroits′felt yä′kôb), or **CJD,** is a human disease also caused by a prion. The incidence of CJD in the United Kingdom is similar to the incidence in other countries. However, by 2005 the United Kingdom reported 154 human cases of CJD caused by a new variant related to the bovine disease (see Chapter 22).

Escherichia coli is a normal inhabitant of the large intestine of vertebrates, including humans, and its presence is beneficial because it helps produce certain vitamins and breaks down otherwise undigestible foodstuffs (see Chapter 25). However, a strain called *E. coli* O157:H7 causes bloody **diarrhea** when it grows in the intestines. This strain was first recognized in 1982 and since then has emerged as a public health problem. It is now one of the leading causes of diarrhea worldwide. In 1996, some 9000 people in Japan became ill, and 7 died, as a result of infec-

tion by *E. coli* O157:H7. The recent outbreaks of *E. coli* O157:H7 in the United States, associated with contamination of undercooked meat and unpasteurized beverages, have made public health officials aware that new methods of testing for bacteria in food must be developed.

In 1995, infections of so-called **flesh-eating bacteria** were reported on the front pages of major newspapers. The bacteria are more correctly named invasive group A *Streptococcus* (strep-tō-kok′kus), or IGAS. There has been a trend toward increasing rates of IGAS in the United States, Scandinavia, England, and Wales.

In 1995, a hospital laboratory technician in Congo (Zaire) who had fever and bloody diarrhea underwent surgery for a suspected perforated bowel. Subsequent to surgery, he started hemorrhaging, and his blood began clotting in his blood vessels. A few days later, health care workers in the hospital where he was staying developed similar symptoms. One of them was transferred to a hospital in a different city; personnel in the second hospital who cared for this patient also developed symptoms. By the time the epidemic was over, 315 people had contracted **Ebola hemorrhagic fever** (hem-o-raj′ik), or **EHF,** and over 75% of them died. The epidemic was controlled when microbiologists instituted training on the use of protective equipment and educational measures in the community. Human-to-human transmission occurs when there is close personal contact with infectious blood or other body fluids or tissue (see Chapter 23).

Microbiologists first isolated Ebola viruses from humans during earlier outbreaks in Congo in 1976. (The virus is named after Congo's Ebola River.) In 1994, a single case of infection from a newly described Ebola virus occurred in Ivory Coast. In 1989 and 1996, outbreaks among monkeys imported into the United States from the Philippines were caused by another Ebola virus but were not associated with human disease.

Recorded cases of **Marburg virus,** another hemorrhagic fever virus, are rare. The first cases were laboratory workers in Europe who handled African green monkeys from Uganda. Four outbreaks were identified in Africa between 1975 and 1998, involving 2 to 123 people. In 2004, an outbreak killed 117 people. Microbiologists have been studying many animals but have not yet discovered the natural reservoir (source) of EHF and Marburg viruses.

In 1993, an outbreak of **cryptosporidiosis** (krip-tō-spô-rid-ē-ō′sis) transmitted through the public water supply in Milwaukee, Wisconsin, resulted in diarrheal illness in an estimated 403,000 persons. The microorganism responsible for this outbreak was the protozoan *Cryptosporidium* (krip-tō-spô-ri′dē-um). First reported as a cause of human disease in 1976, it is responsible for up to

30% of the diarrheal illness in developing countries. In the United States, transmission has occurred via drinking water, swimming pools, and contaminated hospital supplies.

AIDS (acquired immunodeficiency syndrome) first came to public attention in 1981 with reports from Los Angeles that a few young homosexual men had died of a previously rare type of pneumonia known as *Pneumocystis* (nü-mō-sis′tis) pneumonia. These men had experienced a severe weakening of the immune system, which normally fights infectious diseases. Soon these cases were correlated with an unusual number of occurrences of a rare form of cancer, Kaposi's sarcoma, among young homosexual men. Similar increases in such rare diseases were found among hemophiliacs and intravenous drug users.

By the end of 2004, over one million people in the United States had been diagnosed as having AIDS, and over 50% of them had died as a result of the disease. A great many more people had tested positive for the presence of the AIDS virus in their blood. As of 2004, health officials estimated that 1.2 million Americans have HIV infection. In 2004, the World Health Organization (WHO) estimated that over 44 million people worldwide are living with HIV/AIDS and that 14,000 new infections occur every day.

Researchers quickly discovered that the cause of AIDS was a previously unknown virus (see Figure 1.1e). The virus, now called **human immunodeficiency virus (HIV),** destroys certain white blood cells of the immune system called CD4 lymphocytes, one of the cell types of the body's defense. Sickness and death result from microorganisms or cancerous cells that might otherwise have been defeated by the body's natural defenses. So far, the disease has been inevitably fatal once symptoms develop.

By studying disease patterns, medical researchers found that HIV could be spread through sexual intercourse, by contaminated needles, from infected mothers to their newborns via breast milk, and by blood transfusions—in short, by the transmission of body fluids from one person to another. Since 1985, blood used for transfusions has been carefully checked for the presence of HIV, and it is now quite unlikely that the virus can be spread by this means.

Since 1994, new treatments have extended the life span of people with AIDS; however, approximately 40,000 new cases occur annually in the United States. The majority of individuals with AIDS are in the sexually active age group; and, because heterosexual partners of AIDS sufferers are at high risk of infection, public health officials are concerned that even more women and minorities will contract AIDS. In 1997, HIV diagnoses began increasing among women and minorities. Among the AIDS cases reported in 2003, 30% were females and 75% were African American.

In the months and years to come, microbiological techniques will continue to be applied to help scientists learn more about the structure of the deadly HIV, how it is transmitted, how it grows in cells and causes disease, how drugs can be directed against it, and whether an effective vaccine can be developed. Public health officials have also focused on prevention through education.

AIDS poses one of this century's most formidable health threats, but it is not the first serious epidemic of a sexually transmitted disease. Syphilis was also once a fatal epidemic disease. As recently as 1941, syphilis caused an estimated 14,000 deaths per year in the United States. With few drugs available for treatment and no vaccines to prevent it, efforts to control the disease focused mainly on altering sexual behavior and on the use of condoms. The eventual development of drugs to treat syphilis contributed significantly to preventing the spread of the disease. According to the Centers for Disease Control and Prevention (CDC), reported cases of syphilis dropped from a record high of 575,000 in 1943 to 7352 cases in 2004.

Just as microbiological techniques helped researchers in the fight against syphilis and smallpox, they will help scientists discover the causes of new emerging infectious diseases in the twenty-first century. Undoubtedly there will be new diseases. Ebola virus and *Influenzavirus* are examples of viruses that may be changing their abilities to infect different host species. Emerging infectious diseases will be discussed further in Chapter 14 on page 438.

Infectious diseases may reemerge because of the development of resistance to antibiotics (see the box in Chapter 26) and through the use of microorganisms as weapons. (See the box in Chapter 23.) The breakdown of public health measures for previously controlled infections has resulted in unexpected cases of tuberculosis, whooping cough, and diphtheria (see Chapter 24).

* * *

The diseases we have mentioned are caused by viruses, bacteria, protozoa, and prions—types of microorganisms. This book introduces you to the enormous variety of microscopic organisms. It shows you how microbiologists use specific techniques and procedures to study the microbes that cause such diseases as AIDS and diarrhea—and diseases that have yet to be discovered. You will also learn about the body's responses to microbial infection and the ways certain drugs combat microbial diseases. Finally, you will learn about the many beneficial roles that microbes play in the world around us.

STUDY OUTLINE

MICROBES IN OUR LIVES (p. 2)

1. Living things too small to be seen with the unaided eye are called microorganisms.

2. Microorganisms are important in the maintenance of an ecological balance on Earth.

3. Some microorganisms live in humans and other animals and are needed to maintain good health.

4. Some microorganisms are used to produce foods and chemicals.

5. Some microorganisms cause disease.

NAMING AND CLASSIFYING MICROORGANISMS (pp. 2–6)

NOMENCLATURE (pp. 2–4)

1. In a nomenclature system designed by Carolus Linnaeus (1735), each living organism is assigned two names.

2. The two names consist of a genus and a specific epithet, both of which are underlined or italicized.

TYPES OF MICROORGANISMS (pp. 4–6)

Bacteria (p. 4)

3. Bacteria are unicellular organisms. Because they have no nucleus, the cells are described as prokaryotic.

4. The three major basic shapes of bacteria are bacillus, coccus, and spiral.

5. Most bacteria have a peptidoglycan cell wall; they divide by binary fission, and they may possess flagella.

6. Bacteria can use a wide range of chemical substances for their nutrition.

Archaea (p. 4)

7. Archaea consist of prokaryotic cells; they lack peptidoglycan in their cell walls.

8. Archaea include methanogens, extreme halophiles, and extreme thermophiles.

Fungi (p. 4)

9. Fungi (mushrooms, molds, and yeasts) have eukaryotic cells (with a true nucleus). Most fungi are multicellular.

10. Fungi obtain nutrients by absorbing organic material from their environment.

Protozoa (p. 4)

11. Protozoa are unicellular eukaryotes.

12. Protozoa obtain nourishment by absorption or ingestion through specialized structures.

Algae (p. 4)

13. Algae are unicellular or multicellular eukaryotes that obtain nourishment by photosynthesis.

14. Algae produce oxygen and carbohydrates that are used by other organisms.

Viruses (pp. 4–6)

15. Viruses are noncellular entities that are parasites of cells.

16. Viruses consist of a nucleic acid core (DNA or RNA) surrounded by a protein coat. An envelope may surround the coat.

Multicellular Animal Parasites (p. 6)

17. The principal groups of multicellular animal parasites are flatworms and roundworms, collectively called helminths.

18. The microscopic stages in the life cycle of helminths are identified by traditional microbiological procedures.

CLASSIFICATION OF MICROORGANISMS (p. 6)

19. All organisms are classified into Bacteria, Archaea, and Eukarya. Eukarya include protists, fungi, plants, and animals.

A BRIEF HISTORY OF MICROBIOLOGY (pp. 6–17)

THE FIRST OBSERVATIONS (p. 6)

1. Robert Hooke observed that cork was composed of "little boxes"; he introduced the term *cell* (1665).

2. Hooke's observations laid the groundwork for development of the cell theory, the concept that all living things are composed of cells.

3. Anton van Leeuwenhoek, using a simple microscope, was the first to observe microorganisms (1673).

THE DEBATE OVER SPONTANEOUS GENERATION (pp. 7–9)

4. Until the mid-1880s, many people believed in spontaneous generation, the idea that living organisms could arise from nonliving matter.

5. Francesco Redi demonstrated that maggots appear on decaying meat only when flies are able to lay eggs on the meat (1668).

6. John Needham claimed that microorganisms could arise spontaneously from heated nutrient broth (1745).

7. Lazzaro Spallanzani repeated Needham's experiments and suggested that Needham's results were due to microorganisms in the air entering his broth (1765).

8. Rudolf Virchow introduced the concept of biogenesis: living cells can arise only from preexisting cells (1858).

9. Louis Pasteur demonstrated that microorganisms are in the air everywhere and offered proof of biogenesis (1861).

10. Pasteur's discoveries led to the development of aseptic techniques used in laboratory and medical procedures to prevent contamination by microorganisms.

THE GOLDEN AGE OF MICROBIOLOGY (pp. 9–11)

11. Rapid advances in the science of microbiology were made between 1857 and 1914.

Fermentation and Pasteurization (p. 9)

12. Pasteur found that yeast ferment sugars to alcohol and that bacteria can oxidize the alcohol to acetic acid.

13. A heating process called pasteurization is used to kill bacteria in some alcoholic beverages and milk.

The Germ Theory of Disease (pp. 9–11)

14. Agostino Bassi (1835) and Pasteur (1865) showed a causal relationship between microorganisms and disease.

15. Joseph Lister introduced the use of a disinfectant to clean surgical wounds in order to control infections in humans (1860s).

16. Robert Koch proved that microorganisms cause disease. He used a sequence of procedures, now called Koch's postulates (1876), that are used today to prove that a particular microorganism causes a particular disease.

Vaccination (p. 11)

17. In a vaccination, immunity (resistance to a particular disease) is conferred by inoculation with a vaccine.

18. In 1798, Edward Jenner demonstrated that inoculation with cowpox material provides humans with immunity to smallpox.

19. About 1880, Pasteur discovered that avirulent bacteria could be used as a vaccine for fowl cholera; he coined the word *vaccine*.

20. Modern vaccines are prepared from living avirulent microorganisms or killed pathogens, from isolated components of pathogens, and by recombinant DNA techniques.

THE BIRTH OF MODERN CHEMOTHERAPY: DREAMS OF A "MAGIC BULLET" (pp. 12–13)

21. Chemotherapy is the chemical treatment of a disease.

22. Two types of chemotherapeutic agents are synthetic drugs (chemically prepared in the laboratory) and antibiotics (substances produced naturally by bacteria and fungi to inhibit the growth of other microorganisms).

23. Paul Ehrlich introduced an arsenic-containing chemical called salvarsan to treat syphilis (1910).

24. Alexander Fleming observed that the *Penicillium* fungus inhibited the growth of a bacterial culture. He named the active ingredient penicillin (1928).

25. Penicillin has been used clinically as an antibiotic since the 1940s.

26. Researchers are tackling the problem of drug-resistant microbes.

MODERN DEVELOPMENTS IN MICROBIOLOGY (pp. 13–17)

27. Bacteriology is the study of bacteria, mycology is the study of fungi, and parasitology is the study of parasitic protozoa and worms.

28. Microbiologists are using genomics, the study of all of an organism's genes, to classify bacteria, fungi, and protozoa.

29. The study of AIDS, analysis of the action of interferons, and the development of new vaccines are among the current research interests in immunology.

30. New techniques in molecular biology and electron microscopy have provided tools for advancement of our knowledge of virology.

31. The development of recombinant DNA technology has helped advance all areas of microbiology.

MICROBES AND HUMAN WELFARE (pp. 17–18)

1. Microorganisms degrade dead plants and animals and recycle chemical elements to be used by living plants and animals.

2. Bacteria are used to decompose organic matter in sewage.

3. Bioremediation processes use bacteria to clean up toxic wastes.

4. Bacteria that cause diseases in insects are being used as biological controls of insect pests. Biological controls are specific for the pest and do not harm the environment.

5. Using microbes to make products such as foods and chemicals is called biotechnology.

6. Using recombinant DNA, bacteria can produce important substances such as proteins, vaccines, and enzymes.

7. In gene therapy, viruses are used to carry replacements for defective or missing genes into human cells.

8. Genetically modified bacteria are used in agriculture to protect plants from frost and insects and to improve the shelf life of produce.

MICROBES AND HUMAN DISEASE (pp. 18–21)

1. Everyone has microorganisms in and on the body; these make up the normal microbiota, or flora.

2. The disease-producing properties of a species of microbe and the host's resistance are important factors in determining whether a person will contract a disease.

3. An infectious disease is one in which pathogens invade a susceptible host.

4. An emerging infectious disease (EID) is a new or changing disease showing an increase in incidence in the recent past or a potential to increase in the near future.

STUDY QUESTIONS

Access more review material either online at **The Microbiology Place** (www.microbiologyplace.com) or with **The Microbiology Place CD-ROM** packaged with your new book. There you'll find activities, practice tests, quizzes, flashcards, case studies, and more to help you succeed.

Answers to the Study Questions can be found in Appendix G.

REVIEW

1. How did the idea of spontaneous generation come about?

2. Some proponents of spontaneous generation believed that air is necessary for life. They thought that Spallanzani did not really disprove spontaneous generation because he hermetically sealed his flasks to keep air out. How did Pasteur's experiments address the air question without allowing the microbes in the air to ruin his experiment?

3. Briefly state the role played by microorganisms in each of the following:
   a. biological control of pests
   b. recycling of elements
   c. normal microbiota
   d. sewage treatment
   e. human insulin production
   f. vaccine production

4. Into which field of microbiology would the following scientists best fit?

Researcher Who	Field
_a,c_ Studies biodegradation of toxic wastes	(a) Biotechnology
_h_ Studies the causative agent of Ebola hemorrhagic fever	(b) Immunology
_a,d,f_ Studies the production of human proteins by bacteria	(c) Microbial ecology
_b_ Studies the symptoms of AIDS	(d) Microbial genetics
_e_ Studies the production of toxin by _E. coli_	(e) Microbial physiology
_c_ Studies the life cycle of _Cryptosporidium_	(f) Molecular biology
_b,d_ Develops gene therapy for a disease	(g) Mycology
_g_ Studies the fungus _Candida albicans_	(h) Virology

5. Match the following microorganisms to their descriptions.

   _g_ Archaea   (a) Not composed of cells
   _d_ Algae   (b) Cell wall made of chitin
   _c_ Bacteria   (c) Cell wall made of peptidoglycan
   _b_ Fungi   (d) Cell wall made of cellulose; photosynthetic
   _f_ Helminths
   _e_ Protozoa   (e) Unicellular, complex cell structure lacking a cell wall
   _a_ Viruses   (f) Multicellular animals
       (g) Prokaryote without peptidoglycan cell wall

6. Match the following people to their contribution toward the advancement of microbiology.

   _k_ Avery, MacLeod, and McCarty   (a) Developed vaccine against smallpox
   _n_ Beadle and Tatum   (b) Discovered how DNA controls protein synthesis in a cell
   _o_ Berg
   _q_ Ehrlich   (c) Discovered penicillin
   _c_ Fleming   (d) Discovered that DNA can be transferred from one bacterium to another
   _i_ Hooke
   _j_ Iwanowski
   _b_ Jacob and Monod   (e) Disproved spontaneous generation
   _a_ Jenner
   _l_ Koch   (f) First to characterize a virus
   _r_ Lancefield   (g) First to use disinfectants in surgical procedures
   _d_ Lederberg and Tatum
   _g_ Lister   (h) First to observe bacteria
   _e_ Pasteur   (i) First to observe cells in plant material and name them
   _f_ Stanley
   _h_ van Leeuwenhoek   (j) Observed that viruses are filterable
   _m_ Virchow
   _p_ Weizmann   (k) Proved that DNA is the hereditary material
       (l) Proved that microorganisms can cause disease
       (m) Said living cells arise from preexisting living cells
       (n) Showed that genes code for enzymes
       (o) Spliced animal DNA to bacterial DNA
       (p) Used bacteria to produce acetone
       (q) Used the first synthetic chemotherapeutic agent
       (r) Proposed a classification system for streptococci based on antigens in their cell walls

7. The genus name of a bacterium is "erwinia" and the specific epithet is "amylovora." Write the scientific name of this organism correctly. Using this name as an example, explain how scientific names are chosen.

8. It is possible to purchase the following microorganisms in a retail store. Provide a reason for buying each.
   a. _Bacillus thuringiensis_
   b. _Saccharomyces_

MULTIPLE CHOICE

1. Which of the following is a scientific name?
   a. _Mycobacterium tuberculosis_
   b. Tubercle bacillus

2. Which of the following is _not_ a characteristic of bacteria?
   a. are prokaryotic

b. have peptidoglycan cell walls
c. have the same shape
d. grow by binary fission
e. have the ability to move

3. Which of the following is the most important element of Koch's germ theory of disease? The animal shows disease symptoms when
   a. the animal has been in contact with a sick animal.
   b. the animal has a lowered resistance.
   c. a microorganism is observed in the animal.
   d. a microorganism is inoculated into the animal.
   e. microorganisms can be cultured from the animal.

4. Recombinant DNA is
   a. DNA in bacteria.
   b. the study of how genes work.
   c. the DNA resulting when genes of two different organisms are mixed.
   d. the use of bacteria in the production of foods.
   e. the production of proteins by genes.

5. Which of the following statements is the best definition of biogenesis?
   a. Nonliving matter gives rise to living organisms.
   b. Living cells can only arise from preexisting cells.
   c. A vital force is necessary for life.
   d. Air is necessary for living organisms.
   e. Microorganisms can be generated from nonliving matter.

6. Which of the following is a beneficial activity of microorganisms?
   a. Some microorganisms are used as food for humans.
   b. Some microorganisms use carbon dioxide.
   c. Some microorganisms provide nitrogen for plant growth.
   d. Some microorganisms are used in sewage treatment processes.
   e. all of the above

7. It has been said that bacteria are essential for the existence of life on Earth. Which of the following would be the essential function performed by bacteria?
   a. control insect populations
   b. directly provide food for humans
   c. decompose organic material and recycle elements
   d. cause disease
   e. produce human growth hormones such as insulin

8. Which of the following is an example of bioremediation?
   a. application of oil-degrading bacteria to an oil spill
   b. application of bacteria to a crop to prevent frost damage
   c. fixation of gaseous nitrogen into usable nitrogen
   d. production by bacteria of a human protein such as interferon
   e. all of the above

9. Spallanzani's conclusion about spontaneous generation was challenged because Lavoisier had just shown that oxygen was the vital component of air. Which of the following statements is true?

a. All life requires air.
b. Only disease-causing organisms require air.
c. Some microbes do not require air.
d. Pasteur kept air out of his biogenesis experiments.
e. Lavoisier was mistaken.

10. Which of the following statements about *E. coli* is *not* true?
    a. *E. coli* was the first disease-causing bacterium identified by Koch.
    b. *E. coli* is part of the normal microbiota of humans.
    c. *E. coli* is beneficial in human intestines.
    d. A disease-causing strain of *E. coli* causes bloody diarrhea.
    e. none of the above

CRITICAL THINKING

1. How did the theory of biogenesis lead the way for the germ theory of disease?

2. Even though the germ theory of disease was not demonstrated until 1876, why did Semmelweis (1840) and Lister (1867) argue for the use of aseptic techniques?

3. Find at least three supermarket products made by microorganisms. (*Hint:* The label will state the scientific name of the organism or include the word *culture, fermented,* or *brewed.*)

4. People believed all microbial diseases would be controlled by the 21st century. Name one emerging infections disease. List three reasons why we are identifying new diseases now.

CLINICAL APPLICATIONS

1. The prevalence of arthritis in the United States is 1 in 100,000 children. However, 1 in 10 children in Lyme, Connecticut, developed arthritis between June and September in 1973. Allen Steere, a rheumatologist at Yale University, investigated the cases in Lyme and found that 25% of the patients remembered having a skin rash during their arthritic episode and that the disease was treatable with penicillin. Steere concluded that this was a new infectious disease and did not have an environmental, genetic, or immunologic cause.
   a. What was the factor that caused Steere to reach his conclusion?
   b. What is the disease?
   c. Why was the disease more prevalent between June and September?

2. In 1864, Lister observed that patients recovered completely from simple fractures, but compound fractures had "disastrous consequences." He knew that the application of phenol (carbolic acid) to fields in the town of Carlisle prevented cattle disease. In 1864, Lister treated compound fractures with phenol, and his patients recovered without complications. How was Lister influenced by Pasteur's work? Why was Koch's work still needed?

Chemical Principles

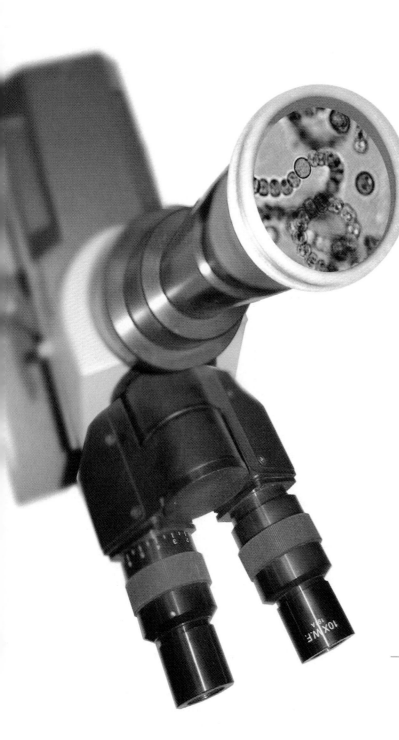

We can see a tree rot and smell milk going sour, but we might not realize what is happening on a microscopic level. In both cases, microbes are conducting chemical operations. The tree rots when microorganisms decompose the wood. Milk turns sour from the production of lactic acid by bacteria. Most of the activities of microorganisms are the result of a series of chemical reactions.

Like all organisms, microorganisms use nutrients to make chemical building blocks for growth and other functions essential to life. For most microorganisms, synthesizing these building blocks requires them to break down nutrient substances and use the energy released to assemble the resulting molecular fragments into new substances.

The chemistry of microbes is one of the most important concerns of microbiologists. Knowledge of chemistry is essential to understanding the roles of microorganisms in nature, how they cause disease, how methods for diagnosing disease are developed, how the body's defenses combat infection, and how antibiotics and vaccines are produced to combat the harmful effects of microbes. To understand the changes that occur in microorganisms and the changes microbes make in the world around us, we need to know how molecules are formed and how they interact.

UNDER THE MICROSCOPE

Cyanobacteria. Light energy drives the oxidation reactions in these photosynthetic bacteria. All organisms use oxidation reactions to get energy.

THE STRUCTURE OF ATOMS

LEARNING OBJECTIVE

• Describe the structure of an atom and its relation to the chemical properties of elements.

All matter—whether air, rock, or a living organism—is made up of small units called **atoms.** Atoms interact with each other in certain combinations to form **molecules.** Living cells are made up of molecules, some of which are very complex. The science of the interaction between atoms and molecules is called **chemistry.**

Atoms are the smallest units of matter that enter into chemical reactions. Every atom has a centrally located **nucleus** and particles called **electrons** that move around the nucleus in patterns known as electronic configurations (Figure 2.1). The nuclei of most atoms are stable—that is, they do not change spontaneously—and nuclei do not participate in chemical reactions. The nucleus is made up of positively (+) charged particles called **protons** and uncharged (neutral) particles called **neutrons.** The nucleus, therefore, bears a net positive charge. Neutrons and protons have approximately the same weight, which is about 1840 times that of an electron. The charge on electrons is negative (−), and in all atoms the number of electrons is equal to the number of protons. Because the total positive charge of the nucleus equals the total negative charge of the electrons, each atom is electrically neutral.

The number of protons in an atomic nucleus ranges from one (in a hydrogen atom) to more than 100 (in the largest atoms known). Atoms are often listed by their **atomic number,** the number of protons in the nucleus. The total number of protons and neutrons in an atom is its approximate **atomic weight.**

TABLE 2.1	The Elements of Life*		
Element	Symbol	Atomic Number	Approximate Atomic Weight
Hydrogen	H	1	1
Carbon	C	6	12
Nitrogen	N	7	14
Oxygen	O	8	16
Sodium	Na	11	23
Magnesium	Mg	12	24
Phosphorus	P	15	31
Sulfur	S	16	32
Chlorine	Cl	17	35
Potassium	K	19	39
Calcium	Ca	20	40
Iron	Fe	26	56
Iodine	I	53	127

*Hydrogen, carbon, nitrogen, and oxygen are the most abundant chemical elements in living organisms.

CHEMICAL ELEMENTS

All atoms with the same number of protons behave the same way chemically and are classified as the same **chemical element.** Each element has its own name and a one- or two-letter symbol, usually derived from the English or Latin name for the element. For example, the symbol for the element hydrogen is H, and the symbol for carbon is C. The symbol for sodium is Na—the first two letters of its Latin name, *natrium*—to distinguish it from nitrogen, N, and from sulfur, S. There are 92 naturally occurring elements. However, only about 26 elements are commonly found in living things. Table 2.1 lists some of the chemical elements found in living organisms, including their atomic numbers and weights. The elements most abundant in living matter are hydrogen, carbon, nitrogen, and oxygen.

Most elements have several **isotopes**—atoms with different numbers of neutrons in their nuclei. All isotopes of an element have the same number of protons in their nuclei, but their atomic weights differ because of the difference in the number of neutrons. For example, in a natural sample of oxygen, all the atoms will contain eight protons. However, 99.76% of the atoms will have eight neutrons, 0.04% will contain nine neutrons, and the remaining 0.2% will contain ten neutrons. Therefore, the three isotopes composing a natural sample of oxygen will

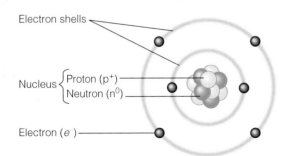

FIGURE 2.1 The structure of an atom. In this simplified diagram of a carbon atom, note the central location of the nucleus. The nucleus contains six neutrons and six protons, although not all are visible in this view. The six electrons move about the nucleus in regions called electron shells, shown here as circles.

Q **What is the atomic number of this atom?**

have atomic weights of 16, 17, and 18, although all will have the atomic number 8. Atomic numbers are written as a subscript to the left of an element's chemical symbol. Atomic weights are written as a superscript above the atomic number. Thus, natural oxygen isotopes are represented as $^{16}_{8}O$, $^{17}_{8}O$, and $^{18}_{8}O$. Isotopes of certain elements are extremely useful in biological research, medical diagnosis, the treatment of some disorders, and in some forms of sterilization.

ELECTRONIC CONFIGURATIONS

In an atom, electrons are arranged in **electron shells,** which are regions corresponding to different **energy levels.** The arrangement is called an **electronic configuration.** Shells are layered outward from the nucleus, and each shell can hold a characteristic maximum number of electrons—two electrons in the innermost shell (lowest energy level), eight electrons in the second shell, and eight electrons in the third shell, if it is the atom's outermost (valence) shell. The fourth, fifth, and sixth electron shells can each accommodate 18 electrons, although there are some exceptions to this generalization. Table 2.2 shows the electronic configurations for atoms of some elements found in living organisms.

There is a tendency for the outermost shell to be filled with the maximum number of electrons. An atom can give up, accept, or share electrons with other atoms to fill this shell. The chemical properties of atoms are largely a function of the number of electrons in the outermost electron shell. When its outer shell is filled, the atom is chemically stable, or inert: it does not tend to react with other atoms. Helium (atomic number 2) and neon (atomic number 10) are examples of atoms of inert gases that have filled outer shells.

When an atom's outer electron shell is only partially filled, the atom is chemically unstable. Such an atom reacts with other atoms, and this reaction depends, in part, on the degree to which the outer energy levels are filled. Notice the number of electrons in the outer energy levels of the atoms in Table 2.2. We will see later how the number correlates with the chemical reactivity of the elements.

HOW ATOMS FORM MOLECULES: CHEMICAL BONDS

LEARNING OBJECTIVE
- Define *ionic bond, covalent bond, hydrogen bond, molecular weight,* and *mole.*

When the outermost energy level of an atom is not completely filled by electrons, you can think of it as having either unfilled spaces or extra electrons in that energy level, depending on whether it is easier for the atom to gain or lose electrons. For example, an atom of oxygen, with two electrons in the first energy level and six in the second, has two unfilled spaces in the second electron shell; an atom of magnesium has two extra electrons in its outermost shell. The most chemically stable configuration for any atom is to have its outermost shell filled, as do the inert gases. Therefore, for these two atoms to attain that state, oxygen must gain two electrons, and magnesium must lose two electrons. All atoms tend to combine so that the extra electrons in the outermost shell of one atom fill the spaces of the outermost shell of the other atom; for example, oxygen and magnesium combine so that the outermost shell of each atom has the full complement of eight electrons.

The **valence,** or combining capacity, of an atom is the number of extra or missing electrons in its outermost electron shell. For example, hydrogen has a valence of 1 (one unfilled space, or one extra electron), oxygen has a valence of 2 (two unfilled spaces), carbon has a valence of 4 (four unfilled spaces, or four extra electrons), and magnesium has a valence of 2 (two extra electrons).

Basically, atoms achieve the full complement of electrons in their outermost energy shells by combining to form molecules, which are made up of atoms of one or more elements. A molecule that contains at least two different kinds of atoms, such as H_2O (the water molecule), is called a **compound.** In H_2O, the subscript 2 indicates that there are two atoms of hydrogen; the absence of a subscript indicates that there is only one atom of oxygen. Molecules hold together because the valence electrons of the combining atoms form attractive forces, called **chemical bonds,** between the atomic nuclei. Therefore, valence may also be viewed as the bonding capacity of an element. Because energy is required for chemical bond formation, each chemical bond possesses a certain amount of potential chemical energy.

In general, atoms form bonds in one of two ways: by either gaining or losing electrons from their outer electron shell, or by sharing outer electrons. When atoms have gained or lost outer electrons, the chemical bond is called an ionic bond. When outer electrons are shared, the bond is called a covalent bond. Although we will discuss ionic and covalent bonds separately, the kinds of bonds actually found in molecules do not belong entirely to either category. Instead, bonds range from the highly ionic to the highly covalent.

IONIC BONDS

Atoms are electrically neutral when the number of positive charges (protons) equals the number of negative charges (electrons). But when an isolated atom gains or loses electrons, this balance is upset. If the atom gains

TABLE 2.2	Electronic Configurations for the Atoms of Some Elements Found in Living Organisms						

Element	First Electron Shell	Second Electron Shell	Third Electron Shell	Diagram	Number of Valence (Outermost) Shell Electrons	Number of Unfilled Spaces	Maximum Number of Bonds Formed
Hydrogen	1	—	—		1	1	1
Carbon	2	4	—		4	4	4
Nitrogen	2	5	—		5	3	3
Oxygen	2	6	—		6	2	2
Magnesium	2	8	2		2	6	2
Phosphorus	2	8	5		5	3	5
Sulfur	2	8	6		6	2	2

FIGURE 2.2 Ionic bond formation. (**a**) A sodium atom (Na), left, loses one electron to an electron acceptor and forms a sodium ion (Na⁺). A chlorine atom (Cl), right, accepts one electron from an electron donor to become a chloride ion (Cl⁻). (**b**) The sodium and chloride ions are attracted because of their opposite charges and are held together by an ionic bond to form a molecule of sodium chloride.

Q **What is an ionic bond?**

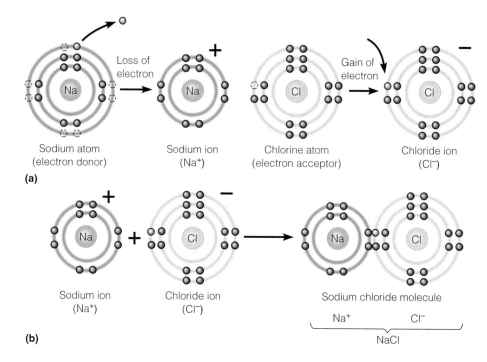

electrons, it acquires an overall negative charge; if the atom loses electrons, it acquires an overall positive charge. Such a negatively or positively charged atom (or group of atoms) is called an **ion.**

Consider the following examples. Sodium (Na) has 11 protons and 11 electrons, with one electron in its outer electron shell. Sodium tends to lose the single outer electron; it is an *electron donor* (Figure 2.2a). When sodium donates an electron to another atom, it is left with 11 protons and only 10 electrons and so has an overall charge of +1. This positively charged sodium atom is called a sodium ion and is written as Na⁺. Chlorine (Cl) has a total of 17 electrons, seven of them in the outer electron shell. Because this outer shell can hold eight electrons, chlorine tends to pick up an electron that has been lost by another atom; it is an *electron acceptor* (see Figure 2.2a). By accepting an electron, chlorine totals 18 electrons. However, it still has only 17 protons in its nucleus. The chloride ion therefore has a charge of −1 and is written as Cl⁻.

The opposite charges of the sodium ion (Na⁺) and chloride ion (Cl⁻) attract each other. The attraction, an ionic bond, holds the two atoms together, and a molecule is formed (Figure 2.2b). The formation of this molecule, called sodium chloride (NaCl) or table salt, is a common example of ionic bonding. Thus, an **ionic bond** is an attraction between ions of opposite charge that holds them together to form a stable molecule. Put another way, an ionic bond is an attraction between atoms in which one atom loses electrons and another atom gains electrons. Strong ionic bonds, such as those that hold Na⁺ and Cl⁻ together in salt crystals, have limited importance in living cells. But the weaker ionic bonds formed in aqueous

(water) solutions are important in biochemical reactions in microbes and other organisms. For example, weaker ionic bonds assume a role in certain antigen–antibody reactions—that is, reactions in which molecules produced by the immune system (antibodies) combine with foreign substances (antigens) to combat infection.

In general, an atom whose outer electron shell is less than half-filled will lose electrons and form positively charged ions, called **cations.** Examples of cations are the potassium ion (K⁺), calcium ion (Ca²⁺), and sodium ion (Na⁺). When an atom's outer electron shell is more than half-filled, the atom will gain electrons and form negatively charged ions, called **anions.** Examples are the iodide ion (I⁻), chloride ion (Cl⁻), and sulfide ion (S²⁻).

COVALENT BONDS

A **covalent bond** is a chemical bond formed by two atoms sharing one or more pairs of electrons. Covalent bonds are stronger and far more common in organisms than are true ionic bonds. In the hydrogen molecule, H₂, two hydrogen atoms share a pair of electrons. Each hydrogen atom has its own electron plus one electron from the other atom (Figure 2.3a). The shared pair of electrons actually orbits the nuclei of both atoms. Therefore, the outer electron shells of both atoms are filled. When only one pair of electrons is shared between atoms, a *single covalent bond* is formed. For simplicity, a single covalent bond is expressed as a single line between the atoms (H—H). When two pairs of electrons are shared between atoms, a *double covalent bond* is formed, expressed as two single lines (=). A *triple covalent bond*, expressed as three single lines (≡), occurs when three pairs of electrons are shared.

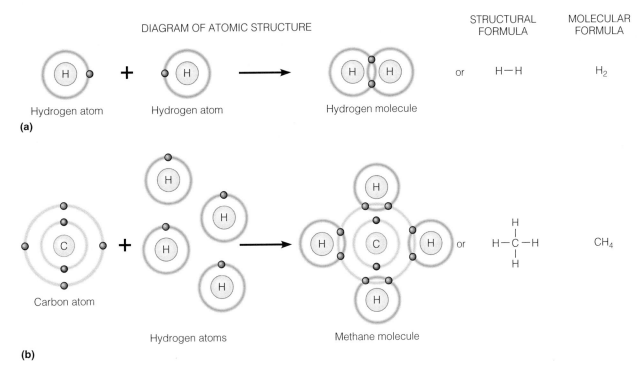

| DIAGRAM OF ATOMIC STRUCTURE | STRUCTURAL FORMULA | MOLECULAR FORMULA |

Hydrogen atom   Hydrogen atom   Hydrogen molecule   or   H—H H_2

(a)

Carbon atom   Hydrogen atoms   Methane molecule   or   H—C—H CH_4

(b)

FIGURE 2.3 Covalent bond formation. (a) A single covalent bond between two hydrogen atoms. **(b)** Single covalent bonds between four hydrogen atoms and a carbon atom, forming a methane molecule. On the right are simpler ways to represent molecules. In structural formulas, each covalent bond is written as a straight line between the symbols for two atoms. In molecular formulas, the number of atoms in each molecule is noted by subscripts.

Q **What is a covalent bond?**

The principles of covalent bonding that apply to atoms of the same element also apply to atoms of different elements. Methane (CH_4) is an example of covalent bonding between atoms of different elements (Figure 2.3b). The outer electron shell of the carbon atom can hold eight electrons but has only four; each hydrogen atom can hold two electrons but has only one. Consequently, in the methane molecule the carbon atom gains four hydrogen electrons to complete its outer shell, and each hydrogen atom completes its pair by sharing one electron from the carbon atom. Each outer electron of the carbon atom orbits both the carbon nucleus and a hydrogen nucleus. Each hydrogen electron orbits both its own nucleus and the carbon nucleus.

Elements such as hydrogen and carbon, whose outer electron shells are half-filled, form covalent bonds quite easily. In fact, in living organisms, carbon almost always forms covalent bonds; it almost never becomes an ion. *Remember:* Covalent bonds are formed by the *sharing* of electrons between atoms. Ionic bonds are formed by *attraction* between atoms that have lost or gained electrons and are therefore positively or negatively charged.

HYDROGEN BONDS

Another chemical bond of special importance to all organisms is the **hydrogen bond,** in which a hydrogen atom that is covalently bonded to one oxygen or nitrogen atom is attracted to another oxygen or nitrogen atom. Such bonds are weak and do not bind atoms into molecules. However, they do serve as bridges between different molecules or between various portions of the same molecule.

When hydrogen combines with atoms of oxygen or nitrogen, the relatively large nucleus of these larger atoms attracts the hydrogen electron more strongly than does the small hydrogen nucleus. Thus, in a molecule of water (H_2O), all the electrons tend to be closer to the oxygen nucleus than to the hydrogen nuclei. The oxygen portion of the molecule thus has a slightly negative charge, and the hydrogen portion of the molecule has a slightly positive charge (Figure 2.4a). When the positively charged end of one molecule is attracted to the negatively charged end of another molecule, a hydrogen bond is formed (Figure 2.4b). This attraction can also occur between hydrogen and other atoms of the same molecule, especially in large molecules. Oxygen and nitrogen are the elements most frequently involved in hydrogen bonding.

Hydrogen bonds are considerably weaker than either ionic or covalent bonds; they have only about 5% of the strength of covalent bonds. Consequently, hydrogen bonds are formed and broken relatively easily. This property accounts for the temporary bonding that occurs between certain atoms of large and complex molecules, such as proteins and nucleic acids. Even though hydrogen bonds are

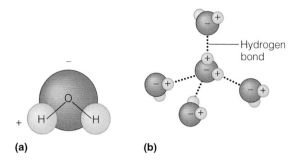

(a)   **(b)**

FIGURE 2.4 Hydrogen bond formation in water.
(**a**) In a water molecule, the electrons of the hydrogen atoms are strongly attracted to the oxygen atom. Therefore, the part of the water molecule containing the oxygen atom has a slightly negative charge, and the part containing hydrogen atoms has a slightly positive charge. (**b**) In a hydrogen bond between water molecules, the hydrogen of one water molecule is attracted to the oxygen of another water molecule. Many water molecules may be attracted to each other by hydrogen bonds (black dots).

Q **Which chemical elements are usually involved in hydrogen bonding?**

relatively weak, large molecules containing several hundred of these bonds have considerable strength and stability.

MOLECULAR WEIGHT AND MOLES

You have seen that bond formation results in the creation of molecules. Molecules are often discussed in terms of units of measure called molecular weight and moles. The **molecular weight** of a molecule is the sum of the atomic weights of all its atoms. To relate the molecular level to the laboratory level, we use a unit called the mole. One **mole** of a substance is its molecular weight expressed in grams. For example, 1 mole of water weighs 18 grams because the molecular weight of H_2O is 18, or $[(2 \times 1) + 16]$.

CHEMICAL REACTIONS

LEARNING OBJECTIVES
• Diagram three basic types of chemical reactions.

As we said earlier, **chemical reactions** involve the making or breaking of bonds between atoms. After a chemical reaction, the total number of atoms remains the same, but there are new molecules with new properties because the atoms have been rearranged.

ENERGY IN CHEMICAL REACTIONS

Some change of energy occurs whenever bonds between atoms are formed or broken during chemical reactions. This energy is called **chemical energy.** When a chemical bond is formed, energy is required. Such a chemical reac-

tion that absorbs more energy than it releases is called an **endergonic reaction** (*endo* = within), meaning that energy is directed inward. When a bond is broken, energy is released. A chemical reaction that releases more energy than it absorbs is called an **exergonic reaction** (*exo* = out), meaning that energy is directed outward.

In this section we will look at three basic types of chemical reactions common to all living cells. By becoming familiar with these reactions, you will be able to understand the specific chemical reactions we will discuss later, particularly in Chapter 5.

SYNTHESIS REACTIONS

When two or more atoms, ions, or molecules combine to form new and larger molecules, the reaction is called a **synthesis reaction.** To synthesize means to put together, and a synthesis reaction *forms new bonds.* Synthesis reactions can be expressed in the following way:

$$\begin{array}{ccc} & & \text{Combine} \\ & & \text{to form} \\ A & + & B & \longrightarrow & AB \end{array}$$

Atom, ion,   Atom, ion,   New molecule
or molecule A   or molecule B   AB

The combining substances, A and B, are called the *reactants;* the substance formed by the combination, AB, is the *product.* The arrow indicates the direction in which the reaction proceeds.

Pathways of synthesis reactions in living organisms are collectively called anabolic reactions, or simply **anabolism** (an-ab'ō-lizm). The combining of sugar molecules to form starch and of amino acids to form proteins are two examples of anabolism.

DECOMPOSITION REACTIONS

The reverse of a synthesis reaction is a **decomposition reaction.** To decompose means to break down into smaller parts, and in a decomposition reaction *bonds are broken.* Typically, decomposition reactions split large molecules into smaller molecules, ions, or atoms. A decomposition reaction occurs in the following way:

$$\begin{array}{ccc} & \text{Breaks} \\ & \text{down into} \\ AB & \longrightarrow & A & + & B \end{array}$$

Molecule AB   Atom, ion,   Atom, ion,
                        or molecule A   or molecule B

Decomposition reactions that occur in living organisms are collectively called catabolic reactions, or simply **catabolism** (ka-tab'ō-lizm). An example of catabolism is the breakdown of sucrose (table sugar) into simpler sugars, glucose and fructose, during digestion. Bacterial decomposition of petroleum is discussed in the box on the facing page.

APPLICATIONS OF MICROBIOLOGY

BIOREMEDIATION—BACTERIA CLEAN UP POLLUTION

Readers of science fiction have long realized that beings from another planet might have quite a different chemical makeup from Earthlings and might be able to eat, drink, and breathe the substances we find deadly. Such aliens could be invaluable in helping clean up pollutants such as crude oil, gasoline, and mercury, which harm plants, animals, and humans alike. Fortunately, however, we need not wait for a visit from outer space to find creatures whose unusual chemistry can be harnessed for environmental cleanup. Although many bacteria have dietary requirements similar to ours—that's why they cause food spoilage—others metabolize (or process chemically) the substances we might expect at a banquet of extraterrestrials: heavy metals, sulfur, nitrogen gas, petroleum, and even polychlorinated biphenyls (PCBs) and mercury.

Bacteria have several advantages as pollution fighters. They can extract pollutants that have combined with soil and

water and hence cannot be simply shoveled away. In addition, they may chemically alter a harmful substance so that it becomes harmless or even beneficial. Bacteria that can degrade many pollutants are naturally present in soil and water; using them to degrade pollutants is called *bioremediation*. However, their small numbers make them inefficient in dealing with large-scale contamination. Scientists are now working to improve the efficiency of natural pollution fighters and, in some cases, are altering organisms by recombinant DNA technology to give them exactly the right chemical appetites.

One of the most promising successes for bioremediation occurred on an Alaskan beach following the *Exxon Valdez* oil spill. Several naturally occurring bacteria in the genus *Pseudomonas* are able to degrade oil for their carbon and energy requirements. In the presence of air, they remove two carbons at a time from a large petroleum molecule (see the figure).

The bacteria degrade the oil too slowly to be helpful in cleaning up an oil spill. However, scientists hit upon a very simple way to speed up the process—with no need for recombinant DNA. They simply dumped ordinary nitrogen and phosphorus plant fertilizers (bioenhancers) onto the beach. The number of oil-degrading bacteria increased compared with that on unfertilized control beaches, and oil was quickly cleared from the test beach.

Another group of bacteria is being investigated for its ability to clean up mercury contamination. Mercury is present in such common substances as discarded leftover paint and can leak into soil and water from garbage dumps. One species

of bacteria that is common in the environment, *Desulfovibrio desulfuricans*, actually makes the mercury more dangerous by adding a methyl group, converting it into the highly toxic substance methyl mercury. Methyl mercury in ponds or marshes sticks to small organisms such as plankton, which are eaten by larger organisms, which in turn are eaten by fish. Fish and human poisonings have been attributed to the ingestion of methyl mercury.

However, other bacteria, such as species of *Pseudomonas*, may offer the solution. To avoid mercury poisoning, these bacteria first convert methyl mercury to mercuric ion:

$$CH_3Hg \rightarrow CH_4 + Hg^{2+}$$

Methyl mercury   Methane   Mercuric ion

Many bacteria can then convert the positively charged mercuric ion to the relatively harmless elemental form by adding electrons, which they take from hydrogen atoms:

$$Hg^{2+} + 2H \xrightarrow{2e^-} Hg + H^+$$

Mercuric ion   Hydrogen atoms   Elemental mercury   Hydrogen ions

These bacteria work too slowly in nature for cleaning up human-made toxic spills, but scientists are experimenting with bioenhancers and other techniques to increase their effectiveness. Unlike some forms of environmental cleanup, in which dangerous substances are removed from one place only to be dumped in another, bacterial cleanup eliminates the toxic substance and often returns a harmless or useful substance to the environment.

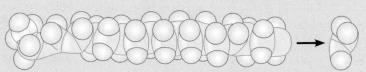

Typical saturated hydrocarbon found in petroleum   Two-carbon unit can be metabolized in cell

EXCHANGE REACTIONS

All chemical reactions are based on synthesis and decomposition. Many reactions, such as **exchange reactions,** are actually part synthesis and part decomposition. An exchange reaction works in the following way:

$$AB + CD \xrightarrow{\text{Recombine to form}} AD + BC$$

First, the bonds between A and B and between C and D are broken in a decomposition process. New bonds are then formed between A and D and between B and C in a synthesis process. For example, an exchange reaction occurs when sodium hydroxide (NaOH) and hydrochloric acid (HCl) react to form table salt (NaCl) and water (H_2O), as follows:

$$NaOH + HCl \longrightarrow NaCl + H_2O$$

THE REVERSIBILITY OF CHEMICAL REACTIONS

All chemical reactions are, in theory, reversible; that is, they can occur in either direction. In practice, however, some reactions do this more easily than others. A chemical reaction that is readily reversible (when the end product can revert to the original molecules) is termed a **reversible reaction** and is indicated by two arrows, as shown here:

$$A + B \underset{\text{Breaks down into}}{\overset{\text{Combines to form}}{\rightleftharpoons}} AB$$

Some reversible reactions occur because neither the reactants nor the end products are very stable. Other reactions will reverse only under special conditions:

$$A + B \underset{\text{Water}}{\overset{\text{Heat}}{\rightleftharpoons}} AB$$

Whatever is written above or below the arrows indicates the special condition under which the reaction in that direction occurs. In this case, A and B react to produce AB only when heat is applied, and AB breaks down into A and B only in the presence of water. See Figure 2.8 as another example.

In Chapter 5 we will examine the various factors that affect chemical reactions.

IMPORTANT BIOLOGICAL MOLECULES

Biologists and chemists divide compounds into two principal classes: inorganic and organic. **Inorganic compounds** are defined as molecules, usually small and structurally simple, that typically lack carbon and in which ionic bonds may play an important role. Inorganic compounds include water, oxygen, carbon dioxide, and many salts, acids, and bases.

Organic compounds always contain carbon and hydrogen and are typically structurally complex. Carbon is a unique element because it has four electrons in its outer shell and four unfilled spaces. It can combine with a variety of atoms, including other carbon atoms, to form straight or branched chains and rings. Carbon chains form the basis of many organic compounds in living cells, including sugars, amino acids, and vitamins. Organic compounds are held together mostly or entirely by covalent bonds. Some organic molecules, such as polysaccharides, proteins, and nucleic acids, are very large and usually contain thousands of atoms. Such giant molecules are called *macromolecules*. In the following section we will discuss inorganic and organic compounds that are essential for cells.

INORGANIC COMPOUNDS

WATER

LEARNING OBJECTIVE
- List several properties of water that are important to living systems.

All living organisms require a wide variety of inorganic compounds for growth, repair, maintenance, and reproduction. Water is one of the most important, as well as one of the most abundant, of these compounds, and it is particularly vital to microorganisms. Outside the cell, nutrients are dissolved in water, which facilitates their passage through cell membranes. And inside the cell, water is the medium for most chemical reactions. In fact, water is by far the most abundant component of almost all living cells. Water makes up at least 5–95% of every cell, the average being between 65% and 75%. Simply stated, no organism can survive without water.

Water has structural and chemical properties that make it particularly suitable for its role in living cells. As we discussed, the total charge on the water molecule is neutral, but the oxygen region of the molecule has a slightly negative charge and the hydrogen region has a slightly positive charge (see Figure 2.4a). Any molecule having such an unequal distribution of charges is called a **polar molecule.** The polar nature of water gives it four characteristics that make it a useful medium for living cells.

First, every water molecule is capable of forming four hydrogen bonds with nearby water molecules (see Figure 2.4b). This property results in a strong attraction between water molecules. Because of this strong attraction, a great deal of heat is required to separate water molecules from each other to form water vapor; thus, water has a relatively high boiling point (100°C). Because water has such a high boiling point, it exists in the liquid state on most of the Earth's surface. Furthermore, the hydrogen bonding between water molecules affects the density of water, depending on whether it occurs as ice or a liquid. For example, the hydrogen bonds in the crystalline structure of water (ice) make ice take up more space. As a result, ice has fewer molecules than an equal volume of liquid water. This makes its crystalline structure less dense than liquid water. For this reason, ice floats and can serve as an insulating layer on the surfaces of lakes and streams that harbor living organisms.

Second, the polarity of water makes it an excellent dissolving medium, or **solvent.** Many polar substances undergo **dissociation,** or separation, into individual molecules in water—that is, they dissolve—because the negative part of the water molecules is attracted to the positive part of the molecules in the **solute,** or dissolving substance, and the positive part of the water molecules is attracted to the negative part of the solute molecules. Substances (such as salts) that are composed of atoms (or groups of atoms) held together by ionic bonds tend to dissociate into separate cations and anions in water. Thus, the polarity of water allows molecules of many different substances to separate and become surrounded by water molecules (Figure 2.5).

Third, polarity accounts for water's characteristic role as a reactant or product in many chemical reactions. Its polarity facilitates the splitting and rejoining of hydrogen ions (H^+) and hydroxide ions (OH^-). Water is a key reactant in the digestive processes of organisms, whereby larger molecules are broken down into smaller ones. Water molecules are also involved in synthetic reactions; water is an important source of the hydrogen and oxygen that are incorporated into numerous organic compounds in living cells.

Finally, the relatively strong hydrogen bonding between water molecules (see Figure 2.4b) makes water an excellent temperature buffer. A given quantity of water

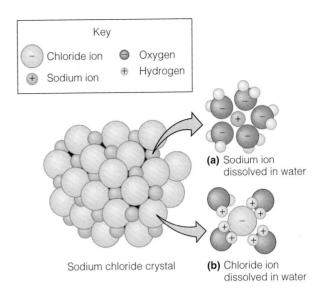

Key

- Chloride ion   ⊖ Oxygen

+ Sodium ion   ⊕ Hydrogen

Sodium chloride crystal

(a) Sodium ion dissolved in water

(b) Chloride ion dissolved in water

FIGURE 2.5 How water acts as a solvent for sodium chloride (NaCl). **(a)** The positively charged sodium ion (Na^+) is attracted to the negative part of the water molecule. **(b)** The negatively charged chloride ion (Cl^-) is attracted to the positive part of the water molecule. In the presence of water molecules, the bonds between the Na^+ and Cl^- are disrupted, and the NaCl dissolves in the water.

Q **What happens during ionization?**

requires a great gain of heat to increase its temperature and a great loss of heat to decrease its temperature, compared with many other substances. Normally, heat absorption by molecules increases their kinetic energy and thus increases their rate of motion and their reactivity. In water, however, heat absorption first breaks hydrogen bonds rather than increasing the rate of motion. Therefore, much more heat must be applied to raise the temperature of water than to raise the temperature of a non–hydrogen-bonded liquid. The reverse is true as water cools. Thus, water more easily maintains a constant temperature than other solvents and tends to protect a cell from fluctuations in environmental temperatures.

ACIDS, BASES, AND SALTS

> **LEARNING OBJECTIVE**
> • Define *acid, base, salt,* and *pH.*

As we saw in Figure 2.5, when inorganic salts such as sodium chloride (NaCl) are dissolved in water, they undergo **ionization** or *dissociation;* that is, they break apart into ions. Substances called acids and bases show similar behavior.

An **acid** can be defined as a substance that dissociates into one or more hydrogen ions (H^+) and one or more negative ions (anions). Thus, an acid can also be defined

FIGURE 2.6 Acids, bases, and salts. (**a**) In water, hydrochloric acid (HCl) dissociates into H^+ and Cl^-. (**b**) Sodium hydroxide (NaOH), a base, dissociates into OH^- and Na^+ in water. (**c**) In water, table salt (NaCl) dissociates into positive ions (Na^+) and negative ions (Cl^-), neither of which are H^+ or OH^-.

Q How do acids and bases differ?

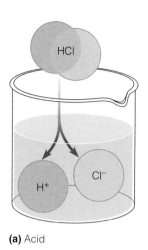

(a) Acid

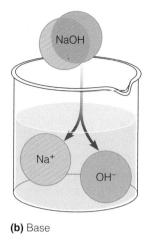

(b) Base

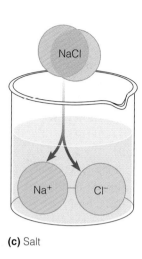

(c) Salt

as a proton (H^+) donor. A **base** dissociates into one or more positive ions (cations) plus one or more negatively charged hydroxide ions (OH^-) that can accept, or combine with, protons. Thus, sodium hydroxide (NaOH) is a base because it dissociates to release OH^-, which has a strong attraction for protons and is among the most important proton acceptors. A **salt** is a substance that dissociates in water into cations and anions, neither of which is H^+ or OH^-. Figure 2.6 shows common examples of each type of compound and how they dissociate in water.

ACID–BASE BALANCE

An organism must maintain a fairly constant balance of acids and bases to remain healthy. For example, if a particular acid or base concentration is too high or too low, enzymes change in shape and no longer function effectively in promoting chemical reactions in a cell. In the aqueous environment within organisms, acids dissociate into hydrogen ions (H^+) and anions. Bases, in contrast, dissociate into hydroxide ions (OH^-) and cations. The more hydrogen ions that are free in a solution, the more acidic the solution is. Conversely, the more hydroxide ions that are free in a solution, the more basic, or alkaline, it is.

Biochemical reactions—that is, chemical reactions in living systems—are extremely sensitive to even small changes in the acidity or alkalinity of the environments in which they occur. In fact, H^+ and OH^- are involved in almost all biochemical processes, and the functions of a cell are modified greatly by any deviation from its narrow band of normal H^+ and OH^- concentrations. For this reason, the acids and bases that are continually formed in an organism must be kept in balance.

It is convenient to express the amount of H^+ in a solution by a logarithmic **pH** scale, which ranges from 0 to 14 (Figure 2.7). The term *pH* means potential of hydrogen.

On a logarithmic scale, a change of one whole number represents a tenfold change from the previous concentration. Thus, a solution of pH 1 has ten times more hydrogen ions than a solution of pH 2 and has 100 times more hydrogen ions than a solution of pH 3.

A solution's pH is calculated as $-\log_{10}[H^+]$, the negative logarithm to the base 10 of the hydrogen ion concentration (denoted by brackets), determined in moles per liter $[H^+]$. For example, if the H^+ concentration of a solution is 1.0×10^{-4} moles/liter, or 10^{-4}, its pH equals $-\log_{10}10^{-4} = -(-4) = 4$; this is about the pH value of wine (see Appendix D). The pH values of some human body fluids and other common substances are also shown in Figure 2.7. In the laboratory, you will usually measure the pH of a solution with a pH meter or with chemical test papers.

Acidic solutions contain more H^+ than OH^- and have a pH lower than 7. If a solution has more OH^- than H^+, it is a basic, or alkaline, solution. In pure water, a small percentage of the molecules are dissociated into H^+ and OH^-, so it has a pH of 7. Because the concentrations of H^+ and OH^- are equal, this pH is said to be the pH of a neutral solution.

Keep in mind that the pH of a solution can be changed. We can increase its acidity by adding substances that will increase the concentration of hydrogen ions. As a living organism takes up nutrients, carries out chemical reactions, and excretes wastes, its balance of acids and bases tends to change, and the pH fluctuates. Fortunately, organisms possess natural pH **buffers,** compounds that help keep the pH from changing drastically. But the pH in our environment's water and soil can be altered by waste products from organisms, pollutants from industry, or fertilizers used in agricultural fields or gardens. When bacteria are grown in a laboratory medium, they excrete waste products such as acids that can alter the pH of the medium. If this

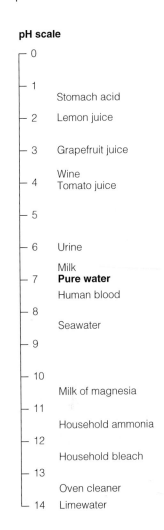

pH scale

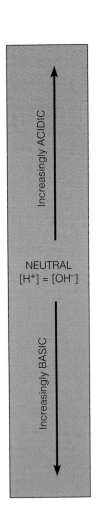

FIGURE 2.7 The pH scale. As pH values decrease from 14 to 0, the H^+ concentration increases. Thus, the lower the pH, the more acidic the solution; the higher the pH, the more basic the solution. If the pH value of a solution is below 7, the solution is acidic; if the pH is above 7, the solution is basic (alkaline). The approximate pH values of some human body fluids and common substances are shown next to the pH scale.

Q At what pH is the concentration of H^+ and OH^- equal?

effect were to continue, the medium would become acidic enough to inhibit bacterial enzymes and cause the death of the bacteria. To prevent this problem, pH buffers are added to the culture medium. One very effective pH buffer for some culture media uses a mixture of K_2HPO_4 and KH_2PO_4 (see Table 6.3, page 169).

Different microbes function best within different pH ranges, but most organisms grow best in environments with a pH value between 6.5 and 8.5. Among microbes, fungi are best able to tolerate acidic conditions, whereas the prokaryotes called cyanobacteria tend to do well in alkaline habitats. *Propionibacterium acnes* (prō-pē-on-ē-bak-ti'rē-um ak'nēz), a bacterium that contributes to acne, has as its natural environment human skin, which tends to be slightly acidic, with a pH of about 4. *Thiobacillus ferrooxidans* (thī-ō-bä-sil'lus fer-rō-oks'i-danz) is a bacterium that metabolizes elemental sulfur and produces sulfuric acid (H_2SO_4). Its pH range for optimum growth is from 1 to 3.5. The sulfuric acid produced by this bacterium in mine water is important in dissolving uranium and copper from low-grade ore (see Chapter 28).

ORGANIC COMPOUNDS

LEARNING OBJECTIVES
- Distinguish between organic and inorganic compounds.
- Define *functional group*.

Inorganic compounds, excluding water, constitute about 1–1.5% of living cells. These relatively simple components, whose molecules have only a few atoms, cannot be used by cells to perform complicated biological functions. Organic molecules, whose carbon atoms can combine in an enormous variety of ways with other carbon atoms and with atoms of other elements, are relatively complex and thus are capable of more complicated biological functions.

STRUCTURE AND CHEMISTRY

In the formation of organic molecules, carbon's four outer electrons can participate in up to four covalent bonds, and carbon atoms can bond to each other to form straight-chain, branched-chain, or ring structures.

In addition to carbon, the most common elements in organic compounds are hydrogen (which can form one bond), oxygen (two bonds), and nitrogen (three bonds). Sulfur (two bonds) and phosphorus (five bonds) appear less often. Other elements are found, but only in a relatively few organic compounds. The elements that are most abundant in living organisms are the same as those that are most abundant in organic compounds (see Table 2.1).

The chain of carbon atoms in an organic molecule is called the **carbon skeleton;** a huge number of combinations is possible for carbon skeletons. Most of these carbons are bonded to hydrogen atoms. The bonding of other elements with carbon and hydrogen forms characteristic **functional groups,** specific groups of atoms that are most commonly involved in chemical reactions and are responsible for most of the characteristic chemical properties and many of the physical properties of a particular organic compound (Table 2.3).

Different functional groups confer different properties on organic molecules. For example, the hydroxyl group of alcohols is hydrophilic (water-loving) and thus attracts water molecules to it. This attraction helps dissolve organic molecules containing hydroxyl groups. Because the carboxyl group is a source of hydrogen ions, molecules containing it have acidic properties. Amino groups, by contrast, function as bases because they readily accept hydrogen ions. The sulfhydryl group helps stabilize the intricate structure of many proteins.

Functional groups help us classify organic compounds. For example, the —OH group is present in each of the following molecules:

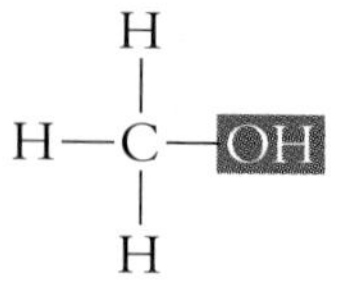

Methanol

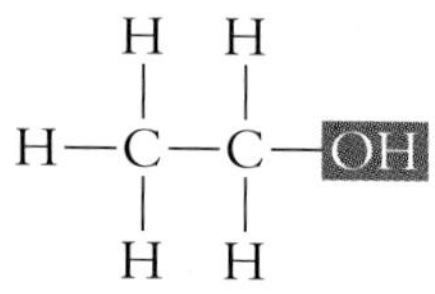

Ethanol

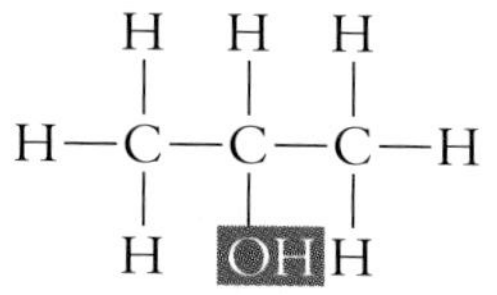

Isopropanol

Because the characteristic reactivity of the molecules is based on the —OH group, they are grouped together in a class called alcohols. The —OH group is called the *hydroxyl group* and is not to be confused with the *hydroxide ion* (OH⁻) of bases. The hydroxyl group of alcohols does not ionize at neutral pH; it is covalently bonded to a carbon atom.

When a class of compounds is characterized by a certain functional group, the letter *R* can be used to stand for

TABLE 2.3	Representative Functional Groups and the Compounds in Which They are Found

Structure	Name of Group	Biological Importance
R—O—H	Alcohol	Lipids, carbohydates
R—C(=O)—H	Aldehyde*	Reducing sugars such as glucose; polysaccharides
R—C(=O)—R	Ketone*	Metabolic intermediates
R—C(H)(H)—H	Methyl	DNA; energy metabolism
R—C(H)(H)—NH₂	Amino	Proteins
R—C(=O)—O—R′	Ester	Bacterial and eukaryotic plasma membranes
R—C(H)(H)—O—C(H)(H)—R′	Ether	Archaeal plasma membranes
R—C(H)(H)—SH	Sulfhydryl	Energy metabolism; protein structure
R—C(=O)—OH	Carboxyl	Organic acids, lipids, proteins
R—O—P(=O)(O⁻)(O⁻)	Phosphate	ATP, DNA

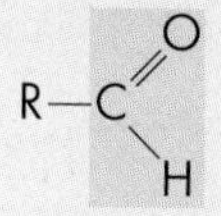

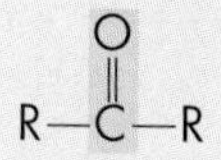

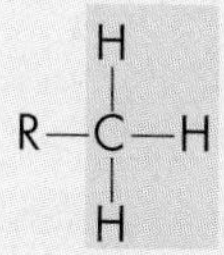

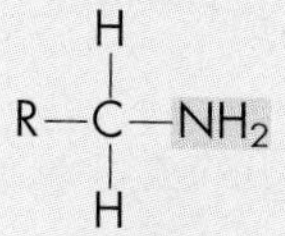

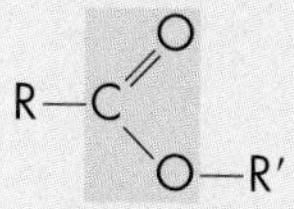

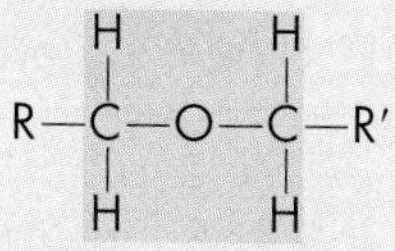

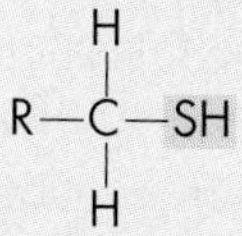

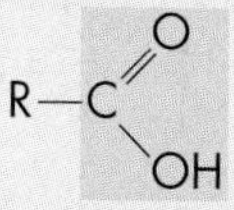

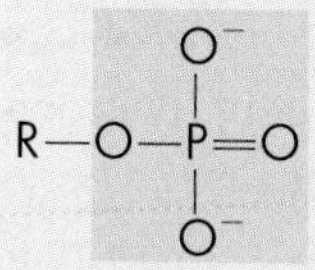

*In an aldehyde, a C=O is at the end of a molecule, in contrast to the internal C=O in a ketone.

the remainder of the molecule. For example, alcohols in general may be written R—OH.

Frequently, more than one functional group is found in a single molecule. For example, an amino acid molecule contains both amino and carboxyl groups. The amino acid glycine has the following structure:

Amino — H   H   O — Carboxyl
group | || group
      H — N — C — C
      | | OH
      H   R

Most of the organic compounds found in living organisms are quite complex; a large number of carbon atoms form the skeleton, and many functional groups are attached. In organic molecules, it is important that each of the four bonds of carbon is satisfied (attached to another atom) and that each of the attaching atoms has its characteristic number of bonds satisfied. Because of this, such molecules are chemically stable.

Small organic molecules can be combined into very large molecules called **macromolecules** (*macro* = large). Macromolecules are usually **polymers** (*poly* = many; *mers* = parts), large molecules formed by covalent bonding of many repeating small molecules called **monomers** (*mono* = one). When two monomers join together, the reaction usually involves the elimination of a hydrogen atom from one monomer and a hydroxyl group from the other; the hydrogen atom and the hydroxyl group combine to produce water:

$$R — \boxed{OH + H} — R' \longrightarrow R — R' + \boxed{H_2O}$$

This type of exchange reaction is called **dehydration synthesis** (*de* = from; *hydro* = water), or a **condensation reaction,** because a molecule of water is released (Figure 2.8a). Such macromolecules as carbohydrates, lipids, proteins, and nucleic acids are assembled in the cell, essentially by dehydration synthesis. However, other molecules must also participate to provide energy for bond formation. ATP, the cell's chief energy provider, is discussed at the end of this chapter.

CARBOHYDRATES

LEARNING OBJECTIVE
- Identify the building blocks of carbohydrates.

The **carbohydrates** are a large and diverse group of organic compounds that includes sugars and starches. Carbohydrates perform a number of major functions in living systems. For instance, one type of sugar (deoxyribose) is a building block of deoxyribonucleic acid (DNA), the molecule that carries hereditary information. Other sugars are needed for the cell walls. Simple carbohydrates are used in the synthesis of amino acids and fats or fatlike substances, which are used to build cell membranes and other structures. Macromolecular carbo-

hydrates function as food reserves. The principal function of carbohydrates, however, is to fuel cell activities with a ready source of energy.

Carbohydrates are made up of carbon, hydrogen, and oxygen atoms. The ratio of hydrogen to oxygen atoms is always 2:1 in simple carbohydrates. This ratio can be seen in the formulas for the carbohydrates ribose ($C_5H_{10}O_5$), glucose ($C_6H_{12}O_6$), and sucrose ($C_{12}H_{22}O_{11}$). Although there are exceptions, the general formula for carbohydrates is $(CH_2O)_n$, where n indicates that there are three or more CH_2O units. Carbohydrates can be classified into three major groups on the basis of size: monosaccharides, disaccharides, and polysaccharides.

MONOSACCHARIDES

Simple sugars are called **monosaccharides** (*sacchar* = sugar); each molecule contains from three to seven carbon atoms. The number of carbon atoms in the molecule of a simple sugar is indicated by the prefix in its name. For example, simple sugars with three carbons are called trioses. There are also tetroses (four-carbon sugars), pentoses (five-carbon sugars), hexoses (six-carbon sugars), and heptoses (seven-carbon sugars). Pentoses and hexoses are extremely important to living organisms. Deoxyribose is a pentose found in DNA. Glucose, a very common hexose, is the main energy-supplying molecule of living cells.

DISACCHARIDES

Disaccharides (*di* = two) are formed when two monosaccharides bond in a dehydration synthesis reaction.* For example, molecules of two monosaccharides, glucose and fructose, combine to form a molecule of the disaccharide sucrose (table sugar) and a molecule of water (Figure 2.8a). Similarly, the dehydration synthesis of the monosaccharides glucose and galactose forms the disaccharide lactose (milk sugar).

It may seem odd that glucose and fructose have the same chemical formula (see Figure 2.8), even though they are different monosaccharides. The positions of the oxygens and carbons differ in the two different molecules, and consequently the molecules have different physical and chemical properties. Two molecules with the same chemical formula but different structures and properties are called **isomers** (*iso* = same).

Disaccharides can be broken down into smaller, simpler molecules when water is added. This chemical reaction, the reverse of dehydration synthesis, is called **hydrolysis**

---

*Carbohydrates composed of 2 to about 20 monosaccharides are called **oligosaccharides** (*oligo* = few). Disaccharides are the most common oligosaccharides.

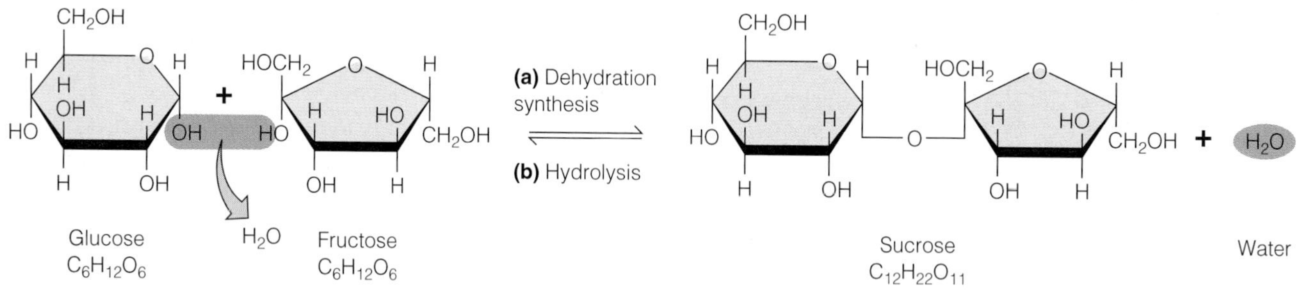

FIGURE 2.8 Dehydration synthesis and hydrolysis. (a) In dehydration synthesis (left to right), the monosaccharides glucose and fructose combine to form a molecule of the disaccharide sucrose. A molecule of water is released in the reaction. **(b)** In hydrolysis (right to left), the sucrose molecule breaks down into the smaller molecules glucose and fructose. For the hydrolysis reaction to proceed, water must be added to the sucrose.

Q **What is the difference between a polymer and a monomer?**

(*hydro* = water; *lysis* = to loosen) (Figure 2.8b). A molecule of sucrose, for example, may be hydrolyzed (digested) into its components of glucose and fructose by reacting with the H^+ and OH^- of water.

As you will see in Chapter 4, the cell walls of bacterial cells are composed of disaccharides and proteins (together called peptidoglycan).

POLYSACCHARIDES

Carbohydrates in the third major group, the **polysaccharides,** consist of tens or hundreds of monosaccharides joined through dehydration synthesis. Polysaccharides often have side chains branching off the main structure and are classified as macromolecules. Like disaccharides, polysaccharides can be split apart into their constituent sugars through hydrolysis. Unlike monosaccharides and disaccharides, however, they usually lack the characteristic sweetness of sugars such as fructose and sucrose and usually are not soluble in water.

One important polysaccharide is *glycogen*, which is composed of glucose subunits and is synthesized as a storage material by animals and some bacteria. *Cellulose*, another important glucose polymer, is the main component of the cell walls of plants and most algae. Although cellulose is the most abundant carbohydrate on Earth, it can be digested by only a few organisms that have the appropriate enzyme. The polysaccharide *dextran*, which is produced as a sugary slime by certain bacteria, is used in a blood plasma substitute. *Chitin* is a polysaccharide that makes up part of the cell wall of most fungi and the exoskeletons of lobsters, crabs, and insects. *Starch* is a polymer of glucose produced by plants and used as food by humans.

Many animals, including humans, produce enzymes called *amylases* that can break the bonds between the glucose molecules in glycogen. However, this enzyme cannot break the bonds in cellulose. Bacteria and fungi that produce enzymes called *cellulases* can digest cellulose. Cellulases from the fungus *Trichoderma* (trik′ō-dėr-mä) are used for a variety of industrial purposes. One of the more unusual uses is producing stone-washed denim. Because washing the fabric with rocks would damage washing machines, cellulase is used to digest, and therefore soften, the cotton. (See the box in Chapter 9, page 267.)

LIPIDS

LEARNING OBJECTIVE

- Differentiate among simple lipids, complex lipids, and steroids.

If lipids were suddenly to disappear from the Earth, all living cells would collapse in a pool of fluid, because lipids are essential to the structure and function of membranes that separate living cells from their environment. **Lipids** (*lip* = fat) are a second major group of organic compounds found in living matter. Like carbohydrates, they are composed of atoms of carbon, hydrogen, and oxygen, but lipids lack the 2:1 ratio between hydrogen and oxygen atoms. Even though lipids are a very diverse group of compounds, they share one common characteristic: they are *nonpolar* molecules so, unlike water, do not have a positive and a negative end (pole). Therefore, most lipids are insoluble in water but dissolve readily in nonpolar solvents, such as ether and chloroform. Lipids function in energy storage and provide the structure of membranes and some cell walls.

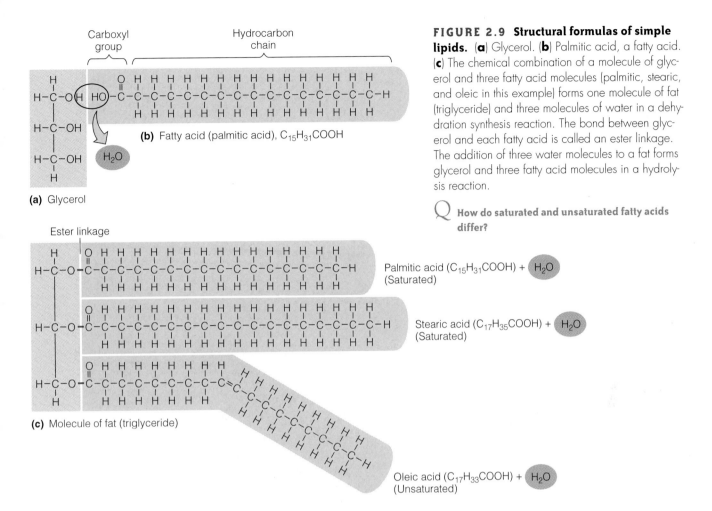

FIGURE 2.9 Structural formulas of simple lipids. (**a**) Glycerol. (**b**) Palmitic acid, a fatty acid. (**c**) The chemical combination of a molecule of glycerol and three fatty acid molecules (palmitic, stearic, and oleic in this example) forms one molecule of fat (triglyceride) and three molecules of water in a dehydration synthesis reaction. The bond between glycerol and each fatty acid is called an ester linkage. The addition of three water molecules to a fat forms glycerol and three fatty acid molecules in a hydrolysis reaction.

Q **How do saturated and unsaturated fatty acids differ?**

SIMPLE LIPIDS

Simple lipids, called *fats* or *triglycerides*, contain an alcohol called *glycerol* and a group of compounds known as *fatty acids*. Glycerol molecules have three carbon atoms to which are attached three hydroxyl (—OH) groups (Figure 2.9a). Fatty acids consist of long hydrocarbon chains (composed only of carbon and hydrogen atoms) ending in a carboxyl (—COOH, organic acid) group (Figure 2.9b). Most common fatty acids contain an even number of carbon atoms.

A fat molecule is formed when a molecule of glycerol combines with one to three fatty acid molecules to form a monoglyceride, diglyceride, or triglyceride (Figure 2.9c). In the reaction, one to three molecules of water are formed (dehydration), depending on the number of fatty acid molecules reacting. The chemical bond formed where the water molecule is removed is called an *ester linkage*. In the reverse reaction, hydrolysis, a fat molecule is broken down into its component fatty acid and glycerol molecules.

Because the fatty acids that form lipids have different structures, there is a wide variety of lipids. For example, three molecules of fatty acid A might combine with a glycerol molecule. Or one molecule each of fatty acids A, B, and C might unite with a glycerol molecule (see Figure 2.9c).

The primary function of lipids is the formation of plasma membranes that enclose cells. A plasma membrane supports the cell and allows nutrients and wastes to pass in and out; therefore, the lipids must maintain the same viscosity, regardless of the temperature of the surroundings. The membrane must be about as viscous as olive oil, without getting too fluid when warmed or too thick when cooled. As everyone who has ever cooked a meal knows, animal fats (such as butter) are usually solid at room temperature, whereas vegetable oils are usually liquid at room temperature. The difference in their respective melting points is caused by the degrees of saturation of the fatty acid chains. A fatty acid is said to be *saturated* when it has no double bonds; then the carbon skeleton contains the maximum number of hydrogen atoms (see Figures 2.9c and 2.10a). Saturated chains become solid more easily because they are relatively straight and are thus able to pack together more closely than unsaturated chains. The double bonds of *unsaturated* chains create kinks in the chain, which keep the chains apart from one another (Figure 2.10b).

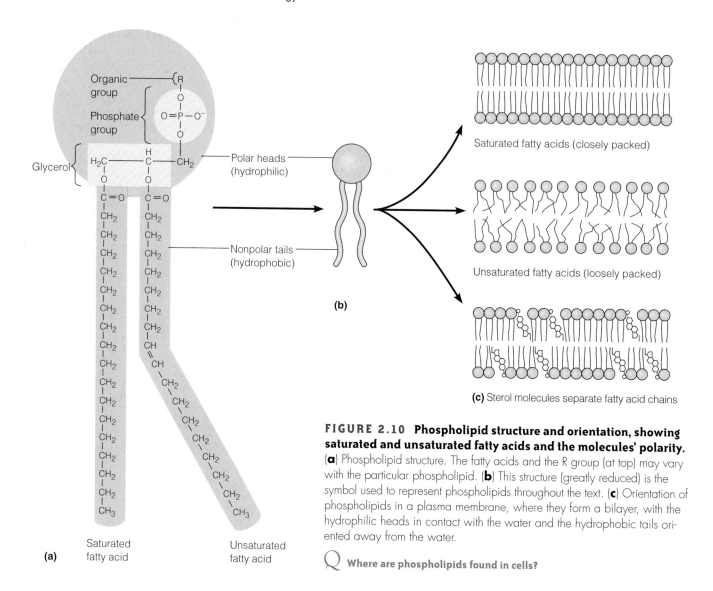

FIGURE 2.10 Phospholipid structure and orientation, showing saturated and unsaturated fatty acids and the molecules' polarity. (**a**) Phospholipid structure. The fatty acids and the R group (at top) may vary with the particular phospholipid. (**b**) This structure (greatly reduced) is the symbol used to represent phospholipids throughout the text. (**c**) Orientation of phospholipids in a plasma membrane, where they form a bilayer, with the hydrophilic heads in contact with the water and the hydrophobic tails oriented away from the water.

Q **Where are phospholipids found in cells?**

COMPLEX LIPIDS

Complex lipids contain such elements as phosphorus, nitrogen, and sulfur, in addition to the carbon, hydrogen, and oxygen found in simple lipids. The complex lipids called *phospholipids* are made up of glycerol, two fatty acids, and, in place of a third fatty acid, a phosphate group bonded to one of several organic groups (see Figure 2.10a). Phospholipids are the lipids that build membranes; they are essential to a cell's survival. Phospholipids have polar as well as nonpolar regions (Figure 2.10a and b; see also Figure 4.13, page 86). When placed in water, phospholipid molecules twist themselves in such a way that all polar (hydrophilic) portions will orient themselves toward the polar water molecules, with which they then form hydrogen bonds. (Recall that *hydrophilic* means water-loving.) This forms the basic structure of a plasma membrane (Figure 2.10c). Polar portions consist of a phosphate group and glycerol. In contrast to the polar regions, all nonpolar (hydrophobic)

parts of the phospholipid make contact only with the nonpolar portions of neighboring molecules. (*Hydrophobic* means water-fearing.) Nonpolar portions consist of fatty acids. This characteristic behavior makes phospholipids particularly suitable to being a major component of the membranes that enclose cells. Phospholipids enable the membrane to act as a barrier that separates the contents of the cell from the water-based environment in which it lives.

Some complex lipids are useful in identifying certain bacteria. For example, the cell wall of *Mycobacterium tuberculosis* (mī-kō-bak-ti'rē-um tü-bėr-kū-lō'sis), the bacterium that causes tuberculosis, is distinguished by its lipid-rich content. The cell wall contains complex lipids such as waxes and glycolipids (lipids with carbohydrates attached) that give the bacterium distinctive staining characteristics. Cell walls rich in such complex lipids are characteristic of all members of the genus *Mycobacterium*.

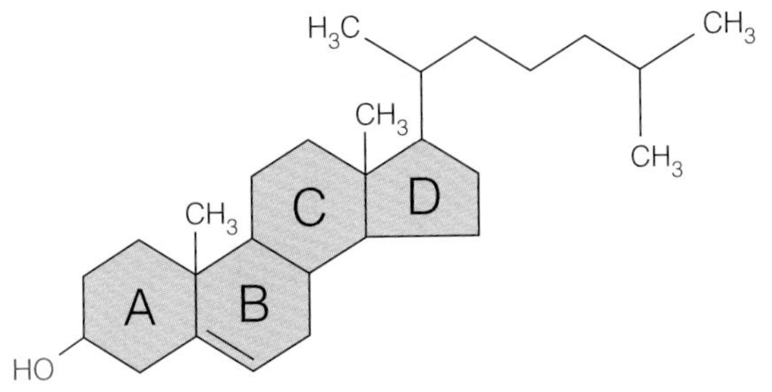

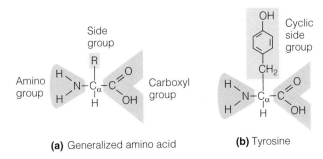

FIGURE 2.11 Cholesterol, a steroid. Note the four "fused" carbon rings (labeled A–D), which are characteristic of steroid molecules. The hydrogen atoms attached to the carbons at the corners of the rings have been omitted. The —OH group (colored red) makes this molecule a sterol.

Q Where are sterols found in cells?

STEROIDS

Steroids are structurally very different from lipids. Figure 2.11 shows the structure of the steroid cholesterol, with the four interconnected carbon rings that are characteristic of steroids. When an —OH group is attached to one of the rings, the steroid is called a *sterol* (an alcohol). Sterols are important constituents of the plasma membranes of animal cells and of one group of bacteria (mycoplasmas), and they are also found in fungi and plants. The sterols separate the fatty acid chains and thus prevent the packing that would harden the plasma membrane at low temperatures (see Figure 2.10c).

PROTEINS

LEARNING OBJECTIVE
• Identify the building blocks and structure of proteins.

Proteins are organic molecules that contain carbon, hydrogen, oxygen, and nitrogen. Some also contain sulfur. If you were to separate and weigh all the groups of organic compounds in a living cell, the proteins would tip the scale. Hundreds of different proteins can be found in any single cell, and together they make up 50% or more of a cell's dry weight.

Proteins are essential ingredients in all aspects of cell structure and function. *Enzymes* are the proteins that speed up biochemical reactions. But proteins have other functions as well. *Transporter proteins* help transport certain chemicals into and out of cells. Other proteins, such as the *bacteriocins* produced by many bacteria, kill other bacteria. Certain *toxins*, called exotoxins, produced by some disease-causing microorganisms are also proteins. Some proteins play a role in the *contraction* of animal

muscle cells and the *movement* of microbial and other types of cells. Other proteins are integral parts of *cell structures* such as walls, membranes, and cytoplasmic components. Still others, such as the *hormones* of certain organisms, have regulatory functions. As we will see in Chapter 17, proteins called *antibodies* play a role in vertebrate immune systems.

AMINO ACIDS

Just as monosaccharides are the building blocks of larger carbohydrate molecules, and fatty acids and glycerol are the building blocks of fats, **amino acids** are the building blocks of proteins. Amino acids contain at least one carboxyl (—COOH) group and one amino (—NH$_2$) group attached to the same carbon atom, called an alpha-carbon (written C$_\alpha$) (Figure 2.12a). Such amino acids are called *alpha-amino acids*. Also attached to the alpha-carbon is a side group (R group), which is the amino acid's distinguishing feature. The side group can be a hydrogen atom, an unbranched or branched chain of atoms, or a ring structure that is cyclic (all carbon) or heterocyclic (when an atom other than carbon is included in the ring). Figure 2.12b shows the structural formula of tyrosine, an amino acid that has a cyclic side group. The side group can contain functional groups, such as the sulfhydryl group (—SH), the hydroxyl group (—OH), or additional carboxyl or amino groups. These side groups and the carboxyl and alpha-amino groups affect the total structure of a protein, described later. The structures and standard abbreviations of the 20 amino acids found in proteins are shown in Table 2.4.

Most amino acids exist in either of two configurations called **stereoisomers,** designated by D and L. These configurations are mirror images, corresponding to "right-handed" (D) and "left-handed" (L) three-dimensional

(a) Generalized amino acid   **(b)** Tyrosine

FIGURE 2.12 Amino acid structure. **(a)** The general structural formula for an amino acid. The alpha-carbon (C$_\alpha$) is shown in the center. Different amino acids have different R groups, also called side groups. **(b)** Structural formula for the amino acid tyrosine, which has a cyclic side group.

Q What distinguishes one amino acid from another?

TABLE 2.4	The 20 Amino Acids Found in Proteins*

Glycine (Gly)

Hydrogen atom

Alanine (Ala)

Unbranched chain

Valine (Val)

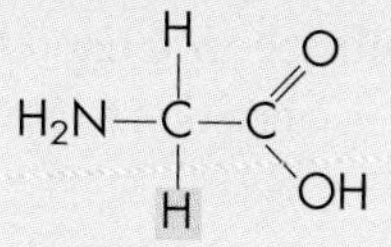

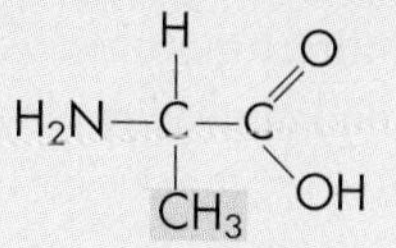

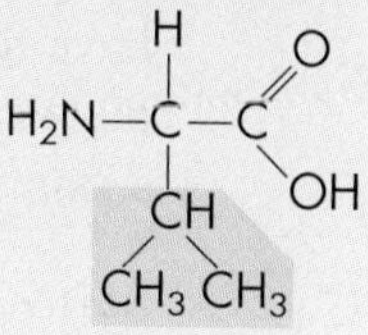

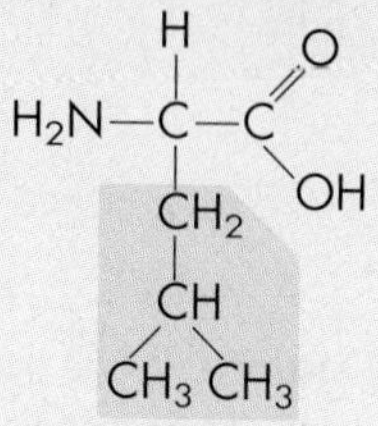

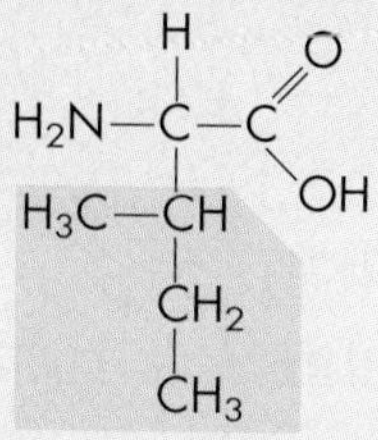

Branched chain

Leucine (Leu)

Branched chain

Isoleucine (Ile)

Branched chain

Serine (Ser)

Hydroxyl (—OH) group

Threonine (Thr)

Hydroxyl (—OH) group

Cysteine (Cys)

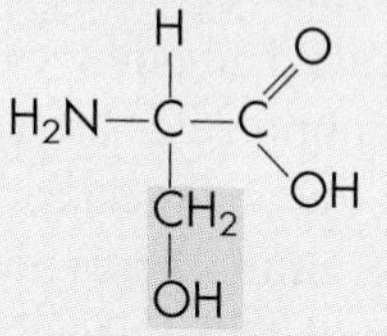

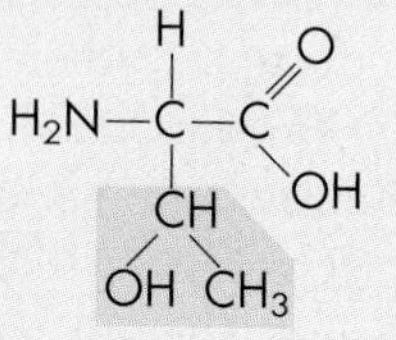

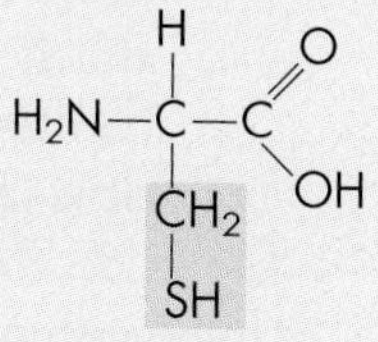

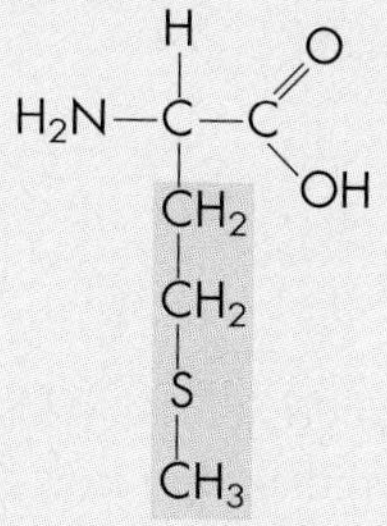

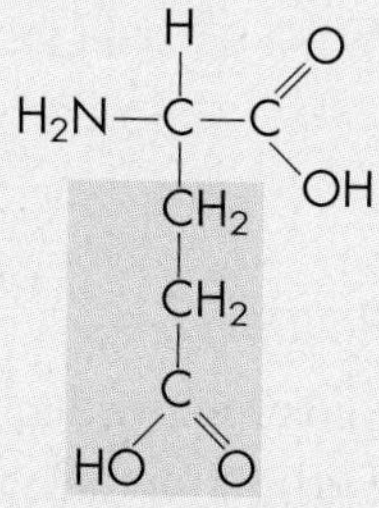

Sulphur-containing (—SH) group

Methionine (Met)

Thioether (SC) group

Glutamic acid (Glu)

Additional carboxyl (—COOH) group, acidic

Aspartic acid (Asp)

Additional carboxyl (—COOH) group, acidic

Lysine (Lys)

Additional amino (—NH₂) group, basic

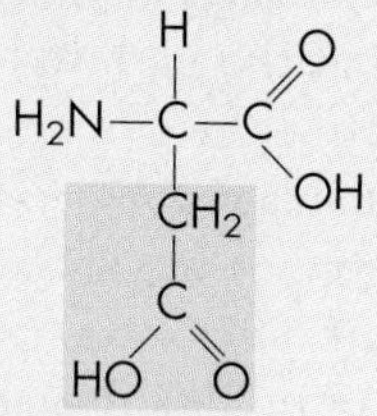

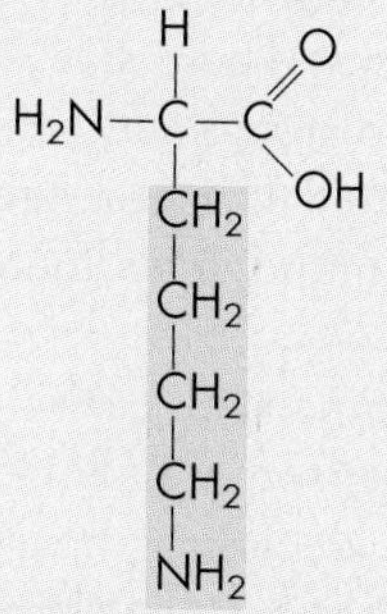

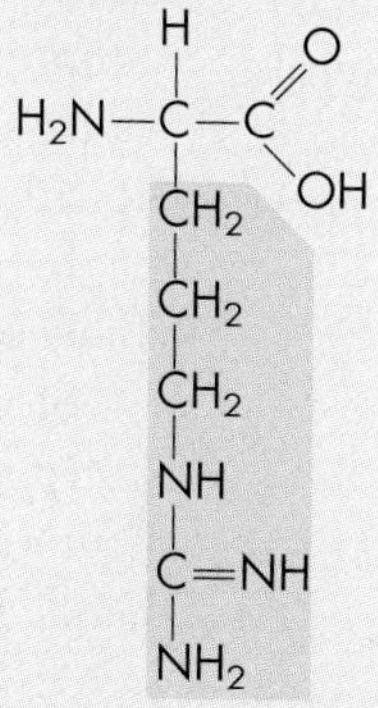

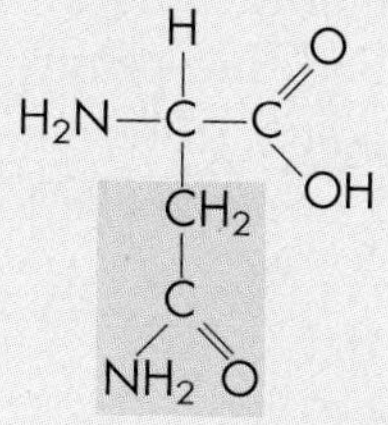

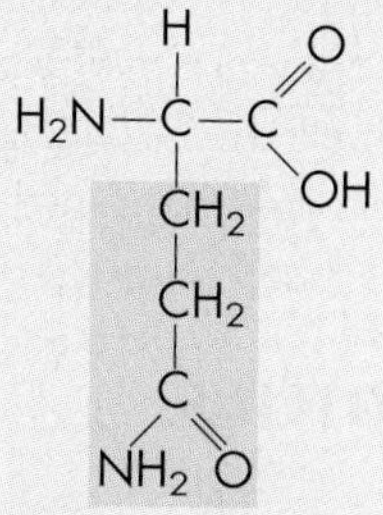

Arginine (Arg)

Additional amino (—NH₂) group, basic

Asparagine (Asn)

Additional amino (—NH₂) group, basic

Glutamine (Gln)

Additional amino (—NH₂) group, basic

Phenylalanine (Phe)

Cyclic

Tyrosine (Tyr)

Cyclic

Histidine (His)

Heterocyclic

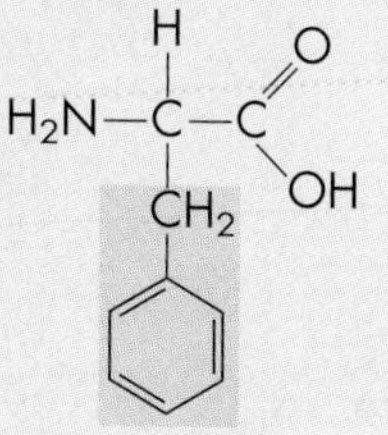

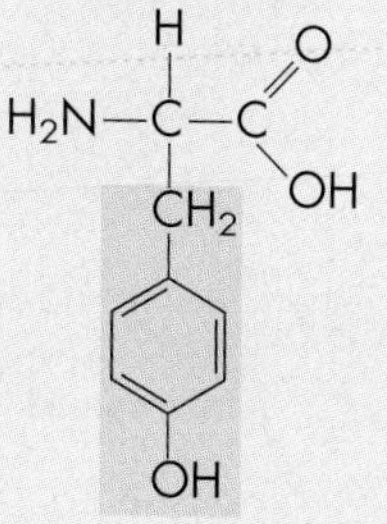

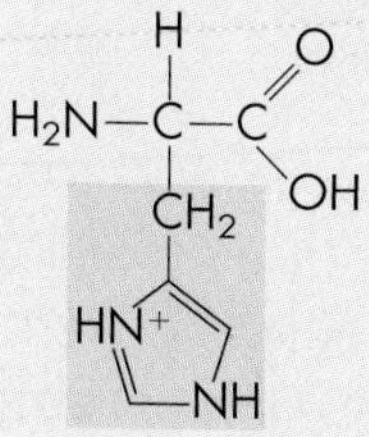

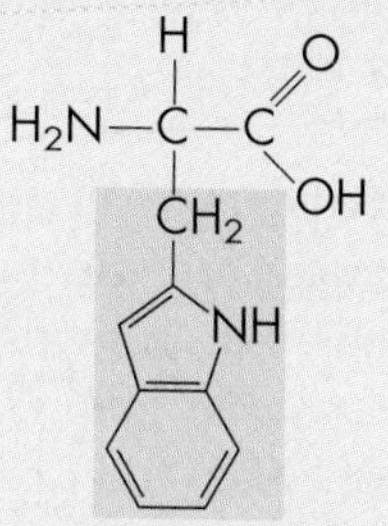

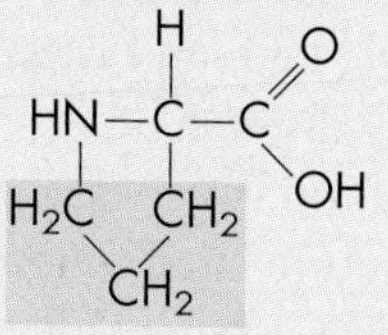

Tryptophan (Trp)

Heterocyclic

Proline (Pro)

Heterocyclic

*Shown are the amino acid names, including the three-letter abbreviation in parentheses (above), their structural formulas (center), and characteristic R group (below). Note that cysteine and methionine are the only amino acids that contain sulfur.

Although only 20 different amino acids occur naturally in proteins, a single protein molecule can contain from 50 to hundreds of amino acid molecules, which can be arranged in an almost infinite number of ways to make proteins of different lengths, compositions, and structures. The number of proteins is practically endless, and every living cell produces many different proteins.

PEPTIDE BONDS

Amino acids bond between the carbon atom of the carboxyl (—COOH) group of one amino acid and the nitrogen atom of the amino (—NH$_2$) group of another (Figure 2.14). The bonds between amino acids are called **peptide bonds.** For the formation of every peptide bond between two amino acids, one water molecule is released; thus, peptide bond formation occurs by dehydration synthesis. In Figure 2.14 the resulting compound is called a *dipeptide* because it consists of two amino acids joined by a peptide bond. Adding another amino acid to a dipeptide would form a *tripeptide.* Further additions of amino acids would produce a long, chainlike molecule called a *peptide* (4–9 amino acids) or *polypeptide* (10–2000 or more amino acids).

LEVELS OF PROTEIN STRUCTURE

Proteins vary tremendously in structure. Different proteins have different architectures and different three-dimensional shapes. This variation in structure is directly related to their diverse functions.

When a cell makes a protein, the polypeptide chain folds spontaneously to assume a certain shape. One reason for folding of the polypeptide is that some parts of a protein are attracted to water and other parts are repelled by it. In practically every case, the function of a protein depends on its ability to recognize and bind to some other molecule. For example, an enzyme binds specifically with its substrate. A hormonal protein binds to a receptor on a cell whose function it will alter. An antibody binds to an antigen (foreign substance) that has invaded the body. The unique shape of each protein permits it to interact with specific other molecules in order to carry out specific functions.

Proteins are described in terms of four levels of organization: primary, secondary, tertiary, and quaternary. The

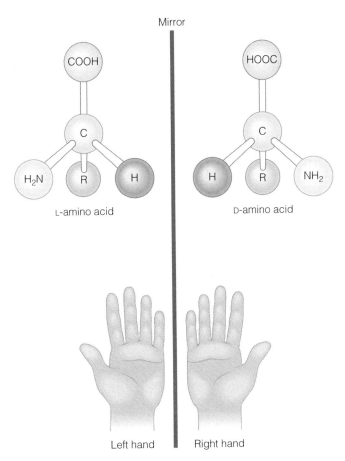

FIGURE 2.13 The L- and D-isomers of an amino acid, shown with ball-and-stick models. The two isomers, like left and right hands, are mirror images of each other and cannot be superimposed on one another. (Try it!)

Q **Which isomer is always found in proteins?**

shapes (Figure 2.13). The amino acids found in proteins are always the L-isomers (except for glycine, the simplest amino acid, which does not have stereoisomers). However, D-amino acids occasionally occur in nature—for example, in certain bacterial cell walls and antibiotics. (Many other kinds of organic molecules also can exist in D and L forms. One example is the sugar glucose, which occurs in nature as D-glucose.)

FIGURE 2.14 Peptide bond formation by dehydration synthesis. The amino acids glycine and alanine combine to form a dipeptide. The newly formed bond between the carbon atom of glycine and the nitrogen atom of alanine is called a peptide bond.

Q **How are amino acids related to proteins?**

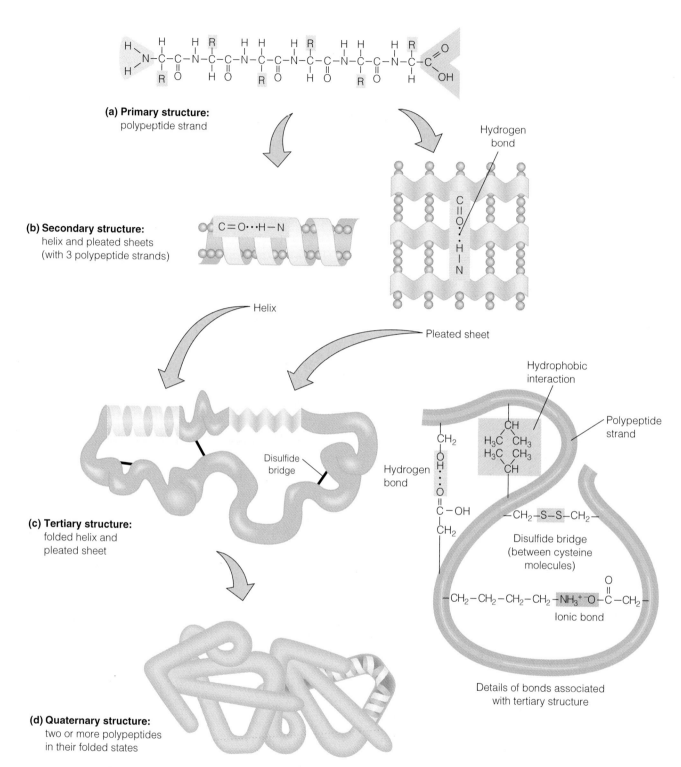

(a) Primary structure:
polypeptide strand

(b) Secondary structure:
helix and pleated sheets
(with 3 polypeptide strands)

Hydrogen bond

C=O···H—N

Helix

Pleated sheet

Hydrophobic interaction

Polypeptide strand

(c) Tertiary structure:
folded helix and
pleated sheet

Disulfide bridge

Hydrogen bond

Disulfide bridge
(between cysteine
molecules)

Ionic bond

Details of bonds associated
with tertiary structure

(d) Quaternary structure:
two or more polypeptides
in their folded states

FIGURE 2.15 Protein structure. (**a**) Primary structure, the amino acid sequence. (**b**) Secondary structures: helix and pleated sheet. (**c**) Tertiary structure, the overall three-dimensional folding of a polypeptide chain. (**d**) Quaternary structure, the relationship between several polypeptide chains that make up a protein. Shown here is the quaternary structure of a hypothetical protein composed of two polypeptide chains.

Q **What property of a protein enables it to carry out specific functions?**

primary structure is the unique sequence in which the amino acids are linked together to form a polypeptide chain (Figure 2.15a). This sequence is genetically determined. Alterations in sequence can have profound metabolic effects. For example, a single incorrect amino acid in a blood protein can produce the deformed hemoglobin molecule characteristic of sickle cell disease. But proteins do not exist as long, straight chains. Each polypeptide chain folds and coils in specific ways into a relatively compact structure with a characteristic three-dimensional shape.

A protein's *secondary structure* is the localized, repetitious twisting or folding of the polypeptide chain. This aspect of a protein's shape results from hydrogen bonds joining the atoms of peptide bonds at different locations along the polypeptide chain. The two types of secondary protein structures are clockwise spirals called helices (singular: *helix*) and pleated sheets, which form from roughly parallel portions of the chain (Figure 2.15b). Both structures are held together by hydrogen bonds between oxygen or nitrogen atoms that are part of the polypeptide's backbone.

Tertiary structure refers to the overall three-dimensional structure of a polypeptide chain (Figure 2.15c). The folding is not repetitive or predictable, as in secondary structure. Whereas secondary structure involves hydrogen bonding between atoms of the amino and carboxyl groups involved in the peptide bonds, tertiary structure involves several interactions between various amino acid side groups in the polypeptide chain. For example, amino acids with nonpolar (hydrophobic) side groups usually interact at the core of the protein, out of contact with water. This *hydrophobic interaction* helps contribute to tertiary structure. Hydrogen bonds between side groups, and ionic bonds between oppositely charged side groups, also contribute to tertiary structure. Proteins that contain the amino acid cysteine form strong covalent bonds called *disulfide bridges*. These bridges form when two cysteine molecules are brought close together by the folding of the protein. Cysteine molecules contain sulfhydryl groups (—SH), and the sulfur of one cysteine molecule bonds to the sulfur on another, forming (by the removal of hydrogen atoms) a disulfide bridge (S—S) that holds parts of the protein together.

Some proteins have a *quaternary structure*, which consists of an aggregation of two or more individual polypeptide chains (subunits) that operate as a single functional unit. Figure 2.15d shows a hypothetical protein consisting of two polypeptide chains. More commonly, proteins have two or more kinds of polypeptide subunits. The bonds that hold a quaternary structure together are basically the same as those that maintain tertiary structure. The overall shape of a protein may be globular (compact and roughly spherical) or fibrous (threadlike).

If a protein encounters a hostile environment in terms of temperature, pH, or salt concentrations, it may unravel and lose its characteristic shape. This process is called **denaturation** (see Figure 5.6, page 121). As a result of denaturation, the protein is no longer functional. This process will be discussed in more detail in Chapter 5 with regard to denaturation of enzymes.

The proteins we have been discussing are *simple proteins*, which contain only amino acids. *Conjugated proteins* are combinations of amino acids with other organic or inorganic components. Conjugated proteins are named by their non–amino acid component. Thus, glycoproteins contain sugars, nucleoproteins contain nucleic acids, metalloproteins contain metal atoms, lipoproteins contain lipids, and phosphoproteins contain phosphate groups. Phosphoproteins are important regulators of activity in eukaryotic cells. Bacterial synthesis of phosphoproteins may be important for the survival of bacteria such as *Legionella pneumophila* that grow inside host cells.

NUCLEIC ACIDS

LEARNING OBJECTIVE
* Identify the building blocks of nucleic acids.

In 1944, three American microbiologists—Oswald Avery, Colin MacLeod, and Maclyn McCarty—discovered that a substance called **deoxyribonucleic acid (DNA)** is the substance of which genes are made. Nine years later, James Watson and Francis Crick, working with molecular models and X-ray information supplied by Maurice Wilkins and Rosalind Franklin, identified the physical structure of DNA. In addition, Crick suggested a mechanism for DNA replication and how it works as the hereditary material. DNA, and another substance called **ribonucleic acid (RNA),** are together referred to as **nucleic acids** because they were first discovered in the nuclei of cells. Just as amino acids are the structural units of proteins, nucleotides are the structural units of nucleic acids.

Each **nucleotide** has three parts: a nitrogen-containing base, a pentose (five-carbon) sugar (either **deoxyribose** or **ribose**), and a phosphate group (phosphoric acid). The nitrogen-containing bases are cyclic compounds made up of carbon, hydrogen, oxygen, and nitrogen atoms. The bases are named adenine (A), thymine (T), cytosine (C), guanine (G), and uracil (U). A and G are double-ring

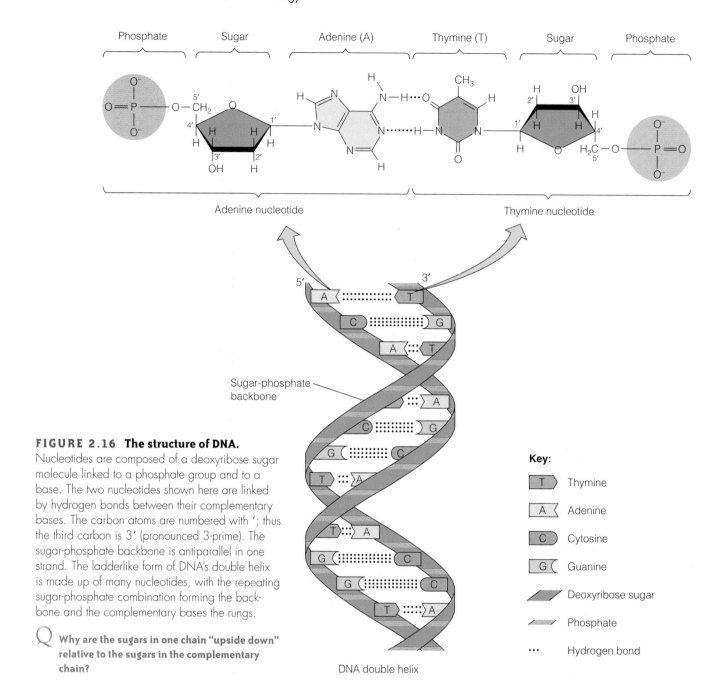

FIGURE 2.16 The structure of DNA.
Nucleotides are composed of a deoxyribose sugar molecule linked to a phosphate group and to a base. The two nucleotides shown here are linked by hydrogen bonds between their complementary bases. The carbon atoms are numbered with ′; thus the third carbon is 3′ (pronounced 3-prime). The sugar-phosphate backbone is antiparallel in one strand. The ladderlike form of DNA's double helix is made up of many nucleotides, with the repeating sugar-phosphate combination forming the backbone and the complementary bases the rungs.

Q **Why are the sugars in one chain "upside down" relative to the sugars in the complementary chain?**

structures called **purines,** whereas T, C, and U are single-ring structures referred to as **pyrimidines.**

Nucleotides are named according to their nitrogen-containing base. Thus, a nucleotide containing thymine is a *thymine nucleotide*, one containing adenine is an *adenine nucleotide*, and so on. The term **nucleoside** refers to the combination of a purine or pyrimidine plus a pentose sugar; it does not contain a phosphate group.

DNA

According to the model proposed by Watson and Crick, a DNA molecule consists of two long strands wrapped around each other to form a **double helix** (Figure 2.16). The double helix looks like a twisted ladder, and each strand is composed of many nucleotides.

Every strand of DNA composing the double helix has a "backbone" consisting of alternating deoxyribose sugar and phosphate groups. The deoxyribose of one nucleotide is joined to the phosphate group of the next. (Refer to Figure 8.3, page 219, to see how nucleotides are bonded.) The nitrogen-containing bases make up the rungs of the ladder. Note that the purine A is always paired with the pyrimidine T and that the purine G is always paired with the pyrimidine C. The bases are held together by hydrogen

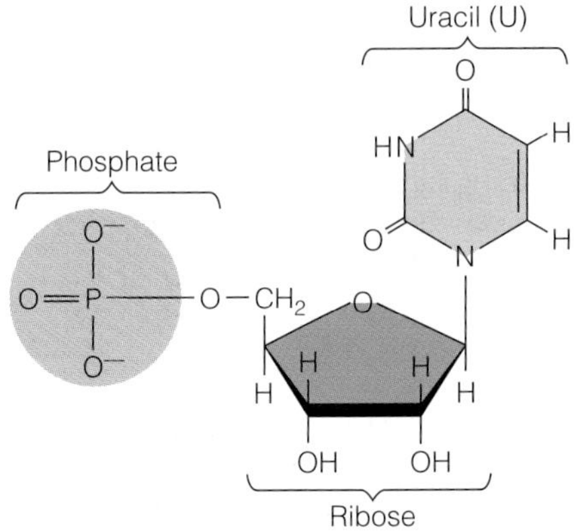

Uracil (U)

Phosphate

FIGURE 2.17 A uracil nucleotide of RNA.

Q **How do DNA and RNA differ in structure?**

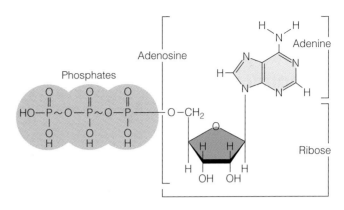

FIGURE 2.18 The structure of ATP. High-energy phosphate bonds are indicated by wavy lines. When ATP breaks down to ADP and inorganic phosphate, a large amount of chemical energy is released for use in other chemical reactions.

Q **How is ATP similar to a nucleotide in RNA? In DNA?**

bonds; A and T are held by two hydrogen bonds, and G and C are held by three. DNA does not contain uracil (U).

The order in which the nitrogen base pairs occur along the backbone is extremely specific and in fact contains the genetic instructions for the organism. Nucleotides form genes, and a single DNA molecule may contain thousands of genes. Genes determine all hereditary traits, and they control all the activities that take place within cells.

One very important consequence of nitrogen-containing base pairing is that if the sequence of bases of one strand is known, then the sequence of the other strand is also known. For example, if one strand has the sequence . . . ATGC . . . , then the other strand has the sequence . . . TACG. . . . Because the sequence of bases of one strand is determined by the sequence of bases of the other, the bases are said to be *complementary*. The actual transfer of information becomes possible because of DNA's unique structure and will be discussed further in Chapter 8.

RNA

RNA, the second principal kind of nucleic acid, differs from DNA in several respects. Whereas DNA is double-stranded, RNA is usually single-stranded. The five-carbon sugar in the RNA nucleotide is ribose, which has one more oxygen atom than deoxyribose. Also, one of RNA's bases is uracil (U) instead of thymine (Figure 2.17). The other three bases (A, G, C) are the same as DNA. Three major kinds of RNA have been identified in cells. These are referred to as **messenger RNA (mRNA), ribosomal RNA (rRNA),** and **transfer RNA (tRNA).** As we will see in Chapter 8, each type of RNA has a specific role in protein synthesis.

ADENOSINE TRIPHOSPHATE (ATP)

LEARNING OBJECTIVE

- Describe the role of ATP in cellular activities.

Adenosine triphosphate (ATP) is the principal energy-carrying molecule of all cells and is indispensable to the life of the cell. It stores the chemical energy released by some chemical reactions, and it provides the energy for reactions that require energy. ATP consists of an adenosine unit, composed of adenine and ribose, with three phosphate groups (abbreviated Ⓟ) attached (Figure 2.18). In other words, it is an adenine nucleotide (also called adenosine monophosphate, or AMP) with two extra phosphate groups. ATP is called a high-energy molecule because it releases a large amount of usable energy when the third phosphate group is hydrolyzed to become **adenosine diphosphate (ADP).** This reaction can be represented as follows:

$$\text{Adenosine} - Ⓟ - Ⓟ - Ⓟ + H_2O \rightleftharpoons$$

Adenosine   Water
triphosphate

$$\text{Adenosine} - Ⓟ - Ⓟ + Ⓟ_i + \text{Energy}$$

Adenosine   Inorganic
diphosphate   phosphate

A cell's supply of ATP at any particular time is limited. Whenever the supply needs replenishing, the reaction goes in the reverse direction; the addition of a phosphate group to ADP and the input of energy produces more ATP. The energy required to attach the terminal phosphate group to ADP is supplied by the cell's various oxidation reactions, particularly the oxidation of glucose. ATP can be stored in every cell, where its potential energy is not released until needed.

STUDY OUTLINE

INTRODUCTION (p. 26)

1. The science of the interaction between atoms and molecules is called chemistry.
2. The metabolic activities of microorganisms involve complex chemical reactions.
3. Nutrients are broken down by microbes to obtain energy and to make new cells.

THE STRUCTURE OF ATOMS (pp. 27–28)

1. Atoms are the smallest units of chemical elements that enter into chemical reactions.
2. Atoms consist of a nucleus, which contains protons and neutrons, and electrons that move around the nucleus.
3. The atomic number is the number of protons in the nucleus; the total number of protons and neutrons is the atomic weight.

CHEMICAL ELEMENTS (pp. 27–28)

4. Atoms with the same number of protons and the same chemical behavior are classified as the same chemical element.
5. Chemical elements are designated by abbreviations called chemical symbols.
6. About 26 elements are commonly found in living cells.
7. Atoms that have the same atomic number (are of the same element) but different atomic weights are called isotopes.

ELECTRONIC CONFIGURATIONS (p. 28)

8. In an atom, electrons are arranged around the nucleus in electron shells.
9. Each shell can hold a characteristic maximum number of electrons.
10. The chemical properties of an atom are due largely to the number of electrons in its outermost shell.

HOW ATOMS FORM MOLECULES: CHEMICAL BONDS (pp. 28–32)

1. Molecules are made up of two or more atoms; molecules consisting of at least two different kinds of atoms are called compounds.
2. Atoms form molecules in order to fill their outermost electron shells.
3. Attractive forces that bind the atomic nuclei of two atoms together are called chemical bonds.

4. The combining capacity of an atom—the number of chemical bonds the atom can form with other atoms—is its valence.

IONIC BONDS (pp. 28–30)

5. A positively or negatively charged atom or group of atoms is called an ion.
6. A chemical attraction between ions of opposite charge is called an ionic bond.
7. To form an ionic bond, one ion is an electron donor, and the other ion is an electron acceptor.

COVALENT BONDS (pp. 30–31)

8. In a covalent bond, atoms share pairs of electrons.
9. Covalent bonds are stronger than ionic bonds and are far more common in organisms.

HYDROGEN BONDS (pp. 31–32)

10. A hydrogen bond exists when a hydrogen atom covalently bonded to one oxygen or nitrogen atom is attracted to another oxygen or nitrogen atom.
11. Hydrogen bonds form weak links between different molecules or between parts of the same large molecule.

MOLECULAR WEIGHT AND MOLES (p. 32)

12. The molecular weight is the sum of the atomic weights of all the atoms in a molecule.
13. A mole of an atom, ion, or molecule is equal to its atomic or molecular weight expressed in grams.

CHEMICAL REACTIONS (pp. 32–34)

1. Chemical reactions are the making or breaking of chemical bonds between atoms.
2. A change of energy occurs during chemical reactions.
3. Endergonic reactions require energy; exergonic reactions release energy.
4. In a synthesis reaction, atoms, ions, or molecules are combined to form a larger molecule.
5. In a decomposition reaction, a larger molecule is broken down into its component molecules, ions, or atoms.
6. In an exchange reaction, two molecules are decomposed, and their subunits are used to synthesize two new molecules.
7. The products of reversible reactions can readily revert to form the original reactants.

IMPORTANT BIOLOGICAL MOLECULES (pp. 34–49)

INORGANIC COMPOUNDS (pp. 34–37)

1. Inorganic compounds are usually small, ionically bonded molecules.
2. Water and many common acids, bases, and salts are examples of inorganic compounds.

WATER (pp. 34–35)

3. Water is the most abundant substance in cells.
4. Because water is a polar molecule, it is an excellent solvent.
5. Water is a reactant in many of the decomposition reactions of digestion.
6. Water is an excellent temperature buffer.

ACIDS, BASES, AND SALTS (pp. 35–36)

7. An acid dissociates into H^+ and anions.
8. A base dissociates into OH^- and cations.
9. A salt dissociates into negative and positive ions, neither of which is H^+ or OH^-.

ACID–BASE, BALANCE (pp. 36–37)

10. The term pH refers to the concentration of H^+ in a solution.
11. A solution of pH 7 is neutral; a pH value below 7 indicates acidity; pH above 7 indicates alkalinity.
12. The pH inside a cell and in culture media is stabilized with pH buffers.

ORGANIC COMPOUNDS (pp. 37–49)

1. Organic compounds always contain carbon and hydrogen.
2. Carbon atoms form up to four bonds with other atoms.
3. Organic compounds are mostly or entirely covalently bonded, and many of them are large molecules.

STRUCTURE AND CHEMISTRY (pp. 37–39)

4. A chain of carbon atoms forms a carbon skeleton.
5. Functional groups of atoms are responsible for most of the properties of organic molecules.
6. The letter R may be used to denote the remainder of an organic molecule.
7. Frequently encountered classes of molecules are R—OH (alcohols) and R—COOH (organic acids).
8. Small organic molecules may combine into very large molecules called macromolecules.
9. Monomers usually bond together by dehydration synthesis, or condensation reactions, that form water and a polymer.

10. Organic molecules may be broken down by hydrolysis, a reaction involving the splitting of water molecules.

CARBOHYDRATES (pp. 39–40)

11. Carbohydrates are compounds consisting of atoms of carbon, hydrogen, and oxygen, with hydrogen and oxygen in a 2:1 ratio.
12. Carbohydrates include sugars and starches.
13. Carbohydrates can be classified as monosaccharides, disaccharides, and polysaccharides.
14. Monosaccharides contain from three to seven carbon atoms.
15. Isomers are two molecules with the same chemical formula but different structures and properties—for example, glucose ($C_6H_{12}O_6$) and fructose ($C_6H_{12}O_6$).
16. Monosaccharides may form disaccharides and polysaccharides by dehydration synthesis.

LIPIDS (pp. 40–43)

17. Lipids are a diverse group of compounds distinguished by their insolubility in water.
18. Simple lipids (fats) consist of a molecule of glycerol and three molecules of fatty acids.
19. A saturated lipid has no double bonds between carbon atoms in the fatty acids; an unsaturated lipid has one or more double bonds. Saturated lipids have higher melting points than unsaturated lipids.
20. Phospholipids are complex lipids consisting of glycerol, two fatty acids, and a phosphate group.
21. Steroids have carbon ring structures; sterols have a functional hydroxyl group.

PROTEINS (pp. 43–47)

22. Amino acids are the building blocks of proteins.
23. Amino acids consist of carbon, hydrogen, oxygen, nitrogen, and sometimes sulfur.
24. Twenty amino acids occur naturally in proteins.
25. By linking amino acids, peptide bonds (formed by dehydration synthesis) allow the formation of polypeptide chains.
26. Proteins have four levels of structure: primary (sequence of amino acids), secondary (helices or pleats), tertiary (overall three-dimensional structure of a polypeptide), and quaternary (two or more polypeptide chains).
27. Conjugated proteins consist of amino acids combined with other organic or inorganic compounds.

NUCLEIC ACIDS (pp. 47–49)

28. Nucleic acids—DNA and RNA—are macromolecules consisting of repeating nucleotides.
29. A nucleotide is composed of a pentose, a phosphate group, and a nitrogen-containing base. A nucleoside is composed of a pentose and a nitrogen-containing base.

30. A DNA nucleotide consists of deoxyribose (a pentose) and one of the following nitrogen-containing bases: thymine or cytosine (pyrimidines) or adenine or guanine (purines).

31. DNA consists of two strands of nucleotides wound in a double helix. The strands are held together by hydrogen bonds between purine and pyrimidine nucleotides: AT and GC.

32. Genes consist of sequences of nucleotides.

33. An RNA nucleotide consists of ribose (a pentose) and one of the following nitrogen-containing bases: cytosine, guanine, adenine, or uracil.

ADENOSINE TRIPHOSPHATE (ATP) (p. 49)

34. ATP stores chemical energy for various cellular activities.

35. When the bond to ATP's terminal phosphate group is hydrolyzed, energy is released.

36. The energy from oxidation reactions is used to regenerate ATP from ADP and inorganic phosphate.

STUDY QUESTIONS

Access more review material either online at **The Microbiology Place** (www.microbiologyplace.com) or with **The Microbiology Place CD-ROM** packaged with your new book. There you'll find activities, practice tests, quizzes, flashcards, case studies, and more to help you succeed. In addition, you'll find the following Interactive Tutorials: Atomic Structure, Ionic and Covalent Bonding, and Hydrogen Bonds and Water.

Answers to the Study Questions can be found in Appendix G.

REVIEW

1. What is a chemical element?

2. Diagram the electronic configuration of a carbon atom.

3. How does $_6^{14}C$ differ from $_6^{12}C$?

4. What type of bonding exists between water molecules?

5. What type of bonds holds the following atoms together?
   a. Li^+ and Cl^- in LiCl
   b. carbon and oxygen atoms in CO_2
   c. oxygen atoms in O_2
   d. a hydrogen atom of one nucleotide to a nitrogen or oxygen atom of another nucleotide in:

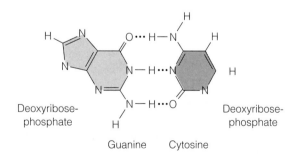

Deoxyribose-phosphate   Deoxyribose-phosphate

Guanine   Cytosine

6. Vinegar, pH 3, is how many times more acidic than pure water, pH 7?

7. Calculate the molecular weight of 1 mole of $C_6H_{12}O_6$.

8. Classify the following types of chemical reactions.
   a. glucose + fructose → sucrose + H_2O
   b. lactose → glucose + galactose
   c. $NH_4Cl + H_2O → NH_4OH + HCl$
   d. ATP → ADP + $\circledP_i$

9. Bacteria use the enzyme urease to obtain nitrogen in a form they can use from urea in the following reaction:

$$CO(NH_2)_2 + H_2O \longrightarrow 2NH_3 + CO_2$$
Urea   Ammonia Carbon dioxide

What purpose does the enzyme serve in this reaction? What type of reaction is this?

10. Classify the following as subunits of either a carbohydrate, lipid, protein, or nucleic acid.
   a. $CH_3-(CH_2)_7-CH=CH-(CH_2)_7-COOH$
   Oleic acid

   b.

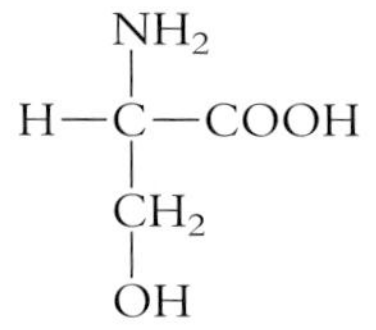

$$
\begin{array}{c}
NH_2 \\
| \\
H-C-COOH \\
| \\
CH_2 \\
| \\
OH
\end{array}
$$

   Serine

   c. $C_6H_{12}O_6$
   d. Thymine nucleotide

11. Add the appropriate functional group(s) to this ethyl group to produce each of the following compounds: ethanol, acetic acid, acetaldehyde, ethanolamine, diethyl ether.

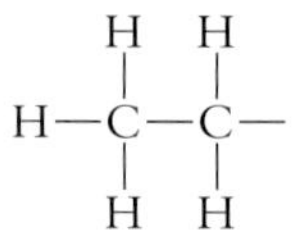

$$
\begin{array}{cc}
H & H \\
| & | \\
H-C-C- \\
| & | \\
H & H
\end{array}
$$

12. The artificial sweetener aspartame, or NutraSweet, is made by joining aspartic acid to methylated phenylalanine, as shown below.

a. What types of molecules are aspartic acid and phenylalanine?
b. What direction is the hydrolysis reaction (left to right or right to left)?
c. What direction is the dehydration synthesis reaction?
d. Circle the atoms involved in the formation of water.
e. Identify the peptide bond.

13. The energy-carrying property of the ATP molecule is due to energy dynamics that favor the breaking of bonds between _____. What type of bonds are these?

14. The following diagram shows a bacteriorhodopsin protein. Indicate the regions of primary, secondary, and tertiary structure. Does this protein have quaternary structure?

15. Draw a simple lipid, and show how it could be modified to a phospholipid.

MULTIPLE CHOICE

Radioisotopes are frequently used to label molecules in a cell. The fate of atoms and molecules in a cell can then be followed. This process is the basis for questions 1–3.

1. Assume *E. coli* are grown in a nutrient medium containing the radioisotope ^{16}N. After a 48-hour incubation period, the ^{16}N would most likely be found in the *E. coli*'s
   a. carbohydrates.   c. proteins.   e. none of the above
   b. lipids.   d. water.

2. If *Pseudomonas* bacteria are supplied with radioactively labeled cytosine, after a 24-hour incubation period this cytosine would most likely be found in the cells'
   a. carbohydrates.   c. lipids.   e. proteins.
   b. DNA.   d. water.

3. If *E. coli* were grown in a medium containing the radioactive isotope ^{32}P, the ^{32}P would be found in all of the following molecules of the cell *except*
   a. ATP.   c. DNA.   e. none of the above
   b. carbohydrates.   d. plasma membrane.

4. A carbonated drink, pH 3, is _____ times more acid than distilled water.
   a. 4   c. 100   e. 10,000
   b. 10   d. 1000

5. The best definition of ATP is that it is:
   a. a molecule stored for food use.
   b. a molecule that supplies energy to do work.
   c. a molecule stored for an energy reserve.
   d. a molecule used as a source of phosphate.

6. Which of the following is an organic molecule?
   a. H_2O (water)
   b. O_2 (oxygen)
   c. $C_{18}H_{29}SO_3$ (Styrofoam)
   d. FeO (iron oxide)
   e. $F_2C=CF_2$ (Teflon)

Classify each of the molecules on the left as an acid, base, or salt. The dissociation products of the molecules are shown to help you.

a 7. $HNO_3 \rightarrow H^+ + NO_3^-$ a. acid
a 8. $H_2SO_4 \rightarrow 2H^+ + SO_4^{2-}$ b. base
b 9. $NaOH \rightarrow Na^+ + OH^-$ c. salt
c 10. $MgSO_4 \rightarrow Mg^{2+} + SO_4^{2-}$

CRITICAL THINKING

1. When you blow bubbles into a glass of water, the following reactions take place:

$$\overset{A}{} \qquad \overset{B}{}$$
$$H_2O + CO_2 \longrightarrow H_2CO_3 \longrightarrow H^+ + HCO_3^-$$

   a. What type of reaction is A?
   b. What does reaction B tell you about the type of molecule H_2CO_3 is?

2. What are the common structural characteristics of ATP and DNA molecules?

3. What happens to the relative amount of unsaturated lipids in its plasma membrane when *E. coli* grown at 25°C are then grown at 37°C?

4. Giraffes, termites, and koalas eat only plant matter. Because animals cannot digest cellulose, how do you suppose these animals get nutrition from the leaves and wood they eat?

CLINICAL APPLICATIONS

1. *Ralstonia* bacteria make poly-β-hydroxybutyrate (PHB), which is used to make a biodegradable plastic. PHB consists of many of the monomers shown below. What type of molecule is PHB? What is the most likely reason a cell would store this molecule?

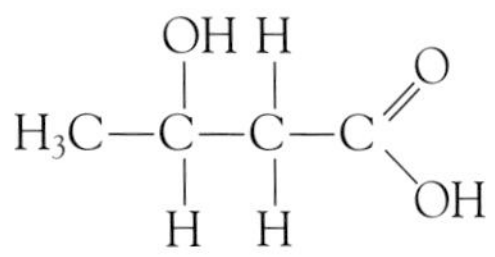

2. *Thiobacillus ferrooxidans* was responsible for destroying buildings in the Midwest by causing changes in the earth. The original rock, which contained lime ($CaCO_3$) and pyrite (FeS_2), expanded as bacterial metabolism caused gypsum ($CaSO_4$) crystals to form. How did *T. ferrooxidans* bring about the change from lime to gypsum?

3. When growing in an animal, *Bacillus anthracis* produces a capsule that is resistant to phagocytosis. The capsule is composed of D-glutamic acid. Why is this capsule resistant to digestion by the host's phagocytes? (Phagocytes are white blood cells that engulf bacteria.)

4. The antibiotic amphotericin B causes leaks in cells by combining with sterols in the plasma membrane. Would you expect to use amphotericin B against a bacterial infection? A fungal infection? Offer a reason why amphotericin B has severe side effects in humans.

5. You can smell sulfur when boiling eggs. What amino acids do you expect in the egg?

3

Observing Microorganisms Through a Microscope

Microorganisms are much too small to be seen with the unaided eye; they must be observed with a microscope. The word microscope is derived from the Latin word *micro*, which means small, and the Greek word *skopos*, to look at. Modern microbiologists use microscopes that produce, with great clarity, magnifications that range from ten to thousands of times greater than those of van Leeuwenhoek's single lens (see Figure 1.2b on page 7). This chapter describes how different types of microscopes function and why one type might be used in preference to another.

Some microbes are more readily visible than others because of their larger size or more easily observable features. Many microbes, however, must undergo several staining procedures before their cell walls, capsules, and other structures lose their colorless natural state. The last part of this chapter will explain some of the more commonly used methods of preparing specimens for examination through a light microscope.

You may wonder how we are going to sort, count, and measure the specimens we will study. To answer these questions, this chapter opens with a discussion of how to use the metric system for measuring microbes.

UNDER THE MICROSCOPE

Paramecium caudatum. Surface structures, including cilia and the oral groove of this *Paramecium*, can be seen with a scanning electron microscope.

TABLE 3.1	Metric Units of Length and U.S. Equivalents		
Metric Unit	Meaning of prefix	Metric Equivalent	U.S. Equivalent
1 kilometer (km)	*kilo* = 1000	1000 m = 10^3 m	3280.84 ft or 0.62 mi; 1 mi = 1.61 km
1 meter (m)		Standard unit of length	39.37 in or 3.28 ft or 1.09 yd
1 decimeter (dm)	*deci* = 1/10	0.1 m = 10^{-1} m	3.94 in
1 centimeter (cm)	*centi* = 1/100	0.01 m = 10^{-2} m	0.394 in; 1 in = 2.54 cm
1 millimeter (mm)	*milli* = 1/1000	0.001 m = 10^{-3} m	
1 micrometer (μm)	*micro* = 1/1,000,000	0.000001 m = 10^{-6} m	
1 nanometer (nm)	*nano* = 1/1,000,000,000	0.000000001 m = 10^{-9} m	

UNITS OF MEASUREMENT

LEARNING OBJECTIVE

- List the metric units of measurement, including their metric equivalents, that are used for microorganisms.

Because microorganisms and their component parts are so very small, they are measured in units that are unfamiliar to many of us in everyday life. When measuring microorganisms, we use the metric system. The standard unit of length in the metric system is the meter (m). A major advantage of the metric system is that the units are related to each other by factors of 10. Thus, 1 m equals 10 decimeters (dm) or 100 centimeters (cm) or 1000 millimeters (mm). Units in the U.S. system of measure do not have the advantage of easy conversion by a single factor of 10. For example, we use 3 feet or 36 inches to equal 1 yard.

Microorganisms and their structural components are measured in even smaller units, such as micrometers and nanometers. A **micrometer (μm)** is equal to 0.000001 m (10^{-6} m). The prefix *micro* indicates that the unit following it should be divided by 1 million, or 10^6 (see the "Exponential Notation" section in Appendix E). A **nanometer (nm)** is equal to 0.000000001 m (10^{-9} m). Angstrom (Å) was previously used for 10^{-10} m, or 0.1 nm.

Table 3.1 presents the basic metric units of length and some of their U.S. equivalents. In Table 3.1, you can compare the microscopic units of measurement with the commonly known macroscopic units of measurement, such as centimeters, meters, and kilometers. If you look ahead to Figure 3.2, you will see the relative sizes of various organisms on the metric scale.

MICROSCOPY: THE INSTRUMENTS

The simple microscope used by van Leeuwenhoek in the seventeenth century had only one lens and was similar to a magnifying glass. However, van Leeuwenhoek was the best lens grinder in the world in his day. His lenses were ground with such precision that a single lens could magnify a microbe 300×. His simple microscopes enabled him to be the first person to see bacteria (see Figure 1.2, page 7).

Contemporaries of van Leeuwenhoek, such as Robert Hooke, built compound microscopes, which have multiple lenses. In fact, a Dutch spectacles maker, Zaccharias Janssen, is credited with making the first compound microscope around 1600. However, these early compound microscopes were of poor quality and could not be used to see bacteria. It was not until about 1830 that a significantly better microscope was developed by Joseph Jackson Lister (the father of Joseph Lister). Various improvements to Lister's microscope resulted in the development of the modern compound microscope, the kind used in microbiology laboratories today. Microscopic studies of live specimens have revealed dramatic interactions between microbes (see the Applications of Microbiology box on the facing page.)

LIGHT MICROSCOPY

LEARNING OBJECTIVES

- Diagram the path of light through a compound microscope.
- Define *total magnification* and *resolution*.
- Identify a use for darkfield, phase-contrast, differential interference contrast (DIC), fluorescence, confocal, and scanning acoustic microscopy, and compare each with brightfield illumination.

Light microscopy refers to the use of any kind of microscope that uses visible light to observe specimens. Here we examine several types of light microscopy.

COMPOUND LIGHT MICROSCOPY

A modern **compound light microscope** has a series of lenses and uses visible light as its source of illumination.

APPLICATIONS OF MICROBIOLOGY

ARE BACTERIA MULTICELLULAR?

The idea that bacteria are unicellular has changed recently. Bacterial cells do not act as unicellular organisms when they are growing in a colony. Instead, the cells interact and exhibit multicellular organization. Researchers cite a number of examples that suggest cells in a colony are not identical and may have different structures and functions. Multicellular organization is indicated where bacteria influence the behavior of neighboring cells (Figure A).

Myxobacteria

Myxobacteria are found in decaying organic material and fresh water throughout the world. Although they are bacteria, many myxobacteria never exist as individual cells. *Myxococcus xanthus* appears to hunt in packs. In its natural aqueous habitat, *M. xanthus* cells form spherical colonies that surround prey bacteria, where they can secrete digestive enzymes and absorb the nutrients.

On solid substrates, other myxobacterial cells glide over a solid surface, leaving slime trails that are followed by other cells. When food is scarce, the cells aggregate to form a mass. Cells within the mass differentiate into a fruiting body that consists of a slime stalk and clusters of spores as shown in Figure B.

Vibrio

Vibrio fischeri is a bioluminescent bacterium that lives as a symbiont in the light-producing organ of squid and certain fish. When free-living, the bacteria are at low concentration and do not give off light. However, when they grow in their host, they are highly concentrated, and each cell is induced to produce the enzyme luciferase, which is used in the chemical pathways of bioluminescence.

V. cholerae bacteria aggregate into biofilms which help them survive in the ocean and may help the bacteria survive stomach acid. After passing through the stomach, individual cells leave the biofilm and colonize the intestine where genes for virulence factors are expressed.

How bacterial group behavior works

Cell density alters gene expression in bacterial cells in a process called quorum sensing. *Quorum sensing* is the ability of bacteria to communicate and coordinate behavior. Bacteria that use quorum sensing produce and secrete a signaling chemical called an *inducer*. As the inducer diffuses into the surrounding medium, other bacterial cells move toward the source and begin producing inducer. The concentration of inducer increases with increasing cell numbers. This, in turn, attracts more cells and initiates synthesis of more inducer.

In law, a quorum is the minimum number of members necessary to conduct the business. Apparently, that's true among bacteria too. *Pseudomonas fluorescens* bacteria that form a pellicle on the surface of a liquid culture medium can use *both* O_2 from the air and nutrient from the broth, a feat that cells could not do individually. *P. aeruginosa* can grow within a human without causing disease until they form a biofilm that overcomes the host's immune system. Biofilm-forming *P. aeruginosa* colonize the lungs of cystic fibrosis patients and are a leading cause of death in these patients (Figure C). Perhaps biofilms that lead to disease can be prevented by new drugs that destroy the inducer.

A. *Paenibacillus.* As one small colony (black dots) moves away from the parent colony, other groups of cells follow the first colony. Soon, all of the other bacteria join the relocation to form this spiraling colony.

SEM   10 µm

B. A fruiting body of a myxobacterium.

C. Intact pellicle of *Pseudomonas aeruginosa.*

40 µm

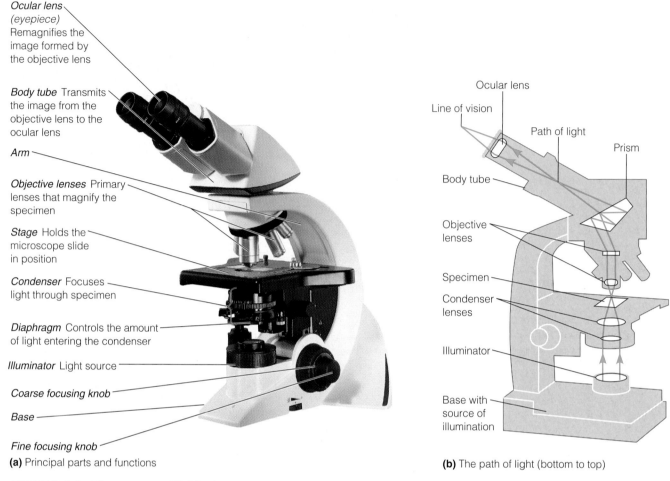

Ocular lens (eyepiece) Remagnifies the image formed by the objective lens

Body tube Transmits the image from the objective lens to the ocular lens

Arm

Objective lenses Primary lenses that magnify the specimen

Stage Holds the microscope slide in position

Condenser Focuses light through specimen

Diaphragm Controls the amount of light entering the condenser

Illuminator Light source

Coarse focusing knob

Base

Fine focusing knob

(a) Principal parts and functions

Ocular lens

Line of vision

Path of light

Prism

Body tube

Objective lenses

Specimen

Condenser lenses

Illuminator

Base with source of illumination

(b) The path of light (bottom to top)

FIGURE 3.1   The compound light microscope.

Q   **How is the total magnification of a compound light microscope calculated?**

(Figure 3.1a). With a compound light microscope, we can examine very small specimens as well as some of their fine detail. A series of finely ground lenses (Figure 3.1b) forms a clearly focused image that is many times larger than the specimen itself. This magnification is achieved when light rays from an **illuminator,** the light source, pass through a **condenser,** which has lenses that direct the light rays through the specimen. From here, light rays pass into the **objective lenses,** the lenses closest to the specimen. The image of the specimen is magnified again by the **ocular lens,** or *eyepiece*.

We can calculate the **total magnification** of a specimen by multiplying the objective lens magnification (power) by the ocular lens magnification (power). Most microscopes used in microbiology have several objective lenses, including 10× (low power), 40× (high power), and 100× (oil immersion, which is described shortly). Most ocular lenses magnify specimens by a factor of 10. Multiplying the magnification of a specific objective lens with that of the ocular, we see that the total magnifications would be 100× for low power, 400× for high power, and 1000× for oil immersion. Some compound light microscopes can achieve a magnification of 2000× with the oil immersion lens.

Resolution (also called *resolving power*) is the ability of the lenses to distinguish fine detail and structure. Specifically, it refers to the ability of the lenses to distinguish between two points a specified distance apart. For example, if a microscope has a resolving power of 0.4 nm, it can distinguish between two points if they are at least 0.4 nm apart. A general principle of microscopy is that the shorter the wavelength of light used in the instrument, the greater the resolution. The white light used in a compound light microscope has a relatively long wavelength and cannot resolve structures smaller than about 0.2 μm. This fact and other practical considerations limit the magnification achieved by even the best compound light microscopes to about 2000×. By comparison, van Leeuwenhoek's microscopes had a resolution of 1 μm.

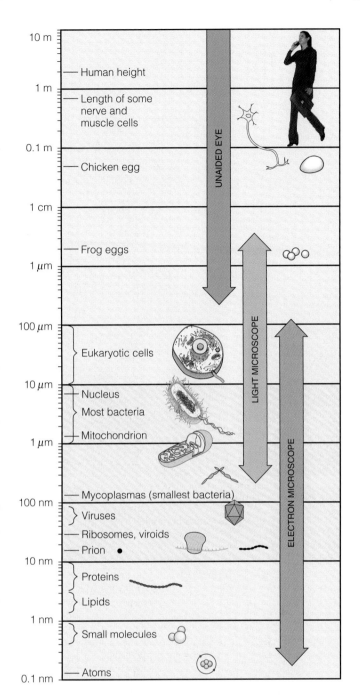

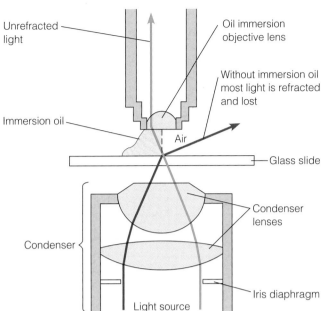

FIGURE 3.3 Refraction in the compound microscope using an oil immersion objective lens. Because the refractive indexes of the glass microscope slide and immersion oil are the same, the light rays do not refract when passing from one to the other when an oil immersion objective lens is used. This method produces images with better resolution at magnifications greater than 900×.

Q **What is meant by** *resolution*?

FIGURE 3.2 Relationships between the sizes of various specimens and the resolution of the human eye, light microscope, and electron microscope. The area in yellow shows the size range of most of the organisms we will be studying in this book.

Q **How many 2 μm bacterial cells fit end-to-end across a 10 μm eukaryotic cell?**

Figure 3.2 shows various specimens that can be resolved by the human eye, light microscope, and electron microscope.

To obtain a clear, finely detailed image under a compound light microscope, specimens must be made to contrast sharply with their *medium* (substance in which they are suspended). To attain such contrast, we must change the

refractive index of specimens from that of their medium. The **refractive index** is a measure of the light-bending ability of a medium. We change the refractive index of specimens by staining them, a procedure we will discuss shortly. Light rays move in a straight line through a single medium. After staining, when light rays pass through the two materials (the specimen and its medium) with different refractive indexes, the rays change direction (refract) from a straight path by bending or changing an angle at the boundary between the materials and increase the image's contrast between the specimen and the medium. As the light rays travel away from the specimen, they spread out and enter the objective lens, and the image is thereby magnified.

To achieve high magnification (1000×) with good resolution, the objective lens must be small. Although we want light traveling through the specimen and medium to refract differently, we do not want to lose light rays after they have passed through the stained specimen. To preserve the direction of light rays at the highest magnification, immersion oil is placed between the glass slide and the oil immersion objective lens (Figure 3.3). The immersion oil has the same refractive index as glass, so the oil becomes part of the optics of the glass of the microscope. Unless immersion oil is used, light rays are refracted as they enter the air from

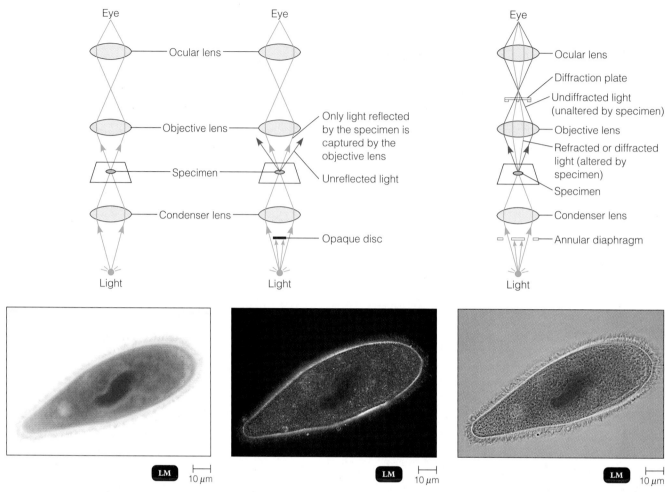

(a) Brightfield. (Top) The path of light in brightfield microscopy, the type of illumination produced by regular compound light microscopes. (Bottom) Brightfield illumination shows internal structures and the outline of the transparent pellicle (external covering).

(b) Darkfield. (Top) The darkfield microscope uses a special condenser with an opaque disc that eliminates all light in the center of the beam. The only light that reaches the specimen comes in at an angle; thus, only light reflected by the specimen (blue rays) reaches the objective lens. (Bottom) Against the black background seen with darkfield microscopy, edges of the cell are bright, some internal structures seem to sparkle, and the pellicle is almost visible.

(c) Phase-contrast. (Top) In phase-contrast microscopy, the specimen is illuminated by light passing through an annular diaphragm. Direct light rays (unaltered by the specimen) travel a different path than light rays that are reflected or diffracted as they pass through the specimen. These two sets of rays are combined at the eye. Reflected or diffracted light rays are indicated in blue; direct rays are red. (Bottom) Phase-contrast microscopy shows greater differentiation of internal structures and clearly shows the pellicle.

FIGURE 3.4 Brightfield, darkfield, and phase-contrast microscopy. The illustrations show the contrasting light pathways of each of these types of microscopy. The photographs compare the same *Paramecium* specimen using these three different microscopy techniques.

Q   **What are the advantages of brightfield, darkfield, and phase-contrast microscopy?**

the slide, and the objective lens would have to be increased in diameter to capture most of them. The oil has the same effect as increasing the objective lens diameter; therefore, it improves the resolving power of the lenses. If oil is not used with an oil immersion objective lens, the image becomes fuzzy, with poor resolution.

Under usual operating conditions, the field of vision in a compound light microscope is brightly illuminated. By focusing the light, the condenser produces a **brightfield illumination** (Figure 3.4a).

It is not always desirable to stain a specimen. However, an unstained cell has little contrast with its surroundings and is therefore difficult to see. Unstained cells are more easily observed with the modified compound microscopes described in the next section.

DARKFIELD MICROSCOPY

A **darkfield microscope** is used for examining live microorganisms that either are invisible in the ordinary light microscope, cannot be stained by standard methods, or are so distorted by staining that their characteristics then cannot be identified. Instead of the normal condenser, a darkfield microscope uses a darkfield condenser that contains an opaque disc. The disc blocks light that would enter the objective lens directly. Only light that is reflected off (turned away from) the specimen enters the objective lens. Because there is no direct background light, the specimen appears light against a black background—the dark field (Figure 3.4b). This technique is frequently used to examine unstained microorganisms suspended in liquid. One use for darkfield microscopy is the examination of very thin spirochetes, such as *Treponema pallidum* (tre-pō-ne'mä pal'li-dum), the causative agent of syphilis.

PHASE-CONTRAST MICROSCOPY

Another way to observe microorganisms is with a **phase-contrast microscope.** Phase-contrast microscopy is especially useful because it permits detailed examination of internal structures in *living* microorganisms. In addition, it is not necessary to fix (attach the microbes to the microscope slide) or stain the specimen—procedures that could distort or kill the microorganisms.

The principle of phase-contrast microscopy is based on the wave nature of light rays and the fact that light rays can be *in phase* (their peaks and valleys match) or *out of phase*. If the wave peak of light rays from one source coincides with the wave peak of light rays from another source, the rays interact to produce *reinforcement* (relative brightness). However, if the wave peak from one light source coincides with the wave trough from another light source, the rays interact to produce *interference* (relative darkness). In a phase-contrast microscope, one set of light rays comes directly from the light source. The other set comes from light that is reflected or diffracted from a particular structure in the specimen. (*Diffraction* is the scattering of light rays as they "touch" a specimen's edge. The diffracted rays are bent away from the parallel light rays that pass farther from the specimen.) When the two sets of light rays—direct rays and reflected or diffracted rays—are brought together, they form an image of the specimen on the ocular lens, containing areas that are relatively light (in phase), through shades of gray, to black (out of phase; Figure 3.4c). In phase-contrast microscopy, the internal structures of a cell become more sharply defined.

DIFFERENTIAL INTERFERENCE CONTRAST (DIC) MICROSCOPY

Differential interference contrast (DIC) microscopy is similar to phase-contrast microscopy in that it uses

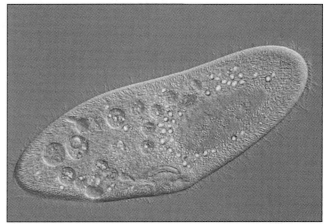

LM | 10 µm

FIGURE 3.5 Differential interference contrast (DIC) microscopy. Like phase-contrast, DIC uses differences in refractive indexes to produce an image, in this case of the protozoan *Paramecium*. The colors in the image are produced by prisms that split the two light beams used in this process.

Q **Why is the resolution of a DIC microscope higher than that of a phase-contrast microscope?**

differences in refractive indexes. However, a DIC microscope uses two beams of light instead of one. In addition, prisms split each light beam, adding contrasting colors to the specimen. Therefore, the resolution of a DIC microscope is higher than that of a standard phase-contrast microscope. Also, the image is brightly colored and appears nearly three-dimensional (Figure 3.5).

FLUORESCENCE MICROSCOPY

Fluorescence microscopy takes advantage of **fluorescence,** the ability of substances to absorb short wavelengths of light (ultraviolet) and give off light at a longer wavelength (visible). Some organisms fluoresce naturally under ultraviolet light; if the specimen to be viewed does not naturally fluoresce, it is stained with one of a group of fluorescent dyes called *fluorochromes*. When microorganisms stained with a fluorochrome are examined under a fluorescence microscope with an ultraviolet or near-ultraviolet light source, they appear as luminescent, bright objects against a dark background.

Fluorochromes have special attractions for different microorganisms. For example, the fluorochrome auramine O, which glows yellow when exposed to ultraviolet light, is strongly absorbed by *Mycobacterium tuberculosis*, the bacterium that causes tuberculosis. When the dye is applied to a sample of material suspected of containing the bacterium, the bacterium can be detected by the appearance of bright yellow organisms against a dark background (Table 3.2). *Bacillus anthracis*, the causative agent of

Text continues on page 64.

TABLE 3.2	A Summary of Various Types of Microscopes		
Microscope Type	Distinguishing Features	Typical Image	Principal Uses
Light Brightfield	Uses visible light as a source of illumination; cannot resolve structures smaller than about 0.2 μm; specimen appears against a bright background. Inexpensive and easy to use.	 LM |—| 10 μm	To observe various stained specimens and to count microbes; does not resolve very small specimens, such as viruses.
Darkfield	Uses a special condenser with an opaque disc that blocks light from entering the objective lens directly; light reflected by specimen enters the objective lens, and the specimen appears light against a black background.	 LM |—| 10 μm	To examine living microorganisms that are invisible in brightfield microscopy, do not stain easily, or are distorted by staining; frequently used to detect *Treponema pallidum* in the diagnosis of syphilis.
Phase-contrast	Uses a special condenser containing an annular (ring-shaped) diaphragm. The diaphragm allows direct light to pass through the condenser, focusing light on the specimen and a diffraction plate in the objective lens. Direct and reflected or diffracted light rays are brought together to produce the image. No staining required.	 LM |—| 10 μm	To facilitate detailed examination of the internal structures of living specimens.
Differential interference contrast (DIC)	Like phase-contrast, uses differences in refractive indexes to produce images. Uses two beams of light separated by prisms; the specimen appears colored as a result of the prism effect. No staining required.	 LM |—| 10 μm	To provide three-dimensional images.
Fluorescence	Uses an ultraviolet or near-ultraviolet source of illumination that causes fluorescent microbes (green-colored) in a specimen to emit light.	 LM |—| 10 μm	For fluorescent-antibody techniques (immunofluorescence) to rapidly detect and identify microbes in tissues or clinical specimens.

TABLE 3.2	*(continued)*		
Microscope Type	Distinguishing Features	Typical Image	Principal Uses
Confocal	Uses laser light to illuminate one plane of a specimen at a time.	 LM 10 μm	To obtain two-and three-dimensional images of cells for biomedical applications.
Scanning acoustic	Uses a sound wave of specific frequency that travels through the specimen with a portion being reflected when it hits an interface within the material.	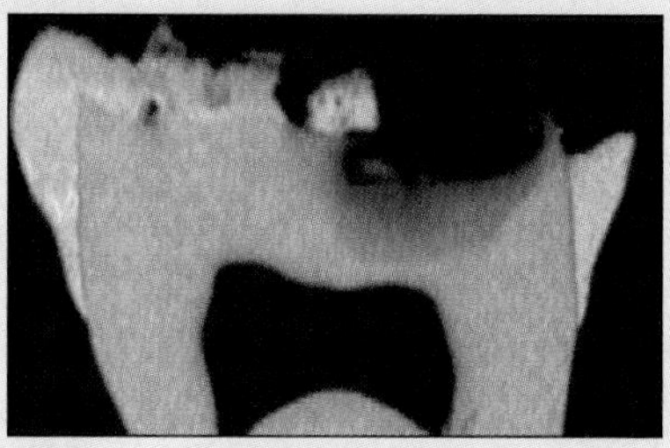 SAM 50 μm	To examine living cells attached to another surface, such as cancer cells, artery plaque, and biofilms.
Electron Transmission	Uses a beam of electrons instead of light; electrons pass through the specimen; because of the shorter wavelength of electrons, structures smaller than 0.2 μm can be resolved. The image produced is two-dimensional.	TEM 10 μm	To examine viruses or the internal ultrastructure in thin sections of cells (usually magnified 10,000–100,000×).
Scanning	Uses a beam of electrons instead of light; electrons are reflected from the specimen; because of the shorter wavelength of electrons, structures smaller than 0.2 μm can be resolved. The image produced appears three-dimensional.	 LM 10 μm	To study the surface features of cells and viruses (usually magnified 1000–10,000×).
Scanned-probe Scanning tunneling	Uses a thin metal probe that scans a specimen and produces an image revealing the bumps and depressions of the atoms on the surface of the specimen. Resolving power is much greater than that of an electron microscope. No special preparation required.	STM 10 nm	Provides very detailed views of molecules inside cells.

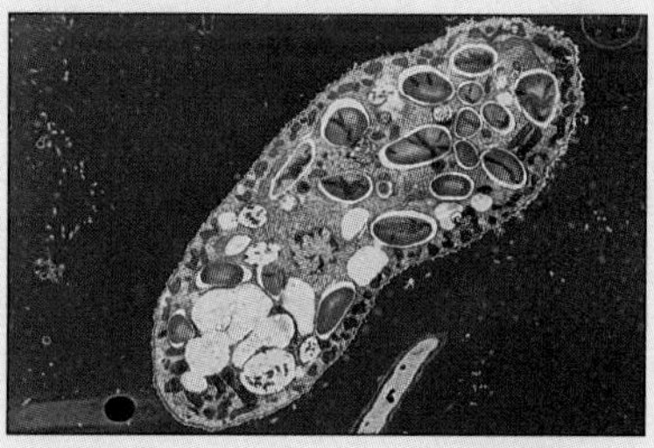

Microscope Type	Distinguishing Features	Typical Image	Principal Uses

TABLE 3.2   **A Summary of Various Types of Microscopes** (continued)

Scanned-probe (continued)

| Atomic force | Uses a metal-and-diamond probe gently forced down along the surface of the specimen. Produces a three-dimensional image. No special preparation required. | 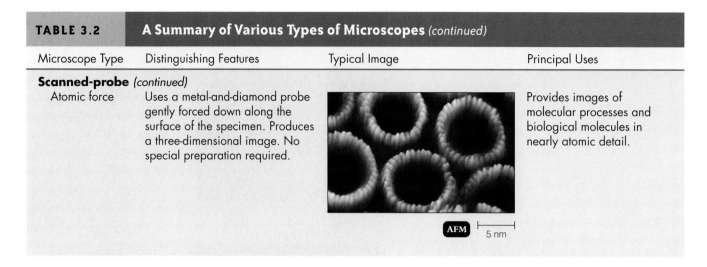 AFM 5 nm | Provides images of molecular processes and biological molecules in nearly atomic detail. |

anthrax, appears apple green when stained with another fluorochrome, fluorescein isothiocyanate (FITC).

The principal use of fluorescence microscopy is a diagnostic technique called the **fluorescent-antibody (FA) technique,** or **immunofluorescence. Antibodies** are natural defense molecules that are produced by humans and many animals in reaction to a foreign substance, or **antigen.** Fluorescent antibodies for a particular antigen are obtained as follows: an animal is injected with a specific antigen, such as a bacterium, and the animal then begins to produce specific antibodies against that antigen. After a sufficient time, the antibodies are removed from the serum of the animal. Next, as shown in Figure 3.6a, a fluorochrome is chemically combined with the antibodies. These fluorescent antibodies are then added to a microscope slide containing an unknown bacterium. If this unknown bacterium is the same bacterium that was injected into the animal, the fluorescent antibodies bind to antigens on the surface of the bacterium, causing it to fluoresce.

This technique can detect bacteria or other pathogenic microorganisms, even within cells, tissues, or other clinical specimens (Figure 3.6b). Of paramount importance, it can be used to identify a microbe in minutes. Immunofluorescence is especially useful in diagnosing syphilis and rabies. We will say more about antigen-antibody reactions and immunofluorescence in Chapter 18.

CONFOCAL MICROSCOPY

A fairly recent development in light microscopy is known as **confocal microscopy.** Like fluorescent microscopy, specimens are stained with fluorochromes so they will emit, or return, light. In confocal microscopy, one plane of a small region of a specimen is illuminated with a laser, which passes the returned light through an aperture aligned with the illuminated region. Each plane corresponds to an image of a fine slice that has been physically cut from a specimen. Successive planes and regions are il-

luminated until the entire specimen has been scanned. Because confocal microscopy uses a pinhole aperture, it eliminates the blurring that occurs with other microscopes. As a result, exceptionally clear two-dimensional images can be obtained, with improved resolution of up to 40% over that of other microscopes.

Most confocal microscopes are used in conjunction with computers to construct three-dimensional images. The scanned planes of a specimen, which resemble a stack of images, are converted to a digital form that can be used by a computer to construct a three-dimensional representation. The reconstructed images can be rotated and viewed in any orientation. This technique has been used to obtain three-dimensional images of entire cells and cellular components (Figure 3.7). In addition, confocal microscopy can be used to evaluate cellular physiology by monitoring the distributions and concentrations of substances such as ATP and calcium ions.

SCANNING ACOUSTIC MICROSCOPY

Scanning acoustic microscopy (SAM) basically consists of interpreting the action of a sound wave sent through a specimen. A sound wave of a specific frequency travels through the specimen, with a portion of it being reflected back every time it hits an interface within the material. The resolution is about 1 μm. SAM is used to study living cells attached to another surface, such as cancer cells, artery plaque, and bacterial biofilms that foul equipment (Figure 3.8).

ELECTRON MICROSCOPY

LEARNING OBJECTIVES
- Explain how electron microscopy differs from light microscopy.
- Identify one use for the TEM, SEM, and scanned-probe microscopes.

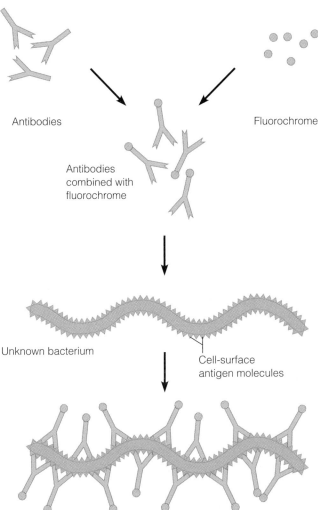

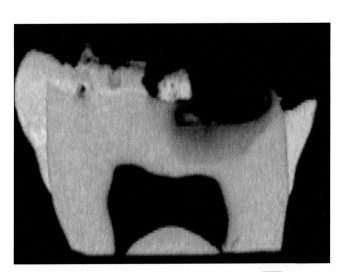

FIGURE 3.7 **Confocal microscopy.** Confocal microscopy produces three-dimensional images and can be used to look inside cells. Shown here are contractile vacuoles in *Paramecium multimicronucleatum*.

Q What feature of confocal microscopy eliminates the blurring that occurs with other microscopes?

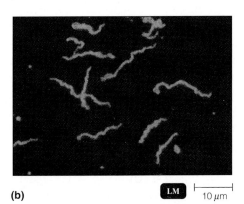

FIGURE 3.6 **The principle of immunofluorescence.**
(**a**) A type of fluorochrome is combined with antibodies against a specific type of bacterium. When the preparation is added to bacterial cells on a microscope slide, the antibodies attach to the bacterial cells, and the cells fluoresce when illuminated with ultraviolet light. (**b**) In the fluorescent treponemal antibody absorption (FTA-ABS) test for syphilis shown here, *Treponema pallidum* shows up as green cells against a darker background.

Q Why won't other bacteria fluoresce in the FTA-ABS test?

Scanning acoustic micrograph of decay on an adult molar tooth

FIGURE 3.8 **Scanning acoustic microscopy (SAM).**
Scanning acoustic microscopy essentially consists of interpreting the action of sound waves through a specimen.

Q What is the principal use of SAM?

Objects smaller than about 0.2 μm, such as viruses or the internal structures of cells, must be examined with an **electron microscope.** In electron microscopy, a beam of electrons is used instead of light. Free electrons travel in waves. The resolving power of the electron microscope is far greater than that of the other microscopes described here so far. The better resolution of electron microscopes is due to the shorter wavelengths of electrons; the wavelengths of electrons are about 100,000 times smaller than the wavelengths of visible light. Thus, electron microscopes are used to examine structures too small to be resolved with light microscopes. Images produced by electron microscopes are always black and white, but they may be colored artificially to accentuate certain details.

Instead of using glass lenses, an electron microscope uses electromagnetic lenses to focus a beam of electrons onto a specimen. There are two types of electron microscopes: the transmission electron microscope and the scanning electron microscope.

TRANSMISSION ELECTRON MICROSCOPY

In the **transmission electron microscope (TEM),** a finely focused beam of electrons from an electron gun passes through a specially prepared, ultrathin section of the specimen (Figure 3.9a). The beam is focused on a small area of the specimen by an electromagnetic condenser lens that performs roughly the same function as the condenser of a light microscope—directing the beam of electrons in a straight line to illuminate the specimen.

Electron microscopes use electromagnetic lenses to control illumination, focus, and magnification. Instead of being placed on a glass slide, as in light microscopes, the specimen is usually placed on a copper mesh grid. The beam of electrons passes through the specimen and then through an electromagnetic objective lens, which magnifies the image. Finally, the electrons are focused by an electromagnetic projector lens (rather than by an ocular lens as in a light microscope) onto a fluorescent screen or photographic plate. The final image, called a *transmission electron micrograph,* appears as many light and dark areas, depending on the number of electrons absorbed by different areas of the specimen.

In practice, the transmission electron microscope can resolve objects as close together as 2.5 nm, and objects are generally magnified 10,000 to 100,000×. Because most microscopic specimens are so thin, the contrast between their ultrastructures and the background is weak. Contrast can be greatly enhanced by using a "stain" that absorbs electrons and produces a darker image in the stained region. Salts of various heavy metals, such as lead, osmium, tungsten, and uranium, are commonly used as stains. These metals can be fixed onto the specimen (*positive staining*) or used to increase the electron opacity of the surrounding field (*negative staining*). Negative staining is useful for the study of the very smallest specimens, such as virus particles, bacterial flagella, and protein molecules.

In addition to positive and negative staining, a microbe can be viewed by a technique called *shadow casting.* In this procedure, a heavy metal such as platinum or gold is sprayed at an angle of about 45° so that it strikes the microbe from only one side. The metal piles up on one side of the specimen, and the uncoated area on the opposite side of the specimen leaves a clear area behind it as a shadow. This gives a three-dimensional effect to the specimen and provides a general idea of the size and shape of the specimen (see Figure 4.6b, page 81).

Transmission electron microscopy has high resolution and is extremely valuable for examining different layers of specimens. However, it does have certain disadvantages. Because electrons have limited penetrating power, only a very thin section of a specimen (about 100 nm) can be studied effectively. Thus, the specimen has no three-dimensional aspect. In addition, specimens must be fixed, dehydrated, and viewed under a high vacuum to prevent electron scattering. These treatments not only kill the specimen, but also cause some shrinkage and distortion, sometimes to the extent that there may appear to be additional structures in a prepared cell. Structures that appear as a result of the method of preparation are called *artifacts.*

SCANNING ELECTRON MICROSCOPY

The **scanning electron microscope (SEM)** overcomes the problem of sectioning associated with a transmission electron microscope. A scanning electron microscope provides striking three-dimensional views of specimens (Figure 3.9b). In scanning electron microscopy, an electron gun produces a finely focused beam of electrons called the primary electron beam. These electrons pass through electromagnetic lenses and are directed over the surface of the specimen. The primary electron beam knocks electrons out of the surface of the specimen, and the secondary electrons thus produced are transmitted to an electron collector, amplified, and used to produce an image on a viewing screen or photographic plate. The image is called a *scanning electron micrograph.* This microscope is especially useful in studying the surface structures of intact cells and viruses. In practice, it can resolve objects as close together as 20 nm, and objects are generally magnified 1000 to 10,000×.

SCANNED-PROBE MICROSCOPY

investigated

Since the early 1980s, several new types of microscopes, called **scanned-probe microscopes,** have been developed. They use various kinds of probes to examine the surface of a specimen at very close range, and they do so without modifying the specimen or exposing it to damaging, high-energy radiation. Such microscopes can be used to

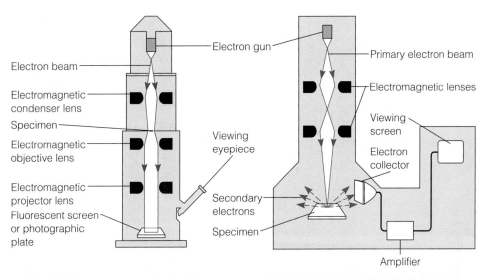

Electron gun

Electron beam

Electromagnetic condenser lens

Specimen

Electromagnetic objective lens

Electromagnetic projector lens

Fluorescent screen or photographic plate

Viewing eyepiece

Primary electron beam

Electromagnetic lenses

Viewing screen

Electron collector

Secondary electrons

Specimen

Amplifier

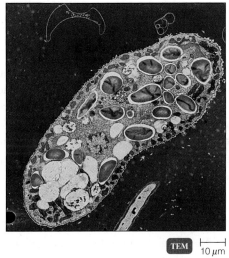

TEM |——| 10 μm

SEM |——| 10 μm

(a) Transmission. (Top) In a transmission electron microscope, electrons pass through the specimen and are scattered. Magnetic lenses focus the image onto a fluorescent screen or photographic plate. (Bottom) This colorized transmission electron micrograph (TEM) shows a thin slice of a *Paramecium*. In this type of microscopy, the internal structures present in the slice can be seen.

(b) Scanning. (Top) In a scanning electron microscope, primary electrons sweep across the specimen and knock electrons from its surface. These secondary electrons are picked up by a collector, amplified, and transmitted onto a viewing screen or photographic plate. (Bottom) In this colorized scanning electron micrograph (SEM), the surface structures of a *Paramecium* can be seen. Note the three-dimensional appearance of this cell, in contrast to the two-dimensional appearance of the transmission electron micrograph in part (a).

FIGURE 3.9 Transmission and scanning electron microscopy. The illustrations show the pathways of electron beams used to create images of the specimens. The photographs show a *Paramecium* viewed with both of these types of electron microscopes. Although electron micrographs are normally black and white, these and other electron micrographs in this book have been artifically colorized for emphasis.

Q **How do TEM and SEM images of the same organism differ?**

map atomic and molecular shapes, to characterize magnetic and chemical properties, and to determine temperature variations inside cells. Among the new scanned-probe microscopes are the scanning tunneling microscope and the atomic force microscope, discussed next.

SCANNING TUNNELING MICROSCOPY

Scanning tunneling microscopy (STM) uses a thin metal (tungsten) probe that scans a specimen and produces an image revealing the bumps and depressions of the

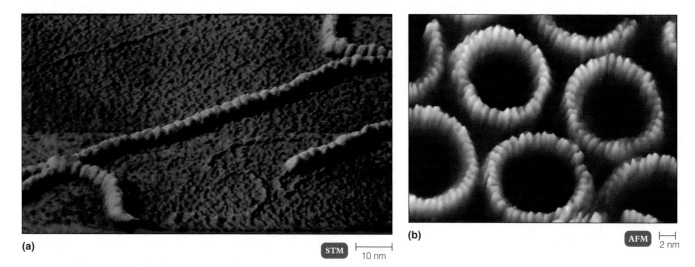

(a)   STM   ⊢—⊣ 10 nm

(b)   AFM   ⊢—⊣ 2 nm

FIGURE 3.10 Scanned-probe microscopy. (**a**) Scanning tunneling microscopy (STM) image of RecA protein from *E.coli*. This protein is involved in repair of DNA. (**b**) Atomic force microscopy (AFM) image of perfringoglysin O toxin from *Clostridium perfringens*. This toxin makes holes in human plasma membranes.

Q **What is the principle employed in scanned-probe microscopy?**

atoms on the surface of the specimen (Figure 3.10a). The resolving power of an STM is much greater than that of an electron microscope; it can resolve features that are only about 1/100 the size of an atom. Moreover, special preparation of the specimen for observation is not needed. STMs are used to provide incredibly detailed views of molecules such as DNA.

ATOMIC FORCE MICROSCOPY

In **atomic force microscopy (AFM),** a metal-and-diamond probe is gently forced down onto a specimen. As the probe moves along the surface of the specimen, its movements are recorded and a three-dimensional image is produced (Figure 3.10b). As with STM, AFM does not require special specimen preparation. AFM is used to image both biological substances (in nearly atomic detail) and molecular processes (such as the assembly of fibrin, a component of a blood clot).

The various types of microscopy just described are summarized in Table 3.2.

PREPARATION OF SPECIMENS FOR LIGHT MICROSCOPY

LEARNING OBJECTIVE
• Differentiate between an acidic dye and a basic dye.

Because most microorganisms appear almost colorless when viewed through a standard light microscope, we often must prepare them for observation. One of the ways

this can be done is by staining (coloring). Next we will discuss several different staining procedures.

PREPARING SMEARS FOR STAINING

Most initial observations of microorganisms are made with stained preparations. **Staining** simply means coloring the microorganisms with a dye that emphasizes certain structures. Before the microorganisms can be stained, however, they must be **fixed** (attached) to the microscope slide. Fixing simultaneously kills the microorganisms and fixes them to the slide. It also preserves various parts of microbes in their natural state with only minimal distortion.

When a specimen is to be fixed, a thin film of material containing the microorganisms is spread over the surface of the slide. This film, called a **smear,** is allowed to air dry. In most staining procedures the slide is then *fixed* by passing it through the flame of a Bunsen burner several times, smear side up, or by covering the slide with methyl alcohol for 1 minute. Stain is applied and then washed off with water; then the slide is blotted with absorbent paper. Without fixing, the stain might wash the microbes off the slide. The stained microorganisms are now ready for microscopic examination.

Stains are salts composed of a positive and a negative ion, one of which is colored and is known as the *chromophore.* The color of so-called **basic dyes** is in the positive ion; in **acidic dyes,** it is in the negative ion. Bacteria are slightly negatively charged at pH 7. Thus, the colored positive ion in a basic dye is attracted to the negatively charged bacterial cell. Basic dyes, which include crystal violet, methylene blue, malachite green, and safranin, are more commonly used than acidic dyes. Acidic dyes are not attracted to most types of bacteria because the dye's negative ions are repelled by the negatively charged bacterial surface, so the stain colors the background instead. Preparing colorless bacteria against a colored background is called **negative staining.** It is valuable in the observation of overall cell shapes, sizes, and capsules because the cells are made highly visible against a contrasting dark background (see Figure 3.13a). Distortions of cell size and shape are minimized because fixing is not necessary and the cells do not pick up the stain. Examples of acidic dyes are eosin, acid fuchsin, and nigrosin.

To apply acidic or basic dyes, microbiologists use three kinds of staining techniques: simple, differential, and special.

SIMPLE STAINS

LEARNING OBJECTIVE

• Explain the purpose of simple staining.

A **simple stain** is an aqueous or alcohol solution of a single basic dye. Although different dyes bind specifically to different parts of cells, the primary purpose of a simple stain is to highlight the entire microorganism so that cellular shapes and basic structures are visible. The stain is applied to the fixed smear for a certain length of time and then washed off, and the slide is dried and examined. Occasionally, a chemical is added to the solution to intensify the stain; such an additive is called a **mordant.** One function of a mordant is to increase the affinity of a stain for a biological specimen; another is to coat a structure (such as a flagellum) to make it thicker and easier to see after it is stained with a dye. Some of the simple stains commonly used in the laboratory are methylene blue, carbolfuchsin, crystal violet, and safranin.

DIFFERENTIAL STAINS

LEARNING OBJECTIVES

• List the steps in preparing a Gram stain, and describe the appearance of gram-positive and gram-negative cells after each step.

• Compare and contrast the Gram stain and the acid-fast stain.

Unlike simple stains, **differential stains** react differently with different kinds of bacteria and thus can be used to distinguish among them. The differential stains most frequently used for bacteria are the Gram stain and the acid-fast stain.

GRAM STAIN

The **Gram stain** was developed in 1884 by the Danish bacteriologist Hans Christian Gram. It is one of the most useful staining procedures because it classifies bacteria into two large groups: gram-positive and gram-negative.

In this procedure (Figure 3.11a),

1 A heat-fixed smear is covered with a basic purple dye, usually crystal violet. Because the purple stain imparts its color to all cells, it is referred to as a **primary stain.**

2 After a short time, the purple dye is washed off, and the smear is covered with iodine, a mordant. When the iodine is washed off, both gram-positive and gram-negative bacteria appear dark violet or purple.

3 Next, the slide is washed with alcohol or an alcohol-acetone solution. This solution is a **decolorizing agent,** which removes the purple from the cells of some species but not from others.

4 The alcohol is rinsed off, and the slide is then stained with safranin, a basic red dye. The smear is washed again, blotted dry, and examined microscopically.

The purple dye and the iodine combine in the cytoplasm of each bacterium and color it dark violet or purple. Bacteria that retain this color after the alcohol has attempted to decolorize them are classified as **gram-positive;** bacteria that lose the dark violet or purple color after decolorization are classified as **gram-negative** (Figure 3.11b). Because gram-negative bacteria are colorless after the alcohol wash, they are no longer visible. This is why the basic dye safranin is applied; it turns the gram-negative bacteria pink. Stains such as safranin that have a contrasting color to the primary stain are called **counterstains.** Because gram-positive bacteria retain the original purple stain, they are not affected by the safranin counterstain.

As you will see in Chapter 4, different kinds of bacteria react differently to the Gram stain because structural differences in their cell walls affect the retention or escape of a combination of crystal violet and iodine, called the crystal violet–iodine (CV–I) complex. Among other differences, gram-positive bacteria have a thicker peptidoglycan (disaccharides and amino acids) cell wall than gram-negative bacteria. In addition, gram-negative bacteria contain a layer of lipopolysaccharide (lipids and polysaccharides) as part of their cell wall (see Figure 4.13, page 86). When applied to both gram-positive and gram-negative cells, crystal violet and then iodine readily enter the cells. Inside the cells, the crystal violet and iodine combine to form CV–I. This complex is larger than the crystal violet molecule that entered the cells, and, because of its size, it cannot be washed out of the intact peptidoglycan layer of

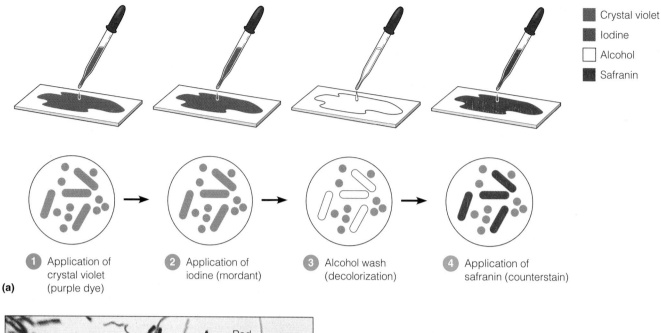

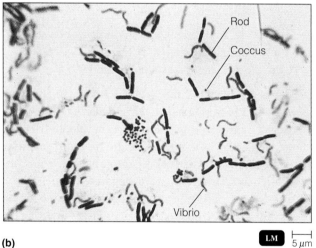

FIGURE 3.11 **Gram staining. (a)** Procedure. **(b)** Micrograph of gram-stained bacteria. The rods and cocci (purple) are gram-positive, and the vibrios (pink) are gram-negative.

Q **How can the Gram reaction be useful in prescribing antibiotic treatment?**

gram-positive cells by alcohol. Consequently, gram-positive cells retain the color of the crystal violet dye. In gram-negative cells, however, the alcohol wash disrupts the outer lipopolysaccharide layer, and the CV–I complex is washed out through the thin layer of peptidoglycan. As a result, gram-negative cells are colorless until counterstained with safranin, after which they are pink.

In summary, gram-positive cells retain the dye and remain purple. Gram-negative cells do not retain the dye; they are colorless until counterstained with a red dye.

The Gram method is one of the most important staining techniques in medical microbiology. But Gram staining results are not universally applicable because some bacterial cells stain poorly or not at all. The Gram reaction is most consistent when it is used on young, growing bacteria.

The Gram reaction of a bacterium can provide valuable information for the treatment of disease. Gram-positive

bacteria tend to be killed easily by penicillins and cephalosporins. Gram-negative bacteria are generally more resistant because the antibiotics cannot penetrate the lipopolysaccharide layer. Some resistance to these antibiotics among both gram-positive and gram-negative bacteria is due to bacterial inactivation of the antibiotics.

ACID-FAST STAIN

Another important differential stain (one that differentiates bacteria into distinctive groups) is the **acid-fast stain,** which binds strongly only to bacteria that have a waxy material in their cell walls. Microbiologists use this stain to identify all bacteria in the genus *Mycobacterium,* including the two important pathogens *Mycobacterium tuberculosis,* the causative agent of tuberculosis, and *Mycobacterium leprae* (lep′rī), the causative agent of leprosy. This stain is also used to identify the pathogenic strains of the genus *Nocardia* (nō-kär′dē-ä).

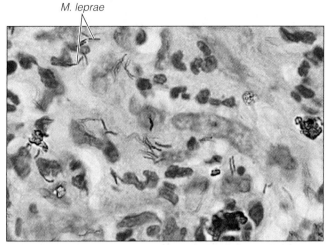

M. leprae

FIGURE 3.12 Acid-fast bacteria. The *Mycobacterium leprae* bacteria that have infected this tissue have been stained red with an acid-fast stain. Non–acid-fast cells are stained with the methylene blue counterstain.

Q **What diseases can be diagnosed using the acid-fast stain?**

In the acid-fast staining procedure, the red dye carbolfuchsin is applied to a fixed smear, and the slide is gently heated for several minutes. (Heating enhances penetration and retention of the dye.) Then the slide is cooled and washed with water. The smear is next treated with acid-alcohol, a decolorizer, which removes the red stain from bacteria that are not acid-fast. The acid-fast microorganisms retain the red color because the carbolfuchsin is more soluble in the cell wall lipids than in the acid-alcohol (Figure 3.12). In non–acid-fast bacteria, whose cell walls lack the lipid components, the carbolfuchsin is rapidly removed during decolorization, leaving the cells colorless. The smear is then stained with a methylene blue counterstain. Non–acid-fast cells appear blue after application of the counterstain.

SPECIAL STAINS

LEARNING OBJECTIVE
- Explain why each of the following is used: capsule stain, endospore stain, flagella stain.

Special stains are used to color and isolate specific parts of microorganisms, such as endospores and flagella, and to reveal the presence of capsules.

NEGATIVE STAINING FOR CAPSULES

Many microorganisms contain a gelatinous covering called a **capsule,** which we will discuss in our examination of the prokaryotic cell in Chapter 4. In medical microbiology, demonstrating the presence of a capsule is a means of

determining the organism's **virulence,** the degree to which a pathogen can cause disease.

Capsule staining is more difficult than other types of staining procedures because capsular materials are soluble in water and may be dislodged or removed during rigorous washing. To demonstrate the presence of capsules, a microbiologist can mix the bacteria in a solution containing a fine colloidal suspension of colored particles (usually India ink or nigrosin) to provide a contrasting background and then stain the bacteria with a simple stain, such as safranin (Figure 3.13a). Because of their chemical composition, capsules do not accept most biological dyes, such as safranin, and thus appear as halos surrounding each stained bacterial cell.

ENDOSPORE (SPORE) STAINING

An **endospore** is a special resistant, dormant structure formed within a cell that protects a bacterium from adverse environmental conditions. Although endospores are relatively uncommon in bacterial cells, they can be formed by a few genera of bacteria. Endospores cannot be stained by ordinary methods, such as simple staining and Gram staining, because the dyes do not penetrate the wall of the endospore.

The most commonly used endospore stain is the *Schaeffer-Fulton endospore stain* (Figure 3.13b). Malachite green, the primary stain, is applied to a heat-fixed smear and heated to steaming for about 5 minutes. The heat helps the stain penetrate the endospore wall. Then the preparation is washed for about 30 seconds with water to remove the malachite green from all of the cells' parts except the endospores. Next, safranin, a counterstain, is applied to the smear to stain portions of the cell other than endospores. In a properly prepared smear, the endospores appear green within red or pink cells. Because endospores are highly refractive, they can be detected under the light microscope when unstained, but they cannot be differentiated from inclusions of stored material without a special stain.

FLAGELLA STAINING

Bacterial **flagella** (singular: **flagellum**) are structures of locomotion too small to be seen with a light microscope without staining. A tedious and delicate staining procedure uses a mordant and the stain carbolfuchsin to build up the diameters of the flagella until they become visible under the light microscope (Figure 3.13c). Microbiologists use the number and arrangement of flagella as diagnostic aids.

* * *

A summary of stains is presented in Table 3.3. In the next chapter we will take a closer look at the structure of microbes and how they protect, nourish, and reproduce themselves. ✳ **Animation: Go to The Microbiology Place website or CD-ROM and click "Animations" to view microscopy.**

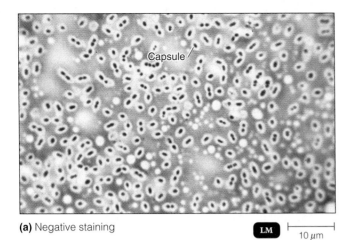

(a) Negative staining LM |———| 10 μm

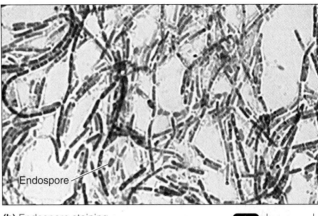

(b) Endospore staining LM |———| 10 μm

FIGURE 3.13 Special staining. (a) Capsule staining provides a contrasting background, so the capsules of these bacteria, *Klebsiella pneumoniae*, show up as light areas surrounding the stained cells. **(b)** Endospores are seen as green ovals in these rod-shaped cells of the bacterium *Bacillus cereus*, using the Schaeffer-Fulton endospore stain. **(c)** Flagella appear as wavy extensions from the ends of these cells of the bacterium *Spirillum volutans*. In relation to the body of the cell, the flagella are much thicker than normal because layers of the stain have accumulated from treatment of the specimen with a mordant.

Q **Of what value are capsules, endospores, and flagella to bacteria?**

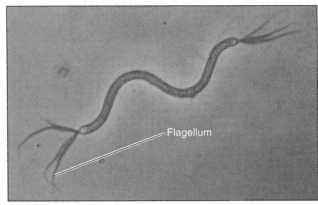

(c) Flagella staining LM |———| 10 μm

TABLE 3.3	A Summary of Various Stains and Their Uses
Stain	**Principal Uses**
Simple (methylene blue, carbolfuchsin, crystal violet, safranin)	Used to highlight microorganisms to determine cellular shapes and arrangements. Aqueous or alcohol solution of a single basic dye stains cells. (Sometimes a mordant is added to intensify the stain.)
Differential Gram	Used to distinguish among different kinds of bacteria. Classifies bacteria into two large groups: gram-positive and gram-negative. Gram-positive bacteria retain the crystal violet stain and appear purple. Gram-negative bacteria do not retain the crystal violet stain; they remain colorless until counterstained with safranin and then appear pink.
Acid-fast	Used to distinguish *Mycobacterium* species and some species of *Nocardia*. Acid-fast bacteria, once stained with carbolfuchsin and treated with acid-alcohol, remain red because they retain the carbolfuchsin stain. Non–acid-fast bacteria, when stained and treated the same way and then stained with methylene blue, appear blue because they lose the carbolfuchsin stain and are then able to accept the methylene blue stain.
Special	Used to color and isolate various structures, such as capsules, endospores, and flagella; sometimes used as a diagnostic aid.
Negative	Used to demonstrate the presence of capsules. Because capsules do not accept most stains, the capsules appear as unstained halos around bacterial cells and stand out against a contrasting background.
Endospore	Used to detect the presence of endospores in bacteria. When malachite green is applied to a heat-fixed smear of bacterial cells, the stain penetrates the endospores and stains them green. When safranin (red) is then applied, it stains the remainder of the cells red or pink.
Flagella	Used to demonstrate the presence of flagella. A mordant is used to build up the diameters of flagella until they become visible microscopically when stained with carbolfuchsin.

STUDY OUTLINE

UNITS OF MEASUREMENT (p. 56)

1. The standard unit of length is the meter (m).
2. Microorganisms are measured in micrometers, μm (10^{-6} m), and in nanometers, nm (10^{-9} m).

MICROSCOPY: THE INSTRUMENTS

(pp. 56–68)

1. A simple microscope consists of one lens; a compound microscope has multiple lenses.

LIGHT MICROSCOPY (pp. 56–68)

Compound Light Microscopy (pp. 56–60)

2. The most common microscope used in microbiology is the compound light microscope (**LM**).
3. The total magnification of an object is calculated by multiplying the magnification of the objective lens by the magnification of the ocular lens.
4. The compound light microscope uses visible light.
5. The maximum resolution, or resolving power (the ability to distinguish between two points) of a compound light microscope is 0.2 μm; maximum magnification is 2000×.
6. Specimens are stained to increase the difference between the refractive indexes of the specimen and the medium.
7. Immersion oil is used with the oil immersion lens to reduce light loss between the slide and the lens.
8. Brightfield illumination is used for stained smears.
9. Unstained cells are more productively observed using darkfield, phase-contrast, or DIC microscopy.

Darkfield Microscopy (p. 61)

10. The darkfield microscope shows a light silhouette of an organism against a dark background.
11. It is most useful for detecting the presence of extremely small organisms.

Phase-Contrast Microscopy (p. 61)

12. A phase-contrast microscope brings direct and reflected or diffracted light rays together (in phase) to form an image of the specimen on the ocular lens.
13. It allows the detailed observation of living organisms.

Differential Interference Contrast (DIC) Microscopy (p. 61)

14. The DIC microscope provides a colored, three-dimensional image of the object being observed.
15. It allows detailed observations of living cells.

Fluorescence Microscopy (pp. 61–64)

16. In fluorescence microscopy, specimens are first stained with fluorochromes and then viewed through a compound microscope by using an ultraviolet light source.

17. The microorganisms appear as bright objects against a dark background.
18. Fluorescence microscopy is used primarily in a diagnostic procedure called fluorescent-antibody (FA) technique, or immunofluorescence.

Confocal Microscopy (p. 64)

19. In confocal microscopy, a specimen is stained with a fluorescent dye and illuminated one plane at a time.
20. Using a computer to process the images, two-dimensional and three-dimensional images of cells can be produced.

SCANNING ACOUSTIC MICROSCOPY (p. 64)

21. Scanning acoustic microscopy (SAM) is based on the interpretation of sound waves through a specimen.
22. It is used to study living cells attached to surfaces such as cancer cells, artery plaque, and biofilms.

ELECTRON MICROSCOPY (pp. 64–66)

23. A beam of electrons, instead of light, is used with an electron microscope.
24. Electromagnets, instead of glass lenses, control focus, illumination, and magnification.
25. Thin sections of organisms can be seen in an electron micrograph produced using a transmission electron microscope (**TEM**). Magnification: 10,000–100,000×. Resolving power: 2.5 nm.
26. Three-dimensional views of the surfaces of whole microorganisms can be obtained with a scanning electron microscope (**SEM**). Magnification: 1000–10,000×. Resolving power: 20 nm.

SCANNED-PROBE MICROSCOPY (pp. 66–68)

27. Scanning tunneling microscopy (STM) and atomic force microscopy (AFM) produce three-dimensional images of the surface of a molecule.

PREPARATION OF SPECIMENS FOR LIGHT MICROSCOPY (pp. 68–72)

PREPARING SMEARS FOR STAINING (pp. 68–69)

1. Staining means coloring a microorganism with a dye to make some structures more visible.
2. Fixing uses heat or alcohol to kill and attach microorganisms to a slide.
3. A smear is a thin film of material used for microscopic examination.

4. Bacteria are negatively charged, and the colored positive ion of a basic dye will stain bacterial cells.

5. The colored negative ion of an acidic dye will stain the background of a bacterial smear; a negative stain is produced.

SIMPLE STAINS (p. 69)

6. A simple stain is an aqueous or alcohol solution of a single basic dye.

7. It is used to make cellular shapes and arrangements visible.

8. A mordant may be used to improve bonding between the stain and the specimen.

DIFFERENTIAL STAINS (pp. 69–71)

9. Differential stains, such as the Gram stain and acid-fast stain, differentiate bacteria according to their reactions to the stains.

10. The Gram stain procedure uses a purple stain (crystal violet), iodine as a mordant, an alcohol decolorizer, and a red counterstain.

11. Gram-positive bacteria retain the purple stain after the decolorization step; gram-negative bacteria do not and thus appear pink from the counterstain.

12. Acid-fast microbes, such as members of the genera *Mycobacterium* and *Nocardia,* retain carbolfuchsin after acid-alcohol decolorization and appear red; non–acid-fast microbes take up the methylene blue counterstain and appear blue.

SPECIAL STAINS (pp. 71–72)

13. Negative staining is used to make microbial capsules visible.

14. The endospore stain and flagella stain are special stains that color only certain parts of bacteria. ✹ **Animation: Microscopy. The Microbiology Place.**

STUDY QUESTIONS

✹ Access more review material either online at **The Microbiology Place** (www.microbiologyplace.com) or with **The Microbiology Place CD-ROM** packaged with your new book. There you'll find activities, practice tests, quizzes, flashcards, case studies, and more to help you succeed.

Answers to the Study Questions can be found in Appendix G.

REVIEW

1. Fill in the following blanks.

   $1 \ \mu m = \underline{10^{-6}} \ m$

   $1 \ \underline{nm} = 10^{-9} \ m$

   $1 \ \mu m = \underline{10^{3}} \ nm$

2. Which type of microscope would be best to use to observe each of the following?
   a. a stained bacterial smear (compound light micr_)
   b. unstained bacterial cells when the cells are small and no detail is needed (Dark field mi_)
   c. unstained live tissue when it is desirable to see some intracellular detail (Phase_contrast mi_)
   d. a sample that emits light when illuminated with ultraviolet light (Fluorescence mi_)
   e. intracellular detail of a cell that is 1 µm long (Electron mi_)
   f. unstained live cells in which intracellular structures are shown in color (D I C mi_)

3. Label the parts of the compound light microscope in the figure below.
   a. Ocular lens   d. Condensor
   b. Objective lens   e. Illuminator
   c. Diaphragm

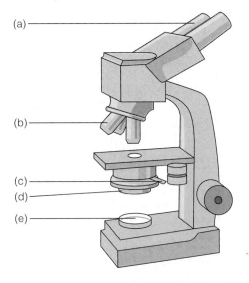

(a)
(b)
(c)
(d)
(e)

4. Calculate the total magnification of the nucleus of a cell being observed through a compound light microscope with a 10× ocular lens and an oil immersion lens. 1000×

5. An electron microscope differs from a light microscope in that _____ focused by _____ is used instead of
   a beam of electrons   magnets

telension-like screen or photograph plate

light, and the image is viewed on _____ instead of through the ocular lenses.

6. The maximum magnification of a compound microscope is *2000 X*; that of an electron microscope, *100,000 X* The maximum resolution of a compound microscope is *0.2 μm*; that of an electron microscope, *0.0025 μm* One advantage of a scanning electron microscope over a transmission electron microscope is *can see 3 dimension*

7. Why do basic dyes stain bacterial cells? Why don't acidic dyes stain bacterial cells?

8. When is it most appropriate to use each of the following?
   a. a simple stain   **c.** a negative stain
   b. a differential stain   **d.** a flagella stain

9. Why is a mordant used in the Gram stain? In the flagella stain?

10. What is the purpose of a counterstain in the acid-fast stain?

11. What is the purpose of a decolorizer in the Gram stain? In the acid-fast stain?

12. Choose from the following terms to fill in the blanks: *counterstain, decolorizer, mordant, primary stain.* In the endospore stain, safranin is the *counter stain* In the Gram stain, safranin is the *counterstain*

13. Fill in the following table regarding the Gram stain:

Steps	Appearance After This Step of	
	Gram-Positive Cells	**Gram-Negative Cells**
Crystal violet	*Purple*	*Purple*
Iodine	*Purple*	*Purple*
Alcohol-acetone	*Purple*	*Colorless*
Safranin	*Purple*	*Red*

MULTIPLE CHOICE

1. Assume you stain *Bacillus* by applying malachite green with heat and then counterstain with safranin. Through the microscope, the green structures are
   a. cell walls.   **d.** flagella.
   b. capsules.   **e.** impossible to identify.
   c. endospores.

3. Carbolfuchsin can be used as a simple stain and a negative stain. As a simple stain, the pH is
   a. 2.
   b. higher than the negative stain.
   c. lower than the negative stain.
   d. the same as the negative stain.

4. Looking at the cell of a photosynthetic microorganism, you observe that the chloroplasts are green in brightfield microscopy and red in fluorescence microscopy. You conclude that
   a. chlorophyll is fluorescent.
   b. the magnification has distorted the image.
   c. you're not looking at the same structure in both microscopes.

d. the stain masked the green color.
   e. none of the above

5. Which of the following is *not* a functionally analogous pair of stains?
   a. nigrosin and malachite green
   b. crystal violet and carbolfuchsin
   c. safranin and methylene blue
   d. ethanol-acetone and acid-alcohol
   e. none of the above

6. Which of the following pairs is mismatched?
   a. capsule—negative stain
   b. cell arrangement—simple stain
   c. cell size—negative stain
   d. Gram stain—bacterial identification
   e. none of the above

7. Assume you stain *Clostridium* by applying a basic stain, carbolfuchsin, with heat, decolorizing with acid-alcohol, and counterstaining with an acidic stain, nigrosin. Through the microscope, the endospores are _____1_____, and the cells are stained _____2_____.
   a. 1—red; 2—black   **d.** 1—red; 2—colorless
   b. 1—black; 2—colorless   **e.** 1—black; 2—red
   c. 1—colorless; 2—black

8. Assume that you are viewing a Gram-stained field of red cocci and blue bacilli through the microscope. You can safely conclude that you have
   a. made a mistake in staining.   **d.** young bacterial cells.
   b. two different species.   **e.** none of the above
   c. old bacterial cells.

9. In 1996, scientists described a new tapeworm parasite that had killed at least one person. The initial examination of the patient's abdominal mass was most likely made using
   a. brightfield microscopy.
   b. darkfield microscopy.
   c. electron microscopy.
   d. phase-contrast microscopy.
   e. fluorescence microscopy.

10. Which of the following is *not* a modification of a compound light microscope?
   a. brightfield microscopy
   b. darkfield microscopy
   c. electron microscopy
   d. phase-contrast microscopy
   e. fluorescence microscopy

CRITICAL THINKING

1. In a Gram stain, one step could be omitted and still allow differentiation between gram-positive and gram-negative cells. What is that one step?

2. Using a good compound light microscope with a resolving power of 0.3 μm, a 10× ocular lens, and a 100× oil immersion lens, would you be able to discern two objects separated by 3 μm? 0.3 μm? 300 nm?

3. Why isn't the Gram stain used on acid-fast bacteria? If you did Gram stain acid-fast bacteria, what would their Gram reaction be? What is the Gram reaction of non–acid-fast bacteria?

4. Endospores can be seen as refractile structures in unstained cells and as colorless areas in Gram-stained cells. Why is it necessary to do an endospore stain to verify the presence of endospores?

CLINICAL APPLICATIONS

1. In 1882, German bacteriologist Paul Erhlich described a method for staining *Mycobacterium* and noted, "It may be that all disinfecting agents which are acidic will be without effect on this [tubercle] bacillus, and one will have to be limited to alkaline agents." How did he reach this conclusion without testing disinfectants?

2. Laboratory diagnosis of *Neisseria gonorrhoeae* infection is based on microscopic examination of Gram-stained pus. Locate the bacteria in this light micrograph. What is the disease?

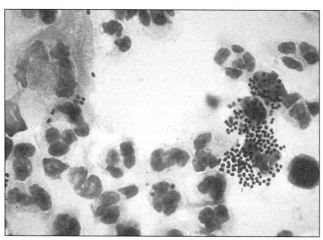

LM $\overline{10\ \mu m}$

3. Assume that you are viewing a Gram-stained sample of vaginal discharge. Large (10 μm) nucleated red cells are coated with small (0.5 μm × 1.5 μm) blue cells on their surfaces. What is the most likely explanation for the red and blue cells?

4. A sputum sample from Calle, a 30-year-old Asian elephant, was smeared onto a slide and air dried. The smear was fixed, covered with carbolfuchsin, and heated for 5 minutes. After washing with water, acid-alcohol was placed on the smear for 30 seconds. Finally, the smear was stained with methylene blue for 30 seconds, washed with water, and dried. On examination at 1000×, the zoo veterinarian saw red rods on the slide. What infections do the results suggest? (Calle was treated and recovered.)

4

Functional Anatomy of Prokaryotic and Eukaryotic Cells

Despite their complexity and variety, all living cells can be classified into two groups, prokaryotes and eukaryotes, based on certain structural and functional characteristics. In general, prokaryotes are structurally simpler and smaller than eukaryotes. The DNA (genetic material) of prokaryotes is usually arranged in a single, circularly arranged chromosome and is not surrounded by a membrane; the DNA of eukaryotes is found in multiple chromosomes in a membrane-enclosed nucleus. Prokaryotes lack membrane-enclosed organelles, specialized structures that carry on various activities. Additional differences are discussed shortly. Plants and animals are entirely composed of eukaryotic cells. In the microbial world, Bacteria and Archaea are prokaryotes. Other cellular microbes—fungi (yeasts and molds), protozoa, and algae—are eukaryotes. Humans are able to exploit the differences between bacterial and human cells to protect themselves from disease. For example, certain drugs kill or inhibit bacteria while not harming human cells, and unique chemicals on the surface of bacteria stimulate the body to mount a defensive response to eliminate them.

Viruses, as noncellular elements, do not fit into any organizational scheme of living cells. They are genetic particles that replicate but are unable to perform the usual chemical activities of living cells. Viral structure and activity will be discussed in Chapter 13. In this chapter we will concentrate on describing prokaryotic and eukaryotic cells.

UNDER THE MICROSCOPE

Bacillus. Cells of the genus *Bacillus* frequently form long chains.

COMPARING PROKARYOTIC AND EUKARYOTIC CELLS: AN OVERVIEW

LEARNING OBJECTIVE
- Compare and contrast the overall cell structure of prokaryotes and eukaryotes.

Prokaryotes and eukaryotes are chemically similar, in the sense that they both contain nucleic acids, proteins, lipids, and carbohydrates. They use the same kinds of chemical reactions to metabolize food, build proteins, and store energy. It is primarily the structure of cell walls and membranes, and the absence of *organelles* (specialized cellular structures that have specific functions), that distinguish prokaryotes from eukaryotes.

The chief distinguishing characteristics of **prokaryotes** (from the Greek words meaning prenucleus) are as follows:

1. Their DNA is not enclosed within a membrane and is usually a singular circularly arranged chromosome. (Some bacteria, such as *Vibrio cholerae*, have two chromosomes, and some bacteria have a linearly arranged chromosome.)

2. Their DNA is not associated with histones (special chromosomal proteins found in eukaryotes); other proteins are associated with the DNA.

3. They lack membrane-enclosed organelles.

4. Their cell walls almost always contain the complex polysaccharide peptidoglycan.

5. They usually divide by **binary fission.** During this process, the DNA is copied and the cell splits into two cells. Binary fission involves fewer structures and processes than eukaryotic cell division.

Eukaryotes (from the Greek words meaning true nucleus) have the following distinguishing characteristics:

1. Their DNA is found in the cell's nucleus, which is separated from the cytoplasm by a nuclear membrane, and the DNA is found in multiple chromosomes.

2. Their DNA is consistently associated with chromosomal proteins called histones and with nonhistones.

3. They have a number of membrane-enclosed organelles, including mitochondria, endoplasmic reticulum, Golgi complex, lysosomes, and sometimes chloroplasts.

4. Their cell walls, when present, are chemically simple.

5. Cell division usually involves mitosis, in which chromosomes replicate and an identical set is distributed into each of two nuclei. This process is guided by the mitotic spindle, a football-shaped assembly of microtubules. Division of the cytoplasm and other organelles follows so that the two cells produced are identical to each other.

Additional differences between prokaryotic and eukaryotic cells are listed in Table 4.2. Next we describe, in detail, the parts of the prokaryotic cell.

THE PROKARYOTIC CELL

The members of the prokaryotic world make up a vast heterogeneous group of very small unicellular organisms. Prokaryotes include bacteria and archaea. The majority of prokaryotes, including the photosynthesizing cyanobacteria, are included in the bacteria. Although bacteria and archaea look similar, they are different in chemical composition, as will be described later. The thousands of species of bacteria are differentiated by many factors, including morphology (shape), chemical composition (often detected by staining reactions), nutritional requirements, biochemical activities, and source of energy (sunlight or chemicals).

THE SIZE, SHAPE, AND ARRANGEMENT OF BACTERIAL CELLS

LEARNING OBJECTIVE
- Identify the three basic shapes of bacteria.

Bacteria come in a great many sizes and several shapes. Most bacteria range from 0.2 to 2.0 μm in diameter and from 2 to 8 μm in length. They have a few basic shapes: spherical **coccus** (plural: *cocci*, meaning berries), rod-shaped **bacillus** (plural: **bacilli,** meaning little staffs), and **spiral.**

Cocci are usually round but can be oval, elongated, or flattened on one side. When cocci divide to reproduce, the cells can remain attached to one another. Cocci that remain in pairs after dividing are called **diplococci;** those that divide and remain attached in chainlike patterns are called **streptococci** (Figure 4.1a). Those that divide in two planes and remain in groups of four are known as **tetrads** (Figure 4.1b). Those that divide in three planes and remain attached in cubelike groups of eight are called **sarcinae** (Figure 4.1c). Those that divide in multiple planes and form grapelike clusters or broad sheets are called **staphylococci** (Figure 4.1d). These group characteristics are frequently helpful in the identification of certain cocci.

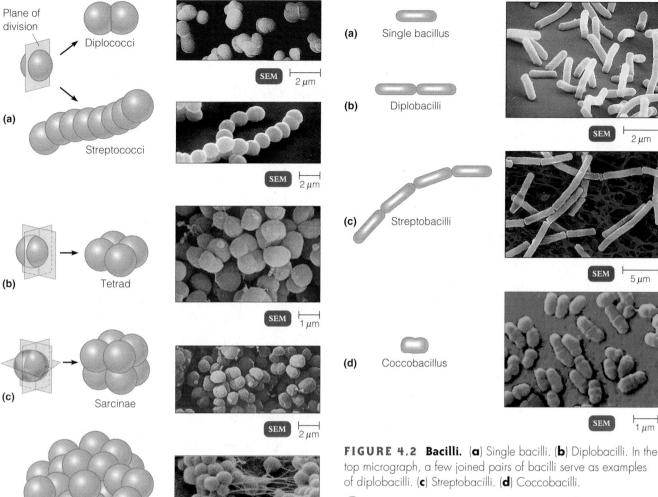

FIGURE 4.1 Arrangements of cocci. (a) Division in one plane produces diplococci and streptococci. **(b)** Division in two planes produces tetrads. **(c)** Division in three planes produces sarcinae, and **(d)** division in multiple planes produces staphylococci.

Q **How do the planes of division determine the arrangement of cells?**

FIGURE 4.2 Bacilli. (a) Single bacilli. **(b)** Diplobacilli. In the top micrograph, a few joined pairs of bacilli serve as examples of diplobacilli. **(c)** Streptobacilli. **(d)** Coccobacilli.

Q **Why don't bacilli form tetrads or clusters?**

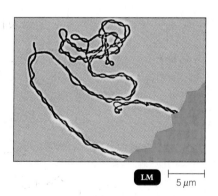

FIGURE 4.3 A double-stranded helix formed by *Bacillus subtilis*.

Q **What is the difference between the term bacillus and *Bacillus*?**

Bacilli divide only across their short axis, so there are fewer groupings of bacilli than of cocci. Most bacilli appear as single rods (Figure 4.2a). **Diplobacilli** appear in pairs after division (Figure 4.2b), and **streptobacilli** occur in chains (Figure 4.2c). Some bacilli look like straws. Others have tapered ends, like cigars. Still others are oval and look so much like cocci that they are called **coccobacilli** (Figure 4.2d).

"Bacillus" has two meanings in microbiology. As we have just used it, bacillus refers to a bacterial shape. When capitalized and italicized, it refers to a specific genus. For example, the bacterium *Bacillus anthracis* is the causative agent of anthrax. Bacillus cells often form long, twisted chains of cells (Figure 4.3).

Spiral bacteria have one or more twists; they are never straight. Bacteria that look like curved rods are called **vibrios** (Figure 4.4a). Others, called **spirilla,** have a helical shape, like a corkscrew, and fairly rigid bodies (Figure 4.4b). Yet another group of spirals are helical and flexible; they are called **spirochetes** (Figure 4.4c). Unlike

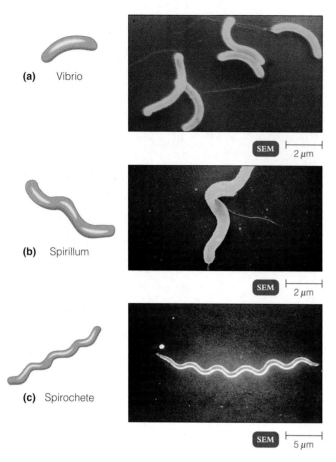

(a) Vibrio

SEM 2 µm

(b) Spirillum

SEM 2 µm

(c) Spirochete

SEM 5 µm

FIGURE 4.4 Spiral bacteria. (**a**) Vibrios. (**b**) Spirillum. (**c**) Spirochete.

Q **What is the distinguishing feature of spirochete bacteria?**

the spirilla, which use whiplike external appendages called flagella to move, spirochetes move by means of axial filaments, which resemble flagella but are contained within a flexible external sheath.

In addition to the three basic shapes, there are star-shaped cells (genus *Stella*); rectangular, flat cells (halophilic archaea) of the genus *Haloarcula* (Figure 4.5); and triangular cells.

The shape of a bacterium is determined by heredity. Genetically, most bacteria are **monomorphic;** that is, they

maintain a single shape. However, a number of environmental conditions can alter that shape. If the shape is altered, identification becomes difficult. Moreover, some bacteria, such as *Rhizobium* (rī-zō′bē-um) and *Corynebacterium* (kô-rī-nē-bak-ti′rē-um), are genetically **pleomorphic,** which means they can have many shapes, not just one.

The structure of a typical prokaryotic cell is shown in Figure 4.6. We will discuss its components according to the following organization: (1) structures external to the cell wall, (2) the cell wall itself, and (3) structures internal to the cell wall.

STRUCTURES EXTERNAL TO THE CELL WALL

LEARNING OBJECTIVE

- Describe the structure and function of the glycocalyx, flagella, axial filaments, fimbriae, and pili.

Among the possible structures external to the prokaryotic cell wall are the glycocalyx, flagella, axial filaments, fimbriae, and pili.

GLYCOCALYX

Many prokaryotes secrete on their surface a substance called glycocalyx. **Glycocalyx** (meaning sugar coat) is the general term used for substances that surround cells. The bacterial glycocalyx is a viscous (sticky), gelatinous polymer that is external to the cell wall and composed of polysaccharide, polypeptide, or both. Its chemical composition varies widely with the species. For the most part, it is made inside the cell and secreted to the cell surface. If the substance is organized and is firmly attached to the cell wall, the glycocalyx is described as a **capsule.** The presence of a capsule can be determined by using negative staining, described in Chapter 3 (see Figure 3.13a, page 72). If the substance is unorganized and only loosely attached to the cell wall, the glycocalyx is described as a **slime layer.**

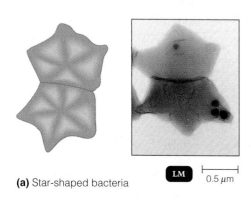

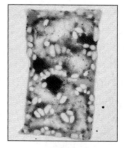

(a) Star-shaped bacteria   LM 0.5 µm

(b) Rectangular bacteria   LM 0.5 µm

FIGURE 4.5 Star-shaped and rectangular prokaryotes. (**a**) *Stella* (star-shaped). (**b**) *Haloarcula*, a genus of halophilic archaea (rectangular cells).

Q **What are the common bacterial shapes?**

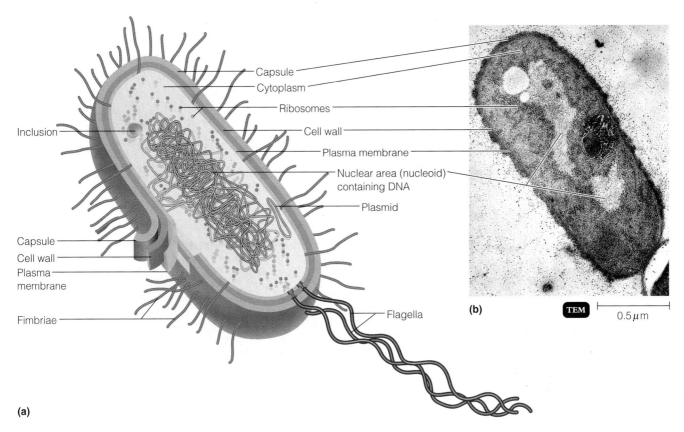

Inclusion

Capsule
Cell wall
Plasma membrane

Fimbriae

Capsule
Cytoplasm
Ribosomes
Cell wall
Plasma membrane
Nuclear area (nucleoid) containing DNA
Plasmid

Flagella

(b)

TEM   0.5 µm

(a)

FIGURE 4.6 A prokaryotic cell showing typical structures. Both the drawing (**a**) and micrograph (**b**) show a bacterium lengthwise to reveal the internal composition.

Q **On what basis are prokaryotic cells distinguished from eukaryotic cells?**

In certain species, capsules are important in contributing to bacterial virulence (the degree to which a pathogen causes disease). Capsules often protect pathogenic bacteria from phagocytosis by the cells of the host. (As you will see later, phagocytosis is the ingestion and digestion of microorganisms and other solid particles.) For example, *Bacillus anthracis* produces a capsule of D-glutamic acid. (Recall from Chapter 2 that the D forms of amino acids are unusual.) Because only encapsulated *B. anthracis* causes anthrax, it is speculated that the capsule may prevent its being destroyed by phagocytosis.

Another example involves *Streptococcus pneumoniae* (strep-tō-kok′kus nü-mō′nē-ī), which causes pneumonia only when the cells are protected by a polysaccharide capsule. Unencapsulated *S. pneumoniae* cells cannot cause pneumonia and are readily phagocytized. The polysaccharide capsule of *Klebsiella* (kleb-sē-el′lä) also prevents phagocytosis and allows the bacterium to adhere to and colonize the respiratory tract. A glycocalyx made of sugars is called an **extracellular polysaccharide (EPS).** The EPS enables a bacterium to survive by attaching to various surfaces in its natural environment in order to survive. Through attachment, bacteria can grow on diverse surfaces such as rocks in fast-moving streams, plant roots,

human teeth, medical implants, water pipes, and even other bacteria. *Streptococcus mutans* (mū′tans), an important cause of dental caries, attaches itself to the surface of teeth by a glycocalyx. *S. mutans* may use its capsule as a source of nutrition by breaking it down and utilizing the sugars when energy stores are low. A glycocalyx also can protect a cell against dehydration, and its viscosity may inhibit the movement of nutrients out of the cell.

FLAGELLA

Some prokaryotic cells have **flagella** (singular: **flagellum,** meaning whip), which are long filamentous appendages that propel bacteria. Bacteria that lack flagella are referred to as **atrichous.** Those that have flagella may have one of four arrangements of flagella (Figure 4.7): **monotrichous** (a single polar flagellum), **amphitrichous** (a tuft of flagella at each end of the cell), **lophotrichous** (two or more flagella at one or both ends of the cell), and **peritrichous** (flagella distributed over the entire cell).

A flagellum has three basic parts (Figure 4.8). The long outermost region, the *filament,* is constant in diameter and contains the globular (roughly spherical) protein *flagellin* arranged in several chains that intertwine and form a helix around a hollow core. In most bacteria, filaments are not

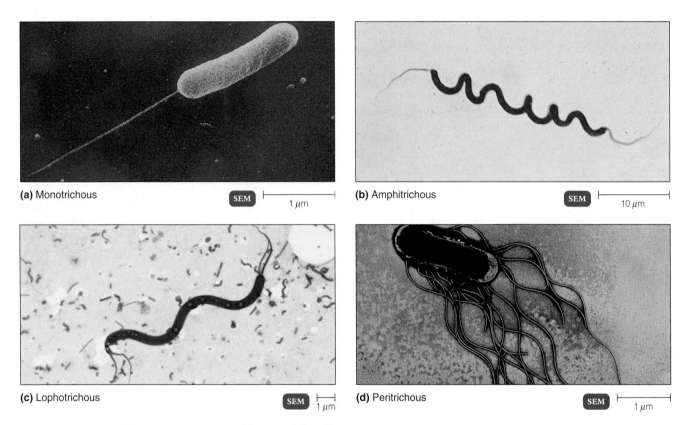

(a) Monotrichous   SEM   ⊢———⊣ 1 μm

(b) Amphitrichous   SEM   ⊢———⊣ 10 μm

(c) Lophotrichous   SEM   ⊢—⊣ 1 μm

(d) Peritrichous   SEM   ⊢———⊣ 1 μm

FIGURE 4.7 Four basic arrangements of bacterial flagella.

Q What are some of the key differences and similarities between flagella and endoflagella?

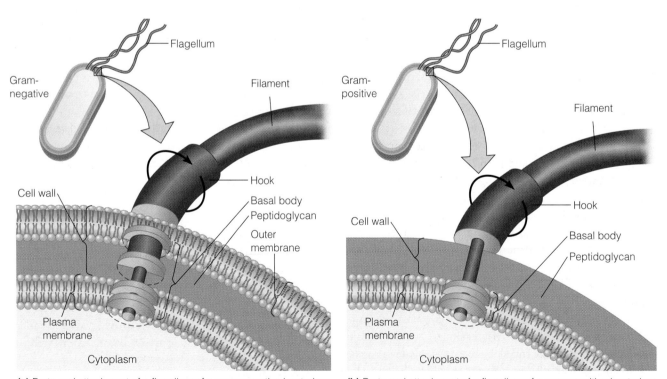

(a) Parts and attachment of a flagellum of a gram-negative bacterium

(b) Parts and attachment of a flagellum of a gram-positive bacterium

FIGURE 4.8 The structure of a prokaryotic flagellum. The parts and attachment of a flagellum of a gram-negative bacterium and gram-positive bacterium are shown in these highly schematic diagrams.

Q How do the basal bodies of gram-negative and gram-positive bacteria differ?

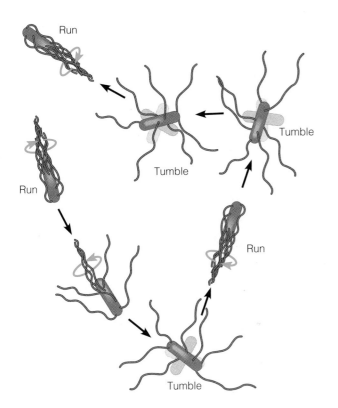

(a) A bacterium running and tumbling. Notice that the direction of flagellar rotation determines which of these movements occurs. Arrows indicate direction of movement.

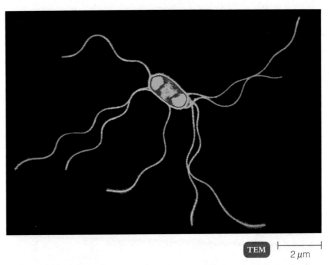

TEM   2 μm

(b) A *Proteus* cell in the swarming stage may have more than 1000 peritrichous flagella.

FIGURE 4.9   Flagella and bacterial motility.

Q Do bacterial flagella push or pull a cell?

covered by a membrane or sheath, as in eukaryotic cells. The filament is attached to a slightly wider *hook*, consisting of a different protein. The third portion of a flagellum is the *basal body*, which anchors the flagellum to the cell wall and plasma membrane.

The basal body is composed of a small central rod inserted into a series of rings. Gram-negative bacteria contain two pairs of rings; the outer pair of rings is anchored to various portions of the cell wall, and the inner pair of rings is anchored to the plasma membrane. In gram-positive bacteria, only the inner pair is present. As you will see later, the flagella (and cilia) of eukaryotic cells are more complex than those of prokaryotic cells.

Each prokaryotic flagellum is a semirigid, helical structure that moves the cell by rotating from the basal body. The rotation of a flagellum is either clockwise or counterclockwise around its long axis. (Eukaryotic flagella, by contrast, undulate in a wavelike motion.) The movement of a prokaryotic flagellum results from rotation of its basal body and is similar to the movement of the shaft of an electric motor. As the flagella rotate, they form a bundle that pushes against the surrounding liquid and propels the bacterium. Flagellar rotation depends on the cell's continuous generation of energy.

Bacterial cells can alter the speed and direction of rotation of flagella and thus are capable of various patterns of

motility, the ability of an organism to move by itself. When a bacterium moves in one direction for a length of time, the movement is called a "run" or "swim." "Runs" are interrupted by periodic, abrupt, random changes in direction called "tumbles." Then, a "run" resumes. "Tumbles" are caused by a reversal of flagellar rotation (Figure 4.9a). Some species of bacteria endowed with many flagella—*Proteus* (prō′tē-us), for example (Figure 4.9b)—can "swarm," or show rapid wavelike movement across a solid culture medium.

One advantage of motility is that it enables a bacterium to move toward a favorable environment or away from an adverse one. The movement of a bacterium toward or away from a particular stimulus is called **taxis.** Such stimuli include chemicals (**chemotaxis**) and light (**phototaxis**). Motile bacteria contain receptors in various locations, such as in or just under the cell wall. These receptors pick up chemical stimuli, such as oxygen, ribose, and galactose. In response to the stimuli, information is passed to the flagella. If the chemotactic signal is positive, called an *attractant*, the bacteria move toward the stimulus with many runs and few tumbles. If the chemotactic signal is negative, called a *repellent*, the frequency of tumbles increases as the bacteria move away from the stimulus.

The flagellar protein called **H antigen** is useful for distinguishing among **serovars,** or variations within a species, of gram-negative bacteria (see page 323). For example, there are at least 50 different H antigens for *E. coli*. Those serovars identified as *E. coli* O157:H7 are associated with foodborne epidemics (see Chapter 1, page 20).

☼ **Animation: Go to the Microbiology Place website or CD-ROM and click on "Animations" to view Bacterial Motility.**

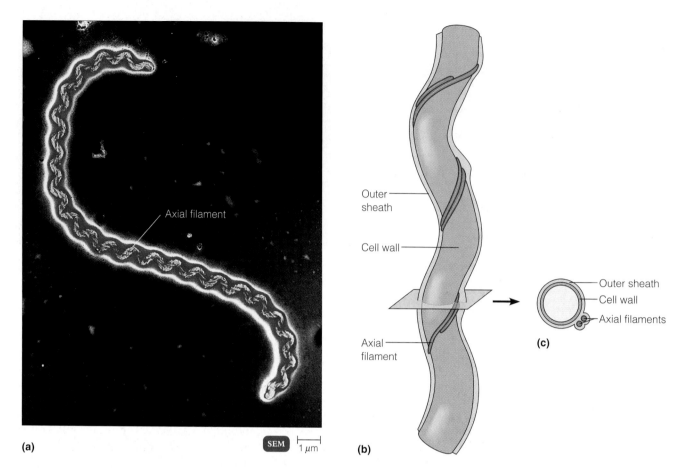

(a)   SEM 1 μm   (b)

FIGURE 4.10 Axial filaments. (**a**) A photomicrograph of the spirochete *Leptospira*, showing an axial filament. (**b**) A diagram of axial filaments wrapping around part of a spirochete. (**c**) A cross-sectional diagram of the spirochete, showing the position of axial filaments.

 How do spirochetes and spirilla differ?

AXIAL FILAMENTS

Spirochetes are a group of bacteria that have unique structure and motility. One of the best-known spirochetes is *Treponema pallidum*, the causative agent of syphilis. Another spirochete is *Borrelia burgdorferi*, the causative agent of Lyme disease. Spirochetes move by means of **axial filaments,** or **endoflagella,** bundles of fibrils that arise at the ends of the cell beneath an outer sheath and spiral around the cell (Figure 4.10).

Axial filaments, which are anchored at one end of the spirochete, have a structure similar to that of flagella. The rotation of the filaments produces a movement of the outer sheath that propels the spirochetes in a spiral motion. This type of movement is similar to the way a corkscrew moves through a cork. This corkscrew motion probably enables a bacterium such as *T. pallidum* to move effectively through body fluids.

FIMBRIAE AND PILI

Many gram-negative bacteria contain hairlike appendages that are shorter, straighter, and thinner than flagella and are used for attachment and transfer of DNA rather than for motility. These structures, which consist of a protein called *pilin* arranged helically around a central core, are divided into two types, fimbriae and pili, having very different functions. (Some microbiologists use the two terms interchangeably to refer to all such structures, but we distinguish between them.)

Fimbriae (singular: **fimbria**) can occur at the poles of the bacterial cell, or they can be evenly distributed over the entire surface of the cell. They can number anywhere from a few to several hundred per cell (Figure 4.11). Like the glycocalyx, fimbriae enable a cell to adhere to surfaces, including the surfaces of other cells. For example, fimbriae attached to the bacterium *Neisseria gonorrhoeae* (nī-se′rē-ä go-nôr-rē′ī), the causative agent of gonorrhea, help the microbe colonize mucous membranes. Once colonization occurs, the bacteria can cause disease. When fimbriae are absent (because of genetic mutation), colonization cannot happen, and no disease ensues.

Pili (singular: **pilus**) are usually longer than fimbriae and number only one or two per cell. Pili join bacterial

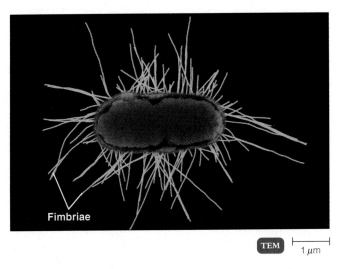

Fimbriae

TEM | 1 μm

FIGURE 4.11 Fimbriae. The fimbriae seem to bristle from this *E. coli* cell, which is beginning to divide.

Q **What is the function of fimbriae?**

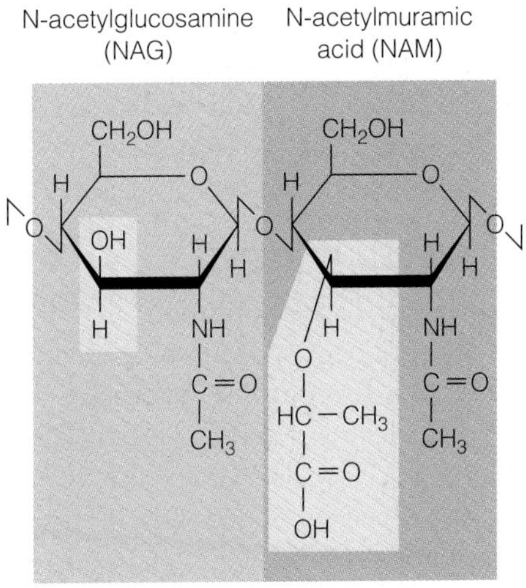

N-acetylglucosamine (NAG)   N-acetylmuramic acid (NAM)

CH₂OH

CH₂OH

OH

H

NH

C=O

CH₃

H

O

HC—CH₃

C=O

OH

NH

C=O

CH₃

FIGURE 4.12 N-acetylglucosamine (NAG) and N-acetylmuramic acid (NAM) joined as in a peptidoglycan. The gold areas show the differences between the two molecules. The linkage between them is called a β-1,4 linkage.

Q **What kind of molecules are these: carbohydrates, lipids, or proteins?**

cells in preparation for the transfer of DNA from one cell to another, a process called *conjugation.* For this reason, they are sometimes also called **conjugation pili** (see page 242).

THE CELL WALL

LEARNING OBJECTIVES

- Compare and contrast the cell walls of gram-positive bacteria, gram-negative bacteria, acid-fast bacteria, archaea, and mycoplasmas.
- Differentiate between *protoplast, spheroplast,* and *L form.*

The **cell wall** of the bacterial cell is a complex, semirigid structure responsible for the shape of the cell. The cell wall surrounds the underlying, fragile plasma (cytoplasmic) membrane and protects it and the interior of the cell from adverse changes in the outside environment (see Figure 4.6). Almost all prokaryotes have cell walls.

The major function of the cell wall is to prevent bacterial cells from rupturing when the water pressure inside the cell is greater than that outside the cell. It also helps maintain the shape of a bacterium and serves as a point of anchorage for flagella. As the volume of a bacterial cell increases, its plasma membrane and cell wall extend as needed. Clinically, the cell wall is important because it contributes to the ability of some species to cause disease and is the site of action of some antibiotics. In addition, the chemical composition of the cell wall is used to differentiate major types of bacteria.

Although the cells of some eukaryotes, including plants, algae, and fungi, have cell walls, their walls differ chemically from those of prokaryotes, are simpler in structure, and are less rigid.

COMPOSITION AND CHARACTERISTICS

The bacterial cell wall is composed of a macromolecular network called **peptidoglycan** (also known as *murein*), which is present either alone or in combination with other substances. Peptidoglycan consists of a repeating disaccharide attached by polypeptides to form a lattice that surrounds and protects the entire cell. The disaccharide portion is made up of monosaccharides called N-acetylglucosamine (NAG) and N-acetylmuramic acid (NAM) (from *murus*, meaning wall), which are related to glucose. The structural formulas for NAG and NAM are shown in Figure 4.12.

The various components of peptidoglycan are assembled in the cell wall (Figure 4.13a). Alternating NAM and NAG molecules are linked in rows of 10 to 65 sugars to form a carbohydrate "backbone" (the glycan portion of peptidoglycan). Adjacent rows are linked by **polypeptides** (the peptide portion of peptidoglycan). Although the structure of the polypeptide link varies, it always includes *tetrapeptide side chains,* which consist of four amino acids attached to NAMs in the backbone. The amino acids occur in an alternating pattern of D and L forms (see Figure 2.13, page 45). This is unique because the amino acids found in other proteins are L forms. Parallel tetrapeptide side chains may be directly bonded to each other or linked by a *peptide cross-bridge,* consisting of a short chain of amino acids.

Penicillin interferes with the final linking of the peptidoglycan rows by peptide cross-bridges (see Figure 4.13a). As a result, the cell wall is greatly weakened and the cell undergoes **lysis,** destruction caused by rupture of the plasma membrane and the loss of cytoplasm.

GRAM-POSITIVE CELL WALLS

In most gram-positive bacteria, the cell wall consists of many layers of peptidoglycan, forming a thick, rigid structure (Figure 4.13b). By contrast, gram-negative cell walls contain only a thin layer of peptidoglycan (Figure 4.13c).

In addition, the cell walls of gram-positive bacteria contain *teichoic acids,* which consist primarily of an alcohol (such as glycerol or ribitol) and phosphate. There are two classes of teichoic acids: *lipoteichoic acid,* which spans the peptidoglycan layer and is linked to the plasma membrane, and *wall teichoic acid,* which is linked to the peptidoglycan layer. Because of their negative charge (from the phosphate groups), teichoic acids may bind and regulate the movement of cations (positive ions) into and out of the cell. They may also assume a role in cell growth, preventing extensive wall breakdown and possible cell lysis. Finally, teichoic acids provide much of the wall's antigenic specificity and thus make it possible to identify bacteria by certain laboratory tests (see Chapter 10). Similarly, the cell walls of gram-positive streptococci are covered with various polysaccharides that allow them to be grouped into medically significant types.

GRAM-NEGATIVE CELL WALLS

The cell walls of gram-negative bacteria consist of one or a very few layers of peptidoglycan and an outer membrane (see Figure 4.13c). The peptidoglycan is bonded to lipoproteins (lipids covalently linked to proteins) in the outer membrane and is in the *periplasm,* a gel-like fluid between the outer membrane and the plasma membrane. The periplasm contains a high concentration of degradative enzymes and transport proteins. Gram-negative cell walls do not contain teichoic acids. Because the cell walls of gram-negative bacteria contain only a small amount of peptidoglycan, they are more susceptible to mechanical breakage.

The *outer membrane* of the gram-negative cell consists of lipopolysaccharides (LPS), lipoproteins, and phospholipids (see Figure 4.13c). The outer membrane has several specialized functions. Its strong negative charge is an important factor in evading phagocytosis and the actions of complement (lyses cells and promotes phagocytosis), two components of the defenses of the host (discussed in detail in Chapter 16). The outer membrane also provides a barrier to certain antibiotics (for example, penicillin), digestive enzymes such as lysozyme, detergents, heavy metals, bile salts, and certain dyes.

However, the outer membrane does not provide a barrier to all substances in the environment because nutrients must pass through to sustain the metabolism of the cell. Part of the permeability of the outer membrane is due to proteins in the membrane, called **porins,** that form channels. Porins permit the passage of molecules such as nucleotides, disaccharides, peptides, amino acids, vitamin B_{12}, and iron.

The LPS component of the outer membrane provides two important characteristics of gram-negative bacteria. First, the polysaccharide portion is composed of sugars, called **O polysaccharides,** that function as antigens and are useful for distinguishing species of gram-negative bacteria. For example, the foodborne pathogen *E. coli* O157:H7 is distinguished from other serovars by certain laboratory tests that test for the specific antigens. This role is comparable to that of teichoic acids in gram-positive cells. Second, the lipid portion of the lipopolysaccharide, called *lipid A,* is referred to as *endotoxin* and is toxic when in the host's bloodstream or gastrointestinal tract. It causes fever and shock. The nature and importance of endotoxins and other bacterial toxins will be discussed in Chapter 15.

CELL WALLS AND THE GRAM STAIN MECHANISM

Now that you have studied the Gram stain (in Chapter 3, page 69) and the chemistry of the bacterial cell wall (in the previous section), it is easier to understand the mechanism of the Gram stain. The mechanism is based on differences in the structure of the cell walls of gram-positive and gram-negative bacteria and how each reacts to the various reagents (substances used for producing a chemical reaction). Crystal violet, the primary stain, stains both gram-positive and gram-negative cells purple because the dye enters the cytoplasm of both types of cells. When iodine (the mordant) is applied, it forms large crystals with the dye that are too large to escape through the cell wall. The application of alcohol dehydrates the peptidoglycan of gram-positive cells to make it more impermeable to the crystal violet-iodine. The effect on gram-negative cells is quite different; alcohol dissolves the outer membrane of gram-negative cells and even leaves small holes in the thin peptidoglycan layer through which crystal violet-iodine diffuse. Because gram-negative bacteria are colorless after the alcohol wash, the addition of safranin (the counterstain) turns the cells pink. Safranin provides a contrasting color to the primary stain (crystal violet). Although gram-positive and gram-negative cells both absorb safranin, the pink color of safranin is masked by the darker purple dye previously absorbed by gram-positive cells.

In any population of cells, some gram-positive cells will give a gram-negative response. These cells are usually

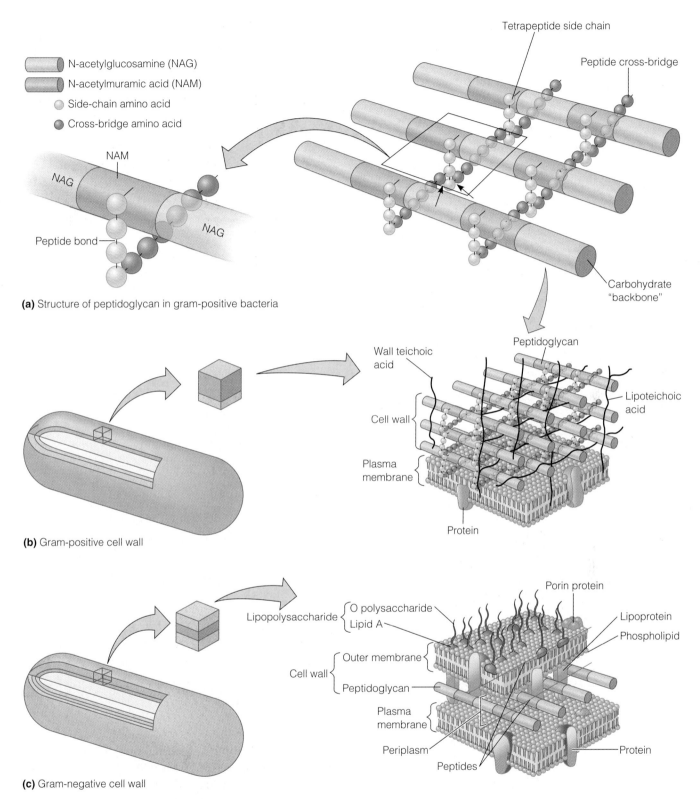

FIGURE 4.13 Bacterial cell walls. (a) The structure of peptidoglycan in gram-positive bacteria. Together the carbohydrate backbone (glycan portion) and tetrapeptide side chains (peptide portion) make up peptidoglycan. The frequency of peptide cross-bridges and the number of amino acids in these bridges vary with the species of bacterium. The small arrows indicate where penicillin interferes with the linkage of peptidoglycan rows by peptide cross-bridges. **(b)** A gram-positive cell wall. **(c)** A gram-negative cell wall.

Q **What are the major structural differences between gram-positive and gram-negative cell walls?**

TABLE 4.1	Some Comparative Characteristics of Gram-Positive and Gram-Negative Bacteria	
Characteristic	Gram-Positive	Gram-Negative
	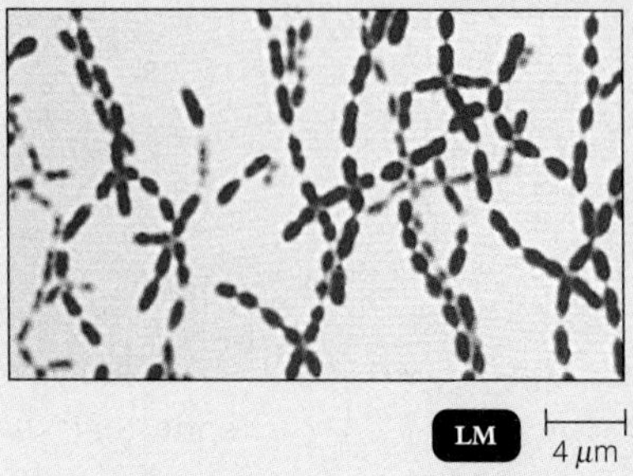 LM ⊢ 4 µm	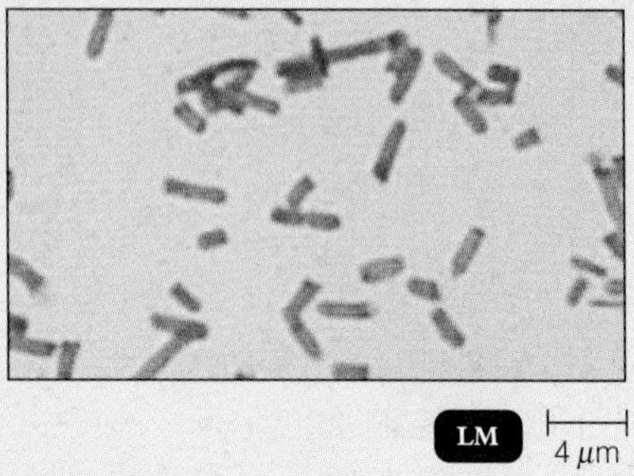 LM ⊢ 4 µm
Gram Reaction	Retain crystal violet dye and stain dark violet or purple	Can be decolorized to accept counterstain (safranin) and stain pink
Peptidoglycan Layer	Thick (multilayered)	Thin (single-layered)
Teichoic Acids	Present in many	Absent
Periplasmic Space	Absent	Present
Outer Membrane	Absent	Present
Lipopolysaccharide (LPS) Content	Virtually none *(in effect)*	High
Lipid and Lipoprotein Content	Low (acid-fast bacteria have lipids linked to peptidoglycan)	High (due to presence of outer membrane)
Flagellar Structure	2 rings in basal body	4 rings in basal body
Toxins Produced	Primarily exotoxins	Primarily endotoxins
Resistance to Physical Disruption	High	Low
Cell Wall Disruption by Lysozyme	High	Low (requires pretreatment to destabilize outer membrane)
Susceptibility to Penicillin and Sulfonamide	High	Low
Susceptibility to Streptomycin, Chloramphenicol, and Tetracycline	Low	High
Inhibition by Basic Dyes	High	Low
Susceptibility to Anionic Detergents	High	Low
Resistance to Sodium Azide	High	Low
Resistance to Drying	High	Low

dead. However, there are a few gram-positive genera that show an increasing number of gram-negative cells as the culture ages. *Bacillus* and *Clostridium* are examples, and are often described as *gram-variable*.

A comparison of some of the characteristics of gram-positive and gram-negative bacteria is presented in Table 4.1.

ATYPICAL CELL WALLS

Among prokaryotes, certain types of cells have no walls or have very little wall material. These include members of the genus *Mycoplasma* (mī-kō-plaz′mä) and related organisms (see Figure 11.19, page 332). Mycoplasmas are the smallest known bacteria that can grow and reproduce outside living host cells. Because of their size and because they have no cell walls, they pass through most bacterial filters and were first mistaken for viruses. Their plasma membranes are unique among bacteria in having lipids called *sterols*, which are thought to help protect them from lysis (rupture).

Archaea may lack walls or may have unusual walls composed of polysaccharides and proteins but not peptidoglycan. These walls do, however, contain a substance similar to peptidoglycan called *pseudomurein*. Pseudomurein contains N-acetyltalosaminuronic acid instead of NAM and lacks the D-amino acids found in bacterial cell walls. Archaea generally cannot be Gram-stained but appear gram-negative because they do not contain peptidoglycan. A representative of the archaea is described in the box in Chapter 5 (page 147).

ACID-FAST CELL WALLS

Recall from Chapter 3 that the acid-fast stain is used to identify all bacteria of the genus *Mycobacterium* and pathogenic species of *Nocardia*. These bacteria contain high concentrations (60%) of a hydrophobic waxy lipid (**mycolic acid**) in their cell wall that prevents the uptake of dyes, including those used in the Gram stain. The mycolic acid forms a layer outside of a thin layer of peptidoglycan. The mycolic acid and peptidoglycan are held together by a polysaccharide. The hydrophobic waxy cell wall causes both cultures of *Mycobacterium* to clump and to stick to the walls of the flask. Acid-fast bacteria can be stained with carbolfuchsin; heating enhances penetration of the stain. The carbolfuchsin penetrates the cell wall, binds to cytoplasm, and resists removal by washing with acid-alcohol. Acid-fast bacteria retain the red color of carbolfuchsin because it is more soluble in the cell wall mycolic acid than in the acid-alcohol. If the mycolic acid layer is removed from the cell wall of acid-fast bacteria, they will stain gram-positive with the Gram stain.

DAMAGE TO THE CELL WALL

Chemicals that damage bacterial cell walls, or interfere with their synthesis, often do not harm the cells of an animal host because the bacterial cell wall is made of chemicals unlike those in eukaryotic cells. Thus, cell wall synthesis is the target for some antimicrobial drugs. One way the cell wall can be damaged is by exposure to the digestive enzyme *lysozyme*. This enzyme occurs naturally in some eukaryotic cells and is a constituent of tears, mucus, and saliva. Lysozyme is particularly active on the major cell wall components of most gram-positive bacteria, making them vulnerable to lysis. Lysozyme catalyzes hydrolysis of the bonds between the sugars in the repeating disaccharide "backbone" of peptidoglycan. This act is analogous to cutting the steel supports of a bridge with a cutting torch: the gram-positive cell wall is almost completely destroyed by lysozyme. The cellular contents that remain surrounded by the plasma membrane may remain intact if lysis does not occur; this wall-less cell is termed a **protoplast.** Typically, a protoplast is spherical and is still capable of carrying on metabolism.

Some members of the genus *Proteus*, as well as other genera, can lose their cell walls and swell into irregularly shaped cells called **L forms,** named for the Lister Institute, where they were discovered. They may form spontaneously or develop in response to penicillin (which inhibits cell wall formation) or lysozyme (which removes the cell wall). L forms can live and divide repeatedly or return to the walled state.

When lysozyme is applied to gram-negative cells, usually the wall is not destroyed to the same extent as in gram-positive cells; some of the outer membrane also remains. In this case, the cellular contents, plasma membrane, and remaining outer wall layer are called a **spheroplast,** also a spherical structure. For lysozyme to exert its effect on gram-negative cells, the cells are first treated with EDTA (ethylenediaminetetraacetic acid). EDTA weakens ionic bonds in the outer membrane and thereby damages it, giving the lysozyme access to the peptidoglycan layer.

Protoplasts and spheroplasts burst in pure water or very dilute salt or sugar solutions because the water molecules from the surrounding fluid rapidly move into and enlarge the cell, which has a much lower internal concentration of water. This rupturing is called **osmotic lysis** and will be discussed in detail shortly.

As noted earlier, certain antibiotics, such as penicillin, destroy bacteria by interfering with the formation of the peptide cross-bridges of peptidoglycan, thus preventing the formation of a functional cell wall. Most gram-negative bacteria are not as susceptible to penicillin as gram-positive bacteria are because the outer membrane of gram-negative bacteria forms a barrier that inhibits the entry of this and other substances, and gram-negative bacteria have fewer peptide cross-bridges. However, gram-negative bacteria are quite susceptible to some β-lactam antibiotics that penetrate the outer membrane better than penicillin. Antibiotics will be discussed in more detail in Chapter 20.

STRUCTURES INTERNAL TO THE CELL WALL

Thus far, we have discussed the prokaryotic cell wall and structures external to it. We will now look inside the prokaryotic cell and discuss the structures and functions of the plasma membrane and components within the cytoplasm of the cell.

THE PLASMA (CYTOPLASMIC) MEMBRANE

LEARNING OBJECTIVES

- Describe the structure, chemistry, and functions of the prokaryotic plasma membrane.
- Define *simple diffusion, facilitated diffusion, osmosis, active transport,* and *group translocation.*

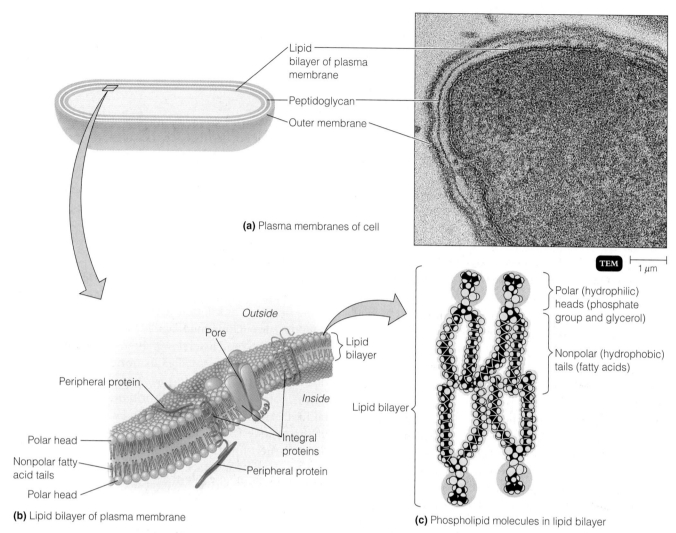

(a) Plasma membranes of cell

Lipid bilayer of plasma membrane

Peptidoglycan

Outer membrane

TEM   1 μm

Outside

Pore

Lipid bilayer

Inside

Lipid bilayer

Peripheral protein

Polar head

Nonpolar fatty acid tails

Polar head

Integral proteins

Peripheral protein

(b) Lipid bilayer of plasma membrane

Polar (hydrophilic) heads (phosphate group and glycerol)

Nonpolar (hydrophobic) tails (fatty acids)

Lipid bilayer

(c) Phospholipid molecules in lipid bilayer

FIGURE 4.14 Plasma membrane. (a) A diagram and micrograph showing the lipid bilayer forming the inner plasma membrane of the gram-negative bacterium *Aquaspirillum serpens*. Layers of the cell wall, including the outer membrane, can be seen outside the inner membrane. **(b)** A portion of the inner membrane showing the lipid bilayer and proteins. The outer membrane of gram-negative bacteria is also a lipid bilayer. **(c)** Space-filling models of several molecules as they are arranged in the lipid bilayer.

Q **What is the action of polymyxins on plasma membranes?**

The **plasma (cytoplasmic) membrane** (or *inner membrane*) is a thin structure lying inside the cell wall and enclosing the cytoplasm of the cell (see Figure 4.6). The plasma membrane of prokaryotes consists primarily of phospholipids (see Figure 2.10, page 42), which are the most abundant chemicals in the membrane, and proteins. Eukaryotic plasma membranes also contain carbohydrates and sterols, such as cholesterol. Because they lack sterols, prokaryotic plasma membranes are less rigid than eukaryotic membranes. One exception is the wall-less prokaryote *Mycoplasma*, which contains membrane sterols.

STRUCTURE

In electron micrographs, prokaryotic and eukaryotic plasma membranes (and the outer membranes of gram-negative bacteria) look like two-layered structures; there are two dark lines with a light space between the lines (Figure 4.14a). The phospholipid molecules are arranged in two parallel rows, called a *lipid bilayer* (Figure 4.14b). As introduced in Chapter 2, each phospholipid molecule contains a polar head, composed of a phosphate group and glycerol that is hydrophilic (water-loving) and soluble in water, and nonpolar tails, composed of fatty acids that are hydrophobic (water-fearing) and insoluble in water (Figure 4.14c).

The polar heads are on the two surfaces of the lipid bilayer, and the nonpolar tails are in the interior of the bilayer.

The protein molecules in the membrane can be arranged in a variety of ways. Some, called *peripheral proteins*, are easily removed from the membrane by mild treatments and lie at the inner or outer surface of the membrane. They may function as enzymes that catalyze chemical reactions, as a "scaffold" for support, and as mediators of changes in membrane shape during movement. Other proteins, called *integral proteins*, can be removed from the membrane only after disrupting the lipid bilayer (by using detergents, for example). Most integral proteins penetrate the membrane completely and are called *transmembrane proteins*. Some integral proteins are channels that have a pore, or hole, through which substances enter and exit the cell.

Many of the proteins and some of the lipids on the outer surface of the plasma membrane have carbohydrates attached to them. Proteins attached to carbohydrates are called **glycoproteins;** lipids attached to carbohydrates are called **glycolipids.** Both glycoproteins and glycolipids help protect and lubricate the cell and are involved in cell-to-cell interactions. For example, glycoproteins play a role in certain infectious diseases. The influenza virus and the toxins that cause cholera and botulism enter their target cells by first binding to glycoproteins on their plasma membranes.

Studies have demonstrated that the phospholipid and protein molecules in membranes are not static but move quite freely within the membrane surface. This movement is most probably associated with the many functions performed by the plasma membrane. Because the fatty acid tails cling together, phospholipids in the presence of water form a self-sealing bilayer, with the result that breaks and tears in the membrane will heal themselves. The membrane must be about as viscous as olive oil which allows membrane proteins to move freely enough to perform their functions without destroying the structure of the membrane. This dynamic arrangement of phospholipids and proteins is referred to as the **fluid mosaic model.**

FUNCTIONS

The most important function of the plasma membrane is to serve as a selective barrier through which materials enter and exit the cell. In this function, plasma membranes have **selective permeability** (sometimes called *semipermeability*). This term indicates that certain molecules and ions pass through the membrane, but others are prevented from passing through it. The permeability of the membrane depends on several factors. Large molecules (such as proteins) cannot pass through the plasma membrane, possibly because these molecules are larger than the pores in integral proteins that function as channels. But smaller molecules (such as water, oxygen, carbon dioxide, and some simple sugars)

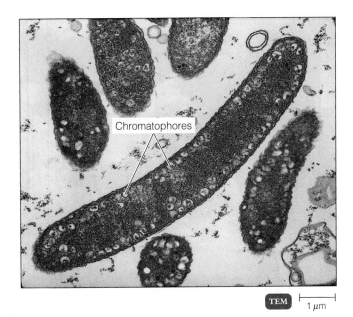

Chromatophores

TEM   1 μm

FIGURE 4.15 Chromatophores. In this micrograph of *Rhodospirillum rubrum,* a purple (nonsulfur) bacterium, the chromatophores are clearly visible.

Q **What is the function of chromatophores?**

usually pass through easily. Ions penetrate the membrane very slowly. Substances that dissolve easily in lipids (such as oxygen, carbon dioxide, and nonpolar organic molecules) enter and exit more easily than other substances because the membrane consists mostly of phospholipids. The movement of materials across plasma membranes also depends on transporter molecules, which will be described shortly.

Plasma membranes are also important to the breakdown of nutrients and the production of energy. The plasma membranes of bacteria contain enzymes capable of catalyzing the chemical reactions that break down nutrients and produce ATP. In some bacteria, pigments and enzymes involved in photosynthesis are found in infoldings of the plasma membrane that extend into the cytoplasm. These membranous structures are called **chromatophores** or **thylakoids** (Figure 4.15).

When viewed with an electron microscope, bacterial plasma membranes often appear to contain one or more large, irregular folds called **mesosomes.** Many functions have been proposed for mesosomes. However, it is now known that they are artifacts, not true cell structures. Mesosomes are believed to be folds in the plasma membrane that develop by the process used for preparing specimens for electron microscopy.

DESTRUCTION OF THE PLASMA MEMBRANE BY ANTIMICROBIAL AGENTS

Because the plasma membrane is vital to the bacterial cell, it is not surprising that several antimicrobial agents exert

FIGURE 4.16 The principle of simple diffusion. (**a**) After a dye pellet is put into a beaker of water, the molecules of dye in the pellet diffuse into the water from an area of high dye concentration to areas of low dye concentration. (**b**) The dye potassium permanganate in the process of diffusing.

Q **What is the distinguishing feature of a passive process?**

(a)   (b)

their effects at this site. In addition to the chemicals that damage the cell wall and thereby indirectly expose the membrane to injury, many compounds specifically damage plasma membranes. These compounds include certain alcohols and quaternary ammonium compounds, which are used as disinfectants. By disrupting the membrane's phospholipids, a group of antibiotics known as the *polymyxins* cause leakage of intracellular contents and subsequent cell death. This mechanism will be discussed in Chapter 20.

THE MOVEMENT OF MATERIALS ACROSS MEMBRANES

Materials move across plasma membranes of both prokaryotic and eukaryotic cells by two kinds of processes: passive and active. In *passive processes*, substances cross the membrane from an area of high concentration to an area of low concentration (move with the concentration gradient, or difference), without any expenditure of energy (ATP) by the cell. In *active processes*, the cell must use energy (ATP) to move substances from areas of low concentration to areas of high concentration (against the concentration gradient).

Passive Processes   Passive processes include simple diffusion, facilitated diffusion, and osmosis.

Simple diffusion is the net (overall) movement of molecules or ions from an area of high concentration to an area of low concentration (Figure 4.16). The movement continues until the molecules or ions are evenly distributed. The point of even distribution is called *equilibrium*. Cells rely on simple diffusion to transport certain small molecules, such as oxygen and carbon dioxide, across their cell membranes.

In **facilitated diffusion,** the substance (glucose, for example) to be transported combines with a plasma membrane protein called a *transporter* (sometimes called a *permease*). In one proposed mechanism for facilitated diffusion, transporters bind a substance on one side of the membrane and, by changing shape, move it to the other side of the membrane, where it is released (Figure 4.17). Facilitated diffusion is similar to simple diffusion in that the cell does not need to expend energy because the substance moves from a high to a low concentration. The process differs from simple diffusion in its use of transporters.

FIGURE 4.17 Facilitated diffusion.
Transporter proteins in the plasma membrane transport molecules across the membrane from an area of high concentration to one of low concentration (with the concentration gradient). The transporter undergoes a change in shape to transport the substance. The process does not require ATP.

Q **How does simple diffusion differ from facilitated diffusion?**

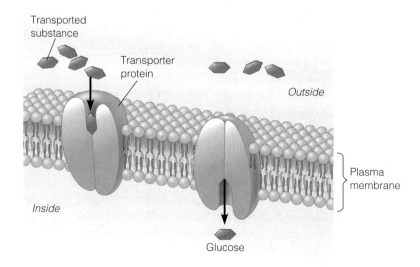

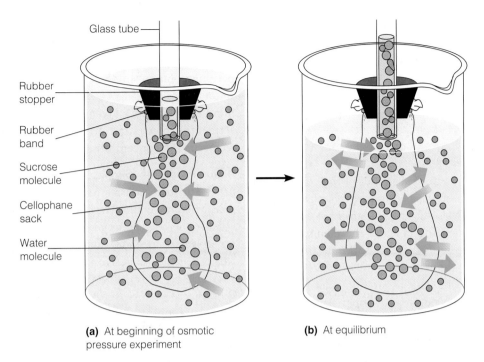

(a) At beginning of osmotic pressure experiment

(b) At equilibrium

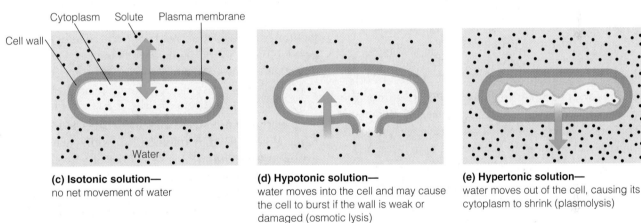

(c) Isotonic solution—
no net movement of water

(d) Hypotonic solution—
water moves into the cell and may cause the cell to burst if the wall is weak or damaged (osmotic lysis)

(e) Hypertonic solution—
water moves out of the cell, causing its cytoplasm to shrink (plasmolysis)

FIGURE 4.18 The principle of osmosis. (a) Setup at the beginning of an osmotic pressure experiment. Water molecules start to move from the beaker into the sack along the concentration gradient. **(b)** Setup at equilibrium. The osmotic pressure exerted by the solution in the sack pushes water molecules from the sack back into the beaker to balance the rate of water entry into the sack. The height of the solution in the glass tube at equilibrium is a measure of the osmotic pressure. **(c)–(e)** The effects of various solutions on bacterial cells.

Q **What is osmosis?**

In some cases, molecules that bacteria need are too large to be transported into the cells by these methods. Most bacteria, however, produce enzymes that can break down large molecules into simpler ones (such as proteins into amino acids, or polysaccharides into simple sugars). Such enzymes, which are released by the bacteria into the surrounding medium, are appropriately called *extracellular enzymes*. Once the enzymes degrade the large molecules, the subunits move into the cell with the help of transporters. For example, specific carriers retrieve DNA bases, such as the purine guanine, from extracellular media and bring them into the cell's cytoplasm.

Osmosis is the net movement of solvent molecules across a selectively permeable membrane from an area with a high concentration of solvent molecules (low concentration of solute molecules) to an area of low concentration of solvent molecules (high concentration of solute molecules). In living systems, the chief solvent is water.

Osmosis may be demonstrated with the apparatus shown in Figure 4.18. A sack constructed from cellophane, which is a selectively permeable membrane, is filled with a solution of 20% sucrose (table sugar). The cellophane sack is placed into a beaker containing distilled water. Initially, the concentrations of water on either side of the membrane are different. Because of the sucrose molecules, the concentration of water is lower inside the cellophane sack. Therefore, water moves from the beaker (where its concentration is higher) into the cellophane sack (where its concentration is lower).

There is no movement of sugar out of the cellophane sack into the beaker, however, because the cellophane is impermeable to molecules of sugar—the sugar molecules are too large to go through the pores of the membrane. As water moves into the cellophane sack, the sugar solution becomes increasingly dilute, and, because the cellophane sack has expanded to its limit as a result of an increased volume of water, water begins to move up the glass tube. In time, the water that has accumulated in the cellophane sack and the glass tube exerts a downward pressure that forces water molecules out of the cellophane sack and back into the beaker. This movement of water through a selectively permeable membrane produces a pressure called osmotic pressure. **Osmotic pressure** is the pressure required to prevent the movement of pure water (water with no solutes) into a solution containing some solutes. In other words, osmotic pressure is the pressure needed to stop the flow of water across the selectively permeable membrane (cellophane). When water molecules leave and enter the cellophane sack at the same rate, equilibrium is reached.

A bacterial cell may be subjected to any of three kinds of osmotic solutions: isotonic, hypotonic, or hypertonic. An **isotonic solution** is a medium in which the overall concentration of solutes equals that found inside a cell (*iso* means equal). Water leaves and enters the cell at the same rate (no net change); the cell's contents are in equilibrium with the solution outside the cell wall (Figure 4.18c).

Earlier we mentioned that lysozyme and certain antibiotics damage bacterial cell walls, causing the cells to rupture, or lyse. Such rupturing occurs because bacterial cytoplasm usually contains such a high concentration of solutes that, when the wall is weakened or removed, additional water enters the cell by osmosis. The damaged (or removed) cell wall cannot constrain the swelling of the cytoplasmic membrane, and the membrane bursts. This is an example of osmotic lysis caused by immersion in a hypotonic solution. A **hypotonic solution** outside the cell is a medium whose concentration of solutes is lower than that inside the cell (*hypo* means under or less). Most bacteria live in hypotonic solutions, and swelling is contained by the cell wall. Cells with weak cell walls, such as gram-negative bacteria, may burst or undergo osmotic lysis as a result of excessive water intake (Figure 4.18d).

A **hypertonic solution** is a medium having a higher concentration of solutes than inside the cell has (*hyper* means above or more). Most bacterial cells placed in a hypertonic solution shrink and collapse or *plasmolyze* because water leaves the cells by osmosis (Figure 4.18e). Keep in mind that the terms *isotonic, hypotonic,* and *hypertonic* describe the concentration of solutions outside the cell *relative to* the concentration inside the cell.

Active Processes Simple diffusion and facilitated diffusion are useful mechanisms for transporting substances into cells when the concentrations of the substances are greater outside the cell. However, when a bacterial cell is in an environment in which nutrients are in low concentration, the cell must use active processes, such as active transport and group translocation, to accumulate the needed substances.

In performing **active transport,** the cell *uses energy* in the form of ATP to move substances across the plasma membrane. The movement of a substance in active transport is usually from outside to inside, even though the concentration might be much higher inside the cell. Like facilitated diffusion, active transport depends on transporter proteins in the plasma membrane (see Figure 4.17). There appears to be a different transporter for each transported substance or group of closely related transported substances.

In active transport, the substance that crosses the membrane is not altered by transport across the membrane. In **group translocation,** a special form of active transport that occurs exclusively in prokaryotes, the substance is chemically altered during transport across the membrane. Once the substance is altered and inside the cell, the plasma membrane is impermeable to it, so it remains inside the cell. This important mechanism enables a cell to accumulate various substances even though they may be in low concentrations outside the cell. Group translocation requires energy supplied by high-energy phosphate compounds, such as phosphoenolpyruvic acid (PEP).

One example of group translocation is the transport of the sugar glucose, which is often used in growth media for bacteria. While a specific carrier protein is transporting the glucose molecule across the membrane, a phosphate group is added to the sugar. This phosphorylated form of glucose, which cannot be transported out, can then be used in the cell's metabolic pathways.

Some eukaryotic cells (those without cell walls) can use two additional active transport processes called phagocytosis and pinocytosis. These processes, which do not occur in bacteria, are explained on p. 102. ❊ **Animation: Go to the Microbiology Place website or CD-ROM and click "Animations" to view Membrane Transport.**

CYTOPLASM

For a prokaryotic cell, the term **cytoplasm** refers to the substance of the cell inside the plasma membrane (see Figure 4.6). Cytoplasm is about 80% water and contains primarily proteins (enzymes), carbohydrates, lipids, inorganic ions, and many low-molecular-weight compounds. Inorganic ions are present in much higher concentrations in cytoplasm than in most media. Cytoplasm is thick, aqueous, semitransparent, and elastic. The major structures in the cytoplasm of prokaryotes are a nuclear area (containing DNA), particles called ribosomes, and reserve deposits called inclusions. Protein filaments in the cytoplasm are most likely responsible for the rod and helical cell shapes of bacteria.

Prokaryotic cytoplasm lacks certain features of eukaryotic cytoplasm, such as a cytoskeleton and cytoplasmic streaming. These features will be described later.

THE NUCLEAR AREA

LEARNING OBJECTIVE

* Identify the functions of the nuclear area, ribosomes, and inclusions.

The nuclear area, or **nucleoid,** of a bacterial cell (see Figure 4.6) usually contains a single long, continuous, and frequently circularly arranged thread of double-stranded DNA called the **bacterial chromosome.** This is the cell's genetic information, which carries all the information required for the cell's structures and functions. Unlike the chromosomes of eukaryotic cells, bacterial chromosomes are not surrounded by a nuclear envelope (membrane) and do not include histones. The nuclear area can be spherical, elongated, or dumbbell-shaped. In actively growing bacteria, as much as 20% of the cell volume is occupied by DNA because such cells presynthesize nuclear material for future cells. The chromosome is attached to the plasma membrane. Proteins in the plasma membrane are believed to be responsible for replication of the DNA and segregation of the new chromosomes to daughter cells during cell division.

In addition to the bacterial chromosome, bacteria often contain small usually circular, double-stranded DNA molecules called **plasmids** (see the F factor in Figure 8.26a, page 243). These molecules are extrachromosomal genetic elements; that is, they are not connected to the main bacterial chromosome, and they replicate independently of chromosomal DNA. Research indicates that plasmids are associated with plasma membrane proteins. Plasmids usually contain from 5 to 100 genes that are generally not crucial for the survival of the bacterium under normal environmental conditions; plasmids may be gained or lost without harming the cell. Under certain conditions,

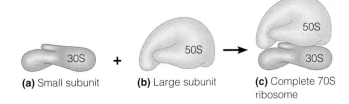

FIGURE 4.19 The prokaryotic ribosome. (a) A small 30S subunit and **(b)** a large 50S subunit make up **(c)** the complete 70S prokaryotic ribosome.

Q **What is the importance of the differences between prokaryotic and eukaryotic ribosomes with regard to antibiotic therapy?**

however, plasmids are an advantage to cells. Plasmids may carry genes for such activities as antibiotic resistance, tolerance to toxic metals, the production of toxins, and the synthesis of enzymes. Plasmids can be transferred from one bacterium to another. In fact, plasmid DNA is used for gene manipulation in biotechnology.

RIBOSOMES

All eukaryotic and prokaryotic cells contain **ribosomes,** which function as the sites of protein synthesis. Cells that have high rates of protein synthesis, such as those that are actively growing, have a large number of ribosomes. The cytoplasm of a prokaryotic cell contains tens of thousands of these very small structures, which give the cytoplasm a granular appearance (see Figure 4.6).

Ribosomes are composed of two subunits, each of which consists of protein and a type of RNA called *ribosomal RNA (rRNA)*. Prokaryotic ribosomes differ from eukaryotic ribosomes in the number of proteins and rRNA molecules they contain; they are also somewhat smaller and less dense than ribosomes of eukaryotic cells. Accordingly, prokaryotic ribosomes are called 70S ribosomes (Figure 4.19), and those of eukaryotic cells are known as 80S ribosomes. The letter S refers to Svedberg units, which indicate the relative rate of sedimentation during ultra-high-speed centrifugation. Sedimentation rate is a function of the size, weight, and shape of a particle. The subunits of a 70S ribosome are a small 30S subunit containing one molecule of rRNA and a larger 50S subunit containing two molecules of rRNA.

Several antibiotics work by inhibiting protein synthesis on prokaryotic ribosomes. Antibiotics such as streptomycin and gentamicin attach to the 30S subunit and interfere with protein synthesis. Other antibiotics, such as erythromycin and chloramphenicol, interfere with protein synthesis by attaching to the 50S subunit. Because of differences in prokaryotic and eukaryotic ribosomes, the microbial cell can be killed by the antibiotic while the eukaryotic host cell remains unaffected.

INCLUSIONS

Within the cytoplasm of prokaryotic cells are several kinds of reserve deposits, known as **inclusions.** Cells may accumulate certain nutrients when they are plentiful and use them when the environment is deficient. Evidence suggests that macromolecules concentrated in inclusions avoid the increase in osmotic pressure that would result if the molecules were dispersed in the cytoplasm. Some inclusions are common to a wide variety of bacteria, whereas others are limited to a small number of species and therefore serve as a basis for identification.

METACHROMATIC GRANULES

Metachromatic granules are large inclusions that take their name from the fact that they sometimes stain red with certain blue dyes such as methylene blue. Collectively they are known as **volutin.** Volutin represents a reserve of inorganic phosphate (polyphosphate) that can be used in the synthesis of ATP. It is generally formed by cells that grow in phosphate-rich environments. Metachromatic granules are found in algae, fungi, and protozoa, as well as in bacteria. These granules are characteristic of *Corynebacterium diphtheriae* (kô-rī-nē-bak-ti′rē-um dif-thi′rē-ī), the causative agent of diphtheria; thus, they have diagnostic significance.

POLYSACCHARIDE GRANULES

Inclusions known as **polysaccharide granules** typically consist of glycogen and starch, and their presence can be demonstrated when iodine is applied to the cells. In the presence of iodine, glycogen granules appear reddish brown and starch granules appear blue.

LIPID INCLUSIONS

Lipid inclusions appear in various species of *Mycobacterium, Bacillus, Azotobacter* (ä-zō-tō-bak′tér), *Spirillum* (spī-ril′lum), and other genera. A common lipid-storage material, one unique to bacteria, is the polymer *poly-β-hydroxybutyric acid.* Lipid inclusions are revealed by staining cells with fat-soluble dyes, such as Sudan dyes.

SULFUR GRANULES

Certain bacteria—for example, the "sulfur bacteria" that belong to the genus *Thiobacillus*—derive energy by oxidizing sulfur and sulfur-containing compounds. These bacteria may deposit **sulfur granules** in the cell, where they serve as an energy reserve.

CARBOXYSOMES

Carboxysomes are inclusions that contain the enzyme ribulose 1,5-diphosphate carboxylase. Photosynthetic bacteria use carbon dioxide as their sole source of carbon and require this enzyme for carbon dioxide fixation. Among

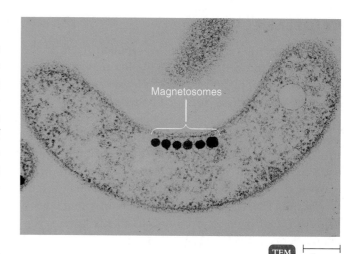

TEM   1 μm

FIGURE 4.20 Magnetosomes. This micrograph of *Magnetospirillum magnetotacticum* shows a chain of magnetosomes. The outer membrane of the gram-negative wall is also visible.

Q **What is the function of magnetosomes?**

the bacteria containing carboxysomes are nitrifying bacteria, cyanobacteria, and thiobacilli.

GAS VACUOLES

Hollow cavities found in many aquatic prokaryotes, including cyanobacteria, anoxygenic photosynthetic bacteria, and halobacteria are called **gas vacuoles.** Each vacuole consists of rows of several individual *gas vesicles,* which are hollow cylinders covered by protein. Gas vacuoles maintain buoyancy so that the cells can remain at the depth in the water appropriate for them to receive sufficient amounts of oxygen, light, and nutrients.

MAGNETOSOMES

Magnetosomes are inclusions of iron oxide (Fe_3O_4), formed by several gram-negative bacteria such as *Magnetospirillum magnetotacticum,* that act like magnets (Figure 4.20). Bacteria may use magnetosomes to move downward until they reach a suitable attachment site. In vitro, magnetosomes can decompose hydrogen peroxide, which forms in cells in the presence of oxygen. Researchers speculate that magnetosomes may protect the cell against hydrogen peroxide accumulation. Industrial microbiologists are developing culture methods to obtain large quantities of magnetite from bacteria to use in the production of magnetic tapes for sound and data recording.

ENDOSPORES

LEARNING OBJECTIVE

• Describe the functions of endospores, sporulation, and endospore germination.

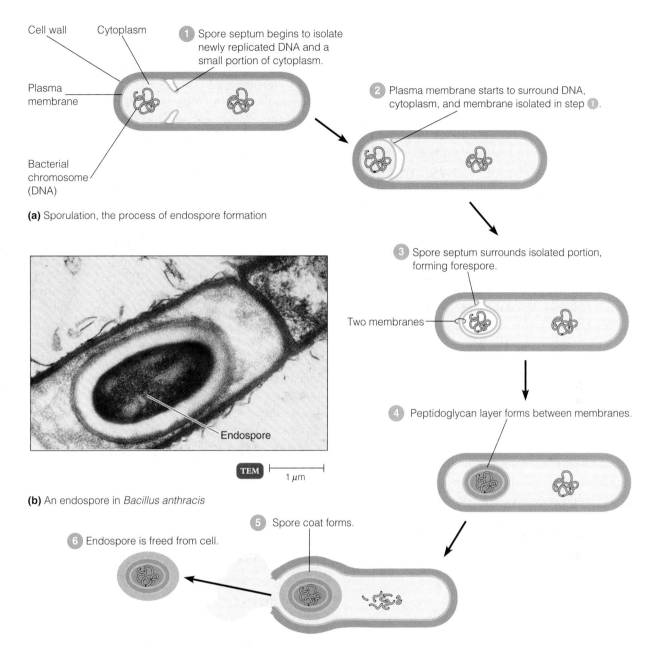

(a) Sporulation, the process of endospore formation

Cell wall
Cytoplasm

1 Spore septum begins to isolate newly replicated DNA and a small portion of cytoplasm.

Plasma membrane

2 Plasma membrane starts to surround DNA, cytoplasm, and membrane isolated in step 1.

Bacterial chromosome (DNA)

3 Spore septum surrounds isolated portion, forming forespore.

Two membranes

4 Peptidoglycan layer forms between membranes.

Endospore

TEM   1 μm

(b) An endospore in *Bacillus anthracis*

5 Spore coat forms.

6 Endospore is freed from cell.

FIGURE 4.21   Formation of endospores by sporulation.

Q **Under what conditions are endospores formed by bacteria?**

When essential nutrients are depleted, certain gram-positive bacteria, such as those of the genera *Clostridium* and *Bacillus*, form specialized "resting" cells called **endospores** (Figure 4.21). As you will see later, some members of the genus *Clostridium* cause diseases such as gangrene, tetanus, botulism, and food poisoning. Some members of the genus *Bacillus* cause anthrax and food poisoning. Unique to bacteria, endospores are highly durable dehydrated cells with thick walls and additional layers. They are formed internal to the bacterial cell membrane.

When released into the environment, they can survive extreme heat, lack of water, and exposure to many toxic chemicals and radiation. For example, 7500-year-old endospores of *Thermoactinomyces vulgaris* (thėr-mō-ak-tin-ō-mī′sēs vul-ga′ris) from the freezing muds of Elk Lake in Minnesota have germinated when rewarmed and placed in a nutrient mēdium, and 25- to 40-million-year-old endospores found in the gut of a stingless bee entombed in amber (hardened tree resin) in the Dominican Republic are reported to have germinated when placed in nutrient media. Although true endospores are found in gram-positive bacteria, one gram-negative species, *Coxiella burnetii* (käks-ē-el′lä bėr-ne′tē-ē), the cause of Q fever, forms endosporelike structures that resist heat and chemicals and can be stained with endospore stains (see Figure 24.15).

The process of endospore formation within a vegetative (parent) cell takes several hours and is known as **sporulation** or **sporogenesis** (Figure 4.21a). Vegetative cells of endospore-forming bacteria begin sporulation when a key nutrient, such as the carbon or nitrogen source, becomes scarce or unavailable. In the first observable stage of sporulation, a newly replicated bacterial chromosome and a small portion of cytoplasm are isolated by an ingrowth of the plasma membrane called a *spore septum*. The spore septum becomes a double-layered membrane that surrounds the chromosome and cytoplasm. This structure, entirely enclosed within the original cell, is called a *forespore*. Thick layers of peptidoglycan are laid down between the two membrane layers. Then a thick *spore coat* of protein forms around the outside membrane; this coat is responsible for the resistance of endospores to many harsh chemicals. The original cell is degraded, and the endospore is released.

The diameter of the endospore may be the same as, smaller than, or larger than the diameter of the vegetative cell. Depending on the species, the endospore might be located *terminally* (at one end), *subterminally* (near one end; Figure 4.21b), or *centrally* inside the vegetative cell. When the endospore matures, the vegetative cell wall ruptures (lyses), killing the cell, and the endospore is freed.

Most of the water present in the forespore cytoplasm is eliminated by the time sporulation is complete, and endospores do not carry out metabolic reactions. The highly dehydrated endospore core contains only DNA, small amounts of RNA, ribosomes, enzymes, and a few important small molecules. The latter include a strikingly large amount of an organic acid called *dipicolinic acid* (found in the cytoplasm), which is accompanied by a large number of calcium ions. These cellular components are essential for resuming metabolism later.

Endospores can remain dormant for thousands of years. An endospore returns to its vegetative state by a process called **germination.** Germination is triggered by physical or chemical damage to the endospore's coat. The endospore's enzymes then break down the extra layers surrounding the endospore, water enters, and metabolism resumes. Because one vegetative cell forms a single endospore, which, after germination, remains one cell, sporulation in bacteria is *not* a means of reproduction. This process does not increase the number of cells. Bacterial endospores differ from spores formed by (prokaryotic) actinomycetes and the eukaryotic fungi and algae, which detach from the parent and develop into another organism and, therefore, represent reproduction.

Endospores are important from a clinical viewpoint and in the food industry because they are resistant to processes that normally kill vegetative cells. Such processes include heating, freezing, desiccation, use of chemicals, and radiation. Whereas most vegetative cells are killed by temperatures above 70°C, endospores can survive in boiling water for several hours or more. Endospores of thermophilic (heat-loving) bacteria can survive in boiling water for 19 hours. Endospore-forming bacteria are a problem in the food industry because they are likely to survive underprocessing, and, if conditions for growth occur, some species produce toxins and disease. Special methods for controlling organisms that produce endospores are discussed in Chapter 7.

* * *

Having examined the functional anatomy of the prokaryotic cell, we will now look at the functional anatomy of the eukaryotic cell.

THE EUKARYOTIC CELL

As mentioned earlier, eukaryotic organisms include algae, protozoa, fungi, plants, and animals. The eukaryotic cell is typically larger and structurally more complex than the prokaryotic cell (Figure 4.22). By comparing the structure of the prokaryotic cell in Figure 4.6 with that of the eukaryotic cell, the differences between the two types of cells become apparent. The principal differences between prokaryotic and eukaryotic cells are summarized in Table 4.2.

The following discussion of eukaryotic cells will parallel our discussion of prokaryotic cells by starting with structures that extend to the outside of the cell.

FLAGELLA AND CILIA

LEARNING OBJECTIVE
- Differentiate between prokaryotic and eukaryotic flagella.

Many types of eukaryotic cells have projections that are used for cellular locomotion or for moving substances along the surface of the cell. These projections contain cytoplasm and are enclosed by the plasma membrane. If the projections are few and are long in relation to the size of the cell, they are called **flagella.** If the projections are numerous and short, they are called **cilia** (singular: **cilium**).

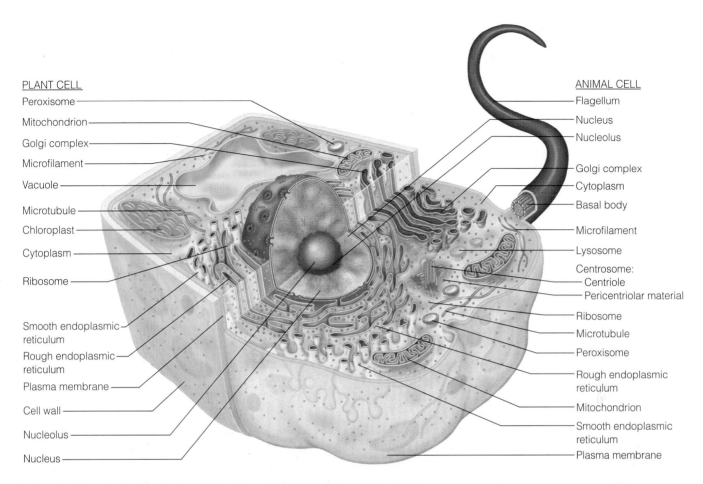

PLANT CELL
Peroxisome
Mitochondrion
Golgi complex
Microfilament
Vacuole
Microtubule
Chloroplast
Cytoplasm
Ribosome
Smooth endoplasmic reticulum
Rough endoplasmic reticulum
Plasma membrane
Cell wall
Nucleolus
Nucleus

ANIMAL CELL
Flagellum
Nucleus
Nucleolus
Golgi complex
Cytoplasm
Basal body
Microfilament
Lysosome
Centrosome:
Centriole
Pericentriolar material
Ribosome
Microtubule
Peroxisome
Rough endoplasmic reticulum
Mitochondrion
Smooth endoplasmic reticulum
Plasma membrane

(a) Highly schematic diagram of a composite eukaryotic cell, half plant and half animal

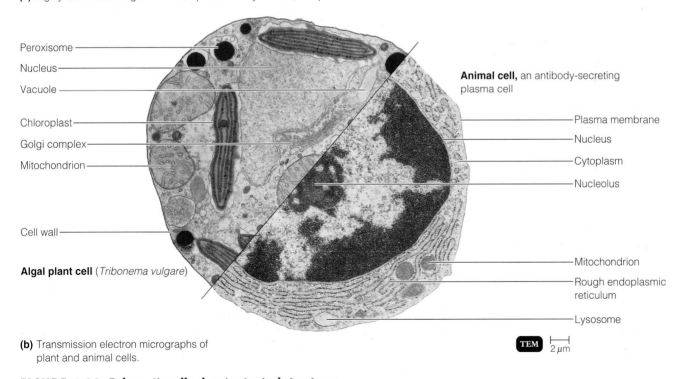

Peroxisome
Nucleus
Vacuole
Chloroplast
Golgi complex
Mitochondrion
Cell wall

Algal plant cell (*Tribonema vulgare*)

Animal cell, an antibody-secreting plasma cell

Plasma membrane
Nucleus
Cytoplasm
Nucleolus
Mitochondrion
Rough endoplasmic reticulum
Lysosome

TEM 2 µm

(b) Transmission electron micrographs of plant and animal cells.

FIGURE 4.22 Eukaryotic cells showing typical structures.

Q **What kingdoms contain eukaryotic organisms?**

TABLE 4.2	Principal Differences Between Prokaryotic and Eukaryotic Cells	
Characteristic	Prokaryotic	Eukaryotic
	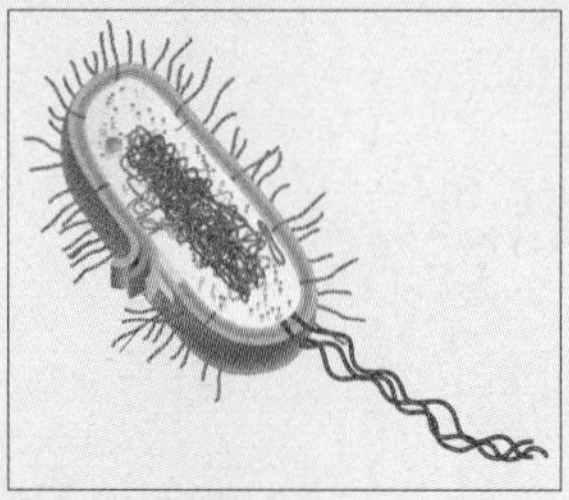	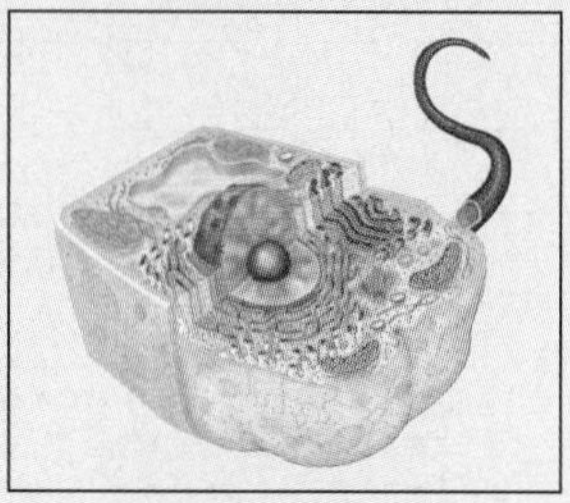
Size of Cell	Typically 0.2–2.0 μm in diameter	Typically 10–100 μm in diameter
Nucleus	No nuclear membrane or nucleoli	True nucleus, consisting of nuclear membrane and nucleoli
Membrane-Enclosed Organelles	Absent	Present; examples include lysosomes, Golgi complex, endoplasmic reticulum, mitochondria, and chloroplasts
Flagella	Consist of two protein building blocks	Complex; consist of multiple microtubules
Glycocalyx	Present as a capsule or slime layer	Present in some cells that lack a cell wall
Cell Wall	Usually present; chemically complex (typical bacterial cell wall includes peptidoglycan)	When present, chemically simple
Plasma Membrane	No carbohydrates and generally lacks sterols	Sterols and carbohydrates that serve as receptors
Cytoplasm	No cytoskeleton or cytoplasmic streaming	Cytoskeleton; cytoplasmic streaming
Ribosomes	Smaller size (70S)	Larger size (80S); smaller size (70S) in organelles
Chromosome (DNA)	Usually single circular chromosome; typically lacks histones	Multiple linear chromosomes with histones
Cell Division	Binary fission	Involves mitosis
Sexual Recombination	None; transfer of DNA fragments only	Involves meiosis

Algae of the genus *Euglena* (ū-glē′na) use a flagellum for locomotion, whereas protozoa, such as *Tetrahymena* (tet-rä-hī′me-nä), use cilia for locomotion (Figure 4.23a and Figure 4.23b). Both flagella and cilia are anchored to the plasma membrane by a basal body, and both consist of nine pairs of microtubules (doublets) arranged in a ring, plus another two microtubules in the center of the ring, an arrangement called a *9 + 2 array* (Figure 4.23c). **Microtubules** are long, hollow tubes made up of a protein called *tubulin*. A prokaryotic flagellum rotates, but a eukaryotic flagellum moves in a wavelike manner (Figure 4.23d). To help keep foreign material out of the lungs, ciliated cells of the human respiratory system move the material along the surface of the cells in the bronchial tubes and trachea toward the throat and mouth (see Figure 16.4, page 478).

THE CELL WALL AND GLYCOCALYX

LEARNING OBJECTIVE

- Compare and contrast prokaryotic and eukaryotic cell walls and glycocalyxes.

Most eukaryotic cells have cell walls, although they are generally much simpler than those of prokaryotic cells. Many algae have cell walls consisting of the polysaccharide *cellulose* (as do all plants); other chemicals may be present as well. Cell walls of some fungi also contain cellulose, but in most fungi the principal structural component of the cell wall is the polysaccharide *chitin*, a polymer of N-acetylglucosamine (NAG) units. (Chitin is also the main structural component of the exoskeleton of crustaceans

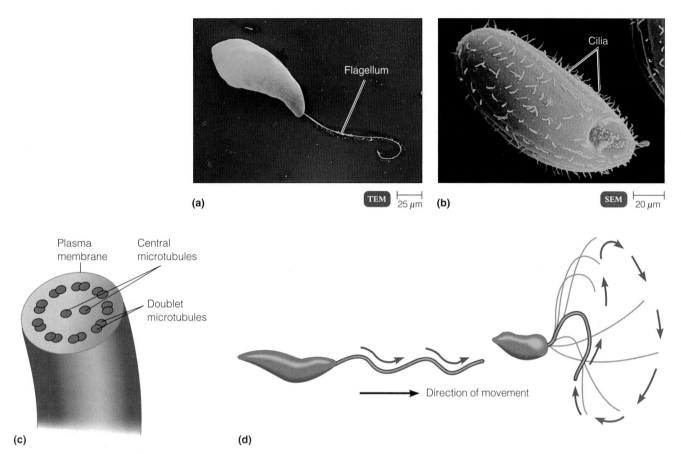

(a)   TEM |—| 25 μm   (b)   SEM |—| 20 μm

(c)   (d)   Direction of movement

FIGURE 4.23 Eukaryotic flagella and cilia. (**a**) A micrograph of *Euglena,* a chlorophyll-containing alga, with its flagellum. (**b**) A micrograph of *Tetrahymena,* a common freshwater protozoan, with cilia. (**c**) The internal structure of a flagellum (or cilium), showing the *9 + 2* arrangement of microtubules. (**d**) The pattern of movement of a eukaryotic flagellum.

Q   **How do prokaryotic and eukaryotic flagella differ?**

and insects.) The cell walls of yeasts contain the polysaccharides *glucan* and *mannan*. In eukaryotes that lack a cell wall, the plasma membrane may be the outer covering; however, cells that have direct contact with the environment may have coatings outside the plasma membrane. Protozoa do not have a typical cell wall; instead, they have a flexible outer protein covering called a *pellicle*.

In other eukaryotic cells, including animal cells, the plasma membrane is covered by a **glycocalyx,** a layer of material containing substantial amounts of sticky carbohydrates. Some of these carbohydrates are covalently bonded to proteins and lipids in the plasma membrane, forming glycoproteins and glycolipids that anchor the glycocalyx to the cell. The glycocalyx strengthens the cell surface, helps attach cells together, and may contribute to cell–cell recognition.

Eukaryotic cells do not contain peptidoglycan, the framework of the prokaryotic cell wall. This is significant medically because antibiotics, such as penicillins and

cephalosporins, act against peptidoglycan and therefore do not affect human eukaryotic cells.

THE PLASMA (CYTOPLASMIC) MEMBRANE

LEARNING OBJECTIVE
• Compare and contrast prokaryotic and eukaryotic plasma membranes.

The **plasma (cytoplasmic) membrane** of eukaryotic and prokaryotic cells is very similar in function and basic structure. There are, however, differences in the types of proteins found in the membranes. Eukaryotic membranes also contain carbohydrates, which serve as attachment sites for bacteria and as receptor sites that assume a role in such functions as cell–cell recognition. Eukaryotic plasma membranes also contain *sterols,* complex lipids not found in prokaryotic plasma membranes (with the exception of *Mycoplasma* cells). Sterols seem to be associated with the ability of the membranes to resist lysis resulting from increased osmotic pressure.

Substances can cross eukaryotic and prokaryotic plasma membranes by simple diffusion, facilitated diffusion, osmosis, or active transport. Group translocation does not occur in eukaryotic cells. However, eukaryotic cells can use a mechanism called **endocytosis.** This occurs when a segment of the plasma membrane surrounds a particle or large molecule, encloses it, and brings it into the cell.

Two very important types of endocytosis are phagocytosis and pinocytosis. During *phagocytosis,* cellular projections called pseudopods engulf particles and bring them into the cell. Phagocytosis is used by white blood cells to destroy bacteria and foreign substances (see Figure 16.8 and further discussion in Chapter 16). In *pinocytosis,* the plasma membrane folds inward, bringing extracellular fluid into the cell, along with whatever substances are dissolved in the fluid. Pinocytosis is one of the ways viruses can enter animal cells (see Figure 13.14a, page 402).

CYTOPLASM

LEARNING OBJECTIVE
• Compare and contrast prokaryotic and eukaryotic cytoplasms.

The **cytoplasm** of eukaryotic cells encompasses the substance inside the plasma membrane and outside the nucleus (see Figure 4.22). The cytoplasm is the substance in which various cellular components are found. (The term **cytosol** refers to the fluid portion of cytoplasm.) A major difference between eukaryotic and prokaryotic cytoplasm is that eukaryotic cytoplasm has a complex internal structure, consisting of exceedingly small rods (*microfilaments* and *intermediate filaments*) and cylinders (*microtubules*). Together, they form the **cytoskeleton.** The cytoskeleton provides support and shape and assists in transporting substances through the cell (and even in moving the entire cell, as in phagocytosis). The movement of eukaryotic cytoplasm from one part of the cell to another, which helps distribute nutrients and move the cell over a surface, is called **cytoplasmic streaming.** Another difference between prokaryotic and eukaryotic cytoplasm is that many of the important enzymes found in the cytoplasmic fluid of prokaryotes are sequestered in the organelles of eukaryotes.

RIBOSOMES

LEARNING OBJECTIVE
• Compare the structure and function of eukaryotic and prokaryotic ribosomes.

Attached to the outer surface of rough endoplasmic reticulum (discussed on page 103) are **ribosomes** (see Figure 4.25), which are also found free in the cytoplasm. As in prokaryotes, ribosomes are the sites of protein synthesis in the cell.

The ribosomes of eukaryotic endoplasmic reticulum and cytoplasm are somewhat larger and denser than those of prokaryotic cells. These eukaryotic ribosomes are 80S ribosomes, each of which consists of a large 60S subunit containing three molecules of rRNA and a smaller 40S subunit with one molecule of rRNA. The subunits are made separately in the nucleolus and, once produced, exit the nucleus and join together in the cytosol. Chloroplasts and mitochondria contain 70S ribosomes, which may indicate their evolution from prokaryotes. (This theory is discussed on page 106) The role of ribosomes in protein synthesis will be discussed in more detail in Chapter 8.

Some ribosomes, called *free ribosomes,* are unattached to any structure in the cytoplasm. Primarily, free ribosomes synthesize proteins used *inside* the cell. Other ribosomes, called *membrane-bound ribosomes,* attach to the nuclear membrane and the endoplasmic reticulum. These ribosomes synthesize proteins destined for insertion in the plasma membrane or for export from the cell. Ribosomes located within mitochondria synthesize mitochondrial proteins. Sometimes 10 to 20 ribosomes join together in a stringlike arrangement called a *polyribosome.*

ORGANELLES

LEARNING OBJECTIVES
• Define *organelle.*
• Describe the functions of the nucleus, endoplasmic reticulum, Golgi complex, lysosomes, vacuoles, mitochondria, chloroplasts, peroxisomes, and centrosomes.

Organelles are structures with specific shapes and specialized functions and are characteristic of eukaryotic cells. They include the nucleus, endoplasmic reticulum, Golgi complex, lysosomes, vacuoles, mitochondria, chloroplasts, peroxisomes, and centrosomes. Not all of the organelles described are found in all cells. Certain cells have their own type and distribution of organelles based on specialization, age, and level of activity.

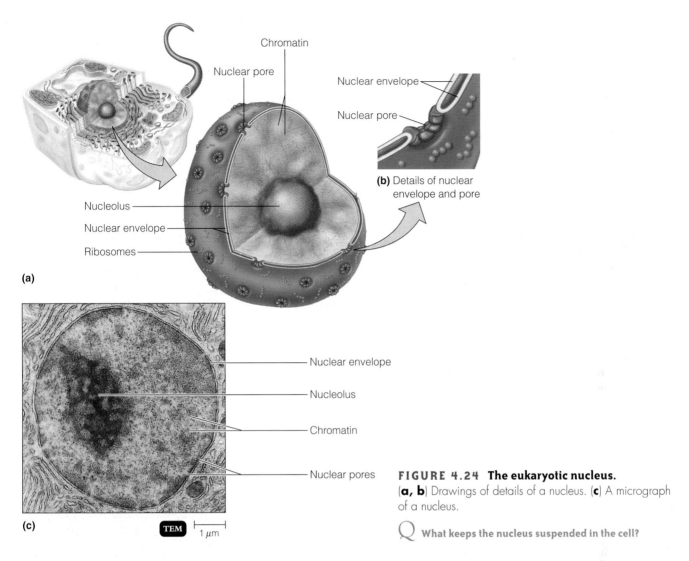

Chromatin

Nuclear pore

Nuclear envelope

Nuclear pore

(b) Details of nuclear envelope and pore

Nucleolus

Nuclear envelope

Ribosomes

(a)

Nuclear envelope

Nucleolus

Chromatin

Nuclear pores

(c)   TEM ⊢─┤ 1 μm

FIGURE 4.24 The eukaryotic nucleus.
(**a, b**) Drawings of details of a nucleus. (**c**) A micrograph of a nucleus.

Q **What keeps the nucleus suspended in the cell?**

THE NUCLEUS

The most characteristic eukaryotic organelle is the nucleus (see Figure 4.22). The **nucleus** (Figure 4.24) is usually spherical or oval, is frequently the largest structure in the cell, and contains almost all of the cell's hereditary information (DNA). Some DNA is also found in mitochondria and in the chloroplasts of photosynthetic organisms.

The nucleus is surrounded by a double membrane called the **nuclear envelope.** Both membranes resemble the plasma membrane in structure. Tiny channels in the membrane called **nuclear pores** allow the nucleus to communicate with the cytoplasm (see Figure 4.24b). Nuclear pores control the movement of substances between the nucleus and cytoplasm. Within the nuclear envelope are one or more spherical bodies called **nucleoli** (singular: **nucleolus**). Nucleoli are actually condensed regions of chromosomes where ribosomal RNA is being synthesized. Ribosomal RNA is an essential component of ribosomes.

The nucleus also contains most of the cell's DNA, which is combined with several proteins, including some

basic proteins called **histones** and nonhistones. The combination of about 165 base pairs of DNA and 9 molecules of histones is referred to as a *nucleosome.* When the cell is not reproducing, the DNA and its associated proteins appear as a threadlike mass called **chromatin.** During nuclear division, the chromatin coils into shorter and thicker rodlike bodies called **chromosomes.** Prokaryotic chromosomes do not undergo this process, do not have histones, and are not enclosed in a nuclear envelope.

Eukaryotic cells require two elaborate mechanisms: mitosis and meiosis to segregate chromosomes prior to cell division. Neither process occurs in prokaryotic cells.

ENDOPLASMIC RETICULUM

Within the cytoplasm of eukaryotic cells is the **endoplasmic reticulum,** or **ER,** an extensive network of flattened membranous sacs or tubules called **cisterns** (Figure 4.25). The ER network is continuous with the nuclear envelope (see Figure 4.22a).

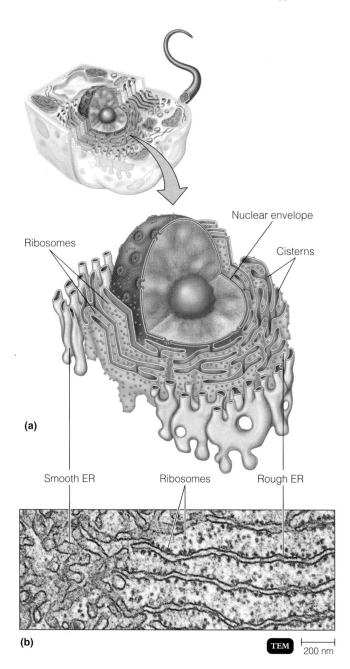

FIGURE 4.25 Rough endoplasmic reticulum and ribosomes. (**a**) A drawing of details of the endoplasmic reticulum. (**b**) A micrograph of the endoplasmic reticulum and ribosomes.

Q **What is the difference between rough ER and smooth ER?**

Most eukaryotic cells contain two distinct, but interrelated, forms of ER that differ in structure and function. The membrane of **rough ER** is continuous with the nuclear membrane and usually unfolds into a series of flattened sacs. The outer surface of rough ER is studded with ribosomes, the sites of protein synthesis. Proteins synthesized by ribosomes that are attached to rough ER enter cisterns within the ER for processing and sorting. In some cases, enzymes within the cisterns attach the proteins to carbohydrates to form glycoproteins. In other cases, enzymes attach the proteins to

phospholipids, also synthesized by rough ER. These molecules may be incorporated into organelle membranes or the plasma membrane. Thus, rough ER is a factory for synthesizing secretory proteins and membrane molecules.

Smooth ER extends from the rough ER to form a network of membrane tubules (see Figure 4.25). Unlike rough ER, smooth ER does not have ribosomes on the outer surface of its membrane. However, smooth ER contains unique enzymes that make it functionally more diverse than rough ER. Although it does not synthesize proteins, smooth ER does synthesize phospholipids, as does rough ER. Smooth ER also synthesizes fats and steroids, such as estrogens and testosterone. In liver cells, enzymes of the smooth ER help release glucose into the bloodstream and inactivate or detoxify drugs and other potentially harmful substances (for example, alcohol). In muscle cells, calcium ions released from the sarcoplasmic reticulum, a form of smooth ER, trigger the contraction process.

GOLGI COMPLEX

Most of the proteins synthesized by ribosomes attached to rough ER are ultimately transported to other regions of the cell. The first step in the transport pathway is through an organelle called the **Golgi complex.** It consists of 3 to 20 cisterns that resemble a stack of pita bread (Figure 4.26). The cisterns are often curved, giving the Golgi complex a cuplike shape.

Proteins synthesized by ribosomes on the rough ER are surrounded by a portion of the ER membrane, which eventually buds from the membrane surface to form a **transport vesicle.** The transport vesicle fuses with a cistern of the Golgi complex, releasing proteins into the cistern. The proteins are modified and move from one cistern to another via **transfer vesicles** that bud from the cisterns' edges. Enzymes in the cisterns modify the proteins to form glycoproteins, glycolipids, and lipoproteins. Some of the processed proteins leave the cisterns in **secretory vesicles,** which detach from the cistern and deliver the proteins to the plasma membrane, where they are discharged by exocytosis. Other processed proteins leave the cisterns in vesicles that deliver their contents to the plasma membrane for incorporation into the membrane. Finally, some processed proteins leave the cisterns in vesicles that are called **storage vesicles.** The major storage vesicle is a lysosome, whose structure and functions are discussed next.

LYSOSOMES

Lysosomes are formed from Golgi complexes and look like membrane-enclosed spheres. Unlike mitochondria, lysosomes have only a single membrane and lack internal structure (see Figure 4.22). But they contain as many as 40 different kinds of powerful digestive enzymes capable of

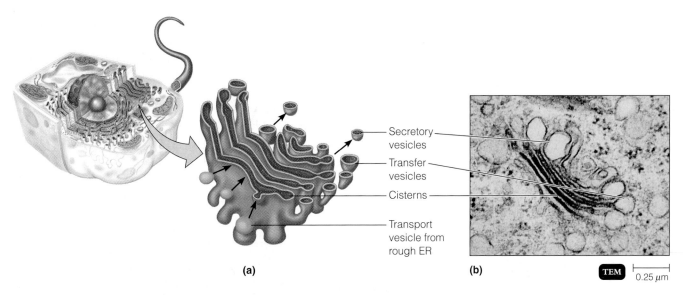

Secretory vesicles

Transfer vesicles

Cisterns

Transport vesicle from rough ER

(a)

(b)

TEM   0.25 μm

FIGURE 4.26 Golgi complex. (**a**) A drawing of details of a Golgi complex. (**b**) A micrograph of a Golgi complex.

Q **What are the functions of the Golgi complex?**

breaking down various molecules. Moreover, these enzymes can also digest bacteria that enter the cell. Human white blood cells, which use phagocytosis to ingest bacteria, contain large numbers of lysosomes.

VACUOLES

A **vacuole** (see Figure 4.22) is a space or cavity in the cytoplasm of a cell that is enclosed by a membrane called a *tonoplast*. In plant cells, vacuoles may occupy 5 to 90% of the cell volume, depending on the type of cell. Vacuoles are derived from the Golgi complex and have several diverse functions. Some vacuoles serve as temporary storage organelles for substances such as proteins, sugars, organic acids, and inorganic ions. Other vacuoles form during endocytosis to help bring food into the cell. Many plant cells also store metabolic wastes and poisons that would otherwise be injurious if they accumulated in the cytoplasm. Finally, vacuoles may take up water, enabling plant cells to increase in size and also providing rigidity to leaves and stems.

MITOCHONDRIA

Spherical or rod-shaped organelles called **mitochondria** (singular: **mitochondrion**) appear throughout the cytoplasm of most eukaryotic cells (see Figure 4.22). The number of mitochondria per cell varies greatly among different types of cells. For example, the protozoan *Giardia* has no mitochondria, whereas liver cells contain 1000 to 2000 per cell. A mitochondrion consists of a double membrane similar in structure to the plasma membrane (Figure 4.27). The outer mitochondrial membrane is smooth, but the

inner mitochondrial membrane is arranged in a series of folds called **cristae** (singular: **crista**). The center of the mitochondrion is a semifluid substance called the **matrix.** Because of the nature and arrangement of the cristae, the inner membrane provides an enormous surface area on which chemical reactions can occur. Some proteins that function in cellular respiration, including the enzyme that makes ATP, are located on the cristae of the inner mitochondrial membrane, and many of the metabolic steps involved in cellular respiration are concentrated in the matrix (see Chapter 5). Mitochondria are often called the "powerhouses of the cell" because of their central role in ATP production.

Mitochondria contain 70S ribosomes and some DNA of their own, as well as the machinery necessary to replicate, transcribe, and translate the information encoded by their DNA. In addition, mitochondria can reproduce more or less on their own by growing and dividing in two.

CHLOROPLASTS

Algae and green plants contain a unique organelle called a **chloroplast** (Figure 4.28), a membrane-enclosed structure that contains both the pigment chlorophyll and the enzymes required for the light-gathering phases of photosynthesis (see Chapter 5). The chlorophyll is contained in flattened membrane sacs called **thylakoids;** stacks of thylakoids are called *grana* (singular: *granum*) (see Figure 4.28).

Like mitochondria, chloroplasts contain 70S ribosomes, DNA, and enzymes involved in protein synthesis. They are capable of multiplying on their own within the cell. The

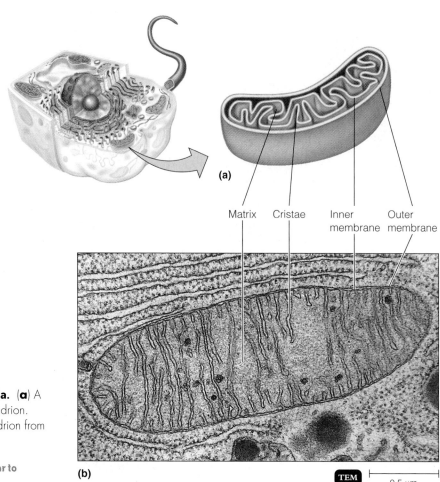

Matrix   Cristae   Inner membrane   Outer membrane

FIGURE 4.27 Mitochondria. (a) A drawing of details of a mitochondrion. **(b)** A micrograph of a mitochondrion from a rat pancreas cell.

 **How are mitochondria similar to prokaryotic cells?**

(b)

TEM   0.5 μm

way both chloroplasts and mitochondria multiply—by increasing in size and then dividing in two—is strikingly reminiscent of bacterial multiplication.

PEROXISOMES

Organelles similar in structure to lysosomes, but smaller, are called **peroxisomes** (see Figure 4.22). Although peroxisomes were once thought to form by budding off the ER, it is now generally agreed that they form by the division of preexisting peroxisomes.

Peroxisomes contain one or more enzymes that can oxidize various organic substances. For example, substances such as amino acids and fatty acids are oxidized in peroxisomes as part of normal metabolism. In addition, enzymes in peroxisomes oxidize toxic substances, such as alcohol. A by-product of the oxidation reactions is hydrogen peroxide (H_2O_2), a potentially toxic compound. However, peroxisomes also contain the enzyme *catalase*, which decomposes H_2O_2 (see Chapter 6, page 167). Because the generation and degradation of H_2O_2 occurs within the same organelle, peroxisomes protect other parts of the cell from the toxic effects of H_2O_2.

CENTROSOME

The **centrosome,** located near the nucleus, consists of two components: the pericentriolar area and centrioles (see Figure 4.22). The *pericentriolar material* is a region of the cytosol composed of a dense network of small protein fibers. This area is the organizing center for the mitotic spindle, which plays a critical role in cell division, and for microtubule formation in nondividing cells. Within the pericentriolar material is a pair of cylindrical structures called *centrioles,* each of which is composed of nine clusters of three microtubules (triplets) arranged in a circular pattern, an arrangement called a 9 + 0 *array.* The 9 refers to the nine clusters of microtubules, and the 0 refers to the absence of microtubules in the center. The long axis of one centriole is at a right angle to the long axis of the other.

THE EVOLUTION OF EUKARYOTES

LEARNING OBJECTIVE

- Discuss evidence that supports the endosymbiotic theory of eukaryotic evolution.

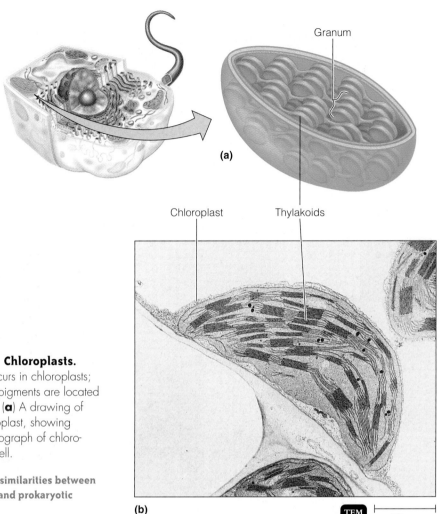

Granum

(a)

Chloroplast   Thylakoids

FIGURE 4.28 Chloroplasts.
Photosynthesis occurs in chloroplasts; the light-trapping pigments are located on the thylakoids. (**a**) A drawing of details of a chloroplast, showing grana. (**b**) A micrograph of chloroplasts in a plant cell.

Q **What are the similarities between chloroplasts and prokaryotic cells?**

(b)   TEM   0.5 µm

Biologists generally believe that life arose on Earth in the form of very simple organisms, similar to prokaryotic cells, about 3.5 to 4 billion years ago. About 2.5 billion years ago, the first eukaryotic cells evolved from prokaryotic cells. Recall that prokaryotes and eukaryotes differ mainly in that eukaryotes contain highly specialized organelles. The theory explaining the origin of eukaryotes from prokaryotes, pioneered by Lynn Margulis, is the **endosymbiotic theory.** According to this theory, larger bacterial cells lost their cell walls and engulfed smaller bacterial cells. This relationship, in which one organism lives within another, is called *endosymbiosis* (*symbiosis* = living together).

According to the endosymbiotic theory, the ancestral eukaryote developed a rudimentary nucleus when the plasma membrane folded around the chromosome (see Figure 10.2, page 286). This cell, called a nucleoplasm, may have ingested aerobic bacteria. Some ingested bacteria lived inside the host nucleoplasm. This arrangement evolved into a symbiotic relationship in which the host nucleoplasm supplied nutrients and the endosymbiotic bacterium produced energy that could be used by the nucleoplasm. Similarly, chloroplasts may be descendants of

photosynthetic prokaryotes ingested by this early nucleoplasm. Eukaryotic flagella and cilia are believed to have originated from symbiotic associations between the plasma membrane of early eukaryotes and motile spiral bacteria called spirochetes. A living example that suggests how flagella developed is described in the box on page 108.

Studies comparing prokaryotic and eukaryotic cells provide evidence for the endosymbiotic theory. For example, both mitochondria and chloroplasts resemble bacteria in size and shape. Further, these organelles contain circular DNA, which is typical of prokaryotes, and the organelles can reproduce independently of their host cell. Moreover, mitochondrial and chloroplast ribosomes resemble those of prokaryotes, and their mechanism of protein synthesis is more similar to that found in bacteria than eukaryotes. Also, the same antibiotics that inhibit protein synthesis on ribosomes in bacteria also inhibit protein synthesis on ribosomes in mitochondria and chloroplasts.

Our next concern is to examine microbial metabolism. In Chapter 5, you will learn about the importance of enzymes to microorganisms and the ways microbes produce and use energy.

APPLICATIONS OF MICROBIOLOGY

WHY MICROBIOLOGISTS STUDY TERMITES

Although termites are famous for their ability to eat wood, causing damage to wooden structures and recycling cellulose in the soil, they are unable to digest the wood that they eat. To break down the cellulose, termites enlist the help of a variety of microorganisms. Some termites, for example, dig tunnels in the wood, then inoculate the tunnels with fungi that grow on the wood. These termites then eat the fungi, not the wood itself.

What microbiologists find more interesting are the termites that contain, within their digestive tracts, symbiotic microorganisms that digest the cellulose that the termites chew and swallow. Even more fascinating to microbiologists is the fact that these microorganisms themselves can survive only because of even smaller symbionts that live on and within them, without which they would not even be able to move. By studying how a single termite survives, microbiologists have begun to gain an entirely new understanding of symbiosis.

The termite's dependence on nitrogen-fixing bacteria to supply its nitrogen and on protozoans such as *Trichonympha*

sphaerica to digest cellulose is an example of endosymbiosis, a symbiotic relationship with an organism that lives inside the body of the host organism (in this case, within the hindgut of the termite).

Theoretically, the protozoan produces the cellulolytic enzymes that digest the cellulose. The picture is more complicated than this, however, for *T. sphaerica* itself is unable to digest cellulose without the aid of bacteria that live within its body: in other words, the protozoan has its own endosymbionts.

Certain hindgut flagellates such as *T. sphaerica* also demonstrate another form of symbiosis—ectosymbiosis, a symbiotic relationship with organisms that live outside its body. Recent advances in microscopy have shown that these flagellates are covered by precise rows consisting of thousands of bacteria, either rods or spirochetes. If these bacteria are killed, the protozoan is unable to move. Instead of using its own flagella, the protozoan relies on the rows of bacteria to row it about like oarsmen in a boat.

The protozoan *Mixotricha*, for example, has rows of spirochetes on its surface. As shown in part (a) of the figure, the end of each spirochete abuts against a swelling known as a bracket. The spirochetes undulate in unison, thereby creating waves of motion along *Mixotricha's* surface.

Rod-shaped bacteria aligned in grooves that cover the surface of devescovinids, another group of termite-hindgut protozoans. Each rod has twelve flagella that overlap the flagella of the adjacent bacteria to form a continuous filament along the groove (see part (b) of the figure). The bacteria rotate their flagella thus creating coordinated waves along all these rows of filaments, which propels the protozoan.

Sid Tamm and his colleagues at Boston University have found that the protozoa cannot control motility of the ectosymbiotics. *Mixotricha* uses its flagella to steer and the bacteria push the protozoa forward—shoving and being shoved by its neighbors, like "bumper cars."

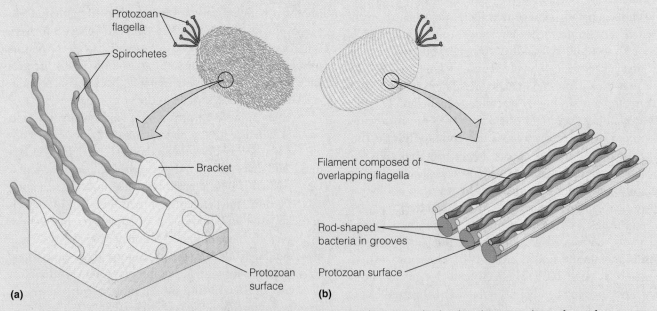

Arrangements of bacteria on the surfaces of two protozoans. **(a)** Spirochetes attached to brackets over the surface of a *Mixotricha* protozoan align themselves and move in unison. **(b)** On this devescovinid protozoan, the flagella from one rod-shaped bacterium overlap the next to form a continuous filament.

STUDY OUTLINE

COMPARING PROKARYOTIC AND EUKARYOTIC CELLS: AN OVERVIEW

(p. 78)

1. Prokaryotic and eukaryotic cells are similar in their chemical composition and chemical reactions.
2. Prokaryotic cells lack membrane-enclosed organelles (including a nucleus).
3. Peptidoglycan is found in prokaryotic cell walls but not in eukaryotic cell walls.
4. Eukaryotic cells have a membrane-bound nucleus and other organelles.

THE PROKARYOTIC CELL (pp. 78–98)

1. Bacteria are unicellular, and most of them multiply by binary fission.
2. Bacterial species are differentiated by morphology, chemical composition, nutritional requirements, biochemical activities, and source of energy.

THE SIZE, SHAPE, AND ARRANGEMENT OF BACTERIAL CELLS (pp. 78–80)

1. Most bacteria are 0.2–2.0 μm in diameter and 2–8 μm in length.
2. The three basic bacterial shapes are coccus (spherical), bacillus (rod-shaped), and spiral (twisted).
3. Pleomorphic bacteria can assume several shapes.

STRUCTURES EXTERNAL TO THE CELL WALL (pp. 80–85)

GLYCOCALYX (pp. 80–81)

1. The glycocalyx (capsule, slime layer, or extracellular polysaccharide) is a gelatinous polysaccharide and/or polypeptide covering.
2. Capsules may protect pathogens from phagocytosis.
3. Capsules enable adherence to surfaces, prevent desiccation, and may provide nutrients.

FLAGELLA (pp. 81–83)

4. Flagella are relatively long filamentous appendages consisting of a filament, hook, and basal body.
5. Prokaryotic flagella rotate to push the cell.

6. Motile bacteria exhibit taxis; positive taxis is movement toward an attractant, and negative taxis is movement away from a repellent.
7. Flagellar (H) protein functions as an antigen. ✳ **Animation: Bacterial Motility. The Microbiology Place.**

AXIAL FILAMENTS (p. 84)

8. Spiral cells that move by means of an axial filament (endoflagellum) are called spirochetes.
9. Axial filaments are similar to flagella, except that they wrap around the cell.

FIMBRIAE AND PILI (pp. 84–85)

10. Fimbriae and pili are short, thin appendages.
11. Fimbriae help cells adhere to surfaces.
12. Pili join cells for the transfer of DNA from one cell to another.

THE CELL WALL (pp. 85–89)

COMPOSITION AND CHARACTERISTICS (pp. 85–87)

1. The cell wall surrounds the plasma membrane and protects the cell from changes in water pressure.
2. The bacterial cell wall consists of peptidoglycan, a polymer consisting of NAG and NAM and short chains of amino acids.
3. Penicillin interferes with peptidoglycan synthesis.
4. Gram-positive cell walls consist of many layers of peptidoglycan and also contain teichoic acids.
5. Gram-negative bacteria have a lipopolysaccharide-lipoprotein-phospholipid outer membrane surrounding a thin peptidoglycan layer.
6. The outer membrane protects the cell from phagocytosis and from penicillin, lysozyme, and other chemicals.
7. Porins are proteins that permit small molecules to pass through the outer membrane; specific channel proteins allow other molecules to move through the outer membrane.
8. The lipopolysaccharide component of the outer membrane consists of sugars (O polysaccharides), which function as antigens, and lipid A, which is an endotoxin.

CELL WALLS AND THE GRAM STAIN MECHANISM (pp. 87–88)

9. The crystal violet–iodine complex combines with peptidoglycan.
10. The decolorizer removes the lipid outer membrane of gram-negative bacteria and washes out the crystal violet.

ATYPICAL CELL WALLS (pp. 88–89)

11. *Mycoplasma* is a bacterial genus that naturally lacks cell walls.

12. Archaea have pseudomurein; they lack peptidoglycan.

13. Acid-fast cell walls have a layer of mycolic acid outside a thin peptidoglycan layer.

DAMAGE TO THE CELL WALL (p. 89)

14. In the presence of lysozyme, gram-positive cell walls are destroyed, and the remaining cellular contents are referred to as a protoplast.

15. In the presence of lysozyme, gram-negative cell walls are not completely destroyed, and the remaining cellular contents are referred to as a spheroplast.

16. L forms are gram-positive or gram-negative bacteria that do not make a cell wall.

17. Antibiotics such as penicillin interfere with cell wall synthesis.

STRUCTURES INTERNAL TO THE CELL WALL (pp. 89–98)

THE PLASMA (CYTOPLASMIC) MEMBRANE (pp. 89–92)

1. The plasma membrane encloses the cytoplasm and is a lipid bilayer with peripheral and integral proteins (the fluid mosaic model).

2. The plasma membrane is selectively permeable.

3. Plasma membranes carry enzymes for metabolic reactions, such as nutrient breakdown, energy production, and photosynthesis.

4. Mesosomes, irregular infoldings of the plasma membrane, are artifacts, not true cell structures.

5. Plasma membranes can be destroyed by alcohols and polymyxins.

THE MOVEMENT OF MATERIALS ACROSS MEMBRANES (pp. 92–94)

6. Movement across the membrane may be by passive processes, in which materials move from areas of higher to lower concentration and no energy is expended by the cell.

7. In simple diffusion, molecules and ions move until equilibrium is reached.

8. In facilitated diffusion, substances are transported by transporter proteins across membranes from areas of high to low concentration.

9. Osmosis is the movement of water from areas of high to low concentration across a selectively permeable membrane until equilibrium is reached.

10. In active transport, materials move from areas of low to high concentration by transporter proteins, and the cell must expend energy.

11. In group translocation, energy is expended to modify chemicals and transport them across the membrane. ❊ **Animation: Membrane Transport. The Microbiology Place.**

CYTOPLASM (p. 95)

12. Cytoplasm is the fluid component inside the plasma membrane.

13. The cytoplasm is mostly water, with inorganic and organic molecules, DNA, ribosomes, and inclusions.

THE NUCLEAR AREA (p. 95)

14. The nuclear area contains the DNA of the bacterial chromosome.

15. Bacteria can also contain plasmids, which are circular, extrachromosomal DNA molecules.

RIBOSOMES (p. 95)

16. The cytoplasm of a prokaryote contains numerous 70S ribosomes; ribosomes consist of rRNA and protein.

17. Protein synthesis occurs at ribosomes; it can be inhibited by certain antibiotics.

INCLUSIONS (p. 96)

18. Inclusions are reserve deposits found in prokaryotic and eukaryotic cells.

19. Among the inclusions found in bacteria are metachromatic granules (inorganic phosphate), polysaccharide granules (usually glycogen or starch), lipid inclusions, sulfur granules, carboxysomes (ribulose 1,5-diphosphate carboxylase), magnetosomes (Fe_3O_4), and gas vacuoles.

ENDOSPORES (pp. 96–98)

20. Endospores are resting structures formed by some bacteria; they allow survival during adverse environmental conditions.

21. The process of endospore formation is called sporulation; the return of an endospore to its vegetative state is called germination.

THE EUKARYOTIC CELL (pp. 98–107)

FLAGELLA AND CILIA (pp. 98–100)

1. Flagella are few and long in relation to cell size; cilia are numerous and short.

2. Flagella and cilia are used for motility, and cilia also move substances along the surface of the cells.

3. Both flagella and cilia consist of an arrangement of nine pairs and two single microtubules.

THE CELL WALL AND GLYCOCALYX (pp. 100–101)

1. The cell walls of many algae and some fungi contain cellulose.

2. The main material of fungal cell walls is chitin.

3. Yeast cell walls consist of glucan and mannan.

4. Animal cells are surrounded by a glycocalyx, which strengthens the cell and provides a means of attachment to other cells.

THE PLASMA (CYTOPLASMIC) MEMBRANE (pp. 101–102)

1. Like the prokaryotic plasma membrane, the eukaryotic plasma membrane is a phospholipid bilayer containing proteins.

2. Eukaryotic plasma membranes contain carbohydrates attached to the proteins and sterols not found in prokaryotic cells (except *Mycoplasma* bacteria).

3. Eukaryotic cells can move materials across the plasma membrane by the passive processes used by prokaryotes and by active transport and endocytosis (phagocytosis and pinocytosis).

CYTOPLASM (p. 102)

1. The cytoplasm of eukaryotic cells includes everything inside the plasma membrane and external to the nucleus.

2. The chemical characteristics of the cytoplasm of eukaryotic cells resemble those of the cytoplasm of prokaryotic cells.

3. Eukaryotic cytoplasm has a cytoskeleton and exhibits cytoplasmic streaming.

RIBOSOMES (p. 102)

1. 80S ribosomes are found in the cytoplasm or attached to the rough endoplasmic reticulum.

ORGANELLES (pp. 102–106)

1. Organelles are specialized membrane-enclosed structures in the cytoplasm of eukaryotic cells.

2. The nucleus, which contains DNA in the form of chromosomes, is the most characteristic eukaryotic organelle.

3. The nuclear envelope is connected to a system of membranes in the cytoplasm called the endoplasmic reticulum (ER).

4. The ER provides a surface for chemical reactions, serves as a transporting network, and stores synthesized molecules. Protein synthesis and transport occur on rough ER; lipid synthesis occurs on smooth ER.

5. The Golgi complex consists of flattened sacs called cisterns. It functions in membrane formation and protein secretion.

6. Lysosomes are formed from Golgi complexes. They store digestive enzymes.

7. Vacuoles are membrane-enclosed cavities derived from the Golgi complex or endocytosis. They are usually found in plant cells that store various substances, help bring food into the cell, increase cell size, and provide rigidity to leaves and stems.

8. Mitochondria are the primary sites of ATP production. They contain 70S ribosomes and DNA, and they multiply by binary fission.

9. Chloroplasts contain chlorophyll and enzymes for photosynthesis. Like mitochondria, they contain 70S ribosomes and DNA and multiply by binary fission.

10. A variety of organic compounds are oxidized in peroxisomes. Catalase in peroxisomes destroys H_2O_2.

11. The centrosome consists of the pericentriolar material and centrioles. Centrioles are 9 triplet microtubules involved in formation of the mitotic spindle and microtubules.

THE EVOLUTION OF EUKARYOTES (pp. 106–107)

1. According to the endosymbiotic theory, eukaryotic cells evolved from symbiotic prokaryotes living inside other prokaryotic cells.

STUDY QUESTIONS

Access more review material either online at **The Microbiology Place** (www.microbiologyplace.com) or with **The Microbiology Place CD-ROM** packaged with your new book. There you'll find activities, practice tests, quizzes, flashcards, case studies, and more to help you succeed.

Answers to the Study Questions can be found in Appendix G.

REVIEW

1. Diagram each of the following flagellar arrangements:
   a. lophotrichous
   b. monotrichous
   c. peritrichous

2. Endospore formation is called _sporogenesis._ It is initiated by _certain adverse environmental env._ Formation of a new cell from an endospore is called _germination._ This process is triggered by _favorable growth condition_

3. Draw the bacterial shapes listed in a, b, and c. Show how d, e, and f are special conditions of a, b, and c, respectively.
   a. spiral
   b. bacillus
   c. coccus
   d. spirochetes
   e. streptobacilli
   f. staphylococci

4. Match the structures to their functions.

   _d_ cell wall   a. attachment to surfaces
   _f_ endospore   b. cell wall formation
   _a_ fimbriae   c. motility
   _c_ flagella   d. protection from osmotic lysis
   _a,e_ glycocalyx   e. protection from phagocytes
   _i_ pili   f. resting
   _b,h_ plasma   g. protein synthesis
       membrane   h. selective permeability
   _g_ ribosomes   i. transfer of genetic material

5. Why is an endospore called a resting structure? Of what advantage is an endospore to a bacterial cell?

6. Compare and contrast the following:
   a. simple diffusion and facilitated diffusion
   b. active transport and facilitated diffusion
   c. active transport and group translocation

7. Why are mycoplasmas resistant to antibiotics that interfere with cell wall synthesis? _Mycoplasma do not have cell wall_

8. Compare and contrast the following:
   a. spheroplast and L form
   b. *Mycoplasma* and L form

9. Answer the following questions using the diagrams provided, which represent cross sections of bacterial cell walls.
   a. Which diagram represents a gram-positive bacterium? How can you tell?

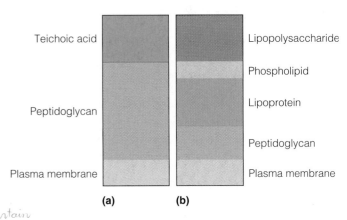

(a)   (b)

(Labels left diagram (a), top to bottom: Teichoic acid, Peptidoglycan, Plasma membrane)

(Labels right diagram (b), top to bottom: Lipopolysaccharide, Phospholipid, Lipoprotein, Peptidoglycan, Plasma membrane)

   b. Explain how the Gram stain works to distinguish between these two types of cell walls.
   c. Why does penicillin have no effect on most gram-negative cells?
   d. How do essential molecules enter cells through each wall?
   e. Which cell wall is toxic to humans?

10. Starch is readily metabolized by many cells, but a starch molecule is too large to cross the plasma membrane. How does a cell obtain the glucose molecules from a starch polymer? How does the cell transport these glucose molecules across the plasma membrane?

11. Match the following characteristics of eukaryotic cells with their functions.

   _c_ pericentriolar material   a. digestive enzyme storage
   _d_ chloroplasts   b. oxidation of fatty acids
   _g_ Golgi complex   c. microtubule formation
   _a_ lysosomes   d. photosynthesis
   _f_ mitochondria   e. protein synthesis
   _b_ peroxisomes   f. respiration
   _e_ rough ER   g. secretion

12. Eukaryotic cells might have evolved from early prokaryotic cells living in close association. What do you know about eukaryotic organelles that would support this theory?

13. What process would a eukaryotic cell use to ingest a prokaryotic cell? To ingest a virus?

14. The antibiotic erythromycin binds with the 50S portion of a ribosome. What effect does this have on a prokaryotic cell? A eukaryotic cell?

MULTIPLE CHOICE

1. Which of the following is *not* a distinguishing characteristic of prokaryotic cells?
   a. They usually have a single, circular chromosome.
   b. They lack membrane-enclosed organelles.
   c. They have cell walls containing peptidoglycan.
   d. Their DNA is not associated with histones.
   e. They lack a plasma membrane.

Use the following choices to answer questions 2–4.
   a. No change will result; the solution is isotonic.
   b. Water will move into the cell.
   c. Water will move out of the cell.
   d. The cell will undergo osmotic lysis.
   e. Sucrose will move into the cell from an area of higher concentration to one of lower concentration.

2. Which statement best describes what happens when a *d* gram-positive bacterium is placed in distilled water and penicillin?

3. Which statement best describes what happens when a *b* gram-negative bacterium is placed in distilled water and penicillin?

4. Which statement best describes what happens when a *a* gram-positive bacterium is placed in an aqueous solution of lysozyme and 10% sucrose?

5. Which of the following statements best describes what happens to a cell exposed to polymyxins that destroy phospholipids?
   a. In an isotonic solution, nothing will happen.
   b. In a hypotonic solution, the cell will lyse.
   c. Water will move into the cell.
   d. Intracellular contents will leak from the cell.
   e. Any of the above might happen.

6. Which of the following is *not* true about fimbriae?
   a. They are composed of protein.
   b. They may be used for attachment.
   c. They are found on gram-negative cells.
   d. They are composed of pilin.
   e. They may be used for motility.

7. Which of the following pairs is mismatched?
   a. glycocalyx—adherence
   b. pili—reproduction
   c. membrane—DNA synthesis
   d. cell wall—protection
   e. plasma membrane—transport

8. Which of the following pairs is mismatched?
   a. metachromatic granules—stored phosphates
   b. polysaccharide granules—stored starch
   c. lipid inclusions—poly-β-hydroxybutyric acid
   d. sulfur granules—energy reserve
   e. ribosomes—protein storage

9. You have isolated a motile, gram-positive cell with no visible nucleus. You can assume this cell has
   a. ribosomes.   **d.** a Golgi complex.
   b. mitochondria.   **e.** all of the above
   c. an endoplasmic
       reticulum.

10. The antibiotic amphotericin B disrupts plasma membranes by combining with sterols; it will affect all of the following cells *except*
   a. animal cells.   **d.** *Mycoplasma* cells.
   b. bacterial cells.   **e.** plant cells.
   c. fungal cells.

CRITICAL THINKING

1. Why can prokaryotic cells be smaller than eukaryotic cells and still carry on all the functions of life?

2. The smallest eukaryotic cell is the motile alga *Micromonas*. What is the minimum number of organelles this alga must have?

3. Two types of prokaryotic cells have been distinguished: bacteria and archaea. How do these cells differ from each other? How are they similar?

4. In 1985, a 0.5-mm cell was discovered in surgeonfish and named *Epulopiscium fishelsoni* (see Figure 11.15, page 320). It was presumed to be a protozoan. In 1993, researchers determined that *Epulopiscium* was actually a gram-positive bacterium. What do you suppose caused the initial identification of this organism as a protozoan? What evidence would change the classification to bacterium?

5. When *E. coli* cells are exposed to a hypertonic solution, the bacteria produce a transporter protein that can move K^+ (potassium ions) into the cell. Of what value is the active transport of K^+, which requires ATP?

CLINICAL APPLICATIONS

1. A child with a bloodborne *Neisseria* infection was treated with gentamicin. After treatment, *Neisseria* could not be cultured from her blood, indicating that the bacteria were killed. However, her symptoms became worse. Annually, nearly half of similar patients die. Explain why antibiotic treatment made her symptoms increase.

2. *Clostridium botulinum* is a strict anaerobe; that is, it is killed by the molecular oxygen (O_2) present in air. Humans can die of botulism from eating foods in which *C. botulinum* is growing. How does this bacterium survive on plants picked for human consumption? Why are home-canned foods most often the source of botulism?

3. Within a 3-day period at a large hospital, five patients undergoing hemodialysis developed fever and chills. *Pseudomonas aeruginosa* and *Klebsiella pneumoniae* were isolated from three of the patients. *P. aeruginosa*, *K. pneumoniae*, and *Enterobacter agglomerans* were isolated from the dialysis system. Why do all three bacteria cause similar symptoms?

4. A South San Francisco child enjoyed bath time at his home because of the colorful orange and red water. The water did not have this rusty color at its source, and the water department could not culture the *Thiobacillus* bacteria responsible for the rusty color from the source. How were the bacteria getting into the household water? What bacterial structures make this possible?

5. Live cultures of *Bacillus thuringiensis* (Dipel) and *B. subtilis* (Kodiak) are sold as pesticides. What bacterial structures make it possible to package and sell these bacteria? For what purpose is each product used? (*Hint:* Refer to Chapter 11.)

5

Microbial Metabolism

Now that you are familiar with the structure of prokaryotic cells, we can discuss the activities that enable these microbes to thrive. The life-support processes of even the most structurally simple organism involve a large number of complex biochemical reactions. Most, although not all, of the biochemical processes of bacteria also occur in eukaryotic microbes and in the cells of multicellular organisms, including humans. However, the reactions that are unique to bacteria are fascinating because they allow microorganisms to do things we cannot do. For example, some bacteria can live on cellulose, while others can live on petroleum. Through their metabolism, bacteria recycle elements after other organisms have used them. Chemoautotrophs can live on diets of such inorganic substances as carbon dioxide, iron, sulfur, hydrogen gas, and ammonia.

This chapter examines some representative chemical reactions that either produce energy (the catabolic reactions) or use energy (the anabolic reactions) in microorganisms. We will also look at how these various reactions are integrated within the cell.

UNDER THE MICROSCOPE

Glaucocystis nostochinearum. This is a colorless (*glauc-*) eukaryotic alga that uses the photosynthesis of endosymbiotic cyanobacteria (*nostoc-*). The cyanobacteria form structures similar to chloroplasts.

CATABOLIC AND ANABOLIC REACTIONS

> **LEARNING OBJECTIVES**
>
> - Define *metabolism*, and describe the fundamental differences between anabolism and catabolism.
> - Identify the role of ATP as an intermediate between catabolism and anabolism.

We use the term **metabolism** to refer to the sum of all chemical reactions within a living organism. Because chemical reactions either release or require energy, metabolism can be viewed as an energy-balancing act. Accordingly, metabolism can be divided into two classes of chemical reactions: those that release energy and those that require energy.

In living cells, the enzyme-regulated chemical reactions that release energy are generally the ones involved in **catabolism,** the breakdown of complex organic compounds into simpler ones. These reactions are called *catabolic*, or *degradative*, reactions. Catabolic reactions are generally *hydrolytic reactions* (reactions that use water and in which chemical bonds are broken), and they are *exergonic* (produce more energy than they consume). An example of catabolism occurs when cells break down sugars into carbon dioxide and water.

The enzyme-regulated energy-requiring reactions are mostly involved in **anabolism,** the building of complex organic molecules from simpler ones. These reactions are called *anabolic*, or *biosynthetic*, reactions. Anabolic processes often involve *dehydration synthesis* reactions (reactions that release water), and they are *endergonic* (consume more energy than they produce). Examples of anabolic processes are the formation of proteins from amino acids, nucleic acids from nucleotides, and polysaccharides from simple sugars. These biosynthetic reactions generate the materials for cell growth.

Catabolic reactions provide building blocks for anabolic reactions and furnish the energy needed to drive anabolic reactions. This coupling of energy-requiring and energy-releasing reactions is made possible through the molecule adenosine triphosphate (ATP). (You can review its structure in Figure 2.18, page 49.) ATP stores energy derived from catabolic reactions and releases it later to drive anabolic reactions and perform other cellular work. Recall from Chapter 2 that a molecule of ATP consists of an adenine, a ribose, and three phosphate groups. When the terminal phosphate group is split from ATP, adenosine diphosphate (ADP) is formed, and energy is released to drive anabolic reactions. Using Ⓟ to represent a phosphate group (Ⓟ$_i$ represents inorganic

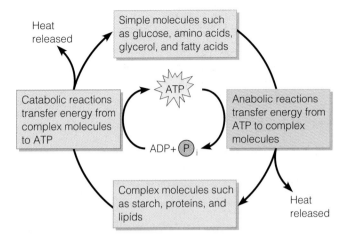

FIGURE 5.1 The role of ATP in coupling anabolic and catabolic reactions. When complex molecules are split apart (catabolism), some of the energy is transferred to and trapped in ATP, and the rest is given off as heat. When simple molecules are combined to form complex molecules (anabolism), ATP provides the energy for synthesis, and again some energy is given off as heat.

Q Which molecule facilitates the coupling of anabolic and catabolic reactions?

phosphate, which is not bound to any other molecule), we write this reaction as:

$$\text{ATP} \rightarrow \text{ADP} + Ⓟ_i + \text{energy}$$

Then, the energy from catabolic reactions is used to combine ADP and a Ⓟ to resynthesize ATP:

$$\text{ADP} + Ⓟ_i + \text{energy} \rightarrow \text{ATP}$$

Thus, anabolic reactions are coupled to ATP breakdown, and catabolic reactions are coupled to ATP synthesis. This concept of coupled reactions is very important; you will see why by the end of this chapter. For now, you should know that the chemical composition of a living cell is constantly changing: some molecules are broken down while others are being synthesized. This balanced flow of chemicals and energy maintains the life of a cell.

The role of ATP in coupling anabolic and catabolic reactions is shown in Figure 5.1. Only part of the energy released in catabolism is actually available for cellular functions because part of the energy is lost to the environment as heat. Because the cell must use energy to maintain life, it has a continuous need for new external sources of energy.

Before we discuss how cells produce energy, let's first consider the principal properties of a group of proteins involved in almost all biologically important chemical reactions. These proteins, the enzymes, were described briefly in Chapter 2. It is important to understand that a cell's

metabolic pathways (sequences of chemical reactions) are determined by its enzymes, which are in turn determined by the cell's genetic makeup. ❊ **Animation: Go to the Microbiology Place website or CD-ROM and click "Animations" to view Metabolic Pathways (Overview).**

ENZYMES

> **LEARNING OBJECTIVES**
> - Identify the components of an enzyme.
> - Describe the mechanism of enzymatic action.
> - List the factors that influence enzymatic activity.
> - Define *ribozyme*.

COLLISION THEORY

We indicated in Chapter 2 that chemical reactions occur when chemical bonds are formed or broken. In order for reactions to take place, atoms, ions, or molecules must collide. The **collision theory** explains how chemical reactions occur and how certain factors affect the rates of those reactions. The basis of the collision theory is that all atoms, ions, and molecules are continuously moving and are thus continuously colliding with one another. The energy transferred by the particles in the collision can disrupt their electron structures enough so that chemical bonds are broken or new bonds are formed.

Several factors determine whether a collision will cause a chemical reaction: the velocities of the colliding particles, their energy, and their specific chemical configurations. Up to a point, the higher the particles' velocities, the greater the probability that their collision will cause a reaction. Also, each chemical reaction requires a specific level of energy. But even if colliding particles possess the minimum energy needed for reaction, no reaction will take place unless the particles are properly oriented toward each other.

Let's assume that molecules of substance AB (the reactant) are to be converted to molecules of substances A and B (the products). In a given population of molecules of substance AB, at a specific temperature, some molecules will possess relatively little energy; the majority of the population will possess an average amount of energy; and a small portion of the population will have high energy. If only the high-energy AB molecules are able to react and be converted to A and B molecules, then only relatively few molecules at any one time possess enough energy to react in a collision. The collision energy required for a chemical reaction is its **activation energy,** which is the amount of energy needed to disrupt the stable electronic configuration of any specific molecule so that the electrons can be rearranged.

The **reaction rate**—the frequency of collisions containing sufficient energy to bring about a reaction—depends on the number of reactant molecules at or above the activation energy level. One way to increase the reaction rate of a substance is to raise its temperature. By causing the molecules to move faster, heat increases both the frequency of collisions and the number of molecules that attain activation energy. The number of collisions also increases when pressure is increased or when the reactants are more concentrated (because the distance between molecules is thereby decreased). In living systems, enzymes increase the reaction rate without raising the temperature.

ENZYMES AND CHEMICAL REACTIONS

Substances that can speed up a chemical reaction without being permanently altered themselves are called **catalysts.** In living cells, **enzymes** serve as biological catalysts. As catalysts, enzymes are specific. Each acts on a specific substance, called the enzyme's **substrate** (or substrates, when there are two or more reactants), and each catalyzes only one reaction. For example, sucrose (table sugar) is the substrate of the enzyme sucrase, which catalyzes the hydrolysis of sucrose to glucose and fructose.

As catalysts, enzymes typically accelerate chemical reactions. The three-dimensional enzyme molecule has an *active site*, a region that will interact with a specific chemical substance (see Figure 5.4).

The enzyme orients the substrate into a position that increases the probability of a reaction. The **enzyme—substrate complex** formed by the temporary binding of enzyme and reactants enables the collisions to be more effective and lowers the activation energy of the reaction (Figure 5.2). The enzyme therefore speeds up the reaction by increasing the number of AB molecules that attain sufficient activation energy to react.

An enzyme's ability to accelerate a reaction without the need for an increase in temperature is crucial to living systems because a significant temperature increase would destroy cellular proteins. The crucial function of enzymes, therefore, is to speed up biochemical reactions at a temperature that is compatible with the normal functioning of the cell.

ENZYME SPECIFICITY AND EFFICIENCY

The specificity of enzymes is made possible by their structures. Enzymes are generally large globular proteins that range in molecular weight from about 10,000 to several million. Each of the thousands of known enzymes has a characteristic three-dimensional shape with a specific

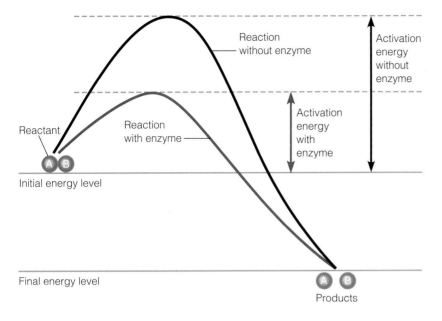

FIGURE 5.2 **Energy requirements of a chemical reaction.** This graph shows the progress of the reaction AB → A + B both without (black line) and with (magenta line) an enzyme. The presence of an enzyme lowers the activation energy of the reaction (see arrows). Thus, more molecules of reactant AB are converted to products A and B because more molecules of reactant AB possess the activation energy needed for the reaction.

Q **How do enzymes speed up chemical reactions?**

surface configuration as a result of its primary, secondary, and tertiary structures (see Figure 2.15, page 46). The unique configuration of each enzyme enables it to "find" the correct substrate from among the large number of diverse molecules in the cell.

Enzymes are extremely efficient. Under optimum conditions, they can catalyze reactions at rates 10^8 to 10^{10} times (up to 10 billion times) higher than those of comparable reactions without enzymes. The **turnover number** (maximum number of substrate molecules an enzyme molecule converts to product each second) is generally between 1 and 10,000 and can be as high as 500,000. For example, the enzyme DNA polymerase I, which participates in the synthesis of DNA, has a turnover number of 15, whereas the enzyme lactate dehydrogenase, which

removes hydrogen atoms from lactic acid, has a turnover number of 1000.

Many enzymes exist in the cell in both active and inactive forms. The rate at which enzymes switch between these two forms is determined by the cellular environment.

NAMING ENZYMES

The names of enzymes usually end in *-ase*. All enzymes can be grouped into six classes, according to the type of chemical reaction they catalyze (Table 5.1). Enzymes within each of the major classes are named according to the more specific types of reactions they assist. For example, the class called oxidoreductases is involved with oxidation-reduction reactions (described shortly). Enzymes in the oxidoreductase class that remove hydrogen from a substrate

TABLE 5.1	Enzyme Classification Based on Type of Chemical Reaction Catalyzed	
Class	Type of Chemical Reaction Catalyzed	Examples
Oxidoreductase	Oxidation-reduction in which oxygen and hydrogen are gained or lost	Cytochrome oxidase, lactate dehydrogenase
Transferase	Transfer of functional groups, such as an amino group, acetyl group, or phosphate group	Acetate kinase, alanine deaminase
Hydrolase	Hydrolysis (addition of water)	Lipase, sucrase
Lyase	Removal of groups of atoms without hydrolysis	Oxalate decarboxylase, isocitrate lyase
Isomerase	Rearrangement of atoms within a molecule	Glucose-phosphate isomerase, alanine racemase
Ligase	Joining of two molecules (using energy usually derived from the breakdown of ATP)	Acetyl-CoA synthetase, DNA ligase

FIGURE 5.3 Components of a holoenzyme. Many enzymes require both an apoenzyme (protein portion) and a cofactor (nonprotein portion) to become active. The cofactor can be a metal ion, or if it is an organic molecule, it is called a coenzyme (as shown here). The apoenzyme and cofactor together make up the holoenzyme, or whole enzyme. The substrate is the reactant acted upon by the enzyme.

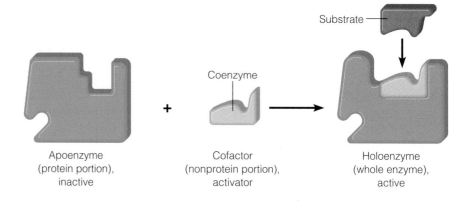

Q Which substances usually function as coenzymes?

are called *dehydrogenases*; those that add molecular oxygen (O_2) are called *oxidases*. As you will see later, dehydrogenase and oxidase enzymes have even more specific names, such as lactate dehydrogenase and cytochrome oxidase, depending on the specific substrates on which they act.

ENZYME COMPONENTS

Although some enzymes consist entirely of proteins, most consist of both a protein portion called an **apoenzyme** and a nonprotein component called a **cofactor.** Ions of iron, zinc, magnesium, or calcium are examples of cofactors. If the cofactor is an organic molecule, it is called a **coenzyme.** Apoenzymes are inactive by themselves; they must be activated by cofactors. Together, the apoenzyme and cofactor form a **holoenzyme,** or whole, active enzyme (Figure 5.3). If the cofactor is removed, the apoenzyme will not function.

Coenzymes may assist the enzyme by accepting atoms removed from the substrate or by donating atoms required by the substrate. Some coenzymes act as electron carriers, removing electrons from the substrate and donating them to other molecules in subsequent reactions. Many coenzymes are derived from vitamins (Table 5.2). Two of the most important coenzymes in cellular metabolism are **nicotinamide adenine dinucleotide (NAD^+)** and **nicotinamide adenine dinucleotide phosphate ($NADP^+$).** Both compounds contain derivatives of the B vitamin niacin (nicotinic acid), and both function as electron carriers. Whereas NAD^+ is primarily involved in catabolic (energy-yielding) reactions, $NADP^+$ is primarily involved in anabolic (energy-requiring) reactions. The flavin coenzymes, such as **flavin mononucleotide (FMN)** and **flavin adenine dinucleotide (FAD),** contain derivatives of the B vitamin riboflavin and are

TABLE 5.2	Selected Vitamins and Their Coenzymatic Functions
Vitamin	Function
Vitamin B$_1$ (thiamine)	Part of coenzyme cocarboxylase; has many functions, including the metabolism of pyruvic acid
Vitamin B$_2$ (riboflavin)	Coenzyme in flavoproteins; active in electron transfers
Niacin (nicotinic acid)	Part of NAD molecule*; active in electron transfers
Vitamin B$_6$ (pyridoxine)	Coenzyme in amino acid metabolism
Vitamin B$_{12}$ (cyanocobalamin)	Coenzyme (methyl cyanocobalamide) involved in the transfer of methyl groups; active in amino acid metabolism
Pantothenic Acid	Part of coenzyme A molecule; involved in the metabolism of pyruvic acid and lipids
Biotin	Involved in carbon dioxide fixation reactions and fatty acid synthesis
Folic Acid	Coenzyme used in the synthesis of purines and pyrimidines
Vitamin E	Needed for cellular and macromolecular syntheses
Vitamin K	Coenzyme used in electron transport (naphthoquinones and quinones)

*NAD = nicotinamide adenine dinucleotide

also electron carriers. Another important coenzyme, **coenzyme A (CoA),** contains a derivative of pantothenic acid, another B vitamin. This coenzyme plays an important role in the synthesis and breakdown of fats and in a series of oxidizing reactions called the Krebs cycle. We will come across all of these coenzymes in our discussion of metabolism later in the chapter.

As noted earlier, some cofactors are metal ions, including iron, copper, magnesium, manganese, zinc, calcium, and cobalt. Such cofactors may help catalyze a reaction by forming a bridge between the enzyme and a substrate. For example, magnesium (Mg^{2+}) is required by many phosphorylating enzymes (enzymes that transfer a phosphate group from ATP to another substrate). The Mg^{2+} can form a link between the enzyme and the ATP molecule. Most trace elements required by living cells are probably used in some such way to activate cellular enzymes.

THE MECHANISM OF ENZYMATIC ACTION

Enzymes lower the activation energy of chemical reactions. The general sequence of events in enzyme action is as follows (Figure 5.4a):

1. The surface of the substrate contacts a specific region of the surface of the enzyme molecule called the **active site.**

2. A temporary intermediate compound forms, called an **enzyme–substrate complex.**

3. The substrate molecule is transformed by the rearrangement of existing atoms, the breakdown of the substrate molecule, or in combination with another substrate molecule.

4. The transformed substrate molecules—the products of the reaction—are released from the enzyme molecule

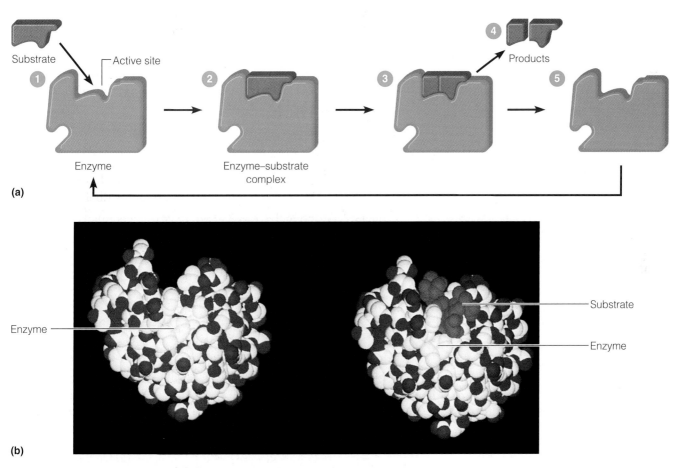

(a)

(b)

FIGURE 5.4 The mechanism of enzymatic action. (a) ❶ The substrate contacts the active site on the enzyme to form ❷ an enzyme–substrate complex. ❸ The substrate is then transformed into products, ❹ the products are released, and ❺ the enzyme is recovered unchanged. In the example shown, the transformation into products involves a breakdown of the substrate into two products. Other transformations, however, may occur. **(b)** Left: A molecular model of the enzyme in step ❶ of part a. The active site of the enzyme can be seen here as a groove on the surface of the protein. Right: As the enzyme and substrate meet in step ❷ of part a, they change shape slightly to fit together more tightly.

Q **What is the function of enzymes in living organisms?**

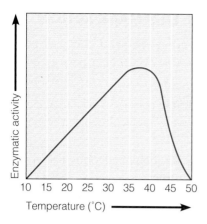

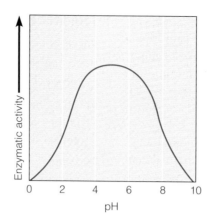

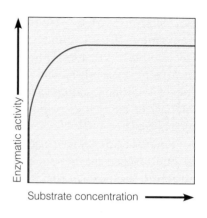

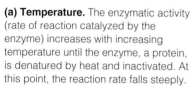

(a) Temperature. The enzymatic activity (rate of reaction catalyzed by the enzyme) increases with increasing temperature until the enzyme, a protein, is denatured by heat and inactivated. At this point, the reaction rate falls steeply.

(b) pH. The enzyme illustrated is most active at about pH 5.0.

(c) Substrate concentration. With increasing concentration of substrate molecules, the rate of reaction increases until the active sites on all the enzyme molecules are filled, at which point the maximum rate of reaction is reached.

FIGURE 5.5 Factors that influence enzymatic activity, plotted for a hypothetical enzyme.

Q **How will this enzyme act at 25°C? At 45°C? At pH 7?**

because they no longer fit in the active site of the enzyme.

⑤ The unchanged enzyme is now free to react with other substrate molecules.

As a result of these events, an enzyme speeds up a chemical reaction.

As mentioned earlier, enzymes have *specificity* for particular substrates. For example, a specific enzyme may be able to hydrolyze a peptide bond only between two specific amino acids. Other enzymes can hydrolyze starch but not cellulose; even though both starch and cellulose are polysaccharides composed of glucose subunits, the orientations of the subunits in the two polysaccharides differ. Enzymes have this specificity because the three-dimensional shape of the active site fits the substrate somewhat as a lock fits with its key (see Figure 5.4b). However, the active site and substrate are flexible, and they change shape somewhat as they meet to fit together more tightly. The substrate is usually much smaller than the enzyme, and relatively few of the enzyme's amino acids make up the active site.

A certain compound can be a substrate for several different enzymes that catalyze different reactions, so the fate of a compound depends on the enzyme that acts upon it. Glucose 6-phosphate, a molecule important in cell metabolism, can be acted upon by at least four different enzymes, and each reaction will yield a different product. �֍ **Animation: Go to The Microbiology Place website or CD-ROM and click "Animations" to view Enzyme–Substrate Interactions.**

FACTORS INFLUENCING ENZYMATIC ACTIVITY

Enzymes are subject to various cellular controls. Two primary types are the control of enzyme *synthesis* (see Chapter 8) and the control of enzyme *activity* (how much enzyme is present versus how active it is).

Several factors influence the activity of an enzyme. Among the more important are temperature, pH, substrate concentration, and the presence or absence of inhibitors.

TEMPERATURE

The rate of most chemical reactions increases as the temperature increases. Molecules move more slowly at lower temperatures than at higher temperatures and so may not have enough energy to cause a chemical reaction. For enzymatic reactions, however, elevation beyond a certain temperature (the optimal temperature) drastically reduces the rate of reaction (Figure 5.5a). The optimal temperature for most disease-producing bacteria in the human body is between 35°C and 40°C. The reduced rate of reaction beyond the optimal temperature is due to the enzyme's **denaturation,** the loss of its characteristic three-dimensional structure (tertiary configuration) (Figure 5.6). Denaturation of a protein involves the breakage of hydrogen bonds and other noncovalent bonds; a common example is the transformation of uncooked egg white (a protein called albumin) to a hardened state by heat.

Denaturation of an enzyme changes the arrangement of the amino acids in the active site, altering its shape and

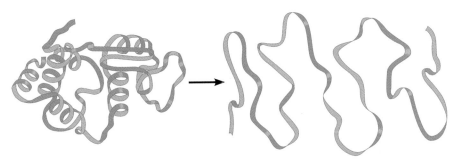

FIGURE 5.6 Denaturation of a protein. Breakage of the noncovalent bonds (such as hydrogen bonds) that hold the active protein in its three-dimensional shape renders the denatured protein nonfunctional.

Q **What factors may cause denaturation?**

Active (functional) protein   Denatured protein

causing the enzyme to lose its catalytic ability. In some cases, denaturation is partially or fully reversible. However, if denaturation continues until the enzyme has lost its solubility and coagulates, the enzyme cannot regain its original properties. Enzymes can also be denatured by concentrated acids, bases, heavy-metal ions (such as lead, arsenic, or mercury), alcohol, and ultraviolet radiation.

pH

Most enzymes have an optimum pH at which their activity is characteristically maximal. Above or below this pH value, enzyme activity, and therefore the reaction rate, decline (see Figure 5.5b). When the H^+ concentration (pH) in the medium is changed drastically, the protein's three-dimensional structure is altered. Extreme changes in pH can cause denaturation. Acids (and bases) alter a protein's three-dimensional structure because the H^+ (and OH^-) compete with hydrogen and ionic bonds in an enzyme, resulting in the enzyme's denaturation.

SUBSTRATE CONCENTRATION

There is a maximum rate at which a certain amount of enzyme can catalyze a specific reaction. Only when the concentration of substrate(s) is extremely high can this maximum rate be attained. Under conditions of high substrate concentration, the enzyme is said to be in **saturation;** that is, its active site is always occupied by substrate or product molecules. In this condition, a further increase in substrate concentration will not affect the reaction rate because all

active sites are already in use (see Figure 5.5c). Under normal cellular conditions, enzymes are not saturated with substrate(s). At any given time, many of the enzyme molecules are inactive for lack of substrate; thus, the rate of reaction is likely to be influenced by the substrate concentration.

INHIBITORS

An effective way to control the growth of bacteria is to control their enzymes. Certain poisons, such as cyanide, arsenic, and mercury, combine with enzymes and prevent them from functioning. As a result, the cells stop functioning and die.

Enzyme inhibitors are classified as either competitive or noncompetitive inhibitors (Figure 5.7). **Competitive inhibitors** fill the active site of an enzyme and compete with the normal substrate for the active site. A competitive inhibitor can do this because its shape and chemical structure are similar to those of the normal substrate (Figure 5.7b). However, unlike the substrate, it does not undergo any reaction to form products. Some competitive inhibitors bind irreversibly to amino acids in the active site, preventing any further interactions with the substrate. Others bind reversibly, alternately occupying and leaving the active site; these slow the enzyme's interaction with the substrate. Reversible competitive inhibition can be overcome by increasing the substrate concentration. As active sites become available, more substrate molecules than competitive inhibitor molecules are available to attach to the active sites of enzymes.

NORMAL BINDING OF SUBSTRATE

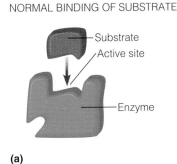

ACTION OF ENZYME INHIBITORS

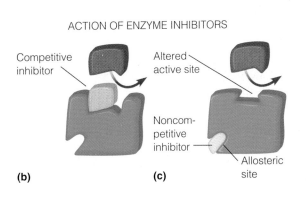

FIGURE 5.7 Enzyme inhibitors. (a) An uninhibited enzyme and its normal substrate. **(b)** A competitive inhibitor. **(c)** One type of noncompetitive inhibitor, causing allosteric inhibition.

Q **How do competitive inhibitors operate?**

(a)   (b)   (c)

One good example of a competitive inhibitor is sulfanilamide (a sulfa drug), which inhibits the enzyme whose normal substrate is *para*-aminobenzoic acid (PABA):

Sulfanilamide   PABA

PABA is an essential nutrient used by many bacteria in the synthesis of folic acid, a vitamin that functions as a coenzyme. When sulfanilamide is administered to bacteria, the enzyme that normally converts PABA to folic acid combines instead with the sulfanilamide. Folic acid is not synthesized, and the bacteria cannot grow. Because human cells do not use PABA to make their folic acid, sulfanilamide kills bacteria but does not harm human cells.

Noncompetitive inhibitors do not compete with the substrate for the enzyme's active site; instead, they interact with another part of the enzyme (see Figure 5.7c). In this process, called **allosteric** ("other space") **inhibition,** the inhibitor binds to a site on the enzyme other than the substrate's binding site, called the **allosteric site.** This binding causes the active site to change its shape, making it nonfunctional. As a result, the enzyme's activity is reduced. This effect can be either reversible or irreversible, depending on whether or not the active site can return to its original shape. In some cases, allosteric interactions can activate an enzyme rather than inhibit it. Another type of noncompetitive inhibition can operate on enzymes that require metal ions for their activity. Certain chemicals can bind or tie up the metal ion activators and thus prevent an enzymatic reaction. Cyanide can bind the iron in iron-containing enzymes, and fluoride can bind calcium or magnesium. Substances such as cyanide and fluoride are sometimes called *enzyme poisons* because they permanently inactivate enzymes. In low concentrations, fluoride kills bacteria in the mouth that can contribute to tooth decay.

FEEDBACK INHIBITION

Allosteric inhibitors play a role in a kind of biochemical control called **feedback inhibition,** or **end-product inhibition.** This control mechanism stops the cell from wasting chemical resources by making more of a substance than it needs. In some metabolic reactions, several steps are required for the synthesis of a particular chemical compound, called the *end-product*. The process is similar to

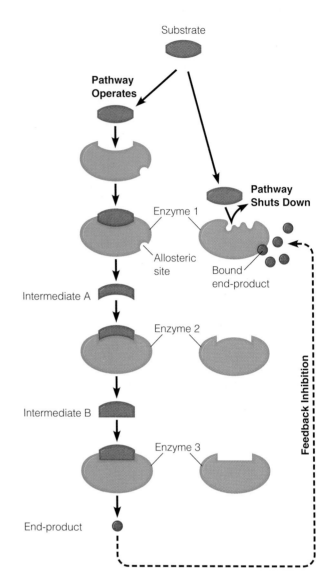

FIGURE 5.8 Feedback inhibition.

Q **What is meant by feedback inhibition?**

an assembly line, with each step catalyzed by a separate enzyme (Figure 5.8). In many anabolic pathways, the final product can allosterically inhibit the activity of one of the enzymes earlier in the pathway. This phenomenon is feedback inhibition.

Feedback inhibition generally acts on the first enzyme in a metabolic pathway (similar to shutting down an assembly line by stopping the first worker). Because the enzyme is inhibited, the product of the first enzymatic reaction in the pathway is not synthesized. Because that unsynthesized product would normally be the substrate for the second enzyme in the pathway, the second reaction stops immediately as well. Thus, even though only the first enzyme in the pathway is inhibited, the entire pathway shuts down and no new end-product is formed. By inhibiting the first enzyme in the pathway, the cell also keeps metabolic intermediates from accumulating. As the existing end-product is used up by the cell, the first enzyme's allosteric site will more often remain unbound, and the pathway will resume activity.

The bacterium *E. coli* can be used to demonstrate feedback inhibition in the synthesis of the amino acid isoleucine, which is required for the cell's growth. In this metabolic pathway, five steps are taken to enzymatically convert the amino acid threonine to isoleucine. If isoleucine is added to the growth medium for *E. coli*, it inhibits the first enzyme in the pathway, and the bacteria stop synthesizing isoleucine. This condition is maintained until the supply of isoleucine is depleted. This type of feedback inhibition is also involved in regulating the cells' production of other amino acids, as well as vitamins, purines, and pyrimidines.

RIBOZYMES

Prior to 1982, it was believed that only protein molecules had enzymatic activity. Researchers working on microbes discovered a unique type of RNA called a **ribozyme.** Like protein enzymes, ribozymes function as catalysts, have active sites that bind to substrates, and are not used up in a chemical reaction. Ribozymes specifically act on strands of RNA by removing sections and splicing together the remaining pieces. In this respect, ribozymes are more restricted than protein enzymes in terms of the diversity of substrates with which they interact.

ENERGY PRODUCTION

> **LEARNING OBJECTIVES**
> - Explain what is meant by *oxidation-reduction.*
> - List and provide examples of three types of phosphorylation reactions that generate ATP.

Nutrient molecules, like all molecules, have energy associated with the electrons that form bonds between their atoms. When it is spread throughout the molecule, this energy is difficult for the cell to use. Various reactions in catabolic pathways, however, concentrate the energy into the bonds of ATP, which serves as a convenient energy carrier. ATP is generally referred to as having "high-energy" bonds. Actually, a better term is probably unstable bonds. Although the amount of energy in these bonds is not exceptionally large, it can be released quickly and easily. In a sense, ATP is similar to a highly flammable liquid such as kerosene. Although a large log might eventually burn to produce more heat than a cup of kerosene, the kerosene is easier to ignite and provides heat more quickly and conveniently. In a similar way, the "high-energy" unstable bonds

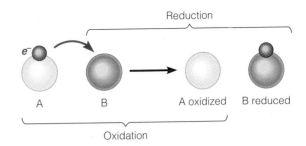

FIGURE 5.9 Oxidation-reduction. An electron is transferred from molecule A to molecule B. In the process, molecule A is oxidized and molecule B is reduced.

Q How do oxidation and reduction differ?

of ATP provide the cell with readily available energy for anabolic reactions.

Before discussing the catabolic pathways, we will consider two general aspects of energy production: the concept of oxidation-reduction and the mechanisms of ATP generation.

OXIDATION-REDUCTION REACTIONS

Oxidation is the removal of electrons (e^-) from an atom or molecule, a reaction that often produces energy. Figure 5.9 shows an example of an oxidation in which molecule A loses an electron to molecule B. Molecule A has undergone oxidation (meaning that it has lost one or more electrons), whereas molecule B has undergone **reduction** (meaning that it has gained one or more electrons).*Oxidation and reduction reactions are always coupled; in other words, each time one substance is oxidized, another is simultaneously reduced. The pairing of these reactions is called **oxidation-reduction** or a **redox reaction.**

In many cellular oxidations, electrons and protons (hydrogen ions, H^+) are removed at the same time; this is equivalent to the removal of hydrogen atoms, because a hydrogen atom is made up of one proton and one electron (see Table 2.2, page 29). Because most biological oxidations involve the loss of hydrogen atoms, they are also called **dehydrogenation** reactions. Figure 5.10 shows an example of a biological oxidation. An organic molecule is

---

*The terms do not seem logical until one considers the history of the discovery of these reactions. When mercury is roasted, it gains weight as mercuric oxide is formed; this was called *oxidation.* Later it was determined that the mercury actually *lost* electrons, and the observed *gain* in oxygen was a direct result of this. Oxidation, therefore, is a *loss* of electrons, and reduction is a *gain* of electrons, but the gain and loss of electrons is not usually apparent as chemical-reaction equations are usually written. For example, in the equations for aerobic respiration on page 134, notice that each carbon in glucose had only one oxygen originally, and later, as carbon dioxide, each carbon now has two oxygens. However, the gain or loss of electrons actually responsible for this is not apparent.

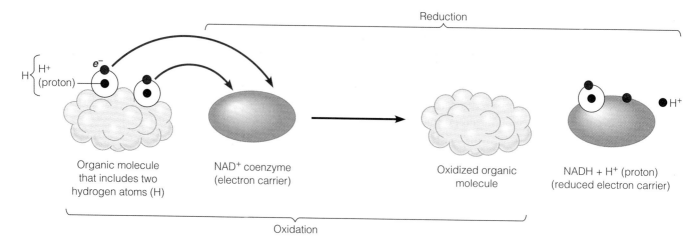

FIGURE 5.10 Representative biological oxidation. Two electrons and two protons (altogether equivalent to two hydrogen atoms) are transferred from an organic substrate molecule to a coenzyme, NAD^+. NAD^+ actually receives one hydrogen atom and one electron, and one proton is released into the medium. NAD^+ is reduced to NADH, which is a more energy-rich molecule.

Q **How do organisms use oxidation-reduction reactions?**

oxidized by the loss of two hydrogen atoms, and a molecule of NAD^+ is reduced. Recall from our earlier discussion of coenzymes that NAD^+ assists enzymes by accepting hydrogen atoms removed from the substrate, in this case the organic molecule. As shown in Figure 5.10, NAD^+ accepts two electrons and one proton. One proton (H^+) is left over and is released into the surrounding medium. The reduced coenzyme, NADH, contains more energy than NAD^+. This energy can be used to generate ATP in later reactions.

An important point to remember about biological oxidation-reduction reactions is that cells use them in catabolism to extract energy from nutrient molecules. Cells take nutrients, some of which serve as energy sources, and degrade them from highly reduced compounds (with many hydrogen atoms) to highly oxidized compounds. For example, when a cell oxidizes a molecule of glucose ($C_6H_{12}O_6$) to CO_2 and H_2O, the energy in the glucose molecule is removed in a stepwise manner and ultimately is trapped by ATP, which can then serve as an energy source for energy-requiring reactions. Compounds such as glucose that have many hydrogen atoms are highly reduced compounds, containing a large amount of potential energy. Thus, glucose is a valuable nutrient for organisms.

THE GENERATION OF ATP

Much of the energy released during oxidation-reduction reactions is trapped within the cell by the formation of ATP. Specifically, a phosphate group, Ⓟ, is added to ADP with the input of energy to form ATP:

$$\underbrace{Adenosine—Ⓟ\sim Ⓟ}_{ADP} + Energy + Ⓟ \rightarrow \underbrace{Adenosine—Ⓟ\sim Ⓟ\sim Ⓟ}_{ATP}$$

The symbol $\sim$ designates a "high-energy" bond—that is, one that can readily be broken to release usable energy. The high-energy bond that attaches the third Ⓟ in a sense contains the energy stored in this reaction. When this Ⓟ is removed, usable energy is released. The addition of Ⓟ to a chemical compound is called **phosphorylation.** Organisms use three mechanisms of phosphorylation to generate ATP from ADP.

SUBSTRATE-LEVEL PHOSPHORYLATION

In **substrate-level phosphorylation,** ATP is usually generated when a high-energy Ⓟ is directly transferred from a phosphorylated compound (a substrate) to ADP. Generally, the Ⓟ has acquired its energy during an earlier reaction in which the substrate itself was oxidized. The following example shows only the carbon skeleton and the Ⓟ of a typical substrate:

$$C—C—C \sim Ⓟ + ADP \rightarrow C—C—C + ATP$$

OXIDATIVE PHOSPHORYLATION

In **oxidative phosphorylation,** electrons are transferred from organic compounds to one group of electron carriers (usually to NAD^+ and FAD). Then, the electrons are passed through a series of different electron carriers to molecules of oxygen (O_2) or other oxidized inorganic and organic molecules. This process occurs in the plasma membrane of prokaryotes and in the inner mitochondrial membrane of eukaryotes. The sequence of electron carriers used in oxidative phosphorylation is called an **electron transport chain (system)** (see Figure 5.14). The transfer of electrons from one electron carrier to the next releases energy, some of which is used to generate ATP from ADP through a process called *chemiosmosis*, to be described on page 132.

PHOTOPHOSPHORYLATION

The third mechanism of phosphorylation, **photophosphorylation,** occurs only in photosynthetic cells, which contain light-trapping pigments such as chlorophylls. In photosynthesis, organic molecules, especially sugars, are synthesized with the energy of light from the energy-poor building blocks carbon dioxide and water. Photophosphorylation starts this process by converting light energy to the chemical energy of ATP and NADPH, which, in turn, are used to synthesize organic molecules. As in oxidative phosphorylation, an electron transport chain is involved.

METABOLIC PATHWAYS OF ENERGY PRODUCTION

LEARNING OBJECTIVE
• Explain the overall function of metabolic pathways.

Organisms release and store energy from organic molecules by a series of controlled reactions rather than in a single burst. If the energy were released all at once, as a large amount of heat, it could not be readily used to drive chemical reactions and would, in fact, damage the cell. To extract energy from organic compounds and store it in chemical form, organisms pass electrons from one compound to another through a series of oxidation-reduction reactions.

As noted earlier, a sequence of enzymatically catalyzed chemical reactions occurring in a cell is called a metabolic pathway. Below is a hypothetical pathway that converts starting material A to end-product F in a series of five steps:

$$NAD^+ \quad NADH + H^+$$

A $\xrightarrow{\;\;\;\;①\;\;\;\;}$ B $\xrightarrow{②}$
Starting material

$$ADP + Ⓟ \quad ATP \qquad\qquad O_2$$

C $\xrightarrow{\;\;\;③\;\;\;}$ D $\rightleftharpoons$ E $\xrightarrow{\;\;\;⑤\;\;\;}$ F
$\qquad\qquad\qquad\qquad ④ \qquad CO_2 \; H_2O$ End-product

The first step is the conversion of molecule A to molecule B. The curved arrow indicates that the reduction of coenzyme NAD^+ to NADH is coupled to that reaction; the electrons and protons come from molecule A. Similarly, the two arrows in ❸ show a coupling of two reactions. As C is converted to D, ADP is converted to ATP; the energy needed comes from C as it transforms into D. The reaction converting D to E is readily reversible, as indicated by the double arrow. In the fifth step, the curved arrow leading from O_2 indicates that O_2 is a reactant in the reaction. The curved arrows leading to CO_2 and H_2O indicate that these substances are secondary products produced in the reaction, in addition to F, the end-product that (presumably) interests us the most. Secondary products such as CO_2 and H_2O shown here are sometimes called "by-products" or "waste products." Keep in mind that almost every reaction in a metabolic pathway is catalyzed by a specific enzyme; sometimes the name of the enzyme is printed near the arrow.

CARBOHYDRATE CATABOLISM

LEARNING OBJECTIVES
• Describe the chemical reactions of glycolysis.
• Explain the products of the Krebs cycle.
• Describe the chemiosmotic model for ATP generation.

Most microorganisms oxidize carbohydrates as their primary source of cellular energy. **Carbohydrate catabolism,** the breakdown of carbohydrate molecules to produce energy, is therefore of great importance in cell metabolism. Glucose is the most common carbohydrate energy source used by cells. Microorganisms can also catabolize various lipids and proteins for energy production (page 137).

To produce energy from glucose, microorganisms use two general processes: *cellular respiration* and *fermentation.* (In discussing cellular respiration, we frequently refer to the process simply as respiration, but it should not be confused with breathing.) Both processes usually start with the same first step, glycolysis, but follow different subsequent pathways (Figure 5.11). Before examining the details of glycolysis, respiration, and fermentation, we will first look at a general overview of the processes.

As shown in Figure 5.11, the respiration of glucose typically occurs in three principal stages: glycolysis, the Krebs cycle, and the electron transport chain (system).

❶ Glycolysis is the oxidation of glucose to pyruvic acid with the production of some ATP and energy-containing NADH.

❷ The Krebs cycle is the oxidation of acetyl CoA (a derivative of pyruvic acid) to carbon dioxide, with

FIGURE 5.11 An overview of respiration and fermentation.

❶ Glycolysis produces ATP and reduces NAD$^+$ to NADH while oxidizing glucose to pyruvic acid. In respiration, the pyruvic acid is converted into the first reactant in ❷ the Krebs cycle, which produces ATP and reduces NAD$^+$ (and another electron carrier called FADH$_2$) while giving off CO$_2$. The NADH from both processes carries electrons to ❸ the electron transport chain, in which their energy is used to produce a great deal of ATP. In fermentation, the pyruvic acid and the electrons carried by NADH from glycolysis are incorporated into fermentation end-products. A small version of this figure will be included in figures throughout the chapter to indicate the relationships of different reactions to the overall processes.

Q **What is the basic difference between respiration and fermentation?**

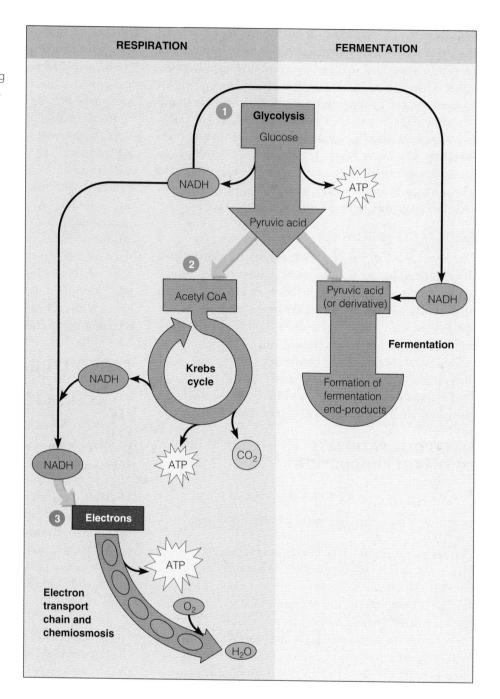

the production of some ATP, energy-containing NADH, and another reduced electron carrier, FADH$_2$ (the reduced form of flavin adenine dinucleotide).

❸ In the electron transport chain (system), NADH and FADH$_2$ are oxidized, contributing the electrons they have carried from the substrates to a "cascade" of oxidation-reduction reactions involving a series of additional electron carriers. Energy from these reactions is used to generate a considerable amount of

ATP. In respiration, most of the ATP is generated in the third step.

Because respiration involves a long series of oxidation-reduction reactions, the entire process can be thought of as involving a flow of electrons from the energy-rich glucose molecule to the relatively energy-poor CO$_2$ and H$_2$O molecules. The coupling of ATP production to this flow is somewhat analogous to the production of electrical power by using energy from a flowing stream. Carrying the analogy further, you could imagine a stream flowing down a gentle slope during glycolysis and the Krebs cycle, supplying energy to turn two old-fashioned waterwheels. Then the stream rushes down a steep slope in the electron transport chain, supplying energy for a large modern power plant. In a similar way, glycolysis and the Krebs cycle generate a small amount of ATP and also supply the electrons that generate a great deal of ATP at the electron transport chain stage.

Typically, the initial stage of fermentation is also glycolysis (see Figure 5.11). However, once glycolysis has taken place, the pyruvic acid is converted into one or more different products, depending on the type of cell. These products might include alcohol (ethanol) and lactic acid. Unlike respiration, there is no Krebs cycle or electron transport chain in fermentation. Accordingly, the ATP yield, which comes only from glycolysis, is much lower.

GLYCOLYSIS

Glycolysis, the oxidation of glucose to pyruvic acid, is usually the first stage in carbohydrate catabolism. Most microorganisms use this pathway; in fact, it occurs in most living cells.

Glycolysis is also called the *Embden-Meyerhof pathway.* The word *glycolysis* means splitting of sugar, and this is exactly what happens. The enzymes of glycolysis catalyze the splitting of glucose, a six-carbon sugar, into two three-carbon sugars. These sugars are then oxidized, releasing energy, and their atoms are rearranged to form two molecules of pyruvic acid. During glycolysis NAD$^+$ is reduced to NADH, and there is a net production of two ATP molecules by substrate-level phosphorylation. Glycolysis does not require oxygen; it can occur whether oxygen is present or not. This pathway is a series of ten chemical reactions, each catalyzed by a different enzyme. The steps are outlined in Figure 5.12; see also Appendix C for a more detailed representation of glycolysis.

To summarize the process, glycolysis consists of two basic stages, a preparatory stage and an energy-conserving stage:

1. First, in the preparatory stage (steps **1**–**4** in Figure 5.12), two molecules of ATP are used as a six-carbon glucose molecule is phosphorylated, restructured, and split into two three-carbon compounds: glyceraldehyde 3-phosphate (GP) and dihydroxyacetone phosphate (DHAP). **5** DHAP is readily converted to GP. (The reverse reaction may also occur.) The conversion of DHAP into GP means that from this point on in glycolysis, two molecules of GP are fed into the remaining chemical reactions.

2. In the energy-conserving stage (steps **6**–**10** in Figure 5.12), the two three-carbon molecules are oxidized in several steps to two molecules of pyruvic acid. In these reactions, two molecules of NAD$^+$ are reduced to NADH, and four molecules of ATP are formed by substrate-level phosphorylation.

Because two molecules of ATP were needed to get glycolysis started and four molecules of ATP are generated by the process, *there is a net gain of two molecules of ATP for each molecule of glucose that is oxidized.* ✳ **Animation: Go to The Microbiology Place website or CD-ROM and click "Animations" to view Glycolysis.**

ALTERNATIVES TO GLYCOLYSIS

Many bacteria have another pathway in addition to glycolysis for the oxidation of glucose. The most common alternative is the pentose phosphate pathway; another alternative is the Entner-Doudoroff pathway.

THE PENTOSE PHOSPHATE PATHWAY

The **pentose phosphate pathway** (or *hexose monophosphate shunt*) operates simultaneously with glycolysis and provides a means for the breakdown of five-carbon sugars (pentoses) as well as glucose (see the detailed figure in Appendix C). A key feature of this pathway is that it produces important intermediate pentoses used in the synthesis of (1) nucleic acids, (2) glucose from carbon dioxide in photosynthesis, and (3) certain amino acids. The pathway is an important producer of the reduced coenzyme NADPH from NADP$^+$. The pentose phosphate pathway yields a net gain of only one molecule of ATP for each molecule of glucose oxidized. Bacteria that use the pentose phosphate pathway include *Bacillus subtilis* (sub′til-us), *E. coli*, *Leuconostoc mesenteroides* (lü-kō-nos′tok mes-en-ter-oi′dēz), and *Enterococcus faecalis* (fē-kāl′is).

THE ENTNER-DOUDOROFF PATHWAY

From each molecule of glucose, the **Entner-Doudoroff pathway** produces two molecules of NADPH and one molecule of ATP for use in cellular biosynthetic reactions (see Appendix C for a more detailed representation). Bacteria that have the enzymes for the Entner-Doudoroff pathway can metabolize glucose without either glycolysis or the pentose phosphate pathway. The Entner-Doudoroff pathway is found in some gram-negative bacteria, including *Rhizobium*, *Pseudomonas* (sū-dō-mō′nas), and *Agrobacterium* (ag-rō-bak-ti′rē-um); it is generally not found among gram-positive bacteria. Tests for the ability to oxidize glucose by this pathway are sometimes used to identify *Pseudomonas* in the clinical laboratory.

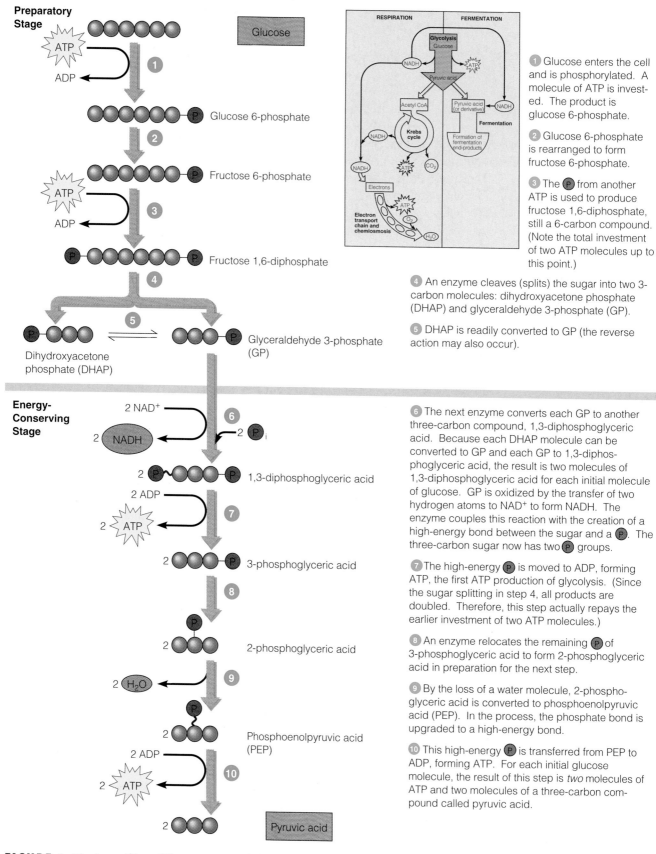

Preparatory Stage

Glucose

① Glucose enters the cell and is phosphorylated. A molecule of ATP is invested. The product is glucose 6-phosphate.

ATP → ADP ①

Glucose 6-phosphate

② Glucose 6-phosphate is rearranged to form fructose 6-phosphate.

② Fructose 6-phosphate

ATP → ADP ③

③ The ⓟ from another ATP is used to produce fructose 1,6-diphosphate, still a 6-carbon compound. (Note the total investment of two ATP molecules up to this point.)

ⓟ Fructose 1,6-diphosphate

④

④ An enzyme cleaves (splits) the sugar into two 3-carbon molecules: dihydroxyacetone phosphate (DHAP) and glyceraldehyde 3-phosphate (GP).

⑤

Dihydroxyacetone phosphate (DHAP)

Glyceraldehyde 3-phosphate (GP)

⑤ DHAP is readily converted to GP (the reverse action may also occur).

Energy-Conserving Stage

2 NAD⁺ ⑥
2 NADH ← 2 ⓟᵢ

⑥ The next enzyme converts each GP to another three-carbon compound, 1,3-diphosphoglyceric acid. Because each DHAP molecule can be converted to GP and each GP to 1,3-diphosphoglyceric acid, the result is two molecules of 1,3-diphosphoglyceric acid for each initial molecule of glucose. GP is oxidized by the transfer of two hydrogen atoms to NAD⁺ to form NADH. The enzyme couples this reaction with the creation of a high-energy bond between the sugar and a ⓟ. The three-carbon sugar now has two ⓟ groups.

2 ⓟ 1,3-diphosphoglyceric acid

2 ADP ⑦
2 ATP

2 3-phosphoglyceric acid

⑦ The high-energy ⓟ is moved to ADP, forming ATP, the first ATP production of glycolysis. (Since the sugar splitting in step 4, all products are doubled. Therefore, this step actually repays the earlier investment of two ATP molecules.)

⑧

2 2-phosphoglyceric acid

⑧ An enzyme relocates the remaining ⓟ of 3-phosphoglyceric acid to form 2-phosphoglyceric acid in preparation for the next step.

2 H₂O ← ⑨

2 Phosphoenolpyruvic acid (PEP)

⑨ By the loss of a water molecule, 2-phosphoglyceric acid is converted to phosphoenolpyruvic acid (PEP). In the process, the phosphate bond is upgraded to a high-energy bond.

2 ADP ⑩
2 ATP

⑩ This high-energy ⓟ is transferred from PEP to ADP, forming ATP. For each initial glucose molecule, the result of this step is *two* molecules of ATP and two molecules of a three-carbon compound called pyruvic acid.

2 Pyruvic acid

FIGURE 5.12 An outline of the reactions of glycolysis (Embden-Meyerhof pathway). The inset indicates the relationship of glycolysis to the overall processes of respiration and fermentation. A more detailed version of glycolysis is presented in Appendix C.

Q **What is glycolysis?**

CELLULAR RESPIRATION

LEARNING OBJECTIVE

• Compare and contrast aerobic and anaerobic respiration.

After glucose has been broken down to pyruvic acid, the pyruvic acid can be channeled into the next step of either fermentation (page 134) or cellular respiration (see Figure 5.11). **Cellular respiration,** or simply **respiration,** is defined as an ATP-generating process in which molecules are oxidized and the final electron acceptor is (almost always) an inorganic molecule. An essential feature of respiration is the operation of an electron transport chain.

There are two types of respiration, depending on whether an organism is an **aerobe,** which uses oxygen, or an **anaerobe,** which does not use oxygen and may even be killed by it. In **aerobic respiration,** the final electron acceptor is O_2; in **anaerobic respiration,** the final electron acceptor is an inorganic molecule other than O_2 or, rarely, an organic molecule. First we will describe respiration as it typically occurs in an aerobic cell.

AEROBIC RESPIRATION

The Krebs Cycle   The **Krebs cycle,** also called the *tricarboxylic acid (TCA) cycle* or *citric acid cycle,* is a series of biochemical reactions in which the large amount of potential chemical energy stored in acetyl CoA is released step by step (see Figure 5.11). In this cycle, a series of oxidations and reductions transfer that potential energy, in the form of electrons, to electron carrier coenzymes, chiefly NAD^+. The pyruvic acid derivatives are oxidized; the coenzymes are reduced.

Pyruvic acid, the product of glycolysis, cannot enter the Krebs cycle directly. In a preparatory step, it must lose one molecule of CO_2 and become a two-carbon compound (Figure 5.13, at top). This process is called **decarboxylation.** The two-carbon compound, called an *acetyl group,* attaches to coenzyme A through a high-energy bond; the resulting complex is known as *acetyl coenzyme A (acetyl CoA).* During this reaction, pyruvic acid is also oxidized and NAD^+ is reduced to NADH.

Remember that the oxidation of one glucose molecule produces two molecules of pyruvic acid, so for each molecule of glucose, two molecules of CO_2 are released in this preparatory step, two molecules of NADH are produced, and two molecules of acetyl CoA are formed. Once the pyruvic acid has undergone decarboxylation and its derivative (the acetyl group) has attached to CoA, the resulting acetyl CoA is ready to enter the Krebs cycle.

As acetyl CoA enters the Krebs cycle, CoA detaches from the acetyl group. The two-carbon acetyl group combines with a four-carbon compound called oxaloacetic acid to form the six-carbon citric acid. This synthesis reaction requires energy, which is provided by the cleavage of the high-energy bond between the acetyl group and CoA. The formation of citric acid is thus the first step in the Krebs cycle. The major chemical reactions of this cycle are outlined in Figure 5.13; a more detailed representation of the Krebs cycle is provided in Appendix C. Keep in mind that each reaction is catalyzed by a specific enzyme.

The chemical reactions of the Krebs cycle fall into several general categories; one of these is decarboxylation. For example, in step ❸ isocitric acid, a six-carbon compound, is decarboxylated to the five-carbon compound called α-ketoglutaric acid. Another decarboxylation takes place in step ❹. Because one decarboxylation has taken place in the preparatory step and two in the Krebs cycle, all three carbon atoms in pyruvic acid are eventually released as CO_2 by the Krebs cycle. This represents the conversion to CO_2 of all six carbon atoms contained in the original glucose molecule.

Another general category of Krebs cycle chemical reactions is oxidation-reduction. For example, in step ❸, two hydrogen atoms are lost during the conversion of the six-carbon isocitric acid to a five-carbon compound. In other words, the six-carbon compound is oxidized. Hydrogen atoms are also released in the Krebs cycle in steps ❹, ❻, and ❽ and are picked up by the coenzymes NAD^+ and FAD. Because NAD^+ picks up two electrons but only one additional proton, its reduced form is represented as NADH; however, FAD picks up two complete hydrogen atoms and is reduced to $FADH_2$.

If we look at the Krebs cycle as a whole, we see that for every two molecules of acetyl CoA that enter the cycle, four molecules of CO_2 are liberated by decarboxylation, six molecules of NADH and two molecules of $FADH_2$ are produced by oxidation-reduction reactions, and two molecules of ATP are generated by substrate-level phosphorylation. Many of the intermediates in the Krebs cycle also play a role in other pathways, especially in amino acid biosynthesis (page 149).

The CO_2 produced in the Krebs cycle is ultimately liberated into the atmosphere as a gaseous by-product of aerobic respiration. (Humans produce CO_2 from the Krebs cycle in most cells of the body and discharge it through the lungs during exhalation.) The reduced coenzymes NADH and $FADH_2$ are the most important products of the Krebs cycle because they contain most of the energy originally stored in glucose. During the next phase of respiration, a series of reductions indirectly transfers the energy stored in those coenzymes to ATP. These reactions are collectively called the electron transport chain.

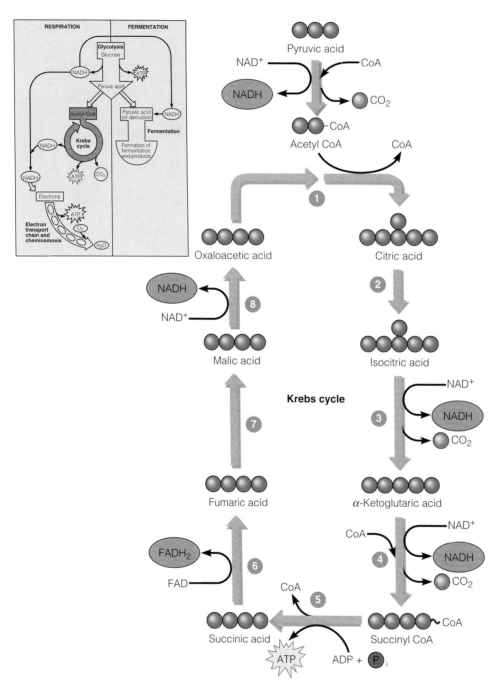

① A turn of the cycle begins as enzymes strip off the CoA portion from acetyl CoA and combine the remaining two-carbon acetyl group with oxaloacetic acid. Adding the acetyl group produces the six-carbon molecule citric acid.

② – ④ Oxidations generate NADH. Step 2 is a rearrangement. Steps 3 and 4 combine oxidations and decarboxylations to dispose of two carbon atoms that came from oxaloacetic acid. The carbons are released as CO_2, and the oxidations generate NADH from NAD^+. During the second oxidation (step 4), CoA is added into the cycle, forming the compound succinyl CoA.

⑤ ATP is produced by substrate-level phosphorylation. CoA is removed from succinyl CoA, leaving succinic acid.

⑥ – ⑧ Enzymes rearrange chemical bonds, producing three different molecules before regenerating oxaloacetic acid. In step 6, an oxidation produces $FADH_2$. In step 8, a final oxidation generates NADH and converts malic acid to oxaloacetic acid, which is ready to enter another round of the Krebs cycle. See Appendix C.

FIGURE 5.13 The Krebs cycle. The inset indicates the relationship of the Krebs cycle to the overall process of respiration.

Q **What are the products of the Krebs cycle?**

The Electron Transport Chain (System) An **electron transport chain (system)** consists of a sequence of carrier molecules that are capable of oxidation and reduction. As electrons are passed through the chain, there is a stepwise release of energy, which is used to drive the chemiosmotic generation of ATP, to be described shortly. The final oxidation is irreversible. In eukaryotic cells, the electron transport chain is contained in the inner membrane of mitochondria; in prokaryotic cells, it is found in the plasma membrane.

There are three classes of carrier molecules in electron transport chains. The first are **flavoproteins.** These proteins contain flavin, a coenzyme derived from riboflavin (vitamin B_2), and are capable of performing alternating oxidations and reductions. One important flavin coenzyme is flavin mononucleotide (FMN). The second class of carrier molecules are **cytochromes,** proteins with an iron-containing group (heme) capable of existing alternately as a reduced form (Fe^{2+}) and an oxidized form (Fe^{3+}). The

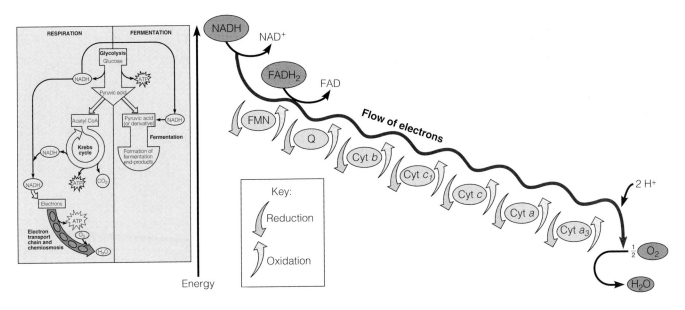

FIGURE 5.14   An electron transport chain (system). The inset indicates the relationship of the electron transport chain to the overall process of respiration. In the mitochondrial electron transport chain shown, the electrons pass along the chain in a gradual and stepwise fashion, so energy is released in manageable quantities. To learn where ATP is formed, see Figure 5.16.

Q **What are the functions of the electron transport chain?**

cytochromes involved in electron transport chains include cytochrome *b* (cyt *b*), cytochrome *c$_1$* (cyt *c$_1$*), cytochrome *c* (cyt *c*), cytochrome *a* (cyt *a*), and cytochrome *a$_3$* (cyt *a$_3$*). The third class is known as **ubiquinones,** or **coenzyme Q,** symbolized Q; these are small nonprotein carriers.

The electron transport chains of bacteria are somewhat diverse, in that the particular carriers used by a bacterium and the order in which they function may differ from those of other bacteria and from those of eukaryotic mitochondrial systems. Even a single bacterium may have several types of electron transport chains. However, keep in mind that all electron transport chains achieve the same basic goal, that of releasing energy as electrons are transferred from higher-energy compounds to lower-energy compounds. Much is known about the electron transport chain in the mitochondria of eukaryotic cells, so this is the chain we will describe.

The first step in the mitochondrial electron transport chain involves the transfer of high-energy electrons from NADH to FMN, the first carrier in the chain (Figure 5.14). This transfer actually involves the passage of a hydrogen atom with two electrons to FMN, which then picks up an additional H^+ from the surrounding aqueous medium. As a result of the first transfer, NADH is oxidized to NAD^+, and FMN is reduced to $FMNH_2$. In the second step in the electron transport chain, $FMNH_2$ passes $2H^+$ to the other side of the mitochondrial membrane (see Figure 5.16) and passes two electrons to Q. As a result,

$FMNH_2$ is oxidized to FMN. Q also picks up an additional $2H^+$ from the surrounding aqueous medium and releases it on the other side of the membrane.

The next part of the electron transport chain involves the cytochromes. Electrons are passed successively from Q to cyt *b*, cyt *c$_1$*, cyt *c*, cyt *a*, and cyt *a$_3$*. Each cytochrome in the chain is reduced as it picks up electrons and is oxidized as it gives up electrons. The last cytochrome, cyt *a$_3$*, passes its electrons to molecular oxygen (O_2), which becomes negatively charged and then picks up protons from the surrounding medium to form H_2O.

Notice that Figure 5.14 shows $FADH_2$, which is derived from the Krebs cycle, as another source of electrons. However, $FADH_2$ adds its electrons to the electron transport chain at a lower level than NADH. Because of this, the electron transport chain produces about one-third less energy for ATP generation when $FADH_2$ donates electrons than when NADH is involved.

An important feature of the electron transport chain is the presence of some carriers, such as FMN and Q, that accept and release protons as well as electrons, and other carriers, such as cytochromes, that transfer electrons only. Electron flow down the chain is accompanied at several points by the active transport (pumping) of protons from the matrix side of the inner mitochondrial membrane to the opposite side of the membrane. The result is a buildup of protons on one side of the membrane. Just as water behind a dam stores energy that can be used to generate electricity, this buildup of protons provides energy for the generation of ATP by the chemiosmotic mechanism.

The Chemiosmotic Mechanism of ATP Generation
The mechanism of ATP synthesis using the electron transport chain is called **chemiosmosis.** To understand chemiosmosis, we need to recall several concepts that were introduced in Chapter 4 as part of the section on the movement of materials across membranes (page 92). Recall that substances diffuse passively across membranes from areas of high concentration to areas of low concentration; this diffusion yields energy. Recall also that the movement of substances *against* such a concentration gradient *requires* energy and that, in such an active transport of molecules or ions across biological membranes, the required energy is usually provided by ATP. In chemiosmosis, the energy released when a substance moves along a gradient is used to *synthesize* ATP. The "substance" in this case refers to protons. In respiration, chemiosmosis is responsible for most of the ATP that is generated. The steps of chemiosmosis are as follows (Figures 5.15 and 5.16):

❶ As energetic electrons from NADH (or chlorophyll) pass down the electron transport chain, some of the carriers in the chain pump—actively transport—

protons across the membrane. Such carrier molecules are called *proton pumps.*

❷ The phospholipid membrane is normally impermeable to protons, so this one-directional pumping establishes a proton gradient (a difference in the concentrations of protons on the two sides of the membrane). In addition to a concentration gradient, there is an electrical charge gradient. The excess H^+ on one side of the membrane makes that side positively charged compared with the other side. The resulting electrochemical gradient has potential energy, called the *proton motive force.*

❸ The protons on the side of the membrane with the higher proton concentration can diffuse across the membrane only through special protein channels that contain an enzyme called *ATP synthase (adenosine triphosphatase).* When this flow occurs, energy is released and is used by the enzyme to synthesize ATP from ADP and ⓟ $_i$.

Figure 5.16 shows in detail how the electron transport chain operates in eukaryotes to drive the chemiosmotic mechanism. ❶ Energetic electrons from NADH pass down the electron transport chains. Within the inner

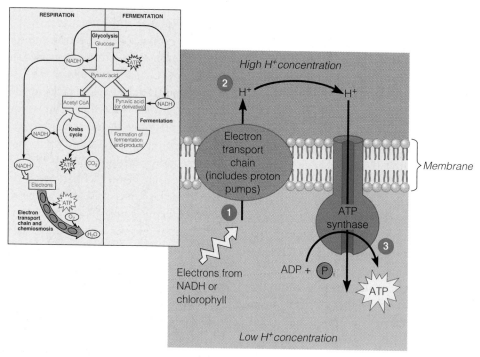

FIGURE 5.15 Chemiosmosis. An overview of the mechanism of chemiosmosis. The membrane shown could be a prokaryotic plasma membrane, a eukaryotic mitochondrial membrane, or a photosynthetic thylakoid. The numbered steps are described in the text.

 What is the proton motive force?

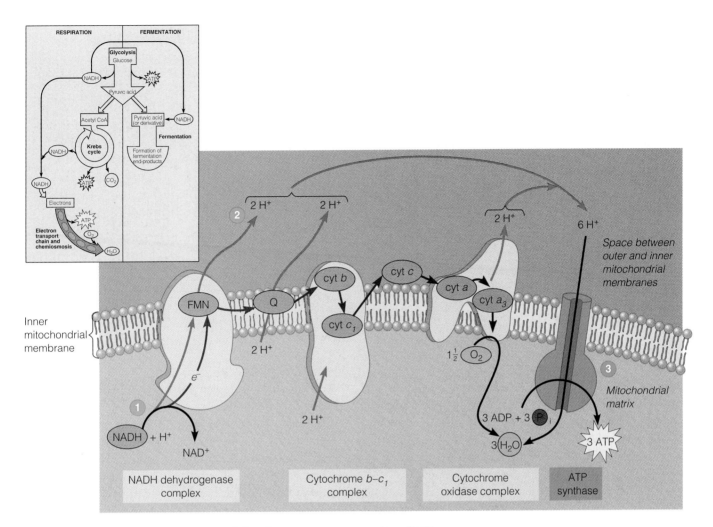

FIGURE 5.16 Electron transport and the chemiosmotic generation of ATP.
In the mitochondrial membrane, electron carriers are organized into three complexes, and protons (H^+) are pumped across the membrane at three points. In a eukaryotic cell, they are pumped from the matrix side of the mitochondrial membrane to the opposite side. In a prokaryotic cell, protons are pumped across the plasma membrane from the cytoplasmic side. The flow of electrons is shown in magenta.

Q **Where does chemiosmosis occur in eukaryotes? In prokaryotes?**

mitochondrial membrane, the carriers of the electron transport chain are organized into three complexes, with Q transporting electrons between the first and second complexes, and cyt c transporting them between the second and third complexes. ❷ Three components of the system pump protons: the first and third complexes and Q. At the end of the chain, electrons join with protons and oxygen (O_2) in the matrix fluid to form water (H_2O). Thus, O_2 is the final electron acceptor.

Both prokaryotic and eukaryotic cells use the chemiosmotic mechanism to generate energy for ATP production. However, in eukaryotic cells, ❸ the inner mitochondrial membrane contains the electron transport carriers and ATP synthase, whereas in most prokaryotic cells, the plasma membrane does so. An electron transport chain

also operates in photophosphorylation and is located in the thylakoid membrane of cyanobacteria and eukaryotic chloroplasts.

A Summary of Aerobic Respiration The electron transport chain regenerates NAD^+ and FAD^+, which can be used again in glycolysis and the Krebs cycle. The various electron transfers in the electron transport chain generate about 34 molecules of ATP from each molecule of glucose oxidized: approximately three from each of the ten molecules of NADH (a total of 30), and approximately two from each of the two molecules of $FADH_2$ (a total of four). To arrive at the total number of ATP molecules generated for each molecule of glucose, the 34 from chemiosmosis are added to those generated by oxidation in glycolysis and the

TABLE 5.3	ATP Yield During Prokaryotic Aerobic Respiration of One Glucose Molecule

Source	ATP Yield (Method)
Glycolysis	
1. Oxidation of glucose to pyruvic acid	2 ATP (substrate-level phosphorylation)
2. Production of 2 NADH	6 ATP (oxidative phosphorylation in electron transport chain)
Preparatory Step	
1. Formation of acetyl CoA produces 2 NADH	6 ATP (oxidative phosphorylation in electron transport chain)
Krebs Cycle 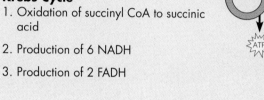	
1. Oxidation of succinyl CoA to succinic acid	2 GTP (equivalent of ATP; substrate-level phosphorylation)
2. Production of 6 NADH	18 ATP (oxidative phosphorylation in electron transport chain)
3. Production of 2 FADH	4 ATP (oxidative phosphorylation in electron transport chain)
	Total: 38 ATP

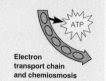

Electron transport chain and chemiosmosis

Krebs cycle. In aerobic respiration among prokaryotes, a total of 38 molecules of ATP can be generated from one molecule of glucose. Note that four of those ATPs come from substrate-level phosphorylation in glycolysis and the Krebs cycle. Table 5.3 provides a detailed accounting of the ATP yield during prokaryotic aerobic respiration.

Aerobic respiration among eukaryotes produces a total of only 36 molecules of ATP. There are fewer ATPs than in prokaryotes because some energy is lost when electrons are shuttled across the mitochondrial membranes that separate glycolysis (in the cytoplasm) from the electron transport chain. No such separation exists in prokaryotes. We can now summarize the overall reaction for aerobic respiration in prokaryotes as follows:

$$C_6H_{12}O_6 + 6O_2 + 38 \text{ ADP} + 38 \text{ ℗}_i \longrightarrow$$
Glucose   Oxygen

$$6 CO_2 + 6 H_2O + 38 \text{ ATP}$$
Carbon   Water
dioxide

A summary of the various stages of aerobic respiration in prokaryotes is presented in Figure 5.17.

ANAEROBIC RESPIRATION

In anaerobic respiration, the final electron acceptor is an inorganic substance other than oxygen (O_2). Some bacteria, such as *Pseudomonas* and *Bacillus*, can use a nitrate ion (NO_3^-) as a final electron acceptor; the nitrate ion is reduced to a nitrite ion (NO_2^-), nitrous oxide (N_2O), or nitrogen gas (N_2). Other bacteria, such as *Desulfovibrio* (dē-sul-fō-vib′rē-ō), use sulfate (SO_4^{2-}) as the final electron

acceptor to form hydrogen sulfide (H_2S). Still other bacteria use carbonate (CO_3^{2-}) to form methane (CH_4). Anaerobic respiration by bacteria using nitrate and sulfate as final acceptors is essential for the nitrogen and sulfur cycles that occur in nature. The amount of ATP generated in anaerobic respiration varies with the organism and the pathway. Because only part of the Krebs cycle operates under anaerobic conditions, and since not all the carriers in the electron transport chain participate in anaerobic respiration, the ATP yield is never as high as in aerobic respiration. Accordingly, anaerobes tend to grow more slowly than aerobes. ❋ **Animations: Go to The Microbiology Place website or CD-ROM and click "Animations" to view Electron Transport Chains and Chemiosmosis, and Krebs Cycle.**

FERMENTATION

> **LEARNING OBJECTIVE**
> • Describe the chemical reactions of, and list some products of, fermentation.

After glucose has been broken down into pyruvic acid, the pyruvic acid can be completely broken down in respiration, as previously described, or it can be converted to an organic product in fermentation, whereupon NAD^+ and $NADP^+$ are regenerated and can enter another round of glycolysis (see Figure 5.11). **Fermentation** can be defined in several ways (see the box, page 137), but we define it here as a process that

1. releases energy from sugars or other organic molecules, such as amino acids, organic acids, purines, and pyrimidines;

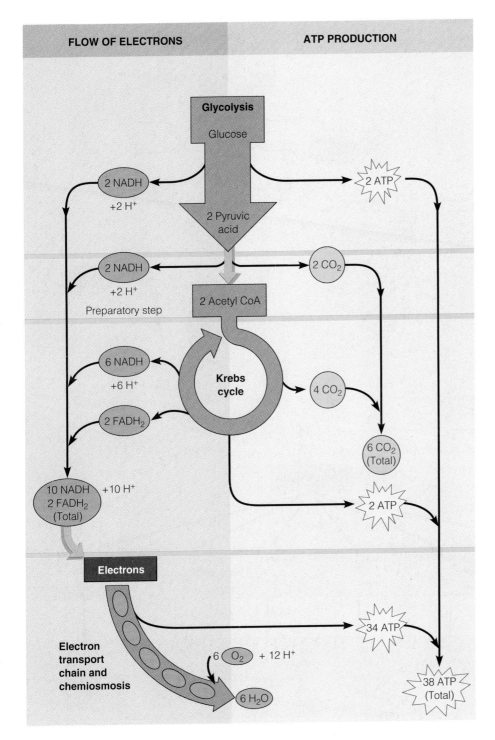

FLOW OF ELECTRONS   **ATP PRODUCTION**

Glycolysis

Glucose

2 NADH
+2 H$^+$

2 ATP

2 Pyruvic
acid

2 NADH
+2 H$^+$

Preparatory step

2 CO$_2$

2 Acetyl CoA

6 NADH
+6 H$^+$

**Krebs
cycle**

2 FADH$_2$

4 CO$_2$

10 NADH
2 FADH$_2$
(Total)   +10 H$^+$

6 CO$_2$
(Total)

2 ATP

Electrons

**Electron
transport
chain and
chemiosmosis**

6 O$_2$ + 12 H$^+$

34 ATP

38 ATP
(Total)

6 H$_2$O

**FIGURE 5.17 A summary of
aerobic respiration in prokaryotes.**
Glucose is broken down completely to
carbon dioxide and water, and ATP is
generated. This process has three major phases: glycolysis, the Krebs cycle,
and the electron transport chain. The
preparatory step is between glycolysis
and the Krebs cycle. The key event in
aerobic respiration is that electrons
are picked up from intermediates of
glycolysis and the Krebs cycle by
NAD$^+$ or FAD and are carried by
NADH or FADH$_2$ to the electron transport chain. NADH is also produced
during the conversion of pyruvic acid
to acetyl CoA. Most of the ATP generated by aerobic respiration is made
by the chemiosmotic mechanism
during the electron transport chain
phase; this is called oxidative
phosphorylation.

Q How do aerobic and anaerobic
respiration differ?

2. does not require oxygen (but sometimes can occur in its presence);

3. does not require the use of the Krebs cycle or an electron transport chain;

4. uses an organic molecule as the final electron acceptor;

5. produces only small amounts of ATP (only one or two ATP molecules for each molecule of starting material) because much of the original energy in glucose

remains in the chemical bonds of the organic end-products, such as lactic acid or ethanol.

During fermentation, electrons are transferred (along with protons) from reduced coenzymes (NADH, NADPH) to pyruvic acid or its derivatives (Figure 5.18a). Those final electron acceptors are reduced to the end-products shown in Figure 5.18b. An essential function of the second stage of fermentation is to ensure a steady supply of

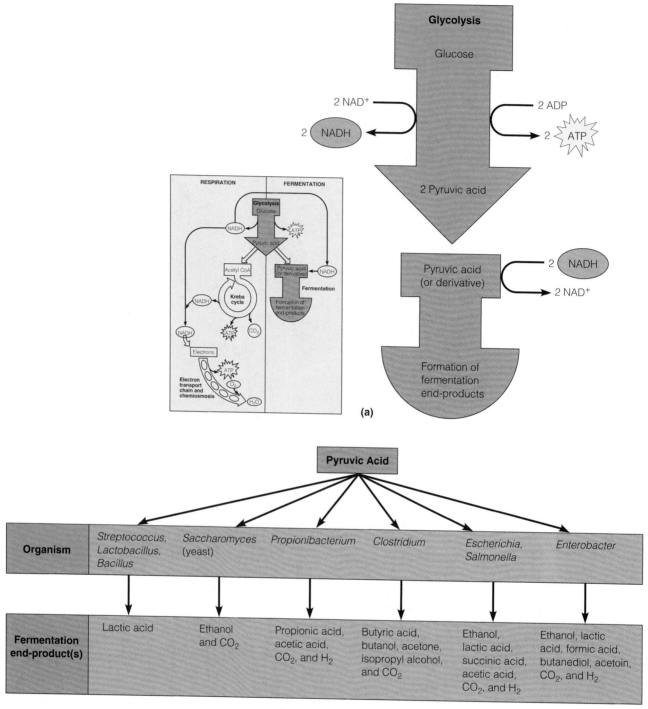

(a)

(b)

FIGURE 5.18 Fermentation. The inset indicates the relationship of fermentation to the overall energy-producing processes. (**a**) An overview of fermentation. The first step is glycolysis, the conversion of glucose to pyruvic acid. In the second step, the reduced coenzymes from glycolysis or its alternatives (NADH, NADPH) donate their electrons and hydrogen ions to pyruvic acid or a derivative to form a fermentation end-product. (**b**) End-products of various microbial fermentations.

Q **During which phase of fermentation is ATP generated?**

APPLICATIONS OF MICROBIOLOGY

WHAT IS FERMENTATION?

To many people, fermentation simply means the production of alcohol: grains and fruits are fermented to produce beer and wine. If a food soured, you might say it was "off" or fermented. Here are some definitions of fermentation. They range from informal, general usage to more scientific definitions. Fermentation is

1. Any spoilage of food by microorganisms (general use).

2. Any process that produces alcoholic beverages or acidic dairy products (general use).

3. Any large-scale microbial process occurring with or without air (common definition used in industry).

4. Any energy-releasing metabolic process that takes place only under anaerobic conditions (becoming more scientific).

5. Any metabolic process that releases energy from a sugar or other organic molecule, does not require oxygen or an electron transport system, and uses an organic molecule as the final electron acceptor. (This is the definition we use in this book.)

NAD^+ and $NADP^+$ so that glycolysis can continue. In fermentation, ATP is generated only during glycolysis.

Microorganisms can ferment various substrates; the end-products depend on the particular microorganism, the substrate, and the enzymes that are present and active. Chemical analyses of these end-products are useful in identifying microorganisms. We next consider two of the more important processes: lactic acid fermentation and alcohol fermentation.

LACTIC ACID FERMENTATION

During glycolysis, which is the first phase of **lactic acid fermentation,** a molecule of glucose is oxidized to two molecules of pyruvic acid (Figure 5.19; see also Figure 5.10). This oxidation generates the energy that is used to form the two molecules of ATP. In the next step, the two molecules of pyruvic acid are reduced by two molecules of NADH to form two molecules of lactic acid (Figure 5.19a). Because lactic acid is the end-product of the reaction, it undergoes no further oxidation, and most of the energy produced by the reaction remains stored in the lactic acid. Thus, this fermentation yields only a small amount of energy.

Two important genera of lactic acid bacteria are *Streptococcus* and *Lactobacillus* (lak-tō-bä-sil′lus). Because these microbes produce only lactic acid, they are referred to as **homolactic** (or *homofermentative*). Lactic acid fermentation can result in food spoilage. However, the process can also produce yogurt from milk, sauerkraut from fresh cabbage, and pickles from cucumbers.

ALCOHOL FERMENTATION

Alcohol fermentation also begins with the glycolysis of a molecule of glucose to yield two molecules of pyruvic

acid and two molecules of ATP. In the next reaction, the two molecules of pyruvic acid are converted to two molecules of acetaldehyde and two molecules of CO_2 (Figure 5.19b). The two molecules of acetaldehyde are next reduced by two molecules of NADH to form two molecules of ethanol. Again, alcohol fermentation is a low-energy-yield process because most of the energy contained in the initial glucose molecule remains in the ethanol, the end-product.

Alcohol fermentation is carried out by a number of bacteria and yeasts. The ethanol and carbon dioxide produced by the yeast *Saccharomyces* (sak-ä-rō-mī′sēs) are waste products for yeast cells but are useful to humans. Ethanol made by yeasts is the alcohol in alcoholic beverages, and carbon dioxide made by yeasts causes bread dough to rise (see the box in Chapter 1, page 3).

Organisms that produce lactic acid as well as other acids or alcohols are known as **heterolactic** (or *heterofermentative*) and often use the pentose phosphate pathway.

Table 5.4 lists some of the various microbial fermentations used by industry to convert inexpensive raw materials into useful end-products. A summary comparison of aerobic respiration, anaerobic respiration, and fermentation is given in Table 5.5.

LIPID AND PROTEIN CATABOLISM

LEARNING OBJECTIVE
• Describe how lipids and proteins undergo catabolism.

Our discussion of energy production has emphasized the oxidation of glucose, the main energy-supplying carbohydrate. However, microbes also oxidize lipids and

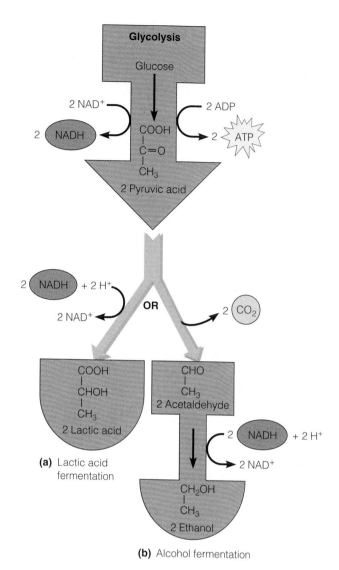

FIGURE 5.19 Types of fermentation.

Q **What is the difference between homolactic and heterolactic fermentation?**

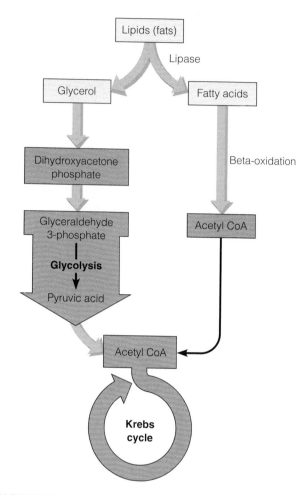

FIGURE 5.20 Lipid catabolism. Glycerol is converted into dihydroxyacetone phosphate (DHAP) and catabolized via glycolysis and the Krebs cycle. Fatty acids undergo beta-oxidation, in which carbon fragments are split off two at a time to form acetyl CoA, which is catabolized via the Krebs cycle.

Q **What is the role of lipases?**

proteins, and the oxidations of all these nutrients are related.

Recall that fats are lipids consisting of fatty acids and glycerol. Microbes produce extracellular enzymes called *lipases* that break fats down into their fatty acid and glycerol components. Each component is then metabolized separately (Figure 5.20). The Krebs cycle functions in the oxidation of glycerol and fatty acids. Many bacteria that hydrolyze fatty acids can use the same enzymes to degrade petroleum products. Although these bacteria are a nuisance when they grow in a fuel storage tank, they are beneficial when they grow in oil spills. Beta-oxidation (the oxidation of fatty acids) of petroleum is illustrated in the box in Chapter 2 (page 33).

Proteins are too large to pass unaided through plasma membranes. Microbes produce extracellular *proteases* and *peptidases*, enzymes that break down proteins into their component amino acids, which can cross the membranes. However, before amino acids can be catabolized, they must be enzymatically converted to other substances that can enter the Krebs cycle. In one such conversion, called **deamination,** the amino group of an amino acid is removed and converted to an ammonium ion (NH_4^+), which can be excreted from the cell. The remaining organic acid can enter the Krebs cycle. Other conversions involve **decarboxylation** (the removal of —COOH) and **dehydrogenation.**

A summary of the interrelationships of carbohydrate, lipid, and protein catabolism is shown in Figure 5.21, page 140.

TABLE 5.4	Some Industrial Uses for Different Types Of Fermentations		
Fermentation End-Product(s)	Industrial or Commercial Use	Starting Material	Microorganism
Ethanol	Beer	Malt extract	*Saccharomyces cerevisiae* (yeast, a fungus)
	Wine	Grape or other fruit juices	*Saccharomyces cerevisiae* var. *ellipsoideus*
	Fuel	Agricultural wastes	*Saccharomyces cerevisiae*
Acetic Acid	Vinegar	Ethanol	*Acetobacter* (bacterium)
Lactic Acid	Cheese, yogurt	Milk	*Lactobacillus, Streptococcus* (bacteria)
	Rye bread	Grain, sugar	*Lactobacillus delbruckii* (bacterium)
	Sauerkraut	Cabbage	*Lactobacillus plantarum* (bacterium)
	Summer sausage	Meat	*Pediococcus* (bacterium)
Propionic Acid and Carbon Dioxide	Swiss cheese	Lactic acid	*Propionibacterium freudenreichii* (bacterium)
Acetone and Butanol	Pharmaceutical, industrial uses	Molasses	*Clostridium acetobutylicum* (bacterium)
Glycerol	Pharmaceutical, industrial uses	Molasses	*Saccharomyces cerevisiae*
Citric Acid	Flavoring	Molasses	*Aspergillus* (fungus)
Methane	Fuel	Acetic acid	*Methanosarcina* (bacterium)
Sorbose	Vitamin C (ascorbic acid)	Sorbitol	*Gluconobacter*

TABLE 5.5	Aerobic Respiration, Anaerobic Respiration, and Fermentation Compared			
Energy-Producing Process	Growth Conditions	Final Hydrogen (Electron) Acceptor	Type of Phosphorylation Used to Generate ATP	ATP Molecules Produced per Glucose Molecule
Aerobic Respiration	Aerobic	Molecular oxygen (O_2)	Substrate-level and oxidative	36(eukaryotes) 38 (prokaryotes)
Anaerobic Respiration	Anaerobic	Usually an inorganic substance (such as NO_3^-, SO_4^{2-}, or CO_3^{2-}) but not molecular oxygen (O_2)	Substrate-level and oxidative	Variable (fewer than 38 but more than 2)
Fermentation	Aerobic or anaerobic	An organic molecule	Substrate-level	2

BIOCHEMICAL TESTS AND BACTERIAL IDENTIFICATION

LEARNING OBJECTIVE

• Describe two examples of the use of biochemical tests to identify bacteria in the laboratory.

Biochemical testing is frequently used to identify bacteria and yeasts because different species produce different enzymes. Such biochemical tests are designed to detect the presence of enzymes. One type of biochemical test is the detection of amino acid catabolizing enzymes involved in decarboxylation and dehydrogenation (discussed on page 138;

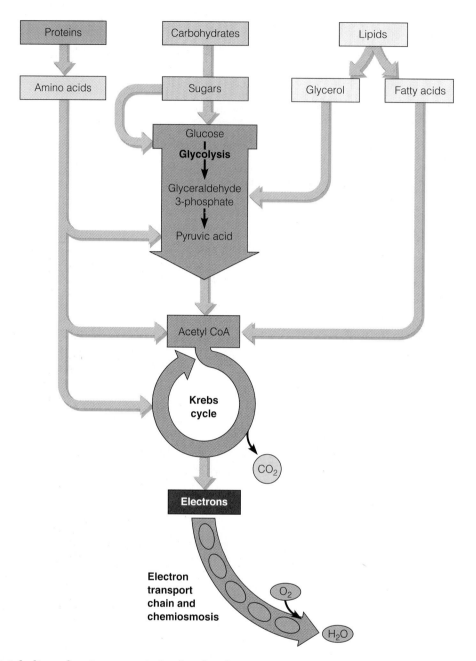

FIGURE 5.21 Catabolism of various organic food molecules. Proteins, carbohydrates, and lipids can all be sources of electrons and protons for respiration. These food molecules enter glycolysis or the Krebs cycle at various points.

Q What are the catabolic pathways through which high-energy electrons from all kinds of organic molecules flow on their energy-releasing pathways?

Figure 5.22). Another is a **fermentation test.** The test medium contains protein, a single carbohydrate, a pH indicator, and an inverted Durham tube, which is used to capture gas (Figure 5.23). Bacteria inoculated into the tube can use the protein or carbohydrate as a carbon and energy source. If they catabolize the carbohydrate and produce acid, the pH indicator changes color. Some organisms produce gas as well as acid from carbohydrate ca-

tabolism. The presence of a bubble in the Durham tube indicates gas formation (Figure 5.23b–d). Another example of the use of biochemical tests is shown in Figure 10.8 on page 296.

Note that in some instances, the waste products of one microorganism can be used as a carbon and energy source by another species. *Acetobacter* (ä-sē-tō-bak′tėr) bacteria oxidize ethanol made by yeast. *Propionibacterium*

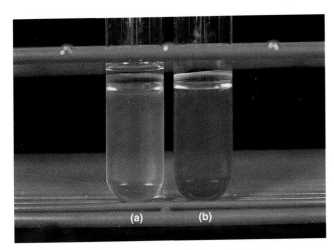

FIGURE 5.22 Detecting amino acid catabolizing enzymes in the lab. Bacteria are inoculated in tubes containing glucose, a pH indicator, and a specific amino acid. (**a**) The pH indicator turns to yellow when bacteria produce acid from glucose. (**b**) Alkaline products from decarboxylation turn the indicator to purple.

Q **What is decarboxylation?**

(prō-pē-on-ē-bak-ti′re-um) can use lactic acid produced by other bacteria. Propionibacteria convert lactic acid to pyruvic acid in preparation for the Krebs cycle. During the Krebs cycle, propionic acid and CO_2 are made. The holes in Swiss cheese are formed by the accumulation of the CO_2 gas.

FIGURE 5.23 A fermentation test. (**a**) An uninoculated fermentation tube containing the carbohydrate mannitol. (**b**) *Staphylococcus epidermidis* grew on the protein but did not use the carbohydrate. This organism is described as mannitol −. (**c**) *Staphylococcus aureus* produced acid but not gas. This species is mannitol +. (**d**) *Escherichia coli* is also mannitol + and produced acid and gas from mannitol. The gas is trapped in the inverted Durham tube.

Q **What is the importance of biochemical testing?**

PHOTOSYNTHESIS

LEARNING OBJECTIVES

- Compare and contrast cyclic and noncyclic photophosphorylation.
- Compare and contrast the light-dependent and light-independent reactions of photosynthesis.
- Compare and contrast oxidative phosphorylation and photophosphorylation.

In all of the metabolic pathways just discussed, organisms obtain energy for cellular work by oxidizing organic compounds. But where do organisms obtain these organic compounds? Some, including animals and many microbes, feed on matter produced by other organisms. For example, bacteria may catabolize compounds from dead plants and animals or may obtain nourishment from a living host.

Other organisms synthesize complex organic compounds from simple inorganic substances. The major mechanism for such synthesis is a process called **photosynthesis,** which is used by plants and many microbes. Essentially, photosynthesis is the conversion of light energy from the sun into chemical energy. The chemical energy is then used to convert CO_2 from the atmosphere to more reduced carbon compounds, primarily sugars. The word *photosynthesis* summarizes the process: *photo* means light, and *synthesis* refers to the assembly of organic compounds. This synthesis of sugars by using carbon atoms from CO_2 gas is also called **carbon fixation.** Continuation of life as we know it on Earth depends on the recycling of carbon in this way (see Figure 27.3). Cyanobacteria, algae, and green plants all contribute to this vital recycling with photosynthesis.

Photosynthesis can be summarized with the following equations:

1. Plants, algae, and cyanobacteria use water as a hydrogen donor, releasing O_2.

$$6\,CO_2 + 12\,H_2O + \text{Light energy} \longrightarrow C_6H_{12}O_6 + 6\,H_2O + 6\,O_2$$

2. Purple sulfur and green sulfur bacteria use H_2S as a hydrogen donor, producing sulfur granules.

$$6\,CO_2 + 12\,H_2S + \text{Light energy} \longrightarrow C_6H_{12}O_6 + 6\,H_2O + 12\,S$$

In the course of photosynthesis, electrons are taken from hydrogen atoms, an energy-poor molecule, and incorporated into sugar, an energy-rich molecule. The energy boost is supplied by light energy, although indirectly.

Photosynthesis takes place in two stages. In the first stage, called the **light-dependent (light) reactions,** light

FIGURE 5.24 Photophosphorylation.
(a) In cyclic photophosphorylation, electrons released from chlorophyll by light return to chlorophyll after passage along the electron transport chain. The energy from electron transfer is converted to ATP. **(b)** In noncyclic photophosphorylation, electrons released from chlorophyll are replaced by electrons from water. The chlorophyll electrons are passed along the electron transport chain to the electron acceptor $NADP^+$. $NADP^+$ combines with electrons and with hydrogen ions from water, forming NADPH.

Q **How are oxidative phosphorylation and photophosphorylation similar?**

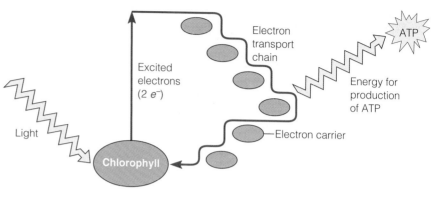

(a) Cyclic photophosphorylation

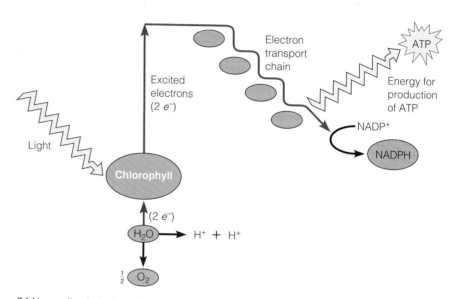

(b) Noncyclic photophosphorylation

energy is used to convert ADP and Ⓟ to ATP. In addition, in the predominant form of the light-dependent reactions, the electron carrier $NADP^+$ is reduced to NADPH. The $NADP^+$ coenzyme NADPH, like NADH, is an energy-rich carrier of electrons. In the second stage, the **light-independent (dark) reactions,** these electrons are used along with energy from ATP to reduce CO_2 to sugar.

THE LIGHT-DEPENDENT REACTIONS: PHOTOPHOSPHORYLATION

Photophosphorylation is one of the three ways ATP is formed, and it occurs only in photosynthetic cells. In this mechanism, light energy is absorbed by chlorophyll molecules in the photosynthetic cell, exciting some of the molecules' electrons. The chlorophyll principally used by green plants, algae, and cyanobacteria is *chlorophyll a.* It is located in the membranous thylakoids of chloroplasts in algae and green plants (see Figure 4.28, page 107) and in

the thylakoids found in the photosynthetic structures of cyanobacteria. Other bacteria use *bacteriochlorophylls.*

The excited electrons jump from the chlorophyll to the first of a series of carrier molecules, an electron transport chain similar to that used in respiration. As electrons are passed along the series of carriers, protons are pumped across the membrane, and ADP is converted to ATP by chemiosmosis. In **cyclic photophosphorylation** the electrons eventually return to chlorophyll (Figure 5.24a). In **noncyclic photophosphorylation,** which is the more common process, the electrons released from chlorophyll do not return to chlorophyll but become incorporated into NADPH (Figure 5.24b). The electrons lost from chlorophyll are replaced by electrons from H_2O. To summarize: the products of noncyclic photophosphorylation are ATP (formed by chemiosmosis using energy released in an electron transport chain), O_2 (from water molecules), and NADPH (in which the hydrogen electrons and protons were derived ultimately from water).

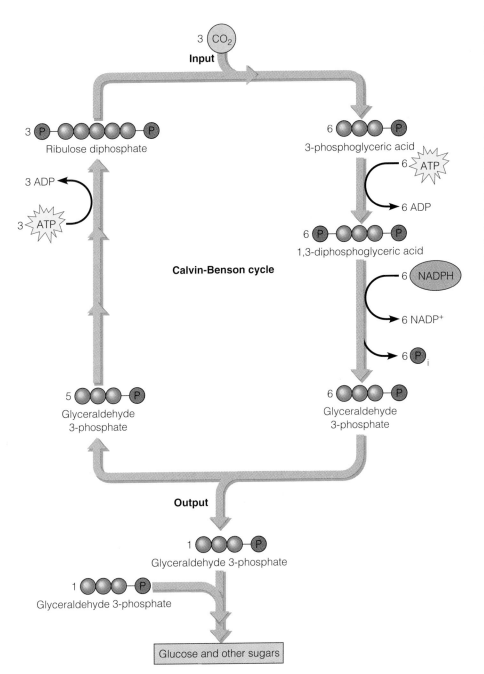

FIGURE 5.25 A simplified version of the Calvin-Benson cycle. This diagram shows three turns of the cycle, in which three molecules of CO_2 are fixed and one molecule of glyceraldehyde 3-phosphate is produced and leaves the cycle. Two molecules of glyceraldehyde 3-phosphate are needed to make one molecule of glucose. Therefore, the cycle must turn six times for each glucose molecule produced, requiring a total investment of 6 molecules of CO_2, 18 molecules of ATP, and 12 molecules of NADPH. A more detailed version of this cycle is presented in Appendix C.

Q **In the Calvin-Benson cycle, which molecule is used to synthesize sugars?**

THE LIGHT-INDEPENDENT REACTIONS: THE CALVIN-BENSON CYCLE

The light-independent (dark) reactions are so named because no light is directly required for them to occur. They include a complex cyclic pathway called the **Calvin-Benson cycle,** in which CO_2 is "fixed"—that is, used to synthesize sugars (Figure 5.25, see also Appendix C). ※ **Animation: Go to The Microbiology Place website or CD-ROM and click "Animations" to view Photosynthesis.**

A SUMMARY OF ENERGY PRODUCTION MECHANISMS

(**LEARNING OBJECTIVE**
• Write a sentence to summarize energy production in cells.

In the living world, energy passes from one organism to another in the form of the potential energy contained in the bonds of chemical compounds. Organisms obtain the energy from oxidation reactions. To obtain energy in a usable form, a cell must have an electron (or hydrogen) donor, which serves as an initial energy source within the cell.

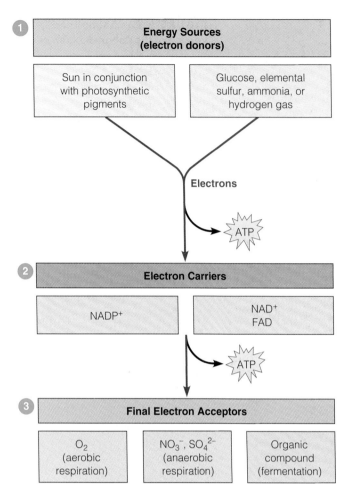

FIGURE 5.26 Requirements of ATP production. The production of ATP requires ❶ an energy source (electron donor), ❷ the transfer of electrons to an electron carrier during an oxidation-reduction reaction, and ❸ the transfer of electrons to a final electron acceptor.

Q **Are energy-generating reactions oxidations or reductions?**

Electron donors can be as diverse as photosynthetic pigments, glucose or other organic compounds, elemental sulfur, ammonia, or hydrogen gas (Figure 5.26). Next, electrons removed from the chemical energy sources are transferred to electron carriers, such as the coenzymes NAD^+, $NADP^+$, and FAD. This transfer is an oxidation-reduction reaction; the initial energy source is oxidized as this first electron carrier is reduced. During this phase, some ATP is produced. In the third stage, electrons are transferred from electron carriers to their final electron acceptors in further oxidation-reduction reactions, producing more ATP.

In aerobic respiration, oxygen (O_2) serves as the final electron acceptor. In anaerobic respiration, inorganic substances other than oxygen, such as nitrate ions (NO_3^-) or sulfate ions (SO_4^{2-}), serve as the final electron acceptors.

In fermentation, organic compounds serve as the final electron acceptors. In aerobic and anaerobic respiration, a series of electron carriers called an electron transport chain releases energy that is used by the mechanism of chemiosmosis to synthesize ATP. Regardless of their energy sources, all organisms use similar oxidation-reduction reactions to transfer electrons and similar mechanisms to use the energy released to produce ATP.

METABOLIC DIVERSITY AMONG ORGANISMS

LEARNING OBJECTIVE

• Categorize the various nutritional patterns among organisms according to carbon source and mechanisms of carbohydrate catabolism and ATP generation.

We have looked in detail at some of the energy-generating metabolic pathways that are used by animals and plants, as well as by many microbes. Microbes are distinguished by their great metabolic diversity, however, and some can sustain themselves on inorganic substances by using pathways that are unavailable to either plants or animals. All organisms, including microbes, can be classified metabolically according to their *nutritional pattern*—their source of energy and their source of carbon.

First considering the energy source, we can generally classify organisms as phototrophs or chemotrophs. **Phototrophs** use light as their primary energy source, whereas **chemotrophs** depend on oxidation-reduction reactions of inorganic or organic compounds for energy. For their principal carbon source, **autotrophs** (self-feeders) use carbon dioxide, and **heterotrophs** (feeders on others) require an organic carbon source. Autotrophs are also referred to as *lithotrophs* (rock eating), and heterotrophs are also referred to as *organotrophs*.

If we combine the energy and carbon sources, we derive the following nutritional classifications for organisms: *photoautotrophs, photoheterotrophs, chemoautotrophs,* and *chemoheterotrophs* (Figure 5.27). Almost all of the medically important microorganisms discussed in this book are chemoheterotrophs. Typically, infectious organisms catabolize substances obtained from the host.

PHOTOAUTOTROPHS

Photoautotrophs use light as a source of energy and carbon dioxide as their chief source of carbon. They include photosynthetic bacteria (green and purple bacteria and cyanobacteria), algae, and green plants. In the photosynthetic reactions of cyanobacteria, algae, and green plants, the hydrogen atoms of water are used to reduce carbon

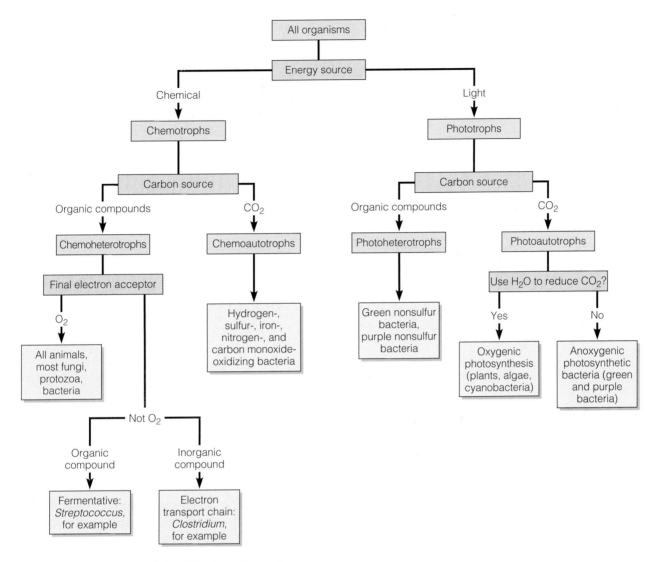

FIGURE 5.27 A nutritional classification of organisms.

Q **What is the basic difference between chemotrophs and phototrophs?**

dioxide, and oxygen gas is given off. Because this photosynthetic process produces O_2, it is sometimes called **oxygenic.**

In addition to the cyanobacteria (see Figure 11.13, page 328), there are several other families of photosynthetic prokaryotes. Each is classified according to the way it reduces CO_2. These bacteria cannot use H_2O to reduce CO_2 and cannot carry on photosynthesis when oxygen is present (they must have an anaerobic environment). Consequently, their photosynthetic process does not produce O_2 and is called **anoxygenic.** The anoxygenic photoautotrophs are the green and purple bacteria. The **green bacteria,** such as *Chlorobium* (klô-rō′bē-um), use sulfur (S), sulfur compounds (such as hydrogen sulfide, H_2S), or hydrogen gas (H_2) to reduce carbon dioxide and form organic compounds. Applying the energy

from light and the appropriate enzymes, these bacteria oxidize sulfide (S^{2-}) or sulfur (S) to sulfate (SO_4^{2-}) or oxidize hydrogen gas to water (H_2O). The **purple bacteria,** such as *Chromatium* (krō-mā′tē-um), also use sulfur, sulfur compounds, or hydrogen gas to reduce carbon dioxide. They are distinguished from the green bacteria by their type of chlorophyll, location of stored sulfur, and ribosomal RNA.

The chlorophylls used by these photosynthetic bacteria are called *bacteriochlorophylls,* and they absorb light at longer wavelengths than that absorbed by chlorophyll *a.* Bacteriochlorophylls of green sulfur bacteria are found in vesicles called *chlorosomes* (or *chlorobium vesicles*) underlying and attached to the plasma membrane. In the purple sulfur bacteria, the bacteriochlorophylls are located in

TABLE 5.6	Photosynthesis Compared in Selected Eukaryotes and Prokaryotes			
Characteristic	Eukaryotes	Prokaryotes		
	Algae, Plants	Cyanobacteria	Green Bacteria	Purple Bacteria
Substance that Reduces CO_2	H atoms of H_2O	H atoms of H_2O	Sulfur, sulfur compounds, H_2 gas	Sulfur, sulfur compounds, H_2 gas
Oxygen Production	Oxygenic	Oxygenic (and anoxygenic)	Anoxygenic	Anoxygenic
Type of Chlorophyll	Chlorophyll *a*	Chlorophyll *a*	Bacteriochlorophyll *a*	Bacteriochlorophyll *a* or *b*
Site of Photosynthesis	Chloroplasts with thylakoids	Thylakoids	Chlorosomes	Intracytoplasmic membrane
Environment	Aerobic	Aerobic (and anaerobic)	Anaerobic	Anaerobic

invaginations of the plasma membrane (*intracytoplasmic membranes*).

Several characteristics that distinguish eukaryotic photosynthesis from prokaryotic photosynthesis are presented in Table 5.6. See the box on the facing page for a discussion of an exceptional photosynthetic system that exists in *Halobacterium*. The system does not use chlorophyll.

PHOTOHETEROTROPHS

Photoheterotrophs use light as a source of energy but cannot convert carbon dioxide to sugar; rather, they use organic compounds, such as alcohols, fatty acids, other organic acids, and carbohydrates, as sources of carbon. They are anoxygenic. The **green nonsulfur bacteria,** such as *Chloroflexus* (klô-rō-flex'us), and **purple nonsulfur bacteria,** such as *Rhodopseudomonas* (rō-dō-sū-dō-mō'nas), are photoheterotrophs.

CHEMOAUTOTROPHS

Chemoautotrophs use the electrons from reduced inorganic compounds as a source of energy, and they use CO_2 as their principal source of carbon (see Figure 27.3, the carbon cycle). Inorganic sources of energy for these organisms include hydrogen sulfide (H_2S) for *Beggiatoa* (bej-jē-ä-tō'ä); elemental sulfur (S) for *Thiobacillus thiooxidans*; ammonia (NH_3) for *Nitrosomonas* (nī-trō-sō-mō'näs); nitrite ions (NO_2^-) for *Nitrobacter* (nī-trō-bak'tėr); hydrogen gas (H_2) for *Hydrogenomonas* (hī-drō-je-nō-mō'näs); ferrous iron (Fe^{2+}) for *Thiobacillus ferrooxidans*; and carbon monoxide (CO) for *Pseudomonas carboxydohydrogena*. The energy derived from the oxidation of these inorganic compounds is

eventually stored in ATP, which is produced by oxidative phosphorylation.

CHEMOHETEROTROPHS

When we discuss photoautotrophs, photoheterotrophs, and chemoautotrophs, it is easy to categorize the energy source and carbon source because they occur as separate entities. However, in chemoheterotrophs, the distinction is not as clear because the energy source and carbon source are usually the same organic compound—glucose, for example. **Chemoheterotrophs** specifically use the electrons from hydrogen atoms in organic compounds as their energy source.

Heterotrophs are further classified according to their source of organic molecules. **Saprophytes** live on dead organic matter, and **parasites** derive nutrients from a living host. Most bacteria, and all fungi, protozoa, and animals, are chemoheterotrophs.

Bacteria and fungi can use a wide variety of organic compounds for carbon and energy sources. This is why they can live in diverse environments. Understanding microbial diversity is scientifically interesting and economically important. In some situations microbial growth is undesirable, such as when rubber-degrading bacteria destroy a gasket or shoe sole. However, these same bacteria might be beneficial if they decomposed discarded rubber products such as tires. *Rhodococcus erythropolis* (rō-dō-kok'kus er-i-throp'ō-lis) is widely distributed in soil and can cause disease in humans and other animals. This same species is able to replace sulfur atoms in petroleum with

APPLICATIONS OF MICROBIOLOGY

BACTERIA MAKE A FASTER, SMARTER COMPUTER

An interesting member of the Archaea, *Halobacterium*, lives where very little else can grow. This bacterium is found in salt lakes, salt licks on ranches, salt flats, or any other environment with a concentration of salt that is five to seven times that of the ocean. Halobacteria are easy to detect because they turn their environment purple (see the photograph). These bacteria cannot ferment carbohydrates and do not contain chlorophyll, so it was assumed that all their energy comes from oxidative phosphorylation. The exciting discovery of a new system of photophosphorylation arose through the study of the plasma membrane of *Halobacterium halobium.*

Researchers found that the plasma membrane of *H. halobium* fragments into two fractions (red and purple) when the cell is broken down and its components are sorted. The red fraction, which constitutes most of the membrane, contains cytochromes, flavoproteins, and other parts of the electron transport chain, which carries out oxidative phosphorylation. The purple fraction is more interesting. This purple membrane occurs in distinct patches of hexagonal lattices within the plasma membrane. The purple color comes from a protein that makes up 75% of the purple membrane. This protein is similar to the retinal pigment in the rod cells of the human eye, rhodopsin, so the protein was named bacteriorhodopsin. At the time it was discovered, its function was not known.

Further studies showed that *H. halobium* can grow in the presence of either light or oxygen but cannot grow when neither is present.

This unexpected result suggested that *Halobacterium* can obtain energy by using either of two systems, one that operates in the presence of oxygen (oxidative phosphorylation) and one that operates in the presence of light (some kind of photophosphorylation). The rate of ATP synthesis by *H. halobium* was found to be highest when the cells receive light that is between 550 and 600 nm in wavelength; this range exactly corresponds to the absorption spectrum of bacteriorhodopsin.

Researchers hypothesized that bacteriorhodopsin, like chlorophyll-containing systems, acts as a proton pump to create a proton gradient across a cell membrane; in this case, the gradient is created across the purple membrane. The proton gradient can do cellular work—can drive the synthesis of ATP or transport solutes.

Halobacterium replaces the silicon chip

At Syracuse University's Center of Molecular Electronics, Robert Birge grows *Halobacterium* for 5 days in 5-liter batches and then extracts bacteriorhodopsin from the cells for a novel use. Birge has developed a computer chip made of a thin layer of bacteriorhodopsin.

Conventional computers store information on thin wafers of silicon. Computers process information by "reading" a series of zeros and ones produced as electrons flow through switches etched in the silicon. Electrons passing through a switch represent a one; a switch that stops the electron flow represents a zero. However, silicon can't hold enough information or process information fast enough for such applications as artificial intelligence or robot vision.

In contrast, the bacteriorhodopsin chip will be able to store more information than a silicon chip and process the information faster, more like a human brain. The bacteriorhodopsin chip works with light, which of course moves at the speed of light, much faster than the flow of electrons. Green light causes the protein to fold; a folded protein is read as a one, whereas an unfolded protein represents a zero. Laser light is used to "see" the configuration of the protein.

At present, the protein chip needs to be stored at −4°C to maintain its structure, but Birge and his coworkers are hopeful they will solve this problem. Russian scientists have made a protein processor for military radar, and the U.S. military is apparently using the protein chips in their combat planes. If such a plane crashes, the cooling system will go off, thus destroying the chip and keeping classified information from being stolen. Eventually, these smaller, faster, and higher-capacity chips will probably make it possible to develop computers that perform functions closer to human intelligence, such as acting as eyes for blind people.

Halobacterium (purple color) grows in the high-salt concentration of solar evaporation ponds used for manufacturing salt around San Francisco Bay.

atoms of oxygen. A Texas company is currently using *R. erythropolis* to produce desulfurized oil.

* * *

We will next consider how cells use ATP pathways for the synthesis of organic compounds such as carbohydrates, lipids, proteins, and nucleic acids.

METABOLIC PATHWAYS OF ENERGY USE

LEARNING OBJECTIVE

• Describe the major types of anabolism and their relationship to catabolism.

Up to now we have been considering energy production. Through the oxidation of organic molecules, organisms produce energy by aerobic respiration, anaerobic respiration, and fermentation. Much of this energy is given off as heat. The complete metabolic oxidation of glucose to carbon dioxide and water is considered a very efficient process, but about 45% of the energy of glucose is lost as heat. Cells use the remaining energy, which is trapped in the bonds of ATP, in a variety of ways. Microbes use ATP to provide energy for the transport of substances across plasma membranes—the process called active transport that we discussed in Chapter 4. Microbes also use some of their energy for flagellar motion (also discussed in Chapter 4). Most of the ATP, however, is used in the production of new cellular components. This production is a continuous process in cells, and, in general, is faster in prokaryotic cells than in eukaryotic cells.

Autotrophs build their organic compounds by fixing carbon dioxide in the Calvin-Benson cycle (see Figure 5.25). This requires both energy (ATP) and electrons (from the oxidation of NADPH). Heterotrophs, by contrast, must have a ready source of organic compounds for biosynthesis—the production of needed cellular components, usually from simpler molecules. The cells use these compounds as both the carbon source and the energy source. We will next consider the biosynthesis of a few representative classes of biological molecules: carbohydrates, lipids, amino acids, purines, and pyrimidines. As we do so, keep in mind that synthesis reactions require a net input of energy.

POLYSACCHARIDE BIOSYNTHESIS

Microorganisms synthesize sugars and polysaccharides. The carbon atoms required to synthesize glucose are derived from the intermediates produced during processes such as glycolysis and the Krebs cycle and from lipids or amino acids. After synthesizing glucose (or other simple

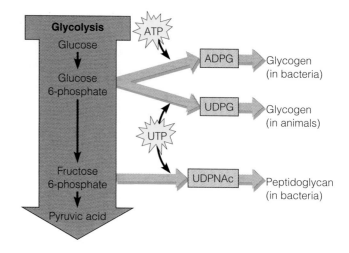

FIGURE 5.28 The biosynthesis of polysaccharides.

Q How are polysaccharides used in cells?

sugars), bacteria may assemble it into more complex polysaccharides such as glycogen. For bacteria to build glucose into glycogen, glucose units must be phosphorylated and linked. The product of glucose phosphorylation is glucose 6-phosphate. Such a process involves the expenditure of energy, usually in the form of ATP. In order for bacteria to synthesize glycogen, a molecule of ATP is added to glucose 6-phosphate to form *adenosine diphosphoglucose (ADPG)* (Figure 5.28). Once ADPG is synthesized, it is linked with similar units to form glycogen.

Using a nucleotide called uridine triphosphate (UTP) as a source of energy and glucose 6-phosphate, animals synthesize glycogen (and many other carbohydrates) from *uridine diphosphoglucose, UDPG* (see Figure 5.28). A compound related to UDPG, called *UDP-N-acetylglucosamine (UDPNAc)*, is a key starting material in the biosynthesis of peptidoglycan, the substance that forms bacterial cell walls. UDPNAc is formed from fructose 6-phosphate, and the reaction also uses UTP.

LIPID BIOSYNTHESIS

Because lipids vary considerably in chemical composition, they are synthesized by a variety of routes. Cells synthesize fats by joining glycerol and fatty acids. The glycerol portion of the fat is derived from dihydroxyacetone phosphate, an intermediate formed during glycolysis. Fatty acids, which are long-chain hydrocarbons (hydrogen linked to carbon), are built up when two-carbon fragments of acetyl CoA are successively added to each other (Figure 5.29). As with polysaccharide synthesis, the building units of fats and other lipids are linked via dehydration synthesis reactions that require energy, not always in the form of ATP.

The most important role of lipids is as structural components of biological membranes, and most membrane

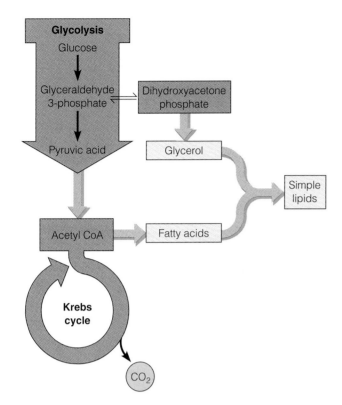

FIGURE 5.29 The biosynthesis of simple lipids.

Q What is the primary use of lipids in cells?

lipids are phospholipids. A lipid of a very different structure, cholesterol, is also found in plasma membranes of eukaryotic cells. Waxes are lipids that are important components of the cell wall of acid-fast bacteria. Other lipids, such as carotenoids, provide the red, orange, and yellow pigments of some microorganisms. Some lipids form portions of chlorophyll molecules. Lipids also function in energy storage. Recall that the breakdown products of lipids after biological oxidation feed into the Krebs cycle.

AMINO ACID AND PROTEIN BIOSYNTHESIS

Amino acids are required for protein biosynthesis. Some microbes, such as *E. coli*, contain the enzymes necessary to use starting materials, such as glucose and inorganic salts, for the synthesis of all the amino acids they need. Organisms with the necessary enzymes can synthesize all amino acids directly or indirectly from intermediates of carbohydrate metabolism (Figure 5.30a). Other microbes require that the environment provide some preformed amino acids.

One important source of the *precursors* (intermediates) used in amino acid synthesis is the Krebs cycle. Adding an amine group to pyruvic acid or to an appropriate organic acid of the Krebs cycle converts the acid into an amino acid. This process is called **amination.** If the amine group comes from a preexisting amino acid, the process is called **transamination** (Figure 5.30b).

Most amino acids within cells are destined to be building blocks for protein synthesis. Proteins play major roles in the cell as enzymes, structural components, and toxins, to name just a few uses. The joining of amino acids to form proteins involves dehydration synthesis and requires energy in the form of ATP. The mechanism of protein synthesis involves genes and is discussed in Chapter 8.

PURINE AND PYRIMIDINE BIOSYNTHESIS

Recall from Chapter 2 that the informational molecules DNA and RNA consist of repeating units called *nucleotides*, each of which consists of a purine or pyrimidine, a pentose (five-carbon sugar), and a phosphate group. The five-carbon sugars of nucleotides are derived from either the pentose phosphate pathway or the Entner-Doudoroff pathway. Certain amino acids—aspartic acid, glycine, and glutamine—made from intermediates produced during glycolysis and in the Krebs cycle participate in the biosyntheses of purines and pyrimidines (Figure 5.31). The carbon and nitrogen atoms derived from these amino acids form the purine and pyrimidine rings, and the energy for synthesis is provided by ATP. DNA contains all the information necessary to determine the specific structures and functions of cells. Both RNA and DNA are required for protein synthesis. In addition, such nucleotides as ATP, NAD^+, and $NADP^+$ assume roles in stimulating and inhibiting the rate of cellular metabolism. The synthesis of DNA and RNA from nucleotides will be discussed in Chapter 8.

THE INTEGRATION OF METABOLISM

LEARNING OBJECTIVE
• Define *amphibolic pathways.*

We have seen thus far that the metabolic processes of microbes produce energy from light, inorganic compounds, and organic compounds. Reactions also occur in which energy is used for biosynthesis. With such a variety of activity, you might imagine that anabolic and catabolic reactions occur independently of each other in space and time. Actually, anabolic and catabolic reactions are joined through a group of common intermediates (identified as key intermediates in Figure 5.32). Both anabolic and catabolic reactions also share some metabolic pathways, such as the Krebs cycle. For example, reactions in the Krebs cycle not only participate in the oxidation of glucose but also produce intermediates that can be converted to amino acids. Metabolic pathways that function in both anabolism and catabolism are called **amphibolic pathways,** meaning that they are dual-purpose.

Amphibolic pathways bridge the reactions that lead to the breakdown and synthesis of carbohydrates, lipids,

FIGURE 5.30 The biosynthesis of amino acids. (**a**) Pathways of amino acid biosynthesis through amination or transamination of intermediates of carbohydrate metabolism from the Krebs cycle, pentose phosphate pathway, and Entner-Doudoroff pathway. (**b**) Transamination, a process by which new amino acids are made with the amine groups from old amino acids. Glutamic acid and aspartic acid are both amino acids; the other two compounds are intermediates in the Krebs cycle.

Q **What is the function of amino acids in cells?**

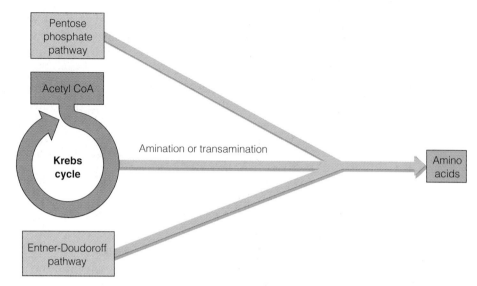

(a) Amino acid biosynthesis

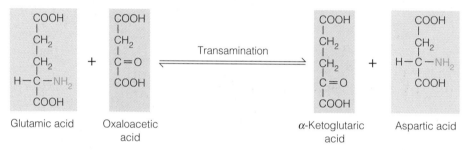

(b) Process of transamination

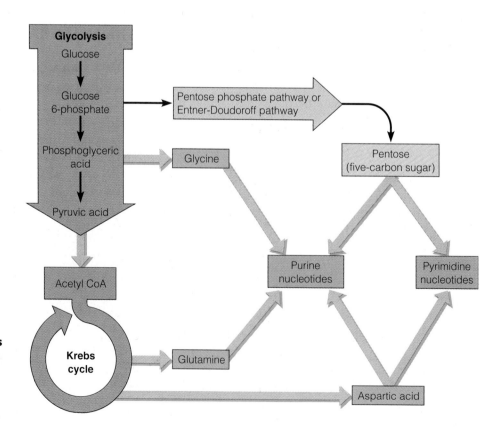

FIGURE 5.31 The biosynthesis of purine and pyrimidine nucleotides.

 Q **What are the functions of nucleotides in a cell?**

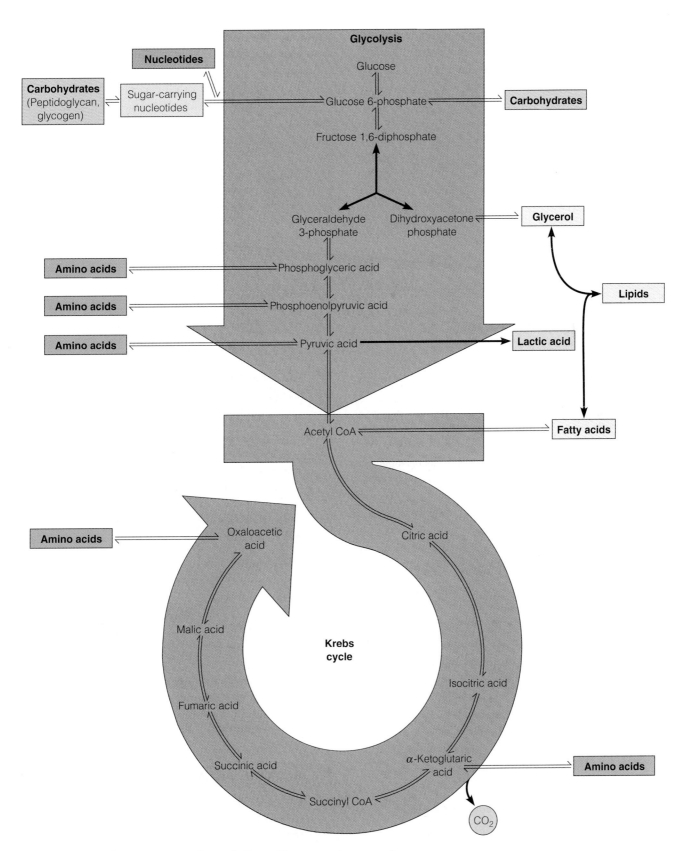

FIGURE 5.32 The integration of metabolism. Key intermediates are shown. Although not indicated in the figure, amino acids and ribose are used in the synthesis of purine and pyrimidine nucleotides (see Figure 5.31). The double arrows indicate amphibolic pathways.

Q **What is an amphibolic pathway?**

proteins, and nucleotides. Such pathways enable simultaneous reactions to occur in which the breakdown product formed in one reaction is used in another reaction to synthesize a different compound, and vice versa. Because various intermediates are common to both anabolic and catabolic reactions, mechanisms exist that regulate synthesis and breakdown pathways and allow these reactions to occur simultaneously. One such mechanism involves the use of different coenzymes for opposite pathways. For example, NAD^+ is involved in catabolic reactions, whereas $NADP^+$ is involved in anabolic reactions. Enzymes can also coordinate anabolic and catabolic reactions by accelerating or inhibiting the rates of biochemical reactions.

The energy stores of a cell can also affect the rates of biochemical reactions. For example, if ATP begins to accumulate, an enzyme shuts down glycolysis; this control helps to synchronize the rates of glycolysis and the Krebs cycle. Thus, if citric acid consumption increases, either because of a demand for more ATP or because anabolic pathways are draining off intermediates of the citric acid cycle, glycolysis accelerates and meets the demand.

STUDY OUTLINE

CATABOLIC AND ANABOLIC REACTIONS (pp. 115–116)

1. The sum of all chemical reactions within a living organism is known as metabolism.

2. Catabolism refers to chemical reactions that result in the breakdown of more complex organic molecules into simpler substances. Catabolic reactions usually release energy.

3. Anabolism refers to chemical reactions in which simpler substances are combined to form more complex molecules. Anabolic reactions usually require energy.

4. The energy of catabolic reactions is used to drive anabolic reactions.

5. The energy for chemical reactions is stored in ATP.
   ❊ **Animation: Metabolic Pathways (Overview). The Microbiology Place.**

ENZYMES (pp. 116–123)

1. Enzymes are proteins, produced by living cells, that catalyze chemical reactions by lowering the activation energy.

2. Enzymes are generally globular proteins with characteristic three-dimensional shapes.

3. Enzymes are efficient, can operate at relatively low temperatures, and are subject to various cellular controls.

NAMING ENZYMES (pp. 117–118)

4. Enzyme names usually end in -ase.

5. The six classes of enzymes are defined on the basis of the types of reactions they catalyze.

ENZYME COMPONENTS (pp. 118–119)

6. Most enzymes are holoenzymes, consisting of a protein portion (apoenzyme) and a nonprotein portion (cofactor).

7. The cofactor can be a metal ion (iron, copper, magnesium, manganese, zinc, calcium, or cobalt) or a complex organic molecule known as a coenzyme (NAD^+, $NADP^+$, FMN, FAD, or coenzyme A).

THE MECHANISM OF ENZYMATIC ACTION (pp. 119–120)

8. When an enzyme and substrate combine, the substrate is transformed, and the enzyme is recovered.

9. Enzymes are characterized by specificity, which is a function of their active sites. ❊ **Animation: Enzyme–Substrate Interactions. The Microbiology Place.**

FACTORS INFLUENCING ENZYMATIC ACTIVITY (pp. 120–122)

10. At high temperatures, enzymes undergo denaturation and lose their catalytic properties; at low temperatures, the reaction rate decreases.

11. The pH at which enzymatic activity is maximal is known as the optimum pH.

12. Within limits, enzymatic activity increases as substrate concentration increases.

13. Competitive inhibitors compete with the normal substrate for the active site of the enzyme. Noncompetitive inhibitors act on other parts of the apoenzyme or on the cofactor and decrease the enzyme's ability to combine with the normal substrate.

FEEDBACK INHIBITION (pp. 122–123)

14. Feedback inhibition occurs when the end-product of a metabolic pathway inhibits an enzyme's activity near the start of the pathway.

RIBOZYMES (p. 123)

15. Ribozymes are enzymatic RNA molecules that cut and splice RNA in eukaryotic cells.

ENERGY PRODUCTION (pp. 123–125)

OXIDATION-REDUCTION REACTIONS (pp. 123–124)

1. Oxidation is the removal of one or more electrons from a substrate. Protons (H^+) are often removed with the electrons.
2. Reduction of a substrate refers to its gain of one or more electrons.
3. Each time a substance is oxidized, another is simultaneously reduced.
4. NAD^+ is the oxidized form; NADH is the reduced form.
5. Glucose is a reduced molecule; energy is released during a cell's oxidation of glucose.

THE GENERATION OF ATP (pp. 124–125)

6. Energy released during certain metabolic reactions can be trapped to form ATP from ADP and Ⓟ (phosphate). Addition of a Ⓟ to a molecule is called phosphorylation.
7. During substrate-level phosphorylation, a high-energy Ⓟ from an intermediate in catabolism is added to ADP.
8. During oxidative phosphorylation, energy is released as electrons are passed to a series of electron acceptors (an electron transport chain) and finally to O_2 or another inorganic compound.
9. During photophosphorylation, energy from light is trapped by chlorophyll, and electrons are passed through a series of electron acceptors. The electron transfer releases energy used for the synthesis of ATP.

METABOLIC PATHWAYS OF ENERGY PRODUCTION (p. 125)

10. A series of enzymatically catalyzed chemical reactions called metabolic pathways store energy in and release energy from organic molecules.

CARBOHYDRATE CATABOLISM

(pp. 125–137)

1. Most of a cell's energy is produced from the oxidation of carbohydrates.
2. Glucose is the most commonly used carbohydrate.
3. The two major types of glucose catabolism are respiration, in which glucose is completely broken down, and fermentation, in which it is partially broken down.

GLYCOLYSIS (p. 127)

4. The most common pathway for the oxidation of glucose is glycolysis. Pyruvic acid is the end-product.
5. Two ATP and two NADH molecules are produced from one glucose molecule.

ALTERNATIVES TO GLYCOLYSIS (p. 127)

6. The pentose phosphate pathway is used to metabolize five-carbon sugars; one ATP and 12 NADPH molecules are produced from one glucose molecule.
7. The Entner-Doudoroff pathway yields one ATP and two NADPH molecules from one glucose molecule.
   ❊ **Animation: Glycolysis. The Microbiology Place.**

CELLULAR RESPIRATION (pp. 129–134)

8. During respiration, organic molecules are oxidized. Energy is generated from the electron transport chain.
9. In aerobic respiration, O_2 functions as the final electron acceptor.
10. In anaerobic respiration, the final electron acceptor is usually an inorganic molecule other than O_2.

Aerobic Respiration (pp. 129–134)

The Krebs Cycle (p. 129)

11. Decarboxylation of pyruvic acid produces one CO_2 molecule and one acetyl group.
12. Two-carbon acetyl groups are oxidized in the Krebs cycle. Electrons are picked up by NAD^+ and FAD for the electron transport chain.
13. From one molecule of glucose, oxidation produces six molecules of NADH, two molecules of $FADH_2$, and two molecules of ATP.
14. Decarboxylation produces six molecules of CO_2.

The Electron Transport Chain (System) (pp. 130–132)

15. Electrons are brought to the electron transport chain by NADH.
16. The electron transport chain consists of carriers, including flavoproteins, cytochromes, and ubiquinones.

The Chemiosmotic Mechanism of ATP Generation (pp. 132–133)

17. Protons being pumped across the membrane generate a proton motive force as electrons move through a series of acceptors or carriers.
18. Energy produced from movement of the protons back across the membrane is used by ATP synthase to make ATP from ADP and Ⓟ.
19. In eukaryotes, electron carriers are located in the inner mitochondrial membrane; in prokaryotes, electron carriers are in the plasma membrane.

A Summary of Aerobic Respiration (pp. 133–134)

20. In aerobic prokaryotes, 38 ATP molecules can be produced from complete oxidation of a glucose molecule in glycolysis, the Krebs cycle, and the electron transport chain.
21. In eukaryotes, 36 ATP molecules are produced from complete oxidation of a glucose molecule.

Anaerobic Respiration (p. 134)

22. The final electron acceptors in anaerobic respiration include NO_3^-, SO_4^{2-}, and CO_3^{2-}.

23. The total ATP yield is less than in aerobic respiration because only part of the Krebs cycle operates under anaerobic conditions. ※ **Animations: Electron Transport Chains and Chemiosmosis, and Krebs Cycle. The Microbiology Place.**

FERMENTATION (pp. 134–137)

24. Fermentation releases energy from sugars or other organic molecules by oxidation.

25. O_2 is not required in fermentation.

26. Two ATP molecules are produced by substrate-level phosphorylation.

27. Electrons removed from the substrate reduce NAD^+.

28. The final electron acceptor is an organic molecule.

29. In lactic acid fermentation, pyruvic acid is reduced by NADH to lactic acid.

30. In alcohol fermentation, acetaldehyde is reduced by NADH to produce ethanol.

31. Heterolactic fermenters can use the pentose phosphate pathway to produce lactic acid and ethanol.

LIPID AND PROTEIN CATABOLISM (pp. 137–138)

1. Lipases hydrolyze lipids into glycerol and fatty acids.

2. Fatty acids and other hydrocarbons are catabolized by beta-oxidation.

3. Catabolic products can be further broken down in glycolysis and the Krebs cycle.

4. Before amino acids can be catabolized, they must be converted to various substances that enter the Krebs cycle.

5. Transamination, decarboxylation, and dehydrogenation reactions convert the amino acids to be catabolized.

BIOCHEMICAL TESTS AND BACTERIAL IDENTIFICATION (pp. 139–141)

1. Bacteria and yeast can be identified by detecting action of their enzymes.

2. Fermentation tests are used to determine whether an organism can ferment a carbohydrate to produce acid and gas.

PHOTOSYNTHESIS (pp. 141–143)

1. Photosynthesis is the conversion of light energy from the sun into chemical energy; the chemical energy is used for carbon fixation.

THE LIGHT-DEPENDENT REACTIONS: PHOTOPHOSPHORYLATION (p. 142)

2. Chlorophyll *a* is used by green plants, algae, and cyanobacteria; it is found in thylakoid membranes.

3. Electrons from chlorophyll pass through an electron transport chain, from which ATP is produced by chemiosmosis.

4. In cyclic photophosphorylation, the electrons return to the chlorophyll.

5. In noncyclic photophosphorylation, the electrons are used to reduce $NADP^+$. The electrons from H_2O or H_2S replace those lost from chlorophyll.

6. When H_2O is oxidized by green plants, algae, and cyanobacteria, O_2 is produced; when H_2S is oxidized by the sulfur bacteria, S granules are produced.

THE LIGHT-INDEPENDENT REACTIONS: THE CALVIN-BENSON CYCLE (p. 143)

7. CO_2 is used to synthesize sugars in the Calvin-Benson cycle. ※ **Animation: Photosynthesis. The Microbiology Place.**

A SUMMARY OF ENERGY PRODUCTION MECHANISMS (pp. 143–144)

1. Sunlight is converted to chemical energy in oxidation-reduction reactions carried on by phototrophs. Chemotrophs can use this chemical energy.

2. In oxidation-reduction reactions, energy is derived from the transfer of electrons.

3. To produce energy, a cell needs an electron donor (organic or inorganic), a system of electron carriers, and a final electron acceptor (organic or inorganic).

METABOLIC DIVERSITY AMONG ORGANISMS (pp. 144–148)

1. Photoautotrophs obtain energy by photophosphorylation and fix carbon from CO_2 via the Calvin-Benson cycle to synthesize organic compounds.

2. Cyanobacteria are oxygenic phototrophs. Green bacteria and purple bacteria are anoxygenic phototrophs.

3. Photoheterotrophs use light as an energy source and an organic compound for their carbon source and electron donor.

4. Chemoautotrophs use inorganic compounds as their energy source and carbon dioxide as their carbon source.

5. Chemoheterotrophs use complex organic molecules as their carbon and energy sources.

METABOLIC PATHWAYS OF ENERGY USE (pp. 148–149)

POLYSACCHARIDE BIOSYNTHESIS (p. 148)

1. Glycogen is formed from ADPG.
2. UDPNAc is the starting material for the biosynthesis of peptidoglycan.

LIPID BIOSYNTHESIS (pp. 148–149)

3. Lipids are synthesized from fatty acids and glycerol.
4. Glycerol is derived from dihydroxyacetone phosphate, and fatty acids are built from acetyl CoA.

AMINO ACID AND PROTEIN BIOSYNTHESIS (p. 149)

5. Amino acids are required for protein biosynthesis.
6. All amino acids can be synthesized either directly or indirectly from intermediates of carbohydrate metabolism, particularly from the Krebs cycle.

PURINE AND PYRIMIDINE BIOSYNTHESIS (p. 149)

7. The sugars composing nucleotides are derived from either the pentose phosphate pathway or the Entner-Doudoroff pathway.
8. Carbon and nitrogen atoms from certain amino acids form the backbones of the purines and pyrimidines.

THE INTEGRATION OF METABOLISM (pp. 149–152)

1. Anabolic and catabolic reactions are integrated through a group of common intermediates.
2. Such integrated metabolic pathways are referred to as amphibolic pathways.

STUDY QUESTIONS

Access more review material either online at **The Microbiology Place** (www.microbiologyplace.com) or with **The Microbiology Place CD-ROM** packaged with your new book. There you'll find activities, practice tests, quizzes, flashcards, case studies, and more to help you succeed. In addition, you'll find the following Interactive Tutorials: Enzyme Inhibitors, Oxidation-Reduction, Glycolysis, Fermentation, Krebs Cycle, and Cellular Respiration.

Answers to the Study Questions can be found in Appendix G.

REVIEW

1. Define *metabolism*.
2. Distinguish between catabolism and anabolism. How are these processes related?
3. Using the diagrams below, show:
   a. where the substrate will bind.
   b. where the competitive inhibitor will bind.
   c. where the noncompetitive inhibitor will bind.
   d. which of the four elements could be the inhibitor in feedback inhibition.

4. What will the effect of the reactions in question 3 be?
5. Why are most enzymes active at one particular temperature? Why are enzymes less active below this temperature? What happens above this temperature?
6. List four compounds that can be made from pyruvic acid by an organism that uses fermentation only.
7. Fill in the following table with the carbon source and energy source of each type of organism.

Organism	Carbon Source	Energy Source
Photoautotroph	CO_2	light
Photoheterotroph	Organic molecules	light
Chemoautotroph	CO_2	inorganic molecule
Chemoheterotroph	Organic molecules	organic molecules

Enzyme   Substrate   Competitive inhibitor   Noncompetitive inhibitor

Use the diagrams for questions 8–16.

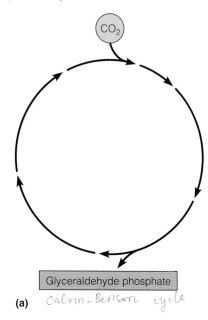

Glyceraldehyde phosphate

(a) *Calvin-Benson cycle*

Glucose
(C—C—C—C—C—C)

Glyceraldehyde
3-phosphate
(C—C—C—Ⓟ)   ⇌   Dihydroxyacetone
phosphate
(C—C—C—Ⓟ)

glycolysis

Two molecules of
pyruvic acid
(two C—C—C)

(b)

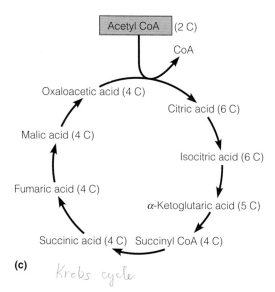

Acetyl CoA (2 C)
CoA
Oxaloacetic acid (4 C)
Malic acid (4 C)
Citric acid (6 C)
Fumaric acid (4 C)
Isocitric acid (6 C)
Succinic acid (4 C)   Succinyl CoA (4 C)
α-Ketoglutaric acid (5 C)

(c) *Krebs cycle*

8. Name the pathways diagrammed in a, b, and c of the figure to the left.

9. Show where glycerol is catabolized and where fatty acids are catabolized.

10. Show where the amino acid glutamic acid is catabolized:

$$HOOC—CH_2—CH_2—\overset{\displaystyle H}{\underset{\displaystyle NH_2}{C}}—COOH$$

11. Show how these pathways are related.

12. Where is ATP required in pathways a and b?

13. Where is CO_2 released in pathways b and c?

14. Show where a long-chain hydrocarbon such as petroleum is catabolized.

15. Where is NADH (or $FADH_2$ or NADPH) used and produced in these pathways?

16. Identify four places where anabolic and catabolic pathways are integrated.

17. There are three mechanisms for the phosphorylation of ADP to produce ATP. Write the name of the mechanism that describes each of the reactions in the following table.

ATP Generated by	Reaction
Photophosphorylation	An electron, liberated from chlorophyll by light, is passed down an electron transport chain.
Oxidative phosphorylation	Cytochrome c passes two electrons to cytochrome a.
Substrate-level phosphorylation	$\underset{\displaystyle COOH}{\overset{\displaystyle CH_2}{\overset{\displaystyle \|}{C}—O\sim Ⓟ}}$ → $\underset{\displaystyle COOH}{\overset{\displaystyle CH_3}{C=O}}$
	Phosphoenolpyruvic acid Pyruvic acid

18. Define oxidation-reduction, and differentiate between the following terms:
    a. aerobic and anaerobic respiration
    b. respiration and fermentation
    c. cyclic and noncyclic photophosphorylation

19. The pentose phosphate pathway produces only one ATP. List four advantages of this pathway for the cell.

20. All of the energy-producing biochemical reactions that occur in cells, such as photophosphorylation and glycolysis, are _oxidation_ reactions.

21. Explain how ATP is a key intermediate in metabolism.

22. An enzyme and substrate are combined. The rate of reaction begins as shown in the following graph. To complete the graph, show the effect of increasing substrate concentration

on a constant enzyme concentration. Show the effect of increasing temperature.

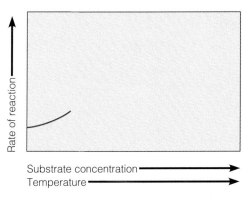

MULTIPLE CHOICE

1. Which substance in the following reaction is being reduced?

$$\underset{\text{Acetaldehyde}}{\overset{H}{\underset{CH_3}{C}}=O + NADH + H^+} \rightarrow \underset{\text{Ethanol}}{\overset{H}{H-\underset{CH_3}{C}-OH}} + NAD^+$$

   a. acetaldehyde
   b. NADH
   c. ethanol
   d. NAD^+

2. Which of the following reactions produces the most molecules of ATP during aerobic metabolism?
   a. glucose → glucose 6-phosphate
   b. phosphoenolpyruvic acid → pyruvic acid
   c. glucose → pyruvic acid
   d. acetyl CoA → $CO_2 + H_2O$
   e. succinic acid → fumaric acid

3. Which of the following processes does not generate ATP?
   a. photophosphorylation
   b. the Calvin-Benson cycle
   c. oxidative phosphorylation
   d. substrate-level phosphorylation
   e. none of the above

4. Which of the following compounds has the greatest amount of energy for a cell?
   a. CO_2 d. O_2
   b. ATP   e. lactic acid
   c. glucose

5. Which of the following is the best definition of the Krebs cycle?
   a. the oxidation of pyruvic acid
   b. the way cells produce CO_2
   c. a series of chemical reactions in which NADH is produced from the oxidation of pyruvic acid
   d. a method of producing ATP by phosphorylating ADP
   e. a series of chemical reactions in which ATP is produced from the oxidation of pyruvic acid.

6. Which of the following is the best definition of respiration?
   a. a sequence of carrier molecules with O_2 as the final electron acceptor
   b. a sequence of carrier molecules with an inorganic molecule as the final electron acceptor
   c. a method of generating ATP
   d. the complete oxidation of glucose to CO_2 and H_2O
   e. a series of reactions in which pyruvic acid is oxidized to CO_2 and H_2O

Use the following choices to answer questions 7–10.
   a. *E. coli* growing in glucose broth at 35°C with O_2 for 5 days
   b. *E. coli* growing in glucose broth at 35°C without O_2 for 5 days
   c. both a and b
   d. neither a nor b

b 7. Which culture produces the most lactic acid?
a 8. Which culture produces the most ATP?
c 9. Which culture uses NAD^+?
b 10. Which culture uses the most glucose?

CRITICAL THINKING

1. Write your own definition of the chemiosmotic mechanism of ATP generation. On Figure 5.16, mark the following using the appropriate letter:
   a. the acidic side of the membrane
   b. the side with a positive electrical charge
   c. potential energy
   d. kinetic energy

2. Explain why, even under ideal conditions, *Streptococcus* grows slowly.

3. Why must NADH be reoxidized? How does this happen in an organism that uses respiration? Fermentation?

4. The following graph shows the normal rate of reaction of an enzyme and its substrate (blue) and the rate when an excess of competitive inhibitor is present (red). Explain why the graph appears as it does.

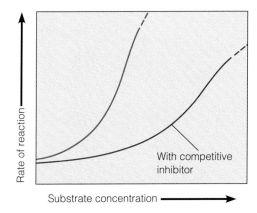

5. Compare and contrast carbohydrate catabolism and energy production in the following bacteria:
   a. *Pseudomonas*, an aerobic chemoheterotroph
   b. *Spirulina*, an oxygenic photoautotroph
   c. *Ectothiorhodospira*, an anoxygenic photoautotroph

6. How much ATP could be obtained from the complete oxidation of one molecule of glucose? From one molecule of butterfat containing one glycerol and three 12-carbon chains?

7. The chemoautotroph *Thiobacillus* can obtain energy from the oxidation of arsenic ($As^{3+} \rightarrow As^{5+}$). How does this reaction provide energy? How can this bacterium be put to use by humans?

CLINICAL APPLICATIONS

1. *Haemophilus influenzae* requires hemin (X factor) to synthesize cytochromes and NAD^+ (V factor) from other cells. For what does it use these two growth factors? What diseases does *H. influenzae* cause?

2. The drug HIVID, also called ddC, inhibits DNA synthesis. It is used to treat HIV infection and AIDS. Compare the following illustration of ddC to Figure 2.16 on page 48. How does this drug work?

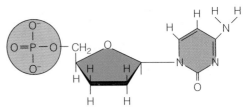

3. The bacterial enzyme streptokinase is used to digest fibrin (blood clots) in patients with atherosclerosis. Why doesn't injection of streptokinase cause a streptococcal infection? How do we know the streptokinase will digest fibrin only and not good tissues?

6

Microbial Growth

When we talk about microbial growth, we are really referring to the *number* of cells, not the *size* of the cells. Microbes that are "growing" are increasing in number, accumulating into *colonies* (groups of cells large enough to be seen without a microscope) of hundreds of thousands of cells, or *populations* of billions of cells. Although individual cells approximately double in size during their lifetime, this change is not very significant compared with the size increases observed during the lifetime of plants and animals.

Microbial populations can become incredibly large in a very short time, as we will see later in this chapter. By understanding the conditions necessary for microbial growth, we can determine how to control the growth of microbes that cause diseases and food spoilage. We can also learn how to encourage the growth of helpful microbes and those we wish to study.

In this chapter we will examine the physical and chemical requirements for microbial growth, the various kinds of culture media, bacterial cell division, the phases of microbial growth, and the methods of measuring microbial growth.

UNDER THE MICROSCOPE

Bacillus licheniformis. *B. licheniformis* is widespread in the soil where it contributes to nutrient recycling. This bacterium is grown industrially for enzymes and antibiotics.

THE REQUIREMENTS FOR GROWTH

The requirements for microbial growth can be divided into two main categories: physical and chemical. Physical aspects include temperature, pH, and osmotic pressure. Chemical requirements include sources of carbon, nitrogen, sulfur, phosphorus, trace elements, oxygen, and organic growth factors.

PHYSICAL REQUIREMENTS

LEARNING OBJECTIVES

* Classify microbes into five groups on the basis of preferred temperature range.
* Identify how and why the pH of culture media is controlled.
* Explain the importance of osmotic pressure to microbial growth.

TEMPERATURE

Most microorganisms grow well at the temperatures favored by humans. However, certain bacteria are capable of growing at extremes of temperature that would certainly hinder the survival of almost all eukaryotic organisms.

Microorganisms are classified into three primary groups on the basis of their preferred range of temperature: **psychrophiles** (cold-loving microbes), **mesophiles** (moderate-temperature–loving microbes), and **thermophiles** (heat-loving microbes). Most bacteria grow only within a limited range of temperatures, and their maximum and minimum growth temperatures are only about 30°C apart. They grow poorly at the high and low temperature extremes within their range.

Each bacterial species grows at particular minimum, optimum, and maximum temperatures. The **minimum growth temperature** is the lowest temperature at which the species will grow. The **optimum growth temperature** is the temperature at which the species grows best. The **maximum growth temperature** is the highest temperature at which growth is possible. By graphing the growth response over a temperature range, we can see that the optimum growth temperature is usually near the top of the range; above that temperature the rate of growth drops off rapidly (Figure 6.1). This happens presumably because the high temperature has inactivated necessary enzymatic systems of the cell.

The ranges and maximum growth temperatures that define bacteria as psychrophiles, mesophiles, or thermophiles are not rigidly defined. Psychrophiles, for example, were originally considered simply to be organisms capable of growing at 0°C. However, there seem to be two fairly distinct groups capable of growth at that temperature. One group, composed of psychrophiles in the strictest sense, can grow at 0°C but has an optimum growth temperature of about 15°C. Most of these organisms are so sensitive to higher temperatures that they will not even grow in a reasonably warm room (25°C). Found mostly in the oceans' depths or in certain polar regions, such organisms seldom cause problems in food preservation. The other group that can grow at 0°C has higher optimum temperatures, usually 20 to 30°C and cannot grow above about 40°C. Organisms of this type are much more common than psychrophiles and are the most likely to be encountered in low-temperature food spoilage because they grow fairly well at refrigerator temperatures. We will use the term **psychrotrophs,** which is favored by

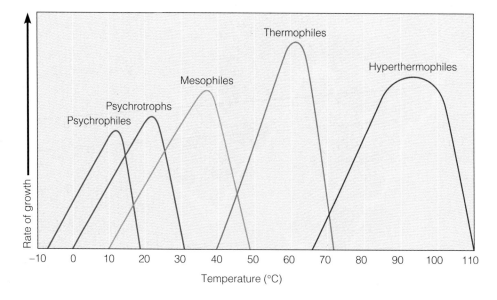

FIGURE 6.1 Typical growth rates of different types of microorganisms in response to temperature. Optimum growth (fastest reproduction) is represented by the peak of the curve. Notice that the reproductive rate drops off very quickly at temperatures only a little above the optimum. At either extreme of the temperature range, the reproductive rate is much lower than the rate at the optimum temperature.

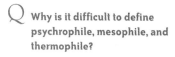 **Why is it difficult to define psychrophile, mesophile, and thermophile?**

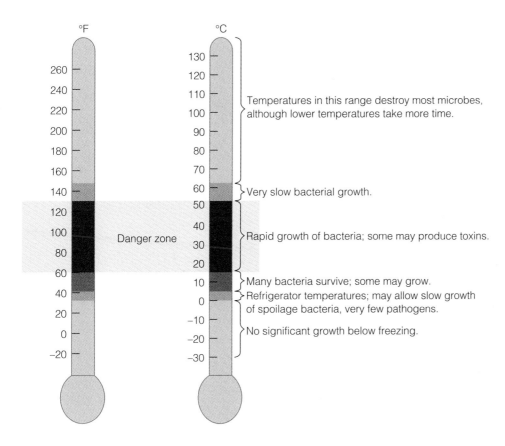

FIGURE 6.2 Food spoilage temperatures. Low temperatures decrease microbial reproduction rates, which is the basic principle of refrigeration. There are always some exceptions to the temperature responses shown here; for example, certain bacteria grow well at temperatures that would kill most bacteria, and a few bacteria can actually grow at temperatures well below freezing.

Q **Which bacterium would theoretically be the more likely to grow at refrigerator temperatures, a human intestinal pathogen or a soil-borne plant pathogen?**

food microbiologists, for this group of spoilage microorganisms. Many environmental microbiologists prefer to call them *moderate psychrophiles* or *facultative psychrophiles* and are dissatisfied with all current attempts to group psychrophilic organisms.

Refrigeration is the most common method of preserving household food supplies. It is based on the principle that microbial reproductive rates decrease at low temperatures. Although microbes usually survive even subfreezing temperatures (they might become entirely dormant), they gradually decline in number. Some species decline faster than others. Psychrotrophs actually do not grow well at low temperatures, except in comparison with other organisms; given time, however, they are able to slowly degrade food. Such spoilage might take the form of mold mycelium, slime on food surfaces, or off-tastes or off-colors in foods. The temperature inside a properly set refrigerator will greatly slow the growth of most spoilage organisms and will entirely prevent the growth of all but a few pathogenic bacteria. Figure 6.2 illustrates the importance of low temperatures for preventing the growth of spoilage and disease organisms. When large amounts of food must be refrigerated, it is important to keep in mind the slow cooling rate of a large quantity of warm food (Figure 6.3).

Mesophiles, with an optimum growth temperature of 25 to 40°C, are the most common type of microbe. Organisms that have adapted to live in the bodies of animals usually

have an optimum temperature close to that of their hosts. The optimum temperature for many pathogenic bacteria is about 37°C, and incubators for clinical cultures are usually set at about this temperature. The mesophiles include most of the common spoilage and disease organisms.

Thermophiles are microorganisms capable of growth at high temperatures. Many of these organisms have an optimum growth temperature of 50 to 60°C, about the temperature of water from a hot water tap. Such temperatures can also be reached in sunlit soil and in thermal waters such as hot springs. Remarkably, many thermophiles cannot grow at temperatures below about 45°C. Endospores formed by thermophilic bacteria are unusually heat resistant and may survive the usual heat treatment given canned goods. Although elevated storage temperatures may cause surviving endospores to germinate and grow, thereby spoiling the food, these thermophilic bacteria are not considered a public health problem. Thermophiles are important in organic compost piles (see Chapter 27), in which the temperature can rise rapidly to 50 to 60°C.

Some microbes, members of the Archaea (page 4), have an optimum growth temperature of 80°C or higher. These organisms are called **hyperthermophiles,** or sometimes **extreme thermophiles.** Most of these organisms live in hot springs associated with volcanic activity; sulfur is usually important in their metabolic activity. The

FIGURE 6.3 **The effect of the amount of food on its cooling rate in a refrigerator and its chance of spoilage.** Notice that in this example the pan of rice with a depth of 5 cm (2 in) cooled through the incubation temperature range of the *Bacillus cereus* in about 1 hour, whereas the pan of rice with a depth of 15 cm (6 in) remained in this temperature range for about 5 hours.

Q **Given a shallow pan and a deep pot with the same volume, which would cool the fastest?**

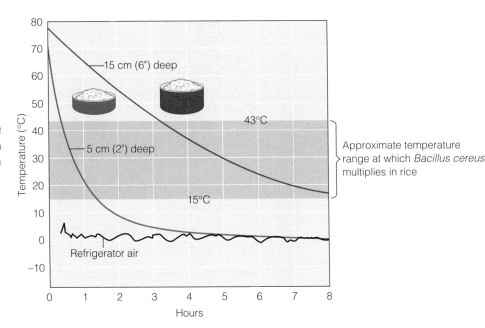

pH

Recall from Chapter 2 (page 36) that pH refers to the acidity or alkalinity of a solution. Most bacteria grow best in a narrow pH range near neutrality, between pH 6.5 and 7.5. Very few bacteria grow at an acidic pH below about pH 4. This is why a number of foods, such as sauerkraut, pickles, and many cheeses, are preserved from spoilage by acids produced by bacterial fermentation. Nonetheless, some bacteria, called **acidophiles,** are remarkably tolerant of acidity. One type of chemoautotrophic bacteria, which is found in the drainage water from coal mines and oxidizes sulfur to form sulfuric acid, can survive at a pH value of 1 (see Chapter 28). Molds and yeasts will grow over a greater pH range than bacteria will, but the optimum pH of molds and yeasts is generally below that of bacteria, usually about pH 5 to 6. Alkalinity also inhibits microbial growth but is rarely used to preserve foods.

When bacteria are cultured in the laboratory, they often produce acids that eventually interfere with their own growth. To neutralize the acids and maintain the proper pH, chemical buffers are included in the growth medium. The peptones and amino acids in some media act as buffers, and many media also contain phosphate salts. Phosphate salts have the advantage of exhibiting their buffering effect in the pH growth range of most bacteria. They are also nontoxic; in fact, they provide phosphorus, an essential nutrient.

OSMOTIC PRESSURE

Microorganisms obtain almost all their nutrients in solution from the surrounding water. Thus, they require water for growth and are made up of 80 to 90% water. High osmotic pressures have the effect of removing necessary water from a cell. When a microbial cell is in a solution that has a higher concentration of solutes than in the cell (the environment is *hypertonic* to the cell), the cellular water passes out through the plasma membrane to the high solute concentration. (See the discussion of osmosis in Chapter 4, pages 93–94, and review Figure 4.18 for the three types of solution environments a cell may encounter.) This osmotic loss of water causes **plasmolysis,** or shrinkage of the cell's cytoplasm (Figure 6.4).

The importance of this phenomenon is that the growth of the cell is inhibited as the plasma membrane pulls away from the cell wall. Thus, the addition of salts (or other solutes) to a solution, and the resulting increase in osmotic pressure, can be used to preserve foods. Salted fish, honey, and sweetened condensed milk are preserved largely by this mechanism; the high salt or sugar concentrations draw water out of any microbial cells that are present and thus prevent their growth. These effects of osmotic pressure are roughly related to the *number* of dissolved molecules and ions in a volume of solution.

Some organisms, called **extreme halophiles,** have adapted so well to high salt concentrations that they actually require them for growth. In this case, they may be termed **obligate halophiles.** Organisms from such saline waters as the Dead Sea often require nearly 30% salt, and the inoculating loop (a device for handling bacteria in the laboratory) used to transfer them must first be dipped into

known record for bacterial growth and replication at high temperatures is about 121°C near deep-sea hydrothermal vent (see the box on page 164). The immense pressure in the ocean depths prevents water from boiling even at temperatures well above 100°C.

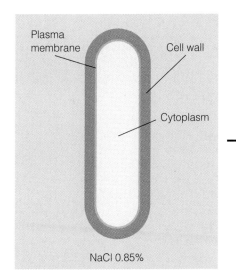

Plasma membrane

Cell wall

Cytoplasm

NaCl 0.85%

(a) Normal cell in isotonic solution. Under these conditions, the osmotic pressure in the cell is equivalent to a solute concentration of 0.85% sodium chloride (NaCl).

H_2O

Plasma membrane

Cytoplasm

NaCl 10%

(b) Plasmolyzed cell in hypertonic solution. If the concentration of solutes such as NaCl is higher in the surrounding medium than in the cell (the environment is hypertonic), water tends to leave the cell. Growth of the cell is inhibited.

FIGURE 6.4 Plasmolysis.

Q **Name a food preserved by high osmotic pressure.**

a saturated salt solution. More common are **facultative halophiles,** which do not require high salt concentrations but are able to grow at salt concentrations up to 2%, a concentration that inhibits the growth of many other organisms. A few species of facultative halophiles can even tolerate 15% salt.

Most microorganisms, however, must be grown in a medium that is nearly all water. For example, the concentration of agar (a complex polysaccharide isolated from marine algae) used to solidify microbial growth media is usually about 1.5%. If markedly higher concentrations are used, the increased osmotic pressure can inhibit the growth of some bacteria.

If the osmotic pressure is unusually low (the environment is *hypotonic*)—such as in distilled water, for example—water tends to enter the cell rather than leave it. Some microbes that have a relatively weak cell wall may be lysed by such treatment.

CHEMICAL REQUIREMENTS

LEARNING OBJECTIVES

- Provide a use for each of the four elements (carbon, nitrogen, sulfur, and phosphorus) needed in large amounts for microbial growth.

- Explain how microbes are classified on the basis of oxygen requirements.

- Identify ways in which aerobes avoid damage by toxic forms of oxygen.

CARBON

Besides water, one of the most important requirements for microbial growth is carbon. Carbon is the structural backbone of living matter; it is needed for all the organic compounds that make up a living cell. Half the dry weight of a typical bacterial cell is carbon. Chemoheterotrophs get most of their carbon from the source of their energy—organic materials such as proteins, carbohydrates, and lipids. Chemoautotrophs and photoautotrophs derive their carbon from carbon dioxide.

NITROGEN, SULFUR, AND PHOSPHORUS

In addition to carbon, other elements are needed by microorganisms for the synthesis of cellular material. For example, protein synthesis requires considerable amounts of nitrogen as well as some sulfur. The syntheses of DNA and RNA also require nitrogen and some phosphorus, as does the synthesis of ATP, the molecule so important for the storage and transfer of chemical energy within the cell. Nitrogen makes up about 14% of the dry weight of a bacterial cell, and sulfur and phosphorus together constitute about another 4%.

Organisms use nitrogen primarily to form the amino group of the amino acids of proteins. Many bacteria meet this requirement by decomposing protein-containing material and reincorporating the amino acids into newly synthesized proteins and other nitrogen-containing compounds. Other bacteria use nitrogen from ammonium ions (NH_4^+), which are already in the reduced form and are usually found in organic cellular material. Still other bacteria are able to

APPLICATIONS OF MICROBIOLOGY

STUDYING HYDROTHERMAL BACTERIA

Until humans explored the deep ocean floor, scientists believed that only a few forms of life could survive in that high-pressure, completely dark, oxygen-poor environment. Then, in 1977, the first manned vehicle capable of penetrating to the bottom of the deepest oceans carried two scientists 2600 meters below the surface at the Galápagos Rift, about 350 km northeast of the Galápagos Islands. There, amid the vast expanse of barren basalt rocks, the scientists found unexpectedly rich oases of life, including mollusks, crustaceans, and worms (see the photograph). How do such creatures survive in these harsh conditions? Many bacterial samples have been taken to the surface and are being studied to answer this question.

Ecosystem of the hydrothermal vents

Life at the surface of the world's oceans depends on photosynthetic organisms, such as bacteria and algae, which harness the sun's energy to fix carbon dioxide (CO_2) to make carbohydrates. At the deep ocean floor, where no light penetrates, photosynthesis is not possible. The scientists found that the primary producers at the ocean floor are chemoautotrophic bacteria. Using chemical energy from hydrogen sulfide (H_2S) as a source of energy to fix CO_2, the chemoautotrophs create an environment that supports higher life forms.

Hydrothermal vents in the seafloor supply the H_2S and CO_2. As superheated water from within the Earth rises through fractures in the Earth's crust called vents, it reacts with surrounding rock and dissolves metal ions, sulfides, and CO_2. The ecosystems of these vents depend on an abundance of sulfur compounds in the hot water. The concentration of sulfide (S^{2-}) is three times greater than the concentration of molecular oxygen in and around the vents. Such high sulfide concentrations are toxic to many organisms but not to the creatures that inhabit this exotic environment.

Mats of bacteria grow along the sides of the vents, where temperatures exceed 100°C. These are the highest temperatures that any organism is known to tolerate. Above the vent, where temperatures are about 30°C, the concentration of bacteria is about four times greater than that in water farther from the vents, and the growth rate of bacteria is equal to that found in productive, sunlit coastal waters. The bacteria in and around the vents create an environment in which bacteria are the producers supporting consumers (such as clams, mussels, and tubeworms) and decomposers.

Biotechnological benefits from hydrothermal vents

Researchers at Oak Ridge National Laboratory in Tennessee have identified two archaea living near deep-sea vents that hold promise for a renewable energy source. *Thermoplasma acidophilus* and *Pyrococcus furiosus* can produce the fuel, hydrogen gas, and an extracellular polysaccharide from glucose. An additional benefit is that the polysaccharide is produced in large quantities and may have industrial applications—for example, as a thickener in foods or other products.

DNA polymerases (enzymes that synthesize DNA) isolated from two archaea living near deep-sea vents are being used in the polymerase chain reaction (PCR), a technique for making many copies of DNA. In PCR, single-stranded DNA is made by heating a chromosome fragment to 98°C and cooling it so that DNA polymerase can copy each strand. DNA polymerase from *Thermococcus litoralis*, called Vent $_R$, and from *Pyrococcus*, called Deep Vent$_R$, are not denatured at 98°C. These enzymes can be used in automatic thermalcyclers to repeat the heating and cooling cycles, allowing many copies of DNA to be made easily and quickly. Vent$_R$ and Deep Vent$_R$ add bases to DNA at a rate of about 1000 bases per minute.

derive nitrogen from nitrates (compounds that dissociate to give the nitrate ion, NO_3^-, in solution).

Some important bacteria, including many of the photosynthesizing cyanobacteria (page 118), use gaseous nitrogen (N_2) directly from the atmosphere. This process is called **nitrogen fixation.** Some organisms that can use this method are free-living, mostly in the soil, but others live cooperatively in **symbiosis** with the roots of legumes such as clover, soybeans, alfalfa, beans, and peas. The nitrogen fixed in the symbiosis is used by both the plant and the bacterium (see Chapter 27).

Sulfur is used to synthesize sulfur-containing amino acids and vitamins such as thiamine and biotin. The box above describes an unusual ecosystem based on a supply of hydrogen sulfide (H_2S) that would be toxic to most organisms. Important natural sources of sulfur include the sulfate ion (SO_4^{2-}), hydrogen sulfide, and the sulfur-containing amino acids.

Phosphorus is essential for the synthesis of nucleic acids and the phospholipids of cell membranes. Among other places, it is also found in the energy bonds of ATP. An important source of phosphorus is the phosphate ion (PO_4^{3-}). Potassium, magnesium, and calcium are also

APPLICATIONS OF MICROBIOLOGY

(continued)

Tubeworms at the Galápagos vent.

elements that microorganisms require, often as cofactors for enzymes (see Chapter 5, pages 118–119).

TRACE ELEMENTS

Microbes require very small amounts of other mineral elements, such as iron, copper, molybdenum, and zinc; these are referred to as **trace elements.** Most are essential for the functions of certain enzymes, usually as cofactors. Although these elements are sometimes added to a laboratory medium, they are usually assumed to be naturally present in tap water and other components of media. Even most distilled waters contain adequate amounts, but tap water is sometimes specified to ensure that these trace minerals will be present in culture media.

OXYGEN

We are accustomed to thinking of molecular oxygen (O_2) as a necessity of life, but it is actually in a sense a poisonous gas. Very little molecular oxygen existed in the atmosphere during most of Earth's history—in fact, it is possible that life could not have arisen had oxygen been present. However, many current forms of life have metabolic systems that require

TABLE 6.1	The Effect of Oxygen on the Growth of Various Types of Bacteria				
	a. Obligate Aerobes	b. Facultative Anaerobes	c. Obligate Anaerobes	d. Aerotolerant Anaerobes	e. Micro-aerophiles
Effect of Oxygen on Growth	Only aerobic growth; oxygen required.	Both aerobic and anaerobic growth; greater growth in presence of oxygen.	Only anaerobic growth; ceases in presence of oxygen.	Only anaerobic growth; but continues in presence of oxygen.	Only aerobic growth; oxygen required in low concentration.
Bacterial Growth in Tube of Solid Growth Medium					
Explanation of Growth Patterns	Growth occurs only where high concentrations of oxygen have diffused into the medium.	Growth is best where most oxygen is present, but occurs throughout tube.	Growth occurs only where there is no oxygen.	Growth occurs evenly; oxygen has no effect.	Growth occurs only where a low concentration of oxygen has diffused into medium.
Explanation of Oxygen's Effects	Presence of enzymes catalase and superoxide dismutase (SOD) allows toxic forms of oxygen to be neutralized; can use oyygen.	Presence of enzymes catalase and SOD allows toxic forms of oxygen to be neutralized; can use oxygen.	Lacks enzymes to neutralize harmful forms of oxygen; cannot tolerate oxygen.	Presence of one enzyme, SOD, allows harmful forms of oxygen to be partially neutralized; tolerates oxygen.	Produce lethal amounts of toxic forms of oxygen if exposed to normal atmospheric oxygen.

oxygen for aerobic respiration. As we have seen, hydrogen atoms that have been stripped from organic compounds combine with oxygen to form water, as shown in Figure 5.14 (page 131). This process yields a great deal of energy while neutralizing a potentially toxic gas—a very neat solution, all in all.

Microbes that use molecular oxygen (aerobes) produce more energy from nutrients than microbes that do not use oxygen (anaerobes). Organisms that require oxygen to live are called **obligate aerobes** (Table 6.1a).

Obligate aerobes are at a disadvantage because oxygen is poorly soluble in the water of their environment. Therefore, many of the aerobic bacteria have developed, or retained, the ability to continue growing in the absence of oxygen. Such organisms are called **facultative anaerobes** (Table 6.1b). In other words, facultative anaerobes can use oxygen when it is present but are able to continue growth by using fermentation or anaerobic respiration when oxygen is not available. However, their efficiency in producing

energy decreases in the absence of oxygen. Examples of facultative anaerobes are the familiar *Escherichia coli* that is found in the human intestinal tract, and many yeasts. Recall from the discussion of anaerobic respiration in Chapter 5 (page 134) that many microbes are able to substitute other electron acceptors, such as nitrate ions, for oxygen.

Obligate anaerobes (Table 6.1c) are bacteria that are unable to use molecular oxygen for energy-yielding reactions. In fact, most are harmed by it. The genus *Clostridium* (klôs-tri′dē-um), which contains the species that cause tetanus and botulism, is the most familiar example. These bacteria do use oxygen atoms present in cellular materials; the atoms are usually obtained from water.

Understanding how organisms can be harmed by oxygen requires a brief discussion of the toxic forms of oxygen:

1. **Singlet oxygen** ($^1O_2^-$) is normal molecular oxygen (O_2) that has been boosted into a higher-energy state and is extremely reactive.

2. **Superoxide free radicals** (O_2^-) are formed in small amounts during the normal respiration of organisms that use oxygen as a final electron acceptor, forming water. In the presence of oxygen, obligate anaerobes also appear to form some superoxide free radicals, which are so toxic to cellular components that all organisms attempting to grow in atmospheric oxygen must produce an enzyme, **superoxide dismutase (SOD)**, to neutralize them. Their toxicity is caused by their great instability, which leads them to steal an electron from a neighboring molecule, which in turn becomes a radical and steals an electron, and so on. Aerobic bacteria, facultative anaerobes growing aerobically, and aerotolerant anaerobes (discussed shortly) produce SOD, with which they convert the superoxide free radical into molecular oxygen (O_2) and hydrogen peroxide (H_2O_2):

$$O_2^- + O_2^- + 2\,H^+ \longrightarrow H_2O_2 + O_2$$

3. The hydrogen peroxide produced in this reaction contains the **peroxide anion** O_2^{2-} and is also toxic. In Chapter 7 (page 205) we will encounter it as the active principle in the antimicrobial agents hydrogen peroxide and benzoyl peroxide. Because the hydrogen peroxide produced during normal aerobic respiration is toxic, microbes have developed enzymes to neutralize it. The most familiar of these is **catalase,** which converts it into water and oxygen:

$$2\,H_2O_2 \longrightarrow 2\,H_2O + O_2$$

Catalase is easily detected by its action on hydrogen peroxide. When a drop of hydrogen peroxide is added to a colony of bacterial cells producing catalase, oxygen bubbles are released. Anyone who has put hydrogen peroxide on a wound will recognize that cells in human tissue also contain catalase. The other enzyme that breaks down hydrogen peroxide is **peroxidase,** which differs from catalase in that its reaction does not produce oxygen:

$$H_2O_2 + 2\,H^+ \longrightarrow 2\,H_2O$$

Another important form of reactive oxygen, **ozone (O_3)**, is also discussed on page 205.

4. The **hydroxyl radical** ($OH\cdot$) is another intermediate form of oxygen and probably the most reactive. It is formed in the cellular cytoplasm by ionizing radiation. Most aerobic respiration produces traces of hydroxyl radicals, but they are transient.

These toxic forms of oxygen are an essential component of one of the body's most important defenses against pathogens, phagocytosis (see page 483 and Figure 16.7). In the phagolysosome of the phagocytic cell, ingested pathogens are killed by exposure to singlet oxygen, superoxide free radicals, peroxide anions of hydrogen peroxide, and hydroxyl radicals and other oxidative compounds.

Obligate anaerobes usually produce neither superoxide dismutase nor catalase. Because aerobic conditions probably lead to an accumulation of superoxide free radicals in their cytoplasm, obligate anaerobes are extremely sensitive to oxygen.

Aerotolerant anaerobes (Table 6.1d) cannot use oxygen for growth, but they tolerate it fairly well. On the surface of a solid medium, they will grow without the use of special techniques (discussed later) required for obligate anaerobes. Many of the aerotolerant bacteria characteristically ferment carbohydrates to lactic acid. As lactic acid accumulates, it inhibits the growth of aerobic competitors and establishes a favorable ecological niche for lactic acid producers. A common example of lactic acid–producing aerotolerant anaerobes is the lactobacilli used in the production of many acidic fermented foods, such as pickles and cheese. In the laboratory, they are handled and grown much like any other bacteria, but they make no use of the oxygen in the air. These bacteria can tolerate oxygen because they possess SOD or an equivalent system that neutralizes the toxic forms of oxygen previously discussed.

A few bacteria are **microaerophiles** (Table 6.1e). They are aerobic; they do require oxygen. However, they grow only in oxygen concentrations lower than those in air. In a test tube of solid nutrient medium, they grow only at a depth where small amounts of oxygen have diffused into the medium; they do not grow near the oxygen-rich surface or below the narrow zone of adequate oxygen. This limited tolerance is probably due to their sensitivity to superoxide free radicals and peroxides, which they produce in lethal concentrations under oxygen-rich conditions.

ORGANIC GROWTH FACTORS

Essential organic compounds an organism is unable to synthesize are known as **organic growth factors;** they must be directly obtained from the environment. One group of organic growth factors for humans is vitamins. Most vitamins function as coenzymes, the organic cofactors required by certain enzymes in order to function. Many bacteria can synthesize all their own vitamins and are not dependent on outside sources. However, some bacteria lack the enzymes needed for the synthesis of certain vitamins, and for them those vitamins are organic growth factors. Other organic growth factors required by some bacteria are amino acids, purines, and pyrimidines.

CULTURE MEDIA

LEARNING OBJECTIVES

• Distinguish between chemically defined and complex media.

• Justify the use of each of the following: anaerobic techniques, living host cells, candle jars, selective and differential media, enrichment medium.

A nutrient material prepared for the growth of microorganisms in a laboratory is called a **culture medium.** Some bacteria can grow well on just about any culture medium; others require special media, and still others cannot grow on any nonliving medium yet developed. When microbes are introduced into a culture medium to initiate growth, they are called an **inoculum.** The microbes that grow and multiply in or on a culture medium are referred to as a **culture.**

Suppose we want to grow a culture of a certain microorganism, perhaps the microbes from a particular clinical specimen. What criteria must the culture medium meet? First, it must contain the right nutrients for the specific microorganism we want to grow. It should also contain sufficient moisture, a properly adjusted pH, and a suitable level of oxygen, perhaps none at all. The medium must initially be **sterile**—that is, it must initially contain no living microorganisms—so that the culture will contain only the microbes (and their offspring) we add to the medium. Finally, the growing culture should be incubated at the proper temperature.

A wide variety of media are available for the growth of microorganisms in the laboratory. Most of these media, which are available from commercial sources, have premixed components and require only the addition of water and then sterilization. Media are constantly being developed or revised for use in the isolation and identification of bacteria that are of interest to researchers in such fields as food, water, and clinical microbiology.

When it is desirable to grow bacteria on a solid medium, a solidifying agent such as agar is added to the medium. A complex polysaccharide derived from a marine alga, **agar** has long been used as a thickener in foods such as jellies and ice cream.

Agar has some very important properties that make it valuable to microbiology, and no satisfactory substitute has ever been found. Few microbes can degrade agar, so it remains solid. Also, agar liquefies at about 100°C (the boiling point of water) and at sea level remains liquid until the temperature drops to about 40°C. For laboratory use, agar is held in water baths at about 50°C. At this temperature, it does not injure most bacteria when it is poured over them (as shown in Figure 6.16a). Once the

TABLE 6.2	A Chemically Defined Medium for Growing a Typical Chemoheterotroph, Such as *Escherichia coli*
Constituent	Amount
Glucose	5.0 g
Ammonium phosphate, monobasic ($NH_4H_2PO_4$)	1.0 g
Sodium chloride (NaCl)	5.0 g
Magnesium sulfate ($MgSO_4 \cdot 7H_2O$)	0.2 g
Potassium phosphate, dibasic (K_2HPO_4)	1.0 g
Water	1 liter

agar has solidified, it can be incubated at temperatures approaching 100°C before it again liquefies; this property is particularly useful when thermophilic bacteria are being grown.

Agar media are usually contained in test tubes or *Petri dishes*. The test tubes are called *slants* when they are allowed to solidify with the tube held at an angle so that a large surface area for growth is available. When the agar solidifies in a vertical tube, it is called a *deep*. Petri dishes, named for their inventor, are shallow dishes with a lid that nests over the bottom to prevent contamination; when filled, they are called *Petri* (or culture) *plates*.

CHEMICALLY DEFINED MEDIA

To support microbial growth, a medium must provide an energy source, as well as sources of carbon, nitrogen, sulfur, phosphorus, and any organic growth factors the organism is unable to synthesize. A **chemically defined medium** is one whose exact chemical composition is known. For a chemoheterotroph, the chemically defined medium must contain organic growth factors that serve as a source of carbon and energy. For example, as shown in Table 6.2, glucose is included in the medium for growing the chemoheterotroph *E. coli*.

As Table 6.3 shows, many organic growth factors must be provided in the chemically defined medium used to cultivate a species of *Neisseria* (page 320). Organisms that require many growth factors are described as "fastidious." Organisms of this type, such as *Lactobacillus* (page 332), are sometimes used in tests that determine the concentration of a particular vitamin in a substance. To perform such a *microbiological assay*, a growth medium is prepared that contains all the growth requirements of the bacterium except the vitamin being assayed. Then the medium, test

TABLE 6.3	A Chemically Defined Medium for Growing a Fastidious Chemoheterotrophic Bacterium, Such as *Neisseria gonorrhoeae*		

Constituent	Amount	Constituent	Amount
Carbon and energy sources		**Amino acids**	
Glucose	9.1 g	Cysteine	1.5 g
Starch	9.1 g	Arginine, proline (each)	0.3 g
Sodium acetate	1.8 g	Glutamic acid, methionine (each)	0.2 g
Sodium citrate	1.4 g	Asparagine, isoleucine, serine (each)	0.2 g
Oxaloacetate	0.3 g	Cystine	0.06 g
Salts		**Organic growth factors**	
Potassium phosphate, dibasic (K_2HPO_4)	12.7 g	Calcium pantothenate	0.02 g
Sodium chloride (NaCl)	6.4 g	Thiamine	0.02 g
Potassium phosphate, monobasic (KH_2PO_4)	5.5 g	Nicotinamide adenine dinucleotide	0.01 g
Sodium bicarbonate ($NaHCO_3$)	1.2 g	Uracil	0.006 g
Potassium sulfate (K_2SO_4)	1.1 g	Biotin	0.005 g
Sodium sulfate (Na_2SO_4)	0.9 g	Hypoxanthine	0.003 g
Magnesium chloride ($MgCl_2$)	0.5 g	**Reducing agent**	
Ammonium chloride (NH_4Cl)	0.4 g	Sodium thioglycolate	0.00003 g
Potassium chloride (KCl)	0.4 g	**Water**	1 liter
Calcium chloride ($CaCl_2$)	0.006 g		
Ferric nitrate [$Fe(NO_3)_3$]	0.006 g		

SOURCE: R. M. Atlas, *Handbook of Microbiological Media*, Ann Arbor, MI: CRC Press, 1993.

substance, and bacterium are combined, and the growth of bacteria is measured. This bacterial growth, which is reflected by the amount of lactic acid produced, will be proportional to the amount of vitamin in the test substance. The more lactic acid, the more the *Lactobacillus* cells have been able to grow, so the more vitamin is present.

COMPLEX MEDIA

Chemically defined media are usually reserved for laboratory experimental work or for the growth of autotrophic bacteria. Most heterotrophic bacteria and fungi, such as you would work with in an introductory lab course, are routinely grown on **complex media,** made up of nutrients including extracts from yeasts, meat, or plants, or digests of proteins from these and other sources. The exact chemical composition varies slightly from batch to batch. Table 6.4 shows one widely used recipe.

In complex media, the energy, carbon, nitrogen, and sulfur requirements of the growing microorganisms are

TABLE 6.4	Composition of Nutrient Agar, a Complex Medium for the Growth of Heterotrophic Bacteria

Constituent	Amount
Peptone (partially digested protein)	5.0 g
Beef extract	3.0 g
Sodium chloride	8.0 g
Agar	15.0 g
Water	1 liter

primarily provided by protein. Protein is a large, relatively insoluble molecule that a minority of microorganisms can utilize directly, but a partial digestion by acids or enzymes reduces protein to shorter chains of amino acids called *peptones.* These small, soluble fragments can be digested by most bacteria.

Vitamins and other organic growth factors are provided by meat extracts or yeast extracts. The soluble vitamins and minerals from the meats or yeasts are dissolved in the extracting water, which is then evaporated so that these factors are concentrated. (These extracts also supplement the organic nitrogen and carbon compounds.) Yeast extracts are particularly rich in the B vitamins. If a complex medium is in liquid form, it is called **nutrient broth.** When agar is added, it is called **nutrient agar.** (This terminology can be confusing; just remember that agar itself is not a nutrient.)

ANAEROBIC GROWTH MEDIA AND METHODS

The cultivation of anaerobic bacteria poses a special problem. Because anaerobes might be killed by exposure to oxygen, special media called **reducing media** must be used. These media contain ingredients, such as sodium thioglycolate, that chemically combine with dissolved oxygen and deplete the oxygen in the culture medium. To routinely grow and maintain pure cultures of obligate anaerobes, microbiologists use reducing media stored in ordinary, tightly capped test tubes. These media are heated shortly before use, to drive off absorbed oxygen.

When the culture must be grown in Petri plates to observe individual colonies, special anaerobic jars are used (Figure 6.5). The culture plates are placed in the jar, and oxygen is removed by the following process: A packet of chemicals (sodium bicarbonate and sodium borohydride) in the jar is moistened with a few milliliters of water, and the jar is sealed. Hydrogen and carbon dioxide are produced by the reaction of the chemicals with the water. A palladium catalyst in the jar combines the oxygen in the jar with the hydrogen produced by the chemical reaction, and water is formed. As a result, the oxygen quickly disappears. Moreover, the carbon dioxide that is produced aids the growth of many anaerobic bacteria.

A relatively new technique to provide an anaerobic environment makes use of an enzyme, oxyrase, that reduces oxygen to water. Oxyrase is a respiratory enzyme derived from the plasma membranes of certain bacteria. When it is added to growth media, it transforms the Petri plate, OxyPlate, into a self-contained anaerobic chamber. This method, which can often avoid the need for more cumbersome apparatus, is now increasingly used in clinical laboratories.

Researchers regularly working with anaerobes use transparent anaerobic chambers equipped with air locks and filled with inert gases (Figure 6.6). Technicians can manipulate the equipment by inserting their hands into airtight rubber gloves called glove ports, which are fitted to the wall of the chamber.

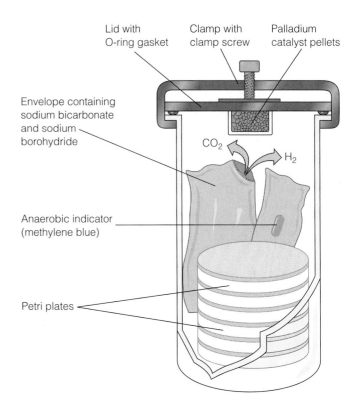

FIGURE 6.5 A jar for cultivating anaerobic bacteria on Petri plates. When water is mixed with the chemical packet containing sodium bicarbonate and sodium borohydride, hydrogen and carbon dioxide are generated. Reacting on the surface of a palladium catalyst in a screened reaction chamber, which may also be incorporated into the chemical packet, the hydrogen and atmospheric oxygen in the jar combine to form water. The oxygen is thus removed. Also in the jar is an anaerobic indicator, containing methylene blue, which is blue when oxidized (as shown here) and turns colorless when the oxygen is removed.

Q   **What is the technical name for bacteria that require a higher-than-atmospheric-concentration of CO_2 for growth?**

SPECIAL CULTURE TECHNIQUES

Many bacteria have never been successfully grown on artificial laboratory media. *Mycobacterium leprae,* the leprosy bacillus, is now usually grown in armadillos, which have a relatively low body temperature that matches the requirements of the microbe. Another example is the syphilis spirochete, although certain nonpathogenic strains of this microbe have been grown on laboratory media. With few exceptions, the obligate intracellular bacteria, such as the rickettsias and the chlamydias, do not grow on artificial media. Like viruses, they can reproduce only in a living host cell. See the discussion of cell culture, page 395.

Many clinical laboratories have special *carbon dioxide incubators* in which to grow aerobic bacteria that require concentrations of CO_2 higher or lower than that found in the atmosphere. Desired CO_2 levels are maintained by

FIGURE 6.6 An anaerobic chamber. The technician is pipetting a bacterial suspension into a flask inside an anaerobic chamber filled with an inert, oxygen-free gas. His arms and hands are encased in glove ports. Organisms and materials enter and leave through the air-lock opening that is visible to the left.

Q In what way would an anaerobic chamber resemble the Space Laboratory orbiting in the vacuum of space?

electronic controls. High CO_2 levels are also obtained with simple *candle jars* (Figure 6.7a). Cultures are placed in a large sealed jar containing a lighted candle, which consumes oxygen. The candle stops burning when the air in the jar has a lowered concentration of oxygen (but one still adequate for the growth of aerobic bacteria). An elevated concentration of CO_2 is also present. Microbes that grow better at high CO_2 concentrations are called **capnophiles.** The low-oxygen, high-CO_2 conditions resemble those found in the intestinal tract, respiratory tract, and other body tissues where pathogenic bacteria grow.

Candle jars are still used occasionally, but more often commercially available chemical packets are used to generate carbon dioxide atmospheres in containers (Figure 6.7b). When only one or two Petri plates of cultures are to be incubated, clinical laboratory investigators often use small plastic bags with self-contained chemical gas generators that are activated by crushing the packet or moistening it with a few milliliters of water. These packets are sometimes specially designed to provide precise concentrations of carbon dioxide (usually higher than can be obtained in candle jars) and oxygen, for culturing organisms such as the microaerophilic *Campylobacter* bacteria (page 327).

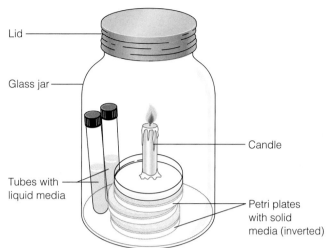

(a) Candle jar. Plates and tubes inoculated with, for example, *Neisseria meningitidis* are placed in a jar with a lighted candle, and the jar is sealed. The burning candle reduces the O_2 concentration to a point where the flame goes out. This will provide a CO_2 atmosphere of approximately 3%.

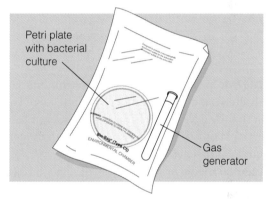

(b) CO_2-generating packet. The packet consists of a bag containing a Petri plate and a CO_2 gas generator. The gas generator is crushed to mix the chemicals it contains and start the reaction that produces CO_2. This gas reduces the O_2 concentration in the bag to about 5% and provides a CO_2 concentration of about 10%.

FIGURE 6.7 Equipment for producing CO_2-rich environments.

Q Which bacterium would be more likely to require a CO_2-rich environment: *Pseudomonas aeruginosa* or *campylobacter jejuni*?

SELECTIVE AND DIFFERENTIAL MEDIA

In clinical and public health microbiology, it is frequently necessary to detect the presence of specific microorganisms associated with disease or poor sanitation. For this task, selective and differential media are used. **Selective media** are designed to suppress the growth of unwanted bacteria and encourage the growth of the desired microbes. For example, bismuth sulfite agar is one medium used to isolate the typhoid bacterium, the gram-negative *Salmonella*

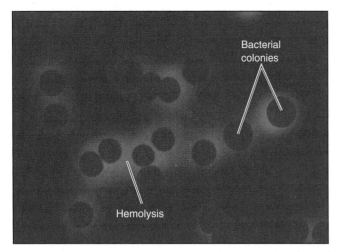

2 mm

FIGURE 6.8 Blood agar, a differential medium containing red blood cells. The bacteria have lysed the red blood cells (beta-hemolysis), causing the clear areas around the colonies.

Q **Can you think of a reason why a pathogen might also produce hemolysins?**

typhi (tī′fē), from feces. Bismuth sulfite inhibits gram-positive bacteria and most gram-negative intestinal bacteria (other than *S. typhi,*) as well. Sabouraud's dextrose agar, which has a pH of 5.6, is used to isolate fungi that outgrow most bacteria at this pH.

Differential media make it easier to distinguish colonies of the desired organism from other colonies growing on the same plate. Similarly, pure cultures of microorganisms have identifiable reactions with differential media in tubes or plates. Blood agar (which contains red blood cells) is a medium that microbiologists often use to identify bacterial species that destroy red blood cells. These species, such as *Streptococcus pyogenes* (pī-äj′en-ēz), the bacterium that causes strep throat, show a clear ring around their colonies (beta-hemolysis, page 333) where they have lysed the surrounding blood cells (Figure 6.8).

Sometimes, selective and differential characteristics are combined in a single medium. Suppose we want to isolate the common bacterium *Staphylococcus aureus,* found in the nasal passages. This organism has a tolerance for high concentrations of sodium chloride; it can also ferment the carbohydrate mannitol to form acid. Mannitol salt agar contains 7.5% sodium chloride, which will discourage the growth of competing organisms and thus *select for* (favor the growth of) *S. aureus.* This salty medium also contains a pH indicator that changes color if the mannitol in the medium is fermented to acid; the mannitol-fermenting colonies of *S. aureus* are thus *differentiated from* colonies of bacteria that do not ferment mannitol. Bacteria that grow

at the high salt concentration *and* ferment mannitol to acid can be readily identified by the color change. These are probably colonies of *S. aureus,* and their identification can be confirmed by additional tests. Figure 6.9 shows the appearance of bacterial colonies on several differential media.

ENRICHMENT CULTURE

Because bacteria present in small numbers can be missed, especially if other bacteria are present in much larger numbers, it is sometimes necessary to use an **enrichment culture.** This is often the case for soil or fecal samples. The medium (enrichment medium) for an enrichment culture is usually liquid and provides nutrients and environmental conditions that favor the growth of a particular microbe but not others. In this sense, it is also a selective medium, but it is designed to increase very small numbers of the desired type of organism to detectable levels.

Suppose we want to isolate from a soil sample a microbe that can grow on phenol and is present in much smaller numbers than other species. If the soil sample is placed in a liquid enrichment medium in which phenol is the only source of carbon and energy, microbes unable to metabolize phenol will not grow. The culture medium is allowed to incubate for a few days, and then a small amount of it is transferred into another flask of the same medium. After a series of such transfers, the surviving population will consist of bacteria capable of metabolizing phenol. The bacteria are given time to grow in the medium between transfers; this is the enrichment stage. (See the box in Chapter 28.) Any nutrients in the original inoculum are rapidly diluted out with the successive transfers. When the last dilution is streaked onto a solid medium of the same composition, only those colonies of organisms capable of using phenol should grow. A remarkable aspect of this particular technique is that phenol is normally lethal to most bacteria.

* * *

Table 6.5 summarizes the purposes of the main types of culture media.

OBTAINING PURE CULTURES

LEARNING OBJECTIVES
- Define *colony.*
- Describe how pure cultures can be isolated by using the streak plate method.

Most infectious materials, such as pus, sputum, and urine, contain several different kinds of bacteria; so do samples of soil, water, or food. If these materials are plated out onto the

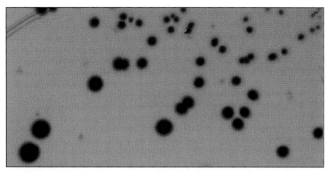

(a) *Staphylococcus aureus* on Tellurite-Glycine medium.

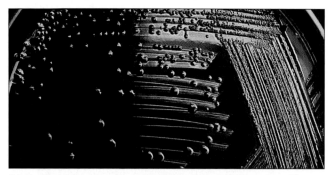

(b) *Escherichia coli* on Eosin Methylene Blue (EMB) medium. The black-centered colonies are surrounded by a characteristic metallic green sheen.

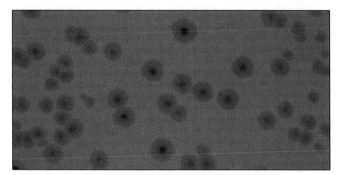

(c) *Enterobacter aerogenes* on EMB medium showing characteristic dark-centered colonies.

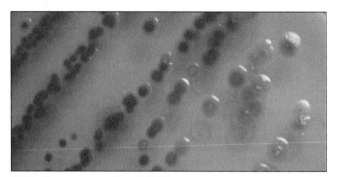

(d) On Pseudomonas Agar P (PSP) medium, *Pseudomonas aeruginosa* produces a blue-green water-soluble pigment.

FIGURE 6.9 Bacterial colonies on several differential media.

Q **Which media pictured here are both selective and differential?**

TABLE 6.5	Culture Media
Type	Purpose
Chemically defined	Growth of chemoautotrophs and photoautotrophs; microbiological assays.
Complex	Growth of most chemoheterotrophic organisms.
Reducing	Growth of obligate anaerobes.
Selective	Suppression of unwanted microbes; encouraging desired microbes.
Differential	Differentiation of colonies of desired microbes from others.
Enrichment	Similar to selective media but designed to increase numbers of desired microbes to detectable levels.

surface of a solid medium, colonies will form that are exact copies of the original organism. A visible **colony** theoretically arises from a single spore or vegetative cell or from a group of the same microorganisms attached to one another in clumps or chains. Microbial colonies often have a distinctive appearance that distinguishes one microbe from another (see Figure 6.9). The bacteria must be distributed widely enough so that the colonies are visibly separated from each other.

Most bacteriological work requires pure cultures, or clones, of bacteria. The isolation method most commonly used to get pure cultures is the **streak plate method** (Figure 6.10). A sterile inoculating loop is dipped into a mixed culture that contains more than one type of microbe and is streaked in a pattern over the surface of the nutrient medium. As the pattern is traced, bacteria are rubbed off the loop onto the medium. The last cells to be rubbed off the loop are far enough apart to grow into isolated colonies. These colonies can be picked up with an inoculating loop and transferred to a test tube of nutrient medium to form a pure culture containing only one type of bacterium.

The streak plate method works well when the organism to be isolated is present in large numbers relative to the

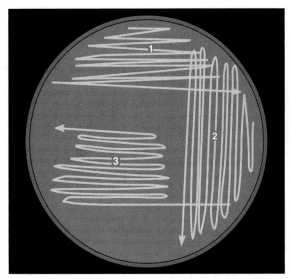

(a) The direction of streaking is indicated by arrows. Streak series 1 is made from the original bacterial culture. The inoculating loop is sterilized following each streak series. In series 2 and 3, the loop picks up bacteria from the previous series, diluting the number of cells each time. There are numerous variants of such patterns.

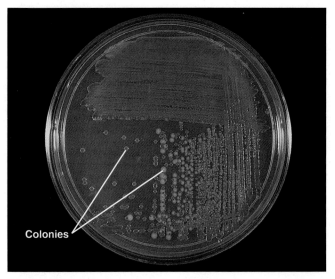

(b) In series 3 of this example, notice that well-isolated colonies of bacteria of two different types, red and yellow, have been obtained.

FIGURE 6.10 The streak plate method for isolating pure bacterial cultures.

 Is a colony formed as a result of streaking a plate always derived from a single bacterium?

total population. However, when the microbe to be isolated is present only in very small numbers, its numbers must be greatly increased by selective enrichment before it can be isolated with the streak plate method.

PRESERVING BACTERIAL CULTURES

LEARNING OBJECTIVE
• Explain how microorganisms are preserved by deep-freezing and lyophilization (freeze-drying).

Refrigeration can be used for the short-term storage of bacterial cultures. Two common methods of preserving microbial cultures for long periods are deep-freezing and lyophilization. **Deep-freezing** is a process in which a pure culture of microbes is placed in a suspending liquid and quick-frozen at temperatures ranging from $-50°$ to $-95°C$. The culture can usually be thawed and cultured even several years later. During **lyophilization (freeze-drying),** a suspension of microbes is quickly frozen at temperatures ranging from $-54°$ to $-72°C$, and the water is removed by a high vacuum (sublimation). While under vacuum, the container is sealed by melting the glass with a high-temperature torch. The remaining powderlike residue that contains the surviving microbes can be stored for years.

The organisms can be revived at any time by hydration with a suitable liquid nutrient medium.

THE GROWTH OF BACTERIAL CULTURES

LEARNING OBJECTIVE
• Define *bacterial growth,* including *binary fission.*

Being able to represent graphically the enormous populations resulting from the growth of bacterial cultures is an essential part of microbiology. It is also necessary to be able to determine microbial numbers, either directly, by counting, or indirectly, by measuring their metabolic activity.

BACTERIAL DIVISION

As we mentioned at the beginning of the chapter, bacterial growth refers to an increase in bacterial numbers, not an increase in the size of the individual cells. Bacteria normally reproduce by **binary fission** (Figure 6.11).

A few bacterial species reproduce by **budding;** they form a small initial outgrowth (a bud) that enlarges until its size approaches that of the parent cell, and then it separates. Some filamentous bacteria (certain actinomycetes)

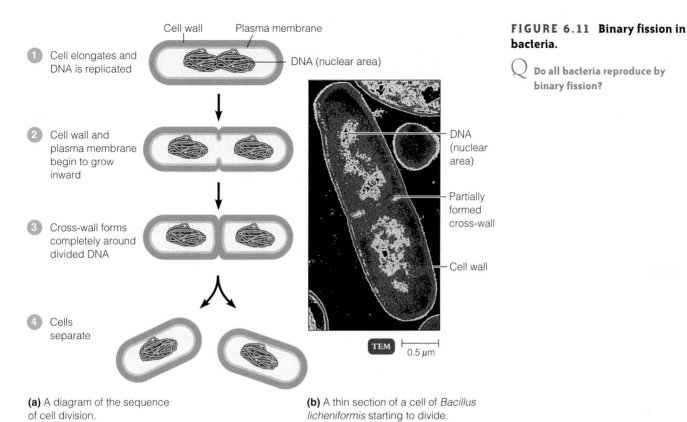

① Cell elongates and DNA is replicated

Cell wall   Plasma membrane

DNA (nuclear area)

② Cell wall and plasma membrane begin to grow inward

③ Cross-wall forms completely around divided DNA

④ Cells separate

DNA (nuclear area)

Partially formed cross-wall

Cell wall

TEM   0.5 μm

FIGURE 6.11 Binary fission in bacteria.

Q **Do all bacteria reproduce by binary fission?**

(a) A diagram of the sequence of cell division.

(b) A thin section of a cell of *Bacillus licheniformis* starting to divide.

reproduce by producing chains of conidiospores carried externally at the tips of the filaments. A few filamentous species simply fragment, and the fragments initiate the growth of new cells. ✷ **Animation: Go to The Microbiology Place Website or CD-Rom and click "Animations" to view Bacterial Growth.**

GENERATION TIME

For purposes of calculating the generation time of bacteria, we will consider only reproduction by binary fission, which is by far the most common method. As you can see in Figure 6.12, one cell's division produces two cells, two cells' divisions produce four cells, and so on. When the number of cells in each generation is expressed as a power of 2, the exponent tells the number of doublings (generations) that have occurred.

The time required for a cell to divide (and its population to double) is called the **generation time.** It varies considerably among organisms and with environmental conditions, such as temperature. Most bacteria have a generation time of 1 to 3 hours; others require more than 24 hours per generation. (The math required to calculate generation times is presented in Appendix D.) If binary fission continues unchecked, an enormous number of cells will be produced. If a doubling occurred every 20 minutes—which is the case for *E. coli* under favorable conditions—after 20 generations a single initial cell would increase to

over 1 million cells. This would require a little less than 7 hours. In 30 generations, or 10 hours, the population would be 1 billion, and in 24 hours it would be a number trailed by 21 zeros. It is difficult to graph population changes of such enormous magnitude by using arithmetic numbers. This is why logarithmic scales are generally used to graph bacterial growth. Understanding logarithmic representations of bacterial populations requires some use of mathematics and is necessary for anyone studying microbiology. (See Appendix D.)

LOGARITHMIC REPRESENTATION OF BACTERIAL POPULATIONS

To illustrate the difference between logarithmic and arithmetic graphing of bacterial populations, let's express 20 bacterial generations both logarithmically and arithmetically. In five generations (2^5), there would be 32 cells; in ten generations (2^{10}), there would be 1024 cells, and so on. (If your calculator has a y^x key and a log key, you can duplicate the numbers in the third column of Figure 6.12.)

In Figure 6.13, notice that the arithmetically plotted line (solid) does not clearly show the population changes in the early stages of the growth curve at this scale. In fact, the first ten generations do not appear to leave the baseline. Furthermore, another one or two arithmetic generations graphed to the same scale would greatly increase the height of the graph and take the line off the page.

Numbers of Cells	Numbers Expressed as a Power of 2	Visual Representation of Numbers
1	2^0	
2	2^1	
4	2^2	
8	2^3	
16	2^4	
32	2^5	

(a) Visual representation of increase in bacterial number over five generations. The number of bacteria doubles in each generation. The superscript indicates the generation, that is, 2^5 = 5 generations.

Generation Number	Number of Cells	Log$_{10}$ of Number of Cells
0	$2^0 = $ 1	0
5	$2^5 = $ 32	1.51
10	$2^{10} = $ 1,024	3.01
15	$2^{15} = $ 32,768	4.52
16	$2^{16} = $ 65,536	4.82
17	$2^{17} = $ 131,072	5.12
18	$2^{18} = $ 262,144	5.42
19	$2^{19} = $ 524,288	5.72
20	$2^{20} = $ 1,048,576	6.02

(b) Conversion of the number of cells in a population into the logarithmic expression of this number. To arrive at the numbers in the center column, use the y^x key on your calculator. Enter 2 on the calculator; press y^x; enter 5; then press the = sign. The calculator will show the number 32. Thus, the fifth-generation population of bacteria will total 32 cells. To arrive at the numbers in the right-hand column, use the log key on your calculator. Enter the number 32; then press the log key. The calculator will show, rounded off, that the log$_{10}$ of 32 is 1.51.

FIGURE 6.12 Cell division.

 If a single bacterium reproduced every 20 minutes, how many would there be in 2 hours?

The dashed line in Figure 6.13 shows how these plotting problems can be avoided by graphing the log$_{10}$ of the population numbers. The log$_{10}$ of the population is plotted at 5, 10, 15, and 20 generations. Notice that a straight line is formed and that a thousand times this population (1,000,000,000, or log$_{10}$ 9.0) could be accommodated in relatively little extra space. However, this advantage is obtained at the cost of distorting our "common sense" perception of the actual situation. We are not accustomed to thinking in logarithmic relationships, but it is necessary for a proper understanding of graphs of microbial populations.

PHASES OF GROWTH

LEARNING OBJECTIVE

• Compare the phases of microbial growth, and describe their relation to generation time.

When a few bacteria are inoculated into a liquid growth medium and the population is counted at intervals, it is possible to plot a **bacterial growth curve** that shows the growth of cells over time (Figure 6.14). There are four basic phases of growth: the lag, log, stationary, and death phases.

THE LAG PHASE

For a while, the number of cells changes very little because the cells do not immediately reproduce in a new medium. This period of little or no cell division is called the **lag phase,** and it can last for 1 hour or several days. During this time, however, the cells are not dormant. The microbial population is undergoing a period of intense metabolic activity involving, in particular, synthesis of enzymes and various molecules. (The situation is analogous to a factory being equipped to produce automobiles; there is considerable

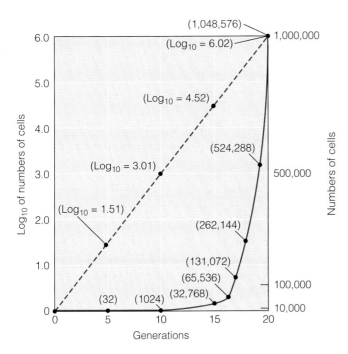

FIGURE 6.13 A growth curve for an exponentially increasing population, plotted logarithmically (dashed line) and arithmetically (solid line).

 If the arithmetic numbers (solid line) were plotted for two more generations, would it still be on the page? Does the line representing generations 1 to 10 perceptibly leave the base line?

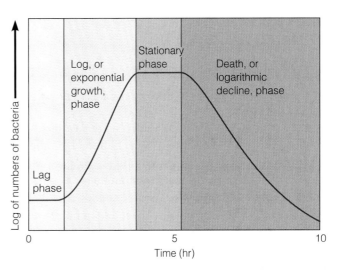

FIGURE 6.14 A bacterial growth curve, showing the four typical phases of growth.

 At what phase of bacterial growth do you think antibiotics are usually the most effective?

tooling-up activity but no immediate increase in the automobile population.)

THE LOG PHASE

Eventually, the cells begin to divide and enter a period of growth, or logarithmic increase, called the **log phase,** or **exponential growth phase.** Cellular reproduction is most active during this period, and generation time reaches a constant minimum. Because the generation time is constant, a logarithmic plot of growth during the log phase is a straight line. The log phase is the time when cells are most active metabolically and is preferred for industrial purposes where, for example, a product needs to be produced efficiently.

However, during their log phase of growth, microorganisms are particularly sensitive to adverse conditions. Radiation and many antimicrobial drugs—for example, the antibiotic penicillin—exert their effect by interfering with some important step in the growth process and are therefore most harmful to cells during this phase.

THE STATIONARY PHASE

If exponential growth continues unchecked, startlingly large numbers of cells could arise. For example, a single bacterium (at a weight of 9.5×10^{-13} g per cell) dividing

every 20 minutes for only 25.5 hours can theoretically produce a population equivalent in weight to that of an 80,000-ton aircraft carrier. In reality, this does not happen. Eventually, the growth rate slows, the number of microbial deaths balances the number of new cells, and the population stabilizes. The metabolic activities of individual surviving cells also slow at this stage. This period of equilibrium is called the **stationary phase.**

What causes exponential growth to stop is not always clear. The exhaustion of nutrients, accumulation of waste products, and harmful changes in pH may all play a role. In a specialized apparatus called a *chemostat*, a population can be kept in the exponential growth phase indefinitely by draining off spent medium and adding fresh medium. This type of *continuous culture* is used in industrial fermentations (Chapter 28).

THE DEATH PHASE

The number of deaths eventually exceeds the number of new cells formed, and the population enters the **death phase,** or **logarithmic decline phase.** This phase continues until the population is diminished to a tiny fraction of the number of cells in the previous phase, or the population dies out entirely. Many bacterial cells often undergo *involution* during this phase, meaning that their morphology changes dramatically, making them difficult to identify. Some species pass through the entire series of phases in only a few days; others retain some surviving cells almost indefinitely. Microbial death will be discussed further in Chapter 7.

DIRECT MEASUREMENT OF MICROBIAL GROWTH

LEARNING OBJECTIVE

• Explain four direct methods of measuring cell growth.

The growth of microbial populations can be measured in a number of ways. Some methods measure cell numbers; other methods measure the population's total mass, which is often directly proportional to cell numbers. Population numbers are usually recorded as the number of cells in a milliliter of liquid or in a gram of solid material. Because bacterial populations are usually very large, most methods of counting them are based on direct or indirect counts of very small samples; calculations then determine the size of the total population. Assume, for example, that a millionth of a milliliter (10^{-6} ml) of sour milk is found to contain 70 bacterial cells. Then there must be 70 times 1 million, or 70 million, cells per milliliter.

However, it is not practical to measure out a millionth of a milliliter of liquid or a millionth of a gram of food. Therefore, the procedure is done indirectly, in a series of dilutions. For example, if we add 1 ml of milk to 99 ml of water, each milliliter of this dilution now has one-hundredth as many bacteria as each milliliter of the original sample had. By making a series of such dilutions, we can readily estimate the number of bacteria in our original sample. To count microbial populations in solid foods (such as hamburger), an homogenate of one part food to nine parts water is finely ground in a food blender. Samples of this initial one-tenth dilution can then be transferred with a pipette for further dilutions or cell counts.

PLATE COUNTS

The most frequently used method of measuring bacterial populations is the **plate count.** An important advantage of this method is that it measures the number of viable cells. One disadvantage may be that it takes some time, usually 24 hours or more, for visible colonies to form. This can be a serious problem in some applications, such as quality control of milk, when it is not possible to hold a particular lot for this length of time.

Plate counts assume that each live bacterium grows and divides to produce a single colony. This is not always true because bacteria frequently grow linked in chains or as clumps (see Figure 4.1, page 79). Therefore, a colony often results, not from a single bacterium, but from short segments of a chain or from a bacterial clump. To reflect this reality, plate counts are often reported as **colony-forming units (CFU).**

When a plate count is performed, it is important that only a limited number of colonies develop in the plate.

When too many colonies are present, some cells are overcrowded and do not develop; these conditions cause inaccuracies in the count. The U.S. Food and Drug Administration convention is to count only plates with 25 to 250 colonies, but many microbiologists prefer plates with 30 to 300 colonies. To ensure that some colony counts will be within this range, the original inoculum is diluted several times in a process called **serial dilution** (Figure 6.15).

Serial Dilutions Let's say, for example, that a milk sample has 10,000 bacteria per milliliter. If 1 ml of this sample were plated out, there would theoretically be 10,000 colonies formed in the Petri plate of medium. Obviously, this would not produce a countable plate. If 1 ml of this sample were transferred to a tube containing 9 ml of sterile water, each milliliter of fluid in this tube would now contain 1000 bacteria. If 1 ml of this sample were inoculated into a Petri plate, there would still be too many potential colonies to count on a plate. Therefore, another serial dilution could be made. One milliliter containing 1000 bacteria would be transferred to a second tube of 9 ml of water. Each milliliter of this tube would now contain only 100 bacteria, and if 1 ml of the contents of this tube were plated out, potentially 100 colonies would be formed—an easily countable number. Learning how to do serial dilutions is an important part of certain experiments in microbiology laboratory classes.

Pour Plates and Spread Plates A plate count is done by either the pour plate method or the spread plate method. The **pour plate method** follows the procedure shown in Figure 6.16a. Either 1.0 ml or 0.1 ml of dilutions of the bacterial suspension is introduced into a Petri dish. The nutrient medium, in which the agar is kept liquid by holding it in a water bath at about 50°C, is poured over the sample, which is then mixed into the medium by gentle agitation of the plate. When the agar solidifies, the plate is incubated. With the pour plate technique, colonies will grow within the nutrient agar (from cells suspended in the nutrient medium as the agar solidifies) as well as on the surface of the agar plate.

This technique has some drawbacks because some relatively heat-sensitive microorganisms may be damaged by the melted agar and will therefore be unable to form colonies. Also, when certain differential media are used, the distinctive appearance of the colony on the surface is essential for diagnostic purposes. Colonies that form beneath the surface of a pour plate are not satisfactory for such tests. To avoid these problems, the **spread plate method** is frequently used instead (Figure 6.16b). A 0.1-ml inoculum is added to the surface of a prepoured, solidified agar medium. The inoculum is then spread uniformly over

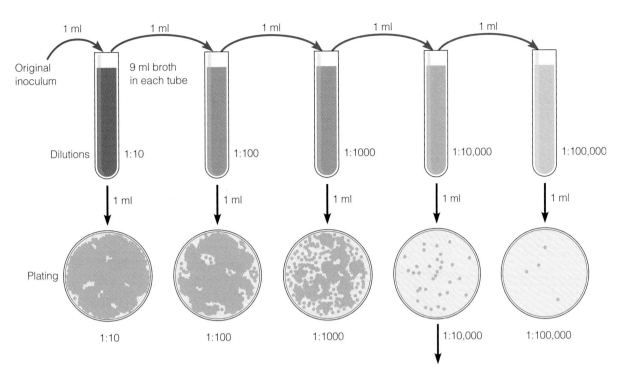

Calculation: Number of colonies on plate × reciprocal of dilution of sample = number of bacteria/ml
(For example, if 32 colonies are on a plate of $^1/_{10,000}$ dilution, then the count is 32 × 10,000 = 320,000 bacteria/ml in sample.)

FIGURE 6.15 Plate counts and serial dilutions. In serial dilutions, the original inoculum is diluted in a series of dilution tubes. In our example, each succeeding dilution tube will have only one-tenth the number of microbial cells as the preceding tube. Then samples of the dilution are used to inoculate Petri plates, on which colonies grow and can be counted. This count is then used to estimate the number of bacteria in the original sample.

Q **Why were the dilutions of 1:1000 and 1:100,000 not counted? Theoretically, how many colonies should appear on the 1:1000 plate?**

the surface of the medium with a specially shaped, sterilized glass rod. This method positions all the colonies on the surface and avoids contact of the cells with melted agar.

FILTRATION

When the quantity of bacteria is very small, as in lakes or relatively pure streams, bacteria can be counted by **filtration** methods (Figure 6.17). In this technique, at least 100 ml of water are passed through a thin membrane filter whose pores are too small to allow bacteria to pass. Thus, the bacteria are filtered out and retained on the surface of the filter. This filter is then transferred to a Petri dish containing a pad soaked in liquid nutrient medium, where colonies arise from the bacteria on the filter's surface. This method is applied frequently to detection and enumeration of coliform bacteria, which are indicators of fecal pollution of food or water (see Chapter 27). The colonies formed by these bacteria are distinctive when a differential nutrient medium is used. (The colonies shown in Figures 6.9b and c are examples of coliforms.)

THE MOST PROBABLE NUMBER (MPN) METHOD

Another method for determining the number of bacteria in a sample is the **most probable number (MPN) method,** illustrated in Figure 6.18. This statistical estimating technique is based on the fact that the greater the number of bacteria in a sample, the more dilution is needed to reduce the density to the point at which no bacteria are left to grow in the tubes in a dilution series. The MPN method is most useful when the microbes being counted will not grow on solid media (such as the chemoautotrophic nitrifying bacteria). It is also useful when the growth of bacteria in a liquid differential medium is used to identify the microbes (such as coliform bacteria, which selectively ferment lactose to acid, in water testing). The MPN is only a statement that there is a 95% chance that the bacterial population falls within a certain range and that the MPN is statistically the most probable number.

DIRECT MICROSCOPIC COUNT

In the method known as the **direct microscopic count,** a measured volume of a bacterial suspension is placed

FIGURE 6.16 Methods of preparing plates for plate counts. (**a**) The pour plate method. (**b**) The spread plate method.

Q **What are the advantages of each method?**

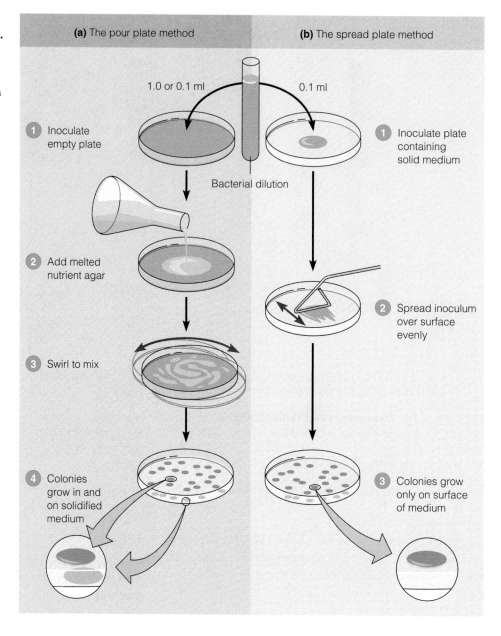

(a) The pour plate method

1.0 or 0.1 ml

1 Inoculate empty plate

2 Add melted nutrient agar

3 Swirl to mix

4 Colonies grow in and on solidified medium

Bacterial dilution

(b) The spread plate method

0.1 ml

1 Inoculate plate containing solid medium

2 Spread inoculum over surface evenly

3 Colonies grow only on surface of medium

within a defined area on a microscope slide. In the *Breed count method*, which is used to count the number of bacteria in milk, for example, a 0.01-ml sample is spread over a marked square centimeter of slide, stain is added so that the bacteria can be seen, and the sample is viewed under the oil immersion objective lens. The area of the viewing field of this objective can be determined. Once the number of bacteria has been counted in several different fields, the average number of bacteria per viewing field can be calculated. From these data, the number of bacteria in the square centimeter over which the sample was spread can also be calculated. Because this area on the slide contained 0.01 ml of sample, the number of bacteria in each milliliter of the suspension is the number of bacteria in the sample times 100.

A specially designed slide called a *Petroff-Hausser cell counter* is also used in direct microscopic counts (Figure 6.19).

Motile bacteria are difficult to count by this method, and, as happens with other microscopic methods, dead cells are about as likely to be counted as live ones. In addition to these disadvantages, a rather high concentration of cells is required to be countable—about 10 million bacteria per milliliter. The chief advantage of microscopic counts is that no incubation time is required, and they are usually reserved for applications in which time is the primary consideration. This advantage also holds for *electronic cell counters*, sometimes known as *Coulter counters*, which automatically count the number of cells in a measured volume of liquid. These instruments are used in some research laboratories and hospitals.

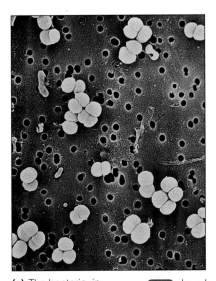

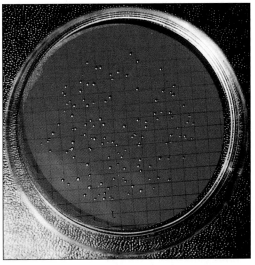

(a) The bacteria in 100 ml of water were sieved out onto the surface of a membrane filter.

SEM ⊢——⊣ 1 μm

(b) A filter such as shown in photo (a), with the bacteria much more widely spaced, was placed on a pad saturated with liquid Endo medium, which is selective for gram-negative bacteria. The individual bacteria grew into visible colonies. One hundred twenty-four colonies are visible, so we would record 124 bacteria per 100 ml of water sample.

FIGURE 6.17 Counting bacteria by Filtration.

Q Could you make a pour plate in the usual Petri dish with a 10 ml inoculum? If not, why not?

Volume of Inoculum for Each Set of Five Tubes	Tubes of Nutrient Medium (Sets of Five Tubes)	Number of Positive Tubes in Set
10 ml		5
1 ml		3
0.1 ml		1

(a) Most probable number (MPN) dilution series. In this example, there are three sets of tubes and five tubes in each set. Each tube in the first set of five tubes receives 10 ml of the inoculum, such as a sample of water. Each tube in the second set of five tubes receives 1 ml of the sample, and the third set, 0.1 ml each. There were enough bacteria in the sample so that all five tubes in the first set showed bacterial growth and were recorded as positive. In the second set, which received only one-tenth as much inoculum, only three tubes were positive. In the third set, which received one-hundredth as much inoculum, only one tube was positive.

Combination of Positives	MPN Index/ 100 ml	95% Confidence Limits	
		Lower	Upper
4-2-0	22	9	56
4-2-1	26	12	65
4-3-0	27	12	67
4-3-1	33	15	77
4-4-0	34	16	80
5-0-0	23	9	86
5-0-1	30	10	110
5-0-2	40	20	140
5-1-0	30	10	120
5-1-1	50	20	150
5-1-2	60	30	180
5-2-0	50	20	170
5-2-1	70	30	210
5-2-2	90	40	250
5-3-0	80	30	250
5-3-1	110	40	300
5-3-2	140	60	360

(b) MPN table. MPN tables enable us to calculate for a sample the microbial numbers that are statistically likely to lead to such a result. The number of positive tubes is recorded for each set: in the shaded example, 5, 3, and 1. If we look up this combination in an MPN table, we find that the MPN index per 100 ml is 110. Statistically, this means that 95% of the water samples that give this result contain 40–300 bacteria, with 110 being the most probable number.

FIGURE 6.18 The most probable number (MPN) method.

Q Under what circumstances is the MPN method used to determine the number of bacteria in a sample?

FIGURE 6.19 Direct microscopic count of bacteria with a Petroff-Hausser cell counter. The average number of cells within a large square multiplied by a factor of 1,250,000 gives the number of bacteria per milliliter.

Q This type of counting, despite its obvious disadvantages, is often used in estimating the bacterial population in dairy products. Why?

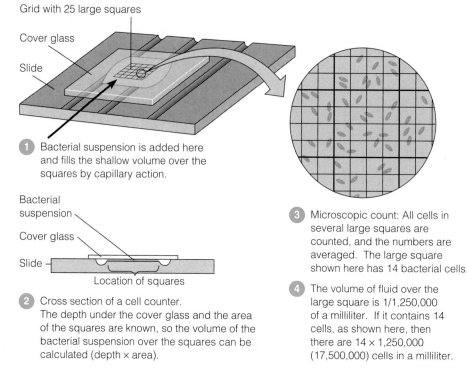

Grid with 25 large squares

Cover glass

Slide

1 Bacterial suspension is added here and fills the shallow volume over the squares by capillary action.

Bacterial suspension

Cover glass

Slide

Location of squares

2 Cross section of a cell counter. The depth under the cover glass and the area of the squares are known, so the volume of the bacterial suspension over the squares can be calculated (depth × area).

3 Microscopic count: All cells in several large squares are counted, and the numbers are averaged. The large square shown here has 14 bacterial cells.

4 The volume of fluid over the large square is 1/1,250,000 of a milliliter. If it contains 14 cells, as shown here, then there are 14 × 1,250,000 (17,500,000) cells in a milliliter.

ESTIMATING BACTERIAL NUMBERS BY INDIRECT METHODS

LEARNING OBJECTIVES

- Differentiate between direct and indirect methods of measuring cell growth.
- Explain three indirect methods of measuring cell growth.

It is not always necessary to count microbial cells to estimate their numbers. In science and industry, microbial numbers and activity are determined by some of the following indirect means as well.

TURBIDITY

For some types of experimental work, estimating **turbidity** is a practical way of monitoring bacterial growth. As bacteria multiply in a liquid medium, the medium becomes turbid, or cloudy with cells.

The instrument used to measure turbidity is a *spectrophotometer* (or colorimeter). In the spectrophotometer, a beam of light is transmitted through a bacterial suspension to a light-sensitive detector (Figure 6.20). As bacterial numbers increase, less light will reach the detector. This change of light will register on the instrument's scale as the *percentage of transmission*. Also printed on the instrument's scale is a logarithmic expression called the *absorbance* (sometimes called *optical density*, or *OD*); a value derived from the percentage of

transmission may also be reported. The absorbance is used to plot bacterial growth. When the bacteria are in logarithmic growth or decline, a graph of absorbance versus time will form an approximately straight line. If absorbance readings are matched with plate counts of the same culture, this correlation can be used in future estimations of bacterial numbers obtained by measuring turbidity.

More than a million cells per milliliter must be present for the first traces of turbidity to be visible. About 10 million to 100 million cells per milliliter are needed to make a suspension turbid enough to be read on a spectrophotometer. Therefore, turbidity is not a useful measure of contamination of liquids by relatively small numbers of bacteria.

METABOLIC ACTIVITY

Another indirect way to estimate bacterial numbers is to measure a population's *metabolic activity*. This method assumes that the amount of a certain metabolic product, such as acid or CO_2, is in direct proportion to the number of bacteria present. An example of a practical application of a metabolic test is the microbiological assay in which acid production is used to determine amounts of vitamins.

DRY WEIGHT

For filamentous bacteria and molds, the usual measuring methods are less satisfactory. A plate count would not

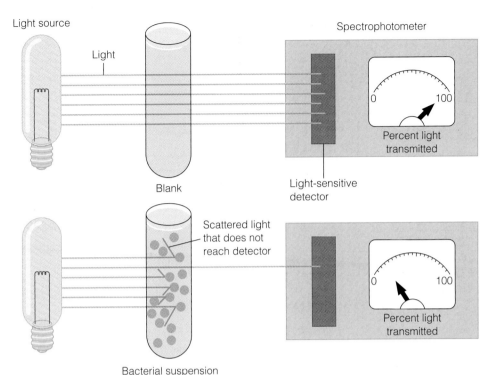

FIGURE 6.20 Turbidity estimation of bacterial numbers. The amount of light striking the light-sensitive detector on the spectrophotometer is inversely proportional to the number of bacteria under standardized conditions. The less light transmitted, the more bacteria in the sample.

Q **Is turbidity a direct or an indirect method of measuring bacterial growth?**

measure this increase in filamentous mass. In plate counts of actinomycetes (see Figure 11.21, page 335) and molds, it is mostly the number of asexual spores that is counted instead. This is not a good measure of growth. One of the better ways to measure the growth of filamentous organisms is by *dry weight*. In this procedure, the fungus is removed from the growth medium, filtered to remove extraneous material, and dried in a desiccator.

It is then weighed. For bacteria, the same basic procedure is followed.

* * *

You now have a basic understanding of the requirements for, and measurements of, microbial growth. In Chapter 7, we will look at how this growth is controlled in laboratories, hospitals, industry, and our homes.

STUDY OUTLINE

THE REQUIREMENTS FOR GROWTH (pp. 160–167)

1. The growth of a population is an increase in the number of cells.

2. The requirements for microbial growth are both physical and chemical.

PHYSICAL REQUIREMENTS (pp. 160–163)

3. On the basis of preferred temperature ranges, microbes are classified as psychrophiles (cold-loving), mesophiles (moderate-temperature–loving), and thermophiles (heat-loving).

4. The minimum growth temperature is the lowest temperature at which a species will grow, the optimum growth temperature is the temperature at which it grows best, and the maximum growth temperature is the highest temperature at which growth is possible.

5. Most bacteria grow best at a pH value between 6.5 and 7.5.

6. In a hypertonic solution, most microbes undergo plasmolysis; halophiles can tolerate high salt concentrations.

CHEMICAL REQUIREMENTS (pp. 163–167)

7. All organisms require a carbon source; chemoheterotrophs use an organic molecule, and autotrophs typically use carbon dioxide.

8. Nitrogen is needed for protein and nucleic acid synthesis. Nitrogen can be obtained from the decomposition of proteins or from NH_4^+ or NO_3^-; a few bacteria are capable of nitrogen (N_2) fixation.

9. On the basis of oxygen requirements, organisms are classified as obligate aerobes, facultative anaerobes, obligate anaerobes, aerotolerant anaerobes, and microaerophiles.

10. Aerobes, facultative anaerobes, and aerotolerant anaerobes must have the enzymes superoxide dismutase ($2 O_2^- + 2 H^+$

$\longrightarrow O_2 + H_2O_2$) and either catalase (2 $H_2O_2 \longrightarrow$ 2 H_2O + O_2) or peroxidase (H_2O_2 + 2 $H^+ \longrightarrow$ 2 H_2O).

11. Other chemicals required for microbial growth include sulfur, phosphorus, trace elements, and, for some microorganisms, organic growth factors.

CULTURE MEDIA (pp. 168–172)

1. A culture medium is any material prepared for the growth of bacteria in a laboratory.
2. Microbes that grow and multiply in or on a culture medium are known as a culture.
3. Agar is a common solidifying agent for a culture medium.

CHEMICALLY DEFINED MEDIA (pp. 168–169)

4. A chemically defined medium is one in which the exact chemical composition is known.

COMPLEX MEDIA (pp. 169–170)

5. A complex medium is one in which the exact chemical composition varies slightly from batch to batch.

ANAEROBIC GROWTH MEDIA AND METHODS (p. 170)

6. Reducing media chemically remove molecular oxygen (O_2) that might interfere with the growth of anaerobes.
7. Petri plates can be incubated in an anaerobic jar, anaerobic chamber, or OxyPlate.

SPECIAL CULTURE TECHNIQUES (pp. 170–171)

8. Some parasitic and fastidious bacteria must be cultured in living animals or in cell cultures.
9. CO_2 incubators or candle jars are used to grow bacteria that require an increased CO_2 concentration.

SELECTIVE AND DIFFERENTIAL MEDIA (pp. 171–172)

10. By inhibiting unwanted organisms with salts, dyes, or other chemicals, selective media allow growth of only the desired microbes.
11. Differential media are used to distinguish among different organisms.

ENRICHMENT CULTURE (p. 172)

12. An enrichment culture is used to encourage the growth of a particular microorganism in a mixed culture.

OBTAINING PURE CULTURES (pp. 172–174)

1. A colony is a visible mass of microbial cells that theoretically arose from one cell.
2. Pure cultures are usually obtained by the streak plate method.

PRESERVING BACTERIAL CULTURES (p. 174)

1. Microbes can be preserved for long periods of time by deep-freezing or lyophilization (freeze-drying).

THE GROWTH OF BACTERIAL CULTURES (pp. 174–183)

BACTERIAL DIVISION (pp. 174–175)

1. The normal reproductive method of bacteria is binary fission, in which a single cell divides into two identical cells.
2. Some bacteria reproduce by budding, aerial spore formation, or fragmentation. ❖ **Animation: Bacterial Growth. The Microbiology Place.**

GENERATION TIME (p. 175)

3. The time required for a cell to divide or a population to double is known as the generation time.

LOGARITHMIC REPRESENTATION OF BACTERIAL POPULATIONS (pp. 175–176)

4. Bacterial division occurs according to a logarithmic progression (two cells, four cells, eight cells, etc.).

PHASES OF GROWTH (pp. 176–177)

5. During the lag phase, there is little or no change in the number of cells, but metabolic activity is high.
6. During the log phase, the bacteria multiply at the fastest rate possible under the conditions provided.
7. During the stationary phase, there is an equilibrium between cell division and death.
8. During the death phase, the number of deaths exceeds the number of new cells formed.

DIRECT MEASUREMENT OF MICROBIAL GROWTH (pp. 178–181)

9. A standard plate count reflects the number of viable microbes and assumes that each bacterium grows into a single colony; plate counts are reported as number of colony-forming units (CFU).
10. A plate count may be done by either the pour plate method or the spread plate method.
11. In filtration, bacteria are retained on the surface of a membrane filter and then transferred to a culture medium to grow and subsequently be counted.
12. The most probable number (MPN) method can be used for microbes that will grow in a liquid medium; it is a statistical estimation.
13. In a direct microscopic count, the microbes in a measured volume of a bacterial suspension are counted with the use of a specially designed slide.

ESTIMATING BACTERIAL NUMBERS BY INDIRECT METHODS (pp. 182–183)

14. A spectrophotometer is used to determine turbidity by measuring the amount of light that passes through a suspension of cells.

15. An indirect way of estimating bacterial numbers is measuring the metabolic activity of the population (for example, acid production or oxygen consumption).

16. For filamentous organisms such as fungi, measuring dry weight is a convenient method of growth measurement.

STUDY QUESTIONS

Access more review material either online at **The Microbiology Place** (www.microbiologyplace.com) or with **The Microbiology Place CD-ROM** packaged with your new book. There you'll find activities, practice tests, quizzes, flashcards, case studies, and more to help you succeed. In addition, you'll find the following Interactive Tutorials: Binary Fission and Growth Curves.

Answers to the Study Questions can be found in Appendix G.

REVIEW

1. Describe binary fission.

2. Draw a typical bacterial growth curve. Label and define each of the four phases.

3. Macronutrients (needed in relatively large amounts) are often listed as CHONPS. What does each of these letters indicate, and why are they needed by the cell?

4. Most bacteria grow best at pH _____.

5. Why can high concentrations of salt or sugar be used to preserve food?

6. Define and explain the importance of each of the following:
   a. catalase   **d.** superoxide free radical
   b. hydrogen peroxide **e.** superoxide dismutase
   c. peroxidase

7. *Clostridium* can be cultured in an anaerobic incubator or in the presence of atmospheric oxygen if thioglycolate is added to the nutrient broth. Compare these two techniques. Using the appropriate terms from question 6, explain why elaborate culture techniques are used for *Clostridium*.

8. Seven methods of measuring microbial growth were explained in this chapter. Categorize each as either a direct or an indirect method.

9. By deep-freezing, bacteria can be stored without harm for extended periods. Why do refrigeration and freezing preserve foods?

10. A pastry chef accidentally inoculated a cream pie with six *S. aureus* cells. If *S. aureus* has a generation time of 60 minutes, how many cells would be in the cream pie after 7 hours?

11. Nitrogen and phosphorus added to beaches following an oil spill encourage the growth of natural oil-degrading bacteria. Explain why the bacteria do not grow if nitrogen and phosphorus are not added.

12. Differentiate between complex and chemically defined media.

13. Draw the following growth curves for *E. coli*, starting with 100 cells with a generation time of 30 minutes at 35°C.
    a. The cells are incubated for 5 hours at 35°C.
    b. After 5 hours, the temperature is changed to 20°C for 2 hours.
    c. After 5 hours at 35°C, the temperature is changed to 5°C for 2 hours followed by 35°C for 5 hours.

MULTIPLE CHOICE

Use the following information to answer questions 1 and 2. Two culture media were inoculated with four different bacteria. After incubation, the following results were obtained:

Organism	Medium 1	Medium 2
Escherichia coli	Red colonies	No growth
Staphylococcus aureus	No growth	Growth
Staphylococcus epidermidis	No growth	Growth
Salmonella enterica	Colorless colonies	No growth

1. Medium 1 is
   a. selective.   **c.** both selective and differential.
   b. differential.

2. Medium 2 is
   a. selective.   **c.** both selective and differential.
   b. differential.

Use the following graph to answer questions 3 and 4.

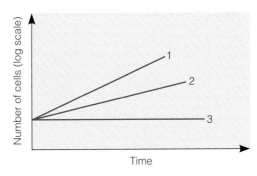

3. Which of the lines best depicts the log phase of a thermophile incubated at room temperature?

4. Which of the lines best depicts the log phase of *Listeria monocytogenes* growing in a human?

5. Assume you inoculated 100 facultatively anaerobic cells onto nutrient agar and incubated the plate aerobically. You then inoculated 100 cells of the same species onto nutrient

agar and incubated the second plate anaerobically. After incubation for 24 hours, you should have

a. more colonies on the aerobic plate.
b. more colonies on the anaerobic plate.
c. the same number of colonies on both plates.

6. The term *trace elements* refers to
a. the elements CHONPS.
b. vitamins.
c. nitrogen, phosphorus, and sulfur.
d. small mineral requirements.
e. toxic substances.

7. Which one of the following temperatures would most likely kill a mesophile?
a. −50°C   c. 9°C   e. 60°C
b. 0°C   d. 37°C

8. All of the following are true about agar *except:*
a. It is a source of nutrients in culture media.
b. It is a polysaccharide.
c. It liquefies at 100°C.
d. It solidifies at approximately 40°C.
e. It is metabolized by few bacteria.

9. Which of the following types of media would *not* be used to culture aerobes?
a. selective media   d. differential media
b. reducing media   e. complex media
c. enrichment media

10. An organism that has peroxidase and superoxide dismutase but lacks catalase is most likely an
a. aerobe.
b. aerotolerant anaerobe.
c. obligate anaerobe.

CRITICAL THINKING

1. *E. coli* was incubated with aeration in a nutrient medium containing two carbon sources, and the following growth curve was made from this culture.
a. Explain what happened at the time marked x.
b. Which substrate provided "better" growth conditions for the bacteria? How can you tell?

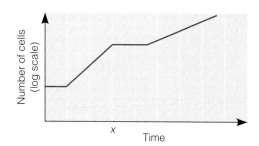

2. *Clostridium* and *Streptococcus* are both catalase-negative. *Streptococcus* grows by fermentation. Why is *Clostridium* killed by oxygen, whereas *Streptococcus* is not?

3. Most laboratory media contain a fermentable carbohydrate and peptone because the majority of bacteria require carbon, nitrogen, and energy sources in these forms. How are these three needs met by glucose–minimal salts medium? (*Hint:* See Table 6.2.)

4. Flask A contains yeast cells in glucose–minimal salts broth incubated at 30°C with aeration. Flask B contains yeast cells in glucose–minimal salts broth incubated at 30°C in an anaerobic jar. The yeasts are facultative anaerobes.
a. Which culture produced more ATP?
b. Which culture produced more alcohol?
c. Which culture had the shorter generation time?
d. Which culture had the greater cell mass?
e. Which culture had the higher absorbance?

CLINICAL APPLICATIONS

1. Assume that after washing your hands, you leave ten bacterial cells on a new bar of soap. You then decide to do a plate count of the soap after it was left in the soap dish for 24 hours. You dilute 1 g of the soap $1:10^6$ and plate it on standard plate count agar. After 24 hours of incubation, there are 168 colonies. How many bacteria were on the soap? How did they get there?

2. Heat lamps are commonly used to maintain foods at about 50°C for as long as 12 hours in cafeteria serving lines. The following experiment was conducted to determine whether this practice poses a potential health hazard.

   Beef cubes were surface-inoculated with 500,000 bacterial cells and incubated at 43–53°C to establish temperature limits for bacterial growth. The following results were obtained from standard plate counts performed on beef cubes at 6 and 12 hours after inoculation:

	Temp. (°C)	Bacteria/Gram of Beef After	
		6 hr	12 hr
S. aureus	43	140,000,000	740,000,000
	51	810,000	59,000
	53	650	300
S. typhimurium	43	3,200,000	10,000,000
	51	950,000	83,000
	53	1,200	300
C. perfringens	43	1,200,000	3,600,000
	51	120,000	3,800
	53	300	300

   Draw the growth curves for each organism. What holding temperature would you recommend? Assuming that cooking kills bacteria in foods, how could these bacteria contaminate the cooked foods? What disease does each organism cause? (*Hint:* See Chapter 25.)

3. The number of bacteria in saliva samples was determined by collecting the saliva, making serial dilutions, and inoculating nutrient agar by the pour plate method. The plates were incubated aerobically for 48 hours at 37°C.

	Bacteria/ml Saliva	
	Before Using Mouthwash	After Using Mouthwash
Mouthwash 1	13.1×10^6	10.9×10^6
Mouthwash 2	11.7×10^6	14.2×10^5
Mouthwash 3	9.3×10^5	7.7×10^5

What can you conclude from these data? Did all the bacteria present in each saliva sample grow?

7

The Control of Microbial Growth

The scientific control of microbial growth began only about 100 years ago. Recall from Chapter 1 that Pasteur's work on microorganisms led scientists to believe that microbes were a possible cause of disease. In the mid-1800s, the Hungarian physician Ignaz Semmelweis and English physician Joseph Lister used this thinking to develop some of the first microbial control practices for medical procedures. These practices included hand washing with microbe-killing chloride of lime and use of the techniques of **aseptic surgery** to prevent microbial contamination of surgical wounds. Until that time, hospital-acquired infections, or *nosocomial infections,* were the cause of death in at least 10% of surgical cases, and deaths of delivering mothers were as high as 25%. Ignorance of microbes was such that, during the American Civil War, a surgeon might have cleaned his scalpel on his bootsole between incisions.

Over the last century, scientists have continued to develop a variety of physical methods and chemical agents to control microbial growth. In Chapter 20 we will discuss methods for the control of microbes once infection has occurred, mainly antibiotic chemotherapy.

UNDER THE MICROSCOPE

Bacteria trapped in a membrane filter. Filtration can be used to remove microorganisms from water and solutions.

THE TERMINOLOGY OF MICROBIAL CONTROL

LEARNING OBJECTIVE

• Define the following key terms related to microbial control: *sterilization, disinfection, antisepsis, degerming, sanitization, biocide, germicide, bacteriostasis,* and *asepsis.*

A word frequently used, and misused, in discussing the control of microbial growth is *sterilization.* **Sterilization** is the removal or destruction of *all forms* of microbial life. Heating is the most common method used for killing microbes, including the most resistant forms such as endospores. A sterilizing agent is called a **sterilant.** Sterilization by removal of microbes from liquids or gases can be done by filtration.

One would think that canned food in the supermarket is completely sterile. In reality, the heat treatment required to ensure absolute sterility would unnecessarily degrade the quality of the food. Instead, food is subjected only to enough heat to destroy the endospores of *Clostridium botulinum,* which can produce a deadly toxin. This limited heat treatment is termed **commercial sterilization.** The endospores of a number of thermophilic bacteria, capable of causing food spoilage but not human disease, are considerably more resistant to heat than C. *botulinum.* If present, they will survive, but their survival is usually of no practical consequence; they will not grow at normal food storage temperatures. If canned foods in a supermarket were incubated at temperatures in the growth range of these thermophiles (above about 45°C), significant food spoilage would occur.

Complete sterilization is often not required in other settings. For example, the body's normal defenses can cope with a few microbes entering a surgical wound. A drinking glass or a fork in a restaurant requires only enough microbial control to prevent the transmission of possibly pathogenic microbes from one person to another.

Control directed at destroying harmful microorganisms is called **disinfection.** It usually refers to the destruction of vegetative (non–endospore-forming) pathogens, which is not the same thing as complete sterility. Disinfection might make use of chemicals, ultraviolet radiation, boiling water, or steam. In practice, the term is most commonly applied to the use of a chemical (a *disinfectant*) to treat an inert surface or substance. When this treatment is directed at living tissue, it is called **antisepsis,** and the chemical is then called an *antiseptic.* Therefore, in practice the same chemical might be called a disinfectant for one use and an antiseptic for another. Of course, many chemicals suitable for swabbing a tabletop would be too harsh to use on living tissue.

There are modifications of disinfection and antisepsis. For example, when someone is about to receive an injection, the skin is swabbed with alcohol—the process of **degerming** (or *degermation*), which mostly results in the mechanical removal, rather than the killing, of most of the microbes in a limited area. Restaurant glassware, china, and tableware are subjected to **sanitization,** which is intended to lower microbial counts to safe public health levels and minimize the chances of disease transmission from one user to another. This is usually accomplished by high-temperature washing or, in the case of glassware in a bar, washing in a sink followed by a dip in a chemical disinfectant.

Table 7.1 summarizes the terminology relating to the control of microbial growth.

TABLE 7.1	Terminology Relating to the Control of Microbial Growth	
	Definition	Comments
Sterilization	Destruction or removal of all forms of microbial life, including endospores.	Usually done by steam under pressure or a sterilizing gas such as ethylene oxide.
Commercial Sterilization	Sufficient heat treatment to kill endospores of *Clostridium botulinum* in canned food.	More-resistant endospores of thermophilic bacteria may survive, but they will not germinate and grow under normal storage conditions.
Disinfection	Destruction of vegetative pathogens.	May make use of physical or chemical methods.
Antisepsis	Destruction of vegetative pathogens on living tissue.	Treatment is almost always by chemical antimicrobials.
Degerming	Removal of microbes from a limited area, such as the skin around an injection site.	Mostly a mechanical removal by an alcohol-soaked swab.
Sanitization	Treatment intended to lower microbial counts on eating and drinking utensils to safe public health levels.	May be done with high-temperature washing or by dipping into a chemical disinfectant.

Names of treatments that cause the outright death of microbes have the suffix *-cide*, meaning kill. A **biocide,** or **germicide,** kills microorganisms (usually with certain exceptions, such as endospores); a *fungicide* kills fungi; a *virucide* inactivates viruses; and so on. Other treatments only inhibit the growth and multiplication of bacteria; their names have the suffix *-stat* or *-stasis*, meaning to stop or to steady, as in **bacteriostasis.** Once a bacteriostatic agent is removed, growth might resume.

Sepsis, from the Greek for decay or putrid, indicates bacterial contamination, as in septic tanks for sewage treatment. (The term is also used to describe a disease condition; see Chapter 23, page 672.) *Aseptic* means that an object or area is free of pathogens. Recall from Chapter 1 that **asepsis** is the absence of significant contamination. Aseptic techniques are important in surgery to minimize contamination from the instruments, operating personnel, and the patient.

THE RATE OF MICROBIAL DEATH

LEARNING OBJECTIVE

• Describe the patterns of microbial death caused by treatments with microbial control agents.

When bacterial populations are heated or treated with antimicrobial chemicals, they usually die at a constant rate. For example, suppose a population of 1 million microbes has been treated for 1 minute, and 90% of the population has died. We are now left with 100,000 microbes. If the population is treated for another minute, 90% of *those* microbes die, and we are left with 10,000 survivors. In other words, for each minute the treatment is applied, 90% of the remaining population is killed (Table 7.2). If

TABLE 7.2	Microbial Death Rate: An Example	
Time (min)	Deaths per Minute	Number of Survivors
0	0	1,000,000
1	900,000	100,000
2	90,000	10,000
3	9000	1000
4	900	100
5	90	10
6	9	1

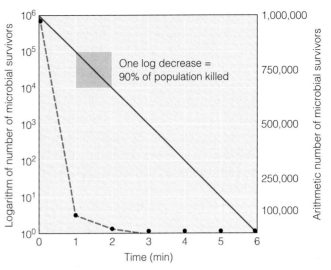

(a) The curve is plotted logarithmically (solid line) and arithmetically (broken line). In this case, the cells are dying at a rate of 90% each minute.

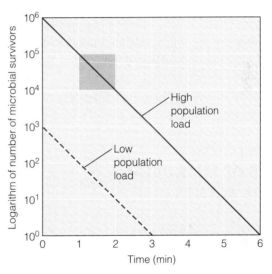

(b) The effect of high or low initial load of microbes. If the rate of killing is the same, it will take longer to kill all members of a larger population than a smaller one. This is true for both heat and chemical treatments.

FIGURE 7.1 A microbial death curve.

Q If the graph in part (a) reflects the experience with a vegetative bacterium, what would the logarithmic curve look like for the endospores of the same bacterium? Less steep? Steeper?

the death curve is plotted logarithmically, the death rate is constant, as shown by the straight line in Figure 7.1a.

Several factors influence the effectiveness of antimicrobial treatments:

• *The number of microbes.* The more microbes there are to begin with, the longer it takes to eliminate the entire population (Figure 7.1b).

- *Environmental influences.* The presence of organic matter often inhibits the action of chemical antimicrobials. In hospitals, the presence of organic matter in blood, vomit, or feces influences the selection of disinfectants. Microbes in surface biofilms, shown in Figure 27.11, are difficult for biocides to reach effectively. Because their activity is due to temperature-dependent chemical reactions, disinfectants work somewhat better under warm conditions. Directions on disinfectant containers frequently specify the use of a warm solution.

    The nature of the suspending medium is also a factor in heat treatment. Fats and proteins are especially protective, and a medium rich in these substances protects microbes, which will then have a higher survival rate. Heat is also measurably more effective under acidic conditions.

- *Time of exposure.* Chemical antimicrobials often require extended exposure for more-resistant microbes or endospores to be affected. In heat treatments, a longer exposure can compensate for a lower temperature, a phenomenon of particular importance to pasteurization of dairy products.

- *Microbial characteristics.* The concluding section of this chapter discusses how microbial characteristics affect chemical and physical control methods.

ACTIONS OF MICROBIAL CONTROL AGENTS

LEARNING OBJECTIVE

- Describe the effects of microbial control agents on cellular structures.

In this section, we examine the ways various agents actually kill or inhibit microbes.

ALTERATION OF MEMBRANE PERMEABILITY

A microorganism's plasma membrane (see Figure 4.14, page 90), located just inside the cell wall, is the target of many microbial control agents. This membrane actively regulates the passage of nutrients into the cell and the elimination of wastes from the cell. Damage to the lipids or proteins of the plasma membrane by antimicrobial agents causes cellular contents to leak into the surrounding medium and interferes with the growth of the cell.

DAMAGE TO PROTEINS AND NUCLEIC ACIDS

Bacteria are sometimes thought of as "little bags of enzymes." Enzymes, which are primarily protein, are vital to all cellular activities. Recall that the functional properties of proteins are the result of their three-dimensional shape (see Figure 2.15, page 46). This shape is maintained by chemical bonds that link adjoining portions of the amino acid chain as it folds back and forth upon itself. Some of those bonds are hydrogen bonds, which are susceptible to breakage by heat or certain chemicals; breakage results in denaturation of the protein. Covalent bonds, which are stronger, are also subject to attack. For example, disulfide bridges, which play an important role in protein structure by joining amino acids with exposed sulfhydryl (—SH) groups, can be broken by certain chemicals or sufficient heat.

The nucleic acids DNA and RNA are the carriers of the cell's genetic information. Damage to these nucleic acids by heat, radiation, or chemicals is frequently lethal to the cell; the cell can no longer replicate, nor can it carry out normal metabolic functions such as the synthesis of enzymes.

PHYSICAL METHODS OF MICROBIAL CONTROL

LEARNING OBJECTIVES

- Compare the effectiveness of moist heat (boiling, autoclaving, pasteurization) and dry heat.
- Describe how filtration, low temperatures, high pressure, desiccation, and osmotic pressure suppress microbial growth.
- Explain how radiation kills cells.

As early as the Stone Age, it is likely that humans were already using some physical methods of microbial control to preserve foods. Drying (desiccation) and salting (osmotic pressure) were probably among the earliest techniques.

When selecting methods of microbial control, consideration must be given to effects on things besides the microbes. For example, certain vitamins or antibiotics in a solution might be inactivated by heat. Many laboratory or hospital materials, such as rubber and latex tubing, are damaged by repeated heating. There are also economic considerations; for example, it may be less expensive to use presterilized, disposable plasticware than to repeatedly wash and resterilize glassware.

HEAT

A visit to any supermarket will demonstrate that heat-preserved canned goods represent one of the most common methods of food preservation. Laboratory media and glassware, and hospital instruments, are also usually sterilized

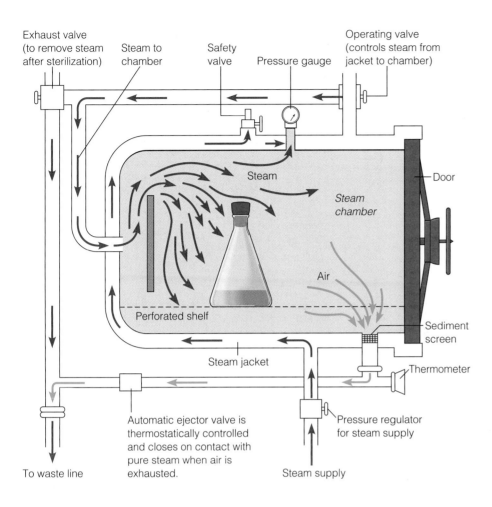

Exhaust valve
(to remove steam
after sterilization)

Steam to
chamber

Safety
valve

Pressure gauge

Operating valve
(controls steam from
jacket to chamber)

Steam

Door

*Steam
chamber*

Air

Perforated shelf

Sediment
screen

Thermometer

Steam jacket

Automatic ejector valve is
thermostatically controlled
and closes on contact with
pure steam when air is
exhausted.

Pressure regulator
for steam supply

To waste line

Steam supply

FIGURE 7.2 An autoclave. The entering steam forces the air out of the bottom (blue arrows). The automatic ejector valve remains open as long as an air-steam mixture is passing out of the waste line. When all the air has been ejected, the higher temperature of the pure steam closes the valve, and the pressure in the chamber increases.

Q **How would an empty, uncapped flask be positioned for sterilization in an autoclave?**

by heat. Heat appears to kill microorganisms by denaturing their enzymes; the resultant changes to the three-dimensional shapes of these proteins inactivate them (see Figure 5.6, page 121).

Heat resistance varies among different microbes; these differences can be expressed through the concept of thermal death point. **Thermal death point (TDP)** is the lowest temperature at which all the microorganisms in a particular liquid suspension will be killed in 10 minutes.

Another factor to be considered in sterilization is the length of time required. This is expressed as **thermal death time (TDT),** the minimal length of time for all bacteria in a particular liquid culture to be killed at a given temperature. Both TDP and TDT are useful guidelines that indicate the severity of treatment required to kill a given population of bacteria.

Decimal reduction time (DRT, or *D value)* is a third concept related to bacterial heat resistance. DRT is the time, in minutes, in which 90% of a population of bacteria at a given temperature will be killed (in Table 7.2 and Figure 7.1a, DRT is 1 minute). In Chapter 28 you can find an important application of DRT in the canning industry; see the discussion of the 12D treatment of canned goods in Chapter 28.

MOIST HEAT

Moist heat kills microorganisms primarily by the coagulation of proteins (denaturation), which is caused by breakage of the hydrogen bonds that hold the proteins in their three-dimensional structure. This coagulation process is familiar to anyone who has watched an egg white frying.

One type of moist heat sterilization is boiling, which kills vegetative forms of bacterial pathogens, almost all viruses, and fungi and their spores within about 10 minutes, usually much faster. Free-flowing (unpressurized) steam is essentially the same temperature as boiling water. Endospores and some viruses, however, are not destroyed this quickly. Some hepatitis viruses, for example, can survive up to 30 minutes of boiling, and some bacterial endospores can resist boiling for more than 20 hours. Boiling is therefore not always a reliable sterilization procedure. However, brief boiling, even at high altitudes, will kill most pathogens. The use of boiling to sanitize baby bottles is a familiar example.

Reliable sterilization with moist heat requires temperatures above that of boiling water. These high temperatures are most commonly achieved by steam under pressure in an **autoclave** (Figure 7.2). Autoclaving is the preferred method of sterilization, unless the material to be sterilized can be damaged by heat or moisture.

TABLE 7.3	The Relationship Between the Pressure and Temperature of Steam at Sea Level*

Pressure (psi in excess of atmospheric pressure)	Temperature (°C)
0	100
5	110
10	116
15	121
20	126
30	135

*At higher altitudes the atmospheric pressure is less, which must be taken into account in operation of an autoclave. For example, in order to reach sterilizing temperatures (121°C) in Denver, Colorado, whose altitude is 5280 feet (1600 meters), the pressure shown on the autoclave gauge would need to be higher than the 15 psi shown in the table.

TABLE 7.4	The Effect of Container Size on Autoclave Sterilization Times for Liquid Solutions*

Container Size	Liquid Volume	Sterilization Time (min)
Test tube: 18 × 150 mm	10 ml	15
Erlenmeyer flask: 125 ml	95 ml	15
Erlenmeyer flask: 2000 ml	1500 ml	30
Fermentation bottle: 9000 ml	6750 ml	70

*Sterilization times in the autoclave include the time for the contents of the containers to reach sterilization temperatures. For smaller containers, this is only 5 min or less, but for a 9000-ml bottle, it might be as much as 70 min. A container is usually not filled past 75% of its capacity.

The higher the pressure in the autoclave, the higher the temperature. For example, when free-flowing steam at a temperature of 100°C is placed under a pressure of 1 atmosphere above sea level pressure—that is, about 15 pounds of pressure per square inch (psi)—the temperature rises to 121°C. Increasing the pressure to 20 psi raises the temperature to 126°C. The relationship between temperature and pressure is shown in Table 7.3.

Sterilization in an autoclave is most effective when the organisms are either contacted by the steam directly or are contained in a small volume of aqueous (primarily water) liquid. Under these conditions, steam at a pressure of about 15 psi (121°C) will kill *all* organisms (but not prions, see page 000) and their endospores in about 15 minutes.

Autoclaving is used to sterilize culture media, instruments, dressings, intravenous equipment, applicators, solutions, syringes, transfusion equipment, and numerous other items that can withstand high temperatures and pressures. Large industrial autoclaves are called *retorts* (see Figure 28.2), but the same principle applies for the common household pressure cooker used in the home canning of foods.

Heat requires extra time to reach the center of solid materials, such as canned meats, because such materials do not develop the efficient heat-distributing convection currents that occur in liquids. Heating large containers also requires extra time. Table 7.4 shows the different time requirements for sterilizing liquids in various container sizes. Unlike sterilizing aqueous solutions, sterilizing the surface of a solid requires that steam actually contact it. To sterilize dry glassware, bandages, and the like, care must be taken to ensure that steam contacts all surfaces.

For example, aluminum foil is impervious to steam and should not be used to wrap dry materials that are to be sterilized; paper should be used instead. Care should also be taken to avoid trapping air in the bottom of a dry container because trapped air will not be replaced by steam, which is lighter than air. The trapped air is the equivalent of a small hot-air oven, which, as we will see shortly, requires a higher temperature and longer time to sterilize materials. Containers that can trap air should be placed in a tipped position so that the steam will force out the air. Products that do not permit penetration by moisture, such as mineral oil or petroleum jelly, are not sterilized by the same methods that would sterilize aqueous solutions.

Several commercially available methods can indicate whether sterilization has been achieved by heat treatment. Some of these are chemical reactions in which an indicator changes color when the proper times and temperatures have been reached (Figure 7.3). In some designs, the word "sterile" or "autoclaved" appears on wrappings or tapes. In another method, a pellet contained within a glass vial melts. A widely used test consists of preparations of specified species of bacterial endospores impregnated into paper strips. After autoclaving, these can then be aseptically inoculated into culture media. Growth in the culture media indicates survival of the endospores and therefore inadequate processing. Other designs use endospore suspensions that can be released, after heating, into a surrounding culture medium within the same vial.

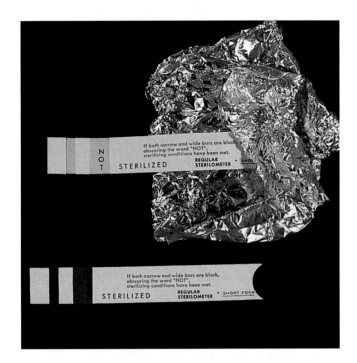

FIGURE 7.3 Examples of sterilization indicators. The strips indicate if the item has been properly sterilized, the word *NOT* appears if heating has been inadequate. In the illustration, the indicator that was wrapped with aluminum foil was not sterilized because steam couldn't penetrate the foil.

Q **What should have been used instead of aluminum foil to wrap the items?**

Steam under pressure fails to sterilize when the air is not completely exhausted. This can happen with the premature closing of the autoclave's automatic ejector valve (see Figure 7.2). The principles of heat sterilization have a direct bearing on home canning. As anyone familiar with home canning knows, the steam must flow vigorously out of the valve in the lid for several minutes to carry with it all the air before the pressure cooker is sealed. If the air is not completely exhausted, the container will not reach the temperature expected for a given pressure. Because of the possibility of botulism, a kind of food poisoning resulting from improper canning methods (see Chapter 22, page 649), people involved in home canning should obtain reliable directions and follow them exactly.

PASTEURIZATION

Recall from Chapter 1 that in the early days of microbiology, Louis Pasteur found a practical method of preventing the spoilage of beer and wine. Pasteur used mild heating, which was sufficient to kill the organisms that caused the particular spoilage problem without seriously damaging the taste of the product. The same principle was later applied to milk to produce what we now call pasteurized milk. The intent of **pasteurization** of milk was to eliminate

pathogenic microbes. It also lowers microbial numbers, which prolongs milk's good quality under refrigeration. Many relatively heat-resistant (**thermoduric**) bacteria survive pasteurization, but these are unlikely to cause disease or cause refrigerated milk to spoil.

Products other than milk, such as ice cream, yogurt, and beer, all have their own pasteurization times and temperatures, which often differ considerably. There are several reasons for these variations. For example, heating is less efficient in foods that are more viscous, and fats in food can have a protective effect on microorganisms. The dairy industry routinely uses a test to determine whether products have been pasteurized: the *phosphatase test* (phosphatase is an enzyme naturally present in milk). If the product has been pasteurized, phosphatase will have been inactivated.

In the classic pasteurization treatment of milk, the milk was exposed to a temperature of about 63°C for 30 minutes. Most milk pasteurization today uses higher temperatures, at least 72°C, but for only 15 seconds. This treatment, known as **high-temperature short-time (HTST) pasteurization,** is applied as the milk flows continuously past a heat exchanger. In addition to killing pathogens, HTST pasteurization lowers total bacterial counts, so the milk keeps well under refrigeration.

Milk can also be sterilized—something quite different from pasteurization—by **ultra-high-temperature (UHT) treatments** so that it can be stored without refrigeration. This is more useful in parts of the world where refrigeration facilities are not always available. In the United States, sterilization is sometimes used on the small containers of coffee creamers found in restaurants. To avoid giving the milk a cooked taste, a UHT system is used in which the liquid milk never touches a surface hotter than the milk itself while being heated by steam. The milk falls in a thin film through a chamber of superheated steam and reaches 140°C in less than a second. It is held for 3 seconds in a holding tube and then cooled in a vacuum chamber, where the steam flashes off. With this process, in less than 5 seconds the milk temperature rises from 74°C to 140°C and drops back to 74°C.

The heat treatments we have just discussed illustrate the concept of **equivalent treatments:** as the temperature is increased, much less time is needed to kill the same number of microbes. For example, the destruction of highly resistant endospores might take 70 minutes at 115°C, whereas only 7 minutes might be needed at 125°C. Both treatments yield the same result. The concept of equivalent treatments also explains why classic pasteurization at 63°C for 30 minutes, HTST treatment at 72°C for 15 seconds, and UHT treatment at 140°C for less than a second can have similar effects.

DRY HEAT STERILIZATION

Dry heat kills by oxidation effects. A simple analogy is the slow charring of paper in a heated oven, even when the temperature remains below the ignition point of paper. One of the simplest methods of dry heat sterilization is direct **flaming.** You will use this procedure many times in the microbiology laboratory when you sterilize inoculating loops. To effectively sterilize the inoculating loop, you heat the wire to a red glow. A similar principle is used in *incineration*, an effective way to sterilize and dispose of contaminated paper cups, bags, and dressings.

Another form of dry heat sterilization is **hot-air sterilization.** Items to be sterilized by this procedure are placed in an oven. Generally, a temperature of about 170°C maintained for nearly 2 hours ensures sterilization. The longer period and higher temperature (relative to moist heat) are required because the heat in water is more readily transferred to a cool body than is the heat in air. For example, imagine the different effects of immersing your hand in boiling water at 100°C (212°F) and of holding it in a hot-air oven at the same temperature for the same amount of time.

FILTRATION

Recall from Chapter 6 that *filtration* is the passage of a liquid or gas through a screenlike material with pores small enough to retain microorganisms (often the same apparatus used for counting; see Figure 6.17, page 181). A vacuum is created in the receiving flask; air pressure then forces the liquid through the filter. Filtration is used to sterilize heat-sensitive materials, such as some culture media, enzymes, vaccines, and antibiotic solutions.

Some operating theaters and rooms occupied by burn patients receive filtered air to lower the numbers of airborne microbes. **High-efficiency particulate air (HEPA) filters** remove almost all microorganisms larger than about 0.3 μm in diameter.

In the early days of microbiology, hollow candle-shaped filters of unglazed porcelain were used to filter liquids. The long and indirect passageways through the walls of the filter adsorbed the bacteria. Unseen pathogens that passed through the filters (causing such diseases as rabies) were called *filterable viruses*.

In recent years, **membrane filters,** composed of such substances as cellulose esters or plastic polymers, have become popular for industrial and laboratory use (Figure 7.4). These filters are only 0.1 mm thick. The pores of membrane filters include, for example, 0.22-μm and 0.45-μm sizes, which are intended for bacteria. Some very flexible bacteria, such as spirochetes, or the wall-less mycoplasma, will sometimes pass through such filters, however. Filters

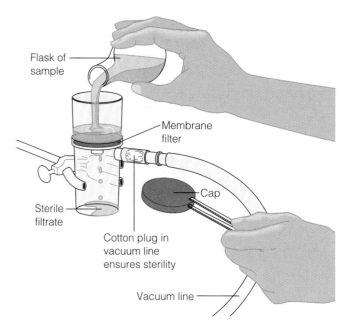

FIGURE 7.4   Filter sterilization with a disposable, presterilized plastic unit.   The sample is placed into the upper chamber and forced through the membrane filter by a vacuum in the lower chamber. Pores in the membrane filter are smaller than the bacteria, so bacteria are retained on the filter. The sterilized sample can then be decanted from the lower chamber. Similar equipment with removable filter disks is used to count bacteria in samples (see Figure 6.17).

Q   **How is a plastic filtration apparatus presterilized? (Assume the plastic cannot be heat sterilized.)**

are available with pores as small as 0.01 μm, a size that will retain viruses and even some large protein molecules.

LOW TEMPERATURES

The effect of low temperatures on microorganisms depends on the particular microbe and the intensity of the application. For example, at temperatures of ordinary refrigerators (0 to 7°C), the metabolic rate of most microbes is so reduced that they cannot reproduce or synthesize toxins. In other words, ordinary refrigeration has a bacteriostatic effect. Yet psychrotrophs do grow slowly at refrigerator temperatures and will alter the appearance and taste of foods after a time. For example, a single microbe reproducing only three times a day would reach a population of more than 2 million within a week. Pathogenic bacteria generally will not grow at refrigerator temperatures, but for at least one important exception, see the discussion of listeriosis in Chapter 22 (page 647).

Surprisingly, some bacteria can grow at temperatures several degrees below freezing. Most foods remain unfrozen until −2°C or lower. Rapidly attained subfreezing temperatures tend to render microbes dormant but do not necessarily kill them. Slow freezing is more harmful to bacteria; the ice crystals that form and grow disrupt the cellular and molecular structure of the bacteria. Thawing, being inherently slower, is actually the more damaging part of a freeze-thaw cycle. Once frozen, one-third of the population of some vegetative bacteria might survive a year, whereas other species might have very few survivors after this time. Many eukaryotic parasites, such as the roundworms that cause human trichinosis, are killed by several days of freezing temperatures. Some important temperatures associated with microorganisms and food spoilage are shown in Figure 6.2 (page 161).

HIGH PRESSURE

High pressure applied to liquid suspensions is transferred instantly and evenly throughout the sample. If the pressure is high enough, the molecular structures of proteins and carbohydrates are altered, resulting in the rapid inactivation of vegetative bacterial cells. Endospores are relatively resistant to high pressure. They can, however, be killed by other techniques, such as combining high pressure with elevated temperatures or by alternating pressure cycles that cause spore germination, followed by pressure-caused death of the resulting vegetative cells. Fruit juices preserved by high-pressure treatments have been marketed in Japan and the United States. An advantage is that these treatments preserve the flavors, colors, and nutrient values of the products.

DESICCATION

In the absence of water, a condition that is known as **desiccation,** microorganisms cannot grow or reproduce but can remain viable for years. Then, when water is made available to them, they can resume their growth and division. This ability is used in the laboratory when microbes are preserved by lyophilization, or freeze-drying, a process described in Chapter 6 (page 174). Certain foods are also freeze-dried (for example, coffee and some fruit additives for dry cereals).

The resistance of vegetative cells to desiccation varies with the species and the organism's environment. For example, the gonorrhea bacterium can withstand dryness for only about an hour, but the tuberculosis bacterium can remain viable for months. Viruses are generally resistant to desiccation, but they are not as resistant as bacterial endospores, some of which have survived for centuries. This ability of certain dried microbes and endospores to remain viable is important in a hospital setting. Dust, clothing, bedding, and dressings might contain infectious microbes in dried mucus, urine, pus, and feces.

OSMOTIC PRESSURE

The use of high concentrations of salts and sugars to preserve food is based on the effects of *osmotic pressure*. High concentrations of these substances create a hypertonic environment that causes water to leave the microbial cell (see Figure 6.4, (page 163). This process resembles preservation by desiccation, in that both methods deny the cell the moisture it needs for growth. The principle of osmotic pressure is used in the preservation of foods. For example, concentrated salt solutions are used to cure meats, and thick sugar solutions are used to preserve fruits.

As a general rule, molds and yeasts are much more capable than bacteria of growing in materials with low moisture or high osmotic pressures. This property of molds, sometimes combined with their ability to grow under acidic conditions, is the reason fruits and grains are spoiled by molds rather than by bacteria. It is also part of the reason molds are able to form mildew on a damp wall or a shower curtain.

RADIATION

Radiation has various effects on cells, depending on its wavelength, intensity, and duration. Radiation that kills microorganisms (sterilizing radiation) is of two types: ionizing and nonionizing.

Ionizing radiation—gamma rays, X rays, or high-energy electron beams—has a wavelength shorter than that of nonionizing radiation, less than about 1 nm. Therefore, it carries much more energy (Figure 7.5). *Gamma rays* are emitted by certain radioactive elements such as cobalt, and electron beams are produced by accelerating electrons to high energies in special machines. *X rays,* which are produced by machines in a manner similar to the production of electron beams, are similar in nature to gamma rays. Gamma rays penetrate deeply but may require hours to sterilize large masses; *high-energy electron beams* have much lower penetrating power but usually require only a few seconds of exposure. The principal effect of ionizing radiation is the ionization of water, which forms highly reactive hydroxyl radicals (see the discussion of toxic forms of oxygen in Chapter 6, pages 166–167). These radicals react with organic cellular components, especially DNA.

The so-called target theory of damage by radiation supposes that ionizing particles, or packets of energy, pass through or close to vital portions of the cell; these constitute "hits." One, or a few, hits may only cause nonlethal mutations, some of them conceivably useful. More hits are likely to cause sufficient mutations to kill the microbe.

The food industry has recently renewed its interest in the use of radiation for food preservation (discussed more fully in Chapter 28). Low-level ionizing radiation, used for years in many countries, has been approved in the United

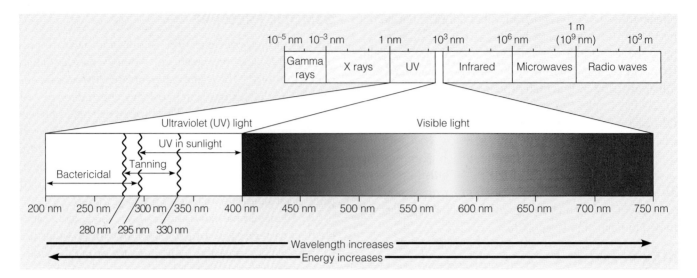

FIGURE 7.5 The radiant energy spectrum. Visible light and other forms of radiant energy radiate through space as waves of various lengths. Ionizing radiation, such as gamma rays and X rays, has a wavelength shorter than 1 nm. Nonionizing radiation, such as ultraviolet (UV) light, has a wavelength between 1 nm and about 380 nm, where the visible spectrum begins.

Q **What effect might increased UV radiation (due to decrease in the ozone layer) have on the Earth's ecosystems?**

States for processing spices and certain meats and vegetables. Ionizing radiation, especially high-energy electron beams, is used for the sterilization of pharmaceuticals and disposable dental and medical supplies, such as plastic syringes, surgical gloves, suturing materials, and catheters. As a protection against bioterrorism, the postal service often uses electron beam radiation to sterilize certain classes of mail.

Nonionizing radiation has a wavelength longer than that of ionizing radiation, usually greater than about 1 nm. The best example of nonionizing radiation is ultraviolet (UV) light. UV light damages the DNA of exposed cells by causing bonds to form between adjacent pyrimidine bases (page 235), usually thymines, in DNA chains (see Figure 8.20). These *thymine dimers* inhibit correct replication of the DNA during reproduction of the cell. The UV wavelengths most effective for killing microorganisms are about 260 nm; these wavelengths are specifically absorbed by cellular DNA. UV radiation is also used to control microbes in the air. A UV, or "germicidal," lamp is commonly found in hospital rooms, nurseries, operating rooms, and cafeterias. UV light is also used to disinfect vaccines and other medical products. A major disadvantage of UV light as a disinfectant is that the radiation is not very penetrating, so the organisms to be killed must be directly exposed to the rays. Organisms protected by solids and such coverings as paper, glass, and textiles are not affected. Another potential problem is that UV light can damage human eyes, and prolonged exposure can cause burns and skin cancer in humans.

Sunlight contains some UV radiation, but the shorter wavelengths—those most effective against bacteria—are screened out by the ozone layer of the atmosphere. The antimicrobial effect of sunlight is due almost entirely to the formation of singlet oxygen in the cytoplasm (see Chapter 6, page 166). Many pigments produced by bacteria provide protection from sunlight.

Microwaves do not have much direct effect on microorganisms, and bacteria can readily be isolated from the interior of recently operated microwave ovens. Moisture-containing foods are heated by microwave action, and the heat will kill most vegetative pathogens. Solid foods heat unevenly because of the uneven distribution of moisture. For this reason, pork cooked in a microwave oven has been responsible for outbreaks of trichinellosis.

* * *

Table 7.5 summarizes the physical methods of microbial control.

CHEMICAL METHODS OF MICROBIAL CONTROL

Chemical agents are used to control the growth of microbes on both living tissue and inanimate objects. Unfortunately, few chemical agents achieve sterility; most

TABLE 7.5	Physical Methods Used to Control Microbial Growth		
Methods	Mechanism of Action	Comment	Preferred Use
Heat			
1. Moist heat			
a. Boiling or flowing steam	Protein denaturation	Kills vegetative bacterial and fungal pathogens and almost all viruses within 10 min; less effective on endospores.	Dishes, basins, pitchers, various equipment
b. Autoclaving	Protein denaturation	Very effective method of sterilization; at about 15 psi of pressure (121°C), all vegetative cells and their endospores are killed in about 15 min.	Microbiological media, solutions, linens, utensils, dressings, equipment, and other items that can withstand temperature and pressure
2. Pasteurization	Protein denaturation	Heat treatment for milk (72°C for about 15 sec) that kills all pathogens and most nonpathogens.	Milk, cream, and certain alcoholic beverages (beer and wine)
3. Dry heat			
a. Direct flaming	Burning contaminants to ashes	Very effective method of sterilization.	Inoculating loops
b. Incineration	Burning to ashes	Very effective method of sterilization.	Paper cups, contaminated dressings, animal carcasses, bags, and wipes
c. Hot-air sterilization	Oxidation	Very effective method of sterilization but requires temperature of 170°C for about 2 hr.	Empty glassware, instruments, needles, and glass syringes
Filtration	Separation of bacteria from suspending liquid	Removes microbes by passage of a liquid or gas through a screen-like material. Most filters in use consist of cellulose acetate or nitrocellulose.	Useful for sterilizing liquids (enzymes, vaccines) that are destroyed by heat
Cold			
1. Refrigeration	Decreased chemical reactions and possible changes in proteins	Has a bacteriostatic effect.	Food, drug, and culture preservation
2. Deep-freezing (see Chapter 6, page 174)	Decreased chemical reactions and possible changes in proteins	An effective method for preserving microbial cultures, in which cultures are quick-frozen between −50° and −95°C.	Food, drug, and culture preservation
3. Lyophilization (see Chapter 6, page 174)	Decreased chemical reactions and possible changes in proteins	Most effective method for longterm preservation of microbial cultures; water removed by high vacuum at low temperature.	Food, drug, and culture preservation
High Pressure	Alteration of molecular structure of proteins and carbohydrates	Preservation of colors, flavors, nutrient values.	Fruit juices
Desiccation	Disruption of metabolism	Involves removing water from microbes; primarily bacteriostatic.	Food preservation
Osmotic Pressure	Plasmolysis	Results in loss of water from microbial cells.	Food preservation
Radiation			
1. Ionizing	Destruction of DNA	Not widespread in routine sterilization.	Used for sterilizing pharmaceuticals and medical and dental supplies
2. Nonionizing	Damage to DNA	Radiation not very penetrating.	Control of closed environment with UV (germicidal) lamp

of them merely reduce microbial populations to safe levels or remove vegetative forms of pathogens from objects. A common problem in disinfection is the selection of an agent. No single disinfectant is appropriate for all circumstances.

PRINCIPLES OF EFFECTIVE DISINFECTION

LEARNING OBJECTIVE
- List the factors related to effective disinfection.

By reading the label, we can learn a great deal about a disinfectant's properties. The label will usually indicate what groups of organisms the disinfectant will be effective against. Remember that the concentration of a disinfectant affects its action, so it should always be diluted exactly as specified by the manufacturer.

Also consider the nature of the material being disinfected. For example, are organic materials present that might interfere with the action of the disinfectant? Similarly, the pH of the medium often has a great effect on a disinfectant's activity.

Another very important consideration is whether the disinfectant will easily make contact with the microbes. An area might need to be scrubbed and rinsed before the disinfectant is applied. In general, disinfection is a gradual process. Thus, to be effective, a disinfectant might need to be left on a surface for several hours.

EVALUATING A DISINFECTANT

LEARNING OBJECTIVE
- Interpret the results of use-dilution tests and the disk-diffusion method.

USE-DILUTION TESTS

There is a need to evaluate the effectiveness of disinfectants and antiseptics. For many years the standard test was the *phenol coefficient test*, which compared the activity of a given disinfectant with that of phenol. However, the current standard is the American Official Analytical Chemist's **use-dilution test.** For most purposes, the three bacteria used in this test are *Salmonella choleraesuis*, *Staphylococcus aureus*, and *Pseudomonas aeruginosa*. Metal carrier rings are dipped into standardized cultures of the test bacteria grown in liquid media, removed, and dried at 37°C for a short time. The dried cultures are then placed into a solution of the disinfectant at the concentration recommended by the manufacturer and left there for 10 minutes at 20°C. Following this exposure, the carrier rings are transferred to a medium that will permit the growth of any surviving bacteria. The effectiveness of the disinfectant can then be determined by the number of cultures that grow.

Variations of this method are used for testing the effectiveness of antimicrobial agents against endospores, mycobacteria that cause tuberculosis, and fungi, because they are difficult to control with chemicals. Also, tests of antimicrobials intended for special purposes, such as dairy utensil disinfection, may substitute other test bacteria. Virucidal chemicals are usually tested against cultures of Newcastle disease virus (a disease of birds and domestic fowl). After exposure to the chemical, the cultures are injected into embryonated chicken eggs; if any viruses survive, they will kill the embryos.

THE DISK-DIFFUSION METHOD

The **disk-diffusion method** is used in teaching laboratories to evaluate the efficacy of a chemical agent. A disk of filter paper is soaked with a chemical and placed on an agar plate that has been previously inoculated and incubated with the test organism. After incubation, if the chemical is effective, a clear zone representing inhibition of growth can be seen around the disk (Figure 7.6).

Disks containing antibiotics are commercially available and used to determine microbial susceptibility to antibiotics (see Figure 20.17, page 601).

TYPES OF DISINFECTANTS

LEARNING OBJECTIVES
- Identify the methods of action and preferred uses of chemical disinfectants.
- Differentiate between halogens used as antiseptics and as disinfectants.
- Identify the appropriate uses for surface-active agents.
- List the advantages of glutaraldehyde over other chemical disinfectants.
- Identify the method of sterilizing plastic labware.

PHENOL AND PHENOLICS

Lister was the first to use **phenol** (carbolic acid) to control surgical infections in the operating room. Its use had been suggested by its effectiveness in controlling odor in sewage. It is now rarely used as an antiseptic or disinfectant because it irritates the skin and has a disagreeable odor. It is often used in throat lozenges for its local anesthetic effect but has little antimicrobial effect at the low concentrations used. At concentrations above 1% (such as in some throat sprays), however, phenol has a significant

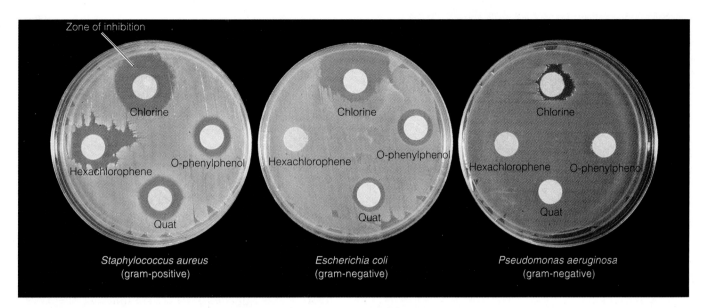

Zone of inhibition

Chlorine

O-phenylphenol

Hexachlorophene

Quat

Staphylococcus aureus
(gram-positive)

Chlorine

O-phenylphenol

Hexachlorophene

Quat

Escherichia coli
(gram-negative)

Chlorine

O-phenylphenol

Hexachlorophene

Quat

Pseudomonas aeruginosa
(gram-negative)

FIGURE 7.6 Evaluation of disinfectants by the disk-diffusion method. In this experiment, paper disks are soaked in a solution of disinfectant and placed on the surface of a nutrient medium on which a culture of test bacteria has been spread to produce uniform growth.

At the top of each plate, the tests show that chlorine (as sodium hypochlorite) was effective against all the test bacteria but was more effective against gram-positive bacteria.

At the bottom row of each plate, the tests show that the quaternary ammonium compound ("quat") was also more effective against the gram-positive bacteria, but it did not affect the pseudomonads at all.

At the left side of each plate, the tests show that hexachlorophene was effective against gram-positive bacteria only.

At the right sides, O-phenylphenol was ineffective against pseudomonads but was almost equally effective against the gram-positive bacteria and the gram-negative bacteria.

All four chemicals worked against the gram-positive test bacteria, but only one of the four chemicals affected pseudomonads.

Q **Which group of bacteria is most resistant to the disinfectants tested?.**

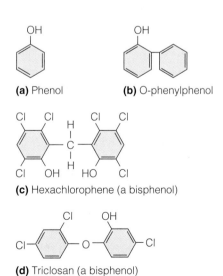

(a) Phenol

(b) O-phenylphenol

(c) Hexachlorophene (a bisphenol)

(d) Triclosan (a bisphenol)

FIGURE 7.7 The structure of phenolics and bisphenols.

Q **Some lozenges intended to alleviate the symptoms of a sore throat contain phenol. Why include this ingredient?**

antibacterial effect. The structure of a phenol molecule is shown in Figure 7.7a.

Derivatives of phenol, called **phenolics,** contain a molecule of phenol that has been chemically altered to reduce its irritating qualities or increase its antibacterial activity in combination with a soap or detergent. Phenolics exert antimicrobial activity by injuring lipid-containing plasma membranes, which results in leakage of cellular contents. The cell wall of mycobacteria, the causes of tuberculosis and leprosy, are rich in lipids, which make them susceptible to phenol derivatives. A useful property of phenolics as disinfectants is that they remain active in the presence of organic compounds, they are stable, and they persist for long periods after application. For these reasons, phenolics are suitable agents for disinfecting pus, saliva, and feces.

One of the most frequently used phenolics is derived from coal tar, a group of chemicals called *cresols*. A very important cresol is *O-phenylphenol* (see Figures 7.6 and 7.7b), the main ingredient in most formulations of Lysol. Cresols are very good surface disinfectants.

BISPHENOLS

Bisphenols are derivatives of phenol that contain two phenolic groups connected by a bridge (*bis* indicates two). One bisphenol, *hexachlorophene* (Figures 7.6 and 7.7c), is an ingredient of a prescription lotion, pHisoHex, used for surgical and hospital microbial control procedures. Gram-positive staphylococci and streptococci, which can cause skin infections in newborns, are particularly susceptible to hexachlorophene, so it is often used to control such infections in nurseries. However, excessive use of this bisphenol, such as bathing infants with it several times a day, can lead to neurological damage.

Another widely used bisphenol is *triclosan* (Figure 7.7d), an ingredient in antibacterial soaps and at least one toothpaste. Triclosan has even been incorporated into kitchen cutting boards and the handles of knives and other plastic kitchenware. Its use is now so widespread that resistant bacteria have been reported, and concerns about its effect on microbes' resistance to certain antibiotics have been raised. Triclosan inhibits an enzyme needed for the biosynthesis of fatty acids (lipids), which mainly affects the integrity of the plasma membrane. It is especially effective against gram-positive bacteria but also works well against fungi and gram-negative bacteria. There are certain exceptions, such as *Pseudomonas aeruginosa*, a gram-negative bacterium that is very resistant to triclosan, as well as to many other antibiotics and disinfectants (see the discussion on pages 321, 436, and 622).

BIGUANIDES

Chlorhexidine is a member of the **biguanide** group with a broad spectrum of activity. It is frequently used for microbial control on skin and mucous membranes. Combined with a detergent or alcohol, chlorhexidine is also used for surgical hand scrubs and preoperative skin preparation in patients. A new product, Avagard, which combines chlorhexidine and ethanol, is persistent for about six hours. It is approved by the FDA as a waterless, scrubless presurgical antiseptic. In such applications, chlorhexidine's strong affinity for binding to the skin or mucous membranes is an advantage, as is its low toxicity. However, contact with the eyes can cause damage. Its killing effect is related to the injury it causes to the plasma membrane by blocking an enzyme needed for lipid synthesis. It is biocidal against most vegetative bacteria and yeasts. Mycobacteria are relatively resistant, and endospores and protozoan cysts are not affected. The only viruses affected are certain enveloped (lipophilic) types (see Chapter 13).

HALOGENS

The **halogens,** particularly iodine and chlorine, are effective antimicrobial agents, both alone and as constituents of inorganic or organic compounds. *Iodine* (I_2) is one of the oldest and most effective antiseptics. It is effective against all kinds of bacteria, many endospores, various fungi, and some viruses. Iodine impairs protein synthesis and alters cell membranes, apparently by forming complexes with amino acids and unsaturated fatty acids.

Iodine is available as a **tincture**—that is, in solution in aqueous alcohol—and as an iodophor. An **iodophor** is a combination of iodine and an organic molecule, from which the iodine is released slowly. Iodophors have the antimicrobial activity of iodine, but they do not stain and are less irritating. The most common commercial preparations is Betadine, which is a *povidone-iodine*. Povidone is a surface-active iodophor that improves the wetting action and serves as a reservoir of free iodine. Iodines are used mainly for skin disinfection and wound treatment. Many campers are familiar with iodine for water treatment. To treat water, iodine tablets are added or the water can be passed through iodine-treated resin filters.

Chlorine (Cl_2), as a gas or in combination with other chemicals, is another widely used disinfectant. Its germicidal action is caused by the hypochlorous acid (HOCl) that forms when chlorine is added to water:

(1)
$$Cl_2 + H_2O \rightleftharpoons H^+ + Cl^- + HOCl$$

Chlorine   Water   Hydrogen   Chloride   Hypochlorous
                                  ion   ion   acid

(2)
$$HOCl \rightleftharpoons H^+ + OCl^-$$

Hypochlorous   Hydrogen   Hypochlorite
    acid   ion   ion

Exactly how hypochlorous acid exerts its killing power is not known. It is a strong oxidizing agent that prevents much of the cellular enzyme system from functioning. Hypochlorous acid is the most effective form of chlorine because it is neutral in electrical charge and diffuses as rapidly as water through the cell wall. Because of its negative charge, the hypochlorite ion (OCl^-) cannot enter the cell freely.

A liquid form of compressed chlorine gas is used extensively for disinfecting municipal drinking water, water in swimming pools, and sewage. Several compounds of chlorine are also effective disinfectants. For example, solutions of *calcium hypochlorite* [$Ca(OCl)_2$] are used to disinfect dairy equipment and restaurant eating utensils. This compound, once called chloride of lime, was used as early as 1825, long before the concept of a germ theory for disease, to soak hospital dressings in Paris hospitals. It was also the disinfectant used in the 1840s by Semmelweis to control hospital infections during childbirth, as mentioned in Chapter 1, page 11. Another chlorine compound, *sodium hypochlorite* (NaOCl; see Figure 7.6), is used as a household disinfectant and bleach (Clorox) and as a disinfectant in dairies, food-processing establishments, and hemodialysis systems. When the quality of drinking water is in question, household bleach can provide a rough equivalent of municipal chlorination. After two drops of bleach are added to a liter of water (four drops if the water is cloudy) and the mixture has sat for 30 minutes, the water is considered safe for drinking under emergency conditions. U.S. military forces in the field are issued tablets (Chlor-Floc) that contain *sodium dichloroisocyanurate,* a form of chlorine combined with an agent that flocculates (coagulates) suspended materials in a water sample, causing them to settle out, thus clarifying it.

Chlorine dioxide (ClO$_2$) is a gaseous form of chlorine, occasionally used for area disinfection, most notably, to kill endospores of the anthrax bacterium.

Another group of chlorine compounds, the *chloramines,* consist of chlorine and ammonia. They are used as disinfectants, antiseptics, or sanitizing agents. Chloramines are very stable compounds that release chlorine over long periods. They are relatively effective in organic matter, but they have the disadvantages of acting more slowly and being less effective purifiers than many other chlorine compounds. Chloramines are used to sanitize glassware and eating utensils and to treat dairy and food-manufacturing equipment. Ammonia is usually mixed with chlorine in municipal water-treatment systems to form chloramines. The chloramines control taste and odor problems caused by the reaction of chlorine with other nitrogenous compounds in the water. Because chloramines are less effective as germicides, sufficient chlorine must be added to ensure a residual of chlorine in the form of HOCl. (Chloramines are toxic to aquarium fish, but pet shops sell chemicals to neutralize them.)

ALCOHOLS

Alcohols effectively kill bacteria and fungi but not endospores and nonenveloped viruses. The mechanism of action of alcohol is usually protein denaturation, but alcohol can also disrupt membranes and dissolve many lipids, including the lipid component of enveloped viruses. Alcohols have the advantage of acting and then evaporating rapidly and leaving no residue. When the skin is swabbed (degermed) before an injection, most of the microbial control activity comes from simply wiping away dirt and microorganisms, along with skin oils. However, alcohols are unsatisfactory antiseptics when applied to wounds. They cause coagulation of a layer of protein under which bacteria continue to grow.

Two of the most commonly used alcohols are ethanol and isopropanol. The recommended optimum concentration of *ethanol* is 70%, but concentrations between 60% and 95% seem to kill as well (Table 7.6). Pure ethanol is less effective than aqueous solutions (ethanol mixed with water) because denaturation requires water. *Isopropanol,*

TABLE 7.6 Biocidal Action of Various Concentrations of Ethanol in Aqueous Solution Against *Streptococcus pyogenes*

Concentration of Ethanol (%)	Time (sec)				
	10	20	30	40	50
100	−	−	−	−	−
95	+	+	+	+	+
90	+	+	+	+	+
80	+	+	+	+	+
70	+	+	+	+	+
60	+	+	+	+	+
50	−	−	+	+	+
40	−	−	−	−	−

NOTE: A minus sign indicates no biocidal action (bacterial growth); a plus sign indicates biocidal action (no bacterial growth). The lighter area represents bacteria killed by biocidal action.

often sold as rubbing alcohol, is slightly superior to ethanol as an antiseptic and disinfectant. Moreover, it is less volatile, less expensive, and more easily obtained than ethanol.

Ethanol and isopropanol are often used to enhance the effectiveness of other chemical agents. For example, an aqueous solution of Zephiran (described on page 203) kills about 40% of the population of a test organism in two minutes, whereas a tincture of Zephiran kills about 85% in the same period. To compare the effectiveness of tinctures and aqueous solutions, see Figure 7.10 on page 204.

HEAVY METALS AND THEIR COMPOUNDS

Several heavy metals can be biocidal or antiseptic, including silver, mercury, and copper. The ability of very small amounts of heavy metals, especially silver and copper, to exert antimicrobial activity is referred to as **oligodynamic action** (*oligo* means few). This action can be seen when we place a coin or other clean piece of metal containing silver or copper on a culture on an inoculated Petri plate. Extremely small amounts of metal diffuse from the coin and inhibit the growth of bacteria for some distance around the coin (Figure 7.8). This effect is produced by the action of heavy metal ions on microbes. When the metal ions combine with the sulfhydryl groups on cellular proteins, denaturation results.

Silver is used as an antiseptic in a 1% *silver nitrate* solution. At one time, many states required that the eyes of

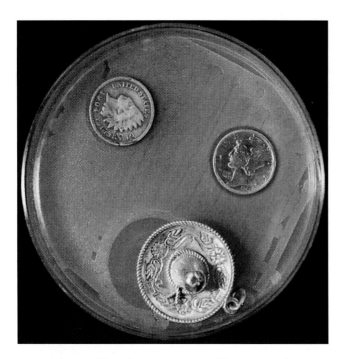

FIGURE 7.8 Oligodynamic action of heavy metals.
Clear zones where bacterial growth has been inhibited are seen around the sombrero charm (pushed aside), the dime, and the penny. The charm and the dime contain silver; the penny contains copper.

Q The coins used in this demonstration were minted many years ago; why were current coins not used?

newborns be treated with a few drops of silver nitrate to guard against an infection of the eyes called gonorrheal ophthalmia neonatorum, which the infants might have contracted as they passed through the birth canal. In recent years, antibiotics have replaced silver nitrate for this purpose.

Recently, there has been renewed interest in silver as an antimicrobial agent. Silver-impregnated dressings that slowly release silver ions have proven especially useful when antibiotic-resistant bacteria are a problem. A combination of silver and the drug sulfadiazine, *silver-sulfadiazine,* is the most common formulation. It is available as a topical cream for use on burns. Silver can also be incorporated into indwelling catheters, which are a common source of hospital infections, and in wound dressings. *Surfacine* is a relatively new antimicrobial for application to surfaces, either animate or inanimate. It contains water-insoluble silver iodide in a polymer carrier and is very persistent, lasting at least 13 days. When a bacterium contacts the surface, the cell's outer membrane is recognized, and a lethal amount of silver ions is released.

Inorganic mercury compounds, such as *mercuric chloride,* have a long history of use as disinfectants. They have a very broad spectrum of activity; their effect is primarily bacteriostatic. However, their use is now limited because of their toxicity, corrosiveness, and ineffectiveness in

organic matter. At present, the primary use of mercurials is to control mildew in paints.

Copper in the form of *copper sulfate* or other copper-containing additives is used chiefly to destroy green algae (algicide) that grow in reservoirs, stock ponds, swimming pools, and fish tanks. If the water does not contain excessive organic matter, copper compounds are effective in concentrations of one part per million of water. To prevent mildew, copper compounds such as copper 8-hydroxy-quinoline are sometimes included in paint.

Another metal used as an antimicrobial is zinc. The effect of trace amounts of zinc can be seen on weathered roofs of buildings down-slope from galvanized (zinc-coated) fittings. The roof is lighter-colored where biological growth is impeded. Copper- and zinc-treated shingles are available. *Zinc chloride* is a common ingredient in mouthwashes, and *zinc oxide* is probably the most widely used antifungal agent in paints, mainly because it is often part of the pigment formulation.

SURFACE-ACTIVE AGENTS

Surface-active agents, or **surfactants,** can decrease surface tension among molecules of a liquid. Such agents include soaps and detergents.

Soaps and Detergents Soap has little value as an antiseptic, but it does have an important function in the mechanical removal of microbes through scrubbing. The skin normally contains dead cells, dust, dried sweat, microbes, and oily secretions from oil glands. Soap breaks the oily film into tiny droplets, a process called *emulsification,* and the water and soap together lift up the emulsified oil and debris and float them away as the lather is washed off. In this sense, soaps are good degerming agents.

Acid-Anionic Sanitizers Acid-anionic surface-active sanitizers are very important in the cleaning of dairy utensils and equipment. Their sanitizing ability is related to the negatively charged portion (anion) of the molecule, which reacts with the plasma membrane. These sanitizers, which act on a wide spectrum of microbes, including troublesome thermoduric bacteria, are nontoxic, noncorrosive, and fast acting.

Quaternary Ammonium Compounds (Quats) The most widely used surface-active agents are the cationic detergents, especially the **quaternary ammonium compounds (quats).** Their cleansing ability is related to the positively charged portion—the cation—of the molecule. Their name is derived from the fact that they are modifications of the four-valence ammonium ion, NH_4^+ (Figure 7.9). Quaternary ammonium compounds are strongly bactericidal against gram-positive bacteria and less active against gram-negative bacteria (see Figure 7.6).

CLINICAL PROBLEM SOLVING

A HOSPITAL-ACQUIRED INFECTION FOLLOWING LIPOSUCTION

As you read through this box you will encounter a series of questions that clinicians ask themselves as they formulate a diagnosis and treatment. Try to answer each question before going on to the next one.

1. During a 17-month period, nine patients in eight hospitals acquired surgical-site infections within 2 months after liposuction. The patients' symptoms included fever, local inflammation, and infected surgical wounds. *What would you do next?*

2. Gram stains and acid-fast stains of pus showed gram-positive, acid-fast rods. *What are possible organisms?*

3. Slow-growing mycobacteria, including *M. tuberculosis* and *M. leprae*, are human pathogens. These infections, however, were caused by rapidly growing mycobacteria (RGM): *Mycobacterium chelonae, M. fortuitum*, and *M. abscessus*.

Where are other species of mycobacteria normally found?

4. RGM are found in soil and water. The liposuction procedure involves making a small surgical wound and using a cannula (needle) for fat suctioning. The cannulae were cleaned with tap water and soap followed by a quat. *What was wrong with this procedure?*

5. RGM can be found in tap water and soap used to remove dust and microbes. Quats do have some disinfecting properties but are considered low-level disinfectants because they are ineffective against endospores and mycobacteria. *Why are mycobacteria resistant to some disinfectants?*

6. The lipid-rich cell wall prevents entry of most biocides into the cells. *How could you find the source of the RGM?*

7. Nosocomial infections associated with contaminated quats that had been used to disinfect patient-care supplies or equipment have been reported. Nosocomial infections have also been associated with RGM growing in a hospital's hot water system. In this case, cultures of the quat and environmental cultures did not yield bacteria or mycobacteria. *What changes in procedures would you recommend?*

8. Surgical instruments used in liposuction are intended to enter normally sterile tissue and should be sterilized between patient procedures. The hospitals modified their procedures by replacing the quat with either a high-level disinfectant using 2% glutaraldehyde or ethylene oxide gas sterilization.

SOURCE: Adapted from *MMWR* 47(49): 1065–1067 (12/18/98).

Quats are also fungicidal, amoebicidal, and virucidal against enveloped viruses. They do not kill endospores or mycobacteria. (See the box above.) Their chemical mode of action is unknown, but they probably affect the plasma membrane. They change the cell's permeability and cause the loss of essential cytoplasmic constituents, such as potassium.

Two popular quats are Zephiran, a brand name of *benzalkonium chloride* (see Figure 7.9), and Cepacol, a brand name of *cetylpyridinium chloride*. They are strongly antimicrobial, colorless, odorless, tasteless, stable, easily diluted, and nontoxic, except at high concentrations. If your mouthwash bottle fills with foam when shaken, the mouthwash probably contains a quat. However, organic matter interferes with their activity, and they are rapidly neutralized by soaps and anionic detergents.

Anyone associated with medical applications of quats should remember that certain bacteria, such as some species of *Pseudomonas*, not only survive in quaternary ammonium compounds but actively grow in them. This resistance occurs not only to the disinfectant solution but also to moistened

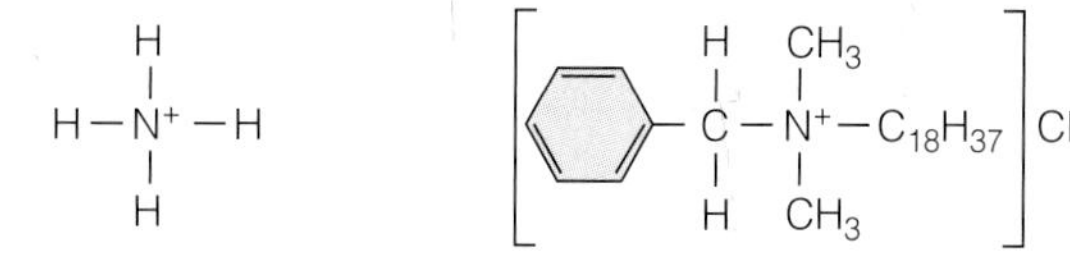

Ammonium ion

Benzalkonium chloride

FIGURE 7.9 The ammonium ion and a quaternary ammonium compound, benzalkonium chloride (Zephiran). Notice how other groups replace the hydrogens of the ammonium ion.

Q **Are quats most effective against gram-positive or gram-negative bacteria?**

gauze and bandages, whose fibers tend to neutralize the quats. Cetylpyridinium chloride, another quat, has recently received FDA (U.S. Food and Drug Administration) approval as an antimicrobial wash during poultry processing.

Before we move on to the next group of chemical agents, refer to Figure 7.10, which compares the effectiveness of some of the antiseptics we have discussed so far.

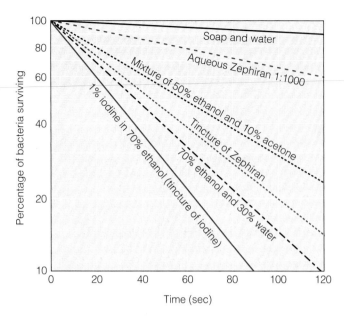

FIGURE 7.10 A comparison of the effectiveness of various antiseptics. The steeper the downward slope of the killing curve of the antiseptic, the more effective it is. A 1% iodine in 70% ethanol solution is the most effective; soap and water are the least effective. Notice that a tincture of Zephiran is more effective than an aqueous solution of the same antiseptic.

Q **Why is the tincture of Zephiran more effective than the aqueous solution?**

CHEMICAL FOOD PRESERVATIVES

Chemical preservatives are frequently added to foods to retard spoilage. *Sulfur dioxide* (SO_2) has long been used as a disinfectant, especially in wine-making. Homer's *Odyssey*, written nearly 2800 years ago, mentions its use. Among the more common additives are sodium benzoate, sorbic acid, and calcium propionate. These chemicals are simple organic acids, or salts of organic acids, which the body readily metabolizes and which are generally judged to be safe in foods. *Sorbic acid,* or its more soluble salt *potassium sorbate,* and *sodium benzoate* prevent molds from growing in certain acidic foods, such as cheese and soft drinks. Such foods, usually with a pH of 5.5 or lower, are most susceptible to mold-type spoilage. *Calcium propionate,* an effective fungistat used in bread, prevents the growth of surface molds and the *Bacillus* bacterium that causes ropy bread. These organic acids inhibit mold growth, not by affecting the pH but by interfering with the mold's metabolism or the integrity of the plasma membrane.

Sodium nitrate and *sodium nitrite* are added to many meat products, such as ham, bacon, hot dogs, and sausage. The active ingredient is sodium nitrite, which certain bacteria in the meats can also produce from sodium nitrate. These bacteria use nitrate as a substitute for oxygen under anaerobic conditions. The nitrite has two main functions: to preserve the pleasing red color of the meat by reacting with blood components in the meat, and to prevent the germination and growth of any botulism endospores that might be present. Nitrite selectively inhibits certain iron-containing enzymes of *Clostridium botulinum.* There has been some concern that the reaction of nitrites with amino acids can form certain carcinogenic products known as **nitrosamines,** and the amount of nitrites added to foods has generally been reduced recently for this reason. However, the use of nitrites continues because of their established value in preventing botulism. Because nitrosamines are formed in the body from other sources, the added risk posed by a limited use of nitrates and nitrites in meats is lower than was once thought.

ANTIBIOTICS

The antimicrobials discussed in this chapter are not useful for ingestion or injection to treat disease. Antibiotics are used for this purpose. The use of antibiotics is highly restricted; however, at least two have considerable use in food preservation. Neither is of value for clinical purposes. *Nisin,* which is often added to cheese to inhibit the growth of certain endospore-forming spoilage bacteria, is an example of a bacteriocin, a protein that is produced by one bacterium and inhibits another (see Chapter 8, page 246). Nisin is present naturally in small amounts in many dairy products. It is tasteless, readily digested, and nontoxic. *Natamycin* (pimaricin) is an antifungal antibiotic approved for use in foods, mostly cheese.

ALDEHYDES

Aldehydes are among the most effective antimicrobials. Two examples are formaldehyde and glutaraldehyde. They inactivate proteins by forming covalent cross-links with several organic functional groups on proteins (—NH$_2$, —OH, —COOH, and—SH). *Formaldehyde gas* is an excellent disinfectant. However, it is more commonly available as *formalin,* a 37% aqueous solution of formaldehyde gas. Formalin was once used extensively to preserve biological specimens and inactivate bacteria and viruses in vaccines.

Glutaraldehyde is a chemical relative of formaldehyde that is less irritating and more effective than formaldehyde. Glutaraldehyde is used to disinfect hospital instruments, including respiratory-therapy equipment. When used in a 2% solution (Cidex), it is bactericidal, tuberculocidal, and virucidal in 10 minutes and sporicidal in 3 to 10 hours. Glutaraldehyde is one of the few liquid chemical disinfectants that can be considered a sterilizing agent. However, 30 minutes is often considered the maximum time allowed for a sporicide to act, which is a criterion glutaraldehyde cannot meet. Both glutaraldehyde and formalin are used by morticians for embalming.

A possible replacement for glutaraldehyde for many uses is *ortho-phthalaldehyde* (OPA), which is more effective against many microbes and has fewer irritating properties.

GASEOUS CHEMOSTERILIZERS

Gaseous chemosterilizers are chemicals that sterilize in a closed chamber (similar to an autoclave). A gas suitable for this method is *ethylene oxide:*

$$H_2C - CH_2$$
$$\diagdown \diagup$$
$$O$$

Its activity depends on the denaturation of proteins: the proteins' labile hydrogens, such as —SH, —COOH, or —OH, are replaced by alkyl groups (*alkylation*), such as —CH₂CH₂OH. Ethylene oxide kills all microbes and endospores but requires a lengthy exposure period of 4 to 18 hours. It is toxic and explosive in its pure form, so it is usually mixed with a nonflammable gas, such as carbon dioxide or nitrogen. One of its advantages is that it is highly penetrating, so much so that ethylene oxide was chosen to sterilize spacecraft sent to land on the moon and Mars. Using heat to sterilize the electronic gear on these vehicles was not practical.

Because of their ability to sterilize without heat, gases like ethylene oxide are also widely used on medical supplies and equipment. Many large hospitals have ethylene oxide chambers, some large enough to sterilize mattresses, as part of their sterilizing equipment. Propylene oxide and β-propiolactone are also used for gaseous sterilization:

$$H_3C - CH - CH_2 \qquad H_2C - CH_2$$
$$\diagdown \diagup \qquad\qquad |\qquad\ |$$
$$O \qquad\qquad\qquad O - C = O$$
Propylene oxide   β-propiolactone

A disadvantage of all these gases is that they are suspected carcinogens, especially β-propiolactone. For this reason, there has been concern about the exposure of hospital workers to ethylene oxide from such sterilizers. Because of this hazard, these gases may eventually be replaced by *plasma gas sterilization*. This makes use of vapors of hydrogen peroxide (discussed shortly) subjected to radio frequencies or microwave radiation to produce reactive free radicals. No by-products toxic to humans are produced, and it is an effective sterilant.

PEROXYGENS (OXIDIZING AGENTS)

Peroxygens exert antimicrobial activity by oxidizing cellular components of the treated microbes. Examples are ozone, hydrogen peroxide, and peracetic acid. *Ozone* (O_3) is a highly reactive form of oxygen that is generated by passing oxygen through high-voltage electrical discharges. It is responsible for the air's rather fresh odor after a lightning storm, in the vicinity of electrical sparking, or around an ultraviolet light. Ozone is often used to supplement chlorine in the disinfection of water because it helps neu-

tralize tastes and odors. Although ozone is a more effective killing agent, its residual activity is difficult to maintain in water, and it is more expensive than chlorine.

Hydrogen peroxide is an antiseptic found in many household medicine cabinets and in hospital supply rooms. It is not a good antiseptic for open wounds; in fact, it may slow wound healing. It is quickly broken down to water and gaseous oxygen by the action of the enzyme catalase, which is present in human cells (see Chapter 6, page 167). However, hydrogen peroxide does effectively disinfect inanimate objects, an application in which it is even sporicidal, especially at elevated temperatures. On a nonliving surface, the normally protective enzymes of aerobic bacteria and facultative anaerobes are overwhelmed by the high concentrations of peroxide used. Because of these factors, the food industry is increasing its use of hydrogen peroxide for aseptic packaging (see Chapter 28). The packaging materials pass through a hot solution of the chemical before being assembled into a container. In addition, many wearers of contact lenses are familiar with disinfection by hydrogen peroxide. After disinfection, a platinum catalyst in the lens-disinfecting kit destroys residual hydrogen peroxide so that it does not persist on the lens, where it might be an irritant.

Oxidizing agents are useful for irrigating deep wounds, where the oxygen released makes an environment that inhibits the growth of anaerobic bacteria.

Benzoyl peroxide is another compound useful for treating wounds infected by anaerobic pathogens, but it is probably more familiar as the main ingredient in over-the-counter medications for acne, which is caused by a type of anaerobic bacterium infecting hair follicles.

Peracetic acid is one of the most effective liquid chemical sporicides available and is considered a sterilant. It is generally effective on endospores and viruses within 30 minutes, and kills vegetative bacteria and fungi in less than 5 minutes. Peracetic acid has many applications in the disinfection of food-processing and medical equipment because it leaves no toxic residues and is minimally affected by the presence of organic matter.

MICROBIAL CHARACTERISTICS AND MICROBIAL CONTROL

LEARNING OBJECTIVE

- Explain how the control of microbial control is affected by the type of microbe.

Many biocides tend to be more effective against gram-positive bacteria, as a group, than against gram-negative bacteria. This is illustrated in Figure 7.11, which presents a simplified hierarchy of relative resistance of major microbial

Most Resistant

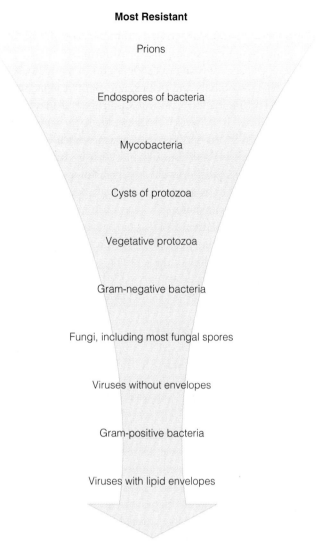

Prions

Endospores of bacteria

Mycobacteria

Cysts of protozoa

Vegetative protozoa

Gram-negative bacteria

Fungi, including most fungal spores

Viruses without envelopes

Gram-positive bacteria

Viruses with lipid envelopes

Least Resistant

FIGURE 7.11 Decreasing order of resistance of microorganisms to chemical biocides.

Q Why are viruses with lipid-containing envelopes relatively susceptible to certain biocides?

TABLE 7.7	The Effectiveness of Chemical Antimicrobials Against Endospores and Mycobaceria	
Chemical Agent	Endospores	Mycobacteria
Mercury	No activity	No activity
Phenolics	Poor	Good
Bisphenols	No activity	No activity
Quaternary ammonium compounds	No activity	No activity
Chlorines	Fair	Fair
Iodine	Poor	Good
Alcohols	Poor	Good
Glutaraldehyde	Fair	Good
Chlorhexidine	No activity	Fair

The mycobacteria are another group of non–endospore-forming bacteria that exhibit greater than normal resistance to chemical biocides. (See the box on page 203.) This group includes *Mycobacterium tuberculosis*, the pathogen that causes tuberculosis. The cell wall of this organism and other members of this genus have a waxy, lipid-rich component. Instruction labels on disinfectants often state whether they are tuberculocidal, indicating if they are effective against mycobacteria. Special tuberculocidal tests have been developed to evaluate the effectiveness of biocides against this bacterial group.

Bacterial endospores are affected by relatively few biocides. (The activity of the major chemical antimicrobial groups against mycobacteria and endospores is summarized in Table 7.7.) The cysts and oocysts of protozoa are also relatively resistant to chemical disinfection.

Viruses are not especially resistant to biocides, but a distinction must be made between those that possess a lipid-containing envelope and those that do not. Antimicrobials that are lipid-soluble are more likely to be effective against enveloped viruses. If so, this will usually be indicated on the label by a statement that they are effective against lipophilic viruses. Nonenveloped viruses, with only a protein coat, are more resistant—fewer biocides are active against them.

A special problem, not yet completely solved, is the reliable killing of prions (see Chapter 13, page 412). Prions are infectious proteins that are the cause of neurological diseases, the spongiform encephalopathies, such as the popularly named mad cow disease (see Chapter 22, page 664). To destroy prions, infected animal carcasses are incinerated. A

groups to biocides. A principal factor in this relative resistance to biocides is the external lipopolysaccharide layer of gram-negative bacteria. Within gram-negative bacteria, members of the genera *Pseudomonas* and *Burkholderia* are of special interest. These closely related bacteria are unusually resistant to biocides (see Figure 7.6) and will even grow actively in some disinfectants and antiseptics, most notably the quaternary ammonium compounds. In Chapter 20, you will see that these bacteria are also resistant to many antibiotics. This resistance to chemical antimicrobials is mostly related to the characteristics of their *porins* (structural openings in the wall of gram-negative bacteria; see Figure 4.13c, page 86). Porins are highly selective of molecules that they permit to enter the cell.

TABLE 7.8	Chemical Agents Used to Control Microbial Growth		
Chemical Agent	Mechanism of Action	Preferred Use	Comment
Phenol and Phenolics			
1. Phenol	Disruption of plasma membrane, denaturation of enzymes.	Rarely used, except as a standard of comparison.	Seldom used as a disinfectant or antiseptic because of its irritating qualities and disagreeable odor.
2. Phenolics	Disruption of plasma membrane, denaturation of enzymes.	Environmental surfaces, instruments, skin surfaces, and mucous membranes.	Derivatives of phenol that are reactive even in the presence of organic material; O-phenylphenol is an example.
3. Bisphenols	Probably disruption of plasma membrane.	Disinfectant hand soaps and skin lotions.	Triclosan is an especially common example of a bisphenol. Broad spectrum, but most effective against gram-positives.
Biguanides (Chlorhexidine)	Disruption of plasma membrane.	Skin disinfection, especially for surgical scrubs.	Bactericidal to gram-positives and gram-negatives; nontoxic, persistent.
Halogens	Iodine inhibits protein function and is a strong oxidizing agent; chlorine forms the strong oxidizing agent hypochlorous acid, which alters cellular components.	Iodine is an effective antiseptic available as a tincture and an iodophor; chlorine gas is used to disinfect water; chlorine compounds are used to disinfect dairy equipment, eating utensils, household items, and glassware.	Iodine and chlorine may act alone or as components of inorganic and organic compounds.
Alcohols	Protein denaturation and lipid dissolution.	Thermometers and other instruments; in swabbing the skin with alcohol before an injection, most of the disinfecting action probably comes from a simple wiping away (degerming) of dirt and some microbes.	Bactericidal and fungicidal, but not effective against endospores or nonenveloped viruses; commonly used alcohols are ethanol and isopropanol.

major problem is the disinfection of surgical instruments exposed to prion contamination. Normal autoclaving has proven to be inadequate. The World Health Organization has recommended the combined use of a solution of sodium hydroxide and autoclaving at 134°C. However, a recent report described another technique, whereby instruments were soaked in a strong solution of sodium hydroxide for an hour followed by an hour of autoclaving at 136°C. Nevertheless, the report described the treatment as only "fairly effective." Recent reports indicate that surgical instruments have been successfully treated to inactivate prions, which are proteins, by addition of protease enzymes to cleaning solution. Surgeons sometimes resort to the use of disposable instruments.

In summary, it is important to remember that microbial control methods, especially biocides, are not uniformly effective against all microbes.

* * *

Table 7.8 summarizes chemical agents used to control microbial growth.

The compounds discussed in this chapter are not generally useful in the treatment of diseases. Because antibiotics are used in chemotherapy, antibiotics and the pathogens against which they are active will be discussed together, in Chapter 20.

| TABLE 7.8 | Chemical Agents Used to Control Microbial Growth *(continued)* |

Chemical Agent	Mechanism of Action	Preferred Use	Comment
Heavy Metals and Their Compounds	Denaturation of enzymes and other essential proteins.	Silver nitrate may be used to prevent gonorrheal ophthalmia neonatorum; silver-sulfadiazine used as a topical cream on burns; copper sulfate is an algicide.	Heavy metals such as silver and mercury are biocidal.
Surface-Active Agents Soaps and Detergents	Mechanical removal of microbes through scrubbing.	Skin degerming and removal of debris.	Many antibacterial soaps contain antimicrobials.
Acid-Anionic Sanitizers	Not certain; may involve enzyme inactivation or disruption.	Sanitizers in dairy and food-processing industries.	Wide spectrum of activity; nontoxic, noncorrosive, fast-acting.
Quaternary Ammonium Compounds (Cationic Detergents)	Enzyme inhibition, protein denaturation, and disruption of plasma membranes.	Antiseptic for skin, instruments, utensils, rubber goods.	Bactericidal, bacteriostatic, fungicidal, and virucidal against enveloped viruses; examples of quats are Zephiran and Cepacol.
Chemical Food Preservatives Organic Acids	Metabolic inhibition, mostly affecting molds; action not related to their acidity.	Sorbic acid and benzoic acid effective at low pH; parabens much used in cosmetics, shampoos; calcium propionate used in bread.	Widely used to control mold and some bacteria in foods and cosmetics.
Nitrates/Nitrites	Active ingredient is nitrite, which is produced by bacterial action on nitrate. Nitrite inhibits certain iron-containing enzymes of anaerobes.	Meat products such as ham, bacon, hot dogs, sausage.	Prevents growth of *Clostridium botulinum* in food; also imparts a red color.
Aldehydes	Protein denaturation.	Glutaraldehyde (Cidex) is less irritating than formaldehyde and is used for disinfection of medical equipment.	Very effective antimicrobials.
Gaseous Chemosterilizers	Protein denaturation.	Excellent sterilizing agent, especially for objects that would be damaged by heat.	Ethylene oxide is the most commonly used.
Peroxygens (Oxidizing Agents)	Oxidation.	Contaminated surfaces; some deep wounds, in which they are very effective against oxygen-sensitive anaerobes.	Ozone is widely used as a supplement for chlorination; hydrogen peroxide is a poor antiseptic but a good disinfectant. Peracetic acid is especially effective.

STUDY OUTLINE

THE TERMINOLOGY OF MICROBIAL CONTROL (pp. 188–189)

1. The control of microbial growth can prevent infections and food spoilage.
2. Sterilization is the process of removing or destroying all microbial life on an object.
3. Commercial sterilization is heat treatment of canned foods to destroy *C. botulinum* endospores.
4. Disinfection is the process of reducing or inhibiting microbial growth on a nonliving surface.
5. Antisepsis is the process of reducing or inhibiting microorganisms on living tissue.
6. The suffix *-cide* means to kill; the suffix *-stat* means to inhibit.
7. Sepsis is bacterial contamination.

THE RATE OF MICROBIAL DEATH
(pp. 189–190)

1. Bacterial populations subjected to heat or antimicrobial chemicals usually die at a constant rate.
2. Such a death curve, when plotted logarithmically, shows this constant death rate as a straight line.
3. The time it takes to kill a microbial population is proportional to the number of microbes.
4. Microbial species and life cycle phases (e.g., endospores) have different susceptibilities to physical and chemical controls.
5. Organic matter may interfere with heat treatments and chemical control agents.
6. Longer exposure to lower heat can produce the same effect as shorter time at higher heat.

ACTIONS OF MICROBIAL CONTROL AGENTS (p. 190)

ALTERATION OF MEMBRANE PERMEABILITY (p. 190)

1. The susceptibility of the plasma membrane is due to its lipid and protein components.
2. Certain chemical control agents damage the plasma membrane by altering its permeability.

DAMAGE TO PROTEINS AND NUCLEIC ACIDS (p. 190)

3. Some microbial control agents damage cellular proteins by breaking hydrogen bonds and covalent bonds.
4. Other agents interfere with DNA and RNA replication and protein synthesis.

PHYSICAL METHODS OF MICROBIAL CONTROL (pp. 190–196)

HEAT (pp. 190–194)

1. Heat is frequently used to kill microorganisms.
2. Moist heat kills microbes by denaturing enzymes.
3. Thermal death point (TDP) is the lowest temperature at which all the microbes in a liquid culture will be killed in 10 minutes.
4. Thermal death time (TDT) is the length of time required to kill all bacteria in a liquid culture at a given temperature.
5. Decimal reduction time (DRT) is the length of time in which 90% of a bacterial population will be killed at a given temperature.
6. Boiling (100°C) kills many vegetative cells and viruses within 10 minutes.
7. Autoclaving (steam under pressure) is the most effective method of moist heat sterilization. The steam must directly contact the material to be sterilized.
8. In HTST pasteurization, a high temperature is used for a short time (72°C for 15 seconds) to destroy pathogens without altering the flavor of the food. Ultra-high-temperature (UHT) treatment (140°C for 3 seconds) is used to sterilize dairy products.
9. Methods of dry heat sterilization include direct flaming, incineration, and hot-air sterilization. Dry heat kills by oxidation.
10. Different methods that produce the same effect (reduction in microbial growth) are called equivalent treatments.

FILTRATION (p. 194)

11. Filtration is the passage of a liquid or gas through a filter with pores small enough to retain microbes.
12. Microbes can be removed from air by high-efficiency particulate air filters.
13. Membrane filters composed of cellulose esters are commonly used to filter out bacteria, viruses, and even large proteins.

LOW TEMPERATURES (pp. 194–195)

14. The effectiveness of low temperatures depends on the particular microorganism and the intensity of the application.
15. Most microorganisms do not reproduce at ordinary refrigerator temperatures (0–7°C).
16. Many microbes survive (but do not grow) at the subzero temperatures used to store foods.

HIGH PRESSURE (p. 195)

17. High pressure denatures proteins in vegetative cells.

DESICCATION (p. 195)

18. In the absence of water, microorganisms cannot grow but can remain viable.

19. Viruses and endospores can resist desiccation.

OSMOTIC PRESSURE (p. 195)

20. Microorganisms in high concentrations of salts and sugars undergo plasmolysis.

21. Molds and yeasts are more capable than bacteria of growing in materials with low moisture or high osmotic pressure.

RADIATION (pp. 195–196)

22. The effects of radiation depend on its wavelength, intensity, and duration.

23. Ionizing radiation (gamma rays, X rays, and high-energy electron beams) has a high degree of penetration and exerts its effect primarily by ionizing water and forming highly reactive hydroxyl radicals.

24. Ultraviolet (UV) radiation, a form of nonionizing radiation, has a low degree of penetration and causes cell damage by making thymine dimers in DNA that interfere with DNA replication; the most effective germicidal wavelength is 260 nm.

25. Microwaves can kill microbes indirectly as materials get hot.

CHEMICAL METHODS OF MICROBIAL CONTROL (pp. 196–205)

1. Chemical agents are used on living tissue (as antiseptics) and on inanimate objects (as disinfectants).

2. Few chemical agents achieve sterility.

PRINCIPLES OF EFFECTIVE DISINFECTION (p. 198)

3. Careful attention should be paid to the properties and concentration of the disinfectant to be used.

4. The presence of organic matter, degree of contact with microorganisms, and temperature should also be considered.

EVALUATING A DISINFECTANT (p. 198)

5. In the use-dilution test, bacterial (*S. choleraesuis*, *S. aureus*, and *P. aeruginosa*) survival in the manufacturer's recommended dilution of a disinfectant is determined.

6. Viruses, endospore-forming bacteria, mycobacteria, and fungi can also be used in the use-dilution test.

7. In the disk-diffusion method, a disk of filter paper is soaked with a chemical and placed on an inoculated agar plate; a zone of inhibition indicates effectiveness.

TYPES OF DISINFECTANTS (pp. 198–205)

Phenol and Phenolics (pp. 198–199)

8. Phenolics exert their action by injuring plasma membranes.

Bisphenols (p. 200)

9. Bisphenols such as triclosan (over the counter) and hexachlorophene (prescription) are widely used in household products.

Biguanides (p. 200)

10. Chlorhexidine damages plasma membranes of vegetative cells.

Halogens (pp. 200–201)

11. Some halogens (iodine and chlorine) are used alone or as components of inorganic or organic solutions.

12. Iodine may combine with certain amino acids to inactivate enzymes and other cellular proteins.

13. Iodine is available as a tincture (in solution with alcohol) or as an iodophor (combined with an organic molecule).

14. The germicidal action of chlorine is based on the formation of hypochlorous acid when chlorine is added to water.

15. Chlorine is used as a disinfectant in gaseous form (Cl_2 or ClO_2) or in the form of a compound, such as calcium hypochlorite, sodium hypochlorite, sodium dichloroisocyanurate, and chloramines.

Alcohols (p. 201)

16. Alcohols exert their action by denaturing proteins and dissolving lipids.

17. In tinctures, they enhance the effectiveness of other antimicrobial chemicals.

18. Aqueous ethanol (60–95%) and isopropanol are used as disinfectants.

Heavy Metals and Their Compounds (pp. 201–202)

19. Silver, mercury, copper, and zinc are used as germicides.

20. They exert their antimicrobial action through oligodynamic action. When heavy metal ions combine with sulfhydryl (—SH) groups, proteins are denatured.

Surface-Active Agents (p. 202)

21. Surface-active agents decrease the surface tension among molecules of a liquid; soaps and detergents are examples.

22. Soaps have limited germicidal action but assist in the removal of microorganisms through scrubbing.

23. Acid-anionic detergents are used to clean dairy equipment.

24. Quats are cationic detergents attached to NH_4^+.

25. By disrupting plasma membranes, they allow cytoplasmic constituents to leak out of the cell.

26. Quats are most effective against gram-positive bacteria.

Chemical Food Preservatives   (p. 204)

27. SO_2, sorbic acid, benzoic acid, and propionic acid inhibit fungal metabolism and are used as food preservatives.
28. Nitrate and nitrite salts prevent germination of *Clostridium botulinum* endospores in meats.

Antibiotics   (p. 204)

29. Nisin and natamycin are anibiotics used to preserve foods, especially cheese.

Aldehydes   (p. 204)

30. Aldehydes such as formaldehyde and glutaraldehyde exert their antimicrobial effect by inactivating proteins.
31. They are among the most effective chemical disinfectants.

Gaseous Chemosterilizers   (p. 205)

32. Ethylene oxide is the gas most frequently used for sterilization.

33. It penetrates most materials and kills all microorganisms by protein denaturation.

Peroxygens (Oxidizing Agents)   (p. 205)

34. Ozone, peroxide, and peracetic acid are used as antimicrobial agents.
35. They exert their effect by oxidizing molecules inside cells.

MICROBIAL CHARACTERISTICS AND MICROBIAL CONTROL (pp. 205–208)

1. Gram-negative bacteria are generally more resistant than gram-positive bacteria to disinfectants and antiseptics.
2. Mycobacteria, endospores, and protozoan cysts and oocysts are very resistant to disinfectants and antiseptics.
3. Nonenveloped viruses are generally more resistant than enveloped viruses to disinfectants and antiseptics.
4. Prions are resistant to disinfection and autoclaving.

STUDY QUESTIONS

Access more review material either online at **The Microbiology Place** (www.microbiologyplace.com) or with **The Microbiology Place CD-ROM** packaged with your new book. There you'll find activities, practice tests, quizzes, flashcards, case studies, and more to help you succeed.

Answers to the Study Questions can be found in Appendix G.

REVIEW

1. Name the cause of cell death resulting from damage to each of the following:
   a. cell wall   c. proteins
   b. plasma membrane   d. nucleic acids
2. The thermal death time for a suspension of *Bacillus subtilis* endospores is 30 minutes in dry heat and less than 10 minutes in an autoclave. Which type of heat is more effective? Why?
3. If pasteurization does not achieve sterilization, why is food treated by pasteurization?
4. Thermal death point is not considered an accurate measure of the effectiveness of heat sterilization. List three factors that can alter thermal death point.
5. The antimicrobial effect of gamma radiation is due to _____. The antimicrobial effect of ultraviolet radiation is due to _____.
6. A bacterial culture was in log phase in the following figure. At time *x*, an antibacterial compound was added to the culture. Which line indicates addition of a bactericidal compound? A bacteriostatic compound? How can you tell?

Explain why the viable count does not immediately drop to zero at *x*.

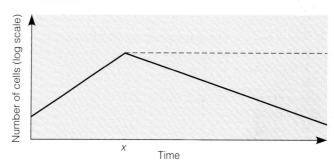

7. Fill in the following table:

Method of Sterilization	Temp.	Time	Type of Heat	Pref'd Use	Types of Action
Autoclaving					
Hot Air					
Pasteurization					

8. How do the examples in question 7 illustrate the concept of equivalent treatments?
9. How do salts and sugars preserve foods? Why are these considered physical rather than chemical methods of microbial control? Name one food that is preserved with sugar and one preserved with salt. How do you account for the occasional growth of *Penicillium* mold in jelly, which is 50% sucrose?
10. List five factors to consider before selecting a disinfectant.

11. Give the method of action and at least one standard use of each of the following types of disinfectants:

    a. phenolics   **e.** heavy metals
    b. iodine   **f.** aldehydes
    c. chlorine   **g.** ethylene oxide
    d. alcohol   **h.** oxidizing agents

12. The use-dilution values for two disinfectants tested under the same conditions are: Disinfectant A—1:2; Disinfectant B—1:10,000. If both disinfectants are designed for the same purpose, which would you select?

13. A large hospital washes burn patients in a stainless steel tub. After each patient, the tub is cleaned with a quat. It was noticed that 14 of 20 burn patients acquired *Pseudomonas* infections after being bathed. Provide an explanation for this high rate of infection.

MULTIPLE CHOICE

1. Which of the following does *not* kill endospores?
    a. autoclaving
    b. incineration
    c. hot-air sterilization
    d. pasteurization
    e. All of the above kill endospores.

2. Which of the following is most effective for sterilizing mattresses and plastic Petri dishes?
    a. chlorine   **d.** autoclaving
    b. ethylene oxide   **e.** nonionizing radiation
    c. glutaraldehyde

3. Which of these disinfectants does *not* act by disrupting the plasma membrane?
    a. phenolics
    b. phenol
    c. quaternary ammonium compounds
    d. halogens
    e. biguanides

4. Which of the following *cannot* be used to sterilize a heat-labile solution stored in a plastic container?
    a. gamma radiation
    b. ethylene oxide
    c. nonionizing radiation
    d. autoclaving
    e. short-wavelength radiation

5. Which of the following is *not* a characteristic of quaternary ammonium compounds?
    a. bactericidal against gram-positive bacteria
    b. sporicidal
    c. amoebicidal
    d. fungicidal
    e. kills enveloped viruses

6. A classmate is trying to determine how a disinfectant might kill cells. You observed that when he spilled the disinfectant in your reduced litmus milk, the litmus turned blue again. You suggest to your classmate that
    a. the disinfectant might inhibit cell wall synthesis.
    b. the disinfectant might oxidize molecules.
    c. the disinfectant might inhibit protein synthesis.
    d. the disinfectant might denature proteins.
    e. he take his work away from yours.

7. Which of the following is most likely to be bactericidal?
    a. membrane filtration
    b. ionizing radiation
    c. lyophilization (freeze-drying)
    d. deep-freezing
    e. all of the above

8. Which of the following is used to control microbial growth in foods?
    a. organic acids
    b. alcohols
    c. aldehydes
    d. heavy metals
    e. all of the above

Use the following information to answer questions 9 and 10. The data were obtained from a use-dilution test comparing four disinfectants against *Salmonella choleraesuis*.

	Bacterial Growth After Exposure to			
Dilution	Disinfectant A	Disinfectant B	Disinfectant C	Disinfectant D
1:2	−	+	−	−
1:4	−	+	−	+
1:8	−	+	+	+
1:16	+	+	+	+

9. Which disinfectant is the most effective? a

10. Which disinfectant(s) is (are) bactericidal?
    a. A, B, C, and D
    b. A, C, and D
    c. A only
    d. B only
    e. none of the above

CRITICAL THINKING

1. The disk-diffusion method was used to evaluate three disinfectants. The results were as follows:

Disinfectant	Zone of Inhibition
X	0 mm
Y	5 mm
Z	10 mm

    a. Which disinfectant was the most effective against the organism?
    b. Can you determine whether compound Y was bactericidal or bacteriostatic?

2. Why is each of the following bacteria often resistant to disinfectants?
    a. *Mycobacterium*
    b. *Pseudomonas*
    c. *Bacillus*

3. A use-dilution test was used to evaluate two disinfectants against *Salmonella choleraesuis*. The results were as follows:

Time of Exposure (min)	Bacterial Growth After Exposures		
	Disinfectant A	Disinfectant B Diluted with Distilled Water	Disinfectant B Diluted with Tap Water
10	+	−	+
20	+	−	−
30	−	−	−

   a. Which disinfectant was the most effective?
   b. Which disinfectant should be used against *Staphylococcus*?

4. To determine the lethal action of microwave radiation, two 10^5 suspensions of *E. coli* were prepared. One cell suspension was exposed to microwave radiation while wet, whereas the other was lyophilized (freeze-dried) and then exposed to radiation. The results are shown in the following figure. Dashed lines indicate the temperature of the samples. What is the most likely method of lethal action of microwave radiation? How do you suppose these data might differ for *Clostridium*?

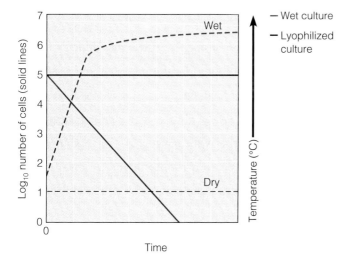

CLINICAL APPLICATIONS

1. *Entamoeba histolytica* and *Giardia lamblia* were isolated from the stool sample of a 45-year-old man, and *Shigella sonnei* was isolated from the stool sample of an 18-year-old woman.

Both patients experienced diarrhea and severe abdominal cramps, and prior to onset of digestive symptoms both had been treated by the same chiropractor. The chiropractor had administered colonic irrigations (enemas) to these patients. The device used for this treatment was a gravity-dependent apparatus using 12 liters of tap water. There were no check valves to prevent backflow, so all parts of the apparatus could have become contaminated with feces during each colonic treatment. The chiropractor provided colonic treatment to four or five patients per day. Between patients, the adaptor piece that is inserted into the rectum was placed in a "hot-water sterilizer."

   What two errors were made by the chiropractor?

2. Between March 9 and April 12, five chronic peritoneal dialysis patients at one hospital became infected with *Pseudomonas aeruginosa*. Four patients developed peritonitis (inflammation of the abdominal cavity), and one developed a skin infection at the catheter insertion site. All patients with peritonitis had low-grade fever, cloudy peritoneal fluid, and abdominal pain. All patients had permanent indwelling peritoneal catheters, which the nurse wiped with gauze that had been soaked with an iodophor solution each time the catheter was connected to or disconnected from the machine tubing. Aliquots of the iodophor were transferred from stock bottles to small in-use bottles. Cultures from the dialysate concentrate and the internal areas of the dialysis machines were negative; iodophor from a small in-use plastic container yielded a pure culture of *P. aeruginosa*.

   What improper technique led to this infection?

3. Eleven patients received injections of methylprednisolone and lidocaine to relieve the pain and inflammation of arthritis at the same orthopedic surgery office. All of them developed septic arthritis caused by *Serratia marcescens*. Unopened bottles of methylprednisolone from the same lot numbers tested sterile; the methylprednisolone was preserved with a quat. Cotton balls were used to wipe multiple-use injection vials before the medication was drawn into a disposable syringe. The site of injection on each patient was also wiped with a cotton ball. The cotton balls were soaked in benzalkonium chloride, and fresh cotton balls were added as the jar was emptied. Opened methylprednisolone containers and the jar of cotton balls contained *S. marcescens*.

   How was the infection transmitted? What part of the routine procedure caused the contamination?

8 Microbial Genetics

Virtually all the microbial traits you have read about in earlier chapters are controlled or influenced by heredity. The inherited traits of microbes include their shape and structural features, their metabolism, their ability to move or behave in various ways, and their ability to interact with other organisms—perhaps causing disease. Individual organisms transmit these characteristics to their offspring through genes.

Researchers are trying to solve the difficult medical problem of microbes developing antibiotic resistance. Microorganisms can become resistant to antibiotics in any of several ways, all of which depend on genetics. The emergence of vancomycin-resistant *Staphylococcus aureus (VRSA)* poses a serious threat to patient care. In this chapter you will see how VRSA acquired this trait.

Emergent diseases provide another example of the importance of understanding genetics. New diseases are the results of genetic changes in some existing organism; for example, *E. coli* O157:H7 acquired the genes for Shiga toxin from *Shigella*.

Currently, microbiologists are using genetics to discover relatedness among organisms and the origins of organisms such as HIV and West Nile virus and to study the potential for avian influenza viruses to infect humans.

UNDER THE MICROSCOPE

Bacterial Chromosome. The single chromosome, normally tightly packed inside a bacterial cell, has burst from an *E. coli* cell after the cell wall and plasma membrane were damaged.

STRUCTURE AND FUNCTION OF THE GENETIC MATERIAL

> **LEARNING OBJECTIVES**
>
> - Define *genetics, genome, chromosome, gene, genetic code, genotype, phenotype,* and *genomics.*
> - Describe how DNA serves as genetic information.

Genetics is the science of heredity; it includes the study of what genes are, how they carry information, how they are replicated and passed to subsequent generations of cells or passed between organisms, and how the expression of their information within an organism determines the particular characteristics of that organism. The genetic information in a cell is called the **genome.** A cell's genome includes its chromosomes and plasmids. **Chromosomes** are structures containing DNA that physically carry hereditary information; the chromosomes contain the genes. **Genes** are segments of DNA (except in some viruses, in which they are made of RNA) that code for functional products. We saw in Chapter 2, on page 48, that DNA is a macromolecule composed of repeating units called *nucleotides.* Recall that each nucleotide consists of a nitrogenous base (adenine, thymine, cytosine, or guanine), deoxyribose (a pentose sugar), and a phosphate group (see Figure 2.16, page 48). The DNA within a cell exists as long strands of nucleotides twisted together in pairs to form a double helix. Each strand has a string of alternating sugar and phosphate groups (its *sugar-phosphate backbone*), and a nitrogenous base is attached to each sugar in the backbone. The two strands are held together by hydrogen bonds between their nitrogenous bases. The **base pairs** always occur in a specific way: adenine always pairs with thymine, and cytosine always pairs with guanine. Because of this specific base pairing, the base sequence of one DNA strand determines the base sequence of the other strand. The two strands of DNA are thus *complementary.* You can think of these complementary DNA sequences as being like a positive photograph and its negative.

The structure of DNA helps explain two primary features of biological information storage. First, the linear sequence of bases provides the actual information. Genetic information is encoded by the sequence of bases along a strand of DNA, in much the same way as our written language uses a linear sequence of letters to form words and sentences. The genetic language, however, uses an alphabet with only four letters—the four kinds of nitrogenous bases in DNA (or RNA). But 1000 of these four bases, the number contained in an average-sized gene, can be arranged in 4^{1000} different ways. This astronomically large number explains how genes can be varied enough to provide all the information a cell needs to grow and perform its functions. The **genetic code,** the set of rules that determines how a nucleotide sequence is converted into the amino acid sequence of a protein, is discussed in more detail later in the chapter.

Second, the complementary structure allows for the precise duplication of DNA during cell division. Again, think of the photograph analogy: if you have a negative, you can always make another copy of the positive print. Likewise with DNA: if you know the sequence of one strand, you also know the sequence of the complementary strand.

Much of cellular metabolism is concerned with translating the genetic message of genes into specific proteins. A gene usually codes for a messenger RNA (mRNA) molecule, which ultimately results in the formation of a protein. Alternatively, the gene product can be a ribosomal RNA (rRNA) or a transfer RNA (tRNA). As we will see, all of these types of RNA are involved in the process of protein synthesis. When the ultimate molecule for which a gene codes (a protein, for example) has been produced, we say that the gene has been *expressed.*

GENOTYPE AND PHENOTYPE

The **genotype** of an organism is its genetic makeup, the information that codes for all the particular characteristics of the organism. The genotype represents *potential* properties, but not the properties themselves. **Phenotype** refers to *actual, expressed* properties, such as the organism's ability to perform a particular chemical reaction. Phenotype, then, is the manifestation of genotype.

In molecular terms, an organism's genotype is its collection of genes, its entire DNA. What constitutes the organism's phenotype in molecular terms? In a sense, an organism's phenotype is its collection of proteins. Most of a cell's properties derive from the structures and functions of its proteins. In microbes, most proteins are either *enzymatic* (catalyze particular reactions) or *structural* (participate in large functional complexes such as membranes or ribosomes). Even phenotypes that depend on structural macromolecules other than proteins (such as lipids or polysaccharides) rely indirectly on proteins. For instance, the structure of a complex lipid or polysaccharide molecule results from the catalytic activities of enzymes that synthesize, process, and degrade those molecules. Thus, although it is not completely accurate to say that phenotypes are due only to proteins, it is a useful simplification.

DNA AND CHROMOSOMES

Bacteria typically have a single circular chromosome consisting of a single circular molecule of DNA with associated proteins. The chromosome is looped and folded (Figure 8.1a)

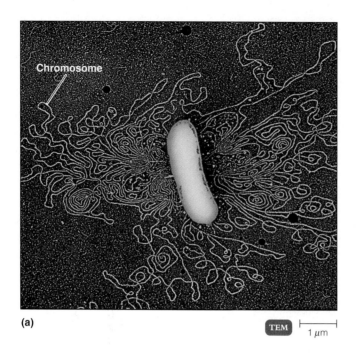

(a)

TEM |⎯⎯⎯⎯⎯|
       1 μm

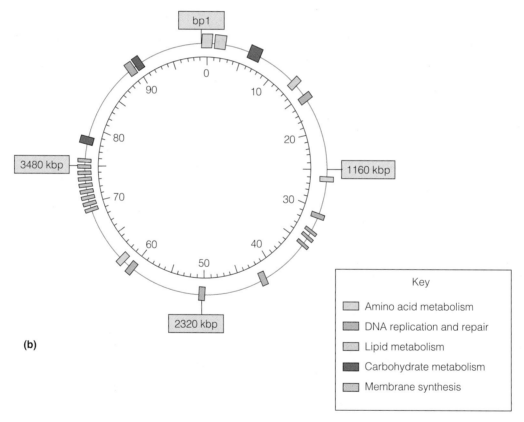

(b)

Key

☐ Amino acid metabolism

☐ DNA replication and repair

☐ Lipid metabolism

■ Carbohydrate metabolism

☐ Membrane synthesis

FIGURE 8.1 Chromosomes. (a) A prokaryotic chromosome. The tangled mass and looping strands of DNA emerging from this disrupted *E. coli* cell are part of its single chromosome. **(b)** A genetic map of the chromosome of *E. coli*. The numbers inside the circle are minutes based on the length of time it takes to transfer the genes during mating between two cells; the numbers in colored boxes are base pairs.

Q **What is a gene? What is an open-reading frame?**

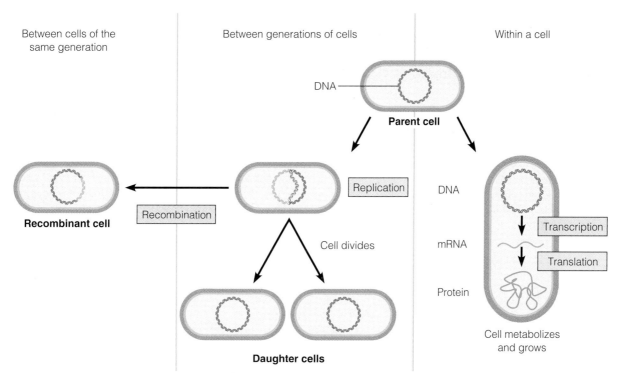

Between cells of the
same generation

Between generations of cells

Within a cell

DNA

Parent cell

Replication

Recombinant cell

Recombination

Cell divides

DNA

mRNA

Protein

Transcription

Translation

Daughter cells

Cell metabolizes
and grows

**FIGURE 8.2   An overview of the flow of genetic
information.** Genetic information can be transferred between
generations of cells, through DNA replication. Occasionally,
genetic information can be transferred between cells of the same
generation through recombination. Genetic information is also
used within a cell to produce the proteins the cell needs to
function (through transcription and translation). The cell represented here is a bacterium with a single circular chromosome. A
small version of this figure will be included in figures throughout
this chapter to indicate the relationships of different processes.

Q   **All of these processes can occur at the same time in a bacterial cell. Which process results in reproduction?**

and attached at one or several points to the plasma membrane. The DNA of *E. coli,* the most-studied bacterial
species, has about 4.6 million base pairs and is about 1 mm
long—1000 times longer than the entire cell. However,
the chromosome takes up only about 10% of the cell's volume because the DNA is twisted, or *supercoiled*—much
like a telephone cord when you put the handset back on
the receiver.

The location of genes on a bacterial chromosome can
be determined by experiments on the transfer of genes
from one cell to another. These processes will be discussed
later in this chapter. The bacterial chromosome map that
results is marked in minutes corresponding to when the
genes are transferred from a donor cell to a recipient cell
(Figure 8.1b).

In recent years, the complete base sequences of several
bacterial chromosomes have been determined. Computers
are used to search for *open-reading frames,* that is, regions of
DNA that are likely to encode a protein. As you will see
later, these are base sequences between start and stop
codons. The sequencing and molecular characterization of
genomes is called **genomics.** The use of genomics to track
West Nile virus is described in the box on the following page.

THE FLOW OF GENETIC INFORMATION

DNA replication makes possible the flow of genetic information from one generation to the next. As shown in
Figure 8.2, the DNA of a cell replicates before cell division
so that each offspring cell receives a chromosome identical
to the parent's. Within each metabolizing cell, the genetic
information contained in DNA also flows in another way:
it is transcribed into mRNA and then translated into protein. We describe the processes of transcription and translation later in this chapter.

DNA REPLICATION

LEARNING OBJECTIVE
• Describe the process of DNA replication.

In DNA replication, one "parental" double-stranded DNA
molecule is converted to two identical "daughter" molecules. The complementary structure of the nitrogenous
base sequences in the DNA molecule is the key to understanding DNA replication. Because the bases along the
two strands of double-helical DNA are complementary,

MORBIDITY & MORTALITY WEEKLY REPORT

TRACKING WEST NILE VIRUS

On August 23, 1999, an infectious disease physician from a hospital in northern Queens contacted the New York City Department of Health (NYCDOH) to report two patients with encephalitis. On investigation, NYCDOH initially identified a cluster of six patients with encephalitis, five of whom had profound muscle weakness and required respiratory support. No bacteria were cultured from the patients' blood or cerebrospinal fluid. Viruses transmitted by mosquitoes are a likely cause of aseptic encephalitis during the summer months. These viruses are called arboviruses. Arboviruses, *arthropod-borne*, are viruses that are maintained in nature through biological transmission between susceptible vertebrate hosts by blood-feeding arthropods such as mosquitoes.

Testing of these initial cases for antibodies to the common North American arboviruses was positive for Saint Louis encephalitis virus (SLE) on September 3 at the CDC. SLE belongs to the family Flaviviridae.

Subsequent nucleic-acid sequencing of these isolates was performed at the CDC on September 23. Comparison of the nucleic acid sequences to databases indicated that the viruses were closely related to West Nile virus (WNV), which had never been isolated in the western hemisphere.

By 2004, WNV had been found in birds in all states except Alaska and Hawaii. The recognition of WNV in the Western Hemisphere in the summer of 1999 marked the first introduction in recent history of an Old World flavivirus into the New World. The United States is not alone, however, in reporting new or heightened activity in humans and other animals. In 2003, WNV caused encephalitis in horses in Mexico, and incursions of flaviviruses into new areas are likely to continue through increasing global commerce and travel.

West Nile virus was first isolated in 1937 in the West Nile district of Uganda. In the early 1950s, scientists

recognized WNV encephalitis outbreaks in humans in Egypt and Israel. Initially considered a minor arbovirus, WNV has recently emerged as a major public health and veterinary concern in southern Europe, the Mediterranean basin, and North America.

Currently researchers are looking at the virus's genome for clues about its path around the world. The flavivirus genome consists of a positive, single-stranded RNA 11,000 to 12,000 nucleotides long. (Positive RNA can act as mRNA and be translated.) The virus has acquired several mutations, and researchers are looking for clues in these mutations to determine the virus's journey.

Using the portions of the genomes (shown below) that encode viral proteins, how similar are these viruses? Can you figure out its movement around the world?

Although genetically related groups or clades can be seen, the actual journey of the virus remains elusive.

SOURCE: Adapted from CDC data.

Portion of the nucleotide base region of the viral envelope protein. Although WNV is an RNA virus, the convention is to write genomes as DNA in the 5′ → 3′ direction.

Australia	A	C	C	C	C	G	T	C	C	A	C	C	C	T	T	T	C	A	A	T	T
Egypt	A	A	T	C	C	C	T	C	C	T	C	T	C	C	T	T	C	G	A	C	T
France	A	A	T	C	C	C	T	C	C	T	C	G	C	C	T	T	C	G	A	C	T
Israel	A	A	C	C	C	C	T	C	C	T	C	T	C	C	T	T	C	G	A	C	T
Italy	A	A	C	C	A	C	T	C	T	T	C	C	C	C	T	A	C	G	A	T	T
Kenya	A	A	C	C	A	C	T	C	T	T	C	C	C	C	T	A	C	G	A	T	T
Mexico	A	A	C	C	C	T	T	C	C	T	C	C	C	C	T	T	C	G	A	T	T
United States	A	A	C	C	C	C	T	C	C	T	C	C	C	C	T	T	C	G	A	T	T
Uganda	A	T	A	C	G	A	T	C	A	T	G	C	T	C	G	T	C	C	A	T	C

one strand can act as a template for the production of the other strand (Figure 8.3a).

DNA replication requires the presence of several cellular proteins that direct a particular sequence of events. Enzymes involved in DNA replication and other processes

are listed in Table 8.1 on page 220. When replication begins, the supercoiling is relaxed by *topoisomerase* or *gyrase* and the two strands of parental DNA are unwound by *helicase* and separated from each other in one small DNA segment after another. Free nucleotides present in the

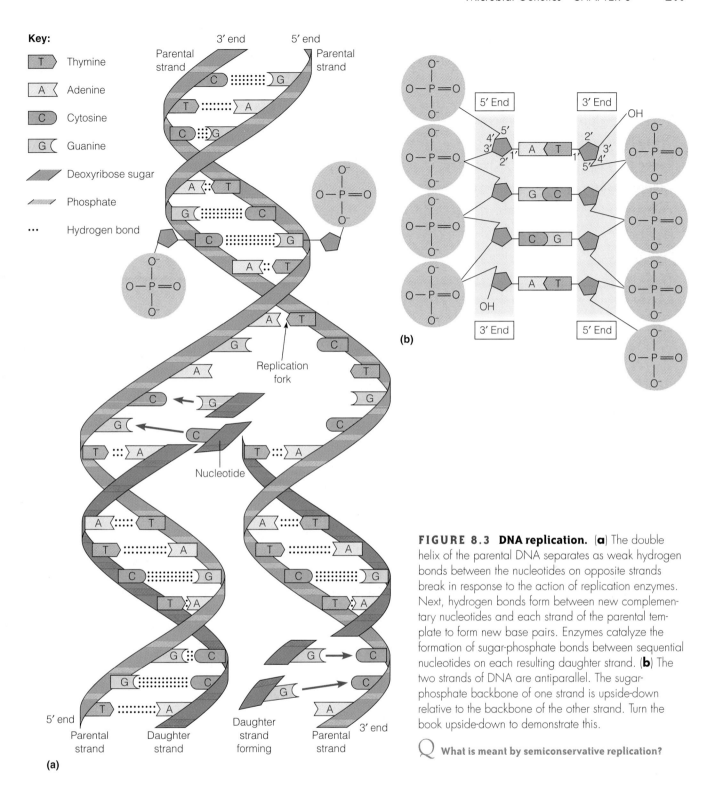

Key:

T	Thymine
A	Adenine
C	Cytosine
G	Guanine
	Deoxyribose sugar
	Phosphate
...	Hydrogen bond

FIGURE 8.3 DNA replication. (a) The double helix of the parental DNA separates as weak hydrogen bonds between the nucleotides on opposite strands break in response to the action of replication enzymes. Next, hydrogen bonds form between new complementary nucleotides and each strand of the parental template to form new base pairs. Enzymes catalyze the formation of sugar-phosphate bonds between sequential nucleotides on each resulting daughter strand. **(b)** The two strands of DNA are antiparallel. The sugar-phosphate backbone of one strand is upside-down relative to the backbone of the other strand. Turn the book upside-down to demonstrate this.

Q **What is meant by semiconservative replication?**

cytoplasm of the cell are matched up to the exposed bases of the single-stranded parental DNA. Where thymine is present on the original strand, only adenine can fit into place on the new strand; where guanine is present on the original strand, only cytosine can fit into place, and so on. Any bases that are improperly base-paired are removed and replaced by replication enzymes. Once aligned, the

newly added nucleotide is joined to the growing DNA strand by an enzyme called **DNA polymerase.** Then the parental DNA is unwound a bit further to allow the addition of the next nucleotides. The point at which replication occurs is called the *replication fork.*

As the replication fork moves along the parental DNA, each of the unwound single strands combines

TABLE 8.1	**Important Enzymes in DNA Replication, Expression, and Repair**
DNA gyrase	Relaxes supercoiling ahead of the replication fork.
DNA ligase	Makes covalent bonds to join DNA strands; joins Okazaki fragments and new segments in excision repair.
DNA polymerase	Synthesizes DNA; proofreads and repairs DNA.
Endonucleases	Cut DNA backbone in a strand of DNA; facilitate repair and insertions.
Exonucleases	Cut DNA from an exposed end of DNA; facilitate repair.
Helicase	Unwinds double-stranded DNA.
Methylase	Adds methyl group to selected bases in newly-made DNA.
Photolyases	Use visible light energy to separate UV-induced pyrimidine dimers.
Primase	Makes RNA primers from a DNA template.
Ribozyme	RNA enzyme that removes introns and splices exons together.
RNA polymerase	Copies RNA from a DNA template.
Topoisomerase	Relaxes supercoiling ahead of the replication fork; separates DNA circles at the end of DNA replication.
Transposase	Cuts DNA backbone leaving single-stranded "sticky ends."

with new nucleotides. The original strand and this newly synthesized daughter strand then rewind. Because each new double-stranded DNA molecule contains one original (conserved) strand and one new strand, the process of replication is referred to as **semiconservative replication.**

Before looking at DNA replication in more detail, let's take a closer look at the structure of DNA (see Figure 2.16, on page 48). It is important to understand the concept that the paired DNA strands are oriented in opposite directions relative to each other. Notice in Figure 2.16 that the carbon atoms of the sugar component of each nucleotide are numbered 1′ (pronounced "one prime") to 5′. In order for the paired bases to be next to each other, the sugar components in one strand are upside-down relative to the other. The end with the hydroxyl attached to the 3′ carbon is called the 3′ end of the DNA strand; the end having a phosphate attached to the 5′ carbon is called the 5′ end. The way in which the two strands fit together dictates that the 5′ → 3′ direction of one strand runs counter to the 5′ → 3′ direction of the other strand (see Figure 8.3b). This structure of DNA affects the replication process because DNA polymerases can add new nucleotides to the 3′ end only. Therefore, as the replication fork moves along the parental DNA, the two new strands must grow in different directions.

DNA replication requires a great deal of energy. The energy is supplied from the nucleotides, which are actually nucleoside triphosphates. You already know about ATP; the only difference between ATP and the adenine nucleotide in DNA is the sugar component. Deoxyribose is the sugar in the nucleosides used to synthesize DNA, and nucleoside triphosphates with ribose are used to synthesize RNA. Two phosphate groups are removed to add the nucleotide to a growing strand of DNA; hydrolysis of the nucleoside is exergonic and provides energy to make the new bonds in the DNA strand (Figure 8.4).

Figure 8.5 provides more detail about the many steps that go into this complex process.

①–② Once the parental DNA is unwound and stabilized, the replication fork forms at a fixed site called the origin of replication.

③ One new DNA strand, called the **leading strand,** is synthesized continuously as the DNA polymerase moves toward the replication fork making DNA in the 5′ → 3′ direction.

④ Remember that DNA polymerase can only add new nucleotides to the 3′ end, so a short piece of RNA called an **RNA primer,** made by *primase*, starts synthesis. DNA polymerase can then add nucleotides to the 3′ end of the RNA. Consequently, the **lagging strand** of

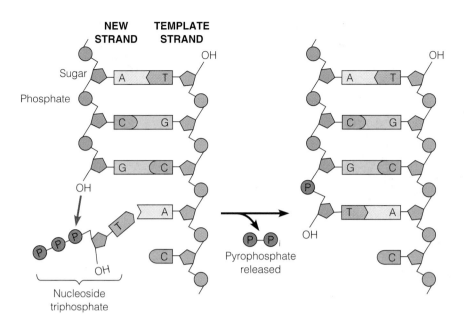

NEW STRAND   TEMPLATE STRAND

Sugar

Phosphate

OH

Nucleoside triphosphate

Pyrophosphate released

FIGURE 8.4 Adding a nucleotide to DNA. When a nucleoside triphosphate bonds to the sugar in a growing DNA strand, it loses two phosphates. Hydrolysis of the phosphate bonds provides the energy for the reaction.

Q **Why is one strand "upside-down" relative to the other strand? Why can't both strands "face" the same way?**

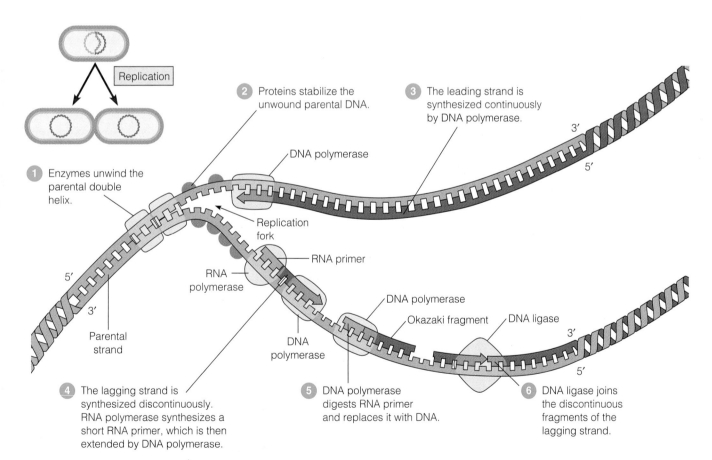

1 Enzymes unwind the parental double helix.

2 Proteins stabilize the unwound parental DNA.

3 The leading strand is synthesized continuously by DNA polymerase.

DNA polymerase

Replication fork

RNA primer

RNA polymerase

5′

3′

Parental strand

DNA polymerase

DNA polymerase

Okazaki fragment

DNA ligase

3′

5′

3′

5′

4 The lagging strand is synthesized discontinuously. RNA polymerase synthesizes a short RNA primer, which is then extended by DNA polymerase.

5 DNA polymerase digests RNA primer and replaces it with DNA.

6 DNA ligase joins the discontinuous fragments of the lagging strand.

Replication

FIGURE 8.5 A summary of events at the DNA replication fork. Enzymes at the replication fork unwind the parental double helix. DNA polymerase synthesizes a continuous strand of new DNA, using one of the parental strands as a template. DNA polymerase also uses the other parental strand as a template, but because the orientation of the sugars is opposite, RNA polymerase starts the synthesis by adding a short stretch of RNA called an RNA primer. The DNA polymerase digests away the RNA as it makes a small piece of DNA. The small units are subsequently joined by DNA ligase.

Q **Why is one strand of DNA synthesized discontinuously?**

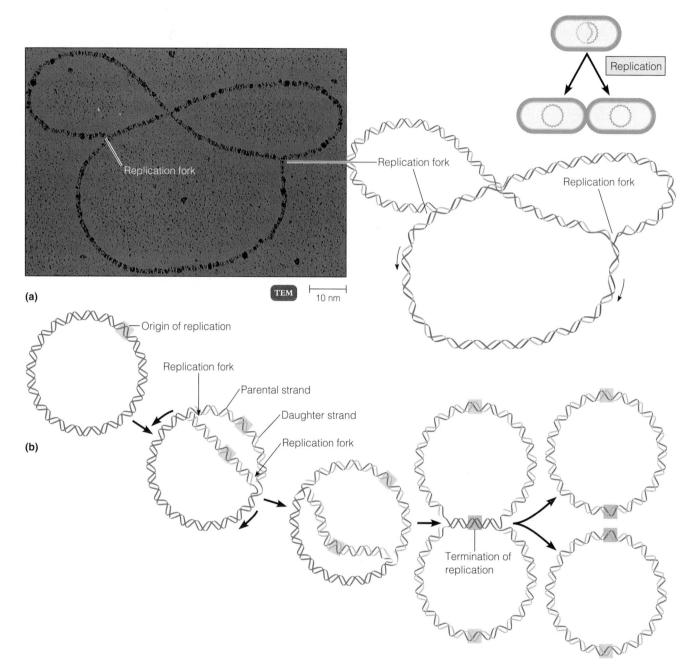

(a)

(b)

FIGURE 8.6 Replication of bacterial DNA. (a) An *E. coli* chromosome in the process of replicating. (In the corresponding diagram at right, the arrows show the direction in which the replication forks are moving.) The chromosome is about one-third replicated. Notice that one of the new helices is perpendicular to the other. **(b)** A diagram of the bidirectional replication of a circular bacterial DNA molecule.

 **What is the origin of replication?**

new DNA is synthesized in pieces consisting of about 1000 nucleotides, called *Okazaki fragments*, as the DNA polymerase moves away from the replication fork.

⑤ DNA polymerase removes the RNA primer, and

⑥ the enzyme **DNA ligase** joins the newly made DNA fragments.

DNA replication by some bacteria, such as *E. coli*, goes *bidirectionally* around the chromosome (Figure 8.6). Two replication forks move in opposite directions away from the origin of replication. Because the bacterial chromosome is a closed loop, the replication forks eventually meet when replication is completed. The two loops must be separated by a topoisomerase. Much evidence shows an association between the bacterial plasma membrane and the origin of replication. After duplication, if each copy of the origin binds to the membrane at opposite poles, then each daughter cell would receive one copy of the DNA molecule—that is, one complete chromosome.

DNA replication is an amazingly accurate process. Typically, mistakes are made at a rate of only 1 in every 10^{10} bases incorporated. Such accuracy is largely due to the *proofreading* capability of DNA polymerase. As each new base is added, the enzyme evaluates whether it forms the proper complementary base-pairing structure. If not, the enzyme excises the improper base and replaces it with the correct one. In this way, DNA replication can be performed very accurately, allowing each daughter chromosome to be virtually identical to the parental DNA. ❋ **Animation: Go to The Microbiology Place website or CD-ROM and click "Animations" to view DNA Replication.**

RNA AND PROTEIN SYNTHESIS

> **LEARNING OBJECTIVE**
>
> • Describe protein synthesis, including transcription, RNA processing, and translation.

How is the information in DNA used to make the proteins that control cell activities? In the process of *transcription,* genetic information in DNA is copied, or transcribed, into a complementary base sequence of RNA. The cell then uses the information encoded in this RNA to synthesize specific proteins through the process of *translation.* We now take a closer look at these two processes as they occur in a bacterial cell.

TRANSCRIPTION

Transcription is the synthesis of a complementary strand of RNA from a DNA template. We will discuss transcription in prokaryotic cells here. Transcription in eukaryotes is discussed on page 226. As mentioned earlier, there are three kinds of RNA in bacterial cells: messenger RNA, ribosomal RNA, and transfer RNA. Ribosomal RNA forms an integral part of ribosomes, the cellular machinery for protein synthesis. Transfer RNA is also involved in protein synthesis, as we will see. **Messenger RNA (mRNA)** carries the coded information for making specific proteins from DNA to ribosomes, where proteins are synthesized.

During transcription, a strand of mRNA is synthesized using a specific gene—a portion of the cell's DNA—as a template. In other words, the genetic information stored in the sequence of nitrogenous bases of DNA is rewritten so that the same information appears in the base sequence of mRNA. As in DNA replication, a G in the DNA template dictates a C in the mRNA being made, a C in the DNA template dictates a G in the mRNA, and a T in the DNA template dictates an A in the mRNA. However, an A in the DNA template dictates a uracil (U) in the mRNA because RNA contains U instead of T. (U has a chemical structure slightly different from T, but it base-pairs in the same way.) If, for example, the template portion of DNA has the base sequence 3'-ATGCAT, the newly synthesized mRNA strand will have the complementary base sequence 5'-UACGUA.

The process of transcription requires both an enzyme called *RNA polymerase* and a supply of RNA nucleotides (Figure 8.7). Transcription begins when

➊ RNA polymerase binds to the DNA at a site called the **promoter.** Only one of the two DNA strands serves as the template for RNA synthesis for a given gene. Like DNA, RNA is synthesized in the 5' → 3' direction.

➋ RNA polymerase assembles free nucleotides into a new chain, using complementary base pairing as a guide.

➌ As the new RNA chain grows, RNA polymerase moves along the DNA.

➍ RNA synthesis continues until RNA polymerase reaches a site on the DNA called the **terminator.**

➎ When this happens, RNA polymerase and the newly formed, single-stranded mRNA are released from the DNA.

The process of transcription allows the cell to produce short-term copies of genes that can be used as the direct source of information for protein synthesis. Messenger RNA acts as an intermediate between the permanent storage form, DNA, and the process that uses the information, translation. ❋ **Animation: Go to The Microbiology Place website or CD-ROM and click "Animations" to view Transcription.**

TRANSLATION

We have seen how the genetic information in DNA is transferred to mRNA during transcription. Now we will see how mRNA serves as the source of information for the synthesis of proteins. Protein synthesis is called **translation** because it involves decoding the "language" of nucleic acids and converting that information into the "language" of proteins.

The language of mRNA is in the form of **codons,** groups of three nucleotides, such as AUG, GGC, or AAA. The sequence of codons on an mRNA molecule determines the sequence of amino acids that will be in the protein being synthesized. Each codon "codes" for a

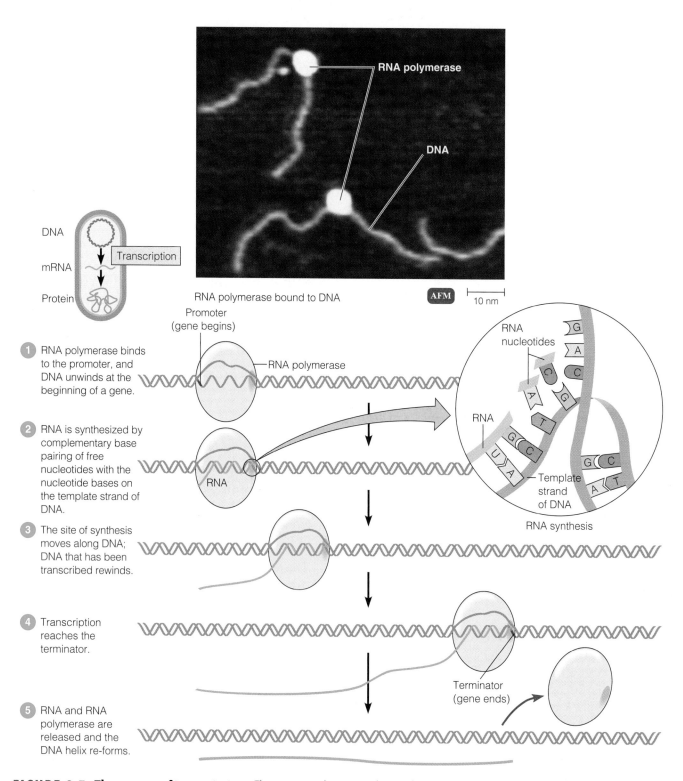

FIGURE 8.7 The process of transcription. The orienting diagram indicates the relationship of transcription to the overall flow of genetic information within a cell.

Q **In transcription, what is copied and what is made?**

particular amino acid. This is the genetic code (Figure 8.8).

Codons are written in terms of their base sequence in mRNA. Notice that there are 64 possible codons but only 20 amino acids. This means that most amino acids are signaled by several alternative codons, a situation referred to as the **degeneracy** of the code. For example, leucine has six codons and alanine has four codons.

Degeneracy allows for a certain amount of change, or mutation, in the DNA without affecting the protein ultimately produced.

Of the 64 codons, 61 are sense codons and 3 are nonsense codons. **Sense codons** code for amino acids, and **nonsense codons** (also called *stop codons*) do not. Rather, the nonsense codons—UAA, UAG, and UGA—signal the end of the protein molecule's synthesis. The start codon that initiates the synthesis of the protein molecule is AUG, which is also the codon for methionine. In bacteria, the start AUG codes for formylmethionine rather than the methionine found in other parts of the protein. The initiating methionine is often removed later, so not all proteins begin with methionine.

The codons of mRNA are converted into protein through the process of translation. The codons of an mRNA are "read" sequentially; and, in response to each codon, the appropriate amino acid is assembled into a growing chain. The site of translation is the ribosome, and **transfer RNA (tRNA)** molecules both recognize the specific codons and transport the required amino acids.

Each tRNA molecule has an **anticodon,** a sequence of three bases that is complementary to a codon. In this way, a tRNA molecule can base-pair with its associated codon. Each tRNA can also carry on its other end the amino acid coded for by the codon that the tRNA recognizes. The functions of the ribosome are to direct the orderly binding of tRNAs to codons and to assemble the amino acids brought there into a chain, ultimately producing a protein.

Figure 8.9 on pages 226–227 shows the details of translation.

1. The necessary components assemble: the two ribosomal subunits, a tRNA with the anticodon UAC, and the mRNA molecule to be translated, along with several additional protein factors. This sets up the initiator codon (AUG) in the proper position to allow translation to begin.
2. The first tRNA binds to the start codon, bringing with it the amino acid methionine.
3. When the tRNA that recognizes the second codon moves into position on the ribosome, the first amino acid is transferred by the ribosome.
4. After the ribosome joins the two amino acids with a peptide bond, the first tRNA molecule leaves the ribosome.
5. The ribosome then moves along the mRNA to the next codon.
6. As the proper amino acids are brought into line one by one, peptide bonds are formed between them, and a polypeptide chain results. (See Figure 2.14, page 45.)

First position	Second position: U		Second position: C		Second position: A		Second position: G		Third position
U	UUU	Phe	UCU	Ser	UAU	Tyr	UGU	Cys	U
U	UUC		UCC		UAC		UGC		C
U	UUA	Leu	UCA		UAA	Stop	UGA	Stop	A
U	UUG		UCG		UAG	Stop	UGG	Trp	G
C	CUU	Leu	CCU	Pro	CAU	His	CGU	Arg	U
C	CUC		CCC		CAC		CGC		C
C	CUA		CCA		CAA	Gln	CGA		A
C	CUG		CCG		CAG		CGG		G
A	AUU	Ile	ACU	Thr	AAU	Asn	AGU	Ser	U
A	AUC		ACC		AAC		AGC		C
A	AUA		ACA		AAA	Lys	AGA	Arg	A
A	AUG	Met/start	ACG		AAG		AGG		G
G	GUU	Val	GCU	Ala	GAU	Asp	GGU	Gly	U
G	GUC		GCC		GAC		GGC		C
G	GUA		GCA		GAA	Glu	GGA		A
G	GUG		GCG		GAG		GGG		G

FIGURE 8.8 The genetic code. The three nucleotides in an mRNA codon are designated, respectively, as the first position, second position, and third position of the codon on the mRNA. Each set of three nucleotides specifies a particular amino acid, represented by a three-letter abbreviation (see Table 2.4, page 44). The codon AUG, which specifies the amino acid methionine, is also the start of protein synthesis. The word *Stop* identifies the nonsense codons that signal the termination of protein synthesis.

Q **Why is the genetic code described as degenerate?**

7. Translation ends when one of the three nonsense codons in the mRNA is reached.
8. When the ribosome arrives at this codon, it comes apart into its two subunits, and the mRNA and newly synthesized polypeptide chain are released. The ribosome, the mRNA, and the tRNAs are then available to be used again.

The ribosome moves along the mRNA in the 5′ → 3′ direction. As a ribosome moves along the mRNA, it will soon allow the start codon to be exposed. Additional ribosomes can then assemble and begin synthesizing protein. In this way, there are usually a number of ribosomes attached to a single mRNA, all at various stages of protein synthesis. In prokaryotic cells, the translation of mRNA into protein can begin even before transcription is complete

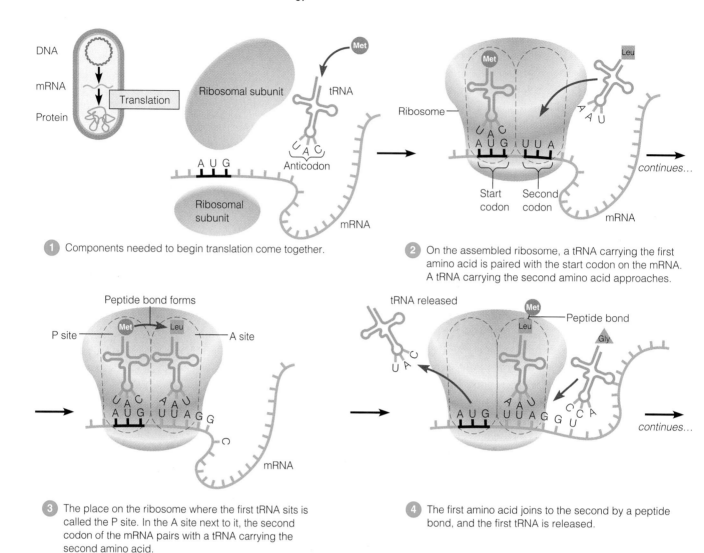

① Components needed to begin translation come together.

② On the assembled ribosome, a tRNA carrying the first amino acid is paired with the start codon on the mRNA. A tRNA carrying the second amino acid approaches.

③ The place on the ribosome where the first tRNA sits is called the P site. In the A site next to it, the second codon of the mRNA pairs with a tRNA carrying the second amino acid.

④ The first amino acid joins to the second by a peptide bond, and the first tRNA is released.

FIGURE 8.9 The process of translation. The overall goal of translation is to produce proteins using mRNAs as the source of biological information. The complex cycle of events illustrated here shows the primary role of tRNA and ribosomes in the decoding of this information. The ribosome acts as the site where the mRNA-encoded information is decoded, as well as the site where individual amino acids are connected into polypeptide chains. The tRNA molecules act as the actual "translators"—one end of each tRNA recognizes a specific mRNA codon, while the other end carries the amino acid coded for by that codon.

Q **Why is this process called translation?**

(Figure 8.10). Because mRNA is produced in the cytoplasm, the start codons of an mRNA being transcribed are available to ribosomes before the entire mRNA molecule is even made.

In eukaryotic cells, transcription takes place in the nucleus. The mRNA must be completely synthesized and moved through the nuclear membrane to the cytoplasm before translation can begin. In addition, the RNA undergoes processing before it leaves the nucleus. In eukaryotic cells the regions of genes that code for proteins are often interrupted by noncoding DNA. Thus, eukaryotic genes are composed of **exons,** the regions of DNA *expressed,* and **introns,** the *intervening* regions of DNA that do not encode protein (Figure 8.11).

① In the nucleus of a eukaryotic cell, RNA polymerase synthesizes an RNA molecule containing exons and introns called an *RNA transcript.*

② This long RNA is then processed by ribozymes, which remove the intron-derived RNA and splice together the exon-derived RNA, producing an mRNA.

③ The resulting mRNA leaves the nucleus to be used by rRNA and tRNA for protein synthesis.

* * *

To summarize, genes are the units of biological information encoded by the sequence of nucleotide bases in DNA. A

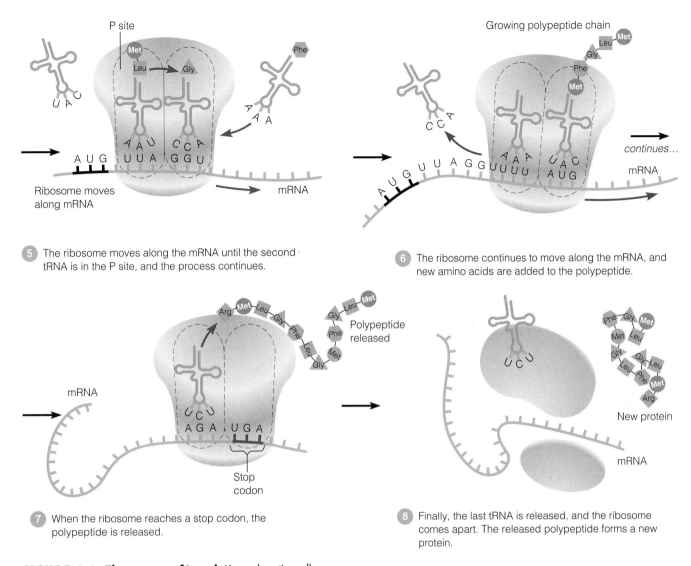

⑤ The ribosome moves along the mRNA until the second tRNA is in the P site, and the process continues.

⑥ The ribosome continues to move along the mRNA, and new amino acids are added to the polypeptide.

⑦ When the ribosome reaches a stop codon, the polypeptide is released.

⑧ Finally, the last tRNA is released, and the ribosome comes apart. The released polypeptide forms a new protein.

FIGURE 8.9 The process of translation. (continued)

gene is expressed, or turned into a product within the cell, through the processes of transcription and translation. The genetic information carried in DNA is transferred to a temporary mRNA molecule by transcription. Then, during translation, the mRNA directs the assembly of amino acids into a polypeptide chain: mRNA attaches to a ribosome, tRNAs deliver the amino acids to the ribosome as directed by the mRNA codon sequence, and the ribosome assembles the amino acids into the chain that will be the newly synthesized protein. ❋ **Animation: Go to The Microbiology Place website or CD-ROM and click "Animations" to view Translation.**

THE REGULATION OF BACTERIAL GENE EXPRESSION

LEARNING OBJECTIVE

• Explain the regulation of gene expression in bacteria by induction, repression, and catabolite repression.

A cell's genetic machinery and its metabolic machinery are integrated and interdependent. Recall from Chapter 5 that the bacterial cell carries out an enormous number of metabolic reactions. The common feature of all metabolic reactions is that they are catalyzed by enzymes. Also recall from Chapter 5 (page 122) that feedback inhibition stops a cell from performing unneeded chemical reactions. Feedback inhibition stops enzymes that have already been synthesized. We will now look at mechanisms to prevent synthesis of enzymes that are not needed.

We have seen that genes, through transcription and translation, direct the synthesis of proteins, many of which serve as enzymes—the very enzymes used for cellular metabolism. Because protein synthesis requires a tremendous expenditure of energy, the regulation of protein synthesis is important to the cell's energy economy. The cell conserves energy by making only those proteins needed at a particular time. We will next look at how chemical reactions are regulated by controlling the synthesis of the enzymes.

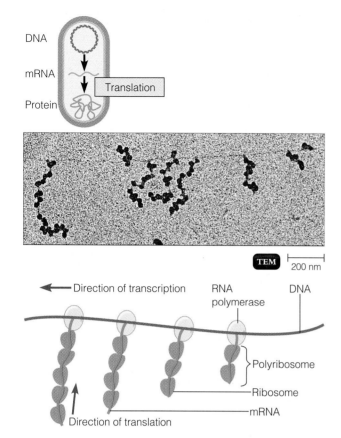

DNA

mRNA

Translation

Protein

TEM   200 nm

Direction of transcription   RNA polymerase   DNA

Polyribosome

Ribosome

mRNA

Direction of translation

Many genes, perhaps 60–80%, are not regulated but are instead *constitutive*, meaning that their products are constantly produced at a fixed rate. Usually these genes, which are effectively turned on all the time, code for enzymes that the cell needs in fairly large amounts for its major life

FIGURE 8.10 Simultaneous transcription and translation in bacteria. The micrograph and diagram show these processes in a single bacterial gene. Many molecules of mRNA are being synthesized simultaneously. The longest mRNA molecules were the first to be transcribed at the promoter. Notice the ribosomes attached to the newly forming mRNA. The newly synthesized polypeptides are not shown.

Q **Why can translation begin before transcription is complete in prokaryotes but not in eukaryotes?**

processes; the enzymes of glycolysis are examples. The production of other enzymes is regulated so that they are present only when needed. *Trypanosoma*, the protozoan parasite that causes African sleeping sickness, has hundreds of genes coding for surface glycoproteins. Each protozoan cell turns on only one glycoprotein gene at a time. As the host's immune system kills parasites with one type of surface molecule, parasites expressing a different surface glycoprotein can continue to grow.

REPRESSION AND INDUCTION

Two genetic control mechanisms known as repression and induction regulate the transcription of mRNA and consequently the synthesis of enzymes from them. These mechanisms control the formation and amounts of enzymes in the cell, not the activities of the enzymes.

REPRESSION

The regulatory mechanism that inhibits gene expression and decreases the synthesis of enzymes is called **repression.**

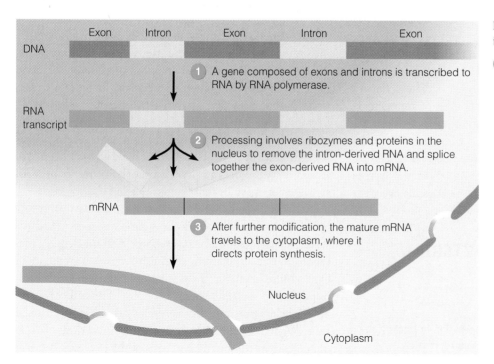

Exon   Intron   Exon   Intron   Exon

DNA

1 A gene composed of exons and introns is transcribed to RNA by RNA polymerase.

RNA transcript

2 Processing involves ribozymes and proteins in the nucleus to remove the intron-derived RNA and splice together the exon-derived RNA into mRNA.

mRNA

3 After further modification, the mature mRNA travels to the cytoplasm, where it directs protein synthesis.

Nucleus

Cytoplasm

FIGURE 8.11 RNA processing in eukaryotic cells.

Q **Why can't the RNA transcript be used for translation?**

Repression is usually a response to the overabundance of an end-product of a metabolic pathway; it causes a decrease in the rate of synthesis of the enzymes leading to the formation of that product. Repression is mediated by regulatory proteins called **repressors,** which block the ability of RNA polymerase to initiate transcription from the repressed genes. The default position of a repressible gene is *on*.

INDUCTION

The process that turns on the transcription of a gene or genes is **induction.** A substance that acts to induce transcription of a gene is called an **inducer,** and enzymes that are synthesized in the presence of inducers are *inducible enzymes*. The genes required for lactose metabolism in *E. coli* are a well-known example of an inducible system. One of these genes codes for the enzyme β-galactosidase, which splits the substrate lactose into two simple sugars, glucose and galactose. (β refers to the type of linkage that joins the glucose and galactose.) If *E. coli* is placed into a medium in which no lactose is present, the organisms contain almost no β-galactosidase; however, when lactose is added to the medium, the bacterial cells produce a large quantity of the enzyme. Lactose is converted in the cell to the related compound allolactose, which is the inducer for these genes; the presence of lactose thus indirectly induces the cells to synthesize more enzyme. The default position of an inducible gene is *off*.

THE OPERON MODEL OF GENE EXPRESSION

Details of the control of gene expression by induction and repression are described by the operon model. François Jacob and Jacques Monod formulated this general model in 1961 to account for the regulation of protein synthesis. They based their model on studies of the induction of the enzymes of lactose catabolism in *E. coli*. In addition to β-galactosidase, these enzymes include lac permease, which is involved in the transport of lactose into the cell, and transacetylase, which metabolizes certain disaccharides other than lactose.

The genes for the three enzymes involved in lactose uptake and utilization are next to each other on the bacterial chromosome and are regulated together (Figure 8.12a). These genes, which determine the structures of proteins, are called **structural genes** to distinguish them from an adjoining control region on the DNA. When lactose is introduced into the culture medium, the *lac* structural genes are all transcribed and translated rapidly and simultaneously. We will now see how this regulation occurs.

In the control region of the *lac* operon are two relatively short segments of DNA. One, the *promoter*, is the region of DNA where RNA polymerase initiates transcription. The other is the **operator,** which is like a traffic light that acts as a go or stop signal for transcription of the structural genes. A set of operator and promoter sites, and the structural genes they control, are what define an **operon;** thus, the combination of the three *lac* structural genes and the adjoining control regions is called the *lac* operon.

1. Near the *lac* operon on the bacterial DNA is a regulatory gene called the *I* gene, which codes for a repressor protein.

2. When lactose is absent, the repressor protein binds tightly to the operator site. This binding prevents RNA polymerase from transcribing the adjacent structural genes; consequently, no mRNA is made and no enzymes are synthesized.

3. But when lactose is present, some of it is transported into the cells and converted into the inducer allolactose. The inducer binds to the repressor protein and alters it so it cannot bind to the operator site. In the absence of an operator-bound repressor protein, RNA polymerase can transcribe the structural genes into mRNA, which is then translated into enzymes. This is why, in the presence of lactose, enzymes are produced. Lactose is said to induce enzyme synthesis, and the *lac* operon is called an inducible operon.

In repressible operons, the structural genes are transcribed until they are turned off, or *repressed* (Figure 8.12b).

1. The genes for the enzymes involved in the synthesis of tryptophan are regulated in this manner.

2. The structural genes are transcribed and translated, leading to tryptophan synthesis.

3. When excess tryptophan is present, the tryptophan acts as a **corepressor** binding to the repressor protein. The repressor protein can now bind to the operator, stopping further tryptophan synthesis.

POSITIVE REGULATION

Regulation of the lactose operon also depends on the level of glucose in the medium, which in turn controls the intracellular level of the small molecule **cyclic AMP (cAMP),** a substance derived from ATP that serves as a cellular alarm signal. Enzymes that metabolize glucose are constitutive, and cells grow at their maximal rate with glucose as their carbon source because they can use it most efficiently (Figure 8.13). When glucose is no longer available, cAMP accumulates in the cell. The cAMP binds to the allosteric site of *catabolic activator protein (CAP)*. CAP then binds to the *lac* promoter, which initiates transcription by making it easier for RNA polymerase to bind to the promoter. Thus transcription of the *lac*

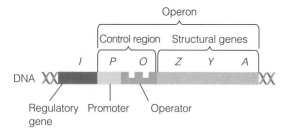

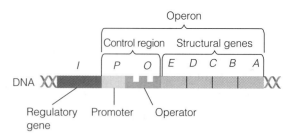

1 **Structure of the operon.** The operon consists of the promoter (*P*) and operator (*O*) sites and structural genes that code for the protein. The operon is regulated by the product of the regulatory gene (*I*).

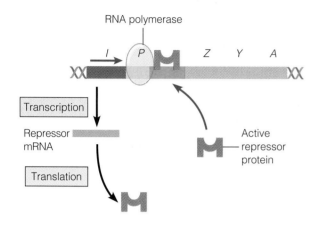

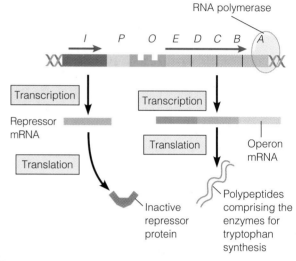

2 **Repressor active, operon off.** The repressor protein binds with the operator, preventing transcription from the operon.

2 **Repressor inactive, operon on.** The repressor is inactive, and transcription and translation proceed, leading to the synthesis of tryptophan.

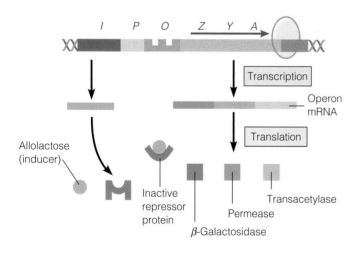

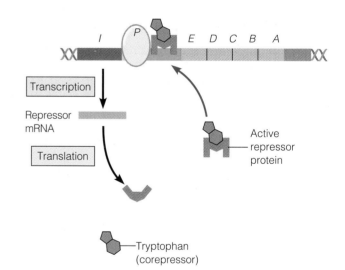

3 **Repressor inactive, operon on.** When the inducer allolactose binds to the repressor protein, the inactivated repressor can no longer block transcription. The structural genes are transcribed, ultimately resulting in the production of the enzymes needed for lactose catabolism.

3 **Repressor active, operon off.** When the corepressor tryptophan binds to the repressor protein, the activated repressor binds with the operator, preventing transcription from the operon.

(a) An inducible operon

(b) A repressible operon

FIGURE 8.12 The operon: regulation of gene expression. (**a**) Lactose is digested by a catabolic pathway catalyzed by inducible enzymes. (**b**) Tryptophan is an amino acid produced by an anabolic pathway catalyzed by repressible enzymes.

Q **How does a repressible enzyme differ from an inducible enzyme?**

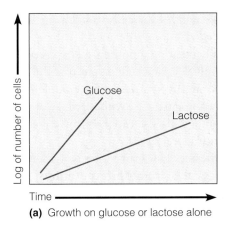

(a) Growth on glucose or lactose alone

(b) Growth on glucose and lactose combined

FIGURE 8.13 The growth rate of *E. coli* on glucose and lactose. The steeper the straight line, the faster the growth. (**a**) Bacteria growing on glucose as the sole carbon source grow faster than on lactose. (**b**) Bacteria growing in a medium containing glucose and lactose first consume the glucose, and then, after a short lag time, the lactose. During the lag time, intracellular cAMP increases, the *lac* operon is transcribed, more lactose is transported into the cell, and β-galactosidase is synthesized to break down lactose.

Q When both glucose and lactose are present, why will cells use glucose first?

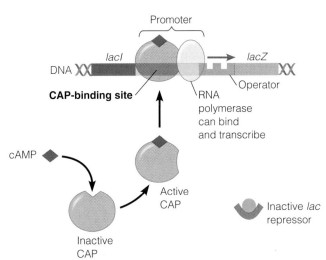

(a) Lactose present, glucose scarce (cAMP level high) If glucose is scarce, the high level of cAMP activates CAP, and the *lac* operon produces large amounts of mRNA for lactose digestion.

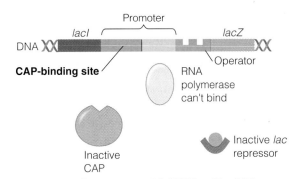

(b) Lactose present, glucose present (cAMP level low) When glucose is present, cAMP is scarce, and CAP is unable to stimulate transcription.

FIGURE 8.14 Positive regulation of the lac operon. (**a**) The enzymes for lactose digestion are synthesized if lactose is present and the cell lacks an energy source. (**b**) The enzymes for lactose digestion are not produced if the cell has sufficient energy.

Q Will transcription of the *lac* operon occur in the presence of lactose and glucose? In the presence of lactose and the absence of glucose? In the presence of glucose and the absence of lactose?

operon requires both the presence of lactose and the absence of glucose (Figure 8.14).

Cyclic AMP is an example of an *alarmone*, a chemical alarm signal that promotes a cell's response to environmental or nutritional stress. (In this case, the stress is the lack of glucose.) The same mechanism involving cAMP allows the cell to grow on other sugars. Inhibition of the metabolism of alternative carbon sources by glucose is termed **catabolite repression** (or the *glucose effect*). When glucose is available, the level of cAMP in the cell is low, and consequently CAP is not bound. ✳ **Animation: Go to The Microbiology Place website or CD-ROM and click "Animations" to view Operons.**

MUTATION: CHANGE IN THE GENETIC MATERIAL

A **mutation** is a change in the base sequence of DNA. Such a change in the base sequence of a gene will sometimes cause a change in the product encoded by that gene.

For example, when the gene for an enzyme mutates, the enzyme encoded by the gene may become inactive or less active because its amino acid sequence has changed. Such a change in genotype may be disadvantageous, or even lethal, if the cell loses a phenotypic trait it needs. However, a mutation can be beneficial if, for instance, the altered enzyme encoded by the mutant gene has a new or enhanced activity that benefits the cell.

Many simple mutations are silent (neutral); the change in DNA base sequence causes no change in the activity of the product encoded by the gene. Silent mutations commonly occur when one nucleotide is substituted for another in the DNA, especially at a location corresponding to the third position of the mRNA codon. Because of the degeneracy of the genetic code, the resulting new codon might still code for the same amino acid. Even if the amino acid is changed, the function of the protein may not change if the amino acid is in a nonvital portion of the protein, or is chemically very similar to the original amino acid.

TYPES OF MUTATIONS

LEARNING OBJECTIVE
- Classify mutations by type, and describe how mutations are prevented or repaired.

The most common type of mutation involving single base pairs is **base substitution** (or *point mutation*), in which a single base at one point in the DNA sequence is replaced with a different base. When the DNA replicates, the result is a substituted base pair (Figure 8.15). For example, AT might be substituted for GC, or CG for GC. If a base substitution occurs within a gene that codes for a protein, the mRNA transcribed from the gene will carry an incorrect base at that position. When the mRNA is translated into protein, the incorrect base may cause the insertion of an incorrect amino acid in the protein. If the base substitution results in an amino acid substitution in the synthesized protein, this change in the DNA is known as a **missense mutation** (Figure 8.16a and b).

The effects of such mutations can be dramatic. For example, sickle cell disease is caused by a single change in the gene for globin, the protein component of hemoglobin. Hemoglobin is primarily responsible for transporting oxygen from the lungs to the tissues. A single missense mutation, a change from an A to a T at a specific site, results in the change from glutamic acid to valine in the protein. The effect of this change is that the shape of the hemoglobin molecule changes under conditions of low oxygen, altering the shape of the red blood cells such that

movement of the cells through small capillaries is greatly impeded.

By creating a nonsense (stop) codon in the middle of an mRNA molecule, some base substitutions effectively prevent the synthesis of a complete functional protein; only a fragment is synthesized. A base substitution resulting in a nonsense codon is thus called a **nonsense mutation** (Figure 8.16c).

Besides base-pair mutations, there are also changes in DNA called **frameshift mutations,** in which one or a few nucleotide pairs are deleted or inserted in the DNA (Figure 8.16d). This mutation can shift the "translational reading frame"—that is, the three-by-three grouping of nucleotides recognized as codons by the tRNAs during translation. For example, deleting one nucleotide pair in the middle of a gene causes changes in many amino acids downstream from the site of the original mutation. Frameshift mutations almost always result in a long stretch of altered amino acids and the production of an inactive protein from the mutated gene. In most cases, a nonsense codon will eventually be encountered and thereby terminate translation.

Occasionally, mutations occur where significant numbers of bases are added to (inserted into) a gene. Huntington's disease, for example, is a progressive neurological disorder caused by extra bases inserted into a particular gene. The reason these insertions occur in this particular gene is still being studied.

Base substitutions and frameshift mutations may occur spontaneously because of occasional mistakes made during DNA replication. These **spontaneous mutations** apparently occur in the absence of any mutation-causing agents. Agents in the environment, such as certain chemicals and radiation, that directly or indirectly bring about mutations are called **mutagens.** Almost any agent that can chemically or physically react with DNA can potentially cause mutations. A wide variety of chemicals, many of which are common in nature or in households, are known to be mutagens. Many forms of radiation, including X rays and ultraviolet light, are also mutagenic, as discussed shortly.

In the microbial world, certain mutations result in resistance to antibiotics (see the box in Chapter 26, page 000) or altered pathogenicity. A mutation in a gene encoding the outer membrane may increase pathogenicity; for example, *Salmonella typhimurium* with an altered outer membrane can survive in phagocytes. A mutation in a capsule-encoding gene may result in decreased pathogenicity because phagocytes can destroy the bacteria, as in the cases of *Streptococcus pneumoniae*, *Haemophilus influenzae*, and *Neisseria meningitidis*.

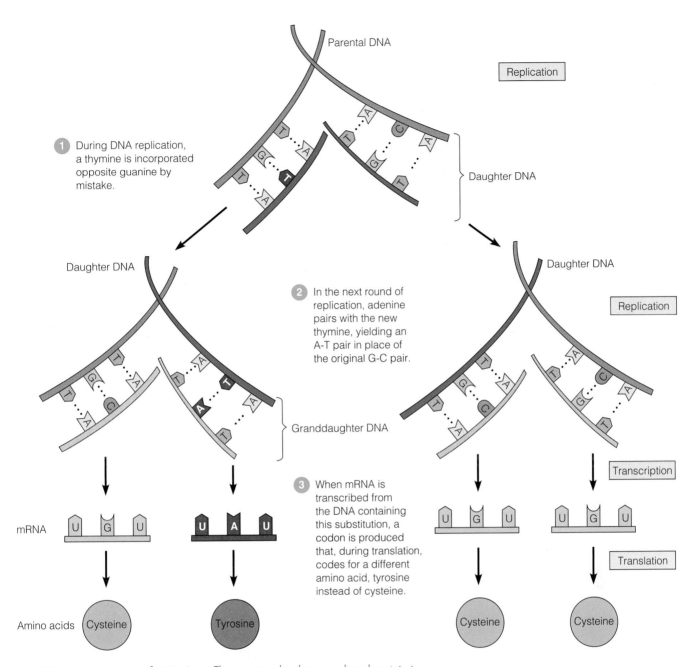

FIGURE 8.15 Base substitution. This mutation leads to an altered protein in a granddaughter cell.

Q **Does a base substitution always result in a different amino acid?**

MUTAGENS

LEARNING OBJECTIVES

• Define *mutagen*.

• Describe two ways mutations can be repaired.

CHEMICAL MUTAGENS

One of the many chemicals known to be a mutagen is nitrous acid. Figure 8.17 shows how exposure of DNA to nitrous acid can convert the base adenine (A) to a form that no longer pairs with thymine (T) but instead pairs with cytosine (C). When DNA containing such modified adenines replicates, one daughter DNA molecule will have a base-pair sequence different from that of the parent DNA. Eventually, some AT base pairs of the parent will have been changed to GC base pairs in a granddaughter cell. Nitrous acid makes a specific base-pair change in DNA. Like all mutagens, it alters DNA at random locations.

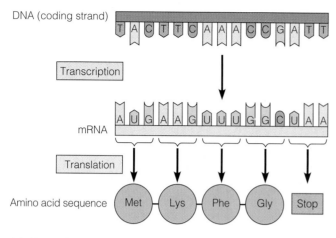

(a) Normal DNA molecule

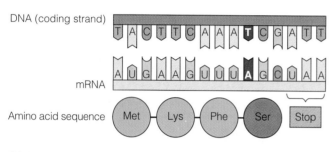

(b) Missense mutation

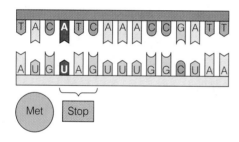

(c) Nonsense mutation

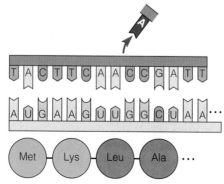

(d) Frameshift mutation

Another type of chemical mutagen is the **nucleoside analog.** These molecules are structurally similar to normal nitrogenous bases, but they have slightly altered base-pairing properties. Examples, 2-aminopurine and 5-bromouracil, are shown in Figure 8.18. When nucleoside analogs are given to growing cells, the analogs are randomly incorporated into cellular DNA in place of the normal bases. Then, during DNA replication, the analogs cause mistakes in base pairing. The incorrectly paired bases will be copied during subsequent replication of the DNA, resulting in base-pair substitutions in the progeny cells. Some antiviral and antitumor drugs are nucleoside analogs, including AZT (azidothymidine), one of the primary drugs used to treat HIV infection.

Still other chemical mutagens cause small deletions or insertions, which can result in frameshifts. For instance, under certain conditions, benzopyrene, which is present in smoke and soot, is an effective *frameshift mutagen.* Aflatoxin—produced by *Aspergillus flavus* (a-spėr-jil′lus flā′vus), a mold that grows on peanuts and grain—is a frameshift mutagen, as are the acridine dyes used experimentally against herpesvirus infections. Frameshift mutagens usually have the right size and chemical properties to slip between the stacked base pairs of the DNA double helix. They may work by slightly offsetting the two strands of DNA, leaving a gap or bulge in one strand or the other. When the staggered DNA strands are copied during DNA synthesis, one or more base pairs can be inserted or deleted in the new double-stranded DNA. Interestingly, frameshift mutagens are often potent carcinogens.

RADIATION

X rays and gamma rays are forms of radiation that are potent mutagens because of their ability to ionize atoms and molecules. The penetrating rays of ionizing radiation cause electrons to pop out of their usual shells (see Chapter 2). These electrons bombard other molecules and cause more damage, and many of the resulting ions and free radicals (molecular fragments with unpaired electrons) are very reactive. Some of these ions can combine with bases in DNA, resulting in errors in DNA replication and repair that produce mutations. An even more serious outcome is the breakage of covalent bonds in the sugar-phosphate backbone of DNA, which causes physical breaks in chromosomes.

FIGURE 8.16 Types of mutations and their effects on the amino acid sequences of proteins.

Q On what basis are missense, nonsense, and frameshift mutations distinguished?

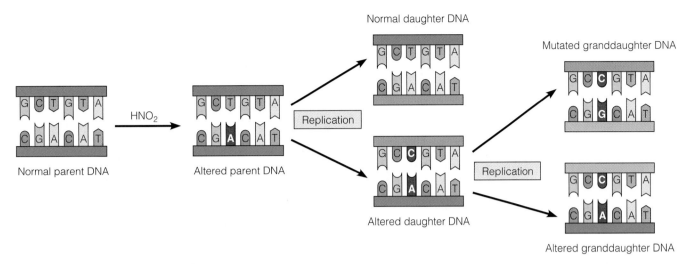

FIGURE 8.17 Nitrous acid (HNO₂) as a mutagen. The nitrous acid alters an adenine in such a way that it pairs with cytosine instead of thymine.

Q **What is a mutagen?**

Another form of mutagenic radiation is ultraviolet (UV) light, a nonionizing component of ordinary sunlight. However, the most mutagenic component of UV light (wavelength 260 nm) is screened out by the ozone layer of the atmosphere. The most important effect of direct UV light on DNA is the formation of harmful covalent bonds between certain bases. Adjacent thymines in a DNA strand can cross-link to form thymine dimers. Such dimers, unless repaired, may cause serious damage or death to the cell because it cannot properly transcribe or replicate such DNA.

Bacteria and other organisms have enzymes that can repair UV-induced damage. **Photolyases,** also known as *light-repair enzymes*, use visible light energy to separate the dimer back to the original two thymines. **Nucleotide excision repair,** shown in Figure 8.19, is not restricted to UV-induced damage; it can repair mutations from other causes as well. Enzymes cut out the incorrect base and fill in the gap with newly synthesized DNA that is complementary to the correct strand. For many years biologists questioned how the incorrect base could be distinguished from the correct base if it was not physically distorted like

NORMAL NITROGENOUS BASE ANALOG

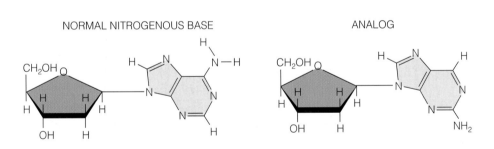

Adenine nucleoside   2-Aminopurine nucleoside

(a) The 2-aminopurine is incorporated into DNA in place of adenine but can pair with cytosine, so an AT pair becomes a CG pair.

Thymine nucleoside   5-Bromouracil nucleoside

(b) 5-bromouracil is used as an anticancer drug because it is mistaken for thymine by cellular enzymes but pairs with cytosine. In the next DNA replication, an AT pair becomes a GC pair.

FIGURE 8.18 Nucleoside analogs and the nitrogenous bases they replace. A nucleoside is phosphorylated and the resulting nucleotide used to synthesize DNA.

Q **Why do these drugs kill cells?**

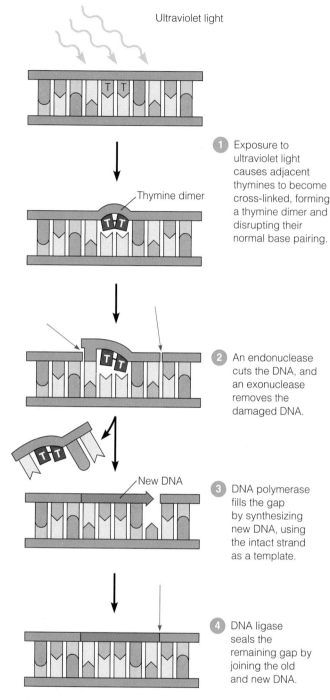

Ultraviolet light

Thymine dimer

1 Exposure to ultraviolet light causes adjacent thymines to become cross-linked, forming a thymine dimer and disrupting their normal base pairing.

2 An endonuclease cuts the DNA, and an exonuclease removes the damaged DNA.

New DNA

3 DNA polymerase fills the gap by synthesizing new DNA, using the intact strand as a template.

4 DNA ligase seals the remaining gap by joining the old and new DNA.

FIGURE 8.19 The creation and repair of a thymine dimer caused by ultraviolet light. After exposure to UV light, adjacent thymines can become cross-linked, forming a thymine dimer. In the absence of visible light, the nucleotide excision repair mechanism is used in a cell to repair the damage.

Q **How do excision repair enzymes "know" which strand is incorrect?**

a thymine dimer. In 1970, Hamilton Smith provided the answer with the discovery of **methylases.** These enzymes add a methyl group to selected bases soon after a DNA strand is made. A repair endonuclease then cuts the non-methylated strand.

Exposure to UV light in humans, such as by excessive suntanning, causes a large number of thymine dimers in skin cells. Unrepaired dimers may result in skin cancers. Humans with xeroderma pigmentosum, an inherited condition that results in increased sensitivity to UV light, have a defect in nucleotide excision repair; consequently, they have an increased risk of skin cancer.

THE FREQUENCY OF MUTATION

LEARNING OBJECTIVE
• Describe the effect of mutagens on the mutation rate.

The **mutation rate** is the probability that a gene will mutate when a cell divides. The rate is usually stated as a power of 10, and because mutations are very rare, the exponent is always a negative number. For example, if there is one chance in 10,000 that a gene will mutate when the cell divides, the mutation rate is 1/10,000, which is expressed as 10^{-4}. Spontaneous mistakes in DNA replication occur at a very low rate, perhaps only once in 10^9 replicated base pairs (a mutation rate of 10^{-9}). Because the average gene has about 10^3 base pairs, the spontaneous rate of mutation is about one in 10^6 (a million) replicated genes.

Mutations usually occur more or less randomly along a chromosome. The occurrence of random mutations at low frequency is an essential aspect of the adaptation of species to their environment, for evolution requires that genetic diversity be generated randomly and at a low rate. For example, in a bacterial population of significant size—say, greater than 10^7 cells—a few new mutant cells will always be produced in every generation. Most mutations either are harmful and likely to be removed from the gene pool when the individual cell dies or are neutral. However, a few mutations may be beneficial. For example, a mutation that confers antibiotic resistance is beneficial to a population of bacteria that is regularly exposed to antibiotics. Once such a trait has appeared through mutation, cells carrying the mutated gene are more likely than other cells to survive and reproduce as long as the environment stays the same. Soon most of the cells in the population will have the gene; an evolutionary change will have occurred, although on a small scale.

A mutagen usually increases the spontaneous rate of mutation, which is about one in 10^6 replicated genes, by a factor of 10–1000 times. In other words, in the presence of a mutagen, the normal rate of 10^{-6} mutations per replicated gene becomes a rate of 10^{-5} to 10^{-3} per replicated gene. Mutagens are used experimentally to enhance the production of mutant cells for research on the genetic properties of microorganisms and for commercial purposes.

IDENTIFYING MUTANTS

> **LEARNING OBJECTIVE**
> • Outline the methods of direct and indirect selection of mutants.

Mutants can be detected by selecting or testing for an altered phenotype. Whether or not a mutagen is used, mutant cells with specific mutations are always rare compared with other cells in the population. The problem is detecting such a rare event.

Experiments are usually performed with bacteria because they reproduce rapidly, so large numbers of organisms (more than 10^9 per milliliter of nutrient broth) can easily be used. Furthermore, because bacteria generally have only one copy of each gene per cell, the effects of a mutated gene are not masked by the presence of a normal version of the gene, as in many eukaryotic organisms.

Positive (direct) selection involves the detection of mutant cells by rejection of the unmutated parent cells. For example, suppose we were trying to find mutant bacteria that are resistant to penicillin. When the bacterial cells are plated on a medium containing penicillin, the mutant can be identified directly. The few cells in the population that are resistant (mutants) will grow and form colonies, whereas the normal, penicillin-sensitive parental cells cannot grow.

To identify mutations in other kinds of genes, **negative (indirect) selection** can be used. This process selects a cell that cannot perform a certain function, using the technique of **replica plating.** For example, suppose we wanted to use replica plating to identify a bacterial cell that has lost the ability to synthesize the amino acid histidine (Figure 8.20). First, about 100 bacterial cells are inoculated onto an agar plate. This plate, called the master plate, contains a medium with histidine on which all cells will grow. After 18 to 24 hours of incubation, each cell reproduces to form a colony. Then a pad of sterile material, such as latex, filter paper, or velvet, is pressed over the master plate, and some of the cells from each colony adhere to the velvet. Next, the velvet is pressed down onto two (or more) sterile plates. One plate contains a medium without histidine, and one contains a medium with histidine on which the original, nonmutant bacteria can grow. Any colony that grows on the medium with histidine on the master plate but that cannot synthesize its own histidine will not be able to grow on the medium without histidine. The mutant colony can then be identified on the master plate. Of course, because mutants are so rare (even those induced by mutagens), many plates must be screened with this technique to isolate a specific mutant.

Replica plating is a very effective means of isolating mutants that require one or more new growth factors. Any mutant microorganism having a nutritional requirement that is absent in the parent is known as an **auxotroph.** For example, an auxotroph may lack an enzyme needed to synthesize a particular amino acid and will therefore require that amino acid as a growth factor in its nutrient medium. ※ **Animation: Go to The Microbiology Place website or CD-ROM and click "Animations" to view Mutations and DNA Repair.**

IDENTIFYING CHEMICAL CARCINOGENS

> **LEARNING OBJECTIVE**
> • Identify the purpose of and outline the procedure for the Ames test.

Many known mutagens have been found to be **carcinogens,** substances that cause cancer in animals, including humans. In recent years, chemicals in the environment, the workplace, and the diet have been implicated as causes of cancer in humans. The usual subjects of tests to determine potential carcinogens are animals, and the testing procedures are time-consuming and expensive. Now there are faster and less expensive procedures for the preliminary screening of potential carcinogens. One of these, called the **Ames test,** uses bacteria as carcinogen indicators.

The Ames test is based on the observation that exposure of mutant bacteria to mutagenic substances may cause new mutations that reverse the effect (the change in phenotype) of the original mutation. These are called back-mutations, or *reversions.* Specifically, the test measures the reversion of histidine auxotrophs of *Salmonella* (his$^-$ cells, mutants that have lost the ability to synthesize histidine) to histidine-synthesizing cells (his$^+$) after treatment with a mutagen (Figure 8.21). Bacteria are incubated in both the presence and absence of the substance being tested. Because many chemicals must be activated (transformed chemically into forms that are chemically reactive) by animal enzymes for mutagenic or carcinogenic activity to appear, the chemical to be tested and the mutant bacteria are incubated together with rat liver extract, a rich source of activation enzymes. If the substance being tested is mutagenic, it will cause the reversion of his$^-$ bacteria to his$^+$ bacteria at a rate higher than the spontaneous reversion rate. The number of observed revertants provides an indication of the degree to which a substance is mutagenic and therefore possibly carcinogenic.

The test can be used in many ways. Several potential mutagens can be qualitatively tested by spotting the

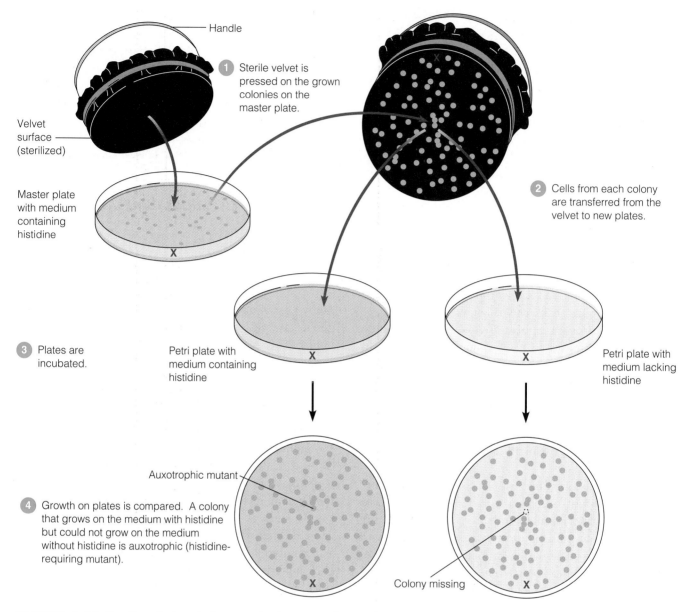

FIGURE 8.20 Replica plating. In this example, the auxotrophic mutant cannot synthesize histidine. The plates must be carefully marked (with an X here) to maintain orientation so that colony positions are known in relation to the original master plate.

Q **What is an auxotroph?**

individual chemicals on small paper disks on a single plate inoculated with bacteria. In addition, mixtures such as wine, blood, smoke condensates, and extracts of foods can also be tested to see whether they contain mutagenic substances.

About 90% of the substances found by the Ames test to be mutagenic have also been shown to be carcinogenic in animals. By the same token, the more mutagenic substances have generally been found to be more carcinogenic.

GENETIC TRANSFER AND RECOMBINATION

LEARNING OBJECTIVES

- Compare the mechanisms of genetic recombination in bacteria.
- Differentiate between horizontal and vertical gene transfer.

Genetic recombination refers to the exchange of genes between two DNA molecules to form new combinations of

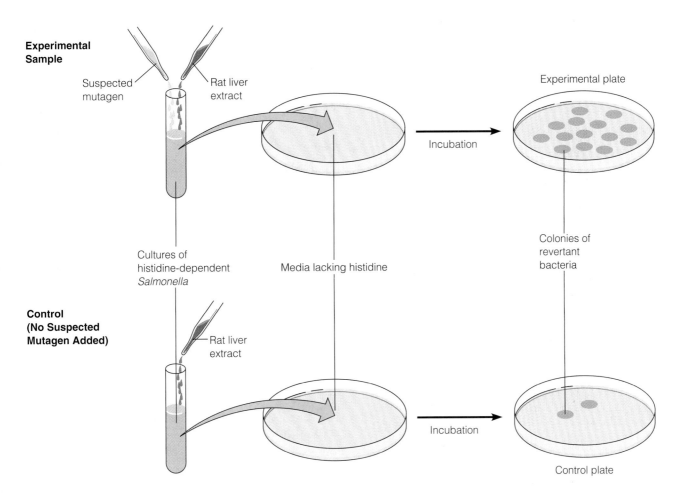

Experimental Sample

Suspected mutagen

Rat liver extract

Cultures of histidine-dependent *Salmonella*

Media lacking histidine

Incubation

Experimental plate

Colonies of revertant bacteria

Control (No Suspected Mutagen Added)

Rat liver extract

Incubation

Control plate

① Two cultures are prepared of *Salmonella* bacteria that have lost the ability to synthesize histidine (histidine-dependent).

② The suspected mutagen is added to the experimental sample only; rat liver extract (an activator) is added to both samples.

③ Each sample is poured onto a plate of medium lacking histidine. The plates are then incubated at 37°C for two days. Only bacteria whose histidine-dependent phenotype has mutated back (reverted) to histidine-synthesizing will grow into colonies.

④ The numbers of colonies on the experimental and control plates are compared. The control plate may show a few spontaneous histidine-synthesizing revertants. The test plates will show an increase in the number of histidine-synthesizing revertants if the test chemical is indeed a mutagen and potential carcinogen. The higher the concentration of mutagen used, the more revertant colonies will result.

FIGURE 8.21   The Ames test.

 **Do all mutagens cause cancer?**

genes on a chromosome. Figure 8.22 shows one type of genetic recombination occurring between two pieces of DNA, which we will call A and B and regard as chromosomes for the sake of simplicity. If these two chromosomes break and rejoin as shown—a process called **crossing over**—some of the genes carried by these chromosomes are shuffled. The original chromosomes have recombined, so that each now carries a portion of the other's genes.

If A and B represent DNA from different individuals, how are they brought close enough together to recombine? In eukaryotes, genetic recombination is an ordered process

that usually occurs as part of the sexual cycle of the organism. Recombination generally takes place during the formation of reproductive cells, such that these cells contain recombinant DNA. In bacteria, genetic recombination can happen in a number of ways, which we will discuss in the following sections.

Like mutation, genetic recombination contributes to a population's genetic diversity, which is the source of variation in evolution. In highly evolved organisms such as present-day microbes, recombination is more likely than mutation to be beneficial because recombination will less

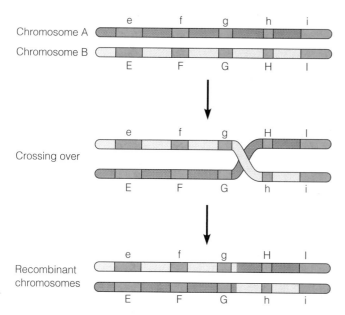

Chromosome A

Chromosome B

Crossing over

Recombinant chromosomes

FIGURE 8.22 Genetic recombination by crossing over between two related chromosomes. Chromosomes A and B each carry one version of genes E through I. The chromosomes cross over by breaking and rejoining, sometimes in more than one location. The result is two recombinant chromosomes, each of which carries genes originating from both chromosomes.

Q **What type of enzyme breaks the DNA? What enzyme rejoins the pieces of DNA?**

likely destroy a gene's function and may bring together combinations of genes that enable the organism to carry out a valuable new function.

The major protein that constitutes the flagella of *Salmonella* is also one of the primary proteins that causes our immune systems to respond. However, these bacteria have the capability of producing two different flagellar proteins. As our immune system mounts a response against those cells containing one form of the flagellar protein, those organisms producing the second are not affected. Which flagellar protein is produced is determined by a recombination event that apparently occurs somewhat randomly within the chromosomal DNA. Thus, by altering the flagellar protein produced, *Salmonella* can better avoid the defenses of the host.

Vertical gene transfer occurs when genes are passed from an organism to its offspring. Plants and animals transmit their genes by vertical transmission. Bacteria can pass their genes not only to their offspring, but also laterally, to other microbes of the same generation. This is known as **horizontal gene transfer.** Horizontal gene transfer between bacteria occurs in several ways. In all of the mechanisms, the transfer involves a **donor cell** that gives a portion of its total DNA to a **recipient cell.** Once transferred, part of the donor's DNA is usually incorporated into the recipient's DNA; the remainder is degraded by cellular

enzymes. The recipient cell that incorporates donor DNA into its own DNA is called a *recombinant.* The transfer of genetic material between bacteria is by no means a frequent event; it may occur in only 1% or less of an entire population. Let's examine in detail the specific types of genetic transfer. ❈ **Animation: Go to The Microbiology Place website or CD-ROM and click "Animations" to view Horizontal Gene Transfer.**

TRANSFORMATION IN BACTERIA

During the process of **transformation,** genes are transferred from one bacterium to another as "naked" DNA in solution. This process was first demonstrated over 70 years ago, although it was not understood at the time. Not only did transformation show that genetic material could be transferred from one bacterial cell to another, but study of this phenomenon eventually led to the conclusion that DNA is the genetic material. The initial experiment on transformation was performed by Frederick Griffith in England in 1928 while he was working with two strains of *Streptococcus pneumoniae.* One, a virulent (pathogenic) strain, has a polysaccharide capsule that prevents phagocytosis. The bacteria grow and cause pneumonia. The other, an avirulent strain, lacks the capsule and does not cause disease.

Griffith was interested in determining whether injections of heat-killed bacteria of the encapsulated strain could be used to vaccinate mice against pneumonia. As he expected, injections of living encapsulated bacteria killed the mouse (Figure 8.23a); injections of live nonencapsulated bacteria (Figure 8.23b) or dead encapsulated bacteria did not kill the mouse (Figure 8.23c). However, when the dead encapsulated bacteria were mixed with live nonencapsulated bacteria and injected into the mice, many of the mice died. In the blood of the dead mice, Griffith found living, encapsulated bacteria. Hereditary material (genes) from the dead bacteria had entered the live cells and changed them genetically so that their progeny were encapsulated and therefore virulent (Figure 8.23d).

Subsequent investigations based on Griffith's research revealed that bacterial transformation could be carried out without mice. A broth was inoculated with live nonencapsulated bacteria. Dead encapsulated bacteria were then added to the broth. After incubation, the culture was found to contain living bacteria that were encapsulated and virulent. The nonencapsulated bacteria had been transformed; they had acquired a new hereditary trait by incorporating genes from the killed encapsulated bacteria.

The next step was to extract various chemical components from the killed cells to determine which component caused the transformation. These crucial experiments were performed in the United States by Oswald T. Avery and his associates Colin M. MacLeod and Maclyn

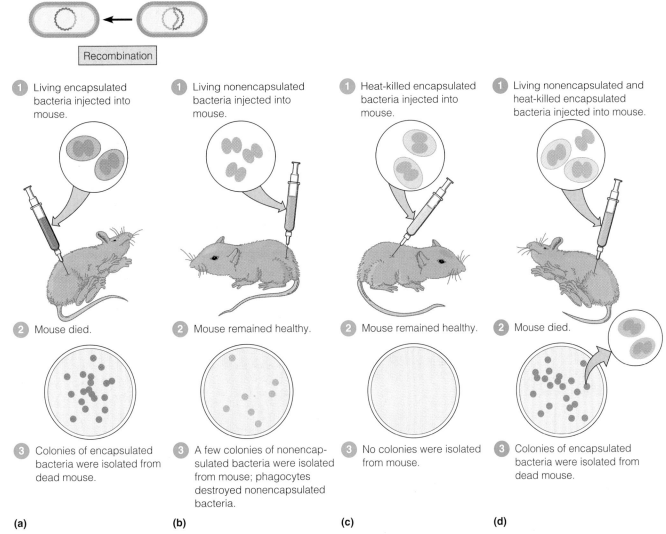

Recombination

(a)

① Living encapsulated bacteria injected into mouse.

② Mouse died.

③ Colonies of encapsulated bacteria were isolated from dead mouse.

(b)

① Living nonencapsulated bacteria injected into mouse.

② Mouse remained healthy.

③ A few colonies of nonencapsulated bacteria were isolated from mouse; phagocytes destroyed nonencapsulated bacteria.

(c)

① Heat-killed encapsulated bacteria injected into mouse.

② Mouse remained healthy.

③ No colonies were isolated from mouse.

(d)

① Living nonencapsulated and heat-killed encapsulated bacteria injected into mouse.

② Mouse died.

③ Colonies of encapsulated bacteria were isolated from dead mouse.

FIGURE 8.23 Griffith's experiment demonstrating genetic transformation. (a) Living encapsulated bacteria caused disease and death when injected into a mouse. **(b)** Living nonencapsulated bacteria are readily destroyed by the phagocytic defenses of the host, so the mouse remained healthy after injection. **(c)** After being killed by heat, encapsulated bacteria lost the ability to cause disease. **(d)** However, the combination of living nonencapsulated bacteria and heat-killed encapsulated bacteria (neither of which alone cause disease) did cause disease. Somehow, the live nonencapsulated bacteria were transformed by the dead encapsulated bacteria so that they acquired the ability to form a capsule and therefore cause disease. Subsequent experiments proved the transforming factor to be DNA.

Q **Why did encapsulated bacteria kill the mouse while nonencapsulated bacteria did not? What killed the mouse in (d)?**

McCarty. After years of research, they announced in 1944 that the component responsible for transforming harmless *S. pneumoniae* into virulent strains was DNA. Their results provided one of the conclusive indications that DNA was indeed the carrier of genetic information.

Since the time of Griffith's experiment, considerable information has been gathered about transformation. In nature, some bacteria, perhaps after death and cell lysis, release their DNA into the environment. Other bacteria can then encounter the DNA and, depending on the particular species and growth conditions, take up fragments of DNA and integrate them into their own chromosomes by recombination.

A protein called RecA (see Figure 3.10a, page 68) binds to the cell's DNA and then to donor DNA causing the exchange of strands. A recipient cell with this new combination of genes is a kind of hybrid, or recombinant cell (Figure 8.24). All the descendants of such a recombinant cell will be identical to it. Transformation occurs naturally among very few genera of bacteria, including *Bacillus*, *Haemophilus* (hē-mä'fi-lus), *Neisseria*, *Acinetobacter* (a-si-ne'tō-bak-tėr), and certain strains of the genera *Streptococcus* and *Staphylococcus*.

Transformation works best when the donor and recipient cells are very closely related. Even though only a small portion of a cell's DNA is transferred to the recipient, the

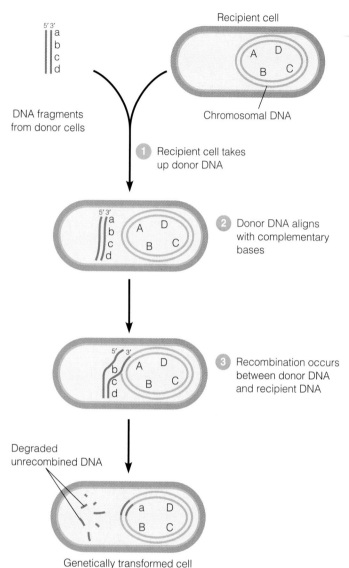

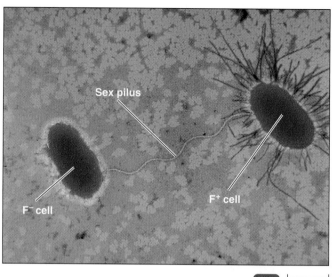

FIGURE 8.25 **Bacterial conjugation.** The sex pilus connecting these cells undergoing conjugation allows the transfer of genetic information. At the actual time of genetic exchange, the cells' connecting bridge contracts and the cells are much closer together. Notice that one cell has numerous fimbriae.

Q **What is an F⁺ cell?**

FIGURE 8.24 **The mechanism of genetic transformation in bacteria.** Some similarity is needed for the donor and recipient to align. Genes *a, b, c,* and *d* may be mutations of genes *A, B, C,* and *D.*

Q **What type of enzyme cuts the donor DNA?**

molecule that must pass through the recipient cell wall and membrane is still very large. When a recipient cell is in a physiological state in which it can take up the donor DNA, it is said to be competent. **Competence** results from alterations in the cell wall that make it permeable to large DNA molecules.

The well-understood and widely used bacterium *E. coli* is not naturally competent for transformation. However, a simple laboratory treatment enables *E. coli* to readily take up DNA. The discovery of this treatment has enabled researchers to use *E. coli* for genetic engineering, discussed in Chapter 9.

CONJUGATION IN BACTERIA

Another mechanism by which genetic material is transferred from one bacterium to another is known as **conjugation.** Conjugation is mediated by one kind of *plasmid,* a circular piece of DNA that replicates independently from the cell's chromosome (discussed on page 245). However, plasmids differ from bacterial chromosomes in that the genes they carry are usually not essential for the growth of the cell under normal conditions. The plasmids responsible for conjugation are transmissible between cells during conjugation.

Conjugation differs from transformation in two major ways. First, conjugation requires direct cell-to-cell contact. Second, the conjugating cells must generally be of opposite mating type; donor cells must carry the plasmid, and recipient cells usually do not. In gram-negative bacteria, the plasmid carries genes that code for the synthesis of *sex pili,* projections from the donor's cell surface that contact the recipient and help bring the two cells into direct contact. Gram-positive bacterial cells produce sticky surface molecules that cause cells to come into direct contact with each other. In the process of conjugation, the plasmid is replicated during the transfer of a single-stranded copy of the plasmid DNA to the recipient, where the complementary strand is synthesized.

Because most experimental work on conjugation has been done with *E. coli,* we will describe the process in this organism. In *E. coli,* the **F factor (fertility factor)** was the first

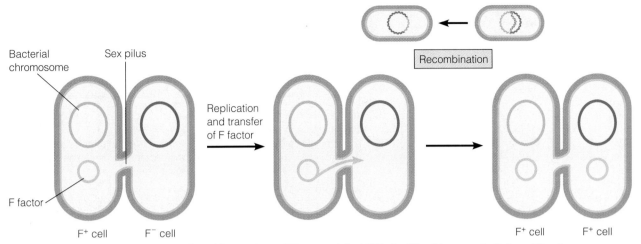

(a) When an F factor (a plasmid) is transferred from a donor (F⁺) to a recipient (F⁻), the F⁻ cell is converted into an F⁺ cell.

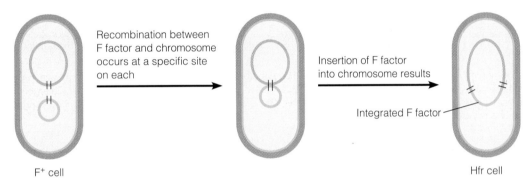

(b) When an F factor becomes integrated into the chromosome of an F⁺ cell, it makes the cell a high frequency of recombination (Hfr) cell.

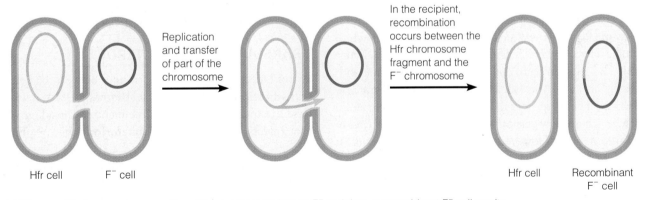

(c) When an Hfr donor passes a portion of its chromosome into an F⁻ recipient, a recombinant F⁻ cell results.

FIGURE 8.26   Conjugation in *E. coli*.

 How does conjugation differ from transformation?

plasmid observed to be transferred between cells during conjugation (Figure 8.25). Donors carrying F factors (F⁺ cells) transfer the plasmid to recipients (F⁻ cells), which become F⁺ cells as a result (Figure 8.26a). In some cells carrying F factors, the factor integrates into the chromosome, converting the F⁺ cell to an **Hfr cell** (high frequency of recombination) (Figure 8.26b). When conjugation occurs between an Hfr cell and an F⁻ cell, the Hfr cell's chromosome (with its integrated F factor) replicates, and a parental strand of the chromosome is transferred to the recipient cell (Figure 8.26c).

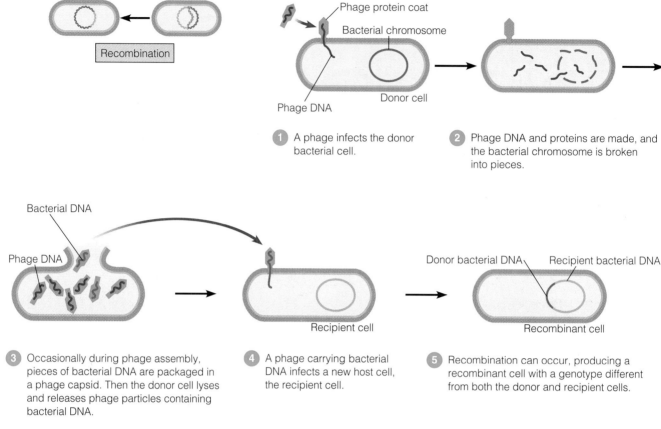

1   A phage infects the donor bacterial cell.

2   Phage DNA and proteins are made, and the bacterial chromosome is broken into pieces.

3   Occasionally during phage assembly, pieces of bacterial DNA are packaged in a phage capsid. Then the donor cell lyses and releases phage particles containing bacterial DNA.

4   A phage carrying bacterial DNA infects a new host cell, the recipient cell.

5   Recombination can occur, producing a recombinant cell with a genotype different from both the donor and recipient cells.

FIGURE 8.27   Transduction by a bacteriophage. Shown here is generalized transduction, in which any bacterial DNA can be transferred from one cell to another.

Q   **What is transduction?**

Replication of the Hfr chromosome begins in the middle of the integrated F factor, and a small piece of the F factor leads the chromosomal genes into the F⁻ cell. Usually, the chromosome breaks before it is completely transferred. Once within the recipient cell, donor DNA can recombine with the recipient's DNA. (Donor DNA that is not integrated is degraded.) Therefore, by conjugation with an Hfr cell, an F⁻ cell may acquire new versions of chromosomal genes (just as in transformation). However, it remains an F⁻ cell because it did not receive a complete F factor during conjugation.

Conjugation is used to map the location of genes on a bacterial chromosome (see Figure 8.1b). The genes for the synthesis of threonine (*thr*) and leucine (*leu*) are first, reading clockwise from 0. Their locations were determined by conjugation experiments. Assume that conjugation is allowed for only 1 minute between an Hfr strain that is his⁺, pro⁺, thr⁺, and leu⁺, and an F⁻ strain that is his⁻, pro⁻, thr⁻, and leu⁻. If the F⁻ acquired the ability to synthesize threonine, then the *thr* gene is located early in the chromosome, between 0 and 1 minute. If after 2 minutes the F⁻ cell now becomes thr⁺ and leu⁺, the order of these two genes on the chromosome must be *thr, leu*. ✳ **Animation:**

Go to The Microbiology Place website or CD-ROM and click "Animations" to view Bacterial Conjugation.

TRANSDUCTION IN BACTERIA

A third mechanism of genetic transfer between bacteria is **transduction.** In this process, bacterial DNA is transferred from a donor cell to a recipient cell inside a virus that infects bacteria, called a **bacteriophage,** or **phage.** (Phages will be discussed further in Chapter 13.)

To understand how transduction works, we will consider the life cycle of one type of transducing phage of *E. coli;* this phage carries out **generalized transduction** (Figure 8.27).

1   In the process of infection, the phage attaches to the donor bacterial cell wall and injects its DNA into the bacterium.

2   The phage DNA acts as a template for the synthesis of new phage DNA and also directs the synthesis of phage protein coats. During phage development inside the infected cell, the bacterial chromosome is broken apart by phage enzymes, and

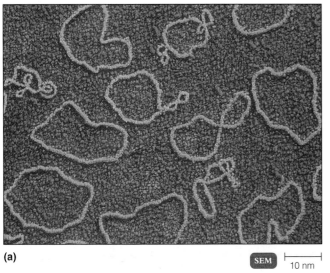

(a)

SEM | 10 nm

(b) Origin of transfer   tet

FIGURE 8.28 R factor, a type of plasmid. (**a**) Plasmids from *Bacteroides fragilis* bacteria that encode resistance to the antibiotic clindamycin. (**b**) A diagram of an R factor, which has two parts: the RTF contains genes needed for plasmid replication and transfer of the plasmid by conjugation, and the r-determinant carries genes for resistance to four different antibiotics and mercury (*sul* = sulfonamide resistance, *str* = streptomycin resistance, *cml* = chloramphenicol resistance, *tet* = tetracycline resistance, *mer* = mercury resistance); numbers are base pairs × 1000.

Q **Why are R factors important in the treatment of infectious diseases?**

❸ at least some pieces of bacterial DNA are mistakenly packaged inside phage protein coats. (Even plasmid DNA or DNA of another virus that is inside the cell can be given phage protein coats.) The resulting phage particles then carry bacterial DNA instead of phage DNA.

❹ When the released phage particles later infect a new population of bacteria, bacterial genes will be transferred to the newly infected recipient cells at low frequency.

❺ Transduction of cellular DNA by a virus can lead to recombination between the DNA of the donor host cell and the DNA of the recipient host cell. The process of generalized transduction is typical of bacteriophages such as phage P1 of *E. coli* and phage P22 of *Salmonella*.

All genes contained within a bacterium infected by a generalized transducing phage are equally likely to be packaged in a phage coat and transferred. In another type of transduction, called **specialized transduction,** only certain bacterial genes are transferred. In one type of specialized transduction, the phage codes for certain toxins produced by their bacterial hosts, such as diptheria toxin for *Corynebacterium diphtheriae* (kôr′-i-nē-bak-ti-rē-um dif-thi′-re-ī), erythrogenic toxin for *Streptococcus pyogenes*, and Shiga toxin for *E. coli* O157:H7. Specialized

transduction will be discussed in Chapter 13 (page 400). In addition to mutation, transformation, and conjugation, transduction is another way bacteria acquire new genotypes.

PLASMIDS AND TRANSPOSONS

LEARNING OBJECTIVE

• Describe the functions of plasmids and transposons.

Plasmids and transposons are genetic elements that provide additional mechanisms for genetic change. They occur in both prokaryotic and eukaryotic organisms, but this discussion focuses on their role in genetic change in prokaryotes.

PLASMIDS

Recall from Chapter 4 (page 95) that plasmids are self-replicating, gene-containing circular pieces of DNA about 1–5% the size of the bacterial chromosome (Figure 8.28a). They are found mainly in bacteria but also in some eukaryotic microorganisms, such as *Saccharomyces cerevisiae*. The F factor is a **conjugative plasmid** that carries genes for sex pili and for the transfer of the plasmid to another cell. Although plasmids are usually dispensable, under certain conditions genes carried by plasmids can be crucial to the survival and growth of the cell. For example,

dissimilation plasmids code for enzymes that trigger the catabolism of certain unusual sugars and hydrocarbons. Some species of *Pseudomonas* can actually use such exotic substances as toluene, camphor, and hydrocarbons of petroleum as primary carbon and energy sources because they have catabolic enzymes encoded by genes carried on plasmids. Such specialized capabilities permit the survival of those microorganisms in very diverse and challenging environments. Because of their ability to degrade and detoxify a variety of unusual compounds, many of them are being investigated for possible use in the cleanup of environmental wastes. (See the box in Chapter 2, page 33.)

Other plasmids code for proteins that enhance the pathogenicity of a bacterium. The strain of *E. coli* that causes infant diarrhea and traveler's diarrhea carries plasmids that code for toxin production and for bacterial attachment to intestinal cells. Without these plasmids, *E. coli* is a harmless resident of the large intestine; with them, it is pathogenic. Other plasmid-encoded toxins include the exfoliative toxin of *Staphylococcus aureus*, *Clostridium tetani* neurotoxin, and toxins of *Bacillus anthracis*. Still other plasmids contain genes for the synthesis of **bacteriocins,** toxic proteins that kill other bacteria. These plasmids have been found in many bacterial genera, and they are useful markers for the identification of certain bacteria in clinical laboratories.

Resistance factors (R factors) are plasmids that have significant medical importance. They were first discovered in Japan in the late 1950s after several dysentery epidemics. In some of these epidemics, the infectious agent was resistant to the usual antibiotic. Following isolation, the pathogen was also found to be resistant to a number of different antibiotics. In addition, other normal bacteria from the patients (such as *E. coli*) proved to be resistant as well. Researchers soon discovered that these bacteria acquired resistance through the spread of genes from one organism to another. The plasmids that mediated this transfer are R factors.

R factors carry genes that confer upon their host cell resistance to antibiotics, heavy metals, or cellular toxins. Many R factors contain two groups of genes. One group is called the **resistance transfer factor (RTF)** and includes genes for plasmid replication and conjugation. The other group, the **r-determinant,** has the resistance genes; it codes for the production of enzymes that inactivate certain drugs or toxic substances (Figure 8.28b). Different R factors, when present in the same cell, can recombine to produce R factors with new combinations of genes in their r-determinants.

In some cases, the accumulation of resistance genes within a single plasmid is quite remarkable. For example, Figure 8.28b shows a genetic map of resistance plasmid R100. Carried on this plasmid are resistance genes for sulfonamides, streptomycin, chloramphenicol, and tetracycline, as well as genes for resistance to mercury. This particular plasmid can be transferred between a number of enteric species, including *Escherichia*, *Klebsiella*, and *Salmonella*.

R factors present very serious problems for the treatment of infectious diseases with antibiotics. The widespread use of antibiotics in medicine and agriculture (many types of animal feed contain antibiotics) has led to the preferential survival (selection) of bacteria that have R factors, so populations of resistant bacteria grow larger and larger. The transfer of resistance between bacterial cells of a population, and even between bacteria of different genera, also contributes to the problem. The ability to reproduce sexually with members of its own species defines a eukaryotic species. However, a bacterial species can conjugate and transfer plasmids to other species. *Neisseria* may have acquired its penicillinase-producing plasmid from *Streptococcus*, and *Agrobacterium* can transfer plasmids to plant cells (see Figure 9.20, page 274). Nonconjugative plasmids may be transferred from one cell to another by inserting themselves into a conjugative plasmid or a chromosome or by transformation when released from a dead cell. Insertion is made possible by an insertion sequence, which will be discussed shortly.

Plasmids are an important tool for genetic engineering, discussed in Chapter 9 (page 258).

TRANSPOSONS

Transposons are small segments of DNA that can move (be "transposed") from one region of a DNA molecule to another. These pieces of DNA are 700–40,000 base pairs long.

In the 1950s, American geneticist Barbara McClintock discovered transposons in corn, but they occur in all organisms and have been studied most thoroughly in microorganisms. They may move from one site to another site on the same chromosome or to another chromosome or plasmid. As you might imagine, the frequent movement of transposons could wreak havoc inside a cell. For example, as transposons move about on chromosomes, they may insert themselves *within* genes, inactivating them. Fortunately, transposition occurs relatively rarely. The frequency of transposition is comparable to the spontaneous mutation rate that occurs in bacteria—that is, from 10^{-5} to 10^{-7} per generation.

All transposons contain the information for their own transposition. As shown in Figure 8.29a, the simplest transposons, also called **insertion sequences (IS),** contain only a gene that codes for an enzyme (*transposase*, which catalyzes the cutting and resealing of DNA that

IS1

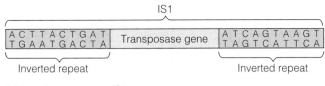

(a) Insertion sequence IS1

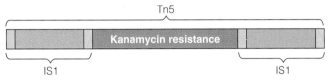

(b) Complex transposon Tn5

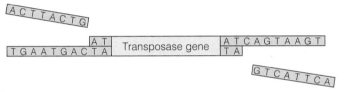

① Transposase cuts DNA, leaving sticky ends.

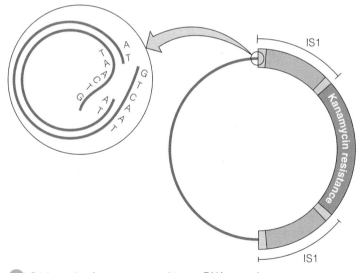

② Sticky ends of transposon and target DNA anneal.

(c) Insertion of a transposon in a plasmid or chromosome (target DNA)

FIGURE 8.29 Transposons and insertion. (a) An insertion sequence (IS), the simplest transposon, contains a gene for transposase, the enzyme that catalyzes transposition. The transposase gene is bounded at each end by inverted repeat sequences that function as recognition sites for the transposon. IS1 is one example of an insertion sequence, shown here with simplified IR sequences. **(b)** Complex transposons carry other genetic material in addition to transposase genes. The example shown here, Tn5, carries the gene for kanamycin resistance and has complete copies of the insertion sequence IS1 at each end. **(c)** Transposition of the transposon Tn5 into R100 plasmid. Notice that a plasmid can acquire IS elements.

Q **Why are transposons sometimes referred to as "jumping genes"?**

antibiotic resistance (Figure 8.29b). Plasmids such as R factors are frequently made up of a collection of transposons (Figure 8.29c).

Transposons with antibiotic resistance genes are of practical interest, but there is no limitation on the kinds of genes that transposons can have. Thus, transposons provide a natural mechanism for the movement of genes from one chromosome to another. Furthermore, because they may be carried between cells on plasmids or viruses, they can also spread from one organism—or even species—to another. For example, vancomycin resistance was transferred from *Enterococcus faecalis* to *Staphylococcus aureus* via a transposon called Tn1546. Transposons are thus a potentially powerful mediator of evolution in organisms. ※ **Animation: Go to The Microbiology Place website or CD-ROM and click "Animations" to view Transposons.**

GENES AND EVOLUTION

LEARNING OBJECTIVE
- Discuss how genetic mutation and recombination provide material for natural selection to act upon.

We have now seen how gene activity can be controlled by the cell's internal regulatory mechanisms and how genes themselves can be altered or rearranged by mutation, transposition, and recombination. All these processes provide diversity in the descendants of cells. Diversity provides the raw material for evolution, and natural selection provides its driving force. Natural selection will act on diverse populations to ensure the survival of those fit for that particular environment. The different kinds of microorganisms that exist today are the result of a long history of evolution. Microorganisms have continually changed by alterations in their genetic properties and acquisition of adaptations to many different habitats.

occurs in transposition) and recognition sites. *Recognition sites* are short inverted repeat sequences of DNA that the enzyme recognizes as recombination sites between the transposon and the chromosome.

Complex transposons also carry other genes not connected with the transposition process. For example, bacterial transposons may contain genes for enterotoxin or for

STUDY OUTLINE

STRUCTURE AND FUNCTION OF THE GENETIC MATERIAL (pp. 215–227)

1. Genetics is the study of what genes are, how they carry information, how their information is expressed, and how they are replicated and passed to subsequent generations or other organisms.

2. DNA in cells exists as a double-stranded helix; the two strands are held together by hydrogen bonds between specific nitrogenous base pairs: AT and CG.

3. A gene is a segment of DNA, a sequence of nucleotides, that codes for a functional product, usually a protein.

4. When a gene is expressed, DNA is transcribed to produce RNA; mRNA is then translated into proteins.

5. The DNA in a cell is duplicated before the cell divides, so each daughter cell receives the same genetic information.

GENOTYPE AND PHENOTYPE (p. 215)

6. Genotype is the genetic composition of an organism, its entire complement of DNA.

7. Phenotype is the expression of the genes: the proteins of the cell and the properties they confer on the organism.

DNA AND CHROMOSOMES (pp. 215–217)

8. The DNA in a chromosome exists as one long double helix associated with various proteins that regulate genetic activity.

9. Bacterial DNA is circular; the chromosome of *E. coli*, for example, contains about 4 million base pairs and is approximately 1000 times longer than the cell.

10. Genomics is the molecular characterization of genomes.

11. Information contained in the DNA is transcribed into RNA and translated into proteins.

DNA REPLICATION (pp. 217–223)

12. During DNA replication, the two strands of the double helix separate at the replication fork, and each strand is used as a template by DNA polymerases to synthesize two new strands of DNA according to the rules of nitrogenous base pairing.

13. The result of DNA replication is two new strands of DNA, each having a base sequence complementary to one of the original strands.

14. Because each double-stranded DNA molecule contains one original and one new strand, the replication process is called semiconservative.

15. DNA is synthesized in one direction designated $5' \rightarrow 3'$. At the replication fork, the leading strand is synthesized continuously and the lagging strand discontinuously.

16. DNA polymerase proofreads new molecules of DNA and removes mismatched bases before continuing DNA synthesis.

17. Each daughter bacterium receives a chromosome that is virtually identical to the parent's. ❊ **Animation: DNA Replication. The Microbiology Place.**

RNA AND PROTEIN SYNTHESIS (pp. 223–227)

18. During transcription, the enzyme RNA polymerase synthesizes a strand of RNA from one strand of double-stranded DNA, which serves as a template.

19. RNA is synthesized from nucleotides containing the bases A, C, G, and U, which pair with the bases of the DNA strand being transcribed.

20. RNA polymerase binds the promoter; transcription begins at AUG; the region of DNA that is the end point of transcription is the terminator; RNA is synthesized in the $5' \rightarrow 3'$ direction. ❊ **Animation: Transcription. The Microbiology Place.**

21. Translation is the process in which the information in the nucleotide base sequence of mRNA is used to dictate the amino acid sequence of a protein.

22. The mRNA associates with ribosomes, which consist of rRNA and protein.

23. Three-base segments of mRNA that specify amino acids are called codons.

24. The genetic code refers to the relationship among the nucleotide base sequence of DNA, the corresponding codons of mRNA, and the amino acids for which the codons code.

25. The genetic code is degenerate; that is, most amino acids are coded for by more than one codon.

26. Of the 64 codons, 61 are sense codons (which code for amino acids), and 3 are nonsense codons (which do not code for amino acids and are stop signals for translation).

27. The start codon, AUG, codes for methionine.

28. Specific amino acids are attached to molecules of tRNA. Another portion of the tRNA has a base triplet called an anticodon.

29. The base pairing of codon and anticodon at the ribosome results in specific amino acids being brought to the site of protein synthesis.

30. The ribosome moves along the mRNA strand as amino acids are joined to form a growing polypeptide; mRNA is read in the $5' \rightarrow 3'$ direction.

31. Translation ends when the ribosome reaches a stop codon on the mRNA.

32. In prokaryotes, translation can begin before transcription is complete. ❊ **Animation: Translation. The Microbiology Place.**

THE REGULATION OF BACTERIAL GENE EXPRESSION (pp. 227–231)

1. Regulating protein synthesis at the gene level is energy-efficient because proteins are synthesized only as they are needed.

2. Constitutive enzymes produce products at a fixed rate. Examples are genes for the enzymes in glycolysis.

3. For these gene regulatory mechanisms, the control is aimed at mRNA synthesis.

REPRESSION AND INDUCTION (pp. 228–229)

4. Repression controls the synthesis of one or several (repressible) enzymes.

5. When cells are exposed to a particular end-product, the synthesis of enzymes related to that product decreases.

6. In the presence of certain chemicals (inducers), cells synthesize more enzymes. This process is called induction.

7. An example of induction is the production of β-galactosidase by *E. coli* in the presence of lactose; lactose can then be metabolized.

THE OPERON MODEL OF GENE EXPRESSION (p. 229)

8. The formation of enzymes is determined by structural genes.

9. In bacteria, a group of coordinately regulated structural genes with related metabolic functions, plus the promoter and operator sites that control their transcription, are called an operon.

10. In the operon model for an inducible system, a regulatory gene codes for the repressor protein.

11. When the inducer is absent, the repressor binds to the operator, and no mRNA is synthesized.

12. When the inducer is present, it binds to the repressor so that it cannot bind to the operator; thus, mRNA is made, and enzyme synthesis is induced.

13. In repressible systems, the repressor requires a corepressor in order to bind to the operator site; thus, the corepressor controls enzyme synthesis.

14. Transcription of structural genes for catabolic enzymes (such as β-galactosidase) is induced by the absence of glucose. Cyclic AMP and CRP must bind to a promoter in the presence of an alternative carbohydrate.

15. The presence of glucose inhibits the metabolism of alternative carbon sources by catabolite repression. ✳ **Animation: Operons. The Microbiology Place.**

MUTATION: CHANGE IN THE GENETIC MATERIAL (pp. 231–238)

1. A mutation is a change in the nitrogenous base sequence of DNA; that change causes a change in the product coded for by the mutated gene.

2. Many mutations are neutral, some are disadvantageous, and others are beneficial.

TYPES OF MUTATIONS (p. 232)

3. A base substitution occurs when one base pair in DNA is replaced with a different base pair.

4. Alterations in DNA can result in missense mutations (which cause amino acid substitutions) or nonsense mutations (which create stop codons).

5. In a frameshift mutation, one or a few base pairs are deleted or added to DNA.

6. Mutagens are agents in the environment that cause permanent changes in DNA.

7. Spontaneous mutations occur without the presence of any mutagen.

MUTAGENS (pp. 233–236)

8. Chemical mutagens include base-pair mutagens, nucleoside analogs, and frameshift mutagens.

9. Ionizing radiation causes the formation of ions and free radicals that react with DNA; base substitutions or breakage of the sugar-phosphate backbone results.

10. Ultraviolet (UV) radiation is nonionizing; it causes bonding between adjacent thymines.

11. Damage to DNA caused by UV radiation can be repaired by enzymes that cut out and replace the damaged portion of DNA.

12. Light-repair enzymes repair thymine dimers in the presence of visible light.

THE FREQUENCY OF MUTATION (p. 236)

13. Mutation rate is the probability that a gene will mutate when a cell divides; the rate is expressed as 10 to a negative power.

14. Mutations usually occur randomly along a chromosome.

15. A low rate of spontaneous mutations is beneficial in providing the genetic diversity needed for evolution.

IDENTIFYING MUTANTS (p. 237)

16. Mutants can be detected by selecting or testing for an altered phenotype.

17. Positive selection involves the selection of mutant cells and the rejection of nonmutated cells.

18. Replica plating is used for negative selection—to detect, for example, auxotrophs that have nutritional requirements not possessed by the parent (nonmutated) cell. ✳ **Animation: Mutations and DNA Repair. The Microbiology Place.**

IDENTIFYING CHEMICAL CARCINOGENS (pp. 237–238)

19. The Ames test is a relatively inexpensive and rapid test for identifying possible chemical carcinogens.

20. The test assumes that a mutant cell can revert to a normal cell in the presence of a mutagen and that many mutagens are carcinogens.

21. Histidine auxotrophs of *Salmonella* are exposed to an enzymatically treated potential carcinogen, and reversions to the nonmutant state are selected.

GENETIC TRANSFER AND RECOMBINATION (pp. 238–247)

1. Genetic recombination, the rearrangement of genes from separate groups of genes, usually involves DNA from different organisms; it contributes to genetic diversity.

2. In crossing over, genes from two chromosomes are recombined into one chromosome containing some genes from each original chromosome.

3. Vertical gene transfer occurs during reproduction when genes are passed from an organism to its offspring.

4. Horizontal gene transfer in bacteria involves a portion of the cell's DNA being transferred from donor to recipient.

5. When some of the donor's DNA has been integrated into the recipient's DNA, the resultant cell is called a recombinant. ✷ **Animation: Horizontal Gene Transfer. The Microbiology Place.**

TRANSFORMATION IN BACTERIA (pp. 240–242)

6. During this process, genes are transferred from one bacterium to another as "naked" DNA in solution.

7. This process was first demonstrated in *Streptococcus pneumoniae* and occurs naturally among a few genera of bacteria.

CONJUGATION IN BACTERIA (pp. 242–244)

8. This process requires contact between living cells.

9. One type of genetic donor cell is an F⁺; recipient cells are F⁻. F cells contain plasmids called F factors; these are transferred to the F⁻ cells during conjugation.

10. When the plasmid becomes incorporated into the chromosome, the cell is called an Hfr (high frequency of recombination) cell.

11. During conjugation, an Hfr cell can transfer chromosomal DNA to an F⁻ cell. Usually, the Hfr chromosome breaks before it is fully transferred. ✷ **Animation: Bacterial Conjugation. The Microbiology Place.**

TRANSDUCTION IN BACTERIA (pp. 244–245)

12. In this process, DNA is passed from one bacterium to another in a bacteriophage and is then incorporated into the recipient's DNA.

13. In generalized transduction, any bacterial genes can be transferred.

PLASMIDS AND TRANSPOSONS (pp. 245–247)

14. Plasmids are self-replicating circular molecules of DNA carrying genes that are not usually essential for the cell's survival.

15. There are several types of plasmids, including conjugative plasmids, dissimilation plasmids, plasmids carrying genes for toxins or bacteriocins, and resistance factors.

16. Transposons are small segments of DNA that can move from one region to another region of the same chromosome or to a different chromosome or a plasmid.

17. Transposons are found in the main chromosomes of organisms, in plasmids, and in the genetic material of viruses. They vary from simple (insertion sequences) to complex.

18. Complex transposons can carry any type of gene, including antibiotic-resistance genes, and are thus a natural mechanism for moving genes from one chromosome to another. ✷ **Animation: Transposons. The Microbiology Place.**

GENES AND EVOLUTION (p. 247)

1. Diversity is the precondition for evolution.

2. Genetic mutation and recombination provide a diversity of organisms, and the process of natural selection allows the growth of those best adapted to a given environment.

STUDY QUESTIONS

✷ Access more review material either online at **The Microbiology Place** (www.microbiologyplace.com) or with **The Microbiology Place CD-ROM** packaged with your new book. There you'll find activities, practice tests, quizzes, flashcards, case studies, and more to help you succeed. In addition, you'll find the following Interactive Tutorials: DNA Replication, Transcription, Translation, Mutation, and Recombination.

Answers to the Study Questions can be found in Appendix G.

REVIEW

1. Briefly describe the components of DNA, and explain its functional relationship to RNA and protein.

2. Draw a diagram showing a portion of a chromosome undergoing replication.
   a. Identify the replication fork.
   b. What is the role of DNA polymerase? Of RNA polymerase?
   c. How does this process represent semiconservative replication?

3. The following is a code for a strand of DNA.

   DNA   3′ A T A T _ _ _ T T T _ _ _ _ _ _ _ _ _ _
         1 2 3 4 5 6 7 8 9 10 11 12 13 14 15 16 17 18 19
   mRNA   C G U   U G A
   tRNA   U G G
   Amino Acid   Met _____   _____   _____   _____
   ATAT = Promoter sequence

   a. Using the genetic code provided in Figure 8.8, fill in the blanks to complete the segment of DNA shown.
   b. Fill in the blanks to complete the sequence of amino acids coded for by this strand of DNA.
   c. Write the code for the complementary strand of DNA completed in part a.
   d. What would be the effect if C were substituted for T at base 10?
   e. What would be the effect if A were substituted for G at base 11?
   f. What would be the effect if G were substituted for T at base 14?
   g. What would be the effect if C were inserted between bases 9 and 10?
   h. How would UV radiation affect this strand of DNA?
   i. Identify a nonsense sequence in this strand of DNA.

4. Describe translation, and be sure to include the following terms: ribosome, rRNA, amino acid activation, tRNA, anticodon, and codon.

5. Explain how you would find an antibiotic-resistant mutant by direct selection and how you would find an antibiotic-sensitive mutant by indirect selection.

6. Match the following examples of mutagens.

   _b_ A mutagen that is incorporated into DNA in place of a normal base
   _d_ A mutagen that causes the formation of highly reactive ions
   _c_ A mutagen that alters adenine so that it base-pairs with cytosine
   _a_ A mutagen that causes insertions
   _e_ A mutagen that causes the formation of pyrimidine dimers

   a. Frameshift mutagen
   b. Nucleoside analog
   c. Base-pair mutagen
   d. Ionizing radiation
   e. Nonionizing radiation

7. Describe the principle of the Ames test for identifying chemical carcinogens.

8. Define *plasmids*, and explain the relationship between F factors and conjugation.

9. Use the following metabolic pathway to answer the questions that follow it.

   Substrate A $\xrightarrow{\text{enzyme } a}$ Intermediate B $\xrightarrow{\text{enzyme } b}$ End-product C

   a. If enzyme a is inducible and is not being synthesized at present, a _____ protein must be bound tightly to the _____ site. When the inducer is present, it will bind to the _____ so that _____ can occur.
   b. If enzyme a is repressible, end-product C, called a _____, causes the _____ to bind to the _____. What causes derepression?
   c. If enzyme a is constitutive, what effect, if any, will the presence of A or C have on it?

10. Identify three ways of preventing mistakes in DNA.

11. Define the following terms:
    a. genotype
    b. phenotype
    c. recombination

12. Which sequence is the best target for damage by UV radiation: AGGCAA, CTTTGA, or GUAAAU? Why aren't all bacteria killed when they are exposed to sunlight?

13. You are provided with cultures with the following characteristics:
    Culture 1: F⁺, genotype A⁺ B⁺ C⁺
    Culture 2: F⁻, genotype A⁻ B⁻ C⁻
    a. Indicate the possible genotypes of a recombinant cell resulting from the conjugation of cultures 1 and 2.
    b. Indicate the possible genotypes of a recombinant cell resulting from conjugation of the two cultures after the F⁺ has become an Hfr cell.

14. Why are semiconservative replication and degeneracy of the genetic code advantageous to the survival of species?

15. Why are mutation and recombination important in the process of natural selection and the evolution of organisms?

MULTIPLE CHOICE

Match the following terms to the definitions in questions 1and 2.
   a. conjugation   d. transformation
   b. transcription   e. translation
   c. transduction

1. The transfer of DNA from a donor to a recipient cell by a bacteriophage.

2. The transfer of DNA from a donor to a recipient as naked DNA in solution.

3. Feedback inhibition differs from repression because feedback inhibition
   a. is less precise.
   b. is slower acting.
   c. stops the action of preexisting enzymes.
   d. stops the synthesis of new enzymes.
   e. all of the above

4. Bacteria can acquire antibiotic resistance by
   a. mutation.
   b. insertion of transposons.
   c. acquiring plasmids.
   d. all of the above
   e. none of the above

5. Suppose you inoculate three flasks of minimal salts broth with *E. coli*. Flask A contains glucose. Flask B contains glucose and lactose. Flask C contains lactose. After a few hours of incubation, you test the flasks for the presence of β-galactosidase. Which flask(s) do you predict will have this enzyme?
   a. A
   b. B
   c. C
   d. A and B
   e. B and C

6. Plasmids differ from transposons because plasmids
   a. become inserted into chromosomes.
   b. are self-replicated outside the chromosome.
   c. move from chromosome to chromosome.
   d. carry genes for antibiotic resistance.
   e. none of the above

Use the following choices to answer questions 7 and 8.
   a. catabolite repression
   b. DNA polymerase
   c. induction
   d. repression
   e. translation

7. The mechanism by which the presence of glucose inhibits the *lac* operon.

8. The mechanism by which lactose controls the *lac* operon.

9. Two daughter cells are most likely to inherit which one of the following from the parent cell?
   a. a change in a nucleotide in mRNA
   b. a change in a nucleotide in tRNA
   c. a change in a nucleotide in rRNA
   d. a change in a nucleotide in DNA
   e. a change in a protein

10. Which of the following is *not* a method of horizontal gene transfer?
   a. binary fission
   b. conjugation
   c. integration of a transposon
   d. transduction
   e. transformation

CRITICAL THINKING

1. Nucleoside analogs and ionizing radiation are used in the treatment of cancer. These mutagens can cause cancer, so how do you suppose they are used to treat the disease?

2. Replication of the *E. coli* chromosome takes 40–45 minutes, but the organism has a generation time of 26 minutes. How does the cell have time to make complete chromosomes for each daughter cell? For each granddaughter cell?

3. *Pseudomonas* has a plasmid containing the *mer* operon, which includes the gene for mercuric reductase. This enzyme catalyzes the reduction of the mercuric ion Hg^{2+} to the uncharged form of mercury, Hg^0. Hg^{2+} is quite toxic to cells; Hg^0 is not.
   a. What do you suppose is the inducer for this operon?
   b. The protein encoded by one of the *mer* genes binds Hg^{2+} in the periplasm and brings it into the cell. Why would a cell bring in a toxin?
   c. What is the value of the *mer* operon to *Pseudomonas*?

CLINICAL APPLICATIONS

1. Ciprofloxacin, erythromycin, and acyclovir are used to treat microbial infections. Ciprofloxacin inhibits DNA gyrase. Erythromycin binds in front of the A site on the 50S subunit of a ribosome. Acyclovir is a guanine analog.
   a. What steps in protein synthesis are inhibited by each drug?
   b. Which drug is more effective against bacteria? Why?
   c. Which drug is more effective against viruses? Why?
   d. Which drugs will have effects on the host's cells? Why?
   e. Use the index to identify the disease for which acyclovir is primarily used. Why is it more effective than erythromycin for treating this disease?

2. HIV, the virus that causes AIDS, was isolated from three individuals, and the amino acid sequences for the viral coat were determined. Of the amino acid sequences shown below, which two of the viruses are most closely related? How can these amino acid sequences be used to identify the source of a virus?

Patient	Viral Amino Acid Sequence
A	Asn Gln Thr Ala Ala Ser Lys Asn Ile Asp Ala Leu
B	Asn Leu His Ser Asp Lys Ile Asn Ile Ile Leu Leu
C	Asn Gln Thr Ala Asp Ser Ile Val Ile Asp Ala Leu

3. Human herpesvirus-8 (HHV-8) is common in certain parts of Africa, the Middle East, and the Mediterranean, but it is rare elsewhere except in AIDS patients. Genetic analyses indicate that the African strain is not changing, whereas the Western strain is accumulating changes. Using the portions of the HHV-8 genomes (shown below) that encode one of the viral proteins, how similar are these two viruses? What mechanism can account for the changes? What disease does HHV-8 cause?
   Western   3′ ATGGAGTTCTTCTGGACAAGA
   African   3′ ATAAACTTTTCTTGACAACG

9

Biotechnology and Recombinant DNA

For thousands of years, people have been consuming foods that are produced by the action of microorganisms. Bread, chocolate, and soy sauce are some of the best known examples. But it was only just over 100 years ago that scientists showed that microorganisms are responsible for these products. This knowledge opened the way for using microorganisms to produce other important products. Since World War I, microbes have been used to produce a variety of chemicals, such as ethanol, acetone, and citric acid. Since World War II, microorganisms have been grown on a large scale to produce antibiotics. More recently, microbes and their enzymes are replacing a variety of chemical processes involved in manufacturing such products as paper, textiles, and fructose. Using microbes or their enzymes instead of chemical syntheses offers several advantages: microbes may use inexpensive, abundant raw materials, such as starch; microbes work at normal temperatures and pressure, thereby avoiding the need for expensive and dangerous pressurized systems; and microbes don't produce toxic, hard-to-treat wastes.

In this chapter you will learn the tools and techniques that are used to research and develop a product. You will also learn how recombinant DNA technology is used to track outbreaks of infectious disease and to provide evidence for courts of law in forensic microbiology.

UNDER THE MICROSCOPE

Escherichia coli. This bacterium has been genetically modified to produce a human protein, gamma interferon. Unlike human cells, it doesn't secrete the protein so the cells will be lysed to harvest the protein.

INTRODUCTION TO BIOTECHNOLOGY

LEARNING OBJECTIVE

- Compare and contrast biotechnology, genetic modification, and recombinant DNA technology.

Biotechnology is the use of microorganisms, cells, or cell components to make a product. Microbes have been used in the commercial production of foods, vaccines, antibiotics, and vitamins for years. Bacteria are also used in mining to extract valuable elements from ore (see Figure 28.14). Additionally, animal cells have been used to produce viral vaccines since the 1950s. Until the 1980s, products made by living cells were all made by naturally occurring cells; the role of scientists was to find the appropriate cell and develop a method for large-scale cultivation of the cells.

Now, microorganisms as well as entire plants are being used as "factories" to produce chemicals that the organisms don't naturally make. The latter is made possible by inserting genes into cells by **recombinant DNA (rDNA) technology,** which is sometimes called *genetic engineering*. The development of rDNA technology is expanding the practical applications of biotechnology almost beyond imagination.

RECOMBINANT DNA TECHNOLOGY

Recall from Chapter 8 that recombination of DNA occurs naturally in microbes. In the 1970s and 1980s, scientists developed artificial techniques for making recombinant DNA.

A gene from a vertebrate animal, including humans, can be inserted into the DNA of a bacterium, or a gene from a virus into a yeast may be used. In many cases, the recipient can then be made to express the gene, which may code for a commercially useful product. Thus, bacteria with genes for human insulin are now being used to produce insulin for treating diabetes, and a vaccine for hepatitis B is being made by yeast carrying a gene for part of the hepatitis virus (the yeast produces a viral coat protein). Scientists hope that such an approach may prove useful in producing vaccines against other infectious agents, thus eliminating the need to use whole organisms, as in conventional vaccines.

The rDNA techniques can also be used to make thousands of copies of the same DNA molecule—to *amplify* DNA, thus generating sufficient DNA for various kinds of experimentation and analysis. This technique has practical application for identifying microbes, such as viruses, that can't be cultured.

AN OVERVIEW OF RECOMBINANT DNA PROCEDURES

LEARNING OBJECTIVE

- Identify the roles of a clone and a vector in making recombinant DNA.

Figure 9.1 presents an overview of some of the procedures typically used for making rDNA, along with some promising applications.

❶–❷ After several preparatory steps,

❸ the gene of interest is inserted into the vector DNA in vitro. In this example, the vector is a plasmid. The DNA molecule chosen as a vector must be a self-replicating type, such as a plasmid or a viral genome.

❹ Next, this recombinant vector DNA is taken up by a cell such as a bacterium, where it can multiply.

❺ The cell containing the recombinant vector is then grown in culture to form a **clone** of many genetically identical cells, each of which carries a copy of the vector. This cell clone therefore contains many copies of the gene of interest. This is why DNA vectors are often called *gene-cloning vectors*, or simply *cloning vectors*. (In addition to referring to a culture of identical cells, the word *clone* is also routinely used as a verb, to describe the entire process, as in "to clone a gene.")

The final step varies according to whether the gene itself or the product of the gene is of interest.

❻Ⓐ From the cell clone, the researcher may isolate ("harvest") large quantities of the gene of interest, which may then be used for a variety of purposes. The gene may even be inserted into another vector for introduction into another kind of cell (such as a plant or animal cell). Alternatively,

❻Ⓑ if the gene of interest is expressed (transcribed and translated) in the cell clone,

❼ its protein product can be harvested and used for a variety of purposes.

The advantages of using recombinant DNA for obtaining such proteins is illustrated by one of its early successes, human growth hormone (hGH). Some individuals do not produce adequate amounts of hGH, and their growth is stunted. In the past, hGH needed to correct this deficiency had to be obtained from human pituitary glands at autopsy. (hGH from other animals is not effective in humans.) This practice was not only expensive but also dangerous because on several occasions neurological diseases were transmitted with the hormone. Human growth hormone produced by

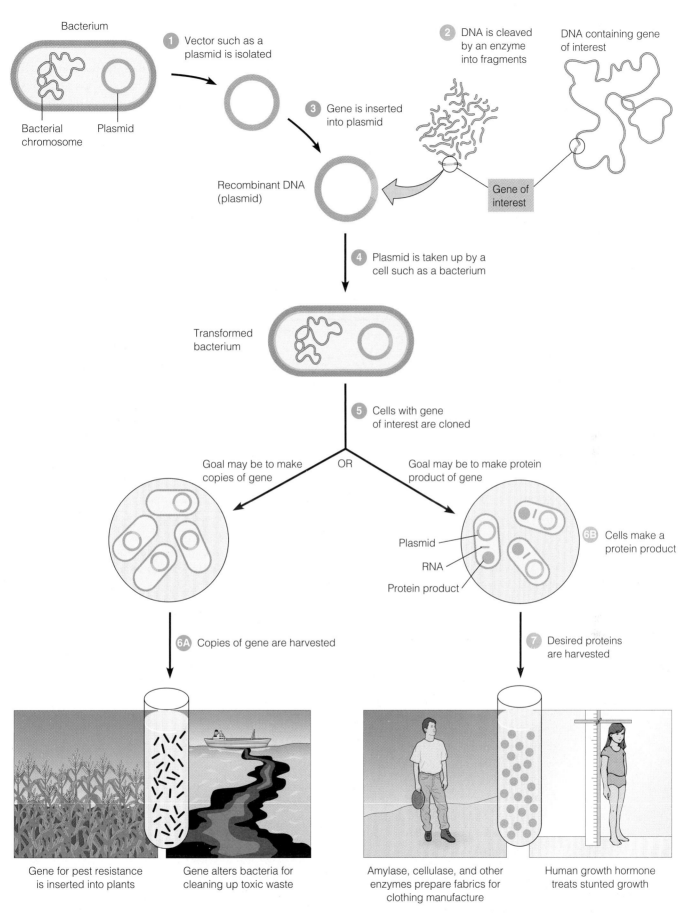

Bacterium

Bacterial chromosome   Plasmid

① Vector such as a plasmid is isolated

② DNA is cleaved by an enzyme into fragments

DNA containing gene of interest

③ Gene is inserted into plasmid

Recombinant DNA (plasmid)

Gene of interest

④ Plasmid is taken up by a cell such as a bacterium

Transformed bacterium

⑤ Cells with gene of interest are cloned

Goal may be to make copies of gene   OR   Goal may be to make protein product of gene

Plasmid

RNA

Protein product

⑥B Cells make a protein product

⑥A Copies of gene are harvested

⑦ Desired proteins are harvested

Gene for pest resistance is inserted into plants

Gene alters bacteria for cleaning up toxic waste

Amylase, cellulase, and other enzymes prepare fabrics for clothing manufacture

Human growth hormone treats stunted growth

FIGURE 9.1 A typical genetic modification procedure, with examples of applications.

Q **What is recombinant DNA?**

genetically modified *E. coli* is a pure and cost-effective product. Recombinant DNA techniques also result in faster production of the hormone than traditional methods might allow.

TOOLS OF BIOTECHNOLOGY

LEARNING OBJECTIVE
• Compare selection and mutation.

Research scientists and technicians isolate bacteria and fungi from natural environments such as soil and water to find, or *select*, the organisms that produce a desired product. The selected organism can be mutated to make more product or to make a better product.

SELECTION

In nature, organisms with characteristics that enhance survival are more likely to survive and reproduce than are variants that lack the desirable traits. This is called *natural selection*. Humans use **artificial selection** to select desirable breeds of animals or strains of plants to cultivate. As microbiologists learned how to isolate and grow microorganisms in pure culture, they were able to select the ones that could accomplish the desired objective, such as brewing beer more efficiently, for example, or producing a new antibiotic. Over 2000 strains of antibiotic-producing bacteria have been discovered by testing soil bacteria and selecting the strains that produce an antibiotic. The box in Chapter 28 describes the selection of a bacterium that converts a waste product into a valuable product.

MUTATION

As we saw in Chapter 8, mutations are responsible for much of the diversity of life. A bacterium with a mutation that confers resistance to an antibiotic will survive and reproduce in the presence of that antibiotic. Biologists working with antibiotic-producing microbes discovered that they could create new strains by exposing microbes to mutagens. After random mutations were created in penicillin-producing *Penicillium* by exposing fungal cultures to radiation, the highest-yielding variant among the survivors was selected for another exposure to a mutagen. Using mutations, biologists increased the amount of penicillin produced by the fungus over 1000 times.

Screening each mutant for penicillin production is a tedious process. **Site-directed mutagenesis** can be used to make a specific change in a gene. Suppose you determine that changing one amino acid will make a laundry enzyme work better in cold water. Using the genetic code (see Figure 8.8, page 225), you could produce the sequence of DNA that encodes that amino acid and insert it into the gene for that enzyme using the techniques described next.

The science of molecular genetics has advanced to such a degree that many routine cloning procedures are performed using prepackaged materials and procedures that are very much like cookbook recipes. Scientists have a grab-bag of methods from which to choose, depending on the ultimate application of their experiments. Next we describe some of the most important tools and techniques, and later we will consider some specific applications.

RESTRICTION ENZYMES

LEARNING OBJECTIVE
• Define *restriction enzymes,* and outline how they are used to make recombinant DNA.

Recombinant DNA technology has its technical roots in the discovery of **restriction enzymes,** a special class of DNA-cutting enzymes that exist in many bacteria. First isolated in 1970, restriction enzymes in nature had actually been observed earlier, when certain bacteriophages were found to have a restricted host range. If these phages were used to infect bacteria other than their usual hosts, restriction enzymes in the new host destroyed almost all the phage DNA. Restriction enzymes protect a bacterial cell by hydrolyzing phage DNA. The bacterial DNA is protected from digestion because the cell **methylates** (adds methyl groups to) some of the cytosines in its DNA. The purified forms of these bacterial enzymes are used in today's laboratories.

What is important for rDNA techniques is that a restriction enzyme recognizes and cuts, or *digests*, only one particular sequence of nucleotide bases in DNA, and it cuts this sequence in the same way each time. Typical restriction enzymes used in cloning experiments recognize four-, six-, or eight-base sequences. Many restriction enzymes make staggered cuts in the two strands of a DNA molecule—cuts that are not directly opposite each other (Figure 9.2).

❶ Notice in this figure that the blue base sequences on the two strands are the same, but they run in opposite directions.

❷ Staggered cuts leave stretches of single-stranded DNA at the ends of the DNA fragments. These are called *sticky ends* because they can "stick" to complementary stretches of single-stranded DNA by base pairing. Hundreds of restriction enzymes are known, each producing DNA fragments with characteristic ends.

❸ If two fragments of DNA from different sources have been produced by the action of the same restriction

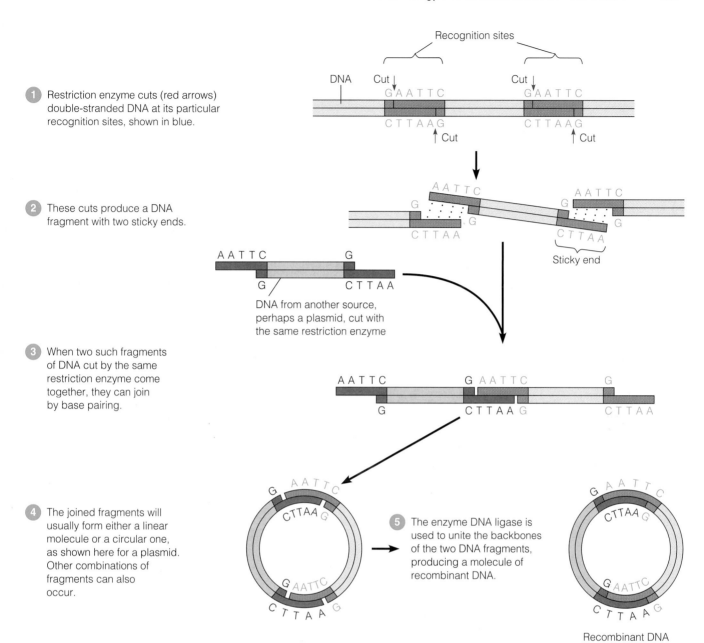

1 Restriction enzyme cuts (red arrows) double-stranded DNA at its particular recognition sites, shown in blue.

2 These cuts produce a DNA fragment with two sticky ends.

DNA from another source, perhaps a plasmid, cut with the same restriction enzyme

3 When two such fragments of DNA cut by the same restriction enzyme come together, they can join by base pairing.

4 The joined fragments will usually form either a linear molecule or a circular one, as shown here for a plasmid. Other combinations of fragments can also occur.

5 The enzyme DNA ligase is used to unite the backbones of the two DNA fragments, producing a molecule of recombinant DNA.

Recombinant DNA

FIGURE 9.2 The role of a restriction enzyme in making recombinant DNA.

Q Why are restriction enzymes used to make recombinant DNA?

enzyme, the two pieces will have identical sets of sticky ends and can be spliced (recombined) in vitro.

4 The sticky ends first join spontaneously by hydrogen bonding (base pairing) in either a linear or a circular form.

5 Then the enzyme DNA ligase is used to covalently link the backbones of the DNA pieces, producing an rDNA molecule.

VECTORS

LEARNING OBJECTIVES

• List the four properties of vectors.

• Describe the use of plasmid and viral vectors.

A great variety of different types of DNA molecules can serve as vectors, provided that they have certain properties. The most important property is self-replication; once

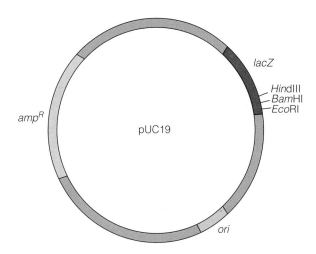

FIGURE 9.3 A plasmid used for cloning. A plasmid vector used for cloning in the bacterium *E. coli* is pUC19. An origin of replication *(ori)* allows the plasmid to be self-replicating. Two genes, one encoding resistance to the antibiotic ampicillin *(amp^R)* and one encoding the enzyme β-galactosidase *(lacZ)*, serve as marker genes. Foreign DNA can be inserted at the restriction enzyme sites.

Q **What is a vector in recombinant DNA technology?**

in a cell, a vector must be capable of replicating. Any DNA that is cloned in the vector will be replicated in the process. Thus, vectors serve as vehicles for the replication of desired DNA sequences.

Vectors also need to be of a size that allows them to be manipulated outside the cell during recombinant DNA procedures. Smaller vectors are more easily manipulated than larger DNA molecules, which tend to be more fragile. Preservation is another important property of vectors. The circular form of DNA molecules is important in protecting the DNA of the vector from destruction by the recipient of the vector. Notice in Figure 9.3 that the DNA of a plasmid is circular. Another preservation mechanism occurs when the DNA of a virus inserts itself quickly into the chromosome of the host (see Chapter 13, page 399).

When it is necessary to retrieve cells containing the vector, a marker gene contained within the vector can often help make selection easy. Common selectable marker genes are for antibiotic resistance or for an enzyme that carries out an easily identified reaction.

Plasmids are one of the primary vectors in use, particularly variants of R factor plasmids. Plasmid DNA can be cut with the same restriction enzymes as the DNA to be cloned, so that all pieces of the DNA will have the same sticky ends. When the pieces are mixed, the DNA to be cloned will become inserted into the plasmid (see Figure 9.3). Note that other possible combinations of fragments can occur as well, including the plasmid re-forming a circle with no DNA inserted.

Some plasmids are capable of existing in several different species. They are called **shuttle vectors** and can be used to move cloned DNA sequences among organisms, such as among bacterial, yeast, and mammalian cells, or among bacterial, fungal, and plant cells. Shuttle vectors can be very useful in the process of genetically modifying multicellular organisms—for example, by trying to insert herbicide resistance genes into plants.

A different kind of vector is viral DNA. This type of vector can usually accept much larger pieces of foreign DNA than plasmids can. After the DNA has been inserted into the viral vector, it can be cloned in the virus's host cells. The choice of a suitable vector depends on many factors, including the organism that will receive the new gene and the size of the DNA to be cloned. Retroviruses, adenoviruses, and herpesviruses are being used to insert corrective genes into human cells that have defective genes. Gene therapy is discussed on page 269.

POLYMERASE CHAIN REACTION

LEARNING OBJECTIVE

• Outline the steps in PCR, and provide an example of its use.

The **polymerase chain reaction (PCR)** is a technique by which small samples of DNA can be quickly amplified, that is, increased to quantities that are large enough for analysis.

Starting with just one gene-sized piece of DNA, PCR can be used to make literally billions of copies in only a few hours. The PCR process is shown in Figure 9.4.

❶ Each strand of the target DNA will serve as a template for DNA synthesis.

❷ To this DNA is added a supply of the four nucleotides (for assembly into new DNA) and the enzyme for catalyzing the synthesis, DNA polymerase (see Chapter 8, page 217). Short pieces of nucleic acid called primers are also added to help start the reaction. The primers are complementary to the ends of the target DNA and

❸ will hybridize to the fragments to be amplified.

❹ Then, the polymerase synthesizes new complementary strands.

❺ After each cycle of synthesis, the DNA is heated to convert all the new DNA into single strands. Each newly synthesized DNA strand serves in turn as a template for more new DNA.

As a result, the process proceeds exponentially. All of the necessary reagents are added to a tube, which is placed

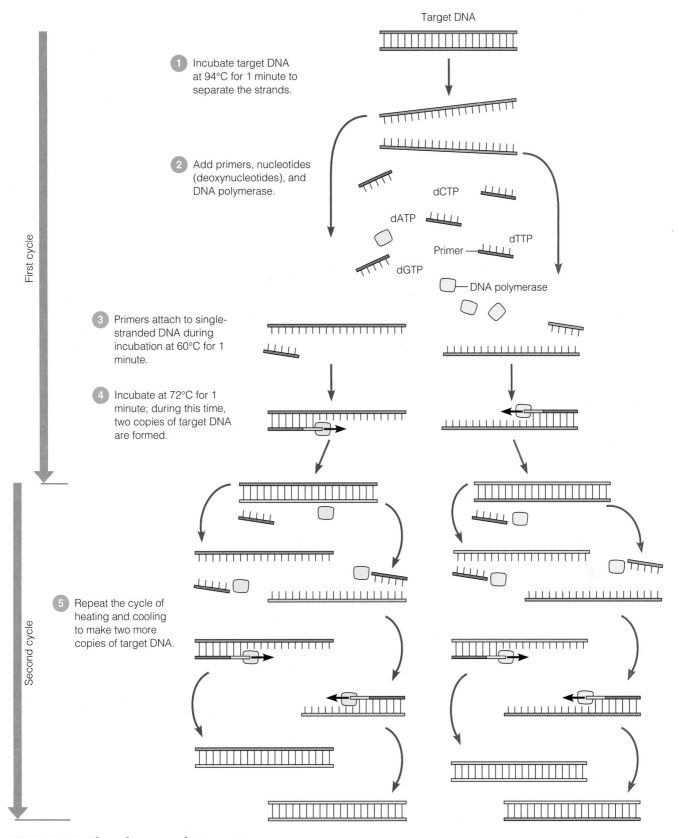

FIGURE 9.4 The polymerase chain reaction.

Q **What enzyme copies DNA in PCR?**

in a *thermalcycler*. The thermalcycler can be set for the desired temperatures, times, and number of cycles. Use of an automated thermalcycler is made possible by the use of DNA polymerase taken from a thermophilic bacterium such as *Thermus aquaticus*; the enzyme from such organisms can survive the heating phase without being destroyed (see the box in Chapter 6, page 164). Thirty cycles, completed in just a few hours, will increase the amount of target DNA by more than a billion times.

Note that PCR can only be used to amplify relatively small, specific sequences of DNA as determined by the choice of primers. It cannot be used to amplify an entire genome.

PCR can be applied to any situation that requires the amplification of DNA. Especially noteworthy are diagnostic tests that use PCR to detect the presence of infectious agents in situations in which they would otherwise be undetectable. ✳ **Animation: Go to The Microbiology Place website or CD-ROM and click "Animations" to view Polymerase Chain Reaction.**

TECHNIQUES OF GENETIC MODIFICATION

INSERTING FOREIGN DNA INTO CELLS

LEARNING OBJECTIVE
● Describe five ways of getting DNA into a cell.

Recombinant DNA procedures require that DNA molecules be manipulated outside the cell and then returned to living cells. There are several ways to introduce DNA into cells. The choice of method is usually determined by the type of vector and host cell being used.

In nature, plasmids are usually transferred between closely related microbes by cell-to-cell contact, such as in conjugation. In genetic engineering, a plasmid must be inserted into a cell by **transformation,** a procedure during which cells can take up DNA from the surrounding environment (see Chapter 8, page 240). Many cell types, including *E. coli*, yeast, and mammalian cells, do not naturally transform; however, simple chemical treatments can make all of these cell types *competent*, or able to take up external DNA. For *E. coli*, the procedure for making cells competent is to soak them in a solution of calcium chloride for a brief period. Following this treatment, the now-competent cells are mixed with the cloned DNA and given a mild heat shock. Some of these cells will then take up the DNA.

There are other ways to transfer DNA to cells. A process called **electroporation** uses an electrical current to form microscopic pores in the membranes of cells; the DNA then enters the cells through the pores. Electroporation is generally applicable to all cells; those with cell walls often must be converted to protoplasts first (see Chapter 4, page 89). **Protoplasts** are produced by enzymatically removing the cell wall, thereby allowing more direct access to the plasma membrane.

The process of **protoplast fusion** also takes advantage of the properties of protoplasts. Protoplasts in solution will fuse at a low but significant rate; the addition of polyethylene glycol increases the frequency of fusion (Figure 9.5a). In the new hybrid cell, the DNA derived from the two "parent" cells may undergo natural recombination. This method is especially valuable in the genetic manipulation of plant and algal cells (Figure 9.5b).

A remarkable way of introducing foreign DNA into plant cells is to literally shoot it directly through the thick cellulose walls using a gene gun (Figure 9.6). Microscopic particles of tungsten or gold are coated with DNA and propelled by a burst of helium through the plant cell walls. Some of the cells express the introduced DNA as though it were their own.

DNA can be introduced directly into an animal cell by **microinjection.** This technique requires the use of a glass micropipette with a diameter that is much smaller than the cell. The micropipette punctures the plasma membrane, and DNA can be injected through it (Figure 9.7).

Thus, there is a great variety of different restriction enzymes, vectors, and methods of inserting DNA into cells. But foreign DNA will survive only if it is either present on a self-replicating vector or incorporated into one of the cell's chromosomes by recombination.

OBTAINING DNA

LEARNING OBJECTIVES
● Describe how a gene library is made.

We have seen how genes can be cloned into vectors by using restriction enzymes, and how genes can be transformed or transferred into a variety of cell types. But how do biologists obtain the genes they are interested in? There are two main sources of genes: (1) gene libraries containing either natural copies of genes or cDNA copies of genes made from mRNA, and (2) synthetic DNA.

GENE LIBRARIES

Isolating specific genes as individual pieces of DNA is seldom practical. Therefore, researchers interested in genes from a particular organism start by extracting the organism's DNA, which can be obtained from cells of any

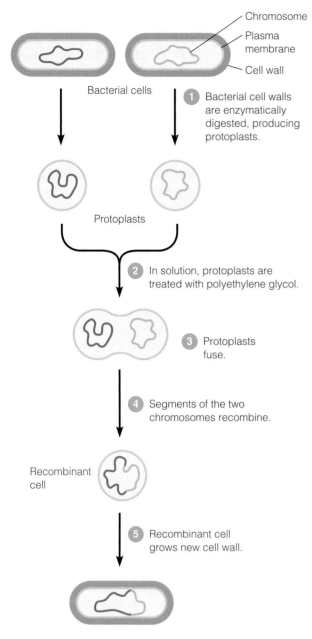

Chromosome

Plasma membrane

Cell wall

Bacterial cells

1 Bacterial cell walls are enzymatically digested, producing protoplasts.

Protoplasts

2 In solution, protoplasts are treated with polyethylene glycol.

3 Protoplasts fuse.

4 Segments of the two chromosomes recombine.

Recombinant cell

5 Recombinant cell grows new cell wall.

(a) Process of protoplast fusion

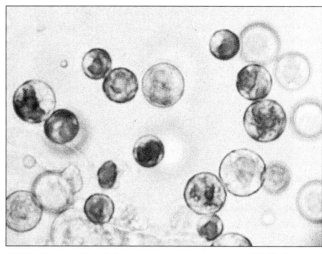

(b) Algal protoplasts fusing

LM   10 μm

FIGURE 9.5 Protoplast fusion. (a) A diagram of protoplast fusion with bacterial cells. **(b)** Protoplast algal cells are shown fusing. Removal of the cell wall left only the delicate plasma membrane to bind the cell contents together, allowing the exchange of DNA.

Q **What is a protoplast?**

organism, whether plant, animal, or microbe, by lysing the cells and precipitating the DNA. This process results in a DNA mass that includes the organism's entire genome. After the DNA is digested by restriction enzymes, the restriction fragments are then spliced into plasmid or phage vectors, and the recombinant vectors are introduced into bacterial cells. The goal is to make a collection of clones

large enough to ensure that at least one clone exists for every gene in the organism. This collection of clones containing different DNA fragments is called a **gene library;** each "book" is a bacterial or phage strain that contains a fragment of the genome (Figure 9.8). Such libraries are essential for maintaining and retrieving DNA clones; they can even be purchased commercially.

Cloning genes from eukaryotic organisms presents a specific problem. Genes of eukaryotic cells generally contain both **exons,** stretches of DNA that code for protein, and **introns,** intervening stretches of DNA that do not code for protein. When the RNA transcript of such a gene is converted to mRNA, the introns are removed (see Figure 8.11 on page 228). In cloning genes of eukaryotic cells, it is desirable to use a version of the gene that lacks introns because a gene that includes introns may be too large to work with easily. In addition, if such a gene is put into a bacterial cell, the bacterium will not usually be able to remove the introns from the RNA transcript, and therefore it will not be able to make the correct protein product. However, an artificial gene that contains only exons can be produced by using an enzyme called **reverse transcriptase** to synthesize **complementary DNA (cDNA)** from an mRNA template (Figure 9.9). This synthesis is the reverse of the normal DNA-to-RNA transcription process. A DNA copy of mRNA is produced by reverse transcriptase. Following this, the mRNA is enzymatically digested away. DNA polymerase then synthesizes a complementary strand of DNA, creating a double-stranded piece of DNA containing the information from the mRNA. Molecules of cDNA produced from a mixture of all the mRNAs from a tissue or cell type can then be cloned to form a cDNA library.

The cDNA method is the most common method of obtaining eukaryotic genes. A difficulty with this method

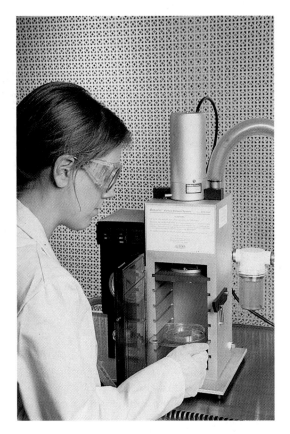

FIGURE 9.6 A gene gun, which can be used to insert DNA-coated "bullets" into a cell.

Q Name four other methods of inserting DNA into a cell.

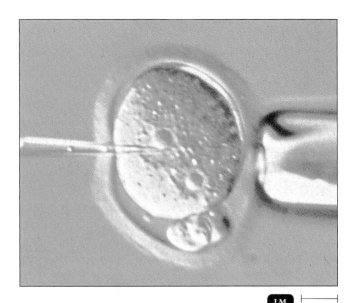

LM | 20 μm

FIGURE 9.7 The microinjection of foreign DNA into a fertilized mouse egg. The egg is first immobilized by applying mild suction to the large, blunt, holding pipette (right). Several hundred copies of the gene of interest are then injected into the nucleus of the cell through the tiny end of the micropipette (left).

Q Why is microinjection impractical for bacterial and fungal cells?

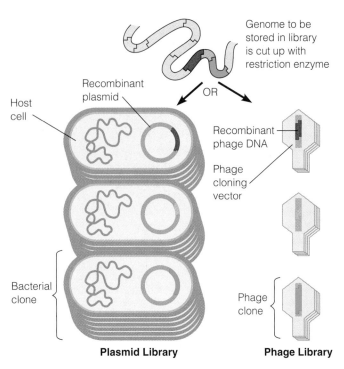

FIGURE 9.8 Gene libraries. Each fragment of DNA, containing about one gene, is carried by a vector, either a plasmid within a bacterial cell or a phage.

Q Differentiate between a RFLP and a gene.

is that long molecules of mRNA may not be completely reverse-transcribed into DNA; the reverse transcription often aborts, forming only parts of the desired gene.

SYNTHETIC DNA

LEARNING OBJECTIVE
• Differentiate cDNA from synthetic DNA.

Under certain circumstances, genes can be made in vitro with the help of DNA synthesis machines (Figure 9.10). A keyboard on the machine is used to enter the desired sequence of nucleotides, much as letters are entered into a word processor to compose a sentence. A microprocessor controls the synthesis of the DNA from stored supplies of nucleotides and the other necessary reagents. A chain of over 120 nucleotides can be synthesized by this method. Unless the gene is very small, at least several chains must be synthesized separately and linked together to form an entire gene.

The difficulty of this approach, of course, is that the sequence of the gene must be known before it can be synthesized. If the gene has not already been isolated, then the only way to predict the DNA sequence is by knowing the amino acid sequence of the protein product of the gene. If this amino acid sequence is known, in principle one can work backward through the genetic code to obtain

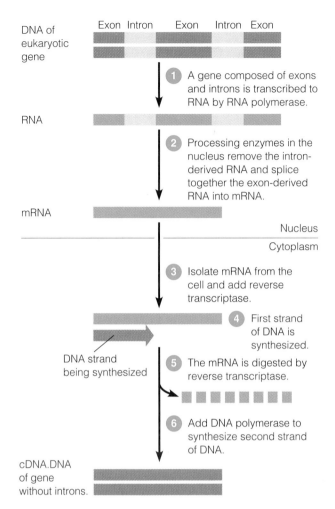

DNA of eukaryotic gene

Exon Intron Exon Intron Exon

1 A gene composed of exons and introns is transcribed to RNA by RNA polymerase.

RNA

2 Processing enzymes in the nucleus remove the intron-derived RNA and splice together the exon-derived RNA into mRNA.

mRNA

Nucleus

Cytoplasm

3 Isolate mRNA from the cell and add reverse transcriptase.

4 First strand of DNA is synthesized.

DNA strand being synthesized

5 The mRNA is digested by reverse transcriptase.

6 Add DNA polymerase to synthesize second strand of DNA.

cDNA.DNA of gene without introns.

FIGURE 9.9 Making complementary DNA (cDNA) for a eukaryotic gene. Reverse transcriptase catalyzes the synthesis of double-stranded DNA from an RNA template.

Q **How does reverse transcriptase differ from DNA polymerase?**

the DNA sequence. Unfortunately, the degeneracy of the code prevents an unambiguous determination; thus, if the protein contains a leucine, for example, which of the six codons for leucine is the one in the gene?

For these reasons, it is rare to clone a gene by synthesizing it directly, although some commercial products such as insulin, interferon, and somatostatin are produced from chemically synthesized genes. Desired restriction sites were added to the synthetic genes so that the genes could be inserted into plasmid vectors for cloning in *E. coli*. Synthetic DNA plays a much more useful role in selection procedures, as we will see.

SELECTING A CLONE

LEARNING OBJECTIVE

- Explain how each of the following is used to locate a clone: antibiotic-resistance genes, DNA probes, gene products.

FIGURE 9.10 A DNA synthesis machine. Short sequences of DNA can be synthesized by instruments such as this one.

Q **What are some of the disadvantages of using a DNA synthesis machine?**

In cloning, it is necessary to select the particular cell that contains the specific gene of interest. This is difficult because out of millions of cells, only a very few cells might contain the desired gene. Here we will examine a typical screening procedure known as *blue-white screening,* from the color of the bacterial colonies formed at the end of the screening process.

The plasmid vector used contains a gene (amp^R) coding for resistance to the antibiotic ampicillin. The host bacterium will not be able to grow on the test medium, which contains ampicillin, unless the vector has transferred the ampicillin-resistance gene. The plasmid vector also contains a second gene, this one for the enzyme β-galactosidase (*lacZ*). Notice in Figure 9.3 that there are several sites in *lacZ* that can be cut by restriction enzymes.

The procedure is shown in Figure 9.11. The two genes, called marker genes, are used so that the insertion of plasmid DNA into the host bacterium can be determined.

1 The plasmid vector and foreign DNA are digested with the same restriction enzyme.

2 The foreign DNA will insert into the gene for β-galactosidase. Therefore, the bacterium receiving the plasmid vector will not produce the enzyme β-galactosidase if foreign DNA has been inserted into the plasmid.

3 The recombinant plasmid is introduced into a culture of ampicillin-sensitive bacteria by transformation.

FIGURE 9.11 One method of selecting recombinant bacteria.

$\mathbb{Q}$ **Why are some colonies blue and others white?**

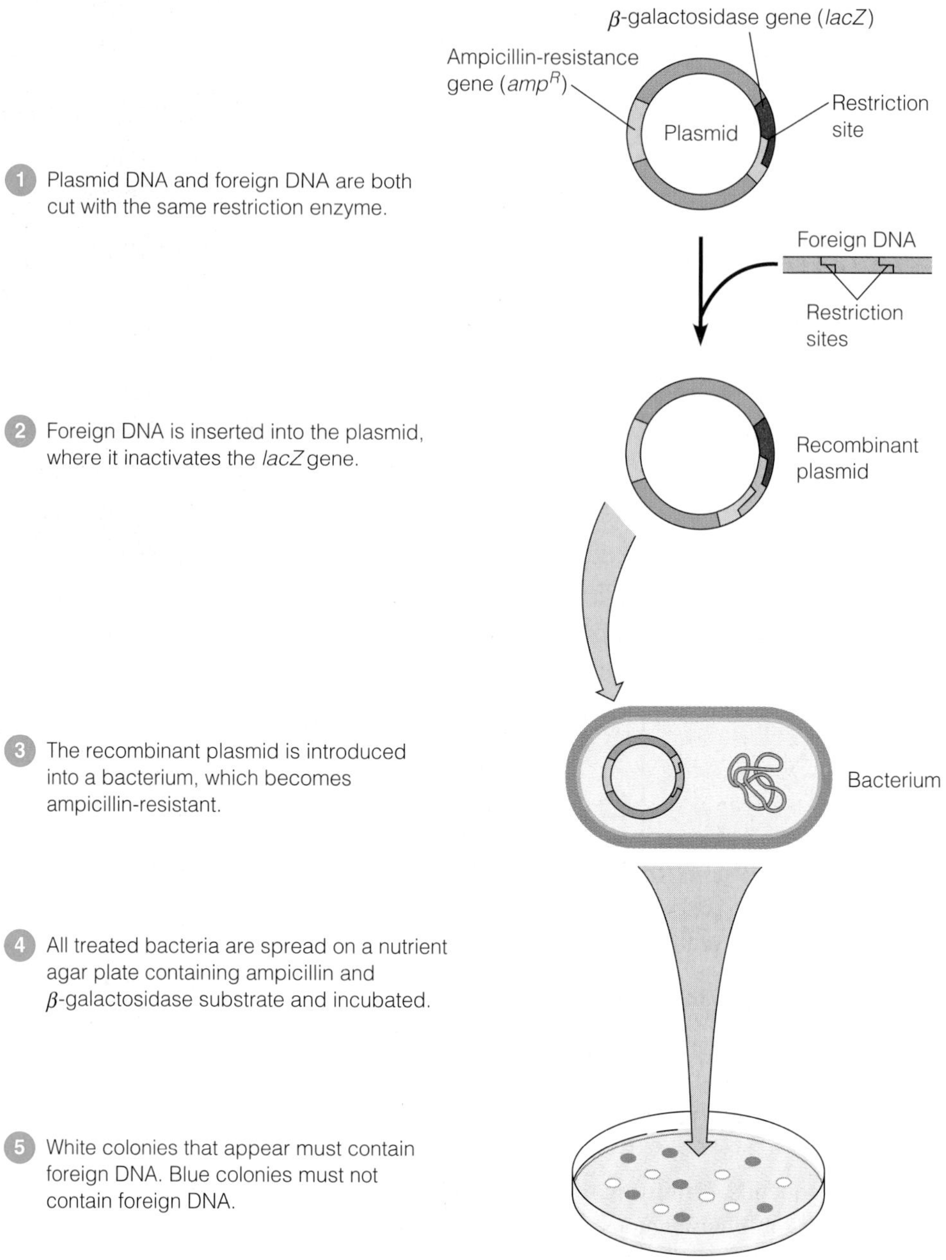

1 Plasmid DNA and foreign DNA are both cut with the same restriction enzyme.

2 Foreign DNA is inserted into the plasmid, where it inactivates the *lacZ* gene.

3 The recombinant plasmid is introduced into a bacterium, which becomes ampicillin-resistant.

4 All treated bacteria are spread on a nutrient agar plate containing ampicillin and β-galactosidase substrate and incubated.

5 White colonies that appear must contain foreign DNA. Blue colonies must not contain foreign DNA.

β-galactosidase gene (*lacZ*)

Ampicillin-resistance gene (*amp^R*)

Plasmid

Restriction site

Foreign DNA

Restriction sites

Recombinant plasmid

Bacterium

4 In the blue-white screening procedure, a library of bacteria is cultured in a medium called X-gal. X-gal contains two essential components other than those necessary to support normal bacterial growth. One is the antibiotic ampicillin, which prevents the growth of any bacterium that has not successfully received the ampicillin-resistance gene from the plasmid. The other, called X-gal, is a substrate for β-galactosidase.

5 Only bacteria that picked up the plasmid will grow—because they are now ampicillin resistant. Bacteria that picked up the recombinant plasmid—in which the new gene was inserted into the *lacZ* gene—will not hydrolyze lactose and will produce white colonies.

If a bacterium received the original plasmid containing the intact *lacZ* gene, the cells will hydrolyze X-gal to produce a blue-colored compound; the colony will be blue.

What remains to be done can still be difficult. The above procedure has isolated white colonies known to contain foreign DNA, but it is still not known whether this is the desired fragment of foreign DNA. A second procedure is needed to identify these bacteria. If the foreign DNA in the plasmid codes for the production of an identifiable product, the bacterial isolate only needs to be grown in culture and tested. However, in some cases the gene itself must be identified in the host bacterium.

Colony hybridization is a common method of identifying cells that carry a specific cloned gene. **DNA probes,** short segments of single-stranded DNA that are complementary to the desired gene, are synthesized. If the DNA probe finds a match, it will adhere to the target gene. The DNA probe is labeled with a radioactive element or fluorescent dye so its presence can be determined. A typical colony hybridization experiment is shown in Figure 9.12.

MAKING A GENE PRODUCT

LEARNING OBJECTIVE

- List one advantage of modifying each of the following: *E. coli*, *Saccharomyces cerevisiae*, mammalian cells, plant cells.

We have just seen how to identify cells carrying a particular gene. The gene products are frequently the objective of genetic modification. Most of the earliest work in genetic modification used *E. coli* to synthesize the gene products. *E. coli* is easily grown, and researchers are very familiar with this bacterium and its genetics. For example, some inducible promoters, such as that of the *lac* operon, have been cloned, and cloned genes can be attached to such promoters. The synthesis of great amounts of the cloned gene product can then be directed by the addition of an inducer. Such a method has been used to produce gamma interferon in *E. coli* (Figure 9.13). However, *E. coli* also has several disadvantages. Like other gram-negative bacteria, it produces endotoxins as part of the outer layer of its cell wall. Because endotoxins cause fever and shock in animals, their accidental presence in products intended for human use would be a serious problem.

Another disadvantage of *E. coli* is that it does not usually secrete protein products. To obtain a product, cells must usually be broken open and the product purified from the resulting "soup" of cell components. Recovering the product from such a mixture is expensive when done on an industrial scale. It is more economical to have an organism secrete the product so that it can be recovered continuously from the growth medium. One approach has been to link the product to a natural *E. coli* protein that the bacterium does secrete. However, gram-positive bacteria, such as *Bacillus subtilis*, are more likely to secrete their products and are often preferred industrially for that reason.

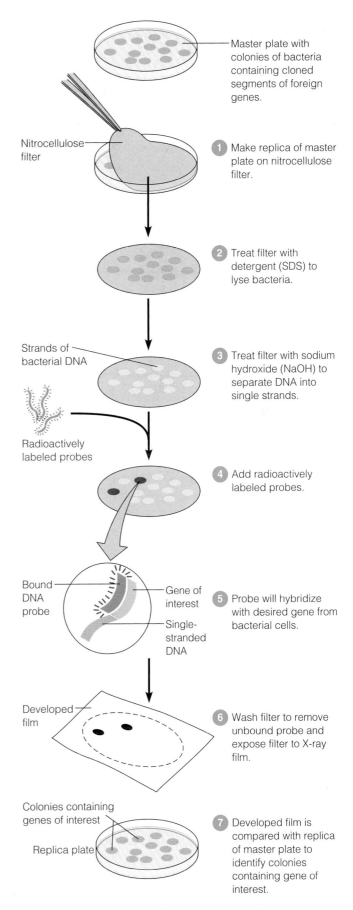

Master plate with colonies of bacteria containing cloned segments of foreign genes.

Nitrocellulose filter

① Make replica of master plate on nitrocellulose filter.

② Treat filter with detergent (SDS) to lyse bacteria.

Strands of bacterial DNA

③ Treat filter with sodium hydroxide (NaOH) to separate DNA into single strands.

Radioactively labeled probes

④ Add radioactively labeled probes.

Bound DNA probe — **Gene of interest** — **Single-stranded DNA**

⑤ Probe will hybridize with desired gene from bacterial cells.

Developed film

⑥ Wash filter to remove unbound probe and expose filter to X-ray film.

Colonies containing genes of interest

Replica plate

⑦ Developed film is compared with replica of master plate to identify colonies containing gene of interest.

FIGURE 9.12 Colony hybridization: using a DNA probe to identify a cloned gene of interest.

 **What is a DNA probe?**

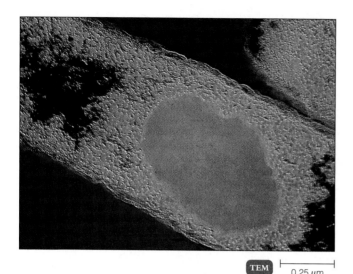

TEM | 0.25 μm

FIGURE 9.13 E. coli genetically engineered to produce gamma interferon, a human protein that promotes an immune response. The product, visible here as an orange colored substance, can be released by lysis of the cell.

Q What is one advantage of using E. coli for genetic engineering? One disadvantage?

Another microbe being used as a vehicle for expressing genetically engineered genes is baker's yeast, *Saccharomyces cerevisiae*. Its genome is only about four times larger than that of *E. coli* and is probably the best understood eukaryotic genome. Yeasts may carry plasmids, and the plasmids are easily transferred into yeast cells after their cell walls have been removed. As eukaryotic cells, yeasts may be more successful in expressing foreign eukaryotic genes than bacteria. Furthermore, yeasts are likely to continuously secrete the product. Because of all these factors, yeasts have become the eukaryotic workhorse of biotechnology.

Mammalian cells in culture, even human cells, can be genetically modified much like bacteria to produce various products. Scientists have developed effective methods of growing certain mammalian cells in culture as hosts for growing viruses (see Chapter 13, page 395). Mammalian cells are often the best suited to making protein products for medical use because the cells secrete their products and there is a low risk of toxins or allergens. Using mammalian cells to make foreign gene products on an industrial scale often requires a preliminary step of cloning the gene in bacteria. Consider the example of colony-stimulating factor (CSF). A protein produced naturally in tiny amounts by white blood cells, CSF is valuable because it stimulates the growth of certain cells that protect against infection. To produce huge amounts of CSF industrially, the gene is first inserted into a plasmid, and bacteria are used to make

multiple copies of the plasmid (see Figure 9.1). The recombinant plasmids are inserted into mammalian cells that are grown in bottles.

Plant cells can also be grown in culture, altered by recombinant DNA techniques, and then used to generate genetically modified plants. Such plants may prove useful as sources of valuable products, such as plant alkaloids (the painkiller codeine, for example), the isoprenoids that are the basis of synthetic rubber, and melanin (the animal skin pigment) for use in sunscreens. Genetically modified plants have many advantages for the production of human therapeutics including vaccines and antibodies. The advantages include large-scale, low-cost production using agriculture and low risk of product contamination by mammalian pathogens or cancer-causing genes. Genetically modifying plants often requires use of a bacterium. We will return to the topic of genetically modified plants later in the chapter (page 273).

APPLICATIONS OF rDNA

> **LEARNING OBJECTIVE**
> - List at least five applications of rDNA Technology.
> - Define RNAi.

We have now described the entire sequence of events in cloning a gene. As indicated earlier, such cloned genes can be applied in a variety of ways. One is to produce useful substances more efficiently and less expensively (see the box on the facing page). Another is to obtain information from the cloned DNA that is useful for either basic research, medicine, or forensics. A third is to use cloned genes to alter the characteristics of cells or organisms. The box in Chapter 27 describes the use of recombinant cells to detect pollutants.

THERAPEUTIC APPLICATIONS

An extremely valuable pharmaceutical product is the hormone insulin, a small protein produced by the pancreas that controls the body's uptake of glucose from blood. For many years, people with insulin-dependent diabetes have controlled their disease by injecting insulin obtained from the pancreases of slaughtered animals. Obtaining this insulin is an expensive process, and the insulin from animals is not as effective as human insulin.

Because of the value of human insulin and the small size of the protein, the production of human insulin by recombinant DNA techniques was an early goal for the pharmaceutical industry. To produce the hormone, synthetic genes were first constructed for each of the two short polypeptide chains that make up the insulin molecule.

APPLICATIONS OF MICROBIOLOGY

DESIGNER JEANS

Denim blue jeans have become increasingly popular ever since they were first made for California gold miners in 1873 by Levi Strauss and Jacob Davis. Now, many companies manufacture jeans for sale around the world. Over the years, very little has changed in the denim-making process.

The denim for jeans is made from cotton. Cotton plants require water, costly fertilizer, and potentially harmful pesticides. The availability of cotton is dependent on weather and the plants' resistance to disease. After harvesting, cotton is made into yarn that is bleached with chlorine. Gaseous chlorine is quite hazardous, but the safer liquid form (hypochlorite) is more expensive. After weaving yarn into cloth, the cloth is treated with starch to hold its shape, and then dyed with indigo. The original indigo was made by fermentation of the indigo plant by *Clostridium*. However, this process was replaced over a century ago by a chemical process that requires mutagenic chemicals and a very high pH.

Now, companies that manufacture blue jeans and other products are returning to microbiology to develop environmentally sound production methods that minimize toxic wastes and the costs associated with treating toxic wastes. Moreover, microbiological methods can reduce their production costs and provide abundant, renewable raw materials.

In the 1980s, a softer denim, called "stone-washed," was introduced. The fabric is not really washed with rocks. Enzymes, called cellulases, from *Trichoderma* fungus are used to digest some of the cellulose in the cotton, thereby softening it. Other microbial enzymes may be able to make the entire product safely. Enzymes catalyze reactions in seconds rather than the days or weeks required of chemical methods, and they usually operate at safe temperatures and pHs as opposed to the high temperatures and extreme pHs of chemical reactions. Because enzymes are proteins, they are readily degraded for removal from wastewater.

Bleaching

Peroxide is a safer bleaching agent than chlorine and can be easily removed from fabric and wastewater by enzymes. Researchers at Novo Nordisk Biotech cloned a mushroom peroxidase gene in yeast and grew the yeasts in washing machine conditions. The yeast that survived the washing machine were selected as the peroxidase producers.

Fabric

Cotton production requires large tracts of land, and the crop yield depends on the weather. Polyester is a synthetic fabric produced from petroleum. Bacteria can produce both cotton and polyester. *Gluconacetobacter xylinus* bacteria make cellulose by attaching glucose units to simple chains in the outer membrane of the bacterial cell wall. The cellulose microfibrils are extruded through pores in the outer membrane, and bundles of microfibrils then twist into ribbons. Polyester is traditionally derived from a by-product of the oil industry, and its production produces toxic waste products. An environmentally friendly polyester can be made by genetically modified bacteria that convert sugar to glycerol using a bacterial enzyme, then glycerol to trimethylene glycol (3G) using a yeast enzyme. The 3G can then be polymerized into polyester. The 3G-producing bacteria can be grown on agricultural waste, and the liquid waste is biodegradable.

Indigo

Chemical synthesis of indigo requires a high pH and produces waste that explodes in contact with air. This is changing because a California biotechnology company, Genencor, developed microbially produced indigo. In the Genencor labs, *E. coli* turned blue when researchers put the gene for conversion of indole to indigo from a soil bacterium, *Pseudomonas putida* into the *E. coli*. Then, using site-directed mutagenesis and modifications of the fermentation medium, Genencor scientists increased *E. coli*'s production of indigo.

Plastic

Microbes can even make plastic zippers and packaging material for the jeans. Over 25 bacteria make polyhydroxyalkanoate (PHA) inclusion granules as a nutrient reserve. PHAs are similar to common plastics and, because they are made by bacteria, they are also readily degraded by many bacteria. PHAs could provide a biodegradable alternative to conventional plastic, which is made from petroleum.

Paper

Labels and even your purchase receipt may soon be printed on paper developed by microbiologists at the University of Texas. The new "e-paper" is made from cellulose produced by *Gluconacetobacter*.

The small size of these chains—only 21 and 30 amino acids long—made it possible to use synthetic genes. Following the procedure described earlier (page 256), each of the two synthetic genes was inserted into a plasmid vector and linked to the end of a gene coding for the bacterial enzyme β-galactosidase, so that the insulin polypeptide was coproduced with the enzyme. Two different *E. coli* bacterial cultures were used, one to produce each of the insulin polypeptide chains. The polypeptides were then recovered from the bacteria, separated from the β-galactosidase, and chemically joined to make human insulin. This accomplishment was one of the early commercial successes of

TABLE 9.1	Some Pharmaceutical Products of Genetic Engineering
Product	Comments
Alpha-interferon	Therapy for leukemia, melanoma, and hepatitis; produced by *Escherichia coli* and *Saccharomyces cerevisiae* (yeast).
Antitrypsin	Assists emphysema patients; produced by genetically modified sheep.
Beta-interferon	Treatment for multiple sclerosis; produced by mammalian cell culture.
Bone morphogenic proteins	Induces new bone formation; useful in healing fractures and reconstructive surgery; produced by mammalian cell culture.
Colony-stimulating factor (CSF)	Counteracts effects of chemotherapy; improves resistance to infectious disease such as AIDS; treatment of leukemia; produced by *E. coli* and *S. cerevisiae*.
Epidermal growth factor (EGF)	Heals wounds, burns, ulcers; produced by *E. coli*.
Erythropoietin (EPO)	Treatment of anemia; produced by mammalian cell culture.
Factor VII	Treatment of hemorrhagic strokes; produced by mammalian cell culture.
Factor VIII	Treatment of hemophilia; improves clotting; produced by mammalian cell culture.
Gamma-interferon	Treatment of chronic granulomatous disease; produced by *E. coli*.
Hepatitis B vaccine	Produced by *S. cerevisiae* that carries hepatitis-virus gene on a plasmid.
Human growth hormone (hGH)	Corrects growth deficiencies in children; produced by *E. coli*.
Human insulin	Therapy for diabetes; better tolerated than insulin extracted from animals; produced by *E. coli*.
Influenza vaccine	Trial vaccine made from *E. coli* or *S. cerevisiae* carrying virus genes.
Interleukins	Regulate the immune system; possible treatment for cancer; produced by *E. coli*.
Monoclonal antibodies	Possible therapy for cancer and transplant rejection; used in diagnostic tests; produced by mammalian cell culture (from fusion of cancer cell and antibody-producing cell).
Orthoclone OKT3 Muromonab-CD3	Monoclonal antibody used in transplant patients to help suppress the immune system, reducing the chance of tissue rejection; produced by mouse cells.
Prourokinase	Anticoagulant; therapy for heart attacks; produced by *E. coli* and yeast.
Pulmozyme (rhDNase)	Enzyme used to break down mucous secretions in cystic fibrosis patients; produced by mammalian cell culture.
Relaxin	Used to ease childbirth; produced by *E. coli*.
Superoxide dismutase (SOD)	Minimizes damage caused by oxygen free radicals when blood is resupplied to oxygen-deprived tissues; produced by *S. cerevisiae* and *K. omagataella pastoris* (yeast).
Taxol	Plant product used for treatment for ovarian cancer; produced in *E. coli*.
Tissue plasminogen activator (Activase)	Dissolves the fibrin of blood clots; therapy for heart attacks; produced by mammalian cell culture.
Tumor necrosis factor (TNF)	Causes disintegration of tumor cells; produced by *E. coli*.

DNA technology, and it illustrates a number of the principles and procedures discussed in this chapter.

Another human hormone that is now being produced commercially by genetic modification of *E. coli* is somatostatin. At one time 500,000 sheep brains were needed to produce 5 mg of animal somatostatin for experimental purposes. By contrast, only 8 liters of a genetically modified bacterial culture are now required to obtain the equivalent amount of the human hormone.

Subunit vaccines, consisting only of a protein portion of a pathogen, are being made by genetically modifying yeasts. Subunit vaccines have been produced for a number of diseases, notably hepatitis B. One of the advantages of a subunit vaccine is that there is no chance of becoming infected from the vaccine. The protein is harvested from genetically modified cells and purified for use as a vaccine. Animal viruses such as vaccinia virus can be genetically modified to carry a gene for another microbe's surface protein. When injected, the virus acts as a vaccine against the other microbe.

DNA vaccines are usually circular plasmids that include a gene encoding a viral protein under the transcriptional control of a promoter region active in human cells. The plasmids are cloned in bacteria. Several trial vaccines against HIV, SARS, influenza, and malaria are being tested. Vaccines are discussed in Chapter 18 (page 528). Table 9.1 lists some other important rDNA products used in medical therapy.

The importance of recombinant DNA technology to medical research cannot be emphasized enough. Artificial blood for use in transfusions can now be prepared using human hemoglobin produced in genetically modified pigs. Sheep have also been genetically modified to produce a number of drugs in their milk. This procedure has no apparent effect upon the sheep, and they provide a ready source of raw material for the product that does not require sacrificing animals.

Gene therapy may eventually provide cures for some genetic diseases. It is possible to imagine removing some cells from a person and transforming them with a normal gene to replace a defective or mutated gene. When these cells are returned to the person, they should function normally. For example, gene therapy has been used to treat hemophilia B and severe combined immunodeficiency. Adenoviruses and retroviruses are used most often to deliver genes; however, some researchers are working with plasmid vectors. The first gene therapy to treat hemophilia in humans was done in 1999. An attenuated retrovirus was used as the vector. Several gene therapy trials are in progress using genetically modified adenovirus carrying human gene *p53* to treat a variety of cancers. The *p53* gene, which encodes a tumor-suppressing protein, is the most frequently mutated gene in cancer cells.

The number of gene therapy trials will increase as technical improvements are made and initial attempts are successful. However, there is a great deal of preliminary work to do, and cures may not be possible for all genetic diseases. Antisense DNA (see page 275) introduced into cells is also being explored to treat hepatitis, cancers, and one type of coronary artery disease.

Gene silencing is a natural process that occurs in a wide variety of organisms and is apparently a defense against viruses and transposons. A new technology called **RNA interference (RNAi)** holds promise for gene therapy and treating cancer and viral infections. Double-stranded RNAs called **short interfering RNAs (siRNA)**

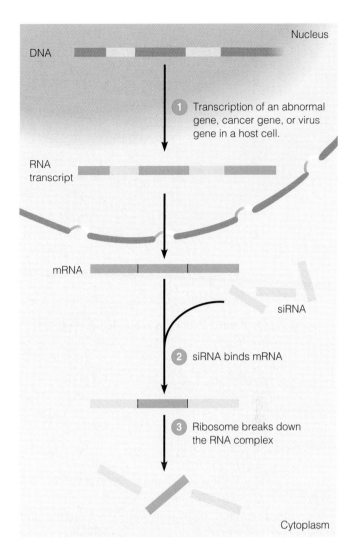

FIGURE 9.14 RNAi could provide treatments for a wide range of diseases.

Q **Does RNAi act during or after transcription?**

that target a particular gene, such as a virus gene, can be introduced into a cell (Figure 9.14). The siRNA molecules bind to mRNA causing its enzymatic destruction, thus *silencing* the expression of a gene. In mice, RNAi has been shown to inhibit hepatitis B virus. The siRNA can be injected into a cell or introduced in a DNA vector. A small DNA insert encoding siRNA against the gene of interest could be cloned into a DNA vector. When transferred into a cell, the cell would produce the desired siRNA.

THE HUMAN GENOME PROJECT

LEARNING OBJECTIVE
• Discuss the value of the Human Genome Project.

One monumental project involving rDNA technology is the Human Genome Project. The Human Genome Project

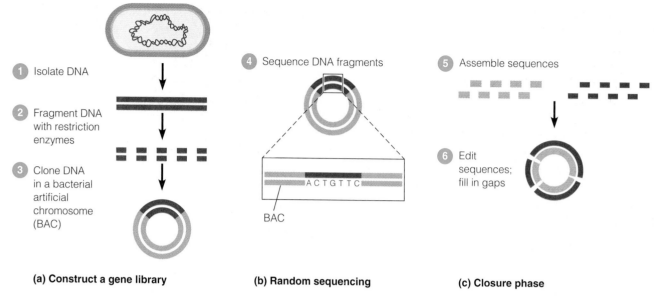

(a) Construct a gene library   **(b) Random sequencing**   **(c) Closure phase**

FIGURE 9.15   Random shotgun sequencing. In this technique a genome is cut into pieces, and each piece is sequenced. Then the pieces are fit together. There may be gaps if a specific DNA fragment was not sequenced.

Q **Does this technique identify genes and their locations?**

was an international 13-year effort, formally begun in October 1990 and completed in 2003. The goal of this project was to sequence the entire human genome, approximately 3 billion nucleotide pairs, comprising 20,000–25,000 genes. Thousands of people in 18 countries participated in this project. Researchers collected blood (female) or sperm (male) samples from a large number of donors. Only a few samples were processed as DNA resources, and the source names are protected so that neither donors nor scientists know whose samples were used. Development of shotgun sequencing (discussed shortly and shown in Figure 9.15) greatly speeded the process, and the genome is nearly complete.

One surprising finding was that less than 2% of the genome encodes a functional product—the other 98% is being called "junk DNA." This includes introns, the chromosome ends (called telomeres), and the transposons (repeating sequences that make up more than half of the human genome; see page 246). Currently, researchers are locating specific genes and determining their functions.

The next goal of researchers is the Human Proteome Project, which will map all the proteins expressed in human cells. Even before it is completed, however, it will yield data that are of immense value to our understanding of biology. It will also eventually be of great medical benefit, especially for the diagnosis and treatment of genetic diseases.

SCIENTIFIC APPLICATIONS

LEARNING OBJECTIVES

- Define the following terms: *random shotgun sequencing, bioinformatics, proteomics.*
- Diagram the Southern blotting procedure, and provide an example of its use.
- Diagram DNA fingerprinting, and provide an example of its use.

Recombinant DNA technology can be used to make products, but this is not its only important application. Because of its ability to produce many copies of DNA, it can serve as a sort of DNA "printing press." Once a large amount of a particular piece of DNA is available, various analytic techniques, discussed in this section, can be used to "read" the information contained in the DNA.

What kind of information can be obtained from cloned DNA? One kind is provided by the process of **DNA sequencing**—the determination of the exact sequence of nucleotide bases in DNA.

One technique for genome sequencing is **random shotgun sequencing.** Small pieces of a genome are sequenced, and the sequences are then assembled using a computer. Any gaps between the pieces then have to be found and sequenced (Figure 9.15). The sequences of entire viral genomes are now relatively easy to obtain. The genomes of *Saccharomyces cerevisiae* yeast, *E. coli*, and over

70 other microbes have been mapped; another 100 are in progress. The Human Genome Project was a massive undertaking resulting in sequencing the human genome (see page 269). The published maps are between 70% and 99% complete, with some gaps remaining to be filled. Most of the gaps are repeated sequences that do not encode genes. For example, the *S. cerevisiae* genome is 93% complete, but the gene encoding regions are 100% complete. Computer applications can also be used to look for protein encoding regions, which can then be "translated" by computer software.

DNA sequencing has produced an enormous amount of information that has spawned the new field of **bioinformatics,** the science of understanding the function of genes through computer-assisted analysis. DNA sequences are stored in web-based databases referred to as GenBank. Genomic information can be searched with computer programs to find specific sequences or to look for similar patterns in the genomes of different organisms. Microbial genes are now being searched to identify molecules that are the virulence factors of pathogens. By comparing genomes, researchers discovered that *Chlamydia trachomatis* (tra-kō′mä-tis) produces a toxin similar to that of *Clostridium difficile* (dif′fi-sē-il). The next goal is to identify proteins encoded by these genes. **Proteomics** is the science of determining all of the proteins expressed in a cell. The Institute for Genomic Research (TIGR) is developing new techniques that can be used to quickly analyze gene function.

An example of the use of human DNA sequencing is the identification and cloning of the mutant gene that causes cystic fibrosis (CF). CF is characterized by the oversecretion of mucus, leading to blocked respiratory passageways. The sequence of the mutated gene can be used as a diagnostic tool in a hybridization technique called **Southern blotting** (Figure 9.16), named for Ed Southern, who developed the technique in 1975.

In this technique,

❶ human DNA is first digested with a restriction enzyme, yielding thousands of fragments of various sizes. The different fragments are then separated by **gel electrophoresis.**

❷ The fragments are put in a well at one end of a layer of agarose gel. Then an electrical current is passed through the gel. While the charge is applied, the different-sized pieces of DNA migrate through the gel at different rates. The fragments are called **RFLPs,** for *restriction fragment length polymorphisms.*

❸–❹ The separated fragments are transferred onto a filter by blotting.

❺ The fragments on the filter are then exposed to a radioactive probe made from the cloned gene of interest, in this case the CF gene. The probe will hybridize to this mutant gene but not to the normal gene.

❻ Fragments to which the probe binds are identified by exposing the filter to X-ray film. With this method, any person's DNA can be tested for the presence of the mutated gene.

This process, called **genetic screening,** can now be used to screen for several hundred genetic diseases. Such screening procedures can be performed on prospective parents and also on fetal tissue. Two of the more commonly screened genes are those associated with inherited forms of breast cancer and the gene responsible for Huntington's disease.

FORENSIC MICROBIOLOGY

Several important diagnostic tools, many of which rely on the technique of hybridization, are now available as a result of rDNA technology. Recall that hybridization enables the identification of a particular DNA sequence among many others. This is exactly what is necessary in most diagnostic situations—the identification of a particular pathogen among many others.

DNA probes such as the ones used to screen gene libraries are promising tools for the rapid identification of microorganisms. For use in medical diagnosis, these probes are derived from the DNA of a pathogenic microbe and are labeled (with a radioactive tag, for example). The probe then serves a diagnostic function by combining with the DNA of the pathogen to reveal its location in body tissue (or perhaps its presence in food). Probes are also being used in nonmedical aspects of microbiology—for example, for locating and identifying specific microbes in soil. We will discuss PCR and DNA probes further in Chapter 10.

For several years, microbiologists have used RFLPs in a method of identification known as **DNA fingerprinting** to identify bacterial or viral pathogens (Figure 9.17). DNA fingerprinting is also used in forensic medicine to determine paternity or to prove that blood on a murder suspect's clothes came from the murder victim. Southern blotting requires substantial amounts of DNA. As mentioned earlier, small samples of DNA can be quickly amplified, or increased to quantities large enough for analysis, by PCR.

DNA chips (see Figure 10.17, page 304) and PCR microarrays that can screen a sample for multiple pathogens at once are being developed. In a PCR microarray, up to 22 primers from different microorganisms can be used to initiate the PCR. The microorganism is identified if DNA is copied from one of the primers.

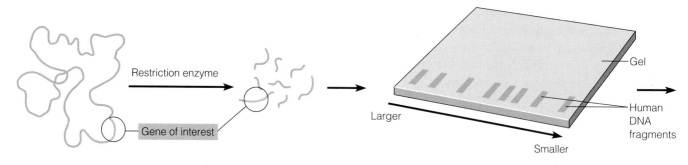

1 DNA containing the gene of interest is extracted from human cells and cut into fragments by restriction enzymes.

2 The fragments are separated according to size by gel electrophoresis. Each band consists of many copies of a particular DNA fragment. The bands are invisible but can be made visible by staining.

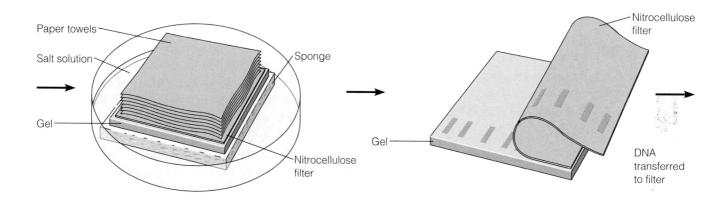

3 The DNA bands are transferred to a nitrocellulose filter by blotting. The solution passes through the gel and filter to the paper towels.

4 This produces a nitrocellulose filter with DNA fragments positioned exactly as on the gel.

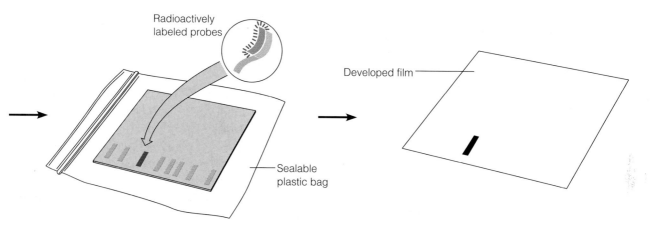

5 The filter is exposed to a radioactively labeled probe for a specific gene. The probe will base-pair (hybridize) with a short sequence present on the gene.

6 The filter is then exposed to X-ray film. The fragment containing the gene of interest is identified by a band on the developed film.

FIGURE 9.16 Southern blotting.

Q **What is the purpose of Southern blotting?**

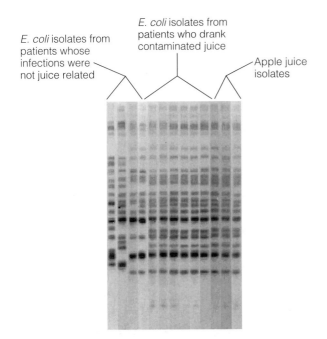

E. coli isolates from patients whose infections were not juice related

E. coli isolates from patients who drank contaminated juice

Apple juice isolates

FIGURE 9.17   DNA fingerprints used to track an infectious disease. This figure shows the relationship among DNA patterns of bacterial isolates from an outbreak of *Escherichia coli* O157:H7. The isolates from apple juice are identical to the patterns of isolates from patients who drank the contaminated juice but different from those from patients whose infections were not juice-related.

Q **What is forensic microbiology?**

The new field of **forensic microbiology** developed because hospitals and food manufacturers can be sued in courts of law and because microorganisms can be used as weapons. The requirements to prove the source of a microbe to a judge are stricter than for the medical community. For example, to prove intent to commit harm requires proper evidence collection and establishing a chain of custody of that evidence. Microbial properties that are unimportant in public health may be important clues in forensic investigations. The American Academy of Microbiology recently proposed professional certification in forensic microbiology.

DNA can often be extracted from preserved and fossilized materials, including mummies and extinct plants and animals. Although such material is very rare, and usually partially degraded, PCR enables researchers to study this genetic material that no longer exists in its natural form. The study of unusual organisms has also led to advances in basic taxonomy; this will be discussed in Chapter 10.

NANOTECHNOLOGY

Nanotechnology deals with the design and manufacture of extremely small electronic circuits and mechanical

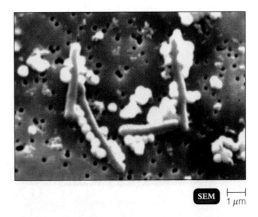

SEM | 1 μm

FIGURE 9.18 *Bacillus* **cells growing on selenium form chains of elemental selenium.**

Q **What might bacteria provide for nanotechnology?**

devices built at the molecular level of matter. Molecule-sized robots or computers can be used to detect contamination in food, diseases in plants, or biological weapons. However, the small machines require small (a nanometer is 10^{-9} meters; 1000 nm fit in 1 μm) wires and components. Bacteria may provide the needed small metals. Researchers at the U.S. Geological Survey have cultured several anaerobic bacteria that reduce toxic selenium, Se^{4+}, to nontoxic elemental Se^{0} which forms into nanospheres (Figure 9.18).

AGRICULTURAL APPLICATIONS

LEARNING OBJECTIVE
- Outline genetic engineering with *Agrobacterium*.

The process of selecting for genetically desirable plants has always been a time-consuming one. Performing conventional plant crosses is laborious and involves waiting for the planted seed to germinate and for the plant to mature. Plant breeding has been revolutionized by the use of plant cells grown in culture. Clones of plant cells, including cells that have been genetically altered by recombinant DNA techniques, can be grown in large numbers. These cells can then be induced to regenerate whole plants, from which seeds can be harvested.

Recombinant DNA can be introduced into plant cells in several ways. Previously we mentioned protoplast fusion and the use of DNA-coated "bullets." The most elegant method, however, makes use of a plasmid called the **Ti plasmid** (Ti stands for tumor-inducing), which occurs naturally in the bacterium *Agrobacterium tumefaciens* (tu′me-fash-enz). This bacterium infects certain plants, in which the Ti plasmid causes the formation of a tumorlike growth called a crown gall (Figure 9.19). A part of the Ti plasmid,

FIGURE 9.19 Crown gall disease on a tomato plant. The tumorlike growth is stimulated by a gene on the Ti plasmid carried by the bacterium *Agrobacterium tumefaciens*, which has infected the plant.

Q **What are some of the agricultural applications of recombinant DNA technology?**

called T-DNA, integrates into the genome of the infected plant. The T-DNA stimulates local cellular growth (the crown gall) and simultaneously causes the production of certain products used by the bacteria as a source of nutritional carbon and nitrogen.

For plant scientists, the attraction of the Ti plasmid is that it provides a vehicle for introducing rDNA into a plant (Figure 9.20). A scientist can insert foreign genes into the T-DNA, put the recombinant plasmid back into the *Agrobacterium* cell, and use the bacterium to insert the recombinant Ti plasmid into a plant cell. The plant cell with the foreign gene can then be used to generate a new plant. With luck, the new plant will express the foreign gene. Unfortunately, *Agrobacterium* does not naturally infect grasses, so it cannot be used to improve grains such as wheat, rice, or corn.

Noteworthy accomplishments of this approach are the introduction into plants of resistance to the herbicide glyphosate, and a *Bacillus thuringiensis*–derived insecticidal toxin (Bt). Normally, the herbicide kills both weeds and useful plants by inhibiting an enzyme necessary for making certain essential amino acids. *Salmonella* bacteria happen to have this enzyme, and some salmonellae have a mutant

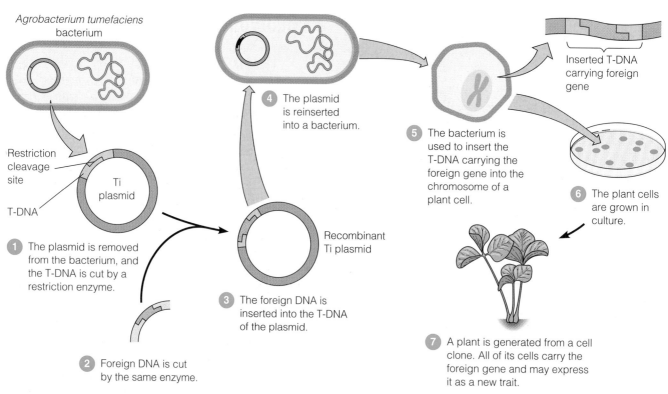

FIGURE 9.20 Using the Ti plasmid as a vector for genetic modification in plants.

Q **Why is the Ti plasmid important to biotechnology?**

enzyme that is resistant to the herbicide. When the DNA for this enzyme is introduced into a crop plant, the crop becomes resistant to the herbicide, which then kills only the weeds. There is now a variety of plants in which different herbicide and pesticide resistances have been engineered. Resistance to drought, viral infection, and several other environmental stresses has also been engineered into crop plants. *Bacillus thuringiensis* bacteria are pathogenic to some insects because they produce a protein called Bt toxin that interferes with the insect digestive tract. The Bt gene has been inserted into a variety of crop plants, including cotton and potatoes, so insects that eat the plants will be killed.

Another example involves MacGregor tomatoes, which stay firm after harvest because the gene for polygalacturonase (PG), the enzyme that breaks down pectin, is suppressed. The suppression was accomplished by **antisense DNA technology.** First, a length of DNA complementary to the PG mRNA is synthesized. This antisense DNA is taken up by the cell and binds to the mRNA to inhibit translation. The DNA-RNA hybrid is broken down by the cell's enzymes, freeing the antisense DNA to disable another mRNA.

Perhaps the most exciting potential use of genetically modified plants concerns nitrogen fixation, the ability to convert the nitrogen gas in the air to compounds that living cells can use (see page 815). The availability of such nitrogen-containing nutrients is usually the main factor limiting crop growth. But in nature, only certain bacteria have genes for carrying out this process. Some plants, such as alfalfa, benefit from a symbiotic relationship with these microbes. Species of the symbiotic bacterium *Rhizobium* have already been genetically modified for enhanced nitrogen fixation. In the future, *Rhizobium* strains may be designed that can colonize such crop plants as corn and wheat, perhaps eliminating their requirement for nitrogen fertilizer. The ultimate goal would be to introduce functioning nitrogen-fixation genes directly into the plants. Although this goal cannot be achieved with our current knowledge, work toward it will continue because of its potential for dramatically increasing the world's food supply.

An example of a genetically modified bacterium now in agricultural use is *Pseudomonas fluorescens* that has been engineered to produce a toxin normally produced by *Bacillus thuringiensis*. This toxin kills certain plant pathogens, such as the European corn borer. The genetically altered *Pseudomonas*, which produces much more toxin than *B. thuringiensis*, can be added to plant seeds and in time will enter the vascular system of the growing plant. Its toxin is ingested by the feeding borer larvae and kills them (but is harmless to humans and other warm-blooded animals).

Animal husbandry has also benefited from rDNA technology. We have seen how one of the early commercial products of rDNA was human growth hormone. By similar methods it is possible to manufacture bovine growth hormone (bGH). When bGH is injected into beef cattle, it increases their weight gain; in dairy cows, it also causes a 10% increase in milk production. Such procedures have met with resistance from consumers, especially in Europe, primarily as a result of as-yet unsubstantiated fears that some of the bGH would be present in the milk or meat of these cattle and might be harmful to humans.

Table 9.2 lists these and several other genetically engineered products used in agriculture and animal husbandry.

SAFETY ISSUES AND THE ETHICS OF USING rDNA

> **LEARNING OBJECTIVE**
> - List the advantages of, and problems associated with, the use of genetic modification techniques.

There will always be concern about the safety of any new technology, and genetic modification and biotechnology are certainly no exceptions. One reason for this concern is that it is nearly impossible to prove that something is entirely safe under all conceivable conditions. People worry that the same techniques that can alter a microbe or plant to make them useful to humans could also inadvertently make them pathogenic to humans or otherwise dangerous to living organisms or could create an ecological nightmare. Therefore, laboratories engaged in rDNA research must meet rigorous standards of control to avoid either accidentally releasing genetically modified organisms into the environment or exposing humans to any risk of infection. To reduce risk further, microbiologists engaged in genetic modification often delete from the microbes' genomes certain genes that are essential for growth in environments outside the laboratory. Genetically modified organisms intended for use in the environment (in agriculture, for example) may be engineered to contain "suicide genes"—genes that eventually turn on to produce a toxin that kills the microbes, thus ensuring that they will not survive in the environment for very long after they have accomplished their task.

The safety issues in agricultural biotechnology are similar to those concerning chemical pesticides: toxicity to humans and to nonpest species. Although not shown to be harmful, genetically modified foods have not been popular with consumers. In 1999, researchers in Ohio noticed that humans may develop allergies to *Bacillus thuringiensis* (Bt) toxin after working in fields sprayed with the insecticide. And an Iowa study showed that the caterpillar stage of monarch butterflies could be killed by ingesting windblown

TABLE 9.2	Some Agriculturally Important Products of Genetic Engineering
Product	**Comments**
Agricultural Products	
Bt cotton and Bt corn	Plants have toxin-producing gene from *Bacillus thuringiensis;* toxin kills insects that eat plants.
MacGregor tomatoes	Antisense gene blocks pectin degradation so fruits have longer shelf life.
Pseudomonas fluorescens bacterium	Has toxin-producing gene from insect pathogen *B. thuringiensis;* toxin kills root-eating insects that ingest bacteria.
Pseudomonas syringae, ice-minus bacterium	Lacks normal protein product that initiates undesirable ice formation on plants.
Rhizobium meliloti bacterium	Modified for enhanced nitrogen fixation.
Round-Up (glyphosate)-resistant crops	Plants have bacterial gene; allows use of herbicide on weeds without damaging crops.
Animal Husbandry Products	
Bovine growth hormone (bGH)	Improves weight gain and milk production in cattle; produced by *E. coli.*
Porcine growth hormone (pGH)	Improves weight gain in swine; produced by *E. coli.*
Transgenic animals	Genetic modification of animals to produce medically useful products in their milk.
Other Food Production Products	
Cellulase	Enzyme that degrades cellulose to make animal feedstocks; produced by *E. coli.*
Rennin	Causes formation of milk curds for dairy products; produced by *Aspergillus niger.*

Bt-carrying pollen that landed on milkweed, the caterpillars' normal food. Crop plants can be genetically modified for herbicide resistance so that fields can be sprayed to eliminate weeds without killing the desired crop. However, if the modified plants pollinate related weed species, weeds could become resistant to herbicides, making it more difficult to control unwanted plants. An unanswered question is whether releasing genetically modified organisms will alter evolution as genes move to wild species.

These developing technologies also raise a variety of moral and ethical issues. If genetic screening for diseases becomes routine, who should have access to this information? Should employers or insurance companies have the right to know the results of such tests? Restricting access to such information will be very difficult, which raises questions concerning the right to privacy. How can we be assured that such information will not be used to discriminate against certain groups?

Applications of genetic screening techniques are not limited to adults. The ability to diagnose a genetic disease in a fetus adds even more controversy to the abortion debate. Genetic counseling, which provides advice and counseling to prospective parents with family histories of genetic disease, is becoming more important in considerations about whether or not to have children. As more is learned about genetic causes for various diseases, such as cancer or Huntington's disease, reproductive choices might become more difficult for some families.

What extra burdens does molecular genetics place on our already overtaxed health care system? Genetic screening and gene therapy are expensive procedures, and we need to consider how they will be delivered to the public as the technology develops. Will a sufficient number of genetic counselors be available to work with people? Will expensive medical cures and treatments be available only to those who can afford them?

There are probably just as many harmful applications of a new technology as there are helpful ones. It is particularly easy to imagine rDNA technology being used to develop new and powerful biological weapons. In addition, because such research efforts are performed under top-secret conditions, it is virtually impossible for the general public to learn of them.

Perhaps more than most new technologies, molecular genetics holds the promise of affecting human life in previously unimaginable ways. It is important that society and individuals be given every opportunity to understand the potential impact of these new developments.

Like the invention of the microscope, the development of rDNA techniques is causing profound changes in science, agriculture, and human health care. With this technology only slightly more than 30 years old, it is difficult to predict exactly what changes will occur. However, it is likely that within another 30 years, many of the treatments and diagnostic methods discussed in this book will have been replaced by far more powerful techniques based on the unprecedented ability to manipulate DNA precisely.

STUDY OUTLINE

INTRODUCTION TO BIOTECHNOLOGY
(pp. 254–256)

1. Biotechnology is the use of microorganisms, cells, or cell components to make a product.

RECOMBINANT DNA TECHNOLOGY (p. 254)

2. Closely related organisms can exchange genes in natural recombination.
3. Genes can be transferred among unrelated species via laboratory manipulation, called recombinant DNA technology.
4. Recombinant DNA is DNA that has been artificially manipulated to combine genes from two different sources.

AN OVERVIEW OF RECOMBINANT DNA TECHNOLOGIES (pp. 254–256)

5. A desired gene is inserted into a DNA vector, such as a plasmid or a viral genome.
6. The vector inserts the DNA into a new cell, which is grown to form a clone.
7. Large quantities of the gene product can be harvested from the clone.

TOOLS OF BIOTECHNOLOGY (pp. 256–260)

SELECTION (p. 256)

1. Microbes with desirable traits are selected for culturing by artificial selection.

MUTATION (p. 256)

2. Mutagens are used to cause mutations that might result in a microbe with desirable traits.
3. Site-directed mutagenesis is used to change a specific codon in a gene.

RESTRICTION ENZYMES (pp. 256–257)

4. Prepackaged kits are available for rDNA techniques.

5. A restriction enzyme recognizes and cuts only one particular nucleotide sequence in DNA.
6. Some restriction enzymes produce sticky ends, short stretches of single-stranded DNA at the ends of the DNA fragments.
7. Fragments of DNA produced by the same restriction enzyme will spontaneously join by base pairing. DNA ligase can covalently link the DNA backbones.

VECTORS (pp. 257–258)

8. Shuttle vectors are plasmids that can exist in several different species.
9. A plasmid containing a new gene can be inserted into a cell by transformation.
10. A virus containing a new gene can insert the gene into a cell.

POLYMERASE CHAIN REACTION (pp. 258–260)

11. The polymerase chain reaction (PCR) is used to make multiple copies of a desired piece of DNA enzymatically.
12. PCR can be used to increase the amounts of DNA in samples to detectable levels. This may allow sequencing of genes, the diagnosis of genetic diseases, or the detection of viruses. ✳ **Animation: Polymerase Chain Reaction. The Microbiology Place.**

TECHNIQUES OF GENETIC MODIFICATION (pp. 260–266)

INSERTING FOREIGN DNA INTO CELLS (p. 260)

1. Cells can take up naked DNA by transformation. Chemical treatments are used to make cells that are not naturally competent take up DNA.
2. Pores made in protoplasts and animal cells by electric current in the process of electroporation can provide entrance for new pieces of DNA.
3. Protoplast fusion is the joining of cells whose cell walls have been removed.

4. Foreign DNA can be introduced into plant cells by shooting DNA-coated particles into the cells.

5. Foreign DNA can be injected into animal cells by using a fine glass micropipette.

OBTAINING DNA (pp. 260–263)

6. Gene libraries can be made by cutting up an entire genome with restriction enzymes and inserting the fragments into bacterial plasmids or phages.

7. Complementary DNA (cDNA) made from mRNA by reverse transcription can be cloned in gene libraries.

8. Synthetic DNA can be made in vitro by a DNA synthesis machine.

SELECTING A CLONE (pp. 263–265)

9. Antibiotic-resistance markers on plasmid vectors are used to identify cells containing the engineered vector by direct selection.

10. In blue-white screening, the vector contains the genes for amp^R and β-galactosidase.

11. The desired gene is inserted into the β-galactosidase gene site, destroying the gene.

12. Clones containing the recombinant vector will be resistant to ampicillin and unable to hydrolyze X-gal (white colonies). Clones containing the vector without the new gene will be blue. Clones lacking the vector will not grow.

13. Clones containing foreign DNA can be tested for the desired gene product.

14. A short piece of labeled DNA called a DNA probe can be used to identify clones carrying the desired gene.

MAKING A GENE PRODUCT (pp. 265–266)

15. *E. coli* is used to produce proteins using rDNA because *E. coli* is easily grown and its genomics are well understood.

16. Efforts must be made to ensure that *E. coli*'s endotoxin does not contaminate a product intended for human use.

17. To recover the product, *E. coli* must be lysed or the gene must be linked to a gene that produces a naturally secreted protein.

18. Yeasts can be genetically modified and are likely to continuously secrete a gene product.

19. Genetically modified mammalian cells can be grown to produce proteins such as hormones for medical use.

20. Genetically modified plant cells can be grown and used to produce plants with new properties.

APPLICATIONS OF rDNA (pp. 266–275)

1. Cloned DNA is used to produce products, study the cloned DNA, and alter the phenotype of an organism.

THERAPEUTIC APPLICATIONS (pp. 266–269)

2. Synthetic genes linked to the β-galactosidase gene (*lacZ*) in a plasmid vector were inserted into *E. coli*, allowing *E. coli* to produce and secrete the two polypeptides used to make human insulin.

3. Cells and viruses can be modified to produce a pathogen's surface protein, which can be used as a vaccine.

4. DNA vaccines consist of rDNA cloned in bacteria.

5. Gene therapy can be used to cure genetic diseases by replacing the defective or missing gene.

6. Recombinant DNA techniques were used to map the human genome through the Human Genome Project.

7. This will provide tools for diagnosis and possibly the repair of genetic diseases.

SCIENTIFIC APPLICATIONS (pp. 270–273)

8. Recombinant DNA techniques can be used to increase understanding of DNA, for genetic fingerprinting, and for gene therapy.

9. DNA sequencing machines are used to determine the nucleotide base sequence of restriction fragments in random shotgun sequencing.

10. Bioinformatics is the use of computer applications to study genetic data; proteomics is the study of a cell's proteins.

11. Southern blotting can be used to locate a gene in a cell.

12. DNA probes can be used to quickly identify a pathogen in body tissue or food.

13. Forensic microbiologists use DNA fingerprinting to identify the source of bacterial or viral pathogens.

AGRICULTURAL APPLICATIONS (pp. 273–275)

14. Cells from plants with desirable characteristics can be cloned to produce many identical cells. These cells can then be used to produce whole plants from which seeds can be harvested.

15. Plant cells can be modified by using the Ti plasmid vector. The tumor-producing T genes are replaced with desired genes, and the recombinant DNA is inserted into *Agrobacterium*. The bacterium naturally transforms its plant hosts.

16. Genes for glyphosate resistance, Bt toxin, and pectinase suppression have been engineered into crop plants.

17. Genetically modified *Rhizobium* has enhanced nitrogen fixation.

18. Genetically modified *Pseudomonas* is a biological insecticide that produces *Bacillus thuringiensis* toxin.

19. Bovine growth hormone is being produced by *E. coli*.

SAFETY ISSUES AND THE ETHICS OF USING rDNA (pp. 275–277)

1. Strict safety standards are used to avoid the accidental release of genetically modified microorganisms.

2. Some microbes used in rDNA cloning have been altered so that they cannot survive outside the laboratory.

3. Microorganisms intended for use in the environment may be modified to contain suicide genes so that the organisms do not persist in the environment.

4. Genetic technology raises ethical questions such as: Should employers and insurance companies have access to a person's genetic records? Will some people be targeted for either breeding or sterilization? Will genetic counseling be available to everyone?

5. Genetically modified crops must be safe for consumption and for release in the environment.

STUDY QUESTIONS

 Access more review material either online at **The Microbiology Place** (www.microbiologyplace.com) or with **The Microbiology Place CD-ROM** packaged with your new book. There you'll find activities, practice tests, quizzes, flashcards, case studies, and more to help you succeed. In addition, you'll find the following Interactive Tutorials: Closing a Gene, Cloning DNA Using PCR, and Identifying Organisms Using PCR.
Answers to the Study Questions can be found in Appendix G.

REVIEW

1. Differentiate recombinant DNA from biotechnology.

2. Compare and contrast the following terms:
   a. *cDNA* and *gene*
   b. *restriction fragment* and *gene*
   c. *DNA probe* and *gene*
   d. *DNA polymerase* and *DNA ligase*
   e. *rDNA* and *cDNA*
   f. *genome* and *proteome*

3. How is each of the following used in rDNA technology?
   a. plasmid
   b. viral genome
   c. antibiotic-resistance genes
   d. restriction enzyme

4. Differentiate between a gene library and synthetic DNA.

5. Differentiate among the following terms. Which one is "hit and miss"—that is, does not add a specific gene to a cell?
   a. protoplast fusion
   b. gene gun
   c. microinjection
   d. electroporation

6. Some commonly used restriction enzymes are listed in the following table. The cutting site is indicated by ↓. Indicate which enzymes produce sticky ends. Of what value are sticky ends in making recombinant DNA?

Enzyme	Bacterial Source	Recognition Sequence
BamHI	Bacillus amyloliquefaciens	G↓G A T C C G C T A G↑G
EcoRI	Escherichia coli	G↓A A T T C C T T A A↑G
HaeIII	Haemophilus aegyptius	G G↓C C C C↑G G
HindIII	Haemophilus influenzae	A↓A G C T T T T C G A↑A

7. Suppose you want multiple copies of a gene you have synthesized. How would you obtain the necessary copies by cloning? By PCR?

8. Describe a recombinant DNA experiment in two or three sentences. Use the following terms: intron, exon, DNA, mRNA, cDNA, RNA polymerase, reverse transcriptase.

9. List at least two examples of the use of rDNA in medicine and in agriculture.

10. You are attempting to insert a gene for saltwater tolerance into a plant by using the Ti plasmid. In addition to the desired gene, you add a gene for tetracycline resistance (tet^R) to the plasmid. What is the purpose of the tet^R gene?

11. How does RNAi "silence" a gene?

MULTIPLE CHOICE

1. Restriction enzymes were first discovered with the observation that
   a. DNA is restricted to the nucleus.
   b. phage DNA is destroyed in a host cell.
   c. foreign DNA is kept out of a cell.
   d. foreign DNA is restricted to the cytoplasm.
   e. all of the above

2. The DNA probe, 3′-GGCTTA, will hybridize with which of the following?

a. 5′-CCGUUA
b. 5′-CCGAAT
c. 5′-GGCTTA
d. 3′-CCGAAT
e. 3′-GGCAAU

3. Which of the following is the fourth basic step to genetically modify a cell?

a. transformation
b. ligation
c. plasmid cleavage
d. restriction-enzyme digestion of gene
e. isolation of gene

4. *The following enzymes are used to make cDNA. What is the second enzyme used to make cDNA?*

a. reverse transcriptase
b. ribozyme
c. RNA polymerase
d. DNA polymerase

5. If you put a gene in a virus, the next step in genetic modification would be

a. insertion of a plasmid.
b. transformation.
c. transduction.
d. PCR.
e. Southern blotting.

6. You have a small gene that you want replicated by PCR. You add radioactively labeled nucleotides to the PCR thermalcycler. After three replication cycles, what percentage of the DNA single-strands are radioactively labeled?

a. 0%
b. 12.5%
c. 50%
d. 87.5%
e. 100%

Match the following choices to the statements in questions 7 through 10.

a. antisense
b. clone
c. library
d. Southern blot
e. vector

7. Pieces of human DNA stored in yeast cells.

8. A population of cells carrying a desired plasmid.

9. Self-replicating DNA for transmitting a gene from one organism to another.

10. A gene that hybridizes with mRNA.

CRITICAL THINKING

1. Using the following map of plasmid pMICRO, give the number of restriction fragments that would result from digesting pMICRO with *Eco*RI, *Hind*III, and both enzymes

together. Which enzyme makes the smallest fragment containing the tetracycline-resistance gene?

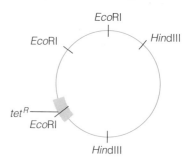

2. Design an experiment using vaccinia virus to make a vaccine against the AIDS virus (HIV).

3. Why did the use of DNA polymerase from the bacterium *Thermus aquaticus* allow researchers to add the necessary reagents to tubes in a preprogrammed heating block?

4. The following picture shows bacterial colonies growing on X-gal plus ampicillin in a blue-white screening test. Which colonies have the recombinant plasmid? The small satellite colonies do not have the plasmid. Why did they start growing on the medium 48 hours after the larger colonies?

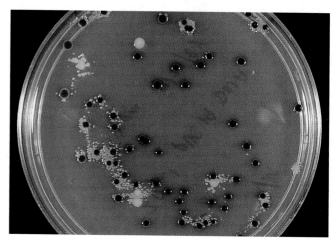

CLINICAL APPLICATIONS

1. PCR has been used to examine oysters for the presence of *Vibrio cholerae*. Oysters from different areas were homogenized, and DNA was extracted from the homogenates. The DNA was digested by the restriction enzyme *Hinc*II. A primer for the hemolysin gene of *V. cholerae* was used for the PCR reaction. After PCR, each sample was electrophoresed and stained with a probe for the hemolysin gene. Which of the oyster samples were (was) positive for *V. cholerae*? How can you tell? Why look for *V. cholerae* in oysters? What is the advantage of PCR over conventional biochemical tests to identify the bacteria?

2. Using the restriction enzyme *Eco*RI, the following gel electrophoresis patterns were obtained from digests of various DNA molecules from a transformation experiment. Can you conclude from these data that transformation occurred? Explain why or why not.

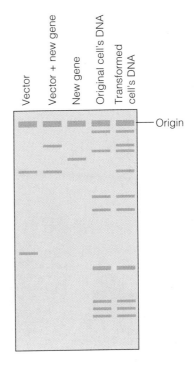

10

Classification of Microorganisms

The science of classification, especially the classification of living forms, is called **taxonomy** (from the Greek for orderly arrangement). The objective of taxonomy is to classify living organisms—that is, to establish the relationships between one group of organisms and another and to differentiate between them. There may be as many as 100 million different living organisms, but fewer than 10% have been discovered, much less classified and identified.

Taxonomy also provides a common reference for identifying organisms already classified. For example, when a bacterium suspected of causing a specific disease is isolated from a patient, characteristics of that isolate are matched to lists of characteristics of previously classified bacteria to identify the isolate (see the box on page 294). Finally, taxonomy is a basic and necessary tool for scientists, providing a universal language of communication.

Modern taxonomy is an exciting and dynamic field. New techniques in molecular biology and genetics are providing new insights into classification and evolution. In this chapter, you will learn the various classification systems, the different criteria used for classification, and tests that are used to identify microorganisms that have already been classified.

UNDER THE MICROSCOPE

Pneumocystis jirovici. This was thought to be a protozoan until DNA analysis showed it is a fungus. *P. jirovici* causes pneumonia in immunocompromised people.

THE STUDY OF PHYLOGENETIC RELATIONSHIPS

LEARNING OBJECTIVES

- Define *taxonomy, taxon,* and *phylogeny.*
- Discuss the limitations of a two-kingdom classification system.
- Identify the contributions of Linnaeus, von Nägeli, Chatton, Whittaker, and Woese.

In 2001, an international project called the All Species Inventory was launched. The project's purpose is to identify and record every species of life on Earth in the next 25 years. These researchers have undertaken a challenging goal: whereas biologists have identified more than 1.7 million different organisms thus far, it is estimated that the number of living species ranges from 10 to 100 million.

Among these many and diverse organisms, however, are many similarities. For example, all organisms are composed of cells surrounded by a plasma membrane, use ATP for energy, and store their genetic information in DNA. These similarities are the result of evolution, or descent from a common ancestor. In 1859, the English naturalist Charles Darwin proposed that natural selection was responsible for the similarities as well as the differences among organisms. The differences can be attributed to the survival of organisms with traits best suited to a particular environment.

To facilitate research, scholarship, and communication, we arrange organisms into taxonomic categories, or **taxa** (singular: *taxon*), to show degrees of similarities among organisms. These similarities are due to relatedness—all organisms are related through evolution. **Systematics,** or **phylogeny,** is the study of the evolutionary history of organisms. The hierarchy of taxa reflects evolutionary, or *phylogenetic,* relationships.

From the time of Aristotle, living organisms were categorized in just two ways, as either plants or animals. In 1735, the Swedish botanist Carolus Linnaeus introduced a formal system of classification dividing living organisms into two kingdoms—Plantae and Animalia. He used latinized names to provide one common "language" for systematics. As the biological sciences developed, however, biologists began looking for a *natural* classification system—one that groups organisms based on ancestral relationships and allows us to see the order in life. In 1857, Carl von Nägeli, a contemporary of Pasteur, proposed that bacteria and fungi be placed in the plant kingdom. In 1866, Ernst Haeckel proposed the Kingdom Protista, to include bacteria, protozoa, algae, and fungi. Because of disagreements over the definition of protists, for the next 100 years biologists continued to follow von Nägeli's placement of bacteria and fungi in the plant kingdom. It is ironic that recent DNA sequencing places fungi closer to animals than plants. Fungi were placed in their own kingdom in 1959.

With the advent of electron microscopy, the physical differences between cells became apparent. The term *prokaryote* was introduced in 1937 by Edouard Chatton to distinguish cells having no nucleus from the nucleated cells of plants and animals. In 1961, Roger Stanier provided the current definition of prokaryotes: cells in which the nuclear material (nucleoplasm) is not surrounded by a nuclear membrane. In 1968, Robert G.E. Murray proposed the Kingdom Prokaryotae.

In 1969, Robert H. Whittaker founded the five-kingdom system in which prokaryotes were placed in the Kingdom Prokaryotae, or Monera, and eukaryotes comprised the other four kingdoms. The Kingdom Prokaryotae had been based on microscopic observations. Subsequently, new techniques in molecular biology revealed that there are actually two types of prokaryotic cells and one type of eukaryotic cell.

THE THREE DOMAINS

LEARNING OBJECTIVES

- Discuss the advantages of the three-domain system.
- List the characteristics of the Bacteria, Archaea, and Eukarya domains.

The discovery of three cell types was based on the observations that ribosomes are not the same in all cells (see Chapter 4, page 95). Ribosomes provide a method of comparing cells because ribosomes are present in all cells. Comparing the sequences of nucleotides in ribosomal RNA (see page 303) from different kinds of cells shows that there are three distinctly different cell groups: the eukaryotes and two different types of prokaryotes—the bacteria and the archaea.

In 1978, Carl R. Woese proposed elevating the three cell types to a level above kingdom, called domain. Woese believed that the archaea and the bacteria, although similar in appearance, should form their own separate domains on the evolutionary tree (Figure 10.1). In this widely accepted scheme, animals, plants, fungi, and protists are kingdoms in the Domain **Eukarya.** Organisms are classified by cell type in the three domain systems. In addition to differences in rRNA, the three domains differ in membrane lipid structure, transfer RNA molecules, and sensitivity to antibiotics (Table 10.1).

The Domain **Bacteria** includes all of the pathogenic prokaryotes as well as many of the nonpathogenic prokaryotes found in soil and water. The photoautotrophic

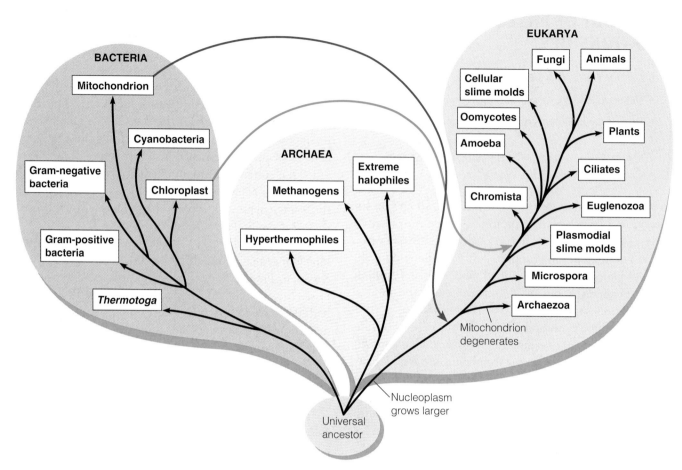

FIGURE 10.1 The three-domain system. The system recognizes the three types of cells. The Domain Eukarya includes the Kingdoms Fungi, Plantae, and Animalia. It also includes the protists discussed on page 289.

Q **What characteristics are used to differentiate the three domains?**

prokaryotes are also in this domain. The Domain **Archaea** includes prokaryotes that do not have peptidoglycan in their cell walls. They often live in extreme environments and carry out unusual metabolic processes. Archaea include three major groups:

1. The methanogens, strict anaerobes that produce methane (CH_4) from carbon dioxide and hydrogen.

2. Extreme halophiles, which require high concentrations of salt for survival.

3. Hyperthermophiles, which normally grow in extremely hot environments.

The evolutionary relationship of the three domains is the subject of current research by biologists. Originally, archaea were thought to be the most primitive group, whereas bacteria were assumed to be more closely related to eukaryotes. However, studies of rRNA indicate that a universal ancestor split into three lineages. That split led to the Archaea, the Bacteria, and what eventually became

the nucleoplasm of the eukaryotes. The oldest known fossils are the remains of prokaryotes that lived more than 3.5 billion years ago. Eukaryotic cells evolved more recently, about 1.4 billion years ago. According to the endosymbiotic theory, eukaryotic cells evolved from prokaryotic cells living inside one another, as endosymbionts (see Chapter 4, page 107). In fact, the similarities between prokaryotic cells and eukaryotic organelles provide striking evidence for this endosymbiotic relationship (Table 10.2).

The original nucleoplasmic cell was prokaryotic. However, infoldings in its plasma membrane may have surrounded the nuclear region to produce a true nucleus (Figure 10.2). Recently, French researchers provided support for this hypothesis with their observations of a true nucleus in *Gemmata* bacteria. Over time, the chromosome of the nucleoplasm may have acquired pieces such as transposons (page 246). In some cells, this large chromosome may have fragmented into smaller linear chromosomes.

TABLE 10.1	Some Characteristics of Archaea, Bacteria, and Eukarya		
	Archaea	Bacteria	Eukarya
	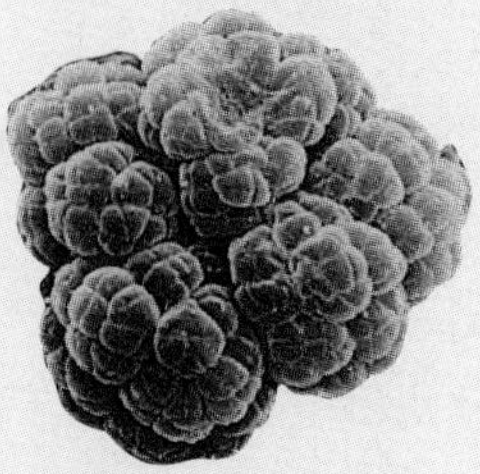 *Methanosarcina* SEM ⊢ 10 µm	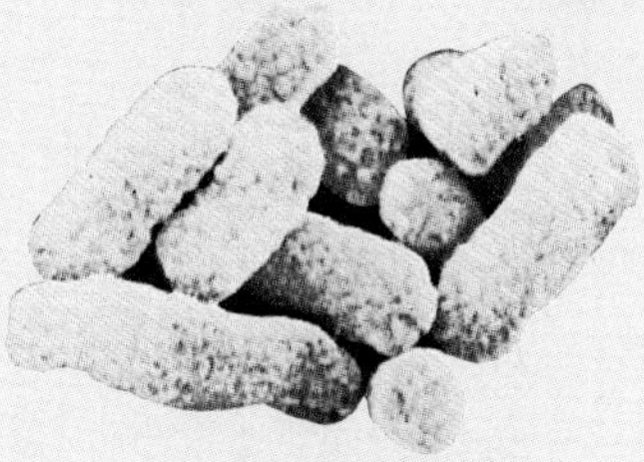 *E. coli* SEM ⊢ 1 µm	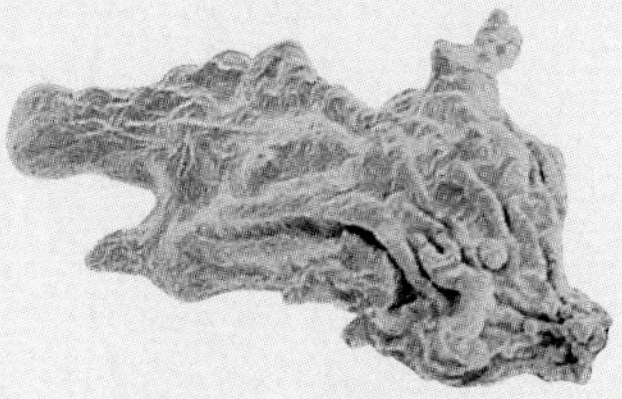 *Amoeba* SEM ⊢ 10 µm
Cell Type	Prokaryotic	Prokaryotic	Eukaryotic
Cell Wall	Varies in composition; contains no peptidoglycan	Contains peptidoglycan	Varies in composition; contains carbohydrates
Membrane Lipids	Composed of branched carbon chains attached to glycerol by ether linkage	Composed of straight carbon chains attached to glycerol by ester linkage	Composed of straight carbon chains attached to glycerol by ester linkage
First Amino Acid in Protein Synthesis	Methionine	Formylmethionine	Methionine
Antibiotic Sensitivity	No	Yes	No
rRNA Loop*	Lacking	Present	Lacking
Common Arm of tRNA†	Lacking	Present	Present

*Binds to ribosomal protein; found in all bacteria.
†A sequence of bases in tRNA found in all eukaryotes and bacteria: guanine-thymine-pseudouridine-cytosine-guanine.

TABLE 10.2	Prokaryotic Cells and Eukaryotic Organelles Compared		
	Prokaryotic Cell	Eukaryotic Cell	Eukaryotic Organelles (Mitochondria and Chloroplasts)
DNA	One circular; some two circular; some linear	Linear	Circular
Histones	In archaea	Yes	No
First Amino Acid in Protein Synthesis	Formylmethionine	Methionine	Formylmethionine
Ribosomes	70S	80S	70S
Growth	Binary fission	Mitosis	Binary fission

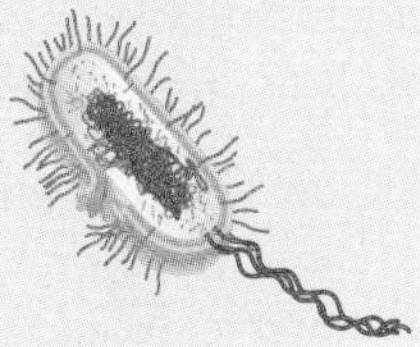

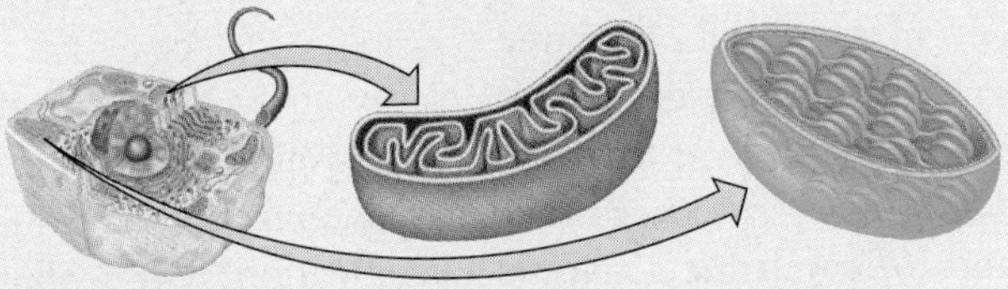

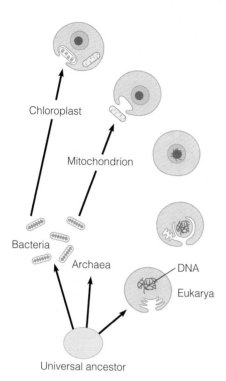

FIGURE 10.2 A model of the origin of eukaryotes.
Invagination of the plasma membrane may have formed the nuclear envelope and endoplasmic reticulum. Similarities, including rRNA sequences, indicate that endosymbiotic prokaryotes gave rise to mitochondria and chloroplasts.

Q **How many membranes make up the nuclear envelope of a eukaryotic cell?**

Perhaps cells with linear chromosomes had an advantage in cell division over those with a large, unwieldy circular chromosome.

That nucleoplasmic cell provided the original host in which endosymbiotic bacteria developed into organisms (see page 107). An example of a modern prokaryote living in a eukaryotic cell is shown in Figure 10.3. The cyanobacterium-like cell and the eukaryotic host require each other for survival.

In sequencing the genome of a prokaryote called *Thermotoga maritima*, microbiologist Karen Nelson has discovered that this species has genes similar to members of both the Domain Bacteria and the Domain Archaea. Her findings suggest that *Thermotoga* is one of the earliest cells. For this reason, *Thermotoga* is referred to as one of the "deeply branching genera," that is, it is near the origin or "root" of the evolutionary tree.

Taxonomy provides tools for clarifying the evolution of organisms, as well as their interrelationships. New organisms are being discovered every day, and taxonomists continue to search for a natural classification system that reflects phylogenetic relationships.

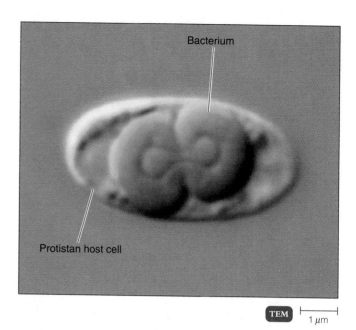

FIGURE 10.3 *Cyanophora paradoxa.* This organism, in which the eukaryotic host and the bacterium require each other for survival, provides a modern example of how eukaryotic cells might have evolved.

Q **What features do chloroplasts, mitochondria, and bacteria have in common?**

A PHYLOGENETIC HIERARCHY

In a phylogenetic hierarchy, grouping organisms according to common properties implies that a group of organisms evolved from a common ancestor; each species retains some of the characteristics of the ancestor. Some of the information used to classify and determine phylogenetic relationships in higher organisms comes from fossils. Bones, shells, or stems that contain mineral matter or have left imprints in rock that was once mud are examples of fossils.

The structures of most microorganisms are not readily fossilized. Some exceptions are the following:

- A marine protist whose fossilized colonies form the White Cliffs of Dover, England.

- Stromatolites, the fossilized remains of filamentous bacteria and sediments that flourished between 0.5 and 2 billion years ago.

- Cyanobacteria-like fossils found in rocks in Western Australia that are 3.0–3.5 billion years old. These are widely believed to be the oldest known fossils. Some fossils of prokaryotes are shown in Figure 10.4.

Because fossil evidence is not available for most prokaryotes, their phylogeny must be based on other evidence. But in one notable exception, scientists may have isolated living bacteria and yeast 25–40 million years old. In 1995,

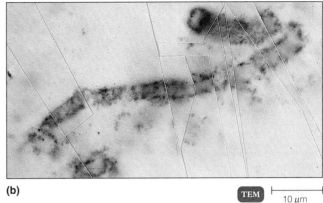

(a) SEM ⊢———⊣ 2 μm   (b) TEM ⊢———⊣ 10 μm

FIGURE 10.4 Fossilized prokaryotes. (a) Coccoid cyanobacteria from the Late Precambrian (850 million years ago) of central Australia. **(b)** Filamentous prokaryotes from the Early Precambrian (3.5 billion years ago) of western Australia.

Q **What evidence is used to determine the phylogeny of prokaryotes?**

the American microbiologist Raul Cano and his colleagues reported growing *Bacillus sphaericus* and other as yet unidentified microorganisms that had survived embedded in amber (fossilized plant resin) for millions of years. If confirmed, this discovery should provide more information about the evolution of microorganisms.

Conclusions from rRNA sequencing and DNA hybridization studies (discussed on page 302) of selected orders and families of eukaryotes are in agreement with the fossil records. This has encouraged workers to use DNA hybridization and rRNA sequencing to gain an understanding of the evolutionary relationships among prokaryotic groups.

CLASSIFICATION OF ORGANISMS

LEARNING OBJECTIVE

• Differentiate among eukaryotic, prokaryotic, and viral species.

Living organisms are grouped according to similar characteristics (classification), and each organism is assigned a unique scientific name. The rules for classifying and naming, which are used by biologists worldwide, are discussed next.

SCIENTIFIC NOMENCLATURE

LEARNING OBJECTIVE

• Explain why scientific names are used.

In a world inhabited by millions of living organisms, biologists must be sure they know exactly which organism is being discussed. We cannot use common names, because the same name is often used for many different organisms in different locales. For example, there are two different organisms with the common name Spanish moss, and neither one is actually a moss. Plus, local languages are used for common names. Because common names can be misleading and are in different languages, a system of scientific names, referred to as *scientific nomenclature*, was developed in the eighteenth century.

Recall from Chapter 1 (page 2) that every organism is assigned two names, or a binomial. These names are the **genus** name and **specific epithet (species),** and both names are printed underlined or italicized. The genus name is always capitalized and is always a noun. The species name is lowercase and is usually an adjective. Because this system gives two names to each organism, the system is called **binomial nomenclature.**

Let's consider some examples. Our own genus and specific epithet are *Homo sapiens* (hō′mō sā′pē-ens). The noun, or genus, means man; the adjective, or specific epithet, means wise. A mold that contaminates bread is called *Rhizopus stolonifer* (rī′zō-pŭs stō′ion-i-fĕr). *Rhizo-* (root) describes rootlike structures on the fungus; *stolo-* (a shoot) describes the long hyphae. Table 10.3 contains more examples.

Binomials are used by scientists worldwide, regardless of their native language, which enables them to share knowledge efficiently and accurately. Several scientific entities are responsible for establishing rules governing the naming of organisms. Rules for assigning names for protozoa and parasitic worms are published in the *International Code of Zoological Nomenclature*. Rules for assigning names for fungi and algae are published in the *International Code of Botanical Nomenclature*. Rules for naming newly classified

TABLE 10.3	Making Scientific Names Familiar

Use the word roots guide inside the book's front and back covers to find out what the name means. The name will not seem so strange if you translate it. When you encounter a new name, practice saying it out loud. The exact pronunciation is not as important as the familiarity you will gain. Guidelines for pronunciation are given in Appendix E.

Following are some examples of microbial names you may encounter in the popular press as well as in the lab.

	Pronunciation	Source of Genus Name	Source of Specific Epithet
Klebsiella pneumoniae (bacterium)	kleb-sē-el′lä nü-mō′nē-ī	Honors bacteriologist Edwin Klebs	The disease it causes
Pfiesteria piscicida (alga)	fes-tėr′ē-ä pi′si-si-dä	Honors dinoflagellate biologist Lois Pfiester	Causes disease in fish (*piscis*)
Salmonella typhimurium (bacterium)	sal-mōn-el′lä tī-fi-mur′ē-um	Honors public health micro-biologist Daniel Salmon	Causes stupor (*typh-*) in mice (*muri-*)
Streptococcus pyogenes (bacterium)	strep-tō-kok′kus pī-äj′en-ēz	Appearance of cells in chains (*strepto-*)	Forms pus (*pyo-*)
Saccharomyces cerevisiae (yeast)	sak-ä-rō-mī′ses se-ri-vis′ē-ī	Fungus (*-myces*) that uses sugar (*saccharo-*)	Makes beer (*cerevisia*)
Penicillium chrysogenum (fungus)	pen-i-sil′lē-um krī-so′jen-um	Tuftlike or paintbrush (*penicill-*) appearance microscopically	Produces a yellow (*chryso-*) pigment
Trypanosoma cruzi (protozoan)	tri-pa-nō-sō′mä krūz′ē	Corkscrew (*trypano-*, borer; *soma-*, body)	Honors epidemiologist Oswaldo Cruz

prokaryotes and for assigning prokaryotes to taxa are established by the International Committee on Systematic Bacteriology and are published in the *Bacteriological Code*. Descriptions of prokaryotes and evidence for their classifications are published in the *International Journal of Systematic Bacteriology* before being incorporated into a reference called *Bergey's Manual*. According to the *Bacteriological Code*, scientific names are to be taken from Latin (a genus name can be taken from Greek) or latinized by the addition of the appropriate suffix. Suffixes for order and family are *-ales* and *-aceae*, respectively.

As new laboratory techniques make more detailed characterizations of microbes possible, two genera may be reclassified as a single genus, or a genus may be divided into two or more genera. For example, the genera "Diplococcus" and *Streptococcus* were combined in 1974; the only diplococcal species is now called *Streptococcus pneumoniae*. (The *Bacteriological Code* states that the older genus name should be retained in such cases.) In 1984, DNA hybridization studies indicated that "Streptococcus faecalis" and "Streptococcus faecium" were only distantly related to the other streptococcal species; consequently, a new genus called *Enterococcus* was created, and these species were renamed *E. faecalis* and *E. faecium* (fē′sē-um) because the rules require that the original specific epithets be retained.

Making the transition to a new name can be confusing, so the old name is often written in parentheses. For example, a physician looking for information on the cause of a patient's pneumonia-like symptoms (meliodosis) would find the bacterial name *Burkholderia (Pseudomonas) pseudomallei* (bėrk′hōld-ér-ē-ä sū-dō-mal′le-ē).

Obtaining the name of the organism is important in order to know what treatment to use; antifungal drugs will not work against bacteria, and antibacterial drugs will not work against viruses.

THE TAXONOMIC HIERARCHY

LEARNING OBJECTIVES

- List the major taxa.
- Differentiate between *culture*, *clone*, and *strain*.

All organisms can be grouped into a series of subdivisions that make up the taxonomic hierarchy. Linnaeus developed this hierarchy for his classification of plants and animals. A **eukaryotic species** is a group of closely related organisms that breed among themselves. (Bacterial species will be discussed shortly.) A genus consists of species that differ from each other in certain ways but are related by descent. For example, *Quercus* (kwer′kus), the genus name for oak, consists of all types of oak trees (white oak, red oak, bur oak, velvet oak, and so on). Even though each species of oak differs from every other species, they are all related genetically. Just as a number of species make up a genus, related genera make up a **family.** A group of similar families constitutes an **order,** and a group of similar orders makes up a **class.** Related classes, in turn, make up a **phylum.** Thus, a particular organism (or species) has a genus name and specific epithet and belongs to a family, order, class, and phylum.

All phyla or divisions that are related to each other make up a **kingdom,** and related kingdoms are grouped into a **domain** (Figure 10.5).

CLASSIFICATION OF PROKARYOTES

The taxonomic classification scheme for prokaryotes is found in *Bergey's Manual of Systematic Bacteriology,* 2nd edition (see Appendix A). The first two volumes have been published, with the remaining three volumes to follow over the next few years. The contents of each volume are shown in Table 10.4. In *Bergey's Manual,* prokaryotes are divided into two domains: Bacteria and Archaea. Each domain is divided into phyla. Remember, the classification is based on similarities in nucleotide sequences in rRNA. Classes are divided into orders; orders, into families; families, into genera; and genera, into species.

A prokaryotic species is defined somewhat differently than a eukaryotic species, which is a group of closely related organisms that can interbreed. Unlike reproduction in eukaryotic organisms, cell division in bacteria is not directly tied to sexual conjugation, which is infrequent and does not always need to be species-specific. A **prokaryotic species,** therefore, is defined simply as a population of cells with similar characteristics. (The types of characteristics will be discussed later in this chapter.) The members of a bacterial species are essentially indistinguishable from each other but are distinguishable from members of other species, usually on the basis of several features. As you know, bacteria grown at a given time in media are called a culture. A pure culture is often a **clone,** that is, a population of cells derived from a single parent cell. All cells in the clone should be identical. However, in some cases, pure cultures of the same species are not identical in all ways. Each such group is called a **strain.** Strains are identified by numbers, letters, or names that follow the specific epithet.

Bergey's Manual provides a reference for identifying bacteria in the laboratory, as well as a classification scheme for bacteria. One scheme for the evolutionary relationships of bacteria is shown in Figure 10.6. Characteristics used to classify and identify bacteria are discussed in Chapter 11.

CLASSIFICATION OF EUKARYOTES

LEARNING OBJECTIVES

- List the major characteristics used to differentiate the three kingdoms of multicellular Eukarya.

- Define *protist.*

Some kingdoms in the Domain Eukarya are shown in Figure 10.1.

In 1969, simple eukaryotic organisms, mostly unicellular, were grouped as the Kingdom **Protista,** a catchall kingdom for a variety of organisms. Historically, eukaryotic organisms that didn't fit into other kingdoms were placed in the Protista. Approximately 200,000 species of protistans have been identified thus far, and these organisms are nutritionally quite diverse—from photosynthetic to obligate intracellular parasite. Ribosomal RNA sequencing is making it possible to divide protists into groups based on their descent from common ancestors. Consequently, for the time being, the organisms once classified as protists are being divided into **clades,** that is, genetically related groups. For convenience, we will continue to use the term *protist* to refer to unicellular eukaryotes and their close relatives. These organisms will be discussed in Chapter 12.

Fungi, plants, and animals make up the three kingdoms of more complex eukaryotic organisms, most of which are multicellular.

The Kingdom **Fungi** includes the unicellular yeasts, multicellular molds, and macroscopic species such as mushrooms. To obtain raw materials for vital functions, a fungus absorbs dissolved organic matter through its plasma membrane. The cells of a multicellular fungus are commonly joined to form thin tubes called *hyphae.* The hyphae are usually divided into multinucleated units by cross-walls that have holes, so that cytoplasm can flow between the cell-like units. Fungi develop from spores or from fragments of hyphae. (See Figure 12.1, page 346.)

The Kingdom **Plantae** (plants) includes some algae and all mosses, ferns, conifers, and flowering plants. All members of this kingdom are multicellular. To obtain energy, a plant uses photosynthesis, the process that converts carbon dioxide and water into organic molecules used by the cell.

The kingdom of multicellular organisms called **Animalia** (animals) includes sponges, various worms, insects, and animals with backbones (vertebrates). Animals obtain nutrients and energy by ingesting organic matter through a mouth of some kind.

CLASSIFICATION OF VIRUSES

Viruses are not classified as part of any of the three domains. Viruses are not composed of cells, and they use the

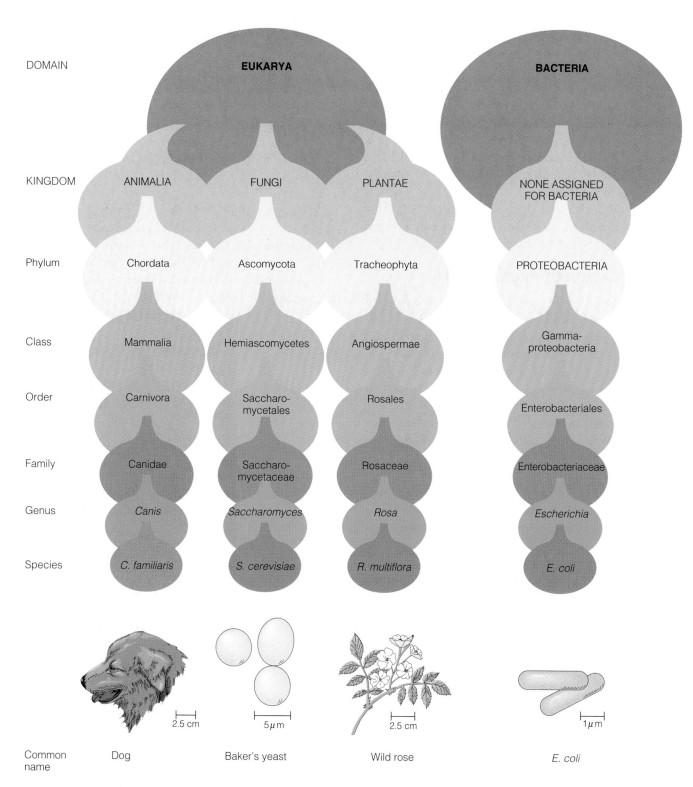

	EUKARYA			BACTERIA
DOMAIN				
KINGDOM	ANIMALIA	FUNGI	PLANTAE	NONE ASSIGNED FOR BACTERIA
Phylum	Chordata	Ascomycota	Tracheophyta	PROTEOBACTERIA
Class	Mammalia	Hemiascomycetes	Angiospermae	Gamma-proteobacteria
Order	Carnivora	Saccharo-mycetales	Rosales	Enterobacteriales
Family	Canidae	Saccharo-mycetaceae	Rosaceae	Enterobacteriaceae
Genus	*Canis*	*Saccharomyces*	*Rosa*	*Escherichia*
Species	*C. familiaris*	*S. cerevisiae*	*R. multiflora*	*E. coli*

	2.5 cm	5 μm	2.5 cm	1 μm
Common name	Dog	Baker's yeast	Wild rose	*E. coli*

FIGURE 10.5 The taxonomic hierarchy. Organisms are grouped according to relatedness. Species that are closely related are grouped into a genus. For example, the baker's yeast belongs to the genus that includes sourdough yeast (*Saccharomyces exiguus*). Related genera, such as *Saccharomyces* and *Candida*, are placed in a family and so on. Each group is more comprehensive. The domain Eukarya includes all organisms with eukaryotic cells.

 What is the biological definition of a family?

TABLE 10.4	Classification of Prokaryotes (*Bergey's Manual of Systematic Bacteriology*, 2nd edition)	
Volume	**Contents**	**Notes**
1	Archaea, cyanobacteria, phototrophs, and deeply branching genera	Includes the Domain Archaea and some gram-negative bacteria
2	Proteobacteria	Phylum Proteobacteria; these are related gram-negative bacteria
3	Low G + C gram-positives	Phylum Firmicutes (gram-positive cell wall); Phylum Mycoplasmas (wall-less)
4	High G + C gram-positives	Includes actinomycetes
5	Chlamydiae, Spirochaetes, Bacteroidetes, and Fusobacteria	These are distinct phyla of bacteria, each with a unique rRNA sequence

anabolic machinery within living host cells to multiply. A viral genome can direct biosynthesis inside a host cell, and some viral genomes can become incorporated into the host genome. The ecological niche of a virus is its specific host cell, so viruses may be more closely related to their hosts than to other viruses. The International Committee on Taxonomy of Viruses defines a **viral species** as a population of viruses with similar characteristics (including morphology, genes, and enzymes) that occupies a particular ecological niche.

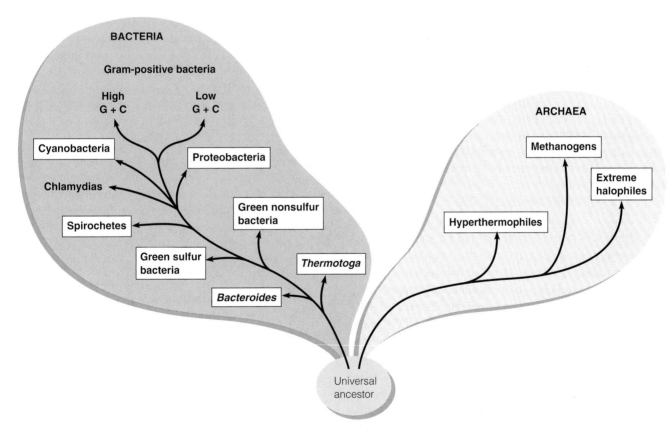

FIGURE 10.6 Phylogenetic relationships of prokaryotes. Arrows indicate major lines of descent of bacterial groups. Selected phyla are indicated by the white boxes.

Q Members of which phylum can be identified by Gram staining?

Viruses are obligatory intracellular parasites, so they must have evolved after a suitable host cell had evolved. There are two hypotheses on the origin of viruses: (1) they arose from independently replicating strands of nucleic acids (such as plasmids), and (2) they developed from degenerative cells that, through many generations, gradually lost the ability to survive independently but could survive when associated with another cell. Viruses will be discussed in Chapter 13.

METHODS OF CLASSIFYING AND IDENTIFYING MICROORGANISMS

> **LEARNING OBJECTIVES**
> • Compare and contrast classification and identification.
> • Explain the purpose of *Bergey's Manual.*

A classification scheme provides a list of characteristics and a means for comparison to aid in the identification of an organism. Once an organism is identified, it can be placed into a previously devised classification scheme. Microorganisms are *identified* for practical purposes—for example, to determine an appropriate treatment for an infection. They are not necessarily identified by the same techniques by which they are *classified.* Most identification procedures are easily performed in a laboratory and use as few procedures or tests as possible. Protozoa, parasitic worms, and fungi can usually be identified microscopically. Most prokaryotic organisms do not have distinguishing morphological features or even much variation in size and shape. Consequently, microbiologists have developed a variety of methods to test metabolic reactions and other characteristics to identify prokaryotes.

Bergey's Manual of Determinative Bacteriology has been a widely used reference since the first edition was published in 1923. *Bergey's Manual of Determinative Bacteriology* (9th ed., 1994) does not classify bacteria according to evolutionary relatedness but instead provides identification (determinative) schemes based on such criteria as cell wall composition, morphology, differential staining, oxygen requirements, and biochemical testing.* The majority of Bacteria and Archaea have not been cultured, and scientists estimate that only 1% of these microbes have been discovered.

Medical microbiology (the branch of microbiology dealing with human pathogens) has dominated the interest in microbes, and this interest is reflected in many identification schemes. However, to put the pathogenic

properties of bacteria in perspective, of the more than 2600 species listed in the *Approved Lists of Bacterial Names,* fewer than 10% are human pathogens.

We next discuss several criteria and methods for the classification of microorganisms and the routine identification of some of those organisms. In addition to properties of the organism itself, the source and habitat of a bacterial isolate is considered as part of the classification and identification processes.

MORPHOLOGICAL CHARACTERISTICS

Morphological (structural) characteristics have helped taxonomists classify organisms for 200 years. Higher organisms are frequently classified according to observed anatomical detail. But many microorganisms look too similar to be classified by their structures. Through a microscope, organisms that might differ in metabolic or physiological properties may look alike. Literally hundreds of bacterial species are small rods or small cocci.

Larger size and the presence of intracellular structures does not always mean easy classification, however. *Pneumocystis* (nü-mō-sis′tis) pneumonia is the most common opportunistic infection in immunosuppressed individuals and is a significant cause of death in AIDS patients. The causative agent of this infection, *P. jiroveci* (ye-rō′vet-zē) [formerly *P. carinii* (kär-i′nē-ē)] was not considered a human pathogen until the 1970s. *Pneumocystis* lacks structures that can be easily used for identification (see Figure 24.22, page 737), and its taxonomic position has been uncertain since its discovery in 1909. Although it was provisionally classified as a protozoan, recent studies comparing its rRNA sequence with those of other protozoa, *Euglena,* cellular slime molds, plants, mammals, and fungi have shown that it is actually a member of the Kingdom Fungi. Researchers have not been able to culture *Pneumocystis,* but they have developed some useful treatments for *Pneumocystis* pneumonia. Perhaps as researchers take into account this organism's relatedness to fungi, appropriate culture methods and treatments will result.

Cell morphology tells us little about phylogenetic relationships. However, morphological characteristics are still useful in identifying bacteria. For example, differences in such structures as endospores or flagella can be helpful.

DIFFERENTIAL STAINING

> **LEARNING OBJECTIVE**
> • Describe how staining and biochemical tests are used to identify bacteria.

Recall from Chapter 3 that one of the first steps in identifying bacteria is differential staining. Most bacteria are

---

*Both *Bergey's Manual of Systematic Bacteriology* (see page 289) and *Bergey's Manual of Determinative Biology* are referred to simply as *Bergey's Manual;* the complete titles are used when the information under discussion is found in one but not the other, for example, an identification table.

MICROBIOLOGY REQUISITION

Lab:

Date, time received:

Date:	Time:	Slip prepared by:
Physician name:	Collected by:	Patient ID#:

▶ DO NOT WRITE BELOW THIS LINE ◀

USE SEPARATE SLIP FOR EACH REQUEST

GRAM STAIN REPORT

- ☐ GRAM POS. COCCI, GROUPS
- ☐ GRAM POS. COCCI, PAIRS/CHAIN
- ☐ GRAM POS. RODS
- ☒ GRAM NEG. COCCI
- ☐ GRAM NEG. RODS
- ☐ GRAM NEG. COCCOBACILLI
- ☐ YEAST
- ☐ OTHER

- ☐ NO GROWTH
- ☐ NO GROWTH IN ___DAYS
- ☐ MIXED MICROBIOTA
- ☐ SPECIMEN IMPROPERLY COLLECTED OR TRANSPORTED
- ☐ ___DIFFERENT TYPES OF ORGANISMS
- ☐ NEGATIVE FOR *SALMONELLA, SHIGELLA,* AND *CAMPYLOBACTER*
- ☐ NO OVA, CYSTS, OR PARASITES SEEN
- ☒ OXIDASE-POSITIVE GRAM-NEGATIVE DIPLOCOCCI
- ☐ PRESUMPTIVE BETA STREP GROUP A BY BACITRACIN

SOURCE OF SPECIMEN

- ☐ BLOOD
- ☐ CEREBROSPINAL FLUID
- ☐ FLUID (Specify Source) _____
- ☐ THROAT
- ☐ SPUTUM, expectorated
- ☐ OTHER Respiratory (Describe) _____
- ☐ URINE, Clean Catch Midstream
- ☐ URINE, Indwelling Catheter
- ☐ URINE, Straight Catheter
- ☐ URINE, Entire First Morning
- ☐ URINE, Other (Describe) _____
- ☐ STOOL
- ☑ GU (Specify Source) _vag._
- ☐ ABSCESS (Specify Source) _____
- ☐ TISSUE (Specify Source) _____
- ☐ ULCER (Specify Source) _____
- ☐ WOUND (Specify Source) _____
- ☐ STERILIZER TEST

TEST(S) REQUESTED

Bacterial
- ☐ **Routine culture;** Gram stain, anaerobic culture, susceptibility testing. Throats done for Gp A Strep.
- ☐ *Legionella* culture
- ☐ *Bartonella*
- ☐ Blood Culture

Other Non-Routine Cultures
- ☐ *E. coli* 0157:H7
- ☐ *Vibrio*
- ☐ *Yersinia*
- ☑ *H. ducreyi*
- ☐ *B. pertussis*
- ☐ Other _____

Screening Cultures
- ☑ Gonococci
- ☐ Group B Strep
- ☐ Group A Strep
- ☐ Other _____

- ☐ **ACID-FAST BACILLI**

- ☐ **FUNGAL**

VIRAL
- ☐ Routine culture
- ☐ Herpes simplex
- ☐ Direct FA for _____

PARASITOLOGY
- ☐ Exam for intestinal ova and parasites
- ☐ *Giardia* immunoassay
- ☐ *Cryptosporidium*
- ☐ Pinworm prep
- ☐ Blood parasites
- ☐ Filaria concentration
- ☐ *Trichomonas*
- ☐ Other _____

TOXIN ASSAY
- ☐ *Clostridium difficile*

DIRECT (Antigen Detection)
- ☐ Cryptococcal antigen-CSF only
- ☐ Bacterial antigens (Specify) _____

SPECIAL
- ☐ Antimicrobial tests (MIC)

Filled out by one person   Filled out by different person

FIGURE 10.7 A clinical microbiology lab report form. In health care, morphology and differential staining are important in determining the proper treatment for microbial diseases. A clinician completes the form to identify the sample and specific tests. In this case, a genitourinary sample will be examined for sexually transmitted diseases. The lab technician reported the Gram stain and culture results. [Minimal inhibitory concentration (MIC) of antibiotics will be discussed in Chapter 20, page 602.]

Q What diseases are suspected if the "acid-fast bacilli" box is checked?

either gram-positive or gram-negative. Other differential stains, such as the acid-fast stain, can be useful for a more limited group of microorganisms. Recall that these stains are based on the chemical composition of cell walls and therefore are not useful in identifying either the wall-less bacteria or the archaea with unusual walls. Microscopic examination of a Gram stain or an acid-fast stain is used to obtain information quickly in the clinical environment. A physician can sometimes get sufficient information from a technician's lab report to begin appropriate treatment (Figure 10.7 and the box in Chapter 21, page 633).

BIOCHEMICAL TESTS

Enzymatic activities are widely used to differentiate bacteria. Even closely related bacteria can usually be separated into distinct species by subjecting them to biochemical

MICROBIOLOGY IN THE NEWS

MASS DEATHS OF MARINE MAMMALS SPUR VETERINARY MICROBIOLOGY

Over the past decade thousands of marine mammals have died unexpectedly all over the world. These deaths occur in outbreaks of a dozen to thousands of mammals, and microbiologists try to determine the cause in each outbreak. In 2004, the deaths of 30 California sea lions were attributed to leptospirosis, and infectious diseases such as toxoplasmosis have been killing California sea otters in increasing numbers. In 2000, nearly a dozen gray whales died and floated ashore in San Francisco Bay. Pneumonia and encephalitis were diagnosed in some of the animals. The current decline in the Southern sea otter population is the result of a 40% mortality rate due to a variety of infectious bacterial diseases. These mortality figures raise concerns that entire populations of marine mammals may ultimately be destroyed. Moreover, alarmed swimmers have asked whether they are also in danger from marine mammal diseases.

Large numbers of opportunistic pathogens, including 55 species of *Vibrio*, were found in dolphins. These bacteria are a part of a dolphin's normal microbiota and the biota of coastal waters.

They can cause disease only if the animals' immune system, their normal defense against infection, has been weakened. To find the ultimate cause of death, scientists must find out what has weakened the immune systems.

One possibility is chemical pollutants. Insecticides and polychlorinated biphenyls (PCBs) have been found in dolphins, manatees, and elephant seals. Daniel Martineau of Cornell University says that PCBs are strong immunosuppressants. Another possibility is a viral infection that affects the immune system.

Phocid distemper virus in seals and the cetacean morbillivirus (CM) were responsible for the deaths of 20,000 marine mammals in European waters and for recurring mortality episodes in bottlenose dolphins along the Atlantic coast of the United States. Evidence suggests that pilot whales may be responsible for transferring the CM virus to other species across wide expanses of ocean.

Information Is Scarce

Such questions are the concern of veterinary microbiology, which until recently has been a neglected branch of medical microbiology. Although the diseases of such animals as cattle, chickens, and mink have been studied, partly because of their availability to researchers, the microbiology of wild animals, especially marine mammals, is a relatively newly emerging field. Gathering samples of animals that live in the open ocean and performing bacteriological analyses on them are very difficult. Currently, the animals being studied are those that live in captivity (see the photograph) and those that come onto the shore to breed, such as the northern fur sea lion.

The scientists are identifying bacteria in marine mammals by using conventional test batteries (see the figure) and genomic data on known species. The bacteria are compared with species described in *Bergey's Manual* to assign names or identify them. New species of bacteria are being found in marine mammals using the FISH technique.

Veterinary microbiologists hope that increased study of the microbiology of wild animals, including marine mammals, will not only promote improved wildlife management but also provide models for the study of human diseases.

---

tests, such as one to determine their ability to ferment an assortment of selected carbohydrates. For one example of the use of biochemical tests to identify bacteria (in this instance, in marine mammals), see the box above. Moreover, biochemical tests can provide insight into a species' niche in the ecosystem. For example, a bacterium that can fix nitrogen gas or oxidize elemental sulfur will provide important nutrients for plants and animals. This will be discussed in Chapter 27.

Enteric, gram-negative bacteria are a large heterogeneous group of microbes whose natural habitat is the intestinal tract of humans and other animals. This family contains several pathogens that cause diarrheal illness. A number of tests have been developed so that technicians can quickly identify the pathogen, a clinician can then provide appropriate treatment, and epidemiologists can

locate the source of an illness. All members of the family Enterobacteriaceae are oxidase-negative. Among the enteric bacteria are members of the genera *Escherichia*, *Enterobacter*, *Shigella*, *Citrobacter*, and *Salmonella*. *Escherichia*, *Enterobacter*, and *Citrobacter*, which ferment lactose to produce acid and gas, can be distinguished from *Salmonella* and *Shigella*, which do not. Further biochemical testing, as represented in Figure 10.8, can differentiate among the genera.

The time needed to identify bacteria can be reduced considerably by the use of selective and differential media or by rapid identification methods. Recall from Chapter 6 (page 171) that selective media contain ingredients that suppress the growth of competing organisms and encourage the growth of desired ones, and that differential media allow the desired organism to form a colony that is somehow distinctive.

MICROBIOLOGY IN THE NEWS

(continued)

Marine mammal researchers examine a Pacific bottlenosed dolphin.

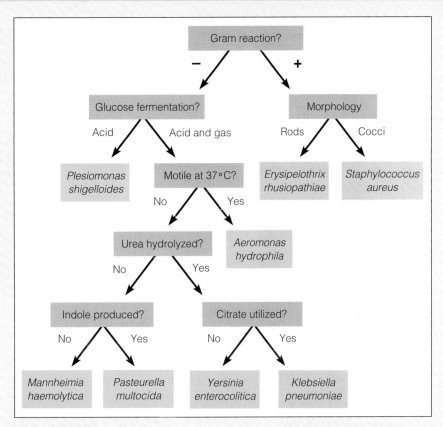

Biochemical tests used to identify selected species of human pathogens isolated from marine mammals.

Q **Assume you isolated a gram-negative rod that produces gas from glucose, is urease negative, citrate negative, and indole positive. What is the bacterium?**

Bergey's Manual does not evaluate the relative importance of each biochemical test and does not always describe strains. In the clinical diagnosis of a disease, a particular species and even a particular strain must be identified in order to proceed with proper treatment. To this end, specific series of biochemical tests have been developed for fast identification in hospital laboratories. Rapid biochemical systems have been developed for yeasts and other fungi, as well as bacteria.

Rapid identification tools are manufactured for groups of medically important bacteria, such as the enterics. Such tools are designed to perform several biochemical tests simultaneously and can identify bacteria within 4–24 hours. This is sometimes called **numerical**

identification because the results of each test are assigned a number. In the simplest form, a positive test would be assigned a value of 1, and a negative is assigned a value of 0. In most commercial testing kits, test results are assigned numbers ranging from 1 to 4 that are based on the relative reliability and importance of each test, and the resulting total is compared to a database of known organisms.

In the example shown in Figure 10.9,

➊ an unknown enteric bacterium is inoculated into a tube designed to perform 15 biochemical tests.

➋ After incubation, results in each compartment are observed.

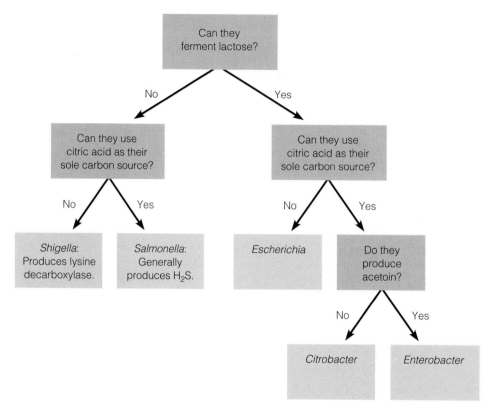

FIGURE 10.8 The use of metabolic characteristics to identify selected genera of enteric bacteria.

Q   Assume you have a gram-negative bacterium that produces acid from lactose and cannot use citric acid as its sole carbon source. What is the bacterium?

❸ Positive results are marked on the scoring form. Notice that each test is assigned a value; the number derived from scoring all the tests is called the ID value. Fermentation of glucose is important, and a positive reaction is valued at 2, compared with the production of acetoin (V–P test, or the Voges-ProsKauer test), which has no value.

❹ A computerized interpretation of the simultaneous test results is essential and is provided by the manufacturer. In this example, the test results indicate that the bacterium is *Enterobacter cloacae* (klō-ā′kē), with the typical test results. A limitation of biochemical testing is that mutations and plasmid acquisition can result in strains with different characteristics. Unless a large number of tests are used, an organism could be incorrectly identified.

SEROLOGY

LEARNING OBJECTIVES

• Differentiate Western blotting from Southern blotting.

• Explain how serological tests and phage typing can be used to identify an unknown bacterium.

Serology is the science that studies serum and immune responses that are evident in serum (see Chapter 18). Microorganisms are antigenic; that is, microorganisms that enter an animal's body stimulate it to form antibodies. Antibodies are proteins that circulate in the blood and combine in a highly specific way with the bacteria that caused their production. For example, the immune system of a rabbit injected with killed typhoid bacteria (antigens) responds by producing antibodies against typhoid bacteria. Solutions of such antibodies used in the identification of many medically important microorganisms are commercially available; such a solution is called an **antiserum** (plural: *antisera*). If an unknown bacterium is isolated from a patient, it can be tested against known antisera and often identified quickly.

In a procedure called a **slide agglutination test,** samples of an unknown bacterium are placed in a drop of saline on each of several slides. Then a different known antiserum is added to each sample. The bacteria agglutinate (clump) when mixed with antibodies that were produced in response to that species or strain of bacterium; a positive test is indicated by the presence of agglutination. Positive and negative slide agglutination tests are shown in Figure 10.10.

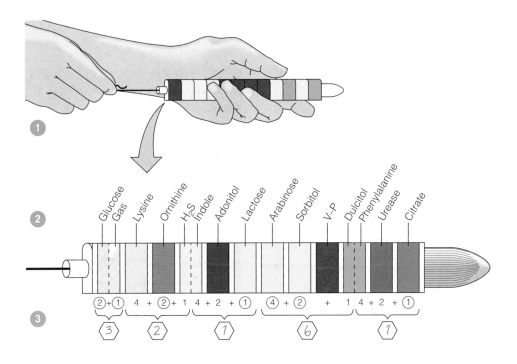

ID Value	Organism	Atypical Test Results	Confirmatory Test
32143	*Enterobacter cloacae*	Sorbitol⁻	–
	Enterobacter sakazakii	Urea⁺	+
32161	*Enterobacter cloacae*	None	V–P⁺
32162	*Enterobacter cloacae*	Citrate⁻	

FIGURE 10.9 One type of rapid identification method for bacteria: Enterotube II from Becton Dickinson. ❶ One tube containing media for 15 tests is inoculated. ❷ After incubation, the tube is observed for results. ❸ The value for each positive test is circled, and the numbers from each group of tests are added to give each portion of the ID value. ❹ Comparing the resultant ID value with a computerized listing shows that the organism in the tube is *Enterobacter cloacae*. This example shows results for a typical strain of *E. cloacae*; however, other strains may produce different test results, which are listed in the Atypical Test Results column. The V–P test is used to confirm an identification.

Q **How can one species have two different ID values?**

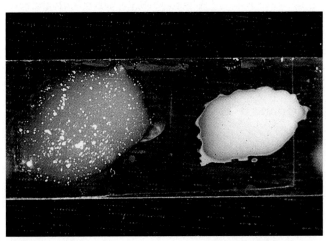

(a) Positive test   **(b)** Negative test

FIGURE 10.10 A slide agglutination test. (a) In a positive test, the grainy appearance is due to the clumping (agglutination) of the bacteria. **(b)** In a negative test, the bacteria are still evenly distributed in the saline and antiserum.

Q Agglutination results when the bacteria are mixed with _____.

Serological testing can differentiate not only among microbial species, but also among strains within species. Strains with different antigens are called **serotypes, serovars,** or **biovars.** See the discussion of *Escherichia* and *Salmonella* serovars on page 323. As mentioned in Chapter 1, Rebecca Lancefield was able to classify

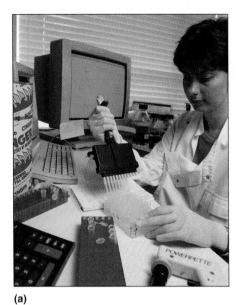

(a)

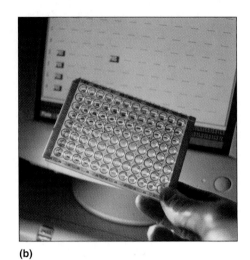

(b)

FIGURE 10.11 Using a computer scanner to read the results of an ELISA test. (**a**) A technician uses a micropipette to add samples to a microplate for an ELISA. (**b**) ELISA results are then read by the computer scanner.

Q What are the similarities between the slide agglutination test and the ELISA test?

serotypes of streptococci by studying serological reactions. She found that the different antigens in the cell walls of various serotypes of streptococci stimulate the formation of different antibodies. In contrast, because closely related bacteria also produce some of the same antigens, serological testing can be used to screen bacterial isolates for possible similarities. If an antiserum reacts with proteins from different bacterial species or strains, these bacteria can be tested further for relatedness.

Serological testing was used to determine whether the increase in number of cases of necrotizing fasciitis in the United States and England since 1987 was due to a common source of the infections. No common source was located, but there has been an increase in two serotypes of *Streptococcus pyogenes* that have been dubbed the flesh-eating bacteria.

A test called the **enzyme-linked immunosorbent assay (ELISA)** is widely used because it is fast and can be read by a computer scanner (Figure 10.11; see Figure 18.14, page 545). In a direct ELISA, known antibodies are placed in (and adhere to) the wells of a microplate, and an unknown type of bacterium is added to each well. A reaction between the known antibodies and the bacteria provides identification of the bacteria. An ELISA is used in AIDS testing to detect the presence of antibodies against human immunodeficiency virus (HIV), the virus that causes AIDS (see Figure 19.12, page 568).

Another serological test, **Western blotting,** is also used to identify antibodies in a patient's serum (Figure 10.12). HIV infection is confirmed by Western blotting, and Lyme disease, caused by *Borrelia burgdorferi,* is often diagnosed by the Western blot.

1 Proteins from a known bacterium or virus are separated by an electric current in electrophoresis. The current causes the proteins to separate according to their molecular weights and charges (see Chapter 9, page 271).

2 The proteins are then transferred to a filter by blotting.

3 Patient's serum is washed over the filter. If the patient has antibodies to one of the proteins in the filter (in this case, *Borrelia* proteins), the antibodies and protein will combine. Anti-human serum linked to an enzyme is then washed over the filter.

4 This will be made visible as a colored band on the filter after addition of the enzyme's substrate.

PHAGE TYPING

Like serological testing, phage typing looks for similarities among bacteria. Both techniques are useful in tracing the origin and course of a disease outbreak. **Phage typing** is a test for determining which phages a bacterium

1 If Lyme disease is suspected in a patient: Electrophoresis is used to separate *B. burgdorferi* proteins in the serum. Proteins move at different rates based on their charge and size when the gel is exposed to an electric current.

2 The bands are transferred to a nitrocellulose filter by blotting. Each band consists of many molecules of a particular protein (antigen). The bands are not visible at this point.

3 The proteins (antigens) are positioned on the filter exactly as they were on the gel. The filter is then washed with patient's serum followed by anti-human antibodies tagged with an enzyme. The patient antibodies that combine with their specific antigen are visible (shown here in red) when the enzyme's substrate is added.

4 The test is read. If the tagged antibodies stick to the filter, evidence of the presence of the microorganism in question—in this case, *Borrelia burgdorferi*—has been found in the patient's serum.

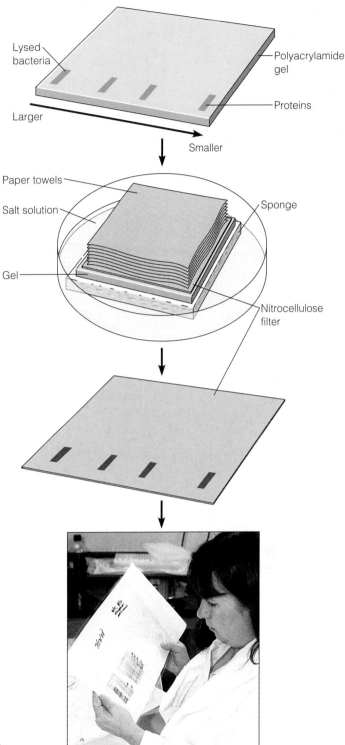

FIGURE 10.12 The Western blot.
Proteins separated by electrophoresis can be detected by their reactions with antibodies.

Q **Name two diseases that may be diagnosed by Western blotting.**

is susceptible to. Recall from Chapter 8 (page 244) that bacteriophages (phages) are bacterial viruses and that they usually cause lysis of the bacterial cells they infect. They are highly specialized, in that they usually infect only members of a particular species, or even particular strains within a species. One bacterial strain might be susceptible to two different phages, whereas another strain of the same species might be susceptible to those two phages plus a third phage. Bacteriophages will be discussed further in Chapter 13.

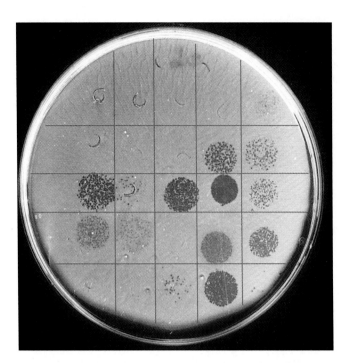

FIGURE 10.13 Phage typing of a strain of *Salmonella enterica.* The tested strain was grown over the entire plate. Plaques, or areas of lysis, were produced by bacteriophages, indicating that the strain was sensitive to infection by these phages. Phage typing is used to distinguish *S. enterica* serotypes and *Staphylococcus aureus* types.

Q   What is being identified in phage typing?

The sources of food-associated infections can be traced by phage typing. One version of this procedure starts with a plate totally covered with bacteria growing on agar. A drop of each different phage type to be used in the test is then placed on the bacteria. Wherever the phages are able to infect and lyse the bacterial cells, clearings in the bacterial growth (called plaques) appear (Figure 10.13). Such a test might show, for instance, that bacteria isolated from a surgical wound have the same pattern of phage sensitivity as those isolated from the operating surgeon or surgical nurses. This establishes that the surgeon or a nurse is the source of infection.

FATTY ACID PROFILES

Bacteria synthesize a wide variety of fatty acids, and in general, these fatty acids are constant for a particular species. Commercial systems have been designed to separate cellular fatty acids to compare them to fatty acid profiles of known organisms. Fatty acid profiles, called **FAME** (*f*atty *a*cid *m*ethyl *e*ster), are widely used in clinical and public health laboratories.

FLOW CYTOMETRY

Flow cytometry can be used to identify bacteria in a sample without culturing the bacteria. In a *flow cytometer*, a moving fluid containing bacteria is forced through a small opening (see Figure 18.12, page 543). The simplest method detects the presence of bacteria by detecting the difference in electrical conductivity between cells and the surrounding medium. If the fluid passing through the opening is illuminated by a laser, the scattering of light provides information about the cell size, shape, density, and surface, which is analyzed by a computer. Fluorescence can be used to detect naturally fluorescent cells, such as *Pseudomonas*, or cells tagged with fluorescent dyes.

Milk can be a vehicle for disease transmission. A proposed test that uses flow cytometry to detect *Listeria* in milk could save time because the bacteria would not need to be cultured for identification. Antibodies against *Listeria* can be labeled with a fluorescent dye and added to the milk to be tested. The milk is passed through the flow cytometer, which records the fluorescence of the antibody-labeled cells.

DNA BASE COMPOSITION

LEARNING OBJECTIVE

- Describe how a newly discovered microbe can be classified by: DNA base composition, DNA fingerprinting, and PCR.

Taxonomists can use an organism's **DNA base composition** to draw conclusions about relatedness. This base composition is usually expressed as the percentage of guanine plus cytosine (G + C). The base composition of a single species is theoretically a fixed property; thus, a comparison of the G + C content in different species can reveal the degree of species relatedness. As we saw in Chapter 8, each guanine (G) in DNA has a complementary cytosine (C). Similarly, each adenine (A) in the DNA has a complementary thymine (T). Therefore, the percentage of DNA bases that are GC pairs also tells us the percentage that are AT pairs (GC + AT = 100%). Two organisms that are closely related and hence have many identical or similar genes will have similar amounts of the various bases in their DNA. However, if there is a difference of more than 10% in their percentage of GC pairs (for example, if one bacterium's DNA contains 40% GC and another bacterium has 60% GC), then these two organisms are probably not related. Of course, two organisms that have the same percentage of

GC are not necessarily closely related; other supporting data are needed to draw conclusions about their phylogenetic relationship.

DNA FINGERPRINTING

Actually determining the entire sequence of bases in an organism's DNA is now possible with modern biochemical methods, but this is currently impractical for laboratory identification because of the great amount of time required. However, the use of restriction enzymes enables researchers to compare the base sequences of different organisms. Restriction enzymes cut a molecule of DNA everywhere a specific base sequence occurs, producing restriction fragments (as discussed in Chapter 9, page 256). For example, the enzyme *Eco*RI cuts DNA at the arrows in the sequence

$$...\text{G}^{\downarrow}\text{A A T T C}...$$

$$...\text{C T T A A}_{\uparrow}\text{G}...$$

In this technique, the DNA from two microorganisms is treated with the same restriction enzyme, and the restriction fragments (RFLPs) produced are separated by electrophoresis (see Figure 9.17, page 273). A comparison of the number and sizes of restriction fragments that are

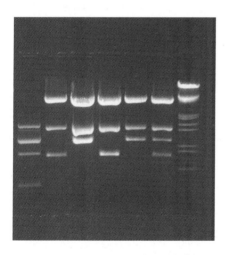

FIGURE 10.14 DNA fingerprints. Plasmids from seven different bacteria were digested with the same restriction enzyme. Each digest was put in a different well (origin) in the agarose gel. An electrical current was then applied to the gel to separate the fragments by size and electrical charge. The DNA was made visible by staining with ethidium bromide, which fluoresces under ultraviolet light. Comparison of the lanes shows that none of the DNA samples (and therefore none of the bacteria) is identical.

Q **What is an RFLP?**

produced from different organisms provides information about their genetic similarities and differences; the more similar the patterns, or DNA *fingerprints*, the more closely related the organisms are expected to be (Figure 10.14).

DNA fingerprinting is used to determine the source of hospital-acquired infections. In one hospital, patients undergoing coronary-bypass surgery developed infections caused by *Rhodococcus bronchialis* (rō-dō-kok′kus bron-kē′al-is). The DNA fingerprints of the patients' bacteria and the bacteria of one nurse were identical. The hospital was thus able to break the chain of transmission of this infection by encouraging this nurse to use aseptic technique. DNA fingerprinting to locate the source of tomato-associated diarrhea is described in the box in Chapter 25 on page 756.

THE POLYMERASE CHAIN REACTION

When a microorganism cannot be cultured by conventional methods, the causative agent of an infectious disease might not be recognized. However, a technique called the **polymerase chain reaction (PCR)** can be used to increase the amount of microbial DNA to levels that can be tested by gel electrophoresis (see Chapter 9, page 258). If a primer for a specific microorganism is used, the presence of amplified DNA indicates that microorganism is present.

In 1992, researchers used PCR to determine the causative agent of Whipple's disease, which was previously an unknown bacterium now named *Tropheryma whippelii* (trō′fer-ē-mä whip-pel′ē-ē). Whipple's disease was first described in 1907 by George Whipple as a gastrointestinal and nervous system disorder caused by an unknown bacillus. No one has been able to culture the bacterium to identify it, and thus PCR provides the only reliable methods of diagnosing and treating the disease.

In recent years, PCR made possible several discoveries. For example, in 1992, Raul Cano used PCR to amplify DNA from *Bacillus* bacteria in amber that was 25–40 million years old. These primers were made from rRNA sequences in living *B. circulans* to amplify DNA coding for rRNA in the amber. These primers will cause amplification of DNA from other *Bacillus* species but do not cause amplification of DNA from other bacteria that might have been present, such as *Escherichia* or *Pseudomonas*. The DNA was sequenced after amplification. This information was used to determine the relationships between the ancient bacteria and modern bacteria.

In 1993, microbiologists identified a *Hantavirus* as the cause of an outbreak of hemorrhagic fever in the American Southwest using PCR. The identification was made in record time—less than 2 weeks. PCR was used in 1994 to identify the causative agent of a new tickborne disease (human granulocytic ehrlichiosis) as the bacterium *Ehrlichia chaffeensis* (ėr'lik-ē-ä chaf'ē-en-sis)(page 687). PCR is used to identify the source of rabies viruses, see the box in Chapter 22 (page 657).

TaqMan is a commercial system that uses PCR to identify pathogenic *E. coli* in food and water. With this system, the newly amplified *E. coli* DNA fluoresces and can be detected using gel electrophoresis.

NUCLEIC ACID HYBRIDIZATION

LEARNING OBJECTIVE

- Describe how microorganisms can be identified by nucleic acid hybridization, Southern blotting, DNA chips, ribotyping, and FISH.

If a double-stranded molecule of DNA is subjected to heat, the complementary strands will separate as the hydrogen bonds between the bases break. If the single strands are then cooled slowly, they will reunite to form a double-stranded molecule identical to the original double strand.

(This reunion occurs because the single strands have complementary sequences.) When this technique is applied to separated DNA strands from two different organisms, it is possible to determine the extent of similarity between the base sequences of the two organisms. This method is known as **nucleic acid hybridization.** The procedure assumes that if two species are similar or related, a major portion of their nucleic acid sequences will also be similar. The procedure measures the ability of DNA strands from one organism to hybridize (bind through complementary base pairing) with the DNA strands of another organism (Figure 10.15). The greater the degree of hybridization, the greater the degree of relatedness.

Similar hybridization reactions can occur between any single-stranded nucleic acid chain: DNA-DNA, RNA-RNA, DNA-RNA. An RNA transcript will hybridize with the separated template DNA to form a DNA-RNA hybrid molecule. Nucleic acid hybridization reactions are the basis of several techniques (described below) that are used to detect the presence of microorganisms and to identify unknown organisms.

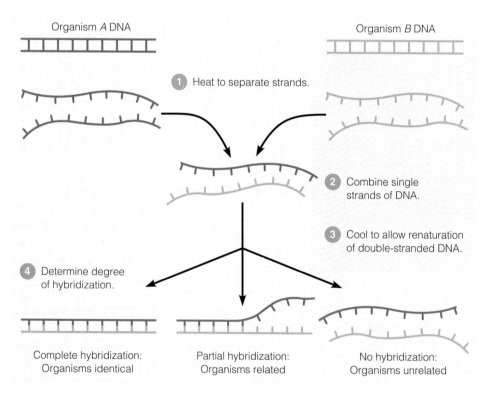

Organism *A* DNA   Organism *B* DNA

1 Heat to separate strands.

2 Combine single strands of DNA.

3 Cool to allow renaturation of double-stranded DNA.

4 Determine degree of hybridization.

Complete hybridization:
Organisms identical

Partial hybridization:
Organisms related

No hybridization:
Organisms unrelated

FIGURE 10.15 DNA-DNA hybridization. The greater the amount of pairing between DNA strands from different organisms (hybridization), the more closely the organisms are related.

Q **What is the principle involved in DNA probes?**

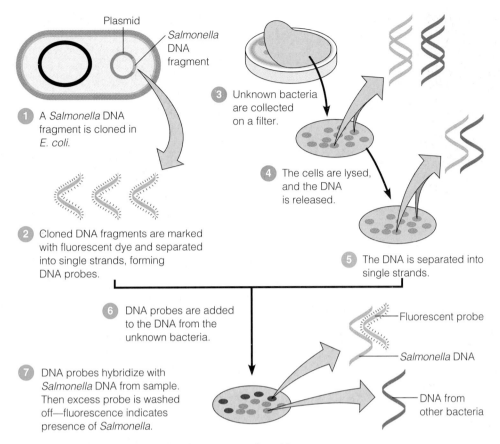

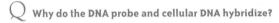

FIGURE 10.16 A DNA probe used to identify bacteria. Southern blotting is used to detect specific DNA. This modification of the Southern blot is used to detect *Salmonella*.

Q **Why do the DNA probe and cellular DNA hybridize?**

SOUTHERN BLOTTING

Nucleic acid hybridization can be used to identify unknown microorganisms by **Southern blotting** (see Figure 9.16, page 272). In addition, rapid identification methods using **DNA probes** are being developed. One method involves breaking DNA extracted from *Salmonella* into fragments with a restriction enzyme, then selecting a specific fragment as the probe for *Salmonella* (Figure 10.16). This fragment must be able to hybridize with the DNA of all *Salmonella* strains, but not with the DNA of closely related enteric bacteria.

DNA CHIPS

An exciting new technology is the **DNA chip,** which will make it possible to quickly detect a pathogen in a host by identifying a gene that is unique to that pathogen (Figure 10.17). The DNA chip is composed of DNA probes. A sample containing DNA from an unknown organism is labeled with a fluorescent dye and added to the chip.

Hybridization between the probe DNA and DNA in the sample is detected by fluoresence.

RIBOTYPING AND RIBOSOMAL RNA SEQUENCING

Ribotyping is currently being used to determine the phylogenetic relationships among organisms. There are several advantages to using rRNA. First, all cells contain ribosomes. Second, RNA genes have undergone few changes over time so all members of a domain, phylum, and, in some cases, a genus, have the same "signature" sequences in their rRNA. The rRNA used most often is a component of the smaller portion of ribosomes. A third advantage of rRNA sequencing is that cells do not have to be cultured in the laboratory.

DNA can be amplified by PCR using an rRNA primer for specific signature sequences. The amplified fragments are subsequently cut with one or more restriction enzymes and separated by electrophoresis. The resulting band patterns can then be compared. Then the rRNA genes in

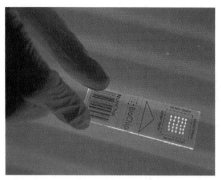

(a) A DNA chip can be manufactured to contain hundreds of thousands of synthetic single-stranded DNA sequences. Assume that each DNA sequence was unique to a different bacterial species.

(b) Unknown DNA from a patient is separated into single strands, enzymatically cut, and labeled with a fluorescent dye.

(c) The unknown DNA is inserted into the chip and allowed to hybridize with the DNA on the chip.

(d) The tagged DNA will bind only to the complementary DNA on the chip. The bound DNA will be detected by its fluorescent dye and analyzed by a computer. The red light is a gene expressed in normal cells; green is a mutated gene expressed in tumor cells; and yellow, in both cells.

FIGURE 10.17 DNA chip technology. DNA chip technology will enable much more rapid, inexpensive analysis using many more probes. Chip technology is new; however, scientists soon hope to use DNA chips to detect microbes in a human or environmental sample, to identify cancer genes, to identify potential suspects in a crime or victims of a crime or catastrophe, to identify endangered species, and to match organ donors. If the DNA chip in (**d**) was constructed from different bacterial DNA, each spot would represent a bacterium, and the spot giving off fluorescent light would identify the bacterium in the sample.

Q **What is on the chip to make it specific for a particular microorganism?**

the amplified fragments can be sequenced to determine evolutionary relationships between organisms. This technique is useful for classifying a newly discovered organism to domain or phylum or to determine the general types of organisms present in one environment. More specific probes (discussed below) are needed to identify individual species, however.

FLUORESCENT IN SITU HYBRIDIZATION (FISH)

Fluorescent dye-labeled RNA or DNA probes are used to specifically stain microorganisms in place, or in situ. This technique is called **fluorescent in situ hybridization,** or **FISH.** Cells are treated so the probe enters the cells and reacts with target ribosome in the cell (in situ). FISH is used to determine the identity, abundance, and relative activity of microorganisms in an environment and can be used to detect bacteria that have not yet been cultured. Using FISH, a tiny bacterium, *Pelagibacter* (pel-aj′ē-bak-tėr), was discovered in the ocean and determined to be related to the rickettsias (page 341). As probes are developed, FISH can be used to detect bacteria in drinking water or bacteria in a patient without the normal 24-hour or longer wait required for culturing the bacteria (Figure 10.18).

PUTTING CLASSIFICATION METHODS TOGETHER

LEARNING OBJECTIVE
• Differentiate a dichotomous key from a cladogram.

Morphological characteristics, differential staining, and biochemical testing were the only identification tools available just a few years ago. Technological advancements are making it possible to use nucleic acid analysis techniques, once reserved for classification, for routine

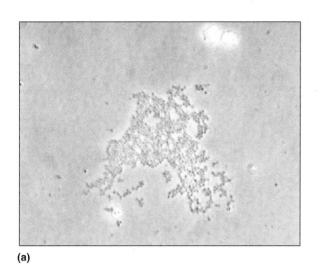

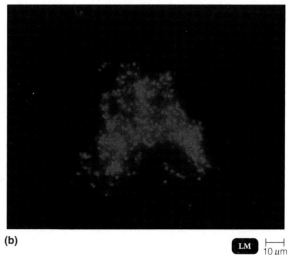

(a)   **(b)**

LM ⊢———⊣ 10 μm

FIGURE 10.18   FISH, or fluorescent in situ hybridization. A DNA or RNA
probe attached to fluorescent dyes is used to identify chromosomes. Bacteria seen with
phase-contrast microscopy (**a**) are identified with a fluorescent-labeled probe that hybridizes with a specific sequence of DNA in *Staphylococcus aureus* (**b**).

Q **What is stained using the FISH technique?**

identification. A summary of taxonomic criteria and methods is provided in Table 10.5.

Information about microbes obtained by these methods is used to identify and classify the organisms. Two
methods of using the information are described below.

DICHOTOMOUS KEYS

Dichotomous keys are widely used for identification. In
a dichotomous key, identification is based on successive
questions, and each question has two possible answers (dichotomous means cut in two). After answering one question, the investigator is directed to another question until
an organism is identified. Although these keys often have
little to do with phylogenetic relationships, they are invaluable for identification. For example, a dichotomous
key for bacteria could begin with an easily determined
characteristic, such as cell shape, and move on to the ability to ferment a sugar. Dichotomous keys are shown in
Figure 10.8 and in the box on page 294.

CLADOGRAMS

Cladograms are maps that show evolutionary relationships among organisms (*clado-* means branch). Cladograms
are shown in Figures 10.1 and 10.6. Each branch point on
the cladogram is defined by a feature shared by various
species on that branch. Historically, cladograms for vertebrates were made using fossil evidence; however, rRNA sequences are now being used to confirm assumptions based

TABLE 10.5	Taxonomic Criteria and Methods for Classifying and Identifying Bacteria	
Criterion or Method	Used for	
	Classification	Identification
Morphological characteristics	No (yes for cyanobacteria)	Yes
Differential Staining	Yes (for cell wall type)	Yes
Biochemical Testing	No	Yes
Serology	No	Yes
Phage Typing	No	Yes
Fatty Acid Profiles	No	Yes
Flow Cytometry	No	Yes
DNA Base Composition	Yes	No
DNA Fingerprinting	Yes	Yes
PCR	Yes	Yes
Nucleic Acid Hybridization Techniques	Yes	Yes
rRNA Sequencing	Yes	No

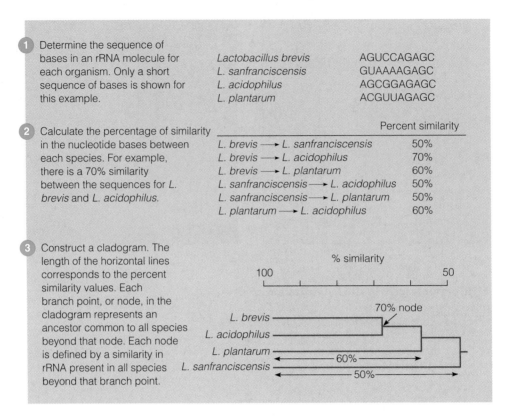

① Determine the sequence of bases in an rRNA molecule for each organism. Only a short sequence of bases is shown for this example.

Lactobacillus brevis	AGUCCAGAGC
L. sanfranciscensis	GUAAAAGAGC
L. acidophilus	AGCGGAGAGC
L. plantarum	ACGUUAGAGC

② Calculate the percentage of similarity in the nucleotide bases between each species. For example, there is a 70% similarity between the sequences for *L. brevis* and *L. acidophilus*.

Percent similarity

L. brevis ⟶ *L. sanfranciscensis*	50%
L. brevis ⟶ *L. acidophilus*	70%
L. brevis ⟶ *L. plantarum*	60%
L. sanfranciscensis ⟶ *L. acidophilus*	50%
L. sanfranciscensis ⟶ *L. plantarum*	50%
L. plantarum ⟶ *L. acidophilus*	60%

③ Construct a cladogram. The length of the horizontal lines corresponds to the percent similarity values. Each branch point, or node, in the cladogram represents an ancestor common to all species beyond that node. Each node is defined by a similarity in rRNA present in all species beyond that branch point.

FIGURE 10.19   Building a cladogram.

Q Why do *L. brevis* and *L. acidophilus* branch from the same node?

on fossils. As we said earlier, most microorganisms do not leave fossils; therefore, rRNA sequencing is primarily used to make cladograms for microorganisms. The small rRNA subunit used has 1500 bases, and computer programs do the calculations. The steps for constructing a cladogram are shown in Figure 10.19.

① Two rRNA sequences are aligned, and

② the percentage of similarity between the sequences is calculated.

③ Then the horizontal branches are drawn in a length proportional to the calculated percent similarity. All species beyond a node (branch point) have similar rRNA sequences, suggesting that they arose from an ancestor at that node. ✳ **Animation: Go to The Microbiology Place website or CD-ROM and click "Animations" to view Dichotomous Keys.**

STUDY OUTLINE

INTRODUCTION (p. 282)

1. Taxonomy is the science of the classification of organisms, with the goal of showing relationships among organisms.

2. Taxonomy also provides a means of identifying organisms.

THE STUDY OF PHYLOGENETIC RELATIONSHIPS (pp. 283–287)

1. Phylogeny is the evolutionary history of a group of organisms.

2. The taxonomic hierarchy shows evolutionary, or phylogenetic, relationships among organisms.

3. Bacteria were separated into the Kingdom Prokaryotae in 1968.
4. Living organisms were divided into five kingdoms in 1969.

THE THREE DOMAINS (pp. 283–286)

5. Living organisms are currently classified into three domains. A domain can be divided into kingdoms.
6. In this system, plants, animals, fungi, and protists belong to the Domain Eukarya.
7. Bacteria (with peptidoglycan) form a second domain.
8. Archaea (with unusual cell walls) are placed in the Domain Archaea.

A PHYLOGENETIC HIERARCHY (pp. 286–287)

9. Organisms are grouped into taxa according to phylogenetic relationships (from a common ancestor).
10. Some of the information for eukaryotic relationships is obtained from the fossil record.
11. Prokaryotic relationships are determined by rRNA sequencing.

CLASSIFICATION OF ORGANISMS
(pp. 287–292)

SCIENTIFIC NOMENCLATURE (pp. 287–288)

1. According to scientific nomenclature, each organism is assigned two names, or a binomial: a genus and a specific epithet, or species.
2. Rules for the assignment of names to bacteria are established by the International Committee on Systematic Bacteriology.
3. Rules for naming fungi and algae are published in the *International Code of Botanical Nomenclature*.
4. Rules for naming protozoa are found in the *International Code of Zoological Nomenclature*.

THE TAXONOMIC HIERARCHY (pp. 288–289)

5. A eukaryotic species is a group of organisms that interbreeds with each other but does not breed with individuals of another species.
6. Similar species are grouped into a genus; similar genera are grouped into a family; families, into an order; orders, into a class; classes, into a phylum; phyla, into a kingdom; and kingdoms, into a domain.

CLASSIFICATION OF PROKARYOTES (p. 289)

7. *Bergey's Manual of Systematic Bacteriology* is the standard reference on bacterial classification.
8. A group of bacteria derived from a single cell is called a strain.
9. Closely related strains constitute a bacterial species.

CLASSIFICATION OF EUKARYOTES (p. 289)

10. Eukaryotic organisms may be classified into the Kingdom Fungi, Plantae, or Animalia.
11. Protists are mostly unicellular organisms; these organisms are currently being assigned to kingdoms.
12. Fungi are absorptive chemoheterotrophs that develop from spores.
13. Multicellular photoautotrophs are placed in the Kingdom Plantae.
14. Multicellular ingestive heterotrophs are classified as Animalia.

CLASSIFICATION OF VIRUSES (pp. 289–292)

15. Viruses are not placed in a kingdom. They are not composed of cells and cannot grow without a host cell.
16. A viral species is a population of viruses with similar characteristics that occupies a particular ecological niche.

METHODS OF CLASSIFYING AND IDENTIFYING MICROORGANISMS
(pp. 292–306)

1. *Bergey's Manual of Determinative Bacteriology* is the standard reference for laboratory identification of bacteria.
2. Morphological characteristics are useful in identifying microorganisms, especially when aided by differential staining techniques.
3. The presence of various enzymes, as determined by biochemical tests, is used in identifying microorganisms.
4. Serological tests, involving the reactions of microorganisms with specific antibodies, are useful in determining the identity of strains and species, as well as relationships among organisms. ELISA and Western blotting are examples of serological tests.
5. Phage typing is the identification of bacterial species and strains by determining their susceptibility to various phages.
6. Fatty acid profiles can be used to identify some organisms.
7. Flow cytometry measures physical and chemical characteristics of cells.
8. The percentage of GC base pairs in the nucleic acid of cells can be used in the classification of organisms.
9. The number and sizes of DNA fragments, or DNA fingerprints, produced by restriction enzymes are used to determine genetic similarities.
10. The polymerase chain reaction (PCR) can be used to amplify a small amount of microbial DNA in a sample. The presence or identification of an organism is indicated by amplified DNA.
11. Single strands of DNA, or of DNA and RNA, from related organisms will hydrogen-bond to form a double-stranded molecule; this bonding is called nucleic acid hybridization.

12. Southern blotting, DNA chips, and FISH are examples of nucleic acid hybridization techniques.

13. The sequence of bases in ribosomal RNA can be used in the classification of organisms.

14. Dichotomous keys are used for the identification of organisms. Cladograms show phylogenetic relationships among organisms. ⁂ **Animation: Dichotomous Keys. The Microbiology Place.**

STUDY QUESTIONS

Access more review material either online at **The Microbiology Place** (www.microbiologyplace.com) or with **The Microbiology Place CD-ROM** packaged with your new book. There you'll find activities, practice tests, quizzes, flashcards, case studies, and more to help you succeed.

Answers to the Study Questions can be found in Appendix G.

REVIEW

1. What is taxonomy?

2. Discuss the evidence that supports classifying organisms into three domains.

3. Explain why a two-, three-, or five-kingdom system is no longer acceptable for classification.

4. List and define the three kingdoms of multicellular eukaryotic organisms.

5. Compare and contrast the following terms:
   a. *archaea* and *bacteria*
   b. *bacteria* and *eukarya*
   c. *archaea* and *eukarya*

6. What is binomial nomenclature?

7. Why is binomial nomenclature preferable to the use of common names?

8. Using *Escherichia coli* and *Entamoeba coli* as examples, explain why the genus name must always be written out the first time you use it in a report.

9. Put the following terms in the correct sequence, from the most general to the most specific: order, class, genus, domain, species, phylum, family.

10. Find the gram-positive bacteria *Staphylococcus* in Appendix A. To what domain, phylum, class, order, and family does it belong? To which bacteria is it most closely related: *Gemella* or *Streptococcus*?

11. Define the following terms: *eukaryotic species*, *bacterial species*, and *viral species*.

12. List the twelve taxonomic criteria and methods discussed in this chapter for the classification of microorganisms. Separate your list into those tests used primarily for taxonomic classification and those used primarily for identifying microorganisms already classified.

13. Higher organisms are arranged into taxonomic groups on the basis of evolutionary relationships. Why is this type of classification only now being developed for bacteria?

14. Can you tell which of the following organisms are most closely related? Are any two the same species? On what did you base your answer?

Characteristic	A	B	C	D
Morphology	Rod	Coccus	Rod	Rod
Gram Reaction	+	−	−	+
Glucose Utilization	Fermentative	Oxidative	Fermentative	Fermentative
Cytochrome Oxidase	Present	Present	Absent	Absent
GC Moles %	48–52	23–40	50–54	49–53

MULTIPLE CHOICE

1. *Bergey's Manual of Systematic Bacteriology* differs from *Bergey's Manual of Determinative Bacteriology* in that the former
   a. groups bacteria into species.
   b. groups bacteria according to phylogenetic relationships.
   c. groups bacteria according to pathogenic properties.
   d. groups bacteria into 19 species.
   e. all of the above

2. *Bacillus* and *Lactobacillus* are not in the same order. This indicates that which one of the following is *not* sufficient to assign an organism to a taxon?
   a. biochemical characteristics
   b. amino acid sequencing
   c. phage typing
   d. serology
   e. morphological characteristics

3. Which of the following is used to classify organisms into the Kingdom Fungi?
   a. ability to photosynthesize; possess a cell wall
   b. unicellular; possess cell wall; prokaryotic
   c. unicellular; lacking cell wall; eukaryotic
   d. absorptive; possess cell wall; eukaryotic
   e. ingestive; lacking cell wall; multicellular; prokaryotic

4. Which of the following is *not* true about scientific nomenclature?
   a. Each name is specific.
   b. Names vary with geographical location.
   c. The names are standardized.
   d. Each name consists of a genus and specific epithet.
   e. It was first designed by Linnaeus.

5. You could identify an unknown bacterium by all of the following *except*
   a. hybridizing a DNA probe from a known bacterium with the unknown's DNA.
   b. making a fatty acid profile of the unknown.
   c. specific antiserum agglutinating the unknown.
   d. ribosomal RNA sequencing.
   e. percentage of guanine + cytosine.

6. The wall-less mycoplasmas are considered to be related to gram-positive bacteria. Which of the following would provide the most compelling evidence for this?
   a. They share common rRNA sequences.
   b. Some gram-positive bacteria and some mycoplasmas produce catalase.
   c. Both groups are prokaryotic.
   d. Some gram-positive bacteria and some mycoplasmas have coccus-shaped cells.
   e. Both groups contain human pathogens.

Use the following choices to answer questions 7 and 8.
   a. Animalia
   b. Fungi
   c. Plantae
   d. Firmicutes (gram-positive bacteria)
   e. Proteobacteria (gram-negative bacteria)

7. Into which group would you place a multicellular organism that has a mouth and lives inside the human liver?

8. Into which group would you place a photosynthetic organism that lacks a nucleus and has a thin peptidoglycan wall surrounded by an outer membrane?

Use the following choices to answer questions 9 and 10.
   1. 9 + 2 flagella
   2. 70S ribosome
   3. fimbria
   4. nucleus
   5. peptidoglycan
   6. plasma membrane

9. Which is (are) found in all three domains?
   a. 2, 6   b. 5
   c. 2, 4, 6   d. 1, 3, 5
   e. all six

10. Which is (are) found *only* in prokaryotes?
   a. 1, 4, 6   b. 3, 5
   c. 1, 2   d. 4
   e. 2, 4, 5

CRITICAL THINKING

1. Here is some additional information on the organisms in review question 14:

Organisms	% DNA Hybridization
A and B	5–15
A and C	5–15
A and D	70–90
B and C	10–20
B and D	2–5

Which of these organisms are most closely related? Compare this answer with your response to review question 14.

2. The GC content of *Micrococcus* is 66–75 moles %, and of *Staphylococcus*, 30–40 moles %. According to this information, would you conclude that these two genera are closely related?

3. Describe the use of a DNA probe and PCR for:
   a. rapid identification of an unknown bacterium.
   b. determining which of a group of bacteria are most closely related.

4. SF medium is a selective medium, developed in the 1940s, to test for fecal contamination of milk and water. Only certain gram-positive cocci can grow in this medium. Why is it named SF? Using this medium, which genus will you culture? (*Hint:* Refer to page 288.)

CLINICAL APPLICATIONS

1. A 55-year-old veterinarian was admitted to a hospital with a 2-day history of fever, chest pain, and cough. Gram-positive cocci were detected in his sputum, and he was treated for lobar pneumonia with penicillin. The next day, another Gram stain of his sputum revealed gram-negative rods, and he was switched to ampicillin and gentamicin. A sputum culture showed biochemically inactive gram-negative rods identified as *Enterobacter agglomerans*. After fluorescent-antibody staining and phage typing, *Yersinia pestis* was identified in the patient's sputum and blood, and chloramphenicol and tetracycline were administered. The patient died 3 days after admission to the hospital. Tetracycline was given to his 220 contacts (hospital personnel, family, and coworkers). What disease did the patient have? Discuss what went wrong in the diagnosis and how his death might have been prevented. Why were the 220 other people treated? (*Hint:* Refer to Chapter 23.)

2. A 6-year-old girl was admitted to a hospital with endocarditis. Blood cultures showed a gram-positive, aerobic rod identified by the hospital laboratory as *Corynebacterium xerosis*. The girl died after 6 weeks of treatment with intravenous penicillin and chloramphenicol. The bacterium was tested by another laboratory and identified as *C. diphtheriae*. The following test results were obtained by each laboratory:

	Hospital Lab	Other lab
Catalase	+	+
Nitrate reduction	+	+
Urea	–	–
Esculin hydrolysis	–	–
Glucose fermentation	+	+
Sucrose fermentation	–	+
Serological test for toxin production	Not done	+

Provide a possible explanation for the incorrect identification. What are the potential public health consequences of misidentifying *C. diphtheriae*? (*Hint:* Refer to Chapter 24.)

3. Use the information in the following table to construct a dichotomous key to these organisms. What is the purpose of a dichotomous key? Look up each genus in Chapter 11, and provide an example of why this organism is of interest to humans.

	Morphology	Gram Reaction	Acid from Glucose	Growth in Air (21% O_2)	Motile by Peritrichous Flagella	Presence of Cytochrome Oxidase	Produce Catalase
Staphylococcus aureus	Coccus	+	+	+	−	−	+
Streptococcus pyogenes	Coccus	+	+	+	−	−	−
Mycoplasma pneumoniae	Coccus	−	+	+ (Colonies < 1mm)	−	−	+
Clostridium botulinum	Rod	+	+	−	+	−	−
Escherichia coli	Rod	−	+	+	+	−	+
Pseudomonas aeruginosa	Rod	−	+	+	−	+	+
Campylobacter fetus	Vibrio	−	−	−	−	+	+
Listeria monocytogenes	Rod	+	+	+	+	−	+

4. Use the additional information shown at right to construct a cladogram for some of the organisms used in question 3. What is the purpose of a cladogram? How does your cladogram differ from a dichotomous key for these organisms?

	Similarity in rRNA Bases
P. aeruginosa—M. pneumoniae	52%
P. aeruginosa—C. botulinum	52%
P. aeruginosa—E. coli	79%
M. pneumoniae—C. botulinum	65%
M. pneumoniae—E. coli	52%
E. coli—C. botulinum	52%

5. Using the following information, create a dichotomous key for distinguishing these organisms.

	Unicellular?	Mitochondria?	Chlorophyll?	Nutritional Type?	Motile?	Cause Human Disease?
Euglena	+	+	+	Both	+	
Giardia				Hetero		
Nosema						
Pfiesteria				Auto		
Trichomonas						
Trypanosoma						

How would you define *protist*?

Using the additional information shown below, create a
dichotomous key for these organisms.

rRNA base #	1	2	3	4	5	6	7	8	9	10	11	12	13	14	15	16	17	18	19	20
Euglena	C	C	A	G	G	U	U	G	U	U	C	C	A	G	U	U	U	U	A	A
Giardia	C	C	A	U	A	U	U	U	U	U	G	A	C	G	A	A	G	G	U	C
Nosema	C	C	A	U	A	U	U	U	U	U	A	A	C	G	A	A	G	G	C	C
Pfiesteria	C	C	A	A	C	U	U	A	U	U	C	C	A	G	U	U	U	C	A	G
Trichomonas	C	C	A	U	A	U	U	U	U	U	G	A	C	G	A	A	G	G	G	C
Trypanosoma	C	C	A	C	G	U	U	G	U	U	C	C	A	G	U	U	U	U	A	A

Do your two keys differ? Explain why. Which key is more useful
for laboratory identification? For classification?

11

The Prokaryotes: Domains Bacteria and Archaea

When biologists first encountered microscopic bacteria, they were puzzled as to how to classify them. Bacteria were clearly not animals or rooted plants. At first, bacteria and fungi were classified as plants, as was discussed in Chapter 10. Attempts to build a taxonomic system for bacteria based on the phylogenetic system developed for plants and animals, failed. Bacteria were grouped by morphology (rod, coccus, helix), staining reactions, presence of endospores, and other obvious features. Although this system had its practical uses, microbiologists were aware that it had many imperfections. In 2001, the first volume of a forthcoming five-volume second edition of *Bergey's Manual* was published. During the years since publication of the first edition, knowledge of bacteria at the molecular level has expanded to such a degree that it is possible to base this new edition on a phylogenetic system. One of the most important expressions of phylogenetic difference is rRNA. Above the species level, rRNA is especially useful. It is slow to change and performs the same functions in all organisms. It is now possible to begin constructing phylogenetic trees in which gene sequences can be compared.

UNDER THE MICROSCOPE

Vibrio cholerae, the cause of cholera, a severe intestinal disease. It is spread through food and water contaminated by human feces and is still common in parts of the world with low sanitary standards.

THE PROKARYOTIC GROUPS

In the second edition of *Bergey's Manual*, the prokaryotes are grouped into two **domains,** the **Archaea** and the **Bacteria.** Both domains consist of prokaryotic cells. Higher organisms are assigned to the Domain Eukarya, which contains all the unicellular and multicellular eukaryotic organisms. The grouping into domains is shown in Figure 10.1 and Table 10.1, and the differences between the two prokaryotic domains are summarized in Figure 10.6. Each domain is divided into phyla, each phylum into classes, and so on. The phyla discussed in this chapter are summarized in Table 11.1 (see also Appendix A).

DOMAIN BACTERIA

Most of us think of bacteria as invisible, potentially harmful little creatures. Actually, relatively few species of bacteria cause disease in humans, animals, plants, or any other organisms. Indeed, once you have completed a course in microbiology, you will realize that without bacteria, much of life as we know it would not be possible. In fact, all organisms made up of eukaryotic cells probably evolved from bacterialike organisms, which were some of the earliest forms of life. In this chapter, you will learn how bacterial groups are differentiated from each other and how important bacteria are in the world of microbiology. Our discussion emphasizes bacteria considered to be of practical importance, those important in medicine, or those that illustrate biologically unusual or interesting principles.

THE PROTEOBACTERIA

The proteobacteria, which includes most of the gram-negative, chemoheterotrophic bacteria, are presumed to have arisen from a common photosynthetic ancestor. They are now the largest taxonomic group of bacteria. However, few are now photosynthetic; other metabolic and nutritional capacities have arisen to replace this characteristic. The phylogenetic relationship in these groups is based upon rRNA studies. The name **Proteobacteria** was taken from the mythological Greek god Proteus, who could assume many shapes. The proteobacteria are separated into five classes designated by Greek letters: alphaproteobacteria, betaproteobacteria, gammaproteobacteria, deltaproteobacteria, and epsilonproteobacteria.

THE ALPHAPROTEOBACTERIA

LEARNING OBJECTIVE

The Learning Objectives throughout this chapter will help you to become familiar with these organisms and to look for similarities and differences between organisms. You will draw a dichotomous key to differentiate the bacteria described in each group. We'll draw the first one to get you started.

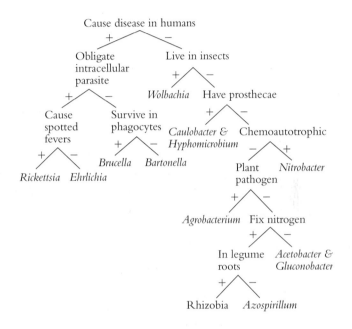

- Make a dichotomous key to distinguish among the alphaproteobacteria described in this chapter.

As a group, the alphaproteobacteria includes most of the proteobacteria that are capable of growth at very low levels of nutrients. Some have unusual morphology, including protrusions such as stalks or buds known as **prosthecae.** The alphaproteobacteria also include agriculturally important bacteria capable of inducing nitrogen fixation in symbiosis with plants, and several plant and human pathogens.

Azospirillum Agricultural microbiologists have been interested in members of the genus *Azospirillum* (ā-zō-spī′ril-lum), a soil bacterium that grows in close association with the roots of many plants, especially tropical grasses. It uses nutrients excreted by the plants and in return fixes nitrogen from the atmosphere. This form of nitrogen fixation is most significant in some tropical grasses and in sugar cane, although the organism can be isolated from the root system of many temperate-climate plants, such as corn. (The prefix *azo-* is frequently encountered in nitrogen-fixing genera of bacteria.

TABLE 11.1	Selected Prokaryotes from *Bergey's Manual of Systematic Bacteriology,* Second Edition*		
Phylum Class	**Order**	**Important Genera**	**Special Features**
DOMAIN BACTERIA **Proteobacteria**			
Alphaproteobacteri	Caulobacterales	*Caulobacter*	Stalked
	Rickettsiales	*Ehrlichia*	Obligately intracellular human pathogens
		Rickettsia	Obligately intracellular human pathogens
		Wolbachia	Symbionts of insects
	Rhizobiales	*Agrobacterium*	Plant pathogens
		Bartonella	Human pathogens
		Beijerinckia	Free-living nitrogen fixers
		Bradyrhizobium	Symbiotic nitrogen fixers
		Brucella	Human pathogens
		Hyphomicrobium	Budding
		Nitrobacter	Nitrifying
		Rhizobium	Symbiotic nitrogen fixers
	Rhodospirillales	*Acetobacter*	Acetic acid production
		Azospirillum	Nitrogen fixers
		Gluconobacter	Acetic acid production
		Rhodospirillum	Photosynthetic, anoxygenic
Betaproteobacteria	Burkholderiales	*Burkholderia*	Opportunistic pathogens
		Bordetella	Human pathogens
		Sphaerotilus	Sheathed
	Hydrogenophilales	*Thiobacillus*	Sulfur oxidizers
	Neisseriales	*Neisseria*	Human pathogens
	Nitrosomonadales	*Nitrosomonas*	Nitrifying
		Spirillum	Found in stagnant freshwater
	Rhodocyclales	*Zoogloea*	Sewage treatment
Gammaproteobacteria	Chromatiales	*Chromatium*	Photosynthetic, anoxygenic
	Thiotrichales	*Beggiatoa*	Sulfur oxiders
		Thiomargarita	Giant bacterium
		Francisella	Human pathogens
	Legionellales	*Legionella*	Human pathogens
		Coxiella	Obligately intracellular human pathogens
	Pseudomonadales	*Azomonas*	Free-living nitrogen fixers
		Azotobacter	Free-living nitrogen fixers
		Moraxella	Human pathogens
		Pseudomonas	Opportunistic pathogens
	Vibrionales	*Vibrio*	Human pathogens
	Enterobacteriales	*Citrobacter*	Opportunistic pathogens
		Enterobacter	Opportunistic pathogens
		Erwinia	Plant pathogens
		Escherichia	Normal intestinal bacteria, some pathogens
		Klebsiella	Opportunistic pathogens
		Proteus	Human intestinal bacteria, occasional pathogens
		Salmonella	Human pathogens

*See Appendix A for a complete taxonomic listing. This table includes prokaryotes mentioned in this text. Descriptions such as "pathogenic" mean that this trait is common in the genus but not that all members of the genus may have this trait.

| TABLE 11.1 | (continued) | | | |

Phylum Class	Order	Important Genera	Special Features
		Serratia	Red pigment, opportunistic pathogens
		Shigella	Human pathogens
		Yersinia	Human pathogens
	Pasteurellales	*Haemophilus*	Human pathogens
		Pasteurella	Human pathogens
Deltaproteobacteria	Bdellovibrionales	*Bdellovibrio*	Parasites of bacteria
	Desulfovibrionales	*Desulfovibrio*	Sulfate reducers
	Myxococcales	*Myxococcus*	Gliding, fruiting
		Stigmatella	Gliding, fruiting
Epsilonproteobacteria	Campylobacterales	*Campylobacter*	Human pathogens
		Helicobacter	Human pathogens, carcinogenic

Nonproteobacteria, Gram-Negative Bacteria

Cyanobacteria

		Anabaena	Photosynthetic, oxygenic
		Gloeocapsa	Photosynthetic, oxygenic

Chlorobi

		Chlorobium	Photosynthetic, anoxygenic

Chloroflexi

		Chloroflexus	Photosynthetic, anoxygenic

Firmicutes (The Low G + C Gram-Positive Bacteria)

	Clostridiales	*Clostridium*	Anaerobes, endospores, some human pathogens
		Epulopiscium	Giant bacterium
		Sarcina	Occur in cubical packets
	Mycoplasmatales[†]	*Mycoplasma*	No cell wall, human pathogens
		Spiroplasma	No cell wall, pleomorphic, plant pathogens
		Ureaplasma	No cell wall, ammonia from urea
	Bacillales	*Bacillus*	Endospores, some pathogens
		Listeria	Human pathogens
		Staphylococcus	Some human pathogens
	Lactobacillales	*Enterococcus*	Opportunistic pathogens
		Lactobacillus	Lactic acid producers
		Streptococcus	Many human pathogens

Actinobacteria (The High G + C Gram-Positive Bacteria)

	Actinomycetales	*Actinomyces*	Filamentous, branching, some human pathogens
		Corynebacterium	Human pathogens
		Frankia	Symbiotic nitrogen-fixers
		Gardnerella	Human pathogens
		Mycobacterium	Acid fast, human pathogens
		Nocardia	Filamentous, branching, opportunistic pathogens
		Propionibacterium	Produce propionic acid
		Streptomyces	Filamentous branching, many produce antibiotics

[†]The bacteria in the order Mycoplasmatales are genetically related to the low G + C gram-positive bacteria, but they lack a cell wall and stain gram-negative.

➤

TABLE 11.1	Selected Prokaryotes from *Bergey's Manual of Systematic Bacteriology*, Second Edition (continued)		
Phylum **Class**	**Order**	**Important** **Genera**	**Special** **Features**
Chlamydiae			
	Chlamydiales	*Chlamydia* *Chlamydophila*	Intracellular parasites, human pathogens Intracellular parasites, human pathogens
Spirochaetes			
	Spirochaetales	*Borrelia* *Leptospira* *Treponema*	Human pathogens Human pathogens Human pathogens
Bacteroidetes			
	Bacteroidales	*Bacteroides* *Prevotella*	Human intestinal tract Human oral cavity
Fusobacteria			
	Fusobacteriales	*Fusobacterium*	Human intestinal tract
DOMAIN ARCHAEA **Crenarchaeota (Gram-Negative)**			
	Desulfurococcales Sulfolobales	*Pyrodictium* *Sulfolobus*	Hyperthermophiles Hyperthermophiles
Euryarchaeota (Gram-Positive to Variable)			
	Methanobacteriales Halobacteriales	*Methanobacterium* *Halobacterium* *Halococcus*	Methanogens Require high salt concentration Require high salt concentration

It is derived from *a* (without) and *zo* (life), in reference to the early days of chemistry when oxygen was removed, by a burning candle, from an experimental atmosphere. Presumably, mostly nitrogen remained, and mammalian life was found to be not possible in this atmosphere. Hence, nitrogen came to be associated with absence of life.)

Acetobacter and Gluconobacter *Acetobacter* (ä′sē-tō-bak-tėr) and *Gluconobacter* (glü′kon-ō-bak-tėr) are industrially important aerobic organisms that convert ethanol into acetic acid (vinegar).

Rickettsia In the first edition of *Bergey's Manual*, the genera *Rickettsia*, *Coxiella*, and *Chlamydia* were grouped closely because they share the common characteristic of being obligate intracellular parasites—that is, they reproduce only within a mammalian cell. In the second edition they are now widely separated. A comparison of rickettsias, chlamydias, and viruses appears in Table 13.1, page 387.

The rickettsias are gram-negative rod-shaped bacteria, or coccobacilli (Figure 11.1a). One distinguishing feature of most rickettsias is that they are transmitted to humans by bites of insects and ticks, as are the *Coxiella* (discussed later with gammaproteobacteria). Rickettsia enter their host cell by inducing phagocytosis. They quickly enter the cytoplasm of the cell and begin reproducing by binary fission (Figure 11.1b). They can usually be cultivated artificially in cell culture or chick embryos (Chapter 13, pages 395–396).

The rickettsias are responsible for a number of diseases known as the spotted fever group. These include epidemic typhus, caused by *Rickettsia prowazekii* (ri-ket′sē-ä prou-wä-ze′kē-ē) and transmitted by lice; endemic murine typhus, caused by *R. typhi* (tī′fē) and transmitted by rat fleas, and Rocky Mountain spotted fever, caused by *R. rickettsii* (ri-ket′sē-ē) and transmitted by ticks. In humans, rickettsial infections damage the permeability of blood capillaries, which results in a characteristic spotted rash.

Ehrlichia Ehrlichiae are gram-negative, rickettsialike bacteria that live obligately within white blood cells.

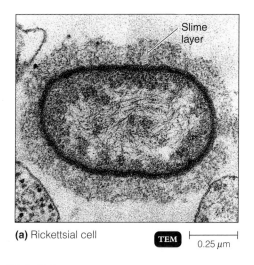

(a) Rickettsial cell   TEM   ├──┤ 0.25 μm

(b) Rickettsials in chicken embryo cell   LM   ├──┤ 50 μm

FIGURE 11.1 Rickettsias.

Q How are rickettsias transmitted from one host to another?

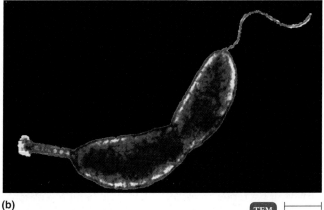

(a)

Flagellated swarmer cell

Flagellum lost

Stalk begins to form

Stalk elongates

Division begins, flagellum forms on new cell

Cell division, the cells continue the division process

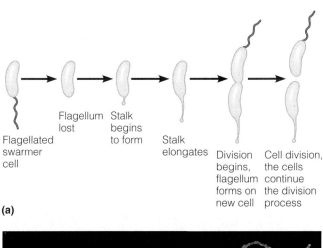

(b)   TEM   ├──┤ 0.5 μm

FIGURE 11.2 *Caulobacter*. (**a**) During its life cycle this bacterium exhibits two forms: a flagellated, swarmer form that provides motility, and a stalked form in which a stalk serves as a "holdfast" by which it can attach to surfaces. (**b**) The photo of a dividing *Caulobacter* bacterium shows one cell with a flagellum and a second with a stalk.

Q What is the competitive advantage provided by attaching to a surface?

that anchor the organs to surfaces (Figure 11.2). This arrangement increases their nutrient uptake because they are exposed to a continuously changing flow of water and because the stalk increases the surface-to-volume ratio of the cell. Also, if the surface to which they anchor is a living host, these bacteria can use the host's excretions as nutrients. When the nutrient concentration is exceptionally low, the size of the stalk increases, evidently to provide an even greater surface area for nutrient absorption.

Budding bacteria do not divide by binary fission into two nearly identical cells. The budding process resembles the asexual reproductive processes of many yeasts (Figure 12.3, page 347). The parent cell retains its identity while the bud increases in size until it separates as a complete new cell. An example is the genus *Hyphomicrobium* (hī-fō-mī-krō'bē-um), as shown in Figure 11.3. These bacteria, like the caulobacteria, are found in low-nutrient aquatic environments and have even been found growing in laboratory water baths. Both *Caulobacter* and *Hyphomicrobium* produce prominent prosthecae.

Ehrlichia (ėr'lik-ē-ä) species are transmitted by ticks to humans and cause ehrlichiosis, a sometimes fatal disease.

Caulobacter and *Hyphomicrobium*

Members of the genus *Caulobacter* (kô-lō-bak'tėr) are found in low-nutrient aquatic environments, such as lakes. They feature stalks

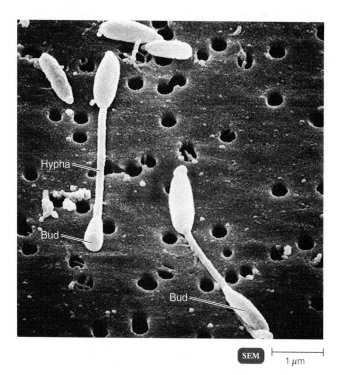

SEM | 1 μm

FIGURE 11.3 *Hyphomicrobium,* **a type of budding bacterium.**

Q Most bacteria do not reproduce by budding; what method do they use?

Rhizobium, Bradyrhizobium, and Agrobacterium The *Rhizobium* (rī-zō′bē-um) and *Bradyrhizobium* (brād-ē-rī-zō′bē-um) are two of the more important genera of a group of agriculturally important bacteria that specifically infect the roots of leguminous plants, such as beans, peas, or clover. For simplicity these bacteria are known by the common name of **rhizobia.** The presence of rhizobia in the roots leads to formation of nodules in which the rhizobia and plant form a symbiotic relationship, resulting in the fixation of nitrogen from the air for use by the plant (see Figure 27.5).

Like rhizobia, the genus *Agrobacterium* (ag′rō-bak-ti′rē-um) has the ability to invade plants. However, they do not induce root nodules or fix nitrogen. Of particular interest is *Agrobacterium tumefaciens.* This is a plant pathogen that causes a disease called crown gall; the crown is the area of the plant where the roots and stem merge. The tumorlike gall is induced when *A. tumefaciens* inserts a plasmid containing bacterial genetic information into the plant's chromosomal DNA (see Figure 9.19, page 274). For this reason, microbial geneticists are very interested in this organism. Plasmids are the most common vector that scientists use to carry new genes into a cell, and the thick wall of plants is especially difficult to penetrate (see Figure 9.20, page 274).

Bartonella The genus *Bartonella* (bär′tō-nel-la) contains several members that are human pathogens. The best known is *Bartonella henselae,* a gram-negative bacillus that causes cat-scratch disease.

Brucella *Brucella* (brü′sel-la) bacteria are small nonmotile coccobacilli. All species of *Brucella* are obligate parasites of mammals and cause the disease brucellosis. Of medical interest is the ability of *Brucella* to survive phagocytosis, an important element of the body's defense against bacteria (see Chapter 16, page 483).

Nitrobacter and Nitrosomonas *Nitrobacter* (ni-trō-bak′tèr) and *Nitrosomonas* (nī-trō-sō-mō′nas) are genera of nitrifying bacteria that are of great importance to the environment and to agriculture. They are chemoautotrophs capable of using inorganic chemicals as energy sources and carbon dioxide as the only source of carbon, from which they synthesize all of their complex chemical makeup. The energy sources of the genera *Nitrobacter* and *Nitrosomonas* (the latter is a member of the betaproteobacteria) are reduced nitrogenous compounds. *Nitrobacter* species oxidize ammonium (NH_4^+) to nitrite (NO_2^-), which is in turn oxidized by *Nitrosomonas* species to nitrates (NO_3^-) in the process of *nitrification.* Nitrate is important to agriculture; it is a nitrogen form that is highly mobile in soil and therefore likely to be encountered and used by plants.

Wolbachia *Wolbachia* (wol-ba′kē-ä) are probably the most common infectious bacterial genus in the world. Even so, little is known about *Wolbachia;* they live only inside the cells of their hosts, usually insects (a relationship known as *endosymbiosis*). Therefore, *Wolbachia* escape detection by the usual culture methods. This fascinating group of bacteria is described further in the box on the facing page.

THE BETAPROTEOBACTERIA

LEARNING OBJECTIVE

- Make a dichotomous key to distinguish among the betaproteobacteria described in this chapter.

There is considerable overlap between the betaproteobacteria and the alphaproteobacteria, for example among the nitrifying bacteria discussed earlier. The betaproteobacteria often use nutrient substances that diffuse away from areas of anaerobic decomposition of organic matter, such as hydrogen gas, ammonia, and methane. Several important pathogenic bacteria are found in this group.

Thiobacillus *Thiobacillus* (thī-ō-bä-sil′lus) species and other sulfur-oxidizing bacteria are important in the sulfur cycle (see Figure 27.7). These chemoautotrophic bacteria are capable of obtaining energy by oxidizing the reduced

APPLICATIONS OF MICROBIOLOGY

BACTERIA AND INSECT SEX

Wolbachia is quite possibly the most common infectious bacterial genus on Earth. Although they were first discovered in 1924, little had been known about them until the 1990s. They escape detection by the usual culture methods because they live as endosymbionts in the cells of insects and other invertebrates.

Wolbachia infect over a million species of insects as well as millipedes, mites, spiders, crustaceans, and nematodes. In all, as many as 75% of species of animals surveyed have been found to carry this bacterium. Wolbachia is essential to the nematodes. If the bacterium is killed with antibiotics, the host worm dies. Pea aphids infected with Wolbachia are not killed by a normally lethal parasitic wasp larva; The bacterium is harmless in the aphid but kills the wasp.

In some insects, Wolbachia destroys males of its host species. Wolbachia can turn males into females by interfering with the male hormone. If Wolbachia does this in an embryo, the developing gonads become ovaries. As shown in the figure, if a male and female insect are uninfected with Wolbachia, they produce offspring normally. If only the male is infected, the insects fail to reproduce. If one or both insects of a mating pair are infected, only the infected females reproduce—and transmit Wolbachia in the cytoplasm of their eggs. Offspring produced without fertilization are female. The result is that the bacteria are transmitted to the next generation. This type of reproduction, called parthenogenesis, has been seen in a variety of insects and in some amphibians and reptiles. Thus a question arises: Is Wolbachia always responsible?

Eukaryotic species are defined as organisms that reproduce only with members of their own species. This reproductive isolation prevents the production of hybrids and thus maintains the uniqueness of each species. In the laboratory, researchers have found that after antibiotic treatment, wasps of one species will produce hybrid offspring with another species. This raises the question about the influence Wolbachia has had on the evolution of insects. Did insects that were not infected successfully reproduce outside of their species?

A virulent strain of Wolbachia called "popcorn" causes host cells to lyse, or "pop," which eventually kills the host insect. The popcorn strain might be used to kill mosquitoes. On the other hand, elimination of Wolbachia from pest insects could result in a lower number of females being produced and thus, reducing population growth. Wolbachia is also being investigated as a potential tool for growing female crabs and shrimps for commercial breeding.

The unique biology of Wolbachia has attracted researchers interested in questions that range from the evolutionary implications of infection to commercial uses of Wolbachia.

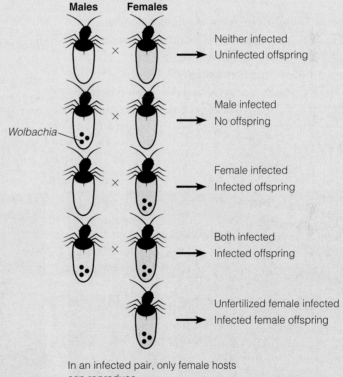

In an infected pair, only female hosts can reproduce.

forms of sulfur, such as hydrogen sulfide (H_2S), or elemental sulfur (S^0), into sulfates (SO_4^{2-}).

Spirillum   The habitat of the genus *Spirillum* (spī-ril′lum) is mainly fresh water. An important morphological difference from the helical spirochetes (discussed on page 338) is that *Spirillum* bacteria are motile by conventional polar flagella, rather than axial filaments. The spirilla are relatively large, gram-negative, aerobic bacteria. *Spirillum volutans* (vō-lū-tans) is often used as a demonstration slide when microbiology students are first introduced to the operation of the microscope (Figure 11.4).

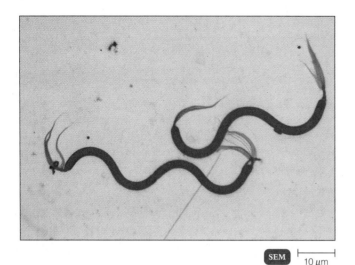

SEM | 10 μm

FIGURE 11.4 *Spirillum volutans.* These large helical bacteria are found in aquatic environments. Note the polar flagella.

Q **Is this bacterium motile? How can you tell?**

Sphaerotilus Sheathed bacteria, which include *Sphaerotilus natans* (sfe-rä′ti-lus na′tans), are found in freshwater and in sewage. These gram-negative bacteria with polar flagella form a hollow, filamentous sheath in which to live (Figure 11.5). Sheaths are protective and also aid in nutrient accumulation. *Sphaerotilus* probably contributes to bulking, an important problem in sewage treatment (see Chapter 27).

Burkholderia The genus *Burkholderia* was formerly grouped with the genus *Pseudomonas*, which is now classified under the gammaproteobacteria. Like the pseudomonads, almost all *Burkholderia* species are motile by a single polar flagellum or tuft of flagella. The best known species is the aerobic, gram-negative rod *Burkholderia cepacia* (berk′hōld-ẽr-ē-ä se-pā′se-ä). It has an extraordinary nutritional spectrum and is capable of degrading more than 100 different organic molecules. This capability is often a factor in the contamination of equipment and drugs in hospitals; these bacteria may actually grow in disinfectant solutions (see the box in Chapter 14, page 445). This bacterium is also a problem for persons with the genetic lung disease cystic fibrosis, in whom it metabolizes accumulated respiratory secretions. *Burkholderia pseudomallei* (sū do-mal′lē-ī) is a resident in moist soils and is the cause of a severe disease (melioidosis) endemic in southeast Asia and northern Australia.

Bordetella Of special importance is the nonmotile, aerobic, gram-negative rod *Bordetella pertussis* (bȯr′de-tel-lä pẽr-tus′sis). This serious pathogen is the cause of pertussis, or whooping cough.

Neisseria Bacteria of the genus *Neisseria* (nī-se′rē-ä) are aerobic, gram-negative cocci that usually inhabit the mucous membranes of mammals. Pathogenic species include the gonococcus bacterium *Neisseria gonorrhoeae* (go-nȯr-rē′ī), the causative agent of gonorrhoea (Figure 11.6 and the box in Chapter 26, page 793.), and *N. meningitidis* (men-nin-ji′ti-dis), the agent of meningococcal meningitis.

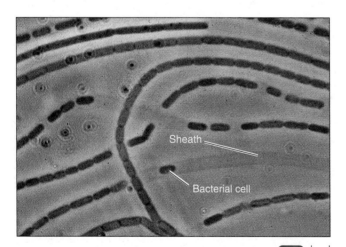

Sheath

Bacterial cell

TEM | 2 μm

FIGURE 11.5 *Sphaerotilus natans.* These sheathed bacteria are found in dilute sewage and aquatic environments. It forms elongated sheaths in which the bacteria live. The bacteria have flagella (not visible here) and can eventually swim free of the sheath.

Q **How does the sheath help the cell?**

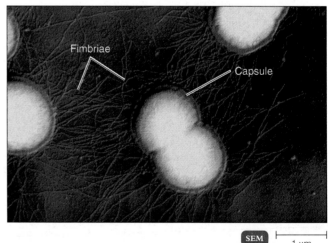

Fimbriae

Capsule

SEM | 1 μm

FIGURE 11.6 The gram-negative coccus *Neisseria gonorrhoeae.* Notice the paired arrangement (diplococci). The fimbriae enabled the organism to attach to mucous membranes and thus contribute to its pathogenicity. *N. gonorrhoeae* causes gonorrhea.

Q **For what are fimbriae used?**

Zoogloea   The genus *Zoogloea* (zō'ō-glē-ä) is important in the context of aerobic sewage-treatment processes, such as the activated sludge system (Chapter 27). As they grow, *Zoogloea* bacteria form fluffy, slimy masses that are essential to the proper operation of such systems.

THE GAMMAPROTEOBACTERIA

LEARNING OBJECTIVE

- Make a dichotomous key to distinguish among the orders of gammaproteobacteria described in this chapter.

The gammaproteobacteria constitute the largest subgroup of the proteobacteria and include a great variety of physiological types. One species that is used in industrial microbiology is described in the box in Chapter 28 on page 848.

Beggiatoa   *Beggiatoa alba* (bej'jē-ä-tō-ä al'ba), the only species of this unusual genus, grows in aquatic sediments at the interface between the aerobic and anaerobic layers. Morphologically, it resembles certain filamentous cyanobacteria (page 328), but it is not photosynthetic. Motility is by gliding. The exact mechanism of gliding motility is uncertain; there may actually be several mechanisms involved. The only common factor among the mechanisms is the production of slime, which attaches to the surface on which movement occurs and also provides lubrication allowing the organism to glide.

Nutritionally *B. alba* uses hydrogen sulfide (H_2S) as an energy source and accumulates internal granules of sulfur. In our discussion of the sulfur cycle in Chapter 27 on page 816, we show how this bacterium was a factor in the discovery of autotrophic metabolism.

Francisella   *Francisella* (fran'sis-el'lä) is a genus of small, pleomorphic bacteria that grow only on complex media enriched with blood or tissue extracts. *Francisella tularensis* (tü'lär-en-sis) causes the disease tularemia. (See the box in Chapter 23, page 677.)

PSEUDOMONADALES

Members of the order Pseudomonadales are gram-negative aerobic rods or cocci. The most important genus in this group is *Pseudomonas*.

Pseudomonas   A very important genus, *Pseudomonas* (sū-dō-mō'nas) consists of aerobic, gram-negative rods that are motile by polar flagella, either single or tufts (Figure 11.7). Pseudomonads are very common in soil and other natural environments.

Many species of pseudomonads excrete extracellular, water-soluble pigments that diffuse into their media. One species, *Pseudomonas aeruginosa* (ā-rü-ji-nō'sä), produces a

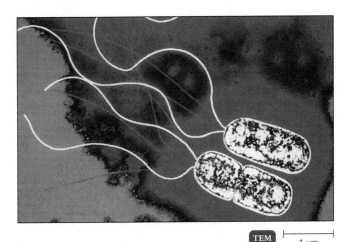

TEM   ⊢——⊣   1 μm

FIGURE 11.7 *Pseudomonas*. This photo of a pair of *Pseudomonas* bacteria shows polar flagella that are a characteristic of the genus. In some species only a single flagellum is present (see Figure 4.7a, page 82).

Q   **How does the nutritional diversity of these bacteria make them a problem in hospitals?**

soluble, blue-green pigmentation. Under certain conditions, particularly in weakened hosts, this organism can infect the urinary tract, burns, and wounds, and can cause blood infections (sepsis), abscesses, and meningitis. Other pseudomonads produce soluble fluorescent pigments that glow when illuminated by ultraviolet light. One species, *P. syringae* (sèr'in-gī), is an occasional plant pathogen. (Some species of *Pseudomonas* have been transferred, based upon rRNA studies, to the genus *Burkholderia*, which was discussed previously with the betaproteobacteria.)

Pseudomonads have almost as much genetic capacity as the eukaryotic yeasts and almost half as much as a fruit fly. While these bacteria are less efficient than some other heterotrophic bacteria in utilizing many of the more common nutrients, they make use of their genetic capacity by compensating for this in other ways. For example, pseudomonads synthesize an unusually large number of enzymes and can metabolize a wide variety of substrates. Therefore, they probably contribute significantly to the decomposition of uncommon chemicals, such as pesticides, that are added to soil.

In hospitals and other places where pharmaceutical agents are prepared, the ability of pseudomonads to grow on minute traces of unusual carbon sources, such as soap residues or cap-liner adhesives found in a solution, has been unexpectedly troublesome. Pseudomonads are even capable of growth in some antiseptics, such as quaternary ammonium compounds. Their resistance to most antibiotics has also been a source of medical concern. This resistance is probably related to the characteristics of the cell wall porins, which control the entrance of molecules through the cell wall (see Chapter 4, page 87). The large genome of pseudomonads also codes for several very efficient efflux pump systems (page 603) that eject antibiotics from the cell before they can function. Pseudomonads are responsible for about one in ten nosocomial infections (hospital-acquired infections; see page 435), especially among infections in burn units. Persons with cystic fibrosis are also especially prone to infections by *Pseudomonas* and the closely related *Burkholderia*.

Although pseudomonads are classified as aerobic, some are capable of substituting nitrate for oxygen as a terminal electron acceptor. This process, anaerobic respiration, yields almost as much energy as aerobic respiration (see Chapter 5). In this way, pseudomonads cause important losses of valuable nitrogen in fertilizer and soil. Nitrate (NO_3^-) is the form of fertilizer nitrogen most easily used by plants. Under anaerobic conditions, as in water-logged soil, pseudomonads eventually convert this valuable nitrate into nitrogen gas (N_2), which is lost to the atmosphere (see Chapter 27).

Many pseudomonads can grow at refrigerator temperatures. This characteristic, combined with their ability to utilize proteins and lipids, makes them an important contributor to food spoilage.

Azotobacter and Azomonas Some nitrogen-fixing bacteria, such as *Azotobacter* (ā-zō-tō-bak′tėr) and *Azomonas* (ā-zō-mō′nas), are free-living in soil. These large, ovoid, heavily capsulated bacteria are frequently used in laboratory demonstrations of nitrogen fixation. However, to fix agriculturally significant amounts of nitrogen, they would require energy sources, such as carbohydrates, that are in limited supply in soil.

Moraxella Members of the genus *Moraxella* (mô-raks-el′lä) are strictly aerobic coccobacilli—that is, intermediate in shape between cocci and rods. *Moraxella lacunata* (la-kü-nä′tä) is implicated in conjunctivitis, an inflammation of the conjunctiva, the membrane that covers the eye and lines the eyelids.

LEGIONELLALES

The genera *Legionella* and *Coxiella* are closely associated in the second edition of *Bergey's Manual*, where both are placed in the same order, Legionellales. Because the *Coxiella* share an intracellular lifestyle with the rickettsial bacteria, they were previously considered rickettsial in nature and grouped with them. *Legionella* bacteria grow readily on suitable artificial media.

Legionella *Legionella* (lē-jä-nel′lä) bacteria were originally isolated during a search for the cause of an outbreak of pneumonia now known as legionellosis. The search was difficult because these bacteria did not grow on the usual laboratory isolation media then available. After intensive effort, special media were developed that enabled researchers to isolate and culture the first *Legionella*. Microbes of this genus are now known to be relatively common in streams, and they colonize such habitats as warm-water supply lines in hospitals and water in the cooling towers of air conditioning systems. (See the box in Chapter 24, page 728.) An ability to survive and reproduce within aquatic amoebae often makes them difficult to eradicate in water systems.

Coxiella *Coxiella burnetii* (käks-ė-el′lä bėr-ne′tē-ē), which causes Q fever, was formerly grouped with the rickettsia. Like them, *Coxiella* bacteria require a mammalian host cell in order to reproduce. Unlike rickettsias, *Coxiella* are not transmitted among humans by insect or tick bites. Although cattle ticks harbor the organism, it is most commonly transmitted by aerosols or contaminated milk. A sporelike body is present in *C. burnetii* (see Figure 24.15b, page 729). This might explain the bacterium's relatively high resistance to the stresses of airborne transmission and heat treatment.

VIBRIONALES

Members of this order are facultatively anaerobic gram-negative rods. Many are slightly curved. They are found mostly in aquatic habitats.

Vibrio Members of the genus *Vibrio* (vib′rē-ō) are rods that are often slightly curved (Figure 11.8). One important pathogen is *Vibrio cholerae* (kol′er-ī), the causative agent of cholera. The disease is characterized by a profuse and watery diarrhea. *V. parahaemolyticus* (pa-rä-hē-mō-li′ti-kus) causes a less serious form of gastroenteritis. Usually inhabiting coastal salt waters, it is transmitted to humans mostly by raw or undercooked shellfish.

ENTEROBACTERIALES

The members of the order Enterobacteriales are facultatively anaerobic, gram-negative rods that are, if motile, peritrichously flagellated. Morphologically, the rods are straight. This is an important bacterial group, often commonly called **enterics.** This reflects the fact that they inhabit the intestinal tracts of humans and other animals. Most enterics are active fermenters of glucose and other carbohydrates.

Because of the clinical importance of enterics, there are many techniques to isolate and identify them. An identification method for some enterics is shown in Figure 10.9 (page 297), which incorporates a modern tool using 15 biochemical tests. Biochemical tests are especially important in clinical laboratory work and in food and water microbiology.

Enterics have fimbriae that help them adhere to surfaces or mucous membranes. Specialized sex pili aid in the exchange of genetic information between cells, which often includes antibiotic resistance (see Figures 8.24 and 8.25, page 242).

Enterics, like many bacteria, produce proteins called bacteriocins that cause the lysis of closely related species of bacteria. Bacteriocins may help maintain the ecological balance of various enterics in the intestines.

Escherichia   The bacterial species *Escherichia coli* is one of the most common inhabitants of the human intestinal tract and is probably the most familiar organism in microbiology. Recall from previous chapters that a great deal is known about the biochemistry and genetics of *E. coli*, and it continues to be an important tool for basic biological research—many researchers consider it almost a laboratory pet. Its presence in water or food is an indication of fecal contamination (see Chapter 27, page 827). *E. coli* is not usually pathogenic. However, it can be a cause of urinary tract infections, and certain strains produce enterotoxins that cause traveler's diarrhea and occasionally cause very serious foodborne disease (see *E. coli* O157:H7 in Chapter 25, page 759).

Salmonella   Almost all members of the genus *Salmonella* (sal'mön-el-lä) are potentially pathogenic. Accordingly, there are extensive biochemical and serological tests to clinically isolate and identify salmonellae. Salmonellae are common inhabitants of the intestinal tracts of many animals, especially poultry and cattle. Under unsanitary conditions, they can contaminate food.

The nomenclature of the genus *Salmonella* is unusual. Instead of multiple species, members of the genus *Salmonella* that are infectious to warm-blooded animals can be considered for practical purposes to be a single species, *Salmonella enterica* (en-ter'i-kä). This species is divided into more than 2400 **serovars,** that is, *serological varieties.* The term **serotype** is often used to mean the same thing. By way of explanation of these terms, when salmonellae are injected into appropriate animals, their flagella, capsules, and cell walls serve as *antigens* that cause the animals to form *antibodies* in their blood that are specific for each of these structures. Thus, *serological* means are used to differentiate the microorganisms. Serology is discussed more fully in Chapter 18, but for now it will be sufficient to state that it can be used to differentiate and identify bacteria.

A serovar such as *Salmonella typhimurium* (tī-fi-mur'ē-um) is not a species and should be more properly written as "*Salmonella enterica* serovar Typhimurium." The convention now used by the Centers for Disease Control and Prevention (CDC) is to spell out the entire name at the first

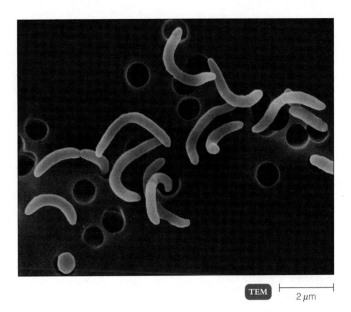

TEM ⊢——⊣ 2 μm

FIGURE 11.8 *Vibrio cholerae.* Notice the slight curvature of these rods, which is a characteristic of the genus.

Q **What disease does *Vibrio cholerae* cause?**

mention and then abbreviate it as, for example, *Salmonella* Typhimurium. For simplicity, we will identify serovars of salmonellae in this text as we would species, that is, *S. typhimurium*, etc.

Specific antibodies, which are available commercially, can be used to differentiate *Salmonella* serovars by a system known as the Kauffmann-White scheme. This scheme designates an organism by numbers and letters that correspond to specific antigens on the organism's capsule, cell wall, and flagella, which are identified by the letters K, O, and H, respectively. For example, the antigenic formula for the bacterium *S. typhimurium* is O1,4,[5],12:H:i,1,2.* Many salmonellae are named only by their antigenic formulas. Serovars can be further differentiated by special biochemical or physiological properties into **biovars,** or **biotypes.**

A recent taxonomic arrangement based upon the latest molecular technology adds another species, *Salmonella bongori* (bon'gôr-ē). This is a resident of "cold-blooded" animals—it was originally isolated from a lizard in the

---

*The letters derive from the original German usage: K represents the German for capsule. (Salmonellae with capsules are identified serologically by a particular capsular antigen named Vi, for virulence.) Colonies that spread in a thin film over the agar surface were described by the German word for film, *hauch.* The motility needed to form a film implied the presence of flagella, and the letter H came to be assigned to the antigens of flagella. Nonmotile bacteria were described as *ohne hauch,* without film, and the O came to be assigned to the cell surface or body antigens. This terminology is also used in the naming of *E. coli* O157:H7, *Vibrio cholerae* O:1 and others.

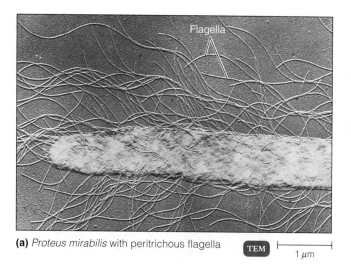

Flagella

(a) *Proteus mirabilis* with peritrichous flagella   TEM   1 µm

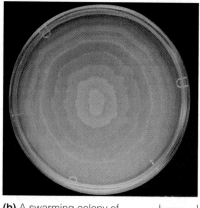

(b) A swarming colony of *Proteus mirabilis*; note concentric rings of growth   20 mm

FIGURE 11.9   *Proteus mirabilis.* Chemical communication between bacterial cells causes changes from cells adapted to swimming in fluid (few flagella) to cells that are able to move on surfaces (numerous flagella). The concentric growth results from periodic synchronized conversion to the highly flagellated form capable of movement on surfaces.

Q **The photo of the *Proteus* cell is probably a swarmer cell. How would you know?**

town of Bongor in the African desert nation of Chad—and is rarely found in humans.

Typhoid fever, caused by *Salmonella typhi* (tī′fē), is the most severe illness caused by any member of the genus *Salmonella*. A less severe gastrointestinal disease caused by other salmonellae is called salmonellosis. Salmonellosis is one of the most common forms of foodborne illness. (See the box in Chapter 25 on page 756.)

Shigella   Species of *Shigella* (shi-gel′lä) are responsible for a disease called bacillary dysentery, or shigellosis. Unlike salmonellae, they are found only in humans. These organisms are second only to *E. coli* as a cause of traveler's diarrhea. Some strains of *Shigella* can cause life-threatening dysentery (see Chapter 25, page 752).

Klebsiella   Members of the genus *Klebsiella* (kleb-sē-el′lä) are commonly found in soil or water. Many isolates are capable of fixing nitrogen from the atmosphere, which has been proposed as being a nutritional advantage in isolated populations with little protein nitrogen in their diet. The species *Klebsiella pneumoniae* (nü-mō′nē-ī) occasionally causes a serious form of pneumonia in humans.

Serratia   *Serratia marcescens* (ser-rä′tē-ä mär-ses′sens) is a bacterial species distinguished by its production of red pigment. In hospital situations the organism can be found on catheters, in saline irrigation solutions, and in other supposedly sterile solutions. Such contamination is probably the cause of many urinary and respiratory tract infections in hospitals.

Proteus   Colonies of *Proteus* (prō′tē-us) bacteria growing on agar exhibit a swarming type of growth. Swarmer cells with many flagella (Figure 11.9a) move outward on the edges of the colony and then revert to normal cells with only a few flagella and reduced motility. Periodically, new generations of highly motile swarmer cells develop and the process is repeated. As a result, a *Proteus* colony has the distinctive appearance of a series of concentric rings (Figure 11.9b). This genus of bacteria is implicated in many infections of the urinary tract and in wounds.

Yersinia   *Yersinia pestis* (yèr-sin′ē-ä pes′tis) causes plague, the Black Death of medieval Europe. Urban rats in some parts of the world and ground squirrels in the American Southwest carry these bacteria. Fleas usually transmit the organisms among animals and to humans, although contact with respiratory droplets from infected animals and people can be involved in transmission.

Erwinia   *Erwinia* (èr-wi′nē-ä) species are primarily plant pathogens; some cause plant soft-rot diseases. These species produce enzymes that hydrolyze the pectin between individual plant cells. This causes the plant cells to separate from each other, a disease that plant pathologists term plant rot.

Enterobacter   Two *Enterobacter* (en-te-rō-bak′tèr) species, *E. cloacae* (klō-ā′kī), and *E. aerogenes* (ā-rä′jen-ēz) can cause urinary tract infections and hospital-acquired infections. They are widely distributed in humans and animals, as well as in water, sewage, and soil.

PASTEURELLALES

The bacteria in this order are nonmotile; they are best known as human and animal pathogens.

Pasteurella   This genus is primarily known as a pathogen of domestic animals. It causes sepsis in cattle, fowl cholera in chickens and other fowl, and pneumonia in several types of animals. The best-known species is *Pasteurella multocida* (pas-tyér-el′lä mul-tō′si-dä), which can be transmitted to humans by dog and cat bites. It is also a prominent member of the microbiota of saliva of the relatively slow-moving Komodo dragon, a large reptile found on an Indonesian island, that bites more mobile prey and waits several days for their death. The Komodo dragon is not venomous, but its prey dies from *P. multocida* introduced by its bite.

Haemophilus   *Haemophilus* (hē-mä′fil-us) is a very important genus of pathogenic bacteria. These organisms commonly inhabit the mucous membranes of the upper respiratory tract, mouth, vagina, and intestinal tract. The best-known species that affects humans is *Haemophilus influenzae* (in-flü-en′zī), named long ago because of the erroneous belief that it was responsible for influenza.

The name *Haemophilus* is derived from the bacteria's requirement for blood in their culture medium (*hemo* = blood). They are unable to synthesize important parts of the cytochrome system needed for respiration, and they obtain these substances from the heme fraction, known as the **X factor,** of blood hemoglobin. The culture medium must also supply the cofactor nicotinamide adenine dinucleotide (from either NAD^+ or $NADP^+$), which is known as **V factor.** Clinical laboratories use tests for the requirement of X and V factors to identify isolates as *Haemophilus* species.

Haemophilus influenzae is responsible for several important diseases. It has been a common cause of meningitis in young children and is a frequent cause of earaches. Other clinical conditions caused by *H. influenzae* include epiglotitis (a life-threatening condition in which the epiglottis becomes infected and inflamed), septic arthritis in children, bronchitis, and pneumonia. *Haemophilus ducreyi* (dü-krā′ē) is the cause of the sexually transmitted disease chancroid.

THE DELTAPROTEOBACTERIA

LEARNING OBJECTIVE

- Make a dichotomous key to distinguish among the deltaproteobacteria described in this chapter.

The deltaproteobacteria are distinctive in that they include some bacteria that are predators on other bacteria. Bacteria in this group are also important contributors to the sulfur cycle.

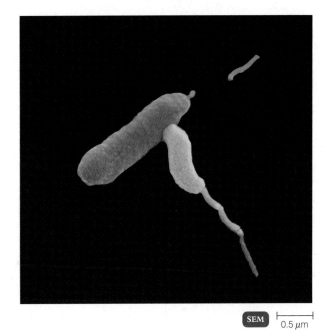

SEM   0.5 µm

FIGURE 11.10 *Bdellovibrio bacteriovorus.* The yellow bacterium is *B. bacteriovorus*. It is attacking a bacterial cell shown in blue.

Q   Would this bacterium attack *Staphylococcus aureus*?

Bdellovibrio   *Bdellovibrio* (del-lō-vib′rē-ō) is a particularly interesting genus. It attacks other gram-negative bacteria. It attaches tightly (*bdella* = leech; Figure 11.10), and after penetrating the outer layer of gram-negative bacteria, it reproduces within the periplasm. There, the cell elongates into a tight spiral, which then fragments almost simultaneously into several individual flagellated cells. The host cell then lyses, releasing the *Bdellovibrio* cells.

DESULFOVIBRIONALES

Members of the order Desulfovibrionales are sulfur reducing bacteria. They are obligately anaerobic bacteria that use oxidized forms of sulfur, such as sulfates (SO_4^{2-}) or elemental sulfur (S^0) rather than oxygen as electron acceptors. The product of this reduction is hydrogen sulfide (H_2S). (Because the H_2S is not assimilated as a nutrient, this type of metabolism is termed *dissimilatory*.) The activity of these bacteria releases millions of tons of H_2S into the atmosphere every year and plays a key part in the sulfur cycle (Figure 27.7). Sulfur-oxidizing bacteria such as *Beggiatoa* are able to use H_2S either as part of photosynthesis or as an autotrophic energy source.

Desulfovibrio   The best studied sulfur-reducing genus is *Desulfovibrio* (dē′sul-fō-vib′rē-ō), which is found in anaerobic sediments and in the intestinal tracts of humans and animals. Sulfur-reducing and sulfate-reducing bacteria use organic compounds such as lactate, ethanol, or fatty acids

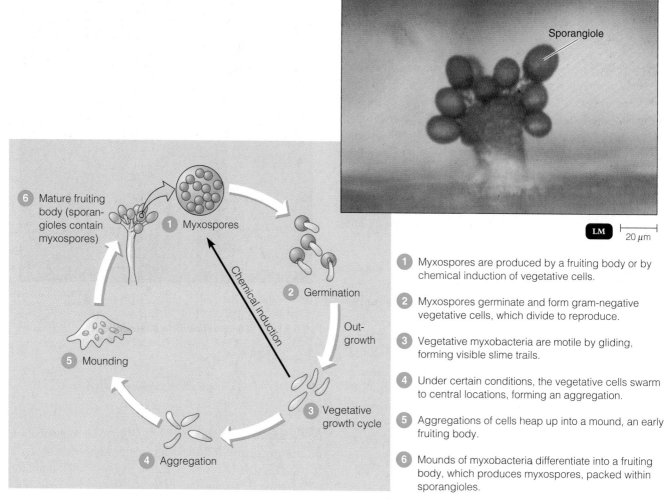

FIGURE 11.11 *Myxococcales.* (**a**) Life cycle of *Myxococcales.* (**b**) A myxobacterium fruiting body; the sporangioles contain myxospores.

Q **What is the feeding stage of this organism?**

---

as electron donors. This reduces sulfur or sulfate to H$_2$S. When H$_2$S reacts with iron it forms insoluble FeS, which is responsible for the black color of many sediments.

MYXOCOCCALES

In the first edition of *Bergey's Manual* the *Myxococcales* were classified among the fruiting and gliding bacteria. They illustrate the most complex life cycle of all bacteria, part of which is predatory upon other bacteria.

Myxococcus Vegetative cells of the myxobacteria (*myxo* = nasal mucus) move by gliding and leave behind a slime trail (Figure 11.10a). *Myxococcus xanthus* (micks-ō-kok′kus zan′thus) and *M. fulvus* (ful′vus) are well-studied representatives of the genus. As they move, their source of nutrition is the bacteria they encounter, enzymatically lyse, and digest. Large numbers of these gram-negative

microbes eventually aggregate (Figure 11.11a). Where the moving cells aggregate, they differentiate and form a macroscopic stalked fruiting body that contains large numbers of resting cells called *myxospores* (Figure 11.11b). Differentiation is usually triggered by low nutrients. Under proper conditions, usually a change in nutrients, the myxospores germinate and form new vegetative gliding cells. You might note the resemblance to the life cycle of the eukaryotic cellular slime molds in Figure 12.22 (page 370).

THE EPSILONPROTEOBACTERIA

LEARNING OBJECTIVE

• Make a dichotomous key to distinguish among the epsilonproteobacteria described in this chapter.

The epsilonproteobacteria are slender gram-negative rods that are helical or vibrioid. *Vibrioid* is a term applied to helical bacteria that do not have a complete turn. We will discuss the two important genera, both of which are motile by means of flagella and are microaerophilic.

Campylobacter Members of the genus *Campylobacter* are microaerophilic vibrios; each cell has one polar flagellum. One species of *Campylobacter*, *C. fetus* (kam′pi-lō-bak-tėr fē′tus), causes spontaneous abortion in domestic animals. Another species, *C. jejuni* (je-ju′ni), is a leading cause of outbreaks of foodborne intestinal disease.

Helicobacter Members of the genus *Helicobacter* are microaerophilic curved rods with multiple flagella. The species *Helicobacter pylori* (hē′lik-ō-bak-tėr pī-lōr′ē) has been identified as the most common cause of peptic ulcers in humans and a cause of stomach cancer (Figure 11.12 and Figure 25.13).

THE NONPROTEOBACTERIA GRAM-NEGATIVE BACTERIA

LEARNING OBJECTIVE

• Make a dichotomous key to distinguish among the gram-negative nonproteobacteria described in this chapter.

There are a number of important gram-negative bacteria that are not closely related to the gram-negative proteobacteria. They include several physiologically and morphologically distinctive photosynthesizing bacteria, such

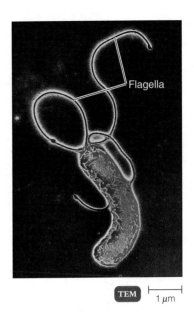

FIGURE 11.12 *Helicobacter pylori,* **an example of a helical bacterium that does not make a complete twist.**

Q How do helical bacteria differ from spirochetes?

as those included in the phyla Cyanobacteria (cyanobacteria), Chlorobi (green sulfur bacteria), and Chloroflexi (green nonsulfur bacteria). The cyanobacteria produce oxygen during photosynthesis (are oxygenic), and the green sulfur and green nonsulfur bacteria do not produce oxygen (are anoxygenic). These groups are summarized in Table 11.2.

TABLE 11.2	Selected Characteristics of Photosynthesizing Bacteria				
Common Name	Example	Phylum	Comments	Electron Donor for CO_2 Reduction	Oxygenic or Anoxygenic
Cyanobacteria	*Anabaena*	Cyanobacteria	Plantlike photosynthesis; some use bacterial photosynthesis under anaerobic conditions	Usually H_2O	Usually oxygenic
Green nonsulfur bacteria	*Chloroflexus*	Chloroflexi	Grow chemoheterotrophically in aerobic environments	Organic compounds	Anoxygenic
Green sulfur bacteria	*Chlorobium*	Chlorobi	Deposit sulfur granules outside cells	Usually H_2S	Anoxygenic
Purple nonsulfur bacteria	*Rhodospirillum*	Proteobacteria	Can grow chemoheterotrophically as well	Organic compounds	Anoxygenic
Purple sulfur bacteria	*Chromatium*	Proteobacteria	Deposit sulfur granules outside cells	Usually H_2S	Anoxygenic

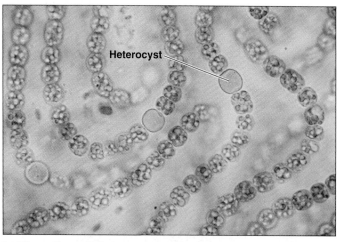

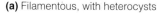

(a) Filamentous, with heterocysts   LM ⊢——⊣ 10 µm

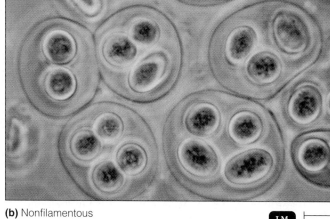

(b) Nonfilamentous   LM ⊢——⊣ 10 µm

(c) Branching filamentous   LM ⊢——⊣ 10 µm

FIGURE 11.13 Cyanobacteria. (**a**) A filamentous cyano-bacterium showing heterocysts, in which nitrogen-fixing activity is located. (**b**) A unicellular, nonfilamentous cyanobacterium, *Gloeocapsa.* Groups of these cells, which divide by binary fission, are held together by the surrounding glycocalyx. (**c**) A branching, filamentous cyanobacterium.

Q **How does the photosynthesis of the cyanobacteria differ from that of the purple sulfur bacteria?**

CYANOBACTERIA (THE OXYGENIC PHOTOSYNTHETIC BACTERIA)

The cyanobacteria, named for their characteristic blue-green (*cyan*) pigmentation, were once called blue-green algae. Although they resemble the eukaryotic algae and often occupy the same environmental niches, this is a misnomer because they are bacteria; algae are not. However, cyanobacteria do carry out oxygen-producing (oxygenic) photosynthesis, as do the eukaryotic plants and algae (see Chapter 12). Many of the cyanobacteria are capable of fixing nitrogen from the atmosphere. In most cases this activity is located in specialized cells called **heterocysts,**

which contain enzymes that fix nitrogen gas (N_2) into ammonium (NH_4^+) that can be used by the growing cell (Figure 11.13a). Species that grow in water usually have gas vacuoles that provide buoyancy, helping the cell float at a favorable environment. Cyanobacteria that move about on solid surfaces use gliding motility.

Cyanobacteria are morphologically varied. They range from unicellular forms that divide by simple binary fission (Figure 11.13b), to colonial forms that divide by multiple fission, to filamentous forms that reproduce by fragmentation of the filaments. The filamentous forms usually exhibit some differentiation of cells that are often bound together within an envelope or sheath (Figure 11.13c).

Evidence indicates that oxygenic cyanobacteria played an important part in the development of life on Earth, which originally had very little free oxygen that would support life as we are familiar with it. Fossil evidence indicates that when cyanobacteria first appeared, the atmosphere contained only about 0.1% free oxygen. When oxygen-producing eukaryotic plants appeared millions of years later, the concentration of oxygen was more than 10%. The increase presumably was a result of photosynthetic activity by cyanobacteria. The atmosphere we breathe today contains about 20% oxygen.

Cyanobacteria, especially those that fix nitrogen, are extremely important to the environment. They occupy environmental niches similar to those occupied by the eukaryotic algae (see Figure 12.11, page 358), but the ability of many of the cyanobacteria to fix nitrogen makes them even more adaptable in nutritionally poor environments. The environmental role of cyanobacteria is presented more fully in Chapter 27, in the discussion of eutrophication (the nutritional overenrichment of bodies of water).

PURPLE AND GREEN PHOTOSYNTHETIC BACTERIA (THE ANOXYGENIC PHOTOSYNTHETIC BACTERIA)

LEARNING OBJECTIVE

• Compare and contrast purple and green photosynthetic bacteria with the cyanobacteria.

The photosynthetic bacteria are taxonomically confusing. The phyla Cyanobacteria, Chlorobi, and Chloroflexi are gram-negative, but they are not genetically included in the proteobacteria. The photosynthetic purple sulfur bacteria and purple nonsulfur bacteria are genetically included in, respectively, the alphaproteobacteria and gammaproteobacteria, although for simplicity we will discuss them at this point. These photosynthetic bacteria, which are not necessarily colored purple or green, are generally anaerobic. Their habitat is usually the deep sediments of lakes and ponds. Like plants, algae, and the cyanobacteria, purple and green bacteria carry out photosynthesis to make carbohydrates (CH_2O). Growing as they do in aquatic depths (see Figure 27.12), these bacteria possess chlorophyll that makes use of parts of the visible spectrum not intercepted by photosynthetic organisms located at higher levels. Also, unlike plantlike photosynthesis, the photosynthesis of purple or green bacteria is *anoxygenic*—it does not produce oxygen.

Cyanobacteria, as well as eukaryotic plants and algae produce oxygen (O_2) from water (H_2O) as they carry out photosynthesis:

$$(1)\ 2H_2O + CO_2 \xrightarrow{\text{light}} (CH_2O) + H_2O + O_2$$

The *purple sulfur* and *green sulfur bacteria* use reduced sulfur compounds, such as hydrogen sulfide (H_2S), instead of water, and they produce granules of sulfur (S^0) rather than oxygen, as follows:

$$(2)\ 2H_2S + CO_2 \xrightarrow{\text{light}} (CH_2O) + H_2O + 2S^0$$

Chromatium (krō-mā′tē-um), shown in Figure 11.14, is a representative genus. At one time, an important question in biology concerned the source of the oxygen produced by

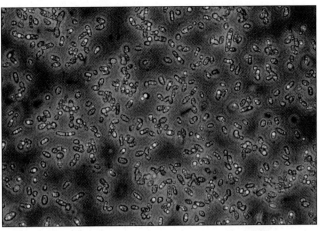

LM |—— 10 μm

FIGURE 11.14 Purple sulfur bacteria. This photomicrograph of cells of the genus *Chromatium* shows the intracellular sulfur granules as multicolored refractile objects. The reason the sulfur accumulates can be surmised from inspection of equation 2 in the discussion.

Q *What does anoxygenic mean?*

plant photosynthesis; was it from CO_2 or from H_2O? Until the introduction of radioisotope tracers, which traced the oxygen in water and carbon dioxide and finally settled the question, comparison of equations 1 and 2 was the best evidence that the oxygen source was from H_2O. It is important, also, to compare these two equations for an understanding of how reduced sulfur compounds, such as H_2S, can substitute for H_2O in photosynthesis. See "Life Without Sunshine" on page 818. (See also the box in Chapter 27 on page 826 and the box in Chapter 6 on page 164.)

Other photoautotrophs, the *purple nonsulfur* and *green nonsulfur bacteria,* use organic compounds, such as acids and carbohydrates, for the photosynthetic reduction of carbon dioxide.

Morphologically, the photosynthetic bacteria are very diverse, with spirals, rods, cocci, and even budding forms.

THE GRAM-POSITIVE BACTERIA

The gram-positive bacteria can be divided into two groups: those that have a high G + C ratio, and those that have a low G + C ratio (see "Nucleic Acids," page 47). To illustrate the variations in G + C ratio, the genus *Streptococcus* has a low G + C content of 33 to 44%; and the genus *Clostridium* has a low content of 21 to 54%. Included with the gram-positive, low G + C bacteria are the mycoplasmas, even though they lack a cell wall and therefore do not have a Gram reaction. Their G + C ratio is 23 to 40%.

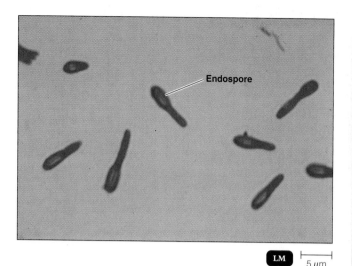

FIGURE 11.15 *Clostridium tetani.* The endospores of clostridia usually distend the cell wall as shown here.

Q What physiological characteristic of *Clostridium* makes it a problem in contamination of deep wounds?

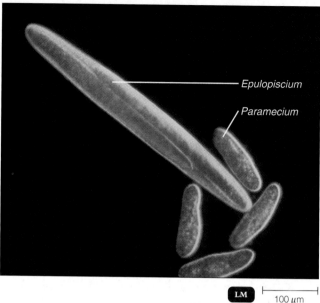

FIGURE 11.16 A giant prokaryote, *Epulopiscium fishelsoni.*

Q Compare the size of this bacterium with the *Paramecium* protozoan, a eukaryotic organism.

By contrast, filamentous actinomycetes of the genus *Streptomyces* have a high G + C content of 69 to 73%. Gram-positive bacteria of a more conventional morphology, such as the genera *Corynebacterium* and *Mycobacterium*, have a G + C content of 51 to 63% and 62 to 70%, respectively.

These bacterial groups are placed into separate phyla, the **Firmicutes** (low G + C ratios) and **Actinobacteria** (high G + C ratios).

FIRMICUTES (LOW G + C GRAM-POSITIVE BACTERIA)

LEARNING OBJECTIVE

• Make a dichotomous key to distinguish among the low G + C gram-positive bacteria described in this chapter.

Low G + C gram-positive bacteria are assigned to the phylum Firmicutes. This group includes important endospore-forming bacteria such as the genera *Clostridium* and *Bacillus*. Also of extreme importance in medical microbiology are the genera *Staphylococcus*, *Enterococcus*, and *Streptococcus*. In industrial microbiology the genus *Lactobacillus*, which produces lactic acid, is well known. The mycoplasma, which do not possess a cell wall, are also found in this phylum.

CLOSTRIDIALES

Clostridium   Members of the genus *Clostridium* (klôs-tri'dē-um) are obligate anaerobes. The rod-shaped cells

contain endospores that usually distend the cell (Figure 11.15). The formation of endospores by bacteria is important to both medicine and the food industry because of the endospore's resistance to heat and many chemicals. Diseases associated with clostridia include tetanus, caused by *C. tetani* (te'tan-e); botulism, caused by *C. botulinum* (bo-tū-lī'num); and gas gangrene, caused by *C. perfringens* (per-frin'jens) and other clostridia. *C. perfringens* is also the cause of a common form of foodborne diarrhea. *C. difficile* (dif'fi-sē-il) is an inhabitant of the intestinal tract that may cause a serious diarrhea. This occurs only when antibiotic therapy alters the normal intestinal microbiota, allowing overgrowth by toxin-producing C. *difficile.*

Epulopiscium   Biologists have long considered bacteria to be small by necessity because they lack the nutrient transport systems used by higher, eukaryotic organisms and because they depend on simple diffusion to obtain nutrients. These characteristics would seem to critically limit size. So, when a cigar-shaped organism living symbiotically in the gut of the Red Sea surgeonfish was first observed in 1985, it was considered to be a protozoan. Certainly, its size suggested this: the organism was as large as 80 μm × 600 μm—over half a millimeter in length—large enough to be seen with the unaided eye (Figure 11.16). Compared to the familiar bacterium *E. coli*, which is about 1 μm × 2 μm, this organism would be about a million times larger in volume.

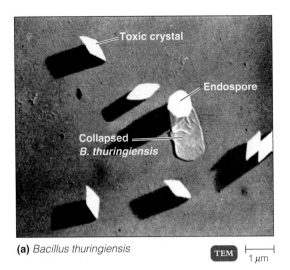

(a) *Bacillus thuringiensis*   TEM   1 µm

FIGURE 11.17 *Bacillus.* (**a**) *Bacillus thuringiensis; the diamond-shaped crystal shown next to the endospore is toxic to insects that ingest it. This electron micrograph was made using the technique of shadow casting described on page 66. (**b**) A germinating cell of Bacillus cereus.*

Q **What structure is made by both *Clostridium* and *Bacillus*?**

Further investigation of the new organism showed that certain external structures thought to resemble the cilia of protozoa were actually similar to bacterial flagella, and it did not have a membrane-enclosed nucleus. Ribosomal RNA analysis conclusively placed *Epulopiscium* (ep′ū-lō-pis-ē-um) with the prokaryotes. (The name means "guest at the banquet of a fish." It is literally bathed in semidigested food.) It most closely resembles grampositive bacteria of the genus *Clostridium*. Strangely, the species *Epulopiscium fishelsoni* (fish-el-sō′nē) does not reproduce by binary fission. Daughter cells formed within the cell are released through a slit opening in the parent cell. This may be related to the evolutionary development of sporulation.

Recently it was discovered that this bacterium does not rely on diffusion to distribute nutrients. Instead, it makes use of its larger genetic capacity—it has 25 times as much DNA as a human cell and as many as 85,000 copies of at least one gene—to manufacture proteins at internal sites where they are needed. *E. fishelsoni* bacteria are large enough for researchers to insert microprobes and directly study such aspects of bacterial physiology, something not possible with conventionally sized bacteria. (We will describe another, more recently discovered giant bacterium, *Thiomargarita*, on page 340.)

BACILLALES

This order includes several important genera of grampositive rods and cocci.

(b) *Bacillus cereus*   TEM   1 µm
germinating

Bacillus Bacteria of the genus *Bacillus* are typically rods that produce endospores. They are common in soil and only a few are pathogenic to humans. Several species produce antibiotics.

Bacillus anthracis (bä-sil′lus an-thrā′sis) causes anthrax, a disease of cattle, sheep, and horses that can be transmitted to humans. It is often mentioned as a possible agent of biological warfare. (See the box in Chapter 23 on page 677.) The anthrax bacillus is a nonmotile facultative anaerobe, often forming chains in culture. The centrally located endospore does not distend the walls. *Bacillus thuringiensis* (thur-in-jē-en′sis) is probably the best-known microbial insect pathogen (Figure 11.17a). It produces intracellular crystals when it sporulates. Commercial preparations containing endospores and crystalline toxin of this bacterium are sold in gardening supply shops to be sprayed on plants. *Bacillus cereus* (se′rē-us) (Figure 11.17b) is a common bacterium in the environment and occasionally is identified as a cause of food poisoning, especially in starchy foods such as rice.

The three species of the genus *Bacillus* that we have just described are dramatically different in important ways, especially their disease-causing properties. However, they are so closely related that taxonomists consider them to be variants of a single species, differing almost entirely in

SEM ⊢—⊣ 1 μm

FIGURE 11.18 ***Staphylococcus aureus.*** Notice the grape-like clusters of these gram-positive cocci.

Q **What is an environmental advantage of a spherical shape?**

genes carried on plasmids, which are easily transferred from one bacterium to another.

Staphylococcus Staphylococci typically occur in grape-like clusters (Figure 11.18). The most important staphylococcal species is *Staphylococcus aureus* (staf-i-lō-kok'kus ô'rē-us), which is named for its yellow-pigmented colonies (*aureus* = golden). Members of this species are facultative anaerobes.

Some characteristics of the staphylococci account for their pathogenicity, which takes many forms. They grow comparatively well under conditions of high osmotic pressure and low moisture, which partially explains why they can grow and survive in nasal secretions (many of us carry the bacteria in our noses) and on the skin. This also explains how *S. aureus* can grow in some foods with high osmotic pressure (such as ham and other cured meats) or in low-moisture foods that tend to inhibit the growth of other organisms. The yellow pigment probably confers some protection from the antimicrobial effects of sunlight.

S. aureus produces many toxins that contribute to the bacterium's pathogenicity by increasing its ability to invade the body or damage tissue. (See the box in Chapter 21, page 633.) The infection of surgical wounds by *S. aureus* is a common problem in hospitals. And its ability to develop resistance quickly to such antibiotics as penicillin contributes to its danger to patients in hospital environments. *S. aureus* produces the toxin responsible for toxic shock syndrome, a severe infection characterized by high fever and vomiting, sometimes even death. *S. aureus* also produces an **enterotoxin** that causes vomiting and nausea when ingested; it is one of the most common causes of food poisoning.

LACTOBACILLALES

Several important genera are found in the order *Lactobacillales.* The genus *Lactobacillus* is a representative of the industrially important lactic acid–producing bacteria.

Most lack a cytochrome system and are unable to use oxygen as an electron acceptor. Unlike most obligate anaerobes, though, they are aerotolerant and capable of growth in the presence of oxygen. But compared to oxygen-utilizing microbes, they grow poorly. However, the production of lactic acid from simple carbohydrates inhibits the growth of competing organisms and allows them to grow competitively in spite of their inefficient metabolism. The genus *Streptococcus* shares the metabolic characteristics of the genus *Lactobacillus.* There are several industrially important species, but the streptococci are best known for their pathogenicity. The genera *Staphylococcus* and *Listeria* are more conventional metabolically. Both are facultative anaerobes and several species are important pathogens.

Lactobacillus In humans, bacteria of the genus *Lactobacillus* (lak-tō-bä-sil'lus) are located in the vagina, intestinal tract, and oral cavity. Lactobacilli are used commercially in the production of sauerkraut, pickles, buttermilk, and yogurt. Typically, a succession of lactobacilli, each more acid tolerant than its predecessor, participates in these lactic acid fermentations.

Streptococcus Members of the genus *Streptococcus* (strep-tō-kok'kus) are spherical, gram-positive bacteria that typically appear in chains (Figure 11.19). They are a taxonomically complex group, probably responsible for more illnesses and causing a greater variety of diseases than any other group of bacteria.

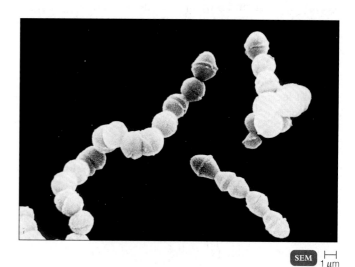

SEM ⊢—⊣ 1 μm

FIGURE 11.19 ***Streptococcus.*** Notice the chains of cells characteristic of most streptococci. Many of the spherical cells are dividing and are somewhat oval in appearance—especially when viewed with a light microscope, which has lower magnification than this electron micrograph.

Q **How does the arrangement of *Streptococcus* differ from *Staphylococcus*?**

Pathogenic streptococci produce several extracellular substances that contribute to their pathogenicity. Among them are products that destroy phagocytic cells that ingest them. Enzymes produced by some streptococci spread infections by digesting connective tissue of the host, which may also result in extensive tissue destruction. (See the discussion on necrotizing fasciitis on page 621). Infections are also allowed to spread from sites of injury by enzymes that lyse the fibrin (a threadlike protein) of blood clots.

A few nonpathogenic species of streptococci are important in the production of dairy products (see Chapter 28, page 847).

Beta-hemolytic streptococci. A useful basis for the classification of some streptococci is their colonial appearance when grown on blood agar. The *beta-hemolytic* species produce a hemolysin that forms a clear zone of hemolysis on blood agar (see Figure 6.8 on page 172). This group includes the principal pathogen of the streptococci, *Streptococcus pyogenes* (pī-äj′en-ēz), also known as the beta-hemolytic group A streptococcus. Group A represents one of an antigenic group (A through G) within the hemolytic streptococci. Among the diseases caused by *S. pyogenes* are scarlet fever, pharyngitis (sore throat), erysipelas, impetigo, and rheumatic fever. The most important virulence factor is the M protein on the bacterial surface (see Figure 21.5, page 620) by which the bacteria avoid phagocytosis. Another member of the beta-hemolytic streptococci is *Streptococcus agalactiae* (ā′gal-acī-ē-ī), in the beta-hemolytic group B. It is the only species with the group B antigen and is the cause of an important disease of the newborn, neonatal sepsis.

Non-beta-hemolytic streptococci. Certain streptococci are not beta-hemolytic, but when grown on blood agar, their colonies are surrounded by a distinctive greening. These are the *alpha-hemolytic* streptococci. The greening represents a partial destruction of the red blood cells caused mostly by the action of bacteria-produced hydrogen peroxide, but it appears only when the bacteria grow in the presence of oxygen. The most important pathogen in this group is *Streptococcus pneumoniae*, the cause of pneumococcal pneumonia. Also included among the alpha-hemolytic streptococci are species of streptococci called *viridans streptococci*. However, not all species form the alpha-hemolytic greening (*virescent* = green) so this is not really a satisfactory group name. Probably the most significant pathogen of the group is *Streptococcus mutans* (mū′tans), the primary cause of dental caries.

Enterococcus The enterococci are adapted to areas of the body that are rich in nutrients but low in oxygen, such as the gastrointestinal tract, vagina, and oral cavity. They are also found in large numbers in human stool. Because they are relatively hardy microbes, they persist as contaminants in a hospital environment, on hands, bedding, and even as a fecal aerosol. In recent years they have become a leading cause of nosocomial infections, especially because of their high resistance to most antibiotics. Two species, *Enterococcus faecalis* (en-te-rō-kok′kus fe-kā′lis) and *Enterococcus faecium* (fē′sē-um), are responsible for much of the infections of surgical wounds and the urinary tract. In medical settings they frequently enter the bloodstream through invasive procedures, such as indwelling catheters.

Listeria The pathogenic species of the genus *Listeria*, *Listeria monocytogenes* (lis-te′rē-ä mo-nō-sī-to′je-nēz), can contaminate food, especially dairy products. Important characteristics of *L. monocytogenes* are that it survives within phagocytic cells and is capable of growth at refrigeration temperatures. If it infects a pregnant woman, the organism poses the threat of stillbirth or serious damage to the fetus.

MYCOPLASMATALES

The mycoplasmas are highly pleomorphic because they lack a cell wall (Figure 11.20) and can produce filaments that resemble fungi, hence their name (*mykes* = fungus, and *plasma* = formed). Cells of the genus *Mycoplasma* (mī-ko-plaz′ma) are very small, ranging in size from 0.1 to 0.25 μm, with a cell volume that is only about 5% of that of a typical bacillus. Because their size and plasticity allowed them to pass through filters that retained bacteria, they were originally considered to be viruses. Mycoplasmas may represent the smallest self-replicating organisms that are capable of a cell-free existence. Studies of their DNA suggest that they are genetically related to the gram-positive bacterial group that includes the genera *Bacillus*, *Streptococcus*, and *Lactobacillus* but have gradually lost genetic material. The term *degenerative evolution* has been used to describe this process.

The most significant human pathogen among the mycoplasmas is *M. pneumoniae* (nu-mō′nē-ī), which is the cause of a common form of mild pneumonia. Other genera in the Order Mycoplasmatales are *Spiroplasma* (spī-rō-plaz′mä), cells with a tight corkscrew morphology that are serious plant pathogens and common parasites of plant-feeding insects, and *Ureaplasma* (ū-rē-ä-plaz′mä), so named because they can enzymatically split the urea in urine and are occasionally associated with urinary tract infections.

Mycoplasmas can be grown on artificial media that provide them with sterols (if necessary) and other special nutritional or physical requirements. Colonies are less than 1 mm in diameter and have a characteristic "fried egg" appearance when viewed under magnification (see

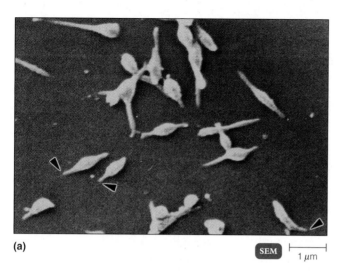

(a)

SEM | 1 μm

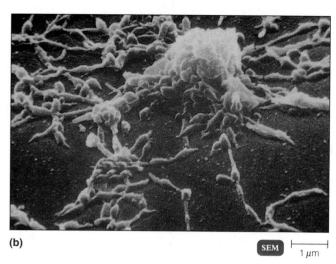

(b)

SEM | 1 μm

FIGURE 11.20 *Mycoplasma pneumoniae.* Bacteria such as *M. pneumoniae* have no cell walls, and their morphology is irregular (pleomorphic). (**a**) Individual cells of *M. pneumoniae.* Arrowheads indicate terminal structures that probably aid in attachment to eukaryotic cells, which then become infected. (**b**) This micrograph shows the filamentous growth of *M. pneumoniae.* Some individual cells can also be seen. The organism reproduces by fragmentation of the filaments at the bulges.

Q **How can the cell structure of mycoplasmas account for their pleomorphism?**

Figure 24.14, page 726). For many purposes, cell culture methods are often more satisfactory. In fact, mycoplasmas grow so well by this method that they are a frequent contamination problem in cell culture laboratories.

ACTINOBACTERIA (HIGH G + C GRAM-POSITIVE BACTERIA)

LEARNING OBJECTIVE

- Make a dichotomous key to distinguish among the high G + C gram-positive bacteria described in this chapter.

High G + C gram-positive bacteria are in the phylum Actinobacteria. Many bacteria in this phylum are highly pleomorphic in their morphology; the genera *Corynebacterium* and *Gardnerella*, for example, and several genera such as *Streptomyces* grow only as extended, often branching filaments. Several important pathogenic genera are found in the Actinobacteria, such as the *Mycobacterium* species causing tuberculosis and leprosy. The genera *Streptomyces*, *Frankia*, *Actinomyces*, and *Nocardia* are often informally called actinomycetes (from the Greek *actino* = ray), because they have a radiate, or starlike, form of growth by reason of their often-branching filaments. Superficially, their morphology resembles that of filamentous fungi; however, the filaments of actinomycetes are prokaryotic cells with a diameter much smaller than that of the eukaryotic molds. Some actinomycetes further resemble molds by their possession of externally carried asexual spores that are used for reproduction. Filamentous bacteria, like filamentous fungi, are very common inhabitants in soil, where a filamentous pattern of growth has advantages. The filamentous organism can bridge water-free gaps between soil particles to move to a new nutritional site. This morphology also gives the organism a much higher surface-to-volume ratio and improves its ability to absorb nutrients in the highly competitive soil environment.

Mycobacterium The mycobacteria are aerobic, non–endospore-forming rods. The name *myco*, meaning fungus-like, was derived from their occasional exhibition of filamentous growth (see Figure 24.9, page 720). Many of the characteristics of mycobacteria, such as acid-fast staining, drug resistance, and pathogenicity, are related to their distinctive cell wall, which is structurally similar to gram-negative bacteria (see Figure 4.13c, page 86). However, the outermost lipopolysaccharide layer in mycobacteria is replaced by mycolic acids, which form a waxy, water-resistant layer. This makes the bacteria resistant to stresses such as drying. Also, few antimicrobial drugs are able to enter the cell. (See the box in Chapter 7 on page 203.) Nutrients enter the cell through this layer very slowly, which is a factor in the slow growth rate of mycobacteria; it sometimes takes weeks for visible colonies to appear. The mycobacteria include the important pathogens

Mycobacterium tuberculosis (mī-kō-bak-ti′rē-um tü-ber-kū-lō′sis), which causes tuberculosis, and M. *leprae* (lep′rī), which causes leprosy. A number of other mycobacteria species are found in soil and water and are occasional pathogens.

Corynebacterium The corynebacteria (*coryne* = club-shaped) tend to be pleomorphic, and their morphology often varies with the age of the cells. The best-known species is *Corynebacterium diphtheriae* (kôr′i-nē-bak-ti-rē-um dif-thi′rē-ī), the causative agent of diphtheria.

Propionibacterium The name of this genus is derived from the organism's ability to form propionic acid; some species are important in the fermentation of Swiss cheese. *Propionibacterium acnes* (prō-pē-on′ē-bak-ti-rē-um ak′nēz) are bacteria that are commonly found on human skin and are implicated as the primary bacterial cause of acne.

Gardnerella *Gardnerella vaginalis* (gard-ne-rel′la va-jin-al′is) is a bacterium that causes one of the most common forms of vaginitis. There has always been some difficulty in assigning a taxonomic position in this species, which is gram-variable, and which exhibits a highly pleomorphic morphology.

Frankia The genus *Frankia* (frank′ē-ä) causes nitrogen-fixing nodules to form in alder tree roots, much as rhizobia cause nodules on the roots of legumes (see Chapter 27).

Streptomyces The genus *Streptomyces* (strep-tō-mī′ses) is the best known of the actinomycetes and is one of the bacteria most commonly isolated from soil (Figure 11.21). The reproductive asexual spores of *Streptomyces* are formed at the ends of aerial filaments. If each spore lands on a suitable substrate, it is capable of germinating into a new colony. These organisms are strict aerobes. They often produce extracellular enzymes that enable them to utilize proteins; polysaccharides, such as starch; cellulose; and many other organic materials found in soil. *Streptomyces* characteristically produce a gaseous compound called *geosmin*, which gives fresh soil its typical musty odor. Species of *Streptomyces* are valuable because they produce most of our commercial antibiotics (see Table 20.1, page 582). This has led to intensive study of the genus—there are nearly 500 described species.

Actinomyces The genus *Actinomyces* (ak-tin-ō-mī′sēs) consists of facultative anaerobes that are found in the mouth and throat of humans and animals. They occasionally form filaments that can fragment (Figure 11.22). One species, *Actinomyces israelii* (is-rā′lē-ē), causes actinomycosis, a tissue-destroying disease usually affecting the head, neck, or lungs.

(a)

(b) SEM 1 μm

FIGURE 11.21 *Streptomyces.* (**a**) Drawing of a typical *Streptomyces* showing filamentous, branching growth with asexual reproductive conidiospores at the filament tips. (**b**) A chain of conidiospores is surrounded by filaments of the streptomycete.

Q **What is an environmental advantage of a filamentous shape?**

Nocardia The genus *Nocardia* (nō-kär′dē-ä) morphologically resemble *Actinomyces*; however, they are aerobic. To reproduce, they form rudimentary filaments, which fragment into short rods. The structure of their cell wall resembles that of the mycobacteria; therefore, they are often acid-fast. *Nocardia* species are common in soil. Some species, such as *Nocardia asteroides* (as′ter-oi-dēz), occasionally cause a chronic, difficult-to-treat pulmonary infection. *N. asteroides* is also one of the causative agents of mycetoma, a localized destructive infection of the feet or hands.

* * *

The fifth and final volume of the second edition of *Bergey's Manual of Systematic Bacteriology* is scheduled to contain

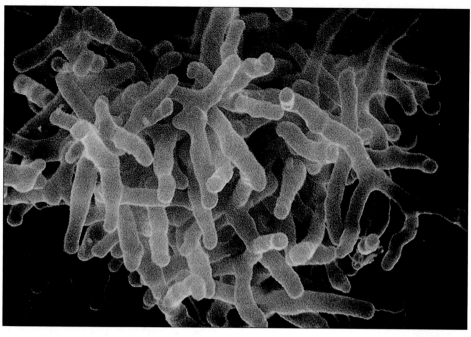

SEM | 1 μm

FIGURE 11.22 *Actinomyces.* Notice the branched filamentous morphology.

Q **Why are these bacteria not classified as fungi?**

a varied assortment of phyla such as Chlamydiae, Spirochaetes, Bacteroidetes, and Fusobacteria. Several important pathogens, such as the genera *Chlamydia*, *Borrelia*, and *Treponema*, are included. Members of the genera *Bacteroides* and *Fusobacterium* are profuse and important inhabitants of the human intestinal tract.

CHLAMYDIAE

LEARNING OBJECTIVE
- Make a dichotomous key to distinguish chlamydias, spirochetes, *Cytophaga*, Bacteroidetes, and Fusobacteria.

Members of the Phylum Chlamydiae are grouped with other genetically similar bacteria that do not contain peptidoglycan in their cell walls. We will discuss only the genera *Chlamydia* and *Chlamydophila*. Earlier editions of *Bergey's Manual of Systematic Bacteriology* grouped these bacteria with the rickettsial bacteria because they all grow intracellularly within host cells. The rickettsia are now classified according to their genetic content with the alphaproteobacteria.

Chlamydia and Chlamydophila These two genera of bacteria, which we will call by the common name of the

chlamydias, have a unique developmental cycle that is perhaps their most distinguishing characteristic (Figure 11.23a). They are gram-negative coccoid bacteria (Figure 11.23b). The **elementary body** shown in Figure 11.23 is the infective agent. Unlike the rickettsias, chlamydias do not require insects or ticks for transmission. They are transmitted to humans by interpersonal contact or by airborne respiratory routes. The chlamydias can be cultivated in laboratory animals, cell cultures, or in the yolk sac of embryonated chicken eggs.

There are three species of the chlamydias that are significant pathogens for humans. *Chlamydia trachomatis* (kla-mi'dē-ä trä-kō'mä-tis) is the best known pathogen of the group and responsible for more than one major disease. These include trachoma, one of the most common causes of blindness in humans in the less developed world. It is also considered to be the primary causative agent of both nongonococcal urethritis, which may be the most common sexually transmitted disease in the United States, and lymphogranuloma venereum, another sexually transmitted disease.

Two members of the genus *Chlamydophila* (kla-mid-o'fil-ä) are well-known pathogens. *Chlamydophila psittaci* (sit'tä-sē) is the causative agent of the respiratory disease psittacosis (ornithosis). *Chlamydophila pneumoniae*

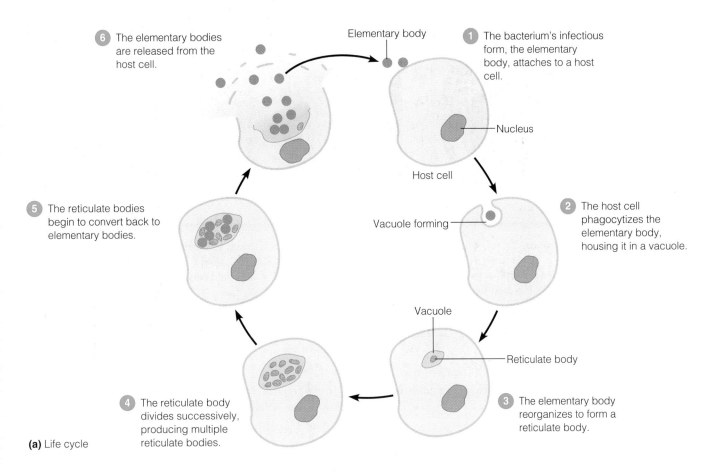

6　The elementary bodies are released from the host cell.

Elementary body

1　The bacterium's infectious form, the elementary body, attaches to a host cell.

Nucleus

Host cell

5　The reticulate bodies begin to convert back to elementary bodies.

2　The host cell phagocytizes the elementary body, housing it in a vacuole.

Vacuole forming

Vacuole

Reticulate body

4　The reticulate body divides successively, producing multiple reticulate bodies.

3　The elementary body reorganizes to form a reticulate body.

(a) Life cycle

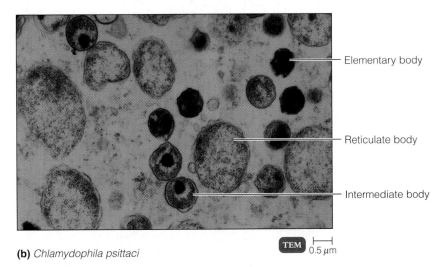

Elementary body

Reticulate body

Intermediate body

TEM ⊢ 0.5 μm

(b) *Chlamydophila psittaci*

FIGURE 11.23 Chlamydias. (a) The generalized life cycle of a chlamydia, which takes about 48 hours to complete. **(b)** A micrograph of *Chlamydophila* in a slice of the cytoplasm of a host cell. The **elementary bodies,** which are the infectious stage, are dense, dark, and relatively small. **Reticulate bodies,** the form in which chlamydia reproduce within the host cell, are larger with a speckled appearance. **Intermediate bodies,** a stage between the change from reticulate to elementary bodies are dark centered.

Q **Which stage of the life cycle is infectious to humans?**

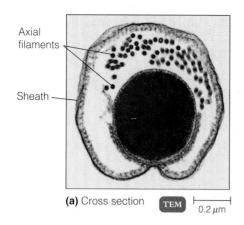

(a) Cross section TEM 0.2 μm

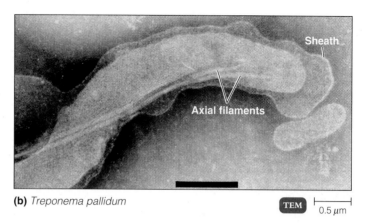

(b) *Treponema pallidum* TEM 0.5 μm

FIGURE 11.24 Spirochetes. Spirochetes are helical and have axial filaments under an outer sheath that enables them to move by a corkscrewlike rotation. **(a)** This section of a spirochete shows numerous axial filaments between the dark cell and the outer sheath. **(b)** This electron micrograph of a portion of *Treponema pallidum* shows the sheath, which has shrunk away from the cell during preparation for microscopy, and two axial filaments, which are attached near one end of the bacterial cell under the sheath.

Q **How does a spirochete's motility differ from that of *Spirillum* (see Figure 11.4)?**

(nü-mō′nē-ī) is the cause of a mild form of pneumonia that is especially prevalent in young adults.

SPIROCHAETES

The spirochetes have a coiled morphology, resembling a metal spring; some are more tightly coiled than others. The most distinctive characteristic of this order, however, is their method of motility, which makes use of two or more **axial filaments** (or *endoflagella*) enclosed in the space between an outer sheath and the body of the cell. One end of each axial filament is attached near a pole of the cell (see Figure 4.10, page 84, and Figure 11.24). By rotating its axial filament, the cell rotates in the opposite direction, like a corkscrew, which is very efficient in moving the organism through liquids. For bacteria, this is more difficult than it might seem. At the scale of a bacterium, water is as viscous as molasses is to a human. However, a bacterium can typically move about 100 times its body length in a second (or about 50 μm/sec), whereas a large fish, such as a tuna, can move only about 10 times its body length in this time.

Many spirochetes are found in the human oral cavity and are probably among the first microorganisms described by van Leeuwenhoek in the 1600s that he found in saliva and tooth scrapings. An extraordinary location for spirochetes is on the surfaces of some of the cellulose-digesting protozoa found in termites, where they may function as substitutes for flagella.

Treponema The spirochetes include a number of important pathogenic bacteria. The best known is the genus *Treponema* (tre-pō-nē′mä), which includes *Treponema pallidum* (pal′li-dum), the cause of syphilis (Figure 11.24b).

Borrelia Members of the genus *Borrelia* (bor′rel-ē-ä) cause relapsing fever and Lyme disease, serious diseases that are usually transmitted by ticks or lice.

Leptospira Leptospirosis is a disease usually spread to humans by water contaminated by *Leptospira* (lep-tō-spī′ra) species. The bacteria are excreted in the urine of such animals as dogs, rats, and swine, so domestic dogs and cats are routinely immunized against leptospirosis. The tightly coiled cells of *Leptospira* are shown in Figure 26.4.

BACTEROIDETES

The phylum Bacteroidetes includes several genera of anaerobic bacteria. Included are the genus *Bacteroides*, a common inhabitant of the human intestinal tract, and the genus *Prevotella*, found in the human mouth. Also included in the phylum Bacteroidetes are the important soil bacteria with gliding motility of the genus *Cytophaga*.

Bacteroides Bacteria of the genus *Bacteroides* (bak-tė-roi′dēz) live in the human intestinal tract in numbers approaching 1 billion per gram of feces. Some *Bacteroides* species also reside in anaerobic habitats such as the gingival crevice (see Figure 25.2) and are also frequently recovered from deep tissue infections. *Bacteroides* organisms are nonmotile and do not form endospores. Infections caused by *Bacteroides* often result from puncture

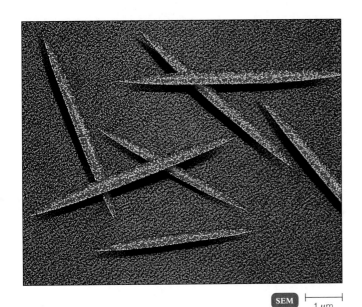

FIGURE 11.25 *Fusobacterium.* This is a common, anaerobic rod found in the human intestine. Notice the characteristic pointed ends.

Q **In what other place in the human body do you often find Fusobacterium?**

wounds or surgery and are a frequent cause of peritonitis, an inflammation resulting from a perforated bowel.

Cytophaga Members of the genus *Cytophaga* (sī-täf′ag-a) are important in the degradation of cellulose and chitin, which are both abundant in soil. Gliding motility places the microbe in close contact with these substrates so that enzymatic action is very efficient.

FUSOBACTERIA

The fusiform bacteria comprise another phylum of anaerobes. These bacteria are often pleomorphic but, as their name suggests, may be spindle-shaped (*fuso* = spindle).

Fusobacterium Members of the genus *Fusobacterium* (fū-sō-bak-ti′rē-um) are long and slender, with pointed rather than blunt ends (Figure 11.25). In humans, they are found most often in the gingival crevice of the gums and may be responsible for some dental abscesses.

DOMAIN ARCHAEA

In the late 1970s, a distinctive type of prokaryotic cell was discovered. Most strikingly, their cell walls lacked the peptidoglycan common to most bacteria. It soon became clear that they also shared many rRNA sequences, and the sequences were different from either those of the Domain Bacteria or the eukaryotic organisms. These differences were so significant that these organisms now constitute a new taxonomic grouping, the Domain Archaea.

DIVERSITY WITHIN THE ARCHAEA

LEARNING OBJECTIVE
• Name a habitat for each group of archaea.

This exceptionally interesting group of prokaryotes is highly diverse. Most archaea are of conventional morphology, that is, rods, cocci, and helixes, but some are of very unusual morphology, as illustrated in Figure 11.26. Some are gram-positive, others gram-negative; some may divide by binary fission, others by fragmentation or budding; a few lack cell walls. Organisms in this domain are physiologically diverse as well, ranging from aerobic, to facultatively anaerobic, to strictly anaerobic. Nutritionally, they

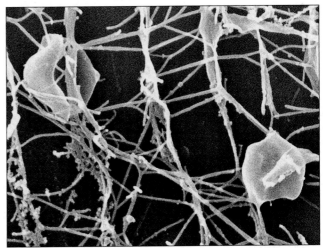

FIGURE 11.26 **Archaea.** *Pyrodictium abyssi,* an unusual member of the archaea found growing in deep ocean sediment at a temperature of 110°C. The cells are disk-shaped with a network of tubules (cannulae). Most archaea are more conventional in their morphology.

Q **Do the terms included in the name, *pyro* and *abyssi,* suggest a basis for the naming of this bacterium?**

include chemoautotrophs, photoautotrophs, and chemo-heterotrophs. Of special interest to microbiologists is the fact that the archaea are frequent inhabitants of exceptionally extreme environments of heat, cold, acidity, and pressure.

Prominent among the archaea are the extreme halophiles, bacteria that survive in very high concentrations of salt, such as the Great Salt Lake and solar evaporating ponds. Examples are *Halobacterium* (hā-lō-bak-ti′rē-um) (see the box in Chapter 5, page 147) and *Halococcus* (hā′lō-kok-kus), which live in high concentrations of sodium chloride (NaCl) and actually require such environmental conditions for growth.

Other archaea thrive in acidic, sulfur-rich hot springs. One such organism is *Sulfolobus* (sul′fō-lō-bus), which has a pH optimum of about 2 and a temperature optimum of more than 70°C. Archaea are also found in ocean depths near hydrothermal vents. Examples of hyperthermophiles, archaea that can thrive in extraordinarily high temperatures, are described in the box in Chapter 6, page 164.

The obligately anaerobic methane-producing members of the archaea, the genus *Methanobacterium* (meth-a-nō-bak-tėr′ē-um), is of considerable economic importance. These bacteria are used in sewage-treatment processes (see Chapter 27) where they derive energy from combining hydrogen (H_2) with carbon dioxide (CO_2) to form methane (CH_4). An essential part of treating sewage is encouraging the growth of these microbes in anaerobic digestion tanks to convert the sludge into CH_4. Methanogens are also part of microbiota of the human colon, vagina, and mouth.

MICROBIAL DIVERSITY

The Earth provides a seemingly infinite number of environmental niches, and novel life forms have evolved to fill them. Many of the microbes that exist in these niches cannot be cultivated by conventional methods on conventional growth media and have remained unknown. In recent years, however, isolation and identification methods have become much more sophisticated and microbes that fill these niches are being identified—many without being cultivated. Particularly interesting are bacteria that test the theoretical limits of size for prokaryotes.

DISCOVERIES ILLUSTRATING THE RANGE OF DIVERSITY

LEARNING OBJECTIVE
• List two factors that contribute to the limits of our knowledge of microbial diversity.

Earlier in this chapter, we described the giant bacterium *Epulopiscium*. In 1999, another giant bacterium was discovered 100 meters deep in the sediments of the coastal waters off Namibia, on the southwestern coast of Africa. Named *Thiomargarita namibiensis* (thī′ō-mär-gär-ē-tà na′mi-bē-ėn-sis), meaning "sulfur pearl of Namibia," these spherical organisms, classified with the gammaproteobacteria, are as large as 750 μm in diameter (Figure 11.27). This is a bit larger than the size of a period at the end of this sentence and even larger than *Epulopiscium*. As we have mentioned, a factor that limits the size of prokaryotic cells is that nutrients must enter the cytoplasm by simple

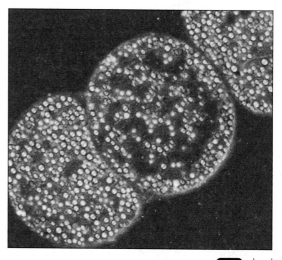

LM ⊢──┤ 100 μm

FIGURE 11.27 *Thiomargarita namibiensis.* *Thiomargarita namibiensis* gets its energy from reduced sulfur compounds such as hydrogen sulfide.

Q **Is a bacterium of this size theoretically possible if the interior were cytoplasm rather than a fluid-filled vacuole?**

diffusion. *T. namibiensis* minimizes this problem by resembling a fluid-filled balloon, the vacuole in the interior being surrounded by a relatively thin outer layer of cytoplasm. This cytoplasm is equal in volume to that of most other prokaryotes. Its energy source is essentially hydrogen sulfide, which is plentiful in the sediments in which it is normally found, and nitrate, which it must extract intermittently from nitrate-rich seawaters when storms stir the loose sediment. The cell's interior vacuole, which makes up about 98% of the bacterium's volume, serves as a storage space to hold the nitrate between recharging of its supply. The cell's energy is derived from the oxidation of hydrogen sulfide; the nitrate, while a source of nutritional nitrogen, primarily serves as an electron acceptor in the absence of oxygen.

The discovery of uniquely large bacteria has raised the question of how large a prokaryotic cell can be and still absorb nutrients by diffusion. At the other extreme, there are reports of bacteria as small as 0.02 to 0.03 μm (**nanobacteria**), found in deep rock formations. Many microbiologists think that nanobacteria are nonliving microscopic artifacts, but even so another question arises: Is there a lower size limit for microorganisms? (See the discussion of the genus *Mycoplasma* on page 333.) Some scientists have used theoretical considerations to calculate that a cell with a significant metabolism would have to have a diameter of at least 0.1 μm. A bacterium that infects plants and insects, *Phytoplasma asteris* (fy′tō-plaz-ma as′zér-is), carries genetic minimalism to an extreme. It uses just 754 genes, many of which are multiple copies—far fewer than any other known organism. It is incapable even of synthesizing ATP.

Until now, microbiologists have described only about 5000 bacterial species, of which about 3000 are listed in *Bergey's Manual*. The true number may be in the millions. Many bacteria in soil or water, or elsewhere in nature, cannot be cultivated with the media and conditions normally used for bacterial growth. Moreover, some bacteria are part of complex food chains and can only grow in the presence of other microbes that supply specific growth requirements. Recently, researchers have been using the polymerase chain reaction (PCR) to make millions of copies of genes found at random in a soil sample. By comparing the genes found in many repetitions of this process, researchers can estimate the different bacterial species in such a sample. One report indicates that a single gram of soil may contain 10,000 or so bacterial types—about twice as many as have ever been described. What may be one of the most abundant life forms on Earth was first observed in seawater only recently. Because it could not be cultivated by conventional methods, its presence was first detected by recovering DNA from seawater and looking for genes that code for certain ribosomal RNA (See the FISH technique on page 304 in Chapter 10). Originally called SAR11 (because it was found in the Sargasso Sea in the mid-Atlantic), it is now named *Pelagibacter ubique* (pel-aj′ē-bak-tér ū′bēk), and methods have been found to cultivate it. An extremely small bacterium, a little over 0.3 μm in diameter, it is barely large enough to be a free-living cell. Estimates are that its populations are so vast that it represents 20% of the prokaryotes in the Earth's seas and perhaps 0.5% of all the prokaryotes on Earth. It apparently metabolizes the waste products of photosynthetic organisms.

In Chapter 27, which discusses environmental microbiology, you will encounter bacteria that live in extraordinary environmental niches kilometers deep in rock and even in clouds. There are also **extremophiles,** bacteria that survive and reproduce under incredible conditions of immense pressures and high temperatures such as found around hydrothermal vents deep in the ocean (see box on page 164) and in conditions of extreme acidity and that are not affected by radiation exposures that would be lethal to any other forms of life.

STUDY OUTLINE

INTRODUCTION (p. 312)

1. *Bergey's Manual* categorizes bacteria into taxa based on rRNA sequences.
2. *Bergey's Manual* lists identifying characteristics such as Gram stain reaction, cellular morphology, oxygen requirements, and nutritional properties.

PROKARYOTIC GROUPS (p. 313)

1. Prokaryotic organisms are classified into two domains: Archaea and Bacteria.

DOMAIN BACTERIA (pp. 313–339)

1. Bacteria are essential to life on Earth.

THE PROTEOBACTERIA (pp. 313–327)

1. Members of the phylum Proteobacteria are gram-negative.
2. Alphaproteobacteria includes nitrogen-fixing bacteria, chemoautotrophs, and chemoheterotrophs.
3. The betaproteobacteria include chemoautotrophs and chemoheterotrophs.

4. Pseudomonadales, Legionellales, Vibrionales, Enterobacteriales, and Pasteurellales are classified as gammaproteobacteria.

5. Purple and green photosynthetic bacteria are photoautotrophs that use light energy and CO_2 and do not produce O_2.

6. *Myxococcus* and *Bdellovibrio* in the deltaproteobacteria prey on other bacteria.

7. Epsilonproteobacteria include *Campylobacter* and *Helicobacter*.

THE NONPROTEOBACTERIA GRAM-NEGATIVE BACTERIA (pp. 327–329)

1. Several phyla of gram-negative bacteria are not related phylogenetically to the Proteobacteria.

2. Cyanobacteria are photoautotrophs that use light energy and CO_2 and do produce O_2.

3. Chemoheterotrophic examples are *Chlamydia*, spirochetes, *Bacteroides*, and *Fusobacterium*.

THE GRAM-POSITIVE BACTERIA (pp. 329–336)

1. In *Bergey's Manual*, gram-positive bacteria are divided into those that have low G + C ratio and those that have high G + C ratio.

2. Low G + C gram-positive bacteria include common soil bacteria, the lactic acid bacteria, and several human pathogens.

3. High G + C gram-positive bacteria include mycobacteria, corynebacteria, and actinomycetes.

DOMAIN ARCHAEA (pp. 339–340)

1. Extreme halophiles, extreme thermophiles, and methanogens are included in the archaea.

MICROBIAL DIVERSITY (pp. 340–341)

1. Few of the total number of different prokaryotes have been isolated and identified.

2. PCR can be used to uncover the presence of bacteria that can't be cultured in the laboratory.

STUDY QUESTIONS

Access more review material either online at **The Microbiology Place** (www.microbiologyplace.com) or with **The Microbiology Place CD-ROM** packaged with your new book. There you'll find activities, practice tests, quizzes, flashcards, case studies, and more to help you succeed.

Answers to the Study Questions can be found in Appendix G.

REVIEW

1. The following outline is a key that can be used to identify medically important bacteria. Fill in a representative genus in the space provided.

Name of Representative Genus

I. Gram-positive
   A. Endospore-forming rod
      1. Obligate anaerobe ___Clostridium___
      2. Not obligate anaerobe ___Bacillus___
   B. Non–endospore-forming
      1. Cells are rods
         a. Produce conidiospores ___Streptomyces___
         b. Acid-fast ___Mycobacterium___
      2. Cells are cocci
         a. Lack cytochrome system ___Streptococcus___
         b. Use aerobic respiration ___Staphylococcus___

II. Gram-negative
   A. Cells are helical or curved
      1. Axial filament ___Treponema___
      2. No axial filament ___Spirillum___
   B. Cells are rods
      1. Aerobic, nonfermenting ___Pseudomonas___
      2. Facultatively anaerobic ___Escherichia___
III. Lack cell walls ___{ Mycoplasma Chlamydia___
IV. Obligate intracellular parasites
   A. Transmitted by ticks ___Rickettsia___
   B. Reticulate bodies in host cells ___Coxiella___

2. Compare and contrast each of the following:
   a. Cyanobacteria and algae
   b. Actinomycetes and fungi
   c. *Bacillus* and *Lactobacillus*
   d. *Pseudomonas* and *Escherichia*
   e. *Leptospira* and *Spirillum*
   f. *Escherichia* and *Bacteroides*
   g. *Rickettsia* and *Chlamydia*
   h. *Ureaplasma* and *Mycoplasma*

3. Matching:
   I. Gram-positive
      A. Nitrogen-fixing ___d___
   II. Gram-negative
      A. Phototrophic
         1. Anoxygenic ___1___
         2. Oxygenic ___a___

   a. Cyanobacteria
   b. Cytophaga
   c. Desulfovibrio
   d. Frankia
   e. Hyphomicrobium
   f. Methanogens

B. Chemoautotrophic
   1. Oxidize NO_2^- _h_
   2. Reduce CO_2 to CH_4 _f_

C. Chemoheterotrophic
   1. Cells inside a sheath _b_
   2. Form myxospores _g_
   3. Reduce sulfate to H_2S
      a. Anaerobic _c_
      b. Thermophilic _k_
   4. Long filaments, found in sewage _j_
   5. Form projections from the cell _e_

g. Myxobacteria
h. *Nitrobacter*
i. Purple bacteria
j. *Sphaerotilus*
k. *Sulfolobus*

MULTIPLE CHOICE

1. If you Gram-stained the bacteria that live in the human intestine, you would expect to find mostly
   a. gram-positive cocci.
   b. gram-negative rods.
   c. gram-positive, endospore-forming rods.
   d. gram-negative, nitrogen-fixing bacteria.
   e. all of the above

2. Which of the following does *not* belong with the others?
   a. Enterobacteriales
   b. Lactobacillales
   c. Legionellales
   d. Pasteurellales
   e. Vibrionales

3. Pathogenic bacteria can be
   a. motile.
   b. rods.
   c. cocci.
   d. anaerobic.
   e. all of the above

4. Which of the following is an intracellular parasite?
   a. *Rickettsia*
   b. *Mycobacterium*
   c. *Bacillus*
   d. *Staphylococcus*
   e. *Streptococcus*

5. Which of the following terms is the most specific?
   a. bacillus
   b. *Bacillus*
   c. gram-positive
   d. endospore-forming rods and cocci
   e. anaerobic

6. Which one of the following does *not* belong with the others?
   a. *Enterococcus*
   b. *Lactobacillus*
   c. *Staphylococcus*
   d. *Streptococcus*
   e. All are grouped together

7. Which of the following pairs is mismatched?
   a. anaerobic endospore-forming gram-positive rods—*Clostridium*
   b. facultatively anaerobic gram-negative rods—*Escherichia*
   c. facultatively anaerobic gram-negative rods—*Shigella*
   d. pleomorphic gram-positive rods—*Corynebacterium*
   e. spirochete—*Helicobacter*

8. *Spirillum* is not classified as a spirochete because spirochetes
   a. do not cause disease.
   b. possess axial filaments.
   c. possess flagella.
   d. are prokaryotes.
   e. none of the above

9. When *Legionella* was newly discovered, it was classified with the pseudomonads because
   a. it is a pathogen.
   b. it is an aerobic gram-negative rod.
   c. it is difficult to culture.
   d. it is found in water.
   e. none of the above

10. Cyanobacteria differ from purple and green phototrophic bacteria because cyanobacteria
   a. produce oxygen during photosynthesis.
   b. do not require light.
   c. use H_2S as an electron donor.
   d. have a membrane-enclosed nucleus.
   e. all of the above

CRITICAL THINKING

1. Place each phylum listed in Table 11.1 in the appropriate category:
   a. typical gram-positive cell wall
   b. typical gram-negative cell wall
   c. no peptidoglycan in cell wall
   d. no cell wall

2. To which of the following is the photosynthetic bacterium *Chromatium* most closely related? Briefly explain why.
   a. cyanobacteria
   b. *Chloroflexus*
   c. *Escherichia*

3. Identify the genus that best fits each of the following descriptions:
   a. This organism can produce a fuel used for home heating and for generating electricity.
   b. This gram-positive genus presents the greatest source of bacterial damage to the beekeeping industry.
   c. This gram-positive rod is used in dairy fermentations.
   d. This gammaproteobacterial genus is well suited to degrade hydrocarbons in an oil spill.

CLINICAL APPLICATIONS

1. After contact with a patient's spinal fluid, a lab technician developed fever, nausea, and purple lesions on her neck and extremities. A throat culture grew gram-negative diplococci. What is the genus of the bacteria?

2. Between April 1 and May 15 of one year, 22 children in three states developed diarrhea, fever, and vomiting. The children had each received pet ducklings. Gram-negative, facultatively anaerobic bacteria were isolated from both the patients' and the ducks' feces; the bacteria were identified as serovar C2. What is the genus of these bacteria?

3. A woman complaining of lower abdominal pain with a temperature of 39°C gave birth soon after to a stillborn baby. Blood cultures from the infant revealed gram-positive rods. The woman had a history of eating unheated hot dogs during her pregnancy. Which organism is most likely involved?

12

The Eukaryotes: Fungi, Algae, Protozoa, and Helminths

Over half of the world's population is infected with eukaryotic pathogens. The World Health Organization (WHO) ranks six parasitic diseases among the top 20 microbial causes of death in the world. Every year, there are over 5 million new cases of malaria, schistosomiasis, amoebiasis, hookworm, African trypanosomiasis, and intestinal parasites reported in developing countries. Emerging eukaryotic pathogens in developed countries include *Pneumocystis,* the leading cause of death in AIDS patients; the protozoan *Cryptosporidium,* which caused disease in 400,000 people in Milwaukee in 1993; a new protozoan parasite, *Cyclospora,* discovered in the United States in 1993; and new and increased poisonings due to algae. Increases in mold-related illness have led to debate about the passage of laws regarding safe levels of exposure to molds.

In this chapter, we examine the eukaryotic microorganisms that affect humans: fungi, algae, protozoa, parasitic helminths, and the arthropods that transmit diseases or worms. (For a comparison of their characteristics, see Table 12.1.)

UNDER THE MICROSCOPE

Penicillium. Fungi such as *Penicillium* are important decomposers that recycle organic matter in the soil.

TABLE 12.1	Major Differences Among Eukaryotic Microorganisms: Fungi, Algae, Protozoa, and Helminths			
	Fungi	Algae	Protozoa	Helminths
Kingdom	Fungi	Protist	Protist	Animalia
Nutritional type	Chemoheterotroph	Photoautotroph	Chemoheterotroph	Chemoheterotroph
Multicellularity	All, except yeasts	Some	None	All
Cellular arrangement	Unicellular, filamentous, fleshy (such as mushrooms)	Unicellular, colonial, filamentous; tissues	Unicellular	Tissues and organs
Food acquisition method	Absorptive	Diffusion	Absorptive; ingestive (cytostome)	Ingestive (mouth); absorptive
Characteristic features	Sexual and asexual spores	Pigments	Motility; some form cysts	Many have elaborate life cycles, including egg, larva, and adult
Embryo formation	None	None	None	All

FUNGI

Over the last 10 years, the incidence of serious fungal infections has been increasing. These infections are occurring as nosocomial infections and in people with compromised immune systems. In addition, thousands of fungal diseases afflict economically important plants, costing more than one billion dollars annually.

Fungi are also beneficial. They are important in the food chain because they decompose dead plant matter, thereby recycling vital elements. Through the use of extracellular enzymes such as cellulases, fungi are the primary decomposers of the hard parts of plants, which cannot be digested by animals. Nearly all plants depend on symbiotic fungi, known as **mycorrhizae,** which help their roots absorb minerals and water from the soil (see Chapter 27). Fungi are also valuable to animals. Fungi-farming ants cultivate fungi that break down cellulose and lignin from

plants, providing glucose that the ants can then digest. Fungi are used by humans for food (mushrooms) and to produce foods (bread and citric acid) and drugs (alcohol and penicillin). Of the more than 100,000 species of fungi, only about 200 are pathogenic to humans and animals.

The study of fungi is called **mycology.** We will first look at the structures that are the basis of fungal identification in a clinical laboratory, then we will explore their life cycles. Recall from Chapter 10 that identification of a pathogen is often required in order to properly treat a disease and to prevent its spread.

We will also examine nutritional needs. All fungi are chemoheterotrophs, requiring organic compounds for energy and carbon. Fungi are aerobic or facultatively anaerobic; only a few anaerobic fungi are known.

Table 12.2 lists the basic differences between fungi and bacteria.

TABLE 12.2	Selected Features of Fungi and Bacteria Compared	
	Fungi	Bacteria
Cell type	Eukaryotic	Prokaryotic
Cell membrane	Sterols present	Sterols absent, except in *Mycoplasma*
Cell wall	Glucans; mannans; chitin (no peptidoglycan)	Peptidoglycan
Spores	Sexual and asexual reproductive spores	Endospores (not for reproduction); some asexual reproductive spores
Metabolism	Limited to heterotrophic; aerobic, facultatively anaerobic	Heterotrophic, autotrophic; aerobic, facultatively anaerobic, anaerobic

FIGURE 12.1 Characteristics of fungal hyphae. (**a**) Septate hyphae have cross-walls, or septa, dividing the hyphae into cell-like units. (**b**) Coenocytic hyphae lack septa. (**c**) Hyphae grow by elongating at the tips.

Q **What is a hypha? A mycelium?**

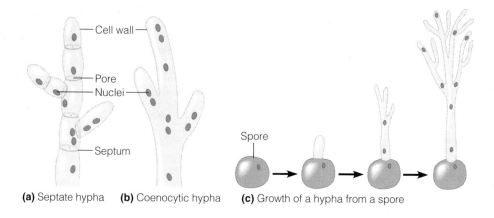

(**a**) Septate hypha　　(**b**) Coenocytic hypha　　(**c**) Growth of a hypha from a spore

CHARACTERISTICS OF FUNGI

> **LEARNING OBJECTIVES**
> • List the defining characteristics of fungi.
> • Differentiate asexual from sexual reproduction, and describe each of these processes in fungi.

Yeast identification, like bacterial identification, involves biochemical tests. However, multicellular fungi are identified on the basis of physical appearance, including colony characteristics and reproductive spores.

VEGETATIVE STRUCTURES

Fungal colonies are described as **vegetative** structures because they are composed of the cells involved in catabolism and growth.

Molds and Fleshy Fungi The **thallus** (body) of a mold or fleshy fungus consists of long filaments of cells joined together; these filaments are called **hyphae** (singular: *hypha*). Hyphae can grow to immense proportions. The hyphae of a single fungus in Oregon extend across 3.5 miles.

In most molds, the hyphae contain cross-walls called **septa** (singular: *septum*), which divide them into distinct, uninucleate (one-nucleus) cell-like units. These hyphae are called **septate hyphae** (Figure 12.1a). In a few classes of fungi, the hyphae contain no septa and appear as long, continuous cells with many nuclei. These are called **coenocytic hyphae** (Figure 12.1b). Even in fungi with septate hyphae, there are usually openings in the septa that make the cytoplasm of adjacent "cells" continuous; these fungi are actually coenocytic organisms, too.

Hyphae grow by elongating at the tips (Figure 12.1c). Each part of a hypha is capable of growth, and when a fragment breaks off, it can elongate to form a new hypha. In the laboratory, fungi are usually grown from fragments obtained from a fungal thallus.

The portion of a hypha that obtains nutrients is called the *vegetative hypha;* the portion concerned with reproduction is the *reproductive* or *aerial hypha,* so named because it projects above the surface of the medium on which the fungus is growing. Aerial hyphae often bear reproductive spores (Figure 12.2a), discussed later. When environmental conditions are suitable, the hyphae grow to form a filamentous mass called a **mycelium,** which is visible to the unaided eye (Figure 12.2b).

Yeasts Yeasts are nonfilamentous, unicellular fungi that are typically spherical or oval. Like molds, yeasts are widely distributed in nature; they are frequently found as a white powdery coating on fruits and leaves. **Budding yeasts,** such as *Saccharomyces* (sak-ä-rō-mi′sēs), divide unevenly.

In budding (Figure 12.3), the parent cell forms a protuberance (bud) on its outer surface. As the bud elongates, the parent cell's nucleus divides, and one nucleus migrates into the bud. Cell wall material is then laid down between the bud and parent cell, and the bud eventually breaks away.

One yeast cell can in time produce up to 24 daughter cells by budding. Some yeasts produce buds that fail to detach themselves; these buds form a short chain of cells called a **pseudohypha.** *Candida albicans* (kan′did-ä al′bi-kanz) attaches to human epithelial cells as a yeast but usually requires pseudohyphae to invade deeper tissues (see Figure 21.17a, page 631).

Fission yeasts, such as *Schizosaccharomyces* (skiz-ō-sak-ä-rō-mī′sēs), divide evenly to produce two new cells. During fission, the parent cell elongates, its nucleus divides, and two daughter cells are produced. Increases in the number of yeast cells on a solid medium produce a colony similar to a bacterial colony.

Yeasts are capable of facultative anaerobic growth. Yeasts can use oxygen or an organic compound as the final electron acceptor; this is a valuable attribute because it allows these fungi to survive in various environments. If

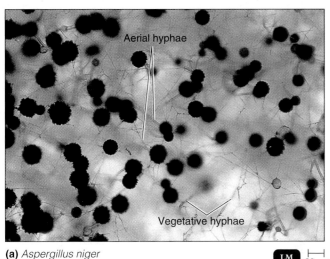

(a) *Aspergillus niger*

LM ⊢——⊣ 20 μm

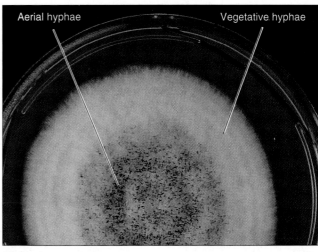

(b) *A. niger* on agar

FIGURE 12.2 Aerial and vegetative hyphae. **(a)** A photomicrograph of aerial hyphae, showing reproductive spores. **(b)** A colony of *Aspergillus niger* grown on a glucose agar plate, showing both vegetative and aerial hyphae.

Q **How do fungal colonies differ from bacterial colonies?**

given access to oxygen, yeasts perform aerobic respiration to metabolize carbohydrates to carbon dioxide and water; denied oxygen, they ferment carbohydrates and produce ethanol and carbon dioxide. This fermentation is used in the brewing, wine-making, and baking industries. *Saccharomyces* species produce ethanol in brewed beverages and carbon dioxide for leavening bread dough (See the box in Chapter 1, page 3).

Dimorphic Fungi   Some fungi, most notably the pathogenic species, exhibit **dimorphism**—two forms of growth.

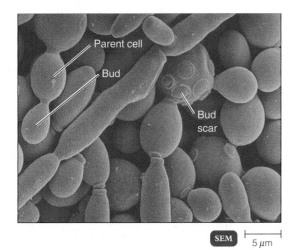

FIGURE 12.3 A budding yeast. A micrograph of *Saccharomyces cerevisiae* in various stages of budding.

Q **How does a bud differ from a spore?**

Such fungi can grow either as a mold or as a yeast. The moldlike forms produce vegetative and aerial hyphae; the yeastlike forms reproduce by budding. Dimorphism in pathogenic fungi is temperature-dependent: at 37°C, the fungus is yeastlike, and at 25°C, it is moldlike. (See Figure 24.17, page 734.) However, the appearance of the dimorphic (in this instance, nonpathogenic) fungus shown in Figure 12.4 changes with CO_2 concentration.

LIFE CYCLE

Filamentous fungi can reproduce asexually by fragmentation of their hyphae. In addition, both sexual and asexual reproduction in fungi occurs by the formation of **spores.** In fact, fungi are usually identified by spore type.

Fungal spores, however, are quite different from bacterial endospores. Bacterial endospores allow a bacterial cell to survive adverse environmental conditions (see Chapter 4). A single vegetative bacterial cell forms one endospore, which eventually germinates to produce a single vegetative bacterial cell. This process is not reproduction because it does not increase the total number of bacterial cells. But after a mold forms a spore, the spore detaches from the parent and germinates into a new mold (see Figure 12.1c). Unlike the bacterial endospore, this is a true reproductive spore; a second organism grows from the spore. Although fungal spores can survive for extended periods in dry or hot environments, most do not exhibit the extreme tolerance and longevity of bacterial endospores.

Spores are formed from aerial hyphae in a number of different ways, depending on the species. Fungal spores can

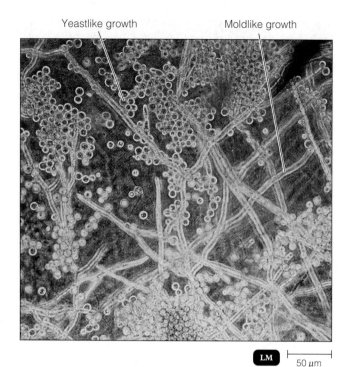

Yeastlike growth   Moldlike growth

LM   50 µm

FIGURE 12.4 Fungal dimorphism. Dimorphism in the fungus *Mucor indicus* depends on CO_2 concentration. On the agar surface, *Mucor* exhibits yeastlike growth, but in the agar it is moldlike.

Q **What is fungal dimorphism?**

be either asexual or sexual. **Asexual spores** are formed by the hyphae of one organism. When these spores germinate, they become organisms that are genetically identical to the parent. **Sexual spores** result from the fusion of nuclei from two opposite mating strains of the same species of fungus. Fungi produce sexual spores less frequently than asexual spores. Organisms that grow from sexual spores will have genetic characteristics of both parental strains. Because spores are of considerable importance in the identification of fungi, we will next look at some of the various types of asexual and sexual spores.

Asexual Spores Asexual spores are produced by an individual fungus through mitosis and subsequent cell division; there is no fusion of the nuclei of cells. Two types of asexual spores are produced by fungi. One type is a **conidiospore,** or **conidium** (plural: *conidia*), a unicellular or multicellular spore that is not enclosed in a sac (Figure 12.5a). Conidia are produced in a chain at the end of a **conidiophore.** Such spores are produced by *Aspergillus*. Conidia formed by the fragmentation of a septate hypha into single, slightly thickened cells are called **arthroconidia** (Figure 12.5b). One species that produces such spores is *Coccidioides immitis* (kok-sid-ē-oi′dēz im′mi-tis) (see Figure 24.19, page 735). Another type of conidium,

blastoconidia, consists of buds coming off the parent cell (Figure 12.5c). Such spores are found in some yeasts, such as *Candida albicans* and *Cryptococcus*. A **chlamydoconidium** is a thick-walled spore formed by rounding and enlargement within a hyphal segment (Figure 12.5d). A fungus that produces chlamydoconidia is the yeast *C. albicans*.

The other type of asexual spore is a **sporangiospore,** formed within a **sporangium,** or sac, at the end of an aerial hypha called a **sporangiophore.** The sporangium can contain hundreds of sporangiospores (Figure 12.5e). Such spores are produced by *Rhizopus*.

Sexual Spores A fungal sexual spore results from sexual reproduction, which consists of three phases:

1. **Plasmogamy.** A haploid nucleus of a donor cell (+) penetrates the cytoplasm of a recipient cell (−).

2. **Karyogamy.** The (+) and (−) nuclei fuse to form a diploid zygote nucleus.

3. **Meiosis.** The diploid nucleus gives rise to haploid nuclei (sexual spores), some of which may be genetic recombinants.

The sexual spores produced by fungi characterize the phyla. In laboratory settings, most fungi exhibit only asexual spores. Consequently, clinical identification is based on microscopic examination of asexual spores.

NUTRITIONAL ADAPTATIONS

Fungi are generally adapted to environments that would be hostile to bacteria. Fungi are chemoheterotrophs, and, like bacteria, they absorb nutrients rather than ingesting them as animals do. However, fungi differ from bacteria in certain environmental requirements and in the following nutritional characteristics:

- Fungi usually grow better in an environment with a pH of about 5, which is too acidic for the growth of most common bacteria.

- Almost all molds are aerobic. Most yeasts are facultative anaerobes.

- Most fungi are more resistant to osmotic pressure than bacteria; most can therefore grow in relatively high sugar or salt concentrations.

- Fungi can grow on substances with a very low moisture content, generally too low to support the growth of bacteria.

- Fungi require somewhat less nitrogen than bacteria for an equivalent amount of growth.

- Fungi are often capable of metabolizing complex carbohydrates, such as lignin (a component of wood), that most bacteria cannot use for nutrients.

(a) Conidia   SEM   5 μm

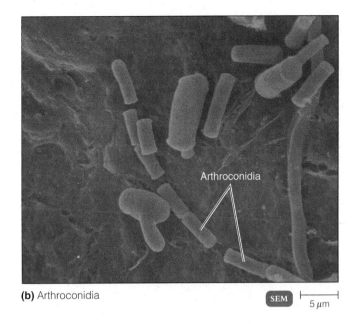

(b) Arthroconidia   SEM   5 μm

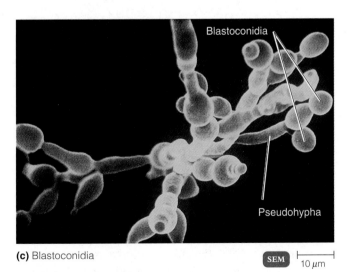

(c) Blastoconidia   SEM   10 μm

(d) Chlamydoconidia   SEM   10 μm

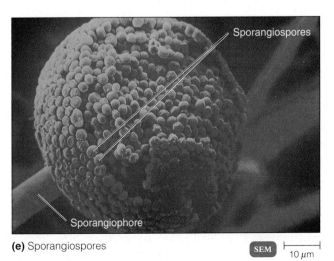

(e) Sporangiospores   SEM   10 μm

FIGURE 12.5 Representative asexual spores.
(a) Conidia are arranged in chains at the end of a conidiophore
on this *Aspergillus flavus*. **(b)** Fragmentation of hyphae results in
the formation of arthroconidia in this *Coccidioides immitis*.
(c) Blastoconidia are formed from the buds of a parent cell of
Candida albicans. **(d)** Chlamydoconidia are thick-walled cells
within hyphae of this *C. albicans*. **(e)** Sporangiospores are
formed within a sporangium (spore sac) of this *Rhizopus*.

Q **What are the green powdery structures on moldy food?**

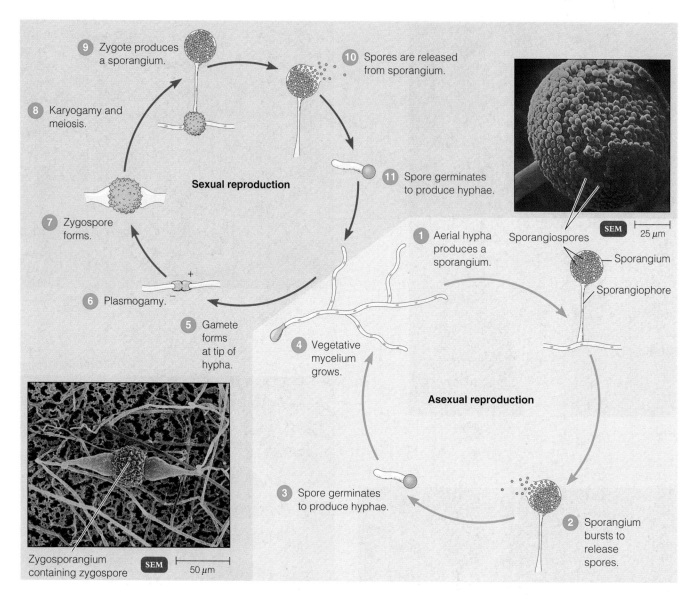

FIGURE 12.6 The life cycle of *Rhizopus*, a zygomycete. This fungus will reproduce asexually most of the time. Two opposite mating strains (designated + and −) are necessary for sexual reproduction.

Q **What is an opportunistic mycosis?**

These characteristics enable fungi to grow on such unlikely substrates as bathroom walls, shoe leather, and discarded newspapers.

MEDICALLY IMPORTANT PHYLA OF FUNGI

> **LEARNING OBJECTIVE**
> • List the defining characteristics of the three phyla of fungi described in this chapter.

This section provides an overview of medically important phyla of fungi. The actual diseases they cause will be studied in Chapters 21 through 26. Note that not all fungi cause disease.

The genera named in the following phyla include many that are readily found as contaminants in foods and in laboratory bacterial cultures. Although these genera are not all of primary medical importance, they are typical examples of their respective groups.

ZYGOMYCOTA

The Zygomycota, or conjugation fungi, are saprophytic molds that have coenocytic hyphae. An example is *Rhizopus stolonifer*, the common black bread mold. The asexual spores of *Rhizopus* are sporangiospores (Figure 12.6, upper right). The dark sporangiospores inside the sporangium give *Rhizopus* its descriptive common name. When the sporangium breaks open, the sporangiospores

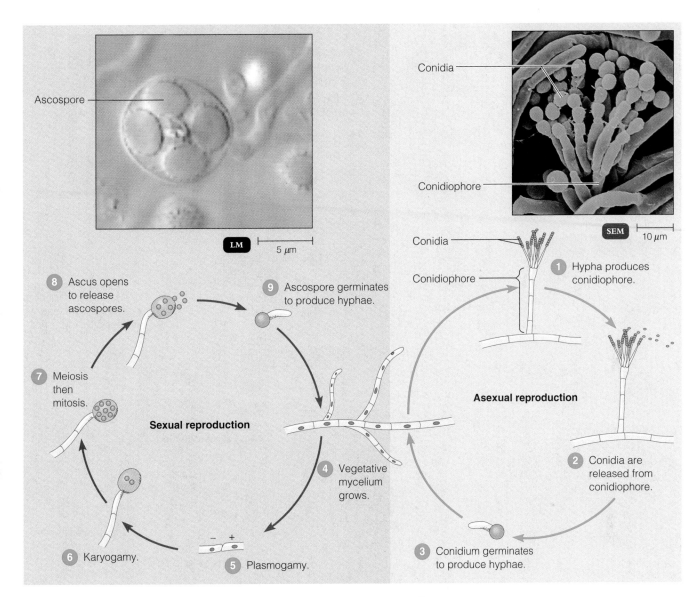

FIGURE 12.7 The life cycle of *Talaromyces*, an ascomycete. Occasionally, when two opposite mating cells from two different strains (+ and −) fuse, sexual reproduction occurs.

Q **Name one ascomycete that can infect humans.**

are dispersed. If they fall on a suitable medium, they will germinate into a new mold thallus.

The sexual spores are zygospores. A **zygospore** is a large spore enclosed in a thick wall (Figure 12.6, lower left). This type of spore results from the fusion of the nuclei of two cells that are morphologically similar to each other.

ASCOMYCOTA

The Ascomycota, or sac fungi, include molds with septate hyphae and some yeasts. Their asexual spores are usually conidia produced in long chains from the conidiophore. The term *conidia* means dust, and these spores freely de-

tach from the chain at the slightest disturbance and float in the air like dust.

An **ascospore** results from the fusion of the nuclei of two cells that can be either morphologically similar or dissimilar. These spores are produced in a saclike structure called an **ascus** (Figure 12.7, upper left). The members of this phylum are called sac fungi because of the ascus.

BASIDIOMYCOTA

The Basidiomycota, or club fungi, also possess septate hyphae. This phylum includes fungi that produce mushrooms. **Basidiospores** are formed externally on a base pedestal called a **basidium** (Figure 12.8). (The common

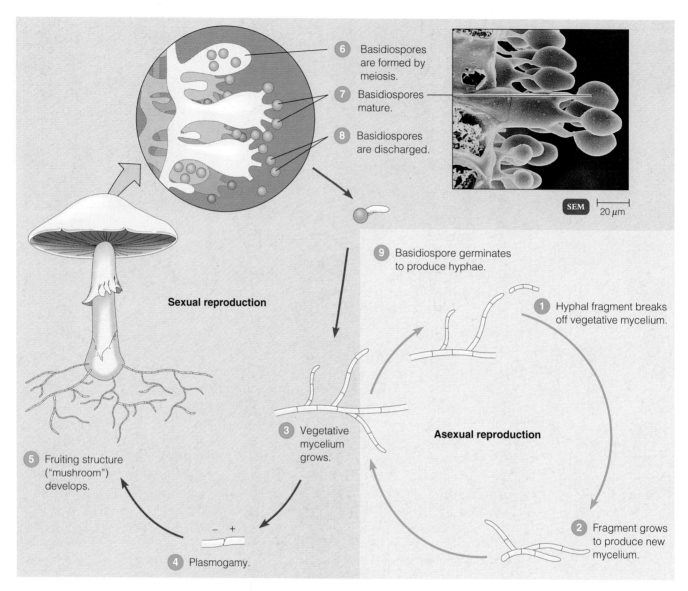

6 Basidiospores are formed by meiosis.

7 Basidiospores mature.

8 Basidiospores are discharged.

SEM 20 μm

9 Basidiospore germinates to produce hyphae.

1 Hyphal fragment breaks off vegetative mycelium.

Sexual reproduction

Asexual reproduction

3 Vegetative mycelium grows.

5 Fruiting structure ("mushroom") develops.

− +

4 Plasmogamy.

2 Fragment grows to produce new mycelium.

FIGURE 12.8 A generalized life cycle of a basidiomycete.
Mushrooms appear after cells from two mating strains (+ and −) have fused.

Q **On what basis are fungi classified into phyla?**

name of the fungus is derived from the club shape of the basidium.) There are usually four basidiospores per basidium. Some of the basidiomycota produce asexual conidiospores. Representative basidiomycetes are shown in Figure 12.9.

The fungi we have looked at thus far are **teleomorphs;** that is, they produce both sexual and asexual spores. Some ascomycetes have lost the ability to reproduce sexually. These asexual fungi are called **anamorphs.** *Penicillium* is an example of an anamorph that arose from a mutation in a teleomorph. Historically, fungi whose sexual cycle had not been observed were put in a "holding category" called *Deuteromycota.* Now, mycologists are using rRNA sequencing to classify these organisms. Most of these previ-

ously unclassified deuteromycetes are anamorph phases of Ascomycota, and a few are basidiomycetes.

Table 12.3 lists some fungi that cause human diseases. Two generic names are given for some of the fungi because medically important fungi that are well known by their anamorph, or asexual, name are often referred to by that name.

FUNGAL DISEASES

Any fungal infection is called a **mycosis.** Mycoses are generally chronic (long-lasting) infections because fungi grow slowly. Mycoses are classified into five groups according to the degree of tissue involvement and mode of entry into

FIGURE 12.9 Representative basidiomycetes. (a) A bird's nest fungus (*Crucibulum vulgare*) growing on a twig. The basidiospores visible in one of the cups will pop out when the cup is hit by a raindrop. **(b)** *Amanita muscaria* grows in close association with plant roots (mycorrhiza) and produces a neurotoxin and a possible antitumor chemical.

Q What is the primary role of fungi in the ecosystem?

(a) *Crucibulum vulgare*

(b) *Amanita muscaria*

the host: systemic, subcutaneous, cutaneous, superficial, or opportunistic. In Chapter 10, we saw that fungi are related to animals. Consequently, drugs that affect fungal cells may also affect animal cells. This fact makes fungal infections of humans and other animals often difficult to treat.

Systemic mycoses are fungal infections deep within the body. They are not restricted to any particular region of the body but can affect a number of tissues and organs. Systemic mycoses are usually caused by fungi that live in the soil. Inhalation of spores is the route of transmission; these infections typically begin in the lungs and then spread to other body tissues. They are not contagious from animal to human or from human to human. Two systemic mycoses, histoplasmosis and coccidioidomycosis, are discussed in Chapter 24.

Subcutaneous mycoses are fungal infections beneath the skin caused by saprophytic fungi that live in soil and on vegetation. Sporotrichosis is a subcutaneous infection acquired by gardeners and farmers (Chapter 21, page 630). Infection occurs by direct implantation of spores or mycelial fragments into a puncture wound in the skin.

Fungi that infect only the epidermis, hair, and nails are called **dermatophytes,** and their infections are called dermatomycoses or **cutaneous mycoses** (see Figure 21.16, page 630). Dermatophytes secrete keratinase, an enzyme that degrades **keratin,** a protein found in hair, skin, and nails. Infection is transmitted from human to human or from animal to human by direct contact or by contact with infected hairs and epidermal cells (as from barber shop clippers or shower room floors).

The fungi that cause **superficial mycoses** are localized along hair shafts and in superficial (surface) epidermal cells. These infections are prevalent in tropical climates.

An **opportunistic pathogen** is generally harmless in its normal habitat but can become pathogenic in a host who is seriously debilitated or traumatized, who is under treatment with broad-spectrum antibiotics, whose immune system is suppressed by drugs or by an immune disorder, or who has a lung disease.

Pneumocystis is an opportunistic pathogen seen in individuals with compromised immune systems and is the most common life-threatening infection in AIDS patients (see Figure 24.22, page 737). It was first classified as a protozoan, but recent studies of its RNA indicate it is a unicellular anamorphic fungus. Another example of an opportunistic pathogen is the fungus *Stachybotrys* (sta′ke-bo-tris), which normally grows on cellulose found in dead plants but in recent years has been found growing on water-damaged walls of homes. See the box in Chapter 27, p. 835. One preliminary study reported that its toxic spores can cause fatal pulmonary hemorrhage in infants. Additional research is being designed to determine whether *Stachybotrys* is a health hazard.

Mucormycosis is an opportunistic mycosis caused by *Rhizopus* and *Mucor* (mū′kôr); the infection occurs mostly in patients with diabetes mellitus, with leukemia, or undergoing treatment with immunosuppressive drugs. Aspergillosis is also an opportunistic mycosis; it is caused by *Aspergillus* (see Figure 12.2). This disease occurs in people who have debilitating lung diseases or cancer and have inhaled *Aspergillus* spores. Opportunistic infections by *Cryptococcus* and *Penicillium* can cause fatal diseases in AIDS patients. These opportunistic fungi may be transmitted from one person to an uninfected person but do not usually infect immunocompetent people. **Yeast infection,** or candidiasis, is most frequently caused by *Candida albicans* and

TABLE 12.3	Characteristics of Some Pathogenic Fungi		
Phylum	Growth Characteristics	Asexual Spore Types	Human Pathogens
Zygomycota	Nonseptate hyphae	Sporangiospores	*Rhizopus* *Mucor*

Ascomycota	Dimorphic	Conidia	*Aspergillus*
			Blastomyces* (Ajellomyces†)
			   dermatitidis
			Histoplasma* (Ajellomyces†)
			   capsulatum

	Septate hyphae, strong affinity for keratin	Conidia Arthroconidia	*Microsporum* *Trichophyton* (Arthroderma†)*
Anamorphs		Conidia	*Epidermophyton*
	Dimorphic	Conidia Arthroconidia	*Sporothrix schenckii, Stachybotrys* *Coccidioides immitis*
	Yeastlike, pseudohyphae	Chlamydoconidia	*Candida albicans*
	Unknown	Unknown	*Pneumocystis*
Basidiomycota	Septate hyphae; includes rusts and smuts, and plant pathogens; yeastlike encapsulated cells	Conidia	*Cryptococcus neoformans** *(Filobasidiella)†* *Malassezia*

*Anamorph name.
†Teleomorph name.

may occur as vulvovaginal candidiasis or thrush, a mucocutaneous candidiasis. Candidiasis frequently occurs in newborns, in people with AIDS, and in people being treated with broad-spectrum antibiotics (see Figure 21.17, page 631).

Some fungi cause disease by producing toxins. These toxins are discussed in Chapter 15.

ECONOMIC EFFECTS OF FUNGI

LEARNING OBJECTIVE
- Identify two beneficial and two harmful effects of fungi.

Fungi have been used in biotechnology for many years. *Aspergillus niger*, for example, has been used to produce citric acid for foods and beverages since 1914. The yeast *Saccharomyces cerevisiae* is used to make bread and wine. It is also genetically modified to produce a variety of proteins, including hepatitis B vaccine. *Saccharomyces* is used as a protein supplement for humans and cattle. *Trichoderma* is used commercially to produce the enzyme cellulase, which is used to remove plant cell walls to produce a clear fruit juice. When the anticancer drug taxol, which is produced by yew trees, was discovered, there was concern that the yew forests of the U.S. Northwest coast would be decimated to harvest the drug. However, in 1993, Andrea and

Donald Stierle saved the yews by discovering that the fungus *Taxomyces* also produces taxol.

Fungi are used as biological controls of pests. In 1990, the fungus *Entomophaga* unexpectedly proliferated and killed gypsy moths that were destroying trees in the eastern United States. Scientists are investigating whether this fungus can be used in place of chemical insecticides. Annually, 25 to 50% of harvested fruits and vegetables are ruined by fungi. Chemical fungicides cannot be used to prevent this decay because of safety and environmental concerns. However, another fungus, *Candida oleophila*, can be and is used to prevent undesirable fungal growth on harvested fruits. This process of biocontrol works because the *C. oleophila* grows on the fruit surface before spoilage fungi grow.

In contrast to these beneficial effects, fungi can have undesirable effects for industry and agriculture because of their nutritional adaptations. As most of us have observed, mold spoilage of fruits, grains, and vegetables is relatively common, but bacterial spoilage of such foods is not. There is little moisture on the unbroken surfaces of such foods, and the interiors of fruits are too acidic for many bacteria to grow there. Jams and jellies also tend to be acidic, and they have a high osmotic pressure from the sugars they contain. These factors all discourage bacterial growth but readily support the growth of molds. A paraffin layer on top of a jar of homemade jelly helps deter mold growth

TABLE 12.3	(continued)		
Habitat	Type of Mycosis	Clinical Notes	Page Reference
Ubiquitous	Systemic	Opportunistic pathogen	737
Ubiquitous	Systemic	Opportunistic pathogen	737
Ubiquitous	Systemic	Opportunistic pathogen	736
Unknown	Systemic	Inhalation	736
Soil	Systemic	Inhalation	734
Soil, animals	Cutaneous	Tinea capitis (ringworm)	629
Soil, animals	Cutaneous	Tinea pedis (athlete's foot)	629
Soil, humans	Cutaneous	Tinea cruris (jock itch), tinea unguium (of fingernails or toenails)	629
Soil	Subcutaneous	Puncture wound	630
Soil	Systemic	Inhalation	735
Human normal microbiota	Cutaneous, systemic, mucocutaneous	Opportunistic pathogen	630
Ubiquitous	Systemic	Opportunistic pathogen	736
Soil, bird feces	Systemic	Inhalation	660
Human skin	Cutaneous	Dandruff; dermatitis	615

because molds are aerobic and the paraffin layer keeps out the oxygen. However, fresh meats and certain other foods are such good substrates for bacterial growth that bacteria not only will outgrow molds but also will actively suppress mold growth in these foods.

The spreading chestnut tree, of which Longfellow wrote, no longer grows in the United States except in a few widely isolated locations; a fungal blight killed virtually all of them. This blight was caused by the ascomycete *Cryphonectria parasitica* (kri-fō-nek′trē-ä par-ä-si′ti-kä), which was introduced from China around 1904. The fungus allows the tree roots to live and put forth shoots regularly, but then it kills the shoots just as regularly. *Cryphonectria*-resistant chestnuts are being developed. Another imported fungal plant disease is Dutch elm disease, caused by *Ceratocystis ulmi* (sē-rä-tō-sis′tis ul′me). Carried from tree to tree by a bark beetle, the fungus blocks the afflicted tree's circulation. The disease has devastated the American elm population.

LICHENS

LEARNING OBJECTIVES

- List the distinguishing characteristics of lichens, and describe their nutritional needs.
- Describe the roles of the fungus and the alga in a lichen.

A **lichen** is a combination of a green alga (or a cyanobacterium) and a fungus. Lichens are placed in the Kingdom Fungi and are classified according to the fungal partner, most often an ascomycete. The two organisms exist in a *mutualistic* relationship, in which each partner benefits. The lichen is very different from either the alga or fungus growing alone, and if the partners are separated, the lichen no longer exists. Approximately 13,500 species of lichens occupy quite diverse habitats. Because they can inhabit areas in which neither fungi nor algae could survive alone, lichens are often the first life forms to colonize newly exposed soil or rock. Lichens secrete organic acids that chemically weather rock, and they accumulate nutrients needed for plant growth. Also found on trees, concrete structures, and rooftops, lichens are some of the slowest-growing organisms on Earth.

Lichens can be grouped into three morphologic categories (Figure 12.10a). *Crustose lichens* grow flush or encrusting onto the substratum, *foliose lichens* are more leaflike, and *fruticose lichens* have fingerlike projections. The lichen's thallus, or body, forms when fungal hyphae grow around algal cells to become the **medulla** (Figure 12.10b). Fungal hyphae project below the lichen body to form **rhizines,** or holdfasts. Fungal hyphae also form a **cortex,** or protective covering, over the algal layer and sometimes under it as well. After incorporation into a lichen thallus, the alga continues to grow, and the growing hyphae can incorporate new algal cells.

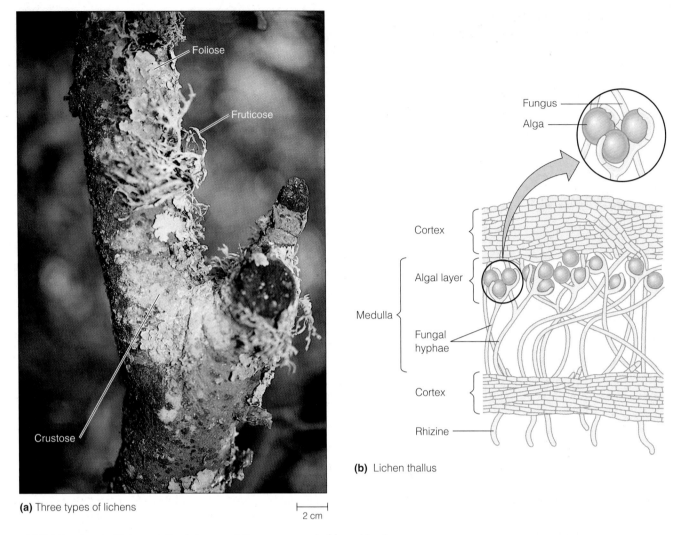

(a) Three types of lichens

2 cm

(b) Lichen thallus

FIGURE 12.10 Lichens. The lichen medulla is composed of fungal hyphae surrounding the algal layer. The protective cortex is a layer of fungal hyphae that covers the surface and sometimes the bottom of the lichen.

Q **In what ways are lichens unique?**

When the algal partner is cultured separately in vitro, about 1% of the carbohydrates produced during photosynthesis are released into the culture medium; however, when the alga is associated with a fungus, the algal plasma membrane is more permeable, and up to 60% of the products of photosynthesis are released to the fungus or are found as end-products of fungus metabolism. The fungus clearly benefits from this association. The alga, while giving up valuable nutrients, is in turn compensated; it receives from the fungus both protection from desiccation (cortex) and attachment (holdfast).

Lichens had considerable economic importance in ancient Greece and other parts of Europe as dyes for clothing. Usnic acid from *Usnea* is used as an antimicrobial agent in China. Erythrolitmin, the dye used in litmus paper to indi-

cate changes in pH, is extracted from a variety of lichens. Some lichens or their acids can cause allergic contact dermatitis in humans.

Populations of lichens readily incorporate cations (positively charged ions) into their thalli. Therefore, the concentrations and types of cations in the atmosphere can be determined by chemical analyses of lichen thalli. In addition, the presence or absence of species that are quite sensitive to pollutants can be used to ascertain air quality. A 1985 study in the Cuyahoga Valley in Ohio revealed that 81% of the 172 lichen species that were present in 1917 were gone. Because this area is severely affected by air pollution, the inference is that air pollutants, primarily sulfur dioxide (the major contributor to acid precipitation), caused the death of sensitive species.

Lichens are the major food for tundra herbivores such as caribou and reindeer. After the 1986 Chernobyl nuclear disaster, 70,000 reindeer in Lapland that had been raised for food had to be destroyed because of high levels of radiation. The lichens on which the reindeer fed had absorbed radioactive cesium-137, which had spread in the air.

ALGAE

Algae are familiar as the large brown kelp in coastal waters, the green scum in a puddle, and the green stains on soil or on rocks. A few algae are responsible for food poisonings. Some algae are unicellular; others form chains of cells (are filamentous); and a few have thalli.

Algae are mostly aquatic, although some are found in soil or on trees when sufficient moisture is available there. Unusual algal habitats include the hair of both the sedentary South American sloth and the polar bear. Water is necessary for physical support, reproduction, and the diffusion of nutrients. Generally, algae are found in cool temperate waters, although the large floating mats of the brown alga *Sargassum* (sär-gas'sum) are found in the subtropical Sargasso Sea. Some species of brown algae grow in antarctic waters.

CHARACTERISTICS OF ALGAE

LEARNING OBJECTIVE
- List the defining characteristics of algae.

Algae are relatively simple eukaryotic photoautotrophs that lack the tissues (roots, stem, and leaves) of plants. The identification of unicellular and filamentous algae requires microscopic examination. Most algae are found in the ocean. Their locations depend on the availability of appropriate nutrients, wavelengths of light, and surfaces on which to grow. Probable locations for representative algae are shown in Figure 12.11a.

VEGETATIVE STRUCTURES

The body of a multicellular alga is called a thallus. Thalli of the larger multicellular algae, those commonly called seaweeds, consist of branched **holdfasts** (which anchor the alga to a rock), stemlike and often hollow **stipes,** and leaflike **blades** (Figure 12.11b). The cells covering the thallus can carry out photosynthesis. The thallus lacks the conductive tissue (xylem and phloem) characteristic of vascular plants; algae absorb nutrients from the water over their entire surface. The stipe is not lignified or woody, so it does not offer the support of a plant's stem; instead, the surrounding water supports the algal thallus; some algae are also buoyed by a floating, gas-filled bladder called a *pneumatocyst.*

LIFE CYCLE

All algae can reproduce asexually. Multicellular algae with thalli and filamentous forms can fragment; each piece is capable of forming a new thallus or filament. When a unicellular alga divides, its nucleus divides (mitosis), and the two nuclei move to opposite parts of the cell. The cell then divides into two complete cells (cytokinesis).

Sexual reproduction occurs in algae (Figure 12.12). In some species, asexual reproduction may occur for several generations and then, under different conditions, the same species reproduce sexually. Other species alternate generations so that the offspring resulting from sexual reproduction reproduce asexually, and the next generation then reproduces sexually.

NUTRITION

Algae is a common name that includes several phyla (Table 12.4, page 359). Most algae are photosynthetic; however, the oomycotes, or fungal-like algae, are chemoheterotrophs. Photosynthetic algae are found throughout the photic (light) zone of bodies of water. Chlorophyll *a* (a light-trapping pigment) and accessory pigments involved in photosynthesis are responsible for the distinctive colors of many algae.

Algae are classified according to their rRNA sequences, structures, pigments, and other qualities (Table 12.4). Following are descriptions of some phyla of algae.

SELECTED PHYLA OF ALGAE

LEARNING OBJECTIVE
- List the outstanding characteristics of the five phyla of algae discussed in this chapter.

The *brown algae,* or kelp, are macroscopic; some reach lengths of 50 m (see Figure 12.11b). Most brown algae are found in coastal waters. Brown algae have a phenomenal growth rate. Some grow at rates exceeding 20 cm per day and therefore can be harvested regularly. **Algin,** a thickener used in many foods (such as ice cream and cake decorations), is extracted from their cell walls. Algin is also used in the production of a wide variety of nonfood goods, including rubber tires and hand lotion. The brown alga *Laminaria japonica* is used to induce vaginal dilation before surgical entry into the uterus through the vagina.

Most *red algae* have delicately branched thalli and can live at greater ocean depths than other algae (see Figure 12.11c). The thalli of a few red algae form crustlike coatings on rocks and shells. The red pigments enable red algae to absorb the blue light that penetrates deepest into the ocean. The agar used in microbiological media is extracted from many red algae. Another gelatinous material, carrageenan, comes from a species of red algae commonly

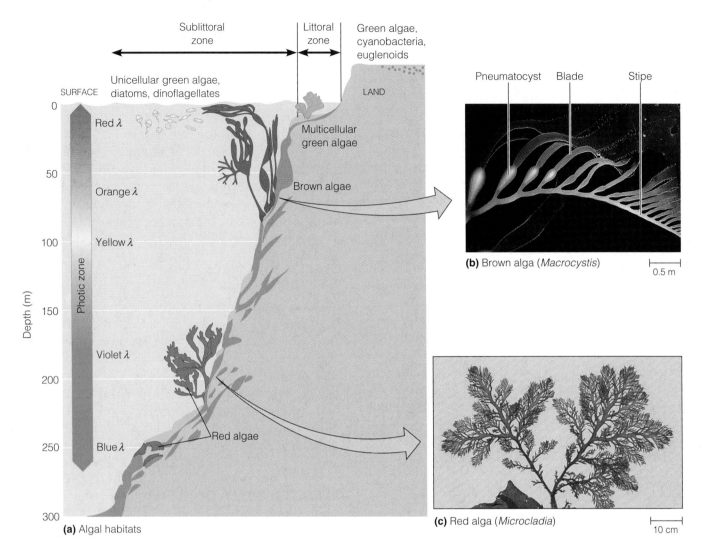

(a) Algal habitats

(b) Brown alga (*Macrocystis*) |— 0.5 m —|

(c) Red alga (*Microcladia*) |— 10 cm —|

FIGURE 12.11 Algae and their habitats. (**a**) Although unicellular and filamentous algae can be found on land, they frequently exist in marine and freshwater environments as plankton. Multicellular green, brown, and red algae require a suitable attachment site, adequate water for support, and light of the appropriate wavelengths. (**b**) *Macrocystis porifera*, a brown alga. The hollow stipe and gas-filled pneumatocysts hold the thallus upright ensuring that sufficient sunlight is received for growth. (**c**) *Microcladia*, a red alga. The delicately branched red algae get their color from phycobiliprotein accessory pigments.

Q **What red alga is toxic for humans?**

called Irish moss. Carrageenan and agar can be a thickening ingredient in evaporated milk, ice cream, and pharmaceutical agents. *Gracilaria* species, which grow in the Pacific Ocean, are used by humans for food. However, members of this genus can produce a lethal toxin.

Green algae have cellulose cell walls, contain chlorophyll *a* and *b*, and store starch, as plants do (see Figure 12.12a). Green algae are believed to have given rise to terrestrial plants. Most green algae are microscopic, although they may be either unicellular or multicellular. Some filamentous kinds form grass-green scum in ponds.

Diatoms, dinoflagellates, and water molds are grouped into the Kingdom Chromista or Stramenopila. *Diatoms* (Figure 12.13) are unicellular or filamentous algae with complex cell walls that consist of pectin and a layer of silica. The two parts of the wall fit together like the halves of a Petri dish. The distinctive patterns of the walls are a useful tool in diatom identification. Diatoms store energy captured through photosynthesis in the form of oil.

The first reported outbreak of a neurological disease caused by diatoms was reported in 1987 in Canada. Affected people ate mussels that had been feeding on diatoms. The diatoms produced *domoic acid*, a toxin that was then concentrated in the mussels. Symptoms included diarrhea and memory loss. The fatality rate was less than 4%. Since 1991, hundreds of marine birds and sea lions have died from the same **domoic acid intoxication** in California.

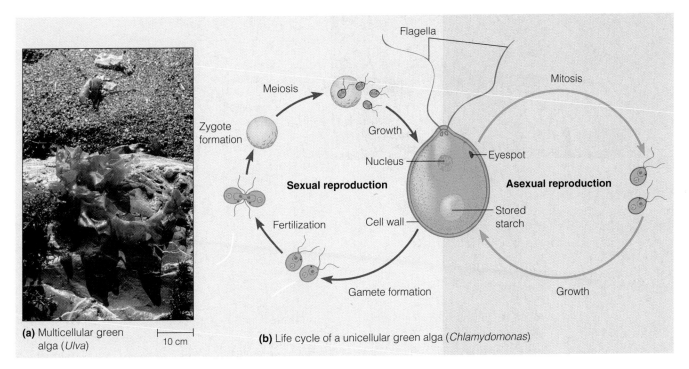

(a) Multicellular green alga (*Ulva*)

10 cm

(b) Life cycle of a unicellular green alga (*Chlamydomonas*)

FIGURE 12.12 Green algae. (**a**) The multicellular green alga *Ulva*. (**b**) The life cycle of the unicellular green alga *Chlamydomonas*. Two whiplike flagella propel this cell.

Q **What is the primary role of algae in the ecosystem?**

TABLE 12.4	Characteristics of Selected Phyla of Algae					
	Brown Algae	Red Algae	Green Algae	Diatoms	Dinoflagellates	Water Molds
Phylum	Phaeophyta	Rhodophyta	Chlorophyta	Bacillariophyta	Dinoflagellata	Oomycota
Color	Brownish	Reddish	Green	Brownish	Brownish	Colorless, white
Cell Wall	Cellulose and alginic acid	Cellulose	Cellulose	Pectin and silica	Cellulose in membrane	Cellulose
Cell Arrangement	Multicellular	Most are multicellular	Unicellular and multicellular	Unicellular	Unicellular	Multicellular
Photosynthetic Pigments	Chlorophyll *a* and *c*, xanthophylls	Chlorophyll *a* and *d*, phycobili- proteins	Chlorophyll *a* and *b*	Chlorophyll *a* and *c*, carotene, xanthophylls	Chlorophyll *a* and *c*, carotene, xanthins	None
Sexual Reproduction	Yes	Yes	Yes	Yes	In a few (?)	Yes (similar to the Zygomycota)
Storage Material	Carbohydrate	Glucose polymer	Glucose polymer	Oil	Starch	None

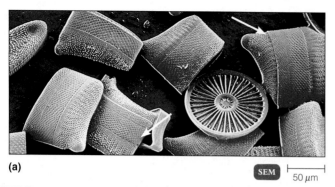

(b) Asexual reproduction of a diatom

(a)

SEM ⊢————⊣ 50 µm

FIGURE 12.13 Diatoms. (a) In this micrograph of *Isthmia nervosa*, notice how the two parts of the cell wall fit together at the arrows. **(b)** Asexual reproduction in a diatom. During mitosis, each daughter cell retains one-half of the cell wall from the parent (green) and must synthesize the remaining half (blue).

Q **What human disease is caused by diatoms?**

Dinoflagellates are unicellular algae collectively called **plankton,** or free-floating organisms (Figure 12.14). Their rigid structure is due to cellulose embedded in the plasma membrane. Some dinoflagellates produce neurotoxins. In the last 20 years, a worldwide increase in toxic marine algae has killed millions of fish, hundreds of marine mammals, and even some humans. When fish swim through large numbers of the dinoflagellate *Gymnodinium breve* (jim′nō-din-ē-um brev′ē), the algae trapped in the gills of the fish release a neurotoxin that stops the fish from breathing. Dinoflagellates in the genus *Alexandrium* (al-eg′zan-drē-um) produce neurotoxins (called **saxitoxins**) that cause **paralytic shellfish poisoning (PSP).** The toxin is concentrated when large numbers of dinoflagellates are eaten by mollusks, such as mussels or clams. Humans who eat these mollusks develop PSP. Large concentrations of *Alexandrium* give the ocean a deep red color, from which the name *red tide* originates (Figure 27.15, page 827). Mollusks should not be harvested for consumption during a red tide. A disease called **ciguatera** occurs when the dinoflagellate *Gambierdiscus toxicus* (gam′bē-er-dis-kus toks′i-kus) passes up the food chain and is concentrated in large fish. Ciguatera is endemic (constantly present) in the south Pacific Ocean and the Caribbean Sea. An emerging disease associated with *Pfiesteria* (fē′ster-ē-ä) is responsible for periodic massive fish deaths along the Atlantic Coast.

Most *water molds*, or *Oomycota*, are decomposers. They form the cottony masses on dead algae and animals, usually in fresh water (Figure 12.15). Asexually, the oomycotes

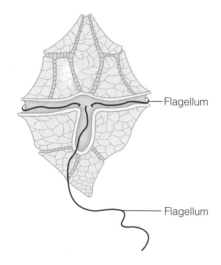

Flagellum

Flagellum

FIGURE 12.14 *Peridinium*, a dinoflagellate. Like some other dinoflagellates, *Peridinium* has two flagella in perpendicular, opposing grooves. When the two flagella beat simultaneously, they cause the cell to spin.

Q **What human diseases are caused by dinoflagellates?**

FIGURE 12.15 An oomycote. Note the fuzzy mass on this fish. The funguslike algae are common decomposers in aquatic environments.

Q **Is this oomycote more closely related to *Penicillium* or to diatoms?**

resemble the zygomycete fungi in that they produce spores in a sporangium (spore sac). However, oomycote spores, called **zoospores,** have two flagella; fungi do not have flagella. Because of their superficial similarity to fungi, oomycotes were previously classified with the fungi. Their cellulose cell walls always raised the question about their relationship to algae, and recent DNA analyses have confirmed that oomycotes are more closely related to diatoms and dinoflagellates than to fungi. Many of the terrestrial oomycotes are plant parasites. The USDA inspects imported plants for white rust and other parasites. Often travelers, or even commercial plant importers, do not realize that one little blossom or seedling could carry a pest that is capable of causing millions of dollars' worth of damage to U.S. agriculture.

In Ireland during the mid-1800s, 1 million people died when the country's potato crop failed. The alga that caused the great potato blight, *Phytophthora infestans* (fī-tof'thô-rä in-fes'tans), was one of the first microorganisms to be associated with a disease. Today, *Phytophthora* infects soybeans, potatoes, and cocoa.

In Australia, *P. cinnamoni* has infected about 20% of one species of *Eucalyptus*. *Phytophthora* was introduced into the United States in the 1990s and caused widespread damage to fruit and vegetable crops. When California oak trees suddenly started dying in 1995, University of California scientists identified the cause of this "sudden oak death" to be a new species, *P. ramorum*. *P. ramorum* also infects redwood trees.

ROLES OF ALGAE IN NATURE

Algae are an important part of any aquatic food chain because they fix carbon dioxide into organic molecules that can be consumed by chemoheterotrophs. Using the energy produced in photophosphorylation, algae convert carbon dioxide in the atmosphere into carbohydrates. Molecular oxygen (O_2) is a by-product of their photosynthesis. The top few meters of any body of water contain planktonic algae. As 75% of the Earth is covered with water, it is estimated that 80% of the Earth's O_2 is produced by planktonic algae.

Seasonal changes in nutrients, light, and temperature cause fluctuations in algal populations; periodic increases in numbers of planktonic algae are called **algal blooms.** Blooms of dinoflagellates are responsible for seasonal red tides. Blooms of a certain few species indicate that the water in which they grow is polluted because these algae thrive in high concentrations of organic materials that exist in sewage or industrial wastes. When algae die, the decomposition of the large numbers of cells associated with an algal bloom depletes the level of dissolved oxygen in the water. (This phenomenon is discussed in Chapter 27.)

Much of the world's petroleum was formed from diatoms and other planktonic organisms that lived several million years ago. When such organisms died and were buried by sediments, the organic molecules they contained did not decompose to be returned to the carbon cycle as CO_2. Heat and pressure resulting from the Earth's geologic movements altered the oil stored in the cells, as well as the cell membranes. Oxygen and other elements were eliminated, leaving a residue of hydrocarbons in the form of petroleum and natural gas deposits.

Many unicellular algae are symbionts in animals. The giant clam *Tridacna* (trī-dak'nä) has evolved special organs that host dinoflagellates. As the clam sits in shallow water, the algae proliferate in these organs when they are exposed to the sun. The algae release glycerol into the clam's bloodstream, thus supplying the clam's carbohydrate requirement. In addition, evidence suggests that the clam gets essential proteins by phagocytizing old algae.

PROTOZOA

Protozoa are unicellular, eukaryotic chemoheterotrophic organisms. Among the protozoa are many variations on this cell structure, as we shall see. Protozoa inhabit water and soil. The feeding and growing stage, or **trophozoite,** feeds upon bacteria and small particulate nutrients. Some protozoa are part of the normal microbiota of animals. *Nosema locustae*, an insect pathogen, is sold commercially as a nontoxic insecticide to kill grasshoppers. Because the protozoa are specific for grasshoppers, they will not affect humans or animals that eat grasshoppers. Of the nearly 20,000 species of protozoa, relatively few cause human disease. Those few, however, have significant health and economic impact. Worldwide, malaria is the fourth leading cause of death.

CHARACTERISTICS OF PROTOZOA

LEARNING OBJECTIVE
• List the defining characteristics of protozoa.

The term *protozoan* means "first animal," which generally describes its animal-like nutrition. In addition to getting food, a protozoan must reproduce, and parasitic species must be able to get from one host to another.

LIFE CYCLE

Protozoa reproduce asexually by fission, budding, or schizogony. **Schizogony** is multiple fission; the nucleus undergoes multiple divisions before the cell divides. After many nuclei are formed, a small portion of cytoplasm concentrates around each nucleus, and then the single cell separates into daughter cells.

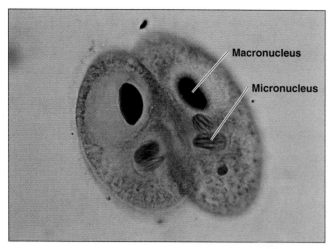

LM   25 μm

FIGURE 12.16 Conjugation in the ciliate protozoan
Paramecium. Sexual reproduction in ciliates is by conjugation.
Each cell has two nuclei: a micronucleus and a macronucleus.
The micronucleus is haploid and is specialized for conjugation.
One micronucleus from each cell will migrate to the other cell dur-
ing conjugation. Both cells will then go on to produce two daugh-
ter cells. Condensed chromosomes are visible in the micronuclei.

Q **Does conjugation result in more cells?**

Sexual reproduction has been observed in some protozoa. The ciliates, such as *Paramecium,* reproduce sexually
by **conjugation** (Figure 12.16), which is very different
from the bacterial process of the same name (see Figure
8.26, page 243). During protozoan conjugation, two
cells fuse, and a haploid nucleus (the micronucleus)
from each cell migrates to the other cell. This haploid
micronucleus fuses with the haploid micronucleus
within the cell. The parent cells separate, each now a
fertilized cell. When the cells later divide, they produce
daughter cells with recombined DNA. Some protozoa
produce **gametes (gametocytes),** haploid sex cells.
During reproduction, two gametes fuse to form a diploid
zygote.

Encystment Under certain adverse conditions, some
protozoa produce a protective capsule called a **cyst.** A cyst
permits the organism to survive when food, moisture, or
oxygen are lacking, when temperatures are not suitable, or
when toxic chemicals are present. A cyst also enables a
parasitic species to survive outside a host. This is important because parasitic protozoa may have to be excreted
from one host in order to get to a new host. The cyst form
in members of the phylum Apicomplexa is called an
oocyst. It is a reproductive structure in which new cells
are produced asexually.

NUTRITION

Protozoa are mostly aerobic heterotrophs, although many
intestinal protozoa are capable of anaerobic growth. Two
chlorophyll-containing groups, dinoflagellates and euglenoids, are often studied with algae.

All protozoa live in areas with a large supply of water.
Some protozoa transport food across the plasma membrane. However, some have a protective covering, or
pellicle, and thus require specialized structures to take in
food. Ciliates take in food by waving their cilia toward a
mouthlike opening called a **cytostome.** Amoebas engulf
food by surrounding it with pseudopods and phagocytizing
it. In all protozoa, digestion takes place in membrane-
enclosed **vacuoles,** and waste may be eliminated through
the plasma membrane or through a specialized **anal pore.**

MEDICALLY IMPORTANT PHYLA OF PROTOZOA

> **LEARNING OBJECTIVES**
> - Describe the outstanding characteristics of the seven
>   phyla of protozoa discussed in this chapter, and give an
>   example of each.
> - Differentiate an intermediate host from a definitive host.

The biology of protozoa is discussed in this chapter. Diseases caused by protozoa are described in Part Four.

Protozoa are a large and diverse group. Current
schemes of classifying protozoan species into phyla are
based on rRNA sequencing. Researchers have begun sorting out groups within the protists based on their evolutionary history; that is, members in a group derived from a
single ancestor. At present, the following groups are phyla
of protists. As more information is obtained, some of these
groups may be classified as kingdoms and others may be
grouped with other kingdoms in the Domain Eukarya.

ARCHAEZOA

The **Archaezoa** are eukaryotes that lack mitochondria.
They do have a unique organelle called a **mitosome.** Mi-
tosomes appear to be a remnant of mitochondria that were
in an ancient ancestor of the archaezoa. Many archaezoans
live as symbionts in the digestive tracts of animals. Archaezoans are typically spindle-shaped, with flagella projecting from the front end (Figure 12.17a). Most have two
or more flagella, which move in a whiplike manner that
pulls the cells through their environment.

An example of an archaezoan that is a human parasite
is *Trichomonas vaginalis* (trik-ō-mōn′as va-jin-al′is), shown
in Figures 12.17b and 26.15. Like some other flagellates,
T. vaginalis has an **undulating membrane,** which consists

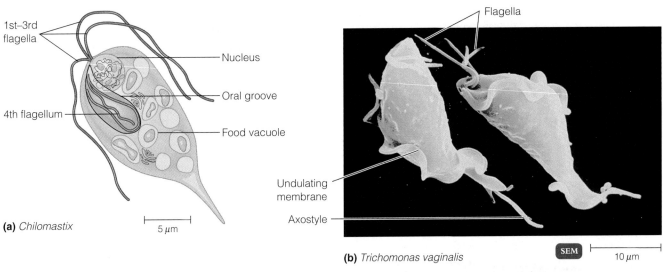

1st–3rd
flagella

Nucleus

Oral groove

4th flagellum

Food vacuole

(a) *Chilomastix*   5 μm

Flagella

Undulating
membrane

Axostyle

(b) *Trichomonas vaginalis*   SEM   10 μm

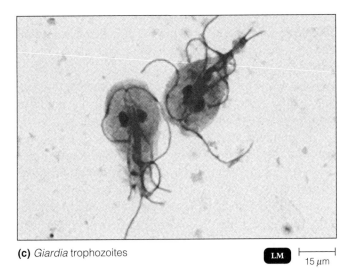

(c) *Giardia* trophozoites   LM   15 μm

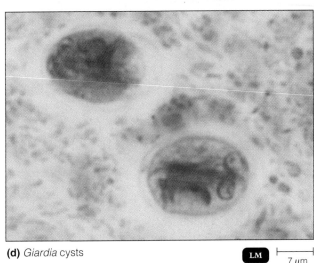

(d) *Giardia* cysts   LM   7 μm

FIGURE 12.17 Archaezoa. (a) *Chilomastix.* This flagellate, found in the human intestine, may be mildly pathogenic. The cysts survive for months outside a human host. The fourth flagellum is used to move food into the oral groove, where food vacuoles are formed. **(b)** *Trichomonas vaginalis.* This flagellate causes urinary and genital tract infections. Notice the small undulating mem-brane. This flagellate does not have a cyst stage. **(c)** *Giardia intestinalis.* The trophozoite of this intestinal parasite has eight flagella and two prominent nuclei, giving it a distinctive appearance. **(d)** The *G. intestinalis* cyst provides protection from the environment before it is ingested by a new host.

Q **How do archaezoans obtain energy without mitochondria?**

of a membrane bordered by a flagellum. *T. vaginalis* does not have a cyst stage and must be transferred from host to host quickly before desiccation occurs. *T. vaginalis* is found in the vagina and in the male urinary tract. It is usually trans-mitted by sexual intercourse but can also be transmitted by toilet facilities or towels.

Another parasitic archaezoan is *Giardia lamblia* (jē-är′d ē-ä lam′lē-ä), also called *G. intestinales* or *G. duode-nalis.* The parasite (Figures 12.17c and 12.18) is found in the small intestine of humans and other mammals. It is excreted in the feces as a cyst (Figure 12.17d) and survives in the environment before being ingested by the next host.

Diagnosis of giardiasis, the disease caused by *G. lamblia,* is often based on the identification of cysts in feces.

MICROSPORA

Microspora, like the Archaezoa, are unusual eukaryotes because they lack mitochondria. Microspora do not have microtubules (see Chapter 4, page 100), and they are obli-gate intracellular parasites. Microsporidial protozoa have been reported since 1984 to be the cause of a number of human diseases, including chronic diarrhea and kerato-conjunctivitis (inflammation of the conjunctiva near the cornea), most notably in AIDS patients.

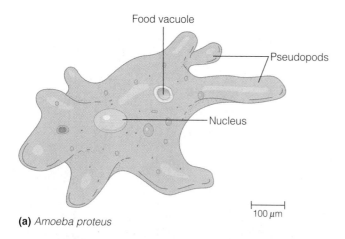

(a) *Amoeba proteus*

100 µm

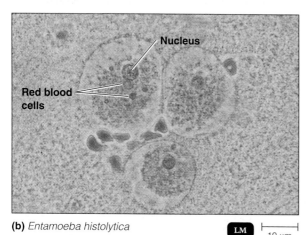

(b) *Entamoeba histolytica*

LM 10 µm

FIGURE 12.18 Amoebozoa. (**a**) To move and to engulf food, amoebas (such as this *Amoeba proteus*) extend cytoplasmic structures called pseudopods. Food vacuoles are created when pseudopods surround food and bring it into the cell. (**b**) *Entamoeba histolytica*. The presence of ingested red blood cells is diagnostic for *Entamoeba*.

Q **How do amoebic dysentery and bacillary dysentery differ?**

AMOEBOZOA

The **Amoebozoans,** or amoebas, move by extending blunt, lobelike projections of the cytoplasm called **pseudopods** (Figure 12.18a). Any number of pseudopods can flow from one side of the amoeba, and the rest of the cell will flow toward the pseudopods.

Entamoeba histolytica (en-ta-me′ba his-to-li′ti-ka) is the only pathogenic amoeba found in the human intestine. As many as 10% of the human population may be colonized by this amoeba. New techniques, including DNA analyses and lectin binding, have revealed that the amoeba thought to be *E. histolytica* are actually two distinct species. The nonpathogenic species, *E. dispar* (dis′par) is most common. The invasive *E. histolytica*

(Figure 12.18b) causes amoebic dysentery. In the human intestine, *E. histolytica* uses proteins called lectins to attach to the galactose of the plasma membrane and causes cell lysis. *E. dispar* does not have galactose-binding lectins. *Entamoeba* is transmitted between humans through ingestion of the cysts that are excreted in the feces of the infected person. *Acanthamoeba* growing in water, including tap water, can infect the cornea and cause blindness.

Since 1990, *Balamuthia* (bal′am-üth-ē-ä) has been reported as the cause of brain abscesses called granulomatous amoebic encephalitis in the United States and other countries. The amoeba most often infects immunocompromised people. Like *Acanthamoeba*, *Balamuthia* is a free-living amoeba found in water and is not transmitted from human to human.

APICOMPLEXA

The **Apicomplexa** are not motile in their mature forms and are obligate intracellular parasites. Apicomplexans are characterized by the presence of a complex of special organelles at the apexes (tips) of their cells (hence the phylum name). The organelles in these apical complexes contain enzymes that penetrate the host's tissues.

Apicomplexans have a complex life cycle that involves transmission between several hosts. An example of an apicomplexan is *Plasmodium* (plaz-mō′dē-um), the causative agent of malaria. Malaria is one of the most common diseases—it affects 10% of the world's population, with 300 million new cases each year. The complex life cycle makes it difficult to develop a vaccine against malaria (see the box in Chapter 18, page 532).

Plasmodium grows by sexual reproduction in the *Anopheles* (an-of′el-ēz) mosquito (Figure 12.19).

❶ When an *Anopheles* carrying the infective stage of *Plasmodium*, called a **sporozoite,** bites a human, sporozoites can be injected into the human. The sporozoites are carried by the blood to the liver.

❷ They undergo schizogony in liver cells and produce thousands of progeny called **merozoites.**

❸ Merozoites enter the bloodstream and infect red blood cells.

❹ The young trophozoite looks like a ring in which the nucleus and cytoplasm are visible. This is called a **ring stage** (see Figure 23.25b, page 697).

❺ The ring stage enlarges and divides repeatedly, and

❻ the red blood cells eventually rupture and release more merozoites. Upon release of the merozoites, their waste products, which cause fever and chills, are also released. Most of the merozoites infect new red blood cells and perpetuate their cycle of asexual reproduction.

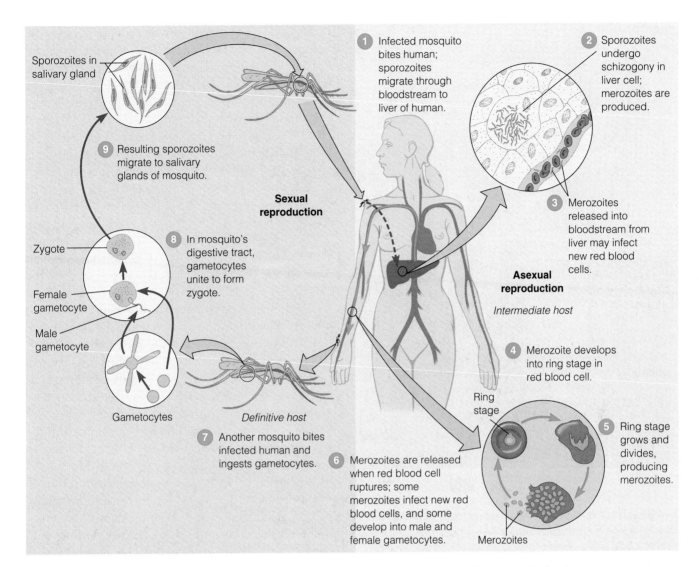

FIGURE 12.19 The life cycle of *Plasmodium vivax*, the apicomplexan that causes malaria. Asexual reproduction (schizogony) of the parasite takes place in the liver and in the red blood cells of a human host. Sexual reproduction occurs in the intestine of an *Anopheles* mosquito after the mosquito has ingested gametocytes.

Q **What is the definitive host for *Plasmodium*?**

However, some develop into male and female sexual forms (gametocytes). Even though the gametocytes themselves cause no further damage,

7 they can be picked up by the bite of another *Anopheles* mosquito; they then enter the mosquito's intestine and begin their sexual cycle.

8 Here the male and female gametocytes unite to form a zygote. The zygote forms an oocyst, in which cell division occurs, and asexual sporozoites are formed.

9 When the oocyst ruptures, the sporozoites migrate to the salivary glands of the mosquito. They can then be injected into a new human host by the biting mosquito.

The mosquito is the **definitive host** because it harbors the sexually reproducing stage of *Plasmodium*. The host in which the parasite undergoes asexual reproduction (in this case, the human) is the **intermediate host.**

Malaria is diagnosed in the laboratory by microscopic observation of thick blood smears for the presence of *Plasmodium* (see Figure 23.25, page 697). A peculiar characteristic of malaria is that the interval between periods of fever caused by the release of merozoites is always the same for a given species of *Plasmodium* and is always a multiple of 24 hours. The reason and mechanism for such precision have intrigued scientists. After all, why should a parasite need a biological clock? *Plasmodium*'s development is regulated by the host's body temperature, which normally

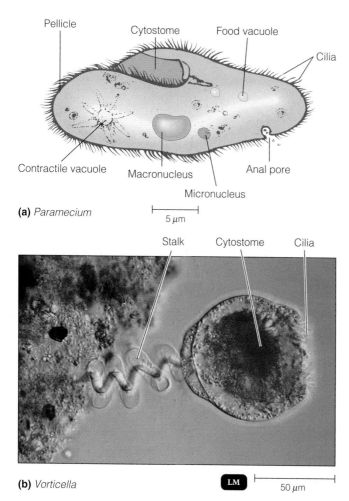

Pellicle
Cytostome
Food vacuole
Cilia
Contractile vacuole
Macronucleus
Micronucleus
Anal pore

(a) *Paramecium*

5 μm

Stalk   Cytostome   Cilia

(b) *Vorticella*

LM   50 μm

FIGURE 12.20 Ciliates. (**a**) *Paramecium* is covered with rows of cilia. It has specialized structures for ingestion (mouth), elimination of wastes (anal pore), and the regulation of osmotic pressure (contractile vacuoles). The macronucleus is involved with protein synthesis and other ongoing cellular activities. The micronucleus functions in sexual reproduction. (**b**) *Vorticella* attaches to objects in water by the base of its stalk. The springlike stalk can expand allowing *Vorticella* to feed in different areas. Cilia surround its cytostome.

Q **What ciliate can cause disease in humans?**

fluctuates over a 24-hour period. The parasite's careful timing ensures that gametocytes are mature at night, when *Anopheles* mosquitoes are feeding, and thereby facilitates transmission of the parasite to a new host.

Another apicomplexan parasite of red blood cells is *Babesia microti* (ba-bē'sē-ä mī-krō'tē). *Babesia* causes fever and anemia in immunosuppressed individuals. In the United States, it is transmitted by the tick *Ixodes scapularis* (iks-ō'-dēs skap-ū-lār'ís).

Toxoplasma gondii (toks- ō-plaz'mä gon'dē-ē) is another apicomplexan intracellular parasite of humans.

The life cycle of this parasite involves domestic cats. The trophozoites, called **tachyzoites,** reproduce sexually and asexually in an infected cat, and **oocysts,** each containing eight sporozoites, are excreted with feces. If the oocysts are ingested by humans or other animals, the sporozoites emerge as trophozoites, which can reproduce in the tissues of the new host (see Figure 23.23, page 695). *T. gondii* is dangerous to pregnant women, as it can cause congenital infections in utero. Tissue examination and observation of *T. gondii* are used for diagnosis. Antibodies may be detected by ELISA and by indirect fluorescent-antibody tests (see Chapter 18).

Cryptosporidium (krip-tō-spô-ri'dē-um) is a newly recognized parasite of humans. In AIDS patients and other immunosuppressed people, *Cryptosporidium* can cause respiratory and gallbladder infections and may be a major cause of death. The organism, which lives inside the cells lining the small intestine, can be transmitted to humans through the feces of cows, rodents, dogs, and cats. Waterborne and nosocomial infections have also been reported. Inside the host cell, each *Cryptosporidium* organism forms four oocysts (see Figure 25.19, page 773), each containing four sporozoites. When the oocyst ruptures, sporozoites may infect new cells in the host or be released with the feces. The disease is diagnosed by acid-fast staining or fluorescent-antibody tests.

During the 1980s, epidemics of waterborne diarrhea were identified on every continent except Antarctica. The causative agent was misidentified as a cyanobacterium because the outbreaks occurred during warm months, and the disease agent looked like a prokaryotic cell. In 1993, the organism was identified as an apicomplexan similar to *Cryptosporidium*. In 2004, the new parasite, named *Cyclospora cayetanensis* (sī'-klō-spô-rä kī'ē-tan-en-sis), was responsible for 300 cases of diarrhea associated with snow peas in the United States and Canada.

CILIOPHORA

Members of the phylum **Ciliophora,** or ciliates, have cilia that are similar to but shorter than flagella. The cilia are arranged in precise rows on the cell (Figure 12.20). They are moved in unison to propel the cell through its environment and to bring food particles to the mouth.

The only ciliate that is a human parasite is *Balantidium coli* (bal-an-tid'ē-um kō'lī), the causative agent of a severe, though rare, type of dysentery. When cysts are ingested by the host, they enter the large intestine, into which the trophozoites are released. The trophozoites produce proteases and other substances that destroy host cells. The

TABLE 12.5	Some Representative Parasitic Protozoa					
Phylum	Human Pathogens	Distinguishing Features	Disease	Source of Human Infections	Figure/ Table Reference	Page Reference
Archaezoa	Giardia lamblia	Two nuclei, eight flagella	Giardial enteritis	Fecal contamination of drinking water	25.18	772
	Trichomonas vaginalis	No encysting stage	Urethritis, vaginitis	Contact with vaginal-urethral discharge	Table 26.1	804
Microspora	Nosema	Unknown	Diarrhea, kerato-conjunctivitis, conjunctivitis	Other animals	—	—
Amoebozoa	Acanthamoeba	Pseudopods	Keratitis	Water	—	—
	Entamoeba histolytica, E. dispar		Amoebic dysentery	Fecal contamination of drinking water	25.20	774
	Balamuthia		Encephalitis	Water	—	—
Apicomplexa	Babesia microti	Complex	Babesiosis	Domestic animals, ticks	—	699
	Cryptosporidium	Life cycles may require more than one host	Diarrhea	Humans, other animals, water	25.19	773
	Cyclospora	—	Diarrhea	Water	Table 25.5	772
	Isospora	—	Coccidiosis	Domestic animals	Table 19.5	572
	Plasmodium	—	Malaria	Bite of Anopheles mosquito	12.19 23.25	365 697
	Toxoplasma gondii	—	Toxoplasmosis	Cats, beef; congenital	23.23	695
Dinoflagellates	Alexandrium, Pfiesteria	Photosynthetic, (see Table 12.4)	Paralytic shellfish poisoning; ciguatera	Ingestion of dinoflagellates in mollusks, fish	27.15	827
Ciliophora	Balantidium coli	Only parasitic ciliate of humans	Balantidial dysentery	Fecal contamination of drinking water	—	—
Euglenozoa	Leishmania	Flagellated form in sand fly; ovoid form in vertebrate host	Leishmaniasis	Bite of sand fly (Phlebotomus)	23.26	699
	Naegleria fowleri	Flagellated and amoeboid forms	Meningo-encephalitis	Water in which people swim	22.17	662
	Trypanosoma cruzi	Undulating membrane	Chagas' disease	Bite of Triatoma (kissing bug)	23.22	693
	T. brucei gambiense, T.b. rhodesiense		African trypanosomiasis	Bite of tsetse fly	22.16	661

trophozoite feeds on host cells and tissue fragments. Its cysts are excreted with feces.

Ciliates, apicomplexans, and dinoflagellates (page 360) may be placed in their own phylum or kingdom, called **Alveolata** because they all have membrane-bound cavities (alveoli) under the cell surface and rRNA sequences in common.

EUGLENOZOA

Two groups of flagellated cells are included in the **Euglenozoa** based on common rRNA sequences, disk-shaped mitochondria, and absence of sexual reproduction.

Euglenoids are photoautotrophs (Figure 12.21). Euglenoids have a semirigid plasma membrane called a pellicle, and they move by means of a flagellum at the anterior

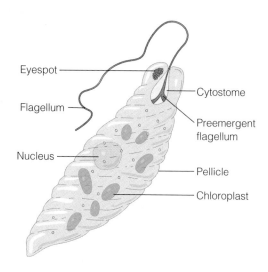

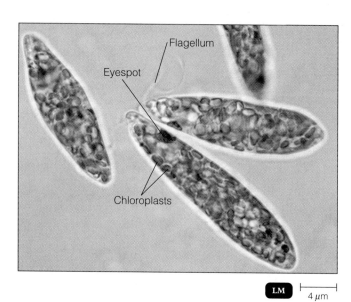

LM | 4 μm

FIGURE 12.21 *Euglena.* Euglenoids are photoautotrophs. Semirigid rings supporting the pellicle allow *Euglena* to change shape.

Q **Why is *Euglena* classified with the hemoflagellates?**

end. Most euglenoids also have a red *eyespot* at the anterior end. This carotenoid-containing organelle senses light and directs the cell in the appropriate direction by using a *preemergent flagellum.* Some euglenoids are facultative chemoheterotrophs. In the dark, they ingest organic matter through a cytostome. Euglenoids are frequently studied with algae because they can photosynthesize.

The **hemoflagellates** (blood parasites) are transmitted by the bites of blood-feeding insects and are found in the circulatory system of the bitten host (see the box on page 371). To survive in this viscous fluid, hemoflagellates usually have long, slender bodies and an undulating membrane. The genus *Trypanosoma* (tri-pa′nō-sō-mä) includes the species that causes African sleeping sickness, *T. brucei gambiense* (brüs′ē gam-bē-ens′), which is transmitted by the tsetse fly. *T. cruzi* (kruz′ē; see Figure 23.22, page 693), the causative agent of Chagas' disease, is transmitted by the "kissing bug," so named because it bites on the face (see Figure 12.33d). After entering the insect, the trypanosome rapidly multiplies by fission. If the insect then defecates while biting a human, it can release trypanosomes that can contaminate the bite wound.

Table 12.5 lists some typical parasitic protozoa and the diseases they cause.

SLIME MOLDS

LEARNING OBJECTIVE

• Compare and contrast cellular slime molds and plasmodial slime molds.

Slime molds have both fungal and amoebal characteristics; they are, however, more closely related to amoeba and placed in the Phylum Amoebozoa. There are two taxa of slime molds: cellular and plasmodial. **Cellular slime molds** are typical eukaryotic cells that resemble amoebas. In the life cycle of cellular slime molds (Figure 12.22),

❶ the amoeboid cells live and grow by ingesting fungi and bacteria by phagocytosis. Cellular slime molds are of interest to biologists who study cellular migration and aggregation, because when conditions are unfavorable,

❷–❸ large numbers of amoeboid cells aggregate to form a single structure. This aggregation occurs because some individual amoebas produce the chemical cyclic AMP (cAMP), toward which the other amoebas migrate.

❹ The aggregated amoebas are enclosed in a slimy sheath called a *slug.* The slug migrates as a unit toward light.

❺ After a period of hours, the slug ceases to migrate and begins to form differentiated structures.

❻ Some of the amoeboid cells form a stalk; others swarm up the stalk to form a spore cap, and

❼ most of these differentiate into spores.

❽ When spores are released under favorable conditions,

❾ they germinate to form single amoebas.

In 1973, a Dallas resident discovered a pulsating red blob in his backyard. The news media claimed that a "new

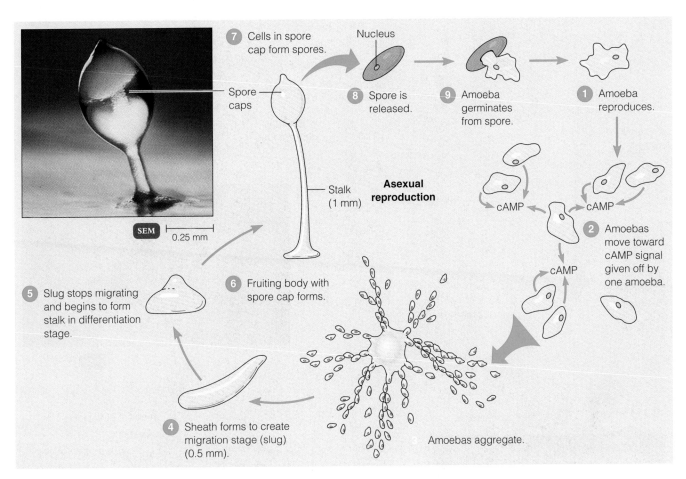

FIGURE 12.22 The generalized life cycle of a cellular slime mold. The micrograph shows a spore cap of *Dictyostelium.*

Q **What characteristics do slime molds share with protozoa? With fungi?**

life form" had been found. For some people, the "creature" evoked spine-chilling recollections of an old science fiction movie. Before imaginations got carried away too far, biologists calmed everyone's worst fears (or highest hopes). The amorphous mass was merely a plasmodial slime mold, they explained. But its unusually large size—46 cm in diameter—startled even scientists.

Plasmodial slime molds were first scientifically reported in 1729. They belong to a separate phylum. A plasmodial slime mold exists as a mass of protoplasm with many nuclei (it is multinucleated). This mass of protoplasm is called a **plasmodium** (Figure 12.23).

① The entire plasmodium moves as a giant amoeba; it engulfs organic debris and bacteria. Biologists have found that musclelike proteins forming microfilaments account for the movement of the plasmodium.

② When plasmodial slime molds are grown in laboratories, a phenomenon called **cytoplasmic streaming** is observed, during which the protoplasm within the plasmodium moves and changes both its speed and direction so that the oxygen and nutrients are evenly distributed. The plasmodium continues to grow as long as there is enough food and moisture for it to thrive.

③ When either is in short supply, the plasmodium separates into many groups of protoplasm;

④ each of these groups forms a stalked sporangium,

⑤ in which spores (a resistant, resting form of the slime mold) develop.

⑥ Nuclei within these spores undergo meiosis and form uninucleate haploid cells.

⑦ The spores are then released.

⑧ When conditions improve, these spores germinate,

⑨ fuse to form diploid cells, and

⑩ develop into a multinucleated plasmodium.

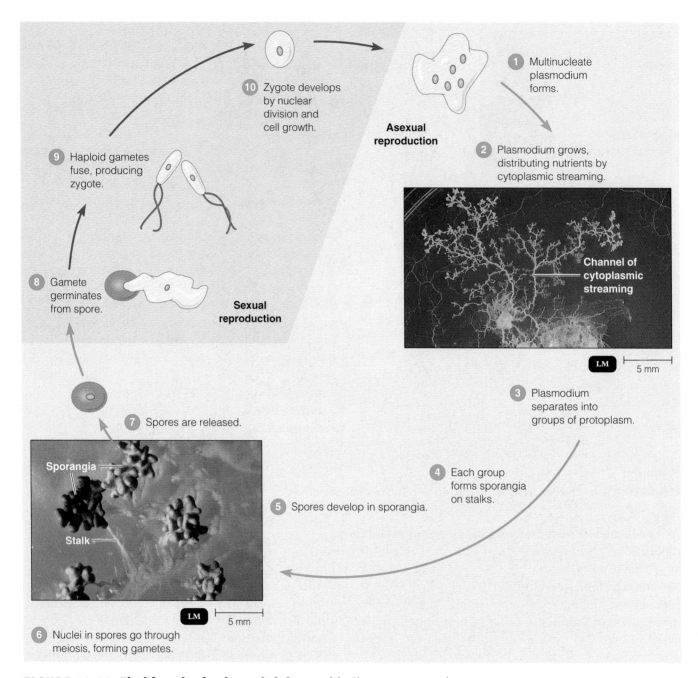

FIGURE 12.23   The life cycle of a plasmodial slime mold. *Physarum* is pictured in the photomicrographs.

Q   How do cellular and acellular slime molds differ?

HELMINTHS

A number of parasitic animals spend part or all of their lives in humans. Most of these animals belong to two phyla: Platyhelminthes (flatworms) and Nematoda (roundworms). These worms are commonly called **helminths.** There are also free-living species in these phyla, but we will limit our discussion to the parasitic species. Diseases caused by parasitic worms are discussed in Part Four.

CHARACTERISTICS OF HELMINTHS

LEARNING OBJECTIVES

- List the distinguishing characteristics of parasitic helminths.
- Provide a rationale for the elaborate life cycle of parasitic worms.

Helminths are multicellular eukaryotic animals that generally possess digestive, circulatory, nervous, excretory, and

CLINICAL PROBLEM SOLVING

A PARASITIC DISEASE

As you read through this box, you will encounter a series of questions that microbiologists ask themselves as they try to diagnose a disease. Try to answer each question before going on to the next one.

1. A previously healthy 30-year-old man first noticed a fever (maximum documented temperature, 40°C [104°F]) in late December. A 10-day course of treatment with Augmentin was prescribed. *For what types of diseases is Augmentin used? (Hint: See page 592.)*

2. Augmentin is an antibiotic that can help rule out a primary bacterial infection and treat any complicating secondary infection. The patient continued to experience fluctuating temperatures and an unintentional weight loss of 25 pounds. *What else do you need to know about this patient?*

3. Fourteen months prior to onset of symptoms while in the U.S. military, the patient had traveled extensively in Afghanistan and had lived and worked with local Afghanis. He had no history of blood transfusions or subsequent travel. *What information do you need about his stay in Afghanistan?*

4. Although he reportedly had used personal protective measures (e.g., permethrin-impregnated bed netting and insect repellent containing 30%–35% DEET [diethyl toluamide]), he had noted multiple insect bites. *What diseases are possible?*

5. While in Afghanistan, he had used mefloquine for prophylaxis against malaria and ciprofloxacin for treatment of occasional diarrheal illnesses. During the course of the illness he developed an enlarged spleen, liver, lymph nodes, and edema. *What tissue samples do you want? How will you examine them?*

6. Light microscopy of white blood cells revealed parasites (see the photo). *Can you identify the parasite.*

The patient recovered with a 28-day course of the antimonial Pentostam.

The patient had classic manifestations of advanced visceral leishmaniasis. Although cutaneous leishmaniasis is common in Afghanistan, including an ongoing epidemic in Kabul with an estimated 200,000 cases, only 21 cases of visceral leishmaniasis attributed to exposures in Afghanistan have been reported previously. Additional cases might have occurred that were not diagnosed or reported. Over 350 cases of cutaneous leishmaniasis have been diagnosed among U.S. military personnel in Iraq. Leishmaniasis is currently prevalent in all continents except Antarctica, with a worldwide prevalence of 12 million cases. The World Health Organization estimates that approximately 500,000 new cases of visceral leishmaniasis occur each year. The annual number of deaths is estimated around 59,000.

SOURCE: Adapted from *MMWR* 52(12) (4/2/2004).

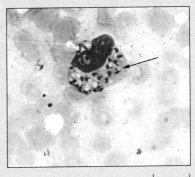

├─────────┤
10 μm

A macrophage practically filled with ovoid cells (arrow), several of which have a clearly visible nucleus and kinetoplast.

---

reproductive systems. Parasitic helminths must be highly specialized to live inside their hosts. The following generalizations distinguish parasitic helminths from their free-living relatives:

1. *They may* lack *a digestive system.* They can absorb nutrients from the host's food, body fluids, and tissues.

2. *Their nervous system is* reduced. They do not need an extensive nervous system because they do not have to search for food or respond much to their environment. The environment within a host is fairly constant.

3. *Their means of locomotion is occasionally* reduced *or completely lacking.* Because they are transferred from host to host, they do not need to search actively for a suitable habitat.

4. *Their reproductive system is often complex.* An individual produces large numbers of eggs, by which a suitable host is infected.

LIFE CYCLE

The life cycle of parasitic helminths can be extremely complex, involving a succession of intermediate hosts for completion of each **larval** (developmental) stage of the parasite and a definitive host for the adult parasite.

Adult helminths may be **dioecious;** male reproductive organs are in one individual, and female reproductive organs are in another. In those species, reproduction occurs only when two adults of the opposite sex are in the same host.

Adult helminths may also be **monoecious,** or **hermaphroditic**—one animal has both male and female

FIGURE 12.24 Infection by a parasitic platyhelminth.
An increase in the trematode *Ribeiroia* in recent years has caused deformed frogs. Frogs with multiple limbs have been found from Minnesota to California. Cercaria of the trematode infect tadpoles. The encysted metacercariae displace developing limb buds, causing abnormal limb development. The increase in the parasite may be due to fertilizer runoff that increases algae, which are food for the parasite's intermediate host snail.

Q **What tailed stage of the parasite lives in a snail?**

reproductive organs. Two hermaphrodites may copulate and simultaneously fertilize each other. A few types of hermaphrodites fertilize themselves.

PLATYHELMINTHS

LEARNING OBJECTIVES

- List the characteristics of the two classes of parasitic platyhelminths, and give an example of each.
- Describe a parasitic infection in which humans serve as a definitive host, as an intermediate host, and as both.

Members of the phylum Platyhelminthes, the **flatworms,** are dorsoventrally flattened. The classes of parasitic flatworms include the trematodes and cestodes. These parasites cause disease or developmental disturbances in a wide variety of animals (Figure 12.24).

TREMATODES

Trematodes, or **flukes,** often have flat, leaf-shaped bodies with a ventral sucker and an oral sucker (Figure

12.25). The suckers hold the organism in place. Flukes obtain food by absorbing it through their nonliving outer covering, called the **cuticle.** Flukes are given common names according to the tissue of the definitive host in which the adults live (for example, lung fluke, liver fluke, blood fluke). The Asian liver fluke *Clonorchis sinensis* (klo-nôr′kis si-nen′sis) is occasionally seen in immigrants in the United States, but it cannot be transmitted because its intermediate hosts are not in the United States.

To exemplify a fluke's life cycle, let's look at the lung fluke, *Paragonimus westermani* (pār-ä-gōn′e-mus we-ster-ma′nē). The intermediate hosts for this fluke, and therefore the fluke itself, occur throughout the world, including the United States and Canada. The adult lung fluke lives in the bronchioles of humans and other mammals and is approximately 6 mm wide and 12 mm long. The hermaphroditic adults liberate eggs into the bronchi. Because sputum that contains eggs is frequently swallowed, the eggs are usually excreted in feces of the definitive host. If the life cycle is to continue, the eggs must reach a body of water. A series of steps occurs that ensure adult flukes can mature in the lungs of a new host. The life cycle is shown in Figure 12.26.

In a laboratory diagnosis, sputum and feces are examined microscopically for fluke eggs. Infection results from eating undercooked crayfish, and the disease can be prevented by thoroughly cooking crayfish.

The cercariae of the blood fluke *Schistosoma* (shis-tō-sō′ma) are not ingested. Instead, they burrow through the skin of the human host and enter the circulatory system. The adults are found in certain abdominal and pelvic veins. The disease schistosomiasis is a major world health problem; it will be discussed further in Chapter 23 (page 701).

CESTODES

Cestodes, or **tapeworms,** are intestinal parasites. Their structure is shown in Figure 12.27. The head, or **scolex** (plural: *scoleces*), has suckers for attaching to the intestinal mucosa of the definitive host; some species also have small hooks for attachment. Tapeworms do not ingest the tissues of their hosts; in fact, they completely lack a digestive system. To obtain nutrients from the small intestine, they absorb food through their cuticle. The body consists of segments called **proglottids.** Proglottids are continually produced by the neck region of the scolex, as long as the scolex is attached and alive. Each mature proglottid contains both male and female reproductive organs. The proglottids farthest away from the scolex are the mature ones containing eggs. Mature proglottids are essentially bags of eggs, each of which is infective to the proper intermediate host.

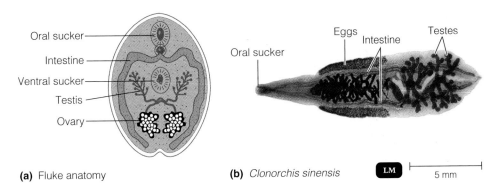

(a) Fluke anatomy

(b) *Clonorchis sinensis*

LM 5 mm

FIGURE 12.25 Flukes. (a) General anatomy of an adult fluke, shown in cross section. The oral and ventral suckers attach the fluke to the host. The mouth is located in the center of the oral sucker. Flukes are hermaphroditic; each animal contains both testes and ovaries. **(b)** The Asian liver fluke *Clonorchis sinensis*. Notice the incomplete digestive system. Heavy infestations may block bile ducts from the liver.

Q **Why is the flatworm digestive system called "incomplete"?**

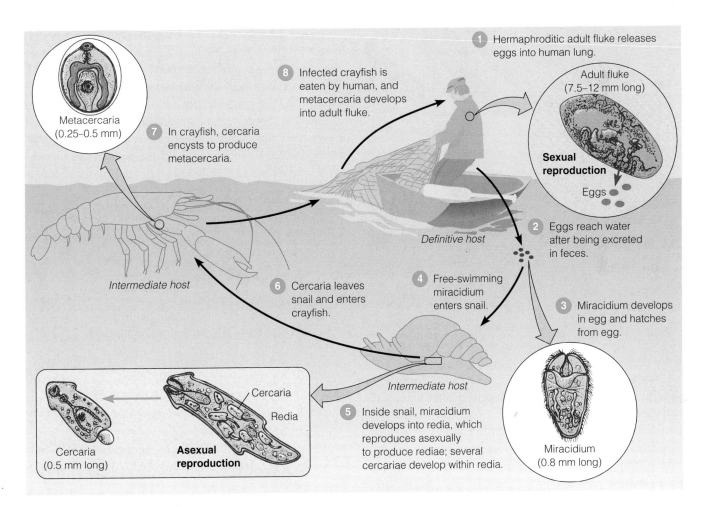

FIGURE 12.26 The life cycle of the lung fluke *Paragonimus westermani*. The trematode reproduces sexually in a human and asexually in a snail, its first intermediate host. Larvae encysted in the second intermediate host, a crayfish, infect humans when ingested. Also see the *Schistosoma* life cycle in Figure 23.27 (page 700).

Q **Of what value is this complex life cycle to *Paragonimus*?**

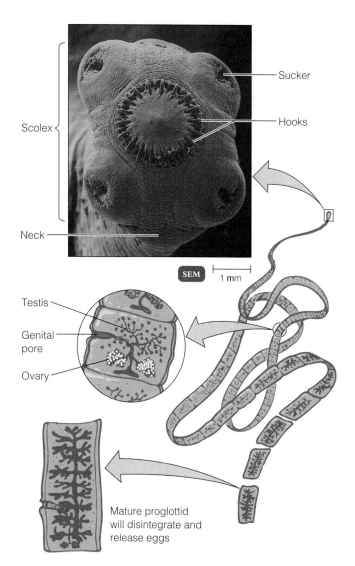

Scolex

Sucker

Hooks

Neck

SEM   1 mm

Testis

Genital pore

Ovary

Mature proglottid will disintegrate and release eggs

FIGURE 12.27   General anatomy of an adult tapeworm. The scolex, shown in the micrograph, consists of suckers and hooks that attach to the host's tissues. The body lengthens as new proglottids form at the neck. Each mature proglottid contains both testes and ovaries.

Q   What are the similarities between tapeworms and flukes?

Humans as Definitive Hosts   The adults of *Taenia saginata* (te′nē-ä sa-ji-nä′tä), the beef tapeworm, live in humans and can reach a length of 6 m. The scolex is about 2 mm long and is followed by a thousand or more proglottids. The feces of an infected human contain mature proglottids, each of which contains thousands of eggs. As the proglottids wriggle away from the fecal material, they increase their chances of being ingested by an animal that is grazing. Upon ingestion by cattle, the larvae hatch from the eggs and bore through the intestinal wall. The larvae migrate to muscle (meat), in which they encyst as **cysticerci.** When the cysticerci are ingested by humans, all

but the scolex is digested. The scolex anchors itself in the small intestine and begins producing proglottids.

Diagnosis of tapeworm infection in humans is based on the presence of mature proglottids and eggs in feces. Cysticerci can be seen macroscopically in meat; their presence is referred to as "measly beef." Inspecting beef that is intended for human consumption for "measly" appearance is one way to prevent infections by beef tapeworm. Another method of prevention is to avoid the use of untreated human sewage as fertilizer in grazing pastures.

Humans are the only known definitive host of the pork tapeworm, *Taenia solium.* Adult worms living in the human intestine produce eggs, which are passed out in feces. When eggs are eaten by pigs, the larval helminth encysts in the pig's muscles; humans become infected when they eat undercooked pork. The human-pig-human cycle of *T. solium* is common in Latin America, Asia, and Africa. In the United States, however, *T. solium* is virtually nonexistent in pigs; the parasite is transmitted from human to human. Eggs shed by one person and ingested by another person hatch, and the larvae encyst in the brain and other parts of the body, causing cysticercosis (see Figure 25.22, page 775). The human hosting *T. solium*'s larvae is serving as an intermediate host. Approximately 7% of the few hundred cases reported in recent years were acquired by people who had never been outside the United States. They may have become infected through household contact with people who were born in or had traveled in other countries.

Humans as Intermediate Hosts   Humans are the intermediate hosts for *Echinococcus granulosus* (ē-kīn-ō-kok′kus gra-nū-lō′sus), shown in Figure 12.28. Dogs and coyotes are the definitive hosts for this minute (2–8 mm) tapeworm.

❶ Eggs are excreted with feces and

❷ are ingested by deer, sheep, or humans. Humans can also become infected by contaminating their hands with dog feces or saliva from a dog that has licked itself.

❸ The eggs hatch in the human's small intestine, and the larvae migrate to the liver or lungs.

❹ The larva develops into a **hydatid cyst.** The cyst contains "brood capsules," from which thousands of scoleces might be produced.

❺ Humans are a dead-end for the parasite, but in the wild, the cysts might be in a deer that is eaten by a wolf.

❻ The scoleces would be able to attach themselves in the wolf's intestine and produce proglottids.

Diagnosis of hydatid cysts is frequently made only on autopsy, although X rays can detect the cysts (see Figure 25.23, page 776).

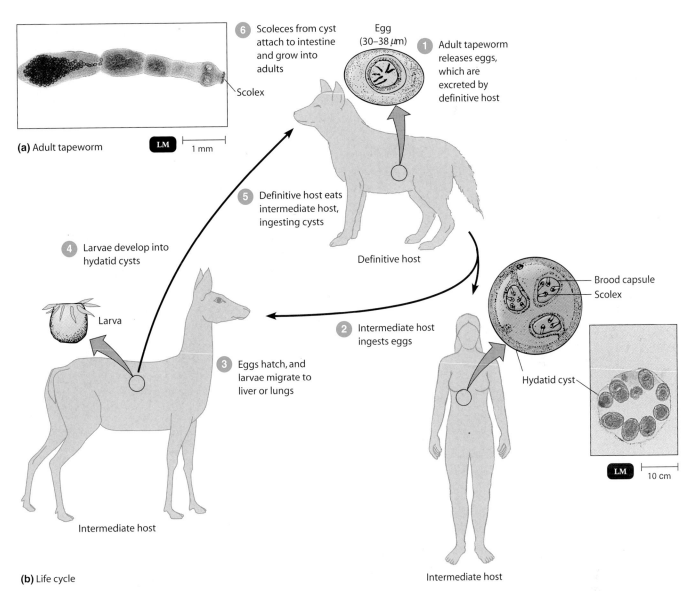

(a) Adult tapeworm   LM |⊢ 1 mm ⊣|

⑥ Scoleces from cyst attach to intestine and grow into adults
Scolex

Egg (30–38 μm)
① Adult tapeworm releases eggs, which are excreted by definitive host

⑤ Definitive host eats intermediate host, ingesting cysts
Definitive host

④ Larvae develop into hydatid cysts
Larva

② Intermediate host ingests eggs

③ Eggs hatch, and larvae migrate to liver or lungs

Intermediate host

Brood capsule
Scolex
Hydatid cyst

LM |⊢ 10 cm ⊣|

(b) Life cycle

Intermediate host

FIGURE 12.28 The tapeworm *Echinococcus granulosus.* This tiny tapeworm is found in the intestines of dogs, wolves, and foxes. (**a**) The adult of the closely related *Echinococcus multilocu-* *laris.* (**b**) Life cycle. The photomicrograph shows a hydatid cyst. The parasite can complete its life cycle only if the cysts are in-gested by a definitive host that eats the intermediate host.

Q **Why isn't being in a human of benefit to *Echinococcus*?**

NEMATODES

LEARNING OBJECTIVES

- List the characteristics of parasitic nematodes, and give an example of infective eggs and infective larvae.
- Compare and contrast platyhelminths and nematodes.

Members of the Phylum Nematoda, the **roundworms,** are cylindrical and tapered at each end. Roundworms have a *complete* digestive system, consisting of a mouth, an intes-tine, and an anus. Most species are dioecious. Males are smaller than females and have one or two hardened

spicules on their posterior ends. Spicules are used to guide sperm to the female's genital pore.

Some species of nematodes are free-living in soil and water, and others are parasites on plants and animals. Some nematodes pass their entire life cycle, from egg to mature adult, in a single host.

Nematode infections of humans can be divided into two categories: those in which the egg is infective, and those in which the larva is infective.

EGGS INFECTIVE FOR HUMANS

The pinworm *Enterobius vermicularis* (en-te-rō'bē-us ver-mi-kū-lar'is) spends its entire life in a human host

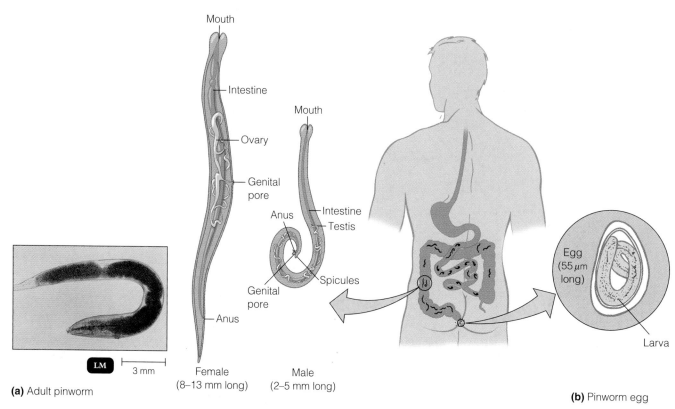

(a) Adult pinworm

(b) Pinworm egg

FIGURE 12.29 The pinworm *Enterobius vermicularis*. (a) Adult pinworms live in the large intestine of humans. Most roundworms are dioecious, and the female (left and photomicrograph) is often distinctly larger than the male (right). **(b)** Pinworm eggs are deposited by the female on the perianal skin at night.

Q **Are humans the definitive or intermediate host for pinworms?**

(Figure 12.29). Adult pinworms are found in the large intestine. From there, the female pinworm migrates to the anus to deposit her eggs on the perianal skin. The eggs can be ingested by the host or by another person exposed through contaminated clothing or bedding. Pinworm infections are diagnosed by the Graham sticky-tape method. A piece of transparent tape is placed on the perianal skin in such a way that the sticky side picks up eggs that were deposited earlier. The tape is then microscopically examined for the presence of eggs.

Ascaris lumbricoides (as'kar-is lum-bri-koi'dēz) is a large nematode (30 cm in length) that infects over one billion people worldwide. (Figure 25.25, page 777). It is dioecious with **sexual dimorphism;** that is, the male and female worms look distinctly different, the male being smaller with a curled tail. The adult *Ascaris* lives in the small intestines of humans exclusively; it feeds primarily on semi-digested food. Eggs, excreted with feces, can survive in the soil for long periods until accidentally ingested by another host. The eggs hatch in the small intestine of the host.

The larvae then burrow out of the intestine and enter the blood. They are carried to the lungs, where they grow. The larvae will then be coughed up, swallowed, and returned to the small intestine, where they mature into adults.

Diagnosis is frequently made when the adult worms are excreted with feces. *Ascaris* eggs remain in soil for 10 years. Children become infected by putting their hands and toys in their mouths. Preventing infection in humans is managed by proper sanitary habits.

LARVAE INFECTIVE FOR HUMANS

Adult hookworms, *Necator americanus* (ne-kā'tôr ä-me-ri-ka'nus) and *Ancylostoma duodenale* (an-sil-os'toma dü'o-den-al-ē), live in the small intestine of humans (Figures 12.30 and 25.24); the eggs are excreted in feces. The larvae hatch in the soil, where they feed on bacteria. A larva enters its host by penetrating the host's skin. It then enters a blood or lymph vessel, which carries it to the lungs. It is coughed up in sputum, swallowed, and finally carried to

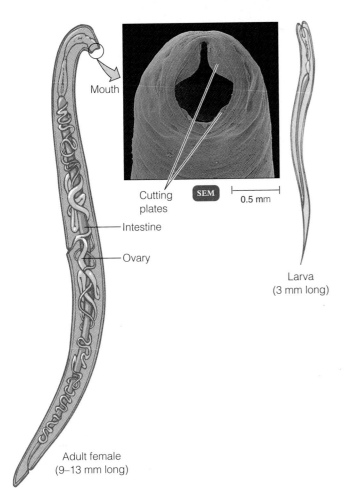

FIGURE 12.30 The hookworm _Necator americanus._
Notice the hook shape of the adult. The cutting plates around the mouth are used for attaching to and removing food from the host tissue. The free-living larvae inhabit the soil and infect their definitive host (humans) by penetrating the skin.

Q **How do roundworms and flatworms differ?**

the small intestine. Diagnosis is based on the presence of eggs in feces. People can avoid hookworm infections by wearing shoes.

Worldwide, _Trichinella spiralis_ (trik-in-el′lä spī-ra′lis) infections, called trichinellosis, are usually acquired by eating encysted larvae in poorly cooked pork. In the United States, trichinellosis is more often acquired from eating game animals, such as bears. In the human digestive tract, the larvae are freed from the cysts. They mature into adults in the small intestine and sexually reproduce there. Eggs develop in the female, and she gives birth to live larvae. The larvae enter lymph and blood vessels in the intestines and migrate from there throughout the body. They encyst in muscles and other tissues and remain there until ingested by another host (see Figure 25.26, page 778).

Diagnosis of trichinellosis is made by microscopic examination for larvae in a muscle biopsy. Trichinellosis can be prevented by thoroughly cooking meat prior to consumption.

Four genera of roundworms called _anisakines_, or wriggly worms, can be transmitted to humans from infected fish and squid. Anisakine larvae are in the fish's intestinal mesenteries and migrate to the muscle when the fish dies. Freezing or thorough cooking will kill the larvae.

Table 12.6 lists representative parasitic helminths of each phylum and class and the diseases they cause.

ARTHROPODS AS VECTORS

> **LEARNING OBJECTIVES**
> - Define _arthropod vector._
> - Differentiate between a tick and a mosquito, and name a disease transmitted by each.

Arthropods are animals characterized by segmented bodies, hard external skeletons, and jointed legs. With nearly 1 million species, this is the largest phylum in the animal kingdom. Although not microbes themselves, we will briefly describe arthropods here because a few suck the blood of humans and other animals and can transmit microbial diseases while doing so. Arthropods that carry pathogenic microorganisms are called **vectors.** Scabies and pediculosis are diseases that are caused by arthropods (see Chapter 21, pages 631–632).

Representative classes of arthropods include the following:

- Arachnida (eight legs): spiders, mites, ticks
- Crustacea (four antennae): crabs, crayfish
- Insecta (six legs): bees, flies, lice

Table 12.7 lists those arthropods that are important vectors, and Figures 12.31, 12.32, and 12.33 (see page 380) illustrate some of them. These insects and ticks reside on an animal only when they are feeding. An exception to this is the louse, which spends its entire life on its host and cannot survive for long away from a host.

Some vectors are just a mechanical means of transport for a pathogen. For example, houseflies lay their eggs on decaying organic matter such as feces. While doing so, a housefly can pick up a pathogen on its feet or body and transport the pathogen to our food.

Some parasites multiply in their vectors. When this happens, the parasites can accumulate in the vector's feces or saliva. Large numbers of parasites can then be deposited on or in the host while the vector is feeding there. The spirochete that causes Lyme disease is transmitted by ticks

TABLE 12.6   Representative Parasitic Helminths

Phylum	Class	Human Parasites	Intermediate Host	Definitive Host Site	Stage Passed to Humans; Method	Disease	Location in Humans	Figure Reference
Platyhel-minthes	Trema-todes	Paragonimus westermani	Freshwater snails and crayfish	Humans; lungs	Metacercaria in crayfish; ingested	Paragonimiasis (lung fluke)	Lungs	12.26
		Schistosoma	Freshwater snails	Humans	Cercariae; through skin	Schistosomiasis	Veins	23.27 23.28
	Ces-todes	Taenia saginata	Cattle	Humans; small intestine	Cysticerci in beef; ingested	Tapeworm	Small intestine	—
		Taenia solium	Humans; pigs	Humans	Eggs; ingested	Neurocysti-cercosis	Brain; any tissue	25.22
		Echinococcus granulosus	Humans	Dogs and other animals; intestines	Eggs from other animals; ingested	Hydatidosis	Lungs, liver, brain	12.28, 25.23
Nematoda		Ascaris lumbricoides	—	Humans; small intestine	Eggs; ingested	Ascariasis	Small intestine	25.25
		Enterobius vermicularis	—	Humans; large intestine	Eggs; ingested	Pinworm	Large intestine	12.29
		Necator americanus	—	Humans; small intestine	Larvae; through skin	Hookworm	Small intestine	12.30
		Ancylostoma duodenale	—	Humans; small intestine	Larvae; through skin	Hookworm	Small Intestine	25.24
		Trichinella spiralis	—	Humans, pigs, and other mammals; small intestine	Larvae; ingested	Trichinellosis	Muscles	25.26
		Anisakines	Marine fish and squid	Marine mammals	Larvae in fish; ingested	Anisakiasis (sashimi worms)	Gastro-intestinal tract	—

TABLE 12.7   Important Arthropod Vectors of Human Diseases

Class	Order	Vector	Disease	Figure Reference
Arachnida	Mites and ticks	Dermacentor (tick)	Rocky Mountain spotted fever	—
		Ixodes (tick)	Lyme disease, babesiosis, ehrlichiosis	12.32
		Ornithodorus (tick)	Relapsing fever	—
Insecta	Sucking lice	Pediculus (human louse)	Epidemic typhus, relapsing fever	12.33a
	Fleas	Xenopsylla (rat flea)	Endemic murine typhus, plague	12.33b
	True flies	Chrysops (deer fly)	Tularemia	12.33c
		Aedes (mosquito)	Dengue fever, yellow fever	—
		Anopheles (mosquito)	Malaria	12.31
		Culex (mosquito)	Arboviral encephalitis	—
		Glossina (tsetse fly)	African trypanosomiasis	—
	True bugs	Triatoma (kissing bug)	Chagas' disease	12.33d

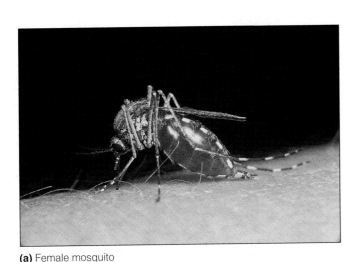

(a) Female mosquito

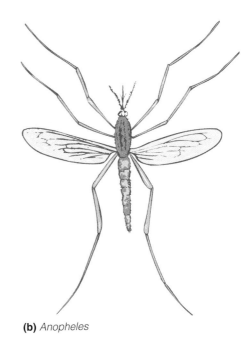

(b) *Anopheles*

FIGURE 12.31 Mosquitoes. **(a)** A female mosquito sucking blood from human skin. Mosquitoes transmit several pathogens from person to person, including the yellow fever and West Nile viruses. **(b)** The *Anopheles* mosquito transmits malaria.

Q **When is a vector also a definitive host?**

LM |—————| 1 mm

FIGURE 12.32 Ticks. *Ixodes pacificus* is the Lyme disease vector on the West Coast.

Q **Why aren't ticks classified as insects?**

in this manner (see Chapter 23, page 685), and the West Nile virus is transmitted in the same way by mosquitoes (see Chapter 22, page 658).

As discussed earlier, *Plasmodium* is an example of a parasite that requires that its vector also be the definitive host. *Plasmodium* can sexually reproduce only in the gut of an *Anopheles* mosquito. *Plasmodium* is introduced into a human host with the mosquito's saliva, which acts as an anticoagulant that keeps blood flowing.

To eliminate vectorborne diseases (such as African sleeping sickness), health workers focus on eradicating the vectors.

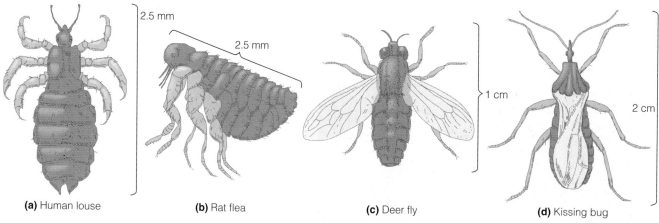

(a) Human louse **(b)** Rat flea **(c)** Deer fly **(d)** Kissing bug

FIGURE 12.33 Arthropod vectors. (a) The human louse, *Pediculus*. **(b)** The rat flea, *Xenopsylla*. **(c)** The deer fly, *Chrysops*. **(d)** The kissing bug, *Triatoma*.

Q **Name one pathogen carried by each of these vectors.**

STUDY OUTLINE

FUNGI (pp. 345–355)

1. Mycology is the study of fungi.

2. The number of serious fungal infections is increasing.

3. Fungi are aerobic or facultatively anaerobic chemo-heterotrophs.

4. Most fungi are decomposers, and a few are parasites of plants and animals.

CHARACTERISTICS OF FUNGI (pp. 346–350)

5. A fungal thallus consists of filaments of cells called hyphae; a mass of hyphae is called a mycelium.

6. Yeasts are unicellular fungi. To reproduce, fission yeasts divide symmetrically, whereas budding yeasts divide asymmetrically.

7. Buds that do not separate from the mother cell form pseudohyphae.

8. Pathogenic dimorphic fungi are yeastlike at 37°C and mold-like at 25°C.

9. Fungi are classified according to rRNA.

10. These spores can be produced asexually: sporangiospores and conidia.

11. Sexual spores are usually produced in response to special circumstances, often changes in the environment.

12. Fungi can grow in acidic, low-moisture, aerobic environments.

13. They are able to metabolize complex carbohydrates.

MEDICALLY IMPORTANT PHYLA OF FUNGI (pp. 350–352)

14. The Zygomycota have coenocytic hyphae and produce sporangiospores and zygospores.

15. The Ascomycota have septate hyphae and produce ascospores and frequently conidia.

16. Basidiomycota have septate hyphae and produce basidiospores; some produce conidiospores.

17. Teleomorphic fungi produce sexual and asexual spores; anamorphic fungi produce asexual spores only.

FUNGAL DISEASES (pp. 352–354)

18. Systemic mycoses are fungal infections deep within the body that affect many tissues and organs.

19. Subcutaneous mycoses are fungal infections beneath the skin.

20. Cutaneous mycoses affect keratin-containing tissues such as hair, nails, and skin.

21. Superficial mycoses are localized on hair shafts and superficial skin cells.

22. Opportunistic mycoses are caused by normal microbiota or fungi that are not usually pathogenic.

23. Opportunistic mycoses include *Pneumocystis* pneumonia; aspergillosis, caused by *Aspergillus*; and candidiasis, caused by *Candida*.

24. Opportunistic mycoses can infect any tissues. However, they are usually systemic.

ECONOMIC EFFECTS OF FUNGI (pp. 354–355)

25. *Saccharomyces* and *Trichoderma* are used in the production of foods.
26. Fungi are used for the biological control of pests.
27. Mold spoilage of fruits, grains, and vegetables is more common than bacterial spoilage of these products.
28. Many fungi cause diseases in plants.

LICHENS (pp. 355–357)

1. A lichen is a mutualistic combination of an alga (or a cyanobacterium) and a fungus.
2. The alga photosynthesizes, providing carbohydrates for the lichen; the fungus provides a holdfast.
3. Lichens colonize habitats that are unsuitable for either the alga or the fungus alone.
4. Lichens may be classified on the basis of morphology as crustose, foliose, or fruticose.
5. Lichens are used for their pigments and as air quality indicators.

ALGAE (pp. 357–361)

1. Algae are unicellular, filamentous, or multicellular (thallic).
2. Most algae live in aquatic environments.

CHARACTERISTICS OF ALGAE (p. 357)

3. Algae are eukaryotic; most are photoautotrophs.
4. The thallus (body) of multicellular algae usually consists of a stipe, a holdfast, and blades.
5. Algae reproduce asexually by cell division and fragmentation.
6. Many algae reproduce sexually.
7. Photoautotrophic algae produce oxygen.
8. Algae are classified according to their structures and pigments.

SELECTED PHYLA OF ALGAE (pp. 357–361)

9. Brown algae (kelp) may be harvested for algin.
10. Red algae grow deeper in the ocean than other algae because their red pigments can absorb the blue light that penetrates to deeper levels.
11. Green algae have cellulose and chlorophyll *a* and *b* and store starch.
12. Diatoms are unicellular and have pectin and silica cell walls; some produce a neurotoxin.
13. Dinoflagellates produce neurotoxins that cause paralytic shellfish poisoning and ciguatera.
14. The oomycotes are heterotrophic; they include decomposers and plant parasites.

ROLES OF ALGAE IN NATURE (p. 361)

15. Algae are the primary producers in aquatic food chains.
16. Planktonic algae produce most of the molecular oxygen in the Earth's atmosphere.
17. Petroleum is the fossil remains of planktonic algae.
18. Unicellular algae are symbionts in such animals as *Tridacna*.

PROTOZOA (pp. 361–368)

1. Protozoa are unicellular, eukaryotic chemoheterotrophs.
2. Protozoa are found in soil and water and as normal microbiota in animals.

CHARACTERISTICS OF PROTOZOA (pp. 361–362)

3. The vegetative form is called a trophozoite.
4. Asexual reproduction is by fission, budding, or schizogony.
5. Sexual reproduction is by conjugation.
6. During ciliate conjugation, two haploid nuclei fuse to produce a zygote.
7. Some protozoa can produce a cyst that provides protection during adverse environmental conditions.
8. Protozoa have complex cells with a pellicle, a cytostome, and an anal pore.

MEDICALLY IMPORTANT PHYLA OF PROTOZOA (pp. 362–368)

9. Archaezoa lack mitochondria and have flagella; they include *Trichomonas* and *Giardia*.
10. Microsporidia lack mitochondria and microtubules; microsporans cause diarrhea in AIDS patients.
11. Amoebozoa are amoeba; they include *Entamoeba* and *Acanthamoeba*.
12. Apicomplexa have apical organelles for penetrating host tissue; they include *Plasmodium* and *Cryptosporidium*.
13. Ciliophora move by means of cilia; *Balantidium coli* is the human parasitic ciliate.
14. Euglenozoa move by means of flagella and lack sexual reproduction; they include *Trypanosoma*.

SLIME MOLDS (pp. 368–370)

1. Cellular slime molds resemble amoebas and ingest bacteria by phagocytosis.
2. Plasmodial slime molds consist of a multinucleated mass of protoplasm that engulfs organic debris and bacteria as it moves.

HELMINTHS (pp. 370–377)

1. Parasitic flatworms belong to the Phylum Platyhelminthes.
2. Parasitic roundworms belong to the Phylum Nematoda.

CHARACTERISTICS OF HELMINTHS (pp. 370–372)

3. Helminths are multicellular animals; a few are parasites of humans.

4. The anatomy and life cycle of parasitic helminths are modified for parasitism.

5. The adult stage of a parasitic helminth is found in the definitive host.

6. Each larval stage of a parasitic helminth requires an intermediate host.

7. Helminths can be monoecious or dioecious.

PLATYHELMINTHS (pp. 372–374)

8. Flatworms are dorsoventrally flattened animals; parasitic flatworms may lack a digestive system.

9. Adult trematodes, or flukes, have an oral and ventral sucker with which they attach to host tissue.

10. Eggs of trematodes hatch into free-swimming miracidia that enter the first intermediate host; two generations of rediae develop in the first intermediate host; the rediae become cercariae that bore out of the first intermediate host and penetrate the second intermediate host; cercariae encyst as metacercariae in the second intermediate host; after they are ingested by the definitive host, the metacercariae develop into adults.

11. A cestode, or tapeworm, consists of a scolex (head) and proglottids.

12. Humans serve as the definitive host for the beef tapeworm, and cattle are the intermediate host.

13. Humans serve as the definitive host and can be an intermediate host for the pork tapeworm.

14. Humans serve as the intermediate host for *Echinococcus granulosus*; the definitive hosts are dogs, wolves, and foxes.

NEMATODES (pp. 375–377)

15. Roundworms have a complete digestive system.

16. The nematodes that infect humans with their eggs are *Enterobius vermicularis* (pinworm) and *Ascaris lumbricoides*.

17. The nematodes that infect humans with their larvae are *Necator americanus*, *Trichinella spiralis*, and anisakine worms.

ARTHROPODS AS VECTORS (pp. 377–379)

1. Jointed-legged animals, including ticks and insects, belong to the Phylum Arthropoda.

2. Arthropods that carry diseases are called vectors.

3. Elimination of vectorborne diseases is best done by the control or eradication of the vectors.

STUDY QUESTIONS

Access more review material either online at **The Microbiology Place** (www.microbiologyplace.com) or with **The Microbiology Place CD-ROM** packaged with your new book. There you'll find activities, practice tests, quizzes, flashcards, case studies, and more to help you succeed.

Answers to the Study Questions can be found in Appendix G.

REVIEW

1. Contrast the mechanisms of conidiospore and ascospore formation by a fungus.

2. Fill in the following table.

Phylum	Spore Type(s) Sexual	Asexual
Zygomycota		
Ascomycota		
Basidiomycota		

3. Following is a list of fungi, their methods of entry into the body, and sites of infections they cause. Categorize each type of mycosis as cutaneous, opportunistic, subcutaneous, superficial, or systemic.

Genus	Method of Entry	Site of Infection	Mycosis
Blastomyces	Inhalation	Lungs	
Sporothrix	Puncture	Ulcerative lesions	
Microsporum	Contact	Fingernails	
Trichosporon	Contact	Hair shafts	
Aspergillus	Inhalation	Lungs	

4. A mixed culture of *Escherichia coli* and *Penicillium chrysogenum* is inoculated onto the following culture media. On which medium would you expect each to grow? Why?
a. 0.5% peptone in tap water
b. 10% glucose in tap water

5. What is the role of the alga in a lichen? What is the role of the fungus?

6. Briefly discuss the importance of lichens in nature. Briefly discuss the importance of algae in nature.

7. Complete the following table.

Phyla	Cell Wall Composition	Special Features/ Importance
Oomycotes		
Dinoflagellates		
Diatoms		
Red algae		
Brown algae		
Green algae		

Indicate which phyla consist primarily of unicellular forms. Which phyla could you include in the plant kingdom? Why?

8. Differentiate between cellular and plasmodial slime molds. How does each survive adverse environmental conditions?

9. Complete the following table.

Phylum	Method of Motility	One Human Parasite
Archaezoa		
Microspora		
Amoebozoa		
Apicomplexa		
Ciliophora		
Euglenozoa		

10. Why is it significant that *Trichomonas* does not have a cyst stage? Name a protozoan parasite that does have a cyst stage.

11. Recall the life cycle of *Plasmodium*. Where does asexual reproduction occur? Where does sexual reproduction occur? Identify the definitive host. Identify the vector.

12. By what means are helminthic parasites transmitted to humans?

13. To what phylum and class does this animal belong?

List two characteristics that put it in this phylum. Name the body parts. What is the name of the encysted larva of this animal?

14. Most nematodes are dioecious. What does this term mean? To what phylum do nematodes belong?

15. Vectors can be divided into three major types, according to the roles they play for the parasite. List the three types of vectors and a disease transmitted by each.

MULTIPLE CHOICE

1. How many phyla are represented in the following list of organisms: *Echinococcus, Cyclospora, Aspergillus, Taenia, Toxoplasma, Trichinella?*
   a. 1   d. 4
   b. 2   e. 5
   c. 3

Use the following choices to answer questions 2 and 3:
   (1) metacercaria   (4) miracidium
   (2) redia   (5) cercaria
   (3) adult

2. Put the above stages in order of development, beginning with the egg.
   a. 5, 4, 1, 2, 3
   b. 4, 2, 5, 1, 3
   c. 2, 5, 4, 3, 1
   d. 3, 4, 5, 1, 2
   e. 2, 4, 5, 1, 3

3. If a snail is the first intermediate host of a parasite with these stages, which stage would be found in the snail?
   a. 1
   b. 2
   c. 3
   d. 4
   e. 5

4. Which of the following statements about yeasts are true?
   (1) Yeasts are fungi.
   (2) Yeasts can form pseudohyphae.
   (3) Yeasts reproduce asexually by budding.
   (4) Yeasts are facultatively anaerobic.
   (5) All yeasts are pathogenic.
   (6) All yeasts are dimorphic.
   a. 1, 2, 3, 4
   b. 3, 4, 5, 6
   c. 2, 3, 4, 5
   d. 1, 3, 5, 6
   e. 2, 3, 4

5. Which of the following events follows cell fusion in an ascomycete?
   a. conidiophore formation   d. ascospore formation
   b. conidiospore germination   e. conidiospore release
   c. ascus opening

6. The definitive host for *Plasmodium vivax* is
   a. human.   c. a sporocyte.
   b. *Anopheles*.   d. a gametocyte.

7. Fleas are the intermediate host for *Dipylidium caninum* tapeworm, and dogs are the definitive host. Which stage of the parasite could be found in the flea?
   a. cysticerus larva   c. scolex
   b. proglottids   d. adult

Use the following choices to answer questions 8–10:
   a. Apicomplexa   c. Dinoflagellates
   b. Ciliophora   d. Microspora

8. These are obligate intracellular parasites that lack mitochondria.

9. These are nonmotile parasites with special organelles for penetrating host tissue.

10. These photosynthetic organisms can cause paralytic shellfish poisoning.

CRITICAL THINKING

1. A generalized life cycle of the liver fluke *Clonorchis sinensis* is shown below. Identify the intermediate host(s). Identify the definitive host(s). To what phylum and class does this animal belong?

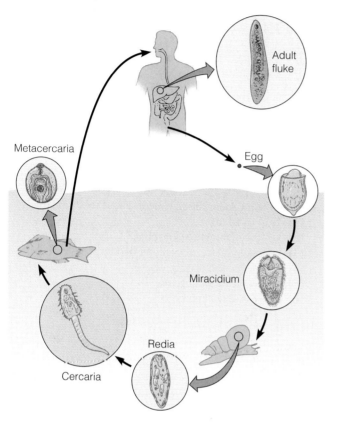

2. The size of a cell is limited by its surface-to-volume ratio; that is, if the volume becomes too great, internal heat cannot be dissipated, and nutrients and wastes cannot be efficiently transported. How do plasmodial slime molds manage to circumvent the surface-to-volume rule?

3. The life cycle of the fish tapeworm *Diphyllobothrium* is similar to that of *Taenia saginata*, except that the intermediate host is fish. Describe the life cycle and method of transmission to humans. Why are freshwater fish more likely to be a source of tapeworm infection than marine fish?

4. *Trypanosoma brucei gambiense* (part a in the figure to the right) is the causative agent of African sleeping sickness. To what phylum does it belong? Part b shows a simplified

life cycle for *T. b. gambiense*. Identify the host and vector of this parasite.

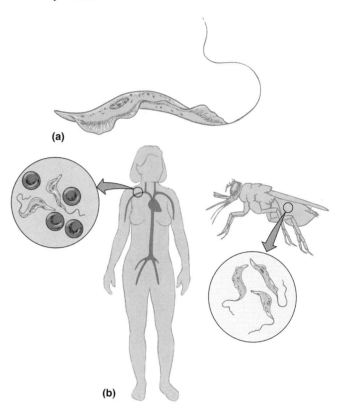

CLINICAL APPLICATIONS

1. A girl developed generalized seizures. A CAT scan revealed a single brain lesion consistent with a tumor. Biopsy of the lesion showed a cysticercus. The patient lived in South Carolina and had never traveled outside the state. What parasite caused her disease? How is this disease transmitted? How might it be prevented?

2. A California farmer developed a low-grade fever, myalgia, and cough. A chest X ray revealed an infiltrate in the lung. Microscopic examination of the sputum revealed round, budding cells. A sputum culture grew mycelia and arthroconidia. What organism is most likely the cause of the symptoms? What is causing the man's disease? How is this disease transmitted? How might it be prevented?

3. A teenaged male in California complained of remittent fever, chills, and headaches. A blood smear revealed ring-shaped cells in his red blood cells. He was successfully treated with primaquine and chloroquine. The patient lives near the San Luis Rey River and has no history of foreign travel, blood transfusion, or intravenous drug use. What is the disease? How was it acquired?

4. Seventeen patients in ten hospitals had cutaneous infections caused by *Rhizopus*. In all 17 patients, Elastoplast bandages were placed over sterile gauze pads to cover wounds. Fourteen of the patients had surgical wounds, two had venous line insertion sites, and one had a bite wound. Lesions present when the bandages were removed ranged from vesiculopustular eruptions to ulcerations and skin necrosis requiring debridement.
   a. How did the wounds most likely get contaminated?
   b. Why is a fungus a more likely contaminant than a bacterium in this instance?

5. In mid-December, a female who had been on prednisone and is an insulin-dependent diabetic fell and received an abrasion on the dorsal side of her right hand. She was placed on penicillin. By the end of January, the ulcer had not healed, and she was referred to a plastic surgeon. On January 30, a swab of the wound was cultured at 35°C on blood agar. On the same day, a smear was made for Gram staining. The Gram stain showed fungal elements. What would you do next? Brownish, waxy colonies grew on the blood agar. Slide cultures set up on February 1 and incubated at 25°C showed septate hyphae and single conidia. What caused the infection? What treatment do you recommend? (See Chapter 20.)

13

Viruses, Viroids, and Prions

Viruses are too small to be seen with a light microscope and cannot be cultured outside their hosts. Therefore, although viral diseases are not new, the viruses themselves could not be studied until the twentieth century. In 1886, the Dutch chemist Adolf Mayer showed that tobacco mosaic disease (TMD) was transmissible from a diseased plant to a healthy plant. In 1892, in an attempt to isolate the cause of TMD, the Russian bacteriologist Dimitri Iwanowski filtered the sap of diseased plants through a porcelain filter that was designed to retain bacteria. He expected to find the microbe trapped in the filter; instead, he found that the infectious agent had passed through the minute pores of the filter. When he infected healthy plants with the filtered fluid, they contracted TMD. The first human disease associated with a filterable agent was yellow fever.

Advances in the molecular biological techniques in the 1980s and 1990s led to the recognition of several new human viruses. Human immunodeficiency virus (HIV), hepatitis C virus, SARS-associated coronavirus, and West Nile virus are a few examples. The diseases caused by these viruses will be discussed in Part Four. In this chapter, we will study the biology of viruses.

UNDER THE MICROSCOPE

Mastadenovirus. This virus causes upper respiratory tract infections in humans.

GENERAL CHARACTERISTICS OF VIRUSES

LEARNING OBJECTIVE

- Differentiate a virus from a bacterium.

One hundred years ago, researchers could not imagine submicroscopic particles, and thus they described the infectious agent as *contagium vivum fluidum*—a contagious fluid. By the 1930s, scientists had begun using the word *virus*, the Latin word for poison, to describe these filterable agents. The nature of viruses, however, remained elusive until 1935, when Wendell Stanley, an American chemist, isolated tobacco mosaic virus, making it possible for the first time to carry out chemical and structural studies on a purified virus. At about the same time, the invention of the electron microscope made it possible to see viruses.

The question of whether viruses are living organisms has an ambiguous answer. Life can be defined as a complex set of processes resulting from the actions of proteins specified by nucleic acids. The nucleic acids of living cells are in action all the time. Because viruses are inert outside living host cells, in this sense they are not considered to be living organisms. However, once viruses enter a host cell, the viral nucleic acids become active, and viral multiplication results. In this sense, viruses are alive when they multiply in the host cells they infect. From a clinical point of view, viruses can be considered alive because they cause infection and disease, just as pathogenic bacteria, fungi, and protozoa do. Depending on one's viewpoint, a virus may be regarded as an exceptionally complex aggregation of nonliving chemicals, or as an exceptionally simple living microorganism.

How, then, do we define a virus? Viruses were originally distinguished from other infectious agents because they are especially small (filterable) and because they are **obligatory intracellular parasites**—that is, they absolutely require living host cells in order to multiply. However, both of these properties are shared by certain small bacteria, such as some rickettsias. Viruses and bacteria are compared in Table 13.1.

The truly distinctive features of viruses are now known to relate to their simple structural organization and their mechanism of multiplication. Accordingly, **viruses** are entities that

- Contain a single type of nucleic acid, either DNA or RNA.
- Contain a protein coat (sometimes itself enclosed by an envelope of lipids, proteins, and carbohydrates) that surrounds the nucleic acid.

TABLE 13.1	Viruses and Bacteria Compared		
	Bacteria		**Viruses**
	Typical Bacteria	Rickettsias/ Chlamydias	
Intracellular parasite	No	Yes	Yes
Plasma membrane	Yes	Yes	No
Binary fission	Yes	Yes	No
Pass through bacteriological filters	No	No/Yes	Yes
Possess both DNA and RNA	Yes	Yes	No
ATP-generating metabolism	Yes	Yes/No	No
Ribosomes	Yes	Yes	No
Sensitive to antibiotics	Yes	Yes	No
Sensitive to interferon	No	No	Yes

- Multiply inside living cells by using the synthesizing machinery of the cell.
- Cause the synthesis of specialized structures that can transfer the viral nucleic acid to other cells.

Viruses have few or no enzymes of their own for metabolism; for example, they lack enzymes for protein synthesis and ATP generation. To multiply, viruses must take over the metabolic machinery of the host cell. This fact has considerable medical significance for the development of antiviral drugs, because most drugs that would interfere with viral multiplication would also interfere with the functioning of the host cell and therefore are too toxic for clinical use. (Antiviral drugs are discussed in Chapter 20.)

HOST RANGE

The **host range** of a virus is the spectrum of host cells the virus can infect. There are viruses that infect invertebrates, vertebrates, plants, protists, fungi, and bacteria. However, most viruses are able to infect specific types of cells of only one host species. In rare cases, viruses cross the host-range barrier, thus expanding their host range. An example is described in the box on page 406. In this chapter we are concerned mainly with viruses that infect

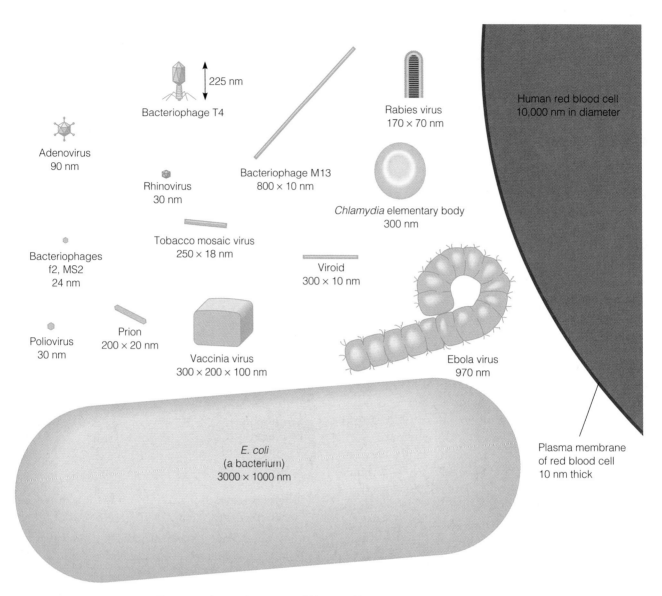

225 nm

Bacteriophage T4

Rabies virus
170 × 70 nm

Human red blood cell
10,000 nm in diameter

Adenovirus
90 nm

Rhinovirus
30 nm

Bacteriophage M13
800 × 10 nm

Chlamydia elementary body
300 nm

Bacteriophages
f2, MS2
24 nm

Tobacco mosaic virus
250 × 18 nm

Viroid
300 × 10 nm

Poliovirus
30 nm

Prion
200 × 20 nm

Vaccinia virus
300 × 200 × 100 nm

Ebola virus
970 nm

Plasma membrane
of red blood cell
10 nm thick

E. coli
(a bacterium)
3000 × 1000 nm

FIGURE 13.1 Virus sizes. The sizes of several viruses (teal blue) and bacteria (pink) are compared with a human red blood cell, shown to the right of the microbes. Dimensions are given in nanometers (nm) and are either diameters or length by width.

Q How do viruses differ from bacteria?

either humans or bacteria. Viruses that infect bacteria are called **bacteriophages,** or **phages.**

The particular host range of a virus is determined by the virus's requirements for its specific attachment to the host cell and the availability within the potential host of cellular factors required for viral multiplication. For the virus to infect the host cell, the outer surface of the virus must chemically interact with specific receptor sites on the surface of the cell. The two complementary components are held together by weak bonds, such as hydrogen bonds. The combination of many attachment and receptor sites leads to a strong association between host cell and virus. For some bacteriophages, the receptor site is part of the cell wall of the host; in other cases, it is part of the fimbriae

or flagella. For animal viruses, the receptor sites are on the plasma membranes of the host cells.

The potential to use viruses to treat diseases is intriguing because of their narrow host range and their ability to kill their host cells. The idea of *phage therapy*—using bacteriophage to treat bacterial infections, has been around for 100 years. Recent advances in our understanding of virus-host interactions have fueled new studies in the field of phage therapy.

Experimentally induced viral infections in cancer patients during the 1920s suggested that viruses might have antitumor activity. These tumor destroying, or *oncolytic,* viruses may selectively infect and kill tumor cells or cause an immune response against tumor cells. Some viruses

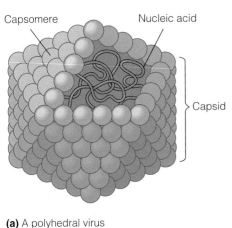

Capsomere   Nucleic acid

Capsid

(a) A polyhedral virus

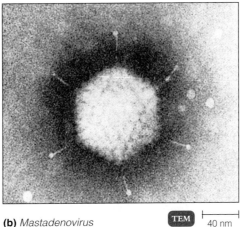

(b) *Mastadenovirus*   TEM | 40 nm

FIGURE 13.2 Morphology of a nonenveloped polyhedral virus.
(**a**) A diagram of a polyhedral (icosahedral) virus. (**b**) A micrograph of the adenovirus *Mastadenovirus*. Individual capsomeres are visible.

Q **What is the chemical composition of a capsid?**

naturally infect tumor cells and other viruses can be genetically modified to infect tumor cells. At present several studies are underway to determine the killing mechanism of oncolytic viruses and the safety of using viral therapy.

VIRAL SIZE

Viral sizes are determined with the aid of electron microscopy. Different viruses vary considerably in size. Although most are quite a bit smaller than bacteria, some of the larger viruses (such as the vaccinia virus) are about the same size as some very small bacteria (such as the mycoplasmas, rickettsias, and chlamydias). Viruses range from 20 to 1000 nm in length. The comparative sizes of several viruses and bacteria are shown in Figure 13.1.

VIRAL STRUCTURE

LEARNING OBJECTIVE

• Describe the chemical and physical structure of both an enveloped and a nonenveloped virus.

A **virion** is a complete, fully developed, infectious viral particle composed of nucleic acid and surrounded by a protein coat that protects it from the environment and is a vehicle of transmission from one host cell to another. Viruses are classified by differences in the structures of these coats.

NUCLEIC ACID

In contrast to prokaryotic and eukaryotic cells, in which DNA is always the primary genetic material (and RNA plays an auxiliary role), a virus can have either DNA or RNA—but never both. The nucleic acid of a virus can be single-stranded or double-stranded. Thus, there are viruses with the familiar double-stranded DNA, with single-stranded DNA, with double-stranded RNA, and with single-stranded RNA. Depending on the virus, the

nucleic acid can be linear or circular. In some viruses (such as the influenza virus), the nucleic acid is in several separate segments.

The percentage of nucleic acid in relation to protein is about 1% for the influenza virus and about 50% for certain bacteriophages. The total amount of nucleic acid varies from a few thousand nucleotides (or pairs) to as many as 250,000 nucleotides. (*E. coli*'s chromosome consists of approximately 4 million nucleotide pairs.)

CAPSID AND ENVELOPE

The nucleic acid of a virus is protected by a protein coat called the **capsid** (Figure 13.2a). The structure of the capsid is ultimately determined by the viral nucleic acid and accounts for most of the mass of a virus, especially of small ones. Each capsid is composed of protein subunits called **capsomeres.** In some viruses, the proteins composing the capsomeres are of a single type; in other viruses, several types of protein may be present. Individual capsomeres are often visible in electron micrographs (see Figure 13.2b for an example). The arrangement of capsomeres is characteristic of a particular type of virus.

In some viruses, the capsid is covered by an **envelope** (Figure 13.3a), which usually consists of some combination of lipids, proteins, and carbohydrates. Some animal viruses are released from the host cell by an extrusion process that coats the virus with a layer of the host cell's plasma membrane; that layer becomes the viral envelope. In many cases, the envelope contains proteins determined by the viral nucleic acid and materials derived from normal host cell components.

Depending on the virus, envelopes may or may not be covered by **spikes,** which are carbohydrate-protein complexes that project from the surface of the envelope. Some viruses attach to host cells by means of spikes. Spikes are such a reliable characteristic of some viruses that they can be used as a means of identification. The ability of certain

FIGURE 13.3 Morphology of an enveloped helical virus.
(**a**) A diagram of an enveloped helical virus. (**b**) A micrograph of *Influenzavirus* A2. Notice the halo of spikes projecting from the outer surface of each envelope (see Chapter 24).

Q What is the nucleic acid in a virus?

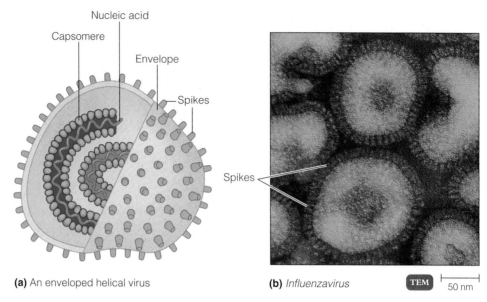

(**a**) An enveloped helical virus

(**b**) *Influenzavirus*

TEM · 50 nm

viruses, such as the influenza virus, to clump red blood cells is associated with spikes. Such viruses bind to red blood cells and form bridges between them. The resulting clumping is called *hemagglutination* and is the basis for several useful laboratory tests. (See Figure 18.7, page 539.)

Viruses whose capsids are not covered by an envelope are known as **nonenveloped viruses** (see Figure 13.2). The capsid of a nonenveloped virus protects the nucleic acid from nuclease enzymes in biological fluids and promotes the virus's attachment to susceptible host cells.

When the host has been infected by a virus, the host immune system is stimulated to produce antibodies (proteins that react with the surface proteins of the virus). This interaction between host antibodies and virus proteins should inactivate the virus and stop the infection. However, some viruses can escape antibodies because regions of the genes that code for these viruses' surface proteins are susceptible to mutations. The progeny of mutant viruses have altered surface proteins, such that the antibodies are not able to react with them. Influenza virus frequently undergoes such changes in its spikes. This is why you can get

influenza more than once. Although you may have produced antibodies to one influenza virus, the virus can mutate and infect you again.

GENERAL MORPHOLOGY

Viruses may be classified into several different morphological types on the basis of their capsid architecture. The structure of these capsids has been revealed by electron microscopy and a technique called X-ray crystallography.

HELICAL VIRUSES

Helical viruses resemble long rods that may be rigid or flexible. The viral nucleic acid is found within a hollow, cylindrical capsid that has a helical structure (Figure 13.4). The viruses that cause rabies and Ebola hemorrhagic fever are helical viruses.

POLYHEDRAL VIRUSES

Many animal, plant, and bacterial viruses are polyhedral, or many-sided, viruses. The capsid of most polyhedral viruses is in the shape of an *icosahedron*, a regular polyhedron with

FIGURE 13.4 Morphology of a helical virus. (**a**) A diagram of a portion of a helical virus. Several rows of capsomeres have been removed to reveal the nucleic acid. (**b**) A micrograph of Ebola virus, a filovirus showing helical rods.

Q What is the chemical composition of a capsomere?

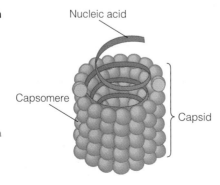

(**a**) A helical virus

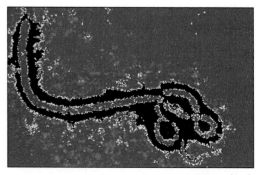

(**b**) Ebola virus

TEM · 100 nm

20 triangular faces and 12 corners (see Figure 13.2a). The capsomeres of each face form an equilateral triangle. An example of a polyhedral virus in the shape of an icosahedron is the adenovirus (shown in Figure 13.2b). Another icosahedral virus is the poliovirus.

ENVELOPED VIRUSES

As noted earlier, the capsid of some viruses is covered by an envelope. Enveloped viruses are roughly spherical. When helical or polyhedral viruses are enclosed by envelopes, they are called *enveloped helical* or *enveloped polyhedral viruses*. An example of an enveloped helical virus is the influenza virus (see Figure 13.3b). An example of an enveloped polyhedral (icosahedral) virus is the herpes simplex virus.

COMPLEX VIRUSES

Some viruses, particularly bacterial viruses, have complicated structures and are called **complex viruses.** One example of a complex virus is a bacteriophage. Some bacteriophages have capsids to which additional structures are attached (Figure 13.5a). In this figure, notice that the capsid (head) is polyhedral and the tail sheath is helical. The head contains the nucleic acid. Later in the chapter we will discuss the functions of the other structures, such as the tail sheath, tail fibers, plate, and pin. Another example of complex viruses are poxviruses, which do not contain clearly identifiable capsids but have several coats around the nucleic acid (Figure 13.5b).

TAXONOMY OF VIRUSES

> **LEARNING OBJECTIVES**
> - Define *viral species.*
> - Give an example of a family, genus, and common name for a virus.

Just as we need taxonomic categories of plants, animals, and bacteria, we need viral taxonomy to help us organize and understand newly discovered organisms. The oldest classification of viruses is based on symptomatology, such as for diseases that affect the respiratory system. This system was convenient but not scientifically acceptable because the same virus may cause more than one disease, depending on the tissue affected. In addition, this system artificially grouped viruses that do not infect humans.

Virologists began addressing the problem of viral taxonomy in 1966 with the formation of the International Committee on Taxonomy of Viruses (ICTV). Since then, the ICTV has been grouping viruses into families based on (1) nucleic acid type, (2) strategy for replication, and (3)

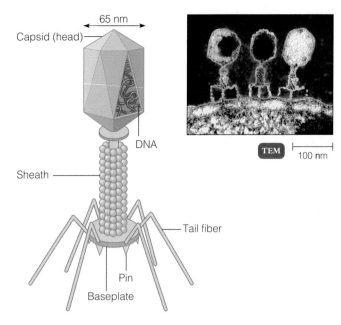

(**a**) A T-even bacteriophage

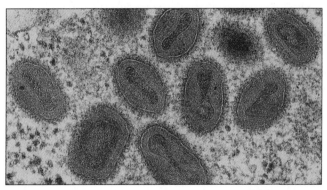

(**b**) *Orthopoxvirus*

FIGURE 13.5 Morphology of complex viruses. (**a**) A diagram and micrograph of a T-even bacteriophage. (**b**) A micrograph of variola virus, a species in the genus *Orthopoxvirus*, which causes smallpox.

 What is the value of a capsid to a virus?

morphology. The suffix *-virus* is used for genus names; family names end in *-viridae;* and order names end in *-ales.* In formal usage, the family and genus names are used in the following manner: Family Herpesviridae, genus *Simplexvirus,* human herpesvirus 2.

A **viral species** is a group of viruses sharing the same genetic information and ecological niche (host range). Specific epithets for viruses are not used. Thus, viral species are designated by descriptive common names, such as human immunodeficiency virus (HIV), with subspecies (if any) designated by a number (HIV-1). Table 13.2 presents a summary of the classification of viruses that infect humans.

| TABLE 13.2 | Families of Viruses That Affect Humans | | |

Characteristics/ Dimensions	Viral Family	Important Genera	Clinical or Special Features
Single-stranded DNA nonenveloped 18–25 nm	Parvoviridae	Human parvovirus B19	Fifth disease; anemia in immunocompromised patients. Refer to Chapter 21.
Double-stranded DNA nonenveloped 70–90 nm	Adenoviridae	*Mastadenovirus*	Medium-sized viruses that cause various respiratory infections in humans; some cause tumors in animals.
40–57 nm	Papovaviridae	*Papillomavirus* (human wart virus) *Polyomavirus*	Small viruses that induce tumors; the human wart virus (papilloma) and certain viruses that produce cancer in animals (polyoma and simian) belong to this family. Refer to Chapters 21 and 26.
Double-stranded DNA enveloped 200–350 nm	Poxviridae	*Orthopoxvirus* (vaccinia and smallpox viruses) *Molluscipoxvirus*	Very large, complex, brick-shaped viruses that cause diseases such as smallpox (variola), molluscum contagiosum (wartlike skin lesion), and cowpox. Refer to Chapter 21.
150–200 nm	Herpesviridae	*Simplexvirus* (HHV-1 and 2) *Varicellovirus* (HHV-3) *Lymphocryptovirus* (HHV-4) *Cytomegalovirus* (HHV-5) *Roseolovirus* (HHV-6) HHV-7 Kaposi's sarcoma (HHV-8)	Medium-sized viruses that cause various human diseases, such as fever blisters, chickenpox, shingles, and infectious mononucleosis; causes a type of human cancer called Burkitt's lymphoma. Refer to Chapters 21, 23, and 26.
42 nm	Hepadnaviridae	*Hepadnavirus* (hepatitis B virus)	After protein synthesis, hepatitis B virus uses reverse transcriptase to produce its DNA from mRNA; causes hepatitis B and liver tumors. Refer to Chapter 25.
Single-stranded RNA, + strand nonenveloped 28–30 nm	Picornaviridae	*Enterovirus* *Rhinovirus* (common cold virus) Hepatitis A virus	At least 70 human enteroviruses are known, including the polio-, coxsackie-, and echoviruses; more than 100 rhinoviruses exist and are the most common cause of colds. Refer to Chapters 22, 24, and 25.
35–40 nm	Caliciviridae	Hepatitis E virus *Norovirus*	Includes causes of gastroenteritis and one cause of human hepatitis. Refer to Chapter 25.
Single-stranded RNA, + strand enveloped 60–70 nm	Togaviridae	*Alphavirus* *Rubivirus* (rubella virus)	Included are many viruses transmitted by arthropods (*Alphavirus*); diseases include eastern equine encephalitis (EEE) and western equine encephalitis (WEE). Rubella virus is transmitted by the respiratory route. Refer to Chapters 21, 22, and 23.

TABLE 13.2	**Families of Viruses That Affect Humans** (continued)		
Characteristics/ Dimensions	Viral Family	Important Genera	Clinical or Special Features
40–50 nm	Flaviviridae 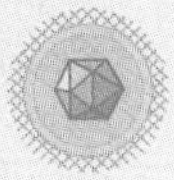	*Flavivirus* *Pestivirus* Hepatitis C virus	Can replicate in arthropods that transmit them; diseases include yellow fever, dengue and St. Louis and West Nile encephalitis. Refer to Chapters 22, 23, and 25.
Nidovirales 80–160 nm	Coronaviridae 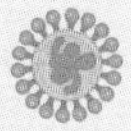	*Coronavirus*	Associated with upper respiratory tract infections and the common cold; SARS virus. Refer to Chapter 24.
Mononegavirales − strand, one strand of RNA 70–180 nm	Rhabdoviridae 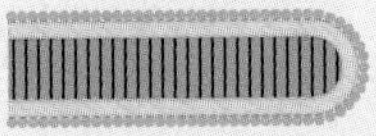	*Vesiculovirus* (vesicular stomatatis virus) *Lyssavirus* (rabies virus)	Bullet-shaped viruses with a spiked envelope; cause rabies and numerous animal diseases. Refer to Chapter 22.
80–14,000 nm	Filoviridae 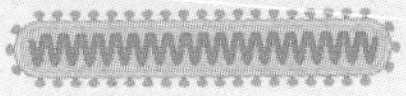	*Filovirus*	Enveloped, helical viruses; Ebola and Marburg viruses are filoviruses. Refer to Chapter 23.
150–300 nm	Paramyxoviridae 	*Paramyxovirus* *Morbillivirus* (measleslike virus)	Paramyxoviruses cause parainfluenza, mumps, and Newcastle disease in chickens. Refer to Chapters 21, 24, and 25.
− strand, one strand of RNA 32 nm	Deltaviridae 	Hepatitis D	Depend on coinfection with hepadnavirus. Refer to Chapter 25.
− strand, multiple strands of RNA 80–200 nm	Orthomyxoviridae 	Influenza virus A, B, and C	Envelope spikes can agglutinate red blood cells. Refer to Chapter 24.
90–120 nm	Bunyaviridae 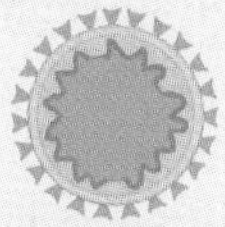	*Bunyavirus* (California encephalitis virus) *Hantavirus*	Hantaviruses cause hemorrhagic fevers such as Korean hemorrhagic fever and *Hantavirus* pulmonary syndrome; associated with rodents. Refer to Chapters 22, 23.
110–130 nm	Arenaviridae 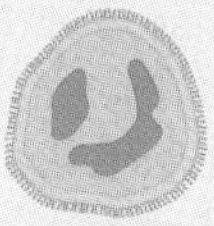	*Arenavirus*	Helical capsids contain RNA-containing granules; cause lymphocytic choriomeningitis, Venezuelan hemorrhagic fever, and Lassa fever. Refer to Chapter 23.
Produce DNA 100–120 nm	Retroviridae 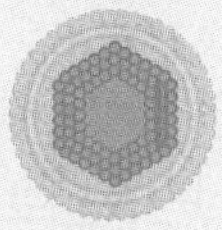	Oncoviruses *Lentivirus* (HIV)	Includes all RNA tumor viruses. Oncoviruses cause leukemia and tumors in animals; the *Lentivirus* HIV causes AIDS. Refer to Chapter 19.
Double-stranded RNA nonenveloped 60–80 nm	Reoviridae 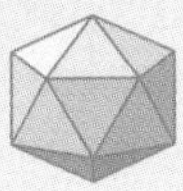	*Reovirus* *Rotavirus*	Involved in mild respiratory infections and gastroenteritis; an unclassified species causes Colorado tick fever. Refer to Chapter 25.

ISOLATION, CULTIVATION, AND IDENTIFICATION OF VIRUSES

The fact that viruses cannot multiply outside a living host cell complicates their detection, enumeration, and identification. Viruses must be provided with living cells instead of a fairly simple chemical medium. Living plants and animals are difficult and expensive to maintain, and pathogenic viruses that grow only in higher primates and human hosts cause additional complications. However, viruses that use bacterial cells as a host (bacteriophages) are rather easily grown on bacterial cultures. This is one reason so much of our understanding of viral multiplication has come from bacteriophages.

GROWING BACTERIOPHAGES IN THE LABORATORY

(**LEARNING OBJECTIVE**
• Describe how bacteriophages are cultured.

Bacteriophages can be grown either in suspensions of bacteria in liquid media or in bacterial cultures on solid media. The use of solid media makes possible the *plaque method* for detecting and counting viruses. A sample of bacteriophage is mixed with host bacteria and melted agar. The agar containing the bacteriophages and host bacteria is then poured into a Petri plate containing a hardened layer of agar growth medium. The virus-bacteria mixture solidifies into a thin top layer that contains a layer of bacteria approximately one cell thick. Each virus infects a bacterium, multiplies, and releases several hundred new viruses. These newly produced viruses infect other bacteria in the immediate vicinity, and more new viruses are produced. Following several viral multiplication cycles, all the bacteria in the area surrounding the original virus are destroyed. This produces a number of clearings, or **plaques,** visible against a lawn of bacterial growth on the surface of the agar (Figure 13.6). While the plaques form, uninfected bacteria elsewhere in the Petri plate multiply rapidly and produce a turbid background.

Each plaque theoretically corresponds to a single virus in the initial suspension. Therefore, the concentrations of viral suspensions measured by the number of plaques are usually given in terms of **plaque-forming units (PFU).**

GROWING ANIMAL VIRUSES IN THE LABORATORY

(**LEARNING OBJECTIVE**
• Describe how animal viruses are cultured.

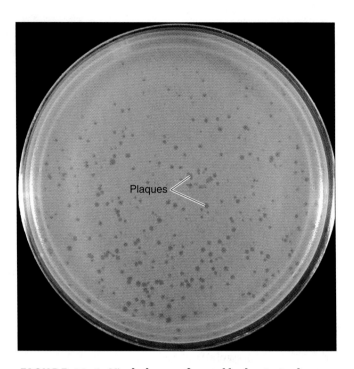

FIGURE 13.6   Viral plaques formed by bacteriophages. Clear viral plaques of various sizes have been formed by bacteriophage λ (lambda) on a lawn of *E. coli.*

Q **What is a plaque-forming unit?**

In the laboratory, three methods are commonly used for culturing animal viruses. These methods involve using living animals, embryonated eggs, or cell cultures.

IN LIVING ANIMALS

Some animal viruses can be cultured only in living animals, such as mice, rabbits, and guinea pigs. Most experiments to study the immune system's response to viral infections must also be performed in virally infected live animals. Animal inoculation may be used as a diagnostic procedure for identifying and isolating a virus from a clinical specimen. After the animal is inoculated with the specimen, the animal is observed for signs of disease or is killed so that infected tissues can be examined for the virus.

Some human viruses cannot be grown in animals or can be grown but do not cause disease. The lack of natural animal models for AIDS has slowed our understanding of its disease process and prevented experimentation with drugs that inhibit growth of the virus in vivo. Chimpanzees can be infected with one subspecies of human immunodeficiency virus (HIV-1, genus *Lentivirus*); but, because they do not show symptoms of the disease, they cannot be used to study the effects of viral growth and disease treatments. AIDS vaccines are presently being tested in humans, but the disease progresses so slowly in humans that it can take years to determine the effectiveness of

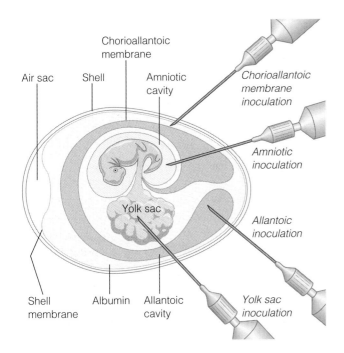

Chorioallantoic
membrane

Air sac   Shell   Amniotic
cavity

*Chorioallantoic
membrane
inoculation*

*Amniotic
inoculation*

Yolk sac

*Allantoic
inoculation*

Shell
membrane   Albumin   Allantoic
cavity

*Yolk sac
inoculation*

FIGURE 13.7 Inoculation of an embryonated egg. The injection site determines the membrane on which the viruses will grow.

Q **Why are viruses grown in eggs and not in culture media?**

these vaccines. In 1986, simian AIDS (an immunodeficiency disease of green monkeys) was reported, followed in 1987 by feline AIDS (an immunodeficiency disease of domestic cats). These diseases are caused by lentiviruses, which are closely related to HIV, and the diseases develop within a few months, thus providing a model for studying viral growth in different tissues. In 1990, a way to infect mice with human AIDS was found when immunodeficient mice were grafted to produce human T cells and human gamma globulin. The mice provide a reliable model for studying viral replication, although they do not provide models for vaccine development.

IN EMBRYONATED EGGS

If the virus will grow in an *embryonated egg*, this can be a fairly convenient and inexpensive form of host for many animal viruses. A hole is drilled in the shell of the embryonated egg, and a viral suspension or suspected virus-containing tissue is injected into the fluid of the egg. There are several membranes in an egg, and the virus is injected near the one most appropriate for its growth (Figure 13.7). Viral growth is signaled by the death of the embryo, by embryo cell damage, or by the formation of typical pocks or lesions on the egg membranes. This method was once the most widely used method of viral isolation and growth, and it is still used to grow viruses for some vaccines. For this reason, you may be asked if you are allergic to eggs before receiving a vaccination, because egg proteins may be present in the viral vaccine preparations. (Allergic reactions will be discussed in Chapter 19.)

IN CELL CULTURES

Cell cultures have replaced embryonated eggs as the preferred type of growth medium for many viruses. Cell cultures consist of cells grown in culture media in the laboratory. Because these cultures are generally rather homogeneous collections of cells and can be propagated and handled much like bacterial cultures, they are more convenient to work with than whole animals or embryonated eggs.

Cell culture lines are started by treating a slice of animal tissue with enzymes that separate the individual cells (Figure 13.8). These cells are suspended in a solution that provides the osmotic pressure, nutrients, and growth factors needed for the cells to grow. Normal cells tend to adhere to the glass or plastic container and reproduce to form a monolayer. Viruses infecting such a monolayer sometimes cause the cells of the monolayer to deteriorate as they multiply. This cell deterioration is called **cytopathic**

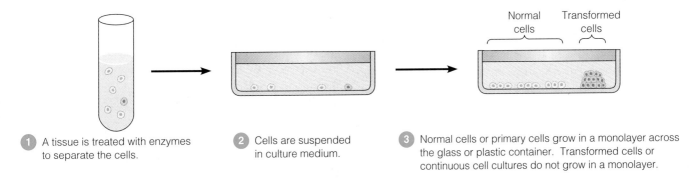

1 A tissue is treated with enzymes to separate the cells.

2 Cells are suspended in culture medium.

Normal   Transformed
cells   cells

3 Normal cells or primary cells grow in a monolayer across the glass or plastic container. Transformed cells or continuous cell cultures do not grow in a monolayer.

FIGURE 13.8 Cell cultures. Transformed cells can be grown indefinitely in laboratory culture.

Q **Why are transformed cells referred to as "immortal"?**

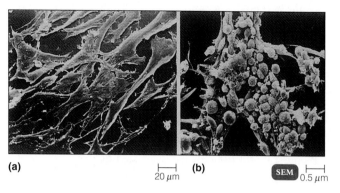

(a)   ⊢——⊣   **(b)**   **SEM** ⊢—⊣
         20 μm   0.5 μm

FIGURE 13.9   The cytopathic effect of viruses. (a) Uninfected mouse cells align next to each other, forming a monolayer. **(b)** The same cells 24 hours after infection with vesicular stomatitis virus (VSV) (see Figure 13.18b). Notice the cells pile up and "round up."

Q How did VSV infection affect the cells?

effect (CPE), illustrated in Figure 13.9. CPE can be detected and counted in much the same way as plaques caused by bacteriophages on a lawn of bacteria and reported as PFU/ml.

Viruses may be grown in primary or continuous cell lines. **Primary cell lines,** derived from tissue slices, tend to die out after only a few generations. Certain cell lines, called **diploid cell lines,** developed from human embryos can be maintained for about 100 generations and are widely used for culturing viruses that require a human host. Cell lines developed from embryonic human cells are used to culture rabies virus for a rabies vaccine called human diploid culture vaccine (see Chapter 22).

When viruses are routinely grown in a laboratory, **continuous cell lines** are used. These are transformed (cancerous) cells that can be maintained through an indefinite number of generations, and they are sometimes called immortal cell lines (see the discussion of transformation on page 411). One of these, the HeLa cell line, was isolated from the cancer of a woman who died in 1951. After years of laboratory cultivation, many such cell lines have lost almost all the original characteristics of the cell, but these changes have not interfered with the use of the cells for viral propagation. In spite of the success of cell culture in viral isolation and growth, there are still some viruses that have never been successfully cultivated in cell culture.

The idea of cell culture dates back to the end of the nineteenth century, but it was not a practical laboratory technique until the development of antibiotics in the years following World War II. A major problem with cell culture is that the cell lines must be kept free of microbial contamination. The maintenance of cell culture lines requires trained technicians with considerable experience working on a full-time basis. Because of these difficulties, most hospital laboratories and many state health laboratories do not isolate and identify viruses in clinical work. Instead, the tissue or serum samples are sent to central laboratories that specialize in such work.

VIRAL IDENTIFICATION

LEARNING OBJECTIVE
• List three techniques used to identify viruses.

Identifying viral isolates is not an easy task. For one thing, viruses cannot be seen at all without the use of an electron microscope. Serological methods, such as Western blotting, are the most commonly used means of identification (see Figure 10.12, page 299). In these tests, the virus is detected and identified by its reaction with antibodies. We will discuss antibodies in detail in Chapter 17 and a number of immunological tests for identifying viruses in Chapter 18. Observation of cytopathic effects, described in Chapter 15 (page 465), is also useful for the identification of a virus.

Virologists can identify and characterize viruses by using such modern molecular methods as use of restriction fragment length polymorphisms (RFLPs) and the polymerase chain reaction (PCR) (Chapter 9, page 258). PCR was used to amplify viral RNA to identify the West Nile virus in 1999 in the United States and the SARS-associated coronavirus in China in 2002.

VIRAL MULTIPLICATION

The nucleic acid in a virion contains only a few of the genes needed for the synthesis of new viruses. These include genes for the virion's structural components, such as the capsid proteins, and genes for a few of the enzymes used in the viral life cycle. These enzymes are synthesized and functional only when the virus is within the host cell. Viral enzymes are almost entirely concerned with replicating or processing viral nucleic acid. Enzymes needed for protein synthesis, ribosomes, tRNA, and energy production are supplied by the host cell and are used for synthesizing viral proteins, including viral enzymes. Although the smallest nonenveloped virions do not contain any preformed enzymes, the larger virions may contain one or a few enzymes, which usually function in helping the virus penetrate the host cell or replicate its own nucleic acid.

Thus, for a virus to multiply, it must invade a host cell and take over the host's metabolic machinery. A single virion can give rise to several or even thousands of similar viruses in a single host cell. This process can drastically change the host cell and usually causes its death. In a few viral infections, cells survive and continue to produce viruses indefinitely.

The multiplication of viruses can be demonstrated with a one-step growth curve (Figure 13.10). The data are obtained by infecting every cell in a culture and then testing the culture medium and cells for virions and viral proteins and nucleic acids.

MULTIPLICATION OF BACTERIOPHAGES

> **LEARNING OBJECTIVES**
> • Describe the lytic cycle of T-even bacteriophages.
> • Describe the lysogenic cycle of bacteriophage lambda.

Although the means by which a virus enters and exits a host cell may vary, the basic mechanism of viral multiplication is similar for all viruses. Bacteriophages can multiply by two alternative mechanisms: the lytic cycle or the lysogenic cycle. The **lytic cycle** ends with the lysis and death of the host cell, whereas the host cell remains alive in the **lysogenic cycle.** Because the *T-even bacteriophages* (T2, T4, and T6) have been studied most extensively, we will describe the multiplication of T-even bacteriophages in their host, *E. coli,* as an example of the lytic cycle.

T-EVEN BACTERIOPHAGES: THE LYTIC CYCLE

The virions of T-even bacteriophages are large, complex, and nonenveloped, with a characteristic head-and-tail structure shown in Figures 13.5a and 13.11. The length of DNA contained in these bacteriophages is only about 6% of that contained in *E. coli,* yet the phage has enough DNA for over 100 genes. The multiplication cycle of these phages, like that of all viruses, occurs in five distinct stages: attachment, penetration, biosynthesis, maturation, and release (see Figure 13.10).

Attachment ❶ After a chance collision between phage particles and bacteria, *attachment,* or *adsorption,* occurs. During this process, an attachment site on the virus attaches to a complementary receptor site on the bacterial cell. This attachment is a chemical interaction in which weak bonds are formed between the attachment and receptor sites. T-even bacteriophages use fibers at the end of the tail as attachment sites. The complementary receptor sites are on the bacterial cell wall.

Penetration ❷ After attachment, the T-even bacteriophage injects its DNA (nucleic acid) into the bacterium. To do this, the bacteriophage's tail releases an enzyme, **phage lysozyme,** which breaks down a portion of the bacterial cell wall. During the process of *penetration,* the tail sheath of the phage contracts, and the tail core is driven through the cell wall. When the tip of the core reaches the plasma membrane, the DNA from the bacteriophage's head passes through the tail core, through the

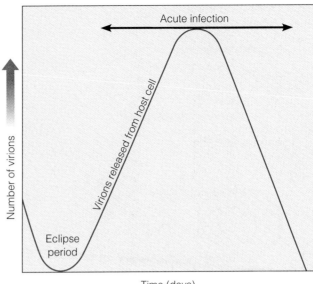

FIGURE 13.10 A viral one-step growth curve. No new infective virions are found in a culture until after biosynthesis and maturation have taken place. Most infected cells die as a result of infection, consequently new virions will not be produced.

Q **What can be found in the cell during biosynthesis and maturation?**

plasma membrane, and enters the bacterial cell. The capsid remains outside the bacterial cell. Therefore, the phage particle functions like a hypodermic syringe to inject its DNA into the bacterial cell.

Biosynthesis ❸ Once the bacteriophage DNA has reached the cytoplasm of the host cell, the biosynthesis of viral nucleic acid and protein occurs. Host protein synthesis is stopped by virus-induced degradation of the host DNA, viral proteins that interfere with transcription, or the repression of translation.

Initially, the phage uses the host cell's nucleotides and several of its enzymes to synthesize many copies of phage DNA. Soon after, the biosynthesis of viral proteins begins. Any RNA transcribed in the cell is mRNA transcribed from phage DNA for the biosynthesis of phage enzymes and capsid proteins. The host cell's ribosomes, enzymes, and amino acids are used for translation. Genetic controls regulate when different regions of phage DNA are transcribed into mRNA during the multiplication cycle. For example, early messages are translated into early phage proteins, the enzymes used in the synthesis of phage DNA. Also, late messages are translated into late phage proteins for the synthesis of capsid proteins.

For several minutes following infection, complete phages cannot be found in the host cell. Only separate components—DNA and protein—can be detected. The

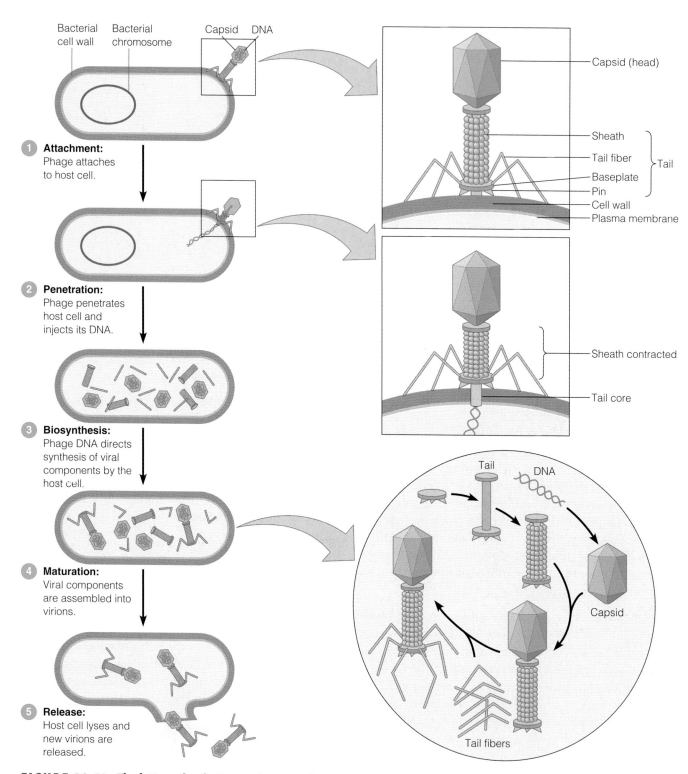

① **Attachment:**
Phage attaches to host cell.

② **Penetration:**
Phage penetrates host cell and injects its DNA.

③ **Biosynthesis:**
Phage DNA directs synthesis of viral components by the host cell.

④ **Maturation:**
Viral components are assembled into virions.

⑤ **Release:**
Host cell lyses and new virions are released.

FIGURE 13.11 The lytic cycle of a T-even bacteriophage.

Q **What is the result of the lytic cycle?**

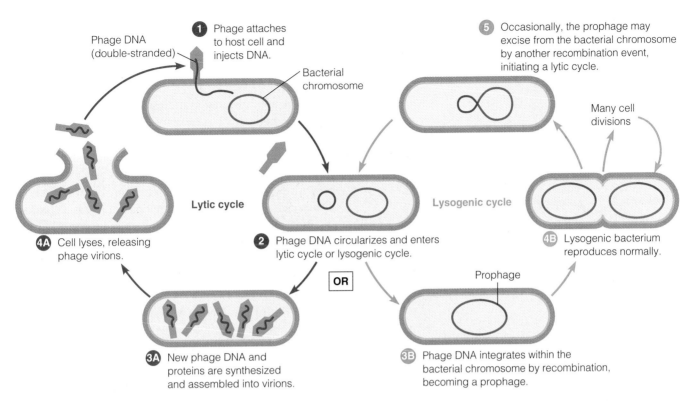

1 Phage attaches to host cell and injects DNA.

Phage DNA (double-stranded)

Bacterial chromosome

5 Occasionally, the prophage may excise from the bacterial chromosome by another recombination event, initiating a lytic cycle.

Many cell divisions

Lytic cycle

Lysogenic cycle

4A Cell lyses, releasing phage virions.

2 Phage DNA circularizes and enters lytic cycle or lysogenic cycle.

4B Lysogenic bacterium reproduces normally.

OR

Prophage

3A New phage DNA and proteins are synthesized and assembled into virions.

3B Phage DNA integrates within the bacterial chromosome by recombination, becoming a prophage.

FIGURE 13.12 The lysogenic cycle of bacteriophage λ in *E. coli.*

Q **How does lysogeny differ from the lytic cycle?**

period during viral multiplication when complete, infective virions are not yet present is called the **eclipse period.**

Maturation ❹ In the next sequence of events, *maturation* occurs. In this process, bacteriophage DNA and capsids are assembled into complete virions. The viral components essentially assemble into a viral particle spontaneously, eliminating the need for many nonstructural genes and gene products. The phage heads and tails are separately assembled from protein subunits, and the head is filled with phage DNA and attached to the tail.

Release ❺ The final stage of viral multiplication is the *release* of virions from the host cell. The term **lysis** is generally used for this stage in the multiplication of T-even phages because in this case, the plasma membrane actually breaks open (lyses). Lysozyme, which is encoded by a phage gene, is synthesized within the cell. This enzyme causes the bacterial cell wall to break down, and the newly produced bacteriophages are released from the host cell. The released bacteriophages infect other susceptible cells in the vicinity, and the viral multiplication cycle is repeated within those cells.

BACTERIOPHAGE LAMBDA (λ): THE LYSOGENIC CYCLE

In contrast to T-even bacteriophages, some viruses do not cause lysis and death of the host cell when they multiply.

These *lysogenic phages* (also called *temperate phages*) may indeed proceed through a lytic cycle, but they are also capable of incorporating their DNA into the host cell's DNA to begin a lysogenic cycle. In **lysogeny,** the phage remains latent (inactive). The participating bacterial host cells are known as *lysogenic cells.*

We will use the bacteriophage λ (lambda), a well-studied lysogenic phage, as an example of the lysogenic cycle (Figure 13.12).

❶ Upon penetration into an *E. coli* cell,

❷ the originally linear phage DNA forms a circle.

3A This circle can multiply and be transcribed,

4A leading to the production of new phage and to cell lysis (the lytic cycle).

3B Alternatively, the circle can recombine with and become part of the circular bacterial DNA (the lysogenic cycle). The inserted phage DNA is now called a **prophage.** Most of the prophage genes are repressed by two repressor proteins that are the products of phage genes. These repressors stop transcription of all the other phage genes by binding to operators. Thus, the phage genes that would otherwise direct the synthesis and release of new virions are turned off, in much the same way that the genes of the *E. coli lac*

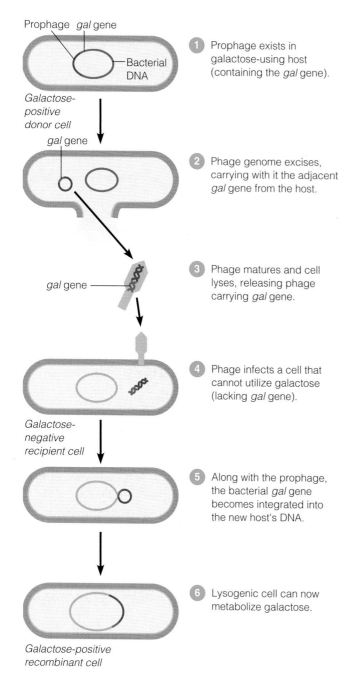

Prophage *gal* gene

Bacterial DNA

Galactose-positive donor cell

gal gene

gal gene

Galactose-negative recipient cell

Galactose-positive recombinant cell

① Prophage exists in galactose-using host (containing the *gal* gene).

② Phage genome excises, carrying with it the adjacent *gal* gene from the host.

③ Phage matures and cell lyses, releasing phage carrying *gal* gene.

④ Phage infects a cell that cannot utilize galactose (lacking *gal* gene).

⑤ Along with the prophage, the bacterial *gal* gene becomes integrated into the new host's DNA.

⑥ Lysogenic cell can now metabolize galactose.

FIGURE 13.13   Specialized transduction. When a prophage is excised from its host chromosome, it can take with it a bit of the adjacent DNA from the bacterial chromosome.

Q   **How does specialized transduction differ from the lytic cycle?**

operon are turned off by the *lac* repressor (Figure 8.12, page 230).

Every time the host cell's machinery replicates the bacterial chromosome,

4B it also replicates the prophage DNA. The prophage remains latent within the progeny cells.

⑤ However, a rare spontaneous event, or the action of UV light or certain chemicals, can lead to the excision (popping-out) of the phage DNA, and to initiation of the lytic cycle.

There are three important results of lysogeny. First, the lysogenic cells are immune to reinfection by the same phage. (However, the host cell is not immune to infection by other phage types.) The second result of lysogeny is **phage conversion;** that is, the host cell may exhibit new properties. For example, the bacterium *Corynebacterium diphtheriae*, which causes diphtheria, is a pathogen whose disease-producing properties are related to the synthesis of a toxin. The organism can produce toxin only when it carries a lysogenic phage, because the prophage carries the gene coding for the toxin. As another example, only streptococci carrying a lysogenic phage are capable of producing the toxin associated with scarlet fever. The toxin produced by *Clostridium botulinum*, which causes botulism, is encoded by a prophage gene, as is the cholera toxin produced by pathogenic strains of *Vibrio cholerae*.

The third result of lysogeny is that it makes **specialized transduction** possible. Recall from Chapter 8 that bacterial genes can be picked up in a phage coat and transferred to another bacterium in a process called generalized transduction (see Figure 8.27, page 244). Any bacterial genes can be transferred by generalized transduction because the host chromosome is broken down into fragments, any of which can be packaged into a phage coat. In specialized transduction, however, only certain bacterial genes can be transferred.

Specialized transduction is mediated by a lysogenic phage, which packages bacterial DNA *along with* its own DNA in the same capsid. When a prophage is excised from the host chromosome, adjacent genes from either side may remain attached to the phage DNA. In Figure 13.13, bacteriophage λ has picked up the *gal* gene for galactose fermentation from its galactose-positive host. The phage carries this gene to a galactose-negative cell, which then becomes galactose-positive.

Certain animal viruses can undergo processes very similar to lysogeny. Animal viruses that can remain latent in cells for long periods without multiplying or causing disease may become inserted into a host chromosome or remain separate from host DNA in a repressed state (as some lysogenic phages). Cancer-causing viruses may also be latent, as will be discussed later in the chapter.

MULTIPLICATION OF ANIMAL VIRUSES

LEARNING OBJECTIVE

• Compare and contrast the multiplication cycle of DNA- and RNA-containing animal viruses.

TABLE 13.3	Bacteriophage and Viral Multiplication Compared	
Stage	Bacteriophages	Animal Viruses

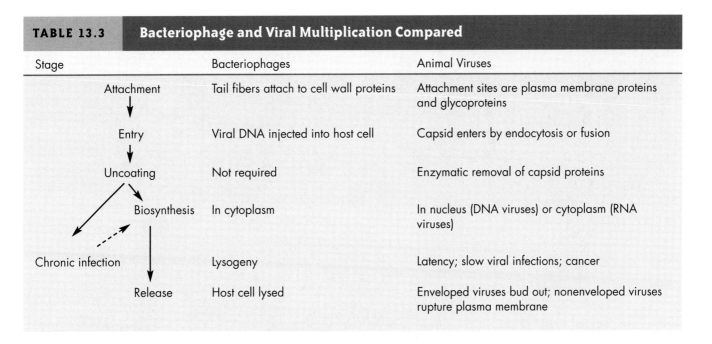

Attachment	Tail fibers attach to cell wall proteins	Attachment sites are plasma membrane proteins and glycoproteins
Entry	Viral DNA injected into host cell	Capsid enters by endocytosis or fusion
Uncoating	Not required	Enzymatic removal of capsid proteins
Biosynthesis	In cytoplasm	In nucleus (DNA viruses) or cytoplasm (RNA viruses)
Chronic infection	Lysogeny	Latency; slow viral infections; cancer
Release	Host cell lysed	Enveloped viruses bud out; nonenveloped viruses rupture plasma membrane

The multiplication of animal viruses follows the basic pattern of bacteriophage multiplication but has several differences, summarized in Table 13.3. Animal viruses differ from phages in their mechanism of entering the host cell. Also, once the virus is inside, the synthesis and assembly of the new viral components are somewhat different, partly because of the differences between prokaryotic cells and eukaryotic cells. Animal viruses may have certain types of enzymes not found in phages. Finally, the mechanisms of maturation and release, and the effects on the host cell, differ in animal viruses and phages.

In the following discussion of the multiplication of animal viruses, we will consider the processes that are shared by both DNA- and RNA-containing animal viruses. These processes are attachment, entry, uncoating, and release. We will also examine how DNA- and RNA-containing viruses differ with respect to their processes of biosynthesis.

ATTACHMENT

Like bacteriophages, animal viruses have attachment sites that attach to complementary receptor sites on the host cell's surface. However, the receptor sites of animal cells are proteins and glycoproteins of the plasma membrane (Figure 13.14a). Moreover, animal viruses do not possess appendages like the tail fibers of some bacteriophages. The attachment sites of animal viruses are distributed over the surface of the virus. The sites themselves vary from one group of viruses to another. In adenoviruses, which are icosahedral viruses, the attachment sites are small fibers at the corners of the icosahedron (see Figure 13.2b). In many of the enveloped viruses, such as influenza virus, the attachment sites are spikes located on the surface of the envelope (see Figure 13.3b). As

soon as one spike attaches to a host receptor, additional receptor sites on the same cell migrate to the virus. Attachment is completed when many sites are bound.

Receptor sites are inherited characteristics of the host. Consequently, the receptor for a particular virus can vary from person to person. This could account for the individual differences in susceptibility to a particular virus. For example, people who lack the cellular receptor (called P antigen) for parvovirus B19, are naturally resistant to infection and do not get fifth disease (see page 628). Understanding the nature of attachment can lead to the development of drugs that prevent viral infections. Monoclonal antibodies (discussed in Chapter 17) that combine with a virus's attachment site or the cell's receptor site may soon be used to treat some viral infections.

ENTRY

Following attachment, entry occurs. Viruses enter into eukaryotic cells by **pinocytosis,** an active cellular process by which nutrients and other molecules are brought into a cell (see Chapter 4, page 102). A cell's plasma membrane continuously folds inward to form vesicles. These vesicles contain elements that originate outside the cell and are brought into the interior of the cell to be digested. If a virion attaches to the plasma membrane of a potential host cell, the host cell will enfold the virion into a fold of plasma membrane, forming a vesicle (Figure 13.14a).

Enveloped viruses can enter by an alternative method called **fusion,** in which the viral envelope fuses with the plasma membrane and releases the capsid into the cell's cytoplasm. For example, HIV penetrates cells by this method (Figure 13.14b).

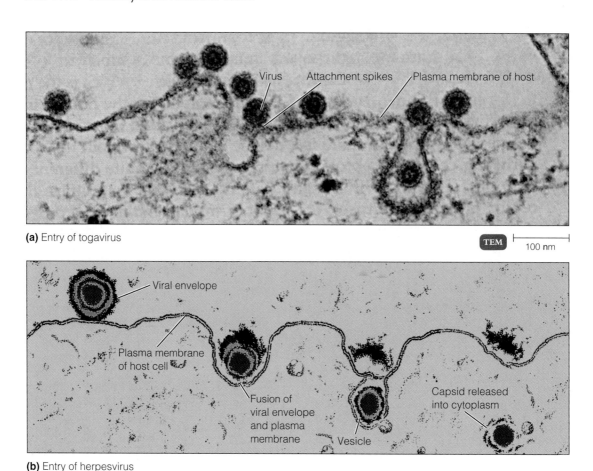

(a) Entry of togavirus

TEM | 100 nm

(b) Entry of herpesvirus

FIGURE 13.14 The entry of viruses into host cells. After attachment, viruses enter cells by (**a**) pinocytosis or (**b**) fusion.

Q **In which process is the cell actively taking in the virus?**

UNCOATING

Viruses disappear during the eclipse period of an infection because they are taken apart inside the cell. **Uncoating** is the separation of the viral nucleic acid from its protein coat once the virion is enclosed within the vesicle. The capsid is digested when the cell attempts to digest the vesicle's contents, or the nonenveloped capsid may be released into the cytoplasm of the host cell. This process varies with the type of virus. Some animal viruses accomplish uncoating by the action of lysosomal enzymes of the host cell. These enzymes degrade the proteins of the viral capsid. The uncoating of poxviruses is completed by a specific enzyme encoded by the viral DNA and synthesized soon after infection. For other viruses, uncoating appears to be exclusively caused by enzymes in the host cell cytoplasm. For at least one virus, the poliovirus, uncoating seems to begin while the virus is still attached to the host cell's plasma membrane.

THE BIOSYNTHESIS OF DNA VIRUSES

Generally, DNA-containing viruses replicate their DNA in the nucleus of the host cell by using viral enzymes, and they synthesize their capsid and other proteins in the cytoplasm

by using host cell enzymes. Then the proteins migrate into the nucleus and are joined with the newly synthesized DNA to form virions. These virions are transported along the endoplasmic reticulum to the host cell's membrane for release. Herpesviruses, papovaviruses, adenoviruses, and hepadnaviruses all follow this pattern of biosynthesis (Table 13.4). Poxviruses are an exception because all of their components are synthesized in the cytoplasm.

As an example of the multiplication of a DNA virus, the sequence of events in papovavirus is shown in Figure 13.15. After

1–2 Following attachment, entry, and uncoating, the viral DNA is released into the nucleus of the host cell.

3 Transcription of a portion of the viral DNA—the "early" genes—occurs next. Translation follows. The products of these genes are enzymes that are required for the multiplication of viral DNA. In most DNA viruses, early transcription is carried out with the host's transcriptase (RNA polymerase); poxviruses, however, contain their own transcriptase.

TABLE 13.4	The Biosynthesis of DNA and RNA Viruses Compared	
Viral Nucleic Acid	Virus Family	Special Features of Biosynthesis
DNA, single-stranded	Parvoviridae	Cellular enzyme transcribes viral DNA in nucleus
DNA, double-stranded	Herpesviridae Papovaviridae	Cellular enzyme transcribes viral DNA in nucleus
	Poxviridae	Viral enzyme transcribes viral DNA in virion, in cytoplasm
DNA, reverse transcriptase	Hepadnaviridae	Cellular enzyme transcribes viral DNA in nucleus; reverse transcriptase copies mRNA to make viral DNA
RNA, + strand	Picornaviridae Togaviridae	Viral RNA functions as a template for synthesis of RNA polymerase which copies − strand RNA to make mRNA in cytoplasm
RNA, − strand	Rhabdoviridae	Viral enzyme copies viral RNA to make mRNA in cytoplasm
RNA, double-stranded	Reoviridae	Viral enzyme copies − strand RNA to make mRNA in cytoplasm
RNA, reverse transcriptase	Retroviridae	Viral enzyme copies viral RNA to make DNA in cytoplasm; DNA moves to nucleus

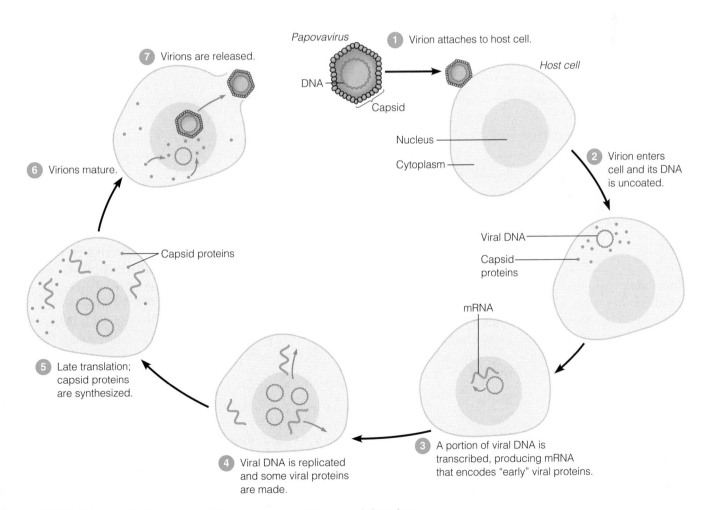

FIGURE 13.15 Multiplication of *Papovavirus*, a DNA-containing virus.

Q **Why is mRNA made?**

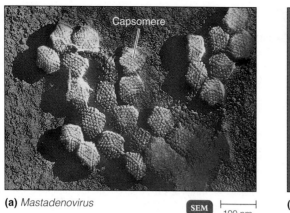

Capsomere

(a) *Mastadenovirus* SEM ⊢——⊣ 100 nm

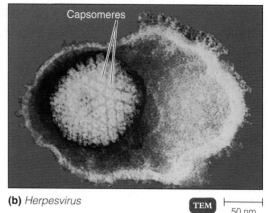

Capsomeres

(b) *Herpesvirus* TEM ⊢——⊣ 50 nm

FIGURE 13.16 DNA-containing animal viruses. (a) Negatively stained adenoviruses that have been concentrated in a centrifuge gradient. The individual capsomeres are clearly visible. **(b)** The envelope around this herpes simplex virus capsid has broken, giving a "fried egg" appearance.

Q **What is the morphology of these viruses?**

④ Sometime after the initiation of DNA replication, transcription and translation of the remaining "late" viral genes occur. Late proteins include capsid and other structural proteins.

⑤ This leads to the synthesis of capsid proteins, which occurs in the cytoplasm of the host cell.

⑥ After the capsid proteins migrate into the nucleus of the host cell, maturation occurs; the viral DNA and capsid proteins assemble to form complete viruses, which

⑦ are then released from the host cell.

Some DNA viruses are described below.

Adenoviridae Named after adenoids, from which they were first isolated, adenoviruses cause acute respiratory diseases—the common cold (Figure 13.16a).

Poxviridae All diseases caused by poxviruses, including smallpox and cowpox, include skin lesions (see Figure 21.9, page 624). *Pox* refers to pus-filled lesions. Viral multiplication is started by viral transcriptase; the viral components are synthesized and assembled in the cytoplasm of the host cell.

Herpesviridae Nearly 100 herpesviruses are known (Figure 13.16b). They are named after the spreading (*herpetic*) appearance of cold sores. Species of human herpesviruses (HHV) include HHV-1 and HHV-2, both in the genus *Simplexvirus*, which cause cold sores; HHV-3, genus *Varicellovirus*, which cause chickenpox; HHV-4, genus *Lymphocryptovirus*, which causes infectious mononucleosis; HHV-5, genus *Cytomegalovirus*, which causes CMV inclusion disease; HHV-6, genus *Roseolovirus*, which

causes roseola; HHV-7, which infects most infants, causing measleslike rashes; and HHV-8, which causes Kaposi's sarcoma, primarily in AIDS patients.

Papovaviridae Papovaviruses are named for *pa*pillomas (warts), *po*lyomas (tumors), and *va*cuolation (cytoplasmic vacuoles produced by some of these viruses). Warts are caused by members of the genus *Papillomavirus*. Some *Papillomavirus* species are capable of transforming cells and causing cancer. Viral DNA is replicated in the host cell's nucleus along with host cell chromosomes. Host cells may proliferate, resulting in a tumor.

Hepadnaviridae Hepadnaviridae are so named because they cause *hepa*titis and contain *DNA* (Figure 25.16, page 765). The only genus in this family causes hepatitis B. (Hepatitis A, C, D, E, F, and G viruses, although not related to each other, are RNA viruses. Hepatitis is discussed in Chapter 25.) Hepadnaviruses differ from other DNA viruses because they synthesize DNA by copying RNA, using viral reverse transcriptase. This enzyme is discussed later with the retroviruses, the only other family with reverse transcriptase.

THE BIOSYNTHESIS OF RNA VIRUSES

The multiplication of RNA viruses is essentially the same as that of DNA viruses, except that several different mechanisms of mRNA formation occur among different groups of RNA viruses (see Table 13.4). Although the details of these mechanisms are beyond the scope of this text, for comparative purposes we will trace the multiplication cycles of the four nucleic acid types of RNA viruses (three of which are shown in Figure 13.17). RNA viruses multiply in the host cell's cytoplasm. The major differences among the

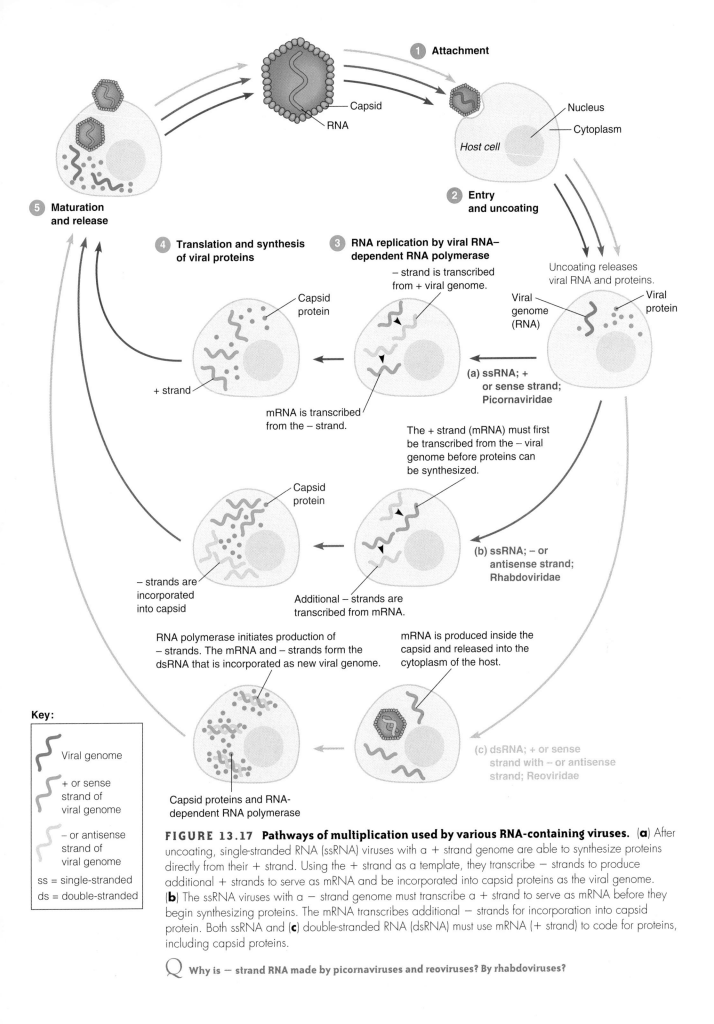

1 Attachment

Capsid

RNA

Nucleus

Cytoplasm

Host cell

2 Entry and uncoating

Uncoating releases viral RNA and proteins.

Viral genome (RNA)

Viral protein

5 Maturation and release

4 Translation and synthesis of viral proteins

3 RNA replication by viral RNA–dependent RNA polymerase

– strand is transcribed from + viral genome.

Capsid protein

mRNA is transcribed from the – strand.

(a) ssRNA; + or sense strand; Picornaviridae

+ strand

The + strand (mRNA) must first be transcribed from the – viral genome before proteins can be synthesized.

Capsid protein

Additional – strands are transcribed from mRNA.

(b) ssRNA; – or antisense strand; Rhabdoviridae

– strands are incorporated into capsid

RNA polymerase initiates production of – strands. The mRNA and – strands form the dsRNA that is incorporated as new viral genome.

mRNA is produced inside the capsid and released into the cytoplasm of the host.

Capsid proteins and RNA-dependent RNA polymerase

(c) dsRNA; + or sense strand with – or antisense strand; Reoviridae

Key:

Viral genome

+ or sense strand of viral genome

– or antisense strand of viral genome

ss = single-stranded
ds = double-stranded

FIGURE 13.17 Pathways of multiplication used by various RNA-containing viruses. (a) After uncoating, single-stranded RNA (ssRNA) viruses with a + strand genome are able to synthesize proteins directly from their + strand. Using the + strand as a template, they transcribe – strands to produce additional + strands to serve as mRNA and be incorporated into capsid proteins as the viral genome. **(b)** The ssRNA viruses with a – strand genome must transcribe a + strand to serve as mRNA before they begin synthesizing proteins. The mRNA transcribes additional – strands for incorporation into capsid protein. Both ssRNA and **(c)** double-stranded RNA (dsRNA) must use mRNA (+ strand) to code for proteins, including capsid proteins.

Q **Why is – strand RNA made by picornaviruses and reoviruses? By rhabdoviruses?**

MORBIDITY & MORTALITY WEEKLY REPORT

INFLUENZA: CROSSING THE SPECIES BARRIER

Influenza A viruses are found in many different animals, including ducks, chickens, pigs, whales, horses, and seals. Sometimes influenza A viruses normally seen in one species can cross over and cause illness in another species. For example, up until 1998, only H1N1 viruses circulated widely in the U.S. pig population. In 1998, H3N2 viruses from humans were introduced into the pig population and caused widespread disease among pigs.

The subtypes differ because of certain proteins on the surface of the virus (hemagglutinin [H] and neuraminidase [N] proteins). There are 16 different H subtypes and 9 different N subtypes of influenza A viruses. Many different combinations of H and N proteins are possible. Each combination is a different subtype.

When we talk about "human flu viruses" we are referring to those subtypes that occur widely in humans. There are only three known subtypes of human influenza viruses (H1N1, H1N2, and H3N2).

All subtypes of influenza virus A can be found in birds. However, when we talk about avian flu viruses, we are referring to those influenza A subtypes that occur mainly in birds. Avian flu viruses do not usually infect humans, although several cases of human infection with bird influenza viruses have occurred since 1997 (see Table A). Avian influenza viruses may be transmitted to humans in two main ways: (1) directly from birds or from avian virus-contaminated environments to people or (2) through an intermediate host, such

as a pig. Pigs are an important carrier because they can be infected with both human and avian flu and also come into contact with both species. The influenza virus genome is composed of eight separate segments. The segmented genome allows viruses from different species to mix and create a new influenza A virus if viruses from two different species infect the same person or animal (see the figure). This type of major change in the influenza A viruses is known as *antigenic shift*.

Thus far avian flu viruses have not caused outbreaks in the human population because they are not infective via human-to-human transmission. Of particular note, however, is the first documented instance of probable human-to-human transmission, in Thailand in 2005.

TABLE A   Recent Cases of Avian Influenza Virus Infections in Humans

Influenza virus	Location	Year	Human cases	Source
H5N1	Thailand, China	2005	130, Sporadic	Poultry outbreaks; two isolated instances of probable limited human-to-human transmission
H5N1	Thailand	2004	12	Poultry outbreak
H5N1	Vietnam	2004	23	Poultry outbreak
H7N3	Canada	2004	Eye infections among poultry workers	Poultry outbreak
H7N2	New York	2003	1	Source unknown
H7N7	Netherlands	2003	89	Started as poultry farm outbreak, later cases reported in pigs and humans
H5N1	China	2003	2	Source unknown
H7N2	Virginia	2002	1	Poultry outbreak
H9N2	China	1999	2	Poultry probable
H5N1	China	1997	18	Poultry

TABLE B   Influenza A Pandemics During the 20th Century

1918–19 H1N1 caused up to 50 million deaths worldwide. The origin of the 1918–19 pandemic virus is not clear.

1957–58 H2N2 caused about 70,000 deaths in the United States. First identified in China in late February 1957. Viruses contained a combination of genes from a human influenza virus and an avian influenza virus.

1968–69 H3N2 caused about 34,000 deaths in the United States. This virus contained genes from a human influenza virus and an avian influenza virus.

MORBIDITY & MORTALITY WEEKLY REPORT

(continued)

During the 20th century, the emergence of new influenza A virus subtypes caused three pandemics, all of which spread around the world within one year of being detected (see Table B). It is likely that some genetic parts of these influenza A strains originally came from birds. Many scientists believe it is only a matter of time until the next influenza pandemic occurs. SOURCE: Adapted from *MMWR* sources.

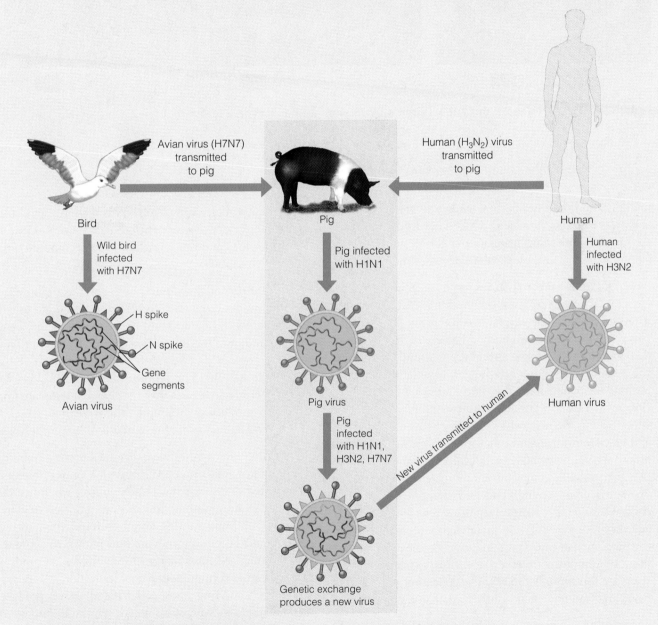

Model for antigenic shift in influenza virus. If a pig were infected with a human influenza virus and an avian influenza virus at the same time, the viruses could reassort and produce a new virus that had most of the genes from the human virus but a hemagglutinin and/or neuraminidase from the avian virus (see the figure). The resulting new virus might then be able to infect humans and spread from person to person, but it would have surface proteins (hemagglutinin and/or neuraminidase) not previously seen in influenza viruses that infect humans.

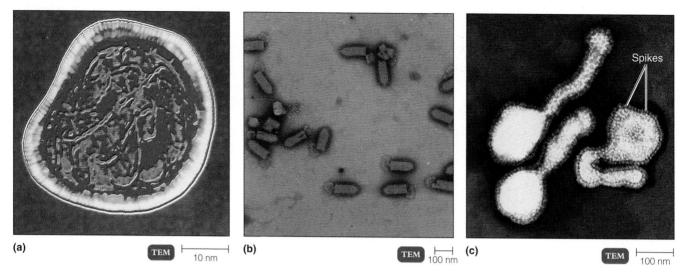

(a) TEM ├──┤ 10 nm **(b)** TEM ├──┤ 100 nm **(c)** TEM ├──┤ 100 nm

Spikes

FIGURE 13.18 RNA-containing animal viruses. (**a**) A single rubella virus (*Rubivirus*), which belongs to the togavirus family. (**b**) Particles of vesicular stomatitis virus (*Vesiculovirus*), a member of the Family Rhabdoviridae. (**c**) Mouse mammary tumor virus, a Retroviridae, causes tumors in mice.

Q Why do viruses with a + strand of RNA make a − strand of RNA?

multiplication processes of these viruses lie in how mRNA and viral RNA are produced. Once viral RNA and viral proteins are synthesized, maturation occurs by similar means among all animal viruses, as will be discussed shortly.

Picornaviridae Picornaviruses, such as poliovirus (see Chapter 22, page 652), are single-stranded RNA viruses. They are the smallest viruses; and the prefix *pico*- (small) plus *RNA* gives these viruses their name. The RNA within the virion is called a **sense strand** (or **+ strand**), because it can act as mRNA. After attachment, penetration, and uncoating are completed, the single-stranded viral RNA (Figure 13.17a) is translated into two principal proteins, which inhibit the host cell's synthesis of RNA and protein and which form an enzyme called *RNA-dependent RNA polymerase*. This enzyme catalyzes the synthesis of another strand of RNA, which is complementary in base sequence to the original infecting strand. This new strand, called an **antisense strand** (or **− strand**), serves as a template to produce additional + strands. The + strands may serve as mRNA for the translation of capsid proteins, may become incorporated into capsid proteins to form a new virus, or may serve as a template for continued RNA multiplication. Once viral RNA and viral protein are synthesized, maturation occurs.

Togaviridae Togaviruses, which include *arthropod-borne* *arbo*viruses or alphaviruses (see Chapter 22, page 658), also contain a single + strand of RNA. Togaviruses are enveloped viruses; their name is from the Latin word for covering, *toga* (Figure 13.18a). Keep in mind that these are not the only enveloped viruses. After a − strand is made from the + strand, two types of mRNA are transcribed from

the − strand. One type of mRNA is a short strand that codes for envelope proteins; the other, longer strand serves as mRNA for capsid proteins and can become incorporated into a capsid.

Rhabdoviridae Rhabdoviruses, such as rabiesvirus (genus *Lyssavirus*; see Chapter 22, page 654), are usually bullet-shaped (Figure 13.18b). *Rhabdo-* is from the Greek word for rod, which is not really an accurate description of their morphology. They contain a single − strand of RNA (see Figure 13.17b). They also contain an *RNA-dependent RNA polymerase* that uses the − strand as a template from which to produce a + strand. The + strand serves as mRNA and as a template for synthesis of new viral RNA.

Reoviridae Reoviruses were named for their habitats: the respiratory and enteric (digestive) systems of humans. They were not associated with any diseases when first discovered, so they were considered orphan viruses. Their name comes from the first letters of *r*espiratory, *e*nteric, and *o*rphan. Three serotypes are now known to cause respiratory tract and intestinal tract infections.

The capsid containing the double-stranded RNA is digested upon entering a host cell. Viral mRNA is produced in the cytoplasm, where it is used to synthesize more viral proteins (see Figure 13.17c). One of the newly synthesized viral proteins acts as *RNA-dependent RNA polymerase* to produce more − strands of RNA. The mRNA + and − strands form the double-stranded RNA that is then surrounded by capsid proteins.

Retroviridae Many retroviruses infect vertebrates (Figure 13.18c). One genus of retrovirus, *Lentivirus*, includes the

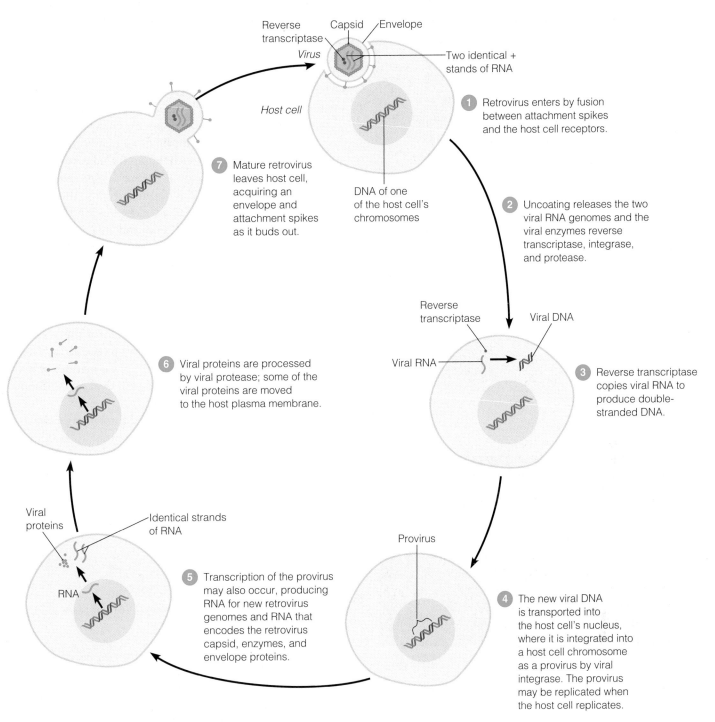

Reverse transcriptase
Capsid
Envelope
Virus
Two identical +
stands of RNA

Host cell

1 Retrovirus enters by fusion between attachment spikes and the host cell receptors.

DNA of one of the host cell's chromosomes

2 Uncoating releases the two viral RNA genomes and the viral enzymes reverse transcriptase, integrase, and protease.

7 Mature retrovirus leaves host cell, acquiring an envelope and attachment spikes as it buds out.

Reverse transcriptase
Viral DNA
Viral RNA

3 Reverse transcriptase copies viral RNA to produce double-stranded DNA.

6 Viral proteins are processed by viral protease; some of the viral proteins are moved to the host plasma membrane.

Viral proteins
Identical strands of RNA

RNA

5 Transcription of the provirus may also occur, producing RNA for new retrovirus genomes and RNA that encodes the retrovirus capsid, enzymes, and envelope proteins.

Provirus

4 The new viral DNA is transported into the host cell's nucleus, where it is integrated into a host cell chromosome as a provirus by viral integrase. The provirus may be replicated when the host cell replicates.

FIGURE 13.19 Multiplication and inheritance processes of the Retroviridae. A retrovirus may become a provirus that replicates in a latent state, and it may produce new retroviruses.

Q How does the biosynthesis of a retrovirus differ from that of other RNA viruses?

subspecies HIV-1 and HIV-2, which cause AIDS (see Chapter 19, page 566–576). The retroviruses that cause cancer will be discussed later in this chapter.

The formation of mRNA and RNA for new retrovirus virions is shown in Figure 13.19. These viruses carry **reverse transcriptase**, which uses the viral RNA as a template to produce complementary double-stranded DNA. This enzyme also degrades the original viral RNA. The name *retrovirus* is derived from the first letters of *reverse transcriptase*. The viral DNA is then integrated into a host cell chromosome as a **provirus**. Unlike a prophage, the provirus never comes out of the chromosome. As a provirus, HIV is protected from the host's immune system and antiviral drugs.

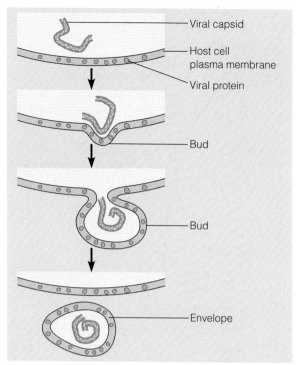

(a) Release by budding

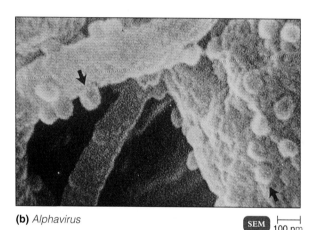

(b) *Alphavirus*

SEM |——| 100 nm

FIGURE 13.20   Budding of an enveloped virus. **(a)** A diagram of the budding process. **(b)** The small "bumps" (at arrows) seen on this freeze-fractured plasma membrane are Sindbis virus (*Alphavirus*) particles caught in the act of budding out from an infected cell.

Q **Of what is a viral envelope composed?**

Sometimes the provirus simply remains in a latent state and replicates when the DNA of the host cell replicates. In other cases, the provirus is expressed and produces new viruses, which may infect adjacent cells. Mutagens such as gamma radiation can induce expression of a provirus. In oncogenic retroviruses, the provirus can also convert the host cell into a tumor cell; possible mechanisms for this phenomenon will be discussed later.

MATURATION AND RELEASE

The first step in viral maturation is the assembly of the protein capsid; this assembly is usually a spontaneous process. The capsids of many animal viruses are enclosed by an envelope consisting of protein, lipid, and carbohydrate, as noted earlier. Examples of such viruses include orthomyxoviruses and paramyxoviruses. The envelope protein is encoded by the viral genes and is incorporated into the plasma membrane of the host cell. The envelope lipid and carbohydrate are encoded by host cell genes and are present in the plasma membrane. The envelope actually develops around the capsid by a process called **budding** (Figure 13.20).

After the sequence of attachment, entry, uncoating, and biosynthesis of viral nucleic acid and protein, the assembled capsid containing nucleic acid pushes through the plasma membrane. As a result, a portion of the plasma membrane, now the envelope, adheres to the virus. This extrusion of a virus from a host cell is one method of release. Budding does not immediately kill the host cell, and in some cases the host cell survives.

Nonenveloped viruses are released through ruptures in the host cell plasma membrane. In contrast to budding, this type of release usually results in the death of the host cell. ❀ **Animation: Go to the Mircobiology Place website or CD-ROM and click "Animations" to view Viral Replication.**

VIRUSES AND CANCER

LEARNING OBJECTIVES

- Define *oncogene* and *transformed cell.*
- Discuss the relationship between DNA- and RNA-containing viruses and cancer.

Several types of cancer are now known to be caused by viruses. Molecular biological research shows that the mechanisms of the diseases are similar, even when a virus does not cause the cancer.

The relationship between cancers and viruses was first demonstrated in 1908, when virologists Wilhelm Ellerman and Olaf Bang, working in Denmark, were trying to isolate the causative agent of chicken leukemia. They found that leukemia could be transferred to healthy chickens by cell-free filtrates that contained viruses. Three years later, F. Peyton Rous, working at the Rockefeller Institute in New York, found that a chicken **sarcoma** (cancer of connective tissue) can be similarly transmitted. Virus-induced **adenocarcinomas** (cancers of glandular epithelial tissue) in mice were discovered in 1936. At that time, it was clearly shown that mouse mammary gland tumors are transmitted from mother to offspring through the mother's milk. A human cancer-causing virus was discovered and isolated in 1972 by American bacteriologist Sarah Stewart.

The viral cause of cancer can often go unrecognized for several reasons. First, most of the particles of some viruses infect cells but do not induce cancer. Second, cancer might not develop until long after viral infection. Third, cancers do not seem to be contagious, as viral diseases usually are.

THE TRANSFORMATION OF NORMAL CELLS INTO TUMOR CELLS

Almost anything that can alter the genetic material of a eukaryotic cell has the potential to make a normal cell cancerous. These cancer-causing alterations to cellular DNA affect parts of the genome called **oncogenes.** Oncogenes were first identified in cancer-causing viruses and were thought to be a part of the normal viral genome. However, American microbiologists J. Michael Bishop and Harold E. Varmus received the 1989 Nobel Prize in Medicine for proving that the cancer-inducing genes carried by viruses are actually derived from animal cells. Bishop and Varmus showed that the cancer-causing *src* gene in avian sarcoma viruses is derived from a normal part of chicken genes.

Oncogenes can be activated to abnormal functioning by a variety of agents, including mutagenic chemicals, high-energy radiation, and viruses. Viruses capable of inducing tumors in animals are called **oncogenic viruses,** or *oncoviruses.* Approximately 10% of cancers are known to be virus-induced. An outstanding feature of all oncogenic viruses is that their genetic material integrates into the host cell's DNA and replicates along with the host cell's chromosome. This mechanism is similar to the phenomenon of lysogeny in bacteria, and it can alter the host cell's characteristics in the same way.

Tumor cells undergo **transformation;** that is, they acquire properties that are distinct from the properties of uninfected cells or from infected cells that do not form tumors. After being transformed by viruses, many tumor cells contain a virus-specific antigen on their cell surface, called **tumor-specific transplantation antigen (TSTA),** or an antigen in their nucleus, called the **T antigen.** Transformed cells tend to be less round than normal cells, and they tend to exhibit certain chromosomal abnormalities, such as unusual numbers of chromosomes and fragmented chromosomes.

DNA ONCOGENIC VIRUSES

Oncogenic viruses are found within several families of DNA-containing viruses. These groups include the Adenoviridae, Herpesviridae, Poxviridae, Papovaviridae, and Hepadnaviridae. Among the papovaviruses, papillomaviruses cause uterine (cervical) cancer.

Virtually all cervical cancers are caused by human papillomavirus (HPV); HPV-16, accounts for about half of all cervical cancers. Recent clinical trials of a vaccine against four HPVs, including HPV-16, have been very promising.

Epstein-Barr (EB) virus was isolated from Burkitt's lymphoma cells in 1964 by Michael Epstein and Yvonne Barr. The proof that EB virus can cause cancer was accidentally demonstrated in 1985 when a 12-year-old boy known only as David received a bone marrow transplant. Several months after the transplant, he died of cancer. An autopsy revealed that the virus had been unwittingly introduced into the boy with the bone marrow transplant.

Another DNA virus that causes cancer is hepatitis B virus (HBV). Many animal studies have been performed that have clearly indicated the causal role of HBV in liver cancer. In one human study, virtually all people with liver cancer had previous HBV infections.

RNA ONCOGENIC VIRUSES

Among the RNA viruses, only the oncoviruses in the family Retroviridae cause cancer. The human T-cell leukemia viruses (HTLV-1 and HTLV-2) are retroviruses that cause adult T-cell leukemia and lymphoma in humans. (T cells are a type of white blood cell involved in the immune response.)

Sarcoma viruses of cats, chickens, and rodents, and the mammary tumor viruses of mice, are also retroviruses. Another retrovirus, feline leukemia virus (FeLV), causes leukemia in cats and is transmissible among cats. There is a test to detect the virus in cat serum.

The ability of retroviruses to induce tumors is related to their production of a reverse transcriptase by the mechanism described earlier (see Figure 13.19). The provirus, which is the double-stranded DNA molecule synthesized from the viral RNA, becomes integrated into the host cell's DNA; new genetic material is thereby introduced into the host's genome, and this is the key reason retroviruses can contribute to cancer. Some retroviruses contain oncogenes; others contain promoters that turn on oncogenes or other cancer-causing factors.

LATENT VIRAL INFECTIONS

LEARNING OBJECTIVE
• Provide an example of a latent viral infection.

A virus can remain in equilibrium with the host and not actually produce disease for a long period, often many years. The oncogenic viruses just discussed are examples of such latent infections. All of the human herpesviruses can remain in host cells throughout the life of an individual. When herpesviruses are reactivated by immunosuppression (for example, AIDS), the resulting infection may be fatal. The classic example of such a **latent infection** in viruses is the infection of the skin by herpes simplex virus, which produces cold sores. This virus inhabits the host's nerve cells but causes no damage until it is activated by a stimulus such as fever or sunburn—hence the term *fever blister*.

In some individuals, viruses are produced, but symptoms never appear. Even though a large percentage of the human population carries the herpes simplex virus, only 10 to 15% of people carrying the virus exhibit the disease. The virus of some latent infections can exist in a lysogenic state within host cells.

The chickenpox virus (genus *Varicellovirus*) can also exist in a latent state. Chickenpox (varicella) is a skin disease that is usually acquired in childhood. The virus gains access to the skin via the blood. From the blood, some viruses may enter nerves, where they remain latent. Later, changes in the immune (T-cell) response can activate these latent viruses, causing shingles (zoster). The shingles rash appears on the skin along the nerve in which the virus was latent. Shingles occurs in 10 to 20% of people who have had chickenpox.

PERSISTENT VIRAL INFECTIONS

LEARNING OBJECTIVE

- Differentiate persistent viral infections from latent viral infections.

A **persistent** or **chronic viral infection** occurs gradually over a long period. Typically, persistent viral infections are fatal.

A number of persistent viral infections have in fact been shown to be caused by conventional viruses. For example, several years after causing measles, the measles virus can be responsible for a rare form of encephalitis called subacute sclerosing panencephalitis (SSPE). A persistent viral infection is apparently different from a latent viral infection in that, in most persistent viral infections, detectable infectious virus gradually builds up over a long period, rather than appearing suddenly (Figure 13.21).

Several examples of latent and persistent viral infections are listed in Table 13.5.

PRIONS

LEARNING OBJECTIVE

- Discuss how a protein can be infectious.

A few infectious diseases are caused by prions. In 1982, American neurobiologist Stanley Prusiner proposed that infectious proteins caused a neurological disease in sheep

TABLE 13.5	Examples of Latent and Persistent Viral Infections in Humans	
Disease	Primary Effect	Causative Virus
Latent		
Cold sores	Skin and mucous membrane lesions; genital lesions	Herpes simplex 1 and 2
Shingles	Skin lesions	*Varicella-zoster virus (Herpesvirus)*
HIV/AIDS	Decreased CD4 cells	HIV-1 and -2 (*Lentivirus*)
Leukemia	Increased white blood cell growth	HTLV-1 and -2
Persistent		
Subacute sclerosing panencephalitis (SSPE)	Mental deterioration	Measles virus
Progressive encephalitis	Rapid mental deterioration	Rubella virus
Cancer	Increased cell growth	EB virus
AIDS-dementia complex	Brain degeneration	HIV (*Lentivirus*)
Persistent enterovirus infection	Mental deterioration associated with AIDS	Echoviruses
Liver cancer	Increased cell growth	Hepatitis B virus
Cervical cancer	Increased cell growth	Human papillomavirus

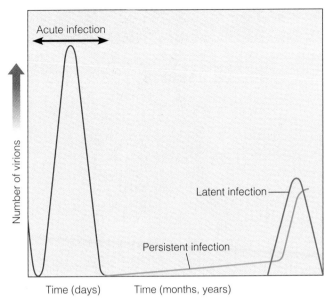

FIGURE 13.21 Latent and persistent viral infections.

Q How do latent and persistent infections differ?

called scrapie. The infectivity of scrapie-infected brain tissue is reduced by treatment with proteases but not by treatment with radiation, suggesting that the infectious agent is pure protein. Prusiner coined the name **prion** for *pro*teinaceous *in*fectious particle.

Nine animal diseases now fall into this category, including the "mad cow disease" that emerged in cattle in Great Britain in 1987. All nine are neurological diseases called spongiform encephalopathies because large vacuoles develop in the brain (Figure 22.18a, page 663). The human diseases are kuru, Creutzfeldt-Jakob disease (CJD), Gerstmann-Sträussler-Scheinker syndrome, and fatal familial insomnia. (Neurological diseases are discussed in Chapter 22.) These diseases run in families, which indicates a possible genetic cause. However, they cannot be purely inherited, because mad cow disease arose from feeding scrapie-infected sheep meat to cattle, and the new (bovine) variant was transmitted to humans who ate undercooked beef from infected cattle (see Chapter 1, page 20). Additionally, CJD has been transmitted with transplanted nerve tissue and contaminated surgical instruments.

These diseases are caused by the conversion of a normal host glycoprotein called PrP^C (for cellular prion protein) into an infectious form called PrP^{Sc} (for scrapie protein). The gene for PrP^C is located on chromosome 20 in humans. Recent evidence suggests that PrP^C is involved in regulating cell death. (See the discussion of apoptosis on page 515.) One hypothesis for how an infectious agent that lacks any nucleic acid can reproduce is shown in Figure 13.22.

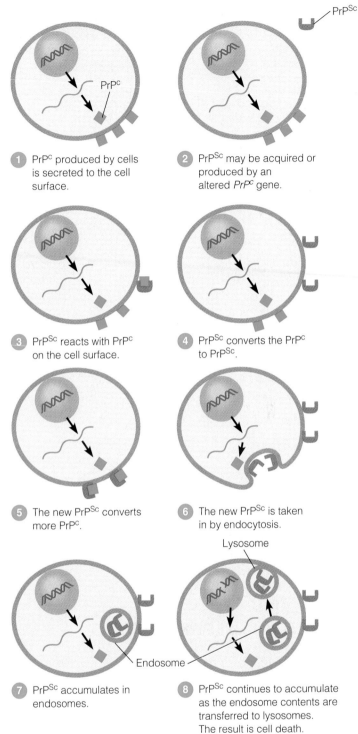

1. PrP^C produced by cells is secreted to the cell surface.

2. PrP^{Sc} may be acquired or produced by an altered *PrP^C* gene.

3. PrP^{Sc} reacts with PrP^C on the cell surface.

4. PrP^{Sc} converts the PrP^C to PrP^{Sc}.

5. The new PrP^{Sc} converts more PrP^C.

6. The new PrP^{Sc} is taken in by endocytosis.

7. PrP^{Sc} accumulates in endosomes.

8. PrP^{Sc} continues to accumulate as the endosome contents are transferred to lysosomes. The result is cell death.

FIGURE 13.22 How a protein can be infectious. If an abnormal prion protein (PrP^{Sc}) enters a cell, it changes a normal prion protein to PrP^{Sc}, which now can change another normal PrP^C, resulting in an accumulation of the abnormal PrP^{Sc}.

Q How do prions differ from viruses?

| | | TABLE 13.6 | **Classification of Some Major Plant Viruses** | | |

Characteristic	Viral Family	Viral Genus or Unclassified Members	Morphology	Method of Transmission
Double-stranded DNA, nonenveloped	Papovaviridae	Cauliflower mosaic virus		Aphids
Single-stranded RNA, + strand, nonenveloped	Picornaviridae	Bean mosaic virus		Pollen
	Tetraviridae	*Tobamovirus*		Wounds
Single-stranded RNA, − strand, enveloped	Rhabdoviridae	Potato yellow dwarf virus		Leafhoppers and aphids
Double-stranded RNA, nonenveloped	Reovirus	Wound tumor virus	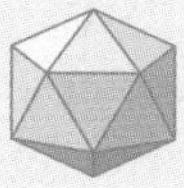	Leafhoppers

The actual cause of cell damage is not known. Fragments of PrP^{Sc} molecules accumulate in the brain, forming plaques; these plaques are used for postmortem diagnosis, but they do not appear to be the cause of cell damage. ✳ **Animation: Go to The Microbiology Place website or CD-ROM and click "Animations" to view Prion Reproduction.**

PLANT VIRUSES AND VIROIDS

LEARNING OBJECTIVES

• Differentiate virus, viroid, and prion.

• Name a virus that causes a plant disease.

Plant viruses resemble animal viruses in many respects: Plant viruses are morphologically similar to animal viruses, and they have similar types of nucleic acid (Table 13.6). In fact, some plant viruses can multiply inside insect cells. Plant viruses cause many diseases of economically important crops, including beans (bean mosaic virus), corn and sugarcane (wound tumor virus), and potatoes (potato yellow dwarf virus). Viruses can cause color change, deformed growth, wilting, and stunted growth in their plant hosts.

Some hosts, however, remain symptomless and only serve as reservoirs of infection.

Plant cells are generally protected from disease by an impermeable cell wall. Viruses must enter through wounds or be assisted by other plant parasites, including nematodes, fungi, and, most often, insects that suck the plant's sap. Once one plant is infected, it can spread infection to other plants in its pollen and seeds.

In laboratories, plant viruses are cultured in protoplasts (plant cells with the cell walls removed) and in insect cell cultures.

Some plant diseases are caused by **viroids,** short pieces of naked RNA, only 300 to 400 nucleotides long, with no protein coat. The nucleotides are often internally paired, so the molecule has a closed, folded, three-dimensional structure that presumably helps protect it from attack by cellular enzymes. The RNA does not code for any proteins. Thus far, viroids have been conclusively identified as pathogens only of plants. Annually, infections by viroids, such as potato spindle tuber viroid, result in losses of millions of dollars from crop damage (Figure 13.23).

Current research on viroids has revealed similarities between the base sequences of viroids and introns. Recall from Chapter 8 (page 226) that introns are sequences of genetic material that do not code for polypeptides. This observation has led to the hypothesis that viroids evolved from introns, leading to speculation that future researchers may discover animal viroids.

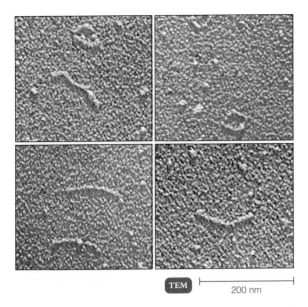

TEM |—————| 200 nm

FIGURE 13.23 Linear and circular potato spindle tuber viroid (PSTV).

Q **How do viroids differ from prions?**

STUDY OUTLINE

GENERAL CHARACTERISTICS OF VIRUSES (pp. 387–389)

1. Depending on one's viewpoint, viruses may be regarded as exceptionally complex aggregations of nonliving chemicals or as exceptionally simple living microbes.
2. Viruses contain a single type of nucleic acid (DNA or RNA) and a protein coat, sometimes enclosed by an envelope composed of lipids, proteins, and carbohydrates.
3. Viruses are obligatory intracellular parasites. They multiply by using the host cell's synthesizing machinery to cause the synthesis of specialized elements that can transfer the viral nucleic acid to other cells.

HOST RANGE (pp. 387–389)

4. Host range refers to the spectrum of host cells in which a virus can multiply.
5. Most viruses infect only specific types of cells in one host species.
6. Host range is determined by the specific attachment site on the host cell's surface and the availability of host cellular factors.

VIRAL SIZE (p. 389)

7. Viral size is ascertained by electron microscopy.
8. Viruses range from 20 to 1000 nm in length.

VIRAL STRUCTURE (pp. 389–391)

1. A virion is a complete, fully developed viral particle composed of nucleic acid surrounded by a coat.

NUCLEIC ACID (p. 389)

2. Viruses contain either DNA or RNA, never both, and the nucleic acid may be single- or double-stranded, linear or circular, or divided into several separate molecules.
3. The proportion of nucleic acid in relation to protein in viruses ranges from about 1% to about 50%.

CAPSID AND ENVELOPE (pp. 389–390)

4. The protein coat surrounding the nucleic acid of a virus is called the capsid.
5. The capsid is composed of subunits, capsomeres, which can be a single type of protein or several types.
6. The capsid of some viruses is enclosed by an envelope consisting of lipids, proteins, and carbohydrates.
7. Some envelopes are covered with carbohydrate-protein complexes called spikes.

GENERAL MORPHOLOGY (pp. 390–391)

8. Helical viruses (for example, Ebola virus) resemble long rods, and their capsids are hollow cylinders surrounding the nucleic acid.

9. Polyhedral viruses (for example, adenovirus) are many-sided. Usually the capsid is an icosahedron.

10. Enveloped viruses are covered by an envelope and are roughly spherical but highly pleomorphic. There are also enveloped helical viruses (for example, influenza virus) and enveloped polyhedral viruses (for example, *Simplexvirus*).

11. Complex viruses have complex structures. For example, many bacteriophages have a polyhedral capsid with a helical tail attached.

TAXONOMY OF VIRUSES (p. 391)

1. Classification of viruses is based on type of nucleic acid, strategy for replication, and morphology.

2. Virus family names end in *-viridae*; genus names end in *-virus*.

3. A viral species is a group of viruses sharing the same genetic information and ecological niche.

ISOLATION, CULTIVATION, AND IDENTIFICATION OF VIRUSES (pp. 394–396)

1. Viruses must be grown in living cells.

2. The easiest viruses to grow are bacteriophages.

GROWING BACTERIOPHAGES IN THE LABORATORY (p. 394)

3. The plaque method mixes bacteriophages with host bacteria and nutrient agar.

4. After several viral multiplication cycles, the bacteria in the area surrounding the original virus are destroyed; the area of lysis is called a plaque.

5. Each plaque originates with a single viral particle; the concentration of viruses is given as plaque-forming units.

GROWING ANIMAL VIRUSES IN THE LABORATORY (pp. 394–396)

6. Cultivation of some animal viruses requires whole animals.

7. Simian AIDS and feline AIDS provide models for studying human AIDS.

8. Some animal viruses can be cultivated in embryonated eggs.

9. Cell cultures are cells growing in culture media in the laboratory.

10. Primary cell lines and embryonic diploid cell lines grow for a short time in vitro.

11. Continuous cell lines can be maintained in vitro indefinitely.

12. Viral growth can cause cytopathic effects in the cell culture.

VIRAL IDENTIFICATION (p. 396)

13. Serological tests are used most often to identify viruses.

14. Viruses may be identified by RFLPs and PCR.

VIRAL MULTIPLICATION (pp. 396–410)

1. Viruses do not contain enzymes for energy production or protein synthesis.

2. For a virus to multiply, it must invade a host cell and direct the host's metabolic machinery to produce viral enzymes and components.

MULTIPLICATION OF BACTERIOPHAGES (pp. 397–400)

3. During the lytic cycle, a phage causes the lysis and death of a host cell.

4. Some viruses can either cause lysis or have their DNA incorporated as a prophage into the DNA of the host cell. The latter situation is called lysogeny.

5. During the attachment phase of the lytic cycle, sites on the phage's tail fibers attach to complementary receptor sites on the bacterial cell.

6. In penetration, phage lysozyme opens a portion of the bacterial cell wall, the tail sheath contracts to force the tail core through the cell wall, and phage DNA enters the bacterial cell. The capsid remains outside.

7. In biosynthesis, transcription of phage DNA produces mRNA coding for proteins necessary for phage multiplication. Phage DNA is replicated, and capsid proteins are produced. During the eclipse period, separate phage DNA and protein can be found.

8. During maturation, phage DNA and capsids are assembled into complete viruses.

9. During release, phage lysozyme breaks down the bacterial cell wall, and the new phages are released.

10. During the lysogenic cycle, prophage genes are regulated by a repressor coded for by the prophage. The prophage is replicated each time the cell divides.

11. Exposure to certain mutagens can lead to excision of the prophage and initiation of the lytic cycle.

12. Because of lysogeny, lysogenic cells become immune to reinfection with the same phage and may undergo phage conversion.

13. A lysogenic phage can transfer bacterial genes from one cell to another through transduction. Any genes can be transferred in generalized transduction, and specific genes can be transferred in specialized transduction.

MULTIPLICATION OF ANIMAL VIRUSES (pp. 400–410)

14. Animal viruses attach to the plasma membrane of the host cell.

15. Entry occurs by endocytosis or fusion.

16. Animal viruses are uncoated by viral or host cell enzymes.

17. The DNA of most DNA viruses is released into the nucleus of the host cell. Transcription of viral DNA and translation produce viral DNA and, later, capsid proteins. Capsid proteins are synthesized in the cytoplasm of the host cell.

18. DNA viruses include members of the families Adenoviridae, Poxviridae, Herpesviridae, Papovaviridae, and Hepadnaviridae.

19. Multiplication of RNA viruses occurs in the cytoplasm of the host cell. RNA-dependent RNA polymerase synthesizes a double-stranded RNA.

20. Picornaviridae + strand RNA acts as mRNA and directs the synthesis of RNA-dependent RNA polymerase.

21. Togaviridae + strand RNA acts as a template for RNA-dependent RNA polymerase, and mRNA is transcribed from a new − RNA strand.

22. Rhabdoviridae − strand RNA is a template for viral RNA-dependent RNA polymerase, which transcribes mRNA.

23. Reoviridae are digested in host cell cytoplasm to release mRNA for viral biosynthesis.

24. Retroviridae reverse transcriptase (RNA-dependent DNA polymerase) transcribes DNA from RNA.

25. After maturation, viruses are released. One method of release (and envelope formation) is budding. Nonenveloped viruses are released through ruptures in the host cell membrane.
 ※ **Animation: Viral Replication. The Microbiology Place.**

VIRUSES AND CANCER (pp. 410–411)

1. The earliest relationship between cancer and viruses was demonstrated in the early 1900s, when chicken leukemia and chicken sarcoma were transferred to healthy animals by cell-free filtrates.

THE TRANSFORMATION OF NORMAL CELLS INTO TUMOR CELLS (p. 411)

2. When activated, oncogenes transform normal cells into cancerous cells.

3. Viruses capable of producing tumors are called oncogenic viruses.

4. Several DNA viruses and retroviruses are oncogenic.

5. The genetic material of oncogenic viruses becomes integrated into the host cell's DNA.

6. Transformed cells lose contact inhibition, contain virus-specific antigens (TSTA and T antigen), exhibit chromosome abnormalities, and can produce tumors when injected into susceptible animals.

DNA ONCOGENIC VIRUSES (p. 411)

7. Oncogenic viruses are found among the Adenoviridae, Herpesviridae, Poxviridae, and Papovaviridae.

8. The EB virus, a herpesvirus, causes Burkitt's lymphoma and nasopharyngeal carcinoma. *Hepadnavirus* causes liver cancer.

RNA ONCOGENIC VIRUSES (p. 411)

9. Among the RNA viruses, only retroviruses seem to be oncogenic.

10. HTLV-1 and HTLV-2 have been associated with human leukemia and lymphoma.

11. The virus's ability to produce tumors is related to the production of reverse transcriptase. The DNA synthesized from the viral RNA becomes incorporated as a provirus into the host cell's DNA.

12. A provirus can remain latent, can produce viruses, or can transform the host cell.

LATENT VIRAL INFECTIONS (pp.411–412)

1. A latent viral infection is one in which the virus remains in the host cell for long periods without producing an infection.

2. Examples are cold sores and shingles.

PERSISTENT VIRAL INFECTIONS (p. 412)

1. Persistent viral infections are disease processes that occur over a long period and are generally fatal.

2. Persistent viral infections are caused by conventional viruses; viruses accumulate over a long period.

PRIONS (pp. 412–414)

1. Prions are infectious proteins first discovered in the 1980s.

2. Prion diseases, such as CJD and mad cow disease, all involve the degeneration of brain tissue.

3. Prion diseases are the result of an altered protein; the cause can be a mutation in the normal gene for PrP^C or contact with an altered protein (PrP^{Sc}). ※ **Animation: Prion Reproduction. The Microbiology Place.**

PLANT VIRUSES AND VIROIDS (pp. 414–415)

1. Plant viruses must enter plant hosts through wounds or with invasive parasites, such as insects.

2. Some plant viruses also multiply in insect (vector) cells.

3. Viroids are infectious pieces of RNA that cause some plant diseases, such as potato spindle tuber viroid disease.

STUDY QUESTIONS

Access more review material either online at **The Microbiology Place** (www.microbiologyplace.com) or with **The Microbiology Place CD-ROM** packaged with your new book. There you'll find activities, practice tests, quizzes, flashcards, case studies, and more to help you succeed. In addition, you'll find the following Interactive Tutorials: Lytic Cycle, Lysogenic Cycle, DNA Viruses, RNA Viruses, and Viruses.
Answers to the Study Questions can by found in Appendix G.

REVIEW

1. Viruses were first detected because they are filterable. What do we mean by the term *filterable*, and how could this property have helped researchers detect viruses before the invention of the electron microscope?

2. Why do we classify viruses as obligatory intracellular parasites?

3. List the four properties that define a virus. What is a virion?

4. Describe the four morphological classes of viruses, then diagram and give an example of each.

5. Describe how bacteriophages are detected and enumerated by the plaque method.

6. Why are continuous cell lines of more practical use than primary cell lines for culturing viruses? What is unique about continuous cell lines?

7. *Vibrio cholerae* produces toxin and is capable of causing cholera only when it is lysogenic. What does this mean?

8. Describe the principal events of attachment, entry, uncoating, biosynthesis, maturation, and release of an enveloped DNA-containing virus.

9. Recall from Chapter 1 that Koch's postulates are used to determine the etiology of a disease. Why is it difficult to determine the etiology of
   a. a viral infection, such as influenza?
   b. cancer?   subacute sclerosing panencephalitis

10. Persistent viral infections such as _____ might be caused by common viruses that are latent in an abnormal tissue

11. The DNA of DNA-containing oncogenic viruses can become integrated into the host DNA. When integrated, the DNA is called a provirus. How does this process result in transformation of the cell? Describe the changes of transformation. How can an RNA-containing virus be oncogenic?

12. Contrast viroids and prions. Name a disease caused by each.

13. Plant viruses cannot penetrate intact plant cells because _____; therefore, they enter cells by _____. Plant viruses can be cultured in _____.

MULTIPLE CHOICE

1. Place the following in the most likely order for biosynthesis of a bacteriophage: (1) phage lysozyme; (2) mRNA; (3) DNA; (4) viral proteins; (5) DNA polymerase.

   a. 5, 4, 3, 2, 1
   b. 1, 2, 3, 4, 5
   c. 5, 3, 4, 2, 1
   d. 3, 5, 2, 4, 1
   e. 2, 5, 3, 4, 1

2. The molecule serving as mRNA can be incorporated in the newly synthesized virus capsids of all of the following *except*
   a. + strand RNA picornaviruses.
   b. + strand RNA togaviruses.
   c. − strand RNA rhabdoviruses.
   d. double-stranded RNA reoviruses.
   e. double-stranded DNA herpesviruses.

3. A virus with RNA-dependent RNA polymerase
   a. synthesizes DNA from an RNA template.
   b. synthesizes double-stranded RNA from an RNA template.
   c. synthesizes double-stranded RNA from a DNA template.
   d. transcribes mRNA from DNA.
   e. none of the above

4. Which of the following would be the first step in the biosynthesis of a virus with reverse transcriptase?
   a. A complementary strand of RNA must be synthesized.
   b. Double-stranded RNA must be synthesized.
   c. A complementary strand of DNA must be synthesized from an RNA template.
   d. A complementary strand of DNA must be synthesized from a DNA template.
   e. none of the above

5. An example of lysogeny in animals could be
   a. slow viral infections.
   b. latent viral infections.
   c. T-even bacteriophages.
   d. infections resulting in cell death.
   e. none of the above

6. The ability of a virus to infect an organism is regulated by
   a. the host species.
   b. the type of cells.
   c. the availability of an attachment site.
   d. cell factors necessary for viral replication.
   e. all of the above

7. Which of the following statements is *not* true?
   a. Viruses contain DNA or RNA.
   b. The nucleic acid of a virus is surrounded by a protein coat.
   c. Viruses multiply inside living cells using viral mRNA, tRNA, and ribosomes.
   d. Viruses cause the synthesis of specialized infectious elements.
   e. Viruses multiply inside living cells.

8. Place the following in the order in which they are found in a host cell: (1) capsid proteins; (2) infective phage particles; (3) phage nucleic acid.
a. 1, 2, 3
b. 3, 2, 1
c. 2, 1, 3
d. 3, 1, 2
e. 1, 3, 2

9. Which of the following does *not* initiate DNA synthesis?
a. a double-stranded DNA virus (Poxviridae)
b. a DNA virus with reverse transcriptase (Hepadnaviridae)
c. an RNA virus with reverse transcriptase (Retroviridae)
d. a single-stranded RNA virus (Togaviridae)
e. none of the above

10. A viral species is not defined on the basis of the disease symptoms it causes. The best example of this is
a. polio.
b. rabies.
c. hepatitis.
d. chickenpox and shingles.
e. measles.

CRITICAL THINKING

1. Discuss the arguments for and against the classification of viruses as living organisms.

2. In some viruses, capsomeres function as enzymes as well as structural supports. Of what advantage is this to the virus?

3. Why was the discovery of simian AIDS and feline AIDS important?

4. Prophages and proviruses have been described as being similar to bacterial plasmids. What similar properties do they exhibit? How are they different?

CLINICAL APPLICATIONS

1. A 40-year-old male who was seropositive for HIV experienced abdominal pain, fatigue, and low-grade fever (38°C) for 2 weeks. A chest X-ray examination revealed lung infiltrates. Gram and acid-fast stains were negative. A viral culture revealed the cause of his symptoms: a large, enveloped polyhedral virus with double-stranded DNA. What is the disease? Which virus causes it? Why was a viral culture done after the Gram and acid-fast stain results were obtained?

2. A newborn female developed extensive vesicular and ulcerative lesions over her face and chest. What is the most likely cause of her symptoms? How would you determine the viral cause of this disease without doing a viral culture?

3. Thirty-two people in the same town reported to their physicians with fever (40°C), jaundice, and tender abdomen. All 32 had eaten an ice-slush beverage purchased from a local convenience store. Liver function tests were abnormal. Over the next several months, the symptoms subsided and liver function returned to normal. What is the disease? This disease could be caused by a member of the Picornaviridae, Hepadnaviridae, or Flaviviridae. Differentiate among these families by method of transmission, morphology, nucleic acid, and type of replication.

14

Principles of Disease and Epidemiology

Now that you have a basic understanding of the structures and functions of microorganisms and some idea of the variety of microorganisms that exist, we can consider how the human body and various microorganisms interact in terms of health and disease.

We all have defenses to keep us healthy. In spite of these, however, we are still susceptible to **pathogens** (disease-causing microorganisms). A rather delicate balance exists between our defenses and the pathogenic mechanisms of microorganisms. When our defenses resist these pathogenic capabilities, we maintain our health—when the pathogen's capability overcomes our defenses, disease results. After the disease has become established, an infected person may recover completely, suffer temporary or permanent damage, or die.

In Part Three we examine some of the principles of infection and disease, the mechanisms by which pathogens cause disease, the body's defenses against disease, and the ways that microbial diseases can be prevented by immunization and controlled by drugs. This first chapter discusses the general principles of disease, starting with a discussion of the meaning and scope of pathology. In the last section of this chapter, "Epidemiology," you will learn how these principles are useful in studying and controlling disease.

UNDER THE MICROSCOPE

Intestinal Bacteria. Millions of bacteria live in the human large intestine and are essential for good health.

PATHOLOGY, INFECTION, AND DISEASE

LEARNING OBJECTIVE
• Define *pathology, etiology, infection,* and *disease.*

Pathology is the scientific study of disease (*pathos* = suffering; *logos* = science). Pathology is first concerned with the cause, or **etiology,** of disease. Second, it deals with **pathogenesis,** the manner in which a disease develops. Third, pathology is concerned with the *structural* and *functional changes* brought about by disease and with their final effects on the body.

Although the terms *infection* and *disease* are sometimes used interchangeably, they differ somewhat in meaning. **Infection** is the invasion or colonization of the body by pathogenic microorganisms; **disease** occurs when an infection results in any change from a state of health. Disease is an abnormal state in which part or all of the body is not properly adjusted or incapable of performing its normal functions. An infection may exist in the absence of detectable disease. For example, the body may be infected with the virus that causes AIDS, but there may be no symptoms of the disease.

The presence of a particular type of microorganism in a part of the body where it is not normally found is also called an infection—and may lead to disease. For example, although large numbers of *E. coli* are normally present in the healthy intestine, their infection of the urinary tract usually results in disease.

Few microorganisms are pathogenic. In fact, the presence of some microorganisms can even benefit the host. Therefore, before we discuss the role of microorganisms in causing disease, let's examine the relationship of the microorganisms to the healthy human body.

NORMAL MICROBIOTA

LEARNING OBJECTIVE
• Define *normal* and *transient microbiota.*

Animals, including humans, are generally free of microbes in utero. At birth, however, normal and characteristic microbial populations begin to establish themselves. Just before a woman gives birth, lactobacilli in her vagina multiply rapidly. The newborn's first contact with microorganisms is usually with these lactobacilli, and they become the predominant organisms in the newborn's intestine. More microorganisms are introduced to the newborn's body from the environment when breathing begins and feeding starts. After birth, *E. coli* and other bacteria acquired from foods begin to inhabit the large intestine. These microorganisms remain there throughout life and, in response to altered environmental conditions, may increase or decrease in number and contribute to disease.

Many other usually harmless microorganisms establish themselves inside other parts of the normal adult body and on its surface. A typical human body contains 1×10^{13} body cells, yet harbors an estimated 1×10^{14} bacterial cells (10 times more bacterial cells than human cells). This gives you an idea of the abundance of microorganisms that normally reside in the human body. The microorganisms that establish more or less permanent residence (colonize) but that do not produce disease under normal conditions are members of the body's **normal microbiota,** or **normal flora** (Figure 14.1). Others, called **transient microbiota,** may be present for several days, weeks, or months and then disappear. Microorganisms are not found throughout the entire human body but are localized in certain regions, as shown in Table 14.1.

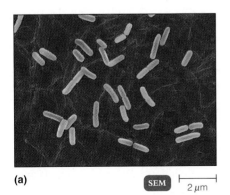
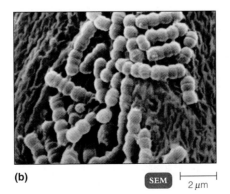
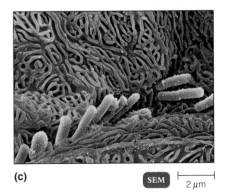

(a) SEM 2 μm   (b) SEM 2 μm   (c) SEM 2 μm

FIGURE 14.1 Representative normal microbiota for different regions of the body. (a) Bacteria on the surface of the skin. **(b)** Bacteria on the surface of the tongue. **(c)** Bacteria on the lining of the esophagus.

Q **Of what value are normal microbiota?**

TABLE 14.1	Representative Members of the Normal Microbiota by Body Region*	
Region	Principal Components	Comments
Skin	*Propionibacterium, Staphylococcus, Corynebacterium, Micrococcus, Acinetobacter, Brevibacterium; Pityrosporum* (fungus), *Candida* (fungus), *Malassezia* (fungus)	Most of the microbes in direct contact with skin do not become residents because secretions from sweat and oil glands have antimicrobial properties. Keratin is a resistant barrier, and the low pH of the skin inhibits many microbes. The skin also has a relatively low moisture content.
Eyes (conjunctiva)	*Staphylococcus epidermidis, S. aureus,* diphtheroids, *Propionibacterium, Corynebacterium,* streptococci, *Micrococcus*	The conjunctiva, a continuation of the skin or mucous membrane, contains basically the same microbiota found on the skin. Tears and blinking also eliminate some microbes or inhibit others from colonizing.

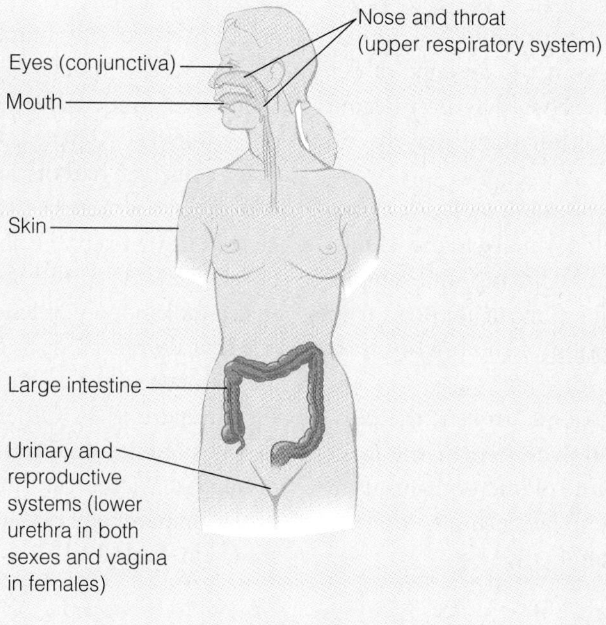

Nose and throat (upper respiratory system)

Eyes (conjunctiva)

Mouth

Skin

Large intestine

Urinary and reproductive systems (lower urethra in both sexes and vagina in females)

There are many factors that determine the distribution and composition of the normal microbiota. Among these are nutrients, physical and chemical factors, defenses of the host, and mechanical factors. Microbes vary with respect to the types of nutrients that they can use as an energy source. Accordingly, microbes can colonize only those body sites that can supply the appropriate nutrients. These nutrients may be derived from secretory and excretory products of cells, substances in body fluids and dead cells, and foods in the gastrointestinal tract.

A number of physical and chemical factors affect the growth of microbes and thus the growth and composition of the normal microbiota. Among these are temperature, pH, available oxygen and carbon dioxide, salinity, and sunlight.

You will learn in Chapters 16 and 17 that the human body has certain defenses against microbes. These defenses include a variety of molecules and activated cells that kill microbes, inhibit their growth, prevent their adhesion to host cell surfaces, and neutralize toxins that microbes produce. Although these defenses are extremely important against pathogens, their role in determining and regulating the normal microbiota is unclear.

Certain regions of the body are subjected to mechanical forces that may affect colonization by the normal microbiota. For example, the chewing actions of the teeth and tongue movements can dislodge microbes attached to tooth and mucosal surfaces. In the gastrointestinal tract, the flow of saliva and digestive secretions and the various

TABLE 14.1	*(continued)*	
Region	Principal Components	Comments
Nose and throat (upper respiratory system)	*Staphylococcus aureus, S. epidermidis,* and aerobic diphtheroids in the nose; *S. epidermidis, S. aureus,* diphtheroids, *Streptococcus pneumoniae, Haemophilus,* and *Neisseria* in the throat	Although some normal microbiota are potential pathogens, their ability to cause disease is reduced by microbial antagonism. Nasal secretions kill or inhibit many microbes, and mucus and ciliary action remove many microbes.
Mouth	*Streptococcus, Lactobacillus, Actinomyces, Bacteroides, Veillonella, Neisseria, Haemophilis, Fusobacterium, Treponema, Staphylococcus, Corynebacterium,* and *Candida* (fungus)	Abundant moisture, warmth, and the constant presence of food make the mouth an ideal environment that supports very large and diverse microbial populations on the tongue, cheeks, teeth, and gums. However, biting, chewing, tongue movements, and salivary flow dislodge microbes. Saliva contains several antimicrobial substances.
Large intestine	*Escherichia coli, Bacteroides, Fusobacterium, Lactobacillus, Enterococcus, Bifidobacterium, Enterobacter, Citrobacter, Proteus, Klebsiella, Candida* (fungus)	The large intestine contains the largest numbers of resident microbiota in the body because of its available moisture and nutrients. Mucus and periodic shedding of the lining prevent many microbes from attaching to the lining of the gastrointestinal tract, and the mucosa produces several antimicrobiol chemicals. Diarrhea also flushes out some of the normal microbiota.
Urinary and reproductive systems	*Staphylococcus, Micrococcus, Enterococcus, Lactobacillus, Bacteroides,* aerobic diphtheroids, *Pseudomonas, Klebsiella,* and *Proteus* in urethra; lactobacilli, aerobic diphtheroids, *Streptococcus, Staphylococcus, Bacteroides, Clostridium, Candida albicans* (fungus), and *Trichomonas vaginalis* (protozoan) in vagina	The lower urethra in both sexes has a resident population; the vagina has its acid-tolerant population of microbes because of the nature of its secretions. Mucus and periodic shedding of the lining prevent microbes from attaching to the lining; urine flow mechanically removes microbes, and the pH of urine and urea are antimicrobial. Cilia and mucus expel microbes from the cervix of the uterus into the vagina, and the acidity of the vagina inhibits or kills microbes.

muscular movements of the throat, esophagus, stomach, and intestines can remove unattached microbes. The flushing action of urine also removes unattached microbes. In the respiratory system, mucus traps microbes, which cilia then propel toward the throat for elimination.

Since there are tremendous variations among humans, the conditions provided by the host at a particular body site vary from one person to another. Among the factors that also affect the normal microbiota are age, nutritional status, diet, health status, disability, hospitalization, emotional state, stress, climate, geography, personal hygiene, living conditions, occupation, and lifestyle.

The principal normal microbiota in different regions of the body and some distinctive features of each region are listed in Table 14.1. Normal microbiota are also discussed more specifically in Part Four.

Animals with no microbiota whatsoever can be reared in the laboratory. Most germfree mammals used in research are obtained by breeding them in a sterile environment. Research with germfree animals has shown that microbes are not absolutely essential to animal life. On the other hand, this research has shown that germfree animals have undeveloped immunity systems and are unusually susceptible to infection and serious disease. Germfree animals also require more calories and vitamins than do normal animals.

RELATIONSHIPS BETWEEN THE NORMAL MICROBIOTA AND THE HOST

LEARNING OBJECTIVE

• Compare commensalism, mutualism, and parasitism, and give an example of each.

Once established, the normal microbiota can benefit the host by preventing the overgrowth of harmful microorganisms. This phenomenon is called **microbial antagonism,** or **competitive exclusion.** Microbial antagonism involves competition among microbes. One consequence of this competition is that the normal microbiota protect the host against colonization by potentially pathogenic microbes by competing for nutrients, producing substances harmful to the invading microbes, and affecting conditions such as pH and available oxygen. When this balance between normal microbiota and pathogenic microbes is upset, disease can result. For example, the normal bacterial microbiota of the adult human vagina maintains a local pH of about 4. The presence of normal microbiota inhibits the overgrowth of the yeast *Candida albicans*, which can grow when the balance between normal microbiota and pathogens is upset and when pH is altered. If the bacterial population is eliminated by antibiotics, excessive douching, or deodorants, the pH of the vagina reverts to nearly neutral, and *C. albicans* can flourish and become the dominant microorganism there. This condition can lead to a form of vaginitis (vaginal infection).

Another example of microbial antagonism is seen in the large intestine. *E. coli* cells produce *bacteriocins*, proteins that inhibit the growth of other bacteria of the same or closely related species, such as pathogenic *Salmonella* and *Shigella*. A bacterium that makes a particular bacteriocin is not killed by that bacteriocin but may be killed by other ones. Bacteriocins are used in medical microbiology to help identify different strains of bacteria. Such identification helps determine whether several outbreaks of an infectious disease are caused by one or more strains of a bacterium.

A final example involves another bacterium, *Clostridium difficile* (dif′-fi-sē-il), also in the large intestine. The normal microbiota of the large intestine effectively inhibit *C. difficile*, possibly by making host receptors unavailable, competing for available nutrients, or producing bacteriocins. However, if the normal microbiota are eliminated (for example, by antibiotics), *C. difficile* can become a problem. This microbe is responsible for nearly all gastrointestinal infections that follow antibiotic therapy, from mild diarrhea to severe or even fatal colitis (inflammation of the colon).

The relationship between the normal microbiota and the host is called **symbiosis,** which means living together

FIGURE 14.2 Symbiosis.

Q **Which type of symbiosis is best represented of the relationship between humans and *E. coli*?**

(Figure 14.2). In the symbiotic relationship called **commensalism,** one of the organisms is benefited, and the other is unaffected. Many of the microorganisms that make up our normal microbiota are commensals; these include the corynebacteria that inhabit the surface of the eye and certain saprophytic mycobacteria that inhabit the ear and external genitals. These bacteria live on secretions and sloughed-off cells, and they bring no apparent benefit or harm to the host.

Mutualism is a type of symbiosis that benefits both organisms. For example, the large intestine contains bacteria, such as *E. coli*, that synthesize vitamin K and some B vitamins. These vitamins are absorbed into the bloodstream and distributed for use by body cells. In exchange, the large intestine provides nutrients used by the bacteria, resulting in their survival.

Recent interest in the importance of bacteria to human health has led to the study of probiotics. **Probiotics** (*pro* = for, *bios* = life) are live microbial cultures applied to or ingested that are intended to exert a beneficial effect. Probiotics may be administered with *prebiotics*, which are chemicals that selectively promote the growth of beneficial bacteria. Several studies have shown that ingestion of certain lactic acid bacteria (LAB) can alleviate diarrhea and prevent colonization by *Salmonella enterica* during antibiotic therapy. If these LAB colonize the large intestine, the lactic acid and bacteriocins they produce can inhibit the growth of certain pathogens. Researchers are also testing the use of LAB to prevent surgical wound infections caused by *Staphylococcus aureus* and vaginal infections caused by *E. coli*. In a Stanford University study, HIV infection was reduced in women treated with a LAB that was genetically modified to produce CD4 protein that binds to HIV.

In still another kind of symbiosis, one organism is benefited by deriving nutrients at the expense of the other; this relationship is called **parasitism.** Many disease-causing bacteria are parasites.

OPPORTUNISTIC MICROORGANISMS

LEARNING OBJECTIVE
- Contrast normal and transient microbiota with opportunistic microorganisms.

Although categorizing symbiotic relationships by type is convenient, keep in mind that under certain conditions the relationship can change. For example, given the proper circumstances, a mutualistic organism, such as *E. coli*, can become harmful. *E. coli* is generally harmless as long as it remains in the large intestine; but if it gains access to other body sites, such as the urinary bladder, lungs, spinal cord, or wounds, it may cause urinary tract infections, pulmonary infections, meningitis, or abscesses, respectively. Microbes such as *E. coli* are called **opportunistic pathogens.** They ordinarily do not cause disease in their normal habitat in a healthy person but may do so in a different environment. For example, microbes that gain access through broken skin or mucous membranes can cause opportunistic infections. Or, if the host is already weakened or compromised by infection, microbes that are usually harmless can cause disease. AIDS is often accompanied by a common opportunistic infection, *Pneumocystis* pneumonia, caused by the opportunistic organism *Pneumocystis jiroveci* (formerly *Pneumocystis carinii*; see Figure 24.22, page 737). This secondary infection can develop in AIDS patients because their immune systems are suppressed. Before the AIDS epidemic, this type of pneumonia was rare. Opportunistic pathogens possess other features that contribute to their ability to cause disease. For example, they are present in or on the body or in the external environment in relatively large numbers. Some opportunistic pathogens may be found in locations in or on the body that are somewhat protected from the body's defenses, and some are resistant to antibiotics.

In addition to the usual symbionts, many people carry other microorganisms that are generally regarded as pathogenic but that may not cause disease in those people. Among the pathogens that are frequently carried in healthy individuals are echoviruses (*echo* comes from *enteric cytopathogenic human orphan*), which can cause intestinal diseases, and adenoviruses, which can cause respiratory diseases. *Neisseria meningitidis*, which often resides benignly in the respiratory tract, can cause meningitis, a disease that inflames the coverings of the brain and spinal cord. *Streptococcus pneumoniae*, a normal resident of the nose and throat, can cause a type of pneumonia.

COOPERATION AMONG MICROORGANISMS

In addition to competition among microbes in causing disease, there are situations in which cooperation among microbes is a factor in causing disease. One example of cooperation among microbes in the development of disease involves oral streptococci that colonize the teeth. Pathogens that cause periodontal disease and gingivitis have been found to have receptors for the streptococci, rather than for the teeth.

THE ETIOLOGY OF INFECTIOUS DISEASES

LEARNING OBJECTIVE
- List Koch's postulates.

Some diseases—such as polio, Lyme disease, and tuberculosis—have a well-known etiology. Some have an etiology that is not completely understood; for example, the relationship between certain viruses and cancer. For still others, such as Alzheimer's disease, the etiology is unknown. Of course, not all diseases are caused by microorganisms. For example, the disease hemophilia is an *inherited (genetic) disease*; osteoarthritis and cirrhosis are considered *degenerative diseases*. There are several other categories of disease, but here we will discuss only *infectious diseases,* those caused by microorganisms. To see how microbiologists determine the etiology of an infectious disease, we will discuss in greater detail the work of Robert Koch, which was introduced in Chapter 1 (page 11).

KOCH'S POSTULATES

In the historical overview of microbiology presented in Chapter 1, we briefly discussed Koch's famous postulates. Recall that Koch was a German physician who played a major role in establishing that microorganisms cause specific diseases. In 1877, he published some early papers on anthrax, a disease of cattle that can also occur in humans. Koch demonstrated that certain bacteria, today known as *Bacillus anthracis*, were always present in the blood of animals that had the disease and were not present in healthy animals. He knew that the mere presence of the bacteria did not prove that they had caused the disease; the bacteria could have been there as a result of the disease. Thus, he experimented further.

He took a sample of blood from a diseased animal and injected it into a healthy one. The second animal developed the same disease and died. He repeated this procedure many times, always with the same results. (A key criterion in the validity of any scientific proof is that experimental results be repeatable.) Koch also cultivated the microorganism in fluids outside the animal's body, and he demonstrated that the bacterium would cause anthrax even after many culture transfers.

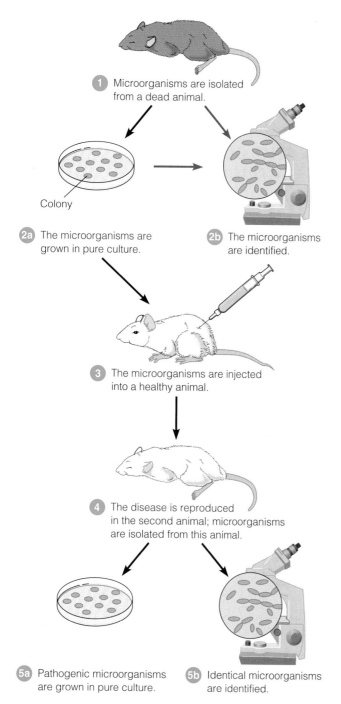

FIGURE 14.3 Application of Koch's postulates.

Q How do Koch's postulates prove the etiology of an infectious disease?

Koch showed that a specific infectious disease (anthrax) is caused by a specific microorganism (*B. anthracis*) that can be isolated and cultured on artificial media. He later used the same methods to show that the bacterium *Mycobacterium tuberculosis* is the causative agent of tuberculosis.

Koch's research provides a framework for the study of the etiology of any infectious disease. Today, we refer to Koch's experimental requirements as **Koch's postulates** (Figure 14.3). They are summarized as follows:

1. The same pathogen must be present in every case of the disease.

2. The pathogen must be isolated from the diseased host and grown in pure culture.

3. The pathogen from the pure culture must cause the disease when it is inoculated into a healthy, susceptible laboratory animal.

4. The pathogen must be isolated from the inoculated animal and must be shown to be the original organism.

EXCEPTIONS TO KOCH'S POSTULATES

Although Koch's postulates are useful in determining the causative agent of most bacterial diseases, there are some exceptions. For example, some microbes have unique culture requirements. The bacterium *Treponema pallidum* is known to cause syphilis, but virulent strains have never been cultured on artificial media. The causative agent of leprosy, *Mycobacterium leprae*, has also never been grown on artificial media. Moreover, many rickettsial and viral pathogens cannot be cultured on artificial media because they multiply only within cells.

The discovery of microorganisms that cannot grow on artificial media has necessitated some modifications of Koch's postulates and the use of alternative methods of culturing and detecting certain microbes. For example, when researchers looking for the microbial cause of legionellosis (Legionnaires' disease) were unable to isolate the microbe directly from a victim, they took the alternative step of inoculating a victim's lung tissue into guinea pigs. These guinea pigs developed the disease's pneumonia-like symptoms, whereas guinea pigs inoculated with tissue from an unafflicted person did not. Then tissue samples from the diseased guinea pigs were cultured in yolk sacs of chick embryos, a method (see Figure 13.7, page 395) that reveals the growth of extremely small microbes. After the embryos were incubated, electron microscopy revealed rod-shaped bacteria in the chick embryos. Finally, modern immunological techniques (which will be discussed in Chapter 18) were used to show that the bacteria in the chick embryos were the same bacteria as those in the guinea pigs and in afflicted humans.

In a number of situations, a human host exhibits certain signs and symptoms that are associated only with a certain pathogen and its disease. For example, the pathogens responsible for diphtheria and tetanus cause distinguishing signs and symptoms (described shortly) that can be produced by no other microbe. They are unequivocally the only organisms that produce their respective diseases. But some infectious diseases are not as clear-cut and provide another exception to Koch's postulates. For example, nephritis (inflammation of the kidneys) can involve any of several different pathogens, all of which cause the same signs and symptoms. (described shortly). Thus, it is often difficult to know which particular microorganism is causing a disease. Other infectious diseases that sometimes have poorly defined etiologies are pneumonia, meningitis, and peritonitis (inflammation of the peritoneum, the membrane that lines the abdomen and covers the organs within them).

Still another exception to Koch's postulates results because some pathogens can cause several disease conditions. *Mycobacterium tuberculosis*, for example, is implicated in diseases of the lungs, skin, bones, and internal organs. *Streptococcus pyogenes* can cause sore throat, scarlet fever, skin infections (such as erysipelas), and osteomyelitis (inflammation of bone), among other diseases. When clinical signs and symptoms are used together with laboratory methods, these infections can usually be distinguished from infections of the same organs by other pathogens.

Ethical considerations may also impose an exception to Koch's postulates. For example, some agents that cause disease in humans have no other known host. An example is human immunodeficiency virus (HIV), the cause of AIDS. This poses the ethical question of whether humans can be intentionally inoculated with infectious agents. In 1721, King George I told condemned prisoners they could be inoculated with smallpox to test a smallpox vaccine (see Chapter 18). He promised them their freedom if they lived. Human experiments with untreatable diseases are not acceptable today. Sometimes accidental inoculation does occur. A contaminated red bone marrow transplant satisfied the third Koch's postulate to prove that a herpesvirus caused cancer (see page 411).

CLASSIFYING INFECTIOUS DISEASES

LEARNING OBJECTIVE
- Differentiate a communicable from a noncommunicable disease.

Every disease that affects the body alters body structures and functions in particular ways, and these alterations are usually indicated by several kinds of evidence. For example, the patient may experience certain **symptoms,** or changes in body function, such as pain and *malaise* (a vague feeling of body discomfort). These *subjective* changes are not apparent to an observer. The patient can also exhibit **signs,** which are *objective* changes the physician can observe and measure. Frequently evaluated signs include lesions (changes produced in tissues by disease), swelling, fever, and paralysis. A specific group of symptoms or signs may always accompany a particular disease; such a group is called a **syndrome.** The diagnosis of a disease is made by evaluation of the signs and symptoms, together with the results of laboratory tests.

Diseases are often classified in terms of how they behave within a host and within a given population. Any disease that spreads from one host to another, either directly or indirectly, is said to be a **communicable disease.** Chickenpox, measles, genital herpes, typhoid fever, and tuberculosis are examples. Chickenpox and measles are also examples of **contagious diseases,** that is, diseases that are *easily* spread from one person to another. A **noncommunicable disease** is not spread from one host to another. These diseases are caused by microorganisms that normally inhabit the body and only occasionally produce disease or by microorganisms that reside outside the body and produce disease only when introduced into the body. An example is tetanus: *Clostridium tetani* produces disease only when it is introduced into the body via abrasions or wounds.

OCCURRENCE OF A DISEASE

LEARNING OBJECTIVE
- Categorize diseases according to frequency of occurrence.

To understand the full scope of a disease, we should know something about its occurrence. The **incidence** of a disease is the number of people in a population who develop a disease during a particular time period. It is an indicator of the spread of the disease. The **prevalence** of a disease is the number of people in a population who develop a disease at a specified time, regardless of when it first appeared. Prevalence takes into account both old and new cases. It's an indicator of how seriously and how long a disease affects a population. For example, the incidence of AIDS in the United States in 2004 was 40,000, while the prevalence in that same year was estimated to be about 900,000. Knowing the incidence and the prevalence of a disease in different populations (for example, in populations representing different geographic regions or different racial groups) enables scientists to estimate the range of the disease's occurrence and its tendency to affect some groups of people more than others.

Frequency of occurrence is another criterion that is used in the classification of diseases. If a particular disease occurs only occasionally, it is called a **sporadic disease;** typhoid fever in the United States is such a disease. A

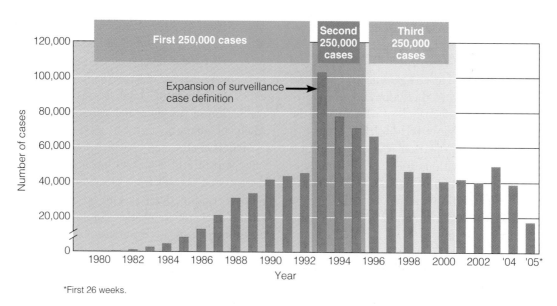

FIGURE 14.4 Reported AIDS cases in the United States. Notice that the first 250,000 cases occurred over a 12-year period, whereas the second and third 250,000 cases in this epidemic occurred in just 3 and 5 years, respectively. Much of the increase shown for 1993 is due to an expanded definition of AIDS cases adopted in that year. SOURCE: CDC.

Q **What was the incidence of AIDS in 2004?**

disease constantly present in a population is called an **endemic disease;** an example of such a disease is the common cold. If many people in a given area acquire a certain disease in a relatively short period, it is called an **epidemic disease;** influenza is an example of a disease that often achieves epidemic status. Figure 14.4 shows the epidemic incidence of AIDS in the United States. Some authorities consider gonorrhea and certain other sexually transmitted diseases to be epidemic at this time as well (see Figures 26.5 and 26.6, pages 790 and 791). An epidemic disease that occurs worldwide is called a **pandemic disease.** We experience pandemics of influenza from time to time. Some authorities also consider AIDS to be pandemic.

SEVERITY OR DURATION OF A DISEASE

LEARNING OBJECTIVES
- Categorize diseases according to severity.
- Define *herd immunity*.

Another useful way of defining the scope of a disease is in terms of its severity or duration. An **acute disease** is one that develops rapidly but lasts only a short time; a good example is influenza. A **chronic disease** develops more slowly, and the body's reactions may be less severe, but the disease is likely to be continual or recurrent for long periods. Infectious mononucleosis, tuberculosis, and hepatitis

B fall into this category. A disease that is intermediate between acute and chronic is described as a **subacute disease;** an example is subacute sclerosing panencephalitis, a rare brain disease characterized by diminished intellectual function and loss of nervous function. A **latent disease** is one in which the causative agent remains inactive for a time but then becomes active to produce symptoms of the disease; an example is shingles, one of the diseases caused by varicella-zoster virus.

The rate at which a disease, or an epidemic, spreads, and the number of individuals involved, are determined in part by the immunity of the population. Vaccination can provide long-lasting and sometimes lifelong protection of an individual against certain diseases. People who are immune to an infectious disease will not be carriers, thereby reducing the occurrence of the disease. Immune individuals act as a barrier to the spread of infectious agents. Even though a highly communicable disease may cause an epidemic, many nonimmune people will be protected because of the unlikelihood of their coming into contact with an infected person. A great advantage of vaccination is that enough individuals in a population will be protected from a disease to prevent its rapid spread to those in the population who are not vaccinated. When many immune people are present in a community, **herd immunity** exists.

EXTENT OF HOST INVOLVEMENT

Infections can also be classified according to the extent to which the host's body is affected. A **local infection** is one in which the invading microorganisms are limited to a relatively small area of the body. Some examples of local infections are boils and abscesses. In a **systemic (generalized) infection,** microorganisms or their products are spread throughout the body by the blood or lymph. Measles is an example of a systemic infection. Very often, agents of a local infection enter a blood or lymphatic vessel and spread to other specific parts of the body, where they are confined to specific areas of the body. This condition is called a **focal infection.** Focal infections can arise from infections in areas such as the teeth, tonsils, or sinuses.

Sepsis is a toxic inflammatory condition arising from the spread of microbes, especially bacteria or their toxins, from a focus of infection. **Septicemia,** also called blood poisoning, is a systemic infection arising from the multiplication of pathogens in the blood. Septicemia is a common example of sepsis. The presence of bacteria in the blood is known as **bacteremia. Toxemia** refers to the presence of toxins in blood (as occurs in tetanus), and **viremia** refers to the presence of viruses in blood.

The state of host resistance also determines the extent of infections. A **primary infection** is an acute infection that causes the initial illness. A **secondary infection** is one caused by an opportunistic pathogen after the primary infection has weakened the body's defenses. Secondary infections of the skin and respiratory tract are common and are sometimes more dangerous than the primary infections. *Pneumocystis* pneumonia as a consequence of AIDS is an example of a secondary infection; streptococcal bronchopneumonia following influenza is an example of a secondary infection that is more serious than the primary infection. A **subclinical (inapparent) infection** is one that does not cause any noticeable illness. Poliovirus and hepatitis A virus, for example, can be carried by people who never develop the illness.

PATTERNS OF DISEASE

A definite sequence of events usually occurs during infection and disease. As you will learn shortly, there must be a reservoir of infection as a source of pathogens for an infectious disease to occur. Next, the pathogen must be transmitted to a susceptible host by direct contact, by indirect contact, or by vectors. Transmission is followed by invasion, in which the microorganism enters the host and multiplies. Following invasion, the microorganism injures the host through a process called pathogenesis (discussed further in the next chapter). The extent of injury depends on the degree to which host cells are damaged, either directly

or by toxins. Despite the effects of all these factors, the occurrence of disease ultimately depends on the resistance of the host to the activities of the pathogen.

PREDISPOSING FACTORS

LEARNING OBJECTIVE
- Identify four predisposing factors for disease.

Certain predisposing factors also affect the occurrence of disease. A **predisposing factor** is one that makes the body more susceptible to a disease and may alter the course of the disease. Gender is sometimes a predisposing factor; for example, females have a higher incidence of urinary tract infections than males, whereas males have higher rates of pneumonia and meningitis. Other aspects of genetic background may play a role as well. For example, sickle cell disease is a severe, life-threatening form of anemia that occurs when the genes for the disease are inherited from both parents. Individuals who carry only one sickle cell gene have a condition called sickle cell trait and are normal unless specially tested. However, they are relatively resistant to the most serious form of malaria. The potential that individuals in a population might inherit a life-threatening disease is more than counterbalanced by protection from malaria among carriers of the gene for sickle cell trait. Of course, in countries where malaria is not present, sickle cell trait is an entirely negative condition.

Climate and weather seem to have some effect on the incidence of infectious diseases. In temperate regions, the incidence of respiratory diseases increases during the winter. This increase may be related to the fact that when people stay indoors, the closer contact with one other facilitates the spread of respiratory pathogens.

Other predisposing factors include inadequate nutrition, fatigue, age, environment, habits, lifestyle, occupation, preexisting illness, chemotherapy, and emotional disturbances. It is often difficult to know the exact relative importance of the various predisposing factors.

DEVELOPMENT OF DISEASE

LEARNING OBJECTIVE
- Put the following in proper sequence, according to the pattern of disease: period of decline, period of convalescence, period of illness, prodromal period, incubation period.

Once a microorganism overcomes the defenses of the host, development of the disease follows a certain sequence that tends to be similar whether the disease is acute or chronic (Figure 14.5).

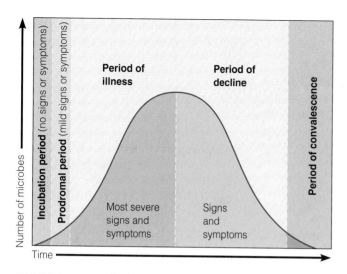

FIGURE 14.5 The stages of a disease.

Q During which periods can a disease be transmitted?

INCUBATION PERIOD

The **incubation period** is the time interval between the initial infection and the first appearance of any signs or symptoms. In some diseases, the incubation period is always the same; in others, it is quite variable. The time of incubation depends on the specific microorganism involved, its virulence (degree of pathogenicity), the number of infecting microorganisms, and the resistance of the host. (See Table 15.1, page 454, for the incubation periods of a number of microbial diseases.)

PRODROMAL PERIOD

The **prodromal period** is a relatively short period that follows the period of incubation in some diseases. The prodromal period is characterized by early, mild symptoms of disease, such as general aches and malaise.

PERIOD OF ILLNESS

During the **period of illness,** the disease is most severe. The person exhibits overt signs and symptoms of disease, such as fever, chills, muscle pain (myalgia), sensitivity to light (photophobia), sore throat (pharyngitis), lymph node enlargement (lymphadenopathy), and gastrointestinal disturbances. During the period of illness, the number of white blood cells may increase or decrease. Generally, the patient's immune response and other defense mechanisms overcome the pathogen, and the period of illness ends. When the disease is not successfully overcome (or successfully treated), the patient dies during this period.

PERIOD OF DECLINE

During the **period of decline,** the signs and symptoms subside. The fever decreases, and the feeling of malaise

diminishes. During this phase, which may take from less than 24 hours to several days, the patient is vulnerable to secondary infections.

PERIOD OF CONVALESCENCE

During the **period of convalescence,** the person regains strength and the body returns to its prediseased state. Recovery has occurred.

We all know that during the period of illness, people can serve as reservoirs of disease and can easily spread infections to other people. However, you should also know that people can spread infection during incubation and convalescence as well. This is especially true of diseases such as typhoid fever and cholera, in which the convalescing person carries the pathogenic microorganism for months or even years.

THE SPREAD OF INFECTION

Now that you have an understanding of normal microbiota, the etiology of infectious diseases, and the types of infectious diseases, we will examine the sources of pathogens and how diseases are transmitted.

RESERVOIRS OF INFECTION

> **LEARNING OBJECTIVES**
> • Define *reservoir of infection.*
> • Contrast human, animal, and nonliving reservoirs, and give one example of each.

For a disease to perpetuate itself, there must be a continual source of the disease organisms. This source can be either a living organism or an inanimate object that provides a pathogen with adequate conditions for survival and multiplication and an opportunity for transmission. Such a source is called a **reservoir of infection.** These reservoirs may be human, animal, or nonliving.

HUMAN RESERVOIRS

The principal living reservoir of human disease is the human body itself. Many people harbor pathogens and transmit them directly or indirectly to others. People with signs and symptoms of a disease may transmit the disease; in addition, some people can harbor pathogens and transmit them to others without exhibiting any signs of illness. These people, called **carriers,** are important living reservoirs of infection. Some carriers have inapparent infections for which no signs or symptoms are ever exhibited. Other people, such as those with latent diseases, carry a disease during its symptom-free stages—during the incubation period (before symptoms appear) or during the convalescent period (recovery). Typhoid Mary is an example of a carrier. Human carriers play an important role in the spread of such diseases as AIDS, diphtheria, typhoid fever, hepatitis, gonorrhea, amoebic dysentery, and streptococcal infections.

ANIMAL RESERVOIRS

Both wild and domestic animals are living reservoirs of microorganisms that can cause human diseases. Diseases that occur primarily in wild and domestic animals and can be transmitted to humans are called **zoonoses** (zō-ō-no′-sēz) (singular: *zoonosis*). Rabies (found in bats, skunks, foxes, dogs, and coyotes), and Lyme disease (found in field mice) are examples of zoonoses. Other representative zoonoses are presented in Table 14.2.

About 150 zoonoses are known. The transmission of zoonoses to humans can occur via one of many routes: by direct contact with infected animals; by direct contact with domestic pet waste (such as cleaning a litter box or bird cage); by contamination of food and water; by air from contaminated hides, fur, or feathers; by consuming infected animal products; or by insect vectors (insects that transmit pathogens).

NONLIVING RESERVOIRS

The two major nonliving reservoirs of infectious disease are soil and water. Soil harbors such pathogens as fungi, which cause mycoses such as ringworm and systemic infections; *Clostridium botulinum*, the bacterium that causes botulism; and *C. tetani*, the bacterium that causes tetanus. Because both species of clostridia are part of the normal intestinal microbiota of horses and cattle, the bacteria are found especially in soil where animal feces are used as fertilizer.

Water that has been contaminated by the feces of humans and other animals is a reservoir for several pathogens, notably those responsible for gastrointestinal diseases. These include *Vibrio cholerae*, which causes cholera, and *Salmonella typhi*, which causes typhoid fever. Other nonliving reservoirs include foods that are improperly prepared or stored. They may be sources of diseases such as trichinellosis and salmonellosis.

TRANSMISSION OF DISEASE

LEARNING OBJECTIVE
• Explain three methods of disease transmission.

The causative agents of disease can be transmitted from the reservoir of infection to a susceptible host by three principal routes: contact, vehicles, and vectors.

CONTACT TRANSMISSION

Contact transmission is the spread of an agent of disease by direct contact, indirect contact, or droplet transmission. **Direct contact transmission,** also known as *person-*

to-person transmission, is the direct transmission of an agent by physical contact between its source and a susceptible host; no intermediate object is involved (Figure 14.6a). The most common forms of direct contact transmission are touching, kissing, and sexual intercourse. Among the diseases that can be transmitted by direct contact are viral respiratory tract diseases (the common cold and influenza), staphylococcal infections, hepatitis A, measles, scarlet fever, and sexually transmitted diseases (syphilis, gonorrhea, and genital herpes). Direct contact is also one way of spreading AIDS and infectious mononucleosis. To guard against person-to-person transmission, health care workers use gloves and other protective measures (Figure 14.6b). Potential pathogens can also be transmitted by direct contact from animals (or animal products) to humans. Examples are the pathogens causing rabies and anthrax.

Indirect contact transmission occurs when the agent of disease is transmitted from its reservoir to a susceptible host by means of a nonliving object. The general term for any nonliving object involved in the spread of an infection is a **fomite.** Examples of fomites are tissues, handkerchiefs, towels, bedding, diapers, drinking cups, eating utensils, toys, money, and thermometers (Figure 14.6c). Contaminated syringes serve as fomites in the transmission of AIDS and hepatitis B. Other fomites may transmit diseases such as tetanus.

Droplet transmission is a third type of contact transmission in which microbes are spread in *droplet nuclei* (mucus droplets) that travel only short distances (Figure 14.6d). These droplets are discharged into the air by coughing, sneezing, laughing, or talking and travel less than 1 meter from the reservoir to the host. In one sneeze, 20,000 droplets may be produced. Disease agents that travel such short distances are not regarded as airborne (airborne transmission is discussed shortly). Examples of diseases spread by droplet transmission are influenza, pneumonia, and pertussis (whooping cough).

VEHICLE TRANSMISSION

Vehicle transmission is the transmission of disease agents by a medium, such as water, food, or air (Figure 14.7 on page 434). Other media include blood and other body fluids, drugs, and intravenous fluids. An outbreak of *Salmonella* infections caused by vehicle transmission is described in the box in Chapter 25 (page 756). Here we will discuss water, food, and air as vehicles of transmission.

In *waterborne transmission*, pathogens are usually spread by water contaminated with untreated or poorly treated sewage. Diseases transmitted via this route include cholera, waterborne shigellosis, and leptospirosis. In *foodborne transmission*, pathogens are generally transmitted in foods that are incompletely cooked, poorly refrigerated, or prepared

TABLE 14.2	Selected Zoonoses			
Disease	Causative Agent	Reservoir	Method of Transmission	Chapter Reference
Viral				
Influenza (some types)	*Influenzavirus*	Swine, birds *pig*	Direct contact	24
Rabies	*Lyssavirus*	Bats, skunks, foxes, dogs, racoons	Direct contact (bite)	22
Western equine encephalitis	*Alphavirus*	Horses, birds	*Culex* mosquito bite	22
Hantavirus pulmonary syndrome (HPS)	*Hantavirus*	Rodents (primarily deer mice)	Direct contact with rodent saliva, feces, or urine	23
Bacterial				
Anthrax	*Bacillus anthracis*	Domestic livestock	Direct contact with contaminated hides or animals; air; food	23
Brucellosis	*Brucella* spp.	Domestic livestock	Direct contact with contaminated milk, meat, or animals	23
Bubonic plague	*Yersinia pestis*	Rodents	Flea bites	23
Cat-scratch disease	*Bartonella henselae*	Domestic cats	Direct contact	23
Ehrlichiosis	*Ehrlichia* spp.	Deer, rodents	Tick bite	23
Leptospirosis	*Leptospira*	Wild mammals, domestic dogs and cats	Direct contact with urine, soil, water	26
Lyme disease	*Borrelia burgdorferi*	Field mice	Tick bites	23
Psittacosis (ornithosis)	*Chlamydia psittaci*	Birds, especially parrots	Direct contact	24
Rocky Mountain spotted fever	*Rickettsia rickettsii*	Rodents	Tick bites	23
Salmonellosis	*Salmonella* spp.	Poultry, rats, reptiles	Ingestion of contaminated food and water and putting hands in mouth	25
Endemic typhus	*Rickettsia typhi*	Rodents	Flea bites	23
Fungal				
Ringworm	*Trichophyton Microsporum Epidermophyton*	Domestic mammals	Direct contact; fomites (nonliving objects)	21
Protozoan				
Malaria	*Plasmodium* spp.	Monkeys	Anopheles mosquito bite	23
Toxoplasmosis	*Toxoplasma gondii*	Cats and other mammals	Ingestion of contaminated meat or by direct contact with infected tissues or fecal matter	23
Helminthic				
Tapeworm (pork)	*Taenia solium*	Pigs	Ingestion of undercooked contaminated pork	25
Trichinellosis	*Trichinella spiralis*	Pigs, bears	Ingestion of undercooked contaminated pork	25

(a)

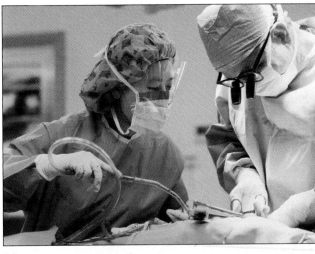

(b)

(d)

(c)

FIGURE 14.6 Contact transmission. (a) Direct contact transmission. **(b)** Preventing direct contact transmission. **(c)** Indirect contact transmission. **(d)** Droplet transmission.

Q **Name a disease transmitted by direct contact, a disease transmitted by indirect contact, and a disease transmitted by droplet transmission.**

under unsanitary conditions. Foodborne pathogens cause diseases such as food poisoning and tapeworm infestation.

Airborne transmission refers to the spread of agents of infection by droplet nuclei in dust that travel more than 1 meter from the reservoir to the host. For example, microbes are spread by droplets, which may be discharged in a fine spray from the mouth and nose during coughing and sneezing (see Figure 14.6d). These droplets are small enough to remain airborne for prolonged periods. The virus that causes measles and the bacterium that causes tuberculosis can be transmitted via airborne droplets. Dust particles can harbor various pathogens. Staphylococci and streptococci can survive on dust and be transmitted by the airborne route. Spores produced by certain fungi are also

transmitted by the airborne route and can cause such diseases as histoplasmosis, coccidioidomycosis, and blastomycosis (see Chapter 24).

VECTORS

Arthropods are the most important group of disease **vectors**—animals that carry pathogens from one host to another. (Insects and other arthropod vectors are discussed in Chapter 12, page 377.) Arthropod vectors transmit disease by two general methods. **Mechanical transmission** is the passive transport of the pathogens on the insect's feet or other body parts (Figure 14.8). If the insect makes contact with a host's food, pathogens can be transferred to the food and later swallowed by the host. Houseflies, for

(a)

(b)

(c)

FIGURE 14.7 Vehicle transmission. (**a**) Water. (**b**) Food. (**c**) Air.

Q How does vehicle transmission differ from contact transmission.

arthropod; these can be passed with feces. If the arthropod defecates or vomits while biting a potential host, the parasite can enter the wound. Other parasites reproduce in the vector's gut and migrate to the salivary gland; these are directly injected into a bite. Some protozoan and helminthic parasites use the vector as a host for a developmental stage in their life cycle.

Table 14.3 lists a few important arthropod vectors and the diseases they transmit.

FIGURE 14.8 Vectors of disease. Mechanical transmission.

Q How do mechanical and biological transmission by vectors differ?

instance, can transfer the pathogens of typhoid fever and bacillary dysentery (shigellosis) from the feces of infected people to food.

Biological transmission is an active process and is more complex. The arthropod bites an infected person or animal and ingests some of the infected blood (see Figure 12.31, page 379). The pathogens then reproduce in the vector, and the increase in the number of pathogens increases the possibility that they will be transmitted to another host. Some parasites reproduce in the gut of the

TABLE 14.3	Representative Arthropod Vectors and the Diseases They Transmit		
Disease	Causative Agent	Arthropod Vector	Chapter Reference
Malaria	*Plasmodium* spp.	*Anopheles* (mosquito)	23
African trypanosomiasis	*Trypanosoma brucei gambiense* and *T. b. rhodesiense*	*Glossina* sp. (tsetse fly)	22
Chagas' disease	*T. cruzi*	*Triatoma* sp. (kissing bug)	23
Yellow fever	*Alphavirus* (yellow fever virus)	*Aedes* (mosquito)	23
Dengue	*Alphavirus* (dengue fever virus)	*A. aegypti* (mosquito)	12, 23
Arthropod-borne encephalitis	*Alphavirus* (encephalitis virus)	*Culex* (mosquito)	22
Ehrlichiosis	*Ehrlichia* spp.	*Ixodes* spp. (tick)	23
Epidemic typhus	*Rickettsia prowazekii*	*Pediculus humanus* (louse)	23
Endemic murine typhus	*R. typhi*	*Xenopsylla cheopis* (rat flea)	23
Rocky Mountain spotted fever	*R. rickettsii*	*Dermacentor andersoni* and other species (tick)	23
Plague	*Yersinia pestis*	*Xenopsylla cheopis* (rat flea)	23
Relapsing fever	*Borrelia* spp.	*Ornithodorus* spp. (soft tick)	23
Lyme disease	*B. burgdorferi*	*Ixodes* spp. (tick)	23

NOSOCOMIAL (HOSPITAL-ACQUIRED) INFECTIONS

LEARNING OBJECTIVE

- Define *nosocomial infections,* and explain their importance.

A **nosocomial** (nōs-ō-kō′mē-al) **infection** is one that does not show any evidence of being present or incubating at the time of admission to a hospital; it is acquired as a result of a hospital stay. (The word *nosocomial* is derived from the Greek word for hospital; the term also includes infections acquired in nursing homes and other health care facilities.) The Centers for Disease Control and Prevention (CDC) estimates that 5 to 15% of all hospital patients acquire some type of nosocomial infection. The work of pioneers in aseptic techniques such as Lister and Semmelweis (Chapter 1, page 11) decreased the rate of nosocomial infections considerably. However, despite modern advances in sterilization techniques and disposable materials, the rate of nosocomial infections has increased 36% during the last 20 years. In the United States, about 2 million people per year contract nosocomial infections, and nearly 20,000 die as a result. Nosocomial infections represent the eighth leading cause of death in the United States behind heart disease, cancer, and strokes.

Nosocomial infections result from the interaction of several factors: (1) microorganisms in the hospital environment, (2) the compromised (or weakened) status of the host, and (3) the chain of transmission in the hospital. Figure 14.9 illustrates that the presence of any one of these

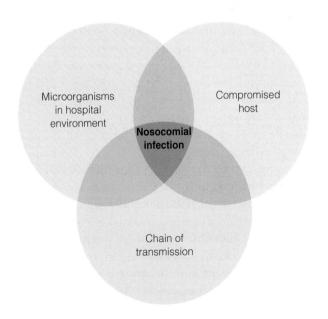

FIGURE 14.9 Nosocomial infections.

Q *What is a nosocomial infection?*

TABLE 14.4	Microorganisms Involved in Most Nosocomial Infections		
Microorganism	Percentage of Total Infections	Percentage Resistant to Antibiotics	Infections Caused
Coagulase-negative staphylococci	25%	89%	Most common cause of sepsis
Staphylococcus aureus	16%	60%	Most frequent cause of pneumonia
Enterococcus	10%	29%	Most common cause of surgical wound infections
Escherichia coli, Pseudomonas aeruginosa, Enterobacter, and *Klebsiella pneumoniae*	23%	5–32%	Pneumonia and surgical wound infections
Clostridium difficile	13%	—	Causes nearly half of all nosocomial diarrhea
Fungi (mostly *Candida albicans*)	6%	—	Urinary tract infections and sepsis
Other gram-negative bacteria (*Acinetobacter, Citrobacter, Haemophilus*)	7%	—	Urinary tract infections and surgical wound infections

Source: Data from CDC, National Nosocomial Infections Surveillance.

factors alone is generally not enough to cause infection; it is the interaction of all three factors that poses a significant risk of nosocomial infection.

MICROORGANISMS IN THE HOSPITAL

Although every effort is made to kill or check the growth of microorganisms in the hospital, the hospital environment is a major reservoir for a variety of pathogens. One reason is that certain normal microbiota of the human body are opportunistic and present a particularly strong danger to hospital patients. In fact, most of the microbes that cause nosocomial infections do not cause disease in healthy people but are pathogenic only for individuals whose defenses have been weakened by illness or therapy (see the boxes on pages 203 and 445).

In the 1940s and 1950s, most nosocomial infections were caused by gram-positive microbes. At one time, the gram-positive *Staphylococcus aureus* was the primary cause of nosocomial infections. In the 1970s, gram-negative rods, such as *E. coli* and *Pseudomonas aeruginosa,* were the most common causes of nosocomial infections. Then, during the 1980s, antibiotic-resistant gram-positive bacteria, *Staphylococcus aureus,* coagulase-negative staphylococci, and *Enterococcus* spp., emerged as nosocomial pathogens. By the 1990s, these gram-positive bacteria accounted for 34% of nosocomial infections and four gram-negative pathogens accounted for 32%. In the 2000s, antibiotic resistance in nosocomial infections is a major concern. The

principal microorganisms involved in nosocomial infections are summarized in Table 14.4.

In addition to being opportunistic, some microorganisms in the hospital become resistant to antimicrobial drugs, which are commonly used there. For example, *P. aeruginosa* and other such gram-negative bacteria tend to be difficult to control with antibiotics because of their R factors, which carry genes that determine resistance to antibiotics (see Chapter 8, page 246). As the R factors recombine, new and multiple resistance factors are produced. These strains become part of the microbiota of patients and hospital personnel and become progressively more resistant to antibiotic therapy. In this way, people become part of the reservoir (and chain of transmission) for antibiotic-resistant strains of bacteria. Usually, if the host's resistance is high, the new strains are not much of a problem. However, if disease, surgery, or trauma has weakened the host's defenses, secondary infections may be difficult to treat.

COMPROMISED HOST

LEARNING OBJECTIVE
• Define *compromised host.*

A **compromised host** is one whose resistance to infection is impaired by disease, therapy, or burns. Two principal conditions can compromise the host: broken skin or mucous membranes, and a suppressed immune system.

TABLE 14.5	Principal Sites of Nosocomial Infections
Type of Infection	Comment
Urinary tract infections	Most common, usually accounts for about 40% of all nosocomial infections. Typically related to urinary catheterization.
Surgical site infections	Ranks second in infection incidence (about 20%). An estimated 5–12% of all surgical patients develop postoperative infections; the percentage can reach 30% for certain surgeries, such as colon surgery and amputations.
Lower respiratory infections	Nosocomial pneumonias account for about 15% and have high mortality rates (13–55%). Most of these pneumonias are related to respiratory devices that aid breathing or administer medications.
Cutaneous infections	Cutaneous infections account for about 8% of nosocomial infections. Newborns have a high rate of susceptibility to skin and eye infections.
Bacteremia, caused primarily by intravenous catheterizations	Bacteremias account for about 6% of nosocomial infections. Intravenous catheterization is implicated in nosocomial infections of the bloodstream, particularly infections caused by bacteria and fungi.
Other	All other infection sites account for about 11% of nosocomial infections.

Source: Data from CDC, National Nosocomial Infection Surveillance.

- Urinary tract infections
- Surgical site infections
- Lower respiratory infections
- Bacteremia caused primarily by IV catheterizations
- Cutaneous infections
- Other

As long as the skin and mucous membranes remain intact, they provide formidable physical barriers against most pathogens. Burns, surgical wounds, trauma (such as accidental wounds), injections, invasive diagnostic procedures, ventilators, intravenous therapy, and urinary catheters (used to drain urine) can all break the first line of defense and make a person more susceptible to disease in hospitals. Burn patients are especially susceptible to nosocomial infections because their skin is no longer an effective barrier to microorganisms.

The risk of infection is also related to other invasive procedures, such as administering anesthesia, which may alter breathing and contribute to pneumonia, and tracheotomy, in which an incision is made into the trachea to assist breathing. Patients who require invasive procedures usually have a serious underlying disease, which further increases susceptibility to infections. Invasive devices provide a pathway for microorganisms in the environment to enter the body; they also help transfer microbes from one part of the body to another. Pathogens can also proliferate on the devices themselves.

In healthy individuals, white blood cells called T cells (T lymphocytes) provide resistance to disease by killing pathogens directly, mobilizing phagocytes and other lymphocytes, and secreting chemicals that kill pathogens. White blood cells called B cells (B lymphocytes), which develop into antibody-producing cells, also protect against infection. Antibodies provide immunity by such actions as neutralizing toxins, inhibiting the attachment of a pathogen to host cells, and helping to lyse pathogens. Drugs, radiation therapy, steroid therapy, burns, diabetes, leukemia, kidney disease, stress, and malnutrition can all adversely affect the actions of T and B cells and compromise the host. In addition, the AIDS virus destroys certain T cells.

A summary of the principal sites of nosocomial infections is presented in Table 14.5.

CHAIN OF TRANSMISSION

Given the variety of pathogens (and potential pathogens) in the hospital and the compromised state of the host, routes of transmission are a constant concern. The principal routes of transmission of nosocomial infections are 1) direct contact transmission from hospital staff to patient and from patient to patient and 2) indirect contact transmission through fomites and the hospital's ventilation system (airborne transmission).

Because hospital personnel are in direct contact with patients, they can often transmit disease. For example, a physician or nurse may transmit microbiota to a patient when changing a dressing, or a kitchen worker who carries *Salmonella* can contaminate a food supply.

Certain areas of a hospital are reserved for specialized care; these include the burn, hemodialysis, recovery, intensive care, and oncology units. Unfortunately, these units also group patients together and provide environments for the epidemic spread of nosocomial infections from patient to patient.

Many diagnostic and therapeutic hospital procedures provide a fomite route of transmission. The urinary catheter used to drain urine from the urinary bladder is a fomite in many nosocomial infections. Intravenous catheters, which pass through the skin and into a vein to provide fluids, nutrients, or medication, can also transmit nosocomial infections. Respiratory aids can introduce contaminated fluids into the lungs. Needles may introduce pathogens into muscle or blood, and surgical dressings can become contaminated and promote disease.

CONTROL OF NOSOCOMIAL INFECTIONS

LEARNING OBJECTIVES

- List several methods of disease transmission in hospitals.
- Explain how nosocomial infections can be prevented.

Control measures aimed at preventing nosocomial infections vary from one institution to another, but certain procedures are generally implemented. It is important to reduce the number of pathogens to which patients are exposed by using aseptic techniques, handling contaminated materials carefully, insisting on frequent and thorough hand-washing, educating staff members about basic infection control measures, and using isolation rooms and wards.

According to the CDC, hand-washing is the single most important means of preventing the spread of infection. Nevertheless, the CDC reports that adherence of health care workers to recommended hand-washing procedures has been poor. On average, health care workers wash their hands before interacting with patients only 40% of the time.

In addition to hand-washing, tubs used to bathe patients should be disinfected between uses so that bacteria from the previous patient will not contaminate the next one. Respirators and humidifiers provide both a suitable growth environment for some bacteria and a method of airborne transmission. These sources of nosocomial infections must be kept scrupulously clean and disinfected, and materials used for bandages and intubation (insertion of tubes into organs, such as the trachea) should be single-use disposable or sterilized before use. Packaging used to maintain sterility should be removed aseptically. Physicians can help improve patients' resistance to infection by prescribing antibiotics only when necessary, avoiding invasive procedures if possible, and minimizing the use of immunosuppressive drugs.

Accredited hospitals should have an infection control committee. Most hospitals have at least an infection control nurse or epidemiologist (an individual who studies disease in populations). The role of these staff members is to identify problem sources, such as antibiotic-resistant strains of bacteria and improper sterilization techniques. The infection control officer should make periodic examinations of hospital equipment to determine the extent of microbial contamination. Samples should be taken from tubing, catheters, respirator reservoirs, and other equipment.

EMERGING INFECTIOUS DISEASES

LEARNING OBJECTIVE

- List several probable reasons for emerging infectious diseases, and name one example for each reason.

As noted in Chapter 1, **emerging infectious diseases (EIDs)** are ones that are new or changing, showing an increase in incidence in the recent past, or a potential to increase in the near future. An emerging disease can be caused by a virus, a bacterium, a fungus, a protozoan, or a helminth. Several criteria are used for identifying an EID. For example, some diseases present symptoms that are clearly distinctive from all other diseases. Some are recognized because improved diagnostic techniques allow the identification of a new pathogen. Others are identified when a local disease becomes widespread, a rare disease becomes common, a mild disease becomes more severe, or an increase in life span permits a slow disease to develop. Examples of emerging infectious diseases are listed in Table 14.6 and described in the boxes in Chapters 13 and 26 (pages 406 and 793).

A variety of factors contribute to the emergence of new infectious diseases:

- New strains, such as *E. coli* O157:H7 and avian influenza (H5N1), may result from genetic recombination between organisms.

- A new serovar, such as *Vibrio cholerae* O139, may result from changes in or the evolution of existing microorganisms.

- The widespread, and sometimes unwarranted, use of antibiotics and pesticides encourages the growth of more resistant populations of microbes and the insects (mosquitoes and lice) and ticks that carry them.

- Global warming and changes in weather patterns may increase the distribution and survival of reservoirs and vectors, resulting in the introduction and dissemination of diseases such as malaria and *Hantavirus* pulmonary syndrome.

| TABLE 14.6 | Emerging Infectious Diseases | | |

Microorganism	Year of Emergence	Disease Caused	Chapter Reference
Bacteria			
Bacillus anthracis	2001	Anthrax	23
Bordetella pertussis	2000	Whooping cough	24
Methicillin-resistant *Staphylococcus aureus*	1997	Bacteremia, pneumonia	20
Vancomycin-resistant *Staphylococcus aureus*	1996	Bacteremia, pneumonia	20
Streptococcus pneumoniae	1995	Antibiotic-resistant pneumonia	24
Streptococcus pyogenes	1995	Streptococcal toxic shock syndrome	21
Corynebacterium diphtheriae	1994	Diphtheria epidemic, eastern Europe	24
Vibrio cholerae O139	1992	New serovar of cholera, Asia	25
Vancomycin-resistant enterococci	1988	Urinary tract infections, bacteremia, endocarditis	26 23
Bartonella henselae	1983	Cat-scratch disease	23
Escherichia coli O157:H7	1982	Hemorrhagic diarrhea	25
Legionella pneumophila	1976	Legionellosis (Legionnaires' disease)	24
Borrelia burgdorferi	1975	Lyme disease	23
Fungi			
Coccidioides immitis	1993	Coccidioidomycosis	24
Pneumocystis jiroveci	1981	Pneumonia in immunocompromised patients	24
Protozoa			
Plasmodium	1986	Malaria in United States	23
Cyclospora cayetanensis	1993	Severe diarrhea and wasting syndrome	25
Cryptosporidium spp.	1976	Cryptosporidiosis	25
Helminths			
Baylisascaris procyonis	2001	Raccoon roundworm encephalitis, in humans	
Viruses			
SARS-associated coronavirus	2002	Severe acute respiratory syndrome (SARS)	24
Ebola virus	2002, 1995, 1975	Ebola hemorrhagic fever	23
West Nile virus	1999	West Nile encephalitis	22
Nipah virus	1998	Encephalitis, Malaysia	22
Avian influenza A (H5N1) virus	1997	Avian influenza	24
Hendra virus	1994	Encephalitis-like symptoms, Australia	24
Hantavirus	1993	*Hantavirus* pulmonary syndrome	23
Venezuelan hemorrhagic fever	1991	Hemorrhagic fever, South America	23

TABLE 14.6	Emerging Infectious Diseases *(continued)*		
Microorganism	Year of Emergence	Disease Caused	Chapter Reference
Hepatitis E virus	1990	Hepatitis	25
Hepatitis C virus	1989	Hepatitis	25
Dengue fever virus	1984	Dengue fever and dengue hemorrhagic fever, South and Central America and the Caribbean	23
HIV	1983	AIDS	19
Prions			
Bovine spongiform encephalitis agent	1996	Mad cow disease, Great Britain	22

- Known diseases, such as cholera and West Nile virus, may spread to new geographic areas by modern transportation. This was less likely 100 years ago when travel took so long that infected travelers either died or recovered during passage.

- Previously unrecognized infections may appear in individuals living or working in regions undergoing ecological changes brought about by natural disaster, construction, wars, and expanding human settlement. In California, the incidence of coccidioidomycosis increased tenfold following the Northridge earthquake of 1994. Workers clearing South American forests are now contracting Venezuelan hemorrhagic fever.

- Even animal control measures may affect the incidence of a disease. The increase in Lyme disease in recent years could be due to rising deer populations resulting from the killing of deer predators.

- Failures in public health measures may be a contributing factor to the emergence of previously controlled infections. For example, the failure of adults to get a diphtheria booster vaccination led to a diphtheria epidemic in the newly independent republics of the former Soviet Union in the 1990s.

The CDC, the National Institutes of Health (NIH), and the World Health Organization (WHO) have developed plans to address issues relating to emerging infectious diseases. Their priorities include the following:

1. To detect, promptly investigate, and monitor emerging infectious pathogens, the diseases they cause, and factors that influence their emergence

2. To expand basic and applied research on ecological and environmental factors, microbial changes and adaptations, and host interactions that influence EIDs

3. To enhance the communication of public health information and the prompt implementation of prevention strategies regarding EIDs

4. To establish plans to monitor and control EIDs worldwide

The importance of emerging infectious diseases to the scientific community resulted in a new publication, *Emerging Infectious Diseases*, devoted exclusively to the topic. It was first published in January 1995.

EPIDEMIOLOGY

LEARNING OBJECTIVE

- Define *epidemiology*, and describe three types of epidemiologic investigations.

In today's crowded, overpopulated world, in which frequent travel and the mass production and distribution of food and other goods are a way of life, diseases can spread rapidly. A contaminated food or water supply, for example, can affect many thousands of people very quickly. Identifying the causative agent is desirable so that a disease can be effectively controlled and treated. It is also desirable to understand the mode of transmission and geographical distribution of the disease. The science that studies when and where diseases occur and how they are transmitted in populations is called **epidemiology** (ep-i-dē-mē-ol'ō-jē).

Modern epidemiology began in the mid-1800s with three now-famous investigations. John Snow, a British physician, conducted a series of investigations related to outbreaks of cholera in London. As the cholera epidemic of 1848 to 1849 raged, Snow analyzed the death records attributed to cholera, gathered information about the victims, and interviewed survivors who lived in the neighborhood. Using the information he compiled, Snow made a map showing that most individuals who died of cholera drank or brought water from the Broad Street pump; those who used other pumps (or drank beer, like the workers at a nearby brewery) did not get cholera. He concluded that contaminated water from the Broad Street pump was the source of the epidemic. When the pump's handle was removed and people could no longer get water from this location, the number of cholera cases dropped significantly.

Between 1846 and 1848, Ignaz Semmelweis meticulously recorded the number of births and maternal deaths at Vienna General Hospital. The First Maternity Clinic had become a source of gossip throughout Vienna because the death rate due to puerperal sepsis ranged between 13% and 18%, four times that of the Second Maternity Clinic. Puerperal sepsis (childbirth fever) is a nosocomial infection that begins in the uterus as a result of childbirth or abortion. It is frequently caused by *Streptococcus pyogenes*. The infection progresses to the abdominal cavity (peritonitis) and in many cases to septicemia (proliferation of microbes in the blood). Wealthy women did not go to the clinic, and poor women had learned they had a better chance of surviving childbirth if they gave birth elsewhere before going to the hospital. Looking at his data, Semmelweis realized that wealthy women and the poor women who had given birth prior to entering the clinic were not examined by the medical students, who had spent their mornings dissecting cadavers. In May 1847, he ordered all medical students to wash their hands with chloride of lime before entering the delivery room, and the mortality rate dropped to under 2%.

Florence Nightingale recorded statistics on epidemic typhus in the English civilian and military populations. In 1858, she published a thousand-page report using statistical comparisons to demonstrate that diseases, poor food, and unsanitary conditions were killing the soldiers. Her work resulted in reforms in the British Army and to her admission to the Statistical Society, their first female member.

These three careful analyses of where and when a disease occurs and how it is transmitted within a population constituted a new approach to medical research and demonstrated the importance of epidemiology. The works of Snow, Semmelweis, and Nightingale resulted in changes that lowered the incidence of diseases even though there was limited knowledge of the causes of infectious disease. Most physicians believed that the symptoms they saw were the causes of the disease, not the result of disease. Koch's work on the germ theory of disease was still 30 years in the future.

An epidemiologist not only determines the etiology of a disease but also identifies other possibly important factors and patterns concerning the people affected. An important part of the epidemiologist's work is assembling and analyzing such data as age, sex, occupation, personal habits,

socioeconomic status, history of immunization, presence of any other diseases, and the common history of affected individuals (such as eating the same food or visiting the same doctor's office). Also important for the prevention of future outbreaks is knowledge of the site at which a susceptible host came into contact with the agent of infection. In addition, the epidemiologist considers the period during which the disease occurs, either on a seasonal basis (to indicate whether the disease is prevalent during the summer or winter) or on a yearly basis (to indicate the effects of immunization or an emerging or reemerging disease).

An epidemiologist is also concerned with various methods for controlling a disease. The strategies controlling diseases include the use of drugs (chemotherapy) and vaccines (immunization). Other methods include the control of human, animal, and nonliving reservoirs of infection, water treatment, proper sewage disposal (enteric diseases), cold storage, pasteurization, food inspection, adequate cooking (foodborne diseases), improved nutrition to bolster host defenses, changes in personal habits, and screening of transfused blood and transplanted organs.

Figure 14.10 contains graphs indicating the incidence of selected diseases. Such graphs provide information about whether disease outbreaks are sporadic or epidemic and, if epidemic, how the disease might have spread. By establishing the frequency of a disease in a population and identifying the factors responsible for its transmission, an epidemiologist can provide physicians with information that is important in determining the prognosis and treatment of a disease. Epidemiologists also evaluate how effectively a disease is being controlled in a community—by a vaccination program, for example. Finally, epidemiologists can provide data to help in evaluating and planning overall health care for a community.

Epidemiologists use three basic types of investigations when analyzing the occurrence of a disease: descriptive, analytical, and experimental. ❖ **Animation: Go to The Microbiology Place website or CD-ROM and click "Animations" to view Epidemiology.**

DESCRIPTIVE EPIDEMIOLOGY

Descriptive epidemiology entails collecting all data that describe the occurrence of the disease under study. Relevant information usually includes information about the affected individuals and the place and period in which the disease occurred. Snow's search for the cause of the cholera outbreak in London is an example of descriptive epidemiology.

Such a study is generally *retrospective* (looking backward after the episode has ended). In other words, the epidemiologist backtracks to the cause and source of the disease (see the Clinical Problem Solving boxes in Chapters 21, 23, 25, and 26). The search for the cause of toxic

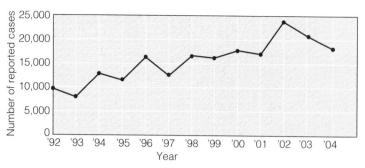

(a) Lyme disease cases, 1992 through 2004

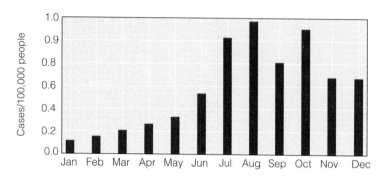

(b) Lyme disease by month, 2004

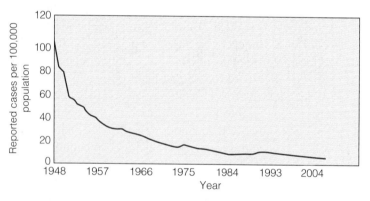

(c) Reported tuberculosis cases, 1948 through 2004

FIGURE 14.10 Epidemiological graphs.
(a) Lyme disease cases, showing the annual occurrence of the disease during the covered period. **(b)** A different perspective of Lyme disease that enabled epidemiologists to begin drawing some conclusions about the disease's epidemiology. This graph records the number of cases per 100,000 people, rather than the total number of cases. **(c)** A graph of the incidence of tuberculosis shows a rapid decrease in the rate of decline from 1948 to 1957. Source: Data from CDC.

 What does graph b indicate about transmission of Lyme disease? What can you conclude from graph c?

shock syndrome is an example of a fairly recent retrospective study. In the initial phase of an epidemiological study, retrospective studies are more common than *prospective* (looking forward) studies, in which an epidemiologist chooses a group of people who are free of a particular disease to study. The group's subsequent disease experiences are then recorded for a given period. Prospective studies were used to test the Salk polio vaccine in 1954 and 1955.

ANALYTICAL EPIDEMIOLOGY

Analytical epidemiology analyzes a particular disease to determine its probable cause. This study can be done in two ways. With the *case control method*, the epidemiologist looks for factors that might have preceded the disease. A

group of people who have the disease is compared with another group of people who are free of the disease. For example, one group with meningitis and one without the disease might be matched by age, sex, socioeconomic status, and location. These statistics are compared to determine which of all the possible factors—genetic, environmental, nutritional, and so forth—might be responsible for the meningitis. Nightingale's work was an example of analytical epidemiology, in which she compared disease in soldiers and civilians. With the *cohort method*, the epidemiologist studies two populations: one that has had contact with the agent causing a disease and another that has not (both groups are called *cohort groups*). For example, a comparison of one group composed of people who have received blood transfusions and one composed of people who have not could reveal an association between blood transfusions and the incidence of hepatitis B virus.

EXPERIMENTAL EPIDEMIOLOGY

Experimental epidemiology begins with a hypothesis about a particular disease; experiments to test the hypothesis are then conducted with a group of people. One such hypothesis could be the assumed effectiveness of a drug. A group of infected individuals is selected and divided randomly so that some receive the drug and others receive a *placebo*, a substance that has no effect. If all other factors are kept constant between the two groups, and if those people who received the drug recover more rapidly than those who received the placebo, it can be concluded that the drug was the experimental factor (variable) that made the difference.

CASE REPORTING

We noted earlier in this chapter that establishing the chain of transmission for a disease is extremely important. Once known, the chain can be interrupted in order to slow down or stop the spread of the disease.

An effective way to establish the chain of transmission is *case reporting*, a procedure that requires health care workers to report specified diseases to local, state, and national health officials. Examples of such diseases are AIDS, measles, gonorrhea, tetanus, and typhoid fever. Case reporting provides epidemiologists with an approximation of the incidence and prevalence of a disease. This information helps officials decide whether or not to investigate a given disease.

Case reporting provided epidemiologists with valuable leads regarding the origin and spread of AIDS. In fact, one of the first clues about AIDS came from reports of young males with Kaposi's sarcoma, formerly a disease of older males. Using these reports, epidemiologists began various studies of the patients. If an epidemiological study shows that a large enough segment of the population is affected by a disease, an attempt is then made to isolate and identify its causative agent. Identification is accomplished by a number of different microbiological methods. Identification of the causative agent often provides valuable information regarding the reservoir for the disease.

Once the chain of transmission is discovered, it is possible to apply control measures to stop the disease from spreading. These might include elimination of the source of infection, isolation and segregation of infected people, the development of vaccines, and, as in the case of AIDS, education.

THE CENTERS FOR DISEASE CONTROL AND PREVENTION (CDC)

LEARNING OBJECTIVES
- Identify the function of the CDC.
- Define the following terms: *morbidity, mortality,* and *notifiable disease.*

Epidemiology is a major concern of state and federal public health departments. The **Centers for Disease Control and Prevention (CDC),** a branch of the U.S. Public Health Service located in Atlanta, Georgia, is a central source of epidemiological information in the United States.

The CDC issues a publication called the *Morbidity and Mortality Weekly Report* (www.cdc.gov). The **MMWR,** as it is called, is read by microbiologists, physicians, and other hospital and public health professionals. The *MMWR* contains data on **morbidity,** the incidence of specific notifiable diseases, and **mortality,** the number of deaths from these diseases. These data are usually organized by state. **Notifiable diseases,** shown in Table 14.7, are those for which physicians are required by law to report cases to the U.S. Public Health Service. As of 2005, a total of 63 infectious diseases were reported at the national level. **Morbidity rate** is the number of people affected by a disease in a given period of time in relation to the total population. **Mortality rate** is the number of deaths resulting from a disease in a population in a given period of time in relation to the total population.

MMWR articles include reports of disease outbreaks, case histories of special interest, and summaries of the status of particular diseases during a recent period. These articles often include recommendations for procedures for diagnosis, immunization, and treatment. Several graphs and other data in this textbook are from the *MMWR,* and the case history boxes are adapted from reports from this publication. See the box on p. 445, for example.

* * *

In the next chapter, we consider the mechanisms of pathogenicity. We will discuss in more detail the methods by which microorganisms enter the body and cause disease, the effects of disease on the body, and the means by which pathogens leave the body.

TABLE 14.7	Nationally Notifiable Diseases, 2005
Acquired immunodeficiency syndrome (AIDS)	Pertussis
Anthrax	Plague
Botulism	Poliomyelitis, paralytic
Brucellosis	Psittacosis
Chancroid	Q fever
Chlamydia trachomatis, genital infections	Rabies, animal
Cholera	Rabies, human
Coccidioidomycosis	Rocky Mountain spotted fever
Cryptosporidiosis	Rubella
Cyclosporiasis	Rubella, congenital syndrome
Diphtheria	Salmonellosis
Ehrlichiosis	Severe acute respiratory syndrome-associated coronavirus (SARS)
Encephalitis, arboviral	Shigellosis
Enterohemorrhagic *Escherichia coli*	Smallpox
Giardiasis	Streptococcal disease, invasive, group A
Gonorrhea	Streptococcal toxic shock syndrome
Haemophilus influenzae, invasive	*Streptococcus pneumoniae,* drug-resistant
Hansen's disease (Leprosy)	*S. pneumoniae,* invasive, children
Hantavirus pulmonary syndrome	Syphilis
Hemolytic uremic syndrome, postdiarrheal	Syphilis, congenital
Hepatitis A	Tetanus
Hepatitis B	Toxic shock syndrome
Hepatitis C (non-A, non-B hepatitis)	Trichinellosis
HIV infection	Tuberculosis
Influenza-associated pediatric mortality	Tularemia
Legionellosis	Typhoid fever
Leprosy (Hansen's disease)	Vancomycin-intermediate resistant *Staphylococcus* aureus (VISA)
Listeriosis	Vancomycin-resistant *Staphylococcus aureus* (VRSA)
Lyme disease	Varicella
Malaria	Yellow fever
Measles	
Meningococcal disease	
Mumps	

CLINICAL PROBLEM SOLVING

A NOSOCOMIAL OUTBREAK

As you read through this box, you will encounter a series of questions that epidemiologists ask themselves as they try to trace an outbreak to its source. Try to answer each question before going to the next one.

1. Over a 7-year period, 361 patients developed bacteremia during their stays at one hospital. All patients had fever (>38°C), chills, and low blood pressure. Blood cultures were grown on brain-heart infusion agar, and the bacteria were identified as motile, aerobic gram-negative rods.
   What organisms are possible? How would you determine that the bacteria were from the same source?

2. *Burkholderia cepacia* was identified by biochemical testing. PCR using primers specifically for *B. cepacia* and restriction enzyme digests of the PCR products were used to determine that the *B. cepacia* were identical. Additionally, all isolates showed resistance to the same antibiotics.
   How would you determine the source?

3. A case control study was conducted using medical records for the 7-year period. Fifty case-patients were compared to two other patients who were hospitalized at the same time, for the same length of time, and within 5 years of age of the case-patient.
   What factors might contribute to infection?

4. Contaminated heparin has been the source of *B. cepacia* bloodstream infections. However, in this outbreak, only 26% of the case-patients received heparin. Fever developed within 36 hours of intravenous (IV) catheter insertion and disappeared within 6 hours of catheter removal.
   What do you want to know now?

5. Insertion sites were cleaned with povidone-iodine purchased from a manufacturer, distributed into plastic squirt bottles, and used on a piece of sterile cotton. The cotton was rubbed in a circular fashion on the insertion site. Alcohol was purchased as 90% ethanol and diluted with water to 70% in a 100-L plastic container. The 70% alcohol was used on cotton that was rubbed in a circular fashion to wipe off the povidone-iodine before insertion.
   Where will you look for B. cepacia?

6. *B. cepacia* was not cultured from the povidone-iodine. Cultures of the hospital water supply showed no growth of *B. cepacia*. Water cultures from taps at nursing stations, in the operating room, and on the dialysis unit were also negative for *B. cepacia*.
   Is there anyplace else to look?

7. Tap water in the pharmacy was used to prepare a new batch of 70% alcohol every two to three days. *B. cepacia* matching the case-patients was isolated from a culture of the inside of the tap and from the 100-L container.
   How can you stop this outbreak?

8. Once the organism was cultured from the pharmacy tap, pharmacy personnel were instructed to use sterile water for dilutions. A policy to use commercially prepared, single-use alcohol and povidone-iodine swabs was implemented hospital-wide.

   B. cepacia is commonly found in liquid reservoirs and moist environments. The *B. cepacia* could have colonized both the water pipe and the 100-L container. Bacteria from the biofilm just above the fluid level could be washed onto the cotton swab and onto the insertion site. *B. cepacia* is a well-known nosocomial pathogen that is responsible for 0.6% of all ventilator-associated pneumonias. Numerous outbreaks of *B. cepacia* infection have been reported among cystic fibrosis patients. Outbreaks involving *B. cepacia* from patients without cystic fibrosis have been traced to mouthwashes, an oral CO_2 blood-gas probe, and heparin.

   Source: Adapted from reports in *MMWR* and *Infection Control and Hospital Epidemiology*, 2004.

	Case-patients	Control-patients
Number of patients	50	100
Risk factor		
Ventilator	11	19
Urinary bladder catheter	24	30
Intravenous (IV) catheter	42	58
More than 2 IV catheters	18	16
Heparin IV to prevent blood clotting	13	7
Chemotherapy for cancer	16	0
Glucose IV for rehydration	31	44
Surgery in past month	13	38

STUDY OUTLINE

INTRODUCTION (p. 420)

1. Disease-causing microorganisms are called pathogens.
2. Pathogenic microorganisms have special properties that allow them to invade the human body or produce toxins.
3. When a microorganism overcomes the body's defenses, a state of disease results.

PATHOLOGY, INFECTION, AND DISEASE (p. 421)

1. Pathology is the scientific study of disease.
2. Pathology is concerned with the etiology (cause), pathogenesis (development), and effects of disease.
3. Infection is the invasion and growth of pathogens in the body.
4. A host is an organism that shelters and supports the growth of pathogens.
5. Disease is an abnormal state in which part or all of the body is not properly adjusted or is incapable of performing normal functions.

NORMAL MICROBIOTA (pp. 421–425)

1. Animals, including humans, are usually germfree in utero.
2. Microorganisms begin colonization in and on the surface of the body soon after birth.
3. Microorganisms that establish permanent colonies inside or on the body without producing disease make up the normal microbiota.
4. Transient microbiota are microbes that are present for various periods and then disappear.

RELATIONSHIPS BETWEEN THE NORMAL MICROBIOTA AND THE HOST (p. 424)

5. The normal microbiota can prevent pathogens from causing an infection; this phenomenon is known as microbial antagonism.
6. Normal microbiota and the host exist in symbiosis (living together).
7. The three types of symbiosis are commensalism (one organism benefits and the other is unaffected), mutualism (both organisms benefit), and parasitism (one organism benefits and one is harmed).

OPPORTUNISTIC MICROORGANISMS (p. 425)

8. Opportunistic pathogens do not cause disease under normal conditions but cause disease under special conditions.

COOPERATION AMONG MICROORGANISMS (p. 425)

9. In some situations, one microorganism makes it possible for another to cause a disease or produce more severe symptoms.

THE ETIOLOGY OF INFECTIOUS DISEASES (pp. 425–427)

KOCH'S POSTULATES (pp. 425–426)

1. Koch's postulates are criteria for establishing that specific microbes cause specific diseases.
2. Koch's postulates have the following requirements: (a) the same pathogen must be present in every case of the disease; (b) the pathogen must be isolated in pure culture; (c) the pathogen isolated from pure culture must cause the same disease in a healthy, susceptible laboratory animal; and (d) the pathogen must be reisolated from the inoculated laboratory animal.

EXCEPTIONS TO KOCH'S POSTULATES (pp. 426–427)

3. Koch's postulates are modified to establish etiologies of diseases caused by viruses and some bacteria, which cannot be grown on artificial media.
4. Some diseases, such as tetanus, have unequivocal signs and symptoms.
5. Some diseases, such as pneumonia and nephritis, may be caused by a variety of microbes.
6. Some pathogens, such as S. pyogenes, cause several different diseases.
7. Certain pathogens, such as HIV, cause disease in humans only.

CLASSIFYING INFECTIOUS DISEASES (pp. 427–429)

1. A patient may exhibit symptoms (subjective changes in body functions) and signs (measurable changes), which a physician uses to make a diagnosis (identification of the disease).
2. A specific group of symptoms or signs that always accompanies a specific disease is called a syndrome.
3. Communicable diseases are transmitted directly or indirectly from one host to another.
4. A contagious disease is one that is easily spread from one person to another.
5. Noncommunicable diseases are caused by microorganisms that normally grow outside the human body and are not transmitted from one host to another.

OCCURRENCE OF A DISEASE (pp. 427–428)

6. Disease occurrence is reported by incidence (number of people contracting the disease) and prevalence (number of cases at a particular time).

7. Diseases are classified by frequency of occurrence: sporadic, endemic, epidemic, and pandemic.

SEVERITY OR DURATION OF A DISEASE (p. 428)

8. The scope of a disease can be defined as acute, chronic, subacute, or latent.

9. Herd immunity is the presence of immunity to a disease in most of the population.

EXTENT OF HOST INVOLVEMENT (p. 429)

10. A local infection affects a small area of the body; a systemic infection is spread throughout the body via the circulatory system.

11. A primary infection is an acute infection that causes the initial illness.

12. A secondary infection can occur after the host is weakened from a primary infection.

13. An inapparent, or subclinical, infection does not cause any signs of disease in the host.

PATTERNS OF DISEASE (pp. 429–430)

PREDISPOSING FACTORS (p. 429)

1. A predisposing factor is one that makes the body more susceptible to disease or alters the course of a disease.

2. Examples include gender, climate, age, fatigue, and inadequate nutrition.

DEVELOPMENT OF DISEASE (pp. 429–430)

3. The incubation period is the time interval between the initial infection and the first appearance of signs and symptoms.

4. The prodromal period is characterized by the appearance of the first mild signs and symptoms.

5. During the period of illness, the disease is at its height, and all disease signs and symptoms are apparent.

6. During the period of decline, the signs and symptoms subside.

7. During the period of convalescence, the body returns to its prediseased state, and health is restored.

THE SPREAD OF INFECTION (pp. 430–435)

RESERVOIRS OF INFECTION (pp. 430–431)

1. A continual source of infection is called a reservoir of infection.

2. People who have a disease or are carriers of pathogenic microorganisms are human reservoirs of infection.

3. Zoonoses are diseases that affect wild and domestic animals and can be transmitted to humans.

4. Some pathogenic microorganisms grow in nonliving reservoirs, such as soil and water.

TRANSMISSION OF DISEASE (pp. 431–434)

5. Transmission by direct contact involves close physical contact between the source of the disease and a susceptible host.

6. Transmission by fomites (inanimate objects) constitutes indirect contact.

7. Transmission via saliva or mucus in coughing or sneezing is called droplet transmission.

8. Transmission by a medium such as water, food, or air is called vehicle transmission.

9. Airborne transmission refers to pathogens carried on water droplets or dust for a distance greater than 1 meter.

10. Arthropod vectors carry pathogens from one host to another by both mechanical and biological transmission.

NOSOCOMIAL (HOSPITAL-ACQUIRED) INFECTIONS (pp. 435–438)

1. A nosocomial infection is any infection that is acquired during the course of stay in a hospital, nursing home, or other health care facility.

2. About 5 to 15% of all hospitalized patients acquire nosocomial infections.

MICROORGANISMS IN THE HOSPITAL (p. 436)

3. Certain normal microbiota are often responsible for nosocomial infections when they are introduced into the body through such medical procedures as surgery and catheterization.

4. Opportunistic, drug-resistant gram-negative bacteria are the most frequent causes of nosocomial infections.

COMPROMISED HOST (pp. 436–437)

5. Patients with burns, surgical wounds, and suppressed immune systems are the most susceptible to nosocomial infections.

CHAIN OF TRANSMISSION (pp. 437–438)

6. Nosocomial infections are transmitted by direct contact between staff members and patients and between patients.

7. Fomites such as catheters, syringes, and respiratory devices can transmit nosocomial infections.

CONTROL OF NOSOCOMIAL INFECTIONS (pp. 438)

8. Aseptic techniques can prevent nosocomial infections.

9. Hospital infection control staff members are responsible for overseeing the proper cleaning, storage, and handling of equipment and supplies.

EMERGING INFECTIOUS DISEASES

(pp. 438–440)

1. New diseases and diseases with increasing incidences are called emerging infectious diseases (EIDs).

2. EIDs can result from the use of antibiotics and pesticides, climatic changes, travel, the lack of vaccinations, and improved case reporting.

3. The CDC, NIH, and WHO are responsible for surveillance and responses to emerging infectious diseases.

EPIDEMIOLOGY (pp. 440–443)

1. The science of epidemiology is the study of the transmission, incidence, and frequency of disease.

2. Modern epidemiology began in in the mid-1800s with the works of Snow, Semmelweis, and Nightingale.
   ✸ **Animation: Epidemiology. The Microbiology Place.**

3. Data about infected people are collected and analyzed in descriptive epidemiology.

4. In analytical epidemiology, a group of infected people is compared with an uninfected group.

5. Controlled experiments designed to test hypotheses are performed in experimental epidemiology.

6. Case reporting provides data on incidence and prevalence to local, state, and national health officials.

7. The Centers for Disease Control and Prevention (CDC) is the main source of epidemiologic information in the United States.

8. The CDC publishes the *Morbidity and Mortality Weekly Report* to provide information on morbidity (incidence) and mortality (deaths).

STUDY QUESTIONS

✸ Access more review material either online at **The Microbiology Place** (www.microbiologyplace.com) or with **The Microbiology Place CD-ROM** packaged with your new book. There you'll find activities, practice tests, quizzes, flashcards, case studies, and more to help you succeed.

Answers to the Study Questions can be found in Appendix G.

REVIEW

1. Differentiate the terms in each of the following pairs:
   a. etiology and pathogenesis
   b. infection and disease
   c. communicable disease and noncommunicable disease

2. What is meant by normal microbiota? How do they differ from transient microbiota?

3. Define *symbiosis*. Differentiate among commensalism, mutualism, and parasitism, and give an example of each.

4. What is a reservoir of infection? Match the following diseases with their reservoirs:

_____b_____ influenza	**a.** nonliving
_____c_____ rabies	**b.** human
_____a_____ botulism	**c.** animal

5. Describe how Koch's postulates establish the etiology of many infectious diseases. Why don't Koch's postulates apply to all infectious diseases?

6. Describe the various ways diseases can be transmitted in each of the following categories. Name one disease transmitted by each method.

a. transmission by direct contact
b. transmission by indirect contact
c. transmission by arthropod vectors
d. droplet transmission
e. vehicle transmission
f. airborne transmission

7. List four predisposing factors to disease.

8. Indicate whether each of the following conditions is typical of subacute, chronic, or acute infections.
   C **a.** The patient experiences a rapid onset of malaise; symptoms last 5 days.
   B **b.** The patient experiences cough and breathing difficulty for months.
   A **c.** The patient has no apparent symptoms and is a known carrier.

9. Of all the hospital patients with infections, one-third do not enter the hospital with an infection. How do they acquire these infections? What is the method of transmission of these infections? What is the reservoir of infection?

10. Differentiate between endemic and epidemic states of infectious disease.

11. What is epidemiology? What is the role of the Centers for Disease Control and Prevention (CDC)?

12. Distinguish between symptoms and signs as signals of disease.

13. How can a local infection become a systemic infection?

14. Why are some organisms that constitute the normal microbiota described as commensals, whereas others are described as mutualistic?

15. Put the following in the correct order to describe the pattern of disease: period of convalescence, prodromal period, period of decline, incubation period, period of illness.

MULTIPLE CHOICE

1. The emergence of new infectious diseases is probably due to all of the following *except*
   a. the need of bacteria to cause disease.
   b. the ability of humans to travel by air.
   c. changing environments (e.g., flood, drought, pollution).
   d. a pathogen crossing the species barrier.
   e. the increasing human population.

2. All members of a group of ornithologists studying barn owls in the wild have had salmonellosis (*Salmonella* gastroenteritis). One birder is experiencing her third infection. What is the most likely source of their infections?
   a. The ornithologists are eating the same food.
   b. They are contaminating their hands while handling the owls and nests.
   c. One of the workers is a *Salmonella* carrier.
   d. Their drinking water is contaminated.

3. Which of the following statements is *not* true?
   a. *E. coli* never causes disease.
   b. *E. coli* provides vitamin K for its host.
   c. *E. coli* often exists in a mutualistic relationship with humans.
   d. *E. coli* gets nutrients from intestinal contents.

4. Which of the following is *not* one of Koch's postulates?
   a. The same pathogen must be present in every case of the disease.
   b. The pathogen must be isolated and grown in pure culture from the diseased host.
   c. The pathogen from pure culture must cause the disease when inoculated into a healthy, susceptible laboratory animal.
   d. The disease must be transmitted from a diseased animal to a healthy, susceptible animal by some form of contact.
   e. The pathogen must be isolated in pure culture from an experimentally infected lab animal.

Use the following information to answer questions 5–7.

On September 6, a 6-year-old boy experienced fever, chills, and vomiting. On September 7, he was hospitalized with diarrhea and swollen lymph nodes under both arms. On September 3, the boy had been scratched and bitten by a cat. The cat was found dead on September 5, and *Yersinia pestis* was isolated from the cat. Chloramphenicol was administered to the boy from September 7, when *Y. pestis* was isolated from him. On September 17, the boy's temperature returned to normal; and on September 22, he was released from the hospital.

5. Identify the incubation period for this case of bubonic plague.
   a. September 3–5.   c. September 6–7.
   b. September 3–6.   d. September 6–17.

6. Identify the prodromal period for this disease.
   a. September 3–5.   c. September 6–7.
   b. September 3–6.   d. September 6–17.

7. Identify the crisis during this disease.
   a. September 6.   c. September 6–17.
   b. September 7.   d. September 17.

Use the following information to answer questions 8–10.

A Maryland woman was hospitalized with dehydration; *Vibrio cholerae* and *Plesiomonas shigelloides* were isolated from the patient. She had neither traveled outside the United States nor eaten raw shellfish during the preceding month. She had attended a party two days before her hospitalization. Two other people at the party had acute diarrheal illness and elevated levels of serum antibodies against *Vibrio*. Everyone at the party ate crabs and rice pudding with coconut milk. Crabs left over from this party were served at a second party. One of the 20 people at the second party had onset of mild diarrhea; specimens from 14 of these people were negative for vibriocidal antibodies.

8. This is an example of
   a. vehicle transmission.   d. direct contact transmission.
   b. airborne transmission.   e. nosocomial transmission.
   c. transmission by fomites.

9. The etiologic agent of the disease is
   a. *Plesiomonas shigelloides*.   d. coconut milk.
   b. crabs.   e. rice pudding.
   c. *Vibrio cholerae*.

10. The source of the disease was
   a. *Plesiomonas shigelloides*.   d. coconut milk.
   b. crabs.   e. rice pudding.
   c. *Vibrio cholerae*.

CRITICAL THINKING

1. Ten years before Robert Koch published his work on anthrax, Anton De Bary showed that potato blight was caused by the alga *Phytophthora infestans*. Why do you suppose we use Koch's postulates instead of something called "De Bary's postulates"?

2. Florence Nightingale gathered the following data in 1855.

Population Sampled	Deaths from Contagious Diseases
Englishmen (in general population)	0.2%
English soldiers (in England)	18.7%
English soldiers (in Crimean War)	42.7%
English soldiers (in Crimean War) after Nightingale's sanitary reforms	2.2%

Discuss how Nightingale used the three basic types of epidemiologic investigation. The contagious diseases were primarily cholera and typhus; how are these diseases transmitted and prevented?

3. Name the method of transmission of each of the following diseases:
   a. malaria   f. mononucleosis
   b. tuberculosis   g. measles
   c. nosocomial infections h. hepatitis A
   d. salmonellosis   i. tetanus
   e. streptococcal   j. hepatitis B
      pharyngitis   k. chlamydial urethritis

4. The following graph shows the incidence of typhoid fever in the United States from 1954 through 2004. Mark the graph to show when this disease occurred sporadically and epidemically. What appears to be the endemic level? What would have to be shown to indicate a pandemic of this disease? How is typhoid fever transmitted?

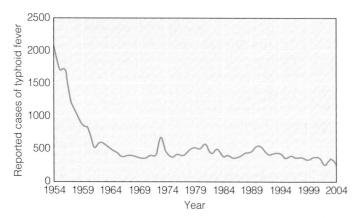

5. Cholera is reemerging in the Western Hemisphere. In 1991, a cholera epidemic caused by *Vibrio cholerae* O1 Inaba was discovered in Peru. The map that follows shows the location of cholera cases in the continental United States caused by *V. cholerae* O1 Inaba (gold), as well as where this strain has been found in coastal waters (blue).

How is cholera transmitted? How did *V. cholerae* O1 Inaba get into the United States? Explain how eight people who did not travel outside the U.S. contracted *V. cholerae* O1 Inaba.

CLINICAL APPLICATIONS

1. Three days before a nurse developed meningococcemia, she assisted with intubation of a patient with a *Neisseria meningitidis* infection. Of the 24 medical personnel involved, only this nurse became ill. The nurse recalled that she was exposed to nasopharyngeal secretions and did not receive antibiotic prophylaxis. What two mistakes did the nurse make? How is meningitis transmitted?

2. Three patients in a large hospital acquired infections of *Pseudomonas cepacia* during their stay. All three patients received cryoprecipitate, which is prepared from blood that has been frozen in a standard plastic blood transfer pack. The transfer pack is then placed in a water bath to thaw. What is the probable origin of the nosocomial infections? What characteristics of *Pseudomonas* would allow it to be involved in this type of infection?

3. Following is a case history of a 49-year-old man. Identify each period in the pattern of disease that he experienced. On February 7, he handled a parakeet with a respiratory illness. On March 9, he experienced intense pain in his legs, followed by severe chills and headaches. On March 16, he had chest pains, cough, and diarrhea, and his temperature was 40°C. Appropriate antibiotics were administered on March 17, and his fever subsided within 12 hours. He continued taking antibiotics for 14 days. (*Note:* The disease is psittacosis. Can you find the etiology?)

4. Twenty-one percent of the patients in a large hospital acquired *Clostridium difficile* diarrhea and colitis during their hospital stay. These patients required longer hospital stays than uninfected patients. Epidemiological studies provided the following information. What is the most likely mode of transmission of this bacterium in hospitals? How can transmission be prevented?

Rate of infection for patients:
Single room	7%
Double room	17%
Triple room	26%

Rate of environmental isolations of *C. difficile*:
Bed rail	10%
Commode	1%
Floor	18%
Call button	6%
Toilet	3%

Hands of hospital personnel after contact with patients that were culture-positive for *C. difficile*:
Used gloves	0%
Did not use gloves	59%
Had *C. difficile* before patient contact	3%
Washed with nondisinfectant soap	40%
Washed with disinfectant soap	3%
Did not wash hands	20%

5. *Mycobacterium avium-intracellulare* is prevalent in AIDS patients. In an effort to determine the source of this infection, hospital water systems were sampled. The water contained chlorine.

Percentage of samples with *M. avium*:

Hot water		Cold water	
February	88%	February	22%
June	50%	June	11%

What is the usual method of transmission for *Mycobacterium*? What is a probable source of infection in hospitals? How can such nosocomial infections be prevented?

15

Microbial Mechanisms of Pathogenicity

Now that you have a basic understanding of how microorganisms cause disease, we will take a look at some of the specific properties of microorganisms that contribute to **pathogenicity,** the ability to cause disease by overcoming the defenses of a host, and **virulence,** the degree or extent of pathogenicity. (As discussed throughout the chapter, the term *host* usually refers to humans.) Microbes don't try to cause disease; the microbial cells are getting food and defending themselves. The box on page 453 describes the balance between survival of the host and microbe. Keep in mind that many of the properties contributing to microbial pathogenicity and virulence are unclear or unknown. We do know, however, that if the microbe overpowers the host defenses, disease results.

UNDER THE MICROSCOPE

Salmonella. *Salmonella* bacteria cause typhoid fever and salmonellosis in humans. These bacteria are transmitted by the fecal-oral route.

HOW MICROORGANISMS ENTER A HOST

To cause disease, most pathogens must gain access to the host, adhere to host tissues, penetrate or evade host defenses, and damage the host tissues. However, some microbes do not cause disease by directly damaging host tissue. Instead, disease is due to the accumulation of microbial waste products. Some microbes, such as those that cause dental caries and acne, can cause disease without penetrating the body. Pathogens can gain entrance to the human body and other hosts through several avenues, which are called **portals of entry.**

PORTALS OF ENTRY

> **LEARNING OBJECTIVE**
> • Identify the principal portals of entry.

The portals of entry for pathogens are mucous membranes, skin, and direct deposition beneath the skin or membranes (the parenteral route).

MUCOUS MEMBRANES

Many bacteria and viruses gain access to the body by penetrating mucous membranes lining the respiratory tract, gastrointestinal tract, genitourinary tract, and conjunctiva, a delicate membrane that covers the eyeballs and lines the eyelids. Most pathogens enter through the mucous membranes of the gastrointestinal and respiratory tracts.

The respiratory tract is the easiest and most frequently traveled portal of entry for infectious microorganisms. Microbes are inhaled into the nose or mouth in drops of moisture and dust particles. Diseases that are commonly contracted via the respiratory tract include the common cold, pneumonia, tuberculosis, influenza, measles, and smallpox.

Microorganisms can gain access to the gastrointestinal tract in food and water and via contaminated fingers. Most microbes that enter the body in these ways are destroyed by hydrochloric acid (HCl) and enzymes in the stomach or by bile and enzymes in the small intestine. Those that survive can cause disease. Microbes in the gastrointestinal tract can cause poliomyelitis, hepatitis A, typhoid fever, amoebic dysentery, giardiasis, shigellosis (bacillary dysentery), and cholera. These pathogens are then eliminated with feces and can be transmitted to other hosts via contaminated water, food, or fingers.

The genitourinary tract is a portal of entry for pathogens that are contracted sexually. Some microbes that cause sexually transmitted diseases (STDs) may penetrate an unbroken mucous membrane. Others require a cut or abrasion of some type. Examples of STDs are HIV infection, genital warts, chlamydia, herpes, syphilis, and gonorrhea.

SKIN

The skin is one of the largest organs of the body in terms of surface area and is an important defense against disease. Unbroken skin is impenetrable by most microorganisms. Some microbes gain access to the body through openings in the skin, such as hair follicles and sweat gland ducts. Larvae of the hookworm actually bore through intact skin, and some fungi grow on the keratin in skin or infect the skin itself.

The conjunctiva is a delicate mucous membrane that lines the eyelids and covers the white of the eyeballs. Although it is a relatively effective barrier against infection, certain diseases such as conjunctivitis, trachoma, and neonatal gonorrheal ophthalmia are acquired through the conjunctiva.

THE PARENTERAL ROUTE

Other microorganisms gain access to the body when they are deposited directly into the tissues beneath the skin or into mucous membranes when these barriers are penetrated or injured. This route is called the **parenteral route.** Punctures, injections, bites, cuts, wounds, surgery, and splitting due to swelling or drying can all establish parenteral routes. HIV, the hepatitis viruses, and bacteria that cause tetanus and gangrene can be transmitted parenterally.

THE PREFERRED PORTAL OF ENTRY

Even after microorganisms have entered the body, they do not necessarily cause disease. The occurrence of disease depends on several factors, only one of which is the portal of entry. Many pathogens have a preferred portal of entry that is a prerequisite to their being able to cause disease. If they gain access to the body by another portal, disease might not occur. For example, the bacteria of typhoid fever, *Salmonella typhi,* produce all the signs and symptoms of the disease when swallowed (preferred route); but, if the same bacteria are rubbed on the skin, no reaction (or only a slight inflammation) occurs. Streptococci that are inhaled (preferred route) can cause pneumonia; those that are swallowed generally do not produce signs or symptoms. Some pathogens, such as *Yersinia pestis,* the microorganism that causes plague and *Bacillus anthracis,* the causative agent of anthrax, can initiate disease from more than one

MICROBIOLOGY IN THE NEWS

HOW HUMAN BEHAVIOR INFLUENCES THE EVOLUTION OF VIRULENCE IN MICROORGANISMS

According to human logic, a parasite that kills its host is harming itself. Thus, it seems reasonable that some parasites have evolved to a sort of commensalism with their human hosts. In this respect, tapeworms might be considered the "perfect parasite." Tapeworms cause no obvious symptoms in most patients, thus ensuring that their hosts will be able to move around and shed tapeworm eggs that will be ingested by intermediate hosts.

However, other diseases such as cholera, antibiotic-resistant tuberculosis, viral cancers, and syphilis sometimes do kill their hosts. Does it make sense for a parasite to kill its host? Remember that nature does not have a plan for evolution; the genetic variations that give rise to evolution are due to random mutations, not logic. However, according to natural selection, organisms best adapted to their environments will survive and reproduce. Coevolution between a parasite and its host seems to occur: the behavior of one influences that of the other.

Some parasites, such as the cholera bacterium, get transmitted before their host dies. *Vibrio cholerae* quickly induces diarrhea, threatening the host's life from a loss of fluids and salts but providing a way to transmit the parasite to another person through the ingestion of contaminated water. This cycle continues in countries where war and poverty have prevented the implementation of water

purification systems. However, the virulent form of cholera has been selected against in countries where water-treatment systems prevent the bacteria from getting to a new host. After water purification was instituted in India, the milder agent, *V. cholerae* eltor, replaced the deadly *V. cholerae*.

When we interfere with a pathogen's survival by using antibiotics, we inadvertently select for antibiotic-resistant mutants. The rapid rise of multidrug-resistant *Mycobacterium tuberculosis*, documented in the popular and scientific media, demonstrates this. In an antibiotic-laden environment, the "fittest" bacterium is one that is resistant to antibiotics. Antibiotic resistance is selected for when chemotherapy is not completed and when antibiotics are not used correctly. For example, tuberculosis (TB) chemotherapy protocols are very long (up to 2 years). In the United States, only 75% of TB patients complete the prescribed chemotherapy. When a person doesn't take the complete regimen of the prescribed antibiotic, the level of the drug in the body decreases, and some resistant bacteria survive. These bacteria reproduce, and some of the new population are able to survive in even higher levels of antibiotics.

Comparisons of human T-cell leukemia virus type 1 (HTLV-1) infections in Japan and Jamaica also suggest that human sexual behavior influences the virulence of sexually transmitted diseases.

HTLV-1 is a sexually transmitted retrovirus that causes a type of adult leukemia. In Japan, the average age of onset of cancers caused by HTLV-1 is 60, whereas in Jamaica, onset occurs at about age 45. In Japan, barrier contraceptives are more widely used than birth control pills, so the virus needs a healthy host for a longer period of time. In Jamaica, barrier contraceptives are not widely used, and HTLV-1 is more virulent—perhaps because it has a greater potential for sexual transmission.

Human behavior may also have influenced the evolution of syphilis. *Treponema pallidum* has a long incubation period and goes into noninfectious latent periods between the primary, secondary, and tertiary stages of the disease. Although it cannot be transmitted during the incubation and latent periods, these periods allow the bacteria to stay alive until the infected host changes sex partners. Partner changes might take several years in a monogamous society.

Paul Ewald of Amherst College reminds us that commensalism is not the inevitable end point of evolution. In order to control diseases, we must modify our behavior to make commensalism the most favorable outcome for the pathogen. Some examples of these modifications are ensuring completion of the prescribed antibiotic protocol for tuberculosis and the use of condoms to prevent the transmission of sexually transmitted diseases.

---

portal of entry. The preferred portals of entry for some common pathogens are listed in Table 15.1.

NUMBERS OF INVADING MICROBES

LEARNING OBJECTIVE
- Define ID_{50} and LD_{50}.

If only a few microbes enter the body, they will probably be overcome by the host's defenses. However, if large numbers of microbes gain entry, the stage is probably set for disease.

Thus, the likelihood of disease increases as the number of pathogens increases.

The virulence of a microbe is often expressed as the ID_{50} (infectious dose for 50% of a sample population). The 50 is not an absolute value; rather, it is used to compare relative virulence under experimental conditions. *Bacillus anthracis* can cause infection via three different portals of entry. The ID_{50} through the skin (cutaneous anthrax) is 10 to 50 endospores; the ID_{50} for inhalation anthrax is inhalation of 10,000 to 20,000 endospores; and the ID_{50} for gastrointestinal anthrax is ingestion of 250,000 to 1,000,000

TABLE 15.1	Portals of Entry for the Pathogens of Some Common Diseases		
Portal of Entry	Pathogen*	Disease	Incubation Period
Mucous Membranes			
Respiratory tract	*Streptococcus pneumoniae*	Pneumococcal pneumonia	Variable
	Mycobacterium tuberculosis[†]	Tuberculosis	Variable
	Bordetella pertussis	Whooping cough (pertussis)	12–20 days
	Influenza virus *(Influenzavirus)*	Influenza	18–36 hours
	Measles virus *(Morbillivirus)*	Measles (rubeola)	11–14 days
	Rubella virus *(Rubivirus)*	German measles (rubella)	2–3 weeks
	Epstein-Barr virus *(Lymphocryptovirus)*	Infectious mononucleosis	2–6 weeks
	Varicella-zoster virus *(Varicellovirus)*	Chickenpox (varicella) (primary infection)	14–16 days
	Histoplasma capsulatum (fungus)	Histoplasmosis	5–18 days
Gastrointestinal tract	*Shigella* spp.	Shigellosis (bacillary dysentery)	1–2 days
	Brucella spp.	Brucellosis (undulant fever)	6–14 days
	Vibrio cholerae	Cholera	1–3 days
	Salmonella enterica	Salmonellosis	7–22 hours
	Salmonella typhi	Typhoid fever	14 days
	Hepatitis A virus *(Hepatovirus)*	Hepatitis A	15–50 days
	Mumps virus *(Rubulavirus)*	Mumps	2–3 weeks
	Trichinella spiralis (helminth)	Trichinellosis	2–28 days
Genitourinary tract	*Neisseria gonorrhoeae*	Gonorrhea	3–8 days
	Treponema pallidum	Syphilis	9–90 days
	Chlamydia trachomatis	Nongonococcal urethritis	1–3 weeks
	Herpes simplex virus type 2	Herpes virus infections	4–10 days
	Human immunodeficiency virus (HIV)[‡]	AIDS	10 years
	Candida albicans (fungus)[‡]	Candidiasis	2–5 days
Skin or Parenteral Route			
	Clostridium perfringens	Gas gangrene	1–5 days
	Clostridium tetani	Tetanus	3–21 days
	Rickettsia rickettsii	Rocky Mountain spotted fever	3–12 days
	Hepatitis B virus *(Hepadnavirus)*[†]	Hepatitis B	6 weeks–6 months
	Rabiesvirus *(Lyssavirus)*	Rabies	10 days–1 year
	Plasmodium spp. (protozoan)	Malaria	2 weeks

*All pathogens are bacteria, unless indicated otherwise. For viruses, the viral species and/or genus name is given.
[†] These pathogens can also cause disease after entering the body via the gastrointestinal tract. Hepatitis B virus can also cause disease after entering the body via the genitourinary tract.
[‡] These pathogens can also cause disease after entering the body via the parenteral route.

endospores. These data show that cutaneous anthrax is significantly easier to acquire than either the inhalation or the gastrointestinal forms. A study of *Vibrio cholerae* showed that the ID_{50} is 10^8 cells; but if stomach acid is neutralized with bicarbonate, the number of cells required to cause an infection decreases significantly.

The potency of a toxin is often expressed as the **LD$_{50}$** (lethal dose for 50% of a sample population). For example, the LD_{50} for botulinum toxin in mice is 0.03 ng/kg; for Shiga toxin, 250 ng/kg; and staphylococcal enterotoxin, 1350 ng/kg. In other words, compared to the other two

toxins, a much smaller dose of botulinum toxin is needed to cause symptoms.

ADHERENCE

LEARNING OBJECTIVE
- Using examples, explain how microbes adhere to host cells.

Almost all pathogens have some means of attaching themselves to host tissues at their portal of entry. For most

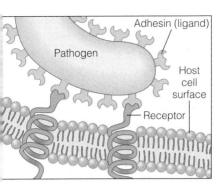

(a) Surface molecules on a pathogen, called adhesins or ligands, bind specifically to complementary surface receptors on cells of certain host tissues.

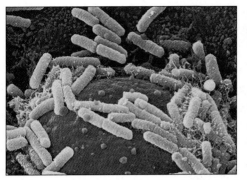

(b) *E. coli* bacteria (yellow-green) on human urinary bladder cells. SEM ⊢—⊣ 1 μm

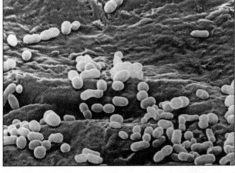

(c) Bacteria (yellow) adhering to human skin. SEM ⊢—⊣ 1 μm

FIGURE 15.1 Adherence.

Q **How are most adhesins classified chemically?**

pathogens, this attachment, called **adherence** (or **adhesion**), is a necessary step in pathogenicity. (Of course, nonpathogens also have structures for attachment.) The attachment between pathogen and host is accomplished by means of surface molecules on the pathogen called **adhesins** or **ligands** that bind specifically to complementary surface **receptors** on the cells of certain host tissues (Figure 15.1). Adhesins may be located on a microbe's glycocalyx or on other microbial surface structures, such as pili, fimbriae, and flagella (see Chapter 4).

The majority of adhesins on the microorganisms studied so far are glycoproteins or lipoproteins. The receptors on host cells are typically sugars, such as mannose. Adhesins on different strains of the same species of pathogen can vary in structure. Different cells of the same host can also have different receptors that vary in structure. If adhesins, receptors, or both can be altered to interfere with adherence, infection can often be prevented (or at least controlled).

The following examples illustrate the diversity of adhesins. *Streptococcus mutans*, a bacterium that plays a key role in tooth decay, attaches to the surface of teeth by its glycocalyx. An enzyme produced by *S. mutans*, called glucosyltransferase, converts glucose (derived from sucrose or table sugar) into a sticky polysaccharide called dextran, which forms the glycocalyx. *Actinomyces* bacterial cells have fimbriae that adhere to the glycocalyx of *S. mutans*. The combination of *S. mutans*, *Actinomyces*, and dextran make up dental plaque and contribute to dental caries (tooth decay; see Chapter 25, page 747).

Microbes have the ability to come together in masses, cling to surfaces, and take in and share available nutrients. These communities, which constitute masses of microbes and their extracellular products that can attach to living and nonliving surfaces, are called **biofilms** (discussed in more detail on page 821). Examples of biofilms include the

dental plaque on teeth, the algae on the walls of swimming pools, and the scum that accumulates on shower doors. A biofilm forms when microbes adhere to a particular surface that is typically moist and contains organic matter. The first microbes to attach are usually bacteria. Once they adhere to the surface, they multiply and secrete a glycocalyx that further attaches the bacteria to each other and to the surface (see Figure 27.11, page 821). In some cases biofilms can be several layers thick and may contain several types of microbes. Biofilms represent another method of adherence and are important because they resist disinfectants and antibiotics. This characteristic is significant, especially when biofilms colonize structures such as teeth, medical catheters, stents, heart valves, hip replacement components, and contact lenses. Dental plaque is actually a biofilm that mineralizes over time. It is estimated that biofilms are involved in 65% of all human bacterial infections.

Enteropathogenic strains of *E. coli* (those responsible for gastrointestinal disease) have adhesins on fimbriae that adhere only to specific kinds of cells in certain regions of the small intestine. After adhering, *Shigella* and *E. coli* induce endocytosis as a vehicle to enter host cells and then multiply within them (see Figure 25.7, page 752). *Treponema pallidum*, the causative agent of syphilis, uses its tapered end as a hook to attach to host cells. *Listeria monocytogenes*, which causes meningitis, spontaneous abortions, and stillbirths, produces an adhesin for a specific receptor on host cells. *Neisseria gonorrhoeae* (nī-se'rē-ä go-nôr-rē'ī), the causative agent of gonorrhea, also has fimbriae containing adhesins, which in this case permit attachment to cells with appropriate receptors in the genitourinary tract, eyes, and pharynx. *Staphylococcus aureus*, which can cause skin infections, binds to skin by a mechanism of adherence that resembles viral attachment (see Chapter 13).

HOW BACTERIAL PATHOGENS PENETRATE HOST DEFENSES

LEARNING OBJECTIVE

• Explain how capsules and cell wall components contribute to pathogenicity.

Although some pathogens can cause damage on the surface of tissues, most must penetrate tissues to cause disease. Here we will consider several factors that contribute to the ability of bacteria to invade a host.

CAPSULES

Recall from Chapter 4 that some bacteria make glycocalyx material that forms capsules around their cell walls; this property increases the virulence of the species. The capsule resists the host's defenses by impairing phagocytosis, the process by which certain cells of the body engulf and destroy microbes (see Chapter 16, page 479). The chemical nature of the capsule appears to prevent the phagocytic cell from adhering to the bacterium. However, the human body can produce antibodies against the capsule, and when these antibodies are present on the capsule surface, the encapsulated bacteria are easily destroyed by phagocytosis.

One bacterium that owes its virulence to the presence of a polysaccharide capsule is *Streptococcus pneumoniae*, the causative agent of pneumococcal pneumonia (see Figure 24.13, page 726). Some strains of this organism have capsules, and others do not. Strains with capsules are virulent, but strains without capsules are avirulent because they are susceptible to phagocytosis. Other bacteria that produce capsules related to virulence are *Klebsiella pneumoniae*, a causative agent of bacterial pneumonia; *Haemophilus influenzae*, a cause of pneumonia and meningitis in children; *Bacillus anthracis*, the cause of anthrax; and *Yersinia pestis*, the causative agent of plague. Keep in mind that capsules are not the only cause of virulence. Many nonpathogenic bacteria produce capsules, and the virulence of some pathogens is not related to the presence of a capsule.

CELL WALL COMPONENTS

The cell walls of certain bacteria contain chemical substances that contribute to virulence. For example, *Streptococcus pyogenes* produces a heat-resistant and acid-resistant protein called **M protein** (see Figure 21.5, page 620). This protein is found on both the cell surface and fimbriae. The M protein mediates attachment of the bacterium to epithelial cells of the host and helps the bacterium resist phagocytosis by white blood cells. The protein thereby increases the virulence of the microorganism. Immunity to *S. pyogenes* depends on the body's production of an antibody specific to M protein. *Neisseria gonorrhoeae* grows inside human epithelial cells and leukocytes. These bacteria use **fimbriae** and an outer membrane protein called **Opa** to attach to host cells. Following attachment by both Opa and fimbriae, the host cells take in the bacteria. (Bacteria that produce Opa form *opaque* colonies in culture media.) The **waxy lipid** (mycolic acid) that makes up the cell wall of *Mycobacterium tuberculosis* also increases virulence by resisting digestion by phagocytes. In fact, *M. tuberculosis* can even multiply inside phagocytes.

ENZYMES

LEARNING OBJECTIVE

• Compare the effects of coagulases, kinases, hyaluronidase, and collagenase.

The virulence of some bacteria is thought to be aided by the production of extracellular enzymes (*exoenzymes*) and related substances. These chemicals can digest materials between cells and form or digest blood clots, among other functions.

Coagulases are bacterial enzymes that coagulate (clot) the fibrinogen in blood. Fibrinogen, a plasma protein produced by the liver, is converted by coagulases into fibrin, the threads that form a blood clot. The fibrin clot may protect the bacterium from phagocytosis and isolate it from other defenses of the host. Coagulases are produced by some members of the genus *Staphylococcus*; they may be involved in the walling-off process in boils produced by staphylococci. However, some staphylococci that do not produce coagulases are still virulent. (Capsules may be more important to their virulence.)

Bacterial **kinases** are bacterial enzymes that break down fibrin and thus digest clots formed by the body to isolate the infection. One of the better-known kinases is *fibrinolysin (streptokinase)*, which is produced by such streptococci as *Streptococcus pyogenes*. Another kinase, *staphylokinase*, is produced by *Staphylococcus aureus*. Injected directly into the blood, streptokinase has been used successfully to remove some types of blood clots in cases of heart attacks due to obstructed coronary arteries.

Hyaluronidase is another enzyme secreted by certain bacteria, such as streptococci. It hydrolyzes hyaluronic acid, a type of polysaccharide that holds together certain cells of the body, particularly cells in connective tissue. This digesting action is thought to be involved in the tissue blackening of infected wounds and to help the microorganism spread from its initial site of infection. Hyaluronidase is also produced by some clostridia that cause gas gangrene. For therapeutic use, hyaluronidase may be mixed with a drug to promote the spread of the drug through a body tissue.

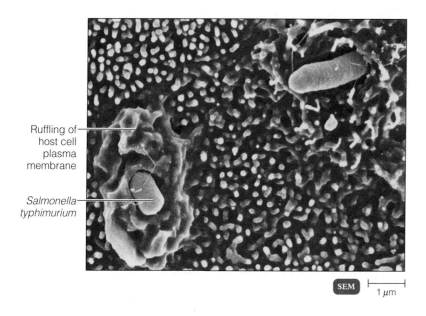

Ruffling of
host cell
plasma
membrane

*Salmonella
typhimurium*

SEM   ⊢——⊣
       1 µm

FIGURE 15.2 *Salmonella* **entering epithelial cells as a result of ruffling.**

Q **What are invasins?**

Another enzyme, **collagenase,** produced by several species of *Clostridium*, facilitates the spread of gas gangrene. Collagenase breaks down the protein collagen, which forms the connective tissue of muscles and other body organs and tissues.

As a defense against adherence of pathogens to mucosal surfaces, the body produces a class of antibodies called IgA antibodies. There are some pathogens with the ability to produce enzymes, called **IgA proteases,** that can destroy these antibodies. *N. gonorrhoeae* has this ability, as does *N. meningitidis* (me-nin-ji′ti-dis), the causative agent of meningococcal meningitis, and other microbes that infect the central nervous system.

ANTIGENIC VARIATION

LEARNING OBJECTIVE

• Define and give an example of *antigenic variation*.

In Chapter 17 you will learn that *adaptive (acquired) immunity* refers to a specific defensive response of the body to an infection or to antigens. In the presence of antigens the body produces proteins called antibodies, which bind to the antigens and inactivate or destroy them. However, some pathogens can alter their surface antigens, by a process called **antigenic variation.** Thus, by the time the body mounts an immune response against a pathogen, the pathogen has already altered its antigens and is unaffected by the antibodies. Some microbes can activate alternative genes, resulting in antigenic changes. For example, *N. gonorrhoeae* has several copies of the Opa-encoding gene, resulting in cells with different antigens and in cells that express different antigens over time.

A wide range of microbes is capable of antigenic variation. Examples include *Influenzavirus*, the causative agent

of influenza (flu); *Neisseria gonorrhoeae*, the causative agent of gonorrhea; and *Trypanosoma brucei gambiense* (tri-pa′nō-sō-mä brüs′ē gam-bē-ens′), the causative agent of African trypanosomiasis (sleeping sickness).

PENETRATION INTO THE HOST CELL CYTOSKELETON

LEARNING OBJECTIVE

• Describe how bacteria use the host cell's cytoskeleton to enter the cell.

As previously noted, microbes attach to host cells by adhesins. The interaction triggers signals in the host cell that activate factors that can result in the entrance of some bacteria. The actual mechanism is provided by the host cell cytoskeleton. Recall from Chapter 4 that eukaryotic cytoplasm has a complex internal structure (the cytoskeleton), consisting of protein filaments called microfilaments, intermediate filaments, and microtubules.

A major component of the cytoskeleton is a protein called actin, which is used by some microbes to penetrate host cells and by others to move through and between host cells.

Salmonella strains and *E. coli* make contact with the host cell plasma membrane. This leads to dramatic changes in the membrane at the point of contact. The microbes produce surface proteins called **invasins** that rearrange nearby actin filaments of the cytoskeleton. For example, when *S. typhimurium* makes contact with a host cell, invasins of the microbe cause the appearance of the host cell plasma membrane to resemble the splash of a drop of a liquid hitting a solid surface. This effect, called *membrane ruffling*, is the result of disruption in the cytoskeleton of the host cell (Figure 15.2). The microbe sinks into the ruffle and is engulfed by the host cell.

Once inside the host cell, certain bacteria such as *Shigella* species and *Listeria* species can actually use actin to propel themselves through the host cell cytoplasm and from one host cell to another. The condensation of actin on one end of the bacteria propels them through the cytoplasm. The bacteria also make contact with membrane junctions that form part of a transport network between host cells. The bacteria use a glycoprotein called *cadherin*, which bridges the junctions, to move from cell to cell.

The study of the numerous interactions between microbes and host cell cytoskeleton is a very intense area of investigation on virulence mechanisms.

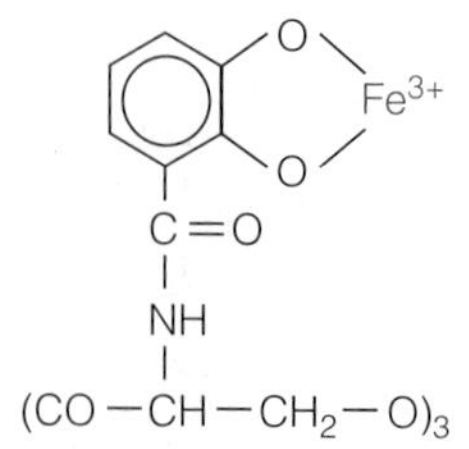

HOW BACTERIAL PATHOGENS DAMAGE HOST CELLS

When a microorganism invades a body tissue, it initially encounters phagocytes of the host. If the phagocytes are successful in destroying the invader, no further damage is done to the host. But if the pathogen overcomes the host's defense, then the microorganism can damage host cells in four basic ways: (1) by using the host's nutrients; (2) by causing direct damage in the immediate vicinity of the invasion; (3) by producing toxins, transported by blood and lymph, that damage sites far removed from the original site of invasion; and (4) by inducing hypersensitivity reactions. This fourth mechanism is considered in detail in Chapter 19. For now, we will discuss only the first three mechanisms.

USING THE HOST'S NUTRIENTS: SIDEROPHORES

LEARNING OBJECTIVE
• Describe the function of siderophores.

Iron is required for the growth of most pathogenic bacteria. However, the concentration of free iron in the human body is fairly low because most of the iron is tightly bound to iron-transport proteins, such as lactoferrin, transferrin, and ferritin, as well as hemoglobin. In order to obtain free iron, some pathogens secrete proteins called **siderophores** (Figure 15.3). When iron is needed by a pathogen, siderophores are released into the medium where they take the iron away from iron-transport proteins by binding the iron even more tightly. Once the iron-siderophore complex is formed, it is taken up by siderophore receptors on the bacterial surface. Then the iron is brought into the

FIGURE 15.3 Structure of enterobactin, one type of bacterial siderophore. Note where the iron (Fe^{3+}) is attached to the siderophore.

Q Of what value are siderophores?

bacterium. In some cases the iron is released from the complex to enter the bacterium; in other cases the iron enters as part of the complex.

As an alternative to iron acquisition by siderophores, some pathogens have receptors that bind directly to iron-transport proteins and hemoglobin. Then these are taken into the bacterium directly along with the iron. Also, it is possible that some bacteria produce toxins (described shortly) when iron levels are low. The toxins kill host cells, releasing their iron, which is then available to the bacteria.

DIRECT DAMAGE

LEARNING OBJECTIVE
• Provide an example of direct damage, and compare this to toxin production.

Once pathogens attach to host cells, they can cause direct damage as the pathogens use the host cell for nutrients and produce waste products. As pathogens metabolize and multiply in cells, the cells usually rupture. Many viruses and some intracellular bacteria and protozoa that grow in host cells are released when the host cell ruptures. Following their release, pathogens that rupture cells can spread to other tissues in even greater numbers. Some bacteria, such as *E. coli*, *Shigella*, *Salmonella*, and *Neisseria gonorrhoeae*, can induce host epithelial cells to engulf them by a process that resembles phagocytosis. These pathogens can disrupt host cells as they pass through and can then be extruded from the host cells by a reverse phagocytosis process, enabling them to enter other host cells. Some bacteria can also penetrate host cells by excreting enzymes and by their own motility; such penetration can itself damage the host cell. Most damage by bacteria, however, is done by toxins.

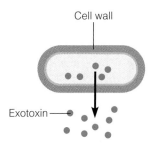

Cell wall

Exotoxin

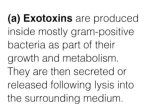

(a) Exotoxins are produced inside mostly gram-positive bacteria as part of their growth and metabolism. They are then secreted or released following lysis into the surrounding medium.

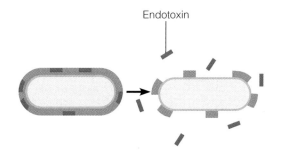

Endotoxin

(b) Endotoxins are part of the outer portion of the cell wall (lipid A; see Figure 4.13c) of gram-negative bacteria. They are liberated when the bacteria die and the cell wall breaks apart.

FIGURE 15.4 Exotoxins and endotoxins.

Q **What are the three principal types of exotoxins based on their mode of action?**

THE PRODUCTION OF TOXINS

LEARNING OBJECTIVES

- Contrast the nature and effects of exotoxins and endotoxins.
- Outline the mechanisms of action of A-B toxins, membrane-disrupting toxins, and superantigens. Classify diphtheria toxin, erythrogenic toxin, botulinum toxin, tetanus toxin, *Vibrio* enterotoxin, and staphylococcal enterotoxin.
- Identify the importance of the LAL assay.

Toxins are poisonous substances that are produced by certain microorganisms. They are often the primary factor contributing to the pathogenic properties of those microbes. The capacity of microorganisms to produce toxins is called **toxigenicity.** Toxins transported by the blood or lymph can cause serious, and sometimes fatal, effects. Some toxins produce fever, cardiovascular disturbances, diarrhea, and shock. Toxins can also inhibit protein synthesis, destroy blood cells and blood vessels, and disrupt the nervous system by causing spasms. Of the 220 or so known bacterial toxins, nearly 40% cause disease by damaging eukaryotic cell membranes. The term **toxemia** refers to the presence of toxins in the blood. Toxins are of two general types, based on their position relative to the microbial cell: exotoxins and endotoxins.

EXOTOXINS

Exotoxins are produced inside some bacteria as part of their growth and metabolism and are secreted by the bacterium into the surrounding medium or released following lysis (Figure 15.4a). Exotoxins are proteins, and many are enzymes that catalyze only certain biochemical reactions. Because of the enzymatic nature of most exotoxins, even small amounts are quite harmful because they can act over and over again. Bacteria that produce exotoxins may be gram-positive or gram-negative. The genes for most (perhaps all) exotoxins are carried on bacterial plasmids or phages. Because exotoxins are soluble in body fluids, they can easily diffuse into the blood and are rapidly transported throughout the body.

Exotoxins work by destroying particular parts of the host's cells or by inhibiting certain metabolic functions. They are highly specific in their effects on body tissues. Exotoxins are among the most lethal substances known. Only 1 mg of the botulinum exotoxin is enough to kill 1 million guinea pigs. Fortunately, only a few bacterial species produce such potent exotoxins.

Diseases caused by bacteria that produce exotoxins are often caused by minute amounts of exotoxins, not by the bacteria themselves. It is the exotoxins that produce the specific signs and symptoms of the disease. Thus, exotoxins are disease-specific. For example, botulism is usually due to ingesting the exotoxin, not a bacterial infection. Likewise, staphylococcal food poisoning is an *intoxication*, not an infection.

The body produces antibodies called **antitoxins** that provide immunity to exotoxins. When exotoxins are inactivated by heat or by formaldehyde, iodine, or other chemicals, they no longer cause the disease but can still stimulate the body to produce antitoxins. Such altered exotoxins are called **toxoids.** When toxoids are injected into the body as a vaccine, they stimulate antitoxin production so that immunity is produced. Diphtheria and tetanus can be prevented by toxoid vaccination.

Naming Exotoxins Exotoxins are named on the basis of several characteristics. One is the type of host cell that is attached. For example, *neurotoxins* attack nerve cells, *cardiotoxins* attack heart cells, *hepatotoxins* attack liver

FIGURE 15.5 The action of an exotoxin.
A proposed model for the mechanism of action of diphtheria toxin.

Q **Why is this called an A-B toxin?**

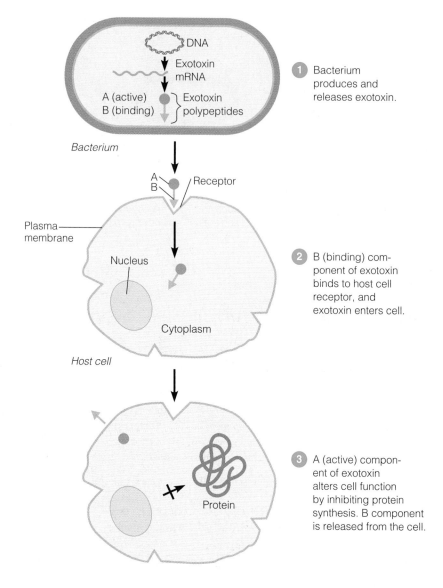

① Bacterium produces and releases exotoxin.

② B (binding) component of exotoxin binds to host cell receptor, and exotoxin enters cell.

③ A (active) component of exotoxin alters cell function by inhibiting protein synthesis. B component is released from the cell.

cells, *leukotoxins* attack leukocytes, *enterotoxins* attack the lining of the gastrointestinal tract, and *cytotoxins* attack a wide variety of cells. Some exotoxins are named for the diseases with which they are associated. Examples include *diphtheria toxin* (cause of diphtheria) and *tetanus toxin* (cause of tetanus). Other exotoxins are named for the specific bacterium that produces them, for example, *botulinum toxin* (*Clostridium botulinum*) and *Vibrio enterotoxin* (*Vibrio cholerae*).

Types of Exotoxins   Exotoxins are divided into three principal types on the basis of their structure and function: (1) A-B toxins, (2) membrane-disrupting toxins, and (3) superantigens.

A-B Toxins A-B toxins were the first toxins to be studied intensively and are so named because they consist of two parts designated A and B, both of which are polypeptides. Most exotoxins are A-B toxins. The A part is the active (enzyme) component, and the B part is the binding component. An example of an A-B toxin is the diphtheria toxin, which is illustrated in Figure 15.5.

① In the first step, the A-B toxin is released from the bacterium.

② The B part binds to a surface receptor on the best cell, usually a carbohydrate. Following binding, the A-B toxin is transported across the plasma membrane into the host cell cytoplasm.

③ Within the cell, the A-B components separate. The A part inhibits protein synthesis and kills the host cell while the B part is released from the cell.

Membrane-Disrupting Toxins Membrane-disrupting toxins cause lysis of host cells by disrupting their plasma membranes. Some do this by forming protein channels in the plasma membrane; others disrupt the phospholipid portion of the membrane. The cell-lysing exotoxin of *Staphylococcus aureus* is an example of an exotoxin that forms protein channels, whereas that of *Clostridium perfringens* is an example of an exotoxin that disrupts the phospholipids. Membrane-disrupting toxins contribute to virulence by killing host cells, especially phagocytes, and by aiding the escape of bacteria from sacs within phagocytes (phagosomes) into the host cell's cytoplasm.

Membrane-disrupting toxins that kill phagocytic leukocytes (white blood cells) are called **leukocidins.** They act by forming protein channels. Leukocidins are also active against macrophages, phagocytes present in tissues. Most leukocidins are produced by staphylococci and streptococci. The damage to phagocytes decreases host resistance. Membrane-disrupting toxins that destroy erythrocytes (red blood cells), also by forming protein channels, are called **hemolysins.** Important producers of hemolysins include staphylococci and streptococci. Hemolysins produced by streptococci are called **streptolysins.** One kind, called *streptolysin O (SLO)*, is so named because it is inactivated by atmospheric oxygen. Another kind of streptolysin is called *streptolysin S (SLS)* because it is stable in an oxygen environment. Both streptolysins can cause lysis not only of red blood cells, but also of white blood cells (whose function is to kill the streptococci) and other body cells.

Superantigens **Superantigens** are antigens that provoke a very intense immune response. They are bacterial proteins. Through a series of interactions with various cells of the immune system, superantigens nonspecifically stimulate the proliferation of immune cells called T cells. These cells are types of white blood cells (lymphocytes) that act against foreign organisms and tissues (transplants) and regulate the activation and proliferation of other cells of the immune system. In response to superantigens, T cells are stimulated to release enormous amounts of chemicals called cytokines (see Chapter 17, page 518). *Cytokines* are small protein hormones produced by various body cells, especially T cells, that regulate immune responses and mediate cell-to-cell communication. The excessively high levels of cytokines released by T cells enter the bloodstream and give rise to a number of symptoms, including fever, nausea, vomiting, diarrhea, and sometimes shock and even death. Bacterial superantigens include the staphylococcal toxins that cause food poisoning and toxic shock syndrome.

Representative Exotoxins

Next we briefly describe a few of the more notable exotoxins (antitoxins will be discussed further in Chapter 18).

Diphtheria Toxin *Corynebacterium diphtheriae* produces the *diphtheria toxin* only when it is infected by a lysogenic phage carrying the *tox* gene. This cytotoxin inhibits protein synthesis in eukaryotic cells. It does this using an A-B toxin mechanism, which is illustrated in Figure 15.5.

Erythrogenic Toxins *Streptococcus pyogenes* has the genetic material to synthesize three types of cytotoxins, designated A, B, and C. These *erythrogenic* (*erythro* = red; *gen* = producing) *toxins* are superantigens that damage the plasma membranes of blood capillaries under the skin and produce a red skin rash. Scarlet fever, caused by *S. pyogenes* exotoxins, is named for this characteristic rash.

Botulinum Toxin *Botulinum toxin* is produced by *Clostridium botulinum*. Although toxin production is associated with the germination of endospores and the growth of vegetative cells, little of the toxin appears in the medium until it is released by lysis late in growth. Botulinum toxin is an A-B neurotoxin; it acts at the neuromuscular junction (the junction between nerve cells and muscle cells) and prevents the transmission of impulses from the nerve cell to the muscle. The toxin accomplishes this by binding to nerve cells and inhibiting the release of a neurotransmitter called acetylcholine. As a result, botulinum toxin causes paralysis in which muscle tone is lacking (flaccid paralysis). *C. botulinum* produces several different types of botulinum toxin, and each possesses a different potency.

Tetanus Toxin *Clostridium tetani* produces tetanus neurotoxin, also known as *tetanospasmin*. This A-B toxin reaches the central nervous system and binds to nerve cells that control the contraction of various skeletal muscles. These nerve cells normally send inhibiting impulses that prevent random contractions and terminate completed contractions. The binding of tetanospasmin blocks this relaxation pathway (see Chapter 22). The result is uncontrollable muscle contractions, producing the convulsive symptoms (spasmodic contractions) of tetanus, or "lockjaw."

Vibrio Enterotoxin *Vibrio cholerae* produces an A-B enterotoxin called *cholera toxin*. Subunit B binds to epithelial cells, and subunit A causes cells to secrete large amounts of fluids and electrolytes (ions). Normal muscular contractions are disturbed, leading to severe diarrhea that may be accompanied by vomiting. *Heat-labile enterotoxin* (so named because it is more sensitive to heat than are most toxins), produced by some strains of *E. coli*, has an action identical to that of *Vibrio* enterotoxin.

Staphylococcal Enterotoxin *Staphylococcus aureus* produces a superantigen that affects the intestines in the same way as *Vibrio* enterotoxin. A strain of *S. aureus* also produces a superantigen that results in the symptoms associated with toxic shock syndrome (see Chapter 21). A summary of diseases produced by exotoxins is shown in Table 15.2.

ENDOTOXINS

Endotoxins differ from exotoxins in several ways. Endotoxins are part of the outer portion of the cell wall of gramnegative bacteria (see Figure 15.4b). Recall from Chapter 4

TABLE 15.2	Diseases Caused by Exotoxins		
Disease	Bacterium	Type of Exotoxin	Mechanism
Botulism	*Clostridium botulinum*	A-B	Neurotoxin prevents the transmission of nerve impulses; flaccid paralysis results.
Tetanus	*Clostridium tetani*	A-B	Neurotoxin blocks nerve impulses to muscle relaxation pathway; results in uncontrollable muscle contractions.
Diphtheria	*Corynebacterium diphtheriae*	A-B	Cytotoxin inhibits protein synthesis, especially in nerve, heart, and kidney cells.
Scalded skin syndrome	*Staphylococcus aureus*	A-B	One exotoxin causes skin layers to separate and slough off (scalded skin).
Cholera	*Vibrio cholerae*	A-B	Enterotoxin causes secretion of large amounts of fluids and electrolytes that result in diarrhea.
Traveler's diarrhea	Enterotoxigenic *Escherichia coli* and *Shigella* spp.	A-B	Enterotoxin causes secretion of large amounts of fluids and electrolytes that result in diarrhea.
Gas gangrene and food poisoning	*Clostridium perfringens* and other species of *Clostridium*	Membrane-disrupting	One exotoxin (cytotoxin) causes massive red blood cell destruction (hemolysis); another exotoxin (enterotoxin) is related to food poisoning and causes diarrhea.
Antibiotic-associated diarrhea	*Clostridium difficile*	Membrane-disrupting	Enterotoxin causes secretion of fluids and electrolytes that results in diarrhea; cytotoxin disrupts host cytoskeleton.
Food poisoning	*Staphylococcus aureus*	Superantigen	Enterotoxin causes secretion of fluids and electrolytes that results in diarrhea.
Toxic shock syndrome (TSS)	*Staphylococcus aureus*	Superantigen	Toxin causes secretion of fluids and electrolytes from capillaries that decreases blood volume and lowers blood pressure.

that gram-negative bacteria have an outer membrane surrounding the peptidoglycan layer of the cell wall. This outer membrane consists of lipoproteins, phospholipids, and lipopolysaccharides (LPSs) (see Figure 4.13c, page 86). The lipid portion of LPS, called **lipid A,** is the endotoxin. Thus, endotoxins are lipopolysaccharides, whereas exotoxins are proteins.

Endotoxins are released when gram-negative bacteria die and their cell walls undergo lysis, thus liberating the endotoxin. (Endotoxins are also released during bacterial multiplication.) Antibiotics used to treat diseases caused by gram-negative bacteria can lyse the bacterial cells; this reaction releases endotoxin and may lead to an immediate worsening of the symptoms, but the condition usually improves as the endotoxin breaks down. Endotoxins exert their effects by stimulating macrophages to release cytokines in very high concentrations. At these levels, cytokines are toxic. All endotoxins produce the same signs and symptoms, regardless of the species of microorganism, although not to the same degree. These include chills, fever, weakness, generalized aches, and, in some cases, shock and even death. Endotoxins can also induce miscarriage.

Another consequence of endotoxins is the activation of blood-clotting proteins, causing the formation of small blood clots. These blood clots obstruct capillaries, and the resulting decreased blood supply induces the death of tissues. This condition is referred to as *disseminated intravascular clotting.*

The fever (pyrogenic response) caused by endotoxins is believed to occur as depicted in Figure 15.6.

❶ When gram-negative bacteria are ingested by phagocytes and

❷ degraded in vacuoles, the LPSs of the bacterial cell wall are released. These endotoxins cause macrophages to produce a cytokine called **interleukin-1 (IL-1),** formerly called *endogenous pyrogen,*

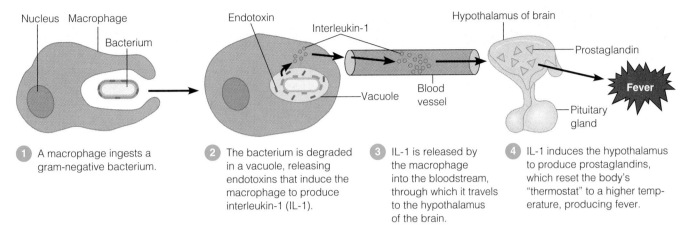

① A macrophage ingests a gram-negative bacterium.

② The bacterium is degraded in a vacuole, releasing endotoxins that induce the macrophage to produce interleukin-1 (IL-1).

③ IL-1 is released by the macrophage into the bloodstream, through which it travels to the hypothalamus of the brain.

④ IL-1 induces the hypothalamus to produce prostaglandins, which reset the body's "thermostat" to a higher temperature, producing fever.

FIGURE 15.6 Endotoxins and the pyrogenic response. The proposed mechanism by which endotoxins cause fever.

Q **What is an endotoxin?**

③ which is carried via the blood to the hypothalamus, a temperature control center in the brain.

④ IL-1 induces the hypothalamus to release lipids called prostaglandins, which reset the thermostat in the hypothalamus at a higher temperature. The result is a fever.

Bacterial cell death caused by lysis or antibiotics can also produce fever by this mechanism. Both aspirin and acetaminophen reduce fever by inhibiting the synthesis of prostaglandins. (The function of fever in the body is discussed in Chapter 16, page 489.)

Shock refers to any life-threatening decrease in blood pressure. Shock caused by bacteria is called **septic shock.** Gram-negative bacteria cause *endotoxic shock.* Like fever, the shock produced by endotoxins is related to the secretion of a cytokine by macrophages. Phagocytosis of gram-negative bacteria causes the phagocytes to secrete a polypeptide called **tumor necrosis factor (TNF),** or *cachectin.* TNF binds to many tissues in the body and alters their metabolism in a number of ways. One effect of TNF is damage to blood capillaries; their permeability is increased, and they lose large amounts of fluid. The result is a drop in blood pressure that results in shock. Low blood pressure has serious effects on the kidneys, lungs, and gastrointestinal tract. Another mechanism that causes septic shock is discussed in the box in Chapter 16 (page 486). In addition, the presence of gram-negative bacteria such as *Haemophilus influenzae* type b in cerebrospinal fluid causes the release of IL-1 and TNF. These, in turn, cause a weakening of the blood–brain barrier that normally protects the central nervous system from infection. The weakened barrier lets phagocytes in, but this also lets more bacteria enter from the bloodstream. In the United States, 750,000 cases of septic shock occur each year. One-third of the patients die within a month, and nearly half die within 6 months.

Endotoxins do not promote the formation of effective antitoxins against the carbohydrate component of an endotoxin. Antibodies are produced, but they tend not to counter the effect of the toxin; sometimes, in fact, they actually enhance its effect.

Representative microorganisms that produce endotoxins are *Salmonella typhi* (the causative agent of typhoid fever), *Proteus* spp. (frequently the causative agents of urinary tract infections), and *Neisseria meningitidis* (the causative agent of meningococcal meningitis).

It is important to have a sensitive test to identify the presence of endotoxins in drugs, medical devices, and body fluids. Materials that have been sterilized may contain endotoxins, even though no bacteria can be cultured from them. One such laboratory test is called the **Limulus amoebocyte lysate (LAL) assay,** which can detect even minute amounts of endotoxin. The hemolymph (blood) of the Atlantic coast horseshoe crab, *Limulus polyphemus,* contains white blood cells called amoebocytes, which have large amounts of a protein (lysate) that causes clotting. In the presence of endotoxin, amoebocytes in the crab hemolymph lyse and liberate their clotting protein. The resulting gel-clot (precipitate) is a positive test for the presence of endotoxin. The degree of the reaction is measured using a spectrophotometer (see Figure 6.20, page 183).

A comparison of exotoxins and endotoxins appears in Table 15.3.

TABLE 15.3	Exotoxins and Endotoxins	
Property	Exotoxin	Endotoxin
Bacterial source	Mostly from gram-positive bacteria	Gram-negative bacteria
Relation to microorganism	Metabolic product of growing cell	Present in LPS of outer membrane of cell wall and released with destruction of cell or during cell division
Chemistry	Proteins, usually with two parts (A-B)	Lipid portion (lipid A) of LPS of outer membrane (lipopolysaccharide).
Pharmacology (effect on body)	Specific for a particular cell structure or function in the host (mainly affects cell functions, nerves, and gastrointestinal tract)	General, such as fever, weaknesses, aches, and shock; all produce the same effects
Heat stability	Unstable; can usually be destroyed at 60–80°C (except staphylococcal enterotoxin)	Stable; can withstand autoclaving (121°C for 1 hour)
Toxicity (ability to cause disease)	High	Low
Fever-producing	No	Yes
Immunology (relation to antibodies)	Can be converted to toxoids to immunize against toxin; neutralized by antitoxin	Not easily neutralized by antitoxin; therefore, effective toxoids cannot be made to immunize against toxin
Lethal dose	Small	Considerably larger
Representative diseases	Gas gangrene, tetanus, botulism, diphtheria, scarlet fever	Typhoid fever, urinary tract infections, and meningococcal meningitis

PLASMIDS, LYSOGENY, AND PATHOGENICITY

LEARNING OBJECTIVE

• Using examples, describe the roles of plasmids and lysogeny in pathogenicity.

Recall from Chapters 4 (page 95) and 8 (page 245) that plasmids are small, circular DNA molecules that are not connected to the main bacterial chromosome and are capable of independent replication. One group of plasmids, called R (resistance) factors, is responsible for the resistance of some microorganisms to antibiotics. In addition, a plasmid may carry the information that determines a microbe's pathogenicity. Examples of virulence factors that are encoded by plasmid genes are tetanus neurotoxin, heat-labile enterotoxin, and staphylococcal enterotoxin. Other examples are dextransucrase, an enzyme produced by *Streptococcus mutans* that is involved in tooth decay;

adhesins and coagulase produced by *Staphylococcus aureus*; and a type of fimbria specific to enteropathogenic strains of *E. coli*.

In Chapter 13, we noted that some bacteriophages (viruses that infect bacteria) can incorporate their DNA into the bacterial chromosome, becoming a prophage, and thus remain latent (do not cause lysis of the bacterium). Such a state is called *lysogeny*, and cells containing a prophage are said to be lysogenic. One outcome of lysogeny is that the host bacterial cell and its progeny may exhibit new properties encoded by the bacteriophage DNA. Such a change in the characteristics of a microbe due to a prophage is called **lysogenic conversion.** As a result of lysogenic conversion, the bacterial cell is immune to infection by the same type of phage. In addition, lysogenic cells are of medical importance because some bacterial pathogenesis is caused by the prophages they contain.

Among the bacteriophage genes that contribute to pathogenicity are the genes for diphtheria toxin, erythrogenic toxins, staphylococcal enterotoxin and pyrogenic toxin, botulinum neurotoxin, and the capsule produced by *Streptococcus pneumoniae*. Pathogenic strains of *Vibrio cholerae* carry lysogenic phages. These phages can transmit the choleratoxin gene to nonpathogenic *V. cholerae* strains, increasing the number of pathogenic bacteria.

PATHOGENIC PROPERTIES OF VIRUSES

LEARNING OBJECTIVE
- List nine cytopathic effects of viral infections.

The pathogenic properties of viruses depend on their gaining access to a host, evading the host's defenses, and then causing damage to or death of the host cell while reproducing themselves.

VIRAL MECHANISMS FOR EVADING HOST DEFENSES

escaping

Viruses have a variety of mechanisms that enable them to evade destruction by the host's immune response (see Chapter 17). For example, viruses can penetrate and grow inside of host cells, where components of the immune system cannot reach them. Viruses gain access to cells because they have attachment sites for receptors on their target cells. When such an attachment site is brought together with an appropriate receptor, the virus can bind to and penetrate the cell. Some viruses gain access to host cells because their attachment sites mimic substances useful to those cells. For example, the attachment sites of rabies virus can mimic the neurotransmitter acetylcholine. As a result, the virus can enter the host cell along with the neurotransmitter.

The AIDS virus (HIV) goes further by hiding its attachment sites from the immune response and by attacking components of the immune system directly. Like most viruses, HIV is cell-specific; that is, it attacks only particular body cells. HIV attacks only those cells that have a surface marker called the CD4 protein, most of which are cells of the immune system called T cells (T lymphocytes). Binding sites on HIV are complementary to the CD4 protein. The surface of the virus is folded to form ridges and valleys, and the HIV binding sites are located on the floors of the valleys. CD4 proteins are long enough and slender enough to reach these binding sites, whereas antibody molecules made against HIV are too large to make contact with the sites. As a result, it is difficult for these antibodies to destroy HIV.

CYTOPATHIC EFFECTS OF VIRUSES

Infection of a host cell by an animal virus usually kills the host cell (see Chapter 13). Death can be caused by the accumulation of large numbers of multiplying viruses, by the effects of viral proteins on the permeability of the host cell's plasma membrane, or by inhibition of host DNA, RNA, or protein synthesis. The visible effects of viral infection are known as **cytopathic effects (CPE).** Those cytopathic effects that result in cell death are called *cytocidal effects;* those that result in cell damage but not cell death are called *noncytocidal effects*. CPEs are used to diagnose many viral infections.

Cytopathic effects vary with the virus. One difference is the point in the viral infection cycle at which the effects occur. Some viral infections result in early changes in the host cell; in other infections, changes are not seen until a much later stage. A virus can produce one or more of the following cytopathic effects:

1. At some stage in their multiplication, cytocidal viruses cause the macromolecular synthesis within the host cell to stop. Some viruses, such as herpes simplex virus, irreversibly stop mitosis.

2. When a cytocidal virus infects a cell, it causes the cell's lysosomes to release their enzymes, resulting in destruction of intracellular contents and host cell death.

3. **Inclusion bodies** are granules found in the cytoplasm or nucleus of some infected cells (Figure 15.7a). These granules are sometimes viral parts—nucleic acids or proteins in the process of being assembled into virions. The granules vary in size, shape, and staining properties, according to the virus. Inclusion bodies are characterized by their ability to stain with an acidic stain (acidophilic) or with a basic stain (basophilic). Other inclusion bodies arise at sites of earlier viral synthesis but do not contain assembled viruses or their components. Inclusion bodies are important because their presence can help identify the virus causing an infection. For example, in most cases, rabies virus produces inclusion bodies (Negri bodies) in the cytoplasm of nerve cells, and their presence in the brain tissue of animals suspected of being rabid has been used as one diagnostic tool for rabies. Diagnostic inclusion bodies are also associated with measles virus, vaccinia virus, smallpox virus, herpesvirus, and adenoviruses.

4. At times, several adjacent infected cells fuse to form a very large multinucleate cell called a **syncytium**

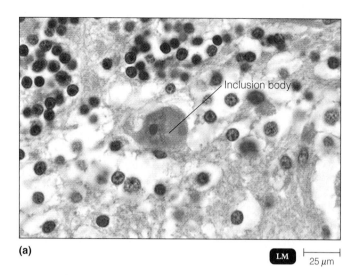

(a)   LM |——| 25 μm

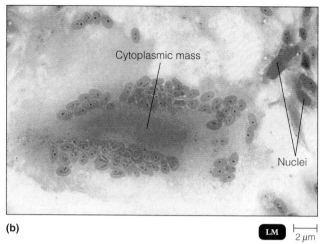

(b)   LM |——| 2 μm

FIGURE 15.7 Some cytopathic effects of viruses. (**a**) Cytoplasmic inclusion body in a brain tissue from a person who died of rabies. (**b**) Portion of a syncytium (giant cell) formed in a cell infected with measles virus. The cytoplasmic mass is probably the Golgi complexes of fused cells.

Q **What are cytopathic effects?**

(Figure 15.7b). Such giant cells are produced from infections by viruses that cause diseases, such as measles, mumps, and the common cold.

5. Some viral infections result in changes in the host cell's functions with no visible changes in the infected cells. For example, when measles virus attaches to its receptor called CD46, the CD46 prompts the cell to reduce production of an immune substance called IL-12, reducing the host's ability to fight the infection.

6. Some virus-infected cells produce substances called **interferons.** Viral infection induces cells to produce interferons, but the host cell's DNA actually codes for the interferon. This protects neighboring uninfected cells from viral infection. (Interferons will be discussed further in Chapter 16, page 494.)

7. Many viral infections induce antigenic changes on the surface of the infected cells. These antigenic changes elicit a host antibody response against the infected cell, and thus they target the cell for destruction by the host's immune system.

8. Some viruses induce chromosomal changes in the host cell. For example, some viral infections result in chromosomal damage to the host cell, most often chromosomal breakage. Frequently oncogenes (cancer-causing genes) may be contributed or activated by a virus.

9. Most normal cells cease growing in vitro when they come close to another cell, a phenomenon known as **contact inhibition.** Viruses capable of causing cancer *transform* host cells, as discussed in Chapter 13. Trans-

formation results in an abnormal, spindle-shaped cell that does not recognize contact inhibition (Figure 15.8). The loss of contact inhibition results in unregulated cell growth.

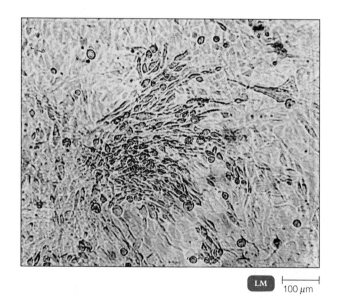

LM |——| 100 μm

FIGURE 15.8 Transformed cells in culture. In the center of this photomicrograph is a cluster of chick embryo cells transformed by Rous sarcoma virus. Such a cluster results from the multiplication of a single cell infected with a transforming virus. Notice how the transformed cells appear dark, in contrast to the monolayer of light, flat, normal cells around them. This appearance is caused by their spindle shapes and their uninhibited growth due to the absence of contact inhibition.

Q **What is contact inhibition?**

TABLE 15.4	Cytopathic Effects of Selected Viruses
Virus (Genus)	**Cytopathic Effect**
Poliovirus *(Enterovirus)*	Cytocidal (cell death)
Papovavirus (family Papovaviridae)	Acidophilic inclusion bodies in nucleus
Adenovirus *(Mastadenovirus)*	Basophilic inclusion bodies in nucleus
Rhabdovirus (family Rhabdoviridae)	Acidophilic inclusion bodies in cytoplasm
Cytomegalovirus	Acidophilic inclusion bodies in nucleus and cytoplasm
Measles virus *(Morbillivirus)*	Cell fusion
Polyomavirus	Transformation
HIV *(Lentivirus)*	Destruction of T cells

Some representative viruses that cause cytopathic effects are presented in Table 15.4. In Part Four we will discuss the pathological properties of viruses in more detail.

PATHOGENIC PROPERTIES OF FUNGI, PROTOZOA, HELMINTHS, AND ALGAE

LEARNING OBJECTIVE
• Discuss the causes of symptoms in fungal, protozoan, helminthic, and algal diseases.

This section describes some general pathological effects of fungi, protozoa, helminths, and algae that cause human disease. Most specific diseases caused by fungi, protozoa, and helminths, along with the pathological properties of these organisms, are discussed in detail in Chapters 21 to 26.

FUNGI

Although fungi cause disease, they do not have a well-defined set of virulence factors. Some fungi have metabolic products that are toxic to human hosts. In such cases, however, the toxin is only an indirect cause of disease, as the fungus is already growing in or on the host. Chronic fungal infections, such as from molds growing in homes, can also provoke an allergic response in the host.

Trichothecenes are fungal toxins that inhibit protein synthesis in eukaryotic cells. Ingestion of these toxins causes headaches, chills, severe nausea, vomiting, and visual disturbances. These toxins are produced by *Fusarium* (fu′săr-ē-um) and *Stachybotrys* (stak′ē-bo-tris) growing on grains and wallboard in homes.

There is evidence that some fungi do have virulence factors. Two fungi that can cause skin infections, *Candida albicans* and *Trichophyton* (trik-ō-fī′ton), secrete proteases. These enzymes may modify host cell membranes to allow attachment of the fungi. *Cryptococcus neoformans* (krip-tō-kok′kus nē-ō-fôr′manz) is a fungus that causes a type of meningitis; it produces a capsule that helps it resist phagocytosis. Some fungi have become resistant to antifungal drugs by decreasing their synthesis of receptors for these drugs.

The disease called ergotism, which was common in Europe during the Middle Ages, is caused by a toxin produced by an ascomycete plant pathogen, *Claviceps purpurea* (kla′vi-seps pŭr-pū-rē′ä), that grows on grains. The toxin is contained in **sclerotia,** highly resistant portions of the mycelia of the fungus that can detach. The toxin itself, **ergot,** is an alkaloid that can cause hallucinations resembling those produced by LSD (lysergic acid diethylamide); in fact, ergot is a natural source of LSD. Ergot also constricts capillaries and can cause gangrene of the limbs by preventing proper blood circulation in the body. Although *C. purpurea* still occasionally occurs on grains, modern milling usually removes the sclerotia.

Several other toxins are produced by fungi that grow on grains or other plants. For example, peanut butter is occasionally recalled because of excessive amounts of **aflatoxin,** a toxin that has carcinogenic properties. Aflatoxin is produced by the growth of the mold *Aspergillus flavus*. When ingested, the toxin might be altered in a human body to a mutagenic compound.

A few mushrooms produce toxins called **mycotoxins** (toxins produced by fungi). Examples are **phalloidin** and **amanitin,** produced by *Amanita phalloides* (am-an-ī′ta fal-loi′dēz), commonly known as the deathcap. These neurotoxins are so potent that ingestion of the *Amanita* mushroom may result in death.

PROTOZOA

The presence of protozoa and their waste products often produces disease symptoms in the host (see Table 12.5, page 369). Some protozoa, such as *Plasmodium*, the causative agent of malaria, invade host cells and reproduce within them, causing their rupture. *Toxoplasma* attaches to macrophages and gains entry by phagocytosis. The parasite prevents normal acidification and digestion; thus, it can grow in the phagocytic vacuole. Other protozoa, such as *Giardia lamblia*, the causative agent of giardiasis, attach to host cells by a sucking disc (see Figure 25.18, page 772) and digest the cells and tissue fluids.

Some protozoa can evade host defenses and cause disease for very long periods of time. For example, *Giardia*, which causes diarrhea, and *Trypanosoma*, which causes African trypanosomiasis (sleeping sickness), both have a mechanism that enables them to stay one step ahead of the host's immune system. The immune system is alerted to recognize foreign substances called antigens; the presence of antigens causes the immune system to produce antibodies designed to destroy them (see Chapter 17). When *Trypanosoma* is introduced into the bloodstream by a tsetse fly, it produces and displays a specific antigen. In response, the body produces antibodies against that antigen. However, within two weeks, the microbe stops displaying the original antigen and instead produces and displays a different one (see Figure 22.16, page 661). Thus, the original antibodies are no longer effective. Because the microbe can make up to 1000 different antigens, such an infection can last for decades.

HELMINTHS

The presence of helminths also often produces disease symptoms in a host (see Table 12.6, page 378). Some of these organisms actually use host tissues for their own growth or produce large parasitic masses; the resulting cellular damage evokes the symptoms. An example is the roundworm *Wuchereria bancrofti* (vū-kėr-ār′ē-ä ban-krof′tē), the causative agent of elephantiasis. This parasite blocks lymphatic circulation, leading to an accumulation of lymph and eventually causing grotesque swelling of the legs and other body parts. Waste products of the metabolism of these parasites can also contribute to the symptoms of a disease.

ALGAE

A few species of algae produce neurotoxins. For example, some genera of dinoflagellates, such as *Alexandrium*, are important medically because they produce a neurotoxin called **saxitoxin**. Although mollusks that feed on the dinoflagellates that produce saxitoxin show no symptoms of disease, people who eat the mollusks develop paralytic shellfish poisoning, with symptoms similar to botulism. Public health agencies frequently prohibit human consumption of mollusks during red tides (see Figure 27.15, page 827).

PORTALS OF EXIT

LEARNING OBJECTIVE
• Compare and contrast portal of entry and portal of exit.

In the beginning of the chapter, you learned how microbes enter the body through a preferred route, or portal of entry.

Microbes also leave the body via specific routes called **portals of exit** in secretions, excretions, discharges, or tissue that has been shed. In general, portals of exit are related to the part of the body that has been infected. Thus, in general, a microbe uses the same portal for entry and exit. By using various portals of exit, pathogens can spread through a population by moving from one susceptible host to another. As you learned in Chapter 14, this type of information about the dissemination of a disease is very important to epidemiologists.

The most common portals of exit are the respiratory and gastrointestinal tracts. For example, many pathogens living in the respiratory tract exit in discharges from the mouth and nose; such discharges are expelled during coughing or sneezing. These microorganisms are found in droplets formed from mucus. Pathogens that cause tuberculosis, whooping cough, pneumonia, scarlet fever, meningococcal meningitis, chickenpox, measles, mumps, smallpox, and influenza are discharged through the respiratory route. Other pathogens exit via the gastrointestinal tract in feces or saliva. Feces may be contaminated with pathogens associated with salmonellosis, cholera, typhoid fever, shigellosis, amoebic dysentery, and poliomyelitis. Saliva can also contain pathogens, such as those that cause rabies, mumps, and infectious mononucleosis.

Another important route of exit is the genitourinary tract. Microbes responsible for sexually transmitted diseases are found in secretions from the penis and vagina. Urine can also contain the pathogens responsible for typhoid fever and brucellosis, which can exit via the urinary tract. Skin or wound infections are other portals of exit. Infections transmitted from the skin include yaws, impetigo, ringworm, herpes simplex, and warts. Drainage from wounds can spread infections to another person directly or by contact with a contaminated fomite. Infected blood can be removed and reinjected by biting insects and contaminated needles and syringes to spread infection within a population. Examples of diseases transmitted by biting insects are yellow fever, plague, tularemia, and malaria. AIDS and hepatitis B may be transmitted by contaminated needles and syringes.

* * *

In the next chapter, we will examine a group of nonspecific defenses of the host against disease. But before proceeding, examine Figure 15.9 carefully. It summarizes some key concepts of the microbial mechanisms of pathogenicity we have discussed in this chapter.

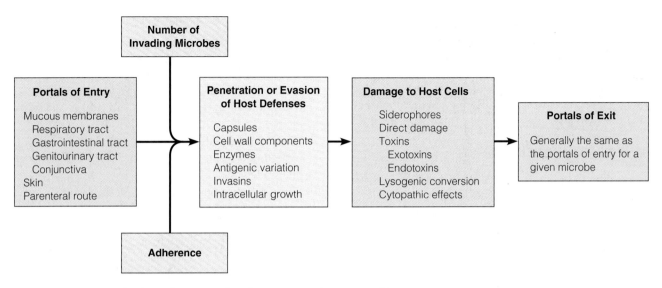

FIGURE 15.9 Microbial mechanisms of pathogenicity. A summary of how microorganisms cause disease.

Q **Which terms in this figure are virulence factors?**

STUDY OUTLINE

INTRODUCTION (p. 451)

1. Pathogenicity is the ability of a pathogen to produce a disease by overcoming the defenses of the host.

2. Virulence is the degree of pathogenicity.

HOW MICROORGANISMS ENTER A HOST (pp. 452–456)

1. The specific route by which a particular pathogen gains access to the body is called its portal of entry.

PORTALS OF ENTRY (p. 452)

2. Many microorganisms can penetrate mucous membranes of the conjunctiva and the respiratory, gastrointestinal, and genitourinary tracts.

3. Microorganisms that are inhaled with droplets of moisture and dust particles gain access to the respiratory tract.

4. The respiratory tract is the most common portal of entry.

5. Microorganisms that gain access via the genitourinary tract can enter the body through mucous membranes.

6. Microorganisms enter the gastrointestinal tract via food, water, and contaminated fingers.

7. Most microorganisms cannot penetrate intact skin; they enter hair follicles and sweat ducts.

8. Some microorganisms can gain access to tissues by inoculation through the skin and mucous membranes in bites, injections, and other wounds. This route of penetration is called the parenteral route.

THE PREFERRED PORTAL OF ENTRY (pp. 452–453)

9. Many microorganisms can cause infections only when they gain access through their specific portal of entry.

NUMBERS OF INVADING MICROBES (pp. 453–454)

10. Virulence can be expressed as LD_{50} (lethal dose for 50% of the inoculated hosts) or ID_{50} (infectious dose for 50% of the inoculated hosts).

ADHERENCE (pp. 454–455)

11. Surface projections on a pathogen called adhesins (ligands) adhere to complementary receptors on the host cells.

12. Adhesins can be glycoproteins or lipoproteins and are frequently associated with fimbriae.

13. Mannose is the most common receptor.

14. Biofilms provide attachment and resistance to antimicrobial agents.

HOW BACTERIAL PATHOGENS PENETRATE HOST DEFENSES (pp. 456–458)

CAPSULES (p. 456)

1. Some pathogens have capsules that prevent them from being phagocytized.

CELL WALL COMPONENTS (p. 456)

2. Proteins in the cell wall can facilitate adherence or prevent a pathogen from being phagocytized.

3. Some microbes can reproduce inside phagocytes.

ENZYMES (pp. 456–457)

4. Leukocidins destroy neutrophils and macrophages.

5. Local infections can be protected in a fibrin clot caused by the bacterial enzyme coagulase.

6. Bacteria can spread from a focal infection by means of kinases (which destroy blood clots), hyaluronidase (which destroys a mucopolysaccharide that holds cells together), and collagenase (which hydrolyzes connective tissue collagen).

7. IgA proteases destroy IgA antibodies.

ANTIGENIC VARIATION (p. 457)

8. Some microbes vary expression of antigens, thus avoiding the host's antibodies.

PENETRATION INTO THE HOST CELL CYTOSKELETON (pp. 457–458)

9. *Salmonella* bacteria produce invasins, proteins that cause the actin of the host cell's cytoskeleton to form a basket that carries the bacteria into the cell.

HOW BACTERIAL PATHOGENS DAMAGE HOST CELLS (pp. 458–465)

USING THE HOST'S NUTRIENTS: SIDEROPHORES (p. 458)

1. Bacteria get iron from the host using siderophores.

DIRECT DAMAGE (p. 458)

2. Host cells can be destroyed when pathogens metabolize and multiply inside the host cells.

THE PRODUCTION OF TOXINS (pp. 459–464)

3. Poisonous substances produced by microorganisms are called toxins; toxemia refers to the presence of toxins in the blood. The ability to produce toxins is called toxigenicity.

4. Exotoxins are produced by bacteria and released into the surrounding medium. Exotoxins, not the bacteria, produce the disease symptoms.

5. Antibodies produced against exotoxins are called antitoxins.

6. Exotoxins occur as A-B toxins, membrane-disrupting toxins, and superantigens.

7. Cytotoxins include diphtheria toxin (which inhibits protein synthesis) and erythrogenic toxins (which damage capillaries).

8. Neurotoxins include botulinum toxin (which prevents nerve transmission) and tetanus toxin (which prevents inhibitory nerve transmission).

9. *Vibrio cholerae* toxin and staphylococcal enterotoxin are enterotoxins, which induce fluid and electrolyte loss from host cells.

10. Endotoxins are lipopolysaccharides (LPS), the lipid A component of the cell wall of gram-negative bacteria.

11. Bacterial cell death, antibiotics, and antibodies may cause the release of endotoxins.

12. Endotoxins cause fever (by inducing the release of interleukin-1) and shock (because of a TNF-induced decrease in blood pressure).

13. Endotoxins allow bacteria to cross the blood–brain barrier.

14. The *Limulus* amoebocyte lysate (LAL) assay is used to detect endotoxins in drugs and on medical devices.

PLASMIDS, LYSOGENY, AND PATHOGENICITY (pp. 464–465)

15. Plasmids may carry genes for antibiotic resistance, toxins, capsules, and fimbriae.

16. Lysogenic conversion can result in bacteria with virulence factors, such as toxins or capsules.

PATHOGENIC PROPERTIES OF VIRUSES (pp. 465–467)

1. Viruses avoid the host's immune response by growing inside cells.

2. Viruses gain access to host cells because they have attachment sites for receptors on the host cell.

3. Visible signs of viral infections are called cytopathic effects (CPE).

4. Some viruses cause cytocidal effects (cell death), and others cause noncytocidal effects (damage but not death).

5. Cytopathic effects include the stopping of mitosis, lysis, the formation of inclusion bodies, cell fusion, antigenic changes, chromosomal changes, and transformation.

PATHOGENIC PROPERTIES OF FUNGI, PROTOZOA, HELMINTHS, AND ALGAE (pp. 467–468)

1. Symptoms of fungal infections can be caused by capsules, toxins, and allergic responses.

2. Symptoms of protozoan and helminthic diseases can be caused by damage to host tissue or by the metabolic waste products of the parasite.

3. Some protozoa change their surface antigens while growing in a host thus avoiding destruction by the host's antibodies.

4. Some algae produce neurotoxins that cause paralysis when ingested by humans.

PORTALS OF EXIT (p. 468)

1. Just as pathogens have preferred portals of entry, they also have definite portals of exit.

2. Three common portals of exit are the respiratory tract via coughing or sneezing, the gastrointestinal tract via saliva or feces, and the genitourinary tract via secretions from the vagina or penis.

3. Arthropods and syringes provide a portal of exit for microbes in blood.

STUDY QUESTIONS

Access more review material either online at **The Microbiology Place** (www.microbiologyplace.com) or with **The Microbiology Place CD-ROM** packaged with your new book. There you'll find activities, practice tests, quizzes, flashcards, case studies, and more to help you succeed. In addition, you'll find the following Interactive Tutorial: Host and Pathogen.

Answers to the Study Questions can be found in Appendix G.

REVIEW

1. List three portals of entry, and describe how microorganisms gain access through each.

2. Compare pathogenicity with virulence.

3. Explain how drugs that bind each of the following would affect pathogenicity:
   a. mannose on human cell membranes
   b. *Neisseria gonorrhoeae* fimbriae
   c. *Streptococcus pyogenes* M protein

4. Define cytopathic effects, and give five examples.

5. Compare and contrast the following aspects of endotoxins and exotoxins: bacterial source, chemistry, toxicity, and pharmacology. Give an example of each toxin.

6. How are capsules and cell wall components related to pathogenicity? Give specific examples.

7. Describe how hemolysins, leukocidins, coagulase, kinases, hyaluronidase, siderophores, and IgA proteases might contribute to pathogenicity.

8. Describe the factors contributing to the pathogenicity of fungi, protozoa, and helminths.

9. Which of the following genera is the most infectious?

Genus	ID$_{50}$	Genus	ID$_{50}$
Legionella	1 cell	*Shigella*	200 cells
Salmonella	10^5 cells	*Treponema*	52 cells

10. The LD$_{50}$ of botulinum toxin is 0.000025 μg. The LD$_{50}$ of *Salmonella* toxin is 200 μg. Which of these is the more potent toxin? How can you tell from the LD$_{50}$ values?

11. Food poisoning can be divided into two categories: food infection and food intoxication. On the basis of toxin produc-

tion by bacteria, explain the difference between these two categories.

12. How can viruses and protozoa avoid being killed by the host's immune response?

MULTIPLE CHOICE

1. The removal of plasmids reduces virulence in which of the following organisms?
   a. *Clostridium tetani*
   b. *Escherichia coli*
   c. *Staphylococcus aureus*
   d. *Streptococcus mutans*
   e. *Clostridium botulinum*

2. What is the LD$_{50}$ for the bacterial toxin tested in the example below?

Dilution (μg/kg)	No. of Animals Died	No. of Animals Survived
a. 6	0	6
b. 12.5	0	6
c. 25	3	3
d. 50	4	2
e. 100	6	0

3. Which of the following is *not* a portal of entry for pathogens?
   a. mucous membranes of the respiratory tract
   b. mucous membranes of the gastrointestinal tract
   c. skin
   d. blood
   e. parenteral route

4. All of the following can occur during bacterial infection. Which would prevent all of the others?
   a. vaccination against fimbriae
   b. phagocytosis
   c. inhibition of phagocytic digestion
   d. destruction of adhesins
   e. alteration of cytoskeleton

5. The ID$_{50}$ for *Campylobacter* sp. is 500 cells; the ID$_{50}$ for *Cryptosporidium* sp. is 100 cells. Which of the following statements is *not* true?

a. Both microbes are pathogens.
b. Both microbes produce infections in 50% of the inoculated hosts.
c. *Cryptosporidium* is more virulent than *Campylobacter*.
d. *Campylobacter* and *Cryptosporidium* are equally virulent; they cause infections in the same number of test animals.
e. The severity of infections caused by *Campylobacter* and *Cryptosporidium* cannot be determined by the information provided.

6. An encapsulated bacterium can be virulent because the capsule
a. resists phagocytosis.
b. is an endotoxin.
c. destroys host tissues.
d. interferes with physiological processes.
e. has no effect; since many pathogens do not have capsules, capsules do not contribute to virulence.

7. A drug that binds to mannose on human cells would prevent
a. the entrance of *Vibrio* enterotoxin.
b. the attachment of pathogenic *E. coli*.
c. the action of botulinum toxin.
d. streptococcal pneumonia.
e. the action of diphtheria toxin.

8. The earliest smallpox vaccines were infected tissue rubbed into the skin of a healthy person. The recipient of such a vaccine usually developed a mild case of smallpox, recovered, and was immune thereafter. The most likely reason this vaccine did not kill more people is:
a. Skin is the wrong portal of entry for smallpox.
b. The vaccine consisted of a mild form of the virus.
c. Smallpox is normally transmitted by skin-to-skin contact.
d. Smallpox is a virus.
e. The virus mutated.

9. Which of the following does *not* represent the same mechanism for avoiding host defenses as the others?
a. Rabiesvirus attaches to the receptor for the neurotransmitter acetylcholine.
b. *Salmonella* attaches to the receptor for epidermal growth factor.
c. Epstein-Barr (EB) virus binds to the host receptor for C3.
d. Surface protein genes in *Neisseria gonorrhoeae* mutate frequently.
e. none of the above

10. Which of the following statements is true?
a. The primary goal of a pathogen is to kill its host.
b. Evolution selects for the most virulent pathogens.
c. A successful pathogen doesn't kill its host before it is transmitted.
d. A successful pathogen never kills its host.

CRITICAL THINKING

1. How can plasmids and lysogeny turn the normally harmless *E. coli* into a pathogen? Using the graph below, which shows confirmed cases of enteropathogenic *E. coli*, what is the usual portal of entry for this bacterium?

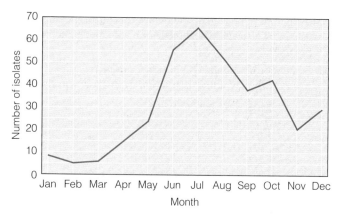

2. The cyanobacterium *Microcystis aeruginosa* produces a peptide that is toxic to humans. According to the graph below, when is this bacterium most toxic?

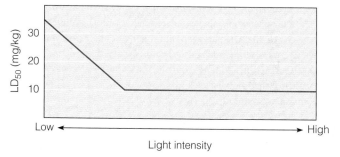

3. When injected into rats, the ID_{50} for *Salmonella typhimurium* is 10^6 cells. If sulfonamides are injected with the salmonellae, the ID_{50} is 35 cells. Explain the change in ID_{50} value.

4. How do each of the following strategies contribute to the virulence of the pathogen? What disease does each organism cause?

Strategy	Pathogen
Changes its cell wall after entry into host	*Yersinia pestis*
Uses urea to produce ammonia	*Helicobacter pylori*
Causes host to make more receptors	*Rhinovirus*
Binds to site for epidermal growth factor on human cells	*Salmonella*

CLINICAL APPLICATIONS

1. On July 8, a woman was given an antibiotic for presumptive sinusitis. However, her condition worsened, and she was unable to eat for 4 days because of severe pain and tightness of the jaw. On July 12, she was admitted to a hospital with severe facial spasms. She reported that on July 5 she had incurred a puncture wound at the base of her big toe; she cleaned the wound but did not seek medical attention. What caused her symptoms? Was her condition due to an infection or an intoxication? Can she transmit this condition to another person?

2. Explain whether each of the following examples is a food infection or intoxication. What is the probable etiological agent in each case?

   a. Eighty-two people who ate shrimp at a dinner in Port Allen, Louisiana, developed diarrhea, cramps, weakness, nausea, chills, headache, and fever from 4 hours to 2 days after eating.

   b. Two people in Vermont who ate barracuda caught in Florida developed malaise, nausea, blurred vision, breathing difficulty, and numbness 3 to 6 hours after eating.

3. Washwater containing *Pseudomonas* was sterilized and used to wash cardiac catheters. Three patients undergoing car-diac catheterization developed fever, chills, and hypoten-sion. The water and catheters were sterile. Why did the pa-tients show these reactions?

4. Cancer patients undergoing chemotherapy are normally *more* susceptible to infections. However, a patient receiving an antitumor drug that inhibited cell division was resistant to *Salmonella*. Provide a possible mechanism for the resis-tance.

16 Innate Immunity: Nonspecific Defenses of the Host

From our discussion to this point, you can see that pathogenic microorganisms are endowed with special properties that enable them to cause disease if given the right opportunity. If microorganisms never encountered resistance from the host, we would constantly be ill and would eventually die of various diseases. In most cases, however, our body's defenses prevent this from happening. Some of these defenses are designed to keep out microorganisms altogether, other defenses remove the microorganisms if they do get in, and still others combat them if they remain inside. Our ability to ward off disease caused by microbes or their products and to protect against environmental agents such as pollen, drugs, foods, chemicals, and animal hair is called **immunity,** or **resistance.** Vulnerability or lack of immunity is referred to as **susceptibility.** We have two lines of defense against pathogens. The first line of defense is our skin and mucous membranes. The second line of defense consists of various defensive cells, inflammation, fever, and antimicrobial substances produced by the body.

UNDER THE MICROSCOPE

A Phagocyte. Phagocytes provide protection against pathogens. This macrophage is engulfing bacteria, which will prevent the bacteria from colonizing and causing disease.

Innate (Nonspecific) Immunity		Adaptive (Acquired) Immunity (Chapter 17)
First line of defense	**Second line of defense**	**Third line of defense**
• Intact skin • Mucous membranes and their secretions • Normal microbiota	• Natural killer cells and phagocytic white blood cells • Inflammation • Fever • Antimicrobial substances	• Specialized lymphocytes: T cells and B cells • Antibodies

FIGURE 16.1 An overview of the body's defenses. Innate immunity involves defenses against any pathogen, regardless of species; adaptive immunity involves defenses against a specific pathogen.

Q **What is the difference between immunity and susceptibility?**

THE CONCEPT OF IMMUNITY

LEARNING OBJECTIVES

• Differentiate between innate and adaptive immunity.

• Define *toll-like receptors*.

When our bodies are challenged by microbes, we defend ourselves by utilizing our various mechanisms of immunity. In general, there are two types of immunity: innate and adaptive (Figure 16.1). **Innate (nonspecific) immunity** refers to defenses that are present at birth. They are always present and available to provide rapid responses to protect us against disease. Innate immunity does not involve specific recognition of a microbe and acts against all microbes in the same way. Further, innate immunity does not have a memory component—that is, it cannot recall a previous contact with a foreign molecule. Among the components of innate immunity are the first line of defense (skin and mucous membranes) and the second line of defense (natural killer cells and phagocytes, inflammation, fever, and antimicrobial substances). Innate immune responses represent immunity's early-warning system and are designed to prevent microbes from gaining access into the body and to help eliminate those that do gain access.

Adaptive (specific) immunity refers to defenses that involve specific recognition of a microbe once it has breached the innate immunity defenses. Adaptive immunity is based on a specific response to a specific microbe. It adapts or adjusts to handle a particular microbe. Unlike innate immunity, adaptive immunity is slower to respond, but it does have a memory component. Adaptive immunity involves lymphocytes (a type of white blood cell) called

T cells (T lymphocytes) and B cells (B lymphocytes) and will be discussed in detail in Chapter 17. Here, we concentrate on innate immunity.

As noted previously the innate immune system responds rapidly to invaders by detecting them and then attempting to eliminate them. It has recently been learned that the responses of the innate system are activated by protein receptors in the plasma membranes of defensive cells; these activators are called **toll-like receptors (TLRs).** These TLRs attach to various components of microbes such as the lipopolysaccharide (LPS) of the outer membrane of gram-negative bacteria, the flagellin in the flagella of motile bacteria, the lipoteichoic acid in the cell wall of gram-positive bacteria, the DNA of bacteria, and the DNA and RNA of viruses. TLRs also attach to components of fungi and parasites. You will learn later in this chapter that two of the defensive cells involved in innate immunity are called macrophages and dendritic cells. When the TLRs on these cells encounter components of microbes, such as the LPS of gram-negative bacteria, the TLRs induce the defensive cells to release chemicals called cytokines. **Cytokines** (*cyto-* = cell; *-kinesis* = motion) are proteins that regulate the intensity and duration of immune responses. One role of cytokines is to recruit other macrophages and dendritic cells, as well as other defensive cells, to isolate and destroy the microbes as part of the inflammatory response. Cytokines can also activate the T cells and B cells involved in adaptive immunity. You will learn more about the different cytokines and their functions in Chapter 17.

FIRST LINE OF DEFENSE: SKIN AND MUCOUS MEMBRANES

LEARNING OBJECTIVES

- Describe the role of the skin and mucous membranes in innate immunity.
- Differentiate physical from chemical factors, and list five examples of each.
- Describe the role of normal microbiota in innate immunity.

The skin and mucous membranes are the body's first line of defense against pathogens. This function results from both physical and chemical factors. Whereas physical factors include barriers to entry or processes that remove microbes from the body's surface, chemical factors include substances made by the body that inhibit microbial growth or destroy them.

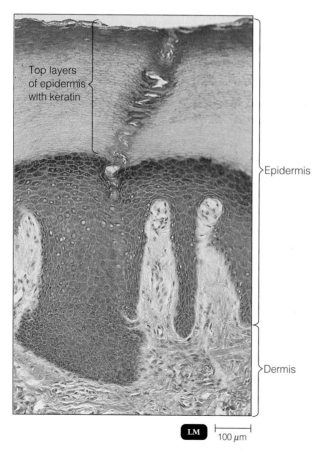

FIGURE 16.2 A section through human skin. The thin layers at the top of this photomicrograph contain keratin. These layers and the darker purple cells beneath them make up the epidermis. The lighter purple material near the bottom is the dermis.

Q **The intact skin and mucous membranes are components of which line of defense against pathogens?**

PHYSICAL FACTORS

The intact **skin** is one of the human body's largest organs in terms of surface area. It consists of two distinct portions: the dermis and the epidermis (Figure 16.2). The **dermis,** the skin's inner, thicker portion, is composed of connective tissue. The **epidermis,** the outer, thinner portion, is in direct contact with the external environment. The epidermis consists of many layers of continuous sheets of tightly packed epithelial cells with little or no material between the cells. The top layer of epidermal cells is dead and contains a protective protein called **keratin.** Because the top layer is shed periodically, this helps remove microbes at the surface. In addition, the dryness of the skin is a major factor in inhibiting microbial growth on the skin. Although normal microbiota and other microbes are present on the entire skin, they are most numerous on moist areas of the skin. When the skin is moist, as in hot, humid climates, skin infections are quite common, especially fungal infections such as athlete's foot. These fungi hydrolyze keratin when water is available.

If we consider the closely packed cells, continuous layering, the presence of keratin, and the dryness and shedding of the skin, we can see why the intact skin provides such a formidable barrier to the entrance of microorganisms. The intact surface of healthy epidermis is rarely, if ever, penetrated by microorganisms. However, when the epithelial surface is broken, a subcutaneous (below-the-skin) infection often develops. The bacteria most likely to cause infection are the staphylococci that normally inhabit the epidermis, hair follicles, and sweat and oil glands of the skin. Infections of the skin and underlying tissues frequently occur as a result of burns, cuts, stab wounds, or other conditions that break the skin.

Epithelial cells that line blood and lymphatic vessels are not closely packed like those of the epidermis. Although this arrangement permits defensive cells to move from blood into tissues during inflammation, it also permits microbes to move into and out of blood and lymph.

Mucous membranes also consist of an epithelial layer and an underlying connective tissue layer. Although mucous membranes inhibit the entrance of many microorganisms, they offer less protection than the skin. Mucous membranes line the entire gastrointestinal, respiratory, and genitourinary tracts. The epithelial layer of a mucous membrane secretes a fluid called **mucus,** a slightly viscous (thick) glycoprotein produced by goblet cells of a mucous membrane. Among other functions, mucus prevents the tracts from drying out. Some pathogens that can thrive on the moist secretions of a mucous membrane are able to penetrate the membrane if the microorganism is present in sufficient numbers. *Treponema pallidum* is such a pathogen. This penetration may be facilitated by toxic substances produced by the microorganism, prior injury by viral infection, or mucosal irritation.

Besides the physical barrier presented by the skin and mucous membranes, several other physical factors help protect certain epithelial surfaces. One such mechanism that protects the eyes is the **lacrimal apparatus,** a group of structures that manufactures and drains away tears (Figure 16.3). The lacrimal glands, located toward the upper, outermost portion of each eye socket, produce the tears and pass them under the upper eyelid. From here, tears pass toward the corner of the eye near the nose and into two small holes that lead through tubes (lacrimal canals) to the nose. The tears are spread over the surface of the eyeball by blinking. Normally, the tears evaporate or pass into the nose as fast as they are produced. This continual washing action helps keep microorganisms from settling on the surface of the eye. If an irritating substance or large numbers of microorganisms come in contact with the eye, the lacrimal glands start to secrete heavily, and the tears accumulate more rapidly than they can be carried away. This excessive production is a protective mechanism because the excess tears dilute and wash away the irritating substance or microorganisms.

In a cleansing action very similar to that of tears, **saliva** is produced by the salivary glands. Saliva helps dilute the numbers of microorganisms and wash them from both the surface of the teeth and the mucous membrane of the mouth. This helps prevent colonization by microbes.

The respiratory and gastrointestinal tracts have many physical forms of defense. Mucus traps many of the microorganisms that enter the respiratory and gastrointestinal tracts. The mucous membrane of the nose also has mucus-coated **hairs** that filter inhaled air and trap microorganisms, dust, and pollutants. The cells of the mucous membrane of the lower respiratory tract are covered with **cilia.** By moving synchronously, these cilia propel inhaled dust and microorganisms that have become trapped in mucus upward toward the throat. This so-called **ciliary escalator** (Figure 16.4) keeps the mucus blanket moving toward the throat at a rate of 1 to 3 cm per hour; coughing and sneezing speed up the escalator. (Some substances in cigarette smoke are toxic to cilia and can seriously impair the functioning of the ciliary escalator by inhibiting or destroying the cilia.) Microorganisms are also prevented from entering the lower respiratory tract by a small lid of cartilage called the **epiglottis,** which covers the larynx (voicebox) during swallowing.

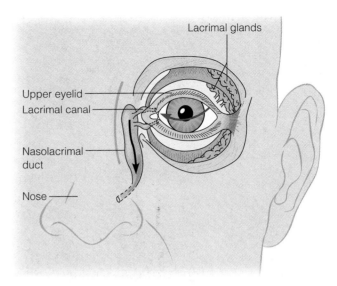

FIGURE 16.3 The lacrimal apparatus. The washing action of the tears is shown by the red arrow passing over the surface of the eyeball. Tears produced by the lacrimal glands pass across the surface of the eyeball into two small holes that convey the tears into the lacrimal canals and the nasolacrimal duct.

Q How does the lacrimal apparatus protect the eyes against infections?

The cleansing of the urethra by the flow of **urine** is another physical factor that prevents microbial colonization in the genitourinary tract. **Vaginal secretions** likewise move microorganisms out of the female body.

Defecation and **vomiting** also expel microbes. For example, in response to microbial toxins, the muscles of the gastrointestinal tract contract vigorously, resulting in vomiting and/or diarrhea, which may rid the body of microbes.

CHEMICAL FACTORS

Physical factors alone do not account for the high degree of resistance of skin and mucous membranes to microbial invasion. Certain chemical factors also play important roles.

Sebaceous (oil) glands of the skin produce an oily substance called **sebum** that prevents hair from drying and becoming brittle. Sebum also forms a protective film over the surface of the skin. One of the components of sebum is unsaturated fatty acids, which inhibit the growth of certain pathogenic bacteria and fungi. The low pH of the skin, between pH 3 and 5, is caused in part by the secretion of fatty acids and lactic acid. The skin's acidity probably discourages the growth of many other microorganisms.

Bacteria that live commensally on the skin decompose sloughed-off skin cells, and the resultant organic molecules and the end-products of their metabolism produce body odor. As we will see in Chapter 21, certain bacteria

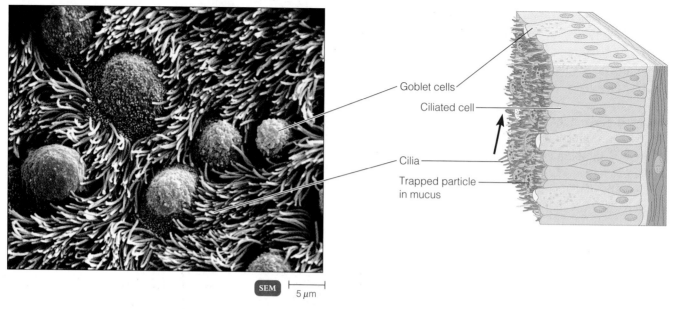

SEM | 5 μm

FIGURE 16.4 The ciliary escalator.

Q **What is the function of the ciliary escalator?**

commonly found on the skin metabolize sebum, and this metabolism forms free fatty acids that cause the inflammatory response associated with acne. Isotretinoin (Accutane), a derivative of vitamin A that prevents sebum formation, is a treatment for a very severe type of acne called cystic acne.

The sweat glands of the skin produce **perspiration,** which helps maintain body temperature, eliminate certain wastes, and flush microorganisms from the surface of the skin. Perspiration also contains **lysozyme,** an enzyme capable of breaking down cell walls of gram-positive bacteria and, to a lesser extent, gram-negative bacteria (see Figure 4.13, page 86). Specifically, lysozyme breaks chemical bonds on peptidoglycan, which destroys the cell walls. Lysozyme is also found in tears, saliva, nasal secretions, and tissue fluids, where it exhibits its antimicrobial activity. Alexander Fleming was studying lysozyme in 1929 when he accidentally discovered the antimicrobial effects of penicillin (see Figure 20.1, page 582).

Gastric juice is produced by the glands of the stomach. It is a mixture of hydrochloric acid, enzymes, and mucus. The very high acidity of gastric juice (pH 1.2–3.0) is sufficient to destroy bacteria and most bacterial toxins, except those of *Clostridium botulinum* and *Staphylococcus aureus.* However, many enteric pathogens are protected by food particles and can enter the intestines via the gastrointestinal tract. The bacterium *Helicobacter pylori* neutralizes stomach acid which allows the bacterium to grow in the stomach. Its growth results in ulcers and gastritis.

NORMAL MICROBIOTA AND INNATE IMMUNITY

In Chapter 14, several relationships between normal microbiota and host cells were described. Some of these relationships assist in preventing the overgrowth of pathogens and thus may be considered components of innate immunity. For example, in microbial antagonism, the normal microbiota prevent pathogens from colonizing the host by competing with them for nutrients (competitive exclusion), by producing substances that are harmful to the pathogens, and by altering conditions that affect the survival of the pathogens, such as pH and oxygen availability. The presence of normal microbiota in the vagina, for example, alters pH thus preventing overpopulation by *Candida albicans,* a pathogenic yeast that causes vaginitis. In the large intestine, *E. coli* produce bacteriocins that inhibit the growth of *Salmonella* and *Shigella.*

In commensalism, one organism uses the body of a larger organism as its physical environment and may make use of the body to obtain nutrients. Thus in commensalism, one organism benefits while the other is unaffected. Most microbes that are part of the commensal microbiota are found on the skin and in the gastrointestinal tract. The majority of such microbes are bacteria that have highly specialized attachment mechanisms and precise environmental requirements for survival. Normally, such microbes are harmless, but they may cause disease if their environmental conditions change. These opportunistic pathogens include *E. coli, Staphylococcus aureus, S. epidermidis, Enterococcus faecalis, Pseudomonas aeruginosa,* and oral streptococci.

SECOND LINE OF DEFENSE

When microbes penetrate the first line of defense, they encounter a second line of defense that includes defensive cells, such as phagocytic cells; inflammation; fever; and antimicrobial substances.

Before we look at the phagocytic cells, it will be helpful to first have an understanding of the cellular components of blood.

FORMED ELEMENTS IN BLOOD

LEARNING OBJECTIVES

- Classify phagocytic cells, and describe the roles of granulocytes and monocytes.
- Define *differential white blood cell count*.

Blood consists of a fluid called **plasma,** which contains **formed elements**—that is, cells and cell fragments (Table 16.1). Of the cells listed in Table 16.1, those that concern us at present are the **leukocytes,** or white blood cells.

During many kinds of infections, especially bacterial infections, the total number of white blood cells increases as a protective response to combat the microbes; this increase is called *leukocytosis*. During the active stage of infection, the leukocyte count might double, triple, or quadruple, depending on the severity of the infection. Diseases that might cause such an elevation in the leukocyte count include meningitis, infectious mononucleosis, appendicitis, pneumococcal pneumonia, and gonorrhea. Other diseases, such as salmonellosis and brucellosis, and some viral and rickettsial infections may cause a *decrease* in the leukocyte count, called *leukopenia*. Leukopenia may be related to either impaired white blood cell production or the effect of increased sensitivity of white blood cell membranes to damage by complement, antimicrobial plasma proteins discussed later in the chapter. Leukocyte increase or decrease can be detected by a **differential white blood cell count,** which is a calculation of the percentage of each kind of white cell in a sample of 100 white blood cells. The percentages in a normal differential white blood cell count are shown in parentheses in the first column of Table 16.1.

Leukocytes are divided into two main categories based on their appearance under a light microscope: granulocytes and agranulocytes. **Granulocytes** owe their name to the presence of large granules in their cytoplasm that can be seen under a light microscope after staining. They are differentiated on the basis of how the granules stain. The granules of **neutrophils** stain pale lilac with a mixture of acidic and basic dyes. Neutrophils are also commonly called *polymorphonuclear leukocytes (PMNs)*, or *polymorphs*. (The term *polymorphonuclear* refers to the fact that the nuclei of neutrophils contain two to five lobes.) Neutrophils, which are highly phagocytic and motile, are active in the initial stages of an infection. They have the ability to leave the blood, enter an infected tissue, and destroy microbes and foreign particles. **Basophils** stain blue-purple with the basic dye methylene blue. Basophils release substances, such as histamine, that are important in inflammation and allergic responses. **Eosinophils** stain red or orange with the acidic dye eosin. Eosinophils are somewhat phagocytic and also have the ability to leave the blood. Their major function is to produce toxic proteins against certain parasites, such as helminths. Although eosinophils are physically too small to ingest and destroy helminths, they can attach to the outer surface of the parasites and discharge peroxide ions that destroy them (see Figure 17.14, page 518). Their number increases significantly during certain parasitic worm infections and hypersensitivity (allergy) reactions.

Also classified with granulocytes are **dendritic cells.** They have long extensions that resemble the dendrites of nerve cells, thus their name. Dendritic cells are especially abundant in the epidermis of the skin, mucous membranes, the thymus, and lymph nodes. The function of dendritic cells is to destroy microbes by phagocytosis and to initiate adaptive immunity responses (see Chapter 17, page 576).

Agranulocytes also have granules in their cytoplasm, but the granules are not visible under the light microscope after staining. **Monocytes** are not actively phagocytic until they leave circulating blood, enter body tissues, and mature into **macrophages.** In fact, the maturation and proliferation of macrophages (along with lymphocytes) is one factor responsible for the swelling of lymph nodes during an infection. As blood and lymph that contain microorganisms pass through organs with macrophages, the microorganisms are removed by phagocytosis. Macrophages also dispose of worn out blood cells.

Lymphocytes include natural killer cells, T cells, and B cells. **Natural killer (NK) cells** are found in blood and in the spleen, lymph nodes, and red bone marrow. NK cells have the ability to kill a wide variety of infected body cells and certain tumor cells. NK cells attack any body cells that display abnormal or unusual plasma membrane proteins. The binding of NK cells to a target cell, such as an infected human cell, causes the release of

| TABLE 16.1 | **Formed Elements in Blood** |

Type of Cell		Numbers per Microliter (µL) or Cubic mm (mm^3)	Function
Erythrocytes (Red Blood Cells)		4.8–5.4 million	Transport of O_2 and CO_2
Leukocytes (White Blood Cells)		5000–10,000	

A. Granulocytes (stained)
  1. Neutrophils (PMNs)
    (60–70% of leukocytes)
			Phagocytosis

2. Basophils (0.5–1%)			Production of histamine
3. Eosinophils (2–4%)			Production of toxic proteins against certain parasites; some phagocytosis
4. Dendritic cells			Initiation of adaptive immune responses

granules containing toxic substances from NK cells. Some granules contain a protein called **perforin,** which inserts into the plasma membrane of the target cell and creates channels (perforations) in the membrane. As a result, extracellular fluid flows into the target cell and the cell bursts, a process called **cytolysis** (sī-tol′-i-sis; *cyto-* = cell; *-lysis* = loosening). Other granules of NK cells release **granzymes,** which are protein-digesting enzymes that induce the target cell to undergo apoptosis, or self-destruction. This type of attack kills infected cells but

TABLE 16.1	*(continued)*		

Type of Cell		Numbers per Microliter (µL) or Cubic mm (mm^3)	Function
B. Agranulocytes (stained) 1. Monocytes (3–8%)	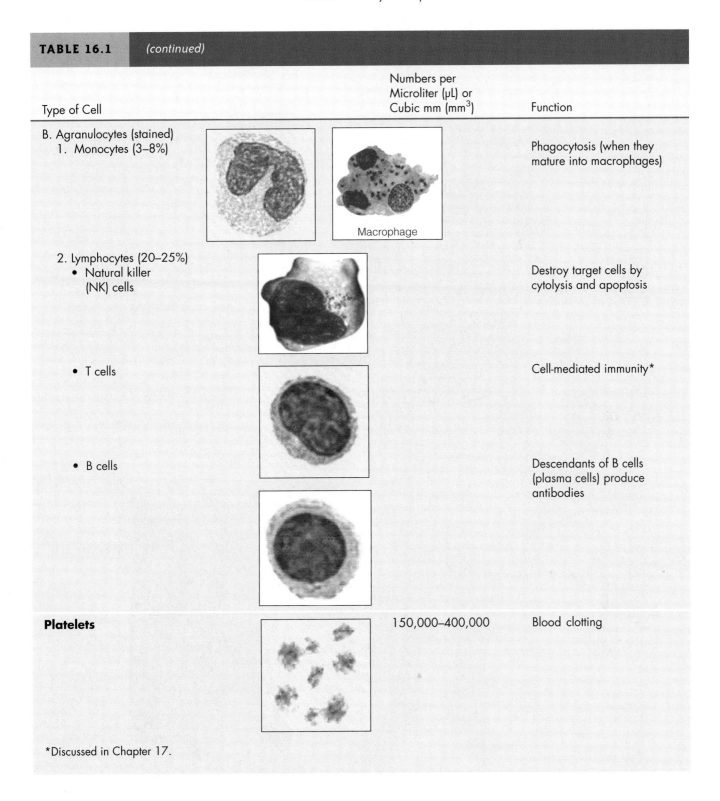 Macrophage		Phagocytosis (when they mature into macrophages)
2. Lymphocytes (20–25%) • Natural killer (NK) cells			Destroy target cells by cytolysis and apoptosis
• T cells			Cell-mediated immunity*
• B cells			Descendants of B cells (plasma cells) produce antibodies
Platelets		150,000–400,000	Blood clotting

*Discussed in Chapter 17.

not the microbes inside the cells; the released microbes, which may or may not be intact, can be destroyed by phagocytes.

T cells and **B cells** are not phagocytic but play a key role in adaptive immunity (see Chapter 17). They occur in lymphoid tissues of the lymphatic system: in the tonsils, spleen, thymus, thoracic duct, red bone marrow, appendix, Peyer's patches of the small intestine, and lymph nodes in the respiratory, gastrointestinal, and reproductive tracts (Figure 16.5). They also circulate in the blood. ✻ **Animation: Go to The Microbiology Place website or CD-ROM and click "Animations" to view Host Defenses.**

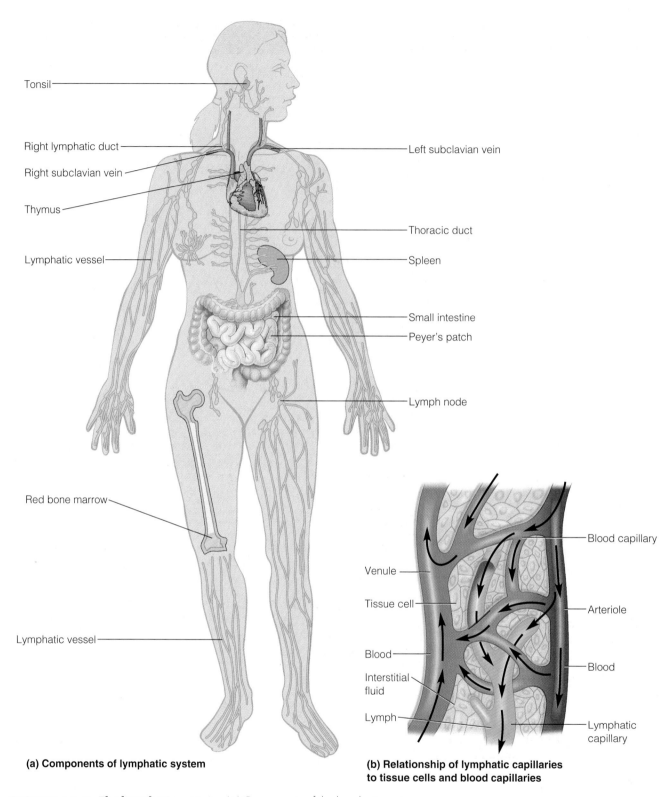

Tonsil

Right lymphatic duct

Right subclavian vein

Thymus

Lymphatic vessel

Left subclavian vein

Thoracic duct

Spleen

Small intestine

Peyer's patch

Lymph node

Red bone marrow

Lymphatic vessel

Venule

Tissue cell

Blood

Interstitial fluid

Lymph

Blood capillary

Arteriole

Blood

Lymphatic capillary

(a) Components of lymphatic system

(b) Relationship of lymphatic capillaries to tissue cells and blood capillaries

FIGURE 16.5 The lymphatic system. (a) Components of the lymphatic system. **(b)** Fluid circulating between tissue cells (interstitial fluid) is picked up by lymphatic capillaries. The fluid is then called lymph. Next, the lymph is passed into lymphatic vessels. Macrophages in lymph nodes at various intervals along lymphatic vessels remove microorganisms from the lymph. Lymph is ultimately collected in the right lymphatic duct and thoracic duct. Eventually, the lymph is returned to the blood from these ducts at the subclavian veins, just before the blood enters the heart.

Q **Why do lymph nodes swell during an infection?**

FIGURE 16.6 A macrophage engulfing rod-shaped bacteria. Macrophages in the mononuclear phagocytic system remove microorganisms after the initial phase of infection.

Q **What are monocytes?**

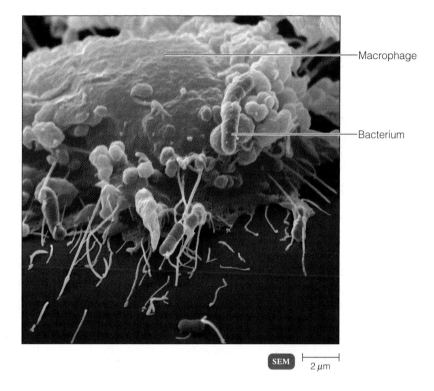

Macrophage

Bacterium

SEM ⊢———⊣ 2 μm

PHAGOCYTES

LEARNING OBJECTIVE

• Define *phagocyte* and *phagocytosis.*

Phagocytosis (from Greek words meaning eat and cell) is the ingestion of a microorganism or other particles (such as debris) by a cell. We have previously mentioned phagocytosis as the method of nutrition of certain protozoa. In this chapter, phagocytosis is discussed as a means by which cells in the human body counter infection—the second line of defense. The cells that perform this function are collectively called **phagocytes,** all of which are types of white blood cells or derivatives of white blood cells. Phagocytes may be activated by components of bacteria such as lipid A or lipopolysaccharides (LPS). Among the more important activators are small protein hormones, called cytokines, which are secreted by phagocytes and other cells involved in adaptive immunity (see Chapter 17).

ACTIONS OF PHAGOCYTIC CELLS

When an infection occurs, both granulocytes (especially neutrophils) and monocytes migrate to the infected area. During this migration, monocytes enlarge and develop into actively phagocytic macrophages (Figure 16.6). These cells leave the blood and migrate into tissues where they enlarge and develop into macrophages. Some macrophages, called **fixed macrophages,** or *histiocytes,* are located in certain tissues and organs of the body. Fixed macrophages are found in the liver (Kupffer's cells), lungs (alveolar macrophages), nervous system (microglial cells), bronchial tubes, spleen (splenic macrophages), lymph nodes, red bone marrow, and the peritoneal cavity surrounding abdominal organs (peritoneal macrophages). Other macrophages are called **wandering macrophages,** which roam the tissues and gather at sites of infection or inflammation. The various macrophages of the body constitute the **mononuclear phagocytic (reticuloendothelial) system.**

During the course of an infection, a shift occurs in the type of white blood cell that predominates in the bloodstream. Granulocytes, especially neutrophils, dominate during the initial phase of bacterial infection, at which time they are actively phagocytic; this dominance is indicated by their increased number in a differential white blood cell count. However, as the infection progresses, the macrophages dominate; they scavenge and phagocytize remaining living bacteria and dead or dying bacteria. The increased number of monocytes (which develop into macrophages) is also reflected in a differential white blood cell count. In viral and fungal infections, macrophages predominate in all phases of defense.

TABLE 16.2	Classification and Functions of Phagocytes	
Type of Phagocyte	Cell Type	Functions
Granulocytes	Neutrophils and eosinophils	Phagocytic against microbes during the initial phase of infection
Agranulocytes	Mononuclear phagocytic system: fixed macrophages and wandering macrophages developed from monocytes	Phagocytic against microbes as infection progresses and against worn-out blood cells as infection subsides; also involved in adaptive immunity (see Chapter 17)

Table 16.2 summarizes the different types of phagocytic cells and their functions.

THE MECHANISM OF PHAGOCYTOSIS

> **LEARNING OBJECTIVE**
> • Describe the process of phagocytosis, and include the stages of adherence and ingestion.

How does phagocytosis occur? For the convenience of study, we will divide phagocytosis into four main phases: chemotaxis, adherence, ingestion, and digestion (Figure 16.7).

CHEMOTAXIS

❶ **Chemotaxis** is the chemical attraction of phagocytes to microorganisms. (The mechanism of chemotaxis is discussed in Chapter 4, page 83.) Among the chemotactic chemicals that attract phagocytes are microbial products, components of white blood cells and damaged tissue cells, and peptides derived from complement, a system of host defense discussed later in the chapter.

ADHERENCE

As it pertains to phagocytosis, **adherence** is the attachment of the phagocyte's plasma membrane to the surface of the microorganism or other foreign material. In some instances, adherence occurs easily, and the microorganism is readily phagocytized. Microorganisms can be more readily phagocytized if they are first coated with certain serum proteins that promote attachment of the microor-

ganisms to the phagocyte. This coating process is called **opsonization.** The proteins that act as *opsonins* include some components of the complement system and antibody molecules (described later in this chapter and in Chapter 17).

INGESTION

❷ Following adherence, **ingestion** occurs. During this process, the plasma membrane of the phagocyte extends projections called **pseudopods** that engulf the microorganism. ❸ Once the microorganism is surrounded, the pseudopods meet and fuse, surrounding the microorganism with a sac called a **phagosome,** or *phagocytic vesicle.* The membrane of a phagosome has enzymes that pump protons (H^+) into the phagosome, reducing the pH to about 4. At this pH, hydrolytic enzymes are activated.

DIGESTION

In this phase of phagocytosis, the phagosome pinches off from the plasma membrane and enters the cytoplasm. Within the cytoplasm, it contacts lysosomes that contain digestive enzymes and bactericidal substances (see Chapter 4, page 104). ❹ Upon contact, the phagosome and lysosome membranes fuse to form a single, larger structure called a **phagolysosome.** ❺ The contents of the phagolysosome take only 10 to 30 minutes to kill most types of bacteria.

Lysosomal enzymes that attack microbial cells directly include lysozyme, which hydrolyzes peptidoglycan in bacterial cell walls. A variety of other enzymes, such as lipases, proteases, ribonuclease, and deoxyribonuclease, hydrolyze other macromolecular components of microorganisms. Lysosomes also contain enzymes that can produce toxic oxygen products such as superoxide radical (O_2^-), hydrogen peroxide (H_2O_2), singlet oxygen ($^1O_2^-$), and hydroxyl radical ($OH\cdot$) (see Chapter 6, page 166). Toxic oxygen products are produced by a process called an *oxidative burst.* Other enzymes can make use of these toxic oxygen products in killing ingested microorganisms. For example, the enzyme myeloperoxidase converts chloride (Cl^-) ions and hydrogen peroxide into highly toxic hypochlorous acid ($HOCl$). The acid contains hypochlorous ions, which are found in household bleach and account for its antimicrobial activity (see Chapter 7, page 200).

❻ After enzymes have digested the contents of the phagolysosome brought into the cell by ingestion, the phagolysosome contains indigestible material and is called a *residual body.* ❼ This residual body then moves toward the cell boundary and discharges its wastes outside the cell. The box on page 486 describes another mechanism by which phagocytes can kill microorganisms and tumor cells.

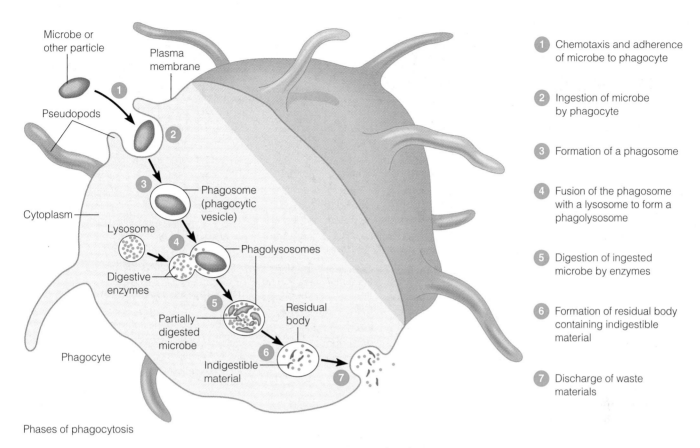

Phases of phagocytosis

1. Chemotaxis and adherence of microbe to phagocyte
2. Ingestion of microbe by phagocyte
3. Formation of a phagosome
4. Fusion of the phagosome with a lysosome to form a phagolysosome
5. Digestion of ingested microbe by enzymes
6. Formation of residual body containing indigestible material
7. Discharge of waste materials

FIGURE 16.7 The mechanism of phagocytosis in a phagocyte. The phases of phagocytosis are chemotaxis, adherence, ingestion, and digestion.

Q **What does *phagocyte* mean?**

MICROBIAL EVASION OF PHAGOCYTOSIS

LEARNING OBJECTIVE
- Identify six methods of avoiding destruction by phagocytes.

Some bacteria have structures that inhibit adherence, such as the M protein and capsules. As mentioned in Chapter 15 (page 456), the M protein of *Streptococcus pyogenes* inhibits the attachment of phagocytes to their surfaces and makes adherence more difficult. Organisms with large capsules include *Streptococcus pneumoniae* and *Haemophilus influenzae* type b. Heavily encapsulated microorganisms like these can be phagocytized only if the phagocyte traps the microorganism against a rough surface, such as a blood vessel, blood clot, or connective tissue fiber, from which the microbe cannot slide away.

Other microbes may be ingested but not killed. For example, *Staphylococcus* produces leukocidins that may kill phagocytes by causing the release of the phagocyte's own lysosomal enzymes into its cytoplasm. Streptolysin released by streptococci has a similar mechanism.

A number of intracellular pathogens secrete membrane attack complexes (described shortly) that lyse phagocyte cell membranes once inside the phagocyte. For example, *Trypanosoma cruzi* (the causative agent of American trypanosomiasis), and *Listeria monocytogenes* (the causative agent of listeriosis), produce membrane attack complexes that lyse phagolysosome membranes and release the microbes into the cytoplasm of the phagocyte, where they propagate. Later the microbes secrete more membrane attack complexes that lyse the plasma membrane and release the microbes from the phagocyte, resulting in lysis of the phagocyte and infection of neighboring cells by the microbe.

Still other microbes have the ability to survive inside phagocytes. *Coxiella burnetti*, the causative agent of Q fever, actually requires the low pH inside a phagolysosome to replicate. *Listeria monocytogenes*, *Shigella* (the causative agent of shigellosis), and *Rickettsia* (the causative agent of Rocky Mountain spotted fever and typhus) have the ability to escape from a phagosome before it fuses with a lysosome. *Mycobacterium tuberculosis* (the causative agent of tuberculosis), HIV (the causative agent of AIDS), *Chlamydia* (the causative agent of trachoma, nongonococcal urethritis, and lymphogranuloma venereum), *Leishmania* (the causative agent of leishmaniasis), and *Plasmodium* (malarial parasites) can prevent both the fusion of a phagosome with a

MICROBIOLOGY IN THE NEWS

MACROPHAGES SAY NO

In 1998, three scientists received the Nobel Prize for Physiology or Medicine for their discovery that nitric oxide (NO) is a powerful signaling molecule in the human body. NO is unusual for a chemical messenger in humans. Chemical messengers such as interferon or epinephrine are large organic molecules, but NO is composed of only two atoms, nitrogen and oxygen ($\cdot N = O$). Nitric oxide is not the same compound as the anesthetic nitrous oxide (N_2O), called laughing gas.

Robert Furchgott of the State University of New York Health Sciences Center, Ferid Murad of the University of Texas, Houston Health Science Center, and Louis Ignarro of the University of California, Los Angeles discovered that NO, released by endothelial cells lining blood vessels, is the molecule that relaxes the smooth muscle of the blood vessel to allow increased blood flow. Murad discovered that nitroglycerin dilates blood vessels because the nitroglycerin releases NO. Since its discovery, NO has been found to be

involved in a wide variety of processes from memory to immunity to impotence to blood pressure regulation.

High nitrate concentrations have been detected in the urine of people with certain types of infections. The reason is that macrophages also produce NO. The macrophage enzyme, NO synthase, produces NO from the amino acid arginine, and the NO is oxidized to nitrates and nitrites, which are excreted in urine.

Macrophages that are activated by gamma-interferon from T cells or bacterial endotoxin produce NO synthase, which makes NO. NO appears to kill microorganisms as well as tumor cells by inhibiting ATP production. Evidence for the role of NO is provided by experiments demonstrating that macrophages can't kill tumor cells or parasites if NO production is inhibited. Although NO is quickly destroyed after it is made, large quantities of NO will dilate blood vessels, which causes blood pressure to drop and the patient to go into shock.

This contributes to the symptoms of gram-negative septic shock. A future treatment for septic shock may involve inhibiting NO synthase.

Exhaled NO is thought to reflect the degree of inflammation in the airways. Several recent studies with asymptomatic asthmatic children have shown that NO in exhaled breath may predict the likelihood of asthma relapse.

NO Suppresses Autoimmunity

Autoimmune diseases, such as lupus erythematosus and rheumatoid arthritis, in which a person makes antibodies against their own molecules, appear to be uncommon in less developed countries. Researchers in Australia suggest that the reason lies in the higher incidence of childhood infections in these countries. We now know that NO production is stimulated by infection and that NO can activate the regulatory cells that are responsible for preventing an immune response to self.

---

lysosome and the proper acidification of digestive enzymes. The microbes then multiply within the phagocyte, almost completely filling it. In most cases, the phagocyte dies and the microbes are released by autolysis to infect other cells. Still other microbes, such as the causative agents of tularemia and brucellosis, can remain dormant within phagocytes for months or years at a time. ✽ **Animation: Go to The Microbiology Place website or CD-ROM and click "Animations" to view Phagocytosis.**

* * *

In addition to providing innate (nonspecific) resistance for the host, phagocytosis plays a role in adaptive (specific) immunity. Macrophages help T and B cells perform vital adaptive immune functions. In Chapter 17, we will discuss in more detail how phagocytosis supports adaptive immunity.

In the next section, we will see how phagocytosis often occurs as part of another innate mechanism of resistance: inflammation.

INFLAMMATION

LEARNING OBJECTIVE
- List the stages of inflammation.

Damage to the body's tissues triggers a defensive response called **inflammation.** The damage can be caused by microbial infection, physical agents (such as heat, radiant energy, electricity, or sharp objects), or chemical agents (acids, bases, and gases). Inflammation is usually characterized by four signs and symptoms: *redness, pain, heat,* and *swelling.* Sometimes a fifth, *loss of function,* is present; its occurrence depends on the site and extent of damage.

If the cause of an inflammation is removed in a relatively short period of time, the inflammatory response is intense and is referred to as an *acute inflammation.* An example is the response to a boil caused by *S. aureus.* If, instead, the cause of an inflammation is difficult or impossible to remove, the inflammatory response is longer lasting but less intense (although overall more destructive).

This type of inflammation is referred to as a *chronic inflammation*. An example is the response to a chronic infection such as tuberculosis, caused by *M. tuberculosis*.

Inflammation has the following functions: (1) to destroy the injurious agent, if possible, and to remove it and its by-products from the body; (2) if destruction is not possible, to limit the effects on the body by confining or walling off the injurious agent and its by-products; and (3) to repair or replace tissue damaged by the injurious agent or its by-products.

During inflammation, there is an activation and increased concentration of a group of proteins in the blood called **acute-phase proteins.** Some acute-phase proteins are produced by the liver; others are present in the blood in an inactive form and are converted to an active form during inflammation. Acute-phase proteins induce both localized and systemic responses and include complement (page 490), the cytokines (Chapter 17, page 518), and several specialized proteins such as fibrinogen for clotting, and kinins for vasodilation.

For purposes of our discussion, we will divide the process of inflammation into three stages: vasodilation and increased permeability of blood vessels, phagocyte migration and phagocytosis, and tissue repair.

VASODILATION AND INCREASED PERMEABILITY OF BLOOD VESSELS

LEARNING OBJECTIVE
• Describe the roles of vasodilation, kinins, prostaglandins, and leukotrienes in inflammation.

Immediately following tissue damage, blood vessels dilate (increase in diameter) in the area of damage, and their permeability increases (Figure 16.8a and b). Dilation of blood vessels, called **vasodilation,** increases blood flow to the damaged area and is responsible for the redness (erythema) and heat associated with inflammation.

Increased permeability permits defensive substances normally retained in the blood to pass through the walls of the blood vessels and enter the injured area. The increase in permeability, which permits fluid to move from the blood into tissue spaces, is responsible for the **edema** (swelling) of inflammation. The pain of inflammation can be caused by nerve damage, irritation by toxins, or the pressure of edema.

❶ Vasodilation and the increase in permeability of blood vessels are caused by chemicals released by damaged cells in response to injury. One such substance is **histamine,** a chemical present in many cells of the body, especially in mast cells in connective tissue, circulating basophils, and blood platelets. Histamine is released in direct response to the injury of cells that contain it; it is also

released in response to stimulation by certain components of the complement system (discussed later). Phagocytic granulocytes attracted to the site of injury can also produce chemicals that cause the release of histamine.

Kinins are another group of substances that cause vasodilation and increased permeability of blood vessels. Kinins are present in blood plasma, and once activated, they play a role in chemotaxis by attracting phagocytic granulocytes, chiefly neutrophils, to the injured area.

Prostaglandins, substances released by damaged cells, intensify the effects of histamine and kinins and help phagocytes move through capillary walls. **Leukotrienes** are substances produced by mast cells (cells especially numerous in the connective tissue of the skin and respiratory system, and in blood vessels) and basophils (one of the types of white blood cells). Leukotrienes cause increased permeability of blood vessels and help attach phagocytes to pathogens. Various components of the complement system stimulate the release of histamine, attract phagocytes, and promote phagocytosis.

Vasodilation and the increased permeability of blood vessels also help deliver clotting elements of blood into the injured area. ❷ The blood clots that form around the site of activity prevent the microbe (or its toxins) from spreading to other parts of the body. ❸ As a result, there may be a localized collection of **pus,** a mixture of dead cells and body fluids, in a cavity formed by the breakdown of body tissues. This focus of infection is called an **abscess.** Common abscesses include pustules and boils.

The next stage in inflammation involves the migration of phagocytes to the injured area.

PHAGOCYTE MIGRATION AND PHAGOCYTOSIS

LEARNING OBJECTIVE
• Describe phagocyte migration.

Generally, within an hour after the process of inflammation is initiated, phagocytes appear on the scene (Figure 16.8c). ❹ As the flow of blood gradually decreases, phagocytes (both neutrophils and monocytes) begin to stick to the inner surface of the endothelium (lining) of blood vessels. This sticking process is called **margination.** ❺ Then the collected phagocytes begin to squeeze between the endothelial cells of the blood vessel to reach the damaged area. This migration, which resembles amoeboid movement, is called **emigration** (*diapedesis*); the migratory process can take as little as 2 minutes. ❻ The phagocytes then begin to destroy invading microorganisms by phagocytosis.

As mentioned earlier, certain chemicals attract neutrophils to the site of injury (chemotaxis). These include

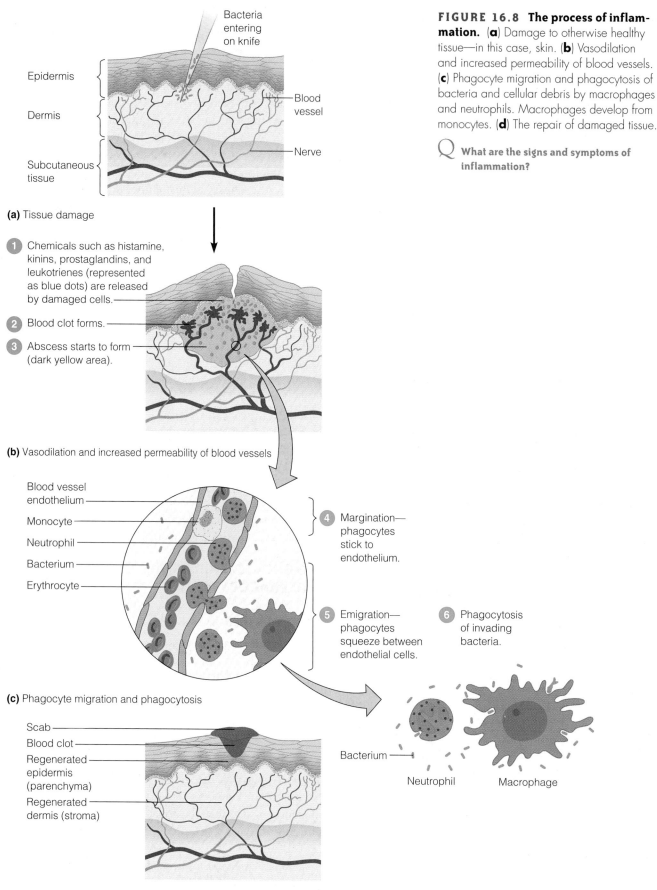

Bacteria
entering
on knife

Epidermis

Dermis

Subcutaneous
tissue

Blood
vessel

Nerve

(a) Tissue damage

**FIGURE 16.8 The process of inflam-
mation.** (**a**) Damage to otherwise healthy
tissue—in this case, skin. (**b**) Vasodilation
and increased permeability of blood vessels.
(**c**) Phagocyte migration and phagocytosis of
bacteria and cellular debris by macrophages
and neutrophils. Macrophages develop from
monocytes. (**d**) The repair of damaged tissue.

**Q   What are the signs and symptoms of
inflammation?**

1 Chemicals such as histamine,
kinins, prostaglandins, and
leukotrienes (represented
as blue dots) are released
by damaged cells.

2 Blood clot forms.

3 Abscess starts to form
(dark yellow area).

(b) Vasodilation and increased permeability of blood vessels

Blood vessel
endothelium

Monocyte

Neutrophil

Bacterium

Erythrocyte

4 Margination—
phagocytes
stick to
endothelium.

5 Emigration—
phagocytes
squeeze between
endothelial cells.

6 Phagocytosis
of invading
bacteria.

(c) Phagocyte migration and phagocytosis

Scab

Blood clot

Regenerated
epidermis
(parenchyma)

Regenerated
dermis (stroma)

Bacterium

Neutrophil

Macrophage

(d) Tissue repair

chemicals produced by microorganisms and even other neutrophils; other chemicals are kinins, leukotrienes, chemokines, and components of the complement system. Chemokines are cytokines that are chemotactic for phagocytes and T cells and thus stimulate both the inflammatory response and an adaptive immune response. The availability of a steady stream of neutrophils is ensured by the production and release of additional granulocytes from red bone marrow.

As the inflammatory response continues, monocytes follow the granulocytes into the infected area. Once the monocytes are contained in the tissue, they undergo changes in biological properties and become wandering macrophages. The granulocytes predominate in the early stages of infection but tend to die off rapidly. Macrophages enter the picture during a later stage of the infection, once granulocytes have accomplished their function. They are several times more phagocytic than granulocytes and are large enough to phagocytize tissue that has been destroyed, granulocytes that have been destroyed, and invading microorganisms.

After granulocytes or macrophages engulf large numbers of microorganisms and damaged tissue, they themselves eventually die. As a result, pus forms, and its formation usually continues until the infection subsides. At times, the pus pushes to the surface of the body or into an internal cavity for dispersal. On other occasions the pus remains even after the infection is terminated. In this case, the pus is gradually destroyed over a period of days and is absorbed by the body.

As effective as phagocytosis is in contributing to nonspecific resistance, there are times when the mechanism becomes less functional in response to certain conditions. For example, some individuals are born with an inability to produce phagocytes. And with age, there is a progressive decline in the efficiency of phagocytosis. Recipients of heart or kidney transplants have impaired nonspecific defenses as a result of receiving drugs that prevent the rejection of the transplant. Radiation treatments can also depress innate immune responses by damaging red bone marrow. Even certain diseases such as AIDS and cancer can cause defective functioning of innate defenses.

TISSUE REPAIR

The final stage of inflammation is tissue repair, the process by which tissues replace dead or damaged cells (Figure 16.8d). Repair begins during the active phase of inflammation, but it cannot be completed until all harmful substances have been removed or neutralized at the site of injury. The ability of a tissue to regenerate, or repair itself, depends on the type of tissue. For example, skin has a high capacity for regeneration, whereas cardiac muscle tissue does not have a high capacity to regenerate.

A tissue is repaired when its stroma or parenchyma produces new cells. The *stroma* is the supporting connective tissue, and the *parenchyma* is the functioning part of the tissue. For example, the capsule around the liver that encloses and protects it is part of the stroma because it is not involved in the functions of the liver; liver cells (hepatocytes) that perform the functions of the liver are part of the parenchyma. If only parenchymal cells are active in repair, a perfect or near-perfect reconstruction of the tissue occurs. A familiar example of perfect reconstruction is a minor skin cut, in which parenchymal cells are more active in repair. However, if repair cells of the stroma of the skin are more active, scar tissue is formed. ❉ **Animation: Go to The Microbiology Place website or CD-ROM and click "Animations" to view Inflammation.**

FEVER

⎧ **LEARNING OBJECTIVE**
⎩ • Describe the cause and effects of fever.

Inflammation is a local response of the body to injury. There are also systemic, or overall, responses; one of the most important is **fever,** an abnormally high body temperature. The most frequent cause of fever is infection from bacteria (and their toxins) or viruses.

Body temperature is controlled by a part of the brain called the hypothalamus. The hypothalamus is sometimes called the body's thermostat, and it is normally set at 37°C (98.6°F). It is believed that certain substances affect the hypothalamus by setting it at a higher temperature. Recall from Chapter 15 that when phagocytes ingest gram-negative bacteria, the lipopolysaccharides (LPS) of the cell wall (endotoxins) are released, causing the phagocytes to release the cytokine interleukin-1 (formerly called endogenous pyrogen). Interleukin-1 causes the hypothalamus to release prostaglandins that reset the hypothalamic thermostat at a higher temperature, thereby causing fever (see Figure 15.6, page 463). Another cytokine called *alpha-tumor necrosis factor*, which is produced by macrophages and mast cells, also induces fever.

Assume that the body is invaded by pathogens and that the thermostat setting is increased to 39°C (102.2°F). To adjust to the new thermostat setting, the body responds with blood vessel constriction, increased rate of metabolism, and **shivering,** all of which raise body temperature. Even though body temperature is climbing higher than normal, the skin remains cold, and shivering occurs. This condition, called a *chill,* is a definite sign that body temperature is rising. When body temperature reaches the setting of the thermostat, the chill disappears. The body will continue to

APPLICATIONS OF MICROBIOLOGY

SERUM COLLECTION

It is common for more than one blood sample to be drawn for laboratory tests. The blood is collected in tubes with different color caps (see the figure). Whole blood may be needed to culture microbes or to type the blood. Serum may be needed to test for enzymes or other chemicals in the blood. Serum is the straw-colored liquid remaining after blood is allow to clot. Plasma is the liquid remaining after formed elements are removed from unclotted blood, for example, by centrifugation.
• Which sample would you use to test for antibodies? To count blood cells?

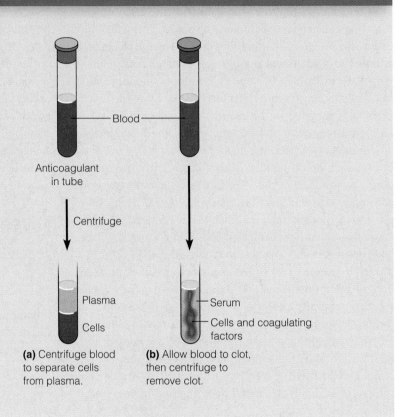

Blood

Anticoagulant in tube

Centrifuge

Plasma

Cells

Serum

Cells and coagulating factors

(a) Centrifuge blood to separate cells from plasma.

(b) Allow blood to clot, then centrifuge to remove clot.

maintain its temperature at 39°C until the IL-1 is eliminated. The thermostat is then reset to 37°C. As the infection subsides, heat-losing mechanisms such as vasodilation and sweating go into operation. The skin becomes warm, and the person begins to sweat. This phase of the fever, called the **crisis,** indicates that body temperature is falling.

Up to a certain point, fever is considered a defense against disease. Interleukin-1 helps step up the production of T cells. High body temperature intensifies the effect of antiviral interferons (page 494) and increases production of transferrins that decrease the iron available to microbes (page 496). Also, because the high temperature speeds up the body's reactions, it may help body tissues repair themselves more quickly.

Among the complications of fever are tachycardia (rapid heart rate), which may compromise elderly persons with cardiopulmonary disease; increased metabolic rate, which may produce acidosis; dehydration; electrolyte imbalances; seizures in young children; and delirium and coma. As a rule, death results if body temperature rises above 44 to 46°C (112–114°F).

ANTIMICROBIAL SUBSTANCES

The body produces certain antimicrobial substances in addition to the chemical factors mentioned earlier. The recently discovered antimicrobial action of nitric oxide is described in the box on page 486. Among the most important of these are the proteins of the complement system and interferons.

THE COMPLEMENT SYSTEM

LEARNING OBJECTIVES
• List the components of the complement system.
• Describe three pathways of activating complement.
• Describe three consequences of complement activation.

The **complement system** is a defensive system consisting of over 30 proteins produced by the liver and found circulating in blood serum (see the box above) and within tissues throughout the body. Together, proteins of the complement system destroy microbes by (1) cytolysis, (2) inflammation, and (3) phagocytosis and also prevent excessive damage to

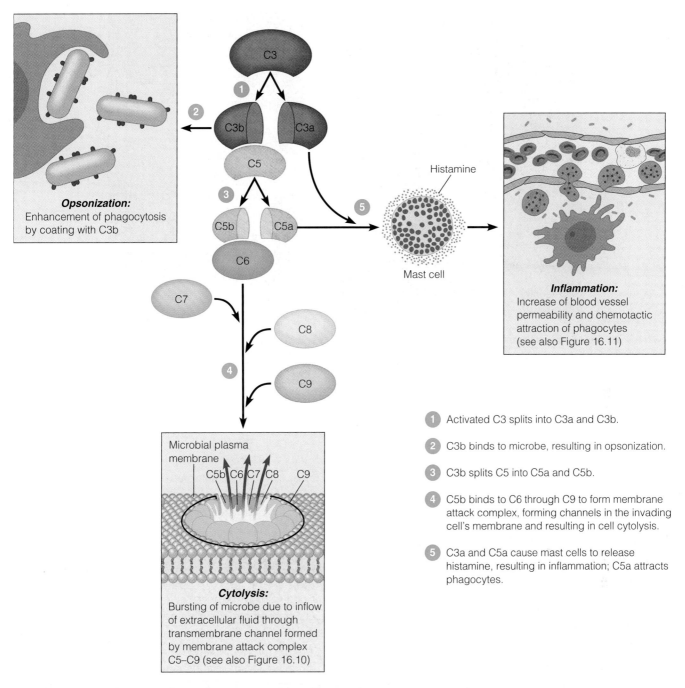

FIGURE 16.9 **Outcomes of complement activation.** The complement system consists of over 30 proteins produced by the liver and found circulating in serum.

Q **What are the outcomes of complement activation?**

host tissues. Complement proteins are usually designated by an uppercase letter C and are inactive until they are split into fragments (products). The proteins are numbered C1 through C9, named for the order in which they were discovered. The fragments are activated proteins and are indicated by the lowercase letters *a* and *b*. For example, inactive complement protein C3 is split into two activated fragments, C3a and C3b. The activated fragments carry out the destructive actions of the C1 through C9 complement proteins.

Complement proteins act in a *cascade*, that is, one reaction triggers another, which in turn triggers another, and so on. Also, as part of the cascade, more product is formed with each succeeding reaction so that the effect is amplified many times as the reactions continue.

THE RESULT OF COMPLEMENT ACTIVATION

The C3 protein can be activated by three mechanisms that will be described shortly. Activation of C3 (Figure 16.9) is

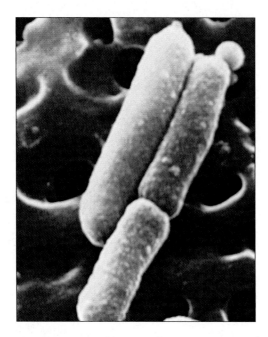

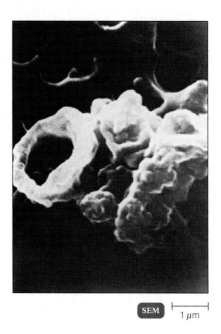

FIGURE 16.10 Cytolysis caused by complement. Micrographs of a rod-shaped bacterium before cytolysis (left) and after cytolysis (right). *Source:* Reprinted from Schreiber, R. D. et al. "Bactericidal Activity of the Alternative Complement Pathway Generated from 11 Isolated Plasma Proteins." *Journal of Experimental Medicine*, 149:870–882, 1979.

Q **How does complement aid in fighting infections?**

very important because it starts a cascade that results in cytolysis, inflammation, and phagocytosis.

❶ When C3 is activated, it splits into C3a and C3b.

❷ C3b binds to the surface of a microbe and receptors on phagocytes attach to the C3b. Thus C3b enhances *phagocytosis* by coating a microbe, a process called *opsonization*, or *immune adherence*. Opsonization promotes attachment of a phagocyte to a microbe.

❸ C3b also initiates a series of reactions that result in cytolysis. First, C3b splits C5. Fragment C5b then binds to C6 and C7, which attach to the invading cell's plasma membrane. Next, C8 and several C9 molecules join the other complement proteins and together form a cylinder-shaped **membrane attack complex (MAC),** which inserts into the membrane.

❹ The MAC creates transmembrane channels (holes) in the membrane that result in *cytolysis*, the bursting of the microbial cell due to the inflow of extracellular fluid through the channels (Figure 16.10).

❺ C3a and C5a bind to mast cells and cause them to release histamine and other chemicals that increase blood vessel permeability during *inflammation* (Figure 16.11). C5a also functions as a very powerful chemotactic factor that attracts phagocytes to the site of an infection.

Host cell plasma membranes contain proteins that protect against lysis by preventing attachment of the MAC proteins to their surfaces. Also, the MAC forms the basis for the complement fixation test used to diagnose some diseases. This is explained in Chapter 18 (see Figure 18.10, page 541). Gram-negative bacteria are more susceptible to cytolysis because they have only one or very few layers of peptidoglycan to protect the plasma membrane from the effects of complement. The many layers of peptidoglycan of gram-positive bacteria limit access of complement to the plasma membrane and thus interfere with cytolysis.

The cascade of complement proteins that occurs during an infection is called **complement activation** and may occur in three pathways.

THE CLASSICAL PATHWAY

The **classical pathway** (Figure 16.12, page 494), so named because it was the first to be discovered, is initiated when antibodies bind to antigens (microbes) and occurs as follows:

❶ Antibodies attach to antigens (for example, proteins or large polysaccharides on the surface of a bacterium or other cell), forming antigen–antibody complexes. The antigen–antibody complexes bind and activate C1.

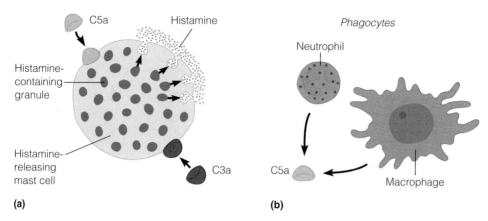

FIGURE 16.11 Inflammation stimulated by complement. (**a**) C3a and C5a bound to mast cells, basophils, and platelets trigger the release of histamine, which increases blood vessel permeability. (**b**) C5a functions as a chemotactic factor that attracts phagocytes to the site of complement activation.

Q **How is complement inactivated?**

❷ Next, activated C1 activates C2 and C4 by splitting them. C2 is split into fragments called C2a and C2b, and C4 is split into fragments called C4a and C4b.

❸ C2a and C4b combine and together they activate C3 by splitting it into C3a and C3b. The C3 fragments then initiate cytolysis, inflammation, and opsonization (see Figure 16.9).

THE ALTERNATIVE PATHWAY

The **alternative pathway** is so named because it was discovered after the classical pathway. Unlike the classical pathway, the alternative pathway does not involve antibodies. The alternative pathway is activated by contact between certain complement proteins and a pathogen. C3 is constantly present in the blood. It combines with complement proteins called factor B, factor D, and factor P (properdin) on the surface of a pathogen (Figure 16.13). The complement proteins are attracted to microbial cell surface material (mostly lipid–carbohydrate complexes of certain bacteria and fungi). Once the complement proteins combine and interact, C3 is split into fragments C3a and C3b. As in the classical pathway, C3a participates in inflammation, and C3b functions in cytolysis and opsonization (see Figure 16.9).

THE LECTIN PATHWAY

The **lectin pathway** is the most recently discovered mechanism for complement activation. When macrophages ingest bacteria, viruses, and other foreign matter by phagocytosis, they release chemicals that stimulate the liver to produce **lectins,** proteins that bind to carbohydrates (Figure 16.14, page 495).

❶ One such lectin, **mannose-binding lectin (MBL),** binds to the carbohydrate mannose. MBL binds to many pathogens because the MBL molecules recognize a distinctive pattern of carbohydrates that includes mannose, which is found in bacterial cell walls and on some viruses. As a result of binding, MBL functions as an opsonin to enhance phagocytosis and

❷ activates C2 and C4;

❸ C2a and C4b activate C3 (see Figure 16.9).

INACTIVATION OF COMPLEMENT

Once complement is activated, its destructive capabilities usually cease very quickly in order to minimize the destruction of host cells. This is accomplished by various regulatory proteins in the host's blood and on certain cells, such as blood cells. The proteins bring about the breakdown of activated complement and function as inhibitors and destructive enzymes.

COMPLEMENT AND DISEASE

In addition to its importance in defense, the complement system assumes a role in causing disease as a result of inherited deficiencies. For example, deficiencies of C1, C2, or C4 cause collagen vascular disorders that result in hypersensitivity (anaphylaxis); deficiency of C3, though rare, results in increased susceptibility to recurrent infections with pyogenic microbes; and C5 through C9 defects result in increased susceptibility to *Neisseria meningitidis* and *N. gonorrhoeae* infections.

EVADING THE COMPLEMENT SYSTEM

Some bacteria evade the complement system by means of their capsules, which prevent complement activation. For

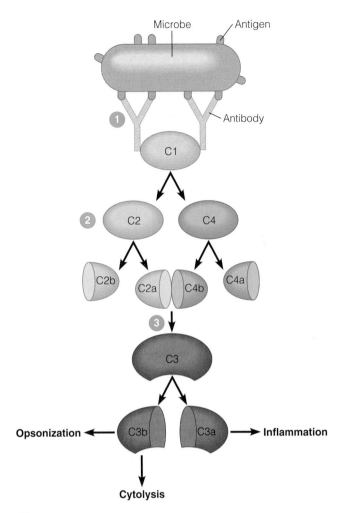

① C1 is activated by binding to antigen–antibody complexes.

② Activated C1 splits C2 into C2a and C2b and C4 into C4a and C4b.

③ C2a and C4b combine and activate C3, splitting it into C3a and C3b (see also Figure 16.9).

FIGURE 16.12 Classical pathway of complement activation. This pathway is initiated by an antigen–antibody reaction. The splitting of C3 into C3a and C3b starts a cascade that results in cytolysis, inflammation, and phagocytosis (see also Figure 16.9).

Q **What happens after C3 is cleaved into C3a and C3b?**

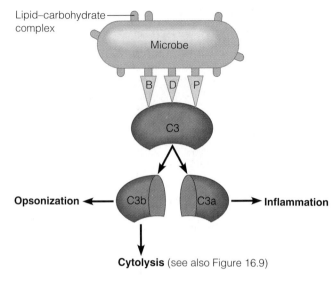

Key: B B factor D D factor P P factor

FIGURE 16.13 Alternative pathway of complement activation. This pathway is initiated by contact between certain complement proteins (C3 and factors B, D, and P) and a pathogen. There are no antibodies involved. Once C3a and C3b are formed, they participate in cytolysis, inflammation, and phagocytosis per the classical pathway (see also Figure 16.12).

Q **How is the alternative pathway similar to the classical pathway?**

INTERFERONS

LEARNING OBJECTIVES

- Define *interferons.*
- Compare and contrast the actions of α-IFN and β-IFN with γ-IFN.

Because viruses depend on their host cells to provide many functions of viral multiplication, it is difficult to inhibit viral multiplication without affecting the host cell itself. One way the infected host counters viral infections is with interferons. **Interferons (IFNs)** are a class of similar antiviral proteins produced by certain animal cells, such as lymphocytes and macrophages, after viral stimulation. One of the principal functions of interferons is to interfere with viral multiplication.

An interesting feature of interferons is that they are host-cell–specific but not virus-specific. Interferons produced by human cells protect human cells but produce little antiviral activity for cells of other species, such as mice or chickens. However, the interferons of a species are active against a number of different viruses.

Just as different animal species produce different interferons, different types of cells in the same animal also

example, some capsules contain large amounts of a substance called sialic acid, which discourages opsonization and MAC formation. Some gram-negative bacteria can lengthen their surface lipid–carbohydrate complexes, which prevents MAC formation. Bacteria that are not killed by the MAC are said to be *serum resistant,* a property of many gram-negative bacteria that cause systemic infections. Gram-positive cocci release an enzyme that breaks down C5a, the fragment that serves as a chemotactic factor that attracts phagocytes. ✳ **Animation: Go to The Microbiology Place website or CD-ROM and click "Animations" to view The Complement System.**

produce different interferons. Human interferons are of three principal types: *alpha-interferon* (α-IFN), *beta-interferon* (β-IFN), and *gamma-interferon* (γ-IFN). There are also various subtypes of interferons within each of the principal groups. In the human body, interferons are produced by fibroblasts in connective tissue and by lymphocytes and other leukocytes. Each of the three types of interferons produced by these cells can have a slightly different effect on the body.

All interferons are small proteins, with molecular weights between 15,000 and 30,000. They are quite stable at low pH and are fairly resistant to heat.

Both α-IFN and β-IFN are produced by virus-infected host cells only in very small quantities and diffuse to uninfected neighboring cells (Figure 16.15). They react with plasma or nuclear membrane receptors, inducing the uninfected cells to manufacture mRNA for the synthesis of **antiviral proteins (AVPs).** These proteins are enzymes that disrupt various stages of viral multiplication. For example, one AVP, called *oligoadenylate synthetase*, degrades viral mRNA. Another, called *protein kinase*, inhibits protein synthesis.

Gamma-interferon is produced by lymphocytes; it induces neutrophils and macrophages to kill bacteria by phagocytosis. Neutrophils and macrophages in individuals with an inherited condition called chronic granulomatous disease (CGD) do not kill bacteria. When these people take a recombinant γ-IFN, their neutrophils and macrophages kill bacteria. Gamma-interferon is not a cure and must be taken for the life of the CGD individual.

Because of their beneficial properties, they would seem to be ideal antiviral substances, but certain problems do exist. For one thing, interferons are effective for only short periods; they do not remain stable for long periods of time in the body. And when injected, interferons have side effects, such as nausea, fatigue, headache, vomiting, weight loss, and fever. High concentrations of interferons are toxic to the heart, liver, kidneys, and red bone marrow. They typically play a major role in infections that are acute and short-term, such as colds and influenza. Another problem is that they have no effect on viral multiplication in cells already infected. Also, some viruses, such as adenoviruses (which cause respiratory infections), have resistance mechanisms that inhibit AVPs. Further, some viruses, such as the hepatitis B virus, do not induce the production of sufficient amounts of interferon in host cells following viral stimulation.

The importance of interferons in protecting the body against viruses, as well as their potential as anticancer agents, has made their production in large quantities a top health priority. Several groups of scientists have successfully applied recombinant DNA technology in inducing

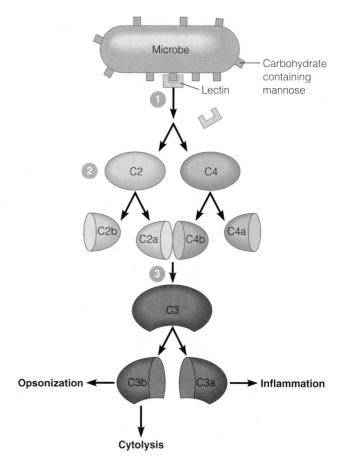

1 Lectin binds to an invading cell.

2 Bound lectin splits C2 and C4.

3 C2a and C4b combine and activate C3 (see also Figure 16.9).

FIGURE 16.14   The lectin pathway of complement activation. When mannose-binding lectin (MBL) binds to mannose on the surface of microbes, MBL functions as an opsin that enhances phagocytosis and activates complement via the classical and alternative pathways. (See Figures 16.12 and 16.13)

Q **How do the lectin and alternative pathways differ from the classical pathway?**

certain species of bacteria to produce interferons. (This technique is described in Chapter 9.) The interferons produced with recombinant DNA techniques, called *recombinant interferons (rIFNs)*, are important for two reasons: they are pure, and they are plentiful.

In clinical trials, IFNs have exhibited no effects against some types of tumors and only limited effects against others. Alpha-interferon (Intron A) is approved in the United States for treating several virus-associated disorders. One is Kaposi's sarcoma, a cancer that often occurs in patients infected with HIV. Other approved uses for α-IFN include treating genital herpes caused by herpesvirus;

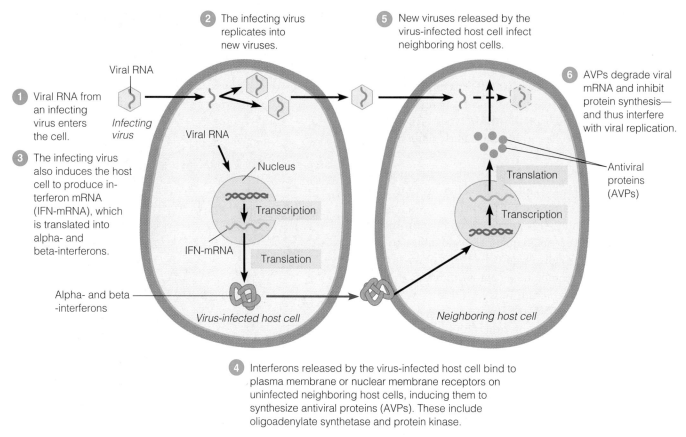

FIGURE 16.15 Antiviral action of alpha- and beta-interferons (IFNs).
Interferons are host-cell–specific but not virus-specific.

Q **How does interferon stop viruses?**

hepatitis B and C, caused by the hepatitis B and C viruses; and hairy cell leukemia. Alpha-interferon also is being tested to see if it can slow the development of AIDS in HIV-infected people. A form of β-IFN (Betaferon) slows the progression of multiple sclerosis (MS) and lessens the frequency and severity of MS attacks.

TRANSFERRINS

LEARNING OBJECTIVE

• Describe the role of transferrins in innate immunity.

Transferrins are iron-binding proteins found in blood, milk, saliva, and tears. They inhibit bacterial growth by reducing the amount of available iron. Iron is not only required for microbial growth (the synthesis of cytochromes and certain enzymes), it also suppresses chemotaxis and phagocytosis. Iron overload increases the risk of infection.

ANTIMICROBIAL PEPTIDES

LEARNING OBJECTIVE

• Describe the role of antimicrobial peptides in innate immunity.

Although newly discovered, **antimicrobial peptides** may be one of the most important elements of innate immunity. These peptides, consisting of about a dozen amino acids, bind to microbial plasma membranes causing cell lysis. Antimicrobial peptides are produced by mucous-membrane cells and phagocytes.

Table 16.3 contains a summary of innate immunity defenses.

* * *

In Chapter 17, we will discuss the principal factors that contribute to adaptive immunity.

TABLE 16.3	Summary of Innate Immunity Defenses
Component	Functions

FIRST LINE OF DEFENSE: SKIN AND MUCOUS MEMBRANES

Physical Factors

Epidermis of skin	Forms a physical barrier to the entrance of microbes.
Mucous membranes	Inhibit the entrance of many microbes, but not as effective as intact skin.
Mucus	Traps microbes in respiratory and gastrointestinal tracts.
Lacrimal apparatus	Tears dilute and wash away irritating substances and microbes.
Saliva	Washes microbes from surfaces of teeth and mucous membranes of mouth.
Hairs	Filter out microbes and dust in nose.
Cilia	Together with mucus, trap and remove microbes and dust from upper respiratory tract.
Epiglottis	Prevents microbes from entering lower respiratory tract.
Urine	Washes microbes from urethra.
Vaginal secretions	Move microbes out of female reproductive tract.
Defecation and vomiting	Expel microbes from body.

Chemical Factors

Sebum	Forms a protective acidic film over the skin surface that inhibits growth of many microbes.
Lysozyme	Antimicrobial substance in perspiration, tears, saliva, nasal secretions, and tissue fluids.
Gastric juice	Destroys bacteria and most toxins in stomach.
Vaginal secretions	Slight acidity discourages bacterial growth.

SECOND LINE OF DEFENSE

Defensive cells

Phagocytes	Phagocytosis by cells such as neutrophils, eosinophils, and macrophages.
Natural killer (NK) cells	Kill infected target cells by releasing granules that contain perforin and granzymes. Phagocytes then kill the infected microbes.
Inflammation	Confines and destroy microbes and initiates tissue repair.
Fever	Intensifies the effects of interferons, inhibits growth of some microbes, and speeds up body reactions that aid repair.

Antimicrobial substances

Complement system	Causes cytolysis of microbes, promotes phagocytosis, and contributes to inflammation.
Interferons	Protect uninfected host cells from viral infection.
Transferrins	Inhibit growth of certain bacteria by reducing the amount of available iron.
Antimicrobial peptides	Cause lysis of bacteria.

STUDY OUTLINE

INTRODUCTION (p. 474)

1. The ability to ward off disease through body defenses is called immunity.
2. Lack of immunity is called susceptibility.

THE CONCEPT OF IMMUNITY (p. 475)

1. Innate immunity refers to all body defenses that protect the body against any kind of pathogen.
2. Adaptive immunity refers to defenses (antibodies) against specific microorganisms.
3. Toll-like receptors are proteins in plasma membranes of macrophages and dendritic cells. TLRs bind to invading microbes.

FIRST LINE OF DEFENSE: SKIN AND MUCOUS MEMBRANES (pp. 476–478)

1. The body's first line of defense against infections is a physical barrier and nonspecific chemicals of the skin and mucous membranes.

PHYSICAL FACTORS (pp. 476–477)

1. The structure of intact skin and the waterproof protein keratin provide resistance to microbial invasion.
2. Some pathogens, if present in large numbers, can penetrate mucous membranes.
3. The lacrimal apparatus protects the eyes from irritating substances and microorganisms.
4. Saliva washes microorganisms from teeth and gums.
5. Mucus traps many microorganisms that enter the respiratory and gastrointestinal tracts; in the lower respiratory tract, the ciliary escalator moves mucus up and out.
6. The flow of urine moves microorganisms out of the urinary tract, and vaginal secretions move microorganisms out of the vagina.

CHEMICAL FACTORS (pp. 477–478)

1. Sebum contains unsaturated fatty acids, which inhibit the growth of pathogenic bacteria. Some bacteria commonly found on the skin can metabolize sebum and cause the inflammatory response associated with acne.
2. Perspiration washes microorganisms off the skin.
3. Lysozyme is found in tears, saliva, nasal secretions, and perspiration.
4. The high acidity (pH 1.2–3.0) of gastric juice prevents microbial growth in the stomach.

NORMAL MICROBIOTA AND INNATE IMMUNITY (pp. 478–479)

1. Normal microbiota change the environment, which can prevent the growth of pathogens.

SECOND LINE OF DEFENSE (pp. 479–496)

1. If a microbe penetrates the first line of defense it encourages production of phagocytes, inflammation, fever, and antimicrobial substances.

FORMED ELEMENTS IN BLOOD (pp. 479–482)

1. Blood consists of plasma (fluid) and formed elements (cells and cell fragments).
2. Leukocytes (white blood cells) are divided into three categories: granulocytes (neutrophils, basophils, eosinophils, and dendritic cells), lymphocytes, and monocytes.
3. During many infections, the number of leukocytes increases (leukocytosis); some infections are characterized by leukopenia (decrease in leukocytes). ❊ **Animation: Host Defenses. The Microbiology Place.**

PHAGOCYTES (pp. 483–486)

1. Phagocytosis is the ingestion of microorganisms or particulate matter by a cell.
2. Phagocytosis is performed by phagocytes, certain types of white blood cells or their derivatives.

ACTIONS OF PHAGOCYTIC CELLS (pp. 483–484)

3. Among the granulocytes, neutrophils are the most important phagocytes.
4. Enlarged monocytes become wandering macrophages and fixed macrophages.
5. Fixed macrophages are located in selected tissues and are part of the mononuclear phagocytic system.
6. Granulocytes predominate during the early stages of infection, whereas monocytes predominate as the infection subsides.

THE MECHANISM OF PHAGOCYTOSIS (p. 484)

7. Chemotaxis is the process by which phagocytes are attracted to microorganisms.
8. The phagocyte then adheres to the microbial cells; adherence may be facilitated by opsonization—coating the microbe with serum proteins.

9. Pseudopods of phagocytes engulf the microorganism and enclose it in a phagocytic vesicle to complete ingestion.

10. Many phagocytized microorganisms are killed by lysosomal enzymes and oxidizing agents.

MICROBIAL EVASION OF PHAGOCYTOSIS
(pp. 485–486)

11. Some microbes are not killed by phagocytes and can even reproduce in phagocytes.

12. Evasion mechanisms include M protein, capsules, leukocidins, membrane attack complexes, and prevention of phagolysosome formation. ❊ **Animation: Phagocytosis. The Microbiology Place.**

INFLAMMATION   (pp. 486–490)

1. Inflammation is a bodily response to cell damage; it is characterized by redness, pain, heat, swelling, and sometimes the loss of function.

2. Acute inflammation is a short, intense response to infection; chronic inflammation is a prolonged response.

VASODILATION AND INCREASED PERMEABILITY OF BLOOD VESSELS   (p. 487)

3. The release of histamine, kinins, and prostaglandins causes vasodilation and increased permeability of blood vessels.

4. Blood clots can form around an abscess to prevent dissemination of the infection.

PHAGOCYTE MIGRATION AND PHAGOCYTOSIS   (pp. 487–489)

5. Phagocytes have the ability to stick to the lining of the blood vessels (margination).

6. They also have the ability to squeeze through blood vessels (emigration).

7. Pus is the accumulation of damaged tissue and dead microbes, granulocytes, and macrophages.

TISSUE REPAIR   (p. 489)

8. A tissue is repaired when the stroma (supporting tissue) or parenchyma (functioning tissue) produces new cells.

9. Stromal repair by fibroblasts produces scar tissue.
   ❊ **Animation: Inflammation. The Microbiology Place.**

FEVER   (pp. 489–490)

1. Fever is an abnormally high body temperature produced in response to a bacterial or viral infection.

2. Bacterial endotoxins, interleukin-1, and alpha-tumor necrosis factor can induce fever.

3. A chill indicates a rising body temperature; crisis (sweating) indicates that the body's temperature is falling.

ANTIMICROBIAL SUBSTANCES   (pp. 490–497)

THE COMPLEMENT SYSTEM   (pp. 490–494)

1. The complement system consists of a group of serum proteins that activate one another to destroy invading microorganisms. Serum is the liquid remaining after blood plasma is clotted.

2. Complement proteins are activated in a cascade.

3. C3 activation can result in cell lysis, inflammation, and opsonization.

4. Complement is activated via the classical pathway, the alternative pathway, and the lectin pathway.

5. Complement is deactivated by host-regulatory proteins.

6. Complement deficiencies can result in an increased susceptibility to disease.

7. Some bacteria evade destruction by complement by means of capsules, surface lipid–carbohydrate complexes, and enzymatic destruction of C5a. ❊ **Animation: The Complement System. The Microbiology Place.**

INTERFERONS   (pp. 494–496)

8. Interferons (IFNs) are antiviral proteins produced in response to viral infection.

9. There are three types of human interferon: α-IFN, β-IFN, and γ-IFN. Recombinant interferons have been produced.

10. The mode of action of α-IFN and β-IFN is to induce uninfected cells to produce antiviral proteins (AVPs) that prevent viral replication.

11. Interferons are host-cell–specific but not virus-specific.

12. Gamma-interferon activates neutrophils and macrophages to kill bacteria.

TRANSFERRINS   (p. 496)

13. Transferrins bind iron.

ANTIMICROBIAL PEPTIDES   (pp. 496–497)

14. Antimicrobial peptides lyse microbial cells.

STUDY QUESTIONS

Access more review material either online at **The Microbiology Place** (www.microbiologyplace.com) or with **The Microbiology Place CD-ROM** packaged with your new book. There you'll find activities, practice tests, quizzes, flashcards, case studies, and more to help you succeed.

Answers to the Study Questions can be found in Appendix G.

REVIEW

1. Define the following terms:
   a. innate immunity
   b. susceptibility
   c. adaptive immunity

2. Identify at least one physical and one chemical factor that prevent microbes from entering the body through each of the following:
   a. skin
   b. eyes
   c. digestive tract
   d. respiratory tract
   e. urinary tract
   f. reproductive tract

3. Describe the six different white blood cells, and name a function for each cell type.

4. Define *phagocytosis*.

5. Compare the structures and functions of granulocytes and monocytes in phagocytosis.

6. How do fixed and wandering macrophages differ?

7. Diagram the following processes that result in phagocytosis: margination, emigration, adherence, and ingestion.

8. Define *inflammation*, and list its characteristics.

9. Why is inflammation beneficial to the body?

10. How is fever related to innate immunity?

11. What causes the periods of chill and crisis during fever?

12. What is complement? List the steps of complement activation via (a) the classical pathway, (b) the alternative pathway, and (c) the lectin pathway.

13. Summarize the major outcomes of complement activation.

14. How is the complement system activated by bacterial endotoxin in the bloodstream? How does endotoxic shock result in massive host cell destruction?

15. What are interferons? Discuss their roles in innate immunity. Why do α-IFN and β-IFN share the same receptor on target cells yet γ-IFN has a different receptor?

16. Are the following involved in innate or adaptive immunity? Identify the role of each in immunity:
   a. TLRs
   b. transferrins
   c. antimicrobial peptides

MULTIPLE CHOICE

1. *Legionella* uses C3b receptors to enter monocytes. This
   a. prevents phagocytosis.
   b. degrades complement.
   c. inactivates complement.
   d. prevents inflammation.
   e. prevents cytolysis.

2. *Chlamydia* can prevent the formation of phagolysosomes, and therefore *Chlamydia* can
   a. avoid being phagocytized.
   b. avoid destruction by complement.
   c. prevent adherence.
   d. avoid being digested.
   e. none of the above

3. If the following are placed in the order of occurrence, which would be the *third* step?
   a. Emigration
   b. Digestion
   c. Formation of a phagosome
   d. Formation of a phagolysosome
   e. Margination

4. If the following are placed in the order of occurrence, which would be the *third* step?
   a. Activation of C5 through C9
   b. Cell lysis
   c. Antigen–antibody reaction
   d. Activation of C3
   e. Activation of C2 through C4

5. A human host can prevent a pathogen from getting enough iron by
   a. reducing dietary intake of iron.
   b. binding iron with transferrin.
   c. binding iron with hemoglobin.
   d. excreting excess iron.
   e. binding iron with siderophores.

6. A decrease in the production of C3 would result in
   a. increased susceptibility to infection.
   b. increased numbers of white blood cells.
   c. increased phagocytosis.
   d. activation of C5 through C9.
   e. none of the above

7. In 1884, Elie Metchnikoff observed blood cells collected around a splinter inserted in a sea star embryo. This was the discovery of
   a. blood cells.
   b. sea stars.
   c. phagocytosis.
   d. immunity.
   e. none of the above

8. *Helicobacter pylori* uses the enzyme urease to counteract a chemical defense in the human organ in which it lives. This chemical defense is
   a. lysozyme.
   b. hydrochloric acid.
   c. superoxide radicals.
   d. sebum.
   e. complement.

9. Which of the following statements about α-IFN is *not* true?
   a. It interferes with viral replication.
   b. It is host-specific.
   c. It is released by fibroblasts.
   d. It is virus-specific.
   e. It is released by lymphocytes.

10. Which of the following does not stimulate phagocytes?
   a. cytokines
   b. γ-IFN
   c. C3b
   d. lipid A
   e. histamine

CRITICAL THINKING

1. Why do serum levels of iron increase during an infection? What can a bacterium do to respond to high levels of transferrin?

2. A variety of drugs with the ability to reduce inflammation are available. Comment on the danger of misuse of these anti-inflammatory drugs.

3. To be a successful parasite, a microbe must avoid destruction by complement. The following list provides examples of complement-evading techniques. For each microbe, identify the disease it causes, and describe how its strategy enables it to avoid destruction by complement.

Pathogen	Strategy
Group A streptococci	C3 does not bind to M protein
Haemophilus influenzae type b	Has a capsule
Pseudomonas aeruginosa	Sheds cell wall polysaccharides
Trypanosoma cruzi	Degrades C1

4. The list below identifies a virulence factor for a selected microorganism. Describe the effect of each factor listed. Name a disease caused by each organism.

Microorganism	Virulence Factor
Influenzavirus	Causes release of lysosomal enzymes
Mycobacterium tuberculosis	Inhibits lysosome fusion
Toxoplasma gondii	Prevents phagosome acidification
Trichophyton	Secretes keratinase
Trypanosoma cruzi	Lyses phagosomal membrane

CLINICAL APPLICATIONS

1. People with *Rhinovirus* infections of the nose and throat have an 80-fold increase in kinins and no increase in histamine. What do you expect for rhinoviral symptoms? What disease is caused by rhinoviruses?

2. A hematologist often performs a differential white blood cell count on a blood sample. Such a count determines the relative numbers of white blood cells. Why are these numbers important? What do you think a hematologist would find in a differential white blood cell count of a patient with mononucleosis? With neutropenia? With eosinophilia?

3. Leukocyte adherence deficiency (LAD) is an inherited disease resulting in the inability of neutrophils to recognize C3b-bound microorganisms. What are the most likely consequences of LAD?

4. The neutrophils of individuals with Chédiak-Higashi syndrome (CHS) have fewer than normal chemotactic receptors and lysosomes that spontaneously rupture. What are the consequences of CHS?

5. Consider the following.
   a. In laboratory experiments, plant lectins can bind to mannose in human plasma membranes. Why doesn't human mannose-binding lectin bind to human cells?
   b. About 4% of the human population have a mannose-binding lectin deficiency. How might this deficiency affect a person?

17

Adaptive Immunity: Specific Defenses of the Host

In Chapter 16, we discussed innate defenses of the body against microorganisms in the environment. These defenses include the skin and mucous membranes, phagocytosis, and inflammation. Humans, and even the simplest animals, have some defenses that are always present which provide instant protection against infection. Taken altogether, these defenses are termed **innate (nonspecific) immunity.** *Immunity* is a term derived from the Latin word *immunis,* meaning to exempt. (For a review of the body's innate defense systems, see Figure 16.1, page 475). Innate immunity seems to have an inherited, genetic component. For example, measles usually has a relatively mild effect on individuals of European ancestry, but the disease devastated the populations of Pacific Islanders when they were first exposed to it by European explorers. The reason probably lies in natural selection. For the Europeans, many generations of exposure to the measles virus presumably led to the selection of genes that conferred some resistance to the virus. Higher animals also are protected by the more specialized adaptive immunity. **Adaptive (specific) immunity** is induced; that is, it adapts to a microbial invader or a foreign substance. Adaptive immunity will be the focus of this chapter.

UNDER THE MICROSCOPE

Dendritic cells. These antigen-presenting cells are named for their long extensions called dendrites.

THE ADAPTIVE IMMUNE SYSTEM

LEARNING OBJECTIVE
- Differentiate between innate and adaptive immunity.

It was recognized long ago that immunity to certain infectious diseases can be acquired during an individual's life. If a person recovered from smallpox or measles, that person was almost always immune to that disease when exposed to it again. In some way they had acquired a memory of the infection; an important factor in adaptive immunity. Eventually, as medicine developed over the centuries, methods were found to mimic the adaptive immunity to disease by deliberately exposing people to harmless versions of pathogens that caused certain diseases, thus rendering them immune; we now call this *vaccination* (see Chapter 18, pages 528–534). Vaccination against smallpox, the first disease for which vaccination was developed, predated by nearly a hundred years any knowledge of microscopic pathogens. However, the systematic advancement of the science of adaptive immunity required the concept of the germ theory of diseases, that is, that specific pathogenic microbes are responsible for specific diseases (see page 425).

DUAL NATURE OF THE ADAPTIVE IMMUNE SYSTEM

LEARNING OBJECTIVE
- Differentiate between humoral and cellular immunity.

A brief review of the development of the theoretical basis of adaptive immunity will introduce many of the terms and concepts important to an understanding of adaptive immunity. It will also serve to emphasize the dual nature of adaptive immunity—that it consists of a humoral and a cellular component.

In 1887, when Louis Pasteur first observed the immunity that developed in chickens when he accidentally injected them with weakened pathogens (page 11), he hypothesized that this was due to the depletion of some essential nutrient that was required by the pathogens in order to multiply. Events moved rapidly in the next few years. Emil von Behring, working with diphtheria and tetanus bacteria, showed that the culture medium in which they were grown contained an apparent *toxin* that was fatal to animals when injected into them. When rabbits were injected with very small amounts of toxin, however, they would often survive. The researchers were surprised to find that when blood serum from surviving animals was injected into mice shortly after they received a

typically lethal amount of toxin that these animals showed no sign at all of receiving the toxin. Apparently, some factor in the serum from the surviving rabbits had neutralized the lethal toxins. They named this factor *antitoxin*. For this work, in 1901, von Behring received the first Nobel Prize awarded in Medicine and Physiology.

Paul Ehrlich, a German physician, found that a certain amount of antitoxin would neutralize a certain amount of toxin. He also determined that the protective nature of antitoxin was highly specific—it neutralized only the toxin that had given rise to it. The scientific community quickly focused attention on this new concept of immunity that clearly showed immense promise in medicine. It was found that protective, similarly derived serum factors (now called by the more general name of *antibodies*) were also produced by exposure to whole bacteria and other pathogens. It was soon learned that antibodies arose against not only microbial pathogens and toxins, but also many other particulate substances, such as plant pollen and red blood cells, that the body recognized as alien, or nonself. Such substances that caused the production of antibodies were called *antigens*—from *anti*body *gen*erators. The combination of an antibody and a particulate antigen caused them to visibly clump together or, in some cases, lyse. It was found that the lysis of a cell bearing an antigen, in response to an antibody against it, required another element naturally found in the blood. This was *complement* (see Chapter 16, page 490), so-called because it complements the action of antibodies.

HUMORAL IMMUNITY

From ancient times until well into the nineteenth century, the medical community believed fervently that health depended upon four different body fluids, or humors: blood, phlegm, black bile, and yellow bile. Deliberate bloodletting as a treatment for disease, which seems so illogical to the modern mind, was intended to correct imbalances in these humors. The new science of immunology adopted the term *humoral immunity* when describing immunity brought about by antibodies.

CELLULAR IMMUNITY

As recently as the mid-1950s, immunology was a relatively simple science. Medical science was familiar with innate immunity largely based upon nonspecific phagocytic cells and a humoral immunity based upon antibodies that specifically targeted certain toxins or particulate antigens. Up to that time the **thymus,** a lymphoid organ found in the upper chest and which slowly atrophies after puberty, had no known function. Similarly, in birds the **bursa of Fabricius,** a structure resembling a lymph node, also had no known function. Experiments designed to determine the function of the thymus or the bursa by removing them had no apparent effect. A breakthrough in the science of immunology occurred in 1956 when an experimenter happened to remove the bursa from very young, rather than mature, birds. There were, as usual, no discernible effects, and the birds were set aside. In an unrelated experiment, nearly a year later, these same birds were selected to produce antibodies against pathogenic *Salmonella* bacteria. This was done by inoculating them with old, presumably weakened, cultures of the bacteria and, in effect, vaccinating them. To the surprise of the experimenter the birds produced no antibodies, and some of them became ill and died. Following this lead, it was found that birds that had their bursa removed as adults had a normal immune response—they produced antibodies when injected with bacteria. However, when the bursa was removed from very young birds, they developed into adult birds with defective immune systems. This clearly indicated that the bursa was needed for *maturation* of the immune system and the eventual ability to produce antibodies.

This important work was obscurely published as a two-page note in the journal, *Poultry Science*. There it might have remained, unnoticed by researchers interested in the immunology of humans. Someone, however, happened to call it to the attention of another immunologist who was also attempting to determine the function of the thymus by removing it from his experimental mice. Following the direction suggested by the article in *Poultry Science*, this researcher then tried removing the thymus from newborn rather than mature mice. These mice, like the chickens with their bursa removed, matured into mice that produced no antibodies and, importantly, were also very slow in rejecting skin transplants from other mice.

This was a revolutionary development, showing that there was more to the immune system than the antibody-based immunity that had been taught for the past 50 years. Ehrlich had proposed that there were certain specialized cells that made an antibody in response to contact with an antigen—which then served as a pattern. This new work seemed to point at the need for at least two cells to produce an antibody. One type of cell recognized the antigen as being foreign and passed this information on to a second type of cell that actually produced the antibodies. The identity of these cells was unknown, but it was thought that they probably were white blood cells called lymphocytes (see Table 16.1, page 480), which were plentiful in lymphoid tissue such as the thymus and the bursa of Fabricius.

Before this time the function of lymphocytes had been mysterious, but by the 1960s the technology required to isolate and culture lymphocytes in the laboratory was in place. The source of lymphocytes was determined to be in the liver during the first few weeks of development. By about the third month, the bone marrow replaces the liver as the source of lymphocytes. From the red bone marrow, stem cells produce lymphocytes that begin their 'schooling'—that is, they differentiate. Some mature in the bone marrow and become **B cells** (named for the bursa of Fabricius) that recognize antigens and make antibodies against them. They recognize different antigens by *immunoglobulins* (another name for antibodies; see page 505) that coat their surface with receptors to the antigens. Other lymphocytes do not have surface immunoglobulins but mature under the influence of the thymus and are therefore called T lymphocytes, or **T cells,** the basis of **cellular immunity.** Both T cells and B cells are found primarily in blood and lymphoid organs. The T cells, like B cells, respond to antigens by means of receptors on their surface—**T-cell receptors (TCRs).** Contact with an antigen complementary to a TCR can cause certain types of T cells to proliferate and secrete *cytokines* rather than antibodies. These are chemical messengers that impart instructions to other cells to perform certain functions (see page 518).

ANTIGENS AND ANTIBODIES

Antigens and antibodies play key roles in the response of the immune system. Antigens provoke a highly specific immune response that, in humoral immunity, results in the production of antibodies that are capable of recognizing the antigen that gave rise to them. Antigens that cause such a response are, therefore, often more descriptively known as *immunogens*.

THE NATURE OF ANTIGENS

LEARNING OBJECTIVE
- Define *antigen*, *epitope*, and *hapten*.

Most **antigens** are either proteins or large polysaccharides. Lipids and nucleic acids are usually antigenic only when combined with proteins and polysaccharides. Antigenic compounds are often components of invading microbes, such as capsules, cell walls, flagella, fimbriae, and toxins of bacteria; the coats of viruses; or the surfaces of other types of microbes. Nonmicrobial antigens include pollen, egg white, blood cell surface molecules, serum proteins from other individuals or species, and surface molecules of transplanted tissues and organs.

Generally, antibodies recognize and interact with specific regions on antigens called **epitopes** or **antigenic determinants,** (Figure 17.1). The nature of this interaction depends on the size, shape, and chemical structure of the binding site on the antibody molecule.

Most antigens have a molecular weight of 10,000 or higher. A foreign substance that has a low molecular

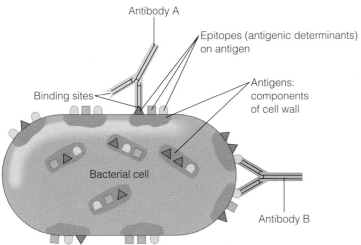

FIGURE 17.1 Epitopes (antigenic determinants). In this illustration the epitopes are components of the antigen—the bacterial cell wall. Each antigen carries more than one epitope. Each Y-shaped antibody molecule has at least two binding sites that can attach to a specific epitope on an antigen. An antibody can also bind to identical epitopes on two different cells at the same time (see Figure 18.5, page 537), which can cause neighboring cells to aggregate.

Q **Which antibody type (IgG or IgM) should be the more efficient in aggregating cells?**

weight is often not antigenic unless it is attached to a carrier molecule. These low molecular weight compounds are called **haptens** (from the Greek *hapto*, to grasp; Figure 17.2). Once an antibody against the hapten has been formed, the hapten will react with the antibody independent of the carrier molecule. Penicillin is a good example of a hapten. This drug is not antigenic by itself, but some people develop an allergic reaction to it. (Allergic

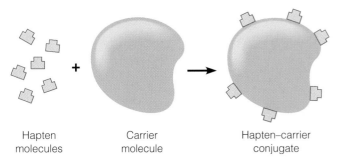

| Hapten molecules | Carrier molecule | Hapten–carrier conjugate |

FIGURE 17.2 Haptens. A hapten is a molecule too small to stimulate antibody formation by itself. However, when the hapten is combined with a larger carrier molecule, usually a serum protein, the hapten and its carrier together form a conjugate that can stimulate an immune response.

Q **How does a hapten differ from an antigen?**

reactions are a type of immune response). In these people, when penicillin combines with serum proteins, the resulting combined molecule initiates an immune response.

THE NATURE OF ANTIBODIES

LEARNING OBJECTIVES

- Explain the function of antibodies, and describe their structural and chemical characteristics.
- Name one function for each of the five classes of antibodies.

Antibodies are globulin proteins (proteins that have a compact, globular form)—therefore, we have come to use the term *immunoglobulins (Ig)* for antibodies. Globulin proteins are relatively soluble. Antibodies are made in response to an antigen and can recognize and bind to the antigen. As was seen in Figure 17.1, a bacterium or virus may have several epitopes that cause the production of different antibodies.

Each antibody has at least two identical sites that bind to epitopes. These sites are known as **antigen-binding sites.** The number of antigen-binding sites on an antibody is called the **valence** of that antibody. For example, most human antibodies have two binding sites; therefore, they are bivalent.

ANTIBODY STRUCTURE

Because a bivalent antibody has the simplest molecular structure, it is called a **monomer.** A typical antibody monomer has four protein chains: two identical *light chains* and two identical *heavy chains.* ("Light" and "heavy" refer to the relative molecular weights.) The chains are joined by disulfide links (see Figure 2.15c, page 46) and other bonds to form a Y-shaped molecule (Figure 17.3). The Y-shaped molecule is flexible and can assume a T shape (notice the hinge region in Figure 17.3a)

The two sections located at the ends of the Y's arms are called *variable (V) regions*. These bind to the epitopes. The amino acid sequences and, therefore, the three-dimensional structure of these two variable regions are identical on any one antibody. Their structure reflects the nature of the antigen for which they are specific—they are specific to the two antigen-binding sites found on each antibody monomer. The stem of the antibody monomer and the lower parts of the arms of the Y are called the *constant (C) regions*. They are the same for a particular class of immunoglobulin. There are five major types of C regions, which account for the five major classes of immunoglobulins (described shortly).

The stem of the Y-shaped antibody monomer is called the *Fc region*, so named because when antibody structure

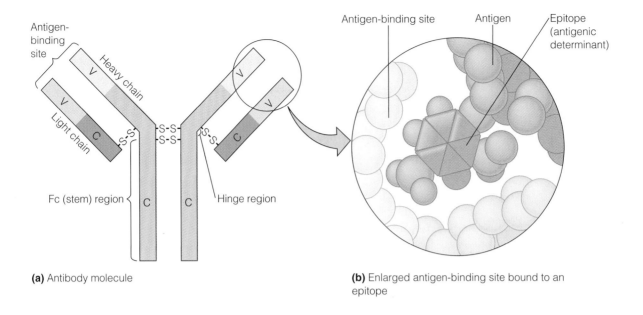

(a) Antibody molecule

(b) Enlarged antigen-binding site bound to an epitope

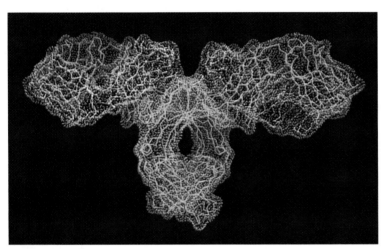

(c) Computer graphic model of an antibody molecule

FIGURE 17.3 The structure of a typical antibody molecule. The Y-shaped molecule is composed of two light chains and two heavy chains linked by disulfide bridges (S—S). Most of the molecule is made up of constant regions (C), which are the same for all antibodies of the same class. The amino acid sequences of the variable regions (V), which form the two antigen-binding sites, differ from molecule to molecule.

Q **What is responsible for the specificity of each different antibody?**

was first being identified, it was a fragment (F) that crystallized (c) in cold storage.

These Fc regions are often important in immunological reactions. If left exposed after both antigen-binding sites attach to an antigen such as a bacterium, the Fc regions of adjacent antibodies can bind complement. This leads to the destruction of the bacterium (see Figure 16.10, page 492). Conversely, the Fc region may bind to a cell, leaving the antigen-binding sites of adjacent antibodies free to react with antigens (see Figure 19.1a, page 552).

IMMUNOGLOBULIN CLASSES

The simplest and most abundant immunoglobulins are monomers, but they can also assume some differences in size and arrangement.

The five classes of Igs are designated IgG, IgM, IgA, IgD, and IgE. Each class has a different role in the immune response. The structures of IgG, IgD, and IgE molecules resemble the structure shown in Figure 17.3a. Molecules of IgA and IgM are aggregates of two or five monomers, respectively, that are joined together. The structures and characteristics of the immunoglobulin classes are summarized in Table 17.1.

IgG.   **IgG** (the name is derived from the blood fraction, gamma globulin; see Figure 17.17, page 521) accounts for about 80% of all antibodies in serum. In regions of inflammation these monomer antibodies readily cross the walls of blood vessels and enter tissue fluids. Maternal IgG antibodies, for example, can cross the placenta and confer passive

TABLE 17.1	A Summary of Immunoglobulin Classes				
Characteristics	IgG	IgM	IgA	IgD	IgE
Structure	Monomer	Pentamer	Dimer (with secretory component)	Monomer	Monomer
Percentage of total serum antibody	80%	5–10%	10–15%*	0.2%	0.002%
Location	Blood, lymph, intestine	Blood, lymph, B cell surface (as monomer)	Secretions (tears, saliva, mucus, intestine, milk), blood, lymph	B cell surface, blood, lymph	Bound to mast and basophil cells throughout body, blood
Molecular weight	150,000	970,000	405,000	175,000	190,000
Half-life in serum	23 days	5 days	6 days	3 days	2 days
Complement fixation	Yes	Yes	No†	No	No
Placental transfer	Yes	No	No	No	No
Known functions	Enhances phagocytosis; neutralizes toxins and viruses; protects fetus and newborn	Especially effective against microorganisms and agglutinating antigens; first antibodies produced in response to initial infection	Localized protection on mucosal surfaces	Serum function not known; presence on B cells functions in initiation of immune response	Allergic reactions; possibly lysis of parasitic worms

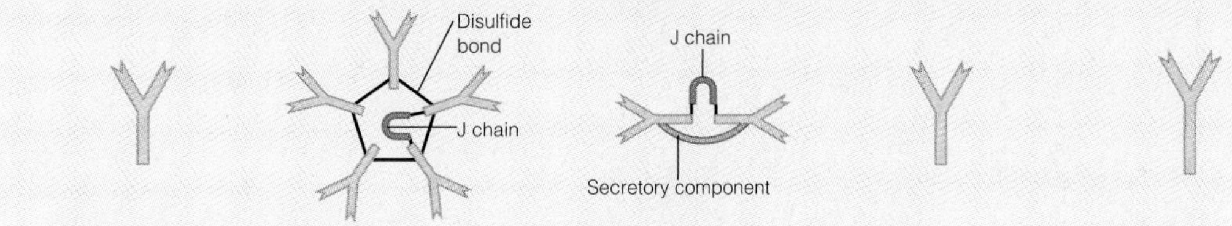

*Percentage in serum only; if mucous membranes and body secretions are included, percentage is much higher.
† May be yes via alternate pathway.

immunity to a fetus (see page 521). IgG antibodies protect against circulating bacteria and viruses, neutralize bacterial toxins, trigger the complement system, and, when bound to antigens, enhance the effectiveness of phagocytic cells.

IgM. Antibodies of the **IgM** (from *macro*, reflecting their large size) class make up 5 to 10% of the antibodies in serum. IgM has a pentamer structure consisting of five monomers held together by a polypeptide called a *J (joining) chain* (see Table 17.1). The large size of the molecule prevents IgM from moving about as freely as IgG does so IgM antibodies generally remain in blood vessels without entering the surrounding tissues.

IgM is the predominant type of antibody involved in the response to the ABO blood group antigens on the surface of red blood cells (see Table 19.2, page 555). It is effective in aggregating antigens and in reactions involving complement, and it can enhance the ingestion of target cells by phagocytic cells, as does IgG.

The fact that IgM appears first in response to a primary infection and is relatively short-lived makes it uniquely valuable in the diagnosis of disease. If high concentrations of IgM against a pathogen are detected in a patient, it is likely that the disease observed is caused by that pathogen. The detection of IgG, which is relatively long-lived, may indicate only that immunity against a particular pathogen was acquired in the more distant past.

IgA. **IgA** accounts for only about 10 to 15% of the antibodies in serum, but it is by far the most common form in mucous membranes and in body secretions such as mucus,

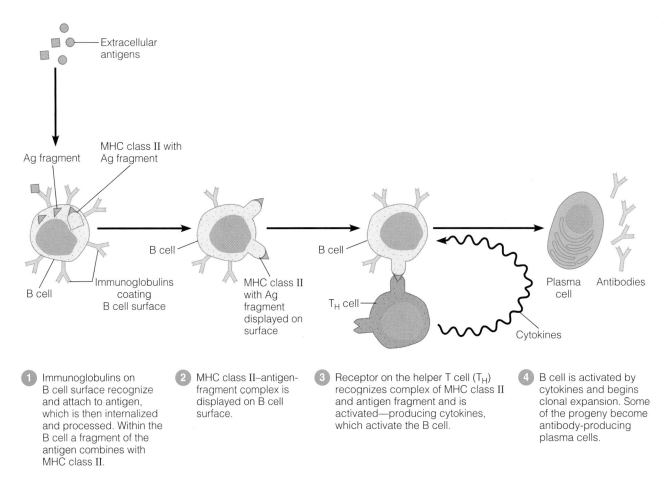

FIGURE 17.4 Activation of B cells to produce antibodies. In this illustration, the B cell is producing antibodies against a T-dependent antigen.

Q **How does activation by T-independent antigens differ from this figure?**

saliva, tears, and breast milk. Taking this into consideration, IgA is the most abundant immunoglobulin in the body (IgG is the most abundant in serum). The form of IgA that circulates in serum, *serum IgA*, is usually in the form of a monomer. The most effective form of IgA, however, consists of two connected monomers that form a *dimer* called *secretory IgA*. It is produced in this form by plasma cells in the mucous membranes. Each dimer then enters and passes through a mucosal cell, where it acquires a polypeptide called a *secretory component* that protects it from enzymatic degradation. The main function of secretory IgA is probably to prevent the attachment of microbial pathogens to mucosal surfaces. This is especially important in resistance to respiratory pathogens. Because IgA immunity is relatively short-lived, the length of immunity to many respiratory infections is correspondingly short. IgA's presence in a mother's milk, especially the colostrum (see page 521) probably helps protect infants from gastrointestinal infections.

IgD. **IgD** antibodies make up only about 0.2% of the total serum antibodies. Their structure resembles that of IgG molecules. IgD antibodies are found in blood, lymph, and particularly on the surfaces of B cells (Figure 17.4). IgD has no well-defined function, although there are indications that it may be involved in deleting B cells that are capable of producing antibodies against host tissue.

IgE. Antibodies of the **IgE** class are slightly larger than IgG molecules, but they constitute only 0.0002% of the total serum antibodies. IgE molecules bind tightly by their Fc (stem) regions to receptors on mast cells and basophils, specialized cells that participate in allergic reactions (see Chapter 19). When an antigen such as pollen cross-links with the IgE antibodies attached to a mast cell or basophil (see Figure 19.la, page 552), that cell releases histamine and other chemical mediators. These chemicals provoke a response—for example, an allergic reaction such as hay fever. However, the response can be protective as well, for it attracts complement and phagocytic cells. This is especially useful when the antibodies bind to parasitic worms. The concentration of IgE is greatly increased during some allergic reactions and parasitic infections, which is often diagnostically useful.

B CELLS AND HUMORAL IMMUNITY

> **LEARNING OBJECTIVES**
> - Compare and contrast T-dependent and T-independent antigens.
> - Differentiate between plasma cell and memory cell.

As we have seen, the humoral (antibody-mediated) response is carried out by antibodies. The production of antibodies is done by a special group of lymphocytes called B cells. The process that leads to the production of antibodies starts when B cells are exposed to *free*, or *extracellular, antigens*.

CLONAL SELECTION OF ANTIBODY-PRODUCING CELLS

> **LEARNING OBJECTIVE**
> - Describe clonal selection.

Each B cell carries immunoglobulins on its surface that are part of its makeup. The majority of the B cell's surface immunoglobulins are IgM and IgD—all of which are specific for the recognition of the same epitope. Ten percent or less of B cells carry other classes of immunoglobulins, but in certain locations their numbers may be high; for example, B cells in the intestinal mucosa are rich in IgA. B cells may carry at least 100,000 identical immunoglobulin molecules embedded in their surface membranes.

When a B cell's immunoglobulins bind to the epitope for which they become specific, the B cell is *activated*. An activated B cell undergoes *clonal expansion*, or proliferation. B cells usually require the assistance of a *helper T cell* (T_H) as shown in Figure 17.4. (T cells will be the subject of a detailed discussion later in this chapter). An antigen that requires a T_H cell for antibody production is known as a **T-dependent antigen.** T-dependent antigens are mainly proteins, such as those found on viruses, bacteria, foreign red blood cells, and haptens with their carrier molecules. For antibodies to be produced in response to a T-dependent antigen, it is necessary that both B and T cells be activated and interact. The process is initiated when the B cell contacts an antigen. It is important to note that the antigen contacts the surface immunoglobulins on the B cell and is enzymatically processed within the B cell and that fragments of it are combined with the

membrane proteins called the **major histocompatibility complex (MHC).** The complex of the antigenic fragments and the MHC are then displayed on the B cell's surface for identification by the TCRs. MHC is a molecule that identifies the host, and its use here prevents the immune system from making antibodies that would be harmful to the host. In this instance, the MHC is of class II, which is found only on the surface of *antigen-presenting cells (APCs)*—in this case, a B cell. We will encounter other APCs later in this chapter.

As shown in Figure 17.4, the T_H cell in contact with the antigenic fragment presented on the surface of the B cell becomes activated and begins producing cytokines. These deliver a message that causes the activation of the B cell. An activated B cell proliferates into a large clone of cells, some of which will differentiate into antibody-producing **plasma cells.** Other clones of the activated B cell become long-lived **memory cells** that are responsible for the enhanced secondary response to an antigen (see Figure 17.15, page 520). This phenomenon, as shown in Figure 17.5, is called **clonal selection.** (A similar process occurs with T cells, as we will see later in the chapter.) The pool of B cells does not contain many that are harmfully reactive against host tissue, or self. These are usually eliminated at the immature lymphocyte stage by the process of **clonal deletion.**

Antigens that stimulate B cells directly without the help of T cells are called **T-independent antigens.** Such antigens are characterized by repeating subunits such as are found in polysaccharides or lipopolysaccharides. Bacterial capsules are often good examples of T-independent antigens. The repeating subunits, as shown in Figure 17.6, can bind to multiple B cell receptors, which is probably why they do not require T cell assistance. T-independent antigens generally provoke a weaker immune response than T-dependent antigens. This response is composed primarily of IgM, and no memory cells are generated. The immune system of infants may not be stimulated by T-independent antigens until about age 2.

THE DIVERSITY OF ANTIBODIES

> **LEARNING OBJECTIVES**
> - Describe how a human can produce different antibodies.

The human immune system is capable of recognizing a mind-boggling number of different antigens—estimates are of a minimum of 10^{15} antigens. The number of genes required for this amount of diversity would seem to require a major part of an individual's inherited DNA. The work of the Japanese immunologist Susumu Tonegawa, for which he received a Nobel prize in 1987, showed how this

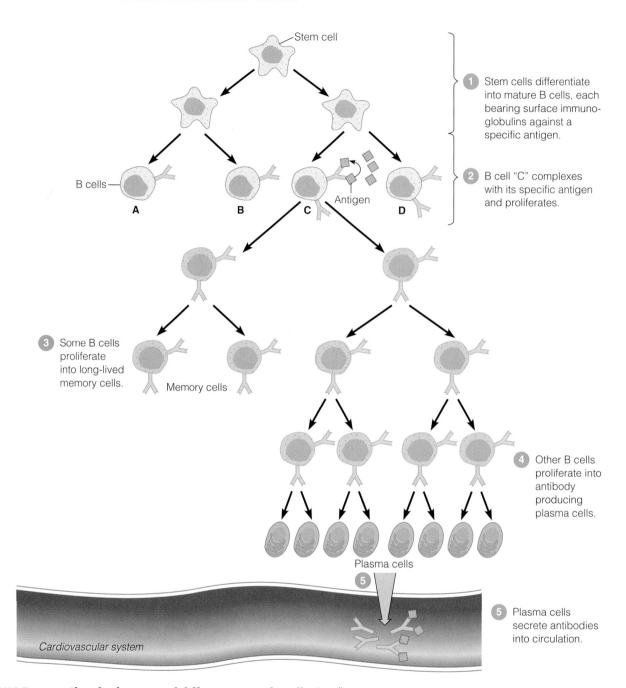

① Stem cells differentiate into mature B cells, each bearing surface immuno-globulins against a specific antigen.

② B cell "C" complexes with its specific antigen and proliferates.

③ Some B cells proliferate into long-lived memory cells.

④ Other B cells proliferate into antibody producing plasma cells.

⑤ Plasma cells secrete antibodies into circulation.

FIGURE 17.5 Clonal selection and differentiation of B cells. B cells can recognize an almost infinite number of antigens, but each particular cell recognizes only one type of antigen. An encounter with a particular antigen triggers the proliferation of a cell that is specific for that antigen (here, B cell "C") into a clone of cells with the same specificity, hence the term *clonal selection*.

Q **What caused cell "C" to respond?**

diversity could be obtained by a set of just hundreds, not billions, of genes. Simplistically, the mechanism is analogous to the generation of huge numbers of words from a limited alphabet. This 'alphabet' is found in the genetic makeup of the variable (V) region genes of the immunoglobulin molecule, which can be linked to various genes in the antibody's constant (C) region (see Figure 17.3). These combinations greatly reduce the amount of genetic information needed, so that a different gene is not needed in order to respond to each antigen. ✳ **Animation: Go to The Microbiology Place website or CD-ROM and click "Animations" to view Humoral Immunity.**

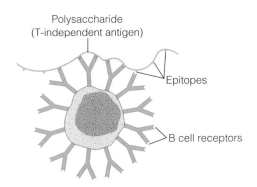

Polysaccharide
(T-independent antigen)

Epitopes

B cell receptors

FIGURE 17.6 T-independent antigens. T-independent antigens have repeating units (epitopes) that can cross-link several antigen receptors on the some B cell. These antigens stimulate the B cell to make antibodies without the aid of helper T cells. The polysaccharides of bacterial capsules ore examples of this type of antigen.

Q How can you differentiate T-dependent from T-independent antigens?

ANTIGEN–ANTIBODY BINDING AND ITS RESULTS

LEARNING OBJECTIVE

- Describe four outcomes of an antigen–antibody reaction.

When an antibody encounters an antigen for which it is specific, an **antigen–antibody complex** rapidly forms. An antibody binds to an antigen such as a bacterium at a specific portion called the *epitope,* or *antigenic determinant* (see Figure 17.1).

The strength of the bond between an antigen and an antibody is called **affinity.** In general, the closer the physical fit between antigen and antibody, the higher the affinity. Antibodies tend to recognize the shape of the antigen's epitope. They also exhibit a capability for **specificity** that is remarkable. They can distinguish between minor differences in the amino acid sequence of a protein and even between two isomers (see Figure 2.13, page 45). Therefore, antibodies can be used to differentiate between the viruses of chickenpox and measles and between bacteria of different species, for example.

The binding of an antibody to an antigen protects the host by tagging foreign cells and molecules for destruction by phagocytes and complement. The antibody molecule itself is not damaging to the antigen. Foreign organisms and toxins are rendered harmless by only a few mechanisms, as summarized in Figure 17.7. These are agglutination, opsonization, neutralization, antibody-dependent cell-mediated cytotoxicity, and the activation of complement

leading to inflammation and cell lysis (see Figure 16.10, page 492).

In **agglutination,** antibodies cause antigens to clump together. For example, the two antigen-binding sites of an IgG antibody can combine with epitopes on two different foreign cells, aggregating the cells into clumps that are more easily ingested by phagocytes. Because of its more numerous binding sites, IgM is more effective at cross-linking and aggregating particulate antigens (see Figure 18.5, page 537). IgG requires 100 to 1000 times as many molecules for the same results. (In Chapter 18, we will see how agglutination is important in the diagnosis of some diseases.)

For **opsonization** (from the Greek *opsonare,* meaning to cater) the antigen, such as a bacterium, is coated with antibodies that enhance its ingestion and lysis by phagocytic cells. **Antibody-dependent cell-mediated cytotoxicity** (see page 517 and Figure 17.14) resembles opsonization in that the target organism becomes coated with antibodies; however, destruction of the target cell is by immune system cells that remain external to the target cell.

In **neutralization** IgG antibodies inactivate viruses by blocking their attachment to host cells, and they neutralize toxins in a similar manner.

Finally, either IgG or IgM antibodies may trigger **activation of the complement system.** For example, inflammation is caused by infection or tissue injury (see Figure 16.8, page 488). One aspect of inflammation is that it will often cause microbes in the inflamed area to become coated with certain proteins. This, in turn, leads to the attachment to the microbe of an antibody–complement complex. This complex lyses the microbe, which then attracts phagocytes and other defensive immune system cells to the area.

As you will see in Chapter 19, the action of antibodies can be harmful. For example, immune complexes of antibody, antigen, and complement can damage host tissue. Antigens combining with IgE on mast cells can initiate allergic reactions, and antibodies can react with host cells and cause autoimmune disorders.

T CELLS AND CELLULAR IMMUNITY

LEARNING OBJECTIVES

- Describe at least one function of each of the following: M cells, T_H1 cells, T_H2 cells, T_C cells, T_R cells, CTL, NK cells.

Humoral antibodies are effective against pathogens such as viruses and bacteria that are circulating freely, where the antibodies can contact them. Intracellular antigens, such as a virus within an infected cell, are not exposed to

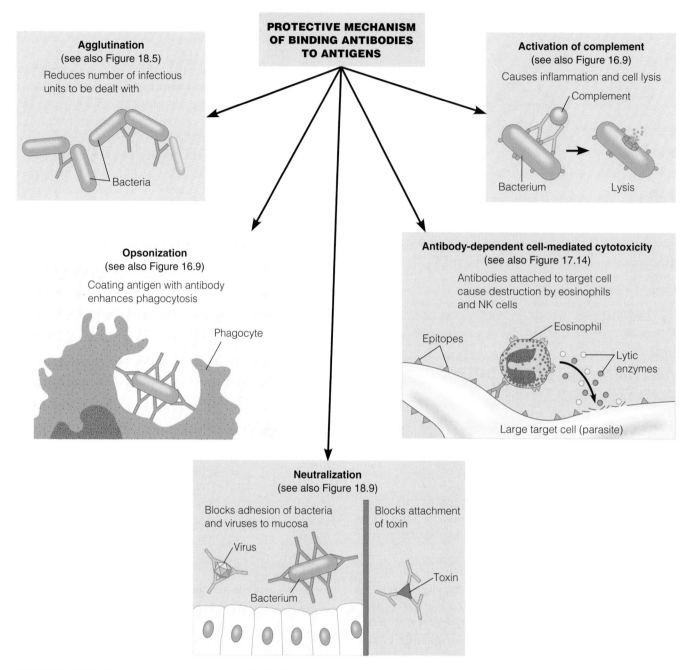

FIGURE 17.7 The results of antigen–antibody binding. The binding of antibodies to antigens to form antigen–antibody complexes tags foreign cells and molecules for destruction by phagocytes and complement.

Q **What are some of the possible outcomes of an antigen–antibody reaction?**

circulating antibodies. Some bacteria and parasites can also invade and live within cells. T cells probably evolved in response to this aspect of pathogenicity—the need to combat intracellular pathogens. They are also the way in which the immune system recognizes cells that are nonself, especially cancer cells.

Like B cells, each T cell is specific for only a certain antigen. Rather than the coating of immunoglobulins that

provide the specificity for B cells, T cells have TCRs. Like B cells and all other cells involved in the immune response, T cells develop from stem cells in the red bone marrow (see Figure 17.8). The precursors of the T cells migrate from the bone marrow and reach maturity in the thymus. Most immature T cells, an estimated 98%, are eliminated in the thymus, which is analogous to clonal deletion in B cells. Next, mature T cells migrate from the thymus by way of

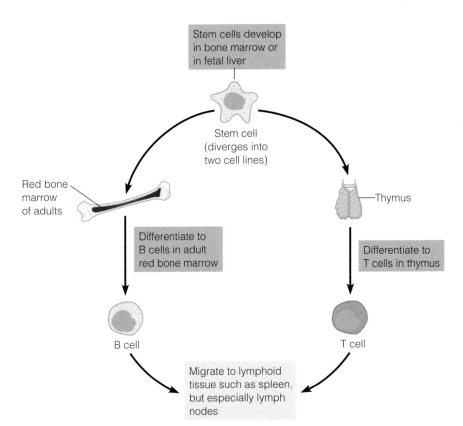

FIGURE 17.8 Differentiation of T cells and B cells. Both B cells and T cells originate from stem cells in adult red bone marrow or in the fetal liver. (Red blood cells, macrophages, neutrophils, and other white blood cells also originate from these same stem cells.) Some cells pass through the thymus and emerge as mature T cells. Other cells probably remain in the red bone marrow and become B cells. Both types of cells then migrate to lymphoid tissues, such as the lymph nodes or spleen.

Q **Which cells, T or B, make antibodies?**

the blood and lymphatic system to various lymphoid tissues (see Figure 16.5, page 482) where they are most likely to encounter antigens.

Most pathogens of the type that the cellular immune system is designed to combat first enter the gastrointestinal tract or lungs, where they encounter a barrier of epithelial cells. Normally, they can pass this barrier in the gastrointestinal tract only by way of a scattered array of gateway cells called **microfold cells,** or **M cells.** (Instead of the myriad of fingerlike microvilli found on the surface of absorptive epithelial cells of the intestinal tract, M cells have microfolds). M cells are located over **Peyer's patches,** which are secondary lymphoid organs located on the intestinal wall. M cells are well adapted to take up antigens from the intestinal tract and allow their transfer to the lymphocytes and antigen-presenting cells of the immune system found throughout the intestinal tract, just under the epithelial-cell layer but especially in the Peyer's patches. It is also here that antibodies, mostly IgA essential for mucosal immunity, are formed and migrate to the mucosal lining.

The recognition of antigens by a T cell requires that they be first processed by specialized **antigen-presenting cells (APC).** This resembles the situation previously discussed in humoral immunity in which a B cell served as the APC (see Figure 17.4). After processing, an antigenic fragment is presented on the APC surface together with a molecule of the MHC. APCs are described fully on page 516; they include activated macrophages and, most importantly, dendritic cells.

The body's ability to make new T cells decreases with age, beginning in late adolescence. Eventually, the T-cell-producing thymus becomes less active, and red bone marrow produces fewer B cells. As a result, the immune system is relatively weak in the elderly. However, sufficient long-lived T and B memory cells survive to make immunization of the elderly effective for such diseases as influenza and pneumococcal pneumonia.

CLASSES OF T CELLS

LEARNING OBJECTIVES
- Differentiate between helper T, cytotoxic T, and regulatory T cells.
- Differentiate between T_H1 and T_H2 cells.
- Define *apoptosis*.

There are classes of T cells that have different functions, rather like the classes of immunoglobulins. For example, helper T cells cooperate with B cells in the production of antibodies, mainly through cytokine signaling (see Figure 17.4). Therefore, helper T cells are an important part of humoral immunity; they are an even more essential element of cellular immunity. In their contributions to cellular immunity, T cells do not contribute to the production of antibodies but interact more directly with antigens. Primarily, the two populations of T cells that concern us here are **helper T cells (T_H) cells** and **cytotoxic T cells (T_C).**

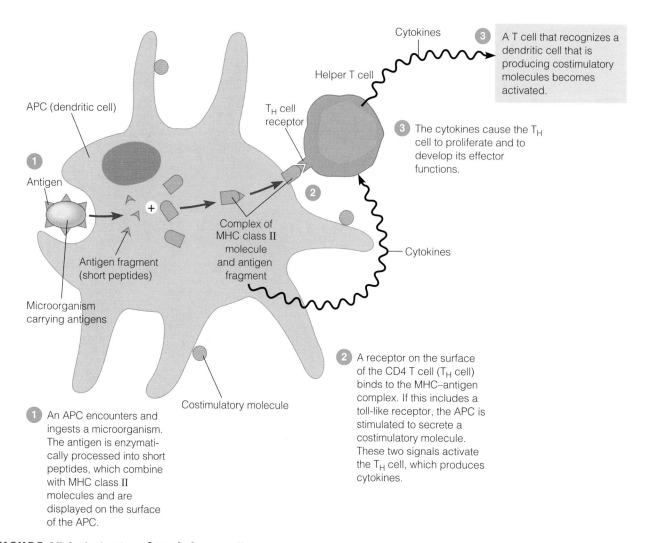

Cytokines

3 A T cell that recognizes a
dendritic cell that is
producing costimulatory
molecules becomes
activated.

Helper T cell

APC (dendritic cell)

T$_H$ cell
receptor

3 The cytokines cause the T$_H$
cell to proliferate and to
develop its effector
functions.

1

Antigen

2

Antigen fragment
(short peptides)

Complex of
MHC class II
molecule
and antigen
fragment

Cytokines

Microorganism
carrying antigens

Costimulatory molecule

2 A receptor on the surface
of the CD4 T cell (T$_H$ cell)
binds to the MHC–antigen
complex. If this includes a
toll-like receptor, the APC is
stimulated to secrete a
costimulatory molecule.
These two signals activate
the T$_H$ cell, which produces
cytokines.

1 An APC encounters and
ingests a microorganism.
The antigen is enzymati-
cally processed into short
peptides, which combine
with MHC class II
molecules and are
displayed on the surface
of the APC.

FIGURE 17.9   Activation of CD4 helper T cells.

Q **What is the role of the macrophage?**

A T$_C$ cell can differentiate into an effector cell called a
cytotoxic T lymphocyte (CTL).

T cells are also classified by certain glycoproteins on
their surface called **clusters of differentiation,** or **CD.**
The T$_H$ cells are classified as **CD4,** which are adhesion
molecules that bind to MHC class II molecules on APCs.
T$_C$ cells are classified as **CD8,** which are adhesion molecules that bind to MHC class I molecules.

HELPER T CELLS

We have seen that an essential part of the body's innate
defenses is phagocytosis by cells such as macrophages.
Macrophages, when functioning as APCs, also are
important in adaptive cellular immunity. T$_H$ cells can
recognize an antigen presented on the surface of a
macrophage and activate the macrophage, making it
more effective in both phagocytosis and in antigen presentation. Even more important as APCs are dendritic cells.

Dendritic cells are especially important in the activation
of CD4 T cells and in developing their effector functions
(Figure 17.9).

In order for a CD4 T cell to become activated, its
TCR recognizes an antigen that has been processed and
is presented as fragments held in a complex with proteins of MHC class II on the surface of the APC. This is
the initial signal for activation; a second signal, the costimulatory signal, which is present on the APC and the
T$_H$ cell, is also required. Because activation should be directed against harmful pathogens, the displayed antigenic fragments should include *toll-like receptors* (see
Chapter 16, page 475), which are signals of a dangerous
microbe. The activated T$_H$ cell begins to proliferate and
to secrete cytokines, which are essential for its effector
functions. The proliferating T$_H$ cell differentiates into
populations of T$_H$1 and T$_H$2 cells; it also forms a population of memory cells.

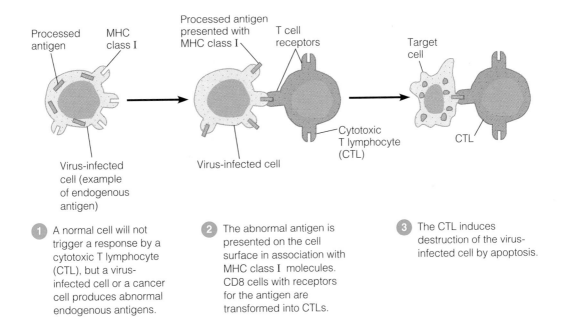

1 A normal cell will not trigger a response by a cytotoxic T lymphocyte (CTL), but a virus-infected cell or a cancer cell produces abnormal endogenous antigens.

2 The abnormal antigen is presented on the cell surface in association with MHC class I molecules. CD8 cells with receptors for the antigen are transformed into CTLs.

3 The CTL induces destruction of the virus-infected cell by apoptosis.

FIGURE 17.10 Activation of CD8 cytotoxic T cells by endogenous antigens.

Q **Differentiate a CD8 from a CD4 cell.**

The cytokines produced by T_H1 **cells** mostly activate cells related to cellular immunity. Among these are macrophages, CD8 T cells, and natural killer cells (see page 517). T_H2 **cells** produce cytokines that are associated with allergic reactions or with responses to certain parasitic infections.

CYTOTOXIC T CELLS

Cytotoxic T cells, despite their name, are not capable of attacking any target cell but are precursors to CTLs. When activated by cytokines from a T_H cell, T_C cells can be transformed into CTLs. A CTL is an effector cell that has the ability to recognize and kill target cells that are considered nonself (Figure 17.10). Primarily, these target cells are self-cells that have been altered by infection with a pathogen, especially viruses. On their surface they carry fragments of **endogenous antigens** that are generally synthesized within the cell and are mostly of viral or parasitic origin. Other important target cells are tumor cells (see Figure 19.10, page 565) and transplanted foreign tissue. Rather than reacting with antigenic fragments presented by an APC in complex with MHC class II molecules, the CD8 T cell recognizes endogenous antigens on the target cell's surface that are in combination with an MHC class I molecule. MHC class I molecules are found on nucleated cells, therefore a CTL can attack almost any cell of the host that has been altered.

In its attack, a CTL attaches to the target cell and releases a pore-forming protein, **perforin.** Pore formation contributes to the subsequent death of the cell and is similar to the action of the complement membrane attack complex described in Chapter 16 (see page 492). **Granzymes,** proteases that induce apoptosis, are now able to enter through the pore. **Apoptosis** (from the Greek for falling away like leaves) is also called programmed cell death. Cells that die from apoptosis first cut their genome into fragments, and the external membranes bulge outward in a manner called *blebbing* (Figure 17.11). Signals are displayed on the cell's surface that attract circulating phagocytes to digest the remains before any significant leakage of contents occurs. This has the advantage of preventing the spread of infectious viruses into other cells. Also, after apoptosis the immune response is rapidly ended, which lessens nonspecific damage to tissue.

REGULATORY T CELLS

Regulatory T cells (T_R), formerly called suppressor T cells, make up about 5 to 10% of the T cell population and typically secrete cytokine IL-10. Recently, improvements in technology have made it easier to isolate and grow these cells. While their function is not clearly defined—and many immunologists question if this category of T cells really exists—they appear to suppress the activity of other T cells. In particular, they modify inflammation

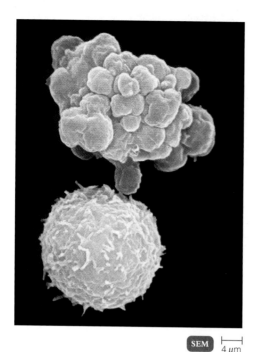

FIGURE 17.11 Apoptosis. A normal B cell is shown at the bottom of the photo. A B cell undergoing apoptosis is at the top. Notice the bubble-like blebs.

Q What is apoptosis?

and regulate the response of the immune system to organ rejection and to autoimmune diseases such as type I (insulin-dependent) diabetes. It is hoped that advances in research on these cells might lead to clinical uses. ✳ **Animations: Go to The Microbiology Place website or CD-ROM and click "Animations" to view Cell-Mediated Immunity, and Antigen Processing and Presentation.**

ANTIGEN-PRESENTING CELLS (APCs)

LEARNING OBJECTIVES
• Define *antigen-presenting cell.*

Although B cells are a form of antigen-presenting cell (APC) that we have already discussed with humoral immunity, we will now consider other APCs associated with cellular immunity. These are the dendritic cells and activated macrophages.

DENDRITIC CELLS

Dendritic cells are characterized by long extensions of the membranes called dendrites (Figure 17.12). They are plentiful in lymph nodes and the spleen, as well as skin. Vaccines injected between the skin layers (intradermally), where there are more dendritic cells, are often more effective than

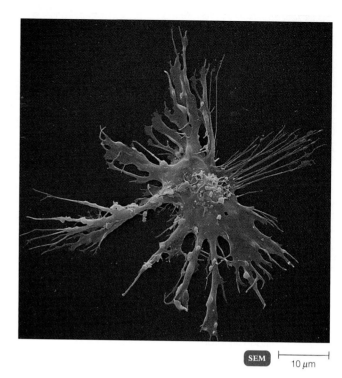

FIGURE 17.12 A dendritic cell. These antigen-presenting cells are named for their long arms, or dendrites. They are especially plentiful in the skin and mucous membranes.

Q What is the role of dendritic cells in immunity?

injections into muscle. Compared to macrophages, dendritic cells are poorly phagocytic; they are much more important in their role as APCs.

MACROPHAGES

Macrophages are cells usually found in a resting state (page 483). We have already discussed the function of macrophages in phagocytosis. They are important for innate immunity and for ridding the body of worn out blood cells and other debris, such as cellular remnants from apoptosis. Their phagocytic capabilities are greatly increased when they are stimulated to become **activated macrophages** (Figure 17.13). This activation can be initiated by ingestion of antigenic material. Other stimuli, such as cytokines produced by an activated helper T cell, can further enhance their capabilities. Once activated, macrophages are more effective as phagocytes and as APCs. Activated macrophages are important factors in the control of cancer cells and such intracellular pathogens as the tubercle bacillus and virus-infected cells. Their appearance becomes recognizably different as well—they are larger and become ruffled.

After taking up an antigen, APCs tend to migrate to lymph nodes or other lymphoid centers on the mucosa where they present the antigen to T cells located there.

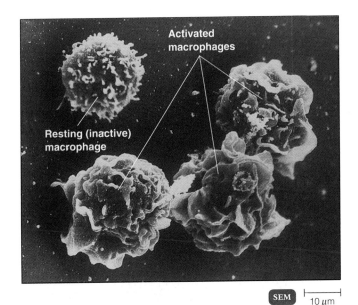

FIGURE 17.13 **Activated macrophages.** When activated, macrophages become larger and ruffled.

Q   **How do macrophages become activated?**

TABLE 17.2	Principal Cells That Function in Cell-Mediated Immunity
Cell	Function
Helper T (T$_H$1) cell	Activates cells related to cell-mediated immunity: macrophages, CD8 T cells, and natural killer cells
Helper T (T$_H$2) cell	Stimulates production of eosinophils, IgM, and IgE
Cytotoxic T lymphocyte (CTL)	Destroys target cells on contact
Regulatory T (T$_R$) cell	Regulates immune response and helps maintain tolerance
Activated macrophage	Enhanced phagocytic activity; attacks cancer cells
Natural killer (NK) cell	Attacks and destroys target cells; participates in antibody-dependent cell-mediated cytotoxicity

T cells carrying receptors that are capable of binding with any specific antigen are present in relatively limited numbers. Migration increases the opportunity for these particular T cells to encounter the antigen for which they are specific.

EXTRACELLULAR KILLING BY THE IMMUNE SYSTEM

LEARNING OBJECTIVE

- Describe the function of natural killer cells.

We have seen how the action of a CTL can lead to the destruction of a target cell. A component of the innate immune system, which has not yet been discussed, can also destroy certain virus-infected cells and tumor cells. These are granular leukocytes called **natural killer (NK) cells.** They can also attack parasites, which are normally much larger than bacteria, as illustrated in Figure 17.14. In contrast to CTLs, NK cells are not immunologically specific; that is, they do not need to be stimulated by an antigen. They must contact the target cell and determine if it expresses MHC class I self-antigens. If it does not—which is often true in the early stages of viral infection and in some viruses that have developed a system of interfering with the usual presentation of antigens on an APC—they kill the target cell by mechanisms similar to that of a CTL. Tumor cells also have a reduced

number of MHC class I molecules on their surfaces. NK cells cause pores to form in the target cell, which leads to either lysis or apoptosis.

The functions of NK cells and the other principal cells involved in cellular immunity are briefly summarized in Table 17.2.

ANTIBODY-DEPENDENT CELL-MEDIATED CYTOTOXICITY

LEARNING OBJECTIVE

- Describe the role of antibodies and natural killer cells in antibody-dependent cell-mediated cytotoxicity.

With the help of antibodies produced by the humoral immune system, the cell-mediated immune system can stimulate natural killer cells (see page 479) and cells of the innate defense system to kill targeted cells. In this way, an organism such as a protozoan or a helminth that is too large to be phagocytized can be attacked by immune system cells. This is referred to as **antibody-dependent cell-mediated cytotoxicity (ADCC).** As is illustrated in Figure 17.14, the target cell is first coated with antibodies. A variety of cells of the immune system bind to the F$_C$ regions of these antibodies, and thus to the target cell. The target cell is then lysed by substances secreted by the attacking cells.

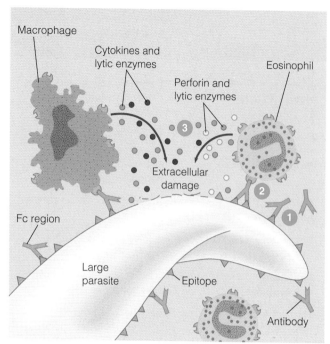

(a) Organisms, such as many parasites, that are too large for ingestion by phagocytic cells must be attacked externally.

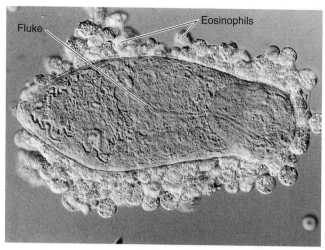

(b) Eosinophils adhering to the larval stage of a parasitic fluke.

FIGURE 17.14 Antibody-dependent cell-mediated cytotoxicity (ADCC). If an organism, such as a parasitic worm, is too large for ingestion and destruction by phagocytosis, it can be attacked by immune system cells that remain external to it.

Q **Why is ADCC important protection against parasitic protozoa and helminths?**

CYTOKINES: CHEMICAL MESSENGERS OF IMMUNE CELLS

LEARNING OBJECTIVES

- Identify at least one function of each of the following: cytokines, interleukins, and chemokines.

The immune response requires complex interactions between different cells. The communication required for this is mediated by chemical messengers called **cytokines.** They are produced by practically all cells of the immune system but especially by helper T cells.

Cytokines that serve as communicators between leukocytes (white blood cells) are now known as **interleukins** (between leukocytes). When enough information, including the amino acid sequence is known, the cytokine is assigned an interleukin number by an international committee. At present, we know of at least 29 interleukins.

One family of cytokines induces the migration of leukocytes into areas of infection or tissue damage. These are called **chemokines,** from *chemotaxis.* The best known of at least 30 identified chemokines is IL-8 (Table 17.3). Certain chemokine receptors are important for infection by the human immunodeficiency virus (HIV; see Chapter 19, page 566).

Interferons that help protect against the viral infection of cells are cytokines. Another named cytokine is tumor

TABLE 17.3	A Summary of Some Important Cytokines
Cytokine	Representative Activity
Interleukin-1 (IL-1)	Stimulates T_H cells in presence of antigens; chemically attracts phagocytes in inflammatory response
Interleukin-2 (IL-2)	Involved in proliferation of antigen-stimulated T_H cells, proliferation and differentiation of B cells, and activation of T_C cells and NK cells
Interleukin-8 (IL-8)	Chemoattractant for immune system cells and phagocytes to site of inflammation
Interleukin-10 (IL-10)	Secreted by T_H2 cells and T_R cells; interferes with activation of T_H1 cells
Interleukin-12 (IL-12)	Mainly involved in differentiation of CD4 T cells
Interferons (IFNs)	
α-IFN and β-IFN	Induces antiviral activity in nucleated cells (interferes with protein synthesis)
γ-IFN	Activates macrophages; improves antigen presentation
Alpha-tumor necrosis factor (α-TNF)	Cytotoxic to tumor cells: enhances activity of phagocytic cells

MICROBIOLOGY IN THE NEWS

IS IL-12 THE NEXT "MAGIC BULLET"?

Worldwide, HIV/AIDS-related illnesses kill 3 million people, and measles kills 1 million people annually. If laboratory tests are any indication of the future, the cytokine IL-12 (interleukin-12) could be the "magic bullet" against these diseases.

Since its discovery in the 1980s, IL-12 proved to be different from the other cytokines. For one thing, it is made from the products of two different genes instead of the usual one gene. IL-12 is released by B cells and macrophages in response to infections. It inhibits the humoral response and activates T_H1 cellular immunity.

Scientists at the National Institute of Allergy and Infectious Diseases (NIAID) have found that treating mice with IL-12 can help activate phagocytes to kill *Histoplasma* fungi, *Leishmania* protozoa, *Cryptosporidium* protozoa, *Toxoplasma gondii* protozoa, and *Mycobacterium*

avium. The last three are common opportunistic infections in people with late-stage AIDS. The NIAID recently began recruiting people with both AIDS and *Mycobacterium avium* complex into a clinical trial to test IL-12.

Researchers have shown that HIV and measles virus decrease the production of IL-12, which may make the patient more susceptible to secondary infections. When treated with IL-12, however, T_H cells taken from HIV-positive people responded to viruses, including HIV.

IL-12 is known to inhibit about 20 kinds of tumors in mice by inhibiting blood vessel growth to tumors. Several clinical trials are in progress to test the effectiveness of IL-12 in patients with advanced kidney cancer.

Because IL-12 activates the T_H1 pathway, it can cause the symptoms associated with chronic inflammatory diseases, including Crohn's disease, psoriasis, rheumatoid arthritis, and multiple sclerosis. NIAID researchers developed the concept of blocking IL-12 in patients with these diseases. There are two approaches to blocking IL-12: one is monoclonal antibodies that bind secreted IL-12, the other using a small molecule that blocks transcription to prevent production and secretion of IL-12. Clinical trials are currently underway at over 30 medical centers to test these in patients with chronic inflammatory diseases.

Will IL-12 be a panacea? Further studies are needed to determine whether treatment with IL-12 could have adverse effects, such as autoimmune disease, or whether blocking IL-12 could allow growth of cancer cells.

---

necrosis factor (TNF), which is important in inflammatory reactions. TNF was so named because tumor cells were observed to be one of its targets. Cytokines known as colony stimulating factors (CSF) stimulate the formation of various blood cells. Cytokines of this type, for example GM-CSF (granulocyte-macrophage colony stimulating factors), are used to increase the numbers of protective macrophages and granulocytes in patients undergoing red bone marrow transplants.

The role of cytokines in the stimulation of the immune system has suggested their use as therapeutic agents (see the box above).

IMMUNOLOGICAL MEMORY

LEARNING OBJECTIVE
- Distinguish a primary from a secondary immune response.

The intensity of the antibody-mediated humoral response can be reflected by the **antibody titer,** the relative amount of antibody in the serum. After the initial contact with an antigen, the exposed person's serum contains no

detectable antibodies for 4 to 7 days. Then there is a slow rise in antibody titer: first IgM class antibodies are produced, followed by IgG peaking in about 10 to 17 days, after which a gradual decline in antibody titer occurs. This pattern is characteristic of a **primary response** to an antigen.

The antibody-mediated immune responses of the host intensify after a second exposure to an antigen. This **secondary response** is also called the **memory (or anamnestic) response.** As shown in Figure 17.15 this response is comparatively more rapid, reaching a peak in only 2 to 7 days, lasts many days, and is considerably greater in magnitude. By way of explanation, as is shown in Figure 17.5, some activated B cells do not become antibody-producing plasma cells but persist as long-lived but nonproliferating memory cells. Years, or even decades later, if these cells are stimulated by the same antigen, they very rapidly differentiate into antibody-producing plasma cells.

A similar response occurs with T cells, which, as we will see in Chapter 19, is necessary for establishing the lifelong memory for distinguishing between self and nonself.

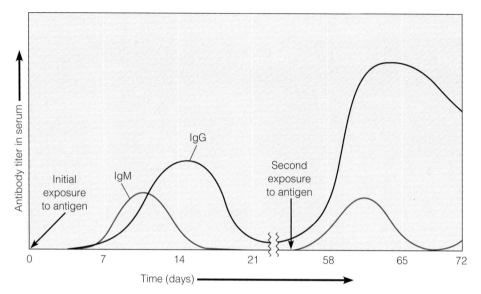

FIGURE 17.15 The primary and secondary immune responses to an antigen. IgM appears first in response to the initial exposure. IgG follows and provides longer-term immunity. The second exposure to the same antigen stimulates the memory cells formed at the time of initial exposure to rapidly produce a large amount of antibody. The antibodies produced in response to this second exposure are mostly IgG.

Q Why do many diseases, such as measles, occur only once in a person, yet others, such as colds, occur more than once?

TYPES OF ADAPTIVE IMMUNITY

LEARNING OBJECTIVE

• Contrast the four types of adaptive immunity.

Adaptive immunity refers to the protection an animal develops against certain specific microbes or foreign substances. The various manifestations of adaptive immunity are summarized in Figure 17.16.

Immunity can be acquired either actively or passively. Immunity is acquired *actively* when a person is exposed to microorganisms or foreign substances and the immune system responds. Immunity is acquired *passively* when antibodies are transferred from one person to another. Passive immunity in the recipient lasts only as long as the antibodies are present—in most cases, a few weeks. Both actively acquired immunity and passively acquired immunity can be obtained by natural or artificial means.

• **Naturally acquired active immunity** develops when a person is exposed to antigens, becomes ill, and then recovers. Once acquired, immunity is lifelong for some diseases, such as measles. For certain other diseases, especially intestinal diseases, the immunity may last for only a few years. *Subclinical infections* or *inapparent infections* (those that produce no

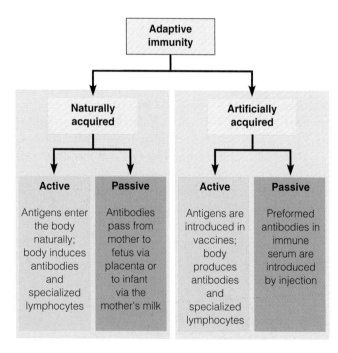

FIGURE 17.16 Types of adaptive immunity.

Q Which type of immunity, active or passive, lasts longer?

noticeable symptoms or signs of illness) can also confer immunity.

- **Naturally acquired passive immunity** involves
the natural transfer of antibodies from a mother to
her infant. Antibodies in a pregnant woman cross
the placenta to her fetus—*transplacental transfer.* If
the mother is immune to diphtheria, rubella, or
polio, for example, the newborn will be temporarily
immune to these diseases as well. Certain antibodies
are also passed from the mother to her nursing
infant in breast milk, especially in the first secretions called *colostrum.* In the infant, this passive
immunity generally lasts only as long as the transmitted antibodies persist—usually a few weeks or
months. These maternal antibodies are essential for
providing immunity to the infant until its own immune system matures. Colostrum is even more important to some other mammals; calves, for example,
do not have antibodies that cross the placenta and
rely on colostrum ingested during the first day of
life. Researchers often specify fetal calf serum for
certain experimental uses because it does not contain maternal antibodies.

- **Artificially acquired active immunity** is the result
of vaccination—which will be discussed in Chapter 18.
Vaccination, also called **immunization,** introduces
vaccines into the body. These are antigens such as
killed or living microorganisms or inactivated bacterial toxins.

- **Artificially acquired passive immunity** involves
the injection of antibodies (rather than antigens) into
the body. These antibodies come from an animal or
person who is already immune to the disease.

Because blood serum is easily obtained (see the box on
page 490) and contains a considerable concentration of
antibodies, **antiserum** has become a generic term for
blood-derived fluids containing antibodies. Hence, the
study of reactions between antibodies and antigens is
called **serology.** As shown in Figure 17.17, when a sample
of serum is subjected to an electrical current in the laboratory during gel electrophoresis (see Chapter 9), the proteins
dissolved within it move at different rates. The globulin
proteins separate into fractions that are termed alpha (α),
beta (β), and gamma (γ) for their relative motility. Because

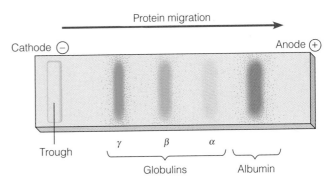

**FIGURE 17.17 The separation of serum proteins by gel
electrophoresis.** In this procedure, serum is placed in a trough
cut into a gel. In response to an electrical current, the negatively
charged proteins of the serum migrate through the gel from the
negatively charged end (cathode) to the positively charged end
(anode).

Q **What serum fraction contains the most antibodies?**

the gamma fraction, called **gamma globulin,** contains
most of the antibodies, it is often used to transfer passive
immunity. As we mentioned in our historical summary, antibodies are also known collectively as immunoglobulins.

When immune serum globulin from an individual who
is immune to a disease is injected into another individual,
it confers an immediate protection against the disease.
However, although artificially acquired passive immunity
is immediate, it is short-lived because antibodies are degraded by the recipient. The half-life of an injected antibody (the time required for half of the antibodies to
disappear) is typically about three weeks.

* * *

This chapter on immunology is intended to provide you
with the general concepts on the subject. It should give
you the information you need to understand the following chapters that emphasize some of the more practical
and clinical aspects of immunology. Immunology can be
a very complex subject; some of you, in pursuit of your academic major, will be taking a full course in immunology
later and will study the subject in much more detail. The
presentation here, you will then find, has been considerably simplified—although you might not have realized it.
Figure 17.18 summarizes the material covered in this
chapter, especially emphasizing the dual nature of immunology.

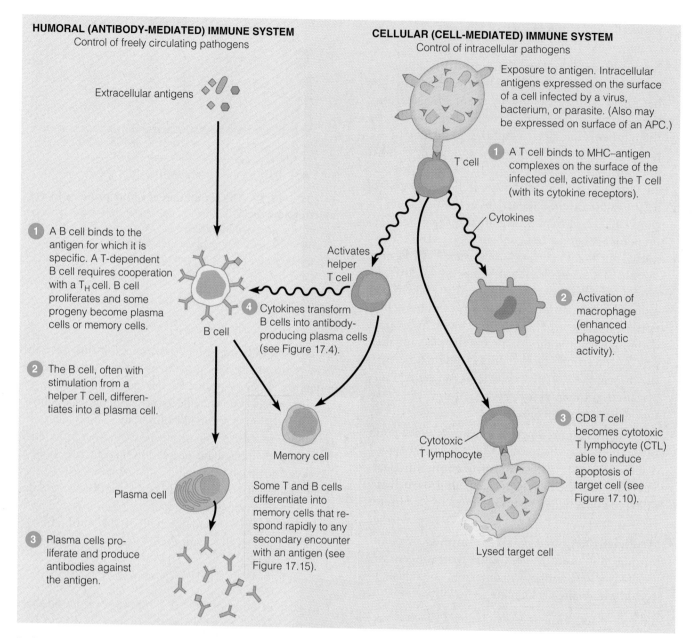

HUMORAL (ANTIBODY-MEDIATED) IMMUNE SYSTEM
Control of freely circulating pathogens

Extracellular antigens

1 A B cell binds to the antigen for which it is specific. A T-dependent B cell requires cooperation with a T$_H$ cell. B cell proliferates and some progeny become plasma cells or memory cells.

B cell

4 Cytokines transform B cells into antibody-producing plasma cells (see Figure 17.4).

2 The B cell, often with stimulation from a helper T cell, differentiates into a plasma cell.

Memory cell

Some T and B cells differentiate into memory cells that respond rapidly to any secondary encounter with an antigen (see Figure 17.15).

Plasma cell

3 Plasma cells proliferate and produce antibodies against the antigen.

CELLULAR (CELL-MEDIATED) IMMUNE SYSTEM
Control of intracellular pathogens

Exposure to antigen. Intracellular antigens expressed on the surface of a cell infected by a virus, bacterium, or parasite. (Also may be expressed on surface of an APC.)

T cell

1 A T cell binds to MHC–antigen complexes on the surface of the infected cell, activating the T cell (with its cytokine receptors).

Cytokines

Activates helper T cell

2 Activation of macrophage (enhanced phagocytic activity).

Cytotoxic T lymphocyte

3 CD8 T cell becomes cytotoxic T lymphocyte (CTL) able to induce apoptosis of target cell (see Figure 17.10).

Lysed target cell

FIGURE 17.18 The dual nature of the immune system. Humoral immunity: pathogens can come into contact with antibodies when they circulate in the blood or spaces outside of cells. Cellular immunity: viruses and some bacterial pathogens and parasites reproduce only inside living cells, where circulating antibodies cannot reach them. Elimination of these intracellular pathogens requires immune responses that depend on T cells, especially cytotoxic T cells.

Q **When does a B cell require stimulation by a T$_H$ cell?**

STUDY OUTLINE

THE ADAPTIVE IMMUNE SYSTEM (p. 503)

1. An individual's genetically predetermined resistance to certain diseases is called innate immunity.
2. Adaptive immunity is the ability of the body to specifically react to a microbial infection.

DUAL NATURE OF THE ADAPTIVE IMMUNE SYSTEM (pp. 503–504)

1. Red bone marrow stem cells produce lymphocytes. Lymphocytes that mature in bone marrow become B cells.
2. Humoral immunity involves antibodies, which are found in serum and lymph and are produced by B cells.
3. Lymphocytes that migrate through the thymus become T cells. Cellular immunity involves T cells.
4. T cell receptors recognize antigens.

ANTIGENS AND ANTIBODIES (pp. 504–509)

THE NATURE OF ANTIGENS (pp. 504–505)

1. An antigen (or immunogen) is a chemical substance that causes the body to produce specific antibodies.
2. As a rule, antigens are proteins or large polysaccharides. Antibodies are formed against specific regions on antigens called epitopes, or antigenic determinants.
3. A hapten is a low-molecular-weight substance that cannot cause the formation of antibodies unless combined with a carrier molecule; haptens react with their antibodies independent of the carrier molecule.

THE NATURE OF ANTIBODIES (pp. 507–509)

4. An antibody, or immunoglobulin, is a protein produced by B cells in response to an antigen and is capable of combining specifically with that antigen.
5. Typical monomers consist of four polypeptide chains: two heavy chains and two light chains.
6. Within each chain is a variable (V) region that binds the epitope and a constant (C) region that distinguishes the different classes of antibodies.
7. An antibody monomer is Y-shaped or T-shaped: the V regions form the tips, the C regions form the base, and F_C (stem) region.
8. The F_C region can attach to a host cell or to complement.
9. IgG antibodies are the most prevalent in serum; they provide naturally acquired passive immunity, neutralize bacterial toxins, participate in complement fixation, and enhance phagocytosis.

10. IgM antibodies consist of five monomers held by a joining chain; they are involved in agglutination and complement fixation.
11. Serum IgA antibodies are monomers; secretory IgA antibodies are dimers that protect mucosal surfaces from invasion by pathogens.
12. IgD antibodies are on B cells; they may delete B cells that produce antibodies against self.
13. IgE antibodies bind to mast cells and basophils and are involved in allergic reactions.

B CELLS AND HUMORAL IMMUNITY (pp. 509–510)

CLONAL SELECTION OF ANTIBODY-PRODUCING CELLS (p. 509)

1. Red bone marrow stem cells give rise to B cells with IgM and IgD on their surfaces, which recognize specific epitopes.
2. For T-independent antigens: a clone of B cells is selected by free antigens.
3. For T-dependent antigens: the B cell's immunoglobulins combine with an antigen, and the antigen fragments combined with MHC class II activate T_H cells. The T_H cells activate a B cell.
4. Activated B cells differentiate into plasma cells and memory cells.
5. B cells that recognize self are eliminated by clonal deletion.

THE DIVERSITY OF ANTIBODIES (p. 509–510)

6. During development, the genes in embryonic B cells recombine so that mature B cells each have different genes for the V region of their antibodies. ❖ **Animation: Humoral Immunity. The Microbiology Place.**

ANTIGEN–ANTIBODY BINDING AND ITS RESULTS (p. 511)

1. An antigen–antibody complex forms when an antibody binds to its specific epitopes on an antigen.
2. Agglutination results when an antibody combines with epitopes on two different cells.
3. Antibodies that attach to viruses or toxins cause neutralization.
4. Opsonization enhances phagocytosis of the antigen.
5. Complement activation results in cell lysis.

T CELLS AND CELLULAR IMMUNITY (pp. 511–516)

1. Red bone marrow stem cells give rise to T cells, which mature in the thymus gland and migrate to lymphoid tissues.
2. An antigen must be processed by an antigen-presenting cell and positioned on the surface of the APC.
3. T cells recognize antigens in association with MHC on an APC.

CLASSES OF T CELLS (pp. 513–514)

4. T cells are classified according to their functions and cell-surface glycoproteins called CDs.

HELPER T CELLS (pp. 514–515)

5. T_H1 cells activate cells involved in cellular immunity.
6. T_H2 cells are associated with allergic reactions and parasitic infections.
7. Helper T cells, or CD4 cells, are activated by MHC class II on APCs. After binding an APC, CD4 cells secrete cytokines that activate other T cells and B cells.

CYTOTOXIC T CELLS (p. 515)

8. Cytotoxic T cells (T_C), or CD8 cells, are activated by endogenous antigens and MHC class I on a target cell and are transformed into a CTL.
9. CTLs lyse the target cell or induce apoptosis in the target cell.

REGULATORY T CELLS (pp. 515–516)

10. Regulatory T cells (T_R) appear to suppress other T cells.
    ❉ **Animations: Cell-Mediated Immunity, and Antigen Processing and Presentation. The Microbiology Place.**

ANTIGEN-PRESENTING CELLS (APCS) (pp. 516–517)

1. APCs include B cells, dendritic cells, and macrophages.

EXTRACELLULAR KILLING BY THE IMMUNE SYSTEM (p. 517)

1. Natural killer (NK) cells lyse virus-infected and tumor cells. They are not immunologically specific.

ANTIBODY-DEPENDENT CELL-MEDIATED CYTOTOXICITY (p. 517)

1. In ADCC, NK cells and macrophages lyse antibody-coated cells.

CYTOKINES: CHEMICAL MESSENGERS OF IMMUNE CELLS (pp. 518–519)

1. Cells of the immune system communicate with each other by means of chemicals called cytokines.
2. Interleukins (IL) are cytokines that serve as communicators between leukocytes.
3. Chemokines cause leukocytes to move to the site of infection.

IMMUNOLOGICAL MEMORY (p. 519)

1. The amount of antibody in serum is called the antibody titer.
2. The response of the body to the first contact with an antigen is called the primary response. It is characterized by the appearance of IgM followed by IgG.
3. Subsequent contact with the same antigen results in a very high antibody titer and is called the secondary, anamnestic, or memory response. The antibodies are primarily IgG.

TYPES OF ADAPTIVE IMMUNITY (pp. 520–522)

1. Immunity resulting from infection is called naturally acquired active immunity; this type of immunity may be long-lasting.
2. Antibodies transferred from a mother to a fetus (transplacental transfer) or to a newborn in colostrum results in naturally acquired passive immunity in the newborn; this type of immunity can last up to a few months.
3. Immunity resulting from vaccination is called artificially acquired active immunity and can be long-lasting.
4. Artificially acquired passive immunity refers to humoral antibodies acquired by injection; this type of immunity can last for a few weeks.
5. Serum containing antibodies is often called antiserum.
6. When serum is separated by gel electrophoresis, antibodies are found in the gamma fraction of the serum and are termed immune serum globulin, or gamma globulin.

STUDY QUESTIONS

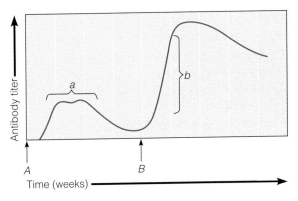

Access more review material either online at **The Microbiology Place** (www.microbiologyplace.com) or with **The Microbiology Place CD-ROM** packaged with your new book. There you'll find activities, practice tests, quizzes, flashcards, case studies, and more to help you succeed. In addition, you'll find the following Interactive Tutorials: B cells and T cells.

Answers to the Study Questions can be found in Appendix G.

REVIEW

1. Contrast the terms in the following pairs:
   a. innate and adaptive immunity
   b. humoral and cellular immunity
   c. active and passive immunity
   d. T_H1 and T_H2 cells
   e. natural and artificial immunity
   f. T-dependent and T-independent antigens
   g. CD4 and CD8
   h. immunoglobulin and TCR

2. Which type of immunity is most effective against viral infections? Why?

3. For what does MHC stand? What is the function of MHC? What types of T cells interact with MHC class I? With MHC class II?

4. Explain what an antigen is. Distinguish an antigen from an epitope and from a hapten.

5. Explain what an antibody is by describing the characteristics of antibodies. Diagram the structure of a typical antibody; label the heavy chain, light chain, and constant, variable, and F_C regions.

6. Discuss the clonal selection mechanism.

7. By means of a diagram, explain the role of T cells and B cells in immunity.

8. Explain a function for the following types of cells: T_C, T_H, and T_R. What is a cytokine?

9. a. In the graph that follows, at time A, the host was injected with tetanus toxoid. At time B, the host was given a booster dose. Explain the meaning of the areas of the curve marked a and b.
   b. Identify in the graph the antibody response of this same individual to exposure to a new antigen indicated at time B.

10. What effect does an antibody have on an antigen?

11. How does a T cell recognize an antigen?

12. What are natural killer cells?

13. How would each of the following prevent infection?
    a. antibodies against *Neisseria gonorrhoeae* fimbriae
    b. antibodies against host cell mannose

14. How can a human make over one billion different antibodies with only 35,000 different genes?

15. Explain why a person who recovers from a disease can attend others with the disease without fear of contracting the disease.

16. Pooled human immune serum globulin is sometimes administered to a patient after exposure to hepatitis A. What is human immune serum globulin? What type of immunity might this confer on the patient?

17. Compare and contrast CTL and complement (p. 490).

MULTIPLE CHOICE

Match the following choices to questions 1–4:
   a. innate resistance
   b. naturally acquired active immunity
   c. naturally acquired passive immunity
   d. artificially acquired active immunity
   e. artificially acquired passive immunity

d 1. The type of protection provided by the injection of diphtheria toxoid

e 2. The type of protection provided by the injection of antirabies serum

b 3. The type of protection resulting from recovery from an infection

c 4. A newborn's immunity to yellow fever

Match the following choices to the statements in questions 5–7:

a. IgA   **d.** IgG
b. IdD   **e.** IgM
c. IgE

d **5.** Antibodies that protect the fetus and newborn

e **6.** The first antibodies synthesized; especially effective against microorganisms

c **7.** Antibodies that are bound to mast cells and involved in allergic reactions

d **8.** Put the following in the correct sequence to elicit an antibody response: (1) T$_H$ cell recognizes B cell; (2) APC contacts antigen; (3) antigen fragment goes to surface of APC; (4) T$_H$ recognizes antigen digest and MHC; (5) B cell proliferates.
  a. 1, 2, 3, 4, 5   **d.** 2, 3, 4, 1, 5
  b. 5, 4, 3, 2, 1   **e.** 4, 5, 3, 1, 2
  c. 3, 4, 5, 1, 2

9. A kidney-transplant patient experienced a cytotoxic rejection of his new kidney. Place the following in order for that rejection: (1) apoptosis occurs; (2) CD8 becomes CTL; (3) granzymes released; (4) MHC class I activates CD8; (5) perforin released.
  a. 1, 2, 3, 4, 5
  b. 5, 4, 3, 2, 1
  c. 4, 2, 5, 3, 1
  d. 3, 4, 5, 1, 2
  e. 2, 3, 4, 1, 5

10. Patients with Chédiak-Higashi syndrome suffer from various types of cancer. These patients are most likely lacking which of the following:
  a. T$_R$ cells   **d.** NK cells
  b. T$_H$1 cells   **e.** T$_H$2 cells
  c. B cells

CRITICAL THINKING

1. Injections of T$_C$ cells completely removed all hepatitis B viruses from infected mice, but they killed only 5% of the infected liver cells. Explain how T$_C$ cells cured the mice.

2. Why is dietary protein deficiency associated with increased susceptibility to infections?

3. A positive tuberculin skin test shows cellular immunity to *Mycobacterium tuberculosis*. How could a person acquire this immunity?

4. On her vacation to Australia, Janet was bitten by a poisonous sea snake. She survived because the emergency room physician injected her with antivenin to neutralize the toxin. What is antivenin? How is it obtained?

CLINICAL APPLICATIONS

1. A woman had life-threatening salmonellosis that was successfully treated with anti-*Salmonella*. Why did this treatment work, when antibiotics and her own immune system failed?

2. A patient with AIDS has a low T$_H$ cell count. Why does this patient have trouble making antibodies? How does this patient make *any* antibodies?

3. A patient with chronic diarrhea was found to lack IgA in his secretions, although he had a normal level of serum IgA. What was this patient found to be unable to produce?

4. Newborns (under 1 year) who contract dengue fever have a higher chance of dying from it if their mothers had dengue fever prior to pregnancy. Explain why.

5. A woman died from a *Capnocytophaga* bacterial infection introduced by a dog bite. *Capnocytophaga* kills only people who lack a spleen. What is the relationship between infection and the spleen?

18

Practical Applications of Immunology

In Chapter 17, we learned the basics of the immune system, in which the body recognizes foreign microbes, toxins, or tissues. In response, the body forms antibodies and activates other immune system cells that are programmed to recognize and neutralize or destroy this foreign material if the body encounters it again. This adaptive immunity is an important element of our defenses against foreign pathogens.

In this chapter, we will discuss some useful tools that have been developed from knowledge of the basics of the immune system. Vaccines were briefly mentioned in the previous chapter; in this chapter we will expand our discussion of this important field of immunology. The diagnosis of disease frequently depends on tests that make use of the specificity of the immune system. Antibodies, especially monoclonal antibodies, are of special value in many of these diagnostic tests.

UNDER THE MICROSCOPE

Immunofluorescence. These streptococci fluoresce under ultraviolet light because dye-tagged antibodies are attached to them.

VACCINES

Long before the invention of vaccines, it was known that people who recovered from certain diseases, such as smallpox, were immune to the disease thereafter. As discussed in Chapter 1 on page 11, Chinese physicians may have been the first to try to exploit this phenomenon to prevent disease when they had children inhale dried smallpox scabs.

In 1717, Lady Mary Montagu reported from her travels in Turkey, an "old woman comes with a nutshell full of the matter of the best sort of smallpox and asks what veins you please to have opened, and puts into the vein as much venom as can lie upon the head of her needle." This practice usually led to a week of mild illness, and the person was subsequently protected from smallpox. Called **variolation,** this procedure became commonplace in England. Unfortunately, however, it sometimes resulted in a serious case of smallpox. In eighteenth-century England, the mortality rate associated with variolation was about 1%, still a significant improvement over the 50% mortality rate that could be expected from smallpox.

One person who received this treatment, at the age of 8, was Edward Jenner. As a physician, Jenner subsequently encountered patients who did not respond with the usual reactions to variolation. Many of them, especially dairymaids, told him that they had no fear of smallpox because they had already had cowpox. Cowpox is a mild disease that causes lesions on cows' udders; dairymaids' hands often became infected during milking. Motivated by his childhood memory of variolation, Jenner began a series of experiments in 1798 in which he deliberately inoculated people with cowpox in an attempt to prevent smallpox. To honor Jenner's work, the term *vaccination* (from the Latin *vacca*, meaning cow) was coined. A **vaccine** is a suspension of organisms or fractions of organisms that is used to induce immunity. Two centuries later, the disease of smallpox has been eliminated worldwide by vaccination, and two other viral diseases, measles and polio, are also targeted for elimination.

PRINCIPLES AND EFFECTS OF VACCINATION

We now know that Jenner's inoculations worked because the cowpox virus, which is not a serious pathogen, is closely related to the smallpox virus. The injection, by skin scratches, provoked a primary immune response in the recipients, leading to the formation of antibodies and long-term memory cells. Later, when the recipient encountered the smallpox virus, the memory cells were stimulated, producing a rapid, intense secondary immune response (see Figure 17.15, page 520). This response mimics the immunity gained by recovering from the disease. The cowpox vaccine was soon replaced by a vaccinia virus vaccine. The vaccinia virus also confers immunity to smallpox, although, strangely, little is known with certainty about the origin of this important virus. It is genetically distinct from cowpox virus and may be a hybrid of an accidental mixing of cowpox and smallpox viruses or perhaps once may have been the cause of a now-extinct disease, horsepox. The development of vaccines based on the model of the smallpox vaccine is the single most important application of immunology.

Many communicable diseases can be controlled by behavioral and environmental methods. For example, proper sanitation can prevent the spread of cholera, and the use of condoms can slow the spread of sexually transmitted diseases. If prevention fails, bacterial diseases can often be treated with antibiotics. Viral diseases, however, often cannot be effectively treated once contracted. Therefore, vaccination is frequently the only feasible method of controlling viral disease. Controlling a disease does not necessarily require that everyone be immune to it. If most of the population is immune, a phenomenon called *herd immunity,* outbreaks are limited to sporadic cases because there are not enough susceptible individuals to support the spread of epidemics.

The principal vaccines used to prevent bacterial and viral diseases in the United States are listed in Tables 18.1 and 18.2. Recommendations for childhood immunizations against some of these diseases are given in Table 18.3. American travelers who might be exposed to cholera, yellow fever, or other diseases not endemic in the United States can obtain current immunization recommendations from the U.S. Public Health Service and local public health agencies.

Experience has shown that vaccines against enteric bacterial pathogens, such as those causing cholera and typhoid, are not nearly as effective or long-lived as those against viral diseases, such as measles and smallpox. The chief reason is that the polysaccharide components of the outer surfaces of these bacterial pathogens do not stimulate production of effective antibodies as well as protein-coated viruses.

TYPES OF VACCINES AND THEIR CHARACTERISTICS

LEARNING OBJECTIVES
• Differentiate among the following, and provide an example of each: attenuated, inactivated, toxoid, subunit, and conjugated vaccines.
• Contrast subunit vaccines and nucleic acid vaccines.

TABLE 18.1	Principal Vaccines Used in the United States to Prevent Bacterial Diseases in Humans		
Disease	Vaccine	Recommendation	Booster
Diphtheria	Purified diphtheria toxoid	See Table 18.3	Every 10 years for adults
Meningococcal meningitis	Purified polysaccharide from *Neisseria meningitidis*	For people with substantial risk of infection. Recommended for college freshmen, especially if living in dormitories.	Need not established
Pertussis (whooping cough)	Killed whole or acellular fragments of *Bordetella pertussis*	Children prior to school age; see Table 18.3	For high-risk adults Available for ages 10–18 years
Pneumococcal pneumonia	Purified polysaccharide from 7 strains of *Streptococcus pneumoniae*	For adults with certain chronic diseases; people over 65; children 2–23 months	
Tetanus	Purified tetanus toxoid	See Table 18.3	Every 10 years for adults
Haemophilus influenzae type b meningitis	Polysaccharide from *Haemophilus influenzae* type b conjugated with protein to enhance effectiveness	Children prior to school age; see Table 18.3	None recommended

There are now several basic types of vaccine. Some of the newer vaccines take full advantage of knowledge and technology developed in recent years.

Attenuated whole-agent vaccines use living but attenuated (weakened) microbes. Live vaccines more closely mimic an actual infection. Lifelong immunity, especially with viruses, is often achieved without booster immunizations, and an effectiveness rate of 95% is not unusual. This long-term effectiveness probably occurs because the attenuated viruses replicate in the body, increasing the original dose and acting as a series of secondary (booster) immunizations.

Examples of attenuated vaccines are the Sabin polio vaccine and those used against measles, mumps, and rubella (MMR). The widely used vaccine against the tuberculosis bacillus and certain of the newly introduced, orally administered typhoid vaccines contain attenuated bacteria. Attenuated microbes are usually derived from mutations accumulated during long-term artificial culture. A danger of such vaccines is that the live microbes can backmutate to a virulent form (discussed on page 654). Attenuated vaccines are not recommended for people whose immune systems are compromised. If available, inactivated vaccines are substituted.

Inactivated whole-agent vaccines use microbes that have been killed, usually by formalin or phenol. Inactivated virus vaccines used in humans include those against rabies (animals sometimes receive a live vaccine considered too hazardous for humans), influenza (Figure 18.1), and polio (the Salk polio vaccine). Inactivated bacterial vaccines include those for pneumococcal pneumonia and cholera. Several long-used inactivated vaccines that are being replaced for most uses by newer, more effective types are those for pertussis (whooping cough) and typhoid.

FIGURE 18.1 Influenza viruses are grown in embryonated eggs. (See Figure 13.7, page 395.) The viruses will be inactivated to make a vaccine.

Q **Could this method of viral cultivation be a problem for people who are allergic to eggs?**

TABLE 18.2	Principal Vaccines Used in the United States to Prevent Viral Diseases in Humans		
Disease	**Vaccine**	**Recommendation**	**Booster**
Influenza	Injected vaccine, inactivated virus (nasally administered vaccine with attenuated virus is now available for some)	For chronically ill, including children over 6 months. Adults over age 65. Healthy children aged 6–23 months (because higher risk of related hospitalizations). Health care workers and others in contact with high risk groups. Healthy persons aged 5–49 years can receive intranasal vaccine.	Annual
Measles	Attenuated virus	For infants age 15 months	See Table 18.3
Mumps	Attenuated virus	For infants age 15 months	(Duration of immunity not known)
Rubella	Attenuated virus	For infants age 15 months; for females of childbearing age who are not pregnant	(Duration of immunity not known)
Chickenpox	Attenuated virus	For infants age 12 months	(Duration of immunity not known)
Poliomyelitis	Killed virus	For children, see Table 18.3; for adults, as risk to exposure warrants	(Duration of immunity not known)
Rabies	Killed virus	For field biologists in contact with wildlife in endemic areas; for veterinarians; for people exposed to rabies virus by bites	Every 2 years
Hepatitis B	Antigenic fragments of virus	For infants and children, see Table 18.3; for adults, especially health care workers, homosexual males, injecting drug users, heterosexual people with multiple partners, and household contacts of hepatitis B carriers	Duration of protection at least 7 years; need for boosters uncertain
Hepatitis A	Inactivated virus	Mostly for travel to endemic areas and protecting contacts during outbreaks	Duration of protection estimated at about 10 years
Smallpox	Live vaccinia virus	Certain military and health care personnel	Duration of protection estimated at about 3 to 5 years

Toxoids, which are inactivated toxins, are vaccines directed at the toxins produced by a pathogen. The tetanus and diphtheria toxoids have long been part of the standard childhood immunization series. They require a series of injections for full immunity, followed by boosters every 10 years. Many older adults have not received boosters; they are likely to have low levels of protection.

Subunit vaccines use only those antigenic fragments of a microorganism that best stimulate an immune response. Subunit vaccines that are produced by genetic modification techniques, meaning that other microbes are programmed to produce the desired antigenic fraction, are called **recombinant vaccines.** For example, the vaccine against the hepatitis B virus consists of a portion of the viral protein coat that is produced by a genetically modified yeast.

Subunit vaccines are inherently safer because they cannot reproduce in the recipient. They also contain little or no extraneous material and therefore tend to produce fewer adverse effects. Similarly, it is possible to separate the fractions of a disrupted bacterial cell, retaining the desired antigenic fractions. The newer **acellular vaccines** for pertussis use this approach.

Conjugated vaccines have been developed in recent years to deal with the poor immune response of children to vaccines based on capsular polysaccharides. As shown in Figure 17.6 (page 511), polysaccharides are T-independent antigens; children's immune systems do not respond well to these antigens until the age of 15 to 24 months. Therefore, the polysaccharides are combined with proteins such as diphtheria toxoid; this approach has led to the very successful vaccine for *Haemophilus influenzae* type b, which gives significant protection even at 2 months.

Nucleic acid vaccines, or DNA vaccines, are among the newest and most promising vaccines, although at this

TABLE 18.3	Schedule of Childhood Immunizations											
Vaccine	**Birth**	**1 mo**	**2 mos**	**4 mos**	**6 mos**	**12 mos**	**15 mos**	**18 mos**	**24 mos**	**4–6 yrs**	**11–12 yrs**	**13–18 yrs**
Hepatitis B (Hep B)*	Hep B 1 (If mother is HBV positive)											
			Hep B 2			Hep B 3					Hep B	
Diphtheria, Tetanus, Pertussis (DTaP)†			DTaP	DTaP	DTaP		DTaP			DTaP	Td	
Haemophilus influenzae type b (Hib)			Hib	Hib	Hib	Hib						
Inactivated polio vaccine (IPV)			IPV	IPV	IPV					IPV		
Pneumococcal conjugate vaccine (PCV)			PCV	PCV	PCV	PCV						
Measles, Mumps, Rubella (MMR)						MMR 1				MMR 2	MMR	
Varicella (Var)						Var					Var	
Hepatitis A (Hep A)‡											Hep A, selected groups	
Influenza§						Influenza (yearly)						

Vaccines are listed under routinely recommended ages. [Bars] indicate range of recommended ages for immunization. Any dose not given at the recommended age should be given as a "catch-up" immunization at any subsequent visit when indicated and feasible. (Ovals) indicate "catch-up" vaccinations.

Note: A vaccine that combines diphtheria, pertussis, tetanus, hepatitis B and polio, administered at 2, 4, and 6 months was approved recently.

*Hepatitis B vaccine (Hep B). Infants born to HBV positive mothers are a special case, Hep B is combined with immune globulin and administered within 12 hours of birth.
†Diphtheria, tetanus, and acellular pertussis vaccine (DTaP) is the preferred vaccine for all doses. (Td vaccines contain only tetanus and diphtheria toxoids.)
‡Hepatitis A vaccine (Hep A) is recommended for use in selected states and regions and for certain high-risk groups.
§Influenza vaccine is recommended annually for healthy children aged 6–23 months; for chronically ill children over 6 months. Healthy children over age 5 can receive intranasal vaccine.

time they have not yet resulted in any commercial vaccine for humans. Experiments with animals show that plasmids of "naked" DNA injected into muscle results in the production of the protein encoded in the DNA. (The "gene gun" method for injecting nucleic acids into plant cells is described in Chapter 9 and Figure 9.6, page 262.) These proteins persist and stimulate an immune response.

A problem with this type of vaccine is that the DNA remains effective only until it is degraded. Indications are that RNA, which could replicate in the recipient, might be a more effective agent.

THE DEVELOPMENT OF NEW VACCINES

LEARNING OBJECTIVES

- Compare and contrast the production of whole-agent vaccines, recombinant vaccines, and DNA vaccines.
- Define *adjuvant*.

An effective vaccine is the most desirable method of disease control. It prevents the targeted disease from ever occurring at all in an individual, and it is generally the most economical. This is especially important in developing parts of the world. A "dream" vaccine would be swallowed instead of injected. It also would give lifelong immunity with a single dose, remain stable without refrigeration, and

MICROBIOLOGY IN THE NEWS

WHY NOT VACCINATE AGAINST EVERYTHING?

Vaccines were responsible for eradicating smallpox and hold promise for eradicating polio and measles within the next 5 to 10 years. In the United States, the chickenpox vaccine reduced the incidence of chickenpox from 4 million cases annually to 18,000 cases in 2004. Vaccines are under development for contraception and cancer, and even a vaccine for cocaine addiction has been proposed. It appears that there is no absolute limit to the number of vaccines that can be given to a person. All standard vaccines recommended for children and adults can be given to the same person, at separate anatomical sites, on the same day or weeks apart, with no risk of interference of one vaccine by another and no potential adverse effects. A few exceptions exist; for example, live vaccines should not be given to immunosuppressed or pregnant individuals, and live bacterial vaccines should not be given within 24 hours after administering antibiotics.

So why not use more vaccines? For example, the vaccines against rabies, botulism, plague, and yellow fever are not routinely used. In the United States in 2004, there were 7 cases of rabies, 3 cases of plague, and 23 cases of botulism; yellow fever has not occurred in the United States since 1924. One might argue that these cases of rabies and plague, few as there were, would have

been prevented by vaccination. However, the plague vaccine affords only limited protection, and society must balance the cost of manufacturing and distributing a vaccine against the overall benefit. The *primary* purpose of vaccination against an infectious disease is to provide herd immunity against it; if herd immunity already exists in a society, it is not usually necessary to undertake efforts at widespread vaccination of all individuals. Why don't we vaccinate against tuberculosis (TB)? Like the plague vaccine, the current TB vaccine affords variable protection. This presents a risk; if people assume they are immune to a disease, they will not practice precautions.

Even in the United States, where the vaccination of children against diseases should be routine, not all children get vaccinated. In 2004, 73% of U.S. children received all the recommended vaccinations, and only 9% were receiving all their vaccinations at the recommended times. More than one in three U.S. children are "under-vaccinated" for more than six months during the first two years of life, a new study finds.

While efforts to vaccinate children in the United States are increasing, the World Health Organization is working to extend protection against diseases preventable with vaccine to children of all nations. Measles still accounts for 10% of

global mortality among children aged less than 5 years (approximately 1 million deaths annually). New vaccines are needed against diseases that affect millions of people in developing nations. Unfortunately, these diseases are not usually candidates for vaccine development. The affected populations cannot afford even minimal costs. Moreover, if electricity and refrigeration are not available, many vaccines cannot be made available.

Recent attempts to develop a malaria vaccine provide an example of the clinical and technical difficulties involved in developing a vaccine. Malaria causes 2 million deaths (75% of them are children) every year throughout the world. Preliminary human trials of a vaccine against a malaria sporozoite antigen showed that human volunteers did develop antibodies against the antigen. However, the antibodies must kill the sporozoites within 30 minutes after the mosquito bites its victim—before the infective sporozoites enter the protection of the liver cells. A separate vaccine is needed to provoke an immune response against the later stage merozoites, which destroy red blood cells and cause the typical symptoms of malaria.

The ultimate vaccine is an affordable, heat-stable, orally administered, multiple-antigen, single immunization given at birth.

be affordable. This dream is currently far from realization, however (see the box above).

Although interest in vaccine development declined with the introduction of antibiotics, it has intensified in recent years. Fear of litigation contributed to the decrease in the development of new vaccines in the United States. However, passage of the National Childhood Vaccine Injury Act in 1986, which limits the liability of vaccine manufacturers, has now helped reverse this trend. Even so, pharmaceutical companies find that the most profitable drugs are those that must be taken daily for extended periods, for example, drugs taken for high blood pressure. In contrast,

a vaccine that is required for a few injections or even only once in a lifetime is inherently less attractive.

Historically, vaccines could be developed only by growing the pathogen in usefully large amounts. The early successful viral vaccines were developed by animal cultivation. The vaccinia virus for smallpox was grown on the shaved bellies of calves, for example.

The introduction of vaccines against polio, measles, mumps, and a number of other viral diseases that would not grow in anything but a living human awaited the development of cell culture techniques. Cell cultures from human sources, or more often from animals such as monkeys that are closely related to humans, enabled growth of these viruses on a large scale. A convenient animal that will grow many viruses is the chick embryo (see Figure 13.7, page 395). Viruses for several vaccines (influenza, for example) are grown this way (see Figure 18.1). Interestingly, the first vaccine against hepatitis B virus used viral antigens extracted from the blood of chronically infected humans because no other source was available.

Recombinant vaccines and DNA vaccines do not need a cell or animal host to grow the vaccine's microbe. This avoids a major problem with certain viruses that so far have not been grown in cell culture—hepatitis B, for example.

Plants are also a potential source for vaccines. They can be genetically altered to produce almost any desired antigen in large amounts, which theoretically would cause production of antibodies against that antigen. Also, the thick cell walls of plants help to protect the cellular contents from degradation by stomach acids. Furthermore, release of the antigens directly into the intestinal tract represents an ideal location for introduction of vaccines against many diseases. These would include the diarrheal diseases, in which the pathogen is ingested and immunity depends primarily on defenses in the intestinal tract (see the discussion of M cells and Peyer's patches on page 513). Diarrheal diseases are a major cause of mortality for infants in the underdeveloped countries, where costs and distribution of vaccines also pose special problems. For example, a vaccine that must be constantly refrigerated can be nearly useless in underdeveloped countries that lack reliable electrical service. Edible, plant-derived vaccines of several types are undergoing clinical trials. Bananas are an especially attractive plant because they are eaten raw, usually relished by children, and can be grown in the tropics, where they are most likely to be needed.

The so-called golden age of immunology occurred from about 1870 to 1910, when most of the basic elements of immunology were discovered and several important vaccines were developed. We may soon be entering another golden age in which new technologies are brought to bear on emerging infectious diseases and problems arising from the decreasing effectiveness of antibiotics. It is remarkable that there are no useful vaccines against chlamydias, fungi, protozoa, or helminthic parasites of humans. Moreover, vaccines for some diseases, such as cholera and tuberculosis, are not reliably protective. At present, vaccines for at least 75 diseases are under development, ranging from those for prominent deadly diseases such as AIDS and malaria to such commonplace conditions as earaches. But we will probably find that the easy vaccines have already been made.

Infectious diseases are not the only possible target of vaccines. Researchers are investigating vaccines' potential for treating and preventing cocaine addiction, Alzheimer's disease, cancer, and for contraception.

Work is underway to improve effectiveness of antigens that may not be very effective when injected by themselves. For example, chemicals added for this purpose, called **adjuvants,** greatly improve the effectiveness of many antigens. Only alum has been approved as an adjuvant for human use; others might be found.

Currently, nearly 20 separate injections are recommended for infants and children, sometimes requiring three or more at one appointment. Developing additional multiple combinations of vaccines would be of some help. The U.S. Food and Drug Administration (FDA) has recently approved such a combination for five childhood diseases. Delivery in ways other than by needle would also be a desirable advance. Many have already received injection by high-pressure "guns," which are commonly used for mass inoculations, and an intranasal spray for influenza is now available. Experiments are also underway to test delivery by skin patches and, as we have discussed, vaccines produced in edible foods and ingested.

SAFETY OF VACCINES

LEARNING OBJECTIVE
- Explain the value of vaccines, and discuss acceptable risks for vaccines.

We have seen how variolation, the first attempt to provide immunity to smallpox, sometimes caused the disease it was intended to prevent. At the time, however, the risk was considered very worthwhile. As you will see later in this book, even today the oral polio vaccine on rare occasions may *cause* the disease. In 1999, a vaccine to prevent infant diarrhea caused by rotaviruses was withdrawn from the market because several recipients developed a life-threatening intestinal obstruction. However, public reaction to such risks has changed. Most parents have never seen a case of polio or measles and therefore tend to view the risk of these diseases as a remote abstraction. Moreover, reports or rumors of harmful effects often lead people to avoid certain vaccines for themselves or their children. In particular, a possible connection between the MMR vaccine and autism has received widespread publicity. Autism is a poorly understood developmental condition that causes a child to withdraw from reality. Because autism is usually diagnosed at the age of 18 to 30 months, about the time vaccine immunization schedules are nearing completion, some people have attempted to make a cause-and-effect connection. Medically, however, most experts agree that autism is a condition with a major genetic component and begins before birth. Extensive scientific surveys have provided no evidence to support a connection between the usual child-hood vaccines and autism or any other disease condition. Some experts even maintain that again introducing the rotavirus vaccine that was withdrawn in the United States would be well justified on a risk-versus-benefit calculus in much of the underdeveloped world. No vaccine will ever be perfectly safe or perfectly effective—neither is any antibiotic or most other drugs for that matter. Nevertheless, vaccines still remain the safest and most effective means of preventing infectious disease in children.

DIAGNOSTIC IMMUNOLOGY

LEARNING OBJECTIVE
- Explain how antibodies are used to diagnose diseases.

Throughout most of history, diagnosing a disease was essentially a matter of observing a patient's signs and symptoms. The writings of ancient and medieval physicians left descriptions of many diseases that are recognizable even today. Essential elements of diagnostic tests are sensitivity and specificity. **Sensitivity** is the probability that the test is reactive if the specimen is a true positive. **Specificity** is the probability that a positive test will not be reactive if a specimen is a true negative.

IMMUNOLOGIC-BASED DIAGNOSTIC TESTS

Knowledge of the high specificity of the immune system soon suggested that this might be used in diagnosing diseases. In fact, it was an accidental observation that led to one of the first diagnostic tests for an infectious disease. More than 100 years ago, Robert Koch was trying to develop a vaccine against tuberculosis. He observed that when guinea pigs with the disease were injected with a suspension of *Mycobacterium tuberculosis*, the site of the injection became red and slightly swollen a day or two later. You may recognize this symptom as a positive result for the widely used tuberculin skin test (see Figure 24.11, page 722)—many colleges and universities require the test as part of admission procedures. Koch, of course, had no idea of the mechanism of cell-mediated immunity that caused this phenomenon, nor did he know of the existence of antibodies.

Since the time of Robert Koch, immunology has given us many other invaluable diagnostic tools, most of which are based on interactions of humoral antibodies with antigens. A known antibody can be used to identify an *unknown* pathogen (antigen) by its reaction with it. This reaction can be reversed, and a *known* pathogen can be used, for example, to determine the presence of an unknown antibody in a person's blood—which would determine if he or she had immunity to the pathogen. One problem that must be overcome in antibody-based diagnostic tests is that antibodies cannot be seen directly. Even at magnifications of well over 100,000×, they appear only as fuzzy, ill-defined particles. Therefore, their presence must be established indirectly. We will describe a number of ingenious solutions to this problem.

Other problems that had to be overcome were that antibodies produced in an animal were mixed with numerous other antibodies produced in that animal and the quantities of any particular antibody were severely limited.

MONOCLONAL ANTIBODIES

LEARNING OBJECTIVE
- Define *monoclonal antibodies,* and identify their advantage over conventional antibody production.

As soon as it was determined that antibodies were produced by specialized cells (B cells), it was understood that these were a potential source of a single type of antibody. If such a B cell producing a single type of antibody could be isolated and cultivated, it would be able to produce the desired antibody in nearly unlimited quantities and without contamination by other antibodies. Unfortunately, a B cell reproduces only a few times under the usual cell culture conditions. This problem was largely solved with the discovery of a method to isolate and indefinitely cultivate B cells capable of producing a single type of antibody. Neils Jerne, Georges Koehler, and Cesar Milstein made this discovery in 1975, for which they were awarded a Nobel Prize.

Scientists have long observed that antibody-producing B cells may become cancerous. In this case, their proliferation is unchecked, and they are called *myelomas*. These cancerous B cells can be isolated and propagated indefinitely in cell culture. Cancer cells, in this sense, are "immortal." The breakthrough came in combining an "immortal" cancerous B cell with an antibody-producing normal B cell. When fused, this combination is termed a **hybridoma.**

When a hybridoma is grown in culture, its genetically identical cells continue to produce the type of antibody characteristic of the ancestral B cell. The importance of the technique is that clones of the antibody-secreting cells now can be maintained indefinitely in cell culture and can produce immense quantities of identical antibody molecules. Because all of these antibody molecules are produced by a single hybridoma clone, they are called **monoclonal antibodies,** or **Mabs** (Figure 18.2).

Monoclonal antibodies are useful for three reasons: they are uniform; they are highly specific; and they can be

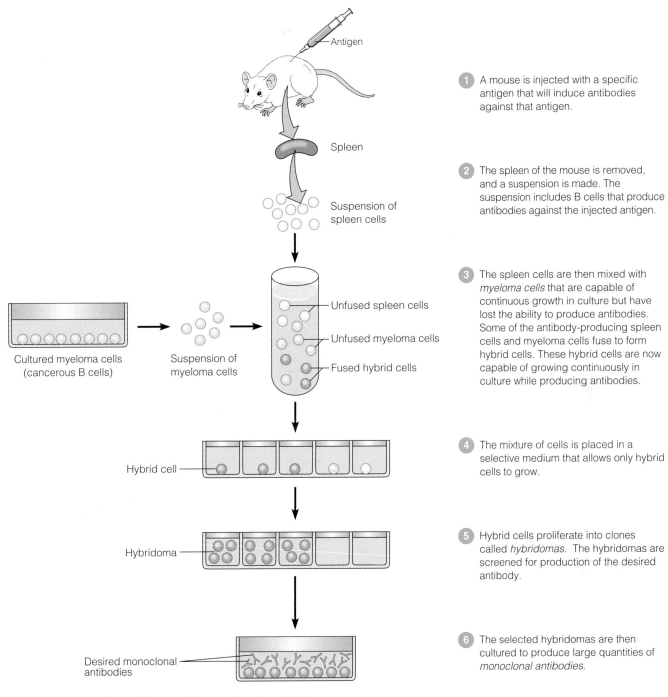

1. A mouse is injected with a specific antigen that will induce antibodies against that antigen.

2. The spleen of the mouse is removed, and a suspension is made. The suspension includes B cells that produce antibodies against the injected antigen.

3. The spleen cells are then mixed with *myeloma cells* that are capable of continuous growth in culture but have lost the ability to produce antibodies. Some of the antibody-producing spleen cells and myeloma cells fuse to form hybrid cells. These hybrid cells are now capable of growing continuously in culture while producing antibodies.

4. The mixture of cells is placed in a selective medium that allows only hybrid cells to grow.

5. Hybrid cells proliferate into clones called *hybridomas*. The hybridomas are screened for production of the desired antibody.

6. The selected hybridomas are then cultured to produce large quantities of *monoclonal antibodies*.

FIGURE 18.2 The production of monoclonal antibodies.

Q **What are the advantages of monoclonal antibodies compared to traditional methods of using animal serum?**

produced readily in large quantities. Because of these qualities, Mabs have assumed enormous importance as diagnostic tools. For instance, commercial kits use Mabs to recognize several bacterial pathogens, and nonprescription pregnancy tests use Mabs to indicate the presence of a hormone excreted only in the urine of a pregnant woman (see Figure 18.13, page 544).

Monoclonal antibodies are also being used therapeutically to overcome unwanted effects of the immune system. For example, *muromonab-CD3 (Orthoclone OKT3)* has been used since 1986 to minimize rejection of kidney transplants. For these purposes, monoclonal antibodies are prepared that react with the T cells that are responsible for rejection of the transplanted tissue. The Mabs suppress the T cell activity.

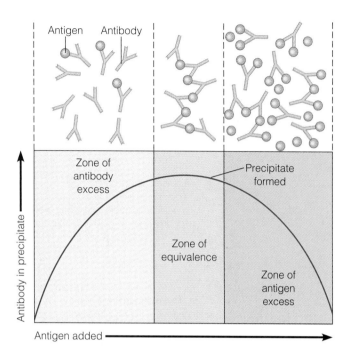

FIGURE 18.3 A precipitation curve. The curve is based on the ratio of antigen to antibody. The maximum amount of precipitate forms in the zone of equivalence, where the ratio is roughly equivalent.

Q **How does precipitation differ from agglutination?**

The use of Mabs is revolutionizing the treatment of many illnesses. The FDA has approved several for the treatment of specific diseases, such as *alemtuzumab (Campath)* for a form of leukemia, *infliximab (Remicade)* for Crohn's disease and rheumatoid arthritis, and *rituximab (Rituxan)* for non-Hodgkin's lymphoma (note that the generic names of those approved to date end in *-mab*). More than 90 other Mabs are in clinical trials.

Cancer is a particularly attractive target for Mab therapy. Theoretically at least, it is possible to combine Mabs that are specifically targeted against cancer cells with a radioactive isotope or a toxin intended to kill the cancer cell and leave the healthy, normal cells untouched. These Mabs are termed **immunotoxins** or **conjugated monoclonal antibodies.** One such product, *trastuzumab (Herceptin)* has proven useful for a particular type of advanced breast cancer that bears a receptor on its cells called HER2.

The therapeutic use of Mabs had been limited because these antibodies once were produced only by mouse (murine) cells. The immune systems of patients reacted against the foreign mouse proteins, leading to rashes, swelling, and even occasional kidney failure, plus the destruction of the Mabs. For example, the success of *muromonab-CD3* in minimizing tissue rejection was severely limited by side effects related to the administration of the foreign (murine) fraction of the monoclonal antibody.

In recognition of this problem, researchers are developing new generations of Mabs that are less likely to cause side effects due to their "foreignness." Essentially, the more human the antibody, the more successful it is likely to be. Researchers are exploring several approaches.

Chimeric monoclonal antibodies use genetically modified mice to make a human–murine hybrid. The variable part of the antibody molecule, including the antigen-binding sites (see Figure 17.3a), is murine. The remainder of the antibody molecule, the constant region, has been derived from a human source. These Mabs are about 66% human. An example is rituximab.

Humanized antibodies are constructed so that the murine portion is limited to the antigen-binding sites. The balance of the variable region and all of the constant region are derived from human sources. Such Mabs are about 90% human. Examples are alemtuzumab and trastuzumab.

The eventual goal is to develop **fully human antibodies.** One approach is to genetically modify mice to contain human antibody genes. The mice would produce antibodies that are fully human; in some cases, it might even be possible to produce an antibody that is an exact match to the patient.

It is also possible that Mab therapies may succeed so well that it would be difficult to produce them in sufficient volumes. Several potential solutions to this problem are under investigation. For example, the use of mice could be avoided entirely by using bacteriophages to insert desired genes into bacteria, which would be able to produce the desired Mabs on an industrial scale. Another approach to the problem is to genetically modify animals that can secrete the Mabs in their milk. Genetic alteration of plants to produce Mabs is another possible avenue to large-scale production.

PRECIPITATION REACTIONS

> **LEARNING OBJECTIVE**
> • Explain how precipitation reactions and immunodiffusion tests work.

Precipitation reactions involve the reaction of soluble antigens with IgG or IgM antibodies to form large, interlocking molecular aggregates called *lattices.*

Precipitation reactions occur in two distinct stages. First, the antigens and antibodies rapidly form small antigen–antibody complexes. This interaction occurs within seconds and is followed by a slower reaction, which may take minutes to hours, in which the antigen–antibody complexes form lattices that precipitate from solution. Precipitation reactions normally occur only when the ratio of antigen to antibody is optimal. Figure 18.3 shows that no visible precipitate forms when either component is in excess. The optimal ratio is produced when separate solutions of antigen and antibody are placed adjacent to each

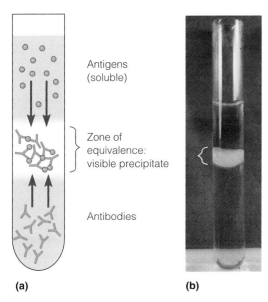

(a)   **(b)**

FIGURE 18.4 The precipitin ring test. (a) This drawing shows the diffusion of antigens and antibodies toward each other in a small-diameter test tube. Where they reach equal proportions, in the zone of equivalence, a visible line or ring of precipitate is formed. **(b)** A photograph of a precipitin ring.

Q **What causes the visible line?**

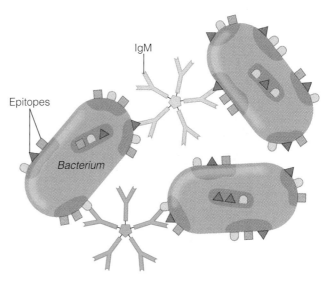

FIGURE 18.5 An agglutination reaction. When antibodies react with epitopes on antigens carried on neighboring cells, such as these bacteria (or red blood cells), the particulate antigens (cells) agglutinate. IgM, the most efficient immunoglobulin for agglutination, is shown here, but IgG also participates in agglutination reactions.

Q **Draw an agglutination reaction involving IgG.**

other and allowed to diffuse together. In a **precipitin ring test** (Figure 18.4), a cloudy line of precipitation (ring) appears in the area in which the optimal ratio has been reached (the *zone of equivalence*).

Immunodiffusion tests are precipitation reactions carried out in an agar gel medium, on either a Petri plate or a microscope slide. A line of visible precipitate develops between the wells at the point where the optimal antigen–antibody ratio is reached.

Other tests use electrophoresis to speed up the movement of antigen and antibody in a gel, sometimes in less than an hour, with this method. The techniques of immunodiffusion and electrophoresis can be combined in a procedure called **immunoelectrophoresis.** The procedure is used in research to separate proteins in human serum and is the basis of certain diagnostic tests. It is an essential part of the Western blot test used in AIDS testing (see Figure 10.12, page 299).

AGGLUTINATION REACTIONS

LEARNING OBJECTIVES

- Differentiate direct from indirect agglutination tests.
- Differentiate agglutination from precipitation tests.
- Define *hemagglutination.*

Whereas precipitation reactions involve *soluble* antigens, agglutination reactions involve either *particulate* antigens

(particles such as cells that carry antigenic molecules) or soluble antigens adhering to particles. These antigens can be linked together by antibodies to form visible aggregates, a reaction called **agglutination** (Figure 18.5). Agglutination reactions are very sensitive, relatively easy to read (see Figure 10.10, page 297), and available in great variety. Agglutination tests are classified as either direct or indirect.

DIRECT AGGLUTINATION TESTS

Direct agglutination tests detect antibodies against relatively large cellular antigens, such as those on red blood cells, bacteria, and fungi. At one time they were carried out in a series of test tubes, but now they are usually done in plastic *microtiter plates*, which have many shallow wells that take the place of the individual test tubes. The amount of particulate antigen in each well is the same, but the amount of serum that contains antibodies is diluted, so that each successive well has half the antibodies of the previous well. These tests are used, for example, to test for brucellosis and to separate *Salmonella* isolates into serovars, types defined by serological means.

Clearly, the more antibody we start with, the more dilutions it will take to lower the amount to the point where there is not enough antibody for the antigen to react with. This is the measure of **titer,** or concentration of serum antibody (Figure 18.6). For infectious diseases in general, the higher the serum antibody titer, the greater

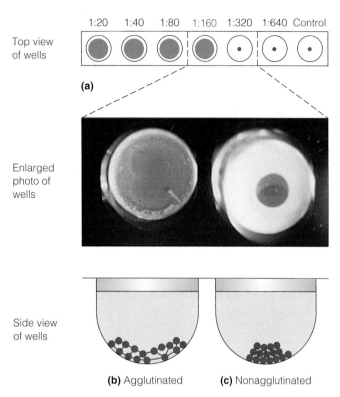

(a) Top view of wells

1:20 1:40 1:80 1:160 1:320 1:640 Control

(a)

Enlarged photo of wells

Side view of wells

(b) Agglutinated   (c) Nonagglutinated

(a) Each well in this microtiter plate contains, from left to right, only half the concentration of serum that is contained in the preceding well. Each well contains the same concentration of particulate antigens, in this instance red blood cells.

(b) In a positive (agglutinated) reaction, sufficient antibodies are present in the serum to link the antigens together, forming a mat of antigen–antibody complexes on the bottom of the well.

(c) In a negative (nonagglutinated) reaction, not enough antibodies are present to cause the linking of antigens. The particulate antigens roll down the sloping sides of the well, forming a pellet at the bottom. In this example, the antibody titer is 160 because the well with a 1:160 concentration is the most dilute concentration that produces a positive reaction.

FIGURE 18.6 Measuring antibody titer with the direct agglutination test.

Q **What is meant by the term *antibody titer*?**

the immunity to the disease. However, the titer alone is of limited use in diagnosing an existing illness. There is no way to know whether the measured antibodies were generated in response to the immediate infection or to an earlier illness. For diagnostic purposes, *a rise in titer* is significant; that is, the titer is higher later in the course of the disease than at its outset. Also, if it can be demonstrated that the person's blood had no antibody titer before the illness but has a significant titer while the disease is progressing, this change, called **seroconversion,** is also diagnostic. This situation is frequently encountered with HIV infections.

Some diagnostic tests specifically identify IgM antibodies. As discussed in Chapter 17, short-lived IgM is more likely to reflect a response to a current disease condition.

INDIRECT (PASSIVE) AGGLUTINATION TESTS

Antibodies against soluble antigens can be detected by agglutination tests if the antigens are adsorbed onto particles such as bentonite clay or, most often, minute latex spheres, each about one-tenth of the diameter of a bacterium. Such tests, known as *latex agglutination tests,* are commonly used for the rapid detection of serum antibodies against many bacterial and viral diseases. In such **indirect (passive)** **agglutination tests,** the antibody reacts with the soluble antigen adhering to the particles (Figure 18.7). The particles then agglutinate with one another, much as particles do in the direct agglutination tests. The same principle can be applied in reverse by using particles coated with antibodies to detect the antigens against which they are specific. This approach is especially common in tests for the streptococci that cause sore throats. A diagnosis can be completed in about 10 minutes.

HEMAGGLUTINATION

When agglutination reactions involve the clumping of red blood cells, the reaction is called **hemagglutination.** These reactions, which involve red blood cell surface antigens and their complementary antibodies, are used routinely in blood typing (see Table 19.2, page 555) and in the diagnosis of infectious mononucleosis.

Certain viruses, such as those causing mumps, measles, and influenza, have the ability to agglutinate red blood cells without an antigen–antibody reaction; this process is called **viral hemagglutination** (Figure 18.8). This type of hemagglutination can be inhibited by antibodies that neutralize the agglutinating virus. Diagnostic tests based on such neutralization reactions are discussed in the next section.

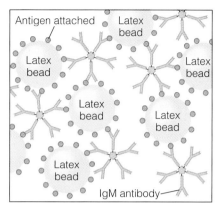

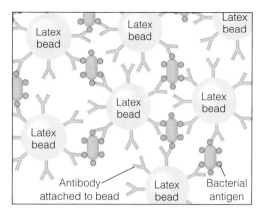

(a) Reaction in a positive indirect test for antibodies. When particles are coated with antigens, agglutination indicates the presence of antibodies, such as the IgM shown here.

(b) Reaction in a positive indirect test for antigens. When particles are coated with monoclonal antibodies, agglutination indicates the presence of antigens.

FIGURE 18.7 Reactions in indirect agglutination tests. These tests are performed using antigens or antibodies coated onto particles such as minute latex spheres.

Q Differentiate direct from indirect agglutination tests.

NEUTRALIZATION REACTIONS

LEARNING OBJECTIVES

• Explain how a neutralization test works.

• Differentiate precipitation from neutralization tests.

Neutralization is an antigen–antibody reaction in which the harmful effects of a bacterial exotoxin or a virus are blocked by specific antibodies. These reactions were first described in 1890, when investigators observed that immune serum could neutralize the toxic substances produced by the diphtheria pathogen, *Corynebacterium diphtheriae*. Such a neutralizing substance, which is called an antitoxin, is a specific antibody produced by a host as it responds to a bacterial exotoxin or its corresponding toxoid (inactivated toxin). The antitoxin combines with the exotoxin to neutralize it (Figure 18.9a). Antitoxins pro-

duced in an animal can be injected into humans to provide passive immunity against a toxin. Antitoxins from horses are routinely used to prevent or treat diphtheria and botulism; tetanus antitoxin is usually of human origin.

These therapeutic uses of neutralization reactions have led to their use as diagnostic tests. Viruses that exhibit their cytopathic (cell-damaging) effects in cell culture or embryonated eggs can be used to detect the presence of neutralizing viral antibodies (see page 465). If the serum to be tested contains antibodies against the particular virus, the antibodies will prevent that virus from infecting cells in the cell culture or eggs, and no cytopathic effects will be seen. Such tests, known as in vitro neutralization tests, can thus be used to both identify a virus and ascertain the viral antibody titer. In vitro neutralization tests are comparatively complex to carry out and are becoming less common in modern clinical laboratories.

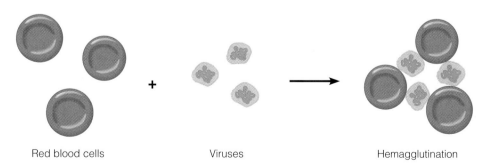

Red blood cells   Viruses   Hemagglutination

FIGURE 18.8 Viral hemagglutination. Viral hemagglutination is not an antigen–antibody reaction.

Q What causes agglutination in viral hemagglutination?

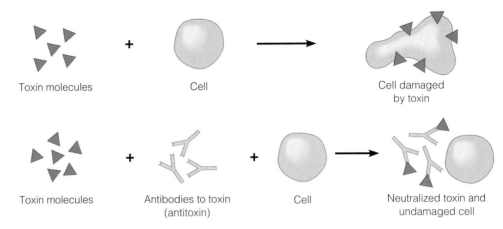

Toxin molecules   +   Cell   →   Cell damaged
                                                        by toxin

Toxin molecules   +   Antibodies to toxin   +   Cell   →   Neutralized toxin and
                            (antitoxin)   undamaged cell

(a) The effects of a toxin on a susceptible cell and neutralization of the toxin by antitoxin.

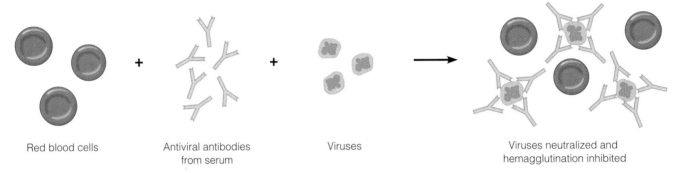

Red blood cells   +   Antiviral antibodies   +   Viruses   →   Viruses neutralized and
                            from serum   hemagglutination inhibited

(b) Viral hemagglutination test to detect antibodies to a virus. These viruses will normally cause hemagglutination when mixed with red blood cells. If antibodies to the virus are present, as shown here, they neutralize and inhibit hemagglutination.

FIGURE 18.9 Reactions in neutralization tests.

Q **Why does hemagglutination indicate that a patient does not have a specific disease?**

A more frequently used neutralization test is the **viral hemagglutination inhibition test.** This test is used in the diagnosis of influenza, measles, mumps, and a number of other infections caused by viruses that can agglutinate red blood cells. If a person's serum contains antibodies against these viruses, these antibodies will react with the viruses and neutralize them (Figure 18.9b). For example, if hemagglutination occurs in a mixture of measles virus and red blood cells but does not occur when the patient's serum is added to the mixture, this result indicates that the serum contains antibodies that have bound to and neutralized the measles virus.

COMPLEMENT-FIXATION REACTIONS

In Chapter 16 (pages 490–494), we discussed a group of serum proteins collectively called complement. During most antigen–antibody reactions, complement binds to the antigen–antibody complex and is used up, or fixed.

This process of **complement fixation** can be used to detect very small amounts of antibody. Antibodies that do not produce a visible reaction, such as precipitation or agglutination, can be demonstrated by the fixing of complement during the antigen–antibody reaction. Complement fixation was once used in the diagnosis of syphilis (Wassermann test) and is still used to diagnose certain viral, fungal, and rickettsial diseases. The complement-fixation test requires great care and good controls, one reason the trend is to replace it with newer, simpler tests. The test is performed in two stages: complement fixation and indicator (Figure 18.10).

FLUORESCENT-ANTIBODY TECHNIQUES

Fluorescent-antibody (FA) techniques can identify microorganisms in clinical specimens and can detect the

FIGURE 18.10 The complement-fixation test. This test is used to indicate the presence of antibodies to a known antigen. Complement will combine (be fixed) with an antibody that is reacting with an antigen. If all the complement is fixed in the complement-fixation stage, then none will remain to cause hemolysis of the red blood cells in the indicator stage.

Q **Why does red blood cell lysis indicate that the patient does not have a specific disease?**

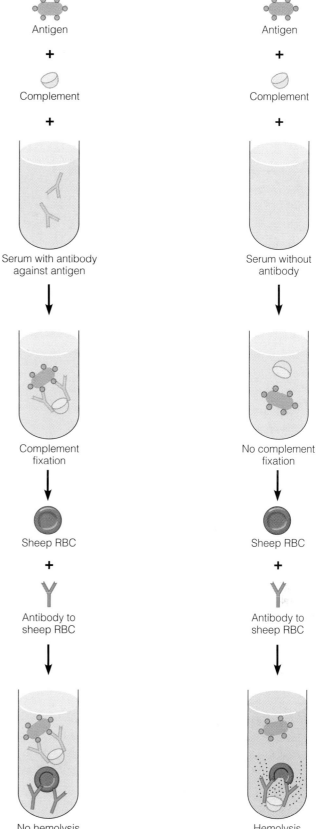

Antigen
+
Complement
+

Serum with antibody against antigen

Complement fixation

Sheep RBC
+
Antibody to sheep RBC

No hemolysis (complement tied up in antigen–antibody reaction)

(a) Positive test. All available complement is fixed by the antigen–antibody reaction; no hemolysis occurs, so the test is positive for the presence of antibodies.

Antigen
+
Complement
+

Serum without antibody

No complement fixation

Sheep RBC
+
Antibody to sheep RBC

Hemolysis (uncombined complement available)

(b) Negative test. No antigen–antibody reaction occurs. The complement remains, and the red blood cells are lysed in the indicator stage, so the test is negative.

Complement-fixation stage

Indicator stage

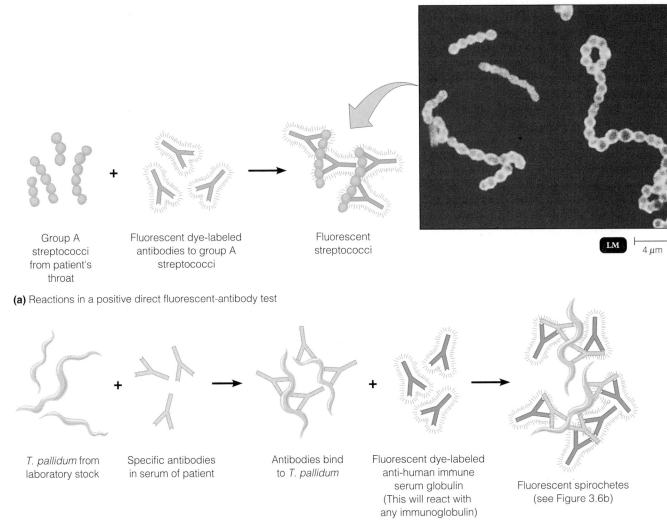

LM |———| 4 μm

Group A
streptococci
from patient's
throat

Fluorescent dye-labeled
antibodies to group A
streptococci

Fluorescent
streptococci

(a) Reactions in a positive direct fluorescent-antibody test

T. pallidum from
laboratory stock

Specific antibodies
in serum of patient

Antibodies bind
to *T. pallidum*

Fluorescent dye-labeled
anti-human immune
serum globulin
(This will react with
any immunoglobulin)

Fluorescent spirochetes
(see Figure 3.6b)

(b) Reactions in a positive indirect fluorescent-antibody test

FIGURE 18.11 Fluorescent-antibody (FA) techniques.
(a) A direct FA test to identify group A streptococci. **(b)** In an indirect FA test such as that used in the diagnosis of syphilis, the fluorescent dye is attached to antihuman immune serum globulin, which reacts with any human immunoglobulin (such as the *Treponema pallidum*–specific antibody) that has previously reacted with the antigen. The reaction is viewed through a fluorescence microscope, and the antigen with which the dye-tagged antibody has reacted fluoresces (glows) in the ultraviolet illumination.

Q **Differentiate a direct from an indirect FA test.**

presence of a specific antibody in serum (Figure 18.11). These techniques combine fluorescent dyes such as fluorescein isothiocyanate (FITC) with antibodies to make them fluoresce when exposed to ultraviolet light (see Figure 3.6, page 65). These procedures are quick, sensitive, and very specific; the FA test for rabies can be performed in a few hours and has an accuracy rate close to 100%.

Fluorescent-antibody tests are of two types, direct and indirect. **Direct FA tests** are usually used to identify a microorganism in a clinical specimen (Figure 18.11a). During this procedure, the specimen containing the antigen to be identified is fixed onto a slide. Fluorescein-labeled

antibodies are then added, and the slide is incubated briefly. Next the slide is washed to remove any antibody not bound to antigen and is then examined under the fluorescence microscope for yellow-green fluorescence.

Indirect FA tests are used to detect the presence of a specific antibody in serum following exposure to a microorganism (Figure 18.11b). They are often more sensitive than direct tests. During this procedure, a known antigen is fixed onto a slide. The test serum is then added, and, if antibody that is specific to that microbe is present, it reacts with the antigen to form a bound complex. In order for the antigen–antibody complex to be seen, fluorescein-labeled

antihuman immune serum globulin (anti-HISG), an antibody that reacts specifically with *any* human antibody, is added to the slide. Anti-HISG will be present only if the specific antibody has reacted with its antigen and is therefore present as well. After the slide has been incubated and washed (to remove unbound antibody), it is examined under a fluorescence microscope. If the known antigen fixed to the slide appears fluorescent, the antibody specific to the test antigen is present.

An especially interesting adaptation of fluorescent antibodies is the **fluorescence-activated cell sorter (FACS).** In Chapter 17, we learned that T cells carry antigenically specific receptors such as CD4 and CD8 on their surface, and these are characteristic of certain groups of T cells. The depletion of CD4 T cells is used to follow the progression of AIDS; their populations can be determined with a FACS.

The FACS is a modification of a *flow cytometer,* in which a suspension of cells leaves a nozzle as droplets containing no more than one cell each (see page 300). A laser beam strikes each cell-containing droplet and is then received by a detector that determines certain characteristics such as size (Figure 18.12). If the cells carry FA markers to identify them as CD4 or CD8 T cells, the detector can measure this fluorescence. As the laser beam detects a cell of a preselected size or fluorescence, an electrical charge, either positive or negative, can be imparted to it. As the charged droplet falls between electrically charged plates, it is attracted to one receiving tube or another, effectively separating cells of different types. Millions of cells can be separated in an hour with this process, all under sterile conditions, which allows them to be used in experimental work.

An interesting application of the flow cytometer is sorting sperm cells to separate male (Y-carrying) and female (X-carrying) sperm. The female sperm (meaning that it will result in a female embryo when it fertilizes the egg) contains more DNA, 2.8% more in humans, 4% in animals. When the sperm is stained with a fluorescent dye specific for DNA, the female sperm glows more brightly when illuminated by the laser beam because it has more DNA and therefore can be separated out. The technique was developed for agricultural purposes. However, it has received medical approval for use in humans where couples carry genes for inherited diseases that affect only boys.

ENZYME-LINKED IMMUNOSORBENT ASSAY (ELISA)

LEARNING OBJECTIVE
- Explain how direct and indirect ELISA tests work.

The **enzyme-linked immunosorbent assay (ELISA)** is the most widely used of a group of tests known as *enzyme*

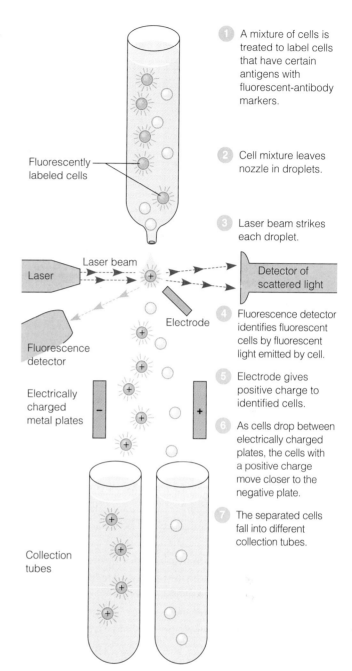

1. A mixture of cells is treated to label cells that have certain antigens with fluorescent-antibody markers.

2. Cell mixture leaves nozzle in droplets.

3. Laser beam strikes each droplet.

4. Fluorescence detector identifies fluorescent cells by fluorescent light emitted by cell.

5. Electrode gives positive charge to identified cells.

6. As cells drop between electrically charged plates, the cells with a positive charge move closer to the negative plate.

7. The separated cells fall into different collection tubes.

Fluorescently labeled cells

Laser Laser beam Detector of scattered light

Fluorescence detector Electrode

Electrically charged metal plates

Collection tubes

FIGURE 18.12 The fluorescence-activated cell sorter (FACS). This technique can be used to separate different classes of T cells. A fluorescence-labeled antibody reacts with, for example, the CD4 receptor on a T cell.

Q **Provide an application of FACS to follow the progress of HIV infection.**

immunoassay (EIA). There are two basic methods. The *direct ELISA* detects antigens, and the *indirect ELISA* detects antibodies. A microtiter plate with numerous shallow wells is used in both procedures (see Figure 10.11a, page 298). Variations of the test exist; for example, the reagents can be bound to tiny latex particles rather than

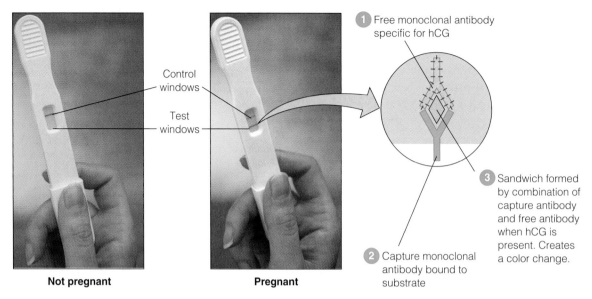

Not pregnant   Pregnant

① Free monoclonal antibody specific for hCG

③ Sandwich formed by combination of capture antibody and free antibody when hCG is present. Creates a color change.

② Capture monoclonal antibody bound to substrate

Control windows

Test windows

FIGURE 18.13 The use of monoclonal antibodies in a home pregnancy test. Home pregnancy tests detect a hormone called human chorionic gonadotropin (hCG) that is excreted only in the urine of a pregnant woman. The test area of the dipstick contains two types of monoclonal antibodies: ① free monoclonal antibodies and ② capture monoclonal antibodies, bound to the substrate in the test window. The free antibodies are specific for hCG and are color-labeled. The capture antibodies are specific for the hCG–free antibody complex. When the dipstick comes in contact with urine that contains hCG, the free antibodies bind the hCG and are then bound in turn by the capture antibodies, forming a "sandwich." ③ The color-labeled antibodies then create a visible color change in the test window. (Other antibodies create the color change in the control window.)

Q **What is the antigen in the home pregnancy test?**

to the surfaces of the microtiter plates. ELISA procedures are popular primarily because they require little interpretive skill to read; the results tend to be clearly positive or clearly negative.

Many ELISA tests are available for clinical use in the form of commercially prepared kits. Procedures are often highly automated, with the results read by a scanner and printed out by computer (see Figure 10.11, page 298). Some tests based on this principle are also available for use by the public; one example is a commonly available home pregnancy test (Figure 18.13).

DIRECT ELISA

The direct ELISA method is shown in Figure 18.14a. A common use of the direct ELISA test is to detect the presence of drugs in urine. For these tests, antibodies specific for the drug are adsorbed to the well on the microtiter plate. (The availability of monoclonal antibodies has been essential to the widespread use of the ELISA test.) When the patient's urine sample is added to the well, any of the drug that it contained would bind to the antibody and is captured. The well is rinsed to remove any unbound drug. In order to make a visible test, more antibodies specific to the drug are now added (these antibodies have an enzyme attached to them—therefore, the term *enzyme-linked*) and

will react with the already-captured drug, forming a "sandwich" of antibody/drug/enzyme-linked antibody. This positive test can be detected by now adding a substrate for the linked enzyme; a visible color is produced by the enzyme reacting with its substrate.

INDIRECT ELISA

The indirect ELISA text, illustrated in Figure 18.14b, detects antibodies in a patient's sample rather than an antigen such as a drug. Indirect ELISA tests are used, for example, to screen blood for antibodies to HIV (see page 571). For such a purpose, the microtiter well contains an antigen, such as the inactivated virus that causes the disease the test is designed to diagnose. A sample of the patient's blood is added to the well; and, if it contains antibodies against the virus, they will react with the virus. The well is rinsed to remove unbound antibodies. If antibodies in the blood and the virus in the well have attached to each other, they will remain in the well—a positive test. In order to make a positive test visible some anti-HISG (an immunoglobulin that will attach to *any* antibody, including the one in the patient's serum that has attached to the virus in the well; see page 543) is added. The anti-HISG is linked to an enzyme. A positive test consists of a "sandwich" or a virus/antibody/enzyme-linked-anti-HISG.

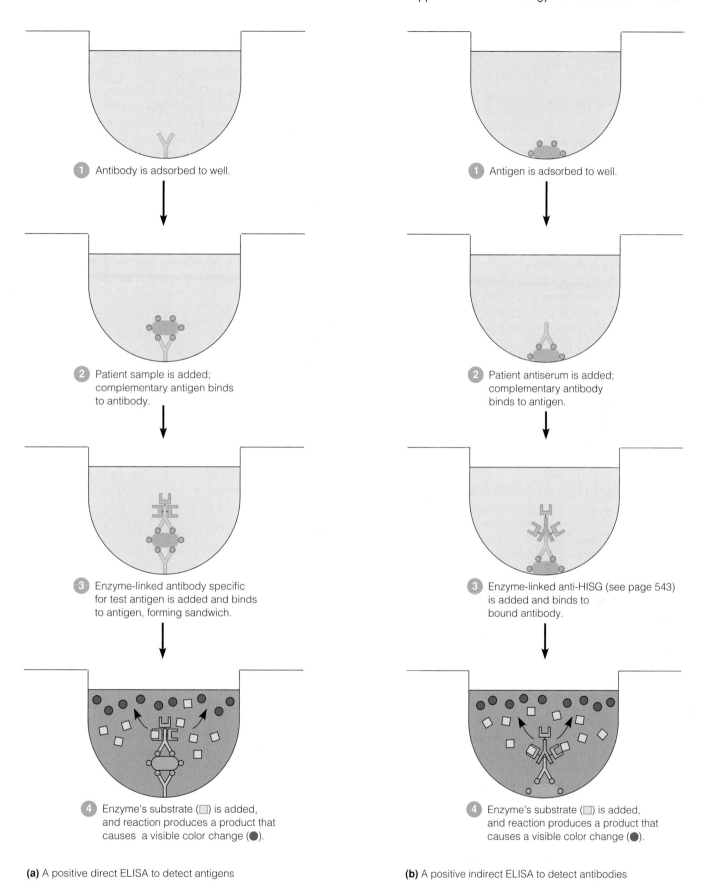

1 Antibody is adsorbed to well.

2 Patient sample is added; complementary antigen binds to antibody.

3 Enzyme-linked antibody specific for test antigen is added and binds to antigen, forming sandwich.

4 Enzyme's substrate (□) is added, and reaction produces a product that causes a visible color change (●).

(a) A positive direct ELISA to detect antigens

1 Antigen is adsorbed to well.

2 Patient antiserum is added; complementary antibody binds to antigen.

3 Enzyme-linked anti-HISG (see page 543) is added and binds to bound antibody.

4 Enzyme's substrate (□) is added, and reaction produces a product that causes a visible color change (●).

(b) A positive indirect ELISA to detect antibodies

FIGURE 18.14 The ELISA method. The components are usually contained in small wells of a microtiter plate.

 Differentiate a direct from an indirect ELISA test.

At this point, the substrate for the enzyme is added, and a positive test is detected by the color change caused by the enzyme linked to the anti-HISG.

THE FUTURE OF DIAGNOSTIC IMMUNOLOGY

LEARNING OBJECTIVE

- Explain the importance of monoclonal antibodies.

The introduction of monoclonal antibodies has revolutionized diagnostic immunology by making available large, economical amounts of specific antibodies. This has led to many newer diagnostic tests that are more sensitive, specific, rapid, and simpler to use. For example, tests to diagnose sexually transmitted chlamydial infections and certain protozoan-caused intestinal parasitic diseases are coming into common use. These tests had previously required relatively difficult culture or microscopic methods for diagnosis. At the same time, the use of many of the classic serological tests, such as complement-fixation tests, is declining. Most newer tests will require less human judgment to read, and they require fewer highly trained personnel.

The use of certain *nonimmunological* tests, such as the PCR and DNA probes that were discussed in Chapter 10 (page 303), is increasing. Some of these tests will become automated to a significant degree. For example, a DNA chip (see Figure 10.17, page 304) containing over 50,000 DNA probes for genetic information expected in possible pathogens can be exposed to a test sample. This chip is scanned and its data automatically analyzed. PCR tests are also becoming highly automated. At the other end of the technical spectrum, efforts are underway to provide low-cost diagnostic methods for developing nations, where the annual medical expenditure per person is only a few dollars. Interest in methods less invasive than blood sampling, such as urinalysis and oral swabs, is also increasing.

The tests described in this chapter are most often used to detect existing disease. In the future, diagnostic testing will probably also be directed at preventing disease. Food and water, for example, are frequently contaminated with pathogens. Controlling diseases spread by ingesting them is a significant public health problem. At present, detecting and identifying such pathogens usually require that the microorganism be selectively isolated and cultivated before diagnostic tests can be applied. This time-consuming process poses significant problems when the source of infection is a perishable consumer product. Much current research in diagnostic methods is now aimed at enabling health workers to detect and identify pathogenic organisms directly and quickly in a food sample without the need for isolation and cultivation.

STUDY OUTLINE

VACCINES (pp. 528–534)

1. Edward Jenner developed the modern practice of vaccination when he inoculated people with cowpox virus to protect them against smallpox.

PRINCIPLES AND EFFECTS OF VACCINATION (p. 528)

2. Herd immunity results when most of a population is immune to a disease.

TYPES OF VACCINES AND THEIR CHARACTERISTICS (pp. 528–531)

3. Attenuated whole-agent vaccines consist of attenuated (weakened) microorganisms; attenuated virus vaccines generally provide lifelong immunity.

4. Inactivated whole-agent vaccines consist of killed bacteria or viruses.

5. Toxoids are inactivated toxins.

6. Subunit vaccines consist of antigenic fragments of a microorganism; these include recombinant vaccines and acellular vaccines.

7. Conjugated vaccines combine the desired antigen with a protein that boosts the immune response.

8. Nucleic acid vaccines, or DNA vaccines, are being developed. These cause the recipient to make the antigenic protein associated with MHC class I.

THE DEVELOPMENT OF NEW VACCINES (pp. 531–533)

9. Viruses for vaccines may be grown in animals, cell cultures, or chick embryos.

10. Recombinant vaccines and nucleic acid vaccines do not need to be grown in cells or animals.

11. Genetically modified plants may someday provide edible vaccines.

12. Adjuvants improve the effectiveness of some antigens.

SAFETY OF VACCINES (pp. 533–534)

13. Vaccines are the safest and most effective means of controlling infectious diseases.

DIAGNOSTIC IMMUNOLOGY (pp. 534–546)

1. Many tests based on the interactions of antibodies and antigens have been developed to determine the presence of antibodies or antigens in a patient.

2. The sensitivity of a diagnostic test is determined by the percentage of positive samples it correctly detects; and its specificity is determined by the percentage of false positive results it gives.

MONOCLONAL ANTIBODIES (pp. 534–536)

3. Hybridomas are produced in the laboratory by fusing a cancerous cell with an antibody-secreting plasma cell.

4. A hybridoma cell culture produces large quantities of the plasma cell's antibodies, called monoclonal antibodies.

5. Monoclonal antibodies are used in serological identification tests, to prevent tissue rejections, and to make immunotoxins to treat cancer.

6. Immunotoxins can be made by combining a monoclonal antibody and a toxin; the toxin will then kill a specific antigen.

PRECIPITATION REACTIONS (pp. 536–537)

7. The interaction of soluble antigens with IgG or IgM antibodies leads to precipitation reactions.

8. Precipitation reactions depend on the formation of lattices and occur best when antigen and antibody are present in optimal proportions. Excesses of either component decrease lattice formation and subsequent precipitation.

9. The precipitin ring test is performed in a small tube.

10. Immunodiffusion procedures are precipitation reactions carried out in an agar gel medium.

11. Immunoelectrophoresis combines electrophoresis with immunodiffusion for the analysis of serum proteins.

AGGLUTINATION REACTIONS (pp. 537–538)

12. The interaction of particulate antigens (cells that carry antigens) with antibodies leads to agglutination reactions.

13. Diseases may be diagnosed by combining the patient's serum with a known antigen.

14. Diseases can be diagnosed by a rising titer or seroconversion (from no antibodies to the presence of antibodies).

15. Direct agglutination reactions can be used to determine antibody titer.

16. Antibodies cause visible agglutination of soluble antigens affixed to latex spheres in indirect or passive agglutination tests.

17. Hemagglutination reactions involve agglutination reactions using red blood cells. Hemagglutination reactions are used in blood typing, the diagnosis of certain diseases, and the identification of viruses.

NEUTRALIZATION REACTIONS (pp. 538–539)

18. In neutralization reactions, the harmful effects of a bacterial exotoxin or virus are eliminated by a specific antibody.

19. An antitoxin is an antibody produced in response to a bacterial exotoxin or a toxoid that neutralizes the exotoxin.

20. In a virus neutralization test, the presence of antibodies against a virus can be detected by the antibodies' ability to prevent cytopathic effects of viruses in cell cultures.

21. Antibodies against certain viruses can be detected by their ability to interfere with viral hemagglutination in viral hemagglutination inhibition tests.

COMPLEMENT-FIXATION REACTIONS (p. 540)

22. Complement-fixation reactions are serological tests based on the depletion of a fixed amount of complement in the presence of an antigen–antibody reaction.

FLUORESCENT-ANTIBODY TECHNIQUES (pp. 540–543)

23. Fluorescent-antibody techniques use antibodies labeled with fluorescent dyes.

24. Direct fluorescent-antibody tests are used to identify specific microorganisms.

25. Indirect fluorescent-antibody tests are used to demonstrate the presence of antibody in serum.

26. A fluorescence-activated cell sorter can be used to detect and count cells labeled with fluorescent antibodies.

ENZYME-LINKED IMMUNOSORBENT ASSAY (ELISA) (pp. 543–546)

27. ELISA techniques use antibodies linked to an enzyme.

28. Antigen–antibody reactions are detected by enzyme activity. If the indicator enzyme is present in the test well, an antigen–antibody reaction has occurred.

29. The direct ELISA is used to detect antigens against a specific antibody bound in a test well.

30. The indirect ELISA is used to detect antibodies against an antigen bound in a test well.

THE FUTURE OF DIAGNOSTIC IMMUNOLOGY (p. 546)

31. The use of monoclonal antibodies will continue to make new diagnostic tests possible.

STUDY QUESTIONS

Access more review material either online at **The Microbiology Place** (www.microbiologyplace.com) or with **The Microbiology Place CD-ROM** packaged with your new book. There you'll find activities, practice tests, quizzes, flashcards, case studies, and more to help you succeed.

Answers to the Study Questions can be found in Appendix G.

REVIEW

1. Classify the following vaccines by type. Which could cause the disease it is supposed to prevent?
   a. attenuated measles virus
   b. dead *Rickettsia prowazekii*
   c. *Vibrio cholerae* toxoid
   d. hepatitis B antigen produced in yeast cells
   e. purified polysaccharides from *Streptococcus pyogenes*
   f. *Haemophilus influenzae* polysaccharide bound to diphtheria toxoid
   g. a plasmid containing genes for influenza A protein

2. Explain the effects of excess antigen and antibody on the precipitation reaction. How is the precipitin ring test different from an immunodiffusion test?

3. How does the antigen in an agglutination reaction differ from that in a precipitation reaction?

4. Define the following terms, and give an example of how each reaction is used diagnostically:
   a. viral hemagglutination
   b. hemagglutination inhibition
   c. passive agglutination

5. Explain the fluorescent-antibody techniques.

6. Describe the direct and indirect FA tests in the following situations. Sketch how the reactions might appear in these tests if the antigens and antibodies could be seen.
   a. Rabies can be diagnosed postmortem by mixing fluorescent-labeled antibodies with brain tissue.
   b. Syphilis can be diagnosed by adding the patient's serum to a slide fixed with *Treponema pallidum*. Antihuman immune serum globulin tagged with a fluorescent dye is added.

7. Explain the ELISA techniques.

8. Describe the direct and indirect ELISA tests in the following situations:
   a. Respiratory secretions to detect respiratory syncytial virus.
   b. Blood to detect human immunodeficiency virus antibodies.

   Which of these tests provides definitive proof of disease?

9. How are monoclonal antibodies produced? What is their advantage over horse serum?

10. Match the following serological tests to the descriptions.
    ___ Precipitation
    ___ Immuno-electrophoresis
    ___ Agglutination
    ___ Complement fixation
    ___ Neutralization
    ___ ELISA

    a. Occurs with particulate antigens
    b. Uses an enzyme for the indicator
    c. Uses red blood cells for the indicator
    d. Uses antihuman immune serum globulin
    e. Occurs with a free soluble antigen
    f. Used to determine the presence of antitoxin

11. Match each of the following tests to its positive reaction.
    ___ Agglutination
    ___ Complement fixation
    ___ ELISA
    ___ FA test
    ___ Neutralization
    ___ Precipitation

    a. Peroxidase activity
    b. Harmful effects of agents not seen
    c. No hemolysis
    d. Cloudy white line
    e. Cell clumping
    f. Fluorescence

MULTIPLE CHOICE

Use the following choices to answer questions 1 and 2:
   a. hemolysis
   b. hemagglutination
   c. hemagglutination-inhibition
   d. no hemolysis
   e. precipitin ring forms

1. Patient's serum, influenza virus, sheep red blood cells, and anti-sheep red blood cells are mixed in a tube. What happens if the patient has antibodies against influenza?

2. Patient's serum, *Chlamydia*, guinea pig complement, sheep red blood cells, and anti-sheep red blood cells are mixed in a tube. What happens if the patient has antibodies against *Chlamydia*?

3. The examples in questions 1 and 2 are
   a. direct tests.
   b. indirect tests.

Use the following choices to answer questions 4 and 5:
   a. anti-*Brucella*
   b. *Brucella*
   c. substrate for the enzyme

4. Which is the third step in a direct ELISA test?

5. Which item is from the patient in an indirect ELISA test?

6. In an immunodiffusion test, a strip of filter paper containing diphtheria antitoxin is placed on a solid culture medium.

Then bacteria are streaked perpendicular to the filter paper. If the bacteria are toxigenic,
   a. the filter paper will turn red.
   b. a line of antigen–antibody precipitate will form.
   c. the cells will lyse.
   d. the cells will fluoresce.
   e. none of the above

Use the following choices to answer questions 7–9.
   a. direct fluorescent antibody
   b. indirect fluorescent antibody
   c. rabies immune globulin
   d. killed rabies virus
   e. none of the above

C **7.** Treatment given to a person bitten by a rabid bat.

a **8.** Test used to identify rabies virus in the brain of a dog.

b **9.** Test used to detect the presence of antibodies in a patient's serum.

10. In an agglutination test, eight serial dilutions to determine antibody titer were set up: Tube 1 contained a 1:2 dilution; tube 2, a 1:4, and so on. If tube 5 is the last tube showing agglutination, what is the antibody titer?
   a. 5
   b. 1:5
   c. 32
   d. 1:32

CRITICAL THINKING

1. What problems are associated with the use of attenuated whole-agent vaccines?

2. The World Health Organization has announced the complete eradication of smallpox and is working toward the eradication of measles and polio. Why would vaccination be more likely to eradicate a viral disease than a bacterial disease?

3. Many of the serological tests require a supply of antibodies against pathogens. For example, to test for *Salmonella*, anti-*Salmonella* antibodies are mixed with the unknown bacterium. How are these antibodies obtained?

4. A test for antibodies against *Treponema pallidum* uses an antigen called cardiolipin and the patient's serum (suspected of having antibodies). Why do the antibodies react with cardiolipin? What is the disease?

CLINICAL APPLICATIONS

1. Which of the following is proof of a disease state? Why doesn't the other situation confirm a disease state? What is the disease?
   a. *Mycobacterium tuberculosis* is isolated from a patient.
   b. Antibodies against M. *tuberculosis* are found in a patient.

2. Streptococcal erythrogenic toxin is injected into a person's skin in the Dick test. What results are expected if a person has antibodies against this toxin? What type of immunological reaction is this? What is the disease?

3. The following data were obtained from FA tests for anti-*Legionella* in four people. What conclusions can you draw? What is the disease?

	Antibody Titer			
	Day 1	**Day 7**	**Day 14**	**Day 21**
Patient A	128	256	512	1024
Patient B	0	0	0	0
Patient C	256	256	256	256
Patient D	0	0	128	512

4. Maria decided against the relatively new chickenpox vaccine and used her parents' method: she wanted her children to get chickenpox in order to develop natural immunity. Her two children did get chickenpox. Her son had slight itching and skin vesicles, but her daughter was hospitalized for months with streptococcal cellulitis and underwent several skin grafts before recovering. Maria's housekeeper contracted chickenpox from the children and subsequently died. Almost half of the deaths due to chickenpox occur in adults.
   a. What responsibilities do parents have for their children's health?
   b. What rights do individuals have? Should vaccination be required by law?
   c. What responsibilities do individuals (e.g., parents) have for the health of society?
   d. Vaccines are given to healthy people, so what risks are acceptable?

19

Disorders Associated with the Immune System

In this chapter, we will see that not all immune system responses produce a desirable result. A familiar example is hay fever, which results from repeated exposure to plant pollen. Most of us also know that a blood transfusion will be rejected if the blood of the donor and the blood of the recipient are not compatible and that rejection is also a potential problem with transplanted organs. One's own tissue may be mistakenly attacked by the immune system, causing diseases we classify as autoimmune. Certain antigens, called superantigens, indiscriminately activate many T-cell receptors at once, resulting in damage to tissue (see Chapter 15, page 461).

Some people are born with a defective immune system, and in all of us the effectiveness of our immune system declines with age. Our immune systems can be deliberately crippled (immunosuppressed) to prevent the rejection of transplanted organs. Disease can also impair the immune system, especially infection by HIV, a virus that specifically attacks the immune system.

UNDER THE MICROSCOPE

Pollen grains such as these are antigens that can cause localized ana-phylaxis; a familiar example is hay fever.

TABLE 19.1	Types of Hypersensitivity		
Type of Reaction	Time Before Clinical Signs	Characteristics	Examples
Type I (anaphylactic)	<30 min	IgE binds to mast cells or basophils; causes degranulation of mast cell or basophil and release of reactive substances such as histamine	Anaphylactic shock from drug injections and insect venom; common allergic conditions, such as hay fever, asthma
Type II (cytotoxic)	5–12 hours	Antigen causes formation of IgM and IgG antibodies that bind to target cell; when combined with action of complement, destroys target cell	Transfusion reactions, Rh incompatibility
Type III (immune complex)	3–8 hours	Antibodies and antigens form complexes that cause damaging inflammation	Arthus reactions, serum sickness
Type IV (delayed cell-mediated, or delayed hypersensitivity)	24–48 hours	Antigens activate T_C that kill target cells.	Rejection of transplanted tissues; contact dermatitis, such as poison ivy; certain chronic diseases, such as tuberculosis

HYPERSENSITIVITY

LEARNING OBJECTIVE

• Define *hypersensitivity.*

The term *hypersensitivity* refers to an antigenic response beyond that which is considered normal; the term *allergy* is more familiar and is essentially synonymous. Hypersensitivity responses occur in individuals who have been *sensitized* by previous exposure to an antigen, which in this context is sometimes called an **allergen.** When an individual who was previously sensitized is exposed to that antigen again, his or her immune system reacts to it in a damaging manner. The four principal types of hypersensitivity reactions, summarized in Table 19.1, are anaphylactic, cytotoxic, immune complex, and cell-mediated (or delayed-type) reactions.

TYPE I (ANAPHYLACTIC) REACTIONS

LEARNING OBJECTIVES

• Describe the mechanism of anaphylaxis.

• Compare and contrast systemic and localized anaphylaxis.

• Explain how allergy skin tests work.

• Define *desensitization* and *blocking antibody.*

Type I, or anaphylactic, reactions often occur within 2 to 30 minutes after a person sensitized to an antigen is re-exposed to that antigen. Anaphylaxis means "the opposite of protected," from the prefix *ana-*, meaning against, and the

Greek *phylaxis*, meaning protection. **Anaphylaxis** is an inclusive term for the reactions caused when certain antigens combine with IgE antibodies. Anaphylactic responses can be *systemic reactions*, which produce shock and breathing difficulties and are sometimes fatal, or *localized reactions*, which include common allergic conditions such as hay fever, asthma, and hives (slightly raised, often itchy and reddened areas of the skin).

The IgE antibodies produced in response to an antigen, such as insect venom or plant pollen, bind to the surfaces of cells such as mast cells and basophils. These two cell types are similar in morphology and in their contribution to allergic reactions. **Mast cells** are especially prevalent in the connective tissue of the skin and respiratory tract and in surrounding blood vessels. The name is from the German word *mastzellen*, meaning "well fed"; they are packed with granules that at one time were mistakenly thought to have been ingested (Figure 19.1a). **Basophils** circulate in the bloodstream, where they constitute fewer than 1% of the leukocytes. Both are filled with granules containing a variety of chemicals called *mediators*.

Mast cells and basophils can have as many as 500,000 sites for IgE attachment. The Fc (stem) region of an IgE antibody (see Figure 17.3, page 506) can attach to one of these specific receptor sites on such a cell, leaving two antigen-binding sites free. Of course, the attached IgE monomers will not all be specific for the same antigen. But when an antigen such as plant pollen encounters two adjacent antibodies of the same appropriate specificity, it can bind to one antigen-binding site on each antibody, bridging the space between

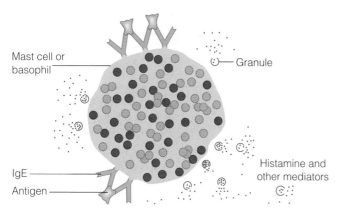

Mast cell or basophil

Granule

IgE

Antigen

Histamine and other mediators

(a) IgE antibodies, produced in response to an antigen, coat mast cells and basophils. When an antigen bridges the gap between two adjacent antibody molecules of the same specificity, the cell undergoes degranulation and releases histamine and other mediators.

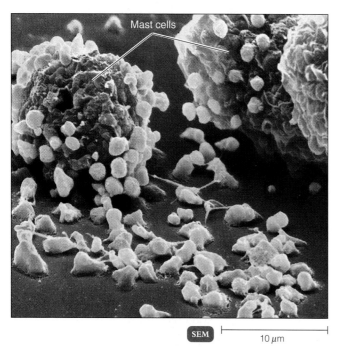

Mast cells

SEM
10 μm

(b) A degranulated mast cell that has reacted with an antigen and released granules of histamine and other reactive mediators.

FIGURE 19.1 The mechanism of anaphylaxis.

Q **What type of cells do IgE antibodies bind to?**

them. This bridge triggers the mast cell or basophil to undergo **degranulation,** which releases the granules inside these cells and also the mediators they contain (Figure 19.1b).

These mediators cause the unpleasant and damaging effects of an allergic reaction. The best-known mediator is **histamine.** The release of histamine increases the permeability and distension of blood capillaries, resulting in edema (swelling) and erythema (redness). Other effects include increased mucus secretion (a runny nose, for example) and smooth muscle contraction, which in the respiratory bronchi results in breathing difficulty.

Other mediators include **leukotrienes** of various types and **prostaglandins.** These mediators are not preformed and stored in the granules but are synthesized by the antigen-triggered cell. Because leukotrienes tend to cause prolonged contractions of certain smooth muscles, their action contributes to the spasms of the bronchial tubes that occur during asthmatic attacks. Prostaglandins affect smooth muscles of the respiratory system and cause increased mucus secretion.

Collectively, all these mediators serve as chemotactic agents that, in a few hours, attract neutrophils and eosinophils to the site of the degranulated cell. They then activate various factors that cause inflammatory symptoms, such as distension of the capillaries, swelling, increased secretion of mucus, and involuntary contractions of smooth muscles.

SYSTEMIC ANAPHYLAXIS

At the turn of the twentieth century, two French biologists studied the responses of dogs to the venom of stinging jellyfish. Large doses of venom usually killed the dogs, but sometimes a few survived the injections. These surviving dogs were used for repeat experiments with the venom, and the results were surprising. Even a very tiny dose of the venom, one that should have been almost harmless, killed the dogs. They suffered difficulty in respiration, entered shock as their cardiovascular systems collapsed, and quickly died. This phenomenon was called *anaphylactic shock.*

Systemic anaphylaxis (or *anaphylactic shock*) can result when an individual sensitized to an antigen is exposed to it again. Injected antigens are more likely to cause a dramatic response than antigens introduced via other portals of entry. The release of mediators causes peripheral blood vessels throughout the body to enlarge, resulting in a drop in blood pressure (shock). This reaction can be fatal within a few minutes. There is very little time to act once someone develops systemic anaphylaxis. Treatment usually involves self-administration with a preloaded syringe of epinephrine, a drug that constricts blood vessels and raises the blood pressure. In the United States, 50 to 60 people die each year from anaphylactic shock caused by insect stings.

You may be acquainted with someone who reacts to penicillin in this way. In these individuals, the penicillin, which is a hapten (it cannot induce antibody formation by itself; see Chapter 17), combines with a carrier serum protein. Only then is penicillin immunogenic. Penicillin allergy probably occurs in about 2% of the population. Skin tests for penicillin sensitivity are available. Patients who have a positive skin test can be desensitized (see page 554) by an orally administered series of increasing doses of

(a) A micrograph of pollen grains   SEM ⊢———⊣ 10 μm

(b) A micrograph of a house dust mite   SEM ⊢———⊣ 500 μm

FIGURE 19.2   Localized anaphylaxis. Inhaled antigens such as these are a common cause of localized anaphylaxis.

Q   **Compare localized and systemic anaphylaxis.**

penicillin V. The interval between doses is only 15 minutes and is completed within 4 hours. The desensitization is valid only for an uninterrupted penicillin series immediately following the procedure. Allergy to penicillin also includes risk from exposure to some related drugs, such as carbapenem (page 592).

LOCALIZED ANAPHYLAXIS

Whereas sensitization to injected antigens is a common cause of systemic anaphylaxis, **localized anaphylaxis** is usually associated with antigens that are ingested (foods) or inhaled (pollen)(Figure 19.2a). The symptoms depend primarily on the route by which the antigen enters the body.

In allergies involving the upper respiratory system, such as hay fever (allergic rhinitis), sensitization usually involves mast cells in the mucous membranes of the upper respiratory tract. Reexposure to the airborne antigen, which might be a common environmental material such as plant pollen, fungal spores, feces of house dust mites (Figure 19.2b), or animal dander.* The typical symptoms are itchy and teary eyes, congested nasal passages, coughing, and sneezing. Antihistamine drugs, which compete for histamine receptor sites, are often used to treat these symptoms.

Asthma is an allergic reaction that mainly affects the lower respiratory system. Symptoms such as wheezing and

shortness of breath are caused by the constriction of smooth muscles in the bronchial tubes.

For unknown reasons, asthma is becoming a near epidemic, affecting about 10% of children in Western society, although they often outgrow it in time. It is speculated that lack of childhood exposure in the developed world to many childhood infections, the so-called *hygiene hypothesis*, is a factor in the increase in the incidence of asthma. Mental or emotional stress can also be a contributing factor in precipitating an attack. Symptoms of asthma are usually controlled by aerosol inhalants, which, unfortunately, may be difficult for very small children to use. Xolair (omalizumab) is a newly available drug but is a very expensive treatment for severe allergic asthma. It blocks IgE.

Antigens that enter the body via the gastrointestinal tract can also sensitize an individual. Many of us may know someone who is allergic to a particular food. Frequently, so-called food allergies may not be related to hypersensitivity at all and are more accurately described as *food intolerances*. For example, many people are unable to digest the lactose in milk because they lack the enzyme that breaks down this disaccharide milk sugar. The lactose enters the intestine, where it osmotically retains fluid, causing diarrhea.

Gastrointestinal upset is a common symptom of food allergies, but it can also result from many other factors. Hives are more characteristic of a true food allergy, and ingestion of the antigen may result in systemic anaphylaxis. Death has even resulted when a person sensitive to fish ate french fries that had been prepared in oil previously used to fry fish. Skin tests are not reliable indicators for the diagnosis of food-related allergies, and completely controlled

---

Dander is a general term for microscopic particles from the fur or skin of animals. A cat, for example, carries about 100 mg of dander on its coat and sheds about 0.1 mg a day. This accumulates in upholstery and carpeting. People with allergies to mice, gerbils, and similar small animals are more likely to be allergic to components of urine accumulating in cages.

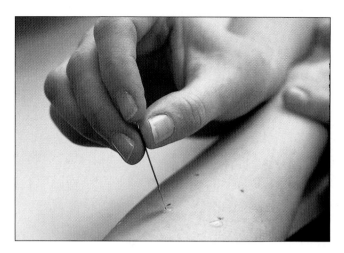

FIGURE 19.3 A skin test to identify allergens. Drops of fluid containing test substances are placed on the skin. A light scratch is made with a needle to allow the substances to penetrate the skin. Reddening and swelling at the site identify the substance as a probable cause of an allergic reaction.

Q **What is scratched into the skin in a skin test?**

tests for hypersensitivity to ingested foods are very difficult to perform. Only eight foods are responsible for 97% of food-related allergies: eggs, peanuts, tree-grown nuts, milk, soy, fish, wheat, and peas. Most children exhibiting allergies to milk, egg, wheat, and soy develop tolerance as they age, but reactions to peanuts, tree nuts, and seafood tend to persist.

An estimated 1.5 million Americans are allergic to peanuts and as many as 100 deaths occur annually. Therefore, considerable research is underway on the problem, ranging from drugs and vaccines to developing less allergenic peanuts. It is interesting that China has a relatively low incidence of peanut allergy, although peanuts are common in Chinese foods. This may be because Chinese cookery involves boiling and lower-temperature frying, whereas dry roasting of peanuts concentrates allergic properties. Children who have a relatively low level of peanut-specific IgE will sometimes outgrow their peanut allergy.

Sulfites, to which many people have an allergy, are a frequent problem. Their use is widespread in foods and beverages; and, although food labels should indicate their presence, these foods may be difficult to avoid in practice. A food product may have come into contact with a food allergen through processing machinery or cookware previously used for other foods. In one U.S. Food and Drug Administration report, 25% of bakery, ice cream, and candy products tested positive for peanut allergens even though peanuts were not listed on the required product labels. In the United States it is estimated that 200 persons a year die of severe allergic reactions to foods.

PREVENTION OF ANAPHYLACTIC REACTIONS

Avoiding contact with the sensitizing antigen is the most obvious way to prevent allergic reactions. Unfortunately, avoidance is not always possible. Some individuals experience an allergic reaction after eating an assortment of foods. In such cases, they may not know exactly what antigen they are sensitive to. In some cases, skin tests might be of use in diagnosis (Figure 19.3). These tests involve inoculating small amounts of the suspected antigen just beneath the epidermis of the skin. Sensitivity to the antigen is indicated by a rapid inflammatory reaction that produces redness, swelling, and itching at the inoculation site. This small affected area is called a *wheal*.

Once the responsible antigen has been identified, the person can either try to avoid contact with it or undergo **desensitization.** This procedure usually consists of a series of gradually increasing dosages of the antigen carefully injected beneath the skin. The objective is to cause the production of IgG rather than IgE antibodies in the hope that the circulating IgG antibodies will act as *blocking antibodies* to intercept and neutralize the antigens before they can react with cell-bound IgE. Desensitization is not a routinely successful procedure, but it is effective in 65 to 75% of individuals whose allergies are induced by inhaled antigens and in a reported 97% for insect venom.

TYPE II (CYTOTOXIC) REACTIONS

LEARNING OBJECTIVES

- Describe the mechanism of cytotoxic reactions and how drugs can induce them.
- Describe the basis of the ABO and Rh blood group systems.
- Explain the relationship between blood groups, blood transfusions, and hemolytic disease of the newborn.

Type II (cytotoxic) reactions generally involve the activation of complement by the combination of IgG or IgM antibodies with an antigenic cell. This activation stimulates complement to lyse the affected cell, which might be either a foreign cell or a host cell that carries a foreign antigenic determinant (such as a drug) on its surface. Additional cellular damage may be caused within 5 to 8 hours by the action of macrophages and other cells that attack antibody-coated cells.

The most familiar cytotoxic hypersensitivity reactions are *transfusion reactions*, in which red blood cells are destroyed as a result of reacting with circulating antibodies. These involve blood group systems that include the ABO and Rh antigens.

TABLE 19.2	The ABO Blood Group System						

Blood Group	Erythrocyte or Red Blood Cell Antigens	Illustration	Plasma Antibodies	Blood That Can Be Received	Frequency (% U.S. Population)		
					White	Black	Asian
AB	A and B	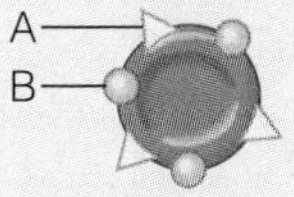	Neither anti-A nor anti-B antibodies	A, B, AB, O (Universal recipient)	3	4	5
B	B	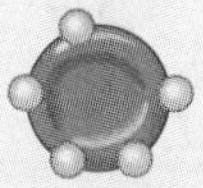	Anti-A	B, O	9	20	27
A	A	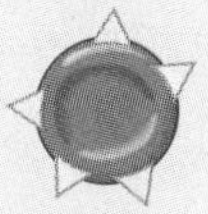	Anti-B	A, O	41	27	28
O	Neither A nor B	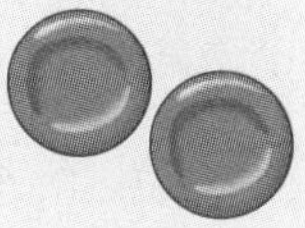	Anti-A and Anti-B	O (Universal donor)	47	49	40

THE ABO BLOOD GROUP SYSTEM

In 1901, Karl Landsteiner discovered that human blood could be grouped into four principal types, which were designated A, B, AB, and O. This method of classification is called the **ABO blood group system.** Since then, other blood group systems, such as the Lewis system and the MN system, have been discovered, but our discussion will be limited to two of the best known, the ABO and the Rh systems. The main features of the ABO blood group system are summarized in Table 19.2.

A person's ABO blood type depends on the presence or absence of carbohydrate antigens located on the cell membranes of red blood cells (RBCs). Cells of blood type O lack both A and B antigens. Table 19.2 shows that the plasma of individuals with a given blood type, such as A, have antibodies against the alternative blood type, anti-B antibody. These antibodies are presumed to arise in response to microorganisms and ingested foodstuffs that have antigenic determinants very similar to blood group antigens. Individuals with type AB cells have plasma with no antibodies to either A or B antigens and therefore can receive cells with either A or B blood without a reaction (are *universal recipients*). Type O individuals have antibodies against both A and B antigens. Lacking antigens, their blood cells can be transfused without difficulty to others

(they are *universal donors*). (The small amounts of anti-A and anti-B antibodies contained in donated type O blood or blood plasma are quickly degraded by the recipient and cause no significant adverse reaction.)

When a transfusion is incompatible, as when type B blood is transfused into a person with type A blood, the antigens on the type B blood cells will react with anti-B antibodies in the recipient's serum. This antigen–antibody reaction activates complement, which in turn causes lysis of the donor's RBCs as they enter the recipient's system.

THE Rh BLOOD GROUP SYSTEM

In the 1930s, researchers discovered the presence of a different surface antigen on human red blood cells. Soon after they injected rabbits with RBCs from rhesus monkeys, the rabbit serum contained antibodies that were directed against the monkey blood cells but that would also agglutinate some human RBCs. This indicated that a common antigen was present on both human and monkey red blood cells. The antigen was named the **Rh factor** (Rh for rhesus monkey). The roughly 85% of the population whose cells possess this antigen are called Rh^+; those lacking this RBC antigen (about 15%) are Rh^-. Antibodies that react with the Rh antigen do not occur naturally in the serum of Rh^- individuals, but exposure to this antigen

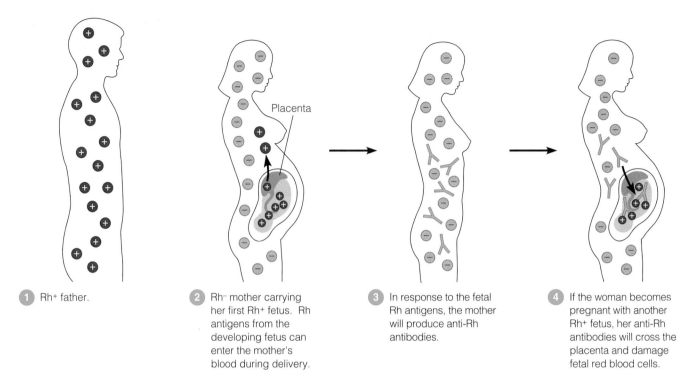

① Rh+ father.

② Rh⁻ mother carrying her first Rh+ fetus. Rh antigens from the developing fetus can enter the mother's blood during delivery.

③ In response to the fetal Rh antigens, the mother will produce anti-Rh antibodies.

④ If the woman becomes pregnant with another Rh+ fetus, her anti-Rh antibodies will cross the placenta and damage fetal red blood cells.

FIGURE 19.4 Hemolytic disease of the newborn.

Q **What type of antibodies cross the placenta?**

can sensitize their immune systems to produce anti-Rh antibodies.

Blood Transfusions and Rh Incompatibility If blood from an Rh⁺ donor is given to an Rh⁻ recipient, the donor's RBCs stimulate the production of anti-Rh antibodies in the recipient. If the recipient then receives Rh⁺ RBCs in a subsequent transfusion, a rapid, serious hemolytic reaction will develop.

Hemolytic Disease of the Newborn Blood transfusions are not the only way in which an Rh⁻ person can become sensitized to Rh⁺ blood. When an Rh⁻ woman and an Rh⁺ man produce a child, there is a 50% chance that the child will be Rh⁺ (Figure 19.4). If the child is Rh⁺, the Rh⁻ mother can become sensitized to this antigen during birth when the placental membranes tear and the fetal Rh⁺ RBCs enter the maternal circulation, causing the mother's body to produce anti-Rh antibodies of the IgG type. If the fetus in a subsequent pregnancy is Rh⁺, her anti-Rh antibodies will cross the placenta and destroy the fetal RBCs. The fetal body responds to this immune attack by producing large numbers of immature RBCs called erythroblasts. Thus, the term *erythroblastosis fetalis* was once used to describe what is now called **hemolytic disease of the newborn (HDNB).** Before the birth of a fetus with this condition, the maternal circulation removes most of the toxic by-

products of fetal RBC disintegration. After birth, however, the fetal blood is no longer purified by the mother, and the newborn develops jaundice and severe anemia.

HDNB is usually prevented today by passive immunization of the Rh⁻ mother at the time of delivery of any Rh⁺ infant with anti-Rh antibodies, which are available commercially (RhoGAM). These anti-Rh antibodies combine with any fetal Rh⁺ RBCs that have entered the mother's circulation, so it is much less likely that she will become sensitized to the Rh antigen. If the disease is not prevented, the newborn's Rh⁺ blood, contaminated with maternal antibodies, may have to be replaced by transfusion of uncontaminated blood.

DRUG-INDUCED CYTOTOXIC REACTIONS

Blood platelets (thrombocytes) are minute cell-like bodies that are destroyed by drug-induced cytotoxic reactions in the disease called **thrombocytopenic purpura.** The drug molecules are usually haptens because they are too small to be antigenic by themselves; but, in the situation illustrated in Figure 19.5, a platelet has become coated with molecules of a drug (quinine is a familiar example), and the combination is antigenic. Both antibody and complement are needed for lysis of the platelet. Because platelets are necessary for blood clotting, their loss results in hemorrhages that appear on the skin as purple spots (purpura).

Drugs may bind similarly to white or red blood cells, causing local hemorrhaging and yielding symptoms described as "blueberry muffin" skin mottling. Immune-caused destruction of granulocytic white cells is called **agranulocytosis,** and it affects the body's phagocytic defenses. When RBCs are destroyed in the same manner, the condition is termed **hemolytic anemia.**

TYPE III (IMMUNE COMPLEX) REACTIONS

LEARNING OBJECTIVE

• Describe the mechanism of immune complex reactions.

Type III reactions involve antibodies against soluble antigens circulating in the serum. (In contrast, type II immune reactions are directed against antigens located on cell or tissue surfaces.) The antigen–antibody complexes are deposited in organs and cause inflammatory damage.

Immune complexes form only when certain ratios of antigen and antibody occur. The antibodies involved are usually IgG. A significant excess of antibody leads to the formation of complement-fixing complexes that are rapidly removed from the body by phagocytosis. When there is a significant excess of antigen, soluble complexes form that do not fix complement and do not cause inflammation. However, when a certain antigen–antibody ratio exists, usually with a slight excess of antigen, the soluble complexes that form are small and escape phagocytosis.

Figure 19.6 illustrates the consequences. These complexes circulate in the blood, pass between endothelial cells of the blood vessels, and become trapped in the basement membrane beneath the cells. In this location, they may activate complement and cause a transient inflammatory reaction: attracting neutrophils that release enzymes. Repeated introduction of the same antigen can lead to more serious inflammatory reactions, causing damage to the basement membrane's endothelial cells within 2 to 8 hours.

Glomerulonephritis is an immune complex condition, usually resulting from an infection, that causes inflammatory damage to the kidney glomeruli, which are sites of blood filtration.

TYPE IV (DELAYED CELL-MEDIATED) REACTIONS

LEARNING OBJECTIVE

• Describe the mechanism of delayed cell-mediated reactions, and name two examples.

Up to this point we have discussed humoral immune responses involving IgE, IgG, or IgM. Type IV reactions involve cell-mediated immune responses and are caused

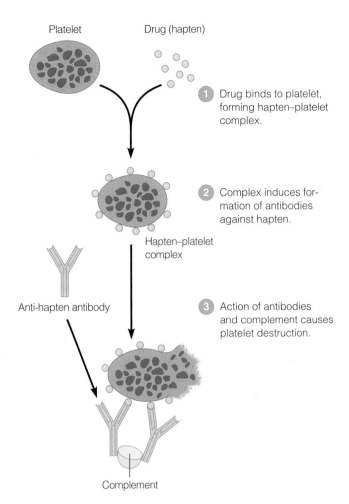

FIGURE 19.5 Drug-induced thrombocytopenic purpura. Molecules of a drug such as quinine accumulate on the surface of a platelet and stimulate an immune response that destroys the platelet.

Q **What actually destroys the platelets in thrombocytopenic purpura?**

mainly by T cells. Instead of occurring within a few minutes or hours after a sensitized individual is again exposed to an antigen, these **delayed cell-mediated reactions,** (or **delayed hypersensitivity**) are not apparent for a day or more. A major factor in the delay is the time required for the participating T cells and macrophages to migrate to and accumulate near the foreign antigens. Transplant rejection is most commonly mediated by CTLs (page 515), but other mechanisms are by ADCC (page 517) or complement-mediated lysis (page 490).

CAUSES OF DELAYED CELL-MEDIATED REACTIONS

Sensitization for delayed hypersensitivity reactions occurs when certain foreign antigens, particularly those that bind to tissue cells, are phagocytized by macrophages and then presented to receptors on the T-cell surface. Contact between the antigenic determinant sites and the appropriate

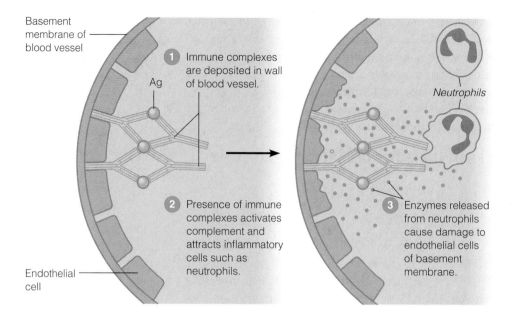

FIGURE 19.6 Immune complex–mediated hypersensitivity.

Q **Name one immune complex disease.**

Basement membrane of blood vessel

Ag

1 Immune complexes are deposited in wall of blood vessel.

2 Presence of immune complexes activates complement and attracts inflammatory cells such as neutrophils.

Endothelial cell

Neutrophils

3 Enzymes released from neutrophils cause damage to endothelial cells of basement membrane.

T cell causes the T cell to proliferate into mature differentiated T cells and memory cells.

When a person sensitized in this way is reexposed to the same antigen, a delayed hypersensitivity reaction might result. Memory cells from the initial exposure activate T cells, which release destructive cytokines in their interaction with the target antigen. In addition, some cytokines contribute to the inflammatory reaction to the foreign antigen by attracting macrophages to the site and activating them.

DELAYED CELL-MEDIATED HYPERSENSITIVITY REACTIONS OF THE SKIN

We have seen that hypersensitivity symptoms are frequently displayed on the skin. One delayed hypersensitivity reaction that involves the skin is the familiar skin test for tuberculosis. Because *Mycobacterium tuberculosis* is often located within macrophages, this organism can stimulate a delayed cell-mediated immune response. As a screening test, protein components of the bacteria are injected into the skin. If the recipient has (or has had) a previous infection by tuberculosis bacteria, an inflammatory reaction to the injection of these antigens will appear on the skin in 1 to 2 days (see Figure 24.11, page 722); this interval is typical of delayed hypersensitivity reactions.

Allergic contact dermatitis, another common manifestation of delayed cell-mediated hypersensitivity, is usually caused by haptens that combine with proteins (particularly the amino acid lysine) in the skin of some people to produce an immune response. Reactions to poison ivy (Figure 19.7), cosmetics, and the metals in jewelry (especially nickel) are familiar examples of these allergies.

The increasing exposure to latex in condoms, in certain catheters, and in gloves used by health care workers has led to a greater awareness of hypersensitivity to latex. Death from anaphylactic shock can also occur. During a recent 4-year period, 15 fatalities among patients were caused by exposure to latex tubing used for enemas or, during abdominal surgery, to the latex gloves of the surgeon. Many hospitals now even restrict entry of latex balloons.

Among physicians and nurses, 5 to 12% report this type of hypersensitivity to latex surgical gloves (Figure 19.8). Delayed hypersensitivity reactions are not to the latex itself but to chemicals used in its manufacture. Because of changes in production methods many current latex products by reputable manufacturers are free of these allergenic chemicals. Allergens tend to be absorbed by the cornstarch powder used in gloves to make them easier to put on. These powders are distributed as aerosols and inhaled. Powder-free gloves minimize this source of sensitization. Non-latex gloves, made of vinyl, have been reported to not provide adequate protection when handling HIV and hepatitis viruses. Many persons who develop an allergy to latex for some reason also have evidence of allergies to certain fruits, most commonly avocado, chestnut, banana, and kiwi. Latex paint, however, does not pose a threat of hypersensitivity reactions. Despite its name, latex paint contains no natural latex, but only synthetic nonallergenic chemical polymers.

The identity of the environmental factor causing the dermatitis can usually be determined by a *patch test*. Samples of suspected materials are taped to the skin; after 48 hours, the area is examined for inflammation.

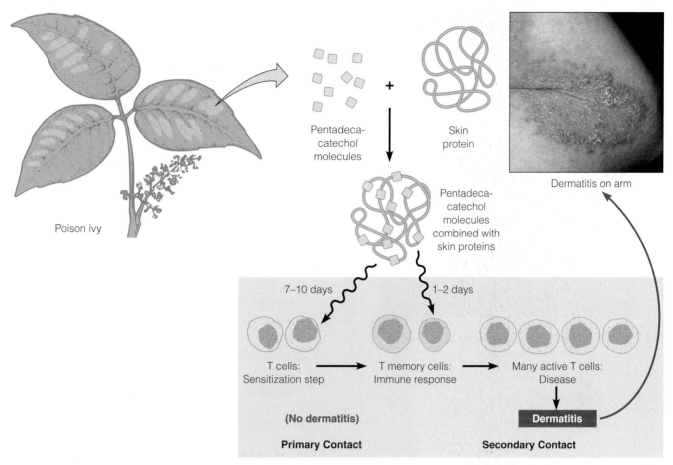

Dermatitis on arm

7–10 days

1–2 days

T cells:
Sensitization step

T memory cells:
Immune response

Many active T cells:
Disease

Dermatitis

(No dermatitis)

Primary Contact

Secondary Contact

Pentadeca-
catechol
molecules

Skin
protein

Pentadeca-
catechol
molecules
combined with
skin proteins

Poison ivy

FIGURE 19.7   The development of an allergy (allergic contact dermatitis) to catechols from the poison ivy plant. Pentadecacatechol is a mixture of catechols, which are oils secreted by the plant that dissolve easily in skin oils and penetrate the skin. In the skin, the catechols function as haptens—that is, they combine with skin proteins to become antigenic and provoke an immune response. The first contact with poison ivy sensitizes the susceptible person and subsequent exposure results in contact dermatitis.

 How does a hapten cause an allergic reaction?

AUTOIMMUNE DISEASES

LEARNING OBJECTIVES

• Describe a mechanism for self-tolerance.

• Give an example of immune complex, cytotoxic, and cell-mediated autoimmune diseases.

When the action of the immune system is in response to self-antigens and causes damage to one's own organs, the result is an **autoimmune disease.** Autoimmune diseases are relatively rare, but overall they affect about 5% of the population in the developed world. One prominent immunologist has observed, "We are all constantly teetering on the brink of autoimmune disease." About 75% of the cases of autoimmune disease selectively affect women.

Treatments for autoimmune diseases are improving as knowledge of the mechanisms controlling immune reactions improves.

Autoimmune diseases occur when there is a loss of **self-tolerance,** the immune system's ability to discriminate self from nonself. In the generally accepted model by which T cells become capable of distinguishing self from nonself, the cells acquire this ability during their passage through the thymus. As we saw in Chapter 17 (page 512), any T cells that will target host cells are either eliminated by *clonal deletion* or inactivated during this period. This requirement makes it unlikely that the T cell will attack its own tissue cells.

In autoimmune diseases, the loss of self-tolerance leads to the production of antibodies or a response by sensitized

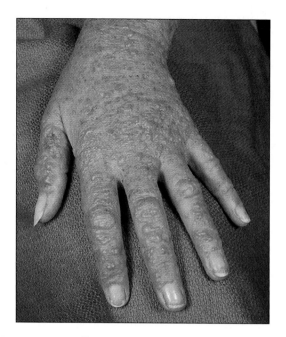

FIGURE 19.8 Allergic contact dermatitis. This person's hand exhibits a severe case of delayed contact dermatitis from wearing latex surgical gloves.

Q What is allergic contact dermatitis?

T cells against a person's own tissue antigens. Autoimmune reactions, and the diseases they cause, can be cytotoxic, immune complex, or cell-mediated in nature.

Autoimmunity involves antibodies that attack self. These antibodies may be made in response to an infectious agent such as a virus, but sequence similarities between viral and self-proteins may result in the antibodies attacking self cells. Hepatitis C virus may be responsible for autoimmune hepatitis by this mechanism.

CYTOTOXIC AUTOIMMUNE REACTIONS

Graves' disease and myasthenia gravis are two examples of disorders caused by cytotoxic autoimmune reactions. Both diseases involve antibody reactions to cell-surface antigens, although there is no cytotoxic destruction of the cells.

Graves' disease is caused by antibodies called long-acting thyroid stimulators. These antibodies attach to receptors on thyroid gland cells that are the normal target cells of the thyroid-stimulating hormone produced by the pituitary gland. The result is that the thyroid gland is stimulated to produce increased amounts of thyroid hormones and becomes greatly enlarged. The most striking signs of the disease are goiter, a disfiguring swelling of the thyroid gland, and markedly bulging, staring eyes.

Myasthenia gravis is a disease in which muscles become progressively weaker. It is caused by antibodies that coat the acetylcholine receptors at the junctions at which

nerve impulses reach the muscles. Eventually, the muscles controlling the diaphragm and the rib cage may fail to receive the necessary nerve signals, and respiratory arrest and death result.

IMMUNE COMPLEX AUTOIMMUNE REACTIONS

Systemic lupus erythematosus is a systemic autoimmune disease, involving immune complex reactions, that mainly affects women. The etiology of the disease is not completely understood, but afflicted individuals produce antibodies directed at components of their own cells, including DNA, which is probably released during the normal breakdown of tissues, especially the skin. The most damaging effects of the disease result from deposits of immune complexes in the kidney glomeruli.

Crippling **rheumatoid arthritis** is a disease in which immune complexes of IgM, IgG, and complement are deposited in the joints. In fact, immune complexes called *rheumatoid factors* may be formed by IgM binding to the Fc region of normal IgG. These factors are found in 70% of individuals suffering from rheumatoid arthritis. The chronic inflammation caused by this deposition eventually leads to severe damage to the cartilage and bone of the joint.

CELL-MEDIATED AUTOIMMUNE REACTIONS

Multiple sclerosis is one of the more common autoimmune diseases, affecting mostly younger adults. Most individuals with multiple sclerosis are whites living in northern latitudes; females are twice as likely to have the disease. It is a neurological disease in which T cells and macrophages attack the myelin sheath of nerves. Symptoms range from only fatigue and weakness to, in some cases, eventual severe paralysis. Progression of the disease is slow, measured over many years. New attacks that worsen the condition are often separated by long periods of remission. There is considerable evidence of genetic susceptibility, probably not from a single gene, but from several genes that interact. The etiology of multiple sclerosis is unknown, but epidemiological evidence indicates that it probably involves some infective agent or agents acquired during early adolescence. The Epstein-Barr virus (page 689) is frequently mentioned as a prime suspect. No cure for the condition exists, but treatments with interferons and several drugs that interfere with immune processes can significantly slow progression of symptoms.

Hashimoto's thyroiditis is a result of the destruction of the thyroid gland, primarily by T cells of the cell-mediated immune system. It is a fairly common disorder and is often found in related family members. **Insulin-dependent diabetes mellitus** is a familiar condition caused by

TABLE 19.3	Diseases Related to Specific HLAs	

Disease	Increased Risk of Occurrence with Specific HLA*	Description
Inflammatory Diseases		
Multiple sclerosis	5 times	Progressive inflammatory disease affecting nervous system
Rheumatic fever	4–5 times	Cross-reaction with antibodies against streptococcal antigen
Endocrine Diseases		
Addison's disease	4–10 times	Deficiency in production of hormones by adrenal gland
Graves' disease	10–12 times	Antibodies attached to certain receptors in the thyroid gland cause it to enlarge and produce excessive hormones
Malignant Disease		
Hodgkin's disease	1.4–1.8 times	Cancer of lymph nodes

*Compared to the general population.

immunological destruction of insulin-secreting cells of the pancreas. T cells are clearly implicated in this disease; animals that are genetically likely to develop diabetes fail to do so when their thymus is removed in infancy.

REACTIONS RELATED TO THE HUMAN LEUKOCYTE ANTIGEN (HLA) COMPLEX

LEARNING OBJECTIVE

- Define *HLA complex,* and explain its importance in disease susceptibility and tissue transplants.

The inherited genetic characteristics of individuals are expressed not only in the color of their eyes and the curl of their hair but also in the composition of the self molecules on their cell surfaces. Some of these are called **histocompatibility antigens.** The genes controlling the production of the most important of these self molecules are known as the **major histocompatibility complex (MHC).** In humans, these genes are called the **human leukocyte antigen (HLA) complex.** We encountered these self molecules in Chapter 17 (page 509), where we saw that most antigens can stimulate an immune reaction only if they are associated with an MHC molecule.

A process called *HLA typing* is used to identify and compare HLAs. Certain HLAs are related to an increased susceptibility to specific diseases; one medical application of HLA typing is to identify such susceptibility. A few of these relationships are summarized in Table 19.3.

Another important medical application of HLA typing is in transplant surgery, in which the donor and the recipient must be matched by *tissue typing.* The serological technique shown in Figure 19.9 is the one most often used. In serological tissue typing, the laboratory uses standardized antisera or monoclonal antibodies that are specific for particular HLAs.

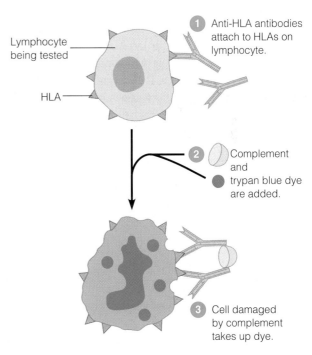

FIGURE 19.9 Tissue typing, a serological method. Lymphocytes from the person being tested are incubated with laboratory test stocks of anti-HLA antibodies specific for a particular HLA. If the antibodies react with the antigens on a lymphocyte, then complement damages the lymphocyte and dye can enter the cell. Such a positive test result indicates that the person has the particular HLA being tested for.

Q **Why is tissue typing done?**

A promising new technique for analyzing HLA is the use of *PCR*, the *polymerase chain reaction*, to amplify the cell's DNA (see Figure 9.4, page 259). If this is done for both donor and recipient, a match between donor DNA and recipient DNA can then be made. Having such a DNA match and matching ABO blood type between the donor and the recipient should result in a much higher success rate in transplant surgery.

Other factors may be involved in the success of a transplant, however. In Chapter 17, we briefly introduced a hypothesis that the body's reaction to transplanted foreign tissue may be a response to surgery-damaged cells. In other words, tissue rejection may result from a learned reaction to the danger signal posed by damaged cells, rather than a learned reaction to nonself.

REACTIONS TO TRANSPLANTATION

> **LEARNING OBJECTIVES**
> * Explain how the rejection of a transplant occurs.
> * Define *privileged site*.
> * Define *autograft*, *isograft*, *allograft*, and *xenotransplant*.
> * Explain how graft-versus-host disease occurs.

In sixteenth-century Italy, crimes were often punished by cutting off the offender's nose. A surgeon of the time, in his attempts to repair this mutilation, observed that if skin was taken from the patient, it healed properly, but if it was taken from another person, it did not. He called this a manifestation of "the force and power of individuality."

We now know the principles behind this phenomenon. Transplants recognized as nonself are rejected—attacked by T cells that directly lyse the grafted cells, by macrophages activated by T cells, and, in certain cases, by antibodies, which activate the complement system and injure blood vessels supplying the transplanted tissue. However, transplants that are not rejected can add many healthy years to a person's life.

Since the first kidney transplant was performed in 1954, this particular type of transplant has become a nearly routine medical procedure. Other types of transplants that are now feasible include bone marrow, lungs, heart, liver, and cornea. Tissues and organs for transplant are usually taken from recently deceased individuals, although one of a pair of organs, such as a kidney, occasionally comes from a living donor.

PRIVILEGED SITES AND PRIVILEGED TISSUE

Some transplants or grafts do not stimulate an immune response. A transplanted cornea, for example, is rarely rejected, mainly because antibodies usually do not circulate into that portion of the eye, which is therefore considered an immunologically **privileged site.** (However, rejections do occur, especially when the cornea has developed many blood vessels from corneal infections or damage.) The brain is also an immunologically privileged site, probably because it does not have lymphatic vessels and because the walls of the blood vessels in the brain differ from blood vessel walls elsewhere in the body (the blood–brain barrier is discussed in Chapter 22). Someday it may even be possible to graft foreign nerves to replace damaged nerves in the brain and spinal cord. Encouraging results have been obtained with these types of grafts in experiments with rats.

It is possible to transplant **privileged tissue** that does not stimulate an immune rejection. An example is replacing a person's damaged heart valve with a valve from a pig's heart. However, privileged sites and tissues are more the exception than the rule.

How animals tolerate pregnancy without rejecting the fetus is only partially understood. The uterus is not a privileged site; yet during pregnancy, the tissues of two genetically different individuals are in direct contact.

STEM CELLS

A development that promises to transform transplantation medicine is the use of **stem cells.** If a stem cell is capable of generating many different types of tissue cells they are called *pluripotent*. A pluripotent cell cannot, however, make all the cells required to make a complete fetus. Examples of pluripotent stem cells are the **embryonic stem (ES) cells** that can be isolated from embryos—usually from very early stage (less than a week after fertilization) embryos created for use for in vitro fertilization attempts. Most ES cell lines now used are derived from this source. Such ES cells can be made to generate many types of tissue cells: nerve cells, muscle cells, liver cells, and the very important hematopoietic stem cells (HSCs) that can form several different kinds of blood and lymphatic cells.

In the scientific and medical community there is great interest in using ES cells in therapy. Theoretically, HSCs could be used to correct genetic immunodeficiency diseases that are responsible for defective immune systems. They might be used to treat blood disorders such as leukemia and to restore the normal blood-making apparatus if it is destroyed by chemical or radiation treatments in attempts to treat cancer. Presently, bone marrow transplantation is used in such attempts to restore the hematopoietic system—essentially, this is a form of stem cell transplantation. Sources of HSCs are umbilical cords (*cord blood*), which are normally discarded, and peripheral blood. These sources are less invasive than bone marrow extraction. It might also be possible to use embryonic stem cells of a host to grow new body parts rather than resorting to transplantation. An example of this might be insulin-producing pancreatic tissue for diabetics and even damaged hearts and other organs.

GRAFTS

When one's own tissue is grafted to another part of the body, as is done in burn treatment or in plastic surgery, the graft is not rejected. Recent technology has made it possible to use a few cells of a burn patient's uninjured skin to culture extensive sheets of new skin. This new skin is an example of an **autograft.** Identical twins have the same genetic makeup; therefore, skin or organs such as kidneys may be transplanted between them without provoking an immune response. Such a transplant is called an **isograft.**

Most transplants, however, are made between people who are not identical twins, and these transplants do trigger an immune response. Attempts are made to match the HLAs of the donor and recipient as closely as possible so that the chances of rejection are reduced. Because HLAs of close relatives are most likely to match, blood relatives, especially siblings, are the preferred donors. Grafts between people who are not identical twins are called **allografts.**

Because of the shortage of available organs, medical researchers hope to increase the success of **xenotransplantation products** (formerly called **xenografts**), which are tissues or organs that have been transplanted from animals. However, the body tends to mount an especially severe immune assault on such transplants. Unsatisfactory attempts have been made to use organs from baboons and other nonhuman primates. Research interest is high in genetically modifying pigs—an animal that is in plentiful supply, is of the right size, and generates relatively little public sympathy—to make acceptable donors of organs. The primary concern about xenotransplantation products is the possibility of transferring harmful animal viruses.

Preliminary research is underway that may eventually allow some bones and organs to be grown from the host's own tissue cells.

In order to be successful, xenotransplantation products must overcome **hyperacute rejection,** caused by the development in early infancy of antibodies against all distantly related animals such as pigs. With the aid of complement, these antibodies attack the transplanted animal tissue and destroy it within an hour. Pigs are now being bred in which the cell-surface antigens that cause hyperacute rejection have been eliminated. These "knockout pigs" are considered an important step toward xenotransplantation. Hyperacute rejection occurs in human-to-

human transplants only when antibodies have been preformed because of previous transfusions, transplantations, or pregnancies.

BONE MARROW TRANSPLANTS

Transplants of bone marrow are frequently in the news. The recipients are usually individuals who lack the capacity to produce B cells and T cells vital for immunity or who are suffering from leukemia. Recall from Chapter 17 that bone marrow stem cells give rise to red blood cells and immune system lymphocytes. The goal of bone marrow transplants is to enable the recipient to produce such vital cells. However, such transplants can result in **graft-versus-host (GVH) disease.** The transplanted bone marrow contains immunocompetent cells that mount primarily a cell-mediated immune response against the tissue into which they have been transplanted. Because the recipients lack effective immunity, GVH disease is a serious complication and can even be fatal.

An extremely promising technique for avoiding this problem is the use of *umbilical cord blood* instead of bone marrow. This blood is harvested from the placenta and umbilical cords of newborns, material that would otherwise be discarded. It is very rich in the stem cells (Figure 17.8, page 513) found in bone marrow. Not only do these cells proliferate into the variety of cells required by the recipient; but, because stem cells from this source are younger and less mature, the "matching" requirements are also less stringent than with bone marrow. As a result, GVH disease is less likely to occur.

IMMUNOSUPPRESSION

LEARNING OBJECTIVE
- Explain how rejection of a transplant is prevented.

To keep the problem of transplant rejection in perspective, it is useful to remember that the immune system is simply doing its job and has no way of recognizing that its attack against the transplant is not helpful. In an attempt to prevent rejection, the recipient of an allograft usually receives treatment to suppress this normal immune response against the graft.

In transplantation surgery, it is generally desirable to suppress cell-mediated immunity, the most important factor in transplant rejection. If humoral (antibody-based) immunity is not suppressed, much of the ability to resist microbial infection will remain. In 1976, the drug *cyclosporine* was isolated from a mold. (Curiously, this fungus has a sexual cycle in a dung beetle that stimulates the beetle to climb high on vegetation and die, ensuring more efficient airborne distribution of fungal spores.) The successful transplantation of organs such as hearts and livers generally dates from the discovery of cyclosporine. The secretion of interleukin-2 (IL-2) is suppressed by cyclosporine, disrupting cell-mediated immunity by cytotoxic T cells. Following the success of this drug, other immunosuppressant drugs soon followed. *Tacrolimus* (FK506) has a mechanism similar to cyclosporine and is a frequent alternative, although both have many serious side effects. Neither cyclosporine nor tacrolimus has much effect on antibody production by the humoral immune system. Both of these drugs remain the mainstay for most regimens to prevent rejection of transplants. Some newer drugs, such as *sirolimus* (Rapamune), are among those that inhibit both cell-mediated and humoral immunity. This can be an advantage if chronic or hyperacute rejection by antibodies are a consideration. Sirolimus is best known for its use in stents (cylindrical meshes) designed to keep blood vessels open after removal of blockages. Drugs such as *mycophenolate mofetil* inhibit the proliferation of T cells and B cells. Some biological agents such as the chimeric monoclonal antibodies (page 534) *basiliximab* and *daclizumab* also block IL-2 and are useful immunosuppressives. Immunosuppressive agents are usually administered in combinations.

THE IMMUNE SYSTEM AND CANCER

LEARNING OBJECTIVE
• Describe the immune responses to cancer and how cells evade immune responses.

Like an infectious disease, cancer represents a failure of the body's defenses, including the immune system. Some of the most promising avenues for effective cancer therapy make use of immunological techniques.

Nearly 100 years ago it was recognized that cancer cells arise frequently in the body and that they are usually eliminated by the immune system much like any other invading cell—the concept of **immune surveillance.** It was postulated that the cell-mediated immune system probably arose to combat cancer cells and that the appearance of a cancerous growth represented a failure of the immune system. This concept has been supported by the observation that cancers occur most often in older adults, whose immune systems are becoming less efficient, or in the very young, whose immune systems may not have developed fully or properly. Also, individuals who are immunosuppressed by either natural or artificial means are more susceptible to certain cancers.

A cell becomes cancerous when it undergoes transformation and begins to proliferate without control (see Chapter 13, page 396). The surfaces of tumor cells acquire tumor-associated antigens that mark them as nonself to the immune system. Figure 19.10 illustrates the attack on such a cancer cell by activated T_C cells (CTLs). Activated macrophages can also destroy cancer cells. While a healthy immune system serves to prevent most cancers, it has limitations. In some cases there is no antigenic epitope for the immune system to target. Tumor cells can even reproduce so rapidly that they exceed the capacity of the immune system to deal with them. Finally, if the tumor cell begins reproducing in tissues and becomes vascularized (connected to the body's blood supply), it usually becomes invisible to the immune system.

IMMUNOTHERAPY FOR CANCER

LEARNING OBJECTIVE
• Give two examples of immunotherapy.

The hypothesis that cancer represents a failure of the immune system has led to the thought that the immune system might be used to prevent or cure cancer—that is, **immunotherapy.**

At the turn of the twentieth century, William B. Coley, a physician at a New York City hospital, observed that if cancer patients contracted typhoid fever, their cancers often diminished noticeably. Following this lead, Coley made mixtures of killed streptococci (gram-positive) and *Serratia marcescens* (gram-negative) bacteria. These so-called Coley's toxins were injected into cancer patients to simulate a bacterial infection. Some of this work was very promising, but its results were inconsistent and advances in surgery and radiation treatment caused it to be nearly forgotten. It is now recognized that endotoxins from such bacteria are powerful stimulants for the production of tumor necrosis factor (TNF) by macrophages. TNF is a small protein that interferes with the blood supply of cancers in animals.

Other research determined many years ago that if animals were injected with dead tumor cells, like a vaccine, they did not develop tumors when injected with live cells from these tumors. Similarly, cancers sometimes undergo spontaneous remission that is probably related to the immune system gaining the advantage. The treatment or prevention of cancer by immunological means is an approach that will probably be increasingly important. An attractive aspect of this approach is that it avoids the damage to healthy cells caused by chemotherapy and radiation treatments. Already, a vaccine for Marek's disease, a cancer of chickens, has been successful. Vaccines to protect cats

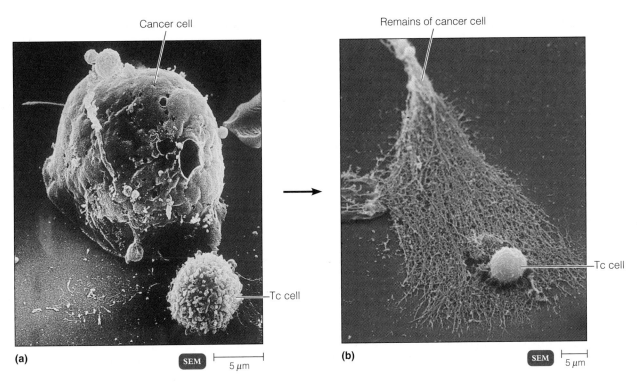

Cancer cell

Remains of cancer cell

Tc cell

Tc cell

(a) SEM 5 μm

(b) SEM 5 μm

FIGURE 19.10 The interaction between a cytotoxic T lymphocyte (CTL) and a cancer cell. (**a**) The small CTL has already made a perforation in the cancer cell. (**b**) The cancer cell has disintegrated.

Q **CTLs can lyse cancer cells. How do they do this?** (*Hint:* **See Figure 17.10.**)

from feline leukemia have also been shown to provide considerable protection.

Cancer vaccines might be either *therapeutic* (used to treat existing cancers) or *prophylactic* (to prevent the development of cancer). Actually, prophylactic vaccines already exist—indirectly. Hepatitis B virus is a common cause of liver cancer and a vaccine against infection by this virus is widely used.

Several therapeutic vaccines are being tested in human trials. There are two primary approaches to making a cancer vaccine: *whole-cell* vaccines and *antigen-type* vaccines. A whole-cell vaccine is prepared from cancer cells taken from a patient's tumor or from tumor cell lines in laboratory collections. These cells need to be altered to inactivate them for safety. Antigen-type vaccines attempt to trigger an immune response to antigens that are found on cancer cells. Such antigens might be from a specific cancer on an individual or to antigens found on several types of cancer.

Another approach for using the immune system against cancer is to try to stimulate the immune system in a more general way; for example, to mix dendritic cells (researchers have found ways to dramatically increase their numbers) with genetic material from a tumor. Hopefully, these dendritic cells will then efficiently present the

cancer antigens to cytotoxic cells of the immune system. Although work with potential cancer vaccines has been in progress for almost a century, it is an area that should be considered to be in only an early stage. Currently, a cancer vaccine would probably be most useful at preventing recurrences after other forms of treatment because the immune system would have to deal with a smaller number of cells.

Monoclonal antibodies are a promising tool for delivering cancer treatment. A humanized monoclonal antibody, *Herceptin* (see Chapter 18, page 536), is currently being used to treat a form of breast cancer. Herceptin specifically neutralizes a genetically determined growth factor, HER2, that promotes the proliferation of the cancer cells. It is expressed in relatively high quantities in about 25 to 30% of breast cancer patients.

Another approach that makes use of monoclonal antibodies is to combine a monoclonal antibody with a toxic agent, forming an **immunotoxin.** Theoretically, an immunotoxin might be used to specifically target and kill cells of a tumor with little damage to healthy cells. There have been some promising results in clinical trials but not with large tumor masses where many of the tumor cells cannot be reached by the immunotoxin.

IMMUNODEFICIENCIES

LEARNING OBJECTIVE

• Compare and contrast congenital and acquired immunodeficiencies.

The absence of a sufficient immune response is called an **immunodeficiency,** which can be either congenital or acquired.

CONGENITAL IMMUNODEFICIENCIES

Some people are born with a defective immune system. Defects in, or the absence of, a number of inherited genes can result in **congenital immunodeficiencies.** For example, individuals with a certain recessive trait, DiGeorge's syndrome, do not have a thymus gland and therefore lack cell-mediated immunity. An animal equivalent, which is extremely valuable for research in transplantation science, is the nude (hairless) mouse (Figure 19.11). These mice have no thymus (the coincidental hairlessness is controlled by the same gene) and therefore do not produce T cells and do not reject transplanted tissue. Even

Site of
M. leprae bacteria

FIGURE 19.11 A nude (hairless) mouse infected with *Mycobacterium leprae* in the hind foot. Nude mice have no thymus and therefore no cell-mediated immunity. The immune response to infection by *M. leprae* (leprosy pathogen) is dependent upon cell-mediated immunity, so these animals play an important role in leprosy research.

Q What is the role of the thymus gland in immunity?

chicken skin, complete with feathers, is readily accepted as a graft.

ACQUIRED IMMUNODEFICIENCIES

A variety of drugs, cancers, or infectious agents can result in **acquired immunodeficiencies.** For example, Hodgkin's disease (a type of cancer) lowers the cell-mediated response. Many viruses are capable of infecting and killing lymphocytes, lowering the immune response. Removal of the spleen decreases humoral immunity. Table 19.4 summarizes several of the better known immune deficiency conditions, including AIDS.

ACQUIRED IMMUNODEFICIENCY SYNDROME (AIDS)

In 1981, a cluster of cases of *Pneumocystis* pneumonia (see page 736) appeared in the Los Angeles area. This extremely rare disease usually only occurred in immunosuppressed individuals. Investigators soon correlated the appearance of this disease with an unusual incidence of a rare form of cancer of the skin and blood vessels called Kaposi's sarcoma. The people affected were all young homosexual men, and all showed a loss of immune function. By 1983, the pathogen causing the loss of immune function had been identified as a virus that selectively infects helper T cells. This virus is now known as human immunodeficiency virus (HIV) (see Figure 1.1e, page 5).

THE ORIGIN OF AIDS

LEARNING OBJECTIVE

• Give two examples of how emerging infectious diseases arise.

HIV is now believed to have arisen by the mutation of a virus that had been endemic in wildlife in some areas of central Africa. Genetic studies of the virus have led to the conclusion that HIV-2 (a type of HIV that is weakly contagious and not found often outside of West Africa) is a mutation of a simian immunodeficiency virus (SIV). Mangabey monkeys in West Africa are naturally and harmlessly infected with this SIV. More recently, studies show that HIV-1 (the primary HIV found worldwide in humans) is genetically related to another SIV that is carried by chimpanzees in Central Africa. The chimpanzee virus is probably a hybrid of two mangabey monkey SIVs. Chimpanzees are predators and eat other monkeys that presumably carried two versions of SIV.

TABLE 19.4	Immunodeficiencies	
Disease	Cells Affected	Comments
Acquired immunodeficiency syndrome (AIDS)	T cells (virus destroys T$_H$ [CD4] cells)	Allows cancer and bacterial, viral, fungal, and protozoan diseases; caused by HIV infection
Selective IgA immunodeficiency	B, T cells	Affects about 1 in 700, causing frequent mucosal infections; specific cause uncertain
Common variable hypogammaglobulinemia	B, T cells (decreased immunoglobulins)	Frequent viral and bacterial infections; second most common immune deficiency, affecting about 1 in 70,000; inherited
Reticular dysgenesis	B, T, and stem cells (a combined immunodeficiency; deficiencies in B and T cells and neutrophils)	Usually fatal in early infancy; very rare; inherited; bone marrow transplant a possible treatment
Severe combined immunodeficiency	B, T, and stem cells (deficiency of both B and T cells)	Affects about 1 in 100,000; allows severe infections; inherited; treated with bone marrow, fetal thymus transplants; gene therapy treatment is promising
Thymic aplasia (DiGeorge's syndrome)	T cells (defective thymus causes deficiency of T cells)	Absence of cell-mediated immunity; usually fatal in infancy from *Pneumocystis* pneumonia or viral or fungal infections; due to failure to develop in embryo
Wiskott-Aldrich syndrome	B, T cells (few platelets in blood, abnormal T cells)	Frequent infections by viruses, fungi, protozoa; eczema, defective blood clotting; usually causes death in childhood; inherited on X chromosome
X-linked infantile (Bruton's) agammaglobulinemia	B cells (decreased immunoglobulins)	Frequent extracellular bacterial infections; affects about 1 in 200,000; the first immunodeficiency disorder recognized (1952); inherited on X chromosome

These SIV infections apparently crossed over relatively recently (well into the twentieth century) into the human population, known to eat "bushmeat." Mathematical models of the supposed evolution of HIV, by Bette Korber of the Los Alamos National Laboratory, calculate that the virus probably made the transition to humans around 1930. The disease may have smoldered with little notice as long as transmission was limited to small villages where rates of sexual promiscuity were lower. The virus could not have killed or incapacitated its hosts quickly; otherwise, it could not have been maintained in the village population. With the sudden end of European colonialism, the social structure of sub-Saharan Africa was disrupted. The population became urbanized; the developments that result from urbanization, such as an increase in prostitution and the growth of highway transportation, which contributed to an increase in sexual promiscuity, are believed to be responsible for the spread of the disease. The earliest documented case of AIDS is from a patient in Leopoldville, Belgian Congo (now Kinshasa, capital of

the Democratic Republic of the Congo). This man died in 1959; preserved samples of his blood contain antibodies to HIV. In the Western world, the first confirmed case of AIDS was the death of a Norwegian sailor in 1976, who probably was infected in 1961 or 1962 by contacts in western Africa.

HIV INFECTION

LEARNING OBJECTIVES

* Explain the attachment of HIV to a host cell.
* List two ways in which HIV avoids the host's antibodies.
* Describe the stages of HIV infection.
* Describe the effects of HIV infection on the immune system.

One of the most common misconceptions is that HIV infection is synonymous with AIDS. AIDS denotes only the final stage of a long infection.

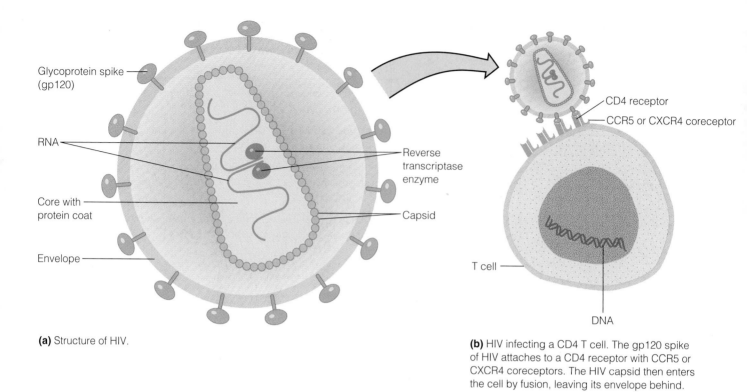

Glycoprotein spike
(gp120)

RNA

Core with
protein coat

Envelope

Reverse
transcriptase
enzyme

Capsid

CD4 receptor

CCR5 or CXCR4 coreceptor

T cell

DNA

(a) Structure of HIV.

(b) HIV infecting a CD4 T cell. The gp120 spike
of HIV attaches to a CD4 receptor with CCR5 or
CXCR4 coreceptors. The HIV capsid then enters
the cell by fusion, leaving its envelope behind.

FIGURE 19.12   HIV structure and attachment to receptors on target T cell.

Q   **Why does HIV preferentially infect CD4 cells?**

THE STRUCTURE OF HIV

HIV, of the genus *Lentivirus*, is a retrovirus (see Figure
13.19, page 409). It has two identical strands of RNA, the
enzyme reverse transcriptase, and an envelope of phospholipid (Figure 19.12a). The envelope has glycoprotein
spikes termed **gp120** (the notation for a glycoprotein with
a molecular weight of 120,000).

THE INFECTIVENESS AND PATHOGENICITY OF HIV

The spikes enable the virus to attach to the CD4 receptor
on the host cells (Figure 19.12b). CD4 receptors are
found on helper T cells, macrophages, and dendritic
cells—the main targets of HIV infection. A CD4 receptor
by itself is not sufficient for HIV infection. Certain coreceptors are also required. The two best known chemokine
coreceptors, originally called *fusins*, are named CCR5 and
CXCR4.*

Attachment of the virus is followed by entry into the
host cell (see Figure 19.12b). In the host cell, viral RNA is

released and transcribed into DNA by the enzyme reverse
transcriptase.

This viral DNA then becomes integrated into the
chromosomal DNA of the host cell. The DNA may control the production of an active infection in which new
viruses bud from the host cell, as shown in Figure 19.13b.

Alternatively, this integrated DNA may not produce
new HIV but remains hidden in the host cell's chromosome as a *provirus* (Figures 19.13a and 19.14a). HIV produced by a host cell is not necessarily released from the cell
but may remain as *latent virions* in vacuoles within the
cell (Figure 19.14b). In fact, a subset of the HIV-infected
cells, instead of being killed, become long-lived memory
T cells in which the reservoir of latent HIV can persist for
decades. This ability of the virus to remain as a provirus or
latent virus within host cells shelters it from the immune
system. Another way HIV evades the immune system is by
cell–cell fusion, by which the virus moves from an infected
cell to an adjacent uninfected cell.

The virus also evades immune defenses by undergoing
rapid antigenic changes. Retroviruses, with the reverse transcriptase enzyme step, have a high mutation rate compared
to DNA viruses. They also lack the corrective "proofreading" capacity of DNA viruses. As a result, a mutation is
probably introduced at every position in the HIV genome
many times each day in an infected person. This may

---

*This nomenclature is based upon the beginning amino acid sequence in
these proteins. The term CCR5 indicates that the beginning sequence consists of cysteines, thus CC. The letter R is a convention representing the balance of the protein molecule, and the number is for identification. If some
other amino acid is located between the first two cysteines, this is shown in
the naming, for example, of CXCR4.

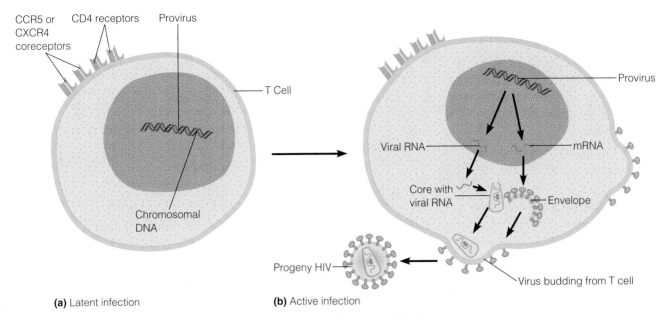

(a) Latent infection   **(b)** Active infection

FIGURE 19.13 Latent and active HIV infection in CD4 T cells. **(a)** Viral DNA is integrated into cellular DNA and forms a *provirus* (page 398) that can later be activated to form an infective virus. **(b)** The provirus is activated, which allows it to control the synthesis of new viruses, which bud from the host cell.

Q **What is a latent infection?**

amount to an accumulation of 1 million variants of the virus in an asymptomatic person and 100 million variants during the final stages of the infection. These dramatic numbers illustrate the potential problems of drug resistance and obstacles to the development of vaccines and diagnostic tests.

CLADES (SUBTYPES) OF HIV

Worldwide, the HIV genome is beginning to separate into distinctive groups called *clades* (Greek for branches), or subtypes. Viruses may vary by 15 to 20% within a clade; between clades, they may vary by 30% or more. Currently, *HIV-1*, the most common major type of HIV, has 11 such clades. In the United States, about 90% of the cases are caused by HIV-1, clade B. Generally, this clade is also the most common elsewhere in the developed Western world. In sub-Saharan Africa, the predominant clade is C; in the Far East and Asia, clade E is the most common.

A second major HIV type, *HIV-2*, is found primarily in western Africa and is rare in the United States. The progression from infection to AIDS is much longer with HIV-2. In fact, most people infected with HIV-2 have a life span that is normal for the area.

THE STAGES OF HIV INFECTION

The CDC classification divides the progress of HIV infection in adults into three clinical stages, or categories (Figure 19.15):

1. *Category A.* At this stage, the infection may be asymptomatic or cause persistent lymphadenopathy (swollen lymph nodes).

2. *Category B.* This stage is characterized by persistent infections by the yeast *Candida albicans*, which can appear in the mouth, throat, or vagina. Other conditions may include shingles, persistent diarrhea and fever, whitish patches on the oral mucosa (hairy leukoplakia), and certain cancerous or precancerous conditions of the cervix.

3. *Category C.* This stage is clinical AIDS. Important AIDS indicator conditions are *Candida albicans* infections of the esophagus, bronchi, and lungs; cytomegalovirus eye infections; tuberculosis; *Pneumocystis* pneumonia; toxoplasmosis of the brain; and Kaposi's sarcoma (caused by human herpesvirus 8).

The CDC also classifies the progress of HIV infections based on T cell populations. The purpose is primarily to furnish guidance for treatment, such as when to administer certain drugs. The normal population of a healthy individual is 800 to 1000 CD4 T cells/mm^3. In the United States, a count below 200/mm^3 is considered diagnostic for AIDS, regardless of the clinical category observed.

The progression from initial HIV infection to AIDS typically takes about 10 years in adults. This figure is typical

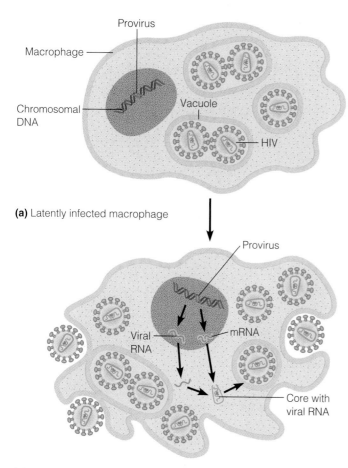

(a) Latently infected macrophage

(b) Activated macrophage

FIGURE 19.14 Latent and active HIV infection in macrophages. (**a**) HIV can persist either as a provirus or as a complete virion in vacuoles. (**b**) New viruses are produced from provirus. Completed virions are either released or persist in the macrophage within vacuoles.

Q **How does an active infection differ from a latent infection?**

in industrialized countries; in Africa, it is often about half this. Cellular warfare on an immense scale occurs during this time. At least 100 billion HIVs are generated every day, each with a remarkably short half-life of about 6 hours. These viruses must be cleared by the body's defenses, which include antibodies, cytotoxic T cells, and macrophages. Almost all HIVs, at least 99%, are produced by infected CD4 T cells, which survive for only about 2 days (T cells normally live for several years). Every day, an average of about 2 billion CD4 T cells are produced in an attempt to compensate for losses. Over time, however, there is a daily net loss of at least 20 million CD4 T cells, one of the main markers for the progression of HIV infection. The most recent studies show that the decrease in CD4 T cells is not due entirely to direct viral destruction of the cells; rather, it is caused primarily by shortened life of the cells and the

body's failure to compensate by increasing production of replacement T cells.

SURVIVAL WITH HIV INFECTION

HIV infection devastates the immune system, which is then unable to respond effectively to pathogens. The diseases or conditions most commonly associated with HIV infection and AIDS are summarized in Table 19.5. Success in treating these conditions has extended the lives of many HIV-infected people.

The age of the infected person can also be an important factor. Older adults are less able to replace CD4 T cell populations. Infants and younger children have an immune system that is not fully developed. They are much more susceptible to opportunistic infections.

Infants born to HIV-positive mothers are not always infected—in fact, only about 20% are. Infants who are most seriously infected survive less than 18 months.

RESISTANCE TO HIV INFECTION

A feature of HIV is that the virus proliferates despite all efforts of the humoral and cellular immune systems. As shown in Figure 19.12, a two-receptor mechanism is required for the virus to infect a CD4 T cell. The glycoprotein spike (gp120) on the virus must attach to a gp120 receptor and either a CCR5 or CXCR4 coreceptor. Although infection stimulates the production of large numbers of humoral antibodies, they have little long-term effect on neutralizing the virus. Furthermore, the gp120 of the virus has evolved in such a way that it resists antibody binding. At the same time the virus is efficiently destroying the CD4 T cells responsible for cellular immunity.

However, despite repeated exposures to HIV, some individuals do not become HIV positive; this population is called "*exposed, but not infected.*" Others, although they are infected and test as HIV positive, do not progress to the low T-cell populations that characterize AIDS. These are the "*long-term nonprogressors.*"

Exposed, But Not Infected Population About 1% of the population has a major deletion in CCR5 and these persons (mostly of European ancestry; the mutation is rare in African and Asian populations) are unusually resistant to repeated exposures to HIV. In this population the CCR5 molecule does not appear on the cell surface, so the infection cannot be completed. (Rare strains of HIV do not require CCR5 and infect the cells anyway).

Another notable population that is resistant to infection by HIV has appeared among certain African prostitutes, who are repeatedly exposed but remain HIV negative. These individuals produce cytotoxic T lymphocytes that are unusually effective in combating HIV.

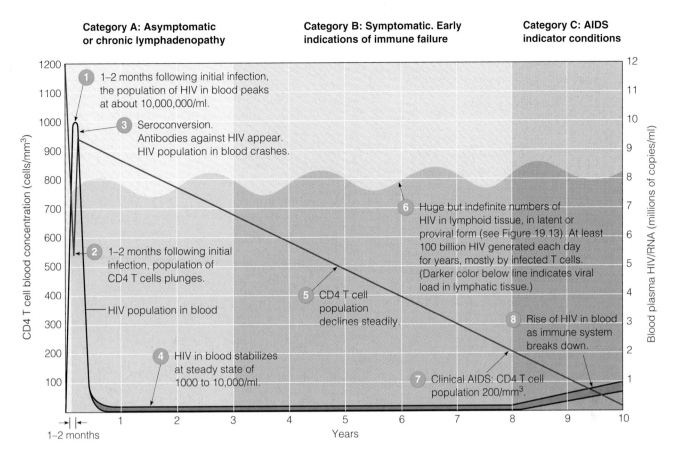

Category A: Asymptomatic or chronic lymphadenopathy

Category B: Symptomatic. Early indications of immune failure

Category C: AIDS indicator conditions

1. 1–2 months following initial infection, the population of HIV in blood peaks at about 10,000,000/ml.

3. Seroconversion. Antibodies against HIV appear. HIV population in blood crashes.

2. 1–2 months following initial infection, population of CD4 T cells plunges.

HIV population in blood

5. CD4 T cell population declines steadily.

6. Huge but indefinite numbers of HIV in lymphoid tissue, in latent or proviral form (see Figure 19.13). At least 100 billion HIV generated each day for years, mostly by infected T cells. (Darker color below line indicates viral load in lymphatic tissue.)

8. Rise of HIV in blood as immune system breaks down.

4. HIV in blood stabilizes at steady state of 1000 to 10,000/ml.

7. Clinical AIDS: CD4 T cell population 200/mm³.

CD4 T cell blood concentration (cells/mm³)

Blood plasma HIV/RNA (millions of copies/ml)

1–2 months

Years

FIGURE 19.15 The progression of HIV infection.

Q What causes the HIV population to drop in the first few months following infection?

Long-Term Nonprogressors Even though infected, these individuals remain free of symptoms and do not progress to the stage of AIDS, and their CD4 cell counts remain stable. Survival exceeding 25 years is predicted. The exact mechanism, or mechanisms, by which the patient successfully combats the infection is still uncertain, although there are several theories. In some cases a genetic factor blocks efficient binding of HIV to the CCR5 coreceptor. There is also evidence that some nonprogressors produce an anti-viral factor that inhibits HIV replication. Interest in these populations is high because a successful search for a vaccine will probably depend upon knowledge of the natural resistance of these populations.

DIAGNOSTIC METHODS

LEARNING OBJECTIVE
- Describe how HIV infection is diagnosed.

Generally speaking, it is simpler and less expensive to detect antibodies against HIV than to detect the virus itself. Therefore, most tests for HIV infection have been developed to detect HIV antibodies. Only a few years ago,

the available tests for HIV antibodies were an ELISA test (see Figure 18.14, page 545) that required confirmation with the Western blot test (see Figure 10.12, page 299). Because several days were usually required to complete the test, as many as a third of patients with a positive test failed to return for their results. Recently, rapid tests that give results within 20 minutes or less have become available. The tests require only a single drop of blood from a finger prick. Although the specificity and sensitivity of these tests is very high, a positive test must also be confirmed with a Western blot. A patient who tests positive with the rapid test is more likely to return for a confirmatory assay, which improves the opportunity for counseling and treatment.

A problem with antibody-type testing is the window of time between infection and the appearance of detectable antibodies, or **seroconversion.** This interval, which can be as long as 3 months, is illustrated in Figure 19.15, where seroconversion follows the peak number of viruses in circulation. Because of this delay, the recipient of an organ transplant or a blood transfusion can become infected with HIV even though antibody tests did not show the presence of the virus. Improvements in testing have gradually narrowed the window to 21 to 25 days.

TABLE 19.5	Some Common Diseases Associated with AIDS
Pathogen or Disease	Disease Description
Protozoa	
Cryptosporidium hominis	Persistent diarrhea
Toxoplasma gondii	Encephalitis
Isospora belli	Gastroenteritis
Viruses	
Cytomegalovirus	Fever, encephalitis, blindness
Herpes simplex virus	Vesicles of skin and mucous membranes
Varicella-zoster virus	Shingles
Bacteria	
Mycobacterium tuberculosis	Tuberculosis
M. avium-intracellulare	May infect many organs; gastroenteritis and other highly variable symptoms
Fungi	
Pneumocystis jiroveci	Life-threatening pneumonia
Histoplasma capsulatum	Disseminated infection
Cryptococcus neoformans	Disseminated, but especially meningitis
Candida albicans	Overgrowth on oral and vaginal mucous membranes (category B stage of HIV infection)
C. albicans	Overgrowth in esophagus, lungs (category C stage of HIV infection)
Cancers or Precancerous Conditions	
Kaposi's sarcoma	Cancer of skin and blood vessels (caused by human herpesvirus 8)
Hairy leukoplakia	Whitish patches on mucous membranes; commonly considered precancerous
Cervical dysplasia	Abnormal cervical growth

In contrast to antibody tests, **plasma viral load tests** can detect and quantify HIV circulating in the blood. Viral load tests detect viral RNA by methods such as PCR (see page 301) or nucleic acid hybridization (see page 302.) Despite their obvious advantages over antibody tests, however, viral load tests are costly and require 48 to 72 hours to complete.

Tests that detect the virus are also useful for instances in which antibody-detection tests cannot be used. For example, newborns of HIV-infected mothers have circulating maternal antibodies that interfere with conventional antibody-detection tests. To ensure safety of the blood supply as much as possible, the American Red Cross has introduced PCR testing for HIV at all their regional collection centers. Pooled samples of 120 donations are tested, and, if HIV is found, each individual sample is tested until the infected sample is found. This will allow blood banks to discontinue antibody-detecting tests and narrow the window of nondetection to about 12 days.

Viral load tests are also used to follow the progress of chemotherapy treatment of HIV infections.

A caution to keep in mind in testing for HIV is that current tests may not reliably detect all of the myriad variants of rapidly mutating HIV, especially subtypes that are not normally present in a population. Furthermore, plasma viral load tests sample only the virions circulating in the blood. Large numbers of HIV in latent form (see Figures 19.13 and 19.14) remain in lymphoid tissue. These are constantly being released into peripheral circulation.

HIV TRANSMISSION

LEARNING OBJECTIVE
• List the routes of HIV transmission.

The transmission of HIV requires the transfer of, or direct contact with, infected body fluids. The most important of these is blood, which contains 1000 to 100,000 infective

MORBIDITY & MORTALITY WEEKLY REPORT

AIDS: THE RISK TO HEALTH CARE WORKERS

With the advent of the AIDS epidemic, health care workers are understandably concerned about the risk of contracting AIDS after exposure to the body fluids of infected patients. However, when precautions are observed, the risk to workers is very small, even for those treating AIDS patients.

Understanding the Risk

The first protection for health care workers is a clear understanding of how HIV can (and cannot) be transmitted in the course of their work. To date, direct inoculation of infected material is the only proven method of transmission in the health care environment. Infected materials that can transmit HIV are blood, semen, vaginal secretions, and breast milk. The most common route of transmission is through accidental needle sticks. However, inoculation is also possible if infected material contacts mucous membranes or a break in the health care worker's skin. There is no evidence of HIV transmission by aerosols, the fecal-oral route, mouth-to-mouth or casual contact, or contact with environmental surfaces such as floors, walls, chairs, and toilets. Although HIV has been detected in fluids such as saliva, tears, cerebrospinal fluid, amniotic fluid, and urine, HIV is unlikely to be transmitted through exposure to these fluids.

CDC is aware of 57 health care workers in the United States who have been documented as having seroconverted to HIV following occupational exposures; twenty-six of them have AIDS.

The CDC estimates that the probability of infection following a needlestick injury with HIV-infected blood is 3 out of 1000 exposures, or 0.3%. The probability of infection from mucous membrane exposure to blood is 0.09%. The probability of acquiring hepatitis B from a needlestick injury with blood containing HBV (hepatitis B virus) is 1 to 31%, depending on the antigens present.

Precautions

Health care workers should be vaccinated against HBV. Avoidance of exposure is the health care worker's first line of defense against HIV. The CDC has developed the strategy of "universal precautions," which should be followed in *all* health care settings. These are described below.

Gloves   Disposable gloves should be used for direct exposure to infected blood, other body fluids, and tissues. Double-gloving is recommended during invasive surgical procedures. Personnel should not work when they have open skin lesions, dermatitis exuding fluids, or cutaneous wounds.

Gowns, Masks, and Goggles

Masks and protective eyewear are recommended when splashes are expected, such as during airway manipulation, endoscopy, and dental procedures, and in the laboratory.

Needles   To minimize the risk of needlesticks, needles should not be resheathed and should be put in a puncture-proof container for sterilization and disposal.

More expensive, safer needle devices that minimize the risk of needlesticks are available.

Disinfection   Routine housekeeping in health care settings should include washing floors, walls, and other areas not normally associated with disease transmissions with a 1:100 dilution of household bleach. A 1:10 dilution is recommended for disinfecting a spill.

Preventive Treatment after Exposure

Accidental exposure cannot always be prevented. Studies indicate that exposed individuals can reduce their risk by the prophylactic use of two or three antiviral drugs for 4 weeks.

The Risk to Patients

Transmission of HBV from health care workers to patients during invasive dental procedures (tooth extractions) has been documented. The only documented cases of HIV transmission from a health care worker to patients involved a dentist in Florida and a surgeon in France. The risk of HIV transmission from an infected health care worker to patients is small and can be reduced with universal precautions.

The risk of HIV transmission from health care workers to patients is greatest during invasive procedures; restriction of patient care by infected workers who perform such procedures is made on an individual basis.
SOURCE: Adapted from MMWR, 2004.

viruses per milliliter, and semen, which contains about 10 to 50 viruses per milliliter. The viruses are often located within cells in these fluids, especially in macrophages. HIV can survive more than 1.5 days inside a cell but only about 6 hours outside a cell.

Routes of HIV transmission include intimate sexual contact, breast milk, transplacental infection of a fetus, blood-contaminated needles, organ transplants, artificial insemination, and blood transfusion. The risk to health care workers is discussed in the box above. Probably the most dangerous form of sexual contact is anal-receptive intercourse. Vaginal intercourse is much more likely to transmit HIV from male to female than vice versa, and transmission either way is much greater when genital

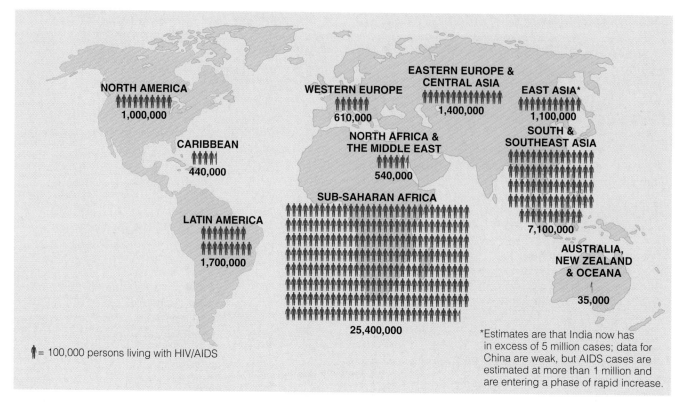

FIGURE 19.16 Distribution of HIV infection and AIDS in regions of the world. Each figure represents 100,000 persons living with HIV infection or AIDS. *Source:* UNAIDS; World Health Organization.

Q Where do you think the most accurate figures would be available?

lesions are present. Although rare, transmission can occur by oral-genital contact.

HIV is not transmitted by insects or casual contact such as hugging or sharing household items. Saliva generally contains less than 1 virus per milliliter, and kissing is not known to transmit the virus. In developed countries, transmission by transfusion is unlikely because blood is tested for HIV or HIV antibodies. However, there will always be a slight risk, as discussed earlier.

AIDS WORLDWIDE

LEARNING OBJECTIVE
• Identify geographic patterns of HIV transmission.

Approximately 25 million people have died from AIDS (Figure 19.16). An estimated 5 million are becoming infected every year. It is the leading cause of death in sub-Saharan Africa. As the disease becomes established in the huge populations of Asia, especially China and India, the incidence of HIV could exceed more than a million new cases a year. Eastern Europe, Russia, and

Central Asia are also areas reporting a steep rise in HIV infections. In Western Europe and the United States the mortality from AIDS has decreased because of the availability of effective antiviral drugs. It is projected that by 2010 there will be a total of more than 100 million HIV-infected individuals in the world; more than 90% of these will be in developing countries. Deaths from HIV-related causes by 2010 will probably exceed 8 million per year.

Certain basic epidemiological patterns have emerged:

• In the United States, Canada, western Europe, Australia, northern Africa, and certain parts of South America, HIV has primarily affected injecting drug users (IDUs) and homosexual and bisexual males. In western Europe and North America, the incidence of heterosexual spread has increased, and women are being infected in at least the same rate as men. The CDC estimates that about 40,000 Americans become infected each year. Certain racial/ethnic populations in the United States have higher rates of HIV infections. In one study, the Hispanic and non-Hispanic black populations constituted 21% of the total population but accounted for 84% of the heterosexually-acquired HIV infections. African-American women and youths now account for about two-thirds of AIDS cases.

- Figure 19.17 illustrates the modes of HIV transmission in the United States. In sub-Saharan Africa, HIV transmission has been almost entirely from heterosexual contact and is now affecting more women than men. In some countries, more than 25% of the population is infected.

- In Asia the epidemic is being spread by IDUs, commercial sex workers, and by young heterosexual males. The numbers of persons infected are expected to eventually be huge because of the immense populations, but the percentage of population infected is not expected to approach that of sub-Saharan Africa.

THE PREVENTION AND TREATMENT OF AIDS

> **LEARNING OBJECTIVE**
> - List the current methods of preventing and treating HIV infection.

At present, for most of the world the only practical means of control is by minimizing transmission. This requires education programs to promote the use of condoms, as well as discouraging sexual promiscuity. In high-income countries, the availability of medications has made HIV infection no longer a certain death sentence. Unfortunately, improvements in managing HIV infection have resulted in a relaxed attitude toward safer-sex practices. The fact tends to be overlooked that the available drugs only delay the progress of the infection, they are not a cure. Attempts to prevent the use of contaminated needles among IDUs is also important. To be effective, educational programs often require fundamental social changes that are not easy to accomplish, but they have slowed the infection rate in some areas, such as Thailand.

HIV VACCINES

An important consideration for vaccine development is the fact that the immune system has not shown much capability in coping with natural infections. Obstacles to the development of a vaccine for HIV are formidable, and the landscape is now littered with unsuccessful vaccine trials. The rapid mutation rate of HIV makes it difficult to develop a vaccine that is effective against all mutational variants of the virus that appear during the course of an infection. Furthermore, the virus has developed clades that

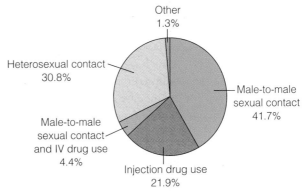

Estimated AIDS cases by transmission mode in 2003

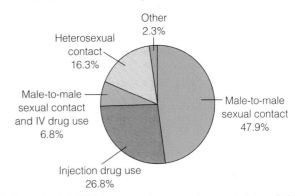

Estimated total AIDS cases by transmission mode through 2003

FIGURE 19.17 Modes of HIV transmission in the United States. In most of the world, transmission is primarily by heterosexual sex. In the United States, this form of transmission is much lower but is growing rapidly. Transmission in western Europe is similar to that in the United States. Data Source: CDC

Q **What is the most common mode of HIV transmission in the United States?**

differ substantially from one geographic area to another, and each would probably require an appropriate vaccine.

Ideally, a vaccine would produce antibodies that would prevent infection. As we have seen, however, the virus resists antibody binding until a last-second exposure just before attachment and entry into the host cell. For persons already infected, a successful cell-mediated type vaccine would be useful in controlling the progression of the disease. To be considered successful, however, the vaccine would have to stimulate the production of CTLs more effective than those produced in response to a natural infection. Cells infected by HIV, it happens, are not very susceptible to attack by CTLs. There is also the problem of a persistent, but immunologically invisible, viral population in the form of proviruses and latent viruses. An effective vaccine would have to protect against transmission by different mucosal routes, which is proving to be an elusive goal in tests with monkeys and SIV. Finally, a vaccine would have to be affordable in regions of the world where subsistence is often marginal. All in all, the development of an HIV vaccine is a formidable task. Some experts believe that no HIV vaccine is possible that will confer nearly complete protection, such as those for smallpox or measles. It is thought that a more practical goal may be to develop a vaccine with more modest goals than "sterilizing immunity." Perhaps a vaccine would stimulate cell-mediated immunity in already infected persons and help the patient's existing immune system to clear the virus. Alternatively, a vaccine might not prevent infection but might greatly moderate the symptoms and severity of the infection.

CHEMOTHERAPY

Much progress has been made in the use of chemotherapy to inhibit HIV infections. Drugs that are available for chemotherapy of HIV are discussed on page 598. The first target of anti-HIV drugs was the enzyme reverse transcriptase (page 261), an enzyme not present in human cells. These *nucleoside reverse transcriptase inhibitors* are analogs of nucleosides and cause the termination of the synthesis of viral DNA by competitive inhibition (page 121). There are other drugs that inhibit reverse transcription but are not analogs of nucleic acids; these are the so-called *non-nucleoside reverse transcriptase inhibitors*.

To reproduce, the virus makes use of certain protease enzymes that cut proteins into pieces, which are then reassembled into the coat of new HIV particles. Drugs called *protease inhibitors* inhibit this enzyme and are now in use.

For infection to take place, the virus must accomplish a series of steps. It must attach to the cell's CD4 receptors, an interplay between the gp120 spike on the virus and the coreceptor (such as CCR5) must occur, and finally there must be a fusion with the cell to allow viral entry into the cell. (Entry into the cell by fusion requires that the host cell membrane fuse with the envelope of the virus, which permits entry into the cell of the nucleus-containing capsid of the virus. The viral envelope is left behind). Some of the newer anti-HIV drugs, grouped as *fusion inhibitors*, target one of the required steps.

After fusion has been completed, reverse transcription produces a double-stranded cDNA version of HIV, which enters the nucleus. Within the nucleus, the complex containing the cDNA must be integrated into the host chromosome to form the HIV provirus. This step requires an enzyme, HIV integrase, which is a target for drugs called *integrase inhibitors*. Several other points of attack for anti-HIV drugs exist, and drugs targeted at these are under investigation or are in clinical trials.

Because of the increasing number of drugs that control reproduction of the virus, at least temporarily, HIV infection is almost at the stage where it can be considered a treatable chronic disease—assuming that the treatment is affordable. The rapid reproductive rate and frequent occurrence of drug-resistant mutations dictates that multiple drugs, given simultaneously, must be used. The current treatment is termed **highly active anti-retroviral therapy (HAART).** This therapy consists of administering drug combinations; one of the most common combinations is two nucleoside analog reverse transcriptase inhibitors plus either a non-nucleoside reverse transcriptase inhibitor or a protease inhibitor. Patients are often required to take as many as 40 pills a day on a complex schedule, which must be adhered to rigorously because the virus is unforgiving. Even so, resistant strains of the virus are likely to emerge. As one scientist put it, drug resistance in HIV is driven by selective pressures that Darwin never imagined. Experience has also shown that eliminating all viruses in latent form in lymphoid tissue is especially difficult. The number of HIVs in circulation is often reduced to fewer than can be detected, but this is not the same as eradication. The more practical goal has been to reduce the viral load to a level where it can be controlled by the immune system. A caution is that a patient with a viral load at an undetectable level might still be infective. Frequent testing for viruses (not antibodies; see the discussion of diagnostic tests on page 571) is required to follow effectiveness of the treatment. A rise in numbers probably indicates development of a resistant population.

One clearly successful application of chemotherapy has been to reduce the chance of HIV transmission from an infected mother to her newborn. The administration of even one nucleoside reverse transcriptase drug alone reduces the incidence.

THE AIDS EPIDEMIC AND THE IMPORTANCE OF SCIENTIFIC RESEARCH

The AIDS epidemic gives clear evidence of the value of basic scientific research. Without the advances in molecular biology of the past few decades, we would have been unable even to identify the causative agent of AIDS. We would not have been able to develop the tests for screening donated blood, to identify points in the viral life cycle for which selectively toxic drugs could be developed, or even to monitor the course of the infection. In the lifetime of most of us, we will have the opportunity to witness medical history being made as the struggle with this deadly and elusive virus continues.

STUDY OUTLINE

INTRODUCTION   (p. 550)

1. Hay fever, transplant rejection, and autoimmunity are examples of harmful immune reactions.

2. Infection and immunosuppression are examples of failure of the immune system.

3. Superantigens activate many T-cell receptors, resulting in the release of excessive amounts of cytokines that can cause adverse host responses.

HYPERSENSITIVITY   (pp. 551–558)

1. Hypersensitivity reactions represent immunological responses to an antigen (allergen) that lead to tissue damage rather than immunity.

2. Hypersensitivity reactions occur when a person has been sensitized to an antigen.

3. Hypersensitivity reactions can be divided into four classes: types I, II, and III are immediate reactions based on humoral immunity, and type IV is a delayed reaction based on cell-mediated immunity.

TYPE I (ANAPHYLACTIC) REACTIONS   (pp. 551–554)

4. Anaphylactic reactions involve the production of IgE antibodies that bind to mast cells and basophils to sensitize the host.

5. The binding of two adjacent IgE antibodies to an antigen causes the target cell to release chemical mediators, such as histamine, leukotrienes, and prostaglandins, which cause the observed allergic reactions.

6. Systemic anaphylaxis may develop in minutes after injection or ingestion of the antigen; this may result in circulatory collapse and death.

7. Localized anaphylaxis is exemplified by hives, hay fever, and asthma.

8. Skin testing is useful in determining sensitivity to an antigen.

9. Desensitization to an antigen can be achieved by repeated injections of the antigen, which leads to the formation of blocking (IgG) antibodies.

TYPE II (CYTOTOXIC) REACTIONS   (pp. 554–557)

10. Type II reactions are mediated by IgG or IgM antibodies and complement.

11. The antibodies are directed toward foreign cells or host cells. Complement fixation may result in cell lysis. Macrophages and other cells may also damage the antibody-coated cells.

The ABO Blood Group System   (p. 555)

12. Human blood may be grouped into four principal types, designated A, B, AB, and O.

13. The presence or absence of two carbohydrate antigens designated A and B on the surface of the red blood cell determines a person's blood type.

14. Naturally occurring antibodies are present in serum against the opposite AB antigen.

15. Incompatible blood transfusions lead to the complement-mediated lysis of the donor red blood cells.

The Rh Blood Group System   (pp. 555–556)

16. Approximately 85% of the human population possesses another blood group antigen, designated the Rh antigen; these individuals are designated Rh^+.

17. The absence of this antigen in certain individuals (Rh^-) can lead to sensitization upon exposure to it.

18. An Rh^+ person can receive Rh^+ or Rh^- blood transfusions.

19. When an Rh^- person receives Rh^+ blood, that person will produce anti-Rh antibodies.

20. Subsequent exposure to Rh^+ cells will result in a rapid, serious hemolytic reaction.

21. An Rh^- mother carrying an Rh^+ fetus will produce anti-Rh antibodies.

22. Subsequent pregnancies involving Rh incompatibility may result in hemolytic disease of the newborn.

23. The disease may be prevented by passive immunization of the mother with anti-Rh antibodies.

Drug-Induced Cytotoxic Reactions   (pp. 556–557)

24. In the disease thrombocytopenic purpura, platelets are destroyed by antibodies and complement.

25. Agranulocytosis and hemolytic anemia result from antibodies against one's own blood cells coated with drug molecules.

TYPE III (IMMUNE COMPLEX) REACTIONS   (p. 557)

26. Immune complex diseases occur when IgG antibodies and soluble antigen form small complexes that lodge in the basement membranes of cells.

27. Subsequent complement fixation results in inflammation.

28. Glomerulonephritis is an immune complex disease.

TYPE IV (DELAYED CELL-MEDIATED) REACTIONS   (pp. 557–558)

29. Delayed cell-mediated hypersensitivity reactions are due primarily to T_D cell proliferation.

30. Sensitized T cells secrete cytokines in response to the appropriate antigen.

31. Cytokines attract and activate macrophages and initiate tissue damage.

32. The tuberculin skin test and allergic contact dermatitis are examples of delayed hypersensitivities.

AUTOIMMUNE DISEASES (pp. 559–561)

1. Autoimmunity results from a loss of self-tolerance.

2. Self-tolerance occurs during fetal development; T cells that will target host cells are eliminated (clonal deletion) or inactivated.

3. Autoimmunity may be due to antibodies against infectious agents.

4. Graves' disease and myasthenia gravis are cytotoxic autoimmune reactions in which antibodies react to cell-surface antigens.

5. Systemic lupus erythematosus and rheumatoid arthritis are immune complex autoimmune reactions in which the deposition of immune complexes results in tissue damage.

6. Multiple sclerosis, Hashimoto's disease, and insulin-dependent diabetes mellitus are cell-mediated autoimmune reactions mediated by T cells.

REACTIONS RELATED TO THE HUMAN LEUKOCYTE ANTIGEN (HLA) COMPLEX (pp. 561–564)

1. Histocompatibility self molecules located on cell surfaces express genetic differences among individuals; these antigens are coded for by MHC or HLA gene complexes.

2. To prevent the rejection of transplants, HLA and ABO blood group antigens of the donor and recipient are matched as closely as possible.

REACTIONS TO TRANSPLANTATION (pp. 562–563)

3. Transplants recognized as foreign antigens may be lysed by T cells and attacked by macrophages and complement-fixing antibodies.

4. Transplantation to a privileged site (such as the cornea) or of a privileged tissue (such as pig heart valves) does not cause an immune response.

5. Pluripotent stem cells differentiate into a variety of tissues that may provide tissues for transplant.

6. Four types of transplants have been defined on the basis of genetic relationships between the donor and the recipient: autografts, isografts, allografts, and xenotransplants.

7. Bone marrow (with immunocompetent cells) can cause graft-versus-host disease.

8. Successful transplant surgery often requires immunosuppressant drugs to prevent an immune response to the transplanted tissue.

THE IMMUNE SYSTEM AND CANCER (pp. 564–565)

1. Cancer cells are normal cells that have undergone transformation, divide uncontrollably, and possess tumor-associated antigens.

2. The response of the immune system to cancer is called immunological surveillance.

3. T_C cells recognize and lyse cancerous cells.

4. Cancer cells can escape detection and destruction by the immune system.

5. Cancer cells may grow faster than the immune system can respond.

IMMUNOTHERAPY FOR CANCER (pp. 564–565)

6. Vaccines consisting of tumor antigens are being tested.

7. Herceptin consists of monoclonal antibodies against a breast cancer growth factor.

8. Immunotoxins are chemical poisons linked to a monoclonal antibody; the antibody selectively locates the cancer cell for release of the poison.

IMMUNODEFICIENCIES (p. 566)

1. Immunodeficiencies can be congenital or acquired.

2. Congenital immunodeficiencies are due to defective or absent genes.

3. A variety of drugs, cancers, and infectious diseases can cause acquired immunodeficiencies.

ACQUIRED IMMUNODEFICIENCY SYNDROME (AIDS) (pp. 566–576)

THE ORIGIN OF AIDS (pp. 566–567)

1. HIV is thought to have originated in central Africa and was brought to other countries by modern transportation and unsafe sexual practices.

HIV INFECTION (pp. 567–571)

2. AIDS is the final stage of HIV infection.

3. HIV is a retrovirus with single-stranded RNA, reverse transcriptase, and a phospholipid envelope with gp120 spikes.

4. HIV spikes attach to CD4 and coreceptors on host cells; the CD4 receptor is found on helper T cells, macrophages, and dendritic cells.

5. Viral RNA is transcribed to DNA by reverse transcriptase. The viral DNA becomes integrated into the host chromosome to direct synthesis of new viruses or to remain latent as a provirus.

6. HIV evades the immune system in latency, in vacuoles, by using cell–cell fusion, and by antigenic change.

7. Genetically distinct groups of HIV are classified into clades.

8. HIV infection is categorized by symptoms: Category A (asymptomatic) and Category B (selected symptoms) are reported as AIDS if CD4 T cells fall below 200 cells/mm^3; Category C (AIDS indicator conditions) is reported as AIDS.

9. HIV infection is also categorized by CD4 T cell numbers: 200 CD4 cells/mm^3 is reported as AIDS.

10. The progression from HIV infection to AIDS takes about 10 years.

11. The life of an AIDS patient can be prolonged by the proper treatment of opportunistic infections.

12. People lacking CCR5 are resistant to HIV infection.

DIAGNOSTIC METHODS   (pp. 571–572)

13. HIV antibodies are detected by ELISA and Western blotting.

14. Plasma viral load tests detect viral nucleic acid and are used to quantify HIV in blood.

HIV TRANSMISSION   (pp. 572–574)

15. HIV is transmitted by sexual contact, breast milk, contaminated needles, transplacental infection, artificial insemination, and blood transfusion.

16. In developed countries, blood transfusions are not a likely source of infection because blood is tested for HIV antibodies.

AIDS WORLDWIDE   (pp. 574–575)

17. In the United States, Canada, western Europe, Australia, northern Africa, and parts of South America, transmission has been by injecting drug users (IDU) and male-to-male sexual contact. Heterosexual transmission is increasing.

18. In sub-Saharan Africa, transmission is by heterosexual contact.

19. In eastern Europe and Asia, transmission is by IDU and heterosexual contact.

THE PREVENTION AND TREATMENT OF AIDS
(pp. 575–576)

20. The use of condoms and sterile needles prevents the transmission of HIV.

21. Vaccine development is difficult because the virus remains inside host cells.

22. Current chemotherapeutic agents target the virus enzymes, including reverse transcriptase and protease.

STUDY QUESTIONS

Access more review material either on line at **The Microbiology Place** (www.microbiologyplace.com) or with **The Microbiology Place CD-ROM** packaged with your new book. There you'll find activities, practice tests, quizzes, flashcards, case studies, and more to help you succeed.

Answers to the Study Questions can be found in Appendix G.

REVIEW

1. Define *hypersensitivity*.

2. List three mediators released in anaphylactic hypersensitivities, and explain their effects.

3. Discuss the roles of antibodies and antigens in an incompatible tissue transplant.

4. What happens to the recipient of an incompatible blood type?

5. Explain how hemolytic disease of the newborn develops and how this disease might be prevented.

6. Explain what happens when a person develops a contact sensitivity to poison oak.
   a. What causes the observed symptoms?
   b. How did the sensitivity develop?
   c. How might this person be desensitized to poison oak?

7. Which type of graft (autograft, isograft, allograft, or xeno-transplantation) is most compatible? Least compatible? How might stem cells help avoid transplant rejections?

8. Which of the following blood transfusions is compatible? Explain your answers.

Donor	Recipient
(a) AB, Rh$^-$	AB, Rh$^+$
(b) B, Rh$^+$	B, Rh$^-$
(c) A, Rh$^+$	O, Rh$^+$

9. Define *autoimmunity*. Propose a hypothesis that could explain autoimmune responses.

10. Differentiate between the three types of autoimmune diseases. Name an example of each type.

11. Summarize the causes of immunodeficiencies. What is the effect of an immunodeficiency?

12. In what ways do tumor cells differ antigenically from normal cells? Explain how tumor cells may be destroyed by the immune system.

13. If tumor cells can be destroyed by the immune system, how does cancer develop? What does immunotherapy involve?

14. Differentiate HIV infection from AIDS. How is HIV transmitted? How can HIV infection be prevented?

MULTIPLE CHOICE

1. Desensitization to prevent an allergic response can be accomplished by injecting small, repeated doses of
   a. IgE antibodies.
   b. the antigen (allergen).
   c. histamine.
   d. IgG antibodies.
   e. antihistamine.

2. In vitro, recipient serum type B will agglutinate donor cells type A; while in vivo, recipient serum type B will lyse donor cells type A. The in vivo response is due to
   a. T cells.
   b. complement.
   c. antibodies.
   d. autoimmunity.
   e. none of the above

3. Cytotoxic autoimmunity differs from immune complex autoimmunity in that cytotoxic reactions
   a. involve antibodies.
   b. do not involve complement.
   c. are caused by T cells.
   d. do not involve IgE antibodies.
   e. none of the above

4. Worldwide, the primary method of transmitting HIV is
   a. homosexual sex.
   b. heterosexual sex.
   c. use of injecting drugs.
   d. blood transfusions.
   e. kissing.

5. Which of the following is *not* the cause of a natural immune deficiency?
   a. a recessive gene resulting in lack of a thymus gland
   b. a recessive gene resulting in few B cells
   c. HIV infection
   d. immunosuppressant drugs
   e. none of the above

6. Which antibodies will be found naturally in the serum of a person with blood type A, Rh$^+$?
   a. anti A, anti B, anti Rh
   b. anti A, anti Rh
   c. anti A
   d. anti B, anti Rh
   e. anti B

Use the following choices to match the type of hypersensitivity to the examples in questions 7 through 10.
   a. type I hypersensitivity
   b. type II hypersensitivity
   c. type III hypersensitivity
   d. type IV hypersensitivity
   e. all of the above

7. Localized anaphylaxis.
8. Allergic contact dermatitis.
9. Due to immune complexes.
10. Reaction to an incompatible blood transfusion.

CRITICAL THINKING

1. When and how does our immune system discriminate between self and nonself antigens?

2. The first preparations used for artificially acquired passive immunity were antibodies in horse serum. A complication that resulted from the therapeutic use of horse serum was immune complex disease. Why did this occur?

3. Do people with AIDS make antibodies? If so, why are they said to have an immune deficiency?

4. What are the methods of action of anti-AIDS drugs?

CLINICAL APPLICATIONS

1. Fungal infections such as athlete's foot are chronic. These fungi degrade skin keratin but are not invasive and do not produce toxins. Why do you suppose that many of the symptoms of a fungal infection are due to hypersensitivity to the fungus?

2. After working in a mushroom farm for several months, a worker develops these symptoms: hives, edema, and swelling lymph nodes.
   a. What do these symptoms indicate?
   b. What mediators cause these symptoms?
   c. How may sensitivity to a particular antigen be determined?
   d. Other employees do not appear to have any immunological reactions. What could explain this?

   (*Hint:* The allergen is conidiospores from molds growing in the mushroom farm.)

3. Physicians administering live, attenuated mumps and measles vaccines prepared in chick embryos are instructed to have epinephrine available. Epinephrine will not treat these viral infections. What is the purpose of keeping this drug on hand?

4. A woman with blood type A+ once received a transfusion of AB+ blood. When she carried a type B+ fetus, the fetus developed hemolytic disease of the newborn. Explain why this fetus developed this condition when another type B+ fetus in a different type A+ mother was normal.

20

Antimicrobial Drugs

When the body's normal defenses cannot prevent or over-come a disease, it is often treated with **chemotherapy.** In this chapter we focus on antimicrobial drugs, the class of chemotherapeutic agents used to treat infectious diseases. Like the disinfectants discussed in Chapter 7, **antimicrobial drugs** act by interfering with the growth of microorganisms. Unlike disinfectants, however, they must often act *within* the host. Therefore, their effects on the cells and tissues of the host are important. The ideal antimicrobial drug kills the harmful microorganism without damaging the host; this is the principle of **selective toxicity.**

Antibiotics were one of the most important discoveries for modern medicine. Within the memory of many people, an abdominal wound or a ruptured appendix represented nearly certain death from infection. There was little that medicine could do to treat diseases such as typhoid fever, tuberculosis, and so-called blood poisoning in which bacteria proliferated uncontrolled in the bloodstream. The use of antimicrobials such as penicillin and sulfanilamide in the treatment of some infections resulted in rapid cures that seemed almost miraculous at the time.

Today, we are seeing these advances threatened by the development of resistance of microbes to these miracle drugs.

UNDER THE MICROSCOPE

This bacterium is lysing because an antibiotic has prevented cell wall synthesis.

THE HISTORY OF CHEMOTHERAPY

LEARNING OBJECTIVES

- Identify the contributions of Paul Ehrlich and Alexander Fleming to chemotherapy.
- Name the microbes that produce most antibiotics.

The birth of modern chemotherapy is credited to the efforts of Paul Ehrlich in Germany during the early part of the twentieth century. While attempting to stain bacteria without staining the surrounding tissue, he speculated about some "magic bullet" that would selectively find and destroy pathogens but not harm the host. This idea provided the basis for *chemotherapy*, a term he coined.

In 1928, Alexander Fleming observed that the growth of the bacterium *Staphylococcus aureus* was inhibited in the area surrounding the colony of a mold that had contaminated a Petri plate (Figure 20.1). The mold was identified as *Penicillium notatum*, and its active compound, which was isolated a short time later, was named penicillin. Similar inhibitory reactions between colonies on solid media are commonly observed in microbiology, and the mechanism of inhibition is called *antibiosis*. From this word comes the term **antibiotic,** a substance produced by microorganisms that in small amounts inhibits another microorganism. Therefore, the wholly synthetic sulfa drugs, for example, technically are not antibiotics. However, this distinction is often ignored in practice.

In 1940, a group of scientists at Oxford University headed by Howard Florey and Ernst Chain succeeded in the first clinical trials of penicillin. Under war time conditions in the United Kingdom, research into the development and large scale production of penicillin was not possible, and this work was transferred to the United States. The original culture of *P. notatum* (later renamed *Penicillium chrysogenum*) was not a very efficient producer of the antibiotic. It was soon replaced by a more prolific strain. This valuable organism was first isolated from a moldy cantaloupe bought at a market in Peoria, Illinois.

Antibiotics are actually rather easy to discover, but few are of medical or commercial value. Some are used commercially other than for treating disease—for example, as a supplement in animal feed (see the box on page 606). Many antibiotics are toxic to humans or lack any advantage over antibiotics already in use.

More than half of our antibiotics are produced by species of *Streptomyces*, filamentous bacteria that commonly inhabit soil. A few antibiotics are produced by endospore-forming bacteria such as *Bacillus*, and others are produced by molds, mostly of the genera *Penicillium* and *Cephalosporium* (sef-ä-lō-spô′rē-um). See Table 20.1 for the sources of

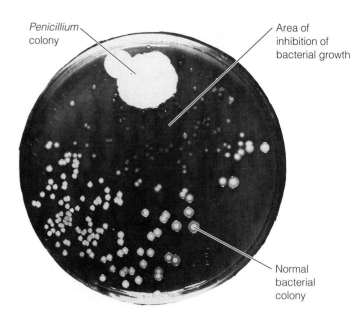

FIGURE 20.1 The discovery of penicillin. Alexander Fleming took this photograph in 1928. The colony of *Penicillium* mold accidentally contaminated the plate and is inhibiting nearby bacterial growth.

Q **Would Fleming have observed this phenomenon if the bacterium on the plate had been gram-negative?**

TABLE 20.1	Representative Sources of Antibiotics
Microorganism	**Antibiotic**
Gram-Positive Rods	
Bacillus subtilis	Bacitracin
Paenibacillus polymyxa	Polymyxin
Actinomycetes	
Streptomyces nodosus	Amphotericin B
Streptomyces venezuelae	Chloramphenicol
Streptomyces aureofaciens	Chlortetracycline and tetracycline
Saccharopolyspora erythraea	Erythromycin
Streptomyces fradiae	Neomycin
Streptomyces griseus	Streptomycin
Micromonospora purpurea	Gentamicin
Fungi	
Cephalosporium spp.	Cephalothin
Penicillium griseofulvum	Griseofulvin
Penicillium chrysogenum	Penicillin

| TABLE 20.2 | The Spectrum of Activity of Antibiotics and Other Antimicrobial Drugs |

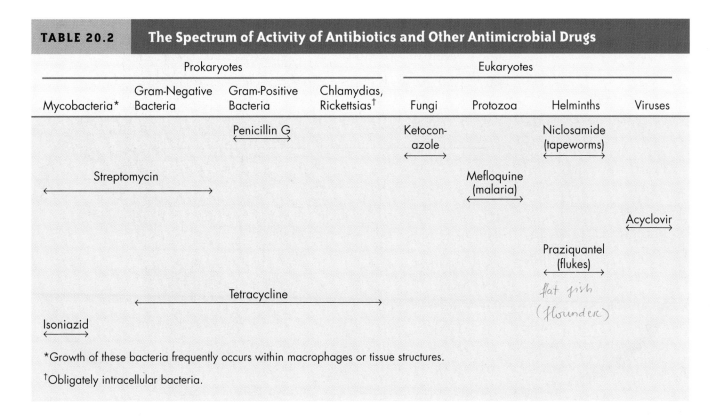

Prokaryotes				Eukaryotes			
Mycobacteria*	Gram-Negative Bacteria	Gram-Positive Bacteria	Chlamydias, Rickettsias†	Fungi	Protozoa	Helminths	Viruses

Penicillin G ← →

Streptomycin ← →

Ketoconazole ← →

Mefloquine (malaria) ← →

Niclosamide (tapeworms) ← →

Acyclovir

Praziquantel (flukes) ← →

flat fish (flounder)

Tetracycline ← →

Isoniazid ← →

*Growth of these bacteria frequently occurs within macrophages or tissue structures.

†Obligately intracellular bacteria.

---

many antibiotics in use today—a surprisingly limited group of organisms. One study screened 400,000 microbial cultures that yielded only three useful drugs. It is especially interesting to note that practically all antibiotic-producing microbes have some sort of sporulation process.

THE SPECTRUM OF ANTIMICROBIAL ACTIVITY

LEARNING OBJECTIVES

- Describe the problems of chemotherapy for viral, fungal, protozoan, and helminthic infections.
- Define the following terms: *spectrum of activity, broad-spectrum antibiotic, superinfection.*

It is comparatively easy to find or develop drugs that are effective against prokaryotic cells and that do not affect the eukaryotic cells of humans. These two cell types differ substantially in many ways, such as in the presence or absence of cell walls, the fine structure of their ribosomes, and details of their metabolism. Thus, selective toxicity has numerous targets. The problem is more difficult when the pathogen is a eukaryotic cell, such as a fungus, protozoan, or helminth. At the cellular level, these organisms resemble the human cell much more closely than a bacterial cell does. We will see that our arsenal against these types of pathogens is much more limited than our arsenal of anti-

bacterial drugs. Viral infections are particularly difficult to treat because the pathogen is within the human host's cells, and the genetic information of the virus is directing the human cell to make viruses rather than to synthesize normal cellular materials.

Some drugs have a narrow **spectrum of microbial activity,** or range of different microbial types they affect. Penicillin G, for example, affects gram-positive bacteria but very few gram-negative bacteria. Antibiotics that affect a broad range of gram-positive or gram-negative bacteria are therefore called **broad-spectrum antibiotics.**

A primary factor involved in the selective toxicity of antibacterial action lies in the lipopolysaccharide outer layer of gram-negative bacteria and the porins that form water-filled channels across this layer (see Figure 4.13c, page 86). Drugs that pass through the porin channels must be relatively small and preferably hydrophilic. Drugs that are lipophilic (having an affinity for lipids) or especially large do not enter gram-negative bacteria readily.

Table 20.2 summarizes the spectrum of activity of a number of chemotherapeutic drugs. Because the identity of the pathogen is not always immediately known, a broad-spectrum drug would seem to have an advantage in treating a disease by saving valuable time. The disadvantage is that many normal microbiota of the host are destroyed by broad-spectrum drugs. The normal microbiota ordinarily compete with and check the growth of pathogens or other microbes. If certain organisms in the normal microbiota

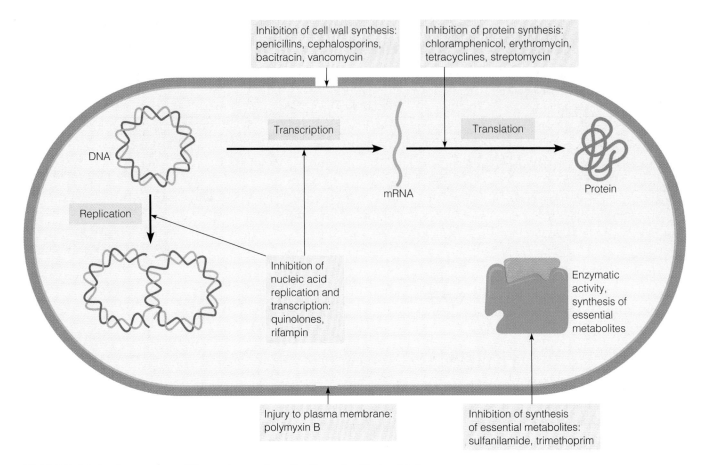

Inhibition of cell wall synthesis:
penicillins, cephalosporins,
bacitracin, vancomycin

Inhibition of protein synthesis:
chloramphenicol, erythromycin,
tetracyclines, streptomycin

Transcription

Translation

DNA

mRNA

Protein

Replication

Inhibition of
nucleic acid
replication and
transcription:
quinolones,
rifampin

Enzymatic
activity,
synthesis of
essential
metabolites

Injury to plasma membrane:
polymyxin B

Inhibition of synthesis
of essential metabolites:
sulfanilamide, trimethoprim

FIGURE 20.2 A summary of the major modes of action of antimicrobial drugs. This illustration shows these actions as they might affect a highly diagrammatic representation of a bacterial cell.

Q **Which modes of action are effective against prokaryotic cells? Against eukaryotic cells?**

are not destroyed by the antibiotic and their competitors are destroyed, the survivors may flourish and become opportunistic pathogens. An example that sometimes occurs is overgrowth by the yeastlike fungus *Candida albicans*, which is not sensitive to bacterial antibiotics. This overgrowth is called a **superinfection,** a term that is also applied to growth of a target pathogen that has developed resistance to the antibiotic. In this situation, such an antibiotic-resistant strain replaces the original sensitive strain, and the infection continues.

THE ACTION OF ANTIMICROBIAL DRUGS

LEARNING OBJECTIVE
• Identify five modes of action of antimicrobial drugs.

Antimicrobial drugs are either **bactericidal** (they kill microbes directly) or **bacteriostatic** (they prevent microbes from growing). In bacteriostasis, the host's own defenses,

such as phagocytosis and antibody production, usually destroy the microorganisms. The major modes of action are summarized in Figure 20.2.

THE INHIBITION OF CELL WALL SYNTHESIS

Recall from Chapter 4 that the cell wall of a bacterium consists of a macromolecular network called peptidoglycan. Peptidoglycan is found only in bacterial cell walls. Penicillin and certain other antibiotics prevent the synthesis of intact peptidoglycan; consequently, the cell wall is greatly weakened, and the cell undergoes lysis (Figure 20.3). Because penicillin targets the synthesis process, only actively growing cells are affected by these antibiotics—and, because human cells do not have peptidoglycan cell walls, penicillin has very little toxicity for host cells.

THE INHIBITION OF PROTEIN SYNTHESIS

Because protein synthesis is a common feature of all cells, whether prokaryotic or eukaryotic, it would seem an unlikely target for selective toxicity. One notable difference

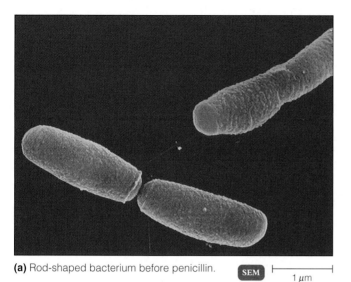

(a) Rod-shaped bacterium before penicillin.   **SEM** ⊢———⊣ 1 μm

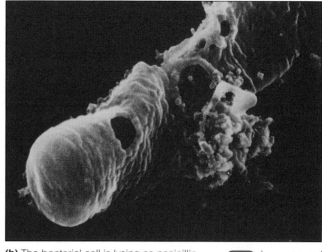

(b) The bacterial cell is lysing as penicillin weakens the cell wall.   **SEM** ⊢———⊣ 1 μm

FIGURE 20.3   The inhibition of bacterial cell synthesis by penicillin.

Q   Why don't penicillins affect the human cell?

between prokaryotes and eukaryotes, however, is the structure of their ribosomes. As discussed in Chapter 4 (page 102), eukaryotic cells have 80S ribosomes; prokaryotic cells have 70S ribosomes. (The 70S ribosome is made up of a 50S and a 30S unit. The S stands for Svedberg unit, which describes the relative rate of sedimentation in a high-speed centrifuge.) The difference in ribosomal structure accounts for the selective toxicity of antibiotics that affect protein synthesis. However, mitochondria (important eukaryotic organelles) also contain 70S ribosomes similar to those of bacteria. Antibiotics targeting the 70S ribosomes can therefore have adverse effects on the cells of the host. Among the antibiotics that interfere with protein synthesis are chloramphenicol, erythromycin, streptomycin, and the tetracyclines (Figure 20.4).

Reacting with the 50S portion of the 70S prokaryotic ribosome, chloramphenicol inhibits the formation of peptide bonds in the growing polypeptide chain. Most drugs that inhibit protein synthesis have a broad spectrum of activity; erythromycin is an exception. Because it does not penetrate the gram-negative cell wall, it affects mostly gram-positive bacteria.

Some other antibiotics react with the 30S portion of the 70S prokaryotic ribosome. The tetracyclines interfere with the attachment of the tRNA carrying the amino acids to the ribosome, preventing the addition of amino acids to the growing polypeptide chain. Tetracyclines do not interfere with mammalian ribosomes because they do not penetrate very well into intact mammalian cells. However, at least small amounts are able to enter the host cell, as is

apparent from the fact that the intracellular pathogenic rickettsias and chlamydias are sensitive to tetracyclines. The selective toxicity of the drug in this case is due to a greater sensitivity of the bacteria at the ribosomal level.

Aminoglycoside antibiotics, such as streptomycin and gentamicin, interfere with the initial steps of protein synthesis by changing the shape of the 30S portion of the 70S prokaryotic ribosome. This interference causes the genetic code on the mRNA to be read incorrectly.

INJURY TO THE PLASMA MEMBRANE

Certain antibiotics, especially polypeptide antibiotics, bring about changes in the permeability of the plasma membrane; these changes result in the loss of important metabolites from the microbial cell. For example, polymyxin B causes disruption of the plasma membrane by attaching to the phospholipids of the membrane.

Some antifungal drugs, such as amphotericin B, miconazole, and ketoconazole, are effective against a considerable range of fungal diseases. Such drugs combine with sterols in the fungal plasma membrane to disrupt the membrane (Figure 20.5). Because bacterial plasma membranes generally lack sterols, these antibiotics do not act on bacteria. However, the plasma membranes of animal cells do contain sterols, and amphotericin B and ketoconazole can be toxic to the host. Fortunately, animal cell membranes have mostly *cholesterol*, and fungal cells have mostly *ergosterol*, against which the drug is most effective, so that the balance of the toxicity is tilted against the fungus.

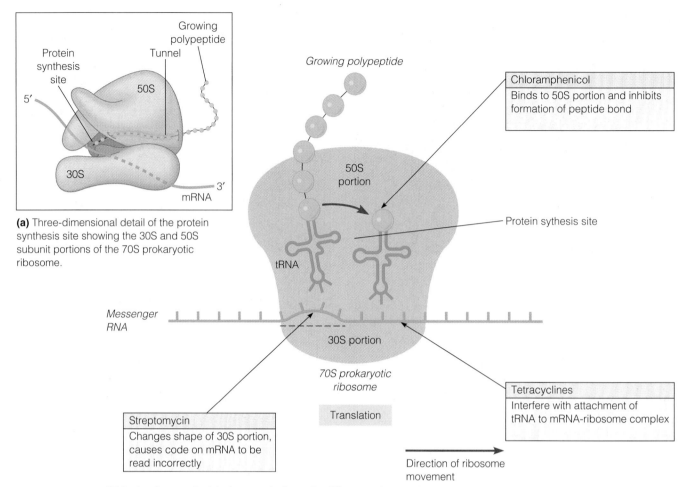

(a) Three-dimensional detail of the protein synthesis site showing the 30S and 50S subunit portions of the 70S prokaryotic ribosome.

Chloramphenicol
Binds to 50S portion and inhibits formation of peptide bond

Protein sythesis site

Streptomycin
Changes shape of 30S portion, causes code on mRNA to be read incorrectly

Tetracyclines
Interfere with attachment of tRNA to mRNA-ribosome complex

Direction of ribosome movement

(b) In the diagram the black arrows indicate the different points at which chloramphenicol, the tetracyclines, and streptomycin exert their activities.

FIGURE 20.4   The inhibition of protein synthesis by antibiotics. (a) The inset shows how the 70S prokaryotic ribosome is assembled from two subunits, 30S and 50S. Note how the growing peptide chain passes through a tunnel in the 50S subunit from the site of protein synthesis. **(b)** The diagram shows the different points at which chloramphenicol, the tetracyclines, and streptomycin exert their activities.

Q **Why do antibiotics that inhibit protein synthesis affect bacteria and not human cells?**

THE INHIBITION OF NUCLEIC ACID SYNTHESIS

A number of antibiotics interfere with the processes of DNA replication and transcription in microorganisms. Some drugs with this mode of action have an extremely limited usefulness because they interfere with mammalian DNA and RNA as well. Others, such as rifampin and the quinolones, are more widely used in chemotherapy because they are more selectively toxic.

INHIBITING THE SYNTHESIS OF ESSENTIAL METABOLITES

In Chapter 5, we mentioned that a particular enzymatic activity of a microorganism can be *competitively inhibited* by

a substance (antimetabolite) that closely resembles the normal substrate for the enzyme (see Figure 5.7, page 121). An example of competitive inhibition is the relationship between the antimetabolite sulfanilamide (a sulfa drug) and *para*-aminobenzoic acid **(PABA).** In many microorganisms, PABA is the substrate for an enzymatic reaction leading to the synthesis of folic acid, a vitamin that functions as a coenzyme for the synthesis of the purine and pyrimidine bases of nucleic acids and many amino acids. In the presence of sulfanilamide, the enzyme that normally converts PABA to folic acid combines with the drug instead of with PABA. This combination prevents folic acid synthesis and stops the growth of the microorganism. Because humans do not produce folic acid from PABA (they obtain it as a vitamin in ingested foods), sulfanilamide

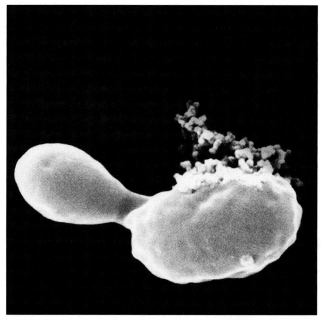

SEM ├─────┤ 10 mm

FIGURE 20.5 Injury to the plasma membrane of a yeast cell caused by an antifungal drug. The cell releases its cytoplasmic contents as the plasma membrane is disrupted by the antifungal drug miconazole.

Q Many antifungal drugs combine with sterols in the plasma membrane. Why don't they combine with sterols in human cell membranes?

exhibits selective toxicity—it affects microorganisms that synthesize their own folic acid but does not harm the human host. Other chemotherapeutic agents that act as antimetabolites are the sulfones and trimethoprim.

A SURVEY OF COMMONLY USED ANTIMICROBIAL DRUGS

LEARNING OBJECTIVE

• Explain why the drugs described in this section are specific for bacteria.

Tables 20.3 and 20.4 summarize the commonly used antimicrobial drugs.

ANTIBACTERIAL ANTIBIOTICS: INHIBITORS OF CELL WALL SYNTHESIS

LEARNING OBJECTIVES

• List the advantages of each of the following over penicillin: semisynthetic penicillins, cephalosporins, and vancomycin.

• Explain why INH and ethambutol are antimycobacterial agents.

TABLE 20.3	Antibacterial Drugs
Drugs by Mode of Action	Comments
Inhibitors of Cell Wall Synthesis	
Natural Penicillins	
Penicillin G	Against gram-positive bacteria, requires injection
Penicillin V	Against gram-positive bacteria, oral administration
Semisynthetic Penicillins	
Oxacillin	Resistant to penicillinase
Ampicillin	Broad spectrum
Amoxicillin	Broad spectrum; combined with inhibitor of penicillinase
Aztreonam	A monobactam; effective for gram-negative bacteria, including *Pseudomonas* spp.
Imipenem	A carbapenem; very broad spectrum
Cephalosporins	
Cephalothin	First-generation cephalosporin; activity similar to penicillin; requires injection
Cefixime	Third-generation cephalosporin; oral administration

➤

TABLE 20.3	**Antibacterial Drugs** *(continued)*

Drugs by Mode of Action	Comments
Polypeptide Antibiotics	
Bacitracin	Against gram-positive bacteria; topical application
Vancomycin	A glycopeptide type; penicillinase-resistant; against gram-positive bacteria
Antimycobacterial Antibiotics	
Isoniazid	Inhibits synthesis of mycolic acid component of cell wall of *Mycobacterium* spp.
Ethambutol	Inhibits incorporation of mycolic acid into cell wall of *Mycobacterium* spp.
Inhibitors of Protein Synthesis	
Chloramphenicol	Broad spectrum, potentially toxic
Aminoglycosides	
Streptomycin	Broad spectrum, including mycobacteria
Neomycin	Topical use, broad spectrum
Gentamicin	Broad spectrum, including *Pseudomonas* spp.
Tetracyclines	
Tetracycline, oxytetracycline, chlortetracycline	Broad spectrum, including chlamydias and rickettsias; animal feed additives
Macrolides	
Erythromycin	Alternative to penicillin
Azithromycin, clarithromycin	Semisynthetic; broader spectrum and better tissue penetration than erythromycin
Telithromycin (Ketek)	New generation of semisynthetic macrolides; used to cope with resistance to other macrolides
Streptogramins	
Quinupristin and dalfopristin (Synercid)	Alternative for treating vancomycin-resistant gram-positive bacteria
Oxazolidinones	
Linezolid (Zyvox)	Useful primarily against penicillin-resistant gram-positive bacteria
Injury to the Plasma Membrane	
Polymyxin B	Topical use, gram-negative bacteria, including *Pseudomonas* spp.
Inhibitors of Nucleic Acid Synthesis	
Rifamycins	
Rifampin (or rifampicin)	Inhibits synthesis of mRNA; treatment of tuberculosis
Quinolones and Fluoroquinolones	
Nalidixic acid, nofloxacin, ciprofloxacin	Inhibit DNA synthesis; broad spectrum; urinary tract infections
Gatifloxacin	Newest generation quinolone; increased potency against gram-positive bacteria
Competitive Inhibitors of the Synthesis of Essential Metabolites	
Sulfonamides	
Trimethoprim-sulfamethoxazole	Broad spectrum; combination is widely used

TABLE 20.4	Antifungal, Antiviral, Antiprotozoan, and Antihelminthic Drugs	
	Mode of Action	Comments
Antifungal Drugs		
Agents Affecting Fungal Sterols (Plasma Membrane)		
Polyenes		
Amphotericin B	Injury to plasma membrane	Systemic fungal infections; fungicidal
Azoles		
Clotrimazole, miconazole	Inhibit synthesis of plasma membrane	Topical use
Ketoconazole	Inhibits synthesis of plasma membrane	Can be taken orally for systemic fungal infections
Voriconazole	Inhibits synthesis of plasma membrane	Can penetrate blood–brain barrier to treat aspergillosis of the central nervous system
Allylamines		
Terbinafine, naftifine	Inhibits synthesis of plasma membrane	New class of antifungals frequently used to treat diseases resistant to azoles
Agents Affecting Fungal Cell Walls		
Echinocandins		
Caspofungin (Cancidas)	New class of antifungals that inhibit synthesis of cell wall	
Agents Inhibiting Nucleic Acids		
Flucytosine	Inhibits synthesis of RNA and therefore protein synthesis	
Other Antifungal Drugs		
Griseofulvin	Inhibition of mitotic microtubules	Fungal infections of the skin
Tolnaftate	Unknown	Athlete's foot
Antiviral Drugs (See also Table 20.5, *Drugs for Chemotherapy of HIV*)		
Nucleoside and Nucleotide Analogs		
Acyclovir, ganciclovir, ribavirin, lamivudine	Inhibit DNA or RNA synthesis	Used primarily against herpesviruses
Cidofovir	Inhibits DNA or RNA synthesis	Cytomegalovirus infections; possibly effective against smallpox
Adefovir dipivoxil (Hepsera)		For resistance against lamivudine
Attachment and Uncoating		
Zanamivir, oseltamivir	Inhibit neuraminidase on influenza virus	Treatment of influenza
Amantadine, zimantadine	Inhibit uncoating	Treatment of influenza
Interferons		
alpha-interferon	Inhibits spread of virus to new cells	Viral hepatitis

TABLE 20.4	Antifungal, Antiviral, Antiprotozoan, and Antihelminthic Drugs *(continued)*	
	Mode of Action	Comments
Antiprotozoan Drugs		
Chloroquine	Inhibits DNA synthesis	Malaria; effective against red blood cell stage only
Diiodohydroxyquin	Unknown	Amoebic infections; amoebicidal
Metronidazole, Tinidazole	Interferes with anaerobic metabolisms	Giardiasis, amebiasis, trichomoniasis
Nitazoxanide	Interferes with anaerobic metabolism	Giardiasis; only drug approved for cryptosporidiosis
Antihelminthic Drugs		
Niclosamide	Prevents ATP generation in mitochondria	Tapeworm infections; kills tapeworms
Praziquantel	Alters permeability of plasma membranes	Tapeworm and fluke infections; kills flatworms
Pyantel pamoate	Neuromuscular block	Intestinal roundworms; kills roundworms
Mebendazole, albendazole	Inhibit absorption of nutrients	Intestinal roundworms
Ivermectin	Paralyzes worm	Intestinal roundworms primarily; occasional use for scabies mite and lice

PENICILLIN

The term *penicillin* refers to a group of over 50 chemically related antibiotics (Figure 20.6). All penicillins have a common core structure containing a β-lactam ring called the nucleus. Penicillin molecules are differentiated by the chemical side chains attached to their nuclei. Penicillins can be produced either naturally or semisynthetically. Penicillins prevent the cross-linking of the peptidoglycans, which interferes with the final stages of the construction of the cell wall (see Figure 4.13a, page 86).

Natural Penicillins Penicillin extracted from cultures of the mold *Penicillium* exists in several closely related forms. These are the so-called **natural penicillins.** The prototype compound of all the penicillins is *penicillin* G. It has a narrow but useful spectrum of activity and is often the drug of choice against most staphylococci, streptococci, and several spirochetes. When injected intramuscularly, penicillin G is rapidly excreted from the body in 3 to 6 hours (Figure 20.7). When taken orally, the acidity of the digestive fluids in the stomach diminishes its concentration. *Procaine penicillin*, a combination of the drugs procaine and penicillin G, is retained at detectable concentrations for up to 24 hours; concentration peaks at about 4 hours. Still longer retention times can be achieved with *benzathine penicillin*, a combination of benzathine and penicillin G. Although retention times of as long as 4 months can be obtained, the concentration of the drug is so low that the organisms must be very sensitive to it. Penicillin V, which is stable in stomach acids and can be taken orally, and penicillin G are the natural penicillins most often used.

Natural penicillins have some disadvantages. Chief among them are their narrow spectrum of activity and their susceptibility to penicillinases. *Penicillinases* are enzymes produced by many bacteria, especially *Staphylococcus* species, that cleave the β-lactam ring of the penicillin molecule (Figure 20.8). Because of this characteristic, penicillinases are sometimes called *β-lactamases*.

Semisynthetic Penicillins A large number of **semisynthetic penicillins** have been developed in attempts to overcome the disadvantages of natural penicillins (see Figure 20.6b). Scientists develop these penicillins in either of two ways. First, they can interrupt synthesis of the molecule by *Penicillium* and obtain only the common penicillin nucleus for use. Second, they can remove the side chains from the completed natural molecules and then chemically add other side chains that make them more resistant to penicillinase, or the scientists can give them an extended spectrum. Thus the term *semisynthetic*: part of the penicillin is produced by the mold, and part is added synthetically.

Penicillinase-Resistant Penicillins The first semisynthetic penicillin designed to evade the action of penicillinases was *methicillin*. Eventually, so many strains of staphylococci

(a) Natural penicillins

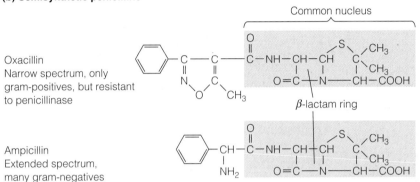

Penicillin G (Requires injection)

Penicillin V (Can be taken orally)

Common nucleus

β-lactam ring

FIGURE 20.6 The structure of penicillins, antibacterial antibiotics. The portion that all penicillins have in common—which contains the β-lactam ring—is shaded in purple. The unshaded portions represent the side chains that distinguish one penicillin from another.

Q What does semisynthetic *mean?*

(b) Semisynthetic penicillins

Oxacillin
Narrow spectrum, only gram-positives, but resistant to penicillinase

Ampicillin
Extended spectrum, many gram-negatives

Common nucleus

β-lactam ring

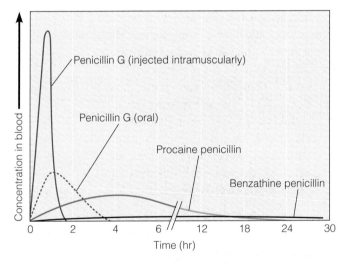

Concentration in blood

Penicillin G (injected intramuscularly)

Penicillin G (oral)

Procaine penicillin

Benzathine penicillin

0   2   4   6   12   18   24   30

Time (hr)

FIGURE 20.7 Retention of penicillin G. Penicillin G is normally injected (solid red line); when administered by this route, the drug is present in high concentrations in the blood but disappears quickly. Taken orally (dotted red line), penicillin G is destroyed by stomach acids and is not very effective. It is possible to improve retention of penicillin G by combining it with compounds such as procaine and benzathine (blue and black lines). However, the blood concentration reached is low, and the target bacterium must be extremely sensitive to the antibiotic.

Q How does a low concentration of penicillin G select for penicillin-resistant bacteria?

developed resistance to methicillin that the abbreviation **MRSA (methicillin-resistant Staphylococcus aureus)** made its appearance. Resistance became so prevalent that methicillin has been discontinued in the United States. Replacement antibiotics similar to methicillin, such as *oxacillin* and *nafcillin*, have been developed.

Extended-Spectrum Penicillins To overcome the problem of the narrow spectrum of activity of natural penicillins, broader-spectrum semisynthetic penicillins have been developed. These new penicillins are effective against many gram-negative bacteria as well as gram-positive ones, although they are not resistant to penicillinases. The first such penicillins were the aminopenicillins, such as *ampicillin* and *amoxicillin*. When bacterial resistance to these became more common, the carboxypenicillins were developed. Members of this group, such as *carbenicillin* and *ticarcillin*, have even greater activity against gram-negative bacteria and have the special advantage of activity against *Pseudomonas aeruginosa*.

Among the more recent additions to the penicillin family are the ureidopenicillins, such as *mezlocillin* and *azlocillin*. These broader-spectrum penicillins are modifications of the structure of ampicillin. The search for even more effective modifications of penicillin continues.

Penicillins Plus β-Lactamase Inhibitors A different approach to the proliferation of penicillinase is to combine penicillins with *potassium clavulanate* (*clavulanic*

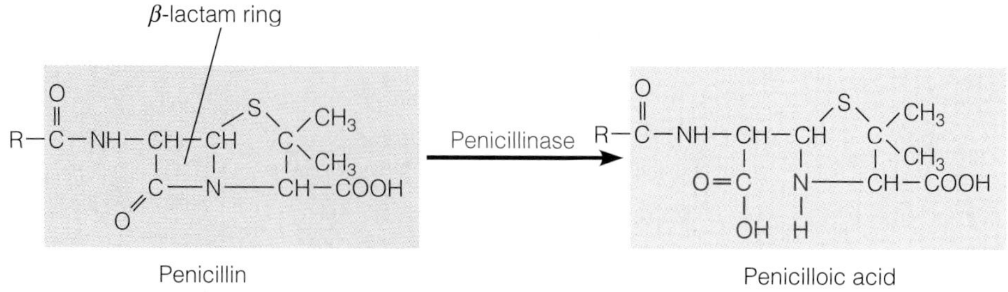

FIGURE 20.8 The effect of penicillinase on penicillins. Bacterial production of this enzyme, which is shown breaking the β-lactam ring, is by far the most common form of resistance to penicillins. R is an abbreviation for the chemical side groups that differentiate similar or otherwise-identical compounds.

β-lactam ring

Penicillinase

Penicillin

Penicilloic acid

Q **What is penicillinase?**

acid), a product of a streptomycete. Potassium clavulanate is a noncompetitive inhibitor of penicillinase with essentially no antimicrobial activity of its own. It has been combined with some new broader-spectrum penicillins, such as *amoxicillin* (the combination is best known by its trade name Augmentin).

Carbapenems The **carbapenems** are a class of β-lactam antibiotics that substitute a carbon atom for a sulfur atom and add a double bond to the penicillin nucleus. These antibiotics, which inhibit cell wall synthesis, have an extremely broad spectrum of activity. Representative of this group is Primaxin, a combination of *imipenem* and *cilastin*. The cilastin has no antimicrobial activity but prevents degradation of the combination in the kidneys. Tests have demonstrated that Primaxin is active against 98% of all organisms isolated from hospital patients.

Monobactams Another method of avoiding the effects of penicillinase is shown by *aztreonam*, which is the first member of a new class of antibiotics. It is a synthetic antibiotic that has only a single ring rather than the conventional β-lactam double ring, and is therefore known as a **monobactam.** Aztreonam's spectrum of activity is remarkable for a penicillin-related compound—this antibiotic, which has unusually low toxicity, affects only certain gram-negative bacteria, including pseudomonads and *E. coli.*

CEPHALOSPORINS

In structure, the nuclei of **cephalosporins** resemble those of penicillin (Figure 20.9). Cephalosporins inhibit cell wall synthesis in essentially the same way as do penicillins. However, cephalosporins differ from penicillin in that they are resistant to penicillinases and are effective against more gram-negative organisms than the natural penicillins. However, the cephalosporins are susceptible to a separate group of β-lactamases.

The number of second-, third-, and even fourth-generation cephalosporins has proliferated in recent years; there are now more than 70 versions. Each generation tends to be more effective against gram-negatives and has a broader spectrum of activity than the previous genera-

tion. Some first-generation cephalosporins are *cephalothin*, *cefamandole*, and *cefotaxime*.

Oral administration is preferred by patients, and several newer cephalosporins allow this. *Cefpodoxime* and *cefixime* are two cephalosporins that have been approved for oral administration.

POLYPEPTIDE ANTIBIOTICS

Bacitracin Bacitracin (the name is derived from its source, a *Bacillus* isolated from a wound on a girl named Tracy) is a polypeptide antibiotic effective primarily against gram-positive bacteria, such as staphylococci and streptococci. Bacitracin inhibits the synthesis of cell walls at an earlier stage than penicillins and cephalosporins. It interferes with the synthesis of the linear strands of the peptidoglycans (see Figure 4.13a, page 86.) Its use is restricted to topical application for superficial infections.

Vancomycin Vancomycin (optimistically named from the word *vanquish*) is one of a small group of glycopeptide antibiotics derived from a species of *Streptomyces* found in the jungles of Borneo. Originally, toxicity of vancomycin was a

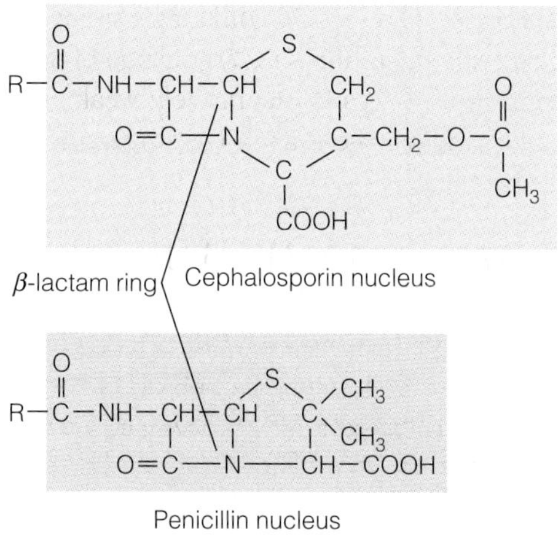

β-lactam ring Cephalosporin nucleus

Penicillin nucleus

FIGURE 20.9 The nuclear structures of cephalosporin and penicillin compared.

Q **Would a β-lactamase effective against penicillin G be likely to affect cephalosporins?**

serious problem, but improved purification procedures in its manufacture have largely corrected this. Although it has a very narrow spectrum of activity, which is based on inhibition of cell wall synthesis, vancomycin has been extremely important in addressing the problem of MRSA (see page 591). Vancomycin has been considered the last line of antibiotic defense for treatment of *Staphylococcus aureus* infections that are resistant to other antibiotics. The widespread use of vancomycin to treat MRSA has led to the selection of **vancomycin-resistant enterococci (VRE).** These are opportunistic, gram-negative pathogens that are particularly troublesome in hospital settings (see page 333 and the box on page 633). This appearance of vancomycin-resistant pathogens, leaving almost no effective alternative, is considered a medical emergency.

ANTIMYCOBACTERIAL ANTIBIOTICS

The cell wall of members of the genus *Mycobacterium* differs from the cell wall of most other bacteria. It incorporates mycolic acids that are a factor in their staining properties, causing them to stain as acid-fast (page 70). The genus includes important pathogens, such as those that cause leprosy and tuberculosis.

Isoniazid (INH) is a very effective synthetic antimicrobial drug against *Mycobacterium tuberculosis*. The primary effect of INH is to inhibit the synthesis of mycolic acids, which are components of cell walls only of the mycobacteria. It has little effect on nonmycobacteria. When used to treat tuberculosis, INH is usually administered simultaneously with other drugs, such as rifampin (also known as rifampicin) or ethambutol. This minimizes development of drug resistance. Because the tubercle bacillus is usually found only within macrophages or walled off in tissue, any antitubercular drug must be able to penetrate into such sites.

Ethambutol is effective only against mycobacteria. The drug apparently inhibits incorporation of mycolic acid into the cell wall. It is a comparatively weak antitubercular drug; its principal use is as the secondary drug to avoid resistance problems.

INHIBITORS OF PROTEIN SYNTHESIS

LEARNING OBJECTIVE
- Describe how each of the following inhibits protein synthesis: aminoglycosides, tetracyclines, chloramphenicol, macrolides.

CHLORAMPHENICOL

Chloramphenicol is a broad-spectrum antibiotic with serious toxicity problems. Because of its relatively simple structure (Figure 20.10), it is less expensive for the pharmaceutical industry to synthesize it chemically than to isolate it from

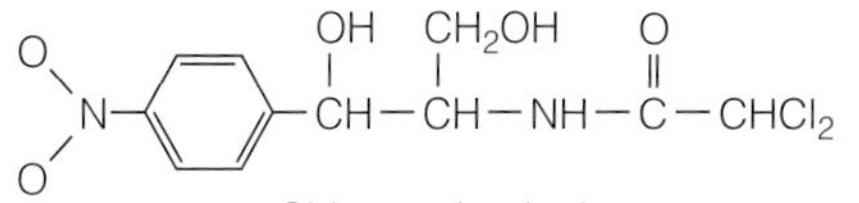

Chloramphenicol

FIGURE 20.10 The structure of the antibacterial antibiotic chloramphenicol. Notice the simple structure, which makes synthesizing this drug less expensive than isolating it from *Streptomyces.*

Q What effect does the binding of chloramphenicol to the 50S portion of the ribosomes have on a cell?

Streptomyces. Because it is relatively inexpensive, chloramphenicol is often used where low cost is essential. Its relatively small molecular size promotes its diffusion into areas of the body that are normally inaccessible to many other drugs. However, chloramphenicol has serious adverse effects; most important is the suppression of bone marrow activity. This suppression affects the formation of blood cells. In about 1 in 40,000 users, the drug appears to cause aplastic anemia, a potentially fatal condition; the normal rate for this condition is only about 1 in 500,000 individuals. Physicians are advised not to use the drug for trivial conditions or ones for which suitable alternatives are available.

AMINOGLYCOSIDES

Aminoglycosides are a group of antibiotics in which amino sugars are linked by glycoside bonds. They were among the first antibiotics to have significant activity against gram-negative bacteria. Probably the best-known aminoglycoside is *streptomycin*, which was discovered in 1944. Streptomycin is still used as an alternative drug in the treatment of tuberculosis, but rapid development of resistance and serious toxic effects have diminished its usefulness.

Aminoglycosides are bactericidal and inhibit protein synthesis. They can affect hearing by causing permanent damage to the auditory nerve, and damage to the kidneys has also been reported. Because of this, their use has been declining. *Neomycin* is present in many nonprescription topical preparations. *Gentamicin* (spelled with an "i" to reflect its source, the filamentous bacterium *Micromonospora*) is especially useful against *Pseudomonas* infections. Pseudomonads are a major problem for persons suffering from cystic fibrosis. The aminoglycoside *tobramycin* is administered in an aerosol to aid in the control of infections that occur in patients with cystic fibrosis.

TETRACYCLINES

Tetracyclines are a group of closely related broad-spectrum antibiotics produced by *Streptomyces* spp. Tetracyclines inhibit protein synthesis. They not only are effective against gram-positive and gram-negative bacteria but also

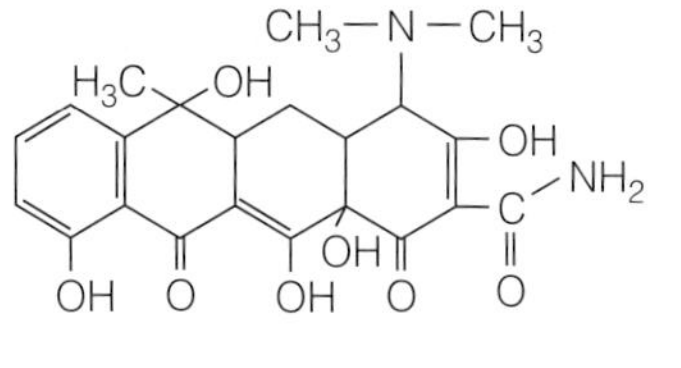

Tetracycline

FIGURE 20.11 The structure of the antibacterial antibiotic tetracycline. Other tetracycline-type antibiotics share the four-cyclic-ring structure of tetracycline and closely resemble it.

Q **How do tetracyclines affect bacteria?**

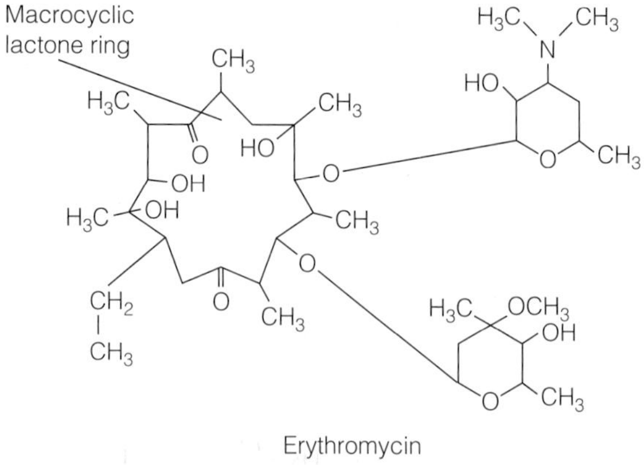

penetrate body tissues well and are especially valuable against the intracellular rickettsias and chlamydias. Three of the more commonly used tetracyclines are *oxytetracycline* (Terramycin), *chlortetracycline* (Aureomycin), and tetracycline itself (Figure 20.11). Some semisynthetic tetracyclines, such as *doxycycline* and *minocycline*, are available. They have the advantage of longer retention in the body.

Tetracyclines are used to treat many urinary tract infections, mycoplasmal pneumonia, and chlamydial and rickettsial infections. They are also frequently used as alternative drugs for such diseases as syphilis and gonorrhea. Tetracyclines often suppress the normal intestinal microbiota because of their broad spectrum, causing gastrointestinal upsets and often leading to superinfections, particularly by the fungus *Candida albicans*. They are not advised for children, who might experience a brownish discoloration of the teeth, or for pregnant women, in whom they might cause liver damage. Tetracyclines are among the most common antibiotics added to animal feeds, where their use results in significantly faster weight gain; however, some human health problems can also result (see the box on page 606).

MACROLIDES

Macrolides are a group of antibiotics named for the presence of a macrocyclic lactone ring. The best-known macrolide in clinical use is *erythromycin* (Figure 20.12). Its mode of action is the inhibition of protein synthesis, apparently by blocking the tunnel shown in Figure 20.4a. However, erythromycin is not able to penetrate the cell walls of most gram-negative bacilli. Its spectrum of activity is therefore similar to that of penicillin G, and it is a frequent alternative drug to penicillin. Because it can be administered orally, an orange-flavored preparation of erythromycin is a frequent penicillin substitute for the treatment of streptococcal and staphylococcal infections in children. Erythromycin is the drug of choice for the treatment of legionellosis, mycoplasmal pneumonia, and several other infections.

Other macrolides now available include *azithromycin* and *clarithromycin*. Compared to erythromycin, they have a broader antimicrobial spectrum and penetrate tissues

Erythromycin

FIGURE 20.12 The structure of the antibacterial antibiotic erythromycin, a representative macrolide. All macrolides have the macrocyclic lactone ring shown here.

Q **How do macrolides affect bacteria?**

better. This is especially important in the treatment of conditions caused by intracellular bacteria such as *Chlamydia*, a frequent cause of sexually transmitted infection.

A new generation of semisynthetic macrolides, the **ketolides,** is being developed to cope with increasing resistance to other macrolides. The prototype of this generation is *telithromycin* (Ketek).

STREPTOGRAMINS

We mentioned previously that the appearance of vancomycin-resistant pathogens constitutes a serious medical problem. One answer may be a unique group of antibiotics, the **streptogramins.** The first of these drugs to be released, Synercid, is a combination of two cyclic peptides, *quinupristin* and *dalfopristin*, which are distantly related to the macrolides. They block protein synthesis by attaching to the 50S portion of the ribosome, as do other antibiotics such as chloramphenicol. Synercid, however, acts at uniquely different points on the ribosome. Dalfopristin blocks an early step in protein synthesis, and quinupristin blocks a later step. The combination causes incomplete peptide chains to be released and is synergistic in its action (see page 605). Synercid is effective against a broad range of gram-positive bacteria that are resistant to other antibiotics. This makes Synercid especially valuable, even though it is expensive and has a high incidence of adverse side effects.

OXAZOLIDINONES

The oxazolidinones are another new class of antibiotics developed in response to vancomycin resistance. When the FDA approved this class of antibiotic in 2001, it represented the first new class of antibiotics approved in 25 years. Like several other antibiotics that inhibit protein synthesis, oxazolidinone antibiotics act on the ribosome (see Figure 20.4, page 586). However, they are unique in their target, binding to the 50S ribosomal subunit close to the point where it interfaces with the 30S subunit. These drugs are totally synthetic, which may make resistance slower to develop. Like vancomycin, they have no usefulness against gram-negative bacteria, but they are active against certain enterococci that are not sensitive to Synercid. One member of this antibiotic group is *linezolid* (Zyvox), used mainly to combat MRSA.

INJURY TO THE PLASMA MEMBRANE

LEARNING OBJECTIVE
- Compare the mode of action of polymyxin B, bacitracin, and neomycin.

Polymyxin B is a bactericidal antibiotic effective against gram-negative bacteria. For many years, it was one of very few drugs used against infections by gram-negative *Pseudomonas*. The mode of action of polymyxin B is to injure plasma membranes. Polymyxin B is seldom used today except in the topical treatment of superficial infections.

Both *bacitracin* and *polymyxin B* are available in nonprescription antiseptic ointments, in which they are usually combined with *neomycin*, a broad-spectrum aminoglycoside. In a rare exception to the rule, these antibiotics do not require a prescription.

Many of the antimicrobial peptides discussed on page 605 target the synthesis of the plasma membrane.

INHIBITORS OF NUCLEIC ACID (DNA/RNA) SYNTHESIS

LEARNING OBJECTIVE
- Describe how rifamycins and quinolones kill bacteria.

RIFAMYCINS

The best known derivative of the **rifamycin** family of antibiotics is *rifampin*. These drugs are structurally related to the macrolides and inhibit the synthesis of mRNA. By far the most important use of rifampin is against mycobacteria in the treatment of tuberculosis and leprosy. A valuable characteristic of rifampin is its ability to penetrate tissues and reach therapeutic levels in cerebrospinal fluid and abscesses. This characteristic is probably an important factor in its antitubercular activity, because the tuberculosis pathogen is usually located inside tissues or macrophages. An unusual side effect of rifampin is the appearance of orange-red urine, feces, saliva, sweat, and even tears.

QUINOLONES AND FLUOROQUINOLONES

In the early 1960s, the synthetic drug *nalidixic acid* was developed—the first of the **quinolone** group of antimicrobials. It exerted a unique bactericidal effect by selectively inhibiting an enzyme (DNA gyrase) needed for the replication of DNA. Although nalidixic acid found only limited use (its only application being for urinary tract infections), it led to the development in the 1980s of a prolific group of synthetic quinolones, the **fluoroquinolones.**

The most widely used fluoroquinolones are *norfloxacin* and *ciprofloxacin*. The latter is better known under its trade name of Cipro and gained widespread publicity for its use against anthrax infections. Although they are relatively safe for adults, fluoroquinolones adversely affect the development of cartilage, and their use is limited among children, adolescents, and pregnant women.

A third generation of fluoroquinolones includes *moaxifloxacin* and *gatifloxacin*. These have a broader antimicrobial spectrum, especially against gram-positives, and can be taken orally. Gatifloxacin is also available in a liquid formulation for treatment of eye infections.

COMPETITIVE INHIBITORS OF THE SYNTHESIS OF ESSENTIAL METABOLITES

LEARNING OBJECTIVE
- Describe how sulfa drugs inhibit microbial growth.

SULFONAMIDES

As noted earlier, **sulfonamides,** or **sulfa drugs,** were among the first synthetic antimicrobial drugs used to treat microbial diseases. Antibiotics have diminished the importance of sulfa drugs in chemotherapy, but they continue to be used to treat certain urinary tract infections and have other specialized uses, as in the combination drug *silver sulfadiazine*, used to control infections in burn patients. Sulfonamides are bacteriostatic; as mentioned, their action is due to their structural similarity to para-aminobenzoic acid (PABA) (see the discussion and formulas on page 122).

Probably the most widely used sulfa drug today is a combination of *trimethoprim* and *sulfamethoxazole* (TMP-SMZ). This combination is an excellent example of drug **synergism.** When used in combination, only 10% of the concentration is needed, compared to when each drug is used alone. The combination also has a broader spectrum of action and greatly reduces the emergence of resistant strains. (Synergism is discussed more fully later in the chapter; see Figure 20.22.)

Figure 20.13 illustrates how the two drugs interfere with different steps of a metabolic sequence leading to the synthesis of precursors of proteins, DNA, and RNA.

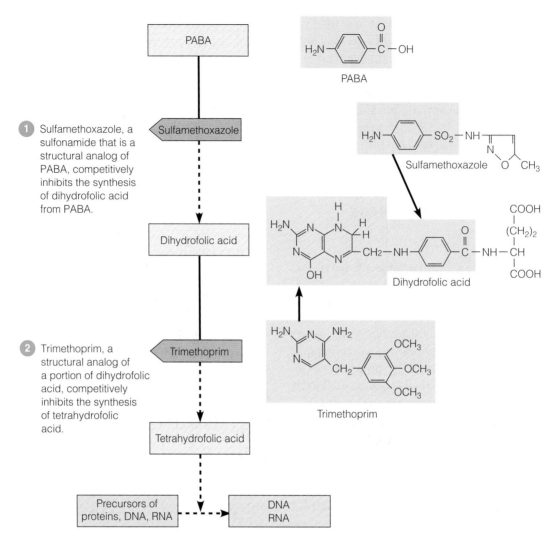

FIGURE 20.13 Actions of the antibacterial synthetics trimethoprim and sulfamethoxazole. TMP-SMZ works by inhibiting different stages in the synthesis of precursors of DNA, RNA, and proteins. Together the drugs are synergistic.

Q Define **synergism.**

ANTIFUNGAL DRUGS

LEARNING OBJECTIVE
- Explain the modes of action of currently used antifungal drugs.

Eukaryotes, such as fungi, use the same mechanisms to synthesize proteins and nucleic acids as higher animals. Therefore it is more difficult to find a point of selective toxicity in eukaryotes than in prokaryotes. Moreover, fungal infections are becoming more frequent because of their role as opportunistic infections in immunosuppressed individuals, especially those with AIDS.

AGENTS AFFECTING FUNGAL STEROLS

Many antifungal drugs target the sterols in the plasma membrane. In fungal membranes, the principal sterol is ergosterol; in animal membranes, cholesterol. When the biosynthesis of ergosterol in a fungal membrane is interrupted, the membrane becomes excessively permeable, killing the cell. Inhibition of ergosterol biosynthesis is the basis for the selective toxicity of many antifungals, which include members of the polyene, azole, and allylamine groups.

Polyenes   *Amphotericin B* is the most commonly used member of the antifungal **polyene antibiotics** (Figure 20.14). For many years amphotericin B, produced by *Streptomyces* species of soil bacteria, has been a mainstay of clinical treatment for systemic fungal diseases such as histoplasmosis, coccidioidomycosis, and blastomycosis. The drug's toxicity, particularly to the kidneys, is a strongly limiting factor in these uses. Administering the drug encapsulated in lipids (liposomes) appears to minimize toxicity.

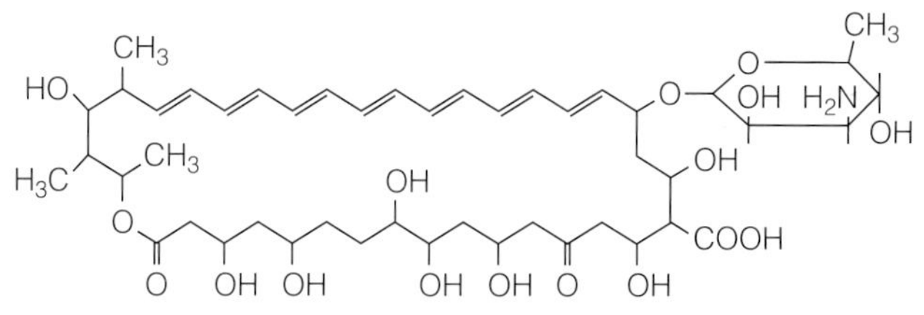

FIGURE 20.14 The structure of the antifungal drug amphotericin B, representative of the polyenes.

Q Why do polyenes injure fungal plasma membranes and not bacterial membranes?

Azoles Some of the most widely used antifungal drugs are represented among the **azole antibiotics.** Before they made their appearance, the only drugs available for systemic fungal infections were amphotericin B and flucytocine (discussed below). The first azoles were **imidazoles,** such as *clotrimazole* and *miconazole* (Figure 20.15), which are now sold without a prescription for topical application for treatment of cutaneous mycoses, such as athlete's foot and vaginal yeast infections. An important addition to this group was *ketoconazole,* which has an unusually broad spectrum of activity among fungi. Ketoconazole, taken orally, is often used as a less toxic alternative to amphotericin B for many systemic fungal infections, although occasional liver damage has been reported. Ketoconazole topical ointments are used to treat dermatomycoses. A promising new broad spectrum antifungal, *voriconazole,* is expected to replace amphotericin B for treatment of many systemic antifungal infections. It has a special advantage in treatment of aspergillosis of the central nervous system because it is able to penetrate the blood–brain barrier (see Figure 22.2, page 644).

The use of ketoconazole diminished sharply upon the introduction of the **triazole antibiotics.** Drugs of this group, such as *fluconazole* and *itraconazole,* are less toxic and have other advantages as well. Unlike ketoconazole, they are very water soluble, which makes their administration for systemic infections easier and more effective.

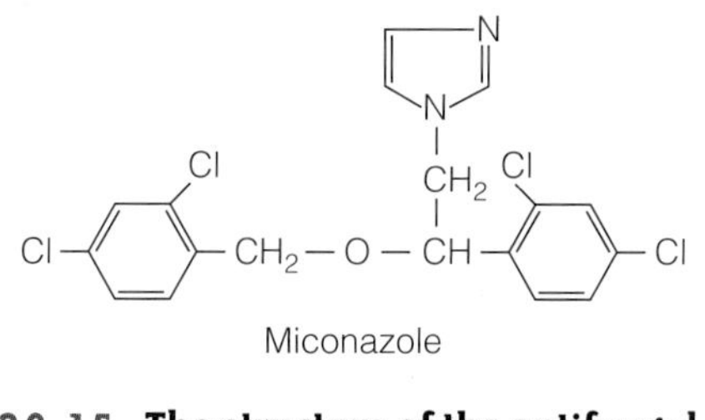

FIGURE 20.15 The structure of the antifungal drug miconazole, representative of the imidazoles.

Q How do azoles affect fungi?

Allylamines The **allylamines** represent a recently developed class of antifungals that inhibit the biosynthesis of ergosterols in a manner that is functionally distinct. *Terbinafine* and *naftifine,* examples of this group, are frequently used when resistance to azole-type antifungals arises.

AGENTS AFFECTING FUNGAL CELL WALLS

The fungal cell wall contains compounds that are unique to these organisms. A primary target for selective toxicity among these compounds is β-glucan. Inhibition of the biosynthesis of this glucan results in an incomplete cell wall and results in lysis of the fungal cell. The first of a new class of antifungal drugs (the first in 40 years) is the **echinocandins,** which inhibit the biosynthesis of glucans. A member of the echinocandin group, *caspofungin (Cancidas)* is now available commercially. This new antifungal agent is expected to become especially valuable for combating systemic *Aspergillus* infections in persons whose immune system is compromised. It is also effective against important fungi such as *Candida* spp. and *Pneumocystis jiroveci,* which causes a pneumonia often seen in AIDS patients.

AGENTS INHIBITING NUCLEIC ACIDS

Flucytosine, an analog of the pyrimidine cytosine, interferes with the biosynthesis of RNA and therefore protein synthesis. The selective toxicity lies in the ability of the fungal cell to convert flucytocine into 5-fluorouracil, which is incorporated into RNA and eventually disrupts protein synthesis. Mammalian cells lack the enzyme to make this conversion of the drug. Flucytosine has a narrow spectrum of activity, and toxicity to the kidneys and bone marrow further limit its use.

OTHER ANTIFUNGAL DRUGS

Griseofulvin is an antibiotic produced by a species of *Penicillium.* It has the interesting property of being active against superficial dermatophytic fungal infections of the hair (tinea capitis, or ringworm) and nails, even though its route of administration is oral. The drug apparently binds selectively to the keratin found in the skin, hair follicles, and nails. Its mode of action is primarily to block microtubule assembly, which interferes with mitosis and thereby inhibits fungal reproduction.

Tolnaftate is a common alternative to miconazole as a topical agent for the treatment of athlete's foot. Its mechanism of action is not known. *Undecylenic acid* is a fatty acid that has antifungal activity against athlete's foot, although it is not as effective as tolnaftate or the imidazoles.

Pentamidine isethionate is used in the treatment of *Pneumocystis* pneumonia. The drug's mode of action is unknown, but it appears to bind DNA.

ANTIVIRAL DRUGS

LEARNING OBJECTIVE

• Explain the modes of action of currently used antiviral drugs.

In developed parts of the world, it is estimated that at least 60% of infectious illnesses are caused by viruses, and about 15% by bacteria. Every year, at least 90% of the U.S. population suffers from a viral disease. Yet relatively few antiviral drugs have been approved in the United States, and they are effective against only an extremely limited group of diseases. Many of the recently developed antiviral drugs are directed against HIV, the pathogen responsible for the pandemic of AIDS. Therefore, as a practical matter the discussion of antivirals is often separated into agents that are directed at chemotherapy of HIV (see page 576 and Table 20.5) and those with more general (non-HIV) applications (see Table 20.4).

Because viruses replicate within the host's cells, very often using the genetic and metabolic mechanisms of the host's own cells, it is relatively difficult to target the virus without damaging the host's cellular machinery. Many of the antivirals in use today are analogs of components of viral DNA or RNA. However, as more becomes known about the reproduction of viruses, more targets suggest themselves for antiviral action.

TABLE 20.5	**Drugs for Chemotherapy of HIV**
Generic Name	Alternative or Brand Name
Nucleoside Analog Reverse Transcriptase Inhibitors	
Abacavir	ABC, Ziagen
Didanosine	ddl, Videx
Emtricitabine	Emtriva
Lamivudine	3TC, Epivir
Stavudine	d4T, Zerit
Zalcitabine	ddC, Hivid
Zidovudine	AZT, ZDV, Retrovir
Nucleotide Analog Reverse Transcriptase Inhibitor	
Tenofovir	Viread
Nonnucleoside Reverse Transcriptase Inhibitors	
Delavirdine	Rescriptor
Efavirenz	Sustiva
Nevirapine	Viramune
Protease Inhibitors	
Atazanavir	Reyataz
Amprenavir	Agenerase
Indinavir	Crixivan
Nelfinavir	Viracept
Ritonavir	Norvir
Saquinavir	Invirase, Fortovase
Lopinavir* plus ritonavir	Kaletra
Fusion Inhibitors	
Enfuvirtide	T-20, Fuzeon

*Available only as this combination.

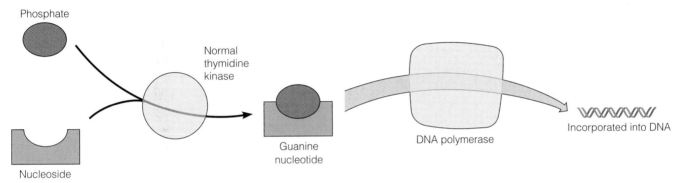

(a) Acyclovir structurally resembles the nucleoside deoxyguanosine.

(b) The enzyme thymidine kinase combines phosphates with nucleosides to form nucleotides, which are then incorporated into DNA.

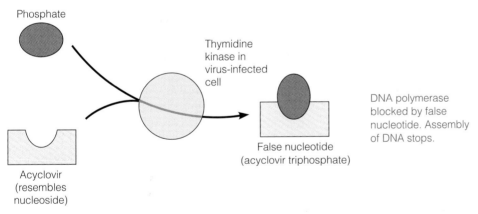

(c) Acyclovir has no effect on a cell not infected by a virus, that is, with normal thymidine kinase. In a virally infected cell, the thymidine kinase is altered and converts the acyclovir (which resembles the nucleoside deoxyguanosine) into a false nucleotide —which blocks DNA synthesis by DNA polymerase.

FIGURE 20.16 The structure and function of the antiviral drug acyclovir.

Q **Why are viral infections generally difficult to treat with chemotherapeutic agents?**

NUCLEOSIDE AND NUCLEOTIDE ANALOGS

Several important antiviral drugs are analogs of nucleosides and nucleotides (page 47). Among the nucleoside analogs, *acyclovir* is the one more widely used (Figure 20.16). While best known for treating genital herpes, it is generally useful for most herpesvirus infections, especially in immunosuppressed individuals. The antiviral drugs *famciclovir*, which can be taken orally, and *ganciclovir* are derivatives of acyclovir and have a similar mode of action. *Ribavirin* resembles the nucleoside guanine and accelerates the already high mutation rate of RNA viruses until the accumulation of errors reaches a crisis point, killing the virus. The nucleoside analog, *lamivudine,* is used to treat hepatitis B. More recently, a nucleotide analog *adefovir dipivoxil (Hepsera)* has been introduced for patients resistant to the nucleoside lamivudine. A nucleoside analog, *cidofovir,* is presently used for treatment of cytomegalovirus infections of the eye, but this drug is especially interesting because it shows promise as a possible treatment of smallpox.

Antiretrovirals   The need for effective chemotherapy of HIV infections has led to the development of a relatively numerous group of antiviral drugs. HIV is an RNA virus, and its reproduction depends on the enzyme reverse transcriptase, which controls the synthesis of RNA from DNA. Analogs of nucleotides or nucleosides are often a basis for drugs to block this essential step. In fact, the term *antiretroviral* currently implies that a drug is used to treat HIV infection. A well-known example of a nucleoside analog is *zidovudine.* Currently, the only nucleotide reverse transcriptase inhibitor is *tenofovir.* Only recently approved, it is often used when other regimens have failed. (See Table 20.5, which summarizes many of the current drugs available for HIV chemotherapy.)

Not all drugs that inhibit reverse transcriptase are nucleoside or nucleotide analogs. For example, several nonnucleoside agents, such as *nevirapine* (see Table 20.5), block RNA synthesis by other mechanisms.

OTHER ENZYME INHIBITORS

Another approach to the control of HIV infections is to inhibit the enzymes that control the last stage of viral reproduction. When the host cell (at the direction of the infecting HIV) makes a new virus, it must begin by cutting up large proteins with protease enzymes. The resulting fragments are used to assemble new viruses. Analogs of amino acid sequences in the large proteins can serve as inhibitors of these proteases by competitively interfering with their activity. The **protease inhibitors** *atazanavir, indinavir,* and *saquinavir* (see Table 20.5) have proved especially effective when combined with inhibitors of reverse transcriptase. Other enzymatic targets of HIV are the enzymes, integrases, that integrate the viral DNA into the host's DNA, forming a provirus—by **integrase inhibitors.** Entry of HIV into the cell by fusion can be blocked by **fusion inhibitors** such as *envuvirtide*—which, however, is dauntingly expensive.

Two inhibitors of the enzyme neuraminidase (page 731) have been introduced for treatment of influenza. These are *zanamivir* (Relenza) and *oseltamivir* (Tamiflu).

INTERFERONS

Cells infected by a virus often produce interferon, which inhibits further spread of the infection. Interferons are classified as cytokines, discussed in Chapter 17. *Alpha-interferon* (see Chapter 16, page 494) is currently a drug of choice for viral hepatitis infections. The production of interferons can be stimulated by a recently introduced antiviral, **imiquimod.** This drug is often prescribed to treat genital warts caused by a herpesvirus.

ANTIPROTOZOAN AND ANTIHELMINTHIC DRUGS

> **LEARNING OBJECTIVE**
> • Explain the modes of action of currently used antiprotozoan and antihelminthic drugs.

For hundreds of years, quinine from the Peruvian cinchona tree was the only drug known to be effective for the treatment of a parasitic infection (malaria). It was first introduced into Europe in the early 1600s and was known as "Jesuit's powder." There are now many antiprotozoan and antihelminthic drugs, although many of them are still considered experimental. This does not preclude their use, however, by qualified physicians. The CDC provides several of them on request when they are not available commercially.

ANTIPROTOZOAN DRUGS

Quinine is still used to control the protozoan disease malaria, but synthetic derivatives, such as *chloroquine,* have largely replaced it. For preventing malaria in areas where the disease has developed resistance to chloroquine, the new drug *mefloquine* is recommended, although serious psychiatric side effects have been reported. *Quinacrine* is the drug of choice for treating the protozoan disease giardiasis. *Diiodohydroxyquin (iodoquinol)* is an important drug prescribed for several intestinal amoebic diseases, but its dosage must be carefully controlled to avoid optic nerve damage. Its mode of action is unknown.

Metronidazole (Flagyl) is one of the most widely used antiprotozoan drugs. It is unique in that it acts not only against parasitic protozoa but also against obligately anaerobic bacteria. For example, as an antiprotozoan agent, it is the drug of choice for vaginitis caused by *Trichomonas vaginalis.* It is also used in the treatment of giardiasis and amoebic dysentery. The mode of action is to interfere with anaerobic metabolism, which incidentally these protozoans share with certain obligately anaerobic bacteria, such as *Clostridium.*

Tinidazole, a drug similar to metronidazole, has only recently been approved for use in the United States—although it has long been used elsewhere under the trade name of Fasigyn. It is effective in the treatment of giardiasis, amebiasis, and trichomoniasis. Another new antiprotozoan agent, and the first to be approved for the chemotheraphy of diarrhea caused by *Cryptosporidium hominis,* is *nitazoxanide.* It is active in treatment of giardiasis and ameboasis. Interestingly, it is also effective in treating several helminthic diseases, as well as having activity against some anaerobic bacteria.

ANTIHELMINTHIC DRUGS

With the increased popularity of sushi, a Japanese specialty often made with raw fish, the CDC began to notice an increased incidence of tapeworm infections. To estimate the incidence, the CDC documents requests for *niclosamide,* which is the usual first choice in treatment. The drug is effective because it inhibits ATP production under aerobic conditions. *Praziquantel* is about equally effective for the treatment of tapeworms; it kills worms by altering the permeability of their plasma membranes. Praziquantel has a broad spectrum of activity and is highly recommended for treating several fluke-caused diseases, especially schistosomiasis. It causes the helminths to undergo muscular spasms and apparently makes them susceptible to attack by the immune system. Apparently, its action exposes surface antigens, which antibodies can then reach.

Mebendazole and *albendazole* are broad-spectrum antihelminthics that have few side effects and have become the drugs of choice for treatment of many intestinal helminthic infections. The mode of action of both drugs is to inhibit the formation of microtubules in the cytoplasm, which interferes with the absorption of nutrients by the parasite. These drugs are also widely used in the livestock industry; for veterinary applications they are relatively more effective in ruminant animals.

Ivermectin is a drug with a wide range of applications. It is known to be produced by only one species of organism, *Streptomyces avermectinius,* which was isolated from the soil near a Japanese golf course. It is effective against many nematodes (roundworms) and several mites (such as scabies), ticks, and insects (such as head lice). (Some mites and insects happen to share certain similar metabolic channels with affected helminths.) Its primary use has been in the livestock industry as a broad-spectrum antihelminthic. Its exact mode of action is uncertain, but the final result is paralysis and death of the helminth without affecting mammalian hosts.

TESTS TO GUIDE CHEMOTHERAPY

LEARNING OBJECTIVE

• Describe two tests for microbial susceptibility to chemotherapeutic agents.

Different microbial species and strains have different degrees of susceptibility to different chemotherapeutic agents. Moreover, the susceptibility of a microorganism can change with time, even during therapy with a specific drug. Thus, a physician must know the sensitivities of the pathogen before treatment can be started. However, physicians often cannot wait for sensitivity tests and must begin treatment based on their "best guess" estimation of the most likely pathogen causing the illness.

Several tests can be used to indicate which chemotherapeutic agent is most likely to combat a specific pathogen. However, if the organisms have been identified—for example, *Pseudomonas aeruginosa,* beta-hemolytic streptococci, or gonococci—certain drugs can be selected without specific testing for susceptibility. Tests are necessary only when susceptibility is not predictable or when antibiotic resistance problems develop.

THE DIFFUSION METHODS

Probably the most widely used, although not necessarily the best, method of testing is the **disk-diffusion method,** also known as the *Kirby–Bauer test* (Figure 20.17). A Petri plate containing an agar medium is inoculated ("seeded") uniformly over its entire surface with a standardized amount of a test organism. Next, filter paper disks impregnated with known concentrations of chemotherapeutic agents are placed on the solidified agar surface. During incubation, the chemotherapeutic agents

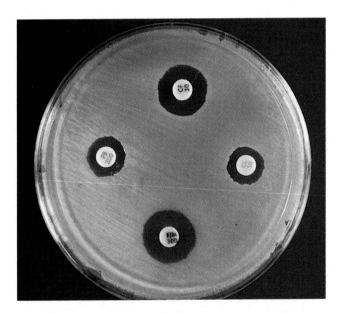

FIGURE 20.17 The disk-diffusion method for determining the activity of antimicrobials. Each disk contains a different chemotherapeutic agent, which diffuses into the surrounding agar. The clear zones indicate inhibition of growth of the microorganism swabbed onto the agar surface.

Q **Which agent is the most effective against the bacterium being tested?**

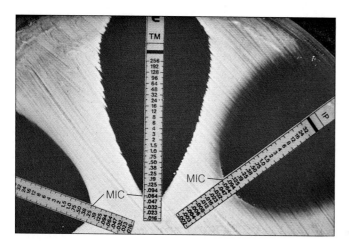

FIGURE 20.18   The E test (for epsilometer), a gradient diffusion method that determines antibiotic sensitivity and estimates minimal inhibitory concentration (MIC). The plastic strip, which is placed on an agar surface inoculated with test bacteria, contains an increasing gradient of the antibiotic. The MIC is clearly shown.

Q   What is the MIC of the central E test?

diffuse from the disks into the agar. The farther the agent diffuses from the disk, the lower its concentration. If the chemotherapeutic agent is effective, a **zone of inhibition** forms around the disk after a standardized incubation. The diameter of the zone can be measured; in general, the larger the zone, the more sensitive the microbe is to the antibiotic. The zone diameter is compared to a standard table for that drug and concentration, and the organism is reported as *sensitive, intermediate,* or *resistant.* For a drug with poor solubility, however, the zone of inhibition indicating that the microbe is sensitive will usually be smaller than for another drug that is more soluble and has diffused more widely. Results obtained by the disk-diffusion method are often inadequate for many clinical purposes. However, the test is simple and inexpensive and is most often used when more sophisticated laboratory facilities are not available.

A more advanced diffusion method, the **E test,** enables a lab technician to estimate the **minimal inhibitory concentration (MIC),** the lowest antibiotic concentration that prevents visible bacterial growth. A plastic-coated strip contains a gradient of antibiotic concentrations, and the MIC can be read from a scale printed on the strip (Figure 20.18).

BROTH DILUTION TESTS

A weakness of the diffusion method is that it does not determine whether a drug is bactericidal and not just bacteriostatic. A **broth dilution test** is often useful in determining the MIC and the **minimal bactericidal concentration (MBC)** of an antimicrobial drug. The MIC is determined by making a sequence of decreasing concentrations of the drug in a broth, which is then inoculated with the test bacteria (Figure 20.19). The wells that do not show growth (higher concentration than the MIC) can be cultured in broth or on agar plates free of the drug. If growth occurs in this broth, the drug was not bactericidal, and the MBC can be determined. Determining the MIC and MBC is important because it avoids the excessive or erroneous use of expensive antibiotics and minimizes the chance of toxic reactions that larger-than-necessary doses might cause.

Dilution tests are often highly automated. The drugs are purchased already diluted into broth in wells formed in a plastic tray. A suspension of the test organism is prepared and inoculated into all the wells simultaneously by a special inoculating device. After incubation, the turbidity may be read visually, although clinical laboratories with high workloads may read the trays with special scanners that enter the data into a computer that provides a printout of the MIC.

Other tests are also useful for the clinician; a determination of the microbe's ability to produce β-lactamase is one example. One popular, rapid method makes use of a cephalosporin that changes color when its β-lactam ring is opened. In addition, a measurement of the *serum concentration* of an antimicrobial is especially important when toxic drugs are used. These assays tend to vary with the drug and may not always be suitable for smaller laboratories.

THE EFFECTIVENESS OF CHEMOTHERAPEUTIC AGENTS

DRUG RESISTANCE

LEARNING OBJECTIVE
• Describe the mechanisms of drug resistance.

Bacteria become resistant to chemotherapeutic agents by four major mechanisms:

1. Destruction or inactivation of the drug (by β-lactamase, for example)

2. Prevention of penetration to the target site within the microbe (a frequent mechanism for tetracycline resistance)

3. Alteration of the drug's target sites (for example, a single amino acid change in the ribosome may be enough

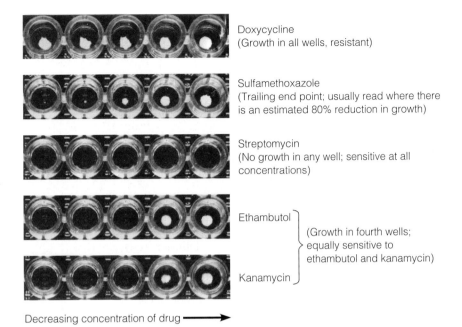

Doxycycline
(Growth in all wells, resistant)

Sulfamethoxazole
(Trailing end point; usually read where there
is an estimated 80% reduction in growth)

Streptomycin
(No growth in any well; sensitive at all
concentrations)

Ethambutol ⎤
          ⎬ (Growth in fourth wells;
          ⎪ equally sensitive to
          ⎪ ethambutol and kanamycin)
Kanamycin ⎦

Decreasing concentration of drug ⟶

FIGURE 20.19 A microdilution, or microtiter, plate used for testing for minimal inhibitory concentration (MIC) of antibiotics. Such plates contain as many as 96 shallow wells that contain measured concentrations of antibiotics. They are usually purchased frozen or freeze dried (page 174). The test microbe is added simultaneously, with a special dispenser, to all the wells in a row of test antibiotics. A button of growth appears if the antibiotic has no effect on the microbe; the microbe is recorded as not sensitive. If there is no growth in a well, the microbe is sensitive to the antibiotic at that concentration. To ensure that the microbe is capable of growth in the absence of the antibiotic, wells that contain no antibiotic are also inoculated (positive control). To ensure against contamination by unwanted microbes, wells that contain nutrient broth but no antibiotics or inoculum are included (negative control).

Q **What is MIC?**

to make the microbe resistant to certain macrolide antibiotics)

4. Rapid efflux (ejection), which pumps the drug out of the cell before it can become effective

Variations on these mechanisms also occur. For example, a microbe could become resistant to trimethoprim by synthesizing very large amounts of the enzyme against which the drug is targeted. Conversely, polyene antibiotics can become less effective when resistant organisms produce smaller amounts of the sterols against which the drug is effective. Of particular concern is the possibility that such *resistant mutants* will increasingly replace the susceptible normal populations (see the box on page 606). Figure 20.20 shows how rapidly bacterial numbers increase during infection as resistance develops.

Hereditary drug resistance is often carried by plasmids, or by small segments of DNA called transposons that can jump from one piece of DNA to another (Chapter 8, page 245). Some plasmids, including those called resistance (R) factors, can be transferred between bacterial cells in a population and between different but closely related bacterial populations (see Figure 8.28, page 245). R factors often contain genes for resistance to several antibiotics.

Antibiotics have been a much misused product, nowhere more so than in the less developed areas of the world. Well-trained personnel are scarce, especially in rural areas, which is perhaps one reason why antibiotics can almost universally be purchased without prescription in these countries. A survey in rural Bangladesh, for example, showed that only 8% of antibiotics had been prescribed by a physician. In much of the world, antibiotics are sold to treat headaches and for other inappropriate uses (Figure 20.21). Even when the use of antibiotics is appropriate, dose regimens are usually shorter than needed to eradicate the infection. This encourages the survival of resistant strains of bacteria. Outdated, adulterated (impure), and even counterfeit antibiotics are common.

The developed world is also contributing to the rise of antibiotic resistance. The CDC estimates that in the United States, 30% of antibiotic prescriptions for ear infections, 100% of prescriptions for the common cold, and 50% of prescriptions for sore throats were unnecessary or not appropriate to treat the probable pathogen. At least half of the more than 100,000 tons of antibiotics consumed in the United States each year are not

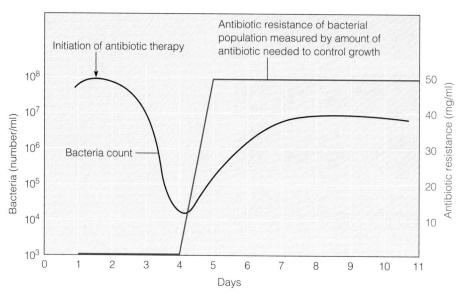

FIGURE 20.20 **The development of an antibiotic-resistant mutant during antibiotic therapy.** The patient, suffering from a chronic kidney infection caused by a gram-negative bacterium, was treated with streptomycin. The red line records the antibiotic resistance of the bacterial population. Until about the fourth day, essentially all of the bacterial population is sensitive to the antibiotic. At this time, resistant mutants that require 50,000 μg/ml of antibiotic (a very high amount) to control them appear, and their numbers increase rapidly. The black line records the bacterial population in the patient. After antibiotic therapy is begun, the population declines until the fourth day. At this time, mutants in the population that are resistant to streptomycin appear. The bacterial population in the patient rises as these resistant mutants replace the sensitive population.

SOURCE: From *Biology of Microorganisms,* 7th Ed. by T. D. Brock, M.T. Madigan, J. M. Martinko, and J. Parker, p. 410 © 1993. Adapted by permission of Prentice-Hall Inc., Upper Saddle River, New Jersey.

Q This test used streptomycin and a gram-negative bacterium. What would the lines have looked like if penicillin G had been the antibiotic?

FIGURE 20.21 **Antibiotics have been sold without prescriptions for many decades in much of the world.**

 How does this practice lead to development of resistant strains of pathogens?

used to treat disease but are used in animal feeds to promote growth—a practice that many feel should be curtailed (see the box on page 606). Patients also contribute to survival of antibiotic-resistant microbes when they fail to finish the full regimen of the prescription or use leftover antibiotics.

Strains of bacteria that are resistant to antibiotics are particularly common among hospital workers, where antibiotics are in constant use. Many hospitals have special monitoring committees to review the use of antibiotics for effectiveness and cost.

ANTIBIOTIC SAFETY

In our discussions of antibiotics, we have occasionally mentioned side effects. These may be potentially serious, such as liver or kidney damage or hearing impairment. Administration of almost any drug involves an assessment of risks against benefits; this is called the *therapeutic index.* Sometimes, the use of another drug can cause toxic effects that do not occur when the drug is taken alone. One drug may also neutralize the intended effects of the other. For example, a few antibiotics have been reported to neutralize the effectiveness of contraceptive pills. Also, some individuals may have hypersensitivity reactions; for example, to penicillins.

A pregnant woman should take only those antibiotics that are classified by the U.S. Food and Drug Administration as presenting no evidence of risk to the fetus.

EFFECTS OF COMBINATIONS OF DRUGS

LEARNING OBJECTIVE
• Compare and contrast synergism and antagonism.

The chemotherapeutic effect of two drugs given simultaneously is sometimes greater than the effect of either given alone (Figure 20.22). This phenomenon, called **synergism,** was introduced earlier. For example, in the treatment of bacterial endocarditis, penicillin and streptomycin are much more effective when taken together than when either drug is taken alone. Damage to bacterial cell walls by penicillin makes it easier for streptomycin to enter.

Other combinations of drugs can show **antagonism.** For example, the simultaneous use of penicillin and tetracycline is often less effective than when either drug is used alone. By stopping the growth of the bacteria, the bacteriostatic drug tetracycline interferes with the action of penicillin, which requires bacterial growth.

THE FUTURE OF CHEMOTHERAPEUTIC AGENTS

LEARNING OBJECTIVE
• Identify three areas of research on new chemotherapeutic agents.

Antibiotics are clearly one of the greatest triumphs of medical science, but they rely upon a limited range of targets. Microbes are now developing resistance to these modes of action. Unfortunately, the problem of antibiotic resistance has now reached a crisis point. Dealing with this crisis will require both intensified educational campaigns to promote the wise use of existing antibiotics and redoubled efforts toward the design of new ones.

A first step will be to continue to modify existing drugs to extend their spectra of use and prevent their destruction by bacterial enzymes of resistance. Research is under way to prevent the resistance strategy of rapid antibiotic efflux. Investigators are also working to identify new targets for antimicrobial activity other than those summarized in Figure 20.2. Most of our present antibiotics are the products of other microorganisms. Interest is now focusing on antibiotics produced by plants and animals.

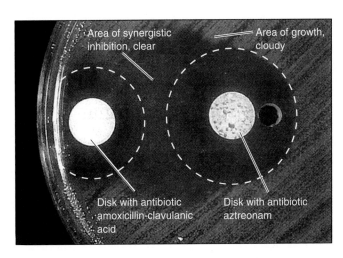

FIGURE 20.22 An example of synergism between two different antibiotics. The photograph shows the surface of a Petri plate seeded with bacteria. The paper disc at the left contains the antibiotic amoxicillin plus clavulanic acid. The disc on the right contains the antibiotic aztreonam. The dashed circles drawn over the photo show the clear areas surrounding each disc where bacterial growth would have been inhibited if there had been no synergy. The additional clear area between these two areas and outside the drawn circles illustrates inhibition of bacterial growth through the effects of synergy.

Q **What would the plate look like if the two antibiotics had been antagonistic?**

ANTIMICROBIAL PEPTIDES

Higher organisms, including humans, often exhibit extraordinary resistance to microbial infections. Especially important are the hundreds of different **antimicrobial peptides** (sometimes called *cationic peptides*) that serve as protective agents. An example of these is *magainin* (from Hebrew for shield) that is found in the skin of certain frogs. Humans also produce antimicrobial peptides, called **defensins.** The most common are alpha-defensins found in neutrophils and the Paneth cells of the intestine. The physical defenses of the intact skin and mucous membranes of our innate defensive system are reinforced by beta-defensins. Peptide antimicrobials work primarily by forming destructive channels in the plasma membrane of microbes, which differ very significantly from mammalian cell membranes. *Nisin*, which we have mentioned for its use as a food preservative (page 204), resembles magainin in its mode of action on membranes.

Considering the time span of exposure, surprisingly little natural resistance has developed—a good reason for medical interest in these products. In order to evade the action of these compounds on the microbial membranes, the membrane would have to be greatly changed, which would be a difficult and metabolically costly revision of a vital

MICROBIOLOGY IN THE NEWS

ANTIBIOTICS IN ANIMAL FEED LINKED TO HUMAN DISEASE

Over 40 years ago, livestock growers began using antibiotics in the feed of closely penned animals that were being fattened for market. The drugs helped reduce the number of bacterial infections and controlled their spread in such ripe conditions. They also unexpectedly accelerated the animals' growth, an effect that led to wider and wider use. The growth-promoting effects are thought to be due to suppression of intestinal bacteria, such as *Clostridium* spp., that produce excessive gas and toxins, which may retard the animals' growth. Today, more than half the antibiotics used worldwide are given to farm animals to promote weight gain and treat infections.

Meat and milk that reach the consumer's table are not heavily laden with antibiotics because the U.S. Food and Drug Administration (FDA) has established limits for antibiotic residues in edible tissues. But the constant presence of antibiotics in these animals puts selective pressure on their normal microbiota, as well as on their pathogens, for drug resistance.

Salmonella

Salmonella can be transferred from animals to humans in meat or milk. The first known case of salmonellosis occurred in Germany in 1888, when 50 people became ill after eating ground beef from one cow. The use of antibiotics in animal feed preferentially allows the growth of strains of bacteria that are resistant to drugs commonly used to treat human infections. In the 1980s, researchers proved that antibiotic-resistant bacteria are being transferred from animals to humans directly in meat or milk. In 1984, Scott Holmberg and his associates at the CDC traced an antibiotic-resistant *Salmonella newport* infection from South Dakota beef cattle to 18 people in four states. Antibiotic-resistant *S. typhimurium* originating from one Illinois dairy infected 16,000 people in seven states in 1985.

In 1985, nearly 1000 cases of infection by *S. newport* resistant to multiple antibiotics were reported in Los Angeles County, California. A radioactive DNA probe was used to identify an identical plasmid carrying multiple antibiotic resistance genes in 99% of the *S. newport* isolates. A plasmid profile analysis was used to track the bacterium from the sick people back to a slaughterhouse, to a meat-deboning plant, and finally to three farms.

Enterococcus

Vancomycin-resistant *Enterococcus* spp. (VRE) were first isolated in France in 1986 and were found in the United States in 1989. Vancomycin and another glycopeptide, avoparcin, are widely used in animal feed in Europe. Veterinary researchers in Denmark cultured vancomycin-resistant enterococci from horses, pigs, and poultry in the eight European countries that use avoparcin in animal feed. They did not find VRE in livestock in Sweden or the United States, which do not use avoparcin. Their conclusion is that use of avoparcin has created a reservoir for VRE in food animals. VRE are frequently present in food produced in European countries, and human VRE carriers are not uncommon.

In 1996, veterinary use of avoparcin was banned in Germany. After the ban, VRE-positive samples decreased from 100% to 25%, and the human carrier rate dropped from 12% to 3%. In 1997, all European Union countries banned the use of avoparcin.

Campylobacter

Campylobacter jejuni is a commensal in the intestines of poultry. Fluoroquinolone (FQ) antibiotics were approved for use in

---

structure. Insects lack a mammalian-type immune system and produce an array of antimicrobial peptides to defend against infecting bacteria and fungi. One such compound is *cecropin*, which is produced by certain moths.

A new class of antibiotics, the **cyclic lipopeptides,** kills bacteria by attacking the microbial cell membrane as we have just described. The first antibiotic of this novel group, *daptomycin*, was approved for use in 2003. Of particular interest is its activity against MRSA and VRE pathogens.

ANTISENSE AGENTS

Another promising new approach is the use of short synthetic strands of DNA, called **antisense agents.** (Because antisense agents involve nucleic acids, persons investigating them whimsically call them "nubiotics.") The principle is to identify sites on the DNA or RNA of the pathogen that are responsible for the pathogenic effects. Segments of DNA are then synthesized that will selectively recognize and bind to the target site, a process that blocks the biosynthesis of the target protein. This approach has a great advantage: it prevents the production of a pathogenic protein, rather than trying to selectively neutralize it once it is made. An antiviral based upon antisense principles, *fomivirsen*, has been approved for treatment of the eye disease cytomegalovirus retinitis.

Mammalian cells have a mechanism that, on occasion, can prevent RNA from giving rise to the proteins for

MICROBIOLOGY IN THE NEWS

(continued)

chicken feed and water to prevent illness in 1995. Fluoroquinolone-resistant *C. jejuni* in humans emerged in the 1990s (see the graph). The emergence corresponds with the presence of FQ-resistant *C. jejuni* in grocery-store purchased chicken meat. FQ-resistant *C. jejuni* could be selected for in patients who had previously taken a fluoroquinolone. However, a study of *Campylobacter* isolates from patients between 1997 and 2001 showed that patients infected with FQ-resistant *C. jejuni* had not taken an FQ prior to their illness and had not traveled out of the United States. Hopefully, FQ-resistant *Campylobacter* will decrease now that FQ is no longer used in poultry feed.

Alternatives to Antibiotics

The U.S. Department of Agriculture (USDA) estimates that bacteria account for 3.6 to 7.1 million foodborne illnesses each year. Among the leading causes are *Salmonella*, *Campylobacter*, and *Escherichia coli* O157:H7, which colonize the intestines of farm animals and subsequently contaminate meat products during processing. *E. coli* O157:H7 colonize most cattle herds at one time or another, and 25% of raw chickens carry *Salmonella*.

Competitive exclusion (CE), the use of bacteria to prevent the colonization of undesirable microorganisms, is being investigated to eliminate undesirable bacteria from animals without using antibiotics. Desirable bacteria can be included with the animals' feed. When ingested, these bacteria will colonize the digestive system, compete for nutrients and attachment sites, and produce antibacterial compounds. Research has

shown that a mixture of three strains of bacteria can inhibit *Campylobacter* from colonizing in poultry.

A variety of approaches may be necessary to reduce the possibility of illness: (1) prevent colonization in the animals at the farm, (2) reduce fecal contamination of meat during processing at the slaughterhouse, and (3) use proper storage and cooking methods.

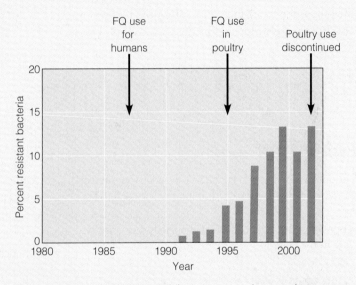

Fluoroquinolone-resistant *Campylobacter jejuni* in the United States, 1982–2001
DATA SOURCE: CDC, *Emerging Infectious Diseases*.

which they are encoded. This has its uses, for example, when an infecting virus tries to take over the cell's metabolism in order to make viral proteins. The mechanism, called **RNA interference (RNAi),** is proposed as the basis of drugs called **short interfering RNAs (siRNAs)** that could selectively block protein synthesis in pathogens (see page 269). This is similar in principle to antisense agents but is much more effective—at least in theory. No resistance mechanisms presently exist in bacteria. Rather than blocking a single messenger RNA, siRNAs serve as catalysts that would repeat their activity indefinitely. The theoretical prospects for this approach to microbial control led *Science* magazine to proclaim RNAi as the "breakthrough of the year" in 2002. There are many difficulties

that will need to be solved before commercially practical drugs appear, however.

Other, even more exotic, approaches to solving the problem of antibiotic resistance are being considered. Even before the age of antibiotics, the use of bacteriophages was considered a possibility for disease therapy. These viruses are highly selective in their infective activity, but experiments with them were not very productive. Russian scientists, in particular, have continued to experiment with so-called phage therapy. Current thinking is that, rather than using intact phage, for which bacteria rapidly develop resistant strains, certain peptides produced by phage that lyse bacteria might be the basis of practical antimicrobials.

Studies of other vertebrates have also proven productive. Compared to mammals, sharks have a rudimentary immune system. The observation that sharks are resistant to infection even in highly contaminated water led to the discovery of an interesting steroid, *squalamine,* that is about as powerful as ampicillin against many pathogens—at least in the laboratory.

Most of these new approaches are still in the investigative stages. Their primary potential targets are viral infections and cancer. There is special interest in such antiviral drugs, as well as in antifungal and antiparasitic drugs, because our arsenal in these categories is especially limited.

A problem with development of antimicrobial agents is that they are not especially profitable. Like vaccines, they are used only at infrequent occasions. Pharmaceutical companies are understandably more interested in developing drugs that treat chronic conditions such as high blood pressure or diabetes for which the patient requires years of regular medication.

* * *

The past few chapters have shown how profoundly science has changed the effects of infectious disease on human mortality and life span. At the turn of the twentieth century, the most common causes of death were infectious diseases. Most of these, including tuberculosis, typhoid fever, and diphtheria, were caused by bacteria. At the beginning of the twenty-first century, viral diseases such as influenza, viral pneumonias, and AIDS, are the only infectious diseases among the ten leading causes of death in the United States. These facts bear testimony to the effectiveness of sanitation, vaccines, and (as discussed in this chapter) the discovery and use of antibiotics. In Chapters 21 through 26, we will see that the struggle against infectious disease continues.

STUDY OUTLINE

INTRODUCTION (p. 581)

1. An antimicrobial drug is a chemical substance that destroys pathogenic microorganisms with minimal damage to host tissues.

2. Chemotherapeutic agents include chemicals that combat disease in the body.

THE HISTORY OF CHEMOTHERAPY (pp. 582–583)

1. Paul Ehrlich developed the concept of chemotherapy to treat microbial diseases; he predicted the development of chemotherapeutic agents, which would kill pathogens without harming the host.

2. Sulfa drugs came into prominence in the late 1930s.

3. Alexander Fleming discovered the first antibiotic, penicillin, in 1929; its first clinical trials were done in 1940.

THE SPECTRUM OF ANTIMICROBIAL ACTIVITY (pp. 583–584)

1. Antibacterial drugs affect many targets in a prokaryotic cell.

2. Fungal, protozoan, and helminthic infections are more difficult to treat because these organisms have eukaryotic cells.

3. Narrow-spectrum drugs affect only a select group of microbes—gram-positive cells, for example; broad-spectrum drugs affect a more diverse range of microbes.

4. Small, hydrophilic drugs can affect gram-negative cells.

5. Antimicrobial agents should not cause excessive harm to normal microbiota.

6. Superinfections occur when a pathogen develops resistance to the drug being used or when normally resistant microbiota multiply excessively.

THE ACTION OF ANTIMICROBIAL DRUGS (pp. 584–587)

1. General action is either by directly killing microorganisms (bactericidal) or by inhibiting their growth (bacteriostatic).

2. Some agents, such as penicillin, inhibit cell wall synthesis in bacteria.

3. Other agents, such as chloramphenicol, tetracyclines, and streptomycin, inhibit protein synthesis by acting on 70S ribosomes.

4. Agents such as polymyxin B cause injury to plasma membranes.

5. Rifampin and the quinolones inhibit nucleic acid synthesis.

6. Agents such as sulfanilamide act as antimetabolites by competitively inhibiting enzyme activity.

A SURVEY OF COMMONLY USED ANTIMICROBIAL DRUGS (pp. 587–601)

ANTIBACTERIAL ANTIBIOTICS: INHIBITORS OF CELL WALL SYNTHESIS (pp. 587–593)

1. All penicillins contain a β-lactam ring.
2. Natural penicillins produced by *Penicillium* are effective against gram-positive cocci and spirochetes.
3. Penicillinases (β-lactamases) are bacterial enzymes that destroy natural penicillins.
4. Semisynthetic penicillins are made in the laboratory by adding different side chains onto the β-lactam ring made by the fungus.
5. Semisynthetic penicillins are resistant to penicillinases and have a broader spectrum of activity than natural penicillins.
6. Carbapenems are broad-spectrum antibiotics that inhibit cell wall synthesis.
7. The monobactam aztreonam affects only gram-negative bacteria.
8. Cephalosporins inhibit cell wall synthesis and are used against penicillin-resistant strains.
9. Polypeptides such as bacitracin inhibit cell wall synthesis primarily in gram-positive bacteria.
10. Vancomycin inhibits cell wall synthesis and may be used to kill penicillinase-producing staphylococci. Streptogramins are bactericidal agents that inhibit protein synthesis and may be used to kill vancomycin-resistant bacteria.
11. Isoniazid (INH) inhibits mycolic acid synthesis in mycobacteria. INH is administered with rifampin or ethambutol to treat tuberculosis.
12. The antimetabolite ethambutol is used with other drugs to treat tuberculosis.

INHIBITORS OF PROTEIN SYNTHESIS (pp. 593–595)

13. Chloramphenicol, aminoglycosides, tetracyclines, macrolides, and streptogramins inhibit protein synthesis at 70S ribosomes.
14. Oxazolidinones prevent formation of 70S ribosomes.

INJURY TO THE PLASMA MEMBRANE (p. 595)

15. Polymyxin B and bacitracin cause damage to plasma membranes.

INHIBITORS OF NUCLEIC ACID (DNA/RNA) SYNTHESIS (p. 595)

16. Rifamycin inhibits mRNA synthesis; it is used to treat tuberculosis.
17. Quinolones and fluoroquinolones inhibit DNA gyrase for treatment of urinary tract infections.

COMPETITIVE INHIBITORS OF THE SYNTHESIS OF ESSENTIAL METABOLITES (p. 595)

18. Sulfonamides competitively inhibit folic acid synthesis.
19. TMP-SMZ competitively inhibits dihydrofolic acid synthesis.

ANTIFUNGAL DRUGS (pp. 596–597)

20. Polyenes, such as nystatin and amphotericin B, combine with plasma membrane sterols and are fungicidal.
21. Azoles and allylamines interfere with sterol synthesis and are used to treat cutaneous and systemic mycoses.
22. Echinocandins interfere with fungal cell wall synthesis.
23. The antifungal agent flucytosine is an antimetabolite of cytosine.
24. Griseofulvin interferes with eukaryotic cell division and is used primarily to treat skin infections caused by fungi.

ANTIVIRAL DRUGS (pp. 597–600)

25. Nucleoside and nucleotide analogs, such as acyclovir and zidovudine, inhibit DNA or RNA synthesis.
26. Inhibitors of viral enzymes are used to treat influenza and HIV infection.
27. Alpha-interferons inhibit the spread of viruses to new cells.

ANTIPROTOZOAN AND ANTIHELMINTHIC DRUGS (pp. 600–601)

28. Chloroquine, quinacrine, diiodohydroxyquin, pentamidine, and metronidazole are used to treat protozoan infections.
29. Antihelminthic drugs include mebendazole, praziquantel, and ivermectin.

TESTS TO GUIDE CHEMOTHERAPY (pp. 601–602)

1. These tests are used to determine which chemotherapeutic agent is most likely to combat a specific pathogen.
2. These tests are used when susceptibility cannot be predicted or when drug resistance arises.

THE DIFFUSION METHODS (pp. 601–602)

3. In this test, also known as the Kirby-Bauer test, a bacterial culture is inoculated on an agar medium, and filter paper disks impregnated with chemotherapeutic agents are overlaid on the culture.
4. After incubation, the absence of microbial growth around a disk is called a zone of inhibition.
5. The diameter of the zone of inhibition, when compared with a standardized reference table, is used to determine whether the organism is sensitive, intermediate, or resistant to the drug.

6. MIC is the lowest concentration of drug capable of preventing microbial growth; MIC can be estimated using the E test.

BROTH DILUTION TESTS (p. 602)

7. In a broth dilution test, the microorganism is grown in liquid media containing different concentrations of a chemotherapeutic agent.

8. The lowest concentration of a chemotherapeutic agent that kills bacteria is called the minimum bactericidal concentration (MBC).

THE EFFECTIVENESS OF CHEMOTHERAPEUTIC AGENTS (pp. 602–608)

DRUG RESISTANCE (pp. 602–604)

1. Resistance may be due to enzymatic destruction of a drug, prevention of penetration of the drug to its target site, or cellular or metabolic changes at target sites.

2. Hereditary drug resistance (R) factors are carried by plasmids and transposons.

3. Resistance can be minimized by the discriminating use of drugs in appropriate concentrations and dosages.

ANTIBIOTIC SAFETY (pp. 604–605)

4. The risk (e.g., side effects) versus the benefit (e.g., curing an infection) must be evaluated prior to use of antibiotics.

EFFECTS OF COMBINATIONS OF DRUGS (p. 605)

5. Some combinations of drugs are synergistic; they are more effective when taken together.

6. Some combinations of drugs are antagonistic; when taken together, both drugs become less effective than when taken alone.

THE FUTURE OF CHEMOTHERAPEUTIC AGENTS (pp. 605–608)

7. Many bacterial diseases, previously treatable with antibiotics, have become resistant to antibiotics.

8. Chemicals produced by plants and animals are providing new antimicrobial agents called antimicrobial peptides.

9. Protein synthesis in pathogens can be blocked by siRNAs.

10. Bacteriophage products may be antimicrobials.

STUDY QUESTIONS

 Access more review material either online at **The Microbiology Place** (www.microbiologyplace.com) or with **The Microbiology Place CD-ROM** packaged with your new book. There you'll find activities, practice tests, quizzes, flashcards, case studies, and more to help you succeed.

Answers to the Study Questions can be found in Appendix G.

1. Fill in the following table:

Antimicrobial Agent	Method of Action	Principal Use
Isoniazid		
Sulfonamides		
Ethambutol		
Trimethoprim		
Fluoroquinolones		
Penicillin, natural		
Penicillin, semisynthetic		
Cephalosporins		
Carbapenems		
Aminoglycosides		
Tetracyclines		
Macrolides		
Polypeptides		
Vancomycin		

Antimicrobial Agent	Method of Action	Principal Use
Rifamycins		
Polyenes		
Griseofulvin		
Amantadine		
Zidovudine		
Niclosamide		
Ketaconazole		

2. Define *chemotherapeutic agent*. Distinguish a synthetic drug from an antibiotic.

3. Discuss the contributions to chemotherapy made by Ehrlich and Fleming.

4. List and explain five criteria used to identify an effective antimicrobial agent.

5. What similar problems are encountered with antiviral, antifungal, antiprotozoan, and antihelminthic drugs?

6. Identify four modes of action of antiviral drugs. Give an example of a currently used antiviral drug for each mode of action.

7. Compare and contrast the broth dilution and disk-diffusion tests. Identify at least one advantage of each.

8. Describe the disk-diffusion test for microbial susceptibility. What information can you obtain from this test?

9. Define *drug resistance*. How is it produced? What measures can be taken to minimize drug resistance?

10. List the advantages of using two chemotherapeutic agents simultaneously to treat a disease. What problem can be encountered using two drugs?

11. Why does a cell die from the following antimicrobial actions?
   a. Colistimethate binds to phospholipids.
   b. Kanamycin binds to 70S ribosomes.

12. How is translation inhibited by each of the following?
   a. chloramphenicol   d. streptomycin
   b. erythromycin   e. oxazolidinone
   c. tetracycline   f. streptogramin

13. Dideoxyinosine (ddI) is an antimetabolite of guanine. The –OH is missing from carbon 3 in ddI. How does ddI inhibit DNA synthesis?

14. Compare the method of action of the following pairs:
   a. penicillin and echinocandin
   b. imidazole and polymyxin B

MULTIPLE CHOICE

1. Which of the following pairs is mismatched?
   a. antihelminthic—inhibition of oxidative phosphorylation
   b. antihelminthic—inhibition of cell wall synthesis
   c. antifungal—injury to plasma membrane
   d. antifungal—inhibition of mitosis
   e. antiviral—inhibition of DNA synthesis

2. All of the following are modes of action of antiviral drugs except
   a. inhibition of protein synthesis at 70S ribosomes.
   b. inhibition of DNA synthesis.
   c. inhibition of RNA synthesis.
   d. inhibition of uncoating.
   e. none of the above

3. Which of the following modes of action would not be fungicidal?
   a. inhibition of peptidoglycan synthesis
   b. inhibition of mitosis
   c. injury to the plasma membrane
   d. inhibition of nucleic acid synthesis
   e. none of the above

4. An antimicrobial agent should meet all of the following criteria except
   a. selective toxicity.
   b. the production of hypersensitivities.
   c. a narrow spectrum of activity.
   d. no production of drug resistance.
   e. none of the above

5. The most selective antimicrobial activity would be exhibited by a drug that
   a. inhibits cell wall synthesis.
   b. inhibits protein synthesis.
   c. injures the plasma membrane.
   d. inhibits nucleic acid synthesis.
   e. all of the above

6. Antibiotics that inhibit translation have side effects
   a. because all cells have proteins.
   b. only in the few cells that make proteins.
   c. because eukaryotic cells have 80S ribosomes.
   d. at the 70S ribosomes in eukaryotic cells.
   e. none of the above

7. Which of the following will *not* affect eukaryotic cells?
   a. inhibition of the mitotic spindle
   b. binding with sterols
   c. binding to 80S ribosomes
   d. binding to DNA
   e. all of the above will affect them.

8. Cell membrane damage causes death because
   a. the cell undergoes osmotic lysis.
   b. cell contents leak out.
   c. the cell plasmolyzes.
   d. the cell lacks a wall.
   e. none of the above

9. A drug that intercalates into DNA has the following effects. Which one leads to the others?
   a. It disrupts transcription.
   b. It disrupts translation.
   c. It interferes with DNA replication.
   d. It causes mutations.
   e. It alters proteins.

10. Chloramphenicol binds to the 50S portion of a ribosome, which will interfere with
   a. transcription in prokaryotic cells.
   b. transcription in eukaryotic cells.
   c. translation in prokaryotic cells.
   d. translation in eukaryotic cells.
   e. DNA synthesis.

CRITICAL THINKING

1. Which of the following can affect human cells? Explain why or why not.
   a. penicillin
   b. indinavir
   c. erythromycin
   d. polymyxin

2. Why is idoxuridine effective if host cells also contain DNA?

3. Some bacteria become resistant to tetracycline because they don't make porins. Why can a porin-deficient mutant be detected by its inability to grow on a medium containing a single carbon source such as succinic acid?

4. The following data were obtained from a disk-diffusion test.

Antibiotic	Zone of Inhibition
A	15 mm
B	0 mm
C	7 mm
D	15 mm

   a. Which antibiotic was most effective against the bacteria being tested?

b. Which antibiotic would you recommend for treatment of a disease caused by this bacterium?

c. Was antibiotic A bactericidal or bacteriostatic? How can you tell?

5. Why do you suppose *Streptomyces griseus* produces an enzyme that inactivates streptomycin? Why is this enzyme produced early in metabolism?

6. The following results were obtained from a broth dilution test for microbial susceptibility.

Antibiotic Concentration	Growth	Growth in Subculture
200 µg/ml	+	+
100 µg/ml	+	+
50 µg/ml	−	+
25 µg/ml	−	−

a. The MIC of this antibiotic is _____.
b. The MBC of this antibiotic is _____.

CLINICAL APPLICATIONS

1. Two patients received corneal transplants of the right eye on the same day, in the same facility, from the same donor. The corneas were kept in a solution of gentamicin and streptomycin, and transplantation was completed within 48 hours of tissue recovery. Both cornea recipients developed eye infections caused by *Clostridium perfringens* and were treated with gentamicin. The eye inflammations persisted. The infections were resolved after treatment was changed to penicillin G. What would you see in a Gram stain of a sample from the infected eyes? Why didn't the first round of treatment cure the infection?

2. A patient with a urinary bladder infection took nalidixic acid, but her condition did not improve. Explain why her infection disappeared when she switched to a sulfonamide.

3. A patient with streptococcal sore throat takes penicillin for two days of a prescribed 10-day regimen. Because he feels better, he then saves the remaining penicillin for some other time. After three more days, he suffers a relapse of the sore throat. Discuss the probable cause of the relapse.

21

Microbial Diseases of the Skin and Eyes

The skin, which covers and protects the body, is the body's first line of defense against pathogens. As a physical barrier, it is almost impossible for pathogens to penetrate it. However, microbes can enter through skin breaks that are not readily apparent, and the larval forms of a few parasites can penetrate intact skin. The skin is an inhospitable place for most microorganisms because the secretions of the skin are acidic and most of the skin contains little moisture. Some parts of the body such as the armpit and the area between the legs have enough moisture, though, to support relatively large bacterial populations. Drier regions, such as the scalp, support rather small numbers of microorganisms. Beyond these ecological factors, the skin contains peptide antibiotics called *defensins* that have a wide spectrum of antimicrobial activity (see page 605). These are also found in mucous membranes, especially those lining the gastrointestinal tract.

UNDER THE MICROSCOPE

Candida albicans. This yeast causes infections of the skin and mucous membranes and is a common cause of nosocomial bloodstream infections.

FIGURE 21.1 The structure of human skin. Notice the passageways between the hair follicle and hair shaft, through which microbes can penetrate the deeper tissues. They can also enter the skin through sweat pores.

Q **What do you perceive from this illustration to be the weak points that would allow penetration of the intact skin to reach the underlying tissue?**

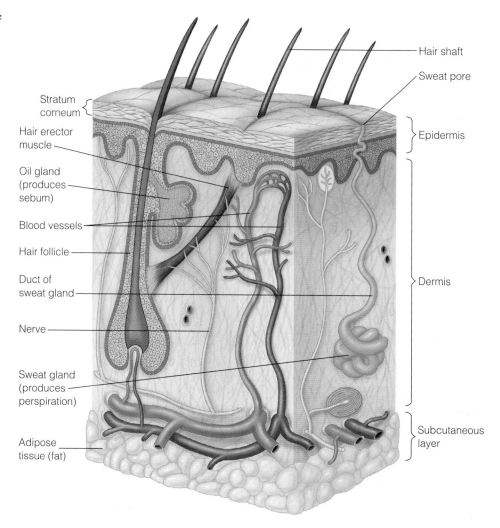

Stratum corneum

Hair erector muscle

Oil gland (produces sebum)

Blood vessels

Hair follicle

Duct of sweat gland

Nerve

Sweat gland (produces perspiration)

Adipose tissue (fat)

Hair shaft

Sweat pore

Epidermis

Dermis

Subcutaneous layer

STRUCTURE AND FUNCTION OF THE SKIN

> **LEARNING OBJECTIVE**
>
> • Describe the structure of the skin and mucous membranes and the ways pathogens can invade the skin.

The skin of an average adult occupies a surface area of about 1.9 m^2 and varies in thickness from 0.05 to 3.0 mm. As we mentioned in Chapter 16, skin consists of two principal parts, the epidermis and the dermis (Figure 21.1). The **epidermis** is the thin outer portion, composed of several layers of epithelial cells. The outermost layer of the epidermis, the *stratum corneum*, consists of dead cells that contain a waterproofing protein called **keratin.** The epidermis, when unbroken, is an effective physical barrier against microorganisms.

The **dermis** is the inner, relatively thick portion of skin, composed mainly of connective tissue. The hair follicles, sweat gland ducts, and oil gland ducts in the dermis provide passageways through which microorganisms can enter the skin and penetrate deeper tissues.

Perspiration provides moisture and some nutrients for microbial growth. However, it contains salt, which inhibits many microorganisms; the enzyme lysozyme, which is capable of breaking down the cell walls of certain bacteria; and, as mentioned previously, antimicrobial peptides.

Sebum, secreted by oil glands, is a mixture of lipids (unsaturated fatty acids), proteins, and salts that prevents skin and hair from drying out. Although the fatty acids inhibit the growth of certain pathogens, sebum, like perspiration, is also nutritive for many microorganisms.

MUCOUS MEMBRANES

In the linings of body cavities, such as those associated with the gastrointestinal, respiratory, urinary, and genital tracts, the outer protective barrier differs from the skin. It consists of sheets of tightly packed *epithelial cells.* These cells are attached at their bases to a layer of extracellular material called the *basement membrane.* Many of these cells secrete mucus—hence the name **mucous membrane,** or **mucosa.** Other mucosal cells have cilia; and, in the respiratory system, the mucous layer traps particles, including microorganisms, which the cilia sweep upward out of the body (see Figure 16.4, page 478). Mucous membranes are often acidic, which tends to limit their microbial populations. Also, the membranes of the eyes are mechanically washed by tears, and the lysozyme in tears destroys the cell walls of certain bacteria. Mucous membranes are often folded to maximize surface area; the total surface area in an average human is about 400 m^2, much more than the surface area of the skin.

NORMAL MICROBIOTA OF THE SKIN

> **LEARNING OBJECTIVE**
> • Provide examples of normal skin microbiota, and state the general locations and ecological roles of its members.

Although the skin is generally inhospitable to most microorganisms, it supports the growth of certain microbes that are established as part of the normal microbiota. On superficial skin surfaces, certain aerobic bacteria produce fatty acids from sebum. These acids inhibit many microbes and allow better-adapted bacteria to flourish.

Microorganisms that find the skin a satisfactory environment are resistant to drying and to relatively high salt concentrations. The skin's normal microbiota contain relatively large numbers of gram-positive bacteria, such as staphylococci and micrococci. Some of these are capable of growth at sodium chloride (table salt) concentrations of 7.5% or more. Scanning electron micrographs show that bacteria on the skin tend to be grouped into small clumps (see Figure 14.1a, page 421). Vigorous washing can reduce their numbers but will not eliminate them. Microorganisms remaining in hair follicles and sweat glands after washing will soon reestablish the normal populations. Areas of the body with more moisture, such as the armpits and between the legs, have higher populations of microbes. These metabolize secretions from the sweat glands and are the main contributors to body odor.

Also part of the skin's normal microbiota are gram-positive pleomorphic rods called *diphtheroids*. Some diphtheroids, such as *Propionibacterium acnes*, are typically anaerobic and inhabit hair follicles. Their growth is supported by secretions from the oil glands (sebum), which, as we will see, makes them a factor in acne. These bacteria produce propionic acid, which helps maintain the low pH of skin, generally between 3 and 5. Other diphtheroids, such as *Corynebacterium xerosis* (ze-rō′sis), are aerobic and occupy the skin surface. A yeast, *Malassezia*

(mal′as-sēz-ē-ä) *furfur* is capable of growing on oily skin secretions and is thought to be responsible for the scaling skin condition known as *dandruff*. Shampoos for treating dandruff contain the antibiotic ketoconazole or zinc pyrithione or selenium sulfide. All are active against this yeast.

MICROBIAL DISEASES OF THE SKIN

Rashes and lesions on the skin do not necessarily indicate an infection of the skin; in fact, many diseases manifested by skin lesions are actually systemic diseases affecting internal organs. Variations in these lesions are often useful in describing the symptoms of the disease. For example, small, fluid-filled lesions are **vesicles** (Figure 21.2a). Vesicles larger than about 1 cm in diameter are termed **bullae** (Figure 21.2b). Flat, reddened lesions are known as **macules** (Figure 21.2c). Raised lesions are called **papules** or, when they contain pus, **pustules** (Figure 21.2d). Although the focus of infection is often elsewhere in the body, it is convenient to classify these diseases by the organ most obviously affected: the skin. A skin rash that arises from disease conditions is called an **exanthem;** on mucous membranes, such as the interior of the mouth, such a rash is called an **enanthem.**

Table 21.1 summarizes the most important diseases associated with the skin.

BACTERIAL DISEASES OF THE SKIN

> **LEARNING OBJECTIVES**
> • Differentiate staphylococci from streptococci, and name several skin infections caused by each.
> • List the causative agent, mode of transmission, and clinical symptoms of *Pseudomonas* dermatitis, otitis externa, and acne.

Two genera of bacteria, *Staphylococcus* and *Streptococcus*, are frequent causes of skin-related diseases and merit special discussion. We will also discuss these bacteria in subsequent chapters in relation to other organs and conditions. Superficial staphylococcal and streptococcal infections of the skin are very common. The bacteria frequently come into contact with the skin and have adapted fairly well to the physiological conditions there. Both genera also produce invasive enzymes and damaging toxins that contribute to the disease process.

STAPHYLOCOCCAL SKIN INFECTIONS

Staphylococci are spherical gram-positive bacteria that form irregular clusters like grapes (see Figure 11.17, page 331). This characteristic occurs because the cells divide at

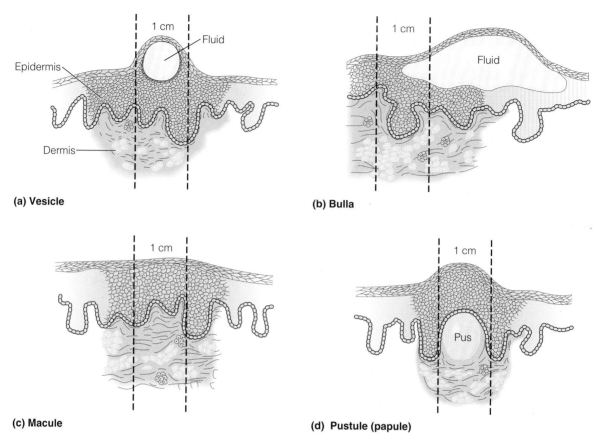

(a) Vesicle

(b) Bulla

(c) Macule

(d) Pustule (papule)

FIGURE 21.2 Skin lesions. (a) Vesicles are small, fluid-filled lesions. **(b)** Bullae are larger fluid-filled lesions. **(c)** Macules are flat lesions that are often reddish. **(d)** Papules are raised lesions; when they contain pus, as shown here, they are called pustules.

Q **Are these skin lesions exanthems or enanthems?**

random points about their circumference, and the daughter cells do not completely separate from each other (see Figure 4.1d, page 79). For almost all clinical purposes, these bacteria can be divided into those that produce **coagulase,** an enzyme that coagulates (clots) fibrin in blood, and those that do not (coagulase-positive and coagulase-negative strains, respectively).

Coagulase-negative strains, such as *Staphylococcus epidermidis*, are very common on the skin, where they may represent 90% of the normal microbiota. They are generally pathogenic only when the skin barrier is broken or is invaded by medical procedures, such as the insertion and removal of catheters into veins. On the surface of the catheter (Figure 21.3), the bacteria are surrounded by a slime layer of capsular material (see discussions of biofilms on pages 57 and 826) that protects them from desiccation and disinfectants.

Staphylococcus aureus is the most pathogenic of the staphylococci. Typically, it forms golden-yellow colonies. Almost all pathogenic strains of *S. aureus* are coagulase-positive. Fibrin clots may protect the microorganisms from phagocytosis and isolate them from other defenses of the

host. There is a high correlation between the bacterium's ability to form coagulase and its production of damaging toxins, several of which may injure tissues. Some staphylococcal toxins, *enterotoxins,* affect the gastrointestinal tract and will be discussed in Chapter 25 with diseases of the digestive system.

S. aureus is often a problem in the hospital environment. Because it is carried by patients, hospital staff members, and visitors, the danger of infection to surgical wounds and other breaks in the skin is great. Such infections are difficult to treat because bacteria in the hospital environment are exposed to so many antibiotics that they quickly become resistant to them. Recent cases of antibiotic-resistant *S. aureus* have been reported on college campuses as well (see the box on page 633).

The nasal passages provide an especially favorable environment for *S. aureus*, which is often present there in large numbers. In fact, its presence on unbroken skin is often the result of transport from the nasal passages. *S. aureus* can enter the body through a natural opening in the skin barrier, the hair follicle's passage through the epidermal layer. Infections of hair follicles, or **folliculitis,** often

TABLE 21.1	**Microbial Diseases of the Skin**				
	Pathogen	Portal of Entry	Symptoms	Method of Transmission	Treatment
Bacterial Diseases. These infections are usually diagnosed by culturing the bacteria.					
Impetigo	*Staphylococcus aureus;* occasionally, *Streptococcus pyogenes*	Skin	Vesicles on skin	Direct contact; fomites	Penicillin (for *Streptococcus* infections only)
Folliculitis	*Staphylococcus aureus*	Hair follicle	Infection of hair follicle	Direct contact; fomites; endogenous* infection	Drain pus; penicillin
Toxic shock syndrome	*Staphylococcus aureus*	Surgical incisions	Fever, rash, shock	Endogenous infection*	Penicillin
Necrotizing fasciitis	*Streptococcus pyogenes*	Skin abrasions	Extensive soft tissue destruction	Direct contact	Surgical removal of tissue; penicillin
Erysipelas	*Streptococcus pyogenes*	Skin; mucous membranes	Reddish patches on skin; often with high fever	Endogenous infection*	Penicillin
Pseudomonas dermatitis	*Pseudomonas aeruginosa*	Skin abrasions	Superficial rash	Swimming water; hot tubs	Usually self-limiting
Otitis externa	*Pseudomonas aeruginosa*	Ear	Superficial infection of external ear canal	Swimming water	Fluoroquinolones
Acne	*Propionibacterium acnes*	Sebum channels	Inflammatory lesions originating with accumulations of sebum that rupture a hair follicle	Direct contact	Benzoyl peroxide, isotretinoin, azelaic acid
Viral Diseases. These infections are usually diagnosed by clinical signs and symptoms and may be confirmed by laboratory tests as indicated below.					
Warts	*Papillomavirus* spp.	Skin	A horny projection of the skin formed by proliferation of cells	Direct contact	May be removed by liquid nitrogen cryotherapy, electrodesiccation, acids, or lasers
Smallpox (variola)	Smallpox (variola) virus	Respiratory tract	Pustules that may be nearly confluent on skin	Aerosol	None
Monkeypox	Monkeypox virus	Respiratory tract	Pustules, similar to smallpox; diagnosis confirmed at the CDC by PCR or serology	Direct contact with or aerosols from infected small mammals	None
Chickenpox (varicella)	Varicella-zoster virus	Respiratory tract	Vesicles in most cases confined to face, throat, and lower back; diagnosis confirmed by fluorescent-antibody test	Aerosol	Acyclovir for immunocompromised patients; preexposure vaccine

*Endogenous infections are those caused by microorganisms already part of the host microbiota.

▶

TABLE 21.1	Microbial Diseases of the Skin (continued)				
	Pathogen	Portal of Entry	Symptoms	Method of Transmission	Treatment
Shingles (herpes-zoster)	Varicella-zoster virus	Endogenous infection of peripheral nerves	Vesicles typically on one side of waist, face and scalp, or upper chest; diagnosis confirmed by virus culture	Recurrence of latent chickenpox infection	Acyclovir for immuno compromised patients; preventive vaccine
Herpes simplex	Herpes simplex virus type 1	Skin; mucous membranes	Vesicles around mouth; can also affect other areas of skin and mucous membranes; diagnosis confirmed by virus culture	Initial infection by direct contact; recurring latent infection	Acyclovir may modify symptoms
Measles (rubeola)	Measles virus	Respiratory tract	Skin rash of reddish macules first appearing on face and spreading to trunk and extremities; diagnosis confirmed by virus culture.	Aerosol	No treatment; pre-exposure vaccine
Rubella (German measles)	Rubella virus	Respiratory tract	Mild macular disease with a rash resembling measles, but less extensive and disappears in 3 days or less; diagnosis is usually based on a rise in antibodies	Aerosol	No treatment; pre-exposure vaccine
Fifth disease (erythema infectiosum)	Human parvovirus B19	Respiratory tract	Mild disease with a macular facial rash; diagnosis confirmed with DNA probe, PCR	Aerosol	None
Roseola	Human herpesvirus 6, human herpesvirus 7	Respiratory tract	High fever followed by macular body rash; diagnosis confirmed with PCR, seroconversion	Aerosol	None
Fungal Diseases					
Ringworm (tinea)	Microsporum, Trichophyton, Epidermophyton spp.	Skin	Skin lesions of highly varied appearance; on scalp may cause local loss of hair; diagnosis confirmed by microscopic examination of skin scrapings	Direct contact; fomites	Griseofulvin (orally), miconazole, clotrimazole (topically)

TABLE 21.1	(continued)				
	Pathogen	Portal of Entry	Symptoms	Method of Transmission	Treatment
Sporotrichosis	*Sporothrix schenckii*	Skin abrasion	Ulcer at site of infection spreading into nearby lymphatic vessels; diagnosis confirmed by culturing the fungus	Soil	Potassium iodide solution (orally)
Candidiasis	*Candida albicans*	Skin; mucous membranes	Infected skin bright red; diagnosis confirmed by Gram staining skin scrapings	Direct contact; endogenous infection	Miconazole, clotrimazole, (topically)
Parasitic Infestations. Diagnosis is confirmed by microscopic examination of parasite.					
Scabies	*Sarcoptes scabiei* (mite)	Skin	Papules	Direct contact	Gamma benzene hexachloride, permethrin (topically)
Pediculosis (lice)	*Pediculus humanus capitis*	Skin	Itching	Primarily direct contact; possible fomites such as bedding, combs	Topical insecticide preparations

occur as pimples. The infected follicle of an eyelash is called a **sty.** A more serious hair follicle infection is the **furuncle (boil),** which is a type of **abscess,** a localized region of pus surrounded by inflamed tissue. Antibiotics do not penetrate well into abscesses, and the infection is therefore difficult to treat. Draining pus from the abscess is frequently a preliminary step to successful treatment.

When the body fails to wall off a furuncle, neighboring tissue can be progressively invaded. The extensive damage is called a **carbuncle,** a hard, round deep inflammation of tissue under the skin. At this stage of infection, the patient usually exhibits the symptoms of generalized illness with fever.

Staphylococci are the primary cause of a very troublesome problem in hospital nurseries, **impetigo of the**

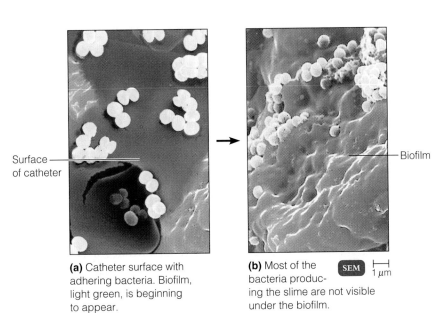

Surface of catheter

Biofilm

(a) Catheter surface with adhering bacteria. Biofilm, light green, is beginning to appear.

(b) Most of the bacteria producing the slime are not visible under the biofilm.

SEM ⊢––⊣ 1 μm

FIGURE 21.3 Coagulase-negative staphylococci. These slime-producing bacteria adhere to surfaces such as the plastic catheter in the photos. Once they have adhered to the surface (**a**), they begin to divide. Eventually (**b**), the entire surface is coated with a biofilm containing the organisms.

Q **What is the most likely source of the bacteria that grow on the catheter?**

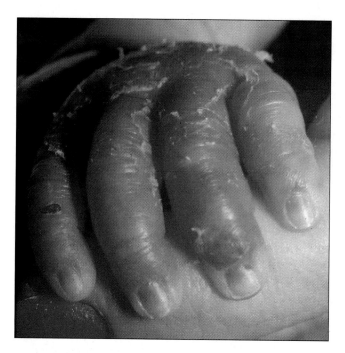

FIGURE 21.4 Lesions of scalded skin syndrome. Some staphylococci produce a toxin that causes the skin to peel off in sheets, as on the hand of this infant. It is especially likely to occur in children under age 2.

Q **What is the name of the toxin that produces this syndrome?**

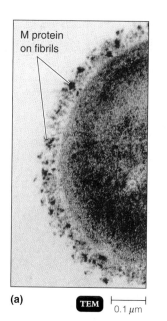 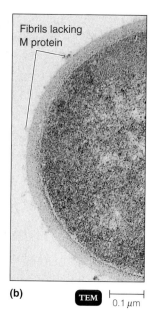

(a)   TEM ⊢─────⊣ 0.1 μm   **(b)**   TEM ⊢─────⊣ 0.1 μm

FIGURE 21.5 The M protein of group A beta-hemolytic streptococci. (**a**) Part of a cell that carries the M protein on a fuzzy layer of surface fibrils. (**b**) Part of a cell that lacks the M protein.

Q **Is the M protein more likely to be antigenic than a polysaccharide capsule?**

newborn. Symptoms of this disease are thin-walled vesicles on the skin that rupture and later crust over. To prevent outbreaks, which can reach epidemic proportions, hexachlorophene-containing skin lotions are commonly prescribed (see Chapter 7).

Staphylococcal infections always carry the risk that the underlying tissue will become infected or that the bacteria will enter the bloodstream, where they produce toxins (see the discussion of sepsis in Chapter 23). A toxin produced by certain staphylococcal strains, the *exfoliative toxin*, is manifested by **scalded skin syndrome.** This condition is first apparent as a lesion around the nose and mouth, which rapidly develops into a bright red area and spreads. Within 48 hours, the skin of the affected areas peels off in sheets when it is touched (Figure 21.4). Scalded skin syndrome is frequently observed in children under age 2, especially in newborns, as a complication of staphylococcal infections. These patients are seriously ill, and vigorous antibiotic therapy is required.

Scalded skin syndrome is also characteristic of the late stages of **toxic shock syndrome (TSS).** In this potentially life-threatening condition, fever, vomiting, and a sunburnlike rash are followed by shock and sometimes organ failure, especially of the kidneys. TSS originally became known as a result of staphylococcal growth associated with the use of a new type of highly absorbent vaginal tampon; the correlation is especially high for cases

in which the tampons remain in place too long. A novel staphylococcal toxin called *toxic shock syndrome toxin 1 (TSST-1)* is formed at the growth site and circulates in the bloodstream. The symptoms are thought to be a result of the superantigenic properties of the toxin (see the discussion of superantigens on page 461).

Today a minority of the cases of TSS are associated with menstruation. Nonmenstrual TSS occurs from staphylococcal infections that follow nasal surgery in which absorbent packing is used, after surgical incisions, and in women who have just given birth.

STREPTOCOCCAL SKIN INFECTIONS

Streptococci are gram-positive spherical bacteria. Unlike staphylococci, streptococcal cells usually grow in chains (see Figure 11.19, page 332). Prior to division, the individual cocci elongate on the axis of the chain, and then the cells divide (see Figure 4.1a, page 79). Streptococci cause a wide range of disease conditions beyond those covered in this chapter, including meningitis, pneumonia, sore throats, otitis media, endocarditis, puerperal fever, and even dental caries.

As streptococci grow, they secrete toxins and enzymes, virulence factors that vary with the different streptococcal species. Among these toxins are *hemolysins*, which lyse red blood cells. Depending on the hemolysin they produce, streptococci are categorized as alpha-hemolytic, beta-

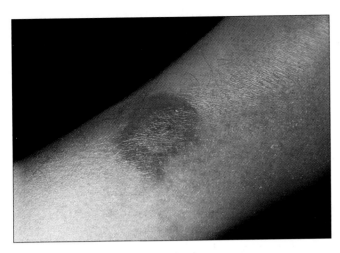

FIGURE 21.6 Lesions of erysipelas, caused by group A beta-hemolytic streptococcal toxins.

 What is the name of the toxin that produces skin reddening?

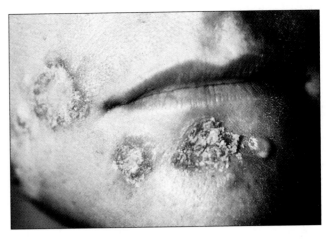

FIGURE 21.7 Lesions of impetigo. This disease is characterized by isolated pustules that become crusted. Impetigo in older children, as depicted here, is usually caused by streptococci.

Q **What other bacteria also cause impetigo?**

hemolytic, and gamma-hemolytic (actually nonhemolytic) streptococci (see Figure 6.8, page 172). Hemolysins can lyse not only red blood cells, but almost any type of cell. It is uncertain, though, just what part they play in streptococcal pathogenicity.

Beta-hemolytic streptococci are often associated with human disease. This group is further differentiated into serological groups, designated A through T, according to antigenic carbohydrates in their cell walls. The group A streptococci (GAS), which are synonymous with the species *Streptococcus pyogenes,* are the most important of the beta-hemolytic streptococci. They are among the most common human pathogens and are responsible for a number of human diseases—some of them deadly. This group of pathogens is divided into over 80 immunological types according to the antigenic properties of the M protein found in some strains (Figure 21.5). This protein is external to the cell wall on a fuzzy layer of fibrils. The M protein prevents the activation of complement and allows the microbe to evade phagocytosis and killing by neutrophils (see page 485). It also appears to aid the bacteria in adhering to and colonizing mucous membranes. Another virulence factor of the GAS is their capsule of hyaluronic acid. Exceptionally virulent strains have a mucoid appearance on blood-agar plates from heavy encapsulation and are rich in M protein. Hyaluronic acid is poorly immunogenic (it resembles human connective tissue) and few antibodies against the capsule are produced.

The GAS produce substances that promote the rapid spread of infection through tissue and by liquefying pus. Among these are *streptokinases* (enzymes that dissolve blood clots), *hyaluronidase* (an enzyme that dissolves the hyaluronic acid in the connective tissue, where it serves to cement the cells together), and *deoxyribonucleases* (enzymes that degrade DNA). These streptococci also pro-

duce certain enzymes, called *streptolysins,* that lyse red blood cells and are toxic to neutrophils.

Streptococcal skin infections are generally localized, but if the bacteria reach deeper tissue, they can be highly destructive.

When *S. pyogenes* infects the dermal layer of the skin, it causes a serious disease, **erysipelas.** In this disease, the skin erupts into reddish patches with raised margins (Figure 21.6). It can progress to local tissue destruction and even enter the bloodstream, causing sepsis (page 673). The infection usually appears first on the face and often has been preceded by a streptococcal sore throat. High fever is common. Fortunately, *S. pyogenes* has remained sensitive to β-lactam-type antibiotics.

S. pyogenes, like the staphylococci, can cause the local infection **impetigo.** This is most common in toddlers and children of grade-school age. Streptococcal impetigo is characterized by isolated pustules that become crusted and rupture (Figure 21.7). The disease is spread mostly by contact, and the bacteria penetrate the skin through some minor abrasion or insect bite. Staphylococci are often found in this type of impetigo, and medical opinions differ as to whether they are the primary cause or only secondary invaders.

Some 15,000 cases of invasive group A streptococcal infection, caused by the so-called flesh-eating bacteria, occur each year in the United States (Figure 21.8). The infection may destroy tissue as rapidly as a surgeon can cut it out, and mortality rates can exceed 40%. The streptococci attack solid tissue (*cellulitis*), muscle (*myositis*), or the muscle covering (*necrotizing fasciitis*). An important factor seems to be an *exotoxin* (exotoxin A), which acts as a superantigen, causing the immune system to contribute to the damage. Myositis or necrotizing fasciitis are often associated with **streptococcal**

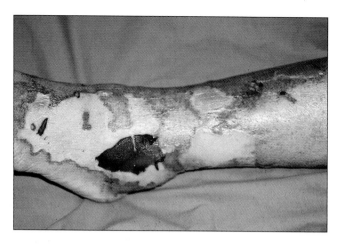

FIGURE 21.8 Necrotizing fasciitis due to group A streptococcus.

Q What is the name of the primary toxin that leads to tissue invasion by the pathogen?

toxic shock syndrome (streptococcal TSS), which resembles staphylococcal TSS, described on page 620. In cases of streptococcal TSS, a rash is less likely to be present, but bacteremia is more likely to occur. M proteins shed from the surfaces of these streptococci form a complex with fibrinogen (see page 456) which binds to neutrophils. This causes the activation of the neutrophils, precipitating the release of damaging enzymes and consequent shock and organ damage.

INFECTIONS BY PSEUDOMONADS

Pseudomonads are aerobic gram-negative rods that are widespread in soil and water. Capable of surviving in any moist environment, they can grow on traces of unusual organic matter, such as soap films or cap liner adhesives, and are resistant to many antibiotics and disinfectants. The most prominent species is *Pseudomonas aeruginosa,* which is considered a model of an opportunistic pathogen (see page 321).

Pseudomonads frequently cause outbreaks of ***Pseudomonas* dermatitis.** This is a self-limiting rash of about two weeks' duration, often associated with swimming pools and pool-type saunas and hot tubs. When many people use these facilities, the alkalinity rises and the chlorines become less effective; at the same time, the concentration of nutrients that support the growth of pseudomonads increases. Hot water causes hair follicles to open wider, facilitating the entry of bacteria. Competition swimmers are often troubled with **otitis externa,** or swimmer's ear, a painful infection of the external ear canal leading to the eardrum that is frequently caused by pseudomonads.

P. aeruginosa produces several exotoxins that account for much of its pathogenicity. It also has an endotoxin. *P. aeruginosa* often grows in dense biofilms (see Figure C, page 57) that contribute to its frequent identification as a cause of nosocomial infections of indwelling medical tubes or devices. This bacterium is also a serious opportunistic pathogen for patients with the genetic lung disease cystic fibrosis; biofilm formation plays a prominent part in this.

P. aeruginosa is also a very common and serious opportunistic pathogen in burn patients, particularly those with second- and third-degree burns. Infection may produce blue-green pus, whose color is caused by the bacterial pigment **pyocyanin.** Of concern in many hospitals is the ease with which *P. aeruginosa* grows in flower vases, mop water, and even dilute disinfectants.

The relative resistance to antibiotics that characterizes pseudomonads is still a problem (see Chapter 20). However, in recent years, several new antibiotics have been developed, and chemotherapy to treat these infections is not as restricted as it once was. The quinolones and the newer, antipseudomonal β-lactam antibiotics are the usual drugs of choice. Silver sulfadiazine is very useful in the treatment of burn infections by *P. aeruginosa.*

ACNE

Acne is probably the most common skin disease in humans, affecting an estimated 17 million people in the United States. More than 85% of all teenagers have the problem to some degree. Acne can be classified by type of lesion into three categories: comedonal acne, inflammatory acne, and nodular cystic acne. They require different treatments.

Acne develops when channels for the passage of sebum to the skin surface are blocked. Normally, skin cells that are shed inside the follicle are able to leave, but acne develops when cells are shed in higher than normal numbers, they combine with sebum, and the mixture clogs the follicle. As sebum accumulates, whiteheads (comedos) form; if the blockage protrudes through the skin, a blackhead (comedone or open comedo) forms. The dark color of blackheads is not due to dirt, but to lipid oxidation and other causes. Topical agents do not affect sebum formation, which is a root cause of acne and is dependent on hormones such as estrogens or androgens. Diet has no known effect on sebum production, but pregnancy, some hormone-based contraceptive methods, and hormonal changes with age do reduce sebum formation.

Comedonal acne is usually treated with topical agents such as azelaic acid (Azelex), salicyclic acid preparations, or retinoids (which are derivatives of vitamin A such as tretinoin, tazarotene [Tazorac], or adapalene [Differin]). These topical agents do not affect sebum formation.

Inflammatory acne arises from bacterial action, especially *Propionibacterium acnes,* an anaerobic diphtheroid commonly found on the skin. *P. acnes* has a nutritional requirement for glycerol in sebum; in metabolizing the sebum, it forms free fatty acids that cause an inflammatory response. Neutrophils that secrete enzymes that damage the wall of the hair follicle are attracted to the site. The resulting inflammation leads to the appearance of pustules and papules. At this stage, therapy is usually focused on preventing formation of sebum; topical agents are not effective for this. A drug that reduces sebum production is isotretinoin (Accutane). Unfortunately, this drug is teratogenic, meaning that it can cause serious damage to the developing fetus in a pregnant woman. About one-third of babies born after exposure will have tragic levels of damage. A side effect of depression in the patient has often been reported, but the relationship remains unproven. Because of these important side effects, isotretinoin is usually reserved for treatment of more severe forms of acne.

Inflammatory acne can also be treated by targeting *P. acnes* with antibiotics. The familiar nonprescription acne treatments containing benzoyl peroxide are effective against some bacteria, especially *P. acnes,* and also cause drying that helps loosen plugged follicles. Benzoyl peroxide is also available as a gel and in products where it is combined with antibiotics such as clindamycin (BenzaClin) and erythromycin (Benzamycin). If a patient is also using tretinoin, it must be applied at different times of the day because tretinoin inactivates benzoyl peroxide. The most recent development for the treatment of acne is the Clear Light system for cases of moderate inflammatory acne. In a series of treatments by a physician the affected skin is exposed to high intensity blue light (405–420 μm) that penetrates skin, generating reactive oxygen damaging to *Propionibacterium acnes* bacteria. This is an alternative for persons worried about side effects of medications, especially Accutane.

Some patients with acne progress to **nodular cystic acne.** Nodular cystic acne is characterized by nodules or cysts, which are inflamed lesions filled with pus deep within the skin. These leave prominent scars on the face and upper body, which often leave psychological scars as well. The most important development in treatment of cystic acne is isotretinoin, which often results in dramatic improvement—but as we have stated, it must be used with extreme caution to avoid administration to a pregnant woman, even for a few days.

VIRAL DISEASES OF THE SKIN

LEARNING OBJECTIVE

- List the causative agent, mode of transmission, and clinical symptoms of these skin infections: warts, smallpox, monkeypox, chickenpox, shingles, cold sores, measles, rubella, fifth disease, and roseola.

Many viral diseases, although systemic in nature and transmitted by respiratory or other routes, are most apparent by their effects on the skin.

WARTS

Warts, or papillomas, are generally benign skin growths caused by viruses. It was long known that warts can be transmitted from one person to another by contact, even sexually, but it was not until 1949 that viruses were identified in wart tissues. More than 50 types of papillomavirus are now known to cause different kinds of warts, often with greatly varying appearances.

After infection, there is an incubation period of several weeks before the warts appear. The most common medical treatments for warts are to apply extremely cold liquid nitrogen (cryotherapy), dry them with an electrical current (electrodesiccation), and burn them with acids. There is evidence that compounds containing salicylic acids are especially effective. Topical application of prescription drugs such as podofilox or imiquimod (Aldara) is often effective; the latter stimulates production of antiviral interferons. Warts that do not respond to any other treatments can be treated with lasers or injected interferon. The use of lasers results in a virus-laden aerosol. Physicians using lasers to remove warts have contracted warts in their nostrils.

Although warts are not a form of cancer, some skin and cervical cancers are associated with papillomaviruses. The incidence of genital warts, discussed separately in Chapter 26, has reached epidemic proportions.

SMALLPOX (VARIOLA)

During the Middle Ages, an estimated 80% of the population of Europe contracted **smallpox*** at some time during their lives. Those who recovered from the disease retained disfiguring scars. The disease, introduced by American colonists, was even more devastating to American Indians, who had had no previous exposure and thus little resistance.

Smallpox is caused by an orthopoxvirus known as the smallpox (variola) virus. There are two basic forms of this disease: **variola major,** with a mortality rate of 20% or higher, and **variola minor,** with a mortality rate of less than 1%. (Variola minor first appeared around 1900.)

Transmitted by the respiratory route, the viruses infect many internal organs before their eventual movement into the bloodstream leads to infection of the skin and the

---

*The origin of the name *smallpox* reportedly arose in the late fifteenth century in France where syphilis had just been introduced. Patients exhibited a severe skin rash called *la grosse verole,* or "the great pox." The rash was compared with that of an endemic disease of the time, which was then referred to as *la petite verole,* or "the small pox." In English, the endemic disease became known as smallpox.

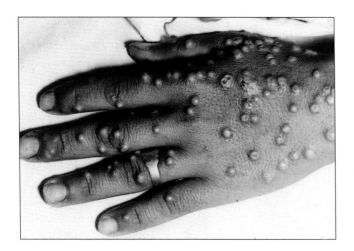

FIGURE 21.9 Smallpox lesions. In some severe cases, the lesions nearly run together (are confluent).

Q **How do these lesions differ from chickenpox?**

production of more recognizable symptoms. The growth of the virus in the epidermal layers of the skin causes lesions that become pustular after 10 days or so (Figure 21.9).

Smallpox was the first disease to which immunity was artificially induced (see pages 11 and 528) and the first to be eradicated from the human population. The last victim of a natural case of smallpox is believed to be an individual who recovered from variola minor in 1977 in Somalia. (However, ten months after this case, there was a smallpox fatality in England caused by escape of the virus from a hospital research laboratory.) The eradication of smallpox was possible because an effective vaccine was developed and because there are no animal host reservoirs for the disease. A concerted worldwide vaccination effort was coordinated by the World Health Organization.

Today, only two sites are known to maintain the smallpox virus, one in the United States and one in Russia. Dates for the destruction of these collections have been set and then postponed.

Smallpox would be an especially dangerous agent for bioterrorism. Vaccination in the United States ended in the early 1970s. People who were vaccinated prior to that time have waning immunity; however, they probably have some remaining protection that would at least moderate the disease. Stocks of smallpox vaccine are being accumulated as a precaution. No general vaccination program of the entire population is contemplated. However, certain groups, among them military and health care workers, may be an exception. Administered to the general population, the vaccine would cause a significant number of deaths, especially among immunosuppressed individuals.

A search is underway for antiviral drugs that would be effective. An intravenous drug, cidofovir, is one candidate

being investigated. The lack of symptomatic smallpox patients prevents really meaningful trials; test tube efficacy does not necessarily transfer to human patients.

With the disappearance of smallpox, there has been some concern with a similar disease, **monkeypox.** This disease first appeared among zoo monkeys that originated in Africa and East Asia and is endemic there in small animals. There are occasional outbreaks among humans in those areas, and one outbreak of more than 50 cases in the United States in 2003 was attributed to contact with pet prairie dogs. These animals apparently were infected by being housed in pet stores with Gambian giant rats imported from western Africa. Monkeypox closely resembles smallpox in symptoms and, while smallpox was endemic, was probably mistaken for it. The mortality rate is typically 1 to 10% in African adults; highest in children. There were no deaths in the U.S. outbreak. The monkeypox virus, like smallpox virus, is an orthopoxvirus, and vaccination for smallpox has a protective effect. Monkeypox is known to jump from animals to humans, but fortunately its transmission from human to human has been very limited. The World Health Organization is monitoring recent outbreaks, especially to see whether human-to-human transmission increases.

CHICKENPOX (VARICELLA) AND SHINGLES (HERPES ZOSTER)

Chickenpox (varicella) is a relatively mild childhood disease. The mortality rate from chickenpox is very low, and is usually from complications such as encephalitis (infection of the brain) or pneumonia. Almost half of such deaths occur in adults.

Chickenpox (Figure 21.10a) is the result of an initial infection with the herpesvirus varicella-zoster. (The official, but less used, name is human herpesvirus 3 [see Chapter 13]. We will encounter other instances in which both the vernacular and the official names of herpesviruses will be stated.) The disease is acquired by entry of the virus into the respiratory system, and the infection localizes in skin cells after about 2 weeks. The infected skin is vesicular for 3 to 4 days. During that time, the vesicles fill with pus, rupture, and form a scab before healing. Lesions are mostly confined to the face, throat, and lower back but can also occur on the chest and shoulders. If varicella infection occurs during early pregnancy, serious fetal damage may occur in about 2% of cases.

Reye's syndrome is an occasional severe complication of chickenpox, influenza, and some other viral diseases. A few days after the initial infection has receded, the patient persistently vomits and exhibits signs of brain dysfunction, such as extreme drowsiness or combative behavior. Coma and death can follow. At one time, the

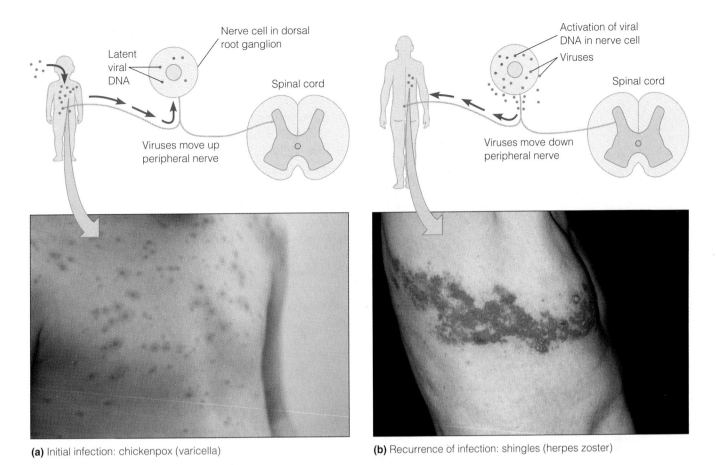

(a) Initial infection: chickenpox (varicella)

(b) Recurrence of infection: shingles (herpes zoster)

FIGURE 21.10 Chickenpox (varicella) and shingles (herpes zoster).
(**a**) Initial infection with the virus, usually during childhood, causes chickenpox. The lesions are vesicles, eventually becoming pustules that rupture and form scabs. The virus then moves to a dorsal root ganglion near the spine where it remains latent indefinitely. (**b**) Later, usually in late adulthood, the latent virus becomes reactivated, causing shingles. Reactivation can be caused by stress or weakening of the immune system. The skin lesions are vesicles.

Q **Does the photo in (a) illustrate an early or late stage of chickenpox?**

death rate of reported cases approached 90%, but this rate has been declining with improved care and is now 30% or lower when the disease is recognized and treated in time. Survivors may show neurological damage, especially if very young. Reye's syndrome affects children and teenagers almost exclusively. The use of aspirin to lower fevers in chickenpox and influenza increases the chances of acquiring Reye's syndrome.

Like all herpesviruses, a characteristic of varicellazoster virus is its ability to remain latent within the body. Following a primary infection, the virus enters the peripheral nerves and moves to a central nerve ganglion (a group of nerve cells lying outside the central nervous system), where it persists as viral DNA (see Figure 21.10a). Humoral antibodies cannot penetrate into the nerve cell, and because no viral antigens are expressed on the surface of the nerve cell, cytotoxic T cells are not activated. Therefore, neither arm of the specific immune system disturbs the latent virus.

Latent varicella-zoster virus is located in the dorsal root ganglion near the spine. Later, perhaps as long as decades later, the virus may be reactivated (Figure 21.10b). The trigger can be stress or simply the lower immune competence associated with aging. The virions produced by the reactivated DNA move along the peripheral nerves to the cutaneous sensory nerves of the skin, where they cause a new outbreak of the virus in the form of **shingles** (herpes zoster).

In shingles, vesicles similar to those of chickenpox occur but are localized in distinctive areas. Typically, they are distributed about the waist (the name *shingles* is derived from the Latin *cingulum* for girdle or belt), although facial

FIGURE 21.11 Cold sores, or fever blisters, caused by herpes simplex virus. Lesions are located mainly at the margin of the red area of the lips.

Q **Why can cold sores reappear, and why do they recur in the same place?**

shingles and infections of the upper chest and back also occur (see Figure 21.10b). The infection follows the distribution of the affected cutaneous sensory nerves and is usually limited to one side of the body at a time because these nerves are unilateral. Occasionally, such nerve infections can result in nerve damage that impairs vision or even causes paralysis. Severe burning or stinging pain is a frequent symptom; occasionally this persists for months or years, a condition called *postherpetic neuralgia.*

Shingles is simply a different expression of the virus that causes chickenpox; it expresses differently because the patient, having had chickenpox, now has partial immunity to the virus. Exposing children to shingles has led to their contracting chickenpox. Shingles seldom occurs in people under age 20, and by far the highest incidence is among older adults. It is unusual for a patient to develop shingles more than once.

Immunocompromised patients are in serious danger from infection by varicella-zoster virus; multiple organs become infected, and a mortality rate of 17% is common. In such cases, the antiviral drugs acyclovir and famciclovir have proven effective.

A live, attenuated varicella vaccine was licensed in 1995. Since then, cases of the disease have declined steadily. There is evidence that the effectiveness of the vaccine, which is about 97% at outset, declines with time. In one study based upon an outbreak in a day-care center, the effectiveness had declined to only 44%. Therefore,

varicella in previously vaccinated persons, called **breakthrough varicella,** is becoming fairly common. Because the vaccine is at least partially effective, it is a relatively mild disease with a rash that does not look much like typical varicella. A booster dose of the vaccine may eventually be needed for complete control of varicella.

Another concern is that waning effectiveness of the vaccine will lead to a population of susceptible adults, for whom the disease tends to be more severe. Also, the latent virus reappears in older adults as shingles. While it is unknown if childhood vaccinations will eventually affect the incidence of shingles, the vaccination of older adults is now proposed for prevention of shingles.

HERPES SIMPLEX

Herpes simplex viruses (HSV) can be separated into two identifiable groups, HSV-1 and HSV-2. The name herpes simplex virus, used here, is the common or vernacular name. The official names are human herpesvirus 1 and 2. HSV-1 is transmitted primarily by oral or respiratory routes, and infection usually occurs in infancy. Serological surveys show that about 90% of the U.S. population has been infected. Frequently, this infection is subclinical, but many cases develop lesions known as **cold sores** or **fever blisters.** These are painful, short-lived vesicles that occur near the outer red margin of the lips (Figure 21.11).

Cold sores, caused by herpesvirus infections, are often confused with **canker sores.** The cause of canker sores is unknown, but their occurrence is often related to stress or menstruation. While similar to cold sores in appearance, canker sores usually appear in different areas. They occur as painful sores on movable mucous membranes, such as those on the tongue, cheeks, and inner surface of the lips. They ordinarily heal in a few days but often recur.

HSV-1 usually remains latent in the trigeminal nerve ganglia communicating between the face and the central nervous system (Figure 21.12). Recurrences can be triggered by events such as excessive exposure to ultraviolet radiation from the sun, emotional upsets, or the hormonal changes associated with menstruation.

HSV-1 infection can be transmitted by skin contact among wrestlers; this is colorfully termed **herpes gladiatorum.** Incidence as high as 3% has been reported among high school wrestlers. Nurses, physicians, and dentists are occupationally susceptible to **herpetic whitlow,** infections of the finger caused by contact with HSV-1 lesions—as are children with herpetic oral ulcers.

A very similar virus, HSV-2, is transmitted primarily by sexual contact. It is the usual cause of genital herpes (see Chapter 26). HSV-2 is differentiated from HSV-1 by its antigenic makeup and by its effect on cells in tissue culture. It is latent in the sacral nerve ganglia found near the base of the spine, a different location than that of HSV-1.

Very rarely, either type of the herpes simplex virus may spread to the brain, causing **herpes encephalitis.** Infections by HSV-2 are more serious, with a fatality rate as high as 70% if untreated. Only about 10% of survivors can expect to lead healthy lives. When administered promptly, acyclovir often cures such encephalitis. Even so, the mortality rate in certain outbreaks was still 28%, and only 38% of the survivors escaped serious neurological damage.

MEASLES (RUBEOLA)

Measles (also called rubeola) is an extremely contagious viral disease that is spread by the respiratory route. Because a person with measles is infectious before symptoms appear, quarantine is not an effective measure of prevention.

Humans are the only reservoir for measles; therefore, it can potentially be eradicated, much as smallpox was. The measles vaccine, now usually administered as the MMR vaccine (measles, mumps, rubella), has almost eliminated measles in the United States. Measles cases have declined from an estimated 5 million cases a year (400,000 were actually reported) to virtual disappearance (Figure 21.13). Worldwide, however, measles still strikes about 30 million persons a year and kills nearly a million. It is the leading cause of vaccine-preventable disease.

Although the vaccine is about 95% effective, cases continue to occur among those who do not develop or retain good immunity. Some of these infections are caused by contact with infected people who come from outside the United States.

An unexpected result of the measles vaccine is that many cases of measles today occur in children under 1 year of age. Measles is especially hazardous to infants; they are more likely to have serious complications. In prevaccination days, measles was rare at this age because infants were protected by maternal antibodies derived from their mothers' recovery from the disease. Unfortunately, maternal antibodies made in response to the vaccine are not as effective in providing protection as are antibodies made in response to the disease. Because the vaccine is not effective when administered in early infancy, the initial vaccination is not made before 12 months. Therefore, the child is vulnerable for a significant time.

The development of measles is similar to that of smallpox and chickenpox. Infection begins in the upper respiratory system. After an incubation period of 10 to 12 days, symptoms develop resembling those of a common cold. Soon, a macular rash appears, beginning on the face and

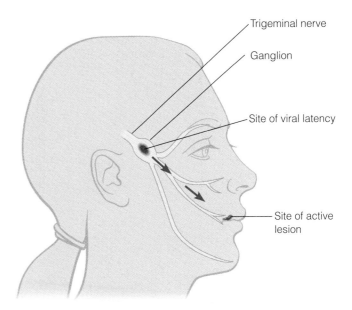

FIGURE 21.12 Site of latency of herpes simplex type 1 in the trigeminal nerve ganglion.

Q Why is this nerve system called *trigeminal*?

spreading to the trunk and extremities (Figure 21.14). Lesions of the oral cavity include *Koplik's spots*, tiny red patches with central white specks, on the oral mucosa opposite the molars. The presence of Koplik's spots is a diagnostic indicator of the disease.

Measles is an extremely dangerous disease, especially in infants and very old people. It is frequently complicated by middle ear infections or pneumonia caused by the virus itself or by a secondary bacterial infection. Encephalitis strikes approximately 1 in 1000 measles victims; its survivors are often left with permanent brain damage. As many as 1 in 3000 cases is fatal, mostly in infants. A rare complication of measles (about 1 in 1,000,000 cases) is **subacute sclerosing panencephalitis.** Occurring mostly in males, it appears about 1 to 10 years after recovery from measles. Severe neurological symptoms result in death within a few years.

RUBELLA

Rubella, or *German measles* (so called because it was first described by German physicians in the eighteenth century), is a much milder viral disease than rubeola (measles) and often goes undetected. A macular rash of small red spots and a light fever are the usual symptoms (Figure 21.15). Complications are rare, especially in children, but encephalitis occurs in about 1 case in 6000, mostly in adults. Transmission is by the respiratory route, and an incubation time of 2 to 3 weeks is the norm. Recovery from clinical or subclinical cases appears to give a firm immunity.

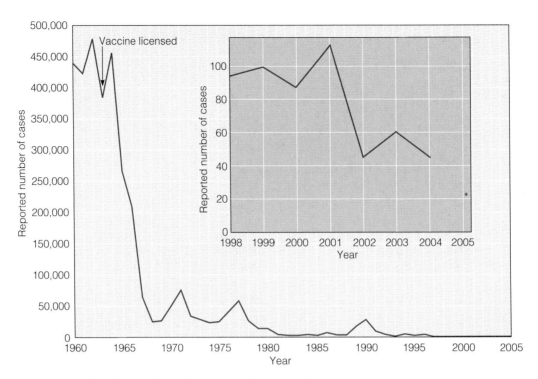

FIGURE 21.13 Reported numbers of measles cases in the United States, 1960–2005. Notice the sharp decline in cases after introduction of the measles vaccine in 1963. **Inset:** Measles cases are low; however, previous periods of low activity have been followed by resurgences.

Note: *Indicates data from the first 26 weeks of 2005.

SOURCES: Date from CDC, *Summary of Notifiable Diseases 2003, MMWR 52 (54) 4/22/05; MMWR 53 (52), 2005.*

 What is the usual cause of measles outbreaks in the United States?

The seriousness of rubella was not appreciated until 1941, when the association was made between certain severe birth defects and maternal infection during the first trimester (3 months) of pregnancy, a condition called **congenital rubella syndrome.** If a pregnant woman contracts the disease during this time, there is about a 35% incidence of serious fetal damage, including deafness, eye cataracts, heart defects, mental retardation, and death. Some 15% of babies with congenital rubella syndrome die during their first year. The last major epidemic of rubella in the United States was during 1964 and 1965. At least 20,000 severely impaired children were born during this epidemic.

It is therefore important to identify women of childbearing age who are not immune to rubella. In some states, the blood test required for a marriage license includes a test for rubella antibodies. Serum antibody can be assayed by a number of commercially available laboratory tests. Accurate diagnosis of immune status always requires such tests; histories alone are unreliable.

In addition to this surveillance, a rubella vaccine was introduced in 1969. Followup studies indicate that more than 90% of vaccinated individuals are protected for at least 15 years. Because of these preventive measures, fewer than 10 annual cases of congenital rubella syndrome are now reported.

The vaccine is not recommended for pregnant women. However, in hundreds of cases in which women were vaccinated 3 months before or 3 months after their presumed date of conception, no case of congenital rubella syndrome defects has occurred. Individuals with an impaired immune system should not receive live vaccine of any disease.

OTHER VIRAL RASHES

Fifth Disease (Erythema Infectiosum) Parents with young children are often baffled by a diagnosis of fifth disease, which they have never heard of before. The name derives from a 1905 list of skin rash diseases: measles, scarlet fever, rubella, Filatov Dukes' disease (a mild form of scarlet fever), and the fifth disease on the list. This **fifth disease,** or **erythema infectiosum,** produces no symptoms at all in about 20% of those infected by the virus (human parvovirus

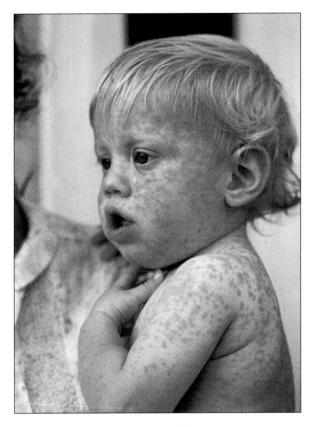

FIGURE 21.14 The rash of small raised spots typical of measles (rubeola). The rash usually begins on the face and spreads to the trunk and extremities.

Q **Why is it potentially possible to eradicate measles?**

B19, first identified in 1989). Symptoms are similar to a mild case of influenza, but there is a distinctive "slapped-cheek" facial rash that slowly fades. In adults who missed an immunizing infection in childhood, the disease may cause anemia, an episode of arthritis, or, rarely, miscarriage.

Roseola **Roseola** is a mild childhood disease that is very common. The child has a high fever for a few days, which is followed by a rash over much of the body lasting for a day or two. Recovery leads to immunity. The pathogens are human herpesviruses 6 (HHV-6) and 7 (HHV-7)—the latter is responsible for 5 to 10% of roseola cases. Both viruses are present in the saliva of most adults.

FUNGAL DISEASES OF THE SKIN AND NAILS

LEARNING OBJECTIVES

- Differentiate cutaneous from subcutaneous mycoses, and provide an example of each.
- List the causative agent of and predisposing factors for candidiasis.

The skin is most susceptible to microorganisms that can resist high osmotic pressure and low moisture. It is not

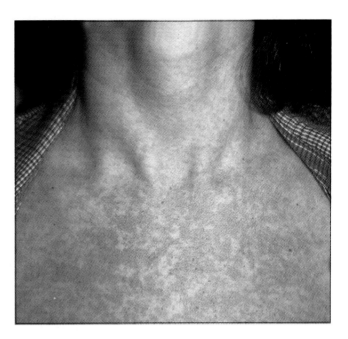

FIGURE 21.15 The rash of red spots characteristic of rubella. The spots are not raised above the surrounding skin.

Q **What is congenital rubella syndrome?**

surprising, therefore, that fungi cause a number of skin disorders. Any fungal infection of the body is called a **mycosis.**

CUTANEOUS MYCOSES

Fungi that colonize the hair, nails, and the outer layer (stratum corneum) of the epidermis (see Figure 21.1) are called **dermatophytes;** they grow on the keratin present in those locations. Termed **dermatomycoses,** these fungal infections are more informally known as *tineas* or *ringworm*. **Tinea capitis,** or ringworm of the scalp, is fairly common among elementary school children and can result in bald patches. This characteristic led to the adoption by the Romans of the name *tinea*, Latin for clothes moth, because the infection resembles the holes left by the wormlike larvae of the moth in wool clothing. The infections tend to expand circularly, hence the term *ringworm* (Figure 21.16a). The infection is usually transmitted by contact with fomites. Dogs and cats are also frequently infected with fungi that cause ringworm in children. Ringworm of the groin, or jock itch, is known as **tinea cruris,** and ringworm of the feet, or athlete's foot, is known as **tinea pedis** (Figure 21.16b). The moisture in such areas favors fungal infections.

Three genera of fungi are involved in cutaneous mycosis. *Trichophyton* (trik-ō-fī′ton) can infect hair, skin, or nails; *Microsporum* (mī-krō-spô′rum) usually involves only the hair or skin; *Epidermophyton* (ep-i-dèr-mō-fī′ton) affects only the skin and nails. The topical drugs available without prescription for tinea infections include

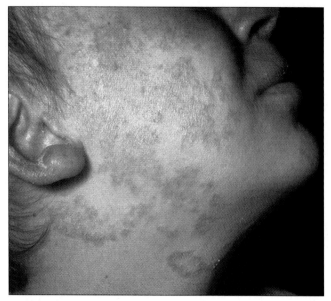

(a) Ringworm

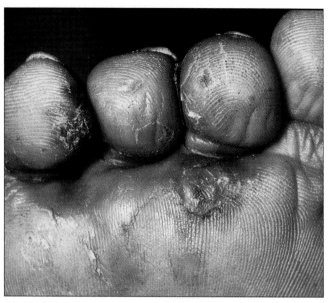

(b) Athlete's foot

FIGURE 21.16 Dermatomycoses.

 Is ringworm caused by a helminth?

miconazole and clotrimazole. When hair is involved, topical treatment is not very effective. An oral antibiotic, griseofulvin, is often useful in such infections because it can localize in keratinized tissue, such as described on page 597. When nails are infected, called **tinea unguium** or *onychomycosis*, oral itraconazole and terbinafine are the drugs of choice, but treatment may require weeks and both must be used with caution because of potential severe side effects. Recently approved topical treatments such as butenafine have become available.

Diagnosis of cutaneous mycosis is usually made by potassium hydroxide (KOH) microscopic examination of scrapings of the affected areas. The KOH dissolves epithelial tissue, allowing a clear view of fungal hyphae.

SUBCUTANEOUS MYCOSES

Subcutaneous mycoses are more serious than cutaneous mycoses. Even when the skin is broken, cutaneous fungi do not seem to be able to penetrate past the stratum corneum, perhaps because they cannot obtain sufficient iron for growth in the epidermis and the dermis. Usually subcutaneous mycoses are caused by fungi that inhabit the soil, especially decaying vegetation, and penetrate the skin through a small wound that allows entry into subcutaneous tissues.

In the United States, the most common disease of this type is **sporotrichosis,** caused by the dimorphic fungus *Sporothrix schenchii* (spô-rō′thriks shen′kē-ē). Most cases occur among gardeners or others working with soil. The infection frequently forms a small ulcer on the hands. The fungus often enters the lymphatic system in the area and there forms similar lesions. The condition is seldom fatal and is effectively treated by ingesting a dilute solution of potassium iodide, even though the organism is not affected in vitro by even a 10% solution of potassium iodide.

CANDIDIASIS

The bacterial microbiota of the mucous membranes in the genitourinary tract and mouth usually suppress the growth of such fungi as *Candida albicans* (Figure 21.17a). Several other species of *Candida*, for example *C. tropicalis* or *C. krusei* (krūs′ā-ē), may also be involved. Because the fungus is not affected by antibacterial drugs, it sometimes overgrows mucosal tissue when antibiotics suppress the normal bacterial microbiota. Changes in the normal mucosal pH may have a similar effect. Such overgrowths by *C. albicans* are called **candidiasis.** Newborn infants, whose normal microbiota have not become established, often suffer from a whitish overgrowth of the oral cavity, called **thrush** (Figure 21.17b). *C. albicans* is also a very common cause of vaginitis (see Chapter 26).

Immunosuppressed individuals, including AIDS patients, are unusually prone to *Candida* infections of the skin and mucous membranes. On people who are obese or diabetic, the areas of the skin with more moisture tend to become infected with this fungus. The infected areas become bright red, with lesions on the borders. Skin and mucosal infections by *C. albicans* are usually treated with topical applications of miconazole, clotrimazole, or nystatin. If

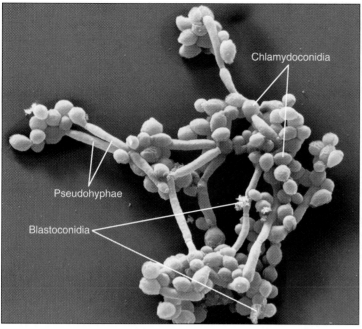

(a) *Candida albicans*

SEM ⊢————⊣ 20 mm

(b) Oral candidiasis, or thrush

FIGURE 21.17 Candidiasis. (**a**) *Candida albicans*. Notice the spherical chlamydoconidia (resting bodies formed from hyphal cells) and the smaller blastoconidia (asexual spores produced by budding) (see Chapter 12). The pseudohyphae (long cells that resemble hyphae) are resistant to phagocytosis and therefore may be a factor in pathogenicity. (**b**) This case of oral candidiasis, or thrush, produced a thick, creamy coating on the tongue.

Q **How can antibacterial drugs lead to candidiasis?**

candidiasis becomes systemic, as can happen in immunosuppressed individuals, fulminating disease (one that appears suddenly and severely) and death can result. Oral ketoconazole is the usual treatment for systemic candidiasis.

PARASITIC INFESTATION OF THE SKIN

LEARNING OBJECTIVE

- List the causative agent, mode of transmission, clinical symptoms, and treatment for scabies and pediculosis.

Parasitic organisms such as some protozoa, helminths, and microscopic arthropods can infest the skin and cause disease conditions. We will describe two examples of common arthropod infestation, scabies and lice.

SCABIES

Probably the first documented connection between a microscopic organism and a disease in humans was **scabies,** which was described by an Italian physician in 1687. The disease involves intense local itching and is caused by the tiny mite *Sarcoptes scabiei* burrowing under the skin to lay

its eggs (Figure 21.18). The burrows are often visible as slightly elevated, serpentine lines about 1 mm in width. However, scabies may appear as a variety of inflammatory skin lesions, many of them secondary infections from scratching. The mite is transmitted by intimate contact, including sexual contact, and is most often seen in family members, nursing home residents, and teenagers infected by children for whom they baby-sit.

About 500,000 people seek treatment for scabies in the United States each year; in developing countries, it is even more prevalent. The mite lives about 25 days, but by that time eggs have hatched and produced a dozen or so progeny. The recommended treatment is a solution of permethrin insecticide or gamma benzene hexachloride (lindane insecticide) applied topically. Oral administration of ivermectin, an antihelminthic drug, is also effective in resistant cases.

PEDICULOSIS (LICE)

Infestations by lice, called **pediculosis,** have afflicted humans for thousands of years. Although usually associated in the public mind with poor sanitation, outbreaks of head lice among middle- and upper-class schoolchildren in the

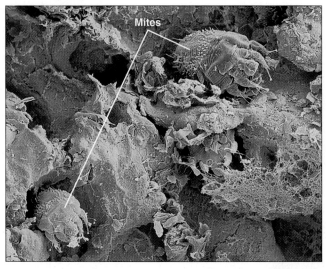

Mites

SEM ⊢——⊣ 1 mm

FIGURE 21.18 Scabies mites in skin.

Q Would it have required a microscope to identify this pathogen?

United States are common. Parents are usually appalled, but head lice are fairly easily transferred by head-to-head contact, such as occurs among children who know each other well. The head louse, *Pediculus humanus capitis*, is

not the same as the body louse, *Pediculus humanus corporis*. These are subspecies of *Pediculus humanus* that have adapted to different areas of the body. Only the body louse spreads diseases, such as epidemic typhus.

Lice (see Figure 12.33a, page 380) require blood from the host and feed several times a day. The victim is often unaware of their silent passengers until itching, which is a result of sensitization to louse saliva, develops several weeks later. Scratching can result in secondary bacterial infections. The head louse has legs especially adapted to grasp scalp hairs (Figure 21.19a). During a life span of a little over a month, the female louse produces several eggs (nits) a day. The eggs are attached to hair shafts close to the scalp (Figure 21.19b) to benefit from a warmer incubation temperature, and they hatch in about a week. The very young stages of the louse are also called nits. Empty egg cases are whitish and more visible. They do not necessarily indicate the presence of live lice. As the hair grows (at the rate of about 1 cm [$\frac{1}{4}$ inch] a month), the attached nit moves away from the scalp.

A point of interest is that the incidence of pediculosis among blacks in the United States is low: in the United States, lice have become adapted to the cylindrical hair shafts found on whites. In Africa, lice have adapted to the noncylindrical hair shafts of blacks.

FIGURE 21.19 Louse and louse egg case. (a) Adult louse grasping hair. **(b)** This egg case (nit) contains the nymphal stage of the louse, which is in the process of exiting through the cap (operculum). It does this by gulping air and forcing it out the anus until it pops free, much like a champagne cork.

Q How is pediculosis transmitted?

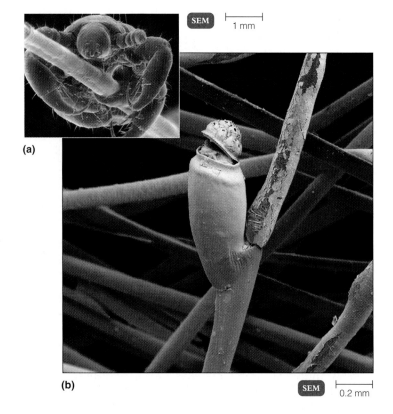

SEM ⊢——⊣ 1 mm

(a)

(b)

SEM ⊢——⊣ 0.2 mm

CLINICAL PROBLEM SOLVING

INFECTIONS IN THE GYM

As you read through this box you will encounter a series of questions that epidemiologists ask themselves as they try to trace on outbreak to its source. Try to answer each question before going on to the next one.

1. A 21-year-old college football player went to the college health center with an 11 cm × 5 cm area of redness on his right thigh. It was swollen and warm and tender when touched. His temperature was normal. He was given sulfamethoxazole-trimethoprim. *What is your diagnosis?*

2. After two days, he returned saying the area was worse. Examination revealed a broader area of redness. He was diagnosed with cellulitis. The pustule was opened and drained. *What do you need to know now?*

3. Results of Gram stains of the pus are shown in Figure a. A coagulase test was also performed on a culture (Figure b). *What is the cause of the infection?*

4. The presence of gram-positive, coagulase-positive cocci indicate *Staphylococcus aureus*. *What treatment is recommended?*

5. The results of sensitivity testing are shown in Figure c. (P = penicillin, M = methicillin, E = erythromycin, V = vancomycin, X = trimethoprim/sulfamethoxazole.) *What treatment is appropriate?*

6. Over a three-month period, ten members of the college football and fencing teams reported to the health center with cellulitis. Seven were hospitalized; one received surgical de-

bridement and skin grafts. *What is the most likely source of the methicillin-resistant* Staphylococcus aureus *(MRSA)?*

Although the investigations described in this report did not determine definitively the roots of MRSA transmission, three factors might have contributed to transmission in these outbreaks. First, abrasions and other skin trauma, which could facilitate entry of pathogens, are likely in some sports. Second, some sports involve frequent physical contact among players (e.g., football and wrestling). *S. aureus* and other skin microbiota can be transmitted easily from person to person with direct contact. Third, sports such as fencing have limited skin-to-skin contact but require multiple pieces of protective clothing. The use of shared equipment or other personal items that are not cleaned or laundered between users could be a vehicle for *S. aureus* transmission.

Investigation of outbreaks of MRSA among professional football (2004) and baseball (2005) players showed that all of the infections occurred at the site of a turf burn and rapidly progressed to large abscesses 5 to 7 cm in diameter that required surgery to drain. MRSA was recovered from whirlpools and taping gel and from 35 of the 84 nasal swabs from players and staff members.

Although outbreaks of MRSA usually have been associated with health care institutions, MRSA is emerging as a cause of skin infections in the community. This report demonstrates that community-acquired MRSA has the potential to spread and cause outbreaks among sports participants. As in this case, patients with recurrent MRSA infections

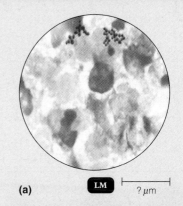

(a) LM |————| ? μm

(b) Negative control   Isolate from patient

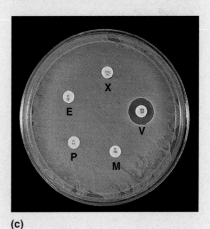

(c)

might make multiple health care visits before a wound culture is obtained. Recurrence of infections might be avoided if physicians obtain cultures more routinely when athletes have infected wounds.
SOURCE: Adapted from *MMWR* 52(33): 793–795 (8/22/03).

Treatments of head lice abound, recalling the medical adage that if there are many treatments for a condition it is probably because none of them are really good. Nonprescription medications such as Nix (permethrin insecticide) and Rid (pyrethrin insecticide) are usually the first choice, but resistance has become common. Other topical

preparations containing insecticides such as malathion and the more toxic lindane are also available. A single-dose treatment with orally administered ivermectin is occasionally used. Combing out the nits with fine-toothed louse combs is another treatment option. This is a difficult, time-consuming procedure that has actually led to the

DISEASES IN FOCUS

MACULAR RASHES

Differential diagnosis is the process of identifying the disease from a list of possible diseases that fit the information derived from examination of a patient. A differential diagnosis is important for providing initial treatment and for laboratory testing. For example, a 9-year-old girl with a history of cough, conjunctivitis, and fever (38.3°C [101°F]) now has a macular rash that starts on her face and neck that is spreading to the rest of her body. Use the table below to identify infections that could cause these symptoms.

	Pathogen	Portal of entry	Symptoms diagnosis	Method of transmission	Treatment
Measles (rubeola)	Measles virus	Respiratory tract	Skin rash of reddish macules first appearing on face and spreading to trunk and extremities; diagnosis confirmed by virus culture	Aerosol	No treatment; pre-exposure vaccine
Rubella (German measles)	Rubella virus	Respiratory tract	Mild macular disease with a rash resembling measles, but less extensive and disappears in 3 days or less; diagnosis is usually based on a rise in antibodies	Aerosol	No treatment; pre-exposure vaccine

appearance of professional removal services in some cities: expensive, but often worth the price to busy mothers.

MICROBIAL DISEASES OF THE EYE

The epithelial cells covering the eye can be considered a continuation of the skin or mucosa. Many microbes can infect the eye, largely through the *conjunctiva*, the mucous membrane that lines the eyelids and covers the outer white surface of the eyeball. It is a transparent layer of living cells replacing the skin.

INFLAMMATION OF THE EYE MEMBRANES: CONJUNCTIVITIS

> **LEARNING OBJECTIVES**
> • Define *conjunctivitis*.
> • List the causative agent, mode of transmission, and clinical symptoms of these eye infections: neonatal gonorrheal ophthalmia, inclusion conjunctivitis, trachoma.

Conjunctivitis is an inflammation of the conjunctiva, often called by the common name **red eye,** or **pinkeye.**

DISEASES IN FOCUS

	Pathogen	Portal of entry	Symptoms diagnosis	Method of transmission	Treatment
Fifth disease (erythema infectiosum)	Human parvovirus B19	Respiratory tract	Mild disease with a macular facial rash; diagnosis confirmed with DNA probe, PCR	Aerosol	None
Roseola	Human herpesvirus 6, human herpes-virus 7	Respiratory tract	High fever followed by macular body rash; diagnosis confirmed with PCR, seroconversion	Aerosol	None
Candidiasis	*Candida albicans*	Skin; mucous membranes	Macular rash; diagnosis confirmed by Gram staining skin scrapings	Direct contact; endogenous infection	Miconazole, clotrimozole, (topically)

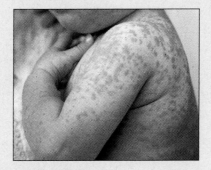

A macular rash

Haemophilus influenzae is the most common bacterial cause; viral conjunctivitis is usually caused by adenoviruses. However, a broad group of bacterial and viral pathogens as well as allergies can cause this condition.

The popularity of contact lenses has been accompanied by an increased incidence of infections of the eye. This is especially true of the soft-lens varieties, which are often worn for extended periods. Among the bacterial pathogens that cause conjunctivitis are pseudomonads, which can cause serious eye damage. (Also see the discussion of *Acanthamoeba* keratitis, page 637.) To prevent infection, contact lens wearers should not use homemade saline solutions, which are a frequent source of infection, and should scrupulously follow the manufacturer's recommendations for cleaning and disinfecting the lenses. The most effective methods for disinfecting contact lenses involve the application of heat; lenses that cannot be heated can be disinfected with hydrogen peroxide, which is then neutralized.

BACTERIAL DISEASES OF THE EYE

The bacterial microorganisms most commonly associated with the eye usually originate from the skin and upper respiratory tract.

NEONATAL GONORRHEAL OPHTHALMIA

Neonatal gonorrheal ophthalmia is a serious form of conjunctivitis caused by *Neisseria gonorrhoeae* (the cause of gonorrhea). Large amounts of pus are formed; if treatment is delayed, ulceration of the cornea will usually result. The disease is acquired as the infant passes through the birth canal, and infection carries a high risk of blindness. Early in the twentieth century, legislation required that the eyes of all newborn infants be treated with a 1% solution of silver nitrate, which proved to be a very effective treatment in preventing this eye infection. Between 1906 and 1959, the percentage of admissions to schools for the blind that could be attributed to neonatal gonorrheal ophthalmia declined from 24% to only 0.3%. Silver nitrate has been almost entirely replaced by antibiotics because of frequent coinfections by gonococci and sexually transmitted chlamydias, and silver nitrate is not effective against chlamydias. In parts of the world where the cost of antibiotics is prohibitive, a dilute solution of povidone-iodine has proven effective.

INCLUSION CONJUNCTIVITIS

Chlamydial conjunctivitis, or **inclusion conjunctivitis,** is quite common today. It is caused by *Chlamydia trachomatis*, a bacterium that grows only as an obligate intracellular parasite. In infants, who acquire it in the birth canal, the condition tends to resolve spontaneously in a few weeks or months, but in rare cases it can lead to scarring of the cornea, much as in trachoma (discussed below). Chlamydial conjunctivitis also appears to spread in the unchlorinated waters of swimming pools; in this context, it is called *swimming pool conjunctivitis*. Tetracycline applied as an ophthalmic ointment is an effective treatment.

TRACHOMA

A serious eye infection, and probably the greatest single cause of blindness by an infectious disease, is **trachoma**—an ancient name derived from the Greek word for rough. It is caused by certain serotypes of *Chlamydia trachomatis* but not the same ones that cause genital infections (see pages 794 and 797). In the arid parts of Africa and Asia, almost all children are infected early in their lives. Worldwide, there are probably 500 million active cases and 7 million blinded victims. Trachoma also occurs occasionally in the southwestern United States, especially among American Indians.

The disease is a conjunctivitis transmitted largely by hand contact or by sharing such personal objects as towels. Flies may also carry the bacteria. Repeated infections cause inflammation (Figure 21.20a), leading to *trichiasis*, an inturning of the eyelashes (Figure 21.20b). Abrasion of the

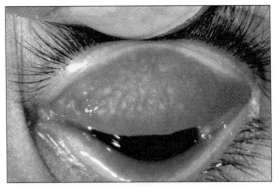

(a) Chronic inflammation of the eyelid

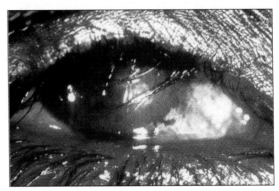

(b) Trichiasis, inturned eyelids, abrading the cornea

FIGURE 21.20 Trachoma (a) Repeated infection with *Chlamydia trachoma* causes chronic inflammation. The eyelid has been pulled back to show the inflammatory nodules that are in contact with the cornea. The abrasion caused by this damages the cornea and makes it susceptible to secondary infections. **(b)** In later stages of trachoma, the eyelashes turn inward (trichiasis) as shown here, further abrading the cornea.

 **How is trachoma transmitted?**

cornea, especially by the eyelashes, eventually causes scarring of the cornea and blindness. Trichiasis can be corrected surgically, a procedure shown in ancient Egyptian papyri. Secondary infections by other bacterial pathogens are also a factor in the disease. Antibiotics to eliminate chlamydia, especially oral azithromycin, are useful in treatment. The disease can be controlled through sanitary practices and health education.

OTHER INFECTIOUS DISEASES OF THE EYE

> **LEARNING OBJECTIVE**
> • List the causative agent, mode of transmission, and clinical symptoms of these eye infections: herpetic keratitis, *Acanthamoeba* keratitis.

Microorganisms such as viruses and protozoa can also cause eye diseases. The diseases discussed here are characterized by inflammation of the cornea, which is called *keratitis*.

TABLE 21.2	Microbial Diseases of the Eye				
	Pathogen	Portal of Entry	Symptoms	Method of Transmission	Treatment
Bacterial Diseases					
Conjunctivitis	*Haemophilus influenzae*	Conjunctiva	Redness	Direct contact; fomites	None
Neonatal gonorrheal ophthalmia	*Neisseria gonorrhoeae*	Conjunctiva	Acute infection with much pus formation	Through birth canal	Silver nitrate, tetracycline, or erythromycin for prevention
Inclusion conjunctivitis	*Chlamydia trachomatis*	Conjunctiva	Swelling of eyelid; mucus and pus formation	Through birth canal; swimming pools	Tetracycline
Trachoma	Chlamydia trachomatis	Conjunctiva	Conjunctivitis	Direct contact; fomites; flies	Azithromycin
Viral Diseases					
Conjunctivitis	Adenoviruses	Conjunctiva	Redness	Direct contact	None
Herpetic keratitis	Herpes simplex type 1 virus	Conjunctiva; cornea	Keratitis	Direct contact; recurring latent infection	Trifluridine may be effective
Protozoan Disease					
Acanthamoeba keratitis	*Acanthamoeba* spp.	Corneal abrasion; soft contact lenses may prevent removal of amoeba by blinking	Keratitis	Contact with freshwater	Topical propamidine isethionate or miconazole; corneal transplant or eye removal surgery may be required

HERPETIC KERATITIS

Herpetic keratitis is caused by the same herpes simplex type 1 virus that causes cold sores and is latent in the trigeminal nerves (see Figure 21.12). The disease is an infection of the cornea, often resulting in deep ulcers, that may be the most common cause of infectious blindness in the United States. The drug trifluridine is often an effective treatment.

ACANTHAMOEBA KERATITIS

The first case of **Acanthamoeba** (a-kan-thä-mē′bä) **keratitis** was reported in 1973 in a Texas rancher. Since then, well over 100 cases have been diagnosed in the United States. This amoeba has been found in fresh water, tap water, hot tubs, and soil. Most recent cases have been associated with the wearing of contact lenses, although any cornea damaged by trauma or infection is susceptible. Contributing factors are inadequate, unsanitary, or faulty disinfecting procedures (only heat will reliably kill the cysts), homemade saline solutions, and wearing the contact lenses overnight and while swimming.

In its early stages, the infection consists of only a mild inflammation, but later stages are often accompanied by severe pain. Damage is often so severe as to require a corneal transplant or even removal of the eye. Diagnosis is confirmed by the presence of trophozoites and cysts in stained scrapings of the cornea.

The diseases of the eye are summarized in Table 21.2.

STUDY OUTLINE

INTRODUCTION (p. 613)

1. The skin is a physical barrier against microorganisms.

2. Moist areas of the skin (such as the armpit) support larger populations of bacteria than dry areas (such as the scalp).

3. Human skin produces antibiotics called defensins.

STRUCTURE AND FUNCTION OF THE SKIN (pp. 614–615)

1. The outer portion of the skin (epidermis) contains keratin, a waterproof coating.

2. The inner portion of the skin, the dermis, contains hair follicles, sweat ducts, and oil glands that provide passageways for microorganisms.

3. Sebum and perspiration are secretions of the skin that can inhibit the growth of microorganisms.

4. Sebum and perspiration provide nutrients for some microorganisms.

5. Body cavities are lined with epithelial cells. When these cells secrete mucus, they constitute the mucous membrane.

NORMAL MICROBIOTA OF THE SKIN (pp. 615–634)

1. Microorganisms that live on skin are resistant to desiccation and high concentrations of salt.

2. Gram-positive cocci predominate on the skin.

3. The normal skin microbiota are not completely removed by washing.

4. Members of the genus *Propionibacterium* metabolize oil from the oil glands and colonize hair follicles.

5. *Malassezia furfur* yeast grows on oily secretions and may be the cause of dandruff.

MICROBIAL DISEASES OF THE SKIN (p. 615–634)

1. Vesicles are small fluid-filled lesions; bullae are vesicles larger than 1 cm; macules are flat, reddened lesions; papules are raised lesions; and pustules are raised lesions containing pus.

BACTERIAL DISEASES OF THE SKIN (pp. 615–623)

Staphylococcal Skin Infections (pp. 615–620)

2. Staphylococci are gram-positive bacteria that often grow in clusters.

3. The majority of skin microbiota consist of coagulase-negative *Staphylococcus epidermidis*.

4. Almost all pathogenic strains of *S. aureus* produce coagulase.

5. Pathogenic *S. aureus* can produce enterotoxins, leukocidins, and exfoliative toxin.

6. Many strains of *S. aureus* produce penicillinase; these are treated with vancomycin.

7. Localized infections (sties, pimples, and carbuncles) result from *S. aureus* entering openings in the skin.

8. Impetigo of the newborn is a highly contagious superficial skin infection caused by *S. aureus*.

9. Toxemia occurs when toxins enter the bloodstream; staphylococcal toxemias include scalded skin syndrome and toxic shock syndrome.

Streptococcal Skin Infections (pp. 620–622)

10. Streptococci are gram-positive cocci that often grow in chains.

11. Streptococci are classified according to their hemolytic enzymes and cell wall antigens.

12. Group A beta-hemolytic streptococci (including *Streptococcus pyogenes*) are the pathogens most important to humans.

13. Group A beta-hemolytic streptococci produce a number of virulence factors: M protein, erythrogenic toxin, deoxyribonuclease, streptokinases, and hyaluronidase.

14. Erysipelas (reddish patches) and impetigo (isolated pustules) are skin infections caused by *S. pyogenes*.

15. Invasive group A beta-hemolytic streptococci cause severe and rapid tissue destruction.

Infections by Pseudomonads (p. 622)

16. Pseudomonads are gram-negative rods. They are aerobes found primarily in soil and water that are resistant to many disinfectants and antibiotics.

17. *Pseudomonas aeruginosa* produces an endotoxin and several exotoxins.

18. Diseases caused by *P. aeruginosa* include otitis externa, respiratory infections, burn infections, and dermatitis.

19. Infections have a characteristic blue-green pus caused by the pigment pyocyanin.

20. Quinolones are useful in treating *P. aeruginosa* infections.

Acne (pp. 622–623)

21. *Propionibacterium acnes* can metabolize sebum trapped in hair follicles.

22. Metabolic end-products (fatty acids) cause inflammatory acne.

23. Tretinoin, benzoyl peroxide, erythromycin, and Accutane are used to treat acne.

VIRAL DISEASES OF THE SKIN (pp. 623–629)

Warts (p. 623)

24. Papillomaviruses cause skin cells to proliferate and produce a benign growth called a wart or papilloma.

25. Warts are spread by direct contact.

26. Warts may regress spontaneously or be removed chemically or physically.

Smallpox (Variola) (pp. 623–624)

27. Variola virus causes two types of skin infections: variola major and variola minor.

28. Smallpox is transmitted by the respiratory route, and the virus is moved to the skin via the bloodstream.

29. The only host for smallpox is humans.

30. Smallpox has been eradicated as a result of a vaccination effort by the WHO.

Chickenpox (Varicella) and Shingles (Herpes Zoster) (pp. 624–626)

31. Varicella-zoster virus is transmitted by the respiratory route and is localized in skin cells, causing a vesicular rash.

32. Complications of chickenpox include encephalitis and Reye's syndrome.

33. After chickenpox, the virus can remain latent in nerve cells and subsequently activate as shingles.

34. Shingles (herpes zoster) is characterized by a vesicular rash along the affected cutaneous sensory nerves.

35. The virus can be treated with acyclovir. An attenuated live vaccine is available.

Herpes Simplex (pp. 626–627)

36. Herpes simplex infection of mucosal cells results in cold sores and occasionally encephalitis.

37. The virus remains latent in nerve cells, and cold sores can recur when the virus is activated.

38. HSV-1 is transmitted primarily by oral and respiratory routes.

39. Herpes encephalitis occurs when herpes simplex viruses infect the brain.

40. Acyclovir has proven successful in treating herpes encephalitis.

Measles (Rubeola) (p. 627)

41. Measles is caused by measles virus and transmitted by the respiratory route.

42. Vaccination provides effective long-term immunity.

43. After the virus has incubated in the upper respiratory tract, macular lesions appear on the skin, and Koplik's spots appear on the oral mucosa.

44. Complications of measles include middle ear infections, pneumonia, encephalitis, and secondary bacterial infections.

Rubella (pp. 627–628)

45. The rubella virus is transmitted by the respiratory route.

46. A red rash and light fever might occur in an infected individual; the disease can be asymptomatic.

47. Congenital rubella syndrome can affect a fetus when a woman contracts rubella during the first trimester of her pregnancy.

48. Damage from congenital rubella syndrome includes stillbirth, deafness, eye cataracts, heart defects, and mental retardation.

49. Vaccination with live rubella virus provides immunity of unknown duration.

Other Viral Rashes (pp. 628–629)

50. Human parvovirus B19 causes fifth disease, and HHV-6 causes roseola.

FUNGAL DISEASES OF THE SKIN AND NAILS (pp. 629–631)

Cutaneous Mycoses (pp. 629–630)

51. Fungi that colonize the outer layer of the epidermis cause dermatomycoses.

52. *Microsporum*, *Trichophyton*, and *Epidermophyton* cause dermatomycoses called ringworm, or tinea.

53. These fungi grow on keratin-containing epidermis, such as hair, skin, and nails.

54. Ringworm and athlete's foot are usually treated with topical antifungal chemicals.

55. Diagnosis is based on the microscopic examination of skin scrapings or fungal culture.

Subcutaneous Mycoses (p. 630)

56. Sporotrichosis results from a soil fungus that penetrates the skin through a wound.

57. The fungi grow and produce subcutaneous nodules along the lymphatic vessels.

Candidiasis (pp. 630–631)

58. *Candida albicans* causes infections of mucous membranes and is a common cause of thrush (in oral mucosa) and vaginitis.

59. *C. albicans* is an opportunistic pathogen that may proliferate when the normal bacterial microbiota are suppressed.

60. Topical antifungal chemicals may be used to treat candidiasis.

PARASITIC INFESTATION OF THE SKIN (pp. 631–634)

61. Scabies is caused by a mite burrowing and laying eggs in the skin.

62. Pediculosis is an infestation by *Pediculus humanus*.

MICROBIAL DISEASES
OF THE EYE (pp. 634–637)

1. The mucous membrane lining the eyelid and covering the eyeball is the conjunctiva.

INFLAMMATION OF THE EYE MEMBRANES:
CONJUNCTIVITIS (pp. 634–635)

2. Conjunctivitis is caused by several bacteria and can be transmitted by improperly disinfected contact lenses.

BACTERIAL DISEASES OF THE EYE (pp. 635–636)

3. Bacterial microbiota of the eye usually originate from the skin and upper respiratory tract.

4. Neonatal gonorrheal ophthalmia is caused by the transmission of *Neisseria gonorrhoeae* from an infected mother to an infant during its passage through the birth canal.

5. All newborn infants are treated with an antibiotic to prevent *Neisseria* and *Chlamydia* infection.

6. Inclusion conjunctivitis is an infection of the conjunctiva caused by *Chlamydia trachomatis*. It is transmitted to infants during birth and is transmitted in unchlorinated swimming water.

7. In trachoma, which is caused by *C. trachomatis*, scar tissue forms on the cornea.

8. Trachoma is transmitted by hands, fomites, and perhaps flies.

OTHER INFECTIOUS DISEASES
OF THE EYE (pp. 636–637)

9. Inflammation of the cornea is called keratitis.

10. Herpetic keratitis causes corneal ulcers. The etiology is HSV-1 that invades the central nervous system and can recur.

11. *Acanthamoeba* protozoa, transmitted via water, can cause a serious form of keratitis.

STUDY QUESTIONS

Access more review material either online at **The Microbiology Place** (www.microbiologyplace.com) or with **The Microbiology Place CD-ROM** packaged with your new book. There you'll find activities, practice tests, quizzes, flashcards, case studies, and more to help you succeed.

Answers to the Study Questions can be found in Appendix G.

REVIEW

1. Discuss the usual mode of entry of bacteria into the skin. Compare bacterial skin infections with those caused by fungi and viruses with respect to mode of entry.

2. What bacteria are identified by a positive coagulase test? What bacteria are characterized as group A beta-hemolytic?

3. Compare and contrast impetigo and erysipelas.

4. Complete the table of epidemiology below.

Disease	Etiologic Agent	Clinical Symptoms	Mode of Transmission
Acne			
Pimples			
Warts			
Chickenpox			
Fever blisters			
Measles			
Rubella			

5. How do sporotrichosis and athlete's foot differ? In what ways are they similar? How is each disease treated?

6. a. Differentiate between conjunctivitis and keratitis.
   b. Select a bacterial and viral eye infection, and discuss the epidemiology of each.

7. Describe the symptoms, etiologic agent, and treatment of candidiasis. How is candidiasis contracted?

8. Why does the blood test prior to a marriage include a test for antibodies against rubella for women?

9. Identify the diseases based on the symptoms in the chart below.

Symptoms	Disease
Koplik's spots	
Macular rash	
Vesicular rash	
Small, spotted rash	
Recurrent "blisters" on oral mucosa	
Corneal ulcer and swelling of lymph nodes	

10. What complications can occur from HSV-1 infections?

11. What is in the MMR vaccine?

12. Explain the relationship between shingles and chickenpox.

13. Why are the eyes of all newborn infants washed with an antiseptic or antibiotic?

14. What is the leading infectious cause of blindness in the world?

15. A patient exhibits inflammatory skin lesions that itch intensely. Microscopic examination of skin scrapings reveals

an eight-legged arthropod. What is your diagnosis? How is the disease treated? What would you conclude if you saw a six-legged arthropod?

MULTIPLE CHOICE

Use the following information to answer questions 1 and 2. A 6-year-old girl was taken to the physician for evaluation of a slowly growing bump on the back of her head. The bump was a raised, scaling lesion 4 cm in diameter. A fungal culture of material from the lesion was positive for a fungus with numerous conidia.

1. The girl's disease was
   a. rubella.
   b. candidiasis.
   c. dermatomycosis.
   d. a cold sore.
   e. none of the above

2. Besides the scalp, this disease can occur on all of the following *except*
   a. feet.
   b. nails.
   c. the groin.
   d. subcutaneous tissue.
   e. none of the above

Use the following information to answer questions 3 and 4. A 12-year-old boy had a fever, rash, headaches, sore throat, and cough. He also had a macular rash on his trunk, face, and arms. A throat culture was negative for *Streptococcus pyogenes*.

3. The boy most likely had
   a. streptococcal sore throat.
   b. measles.
   c. rubella.
   d. smallpox.
   e. none of the above

4. All of the following are complications of this disease *except*
   a. middle ear infections.
   b. pneumonia.
   c. birth defects.
   d. none of the above
   e. encephalitis.

5. A patient has conjunctivitis. If you isolated *Pseudomonas* from the patient's mascara, you would most likely conclude all of the following *except* that
   a. the mascara was the source of the infection.
   b. *Pseudomonas* is causing the infection.
   c. *Pseudomonas* has been growing in the mascara.
   d. the mascara was contaminated by the manufacturer.
   e. none of the above

6. You microscopically examine scrapings from a case of *Acanthamoeba* keratitis. You expect to see
   a. nothing.
   b. viruses.
   c. gram-positive cocci.
   d. eukaryotic cells.
   e. gram-negative cocci.

Use the following choices to answer questions 7 through 9.
   a. *Pseudomonas*
   b. *S. aureus*
   c. scabies
   d. *Sporothrix*
   e. virus

7. Nothing is seen in microscopic examination of a scraping from the patient's rash.

8. Microscopic examination of the patient's ulcer reveals ovoid cells.

9. Microscopic examination of scrapings from the patient's rash shows gram-negative rods.

10. Penicillin would be most effective against
    a. *Chlamydia*.
    b. *Pseudomonas*.
    c. *Candida*.
    d. *Streptococcus*.
    e. human herpesviruses.

CRITICAL THINKING

1. A laboratory test used to determine the identity of *Staphylococcus aureus* is its growth on mannitol salt agar. The medium contains 7.5% sodium chloride (NaCl). Why is it considered a selective medium for *S. aureus*?

2. Is it necessary to treat a patient for warts? Explain briefly.

3. Analyses of nine conjunctivitis cases provided the data in the table below. How were these infections transmitted? How could they be prevented?

No.	Etiology	Isolated from Eye Cosmetics or Contact Lenses
5	S. epidermidis	+
1	Acanthamoeba	+
1	Candida	+
1	P. aeruginosa	+
1	S. aureus	+

4. What factors made the eradication of smallpox possible? What other diseases meet these criteria?

CLINICAL APPLICATIONS

1. A hospitalized patient recovering from surgery develops an infection that has blue-green pus and a grapelike odor. What is the probable etiology? How might the patient have acquired this infection?

2. A 12-year-old diabetic girl using continuous subcutaneous insulin infusion to manage her diabetes developed a fever (39.4°C), low blood pressure, abdominal pain, and erythroderma. She was supposed to change the needle-insertion site every 3 days after cleaning the skin with an iodine solution. Frequently she did not change the insertion site more often than every 10 days. Blood culture was negative, and abscesses at insertion sites were not cultured. What is the probable cause of her symptoms?

3. A teenage male with confirmed influenza was hospitalized when he developed respiratory distress. He had a fever, rash, and low blood pressure. *S. aureus* was isolated from his respiratory secretions. Discuss the relationship between his symptoms and the etiological agent.

22

Microbial Diseases of the Nervous System

Some of the most devastating infectious diseases are those that affect the nervous system, especially the brain and spinal cord. Damage to these areas can lead to deafness, blindness, learning disabilities, paralysis, and death. Because of the crucial importance of the nervous system, it is strongly protected from accident and infection by bone and other structures. Even pathogens that are circulating in the bloodstream usually cannot enter the brain and spinal cord because of the blood–brain barrier (see Figure 22.2). Occasionally, some trauma will disrupt these defenses with serious consequences. It happens that the fluid (cerebrospinal fluid) of the central nervous system is especially vulnerable because it lacks many of the defenses found in the blood, such as phagocytic cells. Pathogens capable of causing diseases of the nervous system often have virulence characteristics of a special nature that enable them to penetrate these defenses. For example, the pathogen can begin replicating in a peripheral nerve and gradually move into the brain and spinal cord.

UNDER THE MICROSCOPE

Naegleria fowleri. Infections by this normally free-living amoeba are associated with swimming. The amoeba crosses the nasal mucosa to enter the central nervous system.

STRUCTURE AND FUNCTION OF THE NERVOUS SYSTEM

LEARNING OBJECTIVES

- Define *central nervous system* and *blood–brain barrier.*
- Differentiate meningitis from encephalitis.

The human nervous system is organized into two divisions: the central nervous system and the peripheral nervous system (Figure 22.1). The **central nervous system (CNS)** consists of the brain and the spinal cord. As the control center for the entire body, the CNS picks up sensory information from the environment, interprets the information, and sends impulses that coordinate the body's activities. The **peripheral nervous system (PNS)** consists of all the nerves that branch off from the brain and spinal cord. These peripheral nerves are the lines of communication between the central nervous system, the various parts of the body, and the external environment.

Both the brain and the spinal cord are covered and protected by three continuous membranes called *meninges* (Figure 22.2). These are the outermost *dura mater,* the middle *arachnoid mater,* and the innermost *pia mater.* Between the pia mater and arachnoid membranes is a space called the *subarachnoid space,* in which an adult has 100 to 160 ml of *cerebrospinal fluid (CSF)* circulating. Because CSF has low levels of complement or circulating antibodies and few phagocytic cells, bacteria can multiply in it with few checks.

Late in the nineteenth century, experiments in which dyes were injected into the body resulted in the staining of all the organs of the body—with the important exception of the brain. On the other hand, when the CSF was injected with dyes, only the brain was stained. These remarkable results were the first evidence of an important feature of anatomy: the **blood–brain barrier.** Certain capillaries permit some substances to pass from the blood into the brain but restrict others. These capillaries are less permeable than others within the body and are therefore more selective in passing materials.

Drugs cannot cross the blood–brain barrier unless they are lipid-soluble. (Glucose and many amino acids are not lipid-soluble, but they can cross the barrier because special transport systems exist for them.) The lipid-soluble antibiotic chloramphenicol enters the brain readily. Penicillin is only slightly lipid-soluble; but, if it is taken in very large doses, enough may cross the barrier to be effective. Inflammations of the brain tend to alter the blood–brain barrier in such a way as to allow antibiotics to cross that would not be able to cross if there were no infection.

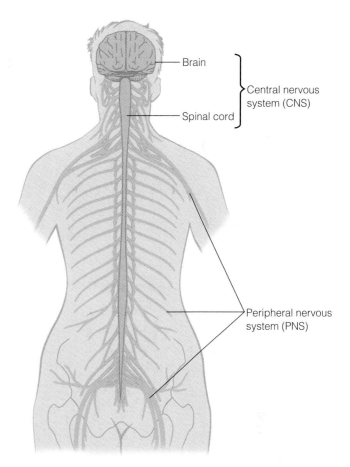

FIGURE 22.1 The human nervous system. This view shows the central and peripheral nervous systems.

Q **Is meningitis an infection of the CNS or the PNS?**

Probably the most common routes of CNS invasion are the bloodstream and lymphatic system (see Chapter 23), when inflammation alters permeability of the blood–brain barrier.

An inflammation of the meninges is called **meningitis.** An inflammation of the brain itself is called **encephalitis.** If both the brain and the meninges are affected, the inflammation is called **meningoencephalitis.**

BACTERIAL DISEASES OF THE NERVOUS SYSTEM

Microbial infections of the central nervous system are infrequent but often have serious consequences. In preantibiotic times, they were almost always fatal.

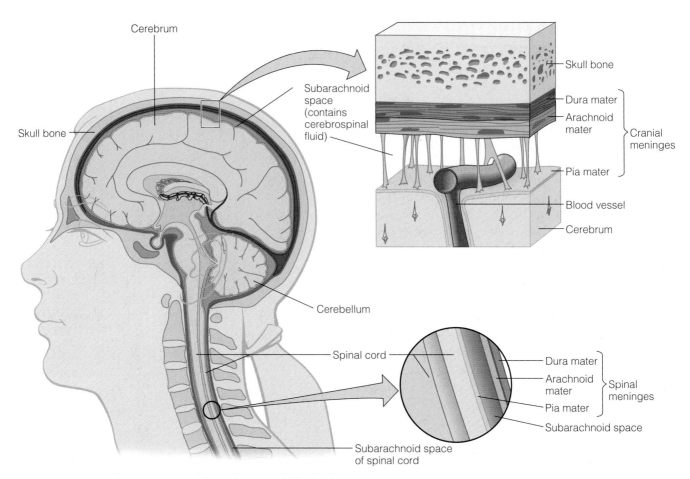

FIGURE 22.2 The meninges and cerebrospinal fluid. The meninges, whether cranial or spinal, consist of three layers: dura mater, arachnoid mater, and pia mater. Between the arachnoid and the pia mater is the subarachnoid space, in which cerebrospinal fluid circulates. Notice that the CSF is vulnerable to contamination by microbes carried in the blood that are able to penetrate the blood–brain barrier at the walls of the blood vessels.

Q **If a patient has meningitis, what barriers would need to be crossed to result in encephalitis?**

BACTERIAL MENINGITIS

> **LEARNING OBJECTIVES**
>
> • Discuss the epidemiology of meningitis caused by *Haemophilus influenzae*, *Neisseria meningitidis*, *Streptococcus pneumoniae*, and *Listeria monocytogenes*.
>
> • Explain how bacterial meningitis is diagnosed and treated.

The initial symptoms of meningitis are not especially alarming: fever, headache, and a stiff neck. Nausea and vomiting often follow. Eventually, meningitis may progress to convulsions and coma. The mortality rate varies with the pathogen but is generally high for an infectious disease today. Many people who survive an attack suffer some degree of neurological damage.

Meningitis can be caused by different types of pathogens, including viruses, bacteria, fungi, and protozoa. Viral meningitis (not to be confused with viral encephalitis, page 658), is probably much more common than bacterial meningitis but tends to be a mild disease.

Only three bacterial species cause more than 70% of the meningitis cases and 70% of the related deaths. These are the gram-positive diplococcus *Streptococcus pneumoniae*, and the gram-negative bacteria *Haemophilus influenzae* and *Neisseria meningitidis*. All three possess a capsule that protects them from phagocytosis as they replicate rapidly in the bloodstream, from which they might enter the cerebrospinal fluid. Death from bacterial meningitis often occurs very quickly, probably from shock and inflammation caused by the release of endotoxins of the gram-negative pathogens or the release of cell wall fragments (peptidoglycans and teichoic acids) of gram-positive bacteria.

Nearly 50 other species of bacteria have been reported to be opportunistic pathogens that occasionally cause meningitis. Especially important are *Listeria monocytogenes*, group B streptococci, staphylococci, and certain gram-negative bacteria.

HAEMOPHILUS INFLUENZAE MENINGITIS

Haemophilus influenzae is an aerobic, gram-negative bacterium that is a common member of the normal throat microbiota. Occasionally, however, it enters the bloodstream and causes several invasive diseases. In addition to causing meningitis, it is also frequently a cause of pneumonia (page 726), otitis media (page 716), and epiglottitis. The carbohydrate capsule of the bacterium is important to its pathogenicity, especially those bacteria with capsular antigens of type b. (Strains that lack a capsule are called *nontypable*). Medically, the bacterium is often referred to by the acronym *Hib*.

The name *Haemophilus influenzae* was given because the microorganism was erroneously thought to be the causative agent of the influenza pandemics of 1889 and World War I. *H. influenzae* was probably only a secondary invader during those virus-caused pandemics. *Haemophilus* refers to the need the microorganism has for factors in blood for growth (*hemo* = blood; *philus* = loving).

Hib-caused meningitis occurs mostly in children under age 4, especially at about 6 months, when antibody protection provided by the mother weakens. The incidence is decreasing, as shown in Figure 22.3, because of the Hib vaccine, which was introduced in 1988. Historically, *H. influenzae* meningitis accounted for most of the cases of reported bacterial meningitis (45%), with a mortality rate of about 6%.

NEISSERIA MENINGITIS
(MENINGOCOCCAL MENINGITIS)

Meningococcal meningitis is caused by *Neisseria meningitidis* (the **meningococcus**). This is an aerobic, gram-negative bacterium with a polysaccharide capsule that is important to its virulence. Like Hib and the pneumococcus, it is frequently present in the nose and throat of carriers without causing disease symptoms (Figure 22.4). These carriers, about 10% of the population, are a reservoir of infection. The symptoms of meningococcal meningitis are mostly caused by an endotoxin which is produced very rapidly and is capable of causing death within just a few hours. The most distinguishing feature is a rash that does not fade when pressed. A case of meningococcal meningitis typically begins with a throat infection, leading to bacteremia and eventually meningitis. It usually occurs in children under 2 years of age; the maternal immunity weakens at about 6 months and leaves them susceptible. Significant numbers of these children have residual damage, such as deafness.

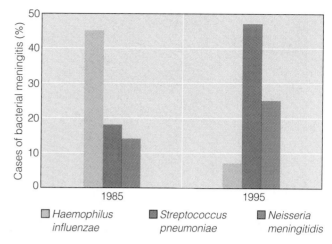

FIGURE 22.3 Changes in the relative incidence of bacterial meningitis in the United States: 1985–1995. Before introduction of the Hib vaccine, *Haemophilus influenzae* accounted for about 45% of the cases of bacterial meningitis. About 70% of these cases occurred in children under age 5. The recent introduction of a pneumococcal conjugate vaccine (PCV) for the childhood vaccination series (see Table 18.3) is expected to cause a decline in diseases caused by *Streptococcus pneumoniae*. Group B streptococci and *Listeria monocytogenes* caused most other cases of bacterial meningitis.
SOURCE: Spach, D. H. "New Issues in Bacterial Meningitis in Adults." *Postgraduate Medicine* 114 (5): 44, 2003.

Q **The percentage of cases of bacterial meningitis caused by *Streptococcus pneumoniae* has increased—is this the same thing as saying that the number of cases of bacterial meningitis caused by this organism has increased?**

In some patients, the bacteria begin to proliferate in the bloodstream, resulting in gram-negative sepsis, a life-threatening condition discussed in Chapter 23 (page 674). Sepsis can lead to extensive tissue destruction and even the need to amputate some of the affected limbs. Death can occur a few hours after the onset of fever; however, antibiotic therapy has helped reduce the mortality rate to about 9 to 12%. Without chemotherapy, mortality rates approach 80%.

The meningococcus occurs in five capsular serotypes. In recent years in the United States meningococcal disease has been caused by several of these serotypes: B (about 35%), C (about 24%), Y (about 34%), and W-135 (about 2%). In Europe, type B predominates, but C also occurs. In arid regions of Africa, China, and the Middle East, type A, and occasionally C and W-135, causes widespread epidemics that coincide with the dry seasons, when the nasal mucous membranes are less resistant to bacterial invasion.

In the United States sporadic meningococcal outbreaks occur among college students, presumably caused

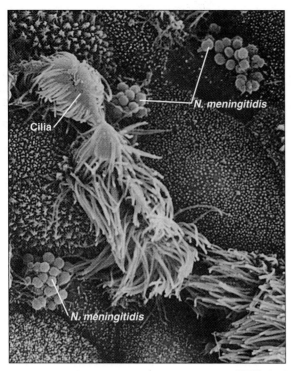

FIGURE 22.4 *Neisseria* **meningitis.** This scanning electron micrograph shows *Neisseria meningitidis* in clusters attached to cells on the mucous membrane in the pharynx.

Q **What would be the effect if the cilia are inactivated by this infection?**

by crowding of susceptible populations in dormitories. At one time, before vaccination was introduced in 1982, this was a major problem in recruit barracks for the U.S. military. This vaccine is directed at the polysaccharide capsules of serotypes A, C, Y, and W-135; it is also recommended for many students entering college and is required by some institutions. No vaccine is available against serotype B, which has a capsule that is not immunogenic in humans.

The effectiveness of polysaccharide vaccines, which are not effective for very young children, can be much improved by conjugation with protein carriers. Such conjugated vaccines are now available for serotypes A and C in some parts of the world. Work on conjugated vaccines for other serotypes is under way. It is hoped that they will have the same degree of success obtained with the Hib conjugated vaccine.

STREPTOCOCCUS PNEUMONIAE MENINGITIS (PNEUMOCOCCAL MENINGITIS)

Streptococcus pneumoniae, like *H. influenzae*, is a common inhabitant of the nasopharyngeal region. About 70% of

the general population are healthy carriers. The pneumococcus, so called because it is best known as a cause of pneumonia (Chapter 24), is a gram-positive, encapsulated diplococcus. It is the leading cause of bacterial meningitis, now that an effective Hib vaccine is in use. In addition to approximately 3000 cases of meningitis, each year *S. pneumoniae* causes 500,000 cases of pneumonia and millions of cases of painful otitis media (earache). Most of the cases of pneumococcal meningitis occur among children between the ages of 1 month and 4 years. For a bacterial disease, the mortality rate is very high: about 30% in children and 80% in the elderly.

A conjugated vaccine, modeled after the Hib vaccine, has been introduced. It is recommended for infants under the age of 2 (see Table 18.3). One side effect of this vaccine is that it results in about a 6 to 7% decrease in cases of otitis media. The large number of serotypes of the pneumococcus will make it difficult to develop vaccines against all of them.

A serious problem with meningitis and other diseases caused by the pneumococcus is the increasing appearance of antibiotic-resistant strains. Currently, nearly 35% are resistant to all β-lactam antibiotics. It is difficult to determine the exact bacterial cause in the early stages of meningitis, a factor that complicates selection of appropriate antibiotics for treatment.

DIAGNOSIS AND TREATMENT OF THE MOST COMMON TYPES OF BACTERIAL MENINGITIS

Bacterial meningitis is life threatening and develops rapidly. Therefore, prompt treatment of any type of bacterial meningitis is essential, and chemotherapy of suspected cases is usually initiated before identification of the pathogen is complete. Broad-spectrum third-generation cephalosporins are usually the first choice of antibiotics; some experts recommend including vancomycin. As soon as identification is confirmed, or perhaps when antibiotic sensitivity has been determined from cultures, the antibiotic treatment may be changed. Antibiotics are also valuable in protecting against the spread of an outbreak when given to patient contacts.

A diagnosis of bacterial meningitis requires a sample of cerebrospinal fluid obtained by a spinal tap. A simple Gram stain is often useful; it will frequently determine the identity of the pathogen with considerable reliability. Cultures are also made from the fluid. For this purpose, prompt and careful handling is required because many of the likely pathogens are very sensitive and will not survive much storage time or even changes in temperature. The most frequently used type of serological tests performed on CSF are latex agglutination tests. Results are available within about 20 minutes. However, a negative result does not eliminate the possibility of less common bacterial pathogens or nonbacterial causes.

LISTERIOSIS

Listeria monocytogenes is a gram-positive rod known to cause stillbirth and neurological disease in animals long before it was recognized as causing human disease. Excreted in animal feces, it is widely distributed in soil and water. The name is derived from the proliferation of monocytes (a type of leukocyte) found in some animals infected by it. In recent years, the disease **listeriosis** has changed from a disease of very limited importance to a major concern for the food industry and health authorities.

The disease appears in two basic forms: in infected adults and as an infection of the fetus and newborn. In adult humans, it is usually a mild, often symptomless disease, but the microbe sometimes invades the CNS, causing meningitis. This is most likely to happen to persons whose immune system is compromised, especially persons with cancer, diabetes, AIDS, or who are taking immunosuppressive medications. Since the introduction of Hib vaccination, listeriosis is now the fourth most common cause of bacterial meningitis. The mortality rate for CNS infections may be as high as 50%. Occasionally, *L. monocytogenes* invades the bloodstream and causes a wide range of disease conditions, especially sepsis. Recovering or apparently healthy individuals often shed the pathogen indefinitely in their feces. An important factor in its virulence is that when *L. monocytogenes* is ingested by phagocytic cells, it is not destroyed; it even proliferates within them, primarily in the liver. It also has the unusual capability of moving directly from one phagocyte to an adjacent one (Figure 22.5).

L. monocytogenes is especially dangerous when it infects a pregnant woman. She usually suffers no more than mild, flulike symptoms. The fetus, however, can be infected via the placenta, often resulting in an abortion or stillborn infant. In some cases the disease is not manifested until a few weeks after birth, usually as meningitis, which can result in significant brain injury or death. The infant mortality rate associated with this type of infection is about 60%.

In human outbreaks, the organism is mostly foodborne. It is frequently isolated from a wide variety of foods; dairy products have been involved in several outbreaks. *L. monocytogenes* is one of the few pathogens capable of growth at refrigerator temperatures, which can lead to an increase in its numbers during a food's shelf life.

Efforts to improve the methods of detecting *L. monocytogenes* in foods are ongoing. Considerable progress has been made with selective growth media and rapid biochemical tests. However, eventually DNA probes and serological tests using monoclonal antibodies are expected to be the most satisfactory (see Chapter 10). Diagnosis in

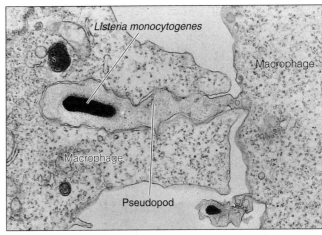

FIGURE 22.5   Cell-to-cell spread of *Listeria monocytogenes*, the cause of listeriosis. Notice that the bacterium has caused the macrophage on the right, in which it resided, to form a pseudopod that is now engulfed by the macrophage on the left. The pseudopod will soon be pinched off and the microbe transferred to the macrophage on the left.

Q How is listeriosis contracted?

humans depends on isolation and culturing of the pathogen, usually from blood or cerebrospinal fluid. Penicillin G is the antibiotic of choice for treatment.

TETANUS

> **LEARNING OBJECTIVE**
>
> • Discuss the epidemiology of tetanus, including mode of transmission, etiology, disease symptoms, and preventive measures.

The causative agent of **tetanus,** *Clostridium tetani,* is an obligately anaerobic, endospore-forming, gram-positive rod. It is especially common in soil contaminated with animal fecal wastes.

The symptoms of tetanus are caused by an extremely potent neurotoxin, *tetanospasmin,* that is released upon death and lysis of the growing bacteria (see Chapter 15). It enters the CNS via peripheral nerves or the blood. An amount of tetanospasmin weighing as much as the ink in one period on this page could kill 30 people. The bacteria themselves do not spread from the infection site, and there is no inflammation.

In a muscle's normal operation, a nerve impulse initiates contraction of the muscle. At the same time, an opposing muscle receives a signal to relax so as not to oppose the contraction. The tetanus neurotoxin blocks the relaxation pathway so that both opposing sets of muscles

FIGURE 22.6 An advanced case of tetanus. A drawing of a British soldier during the Napoleonic wars. These spasms, known as opisthotonos, can actually result in a fractured spine. (Drawing by Charles Bell of the Royal College of Surgeons, Edinburgh.)

Q *What is the name of the toxin that causes opisthotonos?*

contract, resulting in the characteristic muscle spasms. The muscles of the jaw are affected early in the disease, preventing the mouth from opening, a condition known as *lockjaw*. In extreme cases, spasms of the back muscles cause the head and heels to bow backward, a condition called *opisthotonos* (Figure 22.6). Gradually, other skeletal muscles become affected, including those involved in swallowing. Death results from spasms of the respiratory muscles.

Because the microbe is an obligate anaerobe, the wound by which it enters the body must provide anaerobic growth conditions—for example, improperly cleaned deep wounds such as those caused by rusty (and therefore presumably dirt-contaminated) nails. Injecting drug users are at high risk: sanitation during injection is not a priority, and the drugs are often contaminated. However, many cases of tetanus arise from trivial injuries, such as sitting on a tack, that are considered too minor to bring to the attention of a physician.

Effective vaccines for tetanus have been available since the 1940s. But vaccination was not always as common as it is today, where it is part of the standard DTaP (diphtheria, tetanus, and acellular pertussis) childhood vaccine. Currently, about 96% of 6-year-olds in the United States have good immunity, but only about 30% of people age 70 do. The tetanus vaccine is a *toxoid*, an inactivated toxin that stimulates the formation of antibodies that neutralize the toxin produced by the bacteria. A booster is required every 10 years to maintain good immunity, but many people do not obtain these vaccinations. Serological surveys show that at least 50% of the U.S. population does not have adequate protection. In fact, 70% of U.S. tetanus cases occur in individuals over age 50. Some were never immunized at all, and others had lost effective antibody levels over time.

Even so, immunization has made tetanus in the United States a rare disease—typically, fewer than 50 cases a year. In 1903, 406 people died of fireworks-related tetanus injuries alone. (Fireworks explosions drive soil particles deep into human tissue.) Worldwide, there are an estimated 1 million cases annually. At least half occur in newborns. In many parts of the world, the severed umbilical cords of infants are dressed with materials such as soil, clay, and even cow dung. Estimates are that the mortality rate from tetanus is about 50% in developing areas; in the United States, it is about 25%.

When a wound is severe enough to need a physician's attention, the doctor must decide whether it is necessary to provide protection against tetanus. Usually there is not enough time to administer toxoid to produce antibodies and block the progression of the infection, even if given as a booster to a patient who has been immunized. However, temporary immunity can be conferred by *tetanus immune globulin (TIG)*, prepared from the antibody-containing serum of immunized humans. (Prior to World War I, long before tetanus toxoid became available, similar preparations of preformed antibodies called *antisera* were used. Made by inoculating horses, antisera were very effective in lowering the incidence of tetanus in injured people.)

A physician's decision for treatment largely depends on the extent of the deep injuries and the immunization history of the patient, who may not be conscious. People with extensive injures who have previously had three or more doses of toxoid within the past 10 years would be considered protected, requiring no action. For extensive wounds in patients with unknown or low immunity, TIG would be given to provide temporary protection. In addition, the first of a toxoid series would be administered to provide more permanent immunity. When TIG and toxoid are both injected, different sites must be used, to avoid neutralization of the toxoid by the TIG. Adults receive a Td (tetanus and diphtheria) vaccine that also boosts immunity to diphtheria. To minimize the production of more toxin, damaged tissue that provides growth conditions for the pathogen should be removed, a procedure called **debridement,** and antibiotics should be administered. However, once the toxin has attached to the nerves, such therapy is of little use.

BOTULISM

LEARNING OBJECTIVE
* State the causative agent, symptoms, suspect foods, and treatment for botulism.

Botulism, a form of food poisoning, is caused by *Clostridium botulinum,* an obligately anaerobic, endospore-forming gram-positive rod found in soil and many freshwater sediments. Ingesting the endospores usually does no harm, as will be explained shortly. However, in anaerobic environments, such as sealed cans, the microorganism produces an exotoxin that animal assays show to be the most potent of all natural toxins. This neurotoxin is highly specific for the synaptic end of the nerve, where it blocks the release of acetylcholine, a chemical necessary for transmission of nerve impulses across synapses.

Individuals suffering from botulism undergo a progressive *flaccid paralysis* for 1 to 10 days and may die from respiratory and cardiac failure. Nausea, but no fever, may precede the neurological symptoms. The initial neurological symptoms vary, but nearly all sufferers have double or blurred vision. Other symptoms include difficulty swallowing and general weakness. Incubation time varies, but symptoms typically appear within a day or two. As with tetanus, recovery from the disease does not confer immunity because the toxin is usually not present in amounts large enough to be effectively immunogenic.

Botulism was first described as a clinical disease in the early 1800s, when it was known as the sausage disease (*botulus* is the Latin word for sausage). Blood sausage was made by filling a pig stomach with blood and ground meats, tying shut all the openings, boiling it for a short time, and smoking it over a wood fire. The sausage was then stored at room temperature. This attempt at food preservation included most of the requirements for an outbreak of botulism. It killed competing bacteria but allowed the more heat-stable *C. botulinum* endospores to survive, and it provided anaerobic conditions and an incubation period for toxin production.

FIGURE 22.7   Funeral of an Oregon family wiped out by botulism in 1924. The outbreak was caused by home-canned string beans. Altogether there were 12 deaths, but two funerals were held at a different church.

Q **Is such a drastic outcome likely today?**

The botulinal toxin will be destroyed by most ordinary cooking methods that bring the food to a boil. Sausage rarely causes botulism today, largely because of the addition of nitrites. Nitrites prevent *C. botulinum* growth following germination of the endospores.

Botulinal toxin is not formed in acidic foods (below pH 4.7). Such foods as tomatoes can therefore be safely preserved without the use of a pressure cooker. There have been cases of botulism from acidic foods that normally would not have supported the growth of the botulism organisms; however, most of these episodes are related to mold growth, which metabolized enough acid to allow the initiation of growth of *C. botulinum.*

BOTULINAL TYPES
There are several serological types of the botulinal toxin produced by different strains of the pathogen. These differ considerably in their virulence and other factors.

Type A toxin is probably the most virulent. Deaths have resulted from type A toxin when the food was only tasted but not swallowed. It is even possible to absorb lethal doses through skin breaks while handling laboratory samples. In untreated cases, the mortality rate is 60 to 70%. The type A endospore is the most heat-resistant of all *C. botulinum* strains. In the United States, it is found mainly in California, Washington, Colorado, Oregon, and New Mexico. The type A organism is usually proteolytic (the breakdown of proteins by clostridia releases amines with unpleasant odors), but obvious spoilage odor is not always apparent in low-protein foods, such as corn and beans (Figure 22.7).

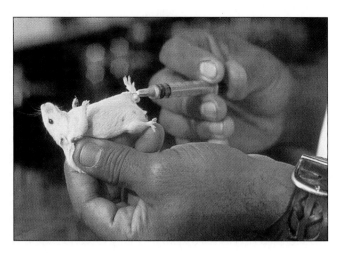

FIGURE 22.8 Diagnosis of botulism by the identification of botulinal toxin type. To determine whether botulinal toxin is present, mice are injected with the liquid portion of food extracts or cell-free cultures. If the mice die within 72 hours, this is evidence of the presence of toxin. To determine the specific type of toxin, groups of mice are passively immunized with antisera specific for *C. botulinum* type A, B, or E. For example, if one group of mice receiving a specific antitoxin lives and the other mice die, the type of toxin in the food or culture has been identified.

Q **What are the symptoms of botulism?**

Type B toxin is responsible for most European outbreaks of botulism and is the most common type in the eastern United States. The mortality rate in cases without treatment is about 25%. Type B botulism organisms occur in both proteolytic and nonproteolytic strains.

Type E toxin is produced by botulism organisms that are often found in marine or lake sediments. Therefore, outbreaks often involve seafood and are especially common in the Pacific Northwest, Alaska, and the Great Lakes area. The endospore of type E botulism is less heat-resistant than that of other strains and is usually destroyed by boiling. Type E is nonproteolytic, so the chance of detecting spoilage by odor in high-protein foods such as fish is minimal. The pathogen is also capable of producing toxin at refrigerator temperatures and requires less strictly anaerobic conditions for growth.

INCIDENCE AND TREATMENT OF BOTULISM

Botulism is not a common disease. Only a few cases are reported each year, but outbreaks from restaurants occasionally involve 20 to 30 cases. About half the cases are type A, and types B and E account about equally for the balance. Alaskan native people probably have the highest rate of botulism in the world, mostly of type E. The problem arises from food preparation methods that reflect a cultural tradition of avoiding the use of scarce fuels for heating or cooking. For example, one food in-

volved in Alaskan outbreaks of botulism is *muktuk*. Muktuk is prepared by slicing the flippers of seals or whales into strips and then drying them for a few days. To tenderize them, they are stored anaerobically in a container of seal oil for several weeks until they approach putrefaction. The 40% mortality rate for type E botulism observed in recent years among Alaskan natives reflects the difficulty in getting prompt treatment for isolated ethnic groups.

Botulism organisms do not seem to be able to compete successfully with the normal intestinal microbiota, so the production of toxin by ingested bacteria almost never causes botulism in adults. However, the intestinal microbiota of infants is not well established, and they may suffer from **infant botulism.** An estimated 250 cases occur in the United States annually, far more than any other form of botulism. Although infants have ample opportunity to ingest soil and other materials contaminated with the endospores of the organism, many reported cases have been associated with honey. Endospores of *C. botulinum* are recovered with some frequency from honey, and a lethal dose may be as few as 2000 bacteria. The recommendation is not to feed honey to infants under 1 year of age; there is no problem with older children or adults who have normal intestinal microbiota.

Botulism is diagnosed by the inoculation of mice with samples from patient serum, stool, or vomitus specimens (Figure 22.8). Different sets of mice are immunized with type A, B, or E antitoxin. All the mice are then inoculated with the test toxin; if, for example, those protected with type A antitoxin are the only survivors, then the toxin is type A. The toxin in food can similarly be identified by mouse inoculation.

The botulism pathogen can also grow in wounds in a manner similar to that of clostridia causing tetanus or gas gangrene (see Chapter 23). Such episodes of **wound botulism** occur occasionally.

The treatment of botulism relies heavily on supportive care. Recovery requires that the nerve endings regenerate; it therefore proceeds slowly. Extended respiratory assistance may be needed, and some neurological impairment may persist for months. Antibiotics are of almost no use because the toxin is preformed. Antitoxins aimed at the neutralization of A, B, and E toxins are available and are usually administered together. This trivalent antitoxin will not affect the toxin already attached to the nerve endings and is probably more effective on type E than on types A and B.

Remarkably, the deadly toxin of botulism has commercial uses as a cosmetic. Local injections of botulism toxin, Botox, at intervals of a few months eliminates forehead wrinkles (worry lines). It also finds a use, requiring expensive twice-yearly injections, in preventing armpit sweating. This

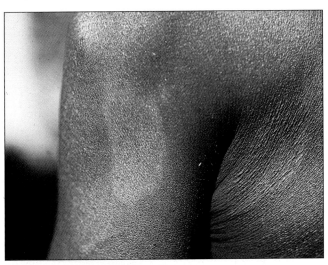

(a) Tuberculoid (neural) leprosy

(b) Lepromatous (progressive) leprosy

FIGURE 22.9 Leprosy lesions. (a) The depigmented area of skin surrounded by a border of nodules is typical of tuberculoid (neural) leprosy. **(b)** If the immune system fails to control the disease, the result is lepromatous (progressive) leprosy. This severely deformed hand shows the progressive tissue damage to the cooler parts of the body typical of this later stage.

Q **Which form of leprosy is more likely to occur in immunosuppressed individuals?**

can be significant to professional models and others who must consider the cost of designer clothing. Botox is used to treat several other more serious conditions caused by excessive muscle contractions, for example, strabismus (cross-eyes) and blepharospasm (inability to keep eyelids raised).

LEPROSY

LEARNING OBJECTIVE

- Discuss the epidemiology of leprosy, including mode of transmission, etiology, disease symptoms, and preventive measures.

Mycobacterium leprae is probably the only bacterium that grows in the peripheral nervous system, although it can also grow in skin cells. It is an acid-fast rod closely related to the tuberculosis pathogen, *Mycobacterium tuberculosis*. The organism was first isolated and identified around 1870 by Gerhard A. Hansen of Norway; his discovery was one of the first links ever made between a specific bacterium and a disease. **Hansen's disease** is the more formal name for **leprosy;** it is sometimes used to avoid the dreaded name leprosy.

The organism has an optimum growth temperature of 30°C and shows a preference for the outer, cooler portions of the human body. It survives ingestion by macrophages and eventually invades cells of the myelin sheath of the peripheral nervous system, where its presence causes nerve damage from a cell-mediated immune response. It is estimated that

M. leprae has a very long generation time, about 12 days. *M. leprae* has never been grown on artificial media. Armadillos have been found to be a useful way to culture the leprosy bacillus; they have a body temperature of 30 to 35°C and are often infected in the wild. Several Texans have actually contracted leprosy from contact with armadillos in that state, where they are common. Probably the most efficient way, however, of culturing *M. leprae* is now the inoculation of the footpads of nude mice (see Figure 19.11, page 566). The ability to grow the bacteria in an animal is invaluable for evaluating chemotherapeutic drugs.

Leprosy occurs in two main forms (although borderline forms are also recognized) that apparently reflect the effectiveness of the host's cell-mediated immune system. The *tuberculoid (neural) form* is characterized by regions of the skin that have lost sensation and are surrounded by a border of nodules (Figure 22.9a). (This disease form is roughly the same as *paucibacillary* in the World Health Organization (WHO) leprosy classification system.) Tuberculoid disease occurs in people with effective immune reactions. Recovery sometimes occurs spontaneously. The disease can be diagnosed by detecting acid-fast rods in the fluids from a slit cut in a cool site, such as an earlobe. The **lepromin test** uses an extract of lepromatous tissue injected into the skin. A visible skin reaction that develops at the injection site indicates that the body has developed an immune response to the leprosy bacillus. This test is negative during the later lepromatous stage of the disease.

In the *lepromatous (progressive) form* of leprosy (which is much the same as *multibacillary* in the WHO system), skin cells are infected, and disfiguring nodules form all over the body. Patients with this type of leprosy have the least effective cell-mediated immune response, and the disease has progressed from the tuberculoid stage. Mucous membranes of the nose tend to become affected, and a lion-faced appearance is associated with this type of leprosy. Deformation of the hand into a clawed form and considerable necrosis of tissue can also occur (Figure 22.9b). The progression of the disease is unpredictable, and remissions may alternate with rapid deterioration.

The exact means of transfer of the leprosy bacillus is uncertain, but patients with lepromatous leprosy shed large numbers in their nasal secretions and in exudates (oozing matter) of their lesions. Most people probably acquire the infection when secretions containing the pathogen contact their nasal mucosa. However, leprosy is not very contagious, and transmission usually occurs only between people in fairly intimate and prolonged contact. The time from infection to the appearance of symptoms is usually measured in years, although children can have a much shorter incubation period. Death is not usually a result of the leprosy itself but of complications, such as tuberculosis.

Much of the public's fear of leprosy can probably be attributed to biblical and historical references to the disease. In the Middle Ages, people with leprosy were rigidly excluded from normal European society and sometimes even wore bells so that people could avoid them. This isolation might have contributed to the near disappearance of the disease in Europe. But patients with leprosy are no longer kept in isolation, because they can be made noncontagious within a few days by the administration of sulfone drugs. The National Leprosy Hospital in Carville, Louisiana, once housed several hundred patients but was closed in 1999. Most patients today are treated at centers on an outpatient basis.

The number of leprosy cases in the United States is gradually increasing. Currently, 100 to 150 cases are reported each year. Most are imported; the disease is usually found in tropical climates. Millions of people, most of them in Asia, Africa, and Brazil, suffer from leprosy today, and over half a million new cases are reported each year.

Dapsone (a sulfone drug), rifampin, and clofazimine, a fat-soluble dye, are the principal drugs used for treatment, usually in combination. A vaccine became commercially available in India in 1998. It is used as an adjunct to chemotherapy. Other vaccines that might be useful in prevention are being tested. One encouraging development is that the BCG vaccine for tuberculosis (also caused by a *Mycobacterium* species) has been found to be somewhat protective against leprosy.

VIRAL DISEASES OF THE NERVOUS SYSTEM

LEARNING OBJECTIVE

- Discuss the epidemiology of poliomyelitis, rabies, and arboviral encephalitis, including mode of transmission, etiology, and disease symptoms.

Most viruses affecting the nervous system enter it by circulation in the blood or lymph. However, some viruses can enter peripheral nerve axons and move along them toward the CNS.

POLIOMYELITIS

LEARNING OBJECTIVE

- Compare the Salk and Sabin polio vaccines.

Poliomyelitis (polio) is best known as a cause of paralysis. However, the paralytic form of poliomyelitis probably affects fewer than 1% of those infected with the poliovirus. The great majority of cases are asymptomatic or exhibit only mild symptoms, such as headache, sore throat, fever, and nausea.

Polio made its first appearance in the United States in an outbreak in Vermont in the summer of 1894. After that, for decades the country was terrified by summertime epidemics. These annual outbreaks increasingly affected adolescents and young adults, and the number of paralytic cases steadily increased. Many victims were killed as their respiratory muscles were paralyzed, and thousands of infants and youths were left with their extremities permanently crippled. Later in the twentieth century, development of the iron lung (Figure 22.10) kept alive thousands with paralyzed respiratory systems.

Why did this disease appear so suddenly? The answer is paradoxical—probably because of improved sanitation. Polioviruses can remain infectious for relatively long periods in water and food. The primary mode of transmission is ingestion of water contaminated with feces containing the virus. Improved sanitation delayed exposure to polioviruses in feces until after the protection provided by maternal antibodies had waned. At one time (and today in parts of the world with poor sanitation), exposure to the poliovirus was frequent. Infants were usually exposed to poliovirus while still protected by maternal antibodies. The result was usually an asymptomatic case of the disease and a lifelong immunity. When infection is delayed until adolescence or early adulthood, the paralytic form of the disease appears more frequently.

Because the infection is initiated by ingestion of the virus, its primary areas of multiplication are the throat and

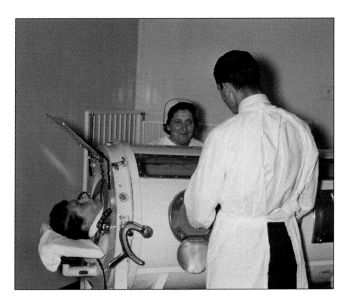

FIGURE 22.10 Polio patient in an iron lung. Many polio patients were able to breathe only with these mechanical aids. A few survivors from these polio epidemics still use these machines, at least part of the time. Others are able to use portable respiratory aids.

Q **What percentage of polio cases resulted in paralysis?**

small intestine. This accounts for the initial sore throat and nausea. Next, the virus invades the tonsils and the lymph nodes of the neck and ileum (the terminal portion of the small intestine). From the lymph nodes, the virus enters the blood, resulting in *viremia*. In most cases the viremia is only transient, the infection does not progress past the lymphatic stage, and clinical disease does not result. If the viremia is persistent, however, the virus penetrates the capillary walls and enters the central nervous system. Once in the CNS, the virus displays a high affinity for nerve cells, particularly motor nerve cells in the upper spinal cord. The virus does not infect the peripheral nerves or the muscles. As the virus multiplies within the cytoplasm of the motor nerve cells, the cells die, and paralysis results. Death can result from respiratory failure.

Diagnosis of polio is usually based on isolation of the virus from feces and throat secretions. Cell cultures can be inoculated, and cytopathic effects on the cells can be observed (see Table 15.4).

The incidence of polio in the United States has decreased markedly since the availability of the polio vaccines (Figure 22.11). The last cases attributed to a wild-type virus were reported in 1979.

There are three different serotypes of the poliovirus, and immunity must be provided for all three. Two vaccines

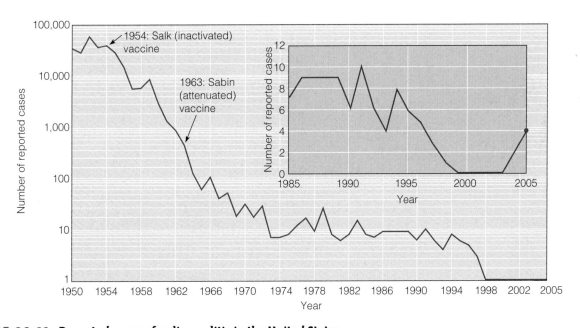

FIGURE 22.11 Reported cases of poliomyelitis in the United States, 1950–2004. Notice the decrease after introduction of the Salk and Sabin vaccines. The inset detail shows the annual cases from 1982 to 2005. The 2005 case occurred in four unvaccinated children.
SOURCES: CDC, *Summary of Notifiable Diseases 2003, MMWR* 52(54) (4/22/05); *MMWR* 53(52) (1/7/05).

Q **Why is it possible to eradicate polio but not tetanus?**

are available. The *Salk vaccine*, developed in 1954, uses viruses that have been inactivated by treatment with formalin. Vaccines of this type, called *inactivated polio vaccines* (IPV), require a series of injections. Their effectiveness rate may be as high as 90% against paralytic polio. The antibody levels decline with time, and booster shots are needed every few years to maintain full immunity. Using only IPV, several European countries have almost eliminated polio from their populations. A newer IPV has been introduced that is produced on human diploid cells. Known as *enhanced inactivated polio vaccine (E-IPV)*, it has replaced the original IPV in the United States.

The *Sabin vaccine*, introduced in 1963, contains three living, attenuated strains of the virus (trivalent) and has been more popular in the United States than the Salk vaccine. It is less expensive to administer, and most people prefer taking a sip of orange-flavored drink containing the virus to having a series of injections. The Sabin vaccine is also called *oral polio vaccine*, or OPV. The immunity achieved with the OPV resembles that acquired by natural infection, and the virus is excreted.

One disadvantage is that, on rare occasions—one in 750,000 first doses, one in about 2.4 million on subsequent doses—one of the attenuated strains of the excreted virus (type 3) may revert to virulence and transmit the disease. These cases often occur in secondary contacts, not in the person who received the vaccine. This has caused a few cases a year but also illustrates how recipients of the Sabin vaccine can infect contacts, leading, in most cases, to immunization.

In 2000 the CDC decided that the advantages of the OPV no longer outweighed the risks. Their recommendation is now to use only IPV for the routine immunization of children. They recommend that the OPV be used only to control widespread outbreaks, for protection of children traveling to areas of high risk, or for children who do not receive all four shots of IPV on schedule.

Immunosuppressed individuals should receive E-IPV to reduce the risk of vaccine-related polio from a live virus.

After the successful worldwide eradication of smallpox, polio was another disease targeted. This promised to be more difficult because the fecal–oral transmission route is still common in less-developed parts of the world and because the disease is often not easy to diagnose. For example, only about one child in 200 develops identifiable paralysis after being infected. Launched in 1988, however, a global campaign to eliminate polio has been very successful. The wild-type polio virus is now circulating in only a few countries in Africa and Asia. This has been the result of huge vaccination campaigns in which, in China and India for example, millions of people were immunized with OPV in a single day.

Because of ease of administration, OPV against all three strains of poliovirus (trivalent OPV) is the most practical vaccine in much of the world. A few immunized persons, however, shed virulent, vaccine-derived mutants for long periods. In several areas where vaccination has eliminated the wild-type virus, the disease has reappeared—caused by vaccine-derived viruses. However, discontinuance of vaccination campaigns would soon leave large populations without immunity.

Therefore, the current successful interruption of transmission of wild-type poliovirus is probably only a first step toward eradication of polio. Stockpiles of the trivalent OPV and the manufacturing facilities will need to be maintained and children routinely vaccinated. The trivalent OPV vaccine is, for several reasons, not as effective in tropical countries with poor sanitation. Even so, of the three types of polio virus, types II and III are considered to be practically eliminated. A monovalent vaccine directed at type I polio, now considered the most likely threat, is being introduced to fight specific outbreaks. It is relatively more potent against this strain of polio virus.

During the 1980s, many middle-aged adults who had had polio as children began showing a muscle weakness now called *postpolio syndrome*. It may be that nerve cells that had survived polio originally are now beginning to die. Fortunately, progress of the disease is extremely slow. Treatment of the condition with exercise therapy is often helpful, and research into drugs that stimulate nerve regeneration is in progress.

RABIES

LEARNING OBJECTIVE
• Compare the preexposure and postexposure treatments for rabies.

Rabies (the word is from the Latin for rage or madness) is a disease that almost always results in fatal encephalitis. The causative agent is the rabies virus, a lyssavirus having a characteristic bullet shape (see Figure 13.18b and discussion on page 408). These are single-stranded RNA viruses with no proofreading capability, and mutant strains develop rapidly. Worldwide, humans usually are infected with the rabies virus from the bite of an infected animal—especially dogs. The virus proliferates in the PNS and moves, fatally, toward the CNS (Figure 22.12). In the United States the most common cause of rabies is a variant of the virus found in silver-haired bats. (Domestic animals have a high rate of vaccination.) This virus has made a unique adaptation and can replicate in epidermal cells and then penetrate to enter a peripheral nerve. Therefore, a lethal dose of the virus can be administered by contact with the unbroken skin. Because deaths from rabies are frequently misdiagnosed, several cases of rabies have been traced to transplanted body tissues, especially corneas.

Rabies is unique in that the incubation period is usually long enough to allow immunity to develop from postexposure vaccination. (The amount of virus introduced into the wound is usually too small to provoke adequate natural immunity in time.) Initially, the virus multiplies in skeletal muscle and connective tissue, where it remains localized for periods ranging from days to months. Then it enters and travels, at the rate of 15 to 100 mm per day, along peripheral nerves to the CNS, where it causes encephalitis. In some extreme cases, incubation periods of as long as 6 years have been reported, but the average is 30 to 50 days. Bites in areas rich in nerve fibers, such as the hands and face, are especially dangerous and the resulting incubation period tends to be short.

Once the virus enters the peripheral nerves, it is not accessible to the immune system until cells of the CNS begin to be destroyed, which triggers a belated immune response.

Preliminary symptoms are mild and varied, resembling several common infections. When the CNS becomes involved, the patient tends to alternate between periods of agitation and intervals of calm. At this time, a frequent symptom is spasms of the muscles of the mouth and pharynx that occur when the patient feels air drafts or swallows liquids. In fact, even the mere sight or thought of water can set off the spasms—thus the common name *hydrophobia* (fear of water). The final stages of the disease result from extensive damage to the nerve cells of the brain and the spinal cord.

Animals with **furious rabies** are at first restless, then become highly excitable and snap at anything within reach. The biting behavior is essential to maintaining the virus in the animal population. Humans also exhibit similar symptoms of rabies, even biting others. When paralysis sets in, the flow of saliva increases as swallowing becomes difficult, and nervous control is progressively lost. The disease is almost always fatal within a few days.

Some animals suffer from **paralytic rabies,** in which there is only minimal excitability. This form is especially common in cats. The animal remains relatively quiet and even unaware of its surroundings, but it might snap irritably if handled. A similar manifestation of rabies occurs in humans and is often misdiagnosed as Guillain-Barré syndrome, a form of paralysis that is usually transient but sometimes fatal, or other neurological conditions. There is some speculation that the two forms of the disease may be caused by slightly different forms of the virus.

Laboratory diagnosis of rabies in humans and animals is based on several findings. When the patient or animal is

5 Virus reaches brain and causes fatal encephalitis.

6 Virus enters salivary glands and other organs of victim.

4 Virus ascends spinal cord.

3 Virus moves up peripheral nervous system to CNS in spinal cord.

2 Virus replicates in muscle near bite.

1 Virus enters tissue from saliva of biting animal.

FIGURE 22.12 Pathology of rabies infection.

Q **What is the postexposure treatment for rabies?**

alive, a diagnosis can sometimes be confirmed by immunofluorescence studies, in which viral antigens are detected in saliva, serum, or cerebrospinal fluid. After death, diagnosis is confirmed by a fluorescent-antibody test performed on brain tissue.

RABIES PREVENTION

Rabies vaccination before known exposure is limited to high-risk individuals such as laboratory workers, animal control professionals, and veterinarians. Any person bitten by an animal that is positive for rabies must undergo *postexposure prophylaxis* (PEP)—meaning a series of antirabies vaccine and immune globulin injections. Another indication for antirabies treatment is any unprovoked bite by a skunk, bat, fox, coyote, bobcat, or raccoon not available for examination. Treatment after a dog or cat bite, if the animal cannot be found, is determined by the prevalence of rabies in the area. The bite of a bat may not be perceptible, and perhaps not even needed for transmission. It may also be impossible to exclude a bite or contact in cases where the bat had access to sleeping persons or small children. Therefore, any significant encounter with a bat is considered by the CDC as a recommendation for PEP—unless the bat can be tested and shown to be negative for rabies.

The original Pasteur treatment, in which the virus was attenuated by drying in the dissected spinal cords of rabies-infected rabbits, has long been replaced in the United

Areas of the United States in which rabies predominates in certain wildlife species. Rabies-infected bats were reported in 47 of the 48 contiguous states. In eastern states in which raccoons are the predominant rabies-infected animal, many cases were also reported in foxes and skunks.

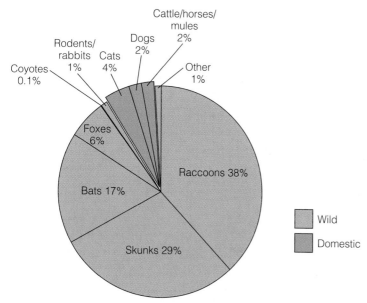

Rabies cases in various wild and domestic animals in the United States in 2003.

FIGURE 22.13 Reported cases of rabies in animals.
Rabies in foxes include different species in different geographical areas.
SOURCE: CDC 2003.

Q **What is the primary reservoir for the rabies virus in your area?**

States by *human diploid cell vaccine* (HDCV), or chick embryo–grown vaccines. These vaccines are administered in a series of five to six injections at intervals during a 28-day period. Passive immunization is provided simultaneously by injecting *human rabies immune globulin* (RIG) that has been harvested from people who are immunized against rabies.

RABIES TREATMENT

Once the symptoms of rabies appear there is very little in the way of effective treatment—only a handful of survivors have been reported. Five survivors had received PEP before the appearance of symptoms. There has been only one reported survival of a patient who had not received PEP. This very recent case was of a 15-year-old girl bitten by a rabid bat. The treatment was primarily to induce an extended coma to minimize excitability while administering antiviral drugs. She survived with some residual neurological symptoms.

DISTRIBUTION OF RABIES

Rabies occurs all over the world, mostly a result of dog bites. Vaccination of pets is prohibitively expensive in most of Africa, Latin America, and Asia. In these areas tens of thousands of deaths by rabies occur annually. In the United States, the vaccination of pets is nearly universal but rabies is widespread among wildlife, predominantly bats, skunks, foxes, and raccoons, although it is also found in domesticated animals (Figure 22.13). As many as 40,000 people are administered postexposure rabies vaccine each year, often as a precaution when the rabies status of the biting animal cannot be determined. Rabies is almost never found in squirrels, rabbits, rats, or mice. The disease has long been endemic in vampire bats of South America. In Europe and North America, there are ongoing experiments to immunize wild animals with live rabies vaccine produced in genetically modified vaccinia viruses that are added to food dropped for the animals to find. In the United States, it has been found that gray foxes prefer dog-food bait flavored with vanilla. Coyotes are not at all choosy. In Europe these campaigns have been highly successful and several countries have been declared free of rabies as a result.

In the United States 7000 to 8000 cases of rabies are diagnosed in animals each year, but in recent years, only one to six cases have been diagnosed in humans annually (see the box on the facing page).

CLINICAL PROBLEM SOLVING

A NEUROLOGICAL DISEASE

You will see questions as you read through this problem. The questions are those that clinicians ask themselves as they proceed through a diagnosis and treatment. Try to answer each question before going on to the next one.

1. In February, a previously healthy man, aged 25 years, from northern Virginia visited his physician with head and body aches, nausea, abdominal pain, chills, fever of 37.2°C to 37.7°C, dry cough, and listlessness. Laboratory findings were normal except a white blood cell (WBC) count of 13.8 × 10³/mm³.
   Compare his WBC count to Table 16.1. What could this indicate?

2. Six days later, the patient awoke disoriented with unsteady gait and slurred speech. He was evaluated in a local emergency department and admitted to the hospital. Upon retrospective questioning, his wife reported that he had showed mild personality changes during the previous days.
   What infections are possible?

3. On the fifth day of hospitalization, the patient was intubated, and twitching on his right side was noted. On day six, he was unresponsive. On day 11, a computerized tomography scan of the head showed diffuse cerebral edema. The patient remained comatose and intermittently febrile. Despite aggressive critical care management, the patient died on the fourteenth hospital day.
   What test(s) would prove the cause of his disease?

4. Tissues were forwarded to CDC for pathologic evaluation for *Naegleria*. Intracytoplasmic inclusions in neurons were seen in several areas of the brain (see the figure).
   What does this suggest? How would you confirm the disease?

5. The diagnosis was confirmed for rabies virus by direct fluorescent-antibody test and reverse-transcriptase PCR.
   How would you treat people who had contact with the patient in January and February?

6. Postexposure prophylaxis (PEP) was administered to 38 people, including 3 health care providers and the pathologist who conducted the autopsy.
   Did the delay in diagnosis affect the outcome of the disease?

7. Early diagnosis cannot save a patient; however, it may help minimize the number of potential exposures and the need for PEP.
   What else must be determined about this case?

8. The nucleotide sequence of the PCR product was used to identify a rabies virus variant associated with raccoons.
   Why is rabies surveillance and case reporting important in the United States?

9. This report describes the first documented case of human rabies associated with a raccoon rabies virus variant. With the isolation of raccoon rabies virus from this patient, human cases have been associated with all of the major reservoirs and vectors of the disease in the United States, including dogs, cats, bats, foxes, skunks, coyotes, and bobcats.

Of the 47 human rabies cases reported in the United States since 1990, four occurred in organ transplant recipients and nine were acquired outside the U.S. Human rabies cases without a definitive history of animal exposure are associated commonly with bat rabies viruses. Challenges to implicating an animal source readily can include failure to seek medical care for perceived minor lesions, non-recognition of the actual exposure event, communication (i.e., language) barriers, and recall bias from memory loss or impaired speech in encephalitic patients. Incubation periods range typically from 1 to 3 months after exposure but in rare cases can exceed 1 year in duration, further complicating collection of an adequate history.

SOURCE: Adapted from *MMWR* 52(45):1102–1103 (11/14/03).

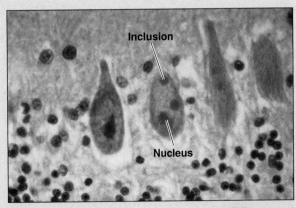

Stained nerve cell.

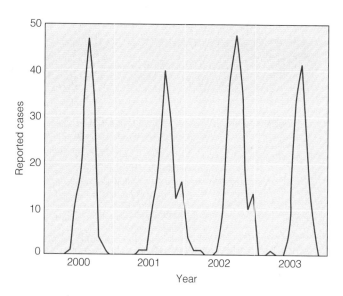

FIGURE 22.14 Arboviral infections of the central nervous system. Cases per month caused by California encephalitis viruses in the United States, 1998–2003. Notice the seasonal occurrence of cases.

SOURCE: CDC, *Summary of Notifiable Diseases*, 2003, *MMWR* 53(54)(4/22/05).

Q **Why do arboviral infections occur during the summer months?**

RELATED LYSSAVIRUS ENCEPHALITIS

In recent years a few fatal cases of encephalitis clinically indistinguishable from classic rabies have occurred in Australia and Scotland—countries considered free of rabies. These cases were found to be caused by genotypes of the genus *Lyssavirus* (see page 408) that are closely related to classic rabies virus: the Australian bat lyssavirus (ABLV) and the European bat lyssavirus (EBLV). Classic rabies is caused by one of seven known genotypes of the genus *Lyssavirus* and is widespread worldwide. Other encephalitis-causing lyssaviruses are indigenous to Europe, Australia, Africa, and Philippines, most commonly in bats.

ARBOVIRAL ENCEPHALITIS

LEARNING OBJECTIVE
• Explain how arboviral encephalitis can be prevented.

Encephalitis caused by mosquito-borne viruses (called arboviruses) is rather common in the United States. (*Arbovirus* is short for *arthropod-borne* virus. This terminology represents a functional grouping; it is not a formal taxonomic term.) Figure 22.14 shows the incidence over a sequence of years. The increase in the summer months coincides with the proliferation of adult mosquitoes during these months. *Sentinel animals*, such as caged rabbits or chickens, are tested periodically for antibodies

to arboviruses. This gives health officials information on the incidence and types of viruses in their area.

A number of clinical types of arboviral encephalitis have been identified; all can cause symptoms ranging from subclinical to severe, including rapid death. Active cases of these diseases are characterized by chills, headache, and fever. As the disease progresses, mental confusion and coma occur. Survivors may suffer from permanent neurological problems.

Horses as well as humans are frequently affected by these viruses; thus, there are strains causing *eastern equine encephalitis (EEE)* and *western equine encephalitis (WEE)*. These two viruses are the most likely to cause severe disease in humans. EEE is the most severe; the mortality rate is 30% or more, and survivors experience a high incidence of brain damage, deafness, and other neurological problems. EEE is uncommon (its main mosquito vector prefers to feed on birds); only about 100 cases a year are reported. WEE has been only rarely reported in recent years and has a mortality rate estimated at about 5%.

St. Louis encephalitis (SLE) acquired its name from the location of an early major outbreak (in which it was originally discovered that mosquitoes are involved in the transmission of these diseases). SLE is distributed from southern Canada to Argentina, but mostly in the central and eastern United States. Fewer than 1% of those infected exhibit symptoms; it can, however, be a severe disease with a mortality rate in symptomatic patients of about 20%.

California encephalitis (CE) was first identified in that state, but most cases occur elsewhere. The La Crosse strain of CE (first isolated in La Crosse, Wisconsin) is the most commonly encountered arbovirus. A relatively mild illness, it is seldom fatal.

A new arbovirus disease, now well known, was introduced into the United States in 1999. First reported in the New York City area, it was quickly identified as being caused by *West Nile virus (WNV)*. The virus infects a considerable number of bird species and has now spread throughout the country. Certain birds such as crows and blue jays are especially susceptible to infection by the virus, and dead birds of these species are often seen before human cases appear. Most human cases of WNV are subclinical or mild, but the disease can cause a polio-like paralysis or fatal encephalitis, especially in the elderly. Patients ill enough for hospitalization have a mortality rate of 4 to 18%. It is suspected that the virus was introduced from the Middle East, although there have also been outbreaks in eastern Europe dating from 1996. See the Diseases in Focus box on the facing page for a summary of the arbovirus-caused diseases of the United States.

The Far East also has endemic arboviral encephalitis. **Japanese B encephalitis** is the best known; it is a serious

DISEASES IN FOCUS

TYPES OF ARBOVIRAL ENCEPHALITIS

Arboviral encephalitis is usually characterized by fever, headache, and altered mental status ranging from confusion to coma. Vector control to decrease contacts between humans and mosquitoes is the best prevention. Mosquito control includes removing standing water and using insect repellent while outdoors.

Disease	Pathogen	Mosquito vector	Reservoir	U.S. distribution	Epidemiology	Mortality
Western equine encephalitis	WEE virus	*Culex*	Birds, horses		Severe disease; frequent neurological damage, especially in infants	5%
Eastern equine encephalitis	EEE virus	*Aedes, Culiseta*	Birds, horses		More severe than WEE; affects mostly young children and younger adults; relatively uncommon in humans	>30%
St. Louis encephalitis	SLE virus	*Culex*	Birds		Mostly urban outbreaks; affects mainly adults over 40	20%
California encephalitis	CE virus	*Aedes*	Small mammals		Affects mostly 4- to 18-year age groups in rural or suburban areas; La Crosse strain medically most important. Rarely fatal; about 10% have neurological damage	1% of those hospitalized
West Nile encephalitis	WN virus	Primarily *Culex* and *Aedes*	Primarily birds, assorted rodents, and large mammals		Most cases asymptomatic, otherwise symptoms vary from mild to severe; likelihood of severe neurological symptoms and fatality increases with age	4–18% of those hospitalized

Culex **mosquito engorged with human blood.**

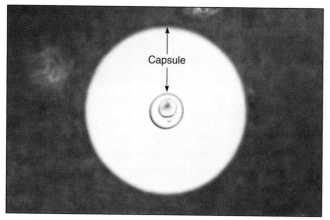

FIGURE 22.15 *Cryptococcus neoformans.* This yeastlike fungus has an unusually thick capsule. In this photomicrograph, the capsule is made visible by suspending the cells in dilute India ink.

Q *What disease does C. neoformans cause?*

public health problem, especially in Japan, Korea, and China. Vaccines are used to control the disease in these countries and are often recommended for visitors.

Diagnosis of arboviral encephalitis is made by serological tests, usually ELISA tests to identify IgM antibodies. The most effective preventive measure is local control of the mosquitoes.

FUNGAL DISEASE OF THE NERVOUS SYSTEM

The central nervous system is seldom invaded by fungi. However, one pathogenic fungus in the genus *Cryptococcus* is well adapted to growth in CNS fluids.

CRYPTOCOCCUS NEOFORMANS MENINGITIS (CRYPTOCOCCOSIS)

LEARNING OBJECTIVE

• Identify the causative agent, reservoir, symptoms, and treatment for cryptococcosis.

Fungi of the genus *Cryptococcus* are spherical cells resembling yeasts; they reproduce by budding and produce polysaccharide capsules, some much thicker than the cells themselves (Figure 22.15). Only one species, *Cryptococcus neoformans*, is pathogenic for humans, causing the disease called **cryptococcosis.** The organism is widely distributed in soil, especially soil contaminated with pigeon droppings. It is also found in pigeon roosts and nests on the window ledges of urban buildings. Most cases of cryptococcosis occur in urban areas, and it is thought to be transmit-

ted by the inhalation of dried infected pigeon or chicken droppings.

Inhalation of *C. neoformans* initially causes infection of the lungs, frequently subclinical, and often the disease does not proceed beyond this stage. However, it can spread through the bloodstream to other parts of the body, including the brain and meninges, especially in immunosuppressed individuals and those receiving steroid treatments for major illnesses. The disease is usually expressed as chronic meningitis, which is often progressive and fatal if untreated.

The best serological diagnostic test is a latex agglutination test to detect cryptococcal antigens in serum or cerebrospinal fluid. The drugs of choice for treatment are amphotericin B and flucytosine in combination. Even so, the mortality rate may approach 30%.

PROTOZOAN DISEASES OF THE NERVOUS SYSTEM

LEARNING OBJECTIVE

• Identify the causative agent, vector, symptoms, and treatment for African trypanosomiasis and amebic meningoencephalitis.

Protozoa capable of invading the CNS are rare. However, those that can reach it cause devastating effects.

AFRICAN TRYPANOSOMIASIS

African trypanosomiasis, or sleeping sickness, is a protozoan disease that affects the nervous system. In 1907, Winston Churchill described Uganda during an epidemic of sleeping sickness as a "beautiful garden of death." Even today, estimates are that as many as half a million Africans are infected, and there are about 100,000 new cases yearly.

The disease is caused by two subspecies of *Trypanosoma brucei* that infect humans: *Trypanosoma brucei gambiense* and *Trypanosoma brucei rhodesiense.* They are morphologically indistinguishable but differ significantly in their epidemiology—that is, in their ability to infect nonhuman hosts. Humans are the only significant reservoir for *T.b. gambiense*, while *T.b. rhodesiense* is a parasite of domestic livestock and many wild animals. These protozoans are flagellates (see Figure 23.22 on page 693 for the appearance of a similar organism) that are spread by tsetse fly vectors. *T.b. gambiense* is transmitted by a tsetse fly species that inhabits stream vegetation, where there are also concentrations of human populations. It is distributed throughout West and Central Africa and is sometimes termed West African trypanosomiasis. More than 97% of reported cases in humans are of this type. Once infected, there are few symptoms for weeks or months. Eventually, fever, headaches, and a variety of other symptoms develop that indicate involvement and deterioration of the CNS. Coma and death are inevitable without effective treatment.

In contrast, infections by *T.b. rhodesiense*, or East African trypanosomiasis, are transmitted by species of tsetse flies that inhabit savannahs (grasslands with scattered trees) of eastern and southern Africa. Wild animals inhabiting these areas are well adapted to the parasite and are little affected, but humans and domestic animals become acutely ill. This has had a profound effect on sub-Saharan Africa; an area nearly the size of the United States. Agricultural development has been practically prohibited because domestic food and working animals become infected. Infections of humans follow a more acute course than that caused by *T.b. gambiense*; symptoms of illness are apparent within a few days or so of infection. Death occurs within weeks or a few months, sometimes from cardiac problems even before the CNS is affected.

There are some moderately effective chemotherapeutic agents, such as suramin and pentamidine, but these do not alter the course of the disease once the CNS is affected. The drug that does alter the disease's course, however, melarsoprol, is very toxic. In 1992 a new drug, eflornithine, was introduced that crosses the blood–brain barrier and blocks an enzyme required for proliferation of the parasite. It requires an extended series of injections but it is so dramatically effective against even late stages of *T.b. gambiense* that it has been called the resurrection drug. (Its effectiveness against *T.b. rhodesiense* is variable; melarsoprol is still recommended). The history of this drug provides a valuable illustration of problems in providing health services in poverty-stricken parts of the world. Because the only populations suffering from African trypanosomiasis were unable to afford it, production was soon discontinued. Happily, it was found that the drug had a profitable use in the industrial world: it reduces growth of unwanted facial hair on women. Because of this, the manufacturer has supplied eflornithine at no cost, but for only a limited time, in many African villages.

The current primary approach in combating the disease is to attempt elimination of the vector, the tsetse fly. A combination of tentlike, insecticide-treated traps that mimic the color and odor of animal hosts of the insect, and large scale releases of sterile males have eliminated the tsetse fly on the offshore island of Zanzibar. (Female tsetse flies mate only once; the release of artificially reared, radiation-sterilized males in vast numbers prevents females that mate with them from producing young). The insect is a weak flyer, and it is hoped to repeat this eradication on selected areas of the mainland.

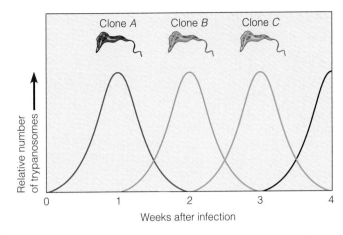

FIGURE 22.16   How trypanosomes evade the immune system. The population of each trypanosome clone drops nearly to zero as the immune system suppresses its members, but a new clone with a different antigenic surface then replaces the previous clone. The black line represents the population of clone *D*.

Q **Can you think of a viral disease that is causing a worldwide pandemic that would make for a similar figure?**

A vaccine is being developed, but a major obstacle is that the trypanosome is able to change protein coats at least 100 times and can thus evade antibodies aimed at only one or a few of the proteins. Each time the body's immune system is successful in suppressing the trypanosome, a new clone of parasites appears with a different antigenic coat (Figure 22.16).

AMEBIC MENINGOENCEPHALITIS

There are two species of protozoa that are a cause of amebic meningoencephalitis, devastating diseases of the nervous system. These protozoa are both found in recreational freshwater. Human exposure to them is apparently widespread, and many in the population carry antibodies—fortunately, symptomatic disease is rare. *Naegleria fowleri* (nī-glē′rē-ä fou′lèr-ē) is a protozoan (amoeba) that causes a neurological disease, **primary amebic meningoencephalitis (PAM)** (Figure 22.17). Although scattered cases of it are reported in most parts of the world, only a few cases are reported in the United States annually. The most common victims are children who swim in ponds or streams. The organism initially infects the nasal mucosa and later penetrates to the brain and proliferates. The fatality rate is nearly 100%, death occurring within a few days after symptoms appear. Diagnosis is typically made at autopsy. The very few known survivors had been treated with the antifungal drug amphotericin B.

A similar neurological disease is **granulomatous amebic encephalitis (GAE).** GAE is caused by a species of *Acanthamoeba* but not the same one that is the cause of

FIGURE 22.17 *Naegleria fowleri.*
This photo shows two vegetative stages of *N. fowleri* beginning to devour a presumably dead amoeba. The suckerlike structures (called amebastomes) function in phagocytic feeding—usually on bacteria or assorted debris that may include host tissue. This protozoan also has a spherical cyst stage and an ovoid flagellated stage (which is most likely to be the infective form) that allows it to swim rapidly in its aquatic habitat.

Q **How is amebic meningoencephalitis transmitted?**

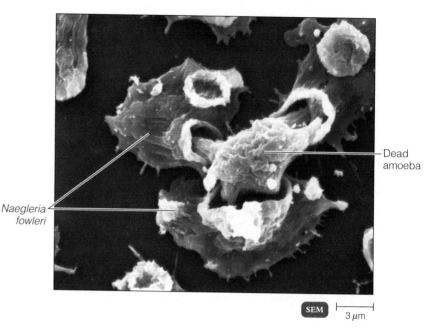

Naegleria fowleri

Dead amoeba

SEM | 3 μm

Acanthamoeba keratitis, a serious disease affecting the eyes. It is chronic, slowly progressive, and fatal in a matter of weeks or months. GAE has an unknown incubation period and months may be necessary before symptoms appear. Granulomas (see Figure 23.28, page 701) form around the organism in response to an immune reaction. The portal of entry is not known but is probably mucous membranes. Multiple lesions are formed in the brain and other organs, especially the lungs. It is probable that many cases of GAE ascribed to *Acanthamoeba* were actually caused by another, similar protozoan, *Balamuthia mandrillaris*, which was first reported in a mandrill baboon in 1990.

NERVOUS SYSTEM DISEASES CAUSED BY PRIONS

LEARNING OBJECTIVE
• List the characteristics of diseases caused by prions.

Several fatal diseases of the human central nervous system are caused by prions. (Prions are abnormally folded proteins that can induce a change in the shape of a normal protein, causing the proteins to clump. See the discussion of prions in Chapter 13, page 412.) Prion-caused diseases have long incubation times, measured in years. CNS damage is insidious and slowly progressive, without the fever and inflammation seen in encephalitis. Autopsies show a characteristic spongiform (porous, like a sponge) degeneration of the brain (Figure 22.18a). Also present in brain tissue are characteristic fibrils (Figure 22.18b). In recent years the study of these diseases, called **transmissible**

spongiform encephalopathies (TSE), has been one of the most interesting areas of medical microbiology.

A typical prion disease in animals is **sheep scrapie,** which has been long known in Great Britain and made its first appearance in the United States in 1947. The infected animal scrapes itself against fences and walls until areas of its body are raw. During a period of several weeks or months, the animal gradually loses motor control and dies. The infection can be experimentally passed to other animals by injection of brain tissue from one animal to the next. Similar conditions are seen in mink, possibly resulting from the animals' being fed mutton. A prion disease, *chronic wasting disease*, affects wild deer and elk in the western United States and Canada. It is invariably fatal, and there are concerns that it might infect humans who eat venison and might eventually infect domestic livestock.

Humans suffer from TSE diseases similar to scrapie; **Creutzfeldt-Jakob disease (CJD)** is an example. CJD is rare (about 200 cases per year in the United States). It often occurs in families, an indication of a genetic component. This form of CJD is sometimes referred to as classic CJD to differentiate it from similar variants that have appeared. There is no doubt that an infective agent is involved because transmission via corneal transplants and accidental scalpel nicks of a surgeon during autopsy have been reported. Several cases have been traced to the injection of a growth hormone derived from human tissue. Boiling and irradiation have no effect, and even routine autoclaving is not reliable. This has led to suggestions that surgeons use disposable instruments where there is a risk of exposure to CJD. To sterilize reusable instruments, the World Health Organization

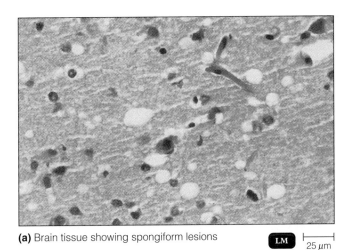

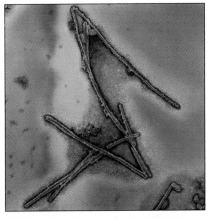

(a) Brain tissue showing spongiform lesions   LM   25 µm

(b) Characteristic fibrils of prion-caused diseases   TEM   0.5 µm

FIGURE 22.18 Spongiform encephalopathies. These diseases, caused by prions, include bovine spongiform encephalopathy, scrapie in sheep, and Creutzfeldt-Jakob disease in humans. All are similar in their pathology. **(a)** Note the clear holes that give this slide of brain tissue a spongy appearance. This pathology is responsible for the term *spongiform*. **(b)** Brain tissue showing characteristic fibrils produced by prion diseases. Prions themselves are not visible by any known technology.

Q **What are prions?**

currently recommends combined use of a strong solution of sodium hydroxide and extended autoclaving at 134°C. However, there are reports that applications of a simple cleaning detergent combined with protease enzymes to disrupt the prions may prove an effective solution to the problem. A similar approach to the disposal of prion-infected animal carcasses, for which incineration is the current primary method, makes use of digestion. The digester tank shown in Figure 22.19 is in use at the Wisconsin Veterinary Diagnostic Laboratory to dispose of deer carcasses suspected of infection with prion-caused chronic wasting disease (for

which incineration was previously used). The animal tissue is subjected to heat and caustic chemicals such as sodium hydroxide or potassium hydroxide. The animal tissue and any microorganisms are reduced to a harmless broth containing only sugars and short peptide chains that can be disposed of in the municipal sanitary sewage system. This process is less expensive than incineration, as well as being more environmentally favorable.

Some tribes in New Guinea have suffered from a TSE disease called **kuru** (a native word for shaking or trembling). Transmission of kuru is apparently related to the

FIGURE 22.19 Tissue digester. (a) The inset photo shows the stainless steel tank of the tissue digester that can be used to reduce prion-infected animals into a noninfectious slurry. **(b)** The technician is dropping a deceased prion-infected sheep into the tissue digester.

Q **Will cooking or freezing meat destroy prions?**

TABLE 22.1	Comparative Characteristics of Classic and Variant CJD	
Characteristic	Classic CJD	Variant CJD
Median age at death (yr)	68 (range 23–97)	28 (range 14–74)
Median duration of illness (mo)	4 to 5	13 to 14
Clinical presentation	Dementia; early neurologic signs	Prominent psychiatric and behavioral symptoms; delayed neurologic signs
Genotype*	Methionine/methionine	Other amino acid combinations

*Victims are homozygous at codon 129, that is, both of their PrP genes (one from each parent) have methionine coded at this position. This is characteristic of only about 37% of Caucasians. Other members of this population have different amino acid combinations at this position—and no one with these genotypes has contracted vCJD to date.

practice of cannibalistic rituals. Carleton Gajdusek received the Nobel Prize for Physiology and Medicine in 1976 for his investigations of kuru. The disease is disappearing as the practice of ritualistic cannibalism dies out.

BOVINE SPONGIFORM ENCEPHALOPATHY AND VARIANT CREUTZFELDT-JAKOB DISEASE

A TSE that is much in the news is **bovine spongiform encephalopathy (BSE).** The disease is better known as *mad cow disease* because of the behavior of the animals. The outbreak that began in 1986 in Great Britain was eventually controlled by drastic culling of herds. The origin of the disease is usually ascribed to feed supplements contaminated with prions from sheep infected with scrapie, a long-endemic neurological disease. As cattle adapted to scrapie, they exhibited the symptoms of BSE. Another hypothesis proposes that BSE resulted from a spontaneous mutation in a cow and that there is no connection with scrapie.

There is an urgent need for reliable tests that will diagnose cases of BSE in early, nonsymptomatic stages in live animals. Currently, the only available tests require postmortem brain tissue and detect only late stages of the disease. In attempts to prevent introduction of BSE into the United States, there are rules prohibiting the use of meat from "downer" animals (fallen and unable to rise and walk) for any purpose, and the use of animal protein as a feed supplement. The U.S. Food and Drug Administration (FDA) has banned for human consumption certain portions of the cattle carcass that are most likely to contain a neurological pathogen. Also, a small percentage of animal carcasses in the United States are tested for BSE—in Europe and Japan, practically all slaughtered animals are tested.

If this disease were to establish itself in domestic cattle in the United States it would be economically devastating.

However, there is another aspect—that the disease could be passed on to humans. In Great Britain and a few other locales around the world, a few cases of apparent classic CJD appeared in relatively young humans. CJD rarely occurs in this age group, and a connection with BSE was feared. Investigation also showed that this variant of CJD (vCJD) also differed in significant ways from classic CJD (Table 22.1). Less than 200 cases have been identified so far. Considering the long incubation times of prion diseases and that an estimated 1 million cattle had been infected with BSE, there was the worrisome possibility that large numbers of vCJD cases might eventually appear. However, this concern has subsided, especially since the number of cases declined from a small peak in 2000 and after it was shown that the affected patients shared a certain limited genetic profile.

DISEASE CAUSED BY UNIDENTIFIED AGENTS

CHRONIC FATIGUE SYNDROME

LEARNING OBJECTIVE
- List some possible causes of chronic fatigue syndrome.

The medical community has long been puzzled by patients who complain of persistent fatigue that prevents them from working and has no apparent cause. They often complain as well of multiple allergies. Called **chronic fatigue syndrome (CFS),** the debilitating condition continues for months or years. There is no effective treatment and not even a universal agreement on the proper diagnosis, although the Centers for Disease Control and Prevention has been working on a diagnostic test that might sometime be universally accepted. Chronic fatigue syndrome is not uncommon in the United States; the prevalence is 0.52% in women and 0.29% in men. There are psychiatric symptoms, and one school of thought contends that CFS is a psychiatric illness. Most experts discount this and believe that some unidentified infectious agent, or several agents, is responsible. For example, CFS often appears after recovery from viral infections such as infectious mononucleosis. Current thinking is that the cause may be an overstimulation of the immune system in response to some infection—probably viral. A recent review of the disease concluded that ". . . CFS is a condition for which every possible etiology has several diverse hypotheses."

* * *

Table 22.2 summarizes the main microbial diseases associated with the nervous system.

TABLE 22.2	**Microbial Diseases of the Nervous System**				
	Pathogen	Portal of Entry	Method of Transmission	Treatment	Prevention
Bacterial Diseases					
Haemophilus influenzae meningitis	*H. influenzae*	Respiratory tract	Endogenous infection; aerosols	Cephalosporin	Capsular Hib vaccine
Meningococcal meningitis	*Neisseria meningitidis*	Respiratory tract	Aerosols	Cephalosporin	Capsular vaccine against A, C, Y, W-135
Pneumococcal meningitis	*Streptococcus pneumoniae*	Respiratory tract	Aerosols	Cephalosporin	Polysaccharide vaccine
Listeriosis	*Listeria monocytogenes*	Mouth	Foodborne infection	Penicillin G	Pasteurizing and cooking food
Tetanus	*Clostridium tetani*	Skin	Puncture wound	Tetanus immune globulin; antibiotics	Toxoid vaccine (DTaP, Td)
Botulism	*Clostridium botulinum*	Mouth	Foodborne intoxication	Antitoxin	Proper canning of foods; infants should not have honey
Leprosy	*Mycobacterium leprae*	Nasal mucosa	Probably prolonged contact with contaminated secretions	Dapsone, riampin, clofaximine	Possibly BCG vaccine
Viral Diseases					
Poliomyelitis	Poliovirus	Mouth	Ingesting contaminated water (fecal–oral route)	Mechanical breathing aid	Inactivated polio vaccine (E-IPV)
Rabies	*Lyssavirus,* including rabies virus	Skin	Animal bite	Postexposure treatment: rabies immunoglobulin + vaccine	Human diploid cell vaccine for high-risk individuals; vaccination of domestic animals
Arboviral encephalitis See Table 22.1	Arboviruses	Skin	Mosquito bite	None	Insect repellent; protective clothing; remove standing water (mosquito breeding)

TABLE 22.2	**Microbial Diseases of the Nervous System** *(continued)*				
	Pathogen	Portal of Entry	Method of Transmission	Treatment	Prevention
Fungal Diseases					
Cryptococcosis	*Cryptococcus neoformans*	Respiratory route	Inhaling soil contaminated with spores	Amphotericin B, flucytosine	None
Protozoan Diseases					
African trypanosomiasis	*Trypanosoma brucei rhodesiense, T. b. gambiense*	Skin	Tsetse fly	Suramin; pentamidine	Vector control
Primary amebic meningoencephalitis	*Naegleria fowleri*	Mucous membranes	Swimming	Amphotericin B	None
Granulomatous amebic encephalitis	*Acanthamoeba* spp.; *Balamethia mandrillaris*	Mucous membranes	Swimming	Amphotericin B	None
Prion Diseases					
Creutzfeldt-Jakob disease	Prion	Injection; mouth	Inherited; ingested; transplants	None	None
Kuru	Prion	Mucous membranes	Contact or ingestion	None	None

STUDY OUTLINE

STRUCTURE AND FUNCTION OF THE NERVOUS SYSTEM (p. 643)

1. The central nervous system (CNS) consists of the brain, which is protected by the skull bones, and the spinal cord, which is protected by the backbone.

2. The peripheral nervous system (PNS) consists of the nerves that branch from the CNS.

3. The CNS is covered by three layers of membranes called meninges: the dura mater, arachnoid mater, and pia mater. Cerebrospinal fluid (CSF) circulates between the arachnoid mater and the pia mater in the subarachnoid space.

4. The blood–brain barrier normally prevents many substances, including antibiotics, from entering the brain.

5. Microorganisms can enter the CNS through trauma, along peripheral nerves, and through the bloodstream and lymphatic system.

6. An infection of the meninges is called meningitis. An infection of the brain is called encephalitis.

BACTERIAL DISEASES OF THE NERVOUS SYSTEM (pp. 643–652)

BACTERIAL MENINGITIS (pp. 644–647)

1. Meningitis can be caused by viruses, bacteria, fungi, and protozoa.

2. The three major causes of bacterial meningitis are *Haemophilus influenzae*, *Streptococcus pneumoniae*, and *Neisseria meningitidis*.

3. Nearly 50 species of opportunistic bacteria can cause meningitis.

Haemophilus influenzae Meningitis (p. 645)

4. *H. influenzae* is part of the normal throat microbiota.

5. *H. influenzae* requires blood factors for growth; there are six types of *H. influenzae* based on capsule differences.

6. *H. influenzae* type b is the most common cause of meningitis in children under 4 years old.

7. A conjugated vaccine directed against the capsular polysaccharide antigen is available.

Neisseria Meningitis
(Meningococcal Meningitis) (pp. 645–646)

8. *N. meningitidis* causes meningococcal meningitis. This bacterium is found in the throats of healthy carriers.

9. The bacteria probably gain access to the meninges through the bloodstream. The bacteria may be found in leukocytes in CSF.

10. Symptoms are due to endotoxin. The disease occurs most often in young children.

11. Purified capsular polysaccharide vaccine against serotypes A, C, Y, and W-135 is available.

Streptococcus pneumoniae Meningitis
(Pneumococcal Meningitis) (p. 646)

12. *S. pneumoniae* is commonly found in the nasopharynx.

13. Hospitalized patients and young children are most susceptible to *S. pneumoniae* meningitis. It is rare but has a high mortality rate.

14. A conjugated vaccine is available.

Diagnosis and Treatment of the Most Common
Types of Bacterial Meningitis (pp. 646–647)

15. Cephalosporins may be administered before identification of the pathogen.

16. Diagnosis is based on Gram stain and serological tests of the bacteria in CSF.

17. Cultures are usually made on blood agar and incubated in an atmosphere containing reduced oxygen levels.

Listeriosis (p. 647)

18. *Listeria monocytogenes* causes meningitis in newborns, the immunosuppressed, pregnant women, and cancer patients.

19. Acquired by ingestion of contaminated food, it may be asymptomatic in healthy adults.

20. *L. monocytogenes* can cross the placenta and cause spontaneous abortion and stillbirth.

TETANUS (pp. 647–649)

21. Tetanus is caused by a localized infection of a wound by *Clostridium tetani*.

22. *C. tetani* produces the neurotoxin tetanospasmin, which causes the symptoms of tetanus: spasms, contraction of muscles controlling the jaw, and death resulting from spasms of respiratory muscles.

23. *C. tetani* is an anaerobe that will grow in deep, unclean wounds and wounds with little bleeding.

24. Acquired immunity results from DPT immunization that includes tetanus toxoid.

25. Following an injury, an immunized person may receive a booster of tetanus toxoid. An unimmunized person may receive (human) tetanus immune globulin.

26. Debridement (removal of tissue) and antibiotics may be used to control the infection.

BOTULISM (pp. 649–651)

27. Botulism is caused by an exotoxin produced by *C. botulinum* growing in foods.

28. Serological types of botulinum toxin vary in virulence, with type A being the most virulent.

29. The toxin is a neurotoxin that inhibits the transmission of nerve impulses.

30. Blurred vision occurs in 1 to 2 days; progressive flaccid paralysis follows for 1 to 10 days, possibly resulting in death from respiratory and cardiac failure.

31. *C. botulinum* will not grow in acidic foods or in an aerobic environment.

32. Endospores are killed by proper canning. The addition of nitrites to foods inhibits growth after endospore germination.

33. The toxin is heat labile and is destroyed by boiling (100°C) for 5 minutes.

34. Infant botulism results from the growth of *C. botulinum* in an infant's intestines.

35. Wound botulism occurs when *C. botulinum* grows in anaerobic wounds.

36. For diagnosis, mice protected with antitoxin are inoculated with toxin from the patient or foods.

LEPROSY (pp. 651–652)

37. *Mycobacterium leprae* causes leprosy, or Hansen's disease.

38. *M. leprae* has never been cultured on artificial media. It can be cultured in armadillos and mouse footpads.

39. The tuberculoid form of the disease is characterized by loss of sensation in the skin surrounded by nodules. The lepromin test is positive.

40. Laboratory diagnosis is based on observations of acid-fast rods in lesions or fluids and the lepromin test.

41. In the lepromatous form, disseminated nodules and tissue necrosis occur. The lepromin test is negative.

42. Leprosy is not highly contagious and is spread by prolonged contact with exudates.

43. Untreated individuals often die of secondary bacterial complications, such as tuberculosis.

44. Patients with leprosy are made noncontagious within 4 to 5 days with sulfone drugs and then treated as outpatients.

45. Leprosy occurs primarily in the tropics.

VIRAL DISEASES OF THE NERVOUS SYSTEM (pp. 652–660)

POLIOMYELITIS (pp. 652–654)

1. The symptoms of poliomyelitis are usually headache, sore throat, fever, stiffness of the back and neck, and occasionally paralysis (fewer than 1% of cases).

2. Poliovirus is transmitted by the ingestion of water contaminated with feces.

3. Poliovirus first invades lymph nodes of the neck and small intestine. Viremia and spinal cord involvement may follow.

4. Diagnosis is based on isolation of the virus from feces and throat secretions.

5. The Salk vaccine (an inactivated polio vaccine, or IPV) involves the injection of formalin-inactivated viruses and boosters every few years. The Sabin vaccine (an oral polio vaccine, or OPV) contains three live, attenuated strains of poliovirus and is administered orally.

6. Polio is a good candidate for elimination through vaccination.

RABIES (pp. 654–658)

7. Rabies virus (a rhabdovirus) causes an acute, usually fatal, encephalitis called rabies.

8. Rabies may be contracted through the bite of a rabid animal, by inhalation of aerosols, or invasion through minute skin abrasions. The virus multiplies in skeletal muscle and connective tissue.

9. Encephalitis occurs when the virus moves along peripheral nerves to the CNS.

10. Symptoms of rabies include spasms of mouth and throat muscles followed by extensive brain and spinal cord damage and death.

11. Laboratory diagnosis may be made by direct FA tests of saliva, serum, and CSF or brain smears.

12. Reservoirs for rabies in the United States include skunks, bats, foxes, and raccoons. Domestic cattle, dogs, and cats may get rabies. Rodents and rabbits seldom get rabies.

13. Current postexposure treatment includes administration of human rabies immune globulin (RIG) along with multiple intramuscular injections of vaccine.

14. Preexposure treatment consists of vaccination.

15. Other genotypes of *Lyssavirus* cause rabies-like diseases.

ARBOVIRAL ENCEPHALITIS (pp. 658–660)

16. Symptoms of encephalitis are chills, headache, fever, and eventually coma.

17. Many types of viruses (called arboviruses) transmitted by mosquitoes cause encephalitis.

18. The incidence of arboviral encephalitis increases in the summer months, when mosquitoes are most numerous.

19. Notifiable arboviral infections are eastern equine encephalitis (EEE), western equine encephalitis (WEE), St. Louis encephalitis (SLE), California encephalitis (CE), and West Nile virus (WNV).

20. Diagnosis is based on serological tests.

21. Control of the mosquito vector is the most effective way to control encephalitis.

FUNGAL DISEASE OF THE NERVOUS SYSTEM (p. 660)

CRYPTOCOCCUS NEOFORMANS MENINGITIS (CRYPTOCOCCOSIS) (p. 660)

1. *Cryptococcus neoformans* is an encapsulated yeastlike fungus that causes cryptococcosis.

2. The disease may be contracted by inhalation of dried infected pigeon or chicken droppings.

3. The disease begins as a lung infection and may spread to the brain and meninges.

4. Immunosuppressed individuals are most susceptible to *Cryptococcus neoformans* meningitis.

5. Diagnosis is based on latex agglutination tests for cryptococcal antigens in serum or CSF.

PROTOZOAN DISEASES OF THE NERVOUS SYSTEM (pp. 660–662)

AFRICAN TRYPANOSOMIASIS (pp. 660–661)

1. African trypanosomiasis is caused by the protozoa *Trypanosoma brucei gambiense* and *T. b. rhodesiense* and transmitted by the bite of the tsetse fly.

2. The disease affects the nervous system of the human host, causing lethargy and eventually coma. It is commonly called sleeping sickness.

3. Vaccine development is hindered by the protozoan's ability to change its surface antigens.

AMEBIC MENINGOENCEPHALITIS (pp. 661–662)

4. Encephalitis caused by the protozoan *Naegleria fowleri* is almost always fatal.

5. Granulomatous amebic encephalitis, caused by *Acanthamoeba* spp. and *Balamuthia mandrillaris*, is a chronic disease.

NERVOUS SYSTEM DISEASES CAUSED BY PRIONS (pp. 662–664)

1. Diseases of the CNS that progress slowly and cause spongiform degeneration are caused by prions.

2. Sheep scrapie and bovine spongiform encephalopathy (BSE) are examples of diseases caused by prions that are transferable from one animal to another.

3. Creutzfeldt-Jakob disease and kuru are human diseases similar to scrapie. They are transmitted between humans.

4. Prions are self-replicating proteins with no detectable nucleic acid.

DISEASE CAUSED BY UNIDENTIFIED AGENTS (pp. 664–665)

CHRONIC FATIGUE SYNDROME (pp. 664–665)

1. Chronic fatigue syndrome (CFS) may be caused by an unidentified infectious agent.

STUDY QUESTIONS

Access more review material either online at **The Microbiology Place** (www.microbiologyplace.com) or with **The Microbiology Place CD-ROM** packaged with your new book. There you'll find activities, practice tests, quizzes, flashcards, case studies, and more to help you succeed.

Answers to the Study Questions can be found in Appendix G.

REVIEW

1. Differentiate meningitis from encephalitis.

2. Fill in the following table:

Causative Agent of Meningitis	Susceptible Population	Mode of Transmission	Treatment
N. meningitidis			
H. influenzae			
S. pneumoniae			
L. monocytogenes			
C. neoformans			

3. Briefly explain the derivation of the name *Haemophilus influenzae*.

4. If *Clostridium tetani* is relatively sensitive to penicillin, why doesn't penicillin cure tetanus?

5. Compare and contrast the Salk and Sabin vaccines with respect to composition, advantages, and disadvantages.

6. What treatment is used against tetanus under the following conditions?
   a. before a person suffers a deep puncture wound
   b. after a person suffers a deep puncture wound

7. Why is the following description used for wounds that are susceptible to *C. tetani* infection: ". . . Improperly cleaned deep puncture wounds . . . ones with little or no bleeding . . ."?

8. List the following information for botulism: etiologic agent, suspect foods, symptoms, treatment, conditions necessary for microbial growth, basis for diagnosis, prevention.

9. Provide the following information on leprosy: etiology, method of transmission, symptoms, treatment, prevention, and susceptible population.

10. Provide the following information on poliomyelitis: etiology, method of transmission, symptoms, prevention. Why aren't the Salk and Sabin vaccines considered treatments for poliomyelitis?

11. Provide the etiology, method of transmission, reservoirs, and symptoms for rabies.

12. Outline the procedures for treating rabies after exposure. Outline the procedures for preventing rabies prior to exposure. What is the reason for the differences in the procedures?

13. Fill in the following table.

Disease	Etiology	Vector	Symptoms	Treatment
Arboviral encephalitis				
African trypanosomiasis				

14. Why are meningitis and encephalitis generally difficult to treat?

15. Provide evidence that Creutzfeldt-Jakob disease is caused by a transmissible agent.

MULTIPLE CHOICE

1. Which of the following is *not* true?
   a. Only puncture wounds by rusty nails result in tetanus.
   b. Rabies is seldom found in rodents (e.g., rats, mice).
   c. Polio is transmitted by the fecal–oral route.
   d. Arboviral encephalitis is rather common in the United States.
   e. All of the above are true.

2. Which of the following does *not* have an animal reservoir or vector?
   a. listeriosis
   b. cryptococcosis
   c. amebic meningoencephalitis
   d. rabies
   e. African trypanosomiasis

3. A 12-year-old girl hospitalized for Guillain-Barré syndrome had a 4-day history of headache, dizziness, fever, sore throat, and weakness of legs. Seizures began 2 weeks later. Bacterial cultures were negative. She died 3 weeks after hospitalization.

An autopsy revealed inclusions in brain cells that tested positive in an immunofluoresence test. She probably had
a. rabies.
b. Creutzfeldt-Jakob disease.
c. botulism.
d. tetanus.
e. leprosy.

4. After receiving a corneal transplant, a woman developed dementia and loss of motor function; she then became comatose and died. Cultures were negative. Serological tests were negative. Autopsy revealed spongiform degeneration of her brain. She most likely had
a. rabies.
b. Creutzfeldt-Jakob disease.
c. botulism.
d. tetanus.
e. leprosy.

5. Endotoxin is responsible for symptoms caused by which of the following organisms?
a. *N. meningitidis*
b. *S. pyogenes*
c. *L. monocytogenes*
d. *C. tetani*
e. *C. botulinum*

6. The increased incidence of encephalitis in the summer months is due to
a. maturation of the viruses.
b. increased temperature.
c. the presence of adult mosquitoes.
d. an increased population of birds.
e. an increased population of horses.

Match the following choices to the statements in questions 7 and 8:
a. antirabies antibodies
b. HDVC

7. Produces the highest antibody titer.

8. Used for passive immunization.

Use the following choices to answer questions 9 and 10:
a. *Cryptococcus*
b. *Haemophilus*
c. *Listeria*
d. *Naegleria*
e. *Neisseria*

9. Microscopic examination of cerebrospinal fluid reveals gram-positive rods.

10. Microscopic examination of cerebrospinal fluid from a person who washes windows on a building in a large city reveals ovoid cells.

CRITICAL THINKING

1. Most of us have been told that a rusty nail causes tetanus. What do you suppose is the origin of this adage?

2. A BCG vaccination will result in positive lepromin and tuberculin tests. What is the relationship between leprosy and tuberculosis?

3. OPV is no longer used for routine vaccination. Provide the rationale for this policy.

CLINICAL APPLICATIONS

1. A 1-year-old infant was lethargic and had a fever. When admitted to the hospital, he had multiple brain abscesses with gram-negative coccobacillary rods. Identify the disease, etiology, and treatment.

2. A 40-year-old bird handler was admitted to the hospital with soreness over his upper jaw, progressive vision loss, and bladder dysfunction. He had been well two months earlier. Within weeks he lost reflexes in his lower extremities and subsequently died. Examination of CSF showed lymphocytes. What etiology do you suspect? What further information do you need?

3. One week after bathing in a hot spring, a 9-year-old girl was hospitalized after a 3-day history of progressive, severe headaches, nausea, lethargy, and stupor. Examination of CSF revealed amoeboid organisms. Identify the disease, etiology, and treatment.

23

Microbial Diseases of the Cardiovascular and Lymphatic Systems

The **cardiovascular system** consists of the heart, blood, and blood vessels. The **lymphatic system** consists of the lymph, lymph vessels, lymph nodes, and the lymphoid organs such as the tonsils, appendix, spleen, and thymus. Fluids in both systems circulate throughout the body, intimately contacting many tissues and organs. Physiologically, the blood and lymph distribute nutrients and oxygen to body tissues and carry away wastes. However, these same qualities make the cardiovascular and lymphatic systems vehicles for the spread of pathogens that enter their circulation. Opportunities for this occur when an insect bite, needle, or wound penetrates the skin. Because of this, many of the body's innate defensive systems are found in the blood and lymph. Especially important for this are circulating phagocytic cells; these are also in fixed locations such as the lymph nodes and spleen. The blood is an important part of our adaptive immune system; antibodies and specialized cells circulate to intercept pathogens introduced into the blood. Occasionally the defensive systems found in the blood are overwhelmed, and pathogens proliferate explosively with disastrous results.

UNDER THE MICROSCOPE

Ebola Virus. This is one of several emerging hemorrhagic fever viruses. A human first contracts the virus, probably a fruit-bat reservoir, then the virus can be spread to other people via blood and secretions from the infected person.

STRUCTURE AND FUNCTION OF THE CARDIOVASCULAR AND LYMPHATIC SYSTEMS

> **LEARNING OBJECTIVE**
>
> • Identify the role of the cardiovascular and lymphatic systems in spreading and eliminating infections.

The center of the cardiovascular system is the heart (Figure 23.1). The function of the cardiovascular system is to circulate blood through the body's tissues so it can deliver certain substances to cells and remove other substances from them.

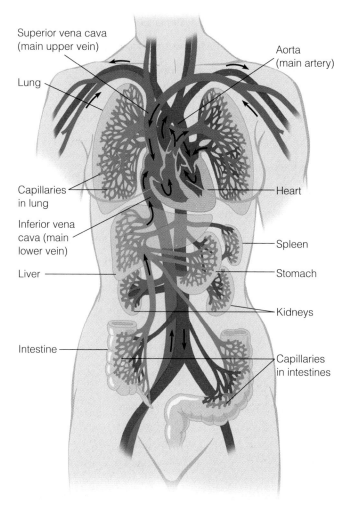

FIGURE 23.1 The human cardiovascular system and related structures. Details of circulation to the head and extremities are not shown in this simplified diagram. The blood circulates from the heart through the arterial system (red) to the capillaries (purple) in the lungs and other parts of the body. From these capillaries, the blood returns through the venous system (blue) to the heart.

Q How can a focal infection become systemic?

Blood is a mixture of formed elements and a liquid called blood plasma (see the box on page 490). The lymphatic system is an essential part of the circulation of blood (Figure 23.2). As the blood circulates, some blood plasma filters out of the blood capillaries into spaces between tissue cells called *interstitial spaces*. The circulating fluid is called *interstitial fluid*. Microscopic lymphatic vessels that surround tissue cells are called *lymph capillaries*. As the interstitial fluid moves around the tissue cells, it is picked up by the lymph capillaries; the fluid is then called *lymph*.

Because lymph capillaries are very permeable, they readily pick up microorganisms or their products. From lymph capillaries, lymph is transported into larger lymph vessels called *lymphatics*, which contain valves that keep the lymph moving toward the heart. Eventually, all the lymph is returned to the blood just before the blood enters the heart. As a result of this circulation, proteins and fluid that have filtered from the plasma are returned to the blood.

At various points in the lymphatic system are oval structures called *lymph nodes,* through which lymph flows. (Also, see Figure 16.5, page 482.) Within the lymph nodes are fixed macrophages that help clear the lymph of infectious microorganisms. At times the lymph nodes themselves get infected and become visibly swollen and tender; swollen lymph nodes are called **buboes** (see Figure 23.10, page 684).

Lymph nodes are also an important component of the body's immune system. Foreign microbes entering lymph nodes encounter two types of lymphocytes: B cells, which are stimulated to become plasma cells that produce humoral antibodies; and T cells, which then differentiate into effector T cells that are essential to the cell-mediated immune system.

BACTERIAL DISEASES OF THE CARDIOVASCULAR AND LYMPHATIC SYSTEMS

Once bacteria gain access to the bloodstream, they become widely disseminated. In some cases, they are also able to reproduce rapidly.

SEPSIS AND SEPTIC SHOCK

> **LEARNING OBJECTIVES**
>
> • List the signs and symptoms of sepsis, and explain the importance of infections that develop into septic shock.
>
> • Differentiate gram-negative sepsis, gram-positive sepsis, and puerperal sepsis.

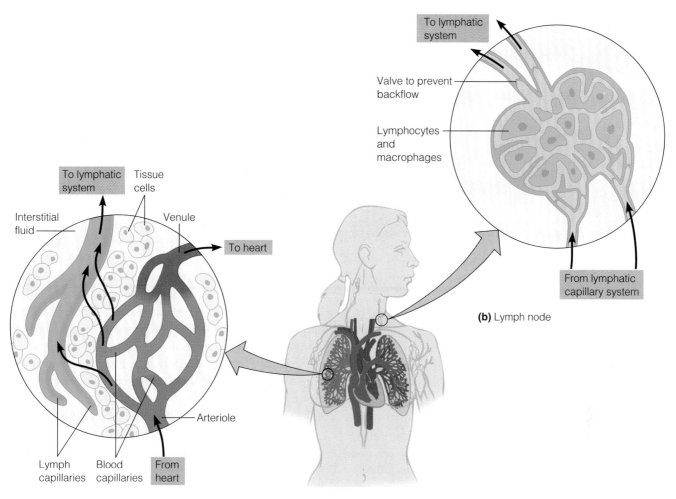

(a) Capillary system in lung

FIGURE 23.2 The relationship between the cardiovascular and lymphatic systems. (**a**) From the blood capillaries, some blood plasma filters into the surrounding tissue and enters the lymph capillaries. This fluid, now called lymph, returns to the heart through the lymphatic circulatory system (green), which channels the lymph to a vein. (**b**) All lymph returning to the heart must pass through at least one lymph node. (See also Figure 16.5.)

Q **What is the role of the lymphatic system in defense against infection?**

Although blood is normally sterile, moderate numbers of microorganisms can enter the bloodstream without causing harm. In hospital conditions, the blood frequently is contaminated as a result of invasive procedures, such as insertion of catheters and intravenous feeding tubes. Blood and lymph contain numerous defensive phagocytic cells. Also, blood is low in iron, which is a requirement for bacterial growth. However, if the defenses of the cardiovascular and lymphatic systems fail, microbes can undergo uncontrolled proliferation in the blood. **Sepsis** is a toxic, inflammatory condition arising from the spread of bacteria or bacterial toxins from a focus of infection. **Septicemia** is sepsis that results from the proliferation of bacterial pathogens in the bloodstream. If these bacteria cause red

blood cells to lyse, the release of iron-containing hemoglobin can result in accelerated microbial growth. Sepsis is often accompanied by the appearance of **lymphangitis,** inflamed lymph vessels visible as red streaks under the skin, running along the arm or leg from the site of the infection (Figure 23.3). Sometimes the streak ends at a lymph node, where fixed phagocytic cells attempt to stop the invading microorganisms.

If the body's defenses do not quickly control the invasion of the bloodstream by pathogenic bacteria, their proliferation can accelerate rapidly, with results that frequently are fatal. The first stage of this progression is **sepsis.** There is evidence of infection and an inflammatory response by the body caused by the release and circulation

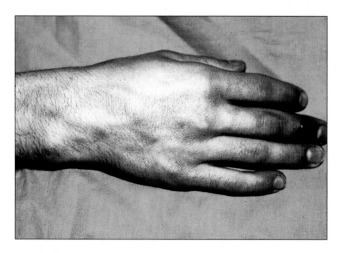

FIGURE 23.3 Lymphangitis, one sign of sepsis. As the infection spreads from its original site along the lymph vessels, the inflamed walls of the vessels become visible as red streaks.

Q Why does the red streak sometimes end at a certain point?

of cytokines. The most obvious signs and symptoms are fever, chills, and accelerated breathing and heart rate. When sepsis results in a drop in blood pressure (*shock*) and dysfunction of at least one organ, it is considered to be **severe sepsis.** Once organs begin to fail, the mortality rate becomes very high. A final stage, when low blood pressure can no longer be controlled by addition of fluids, is **septic shock.**

GRAM-NEGATIVE SEPSIS

Septic shock is most likely to be caused by gram-negative bacteria. Recall that the cell walls of many gram-negative bacteria contain endotoxins that are released upon lysis of the cell. These endotoxins can cause a severe drop in blood pressure with its associated signs and symptoms. Septic shock is often called by the alternative names *gram-negative sepsis* or *endotoxic shock.* Less than one-millionth of a milligram of endotoxin is enough to cause the symptoms. About 750,000 cases of septic shock occur each year in the United States; at least 225,000 are fatal.

An effective treatment for severe sepsis and septic shock has been a medical priority for many years. The early symptoms of sepsis are relatively nonspecific and not especially alarming. Therefore, the antibiotic treatments that might arrest it then are frequently not administered. The progression to lethal stages is rapid and generally impossible to treat effectively. The administration of antibiotics then may even aggravate the condition by causing the lysis of large numbers of bacteria that then release more endotoxins.

The U.S. Food and Drug Administration has recently approved a drug, drotrecogin alfa (Xigris), which is the first to reduce the death rate of sepsis cases. This drug is a

genetically modified version of human activated protein C—a natural anticoagulant found at reduced levels in cases of severe sepsis and septic shock. The drug reduces clotting, which is a factor in organ damage. Xigris is scarcely the sought-for magic bullet to treat sepsis: it is dauntingly expensive and effective only in a minority of cases. Nonetheless, it is expected to be widely prescribed for treatment of gram-negative sepsis and meningococcal meningitis (see page 645).

GRAM-POSITIVE SEPSIS

Gram-positive bacteria are also a frequent source of blood-related disease. Both staphylococci and streptococci produce potent exotoxins that cause toxic shock syndrome, a toxemia discussed in Chapter 21 (page 620). The frequent use of invasive procedures in hospitals allows gram-positive bacteria to enter the bloodstream. Such nosocomial infections are a particular risk for patients who undergo regular dialysis for kidney dysfunction. A vaccine to protect against *Staphylococcus aureus* sepsis in such patients has been tested and conferred partial immunity. The bacterial components that lead to septic shock in gram-positive sepsis are not known with certainty. Possible sources are various fractions of the gram-positive cell wall or even bacterial DNA.

An especially important group of gram-positive bacteria are the enterococci, which are responsible for many nosocomial infections. The enterococci are inhabitants of the human colon and frequently contaminate skin. Once considered relatively harmless, two species in particular, *Enterococcus faecium* and *Enterococcus faecalis,* are now recognized as leading causes of nosocomial infections of wounds and the urinary tract. Enterococci have a natural resistance to penicillin and have rapidly acquired resistance to other antibiotics. What has made them something of a medical emergency is the appearance of vancomycin-resistant strains. Vancomycin (see page 592) was often the only remaining antibiotic to which these bacteria, especially *E. faecium,* were still sensitive. Among isolates of *E. faecium* from nosocomial infections of the bloodstream, almost 90% are now resistant.

To this point our discussion of the streptococci has been focused on serologic group A. There is an emerging awareness of group B streptococci (GBS) and of the enterococci. *S. agalactiae* (ā′gal-act-ē-ī) is the only GBS and is the most common cause of life-threatening neonatal sepsis. The CDC recommends that pregnant women be tested for vaginal GBS and that women with GBS be offered antibiotics during labor.

PUERPERAL SEPSIS

Puerperal sepsis, also called **puerperal fever** and **childbirth fever,** is a nosocomial infection. It begins as an

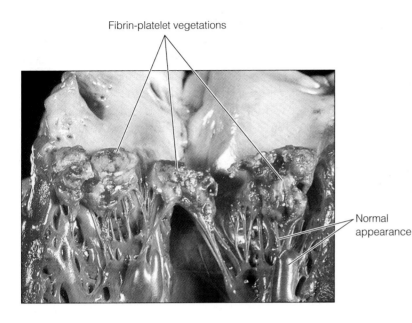

Fibrin-platelet vegetations

Normal appearance

FIGURE 23.4 Bacterial endocarditis. This is a case of subacute endocarditis, meaning that it developed over a period of weeks or months. The heart has been dissected to expose the mitral valve. The cordlike structures connect the heart valve to the operating muscles.

Endocarditis develops as bacteria attach to the surface and multiply, causing damage that promotes the formation of fibrin-platelet vegetations (shown in photo). These vegetations, a biofilm, bury adherent bacteria and allow them to multiply protected from defenses of the host; further deposition of bacteria cause the vegetation to enlarge in layers.

Symptoms usually include fever and a heart murmur from poor mitral valve function that is detectable by echocardiogram. Treatment with antibiotics in high concentrations is often effective.

Q   How is endocarditis contracted?

infection of the uterus as a result of childbirth or abortion. *Streptococcus pyogenes*, a group A beta-hemolytic streptococcus, is the most frequent cause, although other organisms may cause infections of this type.

Puerperal sepsis progresses from an infection of the uterus to an infection of the abdominal cavity (*peritonitis*) and in many cases to sepsis. At a Paris hospital between 1861 and 1864, of the 9886 women who gave birth, 1226 (12%) died of such infections. These deaths were largely unnecessary. Some 20 years before, Oliver Wendell Holmes in the United States and Ignaz Semmelweis in Austria had clearly demonstrated that the disease was transmitted by the hands and instruments of the attending midwives or physicians, and that disinfecting the hands and instruments could prevent such transmission. Antibiotics, especially penicillin, and modern hygienic practices have now made *S. pyogenes* puerperal sepsis an uncommon complication of childbirth.

BACTERIAL INFECTIONS OF THE HEART

LEARNING OBJECTIVE

- Describe the epidemiologies of endocarditis and rheumatic fever.

The wall of the heart consists of three layers. The inner layer, called the *endocardium*, lines the heart muscle itself and covers the valves. An inflammation of the endocardium is called **endocarditis.**

One type of bacterial endocarditis, **subacute bacterial endocarditis** (so named because it develops slowly; Figure 23.4), is characterized by fever, general weakness, and heart murmur. It is usually caused by alpha-hemolytic

streptococci such as are common in the oral cavity, although enterococci or staphylococci are often involved. The condition probably arises from a focus of infection elsewhere in the body, such as in the teeth or tonsils. Microorganisms are released by tooth extractions or tonsillectomies, enter the blood, and find their way to the heart. A more exotic source of infections that have led to cases of endocarditis has been body piercing, especially of the nose, tongue, and even nipples. Normally, such bacteria would be quickly cleared from the blood by the body's defensive mechanisms. However, in people whose heart valves are abnormal, because of either congenital heart defects or such diseases as rheumatic fever and syphilis, the bacteria lodge in the preexisting lesions. Within the lesions, the bacteria multiply and become entrapped in blood clots that protect them from phagocytes and antibodies. As multiplication progresses and the clot gets larger, pieces of the clot break off and can block blood vessels or lodge in the kidneys. In time, the function of the heart valves is impaired. Left untreated by appropriate antibiotics, subacute bacterial endocarditis is fatal within a few months.

A more rapidly progressive type of bacterial endocarditis is **acute bacterial endocarditis,** which is usually caused by *Staphylococcus aureus.* The organisms find their way from the initial site of infection to normal or abnormal heart valves; the rapid destruction of the heart valves is frequently fatal within a few days or weeks if untreated. Penicillin is sometimes used prophylactically to prevent endocarditis during procedures such as tooth extractions and tonsillectomies but with minimal effectiveness. Streptococci can also cause **pericarditis,** inflammation of the sac around the heart (the *pericardium*).

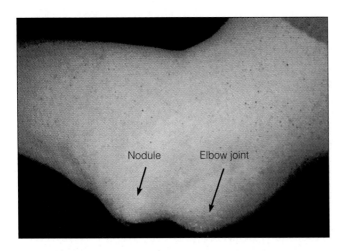

FIGURE 23.5 A nodule caused by rheumatic fever.
Rheumatic fever was named, in part, because of the characteristic subcutaneous nodules that appear at the joints, as shown in this patient's elbow. Infection with group A beta-hemolytic streptococci sometimes leads to this autoimmune complication.

 Is rheumatic fever a bacterial infection?

RHEUMATIC FEVER

Streptococcal infections, such as those caused by *Streptococcus pyogenes*, sometimes lead to **rheumatic fever,** which is generally considered an autoimmune complication. It occurs primarily in people aged 4 to 18 and often follows an episode of streptococcal sore throat. The disease is usually first expressed as a short period of arthritis and fever. Subcutaneous nodules at joints often accompany this stage (Figure 23.5). In about half of persons affected, an inflammation of the heart, probably from a misdirected immune reaction against streptococcal M protein, damages the valves. Reinfection with streptococci renews the immune attack. Damage to heart valves may be serious enough to result in eventual failure and death.

Early in the twentieth century, rheumatic fever killed more school-aged children in the United States than all other diseases combined. The incidence declined steadily in developed countries to the point of becoming rare even before the introduction of effective antimicrobial drugs during the 1930s and 1940s. Many young physicians have never seen a case of the disease, but in much of the underdeveloped world it remains the leading cause of heart disease in the young. The decline of rheumatic fever in the United States is thought to be due to some loss of virulence in the streptococci in circulation. However, since the 1980s there have been a few localized outbreaks of rheumatic fever in the United States that have been related to certain M-protein serotypes. These serotypes had

been prevalent during much earlier epidemics of rheumatic fever but had almost disappeared from circulation. People who have had an episode of rheumatic fever are at risk of renewed immunological damage with repeated streptococcal sore throats. The bacteria have remained sensitive to penicillin, and patients at particular risk, such as these, often receive a monthly preventive injection of long-acting penicillin G benzathine.

As many as 10% of people with rheumatic fever develop **Sydenham's chorea,** an unusual complication known in the Middle Ages as St. Vitus' dance. Several months following an episode of rheumatic fever, the patient (much more likely to be a girl than a boy) exhibits purposeless, involuntary movements during waking hours. Occasionally, sedation is required to prevent self-injury from flailing arms and legs. The condition disappears after a few months.

TULAREMIA

LEARNING OBJECTIVE
• Discuss the epidemiology of tularemia.

Tularemia is an example of a zoonotic disease, that is, a disease transmitted by contact with infected animals, most commonly rabbits and ground squirrels. The name derives from Tulare County, California, where the disease was originally observed in ground squirrels in 1911. The pathogen is *Francisella tularensis*, a small gram-negative bacillus. It can enter humans by several routes. The most common is penetration of the skin at a minor abrasion, where it creates an ulcer at the site. About a week after infection, the regional lymph nodes enlarge; many will contain pockets of pus. (See the box on the facing page.) The bacterium can multiply in macrophages—as much as a thousand-fold. Mortality is normally less than 3%. If not contained, the proliferation of *F. tularensis* can lead to sepsis and infection of multiple organs.

Almost 90% of cases in the United States are related to contact with rabbits, and the disease is often known locally as *rabbit fever.* Tularemia is also transmitted in some areas by ticks and insects and is known there as *deer fly fever.* Respiratory infection, usually by dust contaminated by urine or feces of infected animals, can cause an acute pneumonia with a mortality rate exceeding 30%. The infective dose is very small, and the organism is dangerous to handle if aerosols are likely to be produced.

At one time, so few cases of tularemia were recorded annually in the United States that it was removed from the list of nationally notifiable diseases, However, concern that it might be used as a biological weapon has re-

CLINICAL PROBLEM SOLVING

A SICK CHILD

You will see questions as you read through this problem. The questions are those that primary health care providers ask themselves as they solve a clinical problem. Try to answer each question as a health care provider.

1. On February 15, a 3-year-old boy was seen by his pediatrician for fever, malaise, painful left underarm lymph node, and skin sloughing off his left ring finger. Amoxicillin was prescribed.
What diseases are possible?

2. The child underwent excisional biopsy of the left axillary lymph node when intermittent fever and enlarged lymph node persisted for 49 days. The excised tissue was cultured; a gram-stain of the bacteria that grew is shown in the figure.
What additional tests would you do?

3. Serological tests revealed the following results:

	Antibody titer
Bartonella	0
Ehrlichia	0
Francisella	4,096
CMV	0
Toxoplasma gondii	0

The child improved after treatment with ciprofloxacin.
What is the cause of the infection?
What do you need to know?

4. PCR was used to confirm identification of *Francisella tularensis*. Between January 2 and February 8, the boy's family purchased six hamsters from a pet store. Each hamster died from diarrhea within one week of purchase. One hamster bit the child on the left ring finger.
Where will you look for the source of the infection?

5. Workers at the pet store reported an unusual number of deaths among hamsters but not other animals during January and February. Eight other customers reported that their hamsters died within two weeks of purchase. Available hamsters were negative for *F. tularensis* by serology and culture. One of two cats kept as store pets had a positive serologic test for *F. tularensis* at a titer of 256. The hamsters came from customers who had pets with unanticipated litters.
What is the most likely source of infection?

6. The hamsters came from different sources, thus are probably not the origin of the infection. The positive serological test in a pet cat suggests that infected wild rodents infested the store and spread the infection to hamsters by urinating and defecating through the hamster cages. The infected cat might have had an unrecognized illness after catching or eating an infected wild rodent.

Although tularemia has been associated with hamster hunting in Russia, it has not been associated previously with pet hamsters in the United States. Because proper diagnosis and treatment of tularemia rely on a high index of suspicion and are related to method of acquisition (e.g., ulcer after contact with rabbits), clinicians and public health officials should be aware that pet rodents might be a source of tularemia. Laboratory diagnosis of *F. tularensis* depends on the laboratory being notified that tularemia is a clinical possibility. Identification of the organism is important because it is often resistant to antibiotics commonly used for skin and systemic infections and because it is a potential agent of biological terrorism.

SOURCE: Adapted from reports in *MMWR* 53(52):1202 (1/7/05) and *MMWR* 54(7):170 (2/25/05).

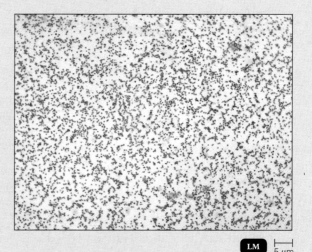

Gram-stained bacteria cultured from lymph node.

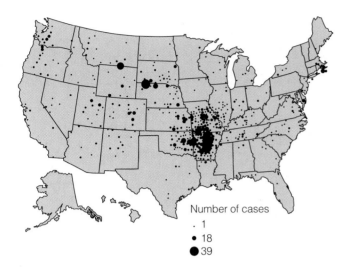

FIGURE 23.6 Tularemia cases in the United States (1990–2000). There were 1347 cases for which a county of residence was reported. The circle sizes are proportional to the number of cases in a county, ranging from 1 to 39.
SOURCE: CDC, *MMWR* 51(9)(3/8/02).

Q **What area reporting tularemia is closest to you?**

Number of cases
· 1
● 18
● 39

cently led to its reinstatement on the list. Figure 23.6 illustrates the geographic distribution of tularemia within the United States. It is also found worldwide in many areas of the Northern Hemisphere.

The intracellular location of the bacterium is a problem in chemotherapy. Tetracycline is the antibiotic of choice. However, this drug is only bacteriostatic, and prolonged administration is necessary to prevent relapses.

BRUCELLOSIS (UNDULANT FEVER)

LEARNING OBJECTIVE
• Discuss the epidemiology of brucellosis.

Like the bacteria that cause tularemia, the *Brucella* spp. that cause **brucellosis** favor intracellular growth and travel to organs via the bloodstream and lymphatic system. These bacteria are very small gram-negative aerobic rods. There are three primary species that cause disease in domestic animals and humans. Depending largely on the infecting bacterial species, brucellosis can range from a mild, self-limiting condition to a life-threatening disease.

The most common *Brucella* species in the United States is *Brucella abortus* (brü-sel'lä ä-bôr'tus), which infects livestock in many parts of the United States and which has a reservoir in the wild in elk and bison. Vaccination of domesticated livestock, including bison, has virtually eliminated the disease in much of the country. However, a point of contention is the presence of the

disease in wild animals, especially in bison in national park herds. Ranchers are concerned that this focus of infection will lead to reintroduction of the disease into their cattle. *Brucella suis* (sü'is) is also found in the United States, mostly in swine, but is emerging as a cause of the disease in cattle. The most common species in the rest of the world is *Brucella melitensis* (me-li-ten'sis), which has a reservoir in goats, sheep, and camels.

Brucella bacteria tend to multiply in the uterus of a susceptible animal, where their growth is favored by the presence of a carbohydrate, *mesoerythritol*, which is produced in the fetus and the membranes surrounding it. In humans, brucellosis usually begins with chills, fever, and malaise. In severe cases, heavy sweating and night-time fevers are common. About 100 or fewer cases of brucellosis are reported in the United States each year; there are few, if any, deaths. The disease caused by *B. abortus* is usually mild and self-limiting, often subclinical. *B. suis* infections are distinguished by the occasional formation of destructive abscesses. The disease caused by *B. melitensis* tends to be severe and often results in disability or death. The fever associated with this bacterium typically spikes to about 40°C (104°F) each evening; that is why the disease is sometimes called *undulant fever* (*undulant* = wavelike).

Mammals excrete the bacteria in their milk, and historically the disease has most commonly been transmitted by unpasteurized milk of cattle or goats. Because of pasteurization, dairy products are involved in only a minority of brucellosis cases today; in the United States, most cases of the disease result from contact with diseased animal tissue by farmers, veterinarians, and meat packers. The microbes apparently enter the human body by passing through minute abrasions in the skin or mucous membranes of the mouth, throat, or gastrointestinal tract. Once in the body, they are ingested by macrophages, in which they multiply and travel via the lymphatic system to the lymph nodes. From there, they can be transported to the liver, spleen, or bone marrow. The ability of the microorganisms to survive and even reproduce within the macrophages largely accounts for their virulence and resistance to antibiotic therapy.

Because the symptoms are difficult to interpret and because the bacteria require an atmosphere rich in carbon dioxide to grow, serological testing is important in the diagnosis of brucellosis. A simple agglutination test used on serum or milk to identify brucellosis-positive cattle has been an important factor in identifying infected herds and establishing a number of brucellosis-free states.

Vaccination has not been an important factor in controlling brucellosis in humans. Treatment with antibiotics must be prolonged because the bacteria grow within phagocytic cells. Frequently, combinations of tetracycline plus streptomycin are used for periods of several weeks.

ANTHRAX

LEARNING OBJECTIVE

• Discuss the epidemiology of anthrax.

In 1877, Robert Koch isolated *Bacillus anthracis,* the bacterium that causes **anthrax** in animals. The endospore-forming bacillus is a large, aerobic, gram-positive microorganism that is apparently able to grow slowly in soil types having specific moisture conditions. The endospores have survived in tests in soil for up to 60 years. The disease strikes primarily grazing animals, such as cattle and sheep. The *B. anthracis* endospores are ingested along with grasses, causing a fulminating, fatal sepsis.

The incidence of human anthrax is now rare in the United States, but occurrences in grazing animals are not uncommon. People at risk are those who handle animals, hides, wool, and other animal products from certain foreign countries. Goat hair and handicrafts containing animal hides from the Middle East have been a repeated source of infection.

Infections by *B. anthracis* are initiated by endospores. Once introduced into the body, they are taken up by macrophages, where they germinate into vegetative cells. These are not killed, but multiply, eventually killing the macrophage. The released bacteria then enter the bloodstream, replicate rapidly, and secrete toxins.

The primary virulence factors of *B. anthracis* are two exotoxins. Both toxins share a third toxic component, a cell receptor–binding protein called the *protective antigen,* that binds the toxins to target cells and permits their entry. One toxin, the *edema toxin,* causes local edema (swelling) and interferes with phagocytosis by macrophages. The other toxin, *lethal toxin,* specifically targets and kills macrophages, which disables an essential defense of the host. Furthermore, the capsule of *B. anthracis* is very unusual. It is not a polysaccharide but rather is composed of amino acid residues, which for some reason do not stimulate a protective response by the immune system. Therefore, once the anthrax bacteria enter the bloodstream, they proliferate without any effective inhibition until there are tens of millions per milliliter. These immense populations of toxin-secreting bacteria ultimately kill the host by precipitating a form of septic shock.

Anthrax affects humans in three forms: cutaneous anthrax, gastrointestinal anthrax, and inhalational (pulmonary) anthrax.

Cutaneous anthrax results from contact with material containing anthrax endospores. Over 90% of naturally occurring cases of anthrax in humans are cutaneous; the endospore enters at some minor skin lesion. A papule appears and then eventually vesicles, which rupture and form

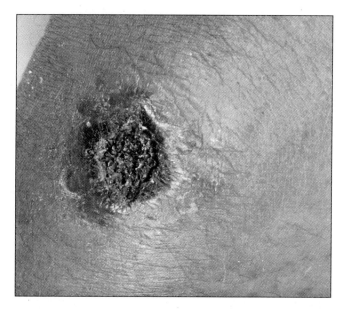

FIGURE 23.7 Anthrax lesion. The swelling and formation of a black scab that forms around the point of infection is a characteristic of cutaneous anthrax.

Q **What are the other types of anthrax?**

a depressed, ulcerated area that is covered by a black eschar (scab) as shown in Figure 23.7. (The name *anthrax* is derived from the Greek word for coal). In most cases the pathogen does not enter the bloodstream, and other symptoms are limited to a low-grade fever and malaise. However, if the bacteria enter the bloodstream, mortality without antibiotic treatment can reach 20%; with antibiotic therapy, mortality is usually less than 1%.

A relatively rare form of anthrax is **gastrointestinal anthrax** caused by ingestion of undercooked food containing anthrax endospores. Symptoms are nausea, abdominal pain, and bloody diarrhea. Ulcerative lesions occur in the gastrointestinal tract ranging from the mouth and throat to, mainly, the intestines. Mortality is usually more than 50%.

The most dangerous form of anthrax in humans is **inhalational (pulmonary) anthrax.** Endospores inhaled into the lungs have a high probability of entering the bloodstream. Symptoms of the first few days of the infection are not especially alarming: mild fever, coughing, and some chest pain. The disease can be arrested at this stage by antibiotics, but unless suspicion of anthrax is high, they are unlikely to be administered. As the bacteria enter the bloodstream and proliferate, the illness progresses in 2 or 3 days into septic shock that usually kills the patient within 24 to 36 hours. The mortality rate is exceptionally high, approaching 100%.

Antibiotics are effective in treating anthrax if they are administered in time. Currently recommended drugs are ciprofloxacin or doxycycline plus one or two additional

MICROBIOLOGY IN THE NEWS

BIOLOGICAL WEAPONS

The idea of biological warfare—that is, the use of living pathogens for hostile purposes—is not new. The earliest recorded use of biological warfare occurred in the fourteenth century, a time predating the discovery of microbes, when people still believed that "demons" caused disease. In 1346, the Tartar army catapulted plague-ridden bodies over the walls of Kaffa (Ukraine). After the fall of Kaffa, survivors escaping the fallen city introduced plague into Europe. Thus began the plague pandemic of 1348–1350.

In 1925, more than 100 countries agreed not to use biological warfare. Not long thereafter, however, airplanes dropped canisters of fleas carrying *Yersinia pestis* on China during the Sino-Japanese War (1937–1945). Japan tested *Francisella tularensis* as a potential weapon during World War II, and the United States and Soviet Union produced large quantities of this bacterium during the Cold War.

The U.S. Epidemic Intelligence Service (EIS) was formed in 1951, just after the start of the Korean War, as an early warning system against biological warfare. Since then EIS officers ("disease detectives") have played an important role in combating epidemics and tracking outbreaks of disease.

During the 1940s through the 1960s, research on biological weapons was conducted in several countries, including the United States and the United Kingdom. Countries that had a "no first use" policy developed weapons as a deterrent to potential aggressors and also did research to develop vaccines or treatments as defenses against the possible use of biological weapons. To determine how a biological weapon would spread, in the 1950s the U.S. Army sprayed *Serratia marcescens* over San Francisco and the Florida communities of Panama City and Key West. One outcome of this test was that 11 people became ill and one man died from *S. marcescens* infections.

In 1972, nearly 100 countries agreed to not even possess biological weapons. Nevertheless, in 1979 *Bacillus anthracis* was being produced in Sverdlovsk (Soviet Union) when an explosion blew the *B. anthracis* into the air; this resulted in 100 deaths in a 2-week period.

Historically, biological weapons have been associated with military action. Toward the end of the twentieth century, however, biological agents began to be used against an unsuspecting civilian population.

- In 1984, a religious cult attacked the people of The Dalles, Oregon, by intentionally contaminating food in restaurants and supermarkets with *Salmonella enterica*.

- In 1996, 15 people developed severe gastroenteritis requiring hospitalization when a laboratory worker intentionally contaminated pastries with *Shigella dysenteriae*.

- In 2001, someone used the U.S. Postal Service to spread *Bacillus anthracis* in New York City and Washington, D.C. This gave rise to a new term, *bioterrorism*, to describe the use of a biological agent to intimidate or coerce a government or group.

One of the problems with biological weapons, or bioweapons, is that they contain living organisms, so their impact is difficult to control or even predict. They could affect the civilians of the target country; moreover, they could be transmitted to civilians in the attacking country by wind, by vectors, or by escaping refugees. They might cause an epidemic that would linger even after hostilities ended; they might spread to neighboring countries, as happened in Europe in 1346; and they might damage livestock and other animals.

Some thought has also been given to the use of bioweapons that target food crops or livestock. The weapon not only might have the immediate effect of debilitating the target country, but also might lead to potentially devastating ecological

agents that are known to be active against the pathogen. People who have been exposed to anthrax endospores can be given preventive doses of antibiotics for a time as a precaution. This time period is usually quite long because experience has shown that up to 60 days can elapse before the inhaled endospores germinate and initiate active disease.

Vaccination of livestock against anthrax is a standard procedure in endemic areas. A single dose of an effective live, attenuated vaccine is used, which is considered unsafe for use in humans. The only vaccine currently approved for use in humans contains an inactivated form of the protective antigen toxin and is designed to prevent entry of the other two toxins into the host's cells. This vaccine requires a series of six injections over a period of 18 months, followed by annual boosters. In view of the recent use of anthrax as a weapon of bioterrorism (see the accompanying box), the need for a more practical human vaccine has become urgent. The target is a vaccine that would require no more than three injections and would work rapidly enough that it could be given *after* exposure to anthrax endospores.

Diagnosis of anthrax has usually consisted of isolation and identification of *B. anthracis* from a clinical specimen—which is too slow for detection of bioterrorism outbreaks. However, the FDA has recently approved a blood test that

MICROBIOLOGY IN THE NEWS

(continued)

consequences if the pathogen escaped to infect wild, native plants or animals.

Clostridium botulinum toxin, although a potent toxin, has limitations as a weapon: It must be delivered via food or water supplies, and it is not communicable. The "ideal" bioweapon is one that is disseminated by aerosol, is spread efficiently from human to human, causes a debilitating disease, and has no readily available treatment. Lists of potential biological weapons usually contain the following organisms:

Bacteria	Viruses
Bacillus anthracis	"Eradicated" polio and measles
Brucella spp.	Encephalitis viruses
Chlamydophila psittaci	Hemorrhagic fever viruses (Ebola, Marburg, Lassa)
Clostridium botulinum toxin	Influenza A (1918 strain)
Coxiella burnetii	Monkeypox
Francisella tularensis	Nipah virus
Rickettsia prowazekii	Smallpox
Shigella spp.	Yellow fever
Vibrio cholerae	
Yersinia pestis	

Early warning systems, such as DNA chips or recombinant cells that fluoresce in the presence of a bioweapon, are being developed. New vaccines are being developed, and existing vaccines are being stockpiled for use where needed. When the use of biological agents is considered a possibility, military personnel and first-responders (health care personnel and others) are vaccinated if a vaccine for the suspected agent exists.

The current plan to protect civilians in the event of an attack with a microbe is illustrated by the smallpox preparedness plan. It is not practical to vaccinate everyone against smallpox. Smallpox vaccination was discontinued when the naturally occurring disease was eradicated because the vaccine does pose a risk of complications, including death. Additionally, some people should not receive the vaccine; for example, people with AIDS, some cancer patients, organ transplant recipients, and people with allergy-induced dermatitis. The U.S. government's current strategy following a confirmed smallpox outbreak includes "ring containment and voluntary vaccination." Ring containment consists of identifying people with the infection and vaccinating everyone who has had contact with them. This procedure was used between 1966 and 1980 during the worldwide campaign to eradicate smallpox. People were vaccinated in villages where smallpox occurred, then people in neighboring villages, and so on. Local voluntary vaccinations can supplement ring containment.

Cultures of microbes have been relatively easy to obtain. The cultures used in your classes were probably purchased from the American Type Culture Collection (ATCC). In contrast, weapons-grade cultures are special preparations and are designed to withstand drying and ultraviolet radiation, with uniform particles between 1 μm and 10 μm for easy distribution through the air. These preparations usually are available only in high-level security labs. It might not be possible to stop all wars, but the public health system is improving its ability to respond to bioweapons. Rapid tests to detect genetic changes in hosts due to bioweapons even before symptoms develop are being investigated.

---

can detect both inhalational and cutaneous cases of anthrax within an hour. Furthermore, locations such as a few mail-sorting facilities are being equipped with automated electronic sensors that can immediately detect anthrax spores.

GANGRENE

LEARNING OBJECTIVE

• Discuss the epidemiology of gas gangrene.

If a wound causes the blood supply to be interrupted, a condition known as **ischemia,** the wound becomes anaerobic.

Ischemia leads to **necrosis,** or death of the tissue. The death of soft tissue resulting from the loss of blood supply is called **gangrene** (Figure 23.8). These conditions can also occur as a complication of diabetes.

Substances released from dying and dead cells provide nutrients for many bacteria. Various species of the genus *Clostridium,* which are gram-positive, endospore-forming anaerobes widely found in soil and in the intestinal tracts of humans and domesticated animals, grow readily in such conditions. *C. perfringens* is the species most commonly involved in gangrene, but other clostridia and several other bacteria can also grow in such wounds.

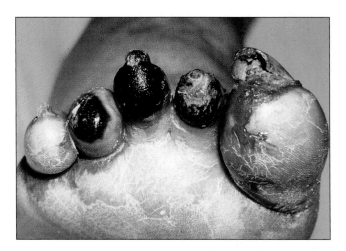

FIGURE 23.8 The toes of a patient with gangrene. This disease is caused by *Clostridium perfringens* and other clostridia. The black, necrotic tissue, resulting from poor circulation or injury, furnishes anaerobic growth conditions for the bacteria, which then progressively destroy adjoining tissue.

Q How can gangrene be prevented?

FIGURE 23.9 Hyperbaric chambers used to treat gas gangrene. A multiplace hyperbaric chamber that can accommodate several patients is shown. Such chambers are usually available at major medical centers. They are also used to treat victims of carbon monoxide poisoning.

Q Name another bacterial disease that might be treated in a hyperbaric chamber.

Once ischemia and the subsequent necrosis caused by impaired blood supply have developed, **gas gangrene** can develop, especially in muscle tissue. As the C. *perfringens* microorganisms grow, they ferment carbohydrates in the tissue and produce gases (carbon dioxide and hydrogen) that swell the tissue. The bacteria produce toxins that move along muscle bundles, killing cells and producing necrotic tissue that is favorable for further growth. Eventually, these toxins and bacteria enter the bloodstream and cause systemic illness. Enzymes produced by the bacteria degrade collagen and proteinaceous tissue, facilitating the spread of the disease. Without treatment, the condition is fatal.

One complication of improperly performed abortions is the invasion of the uterine wall by C. *perfringens*, which resides in the genital tract of about 5% of all women. This infection can lead to gas gangrene and result in a life-threatening invasion of the bloodstream.

The surgical removal of necrotic tissue and amputation are the most common medical treatments for gas gangrene. When gas gangrene occurs in such regions as the abdominal cavity, the patient can be treated in a **hyperbaric chamber,** which contains a pressurized oxygen-rich atmosphere (Figure 23.9). The oxygen saturates the infected tissues and thereby prevents the growth of the obligately anaerobic clostridia. Small chambers are available that can accommodate a gangrenous limb. The prompt cleaning of serious wounds and precautionary antibiotic treatment are the most effective steps in the prevention of gas gangrene. Penicillin is effective against C. *perfringens*.

SYSTEMIC DISEASES CAUSED BY BITES AND SCRATCHES

LEARNING OBJECTIVE

• List three pathogens that are transmitted by animal bites and scratches.

Animal bites can result in serious infections. The frequency of such infections can be estimated by the fact that there are about 4.4 million animal bites in the United States annually, accounting for about 1% of visits to the emergency rooms in hospitals. Most bites are by domestic animals, such as dogs and cats, because they live in close contact with humans.

Dog bites make up at least 80% of reported bite incidents; cat bites are only about 10% of biting incidents. Cat bites are, however, more penetrating, which results in a higher infection rate (30–50%), than the bites of dogs (15–20%). Domestic animals often harbor *Pasteurella multocida* (pas-tyėr-el'lä mul-tō'si-dä), a gram-negative rod similar to the *Yersinia* bacterium that causes plague (page 683). *P. multocida* is primarily a pathogen of animals, and it causes sepsis (hence the name *multocida*, meaning many-killing).

Humans infected with *P. multocida* have varied responses. For example, local infections with severe swelling and pain can develop at the site of the wound. Forms of pneumonia and sepsis may develop and are life-threatening. Penicillin and tetracycline are usually effective in treating these infections.

In addition to *P. multocida,* an assortment of anaerobic bacterial species are often found in infected animal bites, as well as species of *Staphylococcus, Streptococcus,* and *Corynebacterium.* Bites by humans, mostly as a result of fighting, are also prone to serious infections. In fact, before antibiotic therapy became available, nearly 20% of victims of human bites on extremities required amputation—this has been reduced currently to only about 5% of cases.

CAT-SCRATCH DISEASE

Cat-scratch disease, although it receives little attention, is surprisingly common. An estimate is that 22,000 or more cases occur annually in the United States, many more than the well-known Lyme disease. People who own or are closely exposed to cats are at risk. The pathogen is an aerobic, gram-negative bacterium, *Bartonella henselae* (bär′to-nel-lä hen′sel-ī). Microscopy shows that the bacterium can inhabit the interior of some cat red blood cells. It is connected to the exterior of the cell and to the surrounding extracellular fluid by a pore. Resident there, it causes a persistent bacteremia in cats; it is estimated that as many as 50% of domestic and feral (wild) cats carry these bacteria in their blood. A bite or scratch may not be necessary to transmit the disease. It is hypothesized that the bacteria in the cat's saliva can be deposited on the fur and transferred to the pet's owner when the person touches the cat and then rubs his or her eyes. In these instances, conjunctivitis and other symptoms of an eye infection sometimes occur. Fleas may transmit the disease between cats and possibly to humans. However, it is transmitted to humans mostly by a scratch or bite. The initial sign is a papule at the infection site, which appears 3 to 10 days after exposure. Swelling of the lymph nodes and usually malaise and fever follow in a couple of weeks. Cat-scratch disease is ordinarily self-limiting, with a duration of a few weeks, but in severe cases antibiotic therapy may be effective.

RAT-BITE FEVER

In large urban areas (even in the United States), the rat population is not well controlled, and bites from rats are a fairly common occurrence—and may cause the disease of rat bite fever. There are two bacterial pathogens involved. The more common is *Streptobacillus moniliformis;* in this case the disease is called streptobacillary rat-bite fever. After a few days' incubation there is an onset of fever, headache, and muscle aches. The streptobacilli multiply at the bite site, and there is inflammation surrounding a local lesion. The infection can spread in the bloodstream and lead to swollen lymph nodes and a rash. Occasional complications are endocarditis and even a significant mortality rate if the infection remains untreated.

The other bacterial pathogen causing rat bite fever is *Spirillum minor*—in this case the disease is called spirillar fever. It is more likely to occur in bites by wild rodents. The symptoms are similar to streptobacillary rat-bite fever. This gram-negative, spiral-shaped bacterium cannot be cultured.

In a research setting, laboratory rats require handling that sometimes results in bites. Although about half of both wild and laboratory rats are known to carry the bacterial pathogens, only a minority of rat bites (about 10%) result in disease. Treatment by penicillin is usually effective for both forms of rat-bite fever.

VECTOR-TRANSMITTED DISEASES

> **LEARNING OBJECTIVE**
> - Compare and contrast the causative agents, vectors, reservoirs, symptoms, treatments, and preventive measures for plague, Lyme disease and Rocky Mountain spotted fever.
> - Identify the vector, etiology, and symptoms of five diseases transmitted by ticks.
> - Describe the epidemiologies of epidemic typhus, endemic murine typhus, and spotted fevers.

PLAGUE

Few diseases have affected human history more dramatically than **plague,** known in the Middle Ages as the Black Death. This term comes from one of its characteristics, the dark blue areas of skin caused by hemorrhages.

The disease is caused by a gram-negative, rod-shaped bacterium, *Yersinia pestis.* Normally a disease of rats, plague is transmitted from one rat to another by the rat flea, *Xenopsylla cheopis* (ze-nop-sil′lä kē-ō′pis) (see Figure 12.33b, page 380). In the far West and Southwest, the disease is endemic in wild rodents, especially ground squirrels and prairie dogs.

If its host dies, the flea seeks a replacement host, which may be another rodent or a human. It can jump about 3½ inches. A plague-infected flea is hungry for a meal because the growth of the bacteria blocks the flea's digestive tract, and the blood the flea ingests is quickly regurgitated. An arthropod vector is not always necessary for plague transmission. Contact from the skinning of infected animals; scratches, bites, and licks by domestic cats; and similar incidents have been reported to cause infection.

In the United States, exposure to plague is increasing, as residential areas encroach on areas with infected animals. In parts of the world where human proximity to rats is common, infection from this source still prevails.

From the flea bite, bacteria enter the human's bloodstream and proliferate in the lymph and blood. One factor

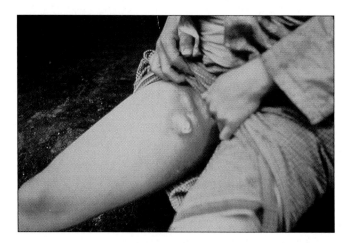

FIGURE 23.10   A case of bubonic plague. Bubonic plague is caused by infection with *Yersinia pestis*. This photograph shows a bubo (swollen lymph node) on the thigh of a patient. Swollen lymph nodes are a common indication of systemic infection.

Q   In what two ways is plague transmitted?

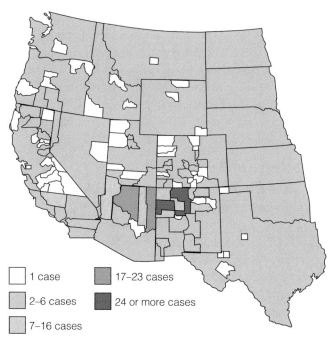

☐ 1 case   ▨ 17–23 cases

▨ 2–6 cases   ■ 24 or more cases

▨ 7–16 cases

FIGURE 23.11   The U.S. geographic distribution of human plague, 1970–2003.
SOURCE: CDC, 2005.

Q   What area reporting plague is closest to you?

in the virulence of the plague bacterium is its ability to survive and proliferate inside phagocytic cells rather than being destroyed by them. An increased number of highly virulent organisms eventually emerges, and an overwhelming infection results. The lymph nodes in the groin and armpit become enlarged, and fever develops as the body's defenses react to the infection. Such swellings, called buboes, account for the name *bubonic plague* (Figure 23.10). The mortality rate of untreated bubonic plague is 50 to 75%. Death, if it occurs, is usually within less than a week after the appearance of symptoms.

A particularly dangerous condition called **septicemic plague** arises when the bacteria enter the blood and proliferate, causing septic shock. Eventually, the bacteria are carried by the blood to the lungs, and a form of the disease called **pneumonic plague** results. The mortality rate for this type of plague is nearly 100%. Even today, this disease can rarely be controlled if it is not recognized within 12 to 15 hours of the onset of fever. People with pneumonic plague usually die within three days.

Pneumonic plague is easily spread by airborne droplets from humans or animals. Great care must be taken to prevent airborne infection of people in contact with patients.

Europe was ravaged by repeated pandemics of plague; from the years 542 to 767, outbreaks occurred repeatedly in cycles of a few years. After a lapse of centuries, the disease reappeared in devastating form in the fourteenth and fifteenth centuries. It is estimated to have killed more than 25% of the population, resulting in lasting effects on the social and economic structure of Europe. A nineteenth-century pandemic primarily affected Asiatic countries; 12 million are estimated to have died in India. The last

major rat-associated urban outbreak in the United States occurred in Los Angeles in 1924 and 1925. Following this, the disease became a rarity until it reappeared in 1965 on the Navajo reservation in the Southwest. Plague, once established in the ground squirrel and prairie dog communities in this area, has gradually spread over much of the western states (Figure 23.11). A peak incidence of 40 cases occurred in 1983. A few cases have also arisen from cats, a new animal reservoir, and one from urban tree squirrels. Awareness of the disease in endemic areas and control measures have again reduced the number of cases—only three were reported in 2004.

Plague has been most commonly diagnosed by isolating the bacterium and then sending it to a laboratory for identification. A recently developed rapid diagnostic test, however, can reliably detect the presence of the capsular antigen of *Y. pestis* in blood and other fluids of patients within 15 minutes even under remote field conditions. The test makes use of monoclonal antibodies coated onto a sampling dipstick. People exposed to infection can be given prophylactic antibiotic protection. A number of antibiotics, including streptomycin and tetracycline, are effective. Recovery from the disease confers reliable immunity. A vaccine is available for people likely to come into contact with infected fleas during field operations or for laboratory workers exposed to the pathogen.

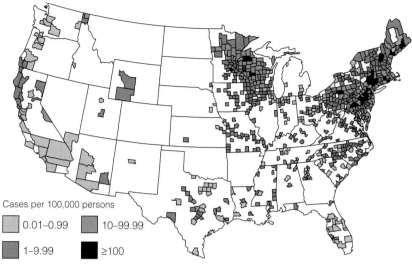

FIGURE 23.12 Lyme disease in the United States, reported cases by county, 2003.
SOURCE: CDC, *MMWR* 53(17): 367 (5/7/04).

Q **What factors are responsible for the geographic distribution of Lyme disease?**

Cases per 100,000 persons

■ 0.01–0.99   ■ 10–99.99

■ 1–9.99   ■ ≥100

*The total number of cases from these counties represented 90% of all cases reported in 2003.

RELAPSING FEVER

Except for the species that causes Lyme disease (discussed below), all members of the spirochete genus *Borrelia* cause **relapsing fever.** In the United States, the disease is transmitted by soft ticks that feed on rodents. The incidence of relapsing fever increases during the summer months, when the activity of rodents and arthropods increases.

The disease is characterized by fever, sometimes in excess of 40.5°C (105°F), jaundice, and rose-colored skin spots. After 3 to 5 days, the fever subsides. Three or four relapses may occur, each shorter and less severe than the initial fever. Each recurrence is caused by a different antigenic type of the spirochete, which evades existing immunity. Diagnosis is made by observing the bacteria in the patient's blood, which is unusual for a spirochete disease. Tetracycline is effective for treatment.

LYME DISEASE (LYME BORRELIOSIS)

In 1975, a cluster of disease cases in young people that was first diagnosed as rheumatoid arthritis was reported near the city of Lyme, Connecticut. The seasonal occurrence (summer months), lack of contagiousness among family members, and descriptions of an unusual skin rash that appeared several weeks before the first symptoms suggested a tickborne disease. The fact that penicillin alleviated the progression of symptoms suggested a bacterial pathogen. In 1983, a spirochete that was later named *Borrelia burgdorferi* was identified as the cause. **Lyme disease** may now be the most common tickborne disease in the United States. In Europe and Asia the disease is usually known as **Lyme borreliosis.** Often, in these locales the tick and *Borrelia* species differ from those in the United States. Tens of thousands of cases are reported annually (Figure 14.10a, page 442). In the United States,

Lyme disease is most prevalent on the Atlantic coast (Figure 23.12).

Field mice are the most important animal reservoir. The nymphal stage of the tick feeds on infected mice and is the most likely to infect humans, even though adult ticks are about twice as likely to carry the bacterial pathogen. This is because nymphal ticks are small and less likely to be noticed before the infection is transmitted. Deer are important in maintenance of the disease because the ticks feed and mate on them. However, they are less likely than mice to carry nymphs or to infect them.

The tick (one of two *Ixodes* species) feeds three times during its life cycle (Figure 23.13a). The first and second feedings, as a larva and then as a nymph, are usually on a field mouse. The third feeding, as an adult, is usually on a deer. These feedings are separated by several months, and the ability of the spirochete to remain viable in the disease-tolerant field mice is crucial to maintaining the disease in the wild.

On humans, the ticks usually attach from a perch on shrubs or grass. They do not feed for about 24 hours, and it usually requires two or three days of attachment before transfer of bacteria and infection occur. Probably only about 1% of tick bites result in Lyme disease.

On the Pacific coast, the tick that transmits Lyme disease is the western black-legged tick *Ixodes pacificus* (iks-ō′dēs pas-i′fi-kus) (see also Figure 12.32). In the rest of the country, *Ixodes scapularis* (scap-ū-lār′is) is most often responsible. This latter tick is so small that it is often missed (Figure 23.13b). On the Atlantic coast, almost all *Ixodes* ticks carry the spirochete (Figure 23.13c); on the Pacific coast, few are infected because that tick feeds on lizards that do not carry the spirochete effectively.

The first symptom of Lyme disease is usually a rash that appears at the bite site. It is a red area that clears in the

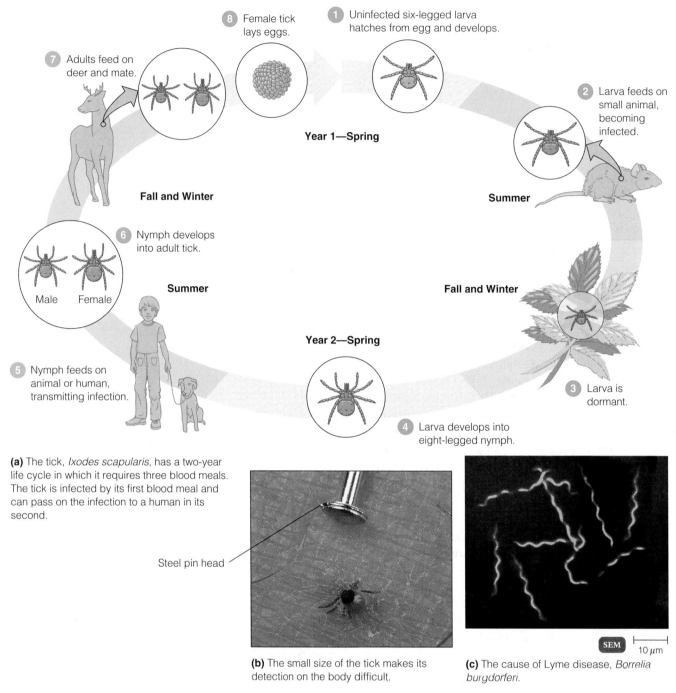

(a) The tick, *Ixodes scapularis*, has a two-year life cycle in which it requires three blood meals. The tick is infected by its first blood meal and can pass on the infection to a human in its second.

Steel pin head

(b) The small size of the tick makes its detection on the body difficult.

(c) The cause of Lyme disease, *Borrelia burgdorferi.*

SEM |——————| 10 µm

FIGURE 23.13   Life cycle of the tick vector of Lyme disease.

 **What other diseases are transmitted by ticks?**

center as it expands to a final diameter of about 15 cm (Figure 23.14). This distinctive rash occurs in about 75% of cases. Flulike symptoms appear in a couple of weeks as the rash fades. Antibiotics taken during this interval are very effective in limiting the disease.

During a second phase, in the absence of effective treatment, there is often evidence that the heart is affected. The heartbeat may become so irregular that a pacemaker is required. Incapacitating, chronic neurological symptoms, such as facial paralysis, meningitis, and encephalitis, may be seen. In a third phase, months or years later, some patients develop arthritis that may affect them for years. Immune responses to the presence of the bacteria are probably the cause of this joint damage. Many of the symptoms of long-term Lyme disease resemble those of the later stages of syphilis, also caused by a spirochete.

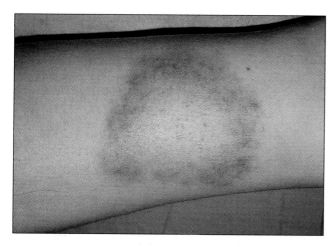

FIGURE 23.14   The common bull's-eye rash of Lyme disease. The rash is not always this obvious.

Q **What symptoms occur once the rash fades?**

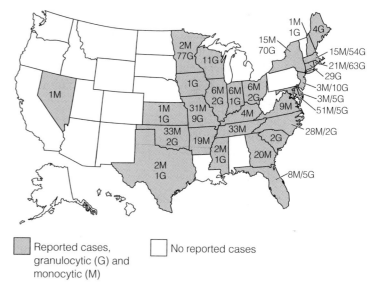

☐ Reported cases, granulocytic (G) and monocytic (M)   ☐ No reported cases

FIGURE 23.15   Ehrlichiosis. Human ehrlichiosis cases (granulocytic (G) and monocytic (M)) in the United States, 2003. SOURCE: CDC, *MMWR* 52(54), 4/22/05.

Q **What factors might contribute to the emergence of this disease?**

Diagnosis of Lyme disease depends partly on the symptoms and an index of suspicion based on the prevalence in the geographic area. Physicians are cautioned that serological tests must be interpreted in conjunction with clinical symptoms and the likelihood of exposure to infection. Some experts contend that although there are tests to detect antibodies against *B. burgdorferi*, there is no real laboratory test for Lyme disease. Also, after effective antibiotic treatment eliminates the bacteria, antibodies—even IgM antibodies—often persist for years and may confuse later attempts at diagnosis.

Several antibiotics are effective in treatment of the disease, although in the later stages, large amounts may be needed.

EHRLICHIOSIS

Ehrlichiosis is caused by *Ehrlichia*, gram-negative, rickettsialike, obligately intracellular bacteria that multiply in white blood cells, such as granulocytes and monocytes. Once considered only a disease of dogs, cattle, and sheep, ehrlichiosis is an emerging disease in much of the United States (Figure 23.15). The first human case was reported in 1987.

There are two distinct forms of ehrlichiosis: **human granulocytic ehrlichiosis (HGE)** and **human monocytic ehrlichiosis (HME).** These are tickborne diseases with a reservoir of infection in animals, especially deer. The vector of HGE is *Ixodes scapularis*, which also transmits Lyme disease in the northeastern and midwestern states. Occasionally coinfections occur with the two diseases. The vector of HME, known familiarly as the Lone Star tick, is common in the southeastern and south central United States. Both forms of ehrlichiosis are usually mild,

flulike diseases, but they have a significant fatality rate (usually less than 5%). They are most dangerous to persons with compromised immune systems. There are serological and PCR-based tests that can detect and differentiate these diseases. The recommended antibiotic for treatment is doxycycline.

TYPHUS

The various typhus diseases are caused by rickettsias, bacteria that are obligate intracellular parasites of eukaryotes. Rickettsias, which are spread by arthropod vectors, infect mostly the endothelial cells of the vascular system and multiply within them. The resulting inflammation causes local blockage and rupture of the small blood vessels.

Epidemic Typhus   Epidemic typhus (louseborne typhus) is caused by *Rickettsia prowazekii* and carried by the human body louse *Pediculus humanus corporis* (ped-ik′ū-lus hü′ma-nus kôr′pô-ris) (see Figure 12.33a, page 380). The pathogen grows in the gastrointestinal tract of the louse and is excreted by it. It is not transmitted directly by the bite of an infected louse; rather, it is transmitted when the feces of the louse are rubbed into the wound when the bitten host scratches the bite. The disease can flourish only in crowded and unsanitary surroundings, when lice can transfer readily from an infected host to a new host. Anne Frank, the teenaged writer of the famed World War II diary, died of typhus contracted in concentration camp conditions.

Epidemic typhus disease produces a high and prolonged fever that lasts at least 2 weeks. Stupor and a rash of small red spots caused by subcutaneous hemorrhaging are characteristic, as the rickettsias invade blood vessel linings. Mortality rates are very high when the disease is untreated.

Tetracycline and chloramphenicol are usually effective against epidemic typhus, but eliminating conditions in which the disease flourishes is more important. The microbe is considered especially hazardous, and attempts to culture it require extreme care. Vaccines are available for military populations, which historically have been highly susceptible to the disease.

Endemic Murine Typhus Endemic murine typhus occurs sporadically rather than in epidemics. The term *murine* (derived from Latin for mouse) refers to the fact that rodents, such as rats and squirrels, are the common hosts for this type of typhus. Endemic murine typhus is transmitted by the rat flea *Xenopsylla cheopis* (see Figure 12.33b), and the pathogen responsible for the disease is *Rickettsia typhi,* a common inhabitant of rats. Texas has had a number of outbreaks of murine typhus in recent years, often associated with campaigns to eliminate rodents, which caused the rat fleas to seek new hosts. With a mortality rate of less than 5%, the disease is considerably less severe than the epidemic form of typhus. Except for the reduced severity of the disease, endemic murine typhus is clinically indistinguishable from epidemic typhus.

Tetracycline and chloramphenicol are effective treatments for endemic murine typhus, and rat control is the best preventive measure.

Spotted Fevers Tickborne typhus, or Rocky Mountain spotted fever, is probably the best-known rickettsial disease in the United States. It is caused by *Rickettsia rickettsii.* Despite its name (it was first recognized in the Rocky Mountain area), it is most common in the southeastern states and Appalachia (Figure 23.16). This rickettsia is a parasite of ticks and is usually passed from one generation of ticks to another through their eggs, a mechanism called *transovarian passage* (Figure 23.17). Surveys show that in endemic areas, perhaps 1 out of every 1000 ticks is infected. In different parts of the United States, different ticks are involved—in the west, the wood tick *Dermacentor andersoni* (dėr-mä-sen′tôr an-dėr-sōn′ē); in the east, the dog tick *Dermacentor variabilis* (vār-ē-a′bil-is).

About a week after the tick bites, a macular rash develops that is sometimes mistaken for measles (Figure 23.18); however, it often appears on palms and soles, which does not occur with viral rashes. The rash is accompanied by fever and headache. Death, which occurs in about 3% of the approximately 1,000 cases reported each year, is usually caused by kidney and heart failure. Antibiotics such as tetracycline and chloramphenicol are very effective if administered early enough. No vaccine is available.

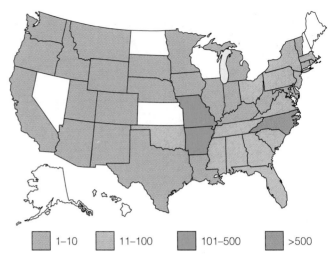

1–10	11–100	101–500	>500

FIGURE 23.16 The U.S. geographic distribution of Rocky Mountain spotted fever (tickborne typhus) by state, 2004.
SOURCE: CDC, *MMWR* 53(52) (1/7/05).

Q What is the nearest area to you that has reported Rocky Mountain spotted fever?

Serological tests do not become positive until late in the illness. Diagnosis before the typical rash appears is difficult; symptoms vary widely. Also, on dark-skinned individuals, the rash is difficult to see. A misdiagnosis can be costly; if treatment is not prompt and correct, the mortality rate is about 20%.

VIRAL DISEASES OF THE CARDIOVASCULAR AND LYMPHATIC SYSTEMS

LEARNING OBJECTIVE

• Describe the epidemiologies of CMV inclusion disease, Burkitt's lymphoma, and infectious mononucleosis.

Viruses are the cause of a number of cardiovascular and lymphatic diseases, prevalent mostly in tropical areas. However, one viral disease of this type, infectious mononucleosis, is an especially familiar infectious disease among American college-aged individuals.

BURKITT'S LYMPHOMA

In the 1950s, Denis Burkitt, an Irish physician working in eastern Africa, noticed the frequent occurrence in children of a fast-growing tumor of the jaw (Figure 23.19). Known as **Burkitt's lymphoma,** this is the most common childhood cancer in Africa. It has a limited geographic distribution similar to that of malaria in central Africa.

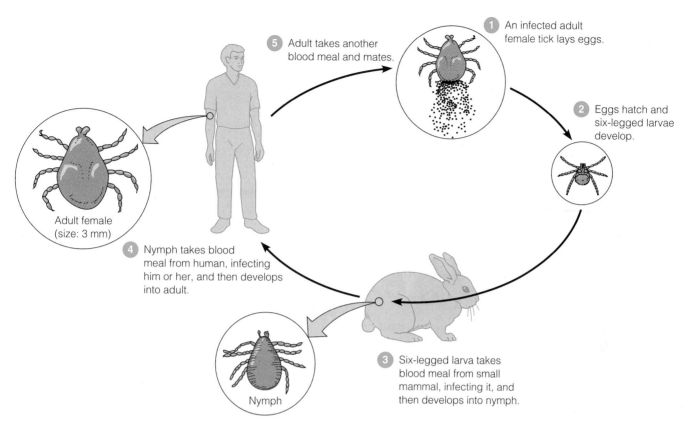

⑤ Adult takes another blood meal and mates.

① An infected adult female tick lays eggs.

② Eggs hatch and six-legged larvae develop.

③ Six-legged larva takes blood meal from small mammal, infecting it, and then develops into nymph.

④ Nymph takes blood meal from human, infecting him or her, and then develops into adult.

Adult female (size: 3 mm)

Nymph

FIGURE 23.17   The life cycle of the tick vector (*Dermacentor* spp.) of Rocky Mountain spotted fever. Mammals are not essential to survival of the pathogen, *Rickettsia rickettsii*, in the tick population; the bacteria may be passed by transovarian passage, so new ticks are infected upon hatching. A blood meal is required for ticks to advance to the next stage in the life cycle.

Q **What is meant by *transovarian passage?***

Burkitt suspected a viral cause of the tumor and a mosquito vector. At that time, there was no known virus that caused human cancer, although several viruses were clearly associated with animal cancers. Intrigued by this possibility, in 1964, British virologist Tony Epstein and his student, Yvonne Barr, performed biopsies on the tumors. A virus was cultured from this material, and the electron microscope showed a herpeslike virus in the culture cells; it was named the *Epstein-Barr virus (EB virus)*. The official name of this virus is human herpesvirus 4.

EB virus is clearly associated with Burkitt's lymphoma, but the mechanism by which it causes the tumor is not understood. Research eventually showed, however, that mosquitoes do not transmit the virus or the disease. Instead, mosquito-borne malarial infections apparently foster the development of Burkitt's lymphoma by impairing the immune response to EB virus, which is almost universally present in human adults worldwide. The virus has, in fact, become so adapted to humans that it is one of our most effective parasites. It establishes a life-long infection in most people (Figure 23.20) that is harmless and rarely causes disease.

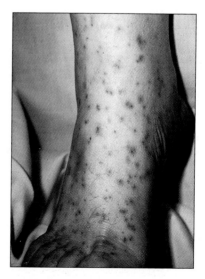

FIGURE 23.18   The rash caused by Rocky Mountain spotted fever. This rash is often mistaken for measles. People with dark skin have a higher mortality rate because the rash is often not recognized early enough for effective treatment.

Q **How can Rocky Mountain spotted fever be prevented?**

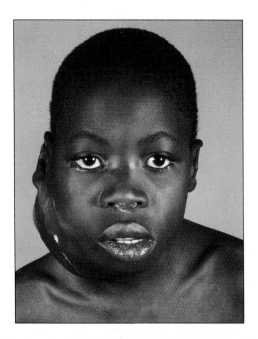

FIGURE 23.19   A child with Burkitt's lymphoma. Cancerous tumors of the jaw caused by Epstein-Barr virus (EB virus) are seen mainly in children. This child was successfully treated.

Q **What is the relationship between malarial areas and areas with Burkitt's lymphoma?**

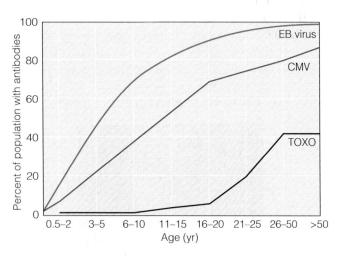

FIGURE 23.20   The typical U.S. prevalence of antibodies against Epstein-Barr virus (EB virus), cytomegalovirus (CMV), and *Toxoplasma gondii* (TOXO) by age.
SOURCE: *Laboratory Management*, June 1987, pp. 23ff.

Q **To judge from this graph, which of these diseases is more likely to result from early-childhood infections?**

In areas without endemic malaria, such as the United States, Burkitt's lymphoma is rare and is usually abdominal. The appearance of the lymphoma in AIDS patients is an indication of the importance of immune surveillance in preventing expression of the disease.

INFECTIOUS MONONUCLEOSIS

The identification of EB virus as the cause of **infectious mononucleosis,** or *mono,* was the result of one of the accidental discoveries that often advance science. A technician in a laboratory investigating EB virus served as a negative control for the virus. While on vacation, she contracted an infection characterized by fever, sore throat, swollen lymph nodes in the neck, and general weakness. The most interesting aspect of the technician's disease was that she now tested serologically positive for EB virus. It was soon confirmed that the same virus that is associated with Burkitt's lymphoma also causes almost all cases of infectious mononucleosis.

In developing parts of the world, infection with EB virus occurs in early childhood, and 90% of the children over age 4 have acquired antibodies. Nearly 20% of adults in the United States carry EB virus in oral secretions. Childhood EB virus infections are usually asymptomatic, but if infection is delayed until young adulthood, as is often the case in the United States, the result is more symptomatic probably because of an intense immunological response. The peak U.S. incidence of the disease occurs at about age 15 to 25 (Figure 23.20). College populations, particularly those in upper socioeconomic levels, have a high incidence of the disease. Most college students have no immunity, and about 15% of them can expect to contract the disease. The disease is generally self-limiting and seldom fatal. A principal cause of the rare deaths is rupture of the enlarged spleen (a common response to a systemic infection) during vigorous activity. Recovery is usually complete in a few weeks, and immunity is permanent.

The usual route of infection is by the transfer of saliva by kissing or, for example, by sharing drinking vessels. It does not spread among casual household contacts, so aerosol transmission is unlikely. The incubation period before appearance of symptoms is 4 to 7 weeks.

EB virus maintains a persistent infection in the oropharynx (mouth and throat), which accounts for its presence in saliva. It is probable that resting memory B cells (see Figure 17.5, page 510) located in lymphoid tissue are the primary site of replication and persistence. Most of the symptoms are attributed to responses of T cells to the infection.

The disease name *mononucleosis* refers to lymphocytes with unusual lobed nuclei that proliferate in the blood during the acute infection. The infected B cells produce nonspecific antibodies called heterophil antibodies. If this test is negative, the symptoms may be caused by cytomegalovirus (see page 691) or several other disease conditions. A fluorescent-antibody test that detects IgM antibodies against EB virus is the most specific diagnostic method. There is no recommended specific therapy for most patients.

OTHER DISEASES AND EPSTEIN-BARR VIRUS

We have just discussed two diseases, Burkitt's lymphoma and infectious mononucleosis, for which there is a clear association with EB virus. There is a lengthy list of diseases for which there is a suspected, but not proven, relationship with EB virus. Some of the more familiar of these include **multiple sclerosis** (autoimmune attack on the nervous system), **Hodgkin's disease** (tumors of the spleen, lymph nodes, or liver), and **nasopharyngeal** (nose and pharynx) **cancer** among certain ethnic groups in southeast Asia and Inuits.

CYTOMEGALOVIRUS INFECTIONS

Almost all of us will become infected with cytomegalovirus (CMV) during our lifetime. The CMV is a very large herpesvirus that, much like the Epstein-Barr virus, remains latent in white blood cells, such as monocytes, neutrophils, and T cells. It is not much affected by the immune system, replicating very slowly and escaping antibody action by moving between cells that are in contact. Carriers of the virus may shed it in body secretions such as saliva, semen, and breast milk. When CMV infects a cell, it causes the formation of distinctive inclusion bodies that are visible by microscopy. When these bodies occur in pairs, they are known as "owl's eyes" and are useful in diagnosis. These inclusion bodies were first reported in 1905 in certain cells of newborn infants affected with congenital abnormalities. The cells were also enlarged (*cytomegaly*), from which the virus eventually received its name. This disease of the newborns was given the name of *cytomegalic inclusion disease* (*CID*). The inclusion bodies were originally thought to be stages in the life cycle of a protozoan, and a viral cause of the disease was not proposed until 1925. The cytomegalovirus was not isolated until some thirty years after that. The official name is human herpesvirus 5.

In the United States, about 8000 infants each year are born suffering symptomatic damage from CID, the most serious of which includes severe mental retardation or hearing loss. The disease can result if a nonimmune mother acquires a primary CMV infection (that is, an infection that is new and not a recurrence) during pregnancy. Tests to determine the immune status of the mother are available, and it is recommended that physicians determine the immune status of female patients of childbearing age. All non-immune women should be informed of the risks of infection during pregnancy.

In healthy adults, acquiring a CMV infection causes either no symptoms or those resembling a mild case of infectious mononucleosis. It has been said that if CMV were accompanied by a skin rash, it would be one of the better-known childhood diseases. It is therefore not surprising, given that 80% of the population of the United States is estimated to carry the virus, that CMV is a common opportunistic pathogen in persons whose immune system has become compromised. Figure 23.20 shows the prevalence of antibodies against CMV, Epstein-Barr virus, and *Toxoplasma gondii* (page 695). In developing parts of the world, infection rates of CMV approach 100%. For immunocompromised individuals, CMV is a frequent cause of a life-threatening pneumonia, but almost any organ can be affected. About 85% of AIDS patients exhibit a CMV-caused eye infection, *cytomegalovirus retinitis*. Without treatment, it results in eventual loss of vision. Several antiviral agents, such as ganciclovir, and the antisense drug fomivirsen are effective, but treatment must be lifelong. Such antiviral drugs are also used in treatment of other CMV-caused diseases.

CMV is transmitted mostly by activities that result in contact with body fluids that contain the virus, such as kissing, and is very common among children in day-care settings. It can also be transmitted sexually, by transfused blood, and by transplanted tissue. Transmission by transfused blood can be eliminated by filtering out the white cells from the blood or by serological testing of the donor for the virus. Transplanted tissue is usually tested for the virus, and products are now available that contain antibodies to neutralize CMV present in donated tissue. Vaccines are under development, but none is currently available.

CLASSIC VIRAL HEMORRHAGIC FEVERS

LEARNING OBJECTIVE

- Compare and contrast the causative agents, vectors, reservoirs, and symptoms of yellow fever, dengue, and dengue hemorrhagic fever.

Most hemorrhagic fevers are zoonotic diseases; they appear in humans only from infectious contact with their normal animal hosts. Some of them have been medically familiar for so long that they are considered "classic" hemorrhagic fevers. First among these is **yellow fever.** The yellow fever virus is injected into the skin by a mosquito, *Aedes aegypti* (ā'ē-dēz ē-jip'tē).

In the early stages of severe cases of the disease, the person experiences fever, chills, and headache, followed by nausea and vomiting. This stage is followed by jaundice, a yellowing of the skin that gave the disease its name. This coloration reflects liver damage, which results in the deposit of bile pigments in the skin and mucous membranes. The mortality rate for yellow fever is high, about 20%.

Yellow fever is still endemic in many tropical areas, such as Central America, tropical South America, and Central Africa. At one time, the disease was endemic in the United States and occurred as far north as Philadelphia. The last U.S. case of yellow fever occurred in Louisiana in 1905 during an outbreak that resulted in about 1000 deaths. Mosquito eradication campaigns initiated by the U.S. Army surgeon Walter Reed were effective in eliminating yellow fever in the United States.

Monkeys are a natural reservoir for the virus, but human-to-human transmission can maintain the disease. Local control of mosquitoes and immunization of the exposed population are effective controls in urban areas.

Diagnosis is usually by clinical signs, but it can be confirmed by a rise in antibody titer or isolation of the virus from the blood. There is no specific treatment for yellow fever. The vaccine is an attenuated live viral strain and yields a very effective immunity, but there recently have been concerns about fatal cases of disease possibly related to receiving the vaccine.

Dengue (den'ghee) is a similar but milder viral disease also transmitted by the *Aedes aegypti* mosquito. This disease is endemic in the Caribbean and other tropical environments, where an estimated 100 million cases occur each year. It is characterized by fever, severe muscle and joint pain, and rash. Except for the painful symptoms, which have led to the name **breakbone fever,** classic dengue fever is a relatively mild disease and is rarely fatal.

The countries surrounding the Caribbean are reporting an increasing number of cases of dengue. In most years, more than 100 cases are imported into the United States, mostly by travelers from the Caribbean and South America. The disease does not appear to have an animal reservoir. The mosquito vector for dengue is common in the Gulf states, and there is some worry that the virus will sooner or later be introduced into this region and become endemic. Health officials are concerned about the American introduction of an Asian mosquito, *Aedes albopictus* (al-bō-pik'tus), an efficient vector for the virus and an aggressive biter. It transmits the virus by transovarian passage and from person to person. The range of this mosquito can potentially cover much of the country. Control measures are directed at eliminating *Aedes* mosquitoes. These are urban mosquitoes that proliferate in sites such as tree holes and discarded plasticware.

A severe form of dengue, **dengue hemorrhagic fever (DHF),** is probably caused when antibodies from a previous infection combine with the virus. DHF can induce shock in the victim (usually a child) and kill in a few hours; it is a leading cause of death among southeast Asian children. Outbreaks have also occurred in Mexico, South America, and the Caribbean.

EMERGING VIRAL HEMORRHAGIC FEVERS

LEARNING OBJECTIVE

• Compare and contrast the causative agents, reservoirs, and symptoms of Ebola hemorrhagic fever and *Hantavirus* pulmonary syndrome.

Certain other hemorrhagic diseases are considered new or "emerging" hemorrhagic fevers. In 1967, 31 people became ill and 7 died after contact with some African monkeys that were imported into Europe. The virus was strangely shaped (in the form of a filament [filoviruses]) and was named for the site of the outbreak in Germany, the **Marburg virus.** The symptoms of infection by hemorrhagic viruses are mild at first; headache and muscle pain. But after a few days the victim suffers from high fever and begins vomiting blood and bleeding profusely, both internally and from external openings such as the nose and eyes. Death comes in a few days from organ failure and shock.

A similar hemorrhagic fever, **Lassa fever,** appeared in Africa in 1969 and was traced to a rodent reservoir. The virus, an arenavirus, is present in the rodent's urine and is the source of human infections. Outbreaks of Lassa fever have killed thousands. Seven years later, outbreaks in Africa of another highly lethal hemorrhagic fever, caused by a filovirus similar to the Marburg virus, caused a disease with a mortality approaching 90%. Named **Ebola virus** for a regional river, this is now a well-publicized disease, the subject of films and books (Figure 23.21).

The natural host reservoir for the Ebola virus is probably a fruit bat, which is used as food and is not acutely affected by the virus it carries. Once a human is infected and shedding blood, the infection is spread by contact with the blood and body fluids and in many cases by the reuse of

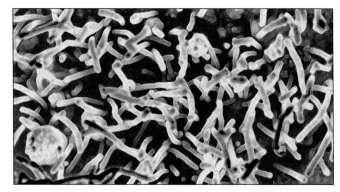

FIGURE 23.21 Ebola hemorrhagic virus. The viruses causing Ebola hemorrhagic fever are shown here. They disrupt the blood clotting system.

Q **Can you see why the Ebola virus is called a filovirus?**

needles used on patients. The local custom of washing the body before burial often triggers new infections.

South America has several hemorrhagic fevers caused by Lassa-like viruses (arenaviruses) that are maintained in the rodent population. **Argentine** and **Bolivian hemorrhagic fevers** are transmitted in rural areas by contact with rodent excretions. A recent handful of deaths in California have been attributed to the **Whitewater Arroyo virus,** an arenavirus with a reservoir in wood rats. These are the first reports of arenavirus-caused hemorrhagic disease in the Northern Hemisphere.

Hantavirus **pulmonary syndrome,** caused by the Sin Nombre virus[*], a bunyavirus, has become well known in the United States because of several outbreaks, mostly in the western states. It manifests itself as a frequently fatal pulmonary infection, in which the lungs fill with fluids. Actually, diseases of this nature have a long history, especially in Asia and Europe. It is best known there as **hemorrhagic fever with renal syndrome** and primarily affects renal (kidney) function. All these related diseases are transmitted by the inhalation of viruses in dried urine and feces from infected small rodents.

The Diseases in Focus box on page 694 describes the various viral hemorrhagic fevers.

PROTOZOAN DISEASES OF THE CARDIOVASCULAR AND LYMPHATIC SYSTEMS

LEARNING OBJECTIVES

- Compare and contrast the causative agents, modes of transmission, reservoirs, symptoms, and treatments for Chagas' disease, toxoplasmosis, malaria, leishmaniasis, and babesiosis.
- Discuss the worldwide effects of these diseases on human health.

Protozoa that cause diseases of the cardiovascular and lymphatic systems often have complex life cycles, and their presence may affect human hosts seriously.

CHAGAS' DISEASE (AMERICAN TRYPANOSOMIASIS)

Chagas' disease, also known as **American trypanosomiasis,** is a protozoan disease of the cardiovascular system. The

SEM |———| 2.5 μm

FIGURE 23.22 *Trypanosoma cruzi,* **the cause of Chagas' disease (American trypanosomiasis).** The trypanosome has an undulating membrane; the flagellum follows the outer margin of the membrane and then projects beyond the body of the trypanosome as a free flagellum. Note the red blood cells in the photo.

Q **Name a common trypanosomal disease that occurs in another part of the world. (Hint: it was discussed in Chapter 22).**

causative agent is *Trypanosoma cruzi* (tri-pa-nō-sō'mä kruz'ē), a flagellated protozoan (Figure 23.22). The protozoan was discovered in its insect vector by the Brazilian microbiologist Carlos Chagas in 1910. He named it for the Brazilian epidemiologist Oswaldo Cruz. The disease occurs in southern Texas, Mexico, Central America, and parts of South America. The disease infects 40 to 50% of the population in some rural areas of South America. An estimated 100,000 infected immigrants carry the disease in the United States.

The reservoir for *T. cruzi* is a wide variety of wild animals, including rodents, opossums, and armadillos. The arthropod vector is the reduviid bug, called the "kissing bug" because it often bites people near the lips (see Figure 12.33d, page 380). The insects live in the cracks and crevices of mud or stone huts with thatched roofs. The trypanosomes, which grow in the gut of the bug, are passed on if the bug defecates while feeding. The bitten human or animal often rubs the feces into the bite wound or other skin abrasions by scratching or into the eye by rubbing. An unusual means of transmission occasionally occurs in remote areas of Mexico, where reduviid bugs are eaten as an aphrodisiac. In Brazil, efforts to control the insect population have met with success; blood transfusions have now become the principal route of infection.

The disease is most dangerous to children, whose death rate may be as high as 10%, primarily because of damage to the heart. If the parasites damage the nerves controlling the peristaltic action of the esophagus or colon, these organs no

---

[*]The virus causing the pulmonary hantavirus outbreak in 1993 in the Four Corners area of the southwestern United States (Arizona, Utah, Colorado, and New Mexico) was originally called the Four Corners virus. Local authorities were concerned about the effect of this name on tourism in the area and complained. The name Sin Nombre, Spanish for no name, was then adopted.

DISEASES IN FOCUS

VIRAL HEMORRHAGIC FEVERS

Viral hemorrhagic fevers (VHFs) refer to a group of illnesses that are caused by several distinct families of viruses. In general, the term *viral hemorrhagic fever* is used to describe a severe syndrome that affects multiple organ systems in the body. Characteristically, the overall vascular system is damaged, and the body's ability to regulate itself is impaired. These symptoms are often accompanied by hemorrhage (bleeding); however, the bleeding is itself rarely life-threatening. While some types of hemorrhagic fever viruses can cause relatively mild illnesses, many of these viruses cause severe, life-threatening disease. These are endemic in tropical countries where, except for dengue, they are found in small mammals. However, increasing international travel has resulted in importation of these viruses into the United States. The CDC's Special Pathogens Branch has specialized containment facilities to confirm diagnosis of these diseases by serology, nucleic acids, and virus culture.

	Pathogen	Portal of entry	Reservoir	Method of transmission	Symptoms	Treatment
Yellow fever	Arbovirus (yellow fever virus)	Skin	Monkeys	*Aedes aegypti*	Fever, chills, headache; jaundice	None Prevention: Vacination; mosquito control
Dengue	Arbovirus (dengue fever virus)	Skin	Humans	*Aedes aegypti; A. albopictus*	Fever, muscle and joint pain, rash	None. Prevention: Mosquito control
Viral hemorrhagic fevers (Marburg, Ebola, Lossa)	Filovirus, arenavirus	Mucous membranes	Possibly fruit bats and other small mammals	Contract with blood	Profuse bleeding.	None
Hantavirus pulmonary syndrome	Bunyavirus (Sin Nombre hantavirus)	Respiratory tract	Field mice	Inhalation	Pneumonia	None

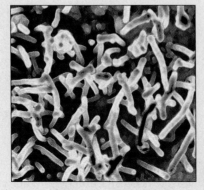

SEM   500 nm

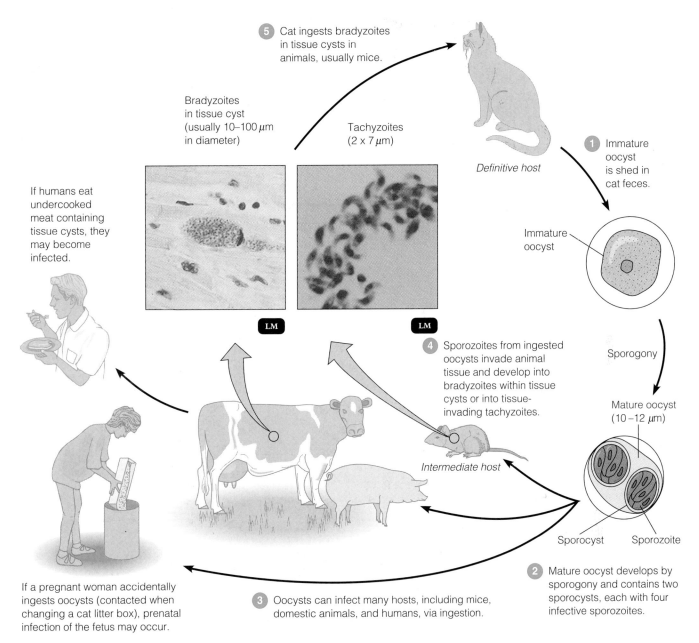

5 Cat ingests bradyzoites in tissue cysts in animals, usually mice.

Bradyzoites in tissue cyst (usually 10–100 μm in diameter)

Tachyzoites (2 x 7 μm)

Definitive host

1 Immature oocyst is shed in cat feces.

Immature oocyst

If humans eat undercooked meat containing tissue cysts, they may become infected.

LM

LM

Sporogony

Mature oocyst (10–12 μm)

4 Sporozoites from ingested oocysts invade animal tissue and develop into bradyzoites within tissue cysts or into tissue-invading tachyzoites.

Intermediate host

Sporocyst   Sporozoite

2 Mature oocyst develops by sporogony and contains two sporocysts, each with four infective sporozoites.

If a pregnant woman accidentally ingests oocysts (contacted when changing a cat litter box), prenatal infection of the fetus may occur.

3 Oocysts can infect many hosts, including mice, domestic animals, and humans, via ingestion.

FIGURE 23.23   The life cycle of *Toxoplasma gondii*, the cause of toxoplasmosis. The domestic cat is the definitive host, in which the protozoa reproduce sexually.

Q   **How do humans contract toxoplasmosis?**

longer transport food and become grossly enlarged, conditions known as *megaesophagus* and *megacolon*, respectively.

Diagnosis in endemic areas is usually based on symptoms. However, serological tests have improved recently and are now required of potential blood donors in Brazil. An extraordinary form of diagnosis, which is still done with some frequency, is *xenodiagnosis*. Reduviid bugs, reared in parasite-free conditions, are allowed to feed on the arm of a suspected patient. The trypanosome is then identified in the intestinal tract of the insect 10 to 20 days later.

Treating Chagas' disease is very difficult when chronic, progressive stages have been reached. The trypanosome

multiplies intracellularly and is difficult to reach chemotherapeutically. Drugs currently available, such as nifurtimox, have toxic side effects and low efficacy.

TOXOPLASMOSIS

Toxoplasmosis, a disease of blood and lymphatic vessels, is caused by the protozoan *Toxoplasma gondii*. *T. gondii* is a spore-forming protozoan, as is the malarial parasite.

Cats are an essential part of the life cycle of *T. gondii* (Figure 23.23). Random tests on urban cats have shown that a large number of them are infected with the organism, which causes no apparent illness in the cat. (A curiosity of the infection in rodents is that it apparently causes them to lose their normal avoidance behavior toward cats, making them more likely to be caught and thus to infect the cat.) The microbe undergoes its only sexual phase in the intestinal tract of the cat. Millions of oocysts are then shed in the cat's feces for 7 to 21 days and contaminate food or water that can be ingested by other animals. The *oocysts* contain *sporozoites* that invade host cells and form trophozoites called *tachyzoites* (about the size of large bacteria, 2 × 7 μm). The intracellular parasite reproduces rapidly (*tachys* is Greek for rapid). The increased numbers cause the rupture of the host cell and the release of more tachyzoites, resulting in a strong inflammatory response.

As the immune system becomes increasingly effective, the disease enters a chronic phase in animals and humans; the infected host cell develops a wall to form a *tissue cyst*. The numerous parasites within such a cyst (in this stage called *bradyzoites; bradys* being Greek for slow) reproduce very slowly, if at all, and persist for years, especially in the brain. These cysts are infective when ingested by intermediate or definitive hosts.

In people with a healthy immune system, toxoplasmosis infection results in only very mild symptoms or none at all. Some surveys have shown that approximately 22 to 40% of the population, without even being aware of it, eventually develops antibodies to *T. gondii* (see Figure 23.20). Humans generally acquire the infection by ingesting undercooked meats containing tachyzoites or tissue cysts, although there is a possibility of contracting the disease more directly by contact with cat feces. The primary danger is congenital infection of a fetus, resulting in stillbirth or a child with severe brain damage or vision problems. This fetal damage occurs only when the initial infection is acquired during pregnancy. As many as 4000 cases are estimated in the United States annually. The problem also affects wildlife. Off the California coast, a fatal encephalitis of sea otters has appeared, caused by *T. gondii*—apparently, they are being infected by oocysts in waste water contaminated from the flushed contents of cat litter boxes. Loss of immune function, AIDS being the best example, allows the inapparent infection to be reactivated from tissue cysts. It often causes severe neurological impairment and may damage vision from the reactivation of tissue cysts in the eye.

Toxoplasmosis can be detected by serological tests, but interpretation is uncertain. This uncertainty is especially important because in some European countries, a person who becomes toxoplasmosis-positive during pregnancy is encouraged to abort the fetus. Recently, PCR tests have become available. If not contaminated, these tests approach an accuracy of 100%, which has revolutionized prenatal

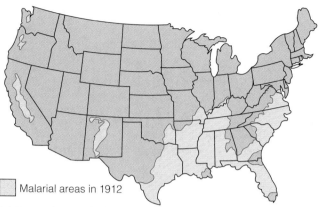

☐ Malarial areas in 1912

(a) Areas where malaria was endemic as recently as 1912

(b) Graph showing reported cases of malaria in the United States, 1967 to (*) the first 26 weeks of 2005

FIGURE 23.24 Malaria in the United States.
SOURCE: CDC, *MMWR* 52(54), 4/22/05; *MMWR* 53(52), 1/7/05.

Q **What factors might contribute to the rise in malaria cases since 1990?**

diagnosis. Toxoplasmosis can be treated with pyrimethamine in combination with sulfadiazine and folinic acid. This does not, however, affect the chronic bradyzoite stage and is quite toxic.

MALARIA

Malaria is characterized by chills and fever and often by vomiting and severe headache. These symptoms typically appear at intervals of two to three days, alternating with asymptomatic periods. Malaria occurs wherever the mosquito vector *Anopheles* is found (see Figure 12.31b, page 375) and there are human hosts for the protozoan parasite *Plasmodium*.

The disease was once widespread in the United States (Figure 23.24a), but effective mosquito control and a reduction in the number of human carriers caused the reported cases to drop below 100 by 1960. In recent years, however, there has been an upward trend in the number of U.S. cases, reflecting a worldwide resurgence of malaria, increased travel to malarial areas, and an increase in immigration from malarial areas. Occasionally, malaria has been transmitted by unsterilized syringes used by drug addicts. Blood transfusions from people who have been in an endemic area are also a potential risk. In tropical Asia, Africa, and Central and South America, malaria is still a serious problem. It is estimated that malaria affects 300 to 500 million people worldwide and causes 2 to 4 million deaths annually. Actually, there are probably more people dying of malaria today than 30 years ago. It is returning to areas where it had been nearly eradicated, such as eastern Europe and central Asia. Africa, where 90% of the mortality from the disease occurs, suffers the most from malaria. It is estimated that it kills an African child every 30 seconds.

There are four major forms of malaria. *Plasmodium vivax* is widely distributed and is the cause of the most prevalent form of malaria. Sometimes referred to as "benign" malaria, the cycle of paroxysms occurs every two days, and the patients generally survive many years even without treatment. *P. ovale* and *P. malariae* also cause a relatively benign malaria, but even so, the victims lack energy. These latter two malarial types are lower in incidence and rather restricted geographically.

The most dangerous malaria is that caused by *P. falciparum*. Perhaps one reason for the virulence of this type of malaria is that humans and the parasite have had less time to become adapted to each other. It is believed that humans have been exposed to this parasite (through contact with birds) only in relatively recent history. Referred to as "malignant" malaria, untreated it eventually kills about half of those infected. The highest mortality rates occur in young children. More red blood cells (RBCs) are infected and destroyed than in other forms of malaria. The resulting anemia severely weakens the victim. Furthermore, the RBCs develop surface knobs that cause them to stick to the walls of the capillary vessels, which become clogged. This clogging prevents the infected RBCs from reaching the spleen, where phagocytic cells would eliminate them. The blocked capillaries and subsequent loss of blood supply leads to death of the tissues. Kidney and liver damage is caused in this fashion. The brain is frequently affected, and *P. falciparum* is the usual cause of cerebral malaria.

The disease of malaria and its symptoms are intimately related to its complex reproductive cycle (see Figure 12.19). Infection is initiated by the bite of a mosquito, which carries the *sporozoite* stage of the *Plasmodium* protozoan in its saliva. The sporozoite enters the bloodstream of the bitten human and within about 30 minutes enters the liver cells. The sporozoites in the liver cells undergo reproductive *schizogony* by a series of steps that finally

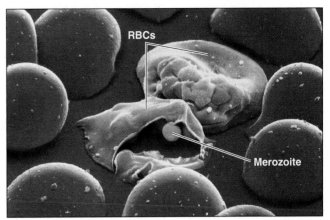

(a) Merozoites being released from lysed RBC.   SEM | 1 μm

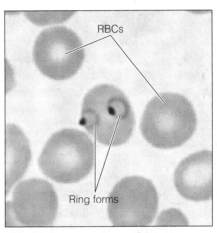

(b) Malarial blood smear; note the ring forms.   LM | 5 μm

FIGURE 23.25 Malaria. (a) Some of the red blood cells (RBCs) are lysing and releasing merozoites that will infect new RBCs. **(b)** Blood smears are used in the diagnosis of malaria; the protozoa can be detected growing in the RBCs. In the early stages, the feeding protozoan resembles a ring within the RBC. The light central area within the circular ring is the food vacuole of the protozoan, and the dark spot on the ring is the nucleus.

Q **Look at the life cycle of the malarial parasite; in Figure 12.19, which of these stages, (a) or (b), actually occurs first?**

results in the release of about 30,000 *merozoite* forms into the bloodstream.

The merozoites infect RBCs. Within the RBCs they again undergo schizogony, and after about 48 hours, the RBCs rupture and each releases about 20 new merozoites (Figure 23.25a). Laboratory diagnosis of malaria is usually made by examining a blood smear (Figure 23.25b) for infected RBCs. With the release of the merozoites there is also a simultaneous release of toxic compounds, which is the cause of the paroxysms (recurrent intensifications of symptoms) of chills and fever that are characteristic of malaria. The fever reaches 40°C (104°F), and a sweating stage begins as the fever subsides. Between paroxysms, the patient feels normal.

Many of the released merozoites infect other RBCs within a few seconds to renew the cycle in the bloodstream. If only 1% of the RBCs contain parasites, an estimated 100,000,000,000 parasites will be in circulation at one time in a typical malaria patient! Some of the merozoites develop into male or female *gametocytes*. When these enter the digestive tract of a feeding mosquito, they pass through a sexual cycle that produces new infective sporozoites. It took the combined labors of several generations of scientists to discover this complex life cycle of the malaria parasite.

People who survive malaria acquire a limited immunity. Although they can be reinfected, they tend to have a less severe form of the disease. This relative immunity almost disappears if the person leaves an endemic area with its periodic reinfections. Malaria is especially dangerous during pregnancy because adaptive immunity is suppressed.

Much effort is being expended on the search for an effective vaccine. Several are being field tested. The sporozoite stage is of primary interest as a target for a vaccine because neutralizing it would prevent the initial infection from becoming well established. A truly global malarial vaccine would have to control not only *P. falciparum* but also the widespread, although milder, *P. vivax*. There are special problems in developing a malarial vaccine. For example, unlike most viral or bacterial pathogens, which are relatively simple genetically and remain much the same during the course of an infection, the malarial parasite has four distinctive stages. In these stages it has as many as 7000 genes that can mutate. The result is that the parasite is very efficient at evading the human immune response.

As we mentioned earlier, the most common diagnostic test for malaria is the blood smear. This requires equipment, such as a microscope, and skill in interpretation. Also, inspection of more than 300 microscopic fields on a slide is recommended, which is time-consuming. The World Health Organization is encouraging development of diagnostic tests for malaria that are rapid, inexpensive, and require minimal skills.

There are two considerations for antimalarial drugs: for treatment and for prophylaxis (prevention). The original drug for the treatment of malaria was quinine, and many drugs used today, such as chloroquine, primaquine, and mefloquine, are derivatives of quinine. Chloroquine is inexpensive, but resistance has developed in many malarial areas. Mefloquine and quinine are often used as an alternative in chloroquine-resistant areas. Travelers to malarial areas are often prescribed prophylactic doses of mefloquine (Lariam) but are cautioned about possible psychological side effects including hallucinations. There is increasing use of several derivatives of artemisinin, the active ingredient of an herb traditionally used by the Chinese to treat fevers. Their action is against the gametocytes, the sexual stage of the parasite that infects feeding mosquitoes. Even the common household antimicrobial, triclosan, has shown promise as an antimalarial drug. This is only a minimal listing of agents available for prophylaxis and chemotherapy of malaria. Cost and drug resistance will be continuing problems. Actually, the most profitable application of antimalarial drugs will probably continue to be prophylaxis of travelers to malarial areas.

Effective control of malaria is not in sight. It will probably require a combination of vector control and chemotherapeutic and immunological approaches. Currently, the most promising control method is the use of insecticide-treated bed nets, because the *Anopheles* mosquito is a night feeder. In malarial areas a sleeping room often will contain hundreds of mosquitoes—1 to 5% of which are infectious. The expense and the need for an effective political organization in malarial areas are probably going to be as important in controlling the disease as are advances in medical research.

LEISHMANIASIS

Leishmaniasis is a widespread and complex disease that exhibits several clinical forms. The protozoan pathogens are of about 20 different species, often categorized into three groups for reasons of simplicity. One group is *Leishmania donovani* (līsh′mā-nē-ä don-ō-van′ē) that causes a visceral leishmaniasis in which parasites invade the internal organs. The *L. tropica* (līsh′mā-nē-ä trop′i-kä) and *L. braziliensis* (brä-sil′ē-en-sis) groups grow preferentially at cooler temperatures and cause lesions of the skin or mucous membranes. Leishmaniasis is transmitted by the bite of female sandflies, about 30 species of which are found in much of the tropical world and around the Mediterranean. These insects are smaller than mosquitoes and often penetrate the mesh of standard netting. Small mammals are an unaffected reservoir of the protozoans. The infective form, the *promastigote*, is in the saliva of the insect. It loses its flagellum when it penetrates the skin of the mammalian victim, becoming an *amastigote* that proliferates in phagocytic cells, mostly in fixed locations in tissue. These amastigotes are then ingested by feeding sandflies, renewing the cycle.

A number of cases of leishmaniasis, mostly cutaneous, occurred among troops fighting the Persian Gulf War. It was once endemic in countries of southern Europe, such as Spain, Italy, Portugal, and the Balkan peninsula. Occasional cases of leishmaniasis as an opportunistic disease of HIV-infected persons are beginning to reappear in these areas.

LEISHMANIA DONOVANI *INFECTION* (*VISCERAL LEISHMANIASIS*)

Leishmania donovani infection occurs in much of the tropical world, although 90% of the cases occur in India, Bangladesh, Sudan, and Brazil. Estimates are that there are about half a million cases per year. (See the box in Chapter 12, page 367.) Known as *kala azar* in India, visceral leishmaniasis is often fatal. Early symptoms, following infection by as long as a year, resemble the chills and sweating of malaria. As the protozoa proliferate in the liver and spleen, these organs enlarge greatly. Eventually, kidney function is also lost as these organs are invaded. This is a debilitating disease that, if untreated, will lead to death within a year or two.

Several inexpensive serological tests that are easy to use have been developed to diagnose visceral leishmaniasis. These have generally replaced microscopic examination of blood and tissues to demonstrate the parasite. PCR tests are very good for confirmation of a diagnosis but usually require a central laboratory.

The primary treatment has long been injected drugs such as sodium stibogluconate that contain the toxic metal, antimony. Development of resistance has made these almost useless where incidence of the disease is highest, but they are still widely used elsewhere. An orally administered drug that was introduced in India in 2002, miltefosine, has been shown to be very effective. However, it is potentially teratogenic and must be used with caution if pregnancy is possible.

LEISHMANIA TROPICA *INFECTION* (*CUTANEOUS LEISHMANIASIS*)

Leishmania tropica infection is a *cutaneous* form of leishmaniasis sometimes called *oriental sore*. A papule appears at the bite site after a few weeks of incubation (Figure 23.26). The papule ulcerates and, after healing, leaves a prominent scar. This form of the disease is found in much of Asia, Africa, and the Mediterranean region. It has been reported in Mexico, Central America, and the northern part of South America.

LEISHMANIA BRAZILIENSIS *INFECTION* (*MUCOCUTANEOUS LEISHMANIASIS*)

Leishmania braziliensis infection is known as *mucocutaneous* leishmaniasis because it affects mucous membranes as well as skin. It causes disfiguring destruction of the tissues of the nose, mouth, and upper throat. This form of leishmaniasis is most commonly found in the Yucatan Peninsula of Mexico and in the rain forest areas of Central and South America; it often affects workers harvesting the chicle sap used for making chewing gum. This disease is often referred to as *American leishmaniasis*.

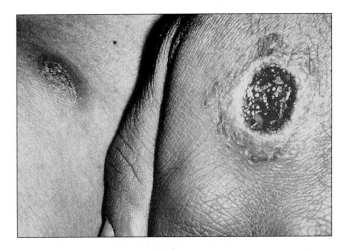

FIGURE 23.26 Cutaneous leishmaniasis. Lesion on the back of the hand of a patient.

Q **Is this case likely to progress to visceral leishmaniasis?**

Diagnosis of cutaneous and mucocutaneous leishmaniasis in the areas where they are endemic usually depends upon its clinical appearance and microscopic examination of the lesion scrapings. Mild cases of cutaneous and mucocutaneous disease will often eventually heal with only local treatment, but antimony compounds are usually effective when required. Recently an alternative treatment with the oral antifungal drug, fluconazole, has been shown to be effective—but requires as long as six weeks of administration.

BABESIOSIS

There have been increased reports of **babesiosis,** a tick-borne disease once thought to be restricted to animals. Rodents are the reservoir in the wild; the tick vectors are most commonly *Ixodes* species. The field of medical entomology largely arose from investigations in the nineteenth century by the American microbiologist Theobald Smith into bovine babesiosis, or tick fever, in Texas cattle. The human disease in the United States is caused by a protozoan, usually of the species *Babesia microti*. The disease resembles malaria in some respects and has been mistaken for it; the parasites replicate in the RBCs and cause a prolonged illness of fever, chills, and night sweats. It can be much more serious, sometimes fatal, in immunocompromised patients. For example, the first human cases were observed in persons who had undergone splenectomy (removal of the spleen). Simultaneous treatment with the drugs atovaquone and azithromycin has been effective.

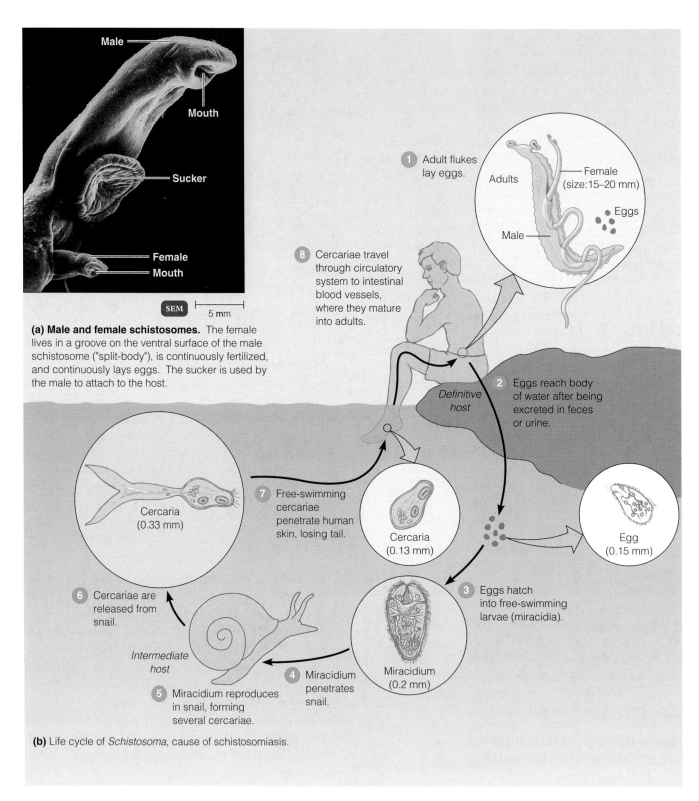

Male
Mouth
Sucker
Female
Mouth

SEM | 5 mm

(a) Male and female schistosomes. The female lives in a groove on the ventral surface of the male schistosome ("split-body"), is continuously fertilized, and continuously lays eggs. The sucker is used by the male to attach to the host.

1 Adult flukes lay eggs.

Adults

Female (size:15–20 mm)

Eggs

Male

8 Cercariae travel through circulatory system to intestinal blood vessels, where they mature into adults.

Definitive host

2 Eggs reach body of water after being excreted in feces or urine.

7 Free-swimming cercariae penetrate human skin, losing tail.

Cercaria (0.33 mm)

Cercaria (0.13 mm)

Egg (0.15 mm)

3 Eggs hatch into free-swimming larvae (miracidia).

6 Cercariae are released from snail.

Miracidium (0.2 mm)

Intermediate host

5 Miracidium reproduces in snail, forming several cercariae.

4 Miracidium penetrates snail.

(b) Life cycle of *Schistosoma*, cause of schistosomiasis.

FIGURE 23.27 Schistosomiasis.

Q What is the role of sanitation and snails in maintaining schistosomiasis in a population?

HELMINTHIC DISEASES OF THE CARDIOVASCULAR AND LYMPHATIC SYSTEMS

LEARNING OBJECTIVES

- Diagram the life cycle of *Schistosoma,* and show where the cycle can be interrupted to prevent human disease.

Many helminths use the cardiovascular system for part of their life cycle. Schistosomes find a home there, shedding eggs that are distributed in the bloodstream.

SCHISTOSOMIASIS

Schistosomiasis is a debilitating disease caused by a small fluke. The symptoms of the disease result from eggs shed by adult schistosomes in the human host. These adult helminths are 15 to 20 mm long, and the slender female lives permanently in a groove in the body of the male, from which is derived the name: *schistosome,* or split-body (Figure 23.27a). The union between the male and female produces a continuing supply of new eggs. Some of these eggs lodge in tissues. Defensive reactions of the human host to these foreign bodies cause local tissue damage called **granulomas** (Figure 23.28). Other eggs are excreted and enter the water to continue the cycle.

The life cycle of *Schistosoma* is depicted in Figure 23.27b. The disease is spread by human feces or urine carrying eggs of the schistosome that enter water supplies with which humans come into contact. In the developed world, sewage and water treatment minimizes the contamination of the water supply. Also, snails of certain species are essential for one stage of the life cycle of the schistosomes. They produce the cercariae that penetrate the skin of a human entering contaminated water. In most areas of the United States, a suitable host snail is not present. Therefore, even though it is estimated that schistosome eggs are being shed by many immigrants, the disease is not being propagated.

There are three primary types of schistosomiasis. The disease caused by *Schistosoma haematobium* (hē'mō-tō-bē-um), sometimes called urinary schistosomiasis, results in inflammation of the urinary bladder wall. Similarly, *S. japonicum* and *S. mansoni* cause intestinal inflammation. Depending on the species, schistosomiasis can cause damage to many different organs when eggs migrate in the bloodstream to different areas; for example, damage to the liver or lungs, urinary bladder cancer, or, when eggs lodge in the brain, neurological symptoms. Geographically, *S. japonicum* is found in east Asia. *S. haematobium* infects many people throughout Africa and the Middle East, most

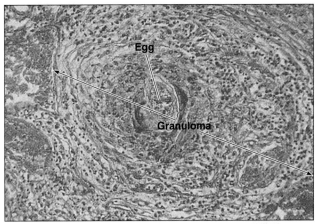

FIGURE 23.28   A granuloma from a patient with schistosomes. Some of the eggs laid by the adult schistosomes lodge in the tissue, and the body responds to the irritant by surrounding it with scarlike tissue, forming a granuloma.

Q **Why is the immune system ineffective against adult schistosomes?**

particularly Egypt. *S. mansoni* has a similar distribution but also is endemic in South America and the Caribbean, including Puerto Rico. It is estimated that more than 250 million of the world's population are affected.

The adult worms appear to be unaffected by the host's immune system. Apparently, they quickly coat themselves with a layer that mimics the host's tissues.

Laboratory diagnosis consists of microscopic identification of the flukes or their eggs in fecal and urine specimens, intradermal tests, and serological tests such as complement-fixation and precipitin tests.

Praziquantel and oxamniquine are approved for use against schistosomes in the United States. Sanitation and elimination of the host snail are also useful forms of control.

SWIMMER'S ITCH

Swimmers in lakes in the northern United States are sometimes troubled by **swimmer's itch.** This is a cutaneous allergic reaction to cercariae, similar to that of schistosomiasis. However, these parasites mature only in wildfowl and not in humans, so infection does not progress beyond penetration of the skin and a local inflammatory response.

* * *

The diseases covered in this chapter are summarized in taxonomic order in Table 23.1.

TABLE 23.1	Microbial Diseases of the Cardiovascular and Lymphatic Systems					
	Pathogen	Portal of Entry	Reservoir	Method of Transmission	Notes	Treatment
Bacterial Diseases The definitive diagnosis for most of these requires isolation of the bacteria.						
Septic shock	Gram-negative enterococci, group B streptococci	Skin	Human body	Injection; catheterization	Fever, chills, increased heart rate, lymphangitis	Xigris (gram-negatives); antibiotics (gram-positives)
Puerperal sepsis	*Streptococcus pyogenes*	Vagina	Human nasopharynx	Nosocomial	Peritonitis; sepsis	Penicillin

Endocarditis
  Subacute bacterial
  Acute bacterial
	Mostly alpha-hemolytic streptococci
	Staphylococcus aureus
		Skin; mucous membranes	Human nasopharynx	From focal infection	Fever, general weakness, heart murmur; damage to heart valves	Antibiotics

Pericarditis	*Streptococcus pyogenes*	Skin; mucous membranes	Human nasopharynx	From focal infection	Fever, general weakness, heart murmur	Antibiotics
Rheumatic fever	Group A beta-hemolytic streptococci	Probably an autoimmune condition; repeated streptococcal infections result in antibodies that damage heart tissue.			Arthritis, fever; damage to heart valves	None Prevention: penicillin to treat streptococcal sore throats
Tularemia	*Francisella tularensis*	Skin; respiratory tract	Rabbits; ground squirrels	Direct contact with infected animals; deer fly bite; inhalation	Local infection; enlarged lymph nodes; pneumonia	Tetracycline
Brucellosis	*Brucella* spp.	Mouth; skin; mucous membranes	Grazing mammals (elk)	Direct contact	Local abscess; undulating fever	Tetracycline, streptomycin
Anthrax	*Bacillus anthracis*	Mouth; skin; respiratory tract	Soil; large grazing mammals	Direct contact; ingestion; inhalation	Papule (cutaneous); bloody diarrhea (gastrointestinal); septic shock (inhalational)	Ciprofloxacin; doxycycline
Gangrene	*Clostridium perfringens*	Skin	Soil	Puncture wound	Tissue necrosis	Surgical removal of necrotic tissue
Animal bites	*Pasteurella multocida*	Skin	Animal mouths	Dog/cat bites	Local infection; sepsis	Penicillin
Rat-bite fever	*Streptobacillus moniliformis, Spirillum minor*	Skin	Rats	Rat bites	Sepsis	Penicillin
Cat-scratch disease	*Bartonella henselae*	Skin	Domestic cat	Cat bites or scratch; fleas	Prolonged fever	Antibiotics
Plague	*Yersinia pestis*	Skin; respiratory tract	Rodents	Fleas; inhalation	Enlarged lymph nodes; septic shock	Streptomycin; tetracycline

TABLE 23.1	*(continued)*					
	Pathogen	Portal of Entry	Reservoir	Method of Transmission	Notes	Treatment
Relapsing fever	*Borrelia* spp.	Skin	Rodents	Soft ticks	Series of fever peaks; diagnosed by microscopic examination of blood for spirochetes	Tetracycline
Lyme disease	*Borrelia burgdorferi*	Skin	Field mice; deer	*Ixodes* ticks	Bull's eye rash; neurologic symptoms; presence of antibodies (Western blot) is used for diagnosis	Antibiotics
Ehrlichiosis	*Ehrlichia* spp.	Skin	Deer	Ticks	Flulike symptoms; diagnosed by serology or PCR	Doxycycline
Epidemic typhus	*Rickettsia prowazekii*	Skin	Squirrels	*Pediculus humanus corporis* louse	Stupor and rash	Tetracycline; chloramphenicol
Endemic murine typhus	*Rickettsia typhi*	Skin	Rodents	*Xenopsylla cheopsis* flea	Fever, headache	Tetracycline; chloramphenicol
Rocky Mountain spotted fever	*Rickettsia rickettsii*	Skin	Ticks; small mammals	*Dermacentor* ticks	Macular rash, fever, headache; diagnosis: serology	Tetracycline, chloramphenicol Prevention: Avoid tick bites

Viral Diseases These diseases are diagnosed by the presence of antibodies.

Burkitt's lymphoma	Epstein-Barr virus (EB virus)	Unknown	Unknown	Unknown	A tumor endemic to central Africa	Surgery
Infectious mononucleosis	EB virus	Mucous membranes of mouth	Humans	Saliva	A mild disease common in young people	None
Cytomegalovirus	Cytomegalovirus	Mucous membranes	Humans	Body fluids	Mostly asymptomatic; if initial infection is acquired during pregnancy, can be damaging to fetus	Ganciclovir, fomivirsen

Viral hemorrhagic fevers (See Diseases in Focus box.)

Protozoan Diseases

Chagas' disease (American trypanosomiasis)	*Trypanosoma cruzi*	Skin	Rodents, opossums	Reduviid bug	Damages heart muscle or peristaltic movement in esophagus and colon; diagnosis with serology	Nifurtimox Prevention: insecticides

➤

TABLE 23.1	Microbial Diseases of the Cardiovascular and Lymphatic Systems *(continued)*					
	Pathogen	Portal of Entry	Reservoir	Method of Transmission	Notes	Treatment
Toxoplasmosis	*Toxoplasma gondii*	Digestive system	Domestic cats	Ingestion	A mild disease in immunocompetent adults; may cause severe fetal damage if initial infection occurs during pregnancy; reactivation in AIDS patients causes serious illness; diagnosis with PCR	Pyrimethamine sulfadiazine, and folinic acid Prevention: sanitary disposal of cat litter
Malaria	*Plasmodium* spp.	Skin	Humans	*Anopheles* mosquito	Fever and chills at intervals; diagnosis with microscopic observation	Chloroquine Prevention: mosquito control
Leishmaniasis	*Leishmania*	Skin	Small mammals	Sandfly	*L. donovani* causes systemic disease of deep body organs; *L. tropica* causes skin sores; *L. braziliensis* causes skin sores and disfiguring damage to mucous membranes of nose, mouth, and so on; diagnosis with PCR or serology	Antimony compounds
Babesiosis	*Babesia microti*	Skin	Rodents	*Ixodes* ticks	Fever and chills at intervals	Atovaquone and azithromycin

Helminthic Diseases These are diagnosed by microscopic observation of the parasite.

	Pathogen	Portal of Entry	Reservoir	Method of Transmission	Notes	Treatment
Schistosomiasis	*Schistosoma* spp.	Skin	Definitive host: humans	Cercariae penetrate skin	Eggs produced by schistosomes lodge in tissue and induce damaging inflammation	Praziquantel; oxamnequine Prevention: sanitation; elimination of host snail
Swimmer's itch	Larvae of schistosomes of nonhuman animals	Skin	Wildfowl	Cercariae penetrate skin	An allergic reaction to parasite in skin	None

STUDY OUTLINE

INTRODUCTION (p. 671)

1. The heart, blood, and blood vessels make up the cardiovascular system.
2. Lymph, lymph vessels, lymph nodes, and lymphoid organs constitute the lymphatic system.

STRUCTURE AND FUNCTION OF THE CARDIOVASCULAR AND LYMPHATIC SYSTEMS (p. 672)

1. The heart circulates substances to and from tissue cells.
2. Blood is a mixture of plasma and cells.
3. Plasma transports dissolved substances. Red blood cells carry oxygen. White blood cells are involved in the body's defense against infection.
4. Fluid that filters out of capillaries into spaces between tissue cells is called interstitial fluid.
5. Interstitial fluid enters lymph capillaries and is called lymph; vessels called lymphatics return lymph to the blood.
6. Lymph nodes contain fixed macrophages, B cells, and T cells.

BACTERIAL DISEASES OF THE CARDIOVASCULAR AND LYMPHATIC SYSTEMS (pp. 672–688)

SEPSIS AND SEPTIC SHOCK (pp. 672–675)

1. Sepsis is an inflammatory response caused by the spread of bacteria or their toxin from a focus of infection septicemia is sepsis that involves proliferation of pathogens in the blood.
2. Gram-negative sepsis can lead to septic shock, characterized by decreased blood pressure. Endotoxin causes the symptoms.
3. Antibiotic-resistant enterococci and group B streptococci cause gram-positive sepsis.
4. Puerperal sepsis begins as an infection of the uterus following childbirth or abortion; it can progress to peritonitis or septicemia.
5. *Streptococcus pyogenes* is the most frequent cause of puerperal sepsis.
6. Oliver Wendell Holmes and Ignaz Semmelweiss demonstrated that puerperal sepsis was transmitted by the hands and instruments of midwives and physicians.
7. Puerperal sepsis is now uncommon because of modern hygienic techniques and antibiotics.

BACTERIAL INFECTIONS OF THE HEART (p. 675)

8. The inner layer of the heart is the endocardium.
9. Subacute bacterial endocarditis is usually caused by alpha-hemolytic streptococci, staphylococci, or enterococci.
10. The infection arises from a focus of infection, such as a tooth extraction.
11. Preexisting heart abnormalities are predisposing factors.
12. Signs include fever, anemia, and heart murmur.
13. Acute bacterial endocarditis is usually caused by *Staphylococcus aureus*.
14. The bacteria cause rapid destruction of heart valves.

RHEUMATIC FEVER (p. 676)

15. Rheumatic fever is an autoimmune complication of streptococcal infections.
16. Rheumatic fever is expressed as arthritis or inflammation of the heart. It can result in permanent heart damage.
17. Antibodies against group A beta-hemolytic streptococci react with streptococcal antigens deposited in joints or heart valves or cross-react with the heart muscle.
18. Rheumatic fever can follow a streptococcal infection, such as streptococcal sore throat. Streptococci might not be present at the time of rheumatic fever.
19. Prompt treatment of streptococcal infections can reduce the incidence of rheumatic fever.
20. Penicillin is administered as a preventive measure against subsequent streptococcal infections.

TULAREMIA (pp. 676–678)

21. Tularemia is caused by *Francisella tularensis*. The reservoir is small wild mammals, especially rabbits.
22. Signs include ulceration at the site of entry, followed by septicemia and pneumonia.

BRUCELLOSIS (UNDULANT FEVER) (p. 678)

23. Brucellosis can be caused by *Brucella abortus, B. melitensis,* and *B. suis.*
24. In the United States, elk and bison constitute the reservoir for *B. abortus.*
25. The bacteria enter through minute breaks in the mucosa or skin, reproduce in macrophages, and spread via lymphatics to liver, spleen, or bone marrow.
26. Signs include malaise and fever that spikes each evening (undulant fever).
27. Diagnosis is based on serological tests.

ANTHRAX (pp. 679–681)

28. *Bacillus anthracis* causes anthrax. In soil, endospores can survive for up to 60 years.

29. Grazing animals acquire an infection after ingesting the endospores.

30. Humans contract anthrax by handling hides from infected animals. The endospores enter through cuts in the skin, respiratory tract, or mouth.

31. Entry through the skin results in a pustule that can progress to sepsis. Entry through the respiratory tract can result in septic shock.

32. Diagnosis is based on isolation and identification of the bacteria.

GANGRENE (pp. 681–682)

33. Soft tissue death from ischemia (loss of blood supply) is called gangrene.

34. Microorganisms grow on nutrients released from gangrenous cells.

35. Gangrene is especially susceptible to the growth of anaerobic bacteria such as *Clostridium perfringens*, the causative agent of gas gangrene.

36. *C. perfringens* can invade the wall of the uterus during improperly performed abortions.

37. Surgical removal of necrotic tissue, hyperbaric chambers, and amputation are used to treat gas gangrene.

SYSTEMIC DISEASES CAUSED BY BITES AND SCRATCHES (pp. 682–683)

38. *Pasteurella multocida*, introduced by the bite of a dog or cat, can cause septicemia.

39. Anaerobic bacteria infect deep animal bites.

40. Rat-bite fever is caused by *Streptobacillus moniliformis*.

41. Cat-scratch disease is caused by *Bartonella henselae*.

VECTOR-TRANSMITTED DISEASES (pp. 683–688)

Plague (pp. 683–684)

42. Plague is caused by *Yersinia pestis*. The vector is usually the rat flea (*Xenopsylla cheopis*).

43. Reservoirs for bubonic plague include European rats and North American rodents.

44. Signs of bubonic plague include bruises on the skin and enlarged lymph nodes (buboes).

45. The bacteria can enter the lungs and cause pneumonic plague.

46. Laboratory diagnosis is based on isolation and identification of the bacteria.

47. Antibiotics are effective in treating plague, but they must be administered promptly after exposure to the disease.

Relapsing Fever (p. 685)

48. Relapsing fever is caused by *Borrelia* species and transmitted by soft ticks.

49. The reservoir for the disease is rodents.

50. Signs include fever, jaundice, and rose-colored spots. Signs recur three or four times after apparent recovery.

51. Laboratory diagnosis is based on the presence of spirochetes in the patient's blood.

Lyme Disease (Lyme Borreliosis) (pp. 685–687)

52. Lyme disease is caused by *Borrelia burgdorferi* and is transmitted by a tick (*Ixodes*).

53. Field mice provide the animal reservoir.

54. Diagnosis is based on serological tests and clinical symptoms.

Ehrlichiosis (p. 687)

55. Human ehrlichioses (HGE and HME) are caused by *Ehrlichia* species and transmitted by *Ixodes* ticks.

Typhus (pp. 687–688)

56. Typhus is caused by rickettsias, obligate intracellular parasites of eukaryotic cells.

Epidemic Typhus (pp. 687–688)

57. The human body louse *Pediculus humanus corporis* transmits *Rickettsia prowazekii* in its feces, which are deposited while the louse is feeding.

58. Epidemic typhus is prevalent in crowded and unsanitary living conditions that allow the proliferation of lice.

59. The signs of typhus are rash, prolonged high fever, and stupor.

60. Tetracyclines and chloramphenicol are used in treatment.

Endemic Murine Typhus (p. 688)

61. Endemic murine typhus is a less severe disease caused by *Rickettsia typhi* and transmitted from rodents to humans by the rat flea.

Spotted Fevers (p. 688)

62. *Rickettsia rickettsii* is a parasite of ticks (*Dermacentor* spp.) in the southeastern United States, Appalachia, and the Rocky Mountain states.

63. The rickettsia may be transmitted to humans, in whom it causes tickborne typhus fever.

64. Chloramphenicol and tetracyclines effectively treat Rocky Mountain spotted fever, or tickborne typhus.

65. Serological tests are used for laboratory diagnosis.

VIRAL DISEASES OF THE CARDIOVASCULAR AND LYMPHATIC SYSTEMS   (pp. 688–693)

BURKITT'S LYMPHOMA   (pp. 688–690)

1. Epstein-Barr virus (EB virus, HHV-4) causes Burkitt's lymphoma.
2. Burkitt's lymphoma tends to occur in patients whose immune system has been weakened; for example, by malaria or AIDS.

INFECTIOUS MONONUCLEOSIS   (pp. 690–691)

3. Infectious mononucleosis is caused by EB virus.
4. The virus multiplies in the parotid glands and is present in saliva. It causes the proliferation of atypical lymphocytes.
5. The disease is transmitted by the ingestion of saliva from infected individuals.
6. Diagnosis is made by an indirect fluorescent-antibody technique.
7. EB virus may cause other diseases, including cancers and multiple sclerosis.

CYTOMEGALOVIRUS INFECTIONS   (p. 691)

8. CMV (HHV-5) causes intranuclear inclusion bodies and cytomegaly of host cells.
9. CMV is transmitted by saliva and other body fluids.
10. CMV inclusion disease can be asymptomatic, a mild disease, or progressive and fatal. Immunosuppressed patients may develop pneumonia.
11. If the virus crosses the placenta, it can cause congenital infection of the fetus, resulting in impaired mental development, neurological damage, and stillbirth.

CLASSIC VIRAL HEMORRHAGIC FEVERS   (pp. 691–692)

12. Yellow fever is caused by the yellow fever virus. The vector is the *Aedes aegypti* mosquito.
13. Signs and symptoms include fever, chills, headache, nausea, and jaundice.
14. Diagnosis is based on the presence of virus-neutralizing antibodies in the host.
15. No treatment is available, but there is an attenuated, live viral vaccine.
16. Dengue is caused by the dengue fever virus and is transmitted by the *Aedes aegypti* mosquito.
17. Signs are fever, muscle and joint pain, and rash.
18. Mosquito abatement is necessary to control the disease.
19. Dengue hemorrhagic fever (DHF) can cause shock.

EMERGING VIRAL HEMORRHAGIC FEVERS   (pp. 692–693)

20. Human diseases caused by Marburg, Ebola, and Lassa fever viruses were first noticed in the late 1960s.
21. Ebola virus is found in fruit bats; Lassa fever viruses are found in rodents.
22. Rodents are the reservoirs for Argentine and Bolivian hemorrhagic fevers.
23. *Hantavirus* pulmonary syndrome and hemorrhagic fever with renal syndrome are caused by hantavirus. The virus is contracted by inhalation of dried rodent urine and feces.

PROTOZOAN DISEASES OF THE CARDIOVASCULAR AND LYMPHATIC SYSTEMS   (pp. 693–699)

CHAGAS' DISEASE (AMERICAN TRYPANOSOMIASIS)   (pp. 693–695)

1. *Trypanosoma cruzi* causes Chagas' disease. The reservoir includes many wild animals. The vector is a reduviid, the "kissing bug."
2. PCR confirms the diagnosis.

TOXOPLASMOSIS   (pp. 695–696)

3. Toxoplasmosis is caused by *Toxoplasma gondii*.
4. *T. gondii* undergoes sexual reproduction in the intestinal tract of domestic cats, and oocysts are eliminated in cat feces.
5. In the host cell, sporozoites reproduce to form either tissue-invading tachyzoites or bradyzoites.
6. Humans contract the infection by ingesting tachyzoites or tissue cysts in undercooked meat from an infected animal or contact with cat feces.
7. Congenital infections can occur. Signs and symptoms include severe brain damage or vision problems.
8. Toxoplasmosis can be identified by serological tests, but interpretation of the results is uncertain.

MALARIA   (pp. 696–698)

9. The signs and symptoms of malaria are chills, fever, vomiting, and headache, which occur at intervals of 2 to 3 days.
10. Malaria is transmitted by *Anopheles* mosquitoes. The causative agent is any one of four species of *Plasmodium*.
11. Sporozoites reproduce in the liver and release merozoites into the bloodstream, where they infect red blood cells and produce more merozoites.
12. New drugs are being developed as the protozoa develop resistance to drugs such as chloroquine.

LEISHMANIASIS (pp. 698–699)

13. *Leishmania* spp., which are transmitted by sandflies, cause leishmaniasis.
14. The protozoa reproduce in the liver, spleen, and kidneys.
15. Antimony compounds are used for treatment.

BABESIOSIS (p. 699)

16. Babesiosis is caused by the protozoan *Babesia microti* and transmitted to humans by ticks.

HELMINTHIC DISEASES OF THE CARDIOVASCULAR AND LYMPHATIC SYSTEMS (p. 701)

SCHISTOSOMIASIS (p. 701)

1. Species of the blood fluke *Schistosoma* cause schistosomiasis.
2. Eggs eliminated with feces hatch into larvae that infect the intermediate host, a snail. Free-swimming cercariae

are released from the snail and penetrate the skin of a human.
3. The adult flukes live in the veins of the liver or urinary bladder in humans.
4. Granulomas are from the host's defense to eggs that remain in the body.
5. Observation of eggs or flukes in feces, skin tests, or indirect serological tests may be used for diagnosis.
6. Chemotherapy is used to treat the disease; sanitation and snail eradication are used to prevent it.

SWIMMER'S ITCH (p. 701)

7. Swimmer's itch is a cutaneous allergic reaction to cercariae that penetrate the skin. The definitive hosts for this fluke are wildfowl.

STUDY QUESTIONS

Access more review material either online at **The Microbiology Place** (www.microbiologyplace.com) or with **The Microbiology Place CD-ROM** packaged with your new book. There you'll find activities, practice tests, quizzes, flashcards, case studies, and more to help you succeed.

Answers to the Study Questions can be found in Appendix G.

REVIEW

1. What are the signs of sepsis? What is septic shock?
2. How can sepsis result from a single focus of infection, such as an abscess?
3. Complete the following table.

Disease	Frequent Causative Agent	Predisposing Condition(s)
Puerperal sepsis		
Subacute bacterial endocarditis		
Acute bacterial endocarditis		

4. Describe the probable cause of rheumatic fever. How is rheumatic fever treated? How is it prevented?
5. Compare and contrast epidemic typhus, endemic murine typhus, and tickborne typhus.

6. Complete the following table.

Disease	Causative Agent	Signs/ Vector	Symptoms	Treatment
Malaria				
Yellow fever				
Dengue				
Relapsing fever				
Leishmaniasis				

7. Complete the following table.

Disease	Causative Agent	Method of Transmission	Reservoir	Signs/ Symptoms	Prevention
Tularemia					
Brucellosis					
Anthrax					
Lyme disease					
Ehrlichiosis					
Cytomegalic inclusion disease					

8. Provide the following information on plague: causative agent, vector, U.S. reservoir, control, treatment, prognosis (probable outcome).

9. List the causative agent, method of transmission, and reservoir for schistosomiasis, toxoplasmosis, and Chagas' disease. Which disease are you most likely to get in the United States? Where are the other diseases endemic?

10. Compare and contrast cat-scratch disease and toxoplasmosis.

11. Why is *Clostridium perfringens* likely to grow in gangrenous wounds?

12. List the causative agents and methods of transmission of infectious mononucleosis.

13. Differentiate between the transmission and symptoms of bubonic plague and pneumonic plague.

MULTIPLE CHOICE

Use the following choices to answer questions 1 through 4:
- **a.** ehrlichiosis
- **b.** Lyme disease
- **c.** septic shock
- **d.** toxoplasmosis
- **e.** viral hemorrhagic fever

1. A patient presents with vomiting, diarrhea, and a history of fever and headache. Bacterial cultures of blood, CSF, and stool are negative. What is your diagnosis?

2. A patient was hospitalized because of continuing fever and progression of symptoms including headache, fatigue, and back pain. Tests for antibodies to *Borrelia burgdorferi* were negative. What is your diagnosis?

3. A patient complained of headache. A CT (computed tomography) scan revealed cysts of varying size in her brain. What is your diagnosis?

4. A patient presents with mental confusion, rapid breathing and heartbeat, and low blood pressure. What is your diagnosis?

5. A patient has a red circular rash on his arm and fever, malaise, and joint pain. The most appropriate treatment is
- **a.** penicillin.
- **b.** chloroquine.
- **c.** anti-inflammatory drugs.
- **d.** rifampin.
- **e.** no treatment.

6. Which of the following is not a tickborne disease?
- **a.** babesiosis
- **b.** ehrlichiosis
- **c.** Lyme disease
- **d.** relapsing fever
- **e.** tularemia

Use the following choices to answer questions 7 and 8:
- **a.** brucellosis
- **b.** malaria
- **c.** relapsing fever
- **d.** Rocky Mountain spotted fever
- **e.** Ebola hemorrhagic fever

7. The patient's fever spikes each evening. Oxidase-positive, gram-negative cocci were isolated from a lesion on his arm. What is your diagnosis?

8. The patient was hospitalized with fever and headache. Spirochetes were observed in her blood. What is your diagnosis?

9. Which of the following diseases has the highest incidence in the United States?
- **a.** brucellosis
- **b.** Ebola hemorrhagic fever
- **c.** malaria
- **d.** plague
- **e.** Rocky Mountain spotted fever

10. Nineteen workers in a slaughterhouse developed fever and chills, with the fever spiking to 40°C each evening. The most likely method of transmission of this disease is
- **a.** a vector.
- **b.** the respiratory route.
- **c.** a puncture wound.
- **d.** an animal bite.
- **e.** water.

CRITICAL THINKING

1. Indirect FA tests on the serum of three 25-year-old women, each of whom is considering pregnancy, provided the information below. Which of these women may have toxoplasmosis? What advice might be given to each woman with regard to toxoplasmosis?

Patient	Antibody Titer Day 1	Day 5	Day 12
Patient A	1024	1024	1024
Patient B	1024	2048	3072
Patient C	0	0	0

2. What is the most effective way to control malaria and dengue?

3. In adults, the second dengue virus infection results in dengue hemorrhagic fever (DHF), which is characterized by bleeding from the skin and mucosa. DHF can be fatal. In infants under 1 year old, the first dengue virus infection results in DHF. Offer an explanation for this.

CLINICAL APPLICATIONS

1. A 19-year-old man went deer hunting. While on the trail, he found a partially dismembered dead rabbit. The hunter picked up the front paws for good luck charms and gave them to another hunter in the party. The rabbit had been handled with bare hands that were bruised and scratched from the hunter's work as an automobile mechanic. Festering sores on his hands, legs, and knees were noted two days later. What infectious disease do you suspect the hunter has? How would you proceed to prove it?

2. On March 30, a 35-year-old veterinarian experienced fever, chills, and vomiting. On March 31, he was hospitalized with diarrhea, left armpit bubo, and secondary bilateral pneumonia. On March 27, he had treated a cat that had labored respiration; an X ray revealed pulmonary infiltrates. The cat died on March 28 and was disposed of. Chloramphenicol was administered to the veterinarian. On April 10, his temperature returned to normal, and on April 20, he

was released from the hospital. Sixty human contacts were given tetracycline. Identify the incubation and prodromal periods for this case. Explain why the 60 contacts were treated. What was the etiologic agent? How would you identify the agent?

3. Three of five patients who underwent heart valve replacement surgery developed bacteremia. The causative agent was *Enterobacter cloacae*. What were the patients' signs and symptoms? How would you identify this bacterium? A manometer used in the operations was culture-positive for *E. cloacae*. What is the most likely source of this contaminant? Suggest a way of preventing such occurrences.

4. In August and September, six people who each at different times spent a night in the same cabin developed the symptoms shown in the graph at the right. Three recovered after tetracycline (TET) therapy, two recovered without therapy, and one was hospitalized with septic shock. What is the disease? What is the incubation period of this disease? How do you account for the periodic temperature changes? What caused septic shock in the sixth patient?

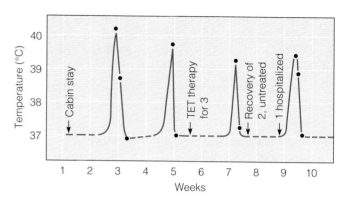

5. A 67-year-old man worked in a textile mill that processed imported goat hair into fabrics. He noticed a painless, slightly swollen pimple on his chin. Two days later he developed a 1-cm ulcer at the pimple site and a temperature of 37.6°C. He was treated with tetracycline. What is the etiology of this disease? Suggest ways to prevent it.

24 Microbial Diseases of the Respiratory System

With every breath, we inhale several microorganisms; therefore, the upper respiratory system is a major portal of entry for pathogens. In fact, respiratory system infections are the most common type of infection—and among the most damaging. Some pathogens that enter via the respiratory route can infect other parts of the body, such as those that cause measles, mumps, and rubella.

The upper respiratory system has several anatomical defenses against airborne pathogens. Coarse hairs in the nose filter large dust particles from the air. The nose is lined with a mucous membrane that contains numerous mucus-secreting cells and cilia. The upper portion of the throat also contains a ciliated mucous membrane. The mucus moistens inhaled air and traps dust and microorganisms. The cilia help remove these particles by moving them toward the mouth for elimination.

At the junction of the nose and throat are masses of lymphoid tissue, the tonsils, which contribute immunity to certain infections. Because the nose and throat are connected to the sinuses, nasolacrimal apparatus, and middle ear, infections commonly spread from one region to another.

UNDER THE MICROSCOPE

Bordetella pertussis cells (orange), the cause of whooping cough, or pertussis, are shown growing on ciliated cells of the respiratory system.

STRUCTURE AND FUNCTION OF THE RESPIRATORY SYSTEM

LEARNING OBJECTIVE

* Describe how microorganisms are prevented from entering the respiratory system.

It is convenient to think of the respiratory system as being composed of two divisions: the upper respiratory system and the lower respiratory system. The **upper respiratory system** consists of the nose, the pharynx (throat), and the structures associated with them, including the middle ear and the auditory (eustachian) tubes (Figure 24.1). Ducts from the sinuses and the nasolacrimal ducts from the lacrimal (tear-forming) apparatus empty into the nasal cavity (see Figure 16.3, page 477). The auditory tubes from the middle ear empty into the upper portion of the throat.

The **lower respiratory system** consists of the larynx (voice box), trachea (windpipe), bronchial tubes, and *alveoli* (Figure 24.2). Alveoli are air sacs that make up the lung tissue; within them, oxygen and carbon dioxide are exchanged between the lungs and blood. Our lungs contain more than 300 million alveoli, with an area for gas exchange of 70 or more square meters in an average adult.

The double-layered membrane enclosing the lungs is the *pleura*, or pleural membranes. A ciliated mucous membrane lines the lower respiratory system down to the smaller bronchial tubes and helps prevent microorganisms from reaching the lungs.

As discussed in Chapter 16, particles trapped in the larynx, trachea, and larger bronchial tubes are moved up toward the throat by a ciliary action called the *ciliary escalator* (see Figure 16.4, page 478). If microorganisms actually reach the lungs, phagocytic cells called *alveolar macrophages* usually locate, ingest, and destroy most of them. IgA antibodies in such secretions as respiratory mucus, saliva, and tears also help protect mucosal surfaces of the respiratory system from many pathogens. Thus, the body has several mechanisms for removing the pathogens that cause airborne infections.

NORMAL MICROBIOTA OF THE RESPIRATORY SYSTEM

LEARNING OBJECTIVE

* Characterize the normal microbiota of the upper and lower respiratory systems.

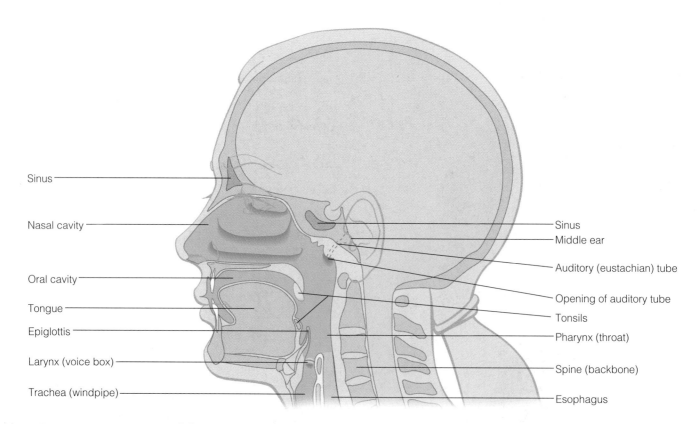

FIGURE 24.1 Structures of the upper respiratory system.

Labels: Sinus; Nasal cavity; Oral cavity; Tongue; Epiglottis; Larynx (voice box); Trachea (windpipe); Sinus; Middle ear; Auditory (eustachian) tube; Opening of auditory tube; Tonsils; Pharynx (throat); Spine (backbone); Esophagus

Q Name the upper respiratory system's defenses against disease.

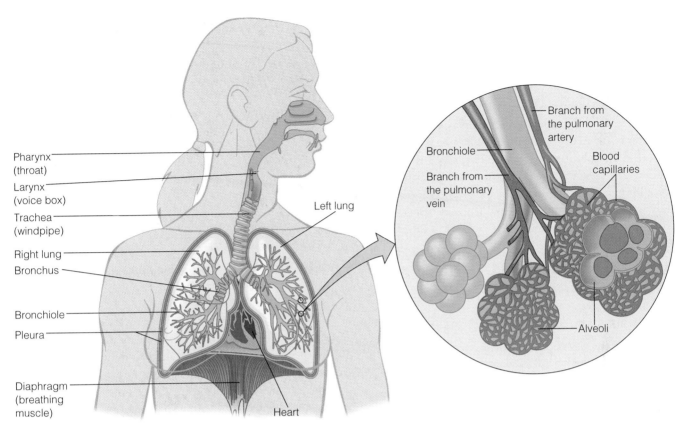

FIGURE 24.2 Structures of the lower respiratory system.

Q Name the lower respiratory system's defenses against disease.

A number of potentially pathogenic microorganisms are part of the normal microbiota in the upper respiratory system. However, they usually do not cause illness because the predominant microorganisms of the normal microbiota suppress their growth by competing with them for nutrients and producing inhibitory substances.

By contrast, the lower respiratory tract is nearly sterile— although the trachea may contain a few bacteria—because of the normally efficient functioning of the ciliary escalator in the bronchial tubes.

MICROBIAL DISEASES OF THE UPPER RESPIRATORY SYSTEM

LEARNING OBJECTIVE
- Differentiate among pharyngitis, laryngitis, tonsillitis, sinusitis, and epiglottitis.

As most of us know from personal experience, the respiratory system is the site of many common infections. We will soon discuss **pharyngitis,** inflammation of the mucous membranes of the throat, or sore throat. When the larynx is the site of infection, we suffer from **laryngitis,** which affects our ability to speak. The microbes that cause pharyngitis also can cause inflamed tonsils, or **tonsillitis.**

The nasal sinuses are cavities in certain cranial bones that open into the nasal cavity. They have a mucous membrane lining that is continuous with that of the nasal cavity. When a sinus becomes infected and there is a heavy nasal discharge of mucus, it is called **sinusitis.** If the opening by which the mucus leaves the sinus becomes blocked, internal pressure can cause pain or a sinus headache. These diseases are almost always *self-limiting,* meaning that recovery will usually occur even without medical intervention.

Probably the most threatening infectious disease of the upper respiratory system is **epiglottitis,** inflammation of the epiglottis. The epiglottis is a flaplike structure of cartilage that prevents ingested material from entering the larynx (see Figure 24.1). Epiglottitis is a rapidly developing disease that can result in death within a few hours. It is caused by opportunistic pathogens, usually *Haemophilus influenzae* type b. The newly introduced Hib vaccine, although directed primarily at meningitis (see Figure 22.3, page 645), has significantly reduced the incidence of epiglottitis in the vaccinated population.

BACTERIAL DISEASES OF THE UPPER RESPIRATORY SYSTEM

> **LEARNING OBJECTIVE**
> - List the causative agent, symptoms, prevention, preferred treatment, and laboratory identification tests for streptococcal pharyngitis, scarlet fever, diphtheria, cutaneous diphtheria, and otitis media.

Airborne pathogens make their first contact with the body's mucous membranes as they enter the upper respiratory system. Many respiratory or systemic diseases initiate infections here.

STREPTOCOCCAL PHARYNGITIS (STREP THROAT)

Streptococcal pharyngitis (strep throat) is an upper respiratory infection caused by group A streptococci (GAS). This gram-positive bacterial group consists solely of *Streptococcus pyogenes,* the same bacterium responsible for many skin and soft tissue infections, such as impetigo, erysipelas, and acute bacterial endocarditis.

The pathogenicity of GAS is enhanced by their resistance to phagocytosis. They are also able to produce special enzymes, called *streptokinases,* which lyse fibrin clots, and *streptolysins,* which are cytotoxic to tissue cells, red blood cells, and protective leukocytes.

At one time, the diagnosis of pharyngitis was based on culturing bacteria from a throat swab. Results took overnight or longer, but, beginning in the early 1980s, rapid antigen detection tests that were capable of detecting GAS directly on throat swabs became available. The first rapid tests used latex indirect agglutination methods (see Figure 18.7, page 539). These have been generally replaced by **enzyme immunoassay (EIA)** tests that are more sensitive and easier to read. Currently, there are a wide range of rapid tests commercially available to evaluate cases of pharyngitis, which reflects the

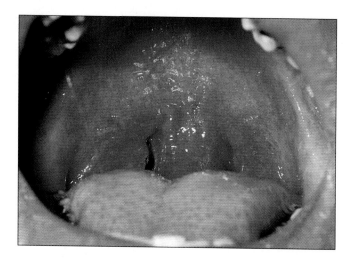

FIGURE 24.3 Streptococcal pharyngitis.

Q **How is strep throat diagnosed?**

fact that millions of patients seek care for it every year. Actually, the majority of patients seen for sore throats do not have a streptococcal infection. Some cases are caused by other bacteria, but many are caused by viruses—for which antibiotic therapy is ineffective. Even the presence of GAS is not a conclusive indication that it is responsible for the sore throat. In areas in which cases of acute rheumatic fever occur, the recommendation is to use both bacterial culture and rapid tests. Fortunately, GAS have remained sensitive to penicillin, although some resistance to erythromycin has appeared.

Pharyngitis is characterized by local inflammation and a fever (Figure 24.3). Frequently, tonsillitis occurs, and the lymph nodes in the neck become enlarged and tender. Another frequent complication is otitis media (infection of the middle ear).

Pharyngitis is now most commonly transmitted by respiratory secretions, but epidemics of streptococcal pharyngitis spread by unpasteurized milk were once frequent.

SCARLET FEVER

When the *Streptococcus pyogenes* strain causing streptococcal pharyngitis produces an *erythrogenic* (reddening) *toxin,* the resulting infection is called **scarlet fever.** When the strain produces this toxin, it has been lysogenized by a bacteriophage (see Figure 13.12, page 399). Recall that this means the genetic information of a bacteriophage (bacterial virus) has been incorporated into the chromosome of the bacterium, so the characteristics of the bacterium have been altered. The toxin causes a pinkish red skin rash, which is probably the skin's hypersensitivity reaction to the circulating toxin, and a high fever. The tongue has a spotted, strawberrylike appearance and then, as it loses its upper membrane, becomes very red

FIGURE 24.4 Scarlet fever rash. One of the characteristic signs of scarlet fever, along with a skin rash, is the "strawberry tongue" shown on this patient.

Q **Compare strep throat and scarlet fever.**

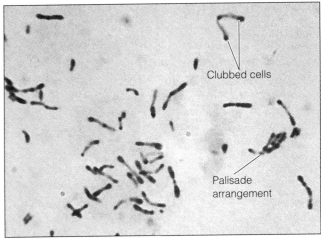

LM |——| 5 μm

FIGURE 24.5 *Corynebacterium diphtheriae*, the cause of diphtheria. This Gram stain shows the club-shaped morphology; the dividing cells are often observed to fold together to form V- and Y-shaped figures. Also notice the side-by-side palisade arrangement.

Q **Are corynebacteria gram-positive or gram-negative?**

and enlarged (Figure 24.4). As the disease runs its course, the affected skin frequently peels off, as if sunburned, very similar to the scalded skin syndrome caused by *Staphylococcus aureus* (see Figure 21.4, page 620). Classically, scarlet fever has been considered to be associated with streptococcal pharyngitis, but it might accompany a streptococcal skin infection.

The incidence of scarlet fever has varied over time in severity and frequency. Between about 1830 and 1880, scarlet fever was a pandemic that was probably the most common cause of death in children, causing more fatalities than measles, diphtheria, or whooping cough. After about 1880, scarlet fever declined steadily in incidence and severity, long before antibiotics were available. Today it is a relatively mild and rare disease. The most likely explanation offered for the pandemic of the 1800s and its decline is spontaneous variations in the virulence of the streptococci. Also, the time of greatest incidence was related to the development of railroad networks that resulted in an unprecedented mixing of populations.

DIPHTHERIA

Another bacterial infection of the upper respiratory system is **diphtheria.** Until 1935, it was the leading infectious killer of children in the United States. The disease begins with a sore throat and fever, followed by general malaise and swelling of the neck. The organism responsible is *Corynebacterium diphtheriae*, a gram-positive, non–endospore-forming rod. Its morphology is pleomorphic, frequently club-shaped, and it stains unevenly (Figure 24.5).

Part of the normal immunization program for children in the United States is the **DTaP vaccine.** The D stands for diphtheria toxoid, an inactivated toxin that causes the body to produce antibodies against the diphtheria toxin.

C. diphtheriae has adapted to a generally immunized population, and relatively nonvirulent strains are found in the throats of many symptomless carriers. The bacterium is well suited to airborne transmission and is very resistant to drying.

Characteristic of diphtheria (from the Greek word for leather) is a tough grayish membrane that forms in the throat in response to the infection (Figure 24.6). It contains

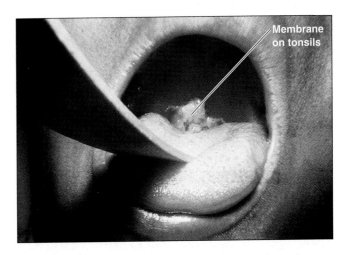

FIGURE 24.6 A diphtheria membrane. In small children, this leatherlike membrane and accompanying swelling of the breathing passages can block the air supply.

Q **What is cutaneous diphtheria?**

fibrin, dead tissue, and bacterial cells and can totally block the passage of air to the lungs.

Although the bacteria do not invade tissues, those that have been lysogenized by a phage can produce a powerful exotoxin. Historically, it was the first disease for which a toxic cause was identified. Circulating in the bloodstream, the toxin interferes with protein synthesis. Only 0.01 mg of this highly virulent toxin is enough to kill a 91-kg (200-lb) person. Thus, if antitoxin therapy is to be effective, it must be administered before the toxin enters the tissue cells. When such organs as the heart and kidneys are affected by the toxin, the disease can rapidly be fatal. In other cases the nerves can be involved, and partial paralysis results.

Laboratory diagnosis by bacterial identification is difficult, requiring several selective and differential media. Identification is complicated by the need to differentiate between toxin-forming isolates and strains that are not toxigenic; both may be found in the same patient.

Even though antibiotics such as penicillin and erythromycin control the growth of the bacteria, they do not neutralize the diphtheria toxin. Thus antibiotics should be used only in conjunction with antitoxin.

The number of diphtheria cases reported in the United States each year is presently five or fewer. In young children, the disease occurs mainly in groups that have not been immunized for religious or other reasons. When diphtheria was more common, repeated contacts with toxigenic strains reinforced the immunity, which otherwise weakens with time. Many adults now lack immunity because routine immunization was less available during their childhood. Some surveys indicate effective immune levels in as few as 20% of the adult population. As an example of what might occur if immunization programs were not maintained, waning levels of immunity in almost all of the countries of the former Soviet Union led to a recent epidemic. In the United States, when any trauma in adults requires tetanus toxoid, it is usually combined with diphtheria toxoid (Td vaccine).

Diphtheria is also expressed as **cutaneous diphtheria.** In this form of the disease, *C. diphtheriae* infects the skin, usually at a wound or similar skin lesion, and there is minimal systemic circulation of the toxin. In cutaneous infections, the bacteria cause slow-healing ulcerations covered by a gray membrane. Cutaneous diphtheria is fairly common in tropical countries. In the United States, it occurs mostly among Native Americans and in adults of low socioeconomic status. It is responsible for most of the reported cases of diphtheria in people over age 30.

In the past, diphtheria was spread mainly to healthy carriers by droplet infection. Respiratory cases have been known to arise from contact with cutaneous diphtheria.

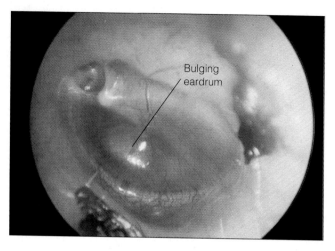

Bulging eardrum

FIGURE 24.7 Acute otitis media, with bulging eardrum.

Q What is the most common bacterium causing middle ear infections?

OTITIS MEDIA

One of the more uncomfortable complications of the common cold, or of any infection of the nose or throat, is infection of the middle ear, **otitis media,** or earache. The pathogens cause the formation of pus, which builds up pressure against the eardrum and causes it to become inflamed and painful (Figure 24.7). The condition is most frequent in early childhood because the auditory tube connecting the middle ear to the throat is small and more horizontal than in adults so it is more easily blocked by infection (see Figure 24.1).

A number of bacteria can cause otitis media. The most commonly isolated pathogen is S. *pneumoniae* (about 35% of cases). Other bacteria frequently involved are nonencapsulated *H. influenzae* (20–30%), *Moraxella catarrhalis* (mô-raks-el'lä ka-tär'al-is) (10–15%), S. *pyogenes* (8–10%), and S. *aureus* (1–2%). In about 3 to 5% of cases, no bacteria can be detected. Viral infections may be responsible in these instances; respiratory syncytial viruses (see page 730) are the most common isolate.

Otitis media affects 85% of children before the age of 3 and accounts for nearly half of office visits to pediatricians—an estimated 8 million cases each year in the United States. Treatment always assumes that bacteria are the cause, and it is estimated that ear infections account for about one-fourth of the prescriptions for antibiotics. Broad-spectrum penicillins, such as amoxicillin, are usually the first choice for children. Many physicians now question the value of antibiotics, uncertain whether these drugs shorten the course of the disease. A conjugate vaccine exists that is intended to prevent pneumonia caused by S. *pneumoniae*.

Because this organism is also a common cause of otitis media, even though different serotypes are often responsible, there was an interest in seeing whether this vaccine has any effect on the incidence of otitis media. Experience so far has shown that the vaccine does have the welcome side effect of reducing the incidence of otitis media by 6 to 7%. This reduction may not sound like much, but it amounts to over a million fewer cases every year. Therefore, the *S. pneumoniae* conjugate vaccine has now been added to the recommended vaccine schedule for children.

VIRAL DISEASES OF THE UPPER RESPIRATORY SYSTEM

LEARNING OBJECTIVE
- List the causative agents and treatments for the common cold.

Probably the most prevalent disease of humans, at least those living in the temperate zones, is a viral disease affecting the upper respiratory system—the common cold.

THE COMMON COLD

A number of different viruses are involved in the etiology of the **common cold.** About 50% of all colds are caused by rhinoviruses. Coronaviruses probably cause another 15 to 20%. About 10% of all colds are caused by one of several other viruses. In about 40% of cases, no causative agent can be identified.

We tend to accumulate immunities against cold viruses during our lifetime, which may be a reason why older people tend to get fewer colds. Immunity is based on the ratio of IgA antibodies to single serotypes and has a reasonably high short-term effectiveness. Isolated populations may develop a group immunity, and their colds disappear until a new set of viruses is introduced. Altogether, probably more than 200 agents cause the common cold. There are at least 113 serotypes of rhinoviruses alone, so a vaccine effective against so many different pathogens does not seem practical.

The symptoms of the common cold are familiar to all of us. They include sneezing, excessive nasal secretion, and congestion. (In ancient times, one school of medical thought held that the nasal discharges were waste products from the brain.) The infection can easily spread from the throat to the sinuses, the lower respiratory system, and the middle ear, leading to complications of laryngitis and otitis media. The uncomplicated cold usually is not accompanied by fever.

Rhinoviruses thrive at a temperature slightly below that of normal body temperature, such as might be found in the upper respiratory system, which is open to the outside environment. No one knows exactly why the number of colds seems to increase with colder weather in temperate zones. It is not known whether closer indoor contact promotes epidemic-type transmission or whether physiological changes increase susceptibility.

A single rhinovirus deposited on the nasal mucosa is often sufficient to cause a cold. However, there is surprisingly little agreement on how the cold virus is transmitted to a site in the nose. One line of experimentation tends to show that cold sufferers deposit the viruses on doorknobs, telephones, and other surfaces, where they remain viable for hours. For whatever reasons, they do not survive nearly as long on tissues or cotton handkerchiefs. Healthy people can transfer these viruses to their hands and then to their nasal passages or eyes, from which the nasal passages are readily reached. This theory was supported by an experiment in which healthy people who used virucidal iodine solutions on their hands had a much reduced incidence of colds.

Another series of experiments involving a group of card players, half with colds and half without colds, supported a different conclusion. Half of the healthy players were restrained so that they could not use their hands to transfer to their noses viruses picked up from the playing cards, and the other half were not restrained. The players who could not touch their noses came down with as many colds as those who could—an argument for airborne transmission. Even when healthy participants in card games were placed in a room separate from anyone suffering from a cold and isolated from their airborne secretions, but handled cards literally soaked in nasal secretions, none developed colds. In a perhaps less repellent series of experiments, researchers required healthy volunteers to kiss cold sufferers for 60 to 90 seconds; only 8% of the volunteers came down with colds.

Because colds are caused by viruses, antibiotics are of no use in treatment. Recovery time is not reported to be affected by nonprescription drugs, such as zinc lozenges or vitamin C. Symptoms can be relieved by cough suppressants and antihistamines, but these medications do not speed recovery. There is still considerable truth in the medical adage that an untreated cold will run its normal course to recovery in a week, whereas with treatment it will take 7 days.

Promising new approaches to shortening the duration of the common cold are, however, being tested. Almost all rhinoviruses, which are among the more common cold-causing viruses, use the same receptor protein on the host's cells to attach and infect the cells lining the nasal passage. Improved insight into the host–virus attachment mechanisms is considered the most likely key to successful therapy for colds.

The diseases affecting the upper respiratory system are summarized in Table 24.1.

TABLE 24.1	**Microbial Diseases of the Upper Respiratory System**		
	Pathogen	Notes	Treatment
Bacterial Diseases			
Epiglottitis	*Haemophilus influenzae*	Inflammation of the epiglottis	Antibiotics; maintain airway Prevention: Hib vaccine
Streptococcal pharyngitis (strep throat)	Streptococci, especially *Streptococcus pyogenes*	Inflamed mucous membranes of the throat; diagnosis by EIA	Penicillin
Scarlet fever	Erythrogenic toxin-producing strains of *Streptococcus pyogenes*	Streptococcal exotoxin causes skin and tongue reddening, peeling of affected skin	Penicillin
Diphtheria	*Corynebacterium diphtheriae*	Bacterial exotoxin interferes with protein synthesis; damages heart, kidneys, and other organs; membrane forms in throat; cutaneous form also occurs; diagnosis by culturing bacteria	Penicillin and antitoxin Prevention: DTaP vaccine
Otitis media	Several agents, especially *Staphylococcus aureus*, *Streptococcus pneumoniae*, and *Haemophilus influenzae*	Accumulations of pus in middle ear build up painful pressure on eardrum	Broad-spectrum antibiotics
Viral Diseases			
Common cold	Coronaviruses, rhinoviruses	Familiar symptoms of coughing, sneezing, runny nose	None

MICROBIAL DISEASES OF THE LOWER RESPIRATORY SYSTEM

The lower respiratory system can be infected by many of the same bacteria and viruses that infect the upper respiratory system. As the bronchi become involved, **bronchitis** or **bronchiolitis** develops (see Figure 24.2). A severe complication of bronchitis is **pneumonia,** in which the pulmonary alveoli become involved.

BACTERIAL DISEASES OF THE LOWER RESPIRATORY SYSTEM

LEARNING OBJECTIVE

- List the causative agent, symptoms, prevention, preferred treatment, and laboratory identification tests for pertussis and tuberculosis.

Bacterial diseases of the lower respiratory system include tuberculosis and the many types of pneumonia caused by

bacteria. Less well-known diseases such as psittacosis and Q fever also fall into this category.

PERTUSSIS (WHOOPING COUGH)

Infection by the bacterium *Bordetella pertussis* results in **pertussis,** or **whooping cough.** *B. pertussis* is a small, obligately aerobic gram-negative coccobacillus. The virulent strains possess a capsule. The bacteria attach specifically to ciliated cells in the trachea, first impeding their ciliary action and then progressively destroying the cells (Figure 24.8). This prevents the movement of mucus by the ciliary escalator system. *B. pertussis* produces several toxins. *Tracheal cytotoxin,* a fixed cell wall fraction of the bacterium, is responsible for damage to the ciliated cells, and *pertussis toxin* enters the bloodstream and is associated with systemic symptoms of the disease.

Primarily a childhood disease, pertussis can be quite severe. The initial stage, called the *catarrhal stage*, resembles a common cold. Prolonged sieges of coughing characterize the *paroxysmal stage*, or second stage. (The name *pertussis* is derived from the Latin *per*, meaning thoroughly, and *tussis*, meaning cough.) When ciliary action is compromised, mucus accumulates, and the infected person desperately attempts to cough up these mucus accumulations. The violence of the coughing in small children can actually result in broken ribs. Gasping for air between coughs causes a whooping sound, hence the informal name of the disease. Coughing episodes occur several times a day for 1 to 6 weeks. The *convalescence stage*, the third stage, may last for months. Because infants are less capable of coping with the effort of coughing to maintain an airway, irreversible damage to the brain occasionally occurs.

Immunity following recovery is good; at least second attacks tend to be very mild. Vaccination rates for pertussis among children in the United States are higher than 80%. The annual number of cases declined from about 270,000 to a low of about 1,000 in 1976. Almost all deaths now occur in infants too young to receive the currently scheduled vaccine (see Table 18.3, page 531). These results were mostly obtained with the whole-cell, heat-killed vaccine introduced in the 1940s (DTP), which was replaced in the 1990s with an acellular vaccine (DTaP) containing an assortment of bacterial virulence factors. Nonetheless, in recent years there has been a steady increase in reported cases, and more than 18,000 occurred in 2004. Even this is probably an underestimate because most of these occurred in adolescents and adults, where it is manifested as a persistent cough that is often misdiagnosed. The reason for this increase in cases is probably that the effectiveness of the vaccination tends to wane after about 12 years—although greater awareness and better diagnostic techniques may be responsible for some of it. In response, the U.S. Food and Drug Administration (FDA) has approved expanded use of pertussis vaccines, frequently in combination with diphtheria and tetanus boosters. These will be marketed to adolescents aged 10 to 18 years. Another proposed cause for the recent surge in pertussis cases in adolescents might be mutational changes in the circulating bacteria; research on this concern is continuing.

Diagnosis of pertussis is primarily based on clinical signs and symptoms. The pathogen can be cultured from a throat swab inserted through the nose on a thin wire and held in the throat while the patient coughs. Culture of the fastidious pathogen requires care for successful results. For alternatives to culture, PCR methods can also be used to test the swabs for presence of the pathogen. Serological tests to detect antibodies against pertussis toxins are under development.

Treatment of pertussis with antibiotics, most commonly erythromycin or other macrolides, is not effective after onset of the paroxysmal coughing stage.

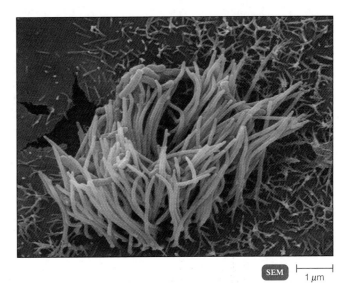

SEM   1 μm

FIGURE 24.8 Ciliated cells of the respiratory system infected with *Bordetella pertussis*. Cells of *B. pertussis* (orange) can be seen growing on the cilia; they will eventually cause the loss of the ciliated cells.

Q What is the name of the toxin produced by *Bordetella pertussis* that causes the loss of cilia?

TUBERCULOSIS

Tuberculosis (TB) is an infectious disease caused by the bacterium *Mycobacterium tuberculosis*, a slender rod and an obligate aerobe. The rods grow slowly (20-hour or longer generation time), sometimes form filaments, and tend to grow in clumps (Figure 24.9). On the surface of liquid media, their growth appears moldlike, which suggested the genus name *Mycobacterium* (*myco* means fungus).

These bacteria are relatively resistant to conventional simple staining procedures. Cells stained with carbolfuchsin dye cannot be decolorized with acid-alcohol and are therefore classified as *acid-fast* (see page 70). This characteristic reflects the unusual composition of the cell wall, which contains large amounts of lipids. These lipids might also be responsible for the resistance of mycobacteria to environmental stresses, such as drying. In fact, these bacteria can survive for weeks in dried sputum and are very resistant to chemical antimicrobials used as antiseptics and disinfectants (see Table 7.7 page 206).

Tuberculosis is a particularly good illustration of the ecological balance between host and parasite in infectious disease. A host is not usually aware of pathogens that invade the body and are defeated. If immune defenses fail, however, the host becomes very much aware of the resulting disease. As might be expected, there is a great synergy between HIV infection and TB—many cases are coinfections. Presently, HIV and TB are the two most common causes of death from infectious diseases.

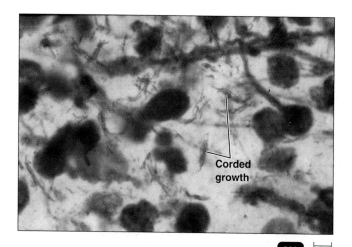

Corded
growth

LM ⊢———⊣
5 μm

FIGURE 24.9 *Mycobacterium tuberculosis*. The filamentous, red-stained funguslike growth shown here in a smear from lung tissue is responsible for the organism's name. Under other conditions, it grows as slender, individual bacilli. A waxy component of the cell, cord factor, is responsible for this ropelike arrangement. An injection of cord factor causes pathogenic effects exactly like that caused by tubercle bacilli.

Q **What characteristic of this bacterium suggests use of the prefix myco-?**

Several factors may affect host resistance levels: the presence of other illness and physiological and environmental factors, such as malnutrition, overcrowding, and stress. A tragic demonstration of individual variation in resistance was the Lübeck disaster in Germany in 1926. By error, 249 babies were inoculated with virulent tuberculosis bacteria instead of the attenuated vaccine strain. Even though all received the same inoculum, there were only 76 deaths, and the remainder did not become seriously ill.

Tuberculosis is most commonly acquired by inhaling the bacillus. Only very fine particles containing one to three bacilli reach the lungs, where they are usually phagocytized by a macrophage in the alveoli (see Figure 24.2). The macrophages of a healthy individual become activated by the presence of the bacilli and usually destroy them.

PATHOGENESIS OF TUBERCULOSIS

Figure 24.10 shows the pathogenesis of TB. The figure depicts the situation in which the body's defenses fail and the disease progresses to a fatal conclusion. However, most healthy people will defeat a potential infection with activated macrophages, especially if the infecting dose is low.

①–② If the infection progresses, the host isolates the pathogens in a walled-off lesion called a *tubercle* (meaning lump or knob), a characteristic that gives the disease its name.

③–④ When the disease is arrested at this point, the lesions slowly heal, becoming calcified. These show up clearly

on X-ray films and are called *Ghon's complexes*. (Computed tomography [CT] is more sensitive than X rays in detecting lesions of TB.)

⑤ If the body's defenses fail at this stage, the tubercle breaks down and releases virulent bacilli into the airways of the lung and then the cardiovascular and lymphatic systems.

The disseminated infection is called *miliary tuberculosis* (the name is derived from the numerous millet seed–sized tubercles formed in the infected tissues). The body's remaining defenses are overwhelmed, and the patient suffers weight loss, coughing (often bringing up blood), and a general loss of vigor. At one time, TB was also known as *consumption*.

TREATMENT OF TUBERCULOSIS

The first effective antibiotic for TB treatment was streptomycin, which was introduced in 1944. It is still in use but now considered a second-line drug. The current treatment for TB recommended by the World Health Organization requires the patient to adhere to a minimum of six months of antibiotic therapy that includes three or four drugs. Many patients fail to follow such a prolonged regimen faithfully, which increases the likelihood of resistance developing. (See the box in Chapter 15, page 453). The two most powerful anti-TB drugs are isoniazid and rifampin (also known as rifampicin). Other FDA approved first-line drugs include pyrazinamide, rifapentine, and ethambutol—in all there are about 10 drugs approved for treatment of TB, many considered secondary choices. The prolonged treatment is necessary because the tubercle bacillus grows very slowly or is dormant (the only drug effective against dormant bacilli is pyrazinamide), and many antibiotics are effective only against growing cells. Also, the bacillus may be hidden for long periods in macrophages or other locations difficult to reach with antibiotics. Multiple-drug therapy is needed to minimize the emergence of resistant strains. Even so, strains of the tubercle bacillus have appeared that are resistant to every anti-tubercular drug available.

There has been little in the way of new anti-TB drugs: the last major anti-tubercular drug, rifampin, was introduced decades ago. There is a serious need for a drug or drugs that would reduce the treatment time to less than three months, kill persistent bacilli that might later reactivate,

FIGURE 24.10 The pathogenesis of tuberculosis. This figure represents the progression of the disease when the defenses of the body fail. In most otherwise healthy individuals, the infection is arrested and fatal tuberculosis does not develop.

Q **Almost a third of the earth's population is infected with *Mycobacterium tuberculosis*—does a study of this figure show why this is not the same as a third of the earth's population *having* tuberculosis?**

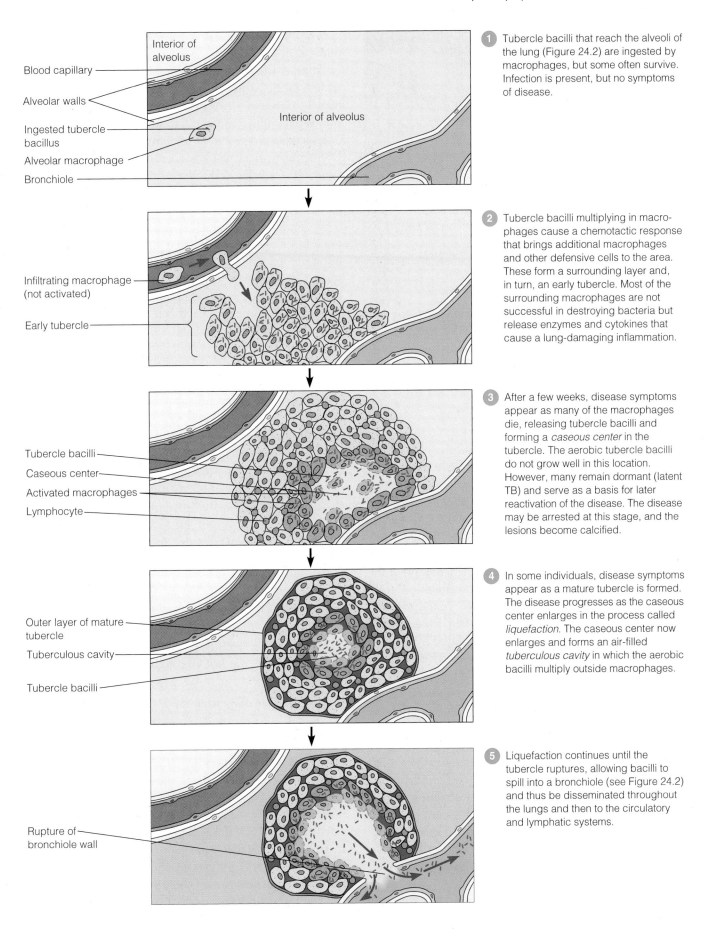

Blood capillary
Alveolar walls
Ingested tubercle bacillus
Alveolar macrophage
Bronchiole
Interior of alveolus
Interior of alveolus

1 Tubercle bacilli that reach the alveoli of the lung (Figure 24.2) are ingested by macrophages, but some often survive. Infection is present, but no symptoms of disease.

Infiltrating macrophage (not activated)
Early tubercle

2 Tubercle bacilli multiplying in macrophages cause a chemotactic response that brings additional macrophages and other defensive cells to the area. These form a surrounding layer and, in turn, an early tubercle. Most of the surrounding macrophages are not successful in destroying bacteria but release enzymes and cytokines that cause a lung-damaging inflammation.

Tubercle bacilli
Caseous center
Activated macrophages
Lymphocyte

3 After a few weeks, disease symptoms appear as many of the macrophages die, releasing tubercle bacilli and forming a *caseous center* in the tubercle. The aerobic tubercle bacilli do not grow well in this location. However, many remain dormant (latent TB) and serve as a basis for later reactivation of the disease. The disease may be arrested at this stage, and the lesions become calcified.

Outer layer of mature tubercle
Tuberculous cavity
Tubercle bacilli

4 In some individuals, disease symptoms appear as a mature tubercle is formed. The disease progresses as the caseous center enlarges in the process called *liquefaction*. The caseous center now enlarges and forms an air-filled *tuberculous cavity* in which the aerobic bacilli multiply outside macrophages.

Rupture of bronchiole wall

5 Liquefaction continues until the tubercle ruptures, allowing bacilli to spill into a bronchiole (see Figure 24.2) and thus be disseminated throughout the lungs and then to the circulatory and lymphatic systems.

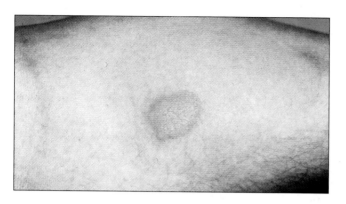

FIGURE 24.11 The tuberculin skin test on an arm.

 What does a positive tuberculin skin test indicate?

and be active against multidrug resistant strains. Currently, experimental interest is focused on a novel diarylquinoline drug. In animal testing it has demonstrated a unique specificity for impeding the synthesis of ATP in mycobacteria and is effective in killing both dormant and actively growing bacilli.

DIAGNOSIS OF TUBERCULOSIS

People infected with tuberculosis respond with cell-mediated immunity against the bacterium. This form of immune response, rather than humoral immunity, develops because the pathogen is located mostly within macrophages. This immunity, involving sensitized T cells, is the basis for the **tuberculin skin test** (Figure 24.11), a screening test for infection. A positive test does not necessarily indicate active disease. In this test, a purified protein derivative of the tuberculosis bacterium, derived by precipitation from broth cultures, is injected cutaneously. If the injected person has been infected with TB in the past, sensitized T cells react with these proteins and a delayed hypersensitivity reaction occurs in about 48 hours. This reaction appears as an induration (hardening) and reddening of the area around the injection site. Probably the most accurate tuberculin test is the *Mantoux test*, in which dilutions of 0.1 ml of antigen are injected, and the reacting area of the skin is measured.

A positive tuberculin test in the very young is a probable indication of an active case of TB. In older individuals, it might indicate only hypersensitivity resulting from a previous infection or vaccination, not a current active case. Nonetheless, it is an indication that further examination is needed, such as a chest X ray or CT examination for the detection of lung lesions, and attempts to isolate the bacterium.

The initial step in laboratory diagnosis of active cases is a microscopic examination of smears, such as sputum. This may be a conventional acid-fast stain or the more specific fluorescent-antibody microscopy. A confirmation of TB by isolation of the bacterium is complicated by the very slow growth of the pathogen. The formation of a colony might take 3 to 6 weeks, with completion of a reliable identification series adding another 3 to 6 weeks. There has been considerable progress in developing rapid diagnostic tests. DNA probes are available to identify cultured isolates (see Figure 10.16, page 303). Several new tests have been introduced that make use of PCR methods that are capable of detecting *M. tuberculosis* directly from sputum or other samples (see pages 301–302). The latest diagnostic test is ELISPOT (Enzyme-Linked ImmunoSpot). Overnight, using a blood sample containing white cells, this test measures the immune messenger called γ interferon (IFN-γ). Large amounts of IFN-γ are produced in response to TB infection. In clinical trials, this test detected many cases that would have been missed with the traditional skin test.

Another mycobacterial species, *Mycobacterium bovis* (bō′vis), is a pathogen mainly of cattle. *M. bovis* is the cause of **bovine tuberculosis,** which is transmitted to humans via contaminated milk or food. Bovine tuberculosis accounts for fewer than 1% of TB cases in the United States. It seldom spreads from human to human; but, before the days of pasteurized milk and the development of control methods such as tuberculin testing of cattle herds, this disease was a frequent form of tuberculosis in humans. *M. bovis* infections cause TB that primarily affects the bones or lymphatic system. At one time, a common manifestation of this type of TB was hunchbacked deformation of the spine.

Other mycobacterial diseases also affect people in the late stages of HIV infection. A majority of the isolates are of a related group of organisms known as the *M. avium-intracellulare* (ā′vē-um in′trä-cel-ū-lä-rē) complex. In the general population, infections by these pathogens are uncommon.

TUBERCULOSIS VACCINES

The **BCG vaccine** is a live culture of *M. bovis* that has been made avirulent by long cultivation on artificial media. (BCG stands for bacillus of Calmette and Guérin, the people who originally isolated the strain.) The BCG vaccine has been available since the 1920s and is one of the most widely used vaccines in the world. In 1990, it was estimated that 70% of the world's schoolchildren received it. In the United States, however, the vaccine is currently recommended only for certain children at high risk who have negative skin tests. People who have received the vaccine show a positive reaction to tuberculin skin tests. This has always been one argument against its widespread use in the United States. Another argument against the universal administration of BCG vaccine is its very uneven effectiveness. Experience has shown that it is fairly effective when given to young children, but for adolescents and adults it sometimes has an effectiveness approaching zero. Recent work indicates that exposure to members of the M. *avium-intracellulare* complex that is often encountered in the environment may interfere with the effectiveness of the BCG vaccine—which might explain why the vaccine is more effective early in life, before much exposure to such environmental mycobacteria. A number of new vaccines are in the experimental pipeline.

WORLDWIDE INCIDENCE OF TUBERCULOSIS

After the introduction of effective antibiotics in the 1950s, TB incidence declined steadily. Currently, about 20,000 new cases are reported in the United States every year (see Figure 14.10c, page 442). The mortality rate, nearly 2000 deaths annually, has continued to decrease. Half of the new annual cases occur among immigrants, especially those from Mexico, the Philippines, and Vietnam. In the United States, certain ethnic groups tend to have much higher rates of TB. For example, African Americans, Native Americans, Asians, and Hispanics account for about two-thirds of cases (Figure 24.12). Most cases in the white population occur among the very elderly.

In the United States, 10 to 12 million people are estimated to be infected by the tubercle bacillus, usually with only latent infections. Worldwide, an estimated one-third of the total population is infected; at least 3 million die of the disease each year.

Globally, the number of TB cases is rising at about 2% a year, and the number of cases that are multidrug-resistant (MDR) is rising even faster. To treat MDR patients in the United States can cost tens of thousands of dollars annually and is economically impractical in most of the world. Tuberculosis remains a major worldwide health problem.

BACTERIAL PNEUMONIAS

> **LEARNING OBJECTIVE**
> • Compare and contrast the seven bacterial pneumonias discussed in this chapter.

The term *pneumonia* is applied to many pulmonary infections, most of which are caused by bacteria. Pneumonia caused by *Streptococcus pneumoniae* is the most common, about two-thirds of cases, and is therefore referred to as *typical pneumonia*. Pneumonias caused by other microorganisms, which can include fungi, protozoa, viruses, and other bacteria, are termed *atypical pneumonias*. This distinction is becoming increasingly blurred in practice.

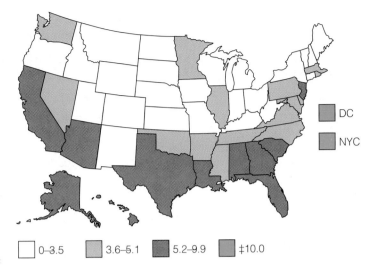

0–3.5   3.6–5.1   5.2–9.9   ‡10.0

(a) Tuberculosis incidence in the United States, per 100,000 population
Source: CDC, 2003

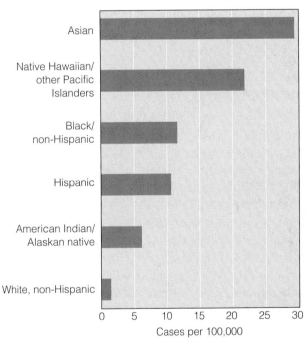

Cases per 100,000

(b) Tuberculosis rates among American ethnic groups in 2003

FIGURE 24.12 U.S. distribution of tuberculosis.
SOURCE: CDC, *Summary of Notifiable Diseases—United States, 2003, MMWR* 52(54), 4/22/05.

Q **How can tuberculosis be eliminated?**

Pneumonias also are named after the portions of the lower respiratory tract they affect. For example, if the lobes of the lungs are infected, it is called *lobar pneumonia*; pneumonias caused by *S. pneumoniae* are usually of this type. *Bronchopneumonia* indicates that the alveoli of the lungs adjacent to the bronchi are infected. *Pleurisy* is often a complication of various pneumonias, in which the pleural membranes become painfully inflamed. (See the box that follows.)

DISEASES IN FOCUS

COMMON BACTERIAL PNEUMONIA

Pneumonia is a leading cause of illness and death among children worldwide and the seventh leading cause of death in the United States. Pneumonia can be caused by a variety of viruses, bacteria, and fungi. Isolation of the bacterium from cultures of blood or, in some cases, lung aspirates is used to prove that a bacterium is the cause. The CDC estimates *S. pneumoniae* causes 40,000 deaths and 500,000 cases of pneumonia annually in the United States.

	Pathogen	Symptoms	Reservoir	Treatment
Pneumococcal pneumonia	*Streptococcus pneumoniae*	Infected alveoli of lung fill with fluids; interferes with oxygen uptake	Humans	Penicillin, fluroquinolones Prevention: pneumococcal vaccine
Haemophilus influenzae pneumonia	*Haemophilus influenzae*	Symptoms resemble pneumococcal pneumonia	Humans	Cephalosporins
Mycoplasmal pneumonia	*Mycoplasma pneumoniae*	Mild but persistent respiratory symptoms; low fever, cough, headache	Humans	Tetracyclines
Legionellosis	*Legionella pneumophila*	Potentially fatal pneumonia that tends to affect older males who drink or smoke heavily	Water	Erythromycin
Psittacosis (ornithosis)	*Chlamydophila psittaci*	Symptoms, it any, are fever, headache, chills	Birds	Tetracyclines

PNEUMOCOCCAL PNEUMONIA

Pneumonia caused by *S. pneumoniae* is called **pneumococcal pneumonia**. *S. pneumoniae* is a gram-positive, ovoid bacterium (Figure 24.13). This microbe is also a common cause of otitis media, meningitis, and sepsis. Because it usually forms cell pairs, the genus was formerly named *Diplococcus pneumoniae*. The cell pairs are surrounded by a dense capsule that makes the pathogen resistant to phagocytosis. These capsules are also the basis of serological differentiation of pneumococci into at least 90 serotypes. Before antibiotic therapy became available, antisera directed at these capsular antigens were used to treat the disease.

Pneumococcal pneumonia involves both the bronchi and the alveoli (see Figure 24.2). Symptoms include high fever, breathing difficulty, and chest pain. (Atypical pneumonias usually have a slower onset and less fever and chest pain.) The lungs have a reddish appearance because blood vessels are dilated. In response to the infection, alveoli fill with some red blood cells, neutrophils (see Table 16.1, pages 480–481), and fluid from surrounding tissues. The

DISEASES IN FOCUS

(continued)

	Pathogen	Symptoms	Reservoir	Treatment
Chlamydial pneumonia	*Chlamydophila pneumoniae*	Mild respiratory illness common in young people; resembles mycoplasmal pneumonia	Humans	Tetracyclines
Q fever	*Coxiella burnetii*	Mild respiratory disease lasting 1–2 weeks; occasional complications such as endocarditis occur	Large mammals; tick vector; can be transmitted via unpasteurized milk	Doxycycline and chloroquine

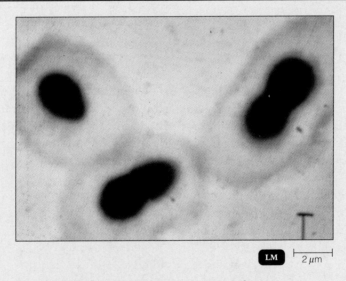

LM   2 μm

Encapsulated *Streptococcus pneumoniae*.

sputum is often rust-colored from blood coughed up from the lungs. Pneumococci can invade the bloodstream, the pleural cavity surrounding the lung, and occasionally the meninges. No bacterial toxin has been clearly related to pathogenicity.

A presumptive diagnosis can be made by isolating the pneumococci from the throat, sputum, and other fluids. Pneumococci can be distinguished from other alpha-hemolytic streptococci by observing the inhibition of growth next to a disk of optochin (ethylhydrocupreine

hydrochloride) or by performing a bile solubility test. They can also be serologically typed.

There are many healthy carriers of the pneumococcus. Virulence of the bacteria seems to be based mainly on the carrier's resistance, which can be lowered by stress. Many illnesses of older adults terminate in pneumococcal pneumonia.

A recurrence of pneumococcal pneumonia is not uncommon, but the serological types are usually different. Before chemotherapy was available, the mortality rate was as high as 25%. This has now been lowered to less than 1% for

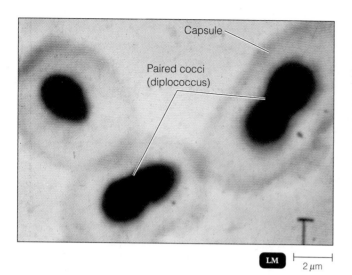

FIGURE 24.13 *Streptococcus pneumoniae*, the cause of pneumococcal pneumonia. Notice the paired arrangement of the cells. The capsule has been made more apparent here by reaction with a specific pneumococcal antiserum that makes it appear to swell.

Q **What component of the cell is the primary antigen?**

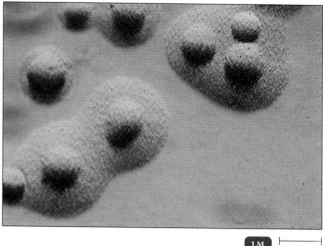

FIGURE 24.14 Colonies of *Mycoplasma pneumoniae*, the cause of mycoplasmal pneumonia.

Q **Could you see these colonies without magnification?**

younger patients treated early in the course of their disease. For elderly patients admitted to a hospital, mortality can approach 20%. Penicillin and fluoroquinolones are the drugs of choice. Antibiotic resistance is an increasing problem, in some areas involving 25% or more of isolates. A vaccine has been developed from the purified capsular material of the 23 types of pneumococci that cause at least 90% of the pneumococcal pneumonias in the United States. This vaccine is used for the groups most susceptible to infection: the elderly and debilitated individuals. A conjugated pneumococcal vaccine has recently been introduced that has also found uses for meningitis and otitis media.

HAEMOPHILUS INFLUENZAE *PNEUMONIA*

Haemophilus influenzae is a gram-negative coccobacillus, and a Gram stain of sputum will differentiate this type of pneumonia from pneumococcal pneumonia. Patients with such conditions as alcoholism, poor nutrition, cancer, or diabetes are especially susceptible. Second-generation cephalosporins are resistant to the β-lactamases produced by many *H. influenzae* strains and are therefore usually the drugs of choice.

MYCOPLASMAL PNEUMONIA

The mycoplasmas, which do not have cell walls, do not grow under the conditions normally used to recover most bacterial pathogens. Because of this characteristic, pneumonias caused by mycoplasmas are often confused with viral pneumonias.

The bacterium *Mycoplasma pneumoniae* is the causative agent of **mycoplasmal pneumonia.** This type of pneu-

monia was first discovered when such atypical infections responded to tetracyclines, indicating that the pathogen was nonviral. Mycoplasmal pneumonia is a common type of pneumonia in young adults and children. It may account for as much as 20% of pneumonias, although it is not a reportable disease. The symptoms, which persist for three weeks or longer, are low-grade fever, cough, and headache. Occasionally, they are severe enough to lead to hospitalization. Other terms for the disease are *primary atypical* (that is, the most common pneumonia not caused by the pneumococcus) and *walking pneumonia.*

When isolates from throat swabs and sputum grow on a medium containing horse serum and yeast extract, they form distinctive colonies with a "fried-egg" appearance (Figure 24.14). The colonies are so small that they must be observed with magnification. The mycoplasmas are highly varied in appearance because they lack cell walls (see Figure 11.20, page 334).

Diagnosis based on recovering the pathogens might not be useful in treatment because as long as three or more weeks may be required for the slow-growing organisms to develop. Diagnostic tests have improved greatly in recent years, however. They include PCR and serological tests that detect IgM antibodies against *M. pneumoniae.*

Treatment with antibiotics such as tetracycline usually hastens the disappearance of symptoms but does not eliminate the bacteria, which the patient continues to carry for several weeks.

LEGIONELLOSIS

Legionellosis, or **Legionnaires' disease,** first received public attention in 1976, when a series of deaths occurred among members of the American Legion who had attended a meeting in Philadelphia. A total of 182 people contracted pulmonary disease, apparently at this meeting, 29 of whom died. Because no obvious bacterial cause could be found, the deaths were attributed to viral pneumonia. Close investigation, mostly with techniques directed at locating a suspected rickettsial agent, eventually identified a previously unknown bacterium, an aerobic gram-negative rod now known as *Legionella pneumophila*, which is capable of replication within macrophages. Over 44 species of *Legionella* have now been identified; not all of them cause disease.

The disease is characterized by a high fever of 40.5°C (105°F), cough, and general symptoms of pneumonia. No person-to-person transmission seems to be involved. Recent studies have shown that the bacterium can be readily isolated from natural waters. In addition, the microbes can grow in the water of air-conditioning cooling towers, which might mean that some epidemics in hotels, urban business districts, and hospitals were caused by airborne transmission. Recent outbreaks have been traced to whirlpool spas, humidifiers, showers, decorative fountains, and even potting soil.

The organism has also been found to inhabit the water lines of many hospitals. Most hospitals keep the temperature of hot water lines relatively low (43–55°C) as a safety measure, and in cooler parts of the system this inadvertently maintains a good growth temperature for *Legionella*. This bacterium is considerably more resistant to chlorine than most other bacteria and can survive for long periods in water with a low level of chlorine. Evidence indicates *Legionella* exist primarily in biofilms that are highly protective. The bacteria are often ingested by waterborne amoebae when these are present but continue to proliferate and may even survive within encysted amoebae. The most successful method for water disinfection in hospitals with a need to control *Legionella* contamination has been installation of copper–silver ionization systems.

The disease appears to have always been fairly common, even if unrecognized. More than 1000 cases are reported each year, but the actual incidence is estimated at over 25,000 annually. Men over age 50 are the most likely to contract legionellosis, especially heavy smokers, alcohol abusers, or the chronically ill. (See the box that follows.)

L. pneumophila is also responsible for **Pontiac fever,** which is essentially another form of legionellosis. Its symptoms include fever, muscular aches, and usually a cough. The condition is mild and self-limiting. During outbreaks of legionellosis both forms may occur.

The best diagnostic method is culture on a selective charcoal–yeast extract medium. Examination of respiratory specimens can be done by fluorescent-antibody methods, and a DNA probe test is available. Erythromycin and other macrolide antibiotics, such as azithromycin, are the drugs of choice for treatment.

PSITTACOSIS (ORNITHOSIS)

The term **psittacosis** is derived from the disease's association with psittacine birds, such as parakeets and other parrots. It was later found that the disease can also be contracted from many other birds, such as pigeons, chickens, ducks, and turkeys. Therefore, the more general term **ornithosis** has come into use.

The causative agent is *Chlamydophila psittaci* (sit'tä-sē), a gram-negative, obligate intracellular bacterium. The taxonomy of this organism has recently been revised. The genus name has been changed from *Chlamydia* to *Chlamydophila*. This taxonomic change has also been made with *C. pneumoniae* (see the discussion of chlamydial pneumonia which follows). We will continue to use the generic terms, chlamydial and chlamydiae. One way chlamydias differ from rickettsias, which are also obligate intracellular bacteria, is that chlamydias form tiny **elementary bodies** as one part of their life cycle (see Figure 11.23, page 337). Unlike most rickettsias, elementary bodies are resistant to environmental stress; therefore, they can be transmitted through air and do not require a bite to transfer the infective agent directly from one host to another.

Psittacosis is a form of pneumonia that usually causes fever, headache, and chills. Subclinical infections are very common, and stress appears to enhance susceptibility to the disease. Disorientation, or even delirium in some cases, indicates that the nervous system can be involved.

The disease is seldom transmitted from one human to another but is usually spread by contact with the droppings and other exudates of fowl. One of the most common modes of transmission is inhalation of dried particles from droppings. The birds themselves usually have diarrhea, ruffled feathers, respiratory illness, and a generally droopy appearance. Parakeets and other parrots sold commercially are usually (but not always) free of the disease. Many birds carry the pathogen in their spleen without symptoms, becoming ill only when stressed. Pet store employees and people involved in raising turkeys are at greatest risk of contracting the disease.

Diagnosis is made by isolating the bacterium in embryonated eggs or by cell culture. Serological tests can be used to identify the isolated organism. No vaccine is available, but tetracyclines are effective antibiotics in treating humans and animals. Effective immunity does not result from recovery, even when high titers of antibody are present in the serum.

Most years, fewer than 100 cases and very few deaths are reported in the United States. The main danger is late diagnosis. Before antibiotic therapy was available, the mortality rate was about 15 to 20%.

CLINICAL PROBLEM SOLVING

OUTBREAK

You will see questions as you read through this problem. The questions are those that epidemiologists ask themselves and each other as they solve a clinical problem. As you read through the problem, try to answer each question as though you were an epidemiologist.

1. A 64-year-old man saw his primary care physician complaining of fever, malaise, and a cough. His vaccinations were up-to-date, including DTaP. His condition worsened over several days; he had difficulty breathing and his temperature rose to 40.4°C (104.7°F). He was hospitalized, and his lungs showed signs of mild inflammation with thin, watery secretion. A Gram stain of bacteria isolated from the patient is shown in the figure. *What diseases are possible?*

2. The same day, a 37-year-old man went to the emergency department because he had shortness of breath, fatigue, and cough. The day before he had had fever and chills, with a maximum body temperature of 38.6°C (101.4°F). *What additional tests would you do on both patients?*

Hotel guests with legionellosis

Age	37–70 yrs (average: 60)
Gender	6 male; 2 female
Number of nights at hotel	1–4 (average: 3)
Diabetes mellitus	4
Immunocompromised	1
Smoker	5
Showered in hotel	8
Used whirlpool spa at hotel	1
Used hotel swimming pool	6

3. Both patients had an antibody titer > 1024 against *Legionella pneumophila* serogroup 1. The local health department (LHD) was contacted because two patients were hospitalized with legionellosis. *What do you need to know now?*

4. One week before hospitalization, both men stayed in the same hotel within one day of each other. The LHD identified six additional cases of legionellosis at other hospitals. The LHD gave a follow-up questionnaire to all eight patients to ascertain travel that preceded the illness, including location, accommodations, dates, and information about exposures to common sources for infection (see the table). *What are likely sources of infection?*

5. Epidemic legionellosis usually results from exposure of susceptible individuals to an aerosol generated by an environmental source of water contaminated with *Legionella*. *Why is it important to identify the source?*

6. Retrospective identification of cases allows further investigation and control and remediation efforts. *L. pneumophila* of the same monoclonal antibody type was recovered from the hot water storage tanks, cooling tower, and showers and faucets in rooms occupied by patients and well guests. *Why didn't other hotel guests get sick?*

7. During outbreaks, attack rates tend to be highest in specific high-risk groups, including the elderly, smokers, and immunocompromised persons. *What are your recommendations for remediation?*

Shower necks and faucets were disinfected with bleach. The spa filter was cleaned, and the potable water system was hyperchlorinated.

Hotels have been common locations for legionellosis outbreaks since the disease was first recognized among hotel guests in Philadelphia in 1976. Active surveillance led to more rapid identification of other cases.

Surveillance data submitted to CDC indicate that approximately 21% of legionellosis cases each year are travel associated. However, identification of travel-associated clusters is hindered because the incubation period is long enough for people to disperse from the point source of infection. Prompt recognition and investigation of clusters can implicate a point source for infection and guide remediation and control efforts. Recognizing the benefits of enhanced surveillance, CDC plans to work with state health departments on new strategies to improve surveillance for travel-associated legionellosis at the national, state, and local levels.

SOURCE: Adapted from reports in *MMWR* 53(52): 1202, 1/7/05, and *MMWR* 54(7): 170, 2/25/05.

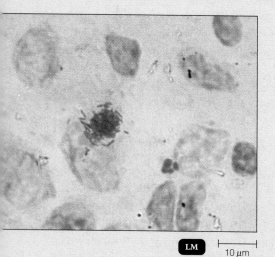

LM   10 μm

Gram stain shows bacteria within a tissue sample.

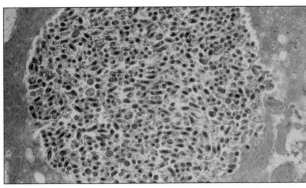

(a) *Coxiella burnetii* growing in a placental cell.   TEM ⊢ 2 μm

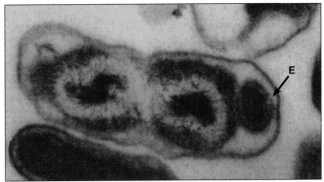

(b) This cell has just divided; notice the endospore-like body (E), which is probably responsible for the relative resistance of the organism.  TEM ⊢ 0.5 μm

FIGURE 24.15 *Coxiella burnetii*, **the cause of Q fever.**

Q **By what two methods is Q fever transmitted?**

CHLAMYDIAL PNEUMONIA

Outbreaks of a respiratory illness in populations of college students were found to be caused by a chlamydial organism. Originally the pathogen was considered a strain of *C. psittaci*, but it has been assigned the species name *Chlamydophila pneumoniae*, and the disease is known as **chlamydial pneumonia.** Clinically, it resembles mycoplasmal pneumonia. (There is also strong evidence of association between *C. pneumoniae* and atherosclerosis, the deposition of fatty deposits that block arteries.)

The disease is apparently transmitted from person to person, probably by the respiratory route, but not as readily as infections such as influenza. Nearly half the U.S. population has antibodies against the organism, an indication that this is a common illness. Several serological tests are useful in diagnosis, but results are complicated by antigenic variation. The most effective antibiotic is tetracycline.

Q FEVER

In Australia during the mid-1930s, a previously unreported flu-like pneumonia made an appearance. In the absence of an obvious cause, the affliction was labeled **Q fever** (for *query*), much as one might say "X fever." The causative agent was subsequently identified as the obligately parasitic, intracellular bacterium *Coxiella burnetii* (käks′ē-el-lä bėr-ne′tē-ē) (Figure 24.15a). Currently, it is classified as a member of the gammaproteobacteria. Along with other bacteria of this group (such as the genera *Franciscella* and *Legionella*), it has the ability to multiply intracellularly. Most intracellular bacteria, such as rickettsia, are not resistant enough to survive airborne transmission, but this microorganism is an exception.

Q fever has a wide range of clinical symptoms, and systematic testing shows that about 60% of cases are not even symptomatic. In cases of *acute Q fever* there are usually symptoms of high fever, headaches, muscle aches, and coughing. A feeling of malaise may persist for months. In France and a few other areas, acute cases of Q fever are often manifested as hepatitis. The heart becomes involved in about 2% of acutely ill patients and is responsible for the rare fatalities. In cases of *chronic Q fever*, the best known manifestation is endocarditis (see page 675). Some 5 to 10 years might elapse between the initial infection and the appearance of endocarditis; and, because these patients show few signs of acute disease, the association with Q fever is often missed. Antibiotic therapy and earlier diagnosis have lowered the mortality rate from chronic Q fever to under 5%.

C. burnetii is a parasite of several arthropods, especially cattle ticks, and it is transmitted among animals by tick bites. Infected animals include cattle, goats, and sheep, as well as most domestic mammalian pets. In animals the infection is usually subclinical. Cattle ticks spread the disease among dairy herds, and the microbes are shed in the feces, milk, and urine of infected cattle. Once the disease is established in a herd, it is maintained by aerosol transmission. The disease is spread to humans by the ingestion of unpasteurized milk and by inhaling aerosols of microbes generated in dairy barns, especially at calving time from placental material, which contains about a billion bacteria per gram.

Inhaling a single pathogen is enough to cause infection, and many dairy workers have acquired at least subclinical infections. Workers in meat- and hide-processing plants are also at risk. The pasteurization temperature of milk, which was originally aimed at eliminating tuberculosis bacilli, was raised slightly in 1956 to ensure the killing of *C. burnetii*. In 1981, an endosporelike body was discovered, which may account for this heat resistance (Figure 24.15b). This resistant body resembles the elementary body of chlamydiae more than typical bacterial endospores.

The pathogen can be identified by isolation and growth in chick embryos in eggs or in cell culture. Laboratory workers testing for *Coxiella*-specific antibodies in a patient's serum can use serological tests.

A disease found worldwide, most cases of Q fever in the United States occur in the western states. The disease is endemic to California, Arizona, Oregon, and Washington. A vaccine for laboratory workers and other high-risk personnel is available. Doxycycline has been recommended for treatment. When growth within macrophages in chronic infections renders C. *burnetii* resistant, the killing activity can be restored by combination of doxycycline with chloroquine, an antimalarial.

MELIOIDOSIS

LEARNING OBJECTIVE
- List the etiology, method of transmission, and symptoms of melioidosis.

In 1911 a new disease was reported among drug addicts in Rangoon, Burma (now Myanmar). The bacterial pathogen, *Burkholderia pseudomallei*, is a gram-negative rod formerly placed in the genus *Pseudomonas*. It closely resembled the bacterium causing glanders, a disease of horses. Therefore, the disease was named **melioidosis** [from the Greek *melis* (distemper of asses) and *eidos* (resemblance)]. It is now recognized as a major infectious disease in southeast Asia and northern Australia, where the pathogen is widely distributed in moist soils. Sporadic cases are reported in Africa, the Caribbean, Central and South America, and the Middle East. Many animal species are also susceptible.

Clinically, melioidosis is most commonly seen as pneumonia. The mortality rate in southeast Asia is about 50% and in Australia approaches 20%. However, it can also appear as abscesses in various body tissues that resemble necrotizing fasciitis (see Figure 21.8, page 622), as severe sepsis, and even as encephalitis. Transmission is primarily by inhalation, but alternative infective routes are by inoculation through puncture wounds and ingestion. About 7% of American soldiers returning from Vietnam showed serological evidence of exposure, which was highest among helicopter crewmen—probably from inhalation. Incubation periods can be very long and occasional delayed-onset cases still surface in this population. Most recently, several cases were reported in Europeans exposed during the Indian Ocean tsunami disaster of 2004.

Diagnosis is usually by isolation of the pathogen from body fluids. Serological tests in endemic areas are problematic because of widespread exposure to a similar, non-pathogenic bacterium. Treatment by antibiotic is uncertain in effectiveness; the most commonly used is ceftazidime, a β-lactam.

VIRAL DISEASES OF THE LOWER RESPIRATORY SYSTEM

LEARNING OBJECTIVE
- List the causative agent, symptoms, prevention, and preferred treatment for viral pneumonia, RSV, and influenza.

For a virus to reach the lower respiratory system and initiate disease, it must pass numerous host defenses designed to trap and destroy it. In 2003 a new disease, SARS (for severe acute respiratory syndrome; see pages 19–20), originated in China and spread rapidly to many countries around the world. It is known to have infected more than 8,000 people and killed at least 774, including several in Canada, before disappearing. Its reservoir apparently was small exotic food animals in Chinese markets, and the virus has recently been found in the Chinese horseshoe bat, where it is latent. This incident illustrates the danger of respiratory viruses, such as influenza, that can originate from animal reservoirs and spread rapidly by modern transportation systems.

VIRAL PNEUMONIA

Viral pneumonia can occur as a complication of influenza, measles, or even chickenpox. A number of enteric and other viruses have been shown to cause viral pneumonia, but viruses are isolated and identified in fewer than 1% of pneumonia-type infections because few laboratories are equipped to properly test clinical samples for viruses. In those cases of pneumonia for which no cause is determined, viral etiology is often assumed if mycoplasmal pneumonia has been ruled out.

RESPIRATORY SYNCYTIAL VIRUS (RSV)

Respiratory syncytial virus (RSV) is probably the most common cause of viral respiratory disease in infants. There are about 4500 deaths from RSV each year in the United States, mostly in infants 2 to 6 months old. It can also cause a life-threatening pneumonia in the elderly, where it is easily misdiagnosed as influenza. Epidemics occur during the winter and early spring. Virtually all children become infected by age 2—of which about 1% would have required hospitalization. We have previously mentioned that RSV is sometimes implicated in cases of otitis media. The name of the virus is derived from its characteristic of causing cell fusion (*syncytium* formation, Figure 15.7b, page 466) when grown in cell culture. The symptoms are coughing and wheezing that last for more than a week. Fever occurs only when there are bacterial complications. Several rapid serological tests are now available that use samples of respiratory secretions to detect both the virus and its antibodies.

Naturally acquired immunity is very poor. An immune globulin product has been approved to protect infants with lung problems that put them at high risk. Protective vaccines are being clinically tested. For chemotherapy in life-threatening situations, where its cost can be justified, the severity of symptoms can sometimes be reduced by aerosol administration of the antiviral drug ribavirin. The most recent approved treatment, usually reserved for high-risk patients, is the humanized monoclonal antibody, palivizumab (Synagis).

INFLUENZA (FLU)

The developed countries of the world are probably more aware of **influenza (flu)** than of any other disease, except for the common cold. The flu is characterized by chills, fever, headache, and general muscular aches. Recovery normally occurs in a few days, and coldlike symptoms appear as the fever subsides. Nonetheless, an estimated 50,000 to 70,000 Americans die annually of the flu, even in nonepidemic years. Diarrhea is not a normal symptom of the disease, and the intestinal discomforts attributed to "stomach flu" are probably from some other cause.

THE INFLUENZA VIRUS

Viruses in the genus *Influenzavirus* consist of eight separate RNA segments of differing lengths enclosed by an inner layer of protein and an outer lipid bilayer (Figures 13.3b, page 390, and 24.16). Embedded in the lipid bilayer are numerous projections that characterize the virus. There are two types of projections: *hemagglutinin (H) spikes* and *neuraminidase (N) spikes*.

The H spikes, of which there are about 500 on each virus, allow the virus to recognize and attach to body cells before infecting them. Antibodies against the influenza virus are directed mainly at these spikes. The term *hemagglutinin* refers to the agglutination of red blood cells (hemagglutination) that occurs when the viruses are mixed with them. This reaction is important in serological tests, such as the hemagglutination inhibition test often used to identify influenza and some other viruses.

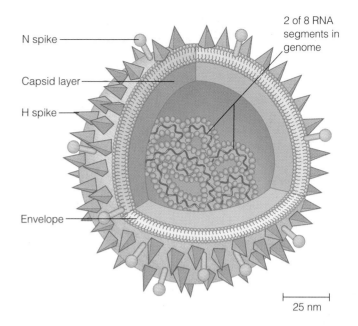

FIGURE 24.16 Detailed structure of the influenza virus. The virus is composed of a protein coat (capsid) that is covered by a lipid bilayer (envelope) and two types of spikes. The genome is composed of eight segments of RNA. Morphologically, under certain environmental conditions, the influenza virus assumes a filamentous form.

Q **What is the primary antigenic structure on the influenza virus?**

The N spikes, of which there are about 100 per virus, differ from the H spikes in appearance and function. Apparently they enzymatically help the virus separate from the infected cell as the virus exits after intracellular reproduction. N spikes also stimulate the formation of antibodies, but these are less important in the body's resistance to the disease than those produced in response to the H spikes.

Viral strains are identified by variation in the H and N antigens. The different forms of the antigens are assigned numbers—for example, H1, H2, H3, N1, and N2. There are 15 subtypes of H and 9 of N. Each number change represents a substantial alteration in the protein makeup of the spike. These changes are called **antigenic shifts,** and they are great enough to evade most of the immunity developed in the human population (see the box in Chapter 13, page 406). This ability is responsible for the outbreaks, including the pandemics of 1918, 1957, and 1968, that are summarized in Table 24.2. Incidentally, influenza viruses were not isolated before 1933, and the antigenic makeup of viruses causing outbreaks before about this time depended on analysis of antibodies taken from persons who had been infected.

Antigenic shifts are probably caused by a major genetic recombination. Because influenza viral RNA occurs as

TABLE 24.2	Human Influenza Viruses*			
Type	Antigenic Subtype		Year	Disease Severity
A	H3N2 (the first "modern" pandemic; originated in southern China)		1889	Moderate
	H1N1 (Spanish)		1918	Severe
	H2N2 (Asian)		1957	Severe
	H3N2 (Hong Kong)		1968	Moderate
	H1N1 (Russian)†		1977	Low
B	None		1940	Moderate
C	None		1947	Very mild

*The conventional wisdom is that H1, H2, and H3 are human-infecting strains; H4, H5, H6, and H7 primarily infect animals, especially swine and poultry. (Avian influenza strains H5N1 and H7N7 have caused human fatalities.)
†Probably escaped from a laboratory. At this time persons over age 20 were mostly immune from similar viruses circulating in the 1950s and earlier in the century.

SOURCE: Adapted from C. Mims, J. Playfair, I. Roitt, D. Wakelin, and R. Williams, *Medical Microbiology*, 2nd ed. London: Mosby International, 1998.

eight segments, recombination is likely in infections caused by more than one strain. Recombination (termed *reassortment*) between the RNA of animal viral strains (found in swine, horses, and birds, for example) and the RNA of human strains might be involved. Swine, ducks, and chickens (especially swine, which can be infected by both human and fowl influenza virus strains) in southern Chinese farming communities have come under suspicion as being the animals most likely to be involved in genetic shifts (and are thus called "mixing vessels"). Wild ducks and other migratory birds then become symptomless carriers that spread the virus over large geographic areas.

In Southeast Asia, poultry is now being produced in huge-scale farms that have become a breeding ground for outbreaks of avian influenza such as H5N1 (which caused a serious outbreak that killed six humans in Hong Kong in 1997). There has been some transmission of the virus from infected birds in these farms to humans. The concern is that mutations could change avian influenza viruses into strains that would allow efficient human-to-human transmission, resulting in a deadly influenza pandemic.

EPIDEMIOLOGY OF INFLUENZA

Between episodes of such major antigenic shifts, there are minor annual variations in the antigenic makeup called

antigenic drift. The virus might still be designated as H3N2, for example, but viral strains arise reflecting minor antigenic changes within the antigenic group. These strains are sometimes assigned names related to the locality in which they were first identified. They usually reflect an alteration of only a single amino acid in the protein makeup of the H or N spike. Such a minor, one-step mutation is probably a response to selective pressure by antibodies (usually IgA in the mucous membranes) that neutralizes all viruses except for those with new mutations. Such mutations can be expected about once in each million multiplications of the virus. High mutation rates are a characteristic of RNA viruses, which lack much of the "proofreading" ability of DNA viruses.

The usual result of antigenic drift is that a vaccine effective against H3, for example, will be less effective against H3 isolates circulating 10 years after the event. There will have been enough drift in that time that the virus can largely evade the antibodies originally stimulated by the earlier strain.

Influenza viruses are also classified into major groups according to the antigens of their protein coats. These groups are A, B, and (rarely) C. The A-type viruses are responsible for the major pandemics. The B-type virus also circulates and mutates, but it is usually responsible for more geographically limited and milder infections.

Almost every year, epidemics of the flu spread rapidly through large populations. The disease is so readily transmissible that epidemics are quickly propagated through populations susceptible to the newly changed strain of virus. The mortality rate from the disease is not high, usually less than 1%, and these deaths are mainly among the very young and the very old. However, so many people are infected in a major epidemic that the total number of deaths is often large.

INFLUENZA VACCINES

Thus far, it has not been possible to make a vaccine for influenza that gives long-term immunity to the general population. Although it is not difficult to make a vaccine for a particular antigenic strain of virus, each new strain of circulating virus must be identified in time, usually about February, for the useful development and distribution of a new vaccine later that year. Strains of the influenza virus are collected in about 100 centers worldwide, then analyzed in central laboratories. This information is then used to decide on the composition of the vaccines to be offered for the next flu season. The vaccines are usually *multivalent*—directed at the three most important strains in circulation at the time. At present, influenza viruses for manufacturing vaccines are grown in embryonated egg cultures. The vaccines are usually 70 to 90% effective, but the duration of protection is probably no more than three years for that strain.

A vaccine that can be administered as a nasal spray was recently introduced and can be used for children aged 1 to 5. (This vaccine may be extended to other age groups). Research into influenza vaccines that are more effective and, especially, more rapidly and easily produced has a high priority. A major problem is that production methods that require growth in egg embryos are clumsy and labor intensive. They also require an unacceptably extended time to respond to the appearance of new viral strains. Furthermore, viruses often grow poorly in egg embryos, and viruses that cause poultry disease usually kill egg embryos. A theoretically promising technology to cope with such problems is to use *reverse genetics*. The viral genome of RNA is converted to DNA and then manipulated to remove the genes causing pathogenicity. The DNA is then converted back to RNA for vaccine production. This procedure results, rather quickly, in a nonpathogenic version of the virus that can be grown in egg embryos or cell culture. It also minimizes the chances of harmful contaminants.

In any discussion of influenza, the great pandemic of 1918–1919 must be mentioned.* Worldwide, more than 20 million people died. No one is sure why it was so unusually lethal. Today, the very young and very old are the principal victims, but in 1918–1919, young adults had the highest mortality rate, often dying within a few hours. The infection is usually restricted to the upper respiratory system, but some change in virulence allowed the virus to invade the lungs and cause viral pneumonia.

Evidence also suggests that the virus was able to infect cells in many organs of the body. In 2005, analysis of material preserved from the lungs of U.S. soldiers killed by the flu and from the exhumed body of a victim buried in permanently frozen soil in Alaska led to the complete genetic sequencing of the 1918 virus. The process of reverse genetics was then used to recreate the virus and grow it in chicken embryos and mice. The conclusion is that the pandemic was caused by an avian virus with ten changes in amino acids. These changes resulted in formation of a lethal gene that allowed the virus to attach itself to human lung cells and multiply. Without the gene, the virus lost its exceptional virulence.

Bacterial complications also frequently accompanied the infection and, in those preantibiotic days, were often fatal. The 1918 viral strain apparently became endemic in the U.S. swine population and may have originated there. (See the box in Chapter 13 on page 406.) Occasionally, influenza is still spread to humans from this reservoir, but the disease has not propagated like the virulent disease of 1918 did.

DIAGNOSIS OF INFLUENZA

Influenza is difficult to diagnose reliably from clinical symptoms, which are shared with numerous respiratory diseases. However, there are now several commercially available techniques that can diagnose influenza A and B within 20 minutes from a swabbed sample in a physician's office. These rapid tests are usually more than 70% sensitive and 90% specific. A central laboratory with sophisticated equipment is required to identify viral strains.

TREATMENT OF INFLUENZA

The antiviral drugs amantadine and rimantadine significantly reduce the symptoms of influenza A if administered promptly. More recently, two drugs for treatment of influenza have been introduced. They are inhibitors of neuraminidase, which the virus uses to separate itself from the host cell after it replicates. These are zanamivir (Relenza), which is inhaled, and oseltamivir phosphate (Tamiflu), which is administered orally. If taken within 30 hours of onset of influenza, these drugs can shorten the duration of symptoms. Neither drug should be considered a substitute for vaccination. The bacterial complications of influenza are amenable to treatment with antibiotics.

FUNGAL DISEASES OF THE LOWER RESPIRATORY SYSTEM

LEARNING OBJECTIVE

- List the causative agent, mode of transmission, preferred treatment, and laboratory identification tests for four fungal diseases of the respiratory system.

Fungi often produce spores that are disseminated through the air. It is therefore not surprising that several serious fungal diseases affect the lower respiratory system. The rate of fungal infections has been increasing in recent years. Opportunistic fungi are able to grow in immunosuppressed patients, and AIDS, transplant drugs, and anticancer drugs have created more immunosuppressed people than ever before.

---

*There will always be uncertainty concerning the origin of this most famous pandemic. The best reliable reports place the first well-documented cases among U.S. Army recruits at Camp Funston, Kansas, in March of 1918. The initial wave of influenza was a relatively mild illness that spread rapidly among the crowded troops and reached France as they were dispatched overseas. There the virus underwent a lethal mutation, seriously incapacitating troops on both sides of the front. Military censorship concealed this, and the first newspaper descriptions were published when the outbreak reached the population of neutral Spain, hence the name assigned to the pandemic: the **Spanish flu.** This second wave of influenza, with its high mortality, soon spread throughout the world and reentered the United States in the autumn and winter of 1918.

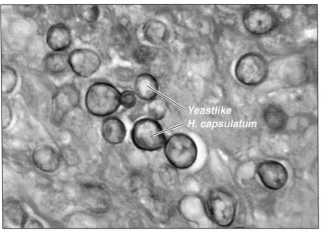

(a) Yeastlike form typical of growth in tissue at 37°C. Notice that one cell near the center is budding.

LM · 5 μm

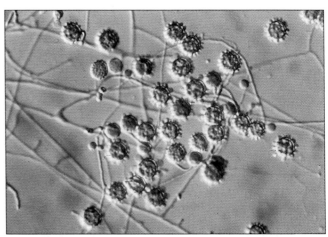

(b) Filamentous, spore-forming phase found in soil or at temperatures below 35°C; the spores are usually the infectious particle.

LM 20 μm

FIGURE 24.17 *Histoplasma capsulatum*, **a dimorphic fungus that causes histoplasmosis.**

Q What does the term *dimorphic* mean?

HISTOPLASMOSIS

Histoplasmosis superficially resembles tuberculosis. In fact, it was first recognized as a widespread disease in the United States when X-ray surveys showed lung lesions in many people who were tuberculin-test negative. Although the lungs are most likely to be initially infected, the pathogens may spread in the blood and lymph, causing lesions in almost all organs of the body.

FIGURE 24.18 Histoplasmosis distribution. Gold indicates the U.S. geographic distribution. SOURCE: CDC.

Q Compared with the disease distribution shown in the map in Figure 24.20, what can you determine about the moisture requirements in the soil for the two fungi involved?

Symptoms are usually poorly defined and mostly subclinical, and the disease passes for a minor respiratory infection. In a few cases, perhaps fewer than 0.1%, histoplasmosis becomes progressive and is a severe, generalized disease. This occurs with an unusually heavy inoculum or upon reactivation, when the infected person's immune system is compromised.

The causative organism, *Histoplasma capsulatum* (histō-plaz′mä kap-su-lä′tum), is a dimorphic fungus; that is, it has a yeastlike morphology in tissue growth (Figure 24.17a), and, in soil or artificial media, it forms a filamentous mycelium carrying reproductive conidia (Figure 24.17b). In the body, the yeastlike form is found intracellularly in macrophages, where it survives and multiplies.

Although histoplasmosis is rather widespread throughout the world, it has a limited geographic range in the United States (Figure 24.18). In general, the disease is found in the states adjoining the Mississippi and Ohio rivers. More than 75% of the population in some of these states have antibodies against the infection. In other states—Maine, for example—a positive test is a rare event. Approximately 50 deaths are reported in the United States each year from histoplasmosis.

Humans acquire the disease from airborne conidia produced under conditions of appropriate moisture and pH levels. These conditions occur especially where droppings from birds and bats have accumulated. Birds themselves, because of their high body temperature, do not carry the disease, but their droppings provide nutrients,

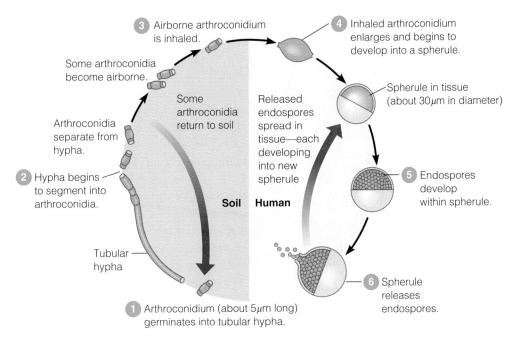

③ Airborne arthroconidium
is inhaled.

④ Inhaled arthroconidium
enlarges and begins to
develop into a spherule.

Some arthroconidia
become airborne.

Some
arthroconidia
return to soil

Released
endospores
spread in
tissue—each
developing
into new
spherule

Spherule in tissue
(about 30μm in diameter)

Arthroconidia
separate from
hypha.

② Hypha begins
to segment into
arthroconidia.

Soil   Human

⑤ Endospores
develop
within spherule.

Tubular
hypha

⑥ Spherule
releases
endospores.

① Arthroconidium (about 5μm long)
germinates into tubular hypha.

**FIGURE 24.19   The life cycle of *Coccidioides immitis*, the cause
of coccidioidomycosis.**

Q   What is the natural habitat of *Coccidioides*?

particularly a source of nitrogen, for the fungus. Bats, which have a lower body temperature than birds, carry the fungus, shed it in their feces, and infect new soil sites.

Clinical signs and history, serological tests, DNA probes, and, most importantly, either isolation of the pathogen or its identification in tissue specimens are necessary for proper diagnosis. Currently, the most effective chemotherapy is with amphotericin B or itraconazole.

COCCIDIOIDOMYCOSIS

Another fungal pulmonary disease, also rather restricted geographically, is **coccidioidomycosis.** The causative agent is *Coccidioides immitis,* a dimorphic fungus. The arthroconidia are found in dry, alkaline soils of the American Southwest and in similar soils of South America and northern Mexico. Because of its frequent occurrence in the San Joaquin Valley of California, it is sometimes known as *Valley fever* or *San Joaquin fever*. In tissues, the organism forms a thick-walled body called a *spherule* filled with endospores (Figure 24.19). In soil, it forms filaments that reproduce by the formation of arthroconidia. The wind carries the arthroconidia to transmit the infection. Arthroconidia are often so abundant that simply driving through an endemic area can result in infection, especially during a dust storm. An estimated 100,000 infections occur each year.

Most infections are not apparent, and almost all patients recover in a few weeks, even without treatment. The symptoms of coccidioidomycosis include chest pain and perhaps fever, coughing, and weight loss. In less than 1% of cases, a progressive disease resembling tuberculosis disseminates throughout the body. A substantial proportion of adults who are long-time residents of areas where the disease is endemic have evidence of prior infection with *C. immitis* by the skin test.

The resemblance to tuberculosis is so close that isolation of the pathogen is necessary to properly diagnose coccidioidomycosis. Diagnosis is most reliably made by identifying the spherules in tissue or fluids. The organism can be cultured from fluids or lesions, but laboratory workers must use great care because of the possibility of infectious aerosols. Several serological tests and DNA probes are available for identifying isolates. A tuberculin-like skin test is used in screening.

The incidence of coccidiodomycosis has been increasing recently in California and Arizona (Figure 24.20). Contributing factors include an increased number of older residents, an increased prevalence of HIV/AIDS, and a severe drought in California that facilitated dustborne transmission. About 50 to 100 deaths occur annually from this disease in the United States.

Amphotericin B has been used to treat serious cases. However, less toxic imidazole drugs, such as ketoconazole and itraconazole, are useful alternatives.

San
Joaquin
Valley

2001
outbreak

Areas where
disease is known
to be endemic

**FIGURE 24.20 The U.S. endemic area for coccid-
ioidomycosis.** The outlined area in California is the San
Joaquin Valley. Because of the very high incidence there, the disease is also sometimes called Valley fever. The small area on the
map in northeastern Utah indicates an outbreak in 2001 in which
ten archeologists working on excavations at the Dinosaur
National Monument were infected.
SOURCE: CDC, 2004.

Q Why does the incidence of coccidioidomycosis increase
after ecological disturbances, such as earthquakes and
construction?

PNEUMOCYSTIS PNEUMONIA

***Pneumocystis* pneumonia** is caused by *Pneumocystis
jiroveci* (ye-rō'vet-zē), formerly *P. carinii* (Figure 24.21).
The taxonomic position of this microbe has been uncertain ever since its discovery in 1909, when it was thought
to be a developmental stage of a trypanosome. Since that
time, there has been no universal agreement about
whether it is a protozoan or a fungus. It has some characteristics of both groups. Analysis of RNA and certain other
structural characteristics indicate that it is closely related
to certain yeasts and it is usually reported as a fungus.

The pathogen is sometimes found in healthy human
lungs. Immunocompetent adults have few or no symptoms,
but newly infected infants occasionally show symptoms of a
lung infection. Persons with compromised immunity are the
most susceptible to symptomatic *Pneumocystis* pneumonia.
This population may also be the reservoir of the organism,
which is not found in the environment, animals, or very often in healthy humans. This portion of the population has
also expanded greatly in recent decades. For example, before

the AIDS epidemic, *Pneumocystis* pneumonia was an uncommon disease; perhaps 100 cases occurred each year. By
1993, it had become a primary indicator of AIDS, with more
than 20,000 annual reported cases. Presumably, the loss of
an effective immune defense allowed the activation of a latent infection. Other groups that are very susceptible to this
disease are people whose immunity is depressed because of
cancer or who are receiving immunosuppressive drugs to
minimize rejection of transplanted tissue.

In the human lung, the microbes are found mostly in
the lining of the alveoli. Diagnosis is usually made from
sputum samples in which cysts are detected. There, they
form a thick-walled cyst in which spherical intracystic
bodies successively divide as part of a sexual cycle. The
mature cyst contains eight such bodies (see Figure 24.21).
Eventually the cyst ruptures and releases them, and each
body develops into a trophozoite. The trophozoite cells
can reproduce asexually by fission, but they may also enter
the encysted sexual stage (Figure 24.22).

The drug of choice for treatment is currently
trimethoprim-sulfamethoxazole, but there are several
alternatives.

BLASTOMYCOSIS
(NORTH AMERICAN BLASTOMYCOSIS)

Blastomycosis is usually called **North American
blastomycosis** to differentiate it from a similar South
American blastomycosis. It is caused by the fungus
Blastomyces dermatitidis (blas-tō-mī'sēz dėr-mä-tit'i-dis),
a dimorphic fungus found most often in the Mississippi
valley, where it probably grows in soil. Approximately 30
to 60 deaths are reported each year, although most infections are asymptomatic.

The infection begins in the lungs and can spread rapidly. Cutaneous ulcers commonly appear, and there is extensive abscess formation and tissue destruction. The
pathogen can be isolated from pus and biopsy specimens.
Amphotericin B is usually an effective treatment.

OTHER FUNGI INVOLVED
IN RESPIRATORY DISEASE

Many other opportunistic fungi may cause respiratory disease, particularly in immunosuppressed hosts or when there
is exposure to massive numbers of spores. **Aspergillosis** is
an important example; it is airborne by the conidia of
Aspergillus fumigatus (fù-mi-gä'tus) and other species of *Aspergillus*, which are widespread in decaying vegetation. Compost piles are ideal sites for growth, and farmers and gardeners
are most often exposed to infective amounts of these conidia.

Similar pulmonary infections sometimes result when individuals are exposed to spores of other mold genera, such as
Rhizopus and *Mucor*. Such diseases can be very dangerous,

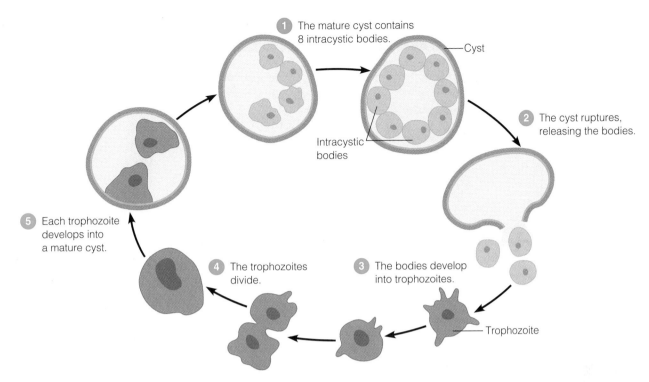

① The mature cyst contains 8 intracystic bodies.

Cyst

② The cyst ruptures, releasing the bodies.

Intracystic bodies

③ The bodies develop into trophozoites.

Trophozoite

④ The trophozoites divide.

⑤ Each trophozoite develops into a mature cyst.

FIGURE 24.21 The life cycle of *Pneumocystis jiroveci*, the cause of *Pneumocystis* pneumonia. Long classified as a protozoan, this organism is now usually considered to be a fungus, but it has characteristics of both groups.

Q **Of what value is proper classification of this organism?**

particularly invasive infections of pulmonary aspergillosis. Predisposing factors include an impaired immune system, cancer, and diabetes. As with most systemic fungal infections, there is only a limited arsenal of antifungal agents available; amphotericin B has proved the most useful.

* * *

Table 24.3 summarizes the microbial respiratory diseases affecting the lower respiratory system discussed in this chapter.

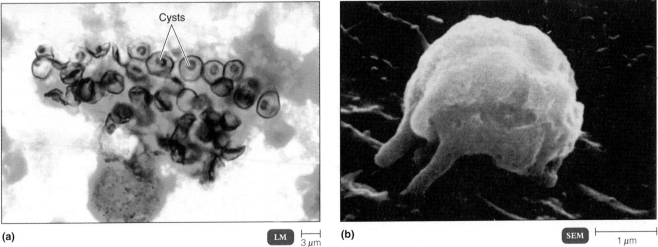

Cysts

(a)   LM ⊢──┤ 3 μm

(b)   SEM ⊢──┤ 1 μm

FIGURE 24.22 *Pneumocystis jiroveci*. **(a)** Cysts of *P. jiroveci* from a smear made from alveolar tissue of a lung. **(b)** Trophozoite stage of *P. jiroveci* adhering to the surface of a chick embryo epithelial lung cell. The tubular extensions are used to extract nutrients from the host.

Q **Does the person with such cysts necessarily have *Pneumocystis* pneumonia?**

TABLE 24.3	Microbial Diseases of the Lower Respiratory System			
	Pathogen	Notes	Reservoir	Treatment
Bacterial Diseases				
Bacterial pneumonia. (See Diseases in Focus, page 725).				
Pertussis (whooping cough)	*Bordetella pertussis*	Cilia in upper respiratory tract inactivated, mucus accumulates, spasms of intense coughing to clear mucus; diagnosis by culturing bacteria	Humans	Erythromycin Prevention: DTaP vaccine
Tuberculosis	*Mycobacterium tuberculosis Mycobacterium bovis*	Tubercle bacilli entering lungs survive phagocytosis, reproduce in macrophages; tubercles formed to isolate pathogen; defenses eventually fail, and infection becomes systemic; diagnosis by acid fast or FA of sputum	Humans, cows; can be transmitted via unpasteurized milk	Multiple-antimycobacterial drugs Prevention: pasteurizing milk; BCG vaccine
Melioidosis	*Burkholderia pseudomallei*	Most common in Southeast Asia and northern Australia. Most frequently presents as pneumonia, but also seen as tissue abscesses and severe sepsis; diagnosis by culturing bacteria	Moist soil	Ceftazidime
Viral Diseases				
Respiratory syncytial virus (RSV) disease	Respiratory syncytial virus	A serious respiratory disease of infants; diagnosis by serologic tests for antibodies or virus	Humans	Palivizumab (if life-threatening)
Influenza	*Influenzavirus;* several serotypes	An illness characterized by chills, fever, headache, and muscular aches; diagnosis by EIA	Humans, pigs, birds	Amantadine, oseltamivir phosphate (Tamiflu)
Fungal Diseases				
Histoplasmosis	*Histoplasma capsulatum*	Resembles tuberculosis; occasionally fatal; diagnosis by isolation of fungus	Soil; widespread in Ohio and Mississippi river valleys	Amphotericin B
Coccidioidomycosis	*Coccidioides immitis*	Fever, coughing, weight loss; occasionally fatal; diagnosis by serological tests	Desert soils of U.S. Southwest	Amphotericin B
Pneumocystis pneumonia	*Pneumocystis jiroveci*	Pneumonia; a common, serious complication of AIDS; no symptoms in people with healthy immune system; diagnosis by microscopy	Unknown; possibly humans or soil	Trimethoprim-sulfamethoxazole
Blastomycosis	*Blastomyces dermatitidis*	Abscesses; extensive tissue damage; diagnosis by isolation of fungus	Soil in Mississippi Valley area	Amphotericin B

STUDY OUTLINE

INTRODUCTION (p. 711)

1. Infections of the upper respiratory system are the most common type of infection.
2. Pathogens that enter the respiratory system can infect other parts of the body.

STRUCTURE AND FUNCTION OF THE RESPIRATORY SYSTEM (p. 712)

1. The upper respiratory system consists of the nose, pharynx, and associated structures, such as the middle ear and auditory tubes.
2. Coarse hairs in the nose filter large particles from air entering the respiratory tract.
3. The ciliated mucous membranes of the nose and throat trap airborne particles and remove them from the body.
4. Lymphoid tissue, tonsils, and adenoids provide immunity to certain infections.
5. The lower respiratory system consists of the larynx, trachea, bronchial tubes, and alveoli.
6. The ciliary escalator of the lower respiratory system helps prevent microorganisms from reaching the lungs.
7. Microbes in the lungs can be phagocytized by alveolar macrophages.
8. Respiratory mucus contains IgA antibodies.

NORMAL MICROBIOTA OF THE RESPIRATORY SYSTEM (pp. 712–713)

1. The normal microbiota of the nasal cavity and throat can include pathogenic microorganisms.
2. The lower respiratory system is usually sterile because of the action of the ciliary escalator.

MICROBIAL DISEASES OF THE UPPER RESPIRATORY SYSTEM (pp. 713–717)

1. Specific areas of the upper respiratory system can become infected to produce pharyngitis, laryngitis, tonsillitis, sinusitis, and epiglottitis.
2. These infections may be caused by several bacteria and viruses, often in combination.
3. Most respiratory tract infections are self-limiting.
4. *H. influenzae* type b can cause epiglottitis.

BACTERIAL DISEASES OF THE UPPER RESPIRATORY SYSTEM (pp. 714–717)

STREPTOCOCCAL PHARYNGITIS (STREP THROAT) (p. 714)

1. This infection is caused by group A beta-hemolytic streptococci, the group that consists of *Streptococcus pyogenes*.
2. Symptoms of this infection are inflammation of the mucous membrane and fever; tonsillitis and otitis media may also occur.
3. Rapid diagnosis is made by enzyme immunoassays.
4. Penicillin is used to treat streptococcal pharyngitis.
5. Immunity to streptococcal infections is type-specific.
6. Strep throat is usually transmitted by droplets.

SCARLET FEVER (pp. 714–715)

7. Strep throat, caused by an erythrogenic toxin-producing *S. pyogenes*, results in scarlet fever.
8. *S. pyogenes* produces erythrogenic toxin when lysogenized by a phage.
9. Symptoms include a red rash, high fever, and a red, enlarged tongue.

DIPHTHERIA (pp. 715–716)

10. Diphtheria is caused by exotoxin-producing *Corynebacterium diphtheriae*.
11. Exotoxin is produced when the bacteria are lysogenized by a phage.
12. A membrane, containing fibrin and dead human and bacterial cells, forms in the throat and can block the passage of air.
13. The exotoxin inhibits protein synthesis, and heart, kidney, or nerve damage may result.
14. Laboratory diagnosis is based on isolation of the bacteria and the appearance of growth on differential media.
15. Antitoxin must be administered to neutralize the toxin, and antibiotics can stop growth of the bacteria.
16. Routine immunization in the United States includes diphtheria toxoid in the DTaP vaccine.
17. Slow-healing skin ulcerations are characteristic of cutaneous diphtheria.
18. There is minimal dissemination of the exotoxin in the bloodstream.

OTITIS MEDIA (pp. 716–717)

20. Earache, or otitis media, can occur as a complication of nose and throat infections.

21. Pus accumulation causes pressure on the eardrum.

22. Bacterial causes include *Streptococcus pneumoniae*, nonencapsulated *Haemophilus influenzae*, *Moraxella catarrhalis*, *Streptococcus pyogenes*, and *Staphylococcus aureus*.

VIRAL DISEASES OF THE UPPER RESPIRATORY SYSTEM (p. 717)

THE COMMON COLD (p. 717)

1. Any one of approximately 200 different viruses can cause the common cold; rhinoviruses cause about 50% of all colds.

2. Symptoms include sneezing, nasal secretions, and congestion.

3. Sinus infections, lower respiratory tract infections, laryngitis, and otitis media can occur as complications of a cold.

4. Colds are most often transmitted by indirect contact.

5. Rhinoviruses grow best slightly below body temperature.

6. The incidence of colds increases during cold weather, possibly because of increased interpersonal indoor contact or physiological changes.

7. Antibodies are produced against the specific viruses.

MICROBIAL DISEASES OF THE LOWER RESPIRATORY SYSTEM (pp. 718–737)

1. Many of the same microorganisms that infect the upper respiratory system also infect the lower respiratory system.

2. Diseases of the lower respiratory system include bronchitis and pneumonia.

BACTERIAL DISEASES OF THE LOWER RESPIRATORY SYSTEM (pp. 718–730)

PERTUSSIS (WHOOPING COUGH) (pp. 718–719)

1. Pertussis is caused by *Bordetella pertussis*.

2. The initial stage of pertussis resembles a cold and is called the catarrhal stage.

3. The accumulation of mucus in the trachea and bronchi causes deep coughs characteristic of the paroxysmal (second) stage.

4. The convalescence (third) stage can last for months.

5. Laboratory diagnosis is based on isolation of the bacteria on enrichment and selective media, followed by serological tests.

6. Regular immunization for children has decreased the incidence of pertussis.

TUBERCULOSIS (pp. 719–723)

7. Tuberculosis is caused by *Mycobacterium tuberculosis*.

8. Large amounts of lipids in the cell wall account for the bacterium's acid-fast characteristic as well as its resistance to drying and disinfectants.

9. *M. tuberculosis* may be ingested by alveolar macrophages; if not killed, the bacteria reproduce in the macrophages.

10. Lesions formed by *M. tuberculosis* are called tubercles; dead macrophages and bacteria form the caseous lesion that might calcify and appear in an X ray as a Ghon's complex.

11. Liquefaction of the caseous lesion results in a tuberculous cavity in which *M. tuberculosis* can grow.

12. New foci of infection can develop when a caseous lesion ruptures and releases bacteria into blood or lymph vessels; this is called miliary tuberculosis.

13. Miliary tuberculosis is characterized by weight loss, coughing, and loss of vigor.

14. Chemotherapy usually involves 3 or 4 drugs taken for at least 6 months; multidrug-resistant *M. tuberculosis* is becoming prevalent.

15. A positive tuberculin skin test can indicate either an active case of TB, prior infection, or vaccination and immunity to the disease.

16. Laboratory diagnosis is based on the presence of acid-fast bacilli and isolation of the bacteria, which requires incubation of up to 8 weeks.

17. *Mycobacterium bovis* causes bovine tuberculosis and can be transmitted to humans by unpasteurized milk.

18. *M. bovis* infections usually affect the bones or lymphatic system.

19. BCG vaccine for tuberculosis consists of a live, avirulent culture of *M. bovis*.

20. *M. avium-intracellulare* complex infects patients in the late stages of HIV infection.

BACTERIAL PNEUMONIAS (pp. 723–730)

21. Typical pneumonia is caused by *S. pneumoniae*.

22. Atypical pneumonias are caused by other microorganisms.

Pneumococcal Pneumonia (pp. 724–726)

23. Pneumococcal pneumonia is caused by encapsulated *Streptococcus pneumoniae*.

24. Symptoms are fever, breathing difficulty, chest pain, and rust-colored sputum.

25. The bacteria can be identified by the production of alpha-hemolysins, inhibition by optochin, bile solubility, and through serological tests.

26. A vaccine consists of purified capsular material from 23 serotypes of *S. pneumoniae*.

Haemophilus influenzae Pneumonia (p. 726)

27. Alcoholism, poor nutrition, cancer, and diabetes are predisposing factors for *H. influenzae* pneumonia.

28. *H. influenzae* is a gram-negative coccobacillus.

Mycoplasmal Pneumonia (p. 726)

29. *Mycoplasma pneumoniae* causes mycoplasmal pneumonia; it is an endemic disease.

30. *M. pneumoniae* produces small "fried-egg" colonies after two weeks' incubation on enriched media containing horse serum and yeast extract.

31. Diagnosis is by PCR or serological tests.

Legionellosis (pp. 726–727)

32. The disease is caused by the aerobic gram-negative rod *Legionella pneumophila*.

33. The bacterium can grow in water, such as air-conditioning cooling towers, and then be disseminated in the air.

34. This pneumonia does not appear to be transmitted from person to person.

35. Bacterial culture, FA tests, and DNA probes are used for laboratory diagnosis.

Psittacosis (Ornithosis) (p. 727)

36. *Chlamydophila psittaci* is transmitted by contact with contaminated droppings and exudates of fowl.

37. Elementary bodies allow the bacteria to survive outside a host.

38. Commercial bird handlers are most susceptible to this disease.

39. The bacteria are isolated in embryonated eggs, mice, or cell culture; identification is based on FA staining.

Chlamydial Pneumonia (p. 729)

40. *Chlamydophila pneumoniae* causes pneumonia; it is transmitted from person to person.

41. Tetracycline is used for treatment.

Q Fever (pp. 729–730)

42. Obligately parasitic, intracellular *Coxiella burnetii* causes Q fever.

43. The disease is usually transmitted to humans through unpasteurized milk or inhalation of aerosols in dairy barns.

44. Laboratory diagnosis is made with the culture of bacteria in embryonated eggs or cell culture.

MELIOIDOSIS (p. 730)

45. Melioidosis, caused by *Burkholderia pseudomallei*, is transmitted by inhalation, ingestion, or through puncture wounds. Symptoms include pneumonia, sepsis, and encephalitis.

VIRAL DISEASES OF THE LOWER RESPIRATORY SYSTEM (pp. 730–733)

VIRAL PNEUMONIA (p. 730)

1. A number of viruses can cause pneumonia as a complication of infections such as influenza.

2. The etiologies are not usually identified in a clinical laboratory because of the difficulty in isolating and identifying viruses.

RESPIRATORY SYNCYTIAL VIRUS (RSV) (pp. 730–731)

3. RSV is the most common cause of pneumonia in infants.

INFLUENZA (FLU) (pp. 731–733)

4. Influenza is caused by *Influenzavirus* and is characterized by chills, fever, headache, and general muscular aches.

5. Hemagglutinin (H) and neuraminidase (N) spikes project from the outer lipid bilayer of the virus.

6. Viral strains are identified by antigenic differences in the H and N spikes; they are also divided by antigenic differences in their protein coats (A, B, and C).

7. Viral isolates are identified by hemagglutination-inhibition tests and immunofluorescence testing with monoclonal antibodies.

8. Antigenic shifts that alter the antigenic nature of the H and N spikes make natural immunity and vaccination of questionable value. Minor antigenic changes are caused by antigenic drift.

9. Deaths during an influenza epidemic are usually from secondary bacterial infections.

10. Multivalent vaccines are available for the elderly and other high-risk groups.

11. Amantadine and rimantadine are effective prophylactic and curative drugs against influenza A virus.

FUNGAL DISEASES OF THE LOWER RESPIRATORY SYSTEM (pp. 733–737)

1. Fungal spores are easily inhaled; they may germinate in the lower respiratory tract.

2. The incidence of fungal diseases has been increasing in recent years.

3. The mycoses in the sections below can be treated with amphotericin B.

HISTOPLASMOSIS (pp. 733–734)

4. *Histoplasma capsulatum* causes a subclinical respiratory infection that only occasionally progresses to a severe, generalized disease.

5. The disease is acquired by inhalation of airborne conidia.

6. Isolation of the fungus or identification of the fungus in tissue samples is necessary for diagnosis.

COCCIDIOIDOMYCOSIS (p. 735)

7. Inhalation of the airborne arthroconidia of *Coccidioides immitis* can result in coccidioidomycosis.

8. Most cases are subclinical, but when there are predisposing factors such as fatigue and poor nutrition, a progressive disease resembling tuberculosis can result.

PNEUMOCYSTIS PNEUMONIA (p. 736)

9. *Pneumocystis jiroveci* is found in healthy human lungs.

10. *P. jiroveci* causes disease in immunosuppressed patients.

BLASTOMYCOSIS (NORTH AMERICAN BLASTOMYCOSIS) (p. 736)

11. *Blastomyces dermatitidis* is the causative agent of blastomycosis.

12. The infection begins in the lungs and can spread to cause extensive abscesses.

OTHER FUNGI INVOLVED IN RESPIRATORY DISEASE (pp. 736–737)

13. Opportunistic fungi can cause respiratory disease in immunosuppressed hosts, especially when large numbers of spores are inhaled.

14. Among these fungi are *Aspergillus*, *Rhizopus*, and *Mucor*.

STUDY QUESTIONS

Access more review material either online at **The Microbiology Place** (www.microbiologyplace.com) or with **The Microbiology Place CD-ROM** packaged with your new book. There you'll find activities, practice tests, quizzes, flashcards, case studies, and more to help you succeed.

Answers to the Study Questions can be found in Appendix G.

REVIEW

1. Respiratory diseases are usually transmitted by _____.

2. Describe how microorganisms are prevented from entering the upper respiratory system. How are they prevented from causing infections in the lower respiratory system?

3. How do the normal microbiota of the respiratory system illustrate microbial antagonism?

4. Compare and contrast mycoplasmal pneumonia and viral pneumonia.

5. How is otitis media contracted? What causes otitis media? Why was otitis media included in a chapter on diseases of the respiratory system?

6. List the causative agent, symptoms, and treatment for four viral diseases of the respiratory system. Separate the diseases according to whether they infect the upper or lower respiratory system.

7. Complete the following table.

Disease	Causative Agent	Symptoms	Treatment
Streptococcal pharyngitis			
Scarlet fever			
Diphtheria			
Whooping cough			
Tuberculosis			
Pneumococcal pneumonia			
H. influenzae pneumonia			
Chlamydial pneumonia			

Disease	Causative Agent	Symptoms	Treatment
Legionellosis			
Psittacosis			
Q fever			
Epiglottitis			
Melioidosis			

8. Under what conditions can the saprophytes *Aspergillus* and *Rhizopus* cause infections?

9. A patient has been diagnosed as having pneumonia. Is this sufficient information to begin treatment with antimicrobial agents? Briefly discuss why or why not.

10. List the causative agent, mode of transmission, and endemic area for the diseases histoplasmosis, coccidioidomycosis, blastomycosis, and *Pneumocystis* pneumonia.

11. Briefly describe the procedures and positive results of the tuberculin test and what is indicated by a positive test.

12. Discuss reasons for the increased incidences of colds and pneumonias during cold weather.

13. Match the bacteria involved in respiratory infections with the following laboratory test results:

Gram-positive cocci

    Catalase-positive _____ S. aureus _____

      Catalase-negative

        Beta-hemolytic, bacitracin inhibition _____ S. pyogenes _____

        Alpha-hemolytic, optochin inhibition _____ C. pneumonia _____

Gram-positive rods

    Not acid-fast _____ C. diphtheria _____

    Acid-fast _____ Mycobacterium tuberculosis _____

Gram-negative cocci _____ Moraxella catarrhalis _____

Gram-negative rods

    Aerobes

      Coccobacilli _____ Bordetella pertussis _____

    Rods

      Grow on nutrient agar _____ Burkholderia pseudom _____

      Require special media _____ Legionella pneumophi _____

Facultative anaerobes

Coccobacilli ___*Haemophilus influenzae*___

Intracellular parasites

Form elementary bodies ___*Chlamydophila psittaci*___

Do not form elementary bodies ___*Coxiella burnetii*___

Wall-less ___*Mycoplasma pneumoniae*___

MULTIPLE CHOICE

1. A patient has fever, difficulty breathing, chest pains, fluid in the alveoli, and a positive tuberculin skin test. Gram-positive cocci are isolated from the sputum. The recommended treatment is
   - **a.** penicillin.
   - **b.** antitoxin.
   - **c.** isoniazid.
   - **d.** tetracyclines.
   - **e.** none of the above

2. No bacterial pathogen can be isolated from the sputum of a patient with pneumonia. Antibiotic therapy has not been successful. The next step should be
   - **a.** culturing for *Mycobacterium tuberculosis*.
   - **b.** culturing for *Mycoplasma pneumoniae*.
   - **c.** culturing for fungi.
   - **d.** a change in antibiotics.
   - **e.** none; nothing more can be done.

Match the following choices to the culture descriptions in questions 3 through 6:
   - **a.** *Chlamydophila*
   - **b.** *Coccidioides*
   - **c.** *Histoplasma*
   - **d.** *Mycobacterium*
   - **e.** *Mycoplasma*

e 3. Your culture from a pneumonia patient appears not to have grown. You do see colonies, however, when the plate is viewed at 100×.

a 4. This pneumonia etiology requires cell culture.

c 5. Microscopic examination of a lung biopsy shows ovoid cells in macrophages. You suspect these are the cause of the patient's symptoms, but your culture grows a filamentous organism.

b 6. Microscopic examination of a lung biopsy shows spherules.

7. In San Francisco, ten animal health care technicians developed pneumonia two weeks after 130 goats were moved to the animal shelter where they worked. Which of the following is *not* true?
   - **a.** Diagnosis is made by a blood agar culture of sputum.
   - **b.** The cause is *Coxiella burnetii*.
   - **c.** The cause is rickettsia.
   - **d.** The disease was transmitted by aerosols.
   - **e.** Diagnosis is made by complement-fixation tests for antibodies.

8. Which of the following leads to all the rest?
   - **a.** catarrhal stage
   - **b.** cough
   - **c.** loss of cilia
   - **d.** mucus accumulation
   - **e.** trachaeal cytotoxin

Match the following choices to the statements in questions 9 and 10:
   - **a.** *Bordetella pertussis*
   - **b.** *Corynebacterium diphtheriae*
   - **c.** *Legionella pneumophila*
   - **d.** *Mycobacterium tuberculosis*
   - **e.** none of the above

9. Causes the formation of a membrane across the throat. *b*

10. Resistant to destruction by phagocytes. *d*

CRITICAL THINKING

1. Differentiate between *S. pyogenes* causing strep throat and *S. pyogenes* causing scarlet fever.

2. Why might the influenza vaccine be less effective than other vaccines?

3. Explain why it would be impractical to include cold and influenza vaccinations in the required childhood vaccinations.

CLINICAL APPLICATIONS

1. In August, a 24-year-old man from Virginia developed difficulty breathing and bilateral lobe infiltrates two months after driving through California. During initial evaluation, typical pneumonia was suspected, and he was treated with antibiotics. Efforts to diagnose the pneumonia were unsuccessful. In October, a laryngeal mass was detected, and laryngeal cancer was suspected; treatment with steroids and bronchodilators did not result in improvement. Lung biopsy and laryngoscopy detected diffuse granular tissue. He was treated with amphotericin B and discharged after five days. What was the disease? What might have been done differently to decrease the patient's recovery time from three months to one week?

2. During a 6-month period, 72 clinic staff members became tuberculin-positive. A case-control study was undertaken to determine the most likely source of *M. tuberculosis* infection among the staff. A total of 16 cases and 34 tuberculin-negative controls were compared. Pentamidine isethionate is not used for TB treatment. What disease was probably being treated with this drug? What is the most likely source of infection?

	Cases	Control
Works ≥ 40 hr/week	100%	62%
In room during aerosolized pentamidine isethionate therapy in TB patients	31	3
Patient contact	94	94
Lunch eaten in staff lounge	38	35
Resident of western Palm Beach	75	65
Female	81	77
Cigarette smoker	6	15
Contact with nurse diagnosed with TB	15	12
In unventilated room during collection of TB-positive sputum samples	13	8

3. In March, six members of a family had fever, anorexia, sore throat, cough, headache, vomiting, and muscle pain. Two people were hospitalized. All six improved after therapy with doxycycline. Convalescent serum samples gave titers of 64 and 32. The family had purchased a cockatiel in mid-February and noticed that the bird was irritable. The bird was euthanized in April. A fluorescent-antibody test for antigens was diagnostic. What is the disease? How was it transmitted?

4. A 27-year-old male with a history of asthma was hospitalized with a 4-day history of progressive cough and two days of spiking fevers. Gram-positive cocci in pairs were cultured from a blood sample. What is the diagnosis?

5. Three weeks after working on the demolition of an abandoned building in Kentucky, a worker was hospitalized for acute respiratory illness. At the time of demolition, a colony of bats inhabited the building. An X-ray examination revealed a lung mass. A purified protein derivative test was negative; a cytological examination for cancer was also negative. The mass was surgically removed. Microscopic examination of the mass revealed ovoid yeast cells. What is the disease? How was it contracted?

25

Microbial Diseases of the Digestive System

Microbial diseases of the digestive system are second only to respiratory diseases as causes of illness in the United States. Most such diseases result from the ingestion of food or water contaminated with pathogenic microorganisms or their toxins. These pathogens usually enter the food or water supply after being shed in the feces of people or animals infected with them. Therefore, microbial diseases of the digestive system are typically transmitted by a **fecal–oral cycle.** This cycle is interrupted by effective sanitation practices in food production and handling. Modern methods of sewage treatment and disinfection of water are essential. There is also an increasing awareness that new tests that will rapidly and reliably detect pathogens in foods (a perishable commodity) will be required.

The CDC estimates that about 76 million cases of foodborne disease resulting in about 5000 deaths occur annually in the United States. As more of our food products—especially fruits and vegetables—are grown in countries with poor sanitation standards, outbreaks of foodborne disease from imported pathogens are expected to increase.

UNDER THE MICROSCOPE

Trichinella spiralis, **adult worm.** These small worms, about 1 mm in length, produce larvae that become encysted in muscle. Large numbers of these larvae in muscle cause the disease trichinellosis.

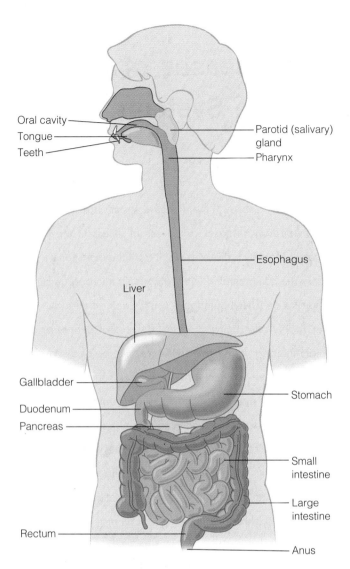

Oral cavity
Tongue
Teeth

Parotid (salivary) gland

Pharynx

Esophagus

Liver

Gallbladder
Duodenum
Pancreas

Stomach

Small intestine

Large intestine

Rectum

Anus

FIGURE 25.1 The human digestive system.

Q **Where are microorganisms normally found in the digestive system?**

STRUCTURE AND FUNCTION OF THE DIGESTIVE SYSTEM

LEARNING OBJECTIVE
• Name the structures of the digestive system that contact food.

The **digestive system** is essentially a tubelike structure, the *gastrointestinal (GI) tract* or *alimentary canal*—mainly the mouth, pharynx (throat), esophagus (food tube leading to the stomach), stomach, and the small and large intestines. It also includes *accessory structures* such as the teeth and tongue. Certain other accessory structures such as the salivary glands, liver, gallbladder, and pancreas lie outside the GI tract and produce secretions that are conveyed by ducts into it (Figure 25.1).

The purpose of the digestive system is to digest foods—that is, to break them down into small molecules that can be taken up and used by body cells. In a process called *absorption*, these end-products of digestion pass from the small intestine into the blood or lymph for distribution to body cells. Then the food moves through the large intestine, where water, vitamins, and nutrients are absorbed from it. Over the course of an average life span, about 25 tons of food pass through the GI tract. The resulting undigested solids, called *feces*, are eliminated from the body through the anus. Intestinal gas, or *flatus*, is a mixture of nitrogen from swallowed air and microbially produced carbon dioxide, hydrogen, and methane. On average, we produce 0.5 to 2.0 liters of flatus every day.

NORMAL MICROBIOTA OF THE DIGESTIVE SYSTEM

LEARNING OBJECTIVE
• List examples of the microbiota for each part of the gastrointestinal tract.

Bacteria heavily populate most of the digestive system. In the mouth, each milliliter of saliva can contain millions of bacteria. The stomach and small intestine have relatively few microorganisms because of the hydrochloric acid produced by the stomach and the rapid movement of food through the small intestine. By contrast, the large intestine has enormous microbial populations, exceeding 100 billion bacteria per gram of feces. (Up to 40% of fecal mass is microbial cell material.) The population of the large intestine is composed mostly of anaerobes and facultative anaerobes. Most of these bacteria assist in the enzymatic breakdown of foods, especially many polysaccharides that would otherwise be indigestible. Some of them synthesize useful vitamins.

It is important to understand that food passing through the tubelike GI tract, although it is in contact with the body, remains outside. Unlike the body's exterior, such as the skin, the GI tract is adapted to absorbing nutrients passing through it. However, at the same time that nutrients are absorbed from the GI tract, harmful microbes ingested in food and water must be kept from invading the body. An important factor in this defense is the high acid content of the stomach, which eliminates many potentially harmful ingested microbes. Curiously, it has been found that many bacteria (such as *E. coli* and *Salmonella*) survive for hours in stomach acid; but when nitric oxide (NO), produced in the stomach by oxidation of ingested nitrates, is combined with stomach acid, it kills these bacteria in less than an hour. This has suggested the possible use of antimicrobial therapies based on nitrate chemistry. (See the box in Chapter 16, page 486).

The small intestine also contains important antimicrobial defenses. Significant among these defenses are millions of specialized, granule-filled cells called *Paneth cells*. These are capable of phagocytizing bacteria, and they also produce antibacterial proteins called *defensins* (see antimicrobial peptides, page 605) and the antibacterial enzyme *lysozyme*.

BACTERIAL DISEASES OF THE MOUTH

LEARNING OBJECTIVE

• Describe the events that lead to dental caries and periodontal disease.

The mouth, which is the entrance to the digestive system, provides an environment that supports a large and varied microbial population.

DENTAL CARIES (TOOTH DECAY)

The teeth are unlike any other exterior surface of the body. They are hard and do not shed surface cells (Figure 25.2). This allows the accumulation of masses of microorganisms and their products. These accumulations, called **dental plaque,** are a type of biofilm (see page 820 in Chapter 27) and are intimately involved in the formation of **dental caries,** or tooth decay.

Oral bacteria convert sucrose and other carbohydrates into lactic acid, which in turn attacks the tooth enamel. The microbial population on and around the teeth is very complex. Based upon ribosomal identification methods (see the discussion of FISH on page 304 in Chapter 10), over 700 species of bacteria have been isolated from the oral cavity. Most of these cannot be cultivated by conventional methods. Probably the most important *cariogenic* (caries-causing) bacterium is *Streptococcus mutans*, a gram-positive coccus that is thought to be capable of metabolizing a wider range of carbohydrates than any other gram-positive organism. Some other species of streptococci are also cariogenic but play a lesser role in initiating caries.

The initiation of caries depends on the attachment of *S. mutans* or other streptococci to the tooth (Figure 25.3). These bacteria do not adhere to a clean tooth, but within minutes a freshly brushed tooth will become coated with a pellicle (thin film) of proteins from saliva. Within a couple of hours, cariogenic bacteria become established on this pellicle and begin to produce a gummy polysaccharide of glucose molecules called *dextran* (Figure 25.3b). In the production of dextran, the bacteria first hydrolyze sucrose into its component monosaccharides, fructose and glucose. The enzyme glucosyltransferase then assembles the glucose

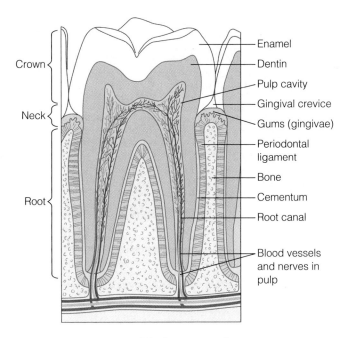

FIGURE 25.2 A healthy human tooth.

Q **Why can a biofilm accumulate on teeth?**

molecules into dextran. The residual fructose is the primary sugar fermented into lactic acid. Accumulations of bacteria and dextran adhering to the teeth make up dental plaque.

The bacterial population of plaque may harbor over 400 bacterial species but is predominantly streptococci and filamentous members of the genus *Actinomyces*. (Older, calcified deposits of plaque are called *dental calculus*, or *tartar*.) *S. mutans* especially favors crevices or other sites on the teeth protected from the shearing action of chewing or the flushing action of the liter or so of saliva produced in the mouth each day. On protected areas of the teeth, plaque accumulations can be several hundred cells thick. Because plaque is not very permeable to saliva, the lactic acid produced by bacteria is not diluted or neutralized, and it breaks down the enamel of the teeth to which the plaque adheres.

Although saliva contains nutrients that encourage the growth of bacteria, it also contains antimicrobial substances, such as lysozyme, that help protect exposed tooth surfaces. Some protection is also provided by *crevicular fluid*, a tissue exudate that flows into the gingival crevice (see Figure 25.2) and is closer in composition to serum than to saliva. It protects teeth by virtue of both its flushing action and its phagocytic cells and immunoglobulin content.

Localized acid production within deposits of dental plaque results in a gradual softening of the external *enamel*. Enamel low in fluoride is more susceptible to the effects of the acid. This is the reason for fluoridation of water and

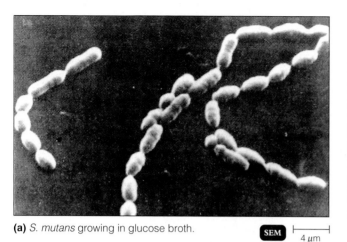

(a) *S. mutans* growing in glucose broth.

SEM | 4 μm

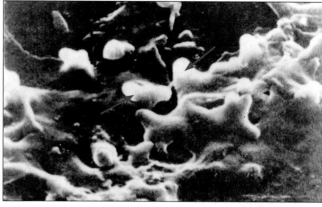

(b) *S. mutans* growing in sucrose broth; note the accumulations of dextran. Arrows point to *S. mutans* cells.

SEM | 4 μm

FIGURE 25.3 The role of *Streptococcus mutans* and sucrose in dental caries.

Q What is dental plaque?

toothpastes, which has been a significant factor in the decline in tooth decay in the United States.

Figure 25.4 shows the stages of tooth decay. If the initial penetration of the enamel by caries remains untreated, bacteria can penetrate into the interior of the tooth. The composition of the bacterial population involved in spreading the decayed area from the enamel into the *dentin* is entirely different from that of the population initiating the decay. The dominant microorganisms are gram-positive rods and filamentous bacteria; *S. mutans* is present in small numbers only. Although once considered the cause of dental caries, *Lactobacillus* spp. actually play no role in initiating the process. However, these very prolific lactic acid producers are important in advancing the front of the decay once it has become established.

The decayed area eventually advances to the *pulp* (see Figure 25.4), which connects with the tissues of the jaw and contains the blood supply and the nerve cells. Almost any member of the normal microbiota of the mouth can be isolated from the infected pulp and roots. Once this stage is reached, root canal therapy is required to remove the infected and dead tissue and to provide access for antimicrobial drugs that suppress renewed infection. If untreated, the infection may advance from the tooth to the soft tissues, producing dental abscesses caused by mixed bacterial populations that contain many anaerobes.

Although dental caries is probably one of the more common infectious diseases in humans today, it was scarce in the Western world until about the seventeenth century. In human remains from older times, only about 10% of the teeth

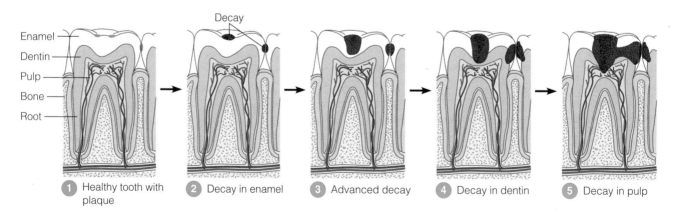

Enamel
Dentin
Pulp
Bone
Root

Decay

1 Healthy tooth with plaque
2 Decay in enamel
3 Advanced decay
4 Decay in dentin
5 Decay in pulp

FIGURE 25.4 The stages of tooth decay. 1 A tooth with plaque accumulation in difficult-to-clean areas. 2 Decay begins as enamel is attacked by acids formed by bacteria. 3 Decay advances through the enamel. 4 Decay advances into the dentin. 5 Decay enters the pulp and may form abscesses in the tissues surrounding the root.

Q Why is the formation of plaque important in tooth decay?

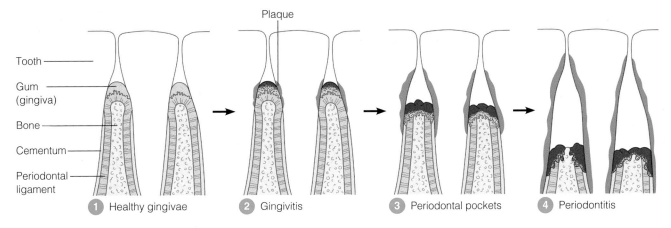

FIGURE 25.5 The stages of periodontal disease.
① Teeth firmly anchored by healthy bone and gum tissue (gingiva). ② Toxins in plaque irritate gums, causing gingivitis. ③ Periodontal pockets form as the tooth separates from the gingiva. ④ Gingivitis progresses to periodontitis. Toxins destroy the gingiva and bone that support the tooth and the cementum that protects the root.

Q **What is the cause of "pink tooth brush"?**

contain caries. The introduction of table sugar, or sucrose, into the diet is highly correlated with our present level of caries in the Western world. Studies have shown that sucrose, a disaccharide composed of glucose and fructose, is much more cariogenic than either glucose or fructose individually (see Figure 25.3). People living on high-starch diets (starch is a polysaccharide of glucose) have a low incidence of tooth decay unless sucrose is also a significant part of their diet. The contribution of bacteria to tooth decay has been shown by experiments with germ-free animals. Such animals do not develop caries even when fed a sucrose-rich diet designed to encourage their formation.

Sucrose is pervasive in the modern Western diet. However, if sucrose is ingested only at regular mealtimes, the protective and repair mechanisms of the body are usually not overwhelmed. It is the sucrose that is ingested between meals that is most damaging to teeth. Sugar alcohols, such as mannitol, sorbitol, and xylitol, are not cariogenic; xylitol appears to inhibit carbohydrate metabolism in *S. mutans*. This is why these sugar alcohols are used to sweeten "sugarless" candies and chewing gum.

The best strategies for preventing dental caries are a minimal ingestion of sucrose; brushing, flossing, and professional cleaning to remove plaque; and the use of fluoride. Professional removal of plaque and tartar at regular intervals lessens the progression to periodontal disease. As for mouthwashes that claim to prevent or reduce plaque, chlorhexidine is probably the most effective. However, proper brushing and flossing are more important. The ancient Chinese were known to use human urine as a mouthwash to improve oral health. Although urine does tend to lower acidity, this preventive measure is not recommended.

PERIODONTAL DISEASE

Even people who avoid tooth decay might, in later years, lose their teeth to **periodontal disease,** a term for a number of conditions characterized by inflammation and degeneration of structures that support the teeth (Figure 25.5). The roots of the tooth are protected by a covering of specialized connective tissue called *cementum*. As the gums recede with age or with overly aggressive brushing, the formation of caries on the cementum becomes more common.

GINGIVITIS

In many cases of periodontal disease, the infection is restricted to the gums, or *gingivae*. This resulting inflammation, called **gingivitis,** is characterized by bleeding of the gums while the teeth are being brushed (see Figure 25.5). This is a condition experienced by at least half of the adult population. It has been shown experimentally that gingivitis will appear in a few weeks if brushing is discontinued and plaque is allowed to accumulate. An assortment of streptococci, actinomycetes, and anaerobic gram-negative bacteria predominate in these infections.

PERIODONTITIS

Gingivitis can progress to a chronic condition called **periodontitis.** This is an insidious condition that generally causes little discomfort. About 35% of adults suffer from periodontitis, which is increasing in incidence as more people retain their teeth into old age. The gums are inflamed and bleed easily. Sometimes pus forms in pockets surrounding the teeth (*periodontal pockets*; see Figure 25.5). As the infection continues, it progresses toward the root

TABLE 25.1	**Bacterial Diseases of the Mouth**			
Disease	Pathogen	Portal of Entry	Treatment	Prevention
Dental caries	Primarily *Streptococcus mutans*	Endogenous infection	Remove decayed area	Brushing, flossing, reduce dietary sucrose
Periodontal disease	Various, primarily *Porphyromonas* spp.	Endogenous infection	Remove damaged area; antibiotics	Plaque removal
Acute necrotizing ulcerative gingivitis	*Prevotella intermedia*	Endogenous infection	Remove damaged area; metronidazole	Brushing, flossing

tips. The bone and tissue that support the teeth are destroyed, leading eventually to loosening and loss of the teeth. Numerous bacteria of many different types, primarily *Porphyromonas* (pôr'fī-rō-mō-nas) species, are found in these infections; the damage to tissue is done by an inflammatory response to the presence of these bacteria. Periodontitis can be treated surgically by eliminating the periodontal pockets. A recent advance in treatment avoids surgery by injecting a gel directly between the gum and the tooth that slowly releases the antibiotic doxycycline.

Acute necrotizing ulcerative gingivitis, also termed **Vincent's disease** or **trench mouth,** is one of the more common serious mouth infections. The disease causes enough pain to make normal chewing difficult. Foul breath (halitosis) also accompanies the infection. Among the bacteria usually associated with this condition is *Prevotella intermedia* (prev'ō-tel-la in'tèr-mē-dē-ä), averaging up to 24% of the isolates. Because these pathogens are usually anaerobic, treatment with oxidizing agents, debridement, and the administration of metronidazole may be temporarily effective. Bacterial diseases of the mouth are summarized in Table 25.1.

BACTERIAL DISEASES OF THE LOWER DIGESTIVE SYSTEM

LEARNING OBJECTIVE

- List the causative agents, suspect foods, signs and symptoms, and treatments for staphylococcal food poisoning, shigellosis, salmonellosis, typhoid fever, cholera, gastroenteritis, and peptic ulcer disease.

Diseases of the digestive system are essentially of two types: infections and intoxications.

An **infection** occurs when a pathogen enters the GI tract and multiplies. Microorganisms can penetrate into the intestinal mucosa and grow there, or they can pass through to other systemic organs. **M cells** (for microfold) are intended to translocate antigens and microorganisms to the other side of the epithelium where they can contact lymphoid tissues (Peyer's patches) to initiate an immune response (see page 513). Pathogens, as shown in Figures 25.8 and 25.9, can exploit this transport system if they have effective weapons to counteract the host's immune system. Infections are characterized by a delay in the appearance of gastrointestinal disturbance while the pathogen increases in numbers or affects invaded tissue. There is also usually a fever, one of the body's general responses to an infective organism.

Some pathogens cause disease by forming toxins that affect the GI tract. An **intoxication** is caused by the ingestion of such a preformed toxin. Most intoxications, such as that caused by *Staphylococcus aureus*, are characterized by a very sudden appearance (usually in only a few hours) of symptoms of a GI disturbance. Fever is less often one of the symptoms.

Both infections and intoxications often cause *diarrhea*, which most of us have experienced. Severe diarrhea accompanied by blood or mucus is called **dysentery.** Both types of digestive system diseases are also frequently accompanied by *abdominal cramps, nausea,* and *vomiting.* Diarrhea and vomiting are both defensive mechanisms designed to rid the body of harmful material.

The general term **gastroenteritis** is applied to diseases causing inflammation of the stomach and intestinal mucosa. Botulism is a special case of intoxication because the ingestion of the preformed toxin affects the nervous system rather than the GI tract (see Chapter 22, page 649).

In developing countries, diarrhea is a major factor in infant mortality. Approximately one child in every four dies of it before the age of 5. It is estimated that mortality from childhood diarrhea could be halved by *oral rehydration therapy* (replacement of lost fluids and electrolytes). This is usually a solution of sodium chloride, potassium chloride, glucose, and sodium bicarbonate to replace lost fluids and electrolytes. These solutions are sold in the infant supply department of many stores. Public health departments often determine the incidence of diarrhea in the population by receiving weekly reports on the sales of oral rehydration preparations.

Diseases of the digestive system are often related to food ingestion.

STAPHYLOCOCCAL FOOD POISONING (STAPHYLOCOCCAL ENTEROTOXICOSIS)

A leading cause of gastroenteritis is **staphylococcal food poisoning,** an intoxication caused by ingesting an enterotoxin produced by S. aureus. Staphylococci are comparatively resistant to environmental stresses, as discussed on page 332. They also have a fairly high resistance to heat; vegetative cells can tolerate 60°C (140°F) for half an hour. Their resistance to drying and radiation helps them survive on skin surfaces. Resistance to high osmotic pressures helps them grow in foods, such as cured ham, in which the high osmotic pressure of salts inhibits the growth of competitors.

S. aureus is often an inhabitant of the nasal passages, from which it contaminates the hands. It is also a frequent cause of skin lesions on the hands. From these sources, it can readily enter food. If the microbes are allowed to incubate in the food, a situation called **temperature abuse,** they reproduce and release enterotoxin into the food. These events, which lead to outbreaks of staphylococcal intoxication, are illustrated in Figure 25.6.

S. aureus produces several toxins that damage tissues or increase the microorganism's virulence. The production of the toxin of serological type A (which is responsible for most cases) is often correlated with the production of an enzyme that coagulates blood plasma. Such bacteria are described as *coagulase-positive*. No direct pathogenic effect can be attributed to the enzyme, but it is useful in the tentative identification of types that are likely to be virulent.

Generally, a population of about 1 million bacteria per gram of food will produce enough enterotoxin to cause illness. The growth of the microbe is facilitated if the competing microorganisms in the food have been eliminated—by cooking, for example. It is also more likely to grow if competing bacteria are inhibited by a higher-than-normal osmotic pressure or by a relatively low moisture level. S. aureus tends to outgrow most competing bacteria under these conditions.

Custards, cream pies, and ham are examples of high-risk foods. Competing microbes are minimized in custards by the high osmotic pressure of sugar and by cooking. In ham they are inhibited by curing agents, such as salts and preservatives. Poultry products can also harbor staphylococci if they are handled and allowed to stand at room

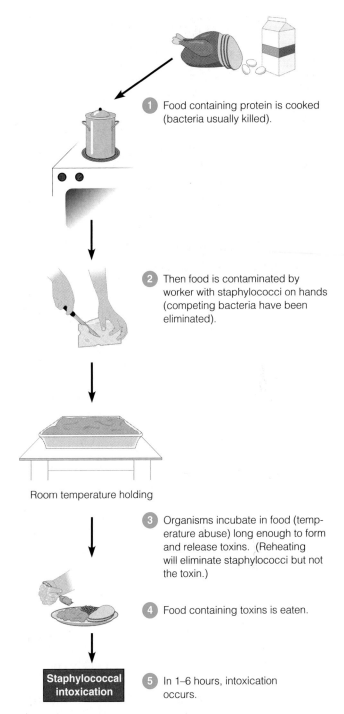

1. Food containing protein is cooked (bacteria usually killed).

2. Then food is contaminated by worker with staphylococci on hands (competing bacteria have been eliminated).

Room temperature holding

3. Organisms incubate in food (temperature abuse) long enough to form and release toxins. (Reheating will eliminate staphylococci but not the toxin.)

4. Food containing toxins is eaten.

Staphylococcal intoxication

5. In 1–6 hours, intoxication occurs.

FIGURE 25.6 The sequence of events in a typical outbreak of staphylococcal food poisoning.

Q How does this differ from foodborne illness caused by a virus?

temperatures. Because staphylococci do not compete well with the large number of microorganisms hamburger contains, it is rarely a factor in this type of food poisoning. Any foods prepared in advance and not kept chilled are a potential source of staphylococcal food poisoning. Because food contamination by human handlers cannot be avoided

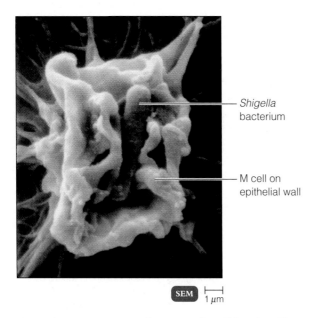

Shigella bacterium

M cell on epithelial wall

SEM ⊢ 1 μm

FIGURE 25.7 Invasion of intestinal wall by *Shigella* bacterium. Note how the membrane ruffles on the M cell of the epithelial wall of the intestine surround the bacterial cell. Invasion by *Salmonella* bacteria is very similar.

Q If a bacterium succeeds in leaving the interior of the intestine, what elements of the immune system would attempt to deal with it?

completely, the most reliable method of preventing staphylococcal food poisoning is adequate refrigeration during storage to prevent toxin formation.

The toxin itself is heat stable and can survive up to 30 minutes of boiling. Therefore, once the toxin is formed, it is not destroyed when the food is reheated, although the bacteria will be killed.

The toxin quickly triggers the brain's vomiting reflex center; abdominal cramps and usually diarrhea then ensue. This reaction is essentially immunological in character; the staphylococcal enterotoxin is a model example of a superantigen (see page 461). Recovery is usually complete within 24 hours.

The mortality rate of staphylococcal food poisoning is almost zero among otherwise healthy people, but it can be significant in weakened individuals, such as residents of nursing homes. No reliable immunity results from recovery. However, there is a great deal of variation in individual susceptibility to the toxin, and it is suspected that immunity acquired from a previous exposure might account for some of this variation.

The diagnosis of staphylococcal food poisoning is usually based on the symptoms, particularly the short incubation time characteristic of intoxication. If the food has not been reheated so that the bacteria are not killed, the pathogen can be recovered and grown. *S. aureus* isolates

can be tested by *phage typing*, a method used in tracing the source of the contamination (see Figure 10.13, page 298). These bacteria grow well in 7.5% sodium chloride, so this concentration is often used in media for their selective isolation. Pathogenic staphylococci usually ferment mannitol, produce hemolysins and coagulase, and form golden-yellow colonies. They cause no obvious spoilage when growing in foods. Detecting the toxin in food samples has always been a problem; there may be only 1 to 2 nanograms in 100 g of food. Reliable serological methods have become commercially available only recently.

SHIGELLOSIS (BACILLARY DYSENTERY)

Bacterial infections, such as salmonellosis and shigellosis, usually have longer incubation periods (12 hours to 2 weeks) than bacterial intoxications, reflecting the time needed for the microorganism to grow in the host. Bacterial infections are often characterized by some fever, indicating the host's response to the infection.

Shigellosis, also known as **bacillary dysentery** to differentiate it from amoebic dysentery (page 773), is a severe form of diarrhea caused by a group of facultatively anaerobic gram-negative rods of the genus *Shigella*. The genus was named for the Japanese microbiologist Kiyoshi Shiga. The bacteria do not have any natural reservoir in animals and spread only from person to person. Outbreaks are most often seen in families, day-care facilities, and similar settings.

There are four species of pathogenic *Shigella*: *S. sonnei* (sōn′ne-ē), *S. dysenteriae* (dis-en-te′rē-ī), *S. flexneri* (fleks′nėr-ē), and *S. boydii* (boi′dē-ē). These bacteria are residents only of the intestinal tract of humans, apes, and monkeys. They are closely related to the pathogenic *E. coli*.

The most common species in the United States is *S. sonnei*; it causes a relatively mild dysentery. Many cases of so-called traveler's diarrhea might be mild forms of shigellosis. At the other extreme, infection with *S. dysenteriae* often results in a severe dysentery and prostration. The toxin responsible is unusually virulent and is known as the **Shiga toxin** (see enterohemorrhagic *E. coli*, page 759). *S. dysenteriae* is the least common species in the United States.

The infective dose required to cause disease is small; the bacteria are not much affected by stomach acidity. They proliferate to immense numbers in the small intestine, but the primary site of disease is the large intestine. There, the bacteria attach to certain epithelial cells. M cells, membranous cellular ruffles surrounding the cell, take the bacterium into the cell (Figure 25.7). The bacteria multiply in the cell and soon spread to neighboring cells, producing Shiga toxin that destroys tissue (Figure 25.8). Dysentery is the result of damage to the intestinal wall.

Shigellosis dysentery can cause as many as 20 bowel movements in one day. Additional symptoms of infection are abdominal cramps and fever. *Shigella* bacteria rarely invade the bloodstream. Macrophages not only fail to kill *Shigella* bacteria that they phagocytose, but they are killed by them. Diagnosis is usually based on recovery of the microbes from rectal swabs.

In recent years, the number of cases reported in the United States has been about 20,000 to 30,000, with 5 to 15 deaths. *S. dysenteriae* has a significant mortality rate, however, and the death rate in tropical areas where it is prevalent can be as high as 20%. Some immunity seems to result from recovery, but a satisfactory vaccine has not yet been developed.

In severe cases of shigellosis, antibiotic therapy and oral rehydration are indicated. At present, fluoroquinolones are the antibiotics of choice.

SALMONELLOSIS (*SALMONELLA* GASTROENTERITIS)

The *Salmonella* bacteria (named for their discoverer, Daniel Salmon) are gram-negative, facultatively anaerobic, non–endospore-forming rods. Their normal habitat is the intestinal tracts of humans and many animals. All salmonellae are considered pathogenic to some degree, causing **salmonellosis,** or *Salmonella* **gastroenteritis.** Pathogenically salmonellae are separated into *typhoidal salmonellae* (see typhoid fever, page 754) and the *nontyphoidal salmonellae*, which cause the milder disease of salmonellosis.

The nomenclature of the *Salmonella* microbes differs from the norm. Rather than recognized species, there are more than 2000 serotypes (or serovars), only about 50 of which are isolated with any frequency in the United States. (For a discussion of the nomenclature of the salmonellae, see page 323.) To summarize, many consider them to belong to only two species, primarily *Salmonella enterica*. Therefore, you might encounter nomenclature such as *S. enterica* serotype Typhimurium, instead of the conventional name *S. typhimurium*.

The salmonellae first invade the intestinal mucosa and multiply there. Occasionally they manage to pass through the intestinal mucosa at M cells to enter the lymphatic and cardiovascular systems, and from there they may spread to eventually affect many organs (Figure 25.9). They replicate readily within macrophages. Salmonellosis has an incubation time of about 12 to 36 hours. There is usually a moderate fever accompanied by nausea, abdominal pain and cramps, and diarrhea. As many as 1 billion salmonellae per gram can be found in an infected person's feces during the acute phase of the illness.

The mortality rate is overall very low, probably less than 1%. However, the death rate is higher in infants and

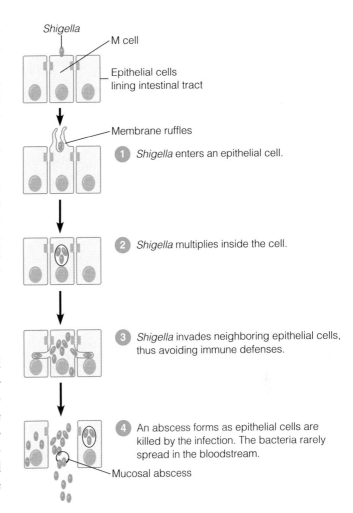

FIGURE 25.8 Shigellosis. This sequence shows the sequence of infection of the intestinal wall. The bacterium attaches to an M cell (see Figure 25.7) of the epithelial wall located over a Peyer's patch (see page 513). This is a region adapted to facilitate transfer of antigens across the intestinal mucosa.

Q **What species of *Shigella* is the most dangerous?**

among the very old; death is usually from septic shock. The severity and incubation time can depend on the number of *Salmonella* ingested. Normally, recovery will be complete in a few days, but many patients will continue to shed the organisms in their feces for up to 6 months. Antibiotic therapy is not useful in treating salmonellosis or, indeed, many diarrheal diseases; treatment consists of oral rehydration therapy.

Salmonellosis is probably greatly underreported. About 40,000 to 50,000 cases are reported each year, but an estimated 2 to 4 million occur, with 500 to 2000 deaths (Figure 25.10). Meat products are particularly susceptible to contamination by *Salmonella*. The sources of the bacteria are the intestinal tract of many animals. Pet reptiles, such as turtles and iguanas, are also a source; their carriage rate is as

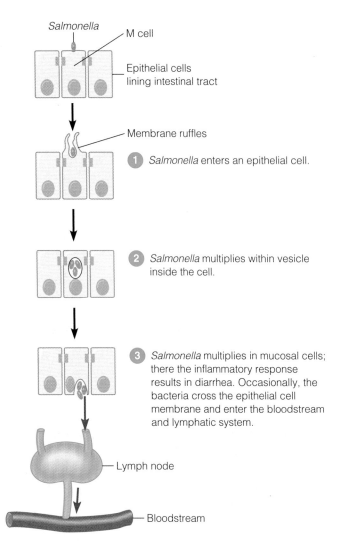

FIGURE 25.9 Salmonellosis. This sequence shows the sequence of infection of the intestinal wall. Compare with Figure 25.8 showing infection with *Shigella*. Note that invasion of the bloodstream, which happens infrequently, can result in septic shock.

Q **Why does salmonellosis have a longer incubation period than a bacterial intoxication?**

high as 90%. *S. enteritidis* and *S. typhimurium* are especially well adapted to commercial chicken production. Hens are highly susceptible to infection, and the bacteria contaminate the eggs. The bacteria have developed the ability to survive in the albumin, which contains natural preservatives such as *lysozyme* (see page 478) and *lactoferrin* (which binds iron the bacteria require). Estimates are that 1 in 20,000 eggs in this country is contaminated by *Salmonella*. Health authorities caution the public to eat only well-cooked eggs. An often unsuspected factor is the presence of inadequately cooked or raw eggs in foods such as hollandaise sauce, cookie batter, and Caesar salad. Surprisingly frequent sources of foodborne illness from ingestion of

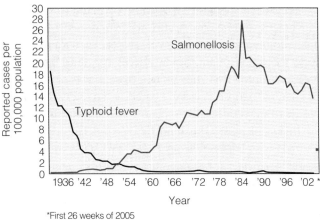

*First 26 weeks of 2005

FIGURE 25.10 The incidence of salmonellosis and typhoid fever. An important factor in comparing the two diseases is that typhoid transmission is almost entirely human, and salmonellosis transmission is primarily between animal products and humans.
SOURCE: CDC. MMWR 52(54), 4/22/05; MMWR 53(52) (1/7/05).

Q **Can you suggest reasons for the change in prevalence of those two diseases?**

Salmonella and *E. coli* O157:H7 have been raw alfalfa sprouts and tomatoes (see the box on page 756).

Prevention also depends on good sanitation practices to deter contamination and on proper refrigeration to prevent increases in bacterial numbers. Recently, it has been possible to offer eggs in which any *Salmonella* have been killed by a special hot water pasteurization procedure that does not cook the eggs. However, these eggs are more expensive. The microbes are generally destroyed by normal cooking. Chicken, for example, should be cooked at temperatures of 76°–82°C (170°–180°F) and ground beef at 71°C (160°F). However, contaminated food can contaminate a surface, such as a cutting board. Although the food first prepared on the board might later be cooked and its bacteria killed, another food subsequently prepared on the board might not be cooked.

Diagnosis usually depends on isolating the pathogen from the patient's stool or from leftover food. Isolation requires specialized selective and differential media; these methods are relatively slow. Also, the small numbers of *Salmonella* generally found in foods present a special problem in detection. The infective dose may be as small as 1000 bacteria. Currently, PCR-based tests are the best for detecting small numbers of *Salmonella* in foods.

TYPHOID FEVER

The most virulent serotype of *Salmonella*, *S. typhi*, causes the bacterial disease **typhoid fever.** Unlike the salmonellae that cause salmonellosis, this pathogen is not found in animals; it is spread only in the feces of other humans. Before the days of proper sewage disposal, water treatment, and food sanitation, typhoid was an extremely common disease. Its incidence has been declining in the United States, whereas that of salmonellosis has been increasing (see Figure 25.10). Typhoid fever is still a frequent cause of death in parts of the world with poor sanitation.

Instead of being destroyed by phagocytic cells, *S. typhi* multiply within them and are disseminated into multiple organs, especially the spleen and liver. Eventually, the phagocytic cells lyse and release *S. typhi* into the bloodstream. The time required for this explains why the incubation period of typhoid fever (2 or 3 weeks) is much longer than for salmonellosis (12 to 36 hours). The patient with typhoid fever suffers from a high fever of about 40°C (104°F) and continual headache. Diarrhea appears only during the second or third week, and the fever then tends to decline. In severe cases, which can be fatal, ulceration and perforation of the intestinal wall can occur. Before antibiotic therapy was available, a mortality rate of 20% was common; with the treatments available today, it is less than 1%.

Substantial numbers of recovered patients, about 1 to 3%, become *chronic carriers*. They harbor the pathogen in the gallbladder and continue to shed bacteria for several months. A number of such carriers continue to shed the organism indefinitely. The classic example of a typhoid carrier was Typhoid Mary. Her name was Mary Mallon; she worked as a cook in New York state in the early part of the twentieth century and was responsible for several outbreaks of typhoid and three deaths. She became well known through the attempts of the state to restrain her from working at her chosen trade.

In recent years there have been about 350 to 500 annual cases of typhoid fever in the United States, of which 70% were acquired during foreign travel. Normally, there are fewer than three deaths each year. Worldwide, at least 16 million annual cases and 600,000 deaths can be attributed to typhoid fever.

When the antibiotic chloramphenicol was introduced in 1948, typhoid became a treatable disease. Although mostly replaced by safer (but more expensive) antibiotics, it is still used in endemic areas in the world—but might require 250 capsules during a course of treatment. The most effective antityphoidal drugs are quinolones or third-generation cephalosporins. Treatment of the carrier state might require weeks of antibiotic therapy. Antibiotic resistance is a frequent problem.

Recovery from typhoid confers lifelong immunity. Vaccines are seldom used in developed countries except for high-risk laboratory or military personnel. The vaccine that has long been in use is a killed-organism type, which

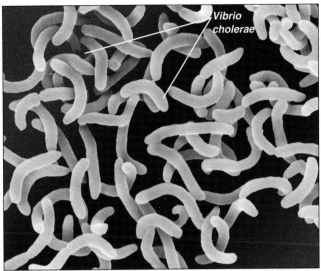

Vibrio cholerae

SEM   ⊢——⊣ 1 μm

FIGURE 25.11 *Vibrio cholerae*, **the cause of cholera.** Notice the slightly curved morphology.

Q **What are the effects of the sudden loss of fluid and electrolytes during infection with *V. cholerae*?**

must be injected. Live, orally ingested conjugated vaccines have become available. Their effectiveness is not much greater than older, killed vaccines (conferring immunity lasting only a few years on 65% of recipients), but they have fewer adverse effects.

CHOLERA

The causative agent of **cholera,** one of the most serious gastrointestinal diseases, is *Vibrio cholerae*, a slightly curved, gram-negative rod with a single polar flagellum (Figure 25.11). Cholera bacilli grow in the small intestine and produce an exotoxin, *cholera toxin* (see Chapter 15, page 461), that causes host cells to secrete water and electrolytes, especially potassium. The result is watery stools containing masses of intestinal mucus and epithelial cells—called "rice water stools" from their appearance. As much as 12 to 20 liters (3 to 5 gallons) of fluids can be lost in a day and the sudden loss of these fluids and electrolytes causes shock, collapse, and often death. The blood, lacking fluids, may become so viscous that vital organs are unable to function properly. Violent vomiting generally also occurs. The microbes are not invasive and a fever is usually not present. The severity of cholera varies greatly, and the number of subclinical cases might be several times the number reported. Untreated cases of cholera may have a mortality rate of 50%, although with proper supportive care it is usually less than 1% today. The diagnosis is based upon symptoms and culturing of *V. cholerae* from feces.

CLINICAL PROBLEM SOLVING

You will see questions as you read through this problem. The questions are those that epidemiologists ask themselves as they solve a clinical problem. Try to answer each question as an epidemiologist.

1. On June 29, a 36-year-old female in Pennsylvania was hospitalized with a 3-day history of nausea, vomiting, and diarrhea. She had a temperature of 39.5°C (103.1°F), and she was dehydrated.
   What smaple is needed from the patient to determine the cause of her signs and symptoms?

2. Stool culture grew gram-negative, non–lactose-fermenting bacteria.
   Can you identify the bacteria?

3. Nine U.S. state and one Canadian provincial health departments were notified of 561 cases of salmonellosis. No deaths occurred, but 30% of the patients were hospitalized. Cultures from the hospitalized cases yielded *Salmonella enterica* serotypes Javiana, Typhimurium, Anatum, Thompson, and Muenchen.
   What information would you try to obtain from these patients?

4. All case-patients had eaten at delicatessen chain A between June 25 and July 13.
   How will you determine the source of the infection?

5. Epidemiologists conducted a case-control study to compare 53 case-patients with 53 well meal companions. All 106 people were asked to complete a questionnaire about foods eaten. The data collected are shown below.

Exposure	Exposed		Not exposed		Relative risk (RR)
	(a) Ill	**(b) Not ill**	**(c) Ill**	**(d) Not ill**	
Chicken salad	47	40	6	13	**1.71**
Cole slaw	32	20	21	33	
Fruit salad	34	30	19	23	
Potato salad	42	39	11	14	
Tomato salad	47	24	6	29	

Cholera bacteria, and other members of the genus *Vibrio* in general, are strongly associated with brackish (salty) waters characteristic of estuaries, although they are also readily spread in contaminated fresh water. They form biofilms and colonize copepods (tiny crustaceans), algae, and other aquatic plants and plankton, which aids their survival. It has even been reported that, because of this growth habit, straining contaminated water through folded layers of finely-woven cloth (such as saris worn by Indian women) often removes these attached bacteria and makes the water safe to drink. Under unfavorable conditions *V. cholerae* may become dormant; the cell shrinks into a nonculturable, spherical state. A favorable change

in the environment causes them to revert rapidly to the culturable form. Both forms are infectious.

Although they survive well in their aquatic environment, cholera bacteria are exceptionally sensitive to stomach acids. Persons with impaired stomach acid secretion or who are taking antacids are at higher risk of infection. Normal individuals may require infective doses on the order of 100 million bacteria to cause severe cholera. Recovery from cholera results in an effective immunity but only to bacterial strains of the same antigenic characteristics. The serogroup O:1 (see the footnote in Chapter 11, page 323), which caused a pandemic in the 1880s, is known as the classical strain. A later pandemic was

CLINICAL PROBLEM SOLVING

(continued)

Relative Risk Calculation using a 2 × 2 table:

	Ill	Not Ill	Relative Risk
Ate _____	(a)	(b)	$(e) = \dfrac{a}{a+b}$
Did not eat _____	(c)	(d)	$(f) = \dfrac{c}{c+d}$
Relative risk =	$\dfrac{e}{f} =$		= Times more likely to become ill by going to this place

Relative risk (RR) is a measure of the probability (risk) that an event will result in disease.

RR is calculated in a 2 × 2 table (shown above). The relative risk must be calculated for each source of exposure. The RR for the chicken salad is given.
Complete the remaining calculations to determine the probable source of the infection.

6. There is a strong association between illness and consumption of (Roma) tomatoes. Delicatessen chain A had purchased presliced Roma tomatoes from a single supplier for all of its 302 stores.
What would you do now?

7. Roma tomatoes were removed from all the stores on July 14 and no further illness occurred. No source of

contamination was identified at the slicing facility. Moreover. *S. javiana* is typically associated with the coastal Southeast; the slicing facility was in the Northeast.
What would you do now?

8. Traceback investigation of tomatoes identified three packing houses in three states as possible sources. Cross-contamination may have occurred in a packing house where substantial numbers of tomatoes passed through a common wash tank. Tomato-associated *Salmonella* outbreaks have increased in frequency and magnitude, causing an estimated 60,000 illnesses during 1990–2004.
What factors might contribute to tomatoes as a vehicle of transmission?

Tomatoes originated in South America, were introduced into Europe in the

sixteenth century, and are now a popular food worldwide. The Roma tomato was developed in the mid-1950s as a firmer and more disease-resistant variety.

In the eastern United States, tomatoes are grown in natural habitats for many known *Salmonella* reservoirs, including birds, amphibians, and reptiles. *Salmonella* can enter tomato plants through roots or flowers and can enter fruit through small cracks in the skin, the stem scar, or the plant itself. Contamination might occur during multiple steps from the tomato seed nursery to the final kitchen. Eradication of *Salmonella* from the interior of the tomato is difficult without cooking, even if treated with highly concentrated chlorine solution.

SOURCE: Adapted from a report in *MMWR* 54(13):325–328 (4/8/05).

caused by a biotype of O:1 named *El Tor* or *eltor* (for the El Tor quarantine camp for pilgrims to Mecca, where it was first isolated). In 1991–1994, an epidemic in South America resulted in over 1 million cases and 9600 deaths (Figure 25.12). This outbreak was traced to seafood contaminated by ballast water picked up in Asia and emptied into harbors in Peru. Until the 1990s it was thought that only *V. cholerae* O:1 caused epidemic cholera, but a widespread epidemic in India and Bangladesh by a new serogroup, O:139, changed this view. There are also nonepidemic strains of *V. cholerae*, non-O:1/O:139, that are only infrequently associated with large-scale outbreaks of cholera. They occasionally cause wound infections or

sepsis, especially in people with liver disease or who are immunosuppressed.

In the United States there have been occasional cases of cholera caused by the O:1 serogroup. These have all been in the Gulf Coast area, and the pathogen may be endemic in these coastal waters. Outbreaks of cholera in this country are limited by high standards of sanitation. This represents the primary means of control and is important because stools may contain 100 million *V. cholerae* per gram. Available oral vaccines provide immunity of relatively short duration and only moderate effectiveness.

Treatment often includes the use of antibiotics such as doxycycline, but the most effective therapy is intravenous

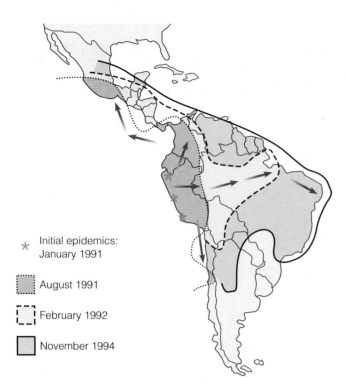

* Initial epidemics:
January 1991

August 1991

February 1992

November 1994

FIGURE 25.12 The extent and progress of a cholera epidemic in Latin America, 1991–1994.
SOURCE: MMWR 44(11) (3/24/95).

Q Note the path of the spread of cholera in February 1992—what major transportation pathway (river) is the disease following?

replacement of the lost fluids and electrolytes. As much as 10% of the patient's body weight within a few hours may be required.

NONCHOLERA VIBRIOS

At least 11 species of *Vibrio*, in addition to *V. cholerae*, can cause human illness. Most are adapted to life in salty coastal waters. *Vibrio parahaemolyticus* (pa-rä-hē-mō-li′ti-kus) is found in salt water estuaries in many parts of the world. It is morphologically similar to *V. cholerae* and the most common cause of gastroenteritis by *Vibrio* spp. in humans. The bacterium is present in coastal waters of the continental United States and Hawaii. Raw oysters and crustaceans, such as shrimp and crabs, have been associated with several outbreaks of *gastroenteritis* in the United States in recent years.

Signs and symptoms, which resemble those of cholera, include abdominal pain, vomiting, a burning sensation in the stomach, and watery stools. Treatment by antibiotics and rehydration is usually effective. The incubation time is normally less than 24 hours. Recovery usually follows in a few days.

Because *V. parahaemolyticus* has a requirement for sodium and a high osmotic pressure, isolation media containing 2 to 4% sodium chloride are used in diagnosing the disease.

Another important pathogenic *Vibrio* is *Vibrio vulnificus*, which is also found in estuaries. It is halophilic and requires 1% sodium chloride in the media used to isolate it. It causes gastroenteritis in only a minority of infections; rather, ingestion can lead to a life-threatening invasion of the bloodstream. People with compromised immune systems are at a higher risk. Anyone suffering from liver disease is also at high risk of sepsis, which in these cases is fatal about 50% of the time. *V. vulnificus* frequently causes very dangerous infections of minor skin lesions incurred in coastal sea waters. Rapidly spreading tissue destruction from these infections may require limb amputation; and, if sepsis occurs, the fatality rate is about 25%. Because these infections are life threatening, they require early antibiotic therapy for successful treatment.

To illustrate the indigenous presence of these bacterial pathogens, after Hurricane Katrina struck the Gulf Coast in 2005 there were over 20 cases of *Vibrio*-caused illness, presumably from contact with contaminated water. Most of these were wound-associated and caused by *V. vulnificus*, although a few were caused by *V. parahaemolyticus*. Five deaths, a mortality rate of 28%, were reported from these infections. Most of the victims had underlying conditions that probably lowered their natural resistance. Several cases of gastroenteritis caused by ingestion of noncholera (non-O1, non-O139) *V. cholerae* were also reported. See the box in Chapter 27, page 835.

ESCHERICHIA COLI GASTROENTERITIS

One of the most prolific microorganisms in the human intestinal tract is *E. coli*. Because it is so common and so easily cultivated, microbiologists often regard it as something of a laboratory pet. However, taxonomists consider the *Shigella* and *E. coli* to be so closely related as to be indistinguishable, and the nomenclature is maintained only because of their clinical significance. *E. coli* are normally harmless, but certain strains can be pathogenic. All pathogenic strains of *E. coli* have specialized fimbriae that allow them to bind to certain intestinal epithelial cells. They also produce toxins that cause gastrointestinal disturbances, collectively termed *E. coli gastroenteritis*.

TRAVELER'S DIARRHEA

It has long been observed that travel tends to broaden the mind and to loosen the bowels, the latter being an affliction with the common name of **traveler's diarrhea**. The probable cause of most cases is either of two strains of *E. coli*. **Enterotoxigenic *E. coli* (ETEC)** is not invasive but produces an enterotoxin that causes a watery diarrhea that resembles a mild case of cholera. **Enteroinvasive *E. coli* (EIEC)** invades the intestinal wall, resulting in inflammation, fever, and sometimes a *Shigella*-like dysentery.

Traveler's diarrhea can also be caused by gastrointestinal pathogens such as *Salmonella* and *Campylobacter*—as well as by various unidentified bacterial pathogens, viruses, and protozoan parasites. In fact, in most cases the causative agent is never identified. Traveler's diarrhea is usually self-limiting, and chemotherapy is not attempted. Once contracted, the best treatment is the usual oral rehydration recommended for all diarrhea. In severe cases, antimicrobial drugs may be necessary. For preventing traveler's diarrhea, reports indicate that there may be some protective effect from taking prescribed antibiotics. Another option is to take bismuth-containing preparations, such as Pepto-Bismol (two tablets, taken four times a day), if the person does not mind the tongue and stool temporarily turning black. But the best advice in risky areas is to prevent infection, as expressed in the saying, "Boil it, peel it, or don't eat it."

SHIGA TOXIN-PRODUCING ESCHERICHIA COLI

In recent years **enterohemorrhagic *E. coli* (EHEC)** strains have become well known in the United States as the cause of several outbreaks of serious disease. The primary virulence factor in these bacteria is the production of Shiga toxin (see page 752), and they are sometimes termed **Shiga-toxin *E. coli* (STEC).** (As we have mentioned, *E. coli* and the genus *Shigella* are closely related.) Another virulence factor is their ability to adhere to the intestinal mucosa. The bacteria cause the destruction of the microvilli and the formation of pedestal-like projections upon which they then rest (Figure 25.13). In the United States, the serotype usually isolated is O157:H7 (see the footnote in Chapter 11, page 323 for an explanation of this nomenclature), but elsewhere other serotypes may predominate. EHEC are found in the intestinal tracts of many animals; in at least 50% of feedlot cattle, for example. The animals do not suffer any obvious symptoms.

The U.S. Department of Agriculture has found that almost 90% of ground meats are contaminated, although usually at a very low level. Such meats, if not cooked well, are a potential source of infection. Poultry and other meats may also be contaminated, but raw alfalfa sprouts and tomatoes have actually been responsible for most cases reported to date. Ingested food is not the only infection source; some cases have been associated with children's visits to farms or petting zoos. The infective dose is estimated to be very small, probably much less than 100 bacteria.

In humans the Shiga toxins often cause only self-limiting diarrhea; but, in about 6% of people infected, it produces an inflammation of the colon (the large intestine above the rectum) with profuse bleeding, called

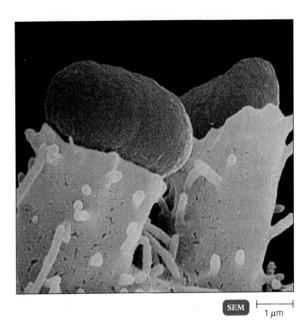

SEM | 1 μm

FIGURE 25.13 Enterohemorrhagic *E. coli* (EHEC) O157:H7. As EHEC bacteria (purple) adhere to the epithelial wall, they destroy the surface microvilli and cause the formation of a pedestal-like projection (yellow) upon which they rest.

Q **Is adhesion a factor in the pathogenicity of a microbe?**

hemorrhagic colitis. Unlike *Shigella* (see Figure 25.8), these *E. coli* do not invade the intestinal wall but release the toxin into the intestinal lumen (space).

Another dangerous complication is *hemolytic uremic syndrome* (HUS), (blood in the urine, often leading to kidney failure), which occurs when the kidneys are affected by the toxin. Some 5 to 10% of small children who have been infected progress to this stage, which has a mortality rate of about 5%. Management of these patients is primarily by intravenous rehydration and careful monitoring of serum electrolytes. Among survivors of HUS, some may require kidney dialysis or even transplants. An estimated 200 to 500 deaths occur annually.

Because of the attention this pathogen has attracted, researchers have been working, with some success, to develop rapid methods of detecting its presence in food without the need for time-consuming culturing methods. For isolation and identification, clinical laboratories use media that differentiate *E. coli* O157:H7 by its inability to ferment sorbitol.

CAMPYLOBACTER GASTROENTERITIS

Campylobacter are gram-negative, microaerophilic, spirally curved bacteria that have emerged as the leading cause of foodborne illness in the United States. They adapt well to the intestinal environment of animal hosts, especially poultry. Culturing *Campylobacter* requires conditions of low oxygen and high carbon dioxide developed in special apparatus (see Figure 6.7b, page 171). Their optimum growth temperature of about 42°C (109°F) approximates that of their animal hosts, but the bacteria do not replicate in food. Almost all retail chicken is contaminated with *Campylobacter*. Nearly 60% of cattle excrete the organism in feces and milk, but retail red meats are less likely to be contaminated.

There are more than an estimated 2 million cases of **Campylobacter gastroenteritis** in the United States annually, usually caused by *C. jejuni*. The infective dose is less than a thousand bacteria. Clinically, it is characterized by fever, cramping abdominal pain, and diarrhea or dysentery. Normally, recovery follows within a week.

An unusual complication of campylobacterial infection is that it is linked, in about 1 in 1000 cases, to the neurological disease Guillain-Barré syndrome, a temporary paralysis. Apparently, a surface molecule of the bacteria resembles a lipid component of nervous tissue and provokes an autoimmune attack.

HELICOBACTER PEPTIC ULCER DISEASE

In 1982, a physician in Australia cultured a spiral-shaped, microaerophilic bacterium observed in the biopsied tissue of stomach ulcer patients. Now named *Helicobacter pylori*, it is accepted that this microbe is responsible for most cases of **peptic ulcer disease.** This syndrome includes gastric and duodenal ulcers. (The duodenum is the first few inches of the small intestine.) About 30 to 50% of the population in the developed world become infected; the infection rate is higher elsewhere. Only about 15% of those infected develop ulcers, so certain host factors are probably involved. For example, people with type O blood are more susceptible, which is also a characteristic of cholera. *H. pylori* is also designated as a carcinogenic bacterium. Gastric cancer develops in about 3% of people infected with these bacteria, but no uninfected persons develop gastric cancer.

The stomach mucosa contains cells that secrete gastric juice containing proteolytic enzymes and hydrochloric acid that activates these enzymes. Other specialized cells produce a layer of mucus that protects the stomach itself from digestion. If this defense is disrupted, an inflammation of the stomach (gastritis) results. This inflammation can then progress to an ulcerated area (Figure 25.14). Through an interesting adaptation, *H. pylori* can grow in the highly acidic environment of the stomach, which is lethal for most microorganisms. *H. pylori* produces large amounts of an especially efficient urease, an enzyme that converts urea to the alkaline compound ammonia, resulting in a locally high pH in the area of growth.

The eradication of *H. pylori* with antimicrobial drugs usually leads to the disappearance of peptic ulcers. Several antibiotics, usually administered in combination, have proven effective. Bismuth subsalicylate (Pepto-Bismol) is also effective and is often part of the drug regimen. When the bacteria are successfully eliminated, the recurrence rate of the ulcer is only about 2 to 4% a year. Reinfection is from many environmental sources but is less likely in areas with high standards of sanitation; in fact, there is some evidence that infection by *H. pylori* is slowly disappearing in developed countries.

The most reliable diagnostic test requires a biopsy of tissue and culture of the organism. An interesting diagnostic approach is the urea breath test. The patient swallows radioactively labeled urea; if the test is positive, within about 30 minutes CO_2 labeled with radioactivity can be detected in the breath. This test is most useful for determining the effectiveness of chemotherapy because a positive test is an indication of live *H. pylori*. Diagnostic tests of stools to detect antigens (not antibodies) for *H. pylori* are suitable for follow-up tests following therapy. They are the noninvasive test of choice, especially for children. Serological tests to detect antibodies are inexpensive but not useful in determining eradication.

YERSINIA GASTROENTERITIS

Other enteric pathogens being identified with increasing frequency are *Yersinia enterocolitica* (en'tér-ō-kōl-it-ik-ä) and *Y. pseudotuberculosis* (sū-dō-tü-bér-kū-lō′sis). These gram-negative bacteria are intestinal inhabitants of many domestic animals and are often transmitted in meat and milk. Both microbes are distinctive in their ability to grow at refrigerator temperatures of 4°C (39°F). This ability increases their numbers in stored refrigerated blood until their endotoxins can result in shock to the blood recipient. *Yersinia* have occasionally been the cause of severe reactions when they contaminate transfused blood.

These pathogens cause **Yersinia gastroenteritis,** or **yersiniosis.** The symptoms are diarrhea, fever, headache, and abdominal pain. The pain is often severe enough to cause a misdiagnosis of appendicitis. Diagnosis requires culturing the organism, which can then be evaluated by serological tests. Adults suffering from yersiniosis usually recover in 1 or 2 weeks; children may take longer. Treatment with antibiotics and oral rehydration may be helpful.

CLOSTRIDIUM PERFRINGENS GASTROENTERITIS

One of the more common, if underrecognized, forms of food poisoning in the United States is caused by *Clostridium perfringens*, a large, gram-positive, endospore-forming,

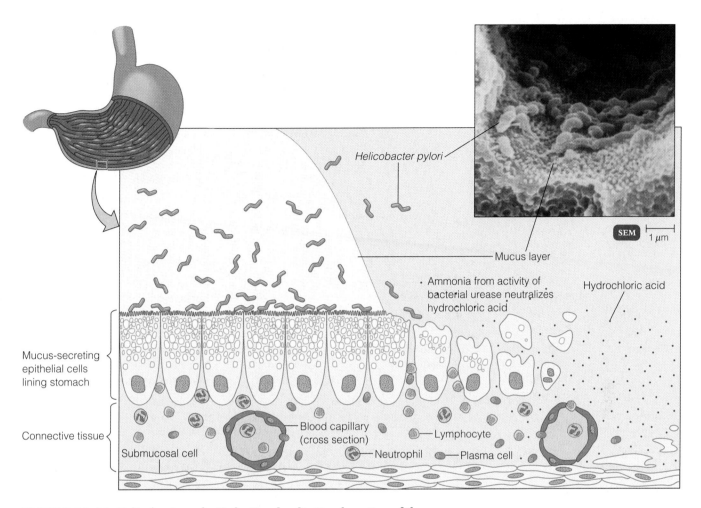

FIGURE 25.14 *Helicobacter pylori* **infection, leading to ulceration of the stomach wall.**

Q **How does the enzyme urease form ammonia?**
   (*Hint*: look up the chemical formula of urea.)

obligately anaerobic rod. This bacterium is also responsible for human gas gangrene (see Chapter 23, page 682).

Most outbreaks of **Clostridium perfringens gastroenteritis** are associated with meats or meat stews contaminated with intestinal contents of the animal during slaughter. The pathogen's nutritional requirement for amino acids is met by such foods, and when the meats are cooked, the oxygen level is lowered enough for clostridial growth. The endospores survive most routine heatings, and the generation time of the vegetative bacterium is less than 20 minutes under ideal conditions. Large populations can therefore build up rapidly when foods are being held for serving or when inadequate refrigeration leads to slow cooling.

The microbe grows in the intestinal tract and produces an exotoxin that causes the typical symptoms of abdominal pain and diarrhea. Most cases are mild and self-limiting and probably are never clinically diagnosed. If treatment is required, oral rehydration is recommended. The symptoms

usually appear 8 to 12 hours after ingestion. Diagnosis is usually based on isolation and identification of the pathogen in stool samples.

CLOSTRIDIUM DIFFICILE–ASSOCIATED DIARRHEA

Every year in the United States there are millions of cases of diarrhea associated with infection by *Clostridium difficile*, a gram-positive endospore-forming bacterium. *C. difficile* produces exotoxins that cause inflammation accompanied by increased fluid secretion and permeability of the intestinal mucosa. The condition, called **Clostridium difficile–associated diarrhea**, occurs mostly in hospitals and nursing homes, where the bacteria or their endospores are a common environmental contaminant. The condition is usually precipitated by the extended use of broad-spectrum antibiotics. The elimination of most competing intestinal bacteria permits the rapid proliferation of the toxin-producing

TABLE 25.2	**Bacterial Diseases of the Lower Digestive System**				
	Pathogen	Symptoms	Intoxication/ Infection	Diagnostic Test	Treatment
Staphylococcal food poisoning	*Staphylococcus aureus*	Nausea, vomiting, and diarrhea	Intoxication (enterotoxin)	Phage typing	None
Shigellosis (bacillary dysentery)	*Shigella* spp.	Tissue damage and dysentery	Infection (endotoxin and Shiga toxin, exotoxin)	Isolation of bacteria on selective media	Quinolones
Salmonellosis	*Salmonella enterica*	Nausea and diarrhea	Infection (endotoxin)	Isolation of bacteria on selective media, serotyping	Oral rehydration
Typhoid fever	*Salmonella typhi*	High fever, significant mortality	Infection (endotoxin)	Isolation of bacteria on selective media, serotyping	Quinolones; cephalosporins
Cholera	*Vibrio cholerae* O:1 and O:139	Diarrhea with large water loss	Cholera toxin (exotoxin)	Isolation of bacteria on selective media	Rehydration; doxycycline
Vibrio parahaemolyticus gastroenteritis	*V. parahaemolyticus*	Cholera-like diarrhea, but generally milder	Infection, enterotoxin	Isolation of bacteria on 2–4% NaCl	Rehydration; antibiotics
V. vulnificus gastroenteritis	*V. vulnificus*	Rapidly spreading tissue destruction	Infection, siderophores	Isolation of bacteria on 1% NaCl	Antibiotics

C. difficile. However, it can cause outbreaks in children in day-care centers that are unrelated to antibiotic use. Caregivers have been known to acquire it from patients. It can be serious; the mortality rate, which is highest in elderly patients, is reported as 1 to 2.5%.

The disease manifests itself in symptoms ranging from only a mild case of diarrhea to life-threatening colitis (inflammation of the colon). The colitis can result in ulceration of the intestinal wall.

A diagnosis of *C. difficile*–associated diarrhea is often suggested by a history of antibiotic use. It can be confirmed by an immunoassay that detects the responsible exotoxins. A more reliable test, but one that is difficult to perform and requires as long as 48 hours for results to be available, is a cytotoxin assay. Treatment, of course, requires discontinuation of the precipitating antibiotic and oral rehydration therapy. Metronidazole, a drug that targets the metabolism of anaerobes, is also part of the usual therapy. In exceptional cases, enemas containing human stools are used in an attempt to restore the normal microbiota.

BACILLUS CEREUS GASTROENTERITIS

Bacillus cereus (se'rē-us) is a large, gram-positive, endospore-forming bacterium that is very common in soil and vegetation and is generally considered harmless. It has, however, been identified as the cause of outbreaks of foodborne illness. Heating the food does not always kill the spores, which germinate as the food cools. Because competing microbes have been eliminated in the cooked food, *B. cereus* grows rapidly and produces toxins. Rice dishes served in Asian restaurants seem especially susceptible.

Some cases of **Bacillus cereus gastroenteritis** resemble *C. perfringens* intoxications and are almost entirely diarrheal in nature (usually appearing 8 to 16 hours after ingestion). Other episodes involve nausea and vomiting (usually 2 to 5 hours after ingestion). It is suspected that different toxins are involved in producing the differing symptoms. Both forms of the disease are self-limiting. The diseases can be differentiated by isolating at least 10^5 *B. cereus* per gram of suspected food.

Bacterial diseases of the GI tract are summarized in Table 25.2

TABLE 25.2	(continued)				
	Pathogen	Symptoms	Intoxication/ Infection	Diagnostic Test	Treatment
Traveler's diarrhea	Enterotoxigenic, enteroinvasive *Escherichia coli*	Watery diarrhea	Infection (endotoxin)	Isolation on selecteve media	Oral rehydration
Shiga toxin– producing *Escherichia coli*	*E. coli* O157:H7	*Shigella*-like dysentery; hemorrhagic colitis (very bloody stools) and hemolytic uremic syndrome (blood in urine, possible kidney failure)	Infection, Shiga toxin (exotoxin)	Isolation, sorbital fermentation	Intravenous rehydration and monitoring of serum electrolytes
Campylobacter gastroenteritis	*Campylobacter jejuni*	Fever, abdominal pain, diarrhea	Infection	Isolation in low O_2, high CO_2	
Helicobacter peptic ulcer disease	*Helicobacter pylori*	Peptic ulcers	Infection	Urea breath test; bacterial culture	Antimocrobial drugs
Yersinia gastroenteritis	*Yersinia enterocolitica*	Abdominal pain and diarrhea, usually mild; may be contused with appendicitis	Infection (endotoxin)	Culture, serotyping	
Clostridium perfringens gastroenteritis	*Clostridium perfringens*	Usually limited to diarrhea	Infection (exotoxin)	Isolation of bacteria	
Clostridium difficile–associated diarrhea	*Clostridium difficile*	Mild diarrhea to colitis; 1–2.5% mortality	Infection (exotoxin)	Cytotoxin assay	Metronidazole
Bacillus cereus gastroenteritis	*Bacillus cereus*	May take form of diarrhea or nausea and vomiting	Intoxication	Isolation of $\geq 10^5$ *B. cereus*/g food	

VIRAL DISEASES OF THE DIGESTIVE SYSTEM

Although viruses do not reproduce within the contents of the digestive system like bacteria, they invade many organs associated with the system.

MUMPS

LEARNING OBJECTIVE

• List the causative agents, modes of transmission, sites of infection, and symptoms for mumps.

The targets of the mumps virus, the parotid glands, are located just below and in front of the ears (see Figure 25.1). Because the parotids are one of the three pairs of salivary glands of the digestive system, it is appropriate to include a discussion of mumps in this chapter.

Mumps typically begins with painful swelling of one or both parotid glands 16 to 18 days after exposure to the virus (Figure 25.15). The virus is transmitted in saliva and respiratory secretions, and its portal of entry is the respiratory tract. An infected person is most infective to others during the first 48 hours before clinical symptoms appear. Once the viruses have begun to multiply in the respiratory tract and local lymph nodes in the neck, they reach the salivary glands via the blood. Viremia (the presence of virus in the blood) begins several days before the onset of mumps symptoms and before the virus appears in saliva. The virus is present in the blood and saliva for 3 to 5 days after the onset of the disease and in the urine after about 10 days.

Mumps is characterized by inflammation and swelling of the parotid glands, fever, and pain during swallowing. About 4 to 7 days after the onset of symptoms, the testes can become inflamed, a condition called *orchitis*. This happens in about 20 to 35% of males past puberty; sterility is a possible

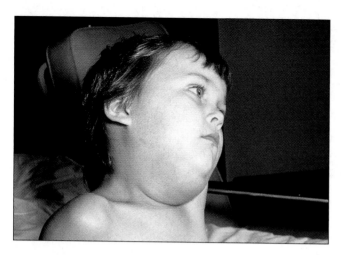

FIGURE 25.15 A case of mumps. This patient shows the typical swelling of mumps.

Q How is the mumps virus transmitted?

but rare consequence. Other possible complications include meningitis, inflammation of the ovaries, and pancreatitis.

An effective attenuated live vaccine is available and is often administered as part of the trivalent measles, mumps, rubella (MMR) vaccine. The number of cases of mumps has dropped sharply since the introduction of the vaccine in 1968. Second attacks are rare, and cases involving only one parotid gland, or subclinical cases (about 30% of those infected), are as effective as bilateral mumps in conferring immunity.

If confirmation of the usual diagnosis based only on symptoms is desired, the virus can be isolated by embryonated egg or cell culture techniques and identified by ELISA tests.

HEPATITIS

> **LEARNING OBJECTIVE**
> • Differentiate among hepatitis A, hepatitis B, hepatitis C, hepatitis D, and hepatitis E.

Hepatitis is an inflammation of the liver. At least five different viruses cause hepatitis, and probably more remain to be discovered or become better known. Hepatitis is also an occasional result of infections by other viruses such as Epstein-Barr virus (EBV) or cytomegalovirus (CMV). Drug and chemical toxicity can also cause acute hepatitis that is clinically identical to viral hepatitis. The characteristics of the various forms of viral hepatitis are summarized in the Diseases in Focus box on page 766.

HEPATITIS A

The *hepatitis A virus (HAV)* is the causative agent of **hepatitis A.** The virus contains single-stranded RNA and lacks an envelope. It can be grown in cell culture.

After a typical entrance via the oral route, HAV multiplies in the epithelial lining of the intestinal tract. Viremia eventually occurs, and the virus spreads to the liver, kidneys, and spleen. The virus is shed in the feces and can also be detected in the blood and urine. The amount of virus excreted is greatest before symptoms appear and then declines rapidly. Therefore, a food handler responsible for spreading the virus might not appear to be ill at the time. The virus can probably survive for several days on such surfaces as cutting boards. Contamination of food or drink by feces is aided by the resistance of HAV to chlorine disinfectants at concentrations ordinarily used in water. Mollusks, such as oysters, that live in contaminated waters are also a source of infection.

At least 50% of infections with HAV are subclinical, especially in children. In clinical cases, the initial symptoms are anorexia (loss of appetite), malaise, nausea, diarrhea, abdominal discomfort, fever, and chills. These symptoms are more likely to appear in adults; they last 2 to 21 days, and the mortality rate is low. Nationwide epidemics occur about every 10 years, mostly in people under 14. In some cases, there is also jaundice (signs are yellowing of the skin and the whites of the eyes) and the dark urine typical of liver infections. In these cases the liver becomes tender and enlarged.

There is no chronic form of hepatitis A, and the virus is usually shed only during the acute stage of disease. The incubation time averages 4 weeks and ranges from 2 to 6 weeks, which makes epidemiological studies for the source of infections difficult. There are no animal reservoirs.

In the United States, the percentage of the population that becomes infected with HAV is much higher among lower socioeconomic groups (72–88%) than among middle and upper socioeconomic groups (18–30%). The 30,000 or more cases reported in the United States each year represent only a fraction of the actual number.

Acute disease is diagnosed by the detection of IgM anti-HAV because these antibodies appear about 4 weeks after infection and disappear about 3 to 4 months after infection. Recovery results in lifelong immunity.

No specific treatment for the disease exists, but people at risk for exposure to hepatitis A can be given immune globulin, which provides protection for several months. Inactivated vaccines are now available and are recommended for travelers to areas of endemic disease and for high-risk groups, such as homosexual men and injecting drug users (IDUs). In 2000 the CDC suggested HAV vaccination for children in certain western states with historically high rates of the disease.

HEPATITIS B

Hepatitis B is caused by the *hepatitis B virus (HBV).* HBV and HAV are completely different viruses: HBV is larger, its

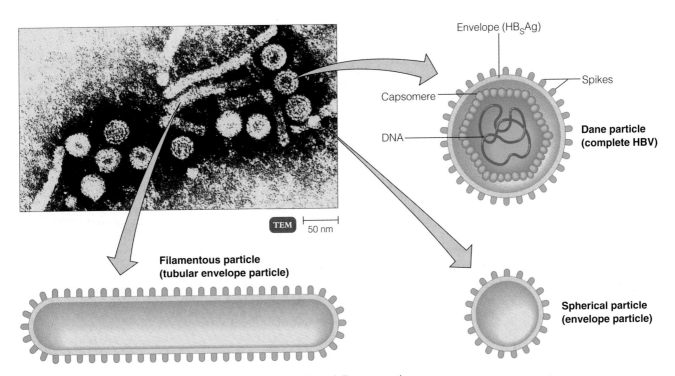

FIGURE 25.16 Hepatitis B virus (HBV). The micrograph and illustrations depict the distinct types of HBV particles discussed in the text.

Q **What are other causes of viral hepatitis?**

genome is double-stranded DNA, and it is enveloped. HBV is a unique DNA virus; instead of replicating its DNA directly, it passes through an intermediate RNA stage resembling a retrovirus. Because HBV has often been transmitted by blood transfusions, this virus has been intensely studied in order to determine how to identify contaminated blood.

The serum from patients with hepatitis B contains three distinct particles. The largest, the *Dane particle*, is the complete virion; it is infectious and capable of replicating. There are also smaller *spherical particles*, about half the size of a Dane particle, and *filamentous particles*, which are tubular particles similar in diameter to the spherical particles but about ten times as long (Figure 25.16). The spherical and filamentous particles are unassembled components of Dane particles without nucleic acids; assembly is evidently not very efficient, and large numbers of these unassembled components accumulate. Fortunately, these numerous unassembled particles contain *hepatitis B surface antigen (HB_sAg)*, which can be detected with antibodies to them. Such antibody tests make convenient screening of blood for HBV possible.

Physicians, nurses, dentists, medical technologists, and others who are in daily contact with blood have a considerably higher incidence of hepatitis B than members of the general population. It is estimated that as many as 10,000 health care workers become infected each year in the United States. Federal regulations require that employees

exposed to blood be offered free vaccinations by their employers. There have also been instances of transmission to patients by surgeons and dentists. Intravenous drug users often share needles and fail to sterilize them properly; as a consequence, they also have a high incidence of hepatitis B. Blood may contain up to a billion viruses per milliliter. Therefore, it is not surprising that it is also present in many body fluids, such as saliva, breast milk, and semen, but not in blood-free feces or urine. Transmission by semen donated for artificial insemination has been documented, and semen has been implicated in transmission between heterosexuals with multiple partners and in male homosexuals. Precautions taken to prevent HIV transmission have also had an effect on the incidence of HBV infections. A mother who is positive for HB_sAg, especially if she is a chronic carrier, may transmit the disease to her infant, usually at birth. In most cases, this type of transmission can be prevented by administering hepatitis B immune globulin (HBIG) to the newborn immediately after birth. These babies should also be vaccinated.

A third of the world's population shows serological evidence of past infection—and HBV is estimated to cause a million deaths every year. An estimated 130,000 Americans, mostly young adults, are infected with HBV each year; only about 10,000 cases are actually reported. About 5000 people die each year of HBV-related liver disease, ranging from cirrhosis (hardening and degeneration) to

DISEASES IN FOCUS

CHARACTERISTICS OF VIRAL HEPATITIS

Hepatitis is an inflammation of the liver. It may be an acute illness with jaundice or elevated serum aminotransferase. Aminotransferases are enzymes found in liver cells and released when the cells are damaged. Chronic hepatitis may be asymptomatic, or there may be evidence of liver disease (including cirrhosis or liver cancer). Hepatitis can be caused by a variety of viruses, alcohol, or drugs; however, it is most often caused by one of the following viruses.

	Pathogen	Method of transmission	Incubation period	Symptoms	Diagnostic test	Antibody prevalence in U.S.	Vaccine
Hepatitis A	Hepatitis A virus (HAV); Picornaviridae	Ingestion	2–6 weeks	Mostly subclinical; fever, headache, in malaise, jaundice severe cases; no chronic disease	IgM antibodies	33%	Inactivated virus Postexposure: immune globulin
Hepatitis B	Hepatitis B virus (HBV); Hepadnaviridae	Parenteral; sexual contact	4–26 weeks	Frequently subclinical; similar to HAV, but no headache; more likely to progress to severe liver damage; chronic disease occurs	IgM antibodies	5–10%	Genetically-modified vaccine produced in yeast
Hepatitis C	Hepatitis C virus; Flaviviridae	Parenteral	2–22 weeks	Similar to HBV; more likely to become chronic	PCR for viral RNA	1.8%	None

cancer. The host's immune response to the virus is primarily responsible for liver damage. The incubation period before the appearance of symptoms averages about 12 weeks: the range is 4 to 26 weeks.

It is important to be able to distinguish between acute and chronic HBV infection. The incubation period for *acute HBV hepatitis* averages about 12 weeks; the range is 4 to 26 weeks. Signs and symptoms are highly variable, and HBV infections cannot be distinguished from other viral hepatitis infections by purely clinical appearance. The patient may have very mild symptoms, such as loss of appetite, low-grade fever, and joint pain. Only a minority of infected infants and small children show any symptoms at all. However, in cases of fulminant HBV hepatitis (rapidly increasing in severity), patients might have fever, nausea, and the typical symptoms of jaundice. At least 90% of acute HBV infections end in complete recovery, and the overall mortality rate for HBV infections is less than 1%. However, although fulminating hepatitis occurs in less than 2% of infections, it has a very high fatality rate.

If HB_sAg persists for more than about 6 months, it is an indication of *chronic HBV hepatitis*; IgM-type antibodies also will have disappeared at about this time. For individuals who were infected 1 to 5 years previously, the risk of developing chronic disease is highest. The risk for infants is about 90%; in children of 1 to 5 years, about 25 to 50%.

DISEASES IN FOCUS

(continued)

	Pathogen	Method of transmission	Incubation period	Symptoms	Diagnostic test	Antibody prevalence in U.S.	Vaccine
Hepatitis D	Hepatitis D virus; Deltaviridae	Parenteral; requires co-infection with hepatitis B	6–26 weeks	Severe liver damage; high mortality rate; chronic disease may occur	IgM antibodies	Unknown	HBV vaccine is protective
Hepatitis E	Hepatitis E virus; Caliciviridae	Ingestion	2–6 weeks	Similar to HAV, but pregnant women may have high mortality; no chronic disease	IgM antibodies; PCR for viral RNA	0.5%	Under development

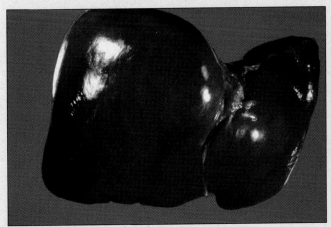

Healthy liver

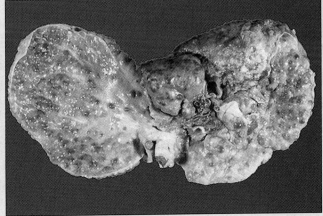

Liver damaged by hepatitis C

Adolescents and young adults have a much lower risk of developing chronic HBV hepatitis: only 6 to 10%.

Overall, up to 10% of patients become chronic carriers. These carriers are reservoirs for transmission of the virus, and they also have a high rate of liver disease. It has been estimated that there are 1.25 million HBV carriers in the United States. A special concern is the strong correlation between the occurrence of liver cancer and the incidence of chronic hepatitis B infections. Chronic carriers are about 200 times more likely to get liver cancer than the general population. Liver cancer is the most prevalent form of cancer in sub-Saharan Africa and the Far East, areas where hepatitis B is extremely common. Worldwide,

the number of HBV carriers is estimated to be 400 million.

Prevention of HBV infection involves several strategies. Important among them are precautions such as disposable needles and syringes and the use of barrier-type contraception. Screening of transfused blood has also been greatly reduced risk. The introduction of HBV vaccines has become widespread worldwide and is now part of the childhood immunization schedule in the United States. The incidence of HBV infections has declined sharply in areas in which the vaccine is in use, and eventual elimination of the disease is conceivable.

It has not been possible to cultivate HBV in cell culture, a step that was necessary for the development of vaccines for polio, mumps, measles, and rubella. The available HBV vaccines use HB$_s$Ag produced by a genetically engineered yeast. Vaccination is recommended for high-risk groups; a partial listing would include health care workers exposed to blood and blood products, people undergoing hemodialysis, patients and staff at mental health care institutions, IDUs, and homosexually active males.

Treatments for chronic HBV infection have been limited and are not curative. Lamivudine (a synthetic nucleoside analogue of cytosine) combined with alpha-interferon (IFN-α) is expensive, but results in improvement in a significant number of patients. Recently, another nucleoside analog, adefovir dipivoxil, has been approved. It is expected to eventually replace the use of IFN-α. Several similar drugs are in clinical trials. Liver transplantation is often a final option in treatment.

HEPATITIS C

In the 1960s, a previously unsuspected form of transfusion-transmitted hepatitis, now called **hepatitis C,** appeared. This new form of hepatitis soon constituted almost all transfusion-transmitted hepatitis—as testing eliminated HBV in the blood supply. Eventually, serological tests to detect hepatitis C virus (HCV) antibodies were developed that similarly reduced the transmission of HCV to very low levels. However, there is a delay of about 70 to 80 days between infection and the appearance of detectable HCV antibodies. The presence of HCV in contaminated blood cannot be detected during this interval and about 1 in 100,000 transfusions can still result in infection. Blood-collecting facilities in the United States can now detect HCV-contaminated blood within 25 days of infection. (See the box on safety of the blood supply, opposite). A PCR test can detect viral RNA within 1 to 2 weeks after infection.

HCV has a single strand of RNA and is enveloped. It is capable of rapid genetic variation to evade the immune system. This characteristic, along with the fact that currently it cannot be cultured in vitro, will complicate the search for an effective vaccine.

Hepatitis C has been described as a silent epidemic, killing more people than AIDS in the United States. It is often clinically inapparent—few people have recognizable symptoms until about 20 years have elapsed. Even today, probably only a minority of infections have been diagnosed. Often, hepatitis C is detected only during some routine testing, such as for insurance or blood donation. A majority of cases, perhaps as high as 85%, progress to chronic hepatitis, a much higher rate than with HBV. Surveys indicated an estimated 1.8%, or nearly 4 million people, of the U.S. population is infected. More than 100,000 people are newly infected each year, and more than 8000 die. About 25% of chronically infected patients develop liver cirrhosis or liver cancer. Hepatitis C is probably the major reason for liver transplantation. Persons infected with HCV should be immunized against both HAV and HBV (a combination vaccine is now available) because they cannot afford the risk of further liver damage.

Prevention of HCV is limited to minimizing exposure—even sharing of items such as razors, toothbrushes, or nail clippers is dangerous. A common source of infection is the sharing of injection equipment among IDUs. At least 80% of this group is infected with HCV. In one exceptional case the disease was transmitted by means of a straw shared for inhaling cocaine. Interestingly, in more than one-third of the cases, a mode of transmission—by contaminated blood, sexual contact, or other means—cannot be identified.

The treatment of choice is a new drug combination, peginterferon alfa-2A (the interferon is conjugated with polyethylene glycol, which has a more sustained concentration in the blood) and ribavirin. Its disadvantages are that it is very expensive and requires a regimen of months. It also has many potentially severe side effects. However, complete eradication of HCV is attained in many cases.

HEPATITIS D (DELTA HEPATITIS)

In 1977, a new hepatitis virus, now known as *hepatitis D virus (HDV)*, was discovered in carriers of HBV in Italy. People who carried this so-called *delta antigen* and were also infected with HBV had a much higher incidence of severe liver damage and a much higher mortality rate than people who had antibodies against HBV alone. With time, it became clearer that **hepatitis D** can occur as either acute (*coinfection form*) or chronic (*superinfection form*) hepatitis. In people with a case of self-limiting acute hepatitis B, coinfection with HDV disappeared as the HBV was cleared from the system, and the condition resembled a typical case of acute hepatitis B. However, if the HBV infection progressed to the chronic stage, superinfection with HDV was often accompanied by progressive liver damage and a fatality rate several times that of people infected with HBV alone.

Epidemiologically, hepatitis D is linked to the epidemiology of hepatitis B. In the United States and northern Europe, the disease occurs predominantly in high-risk groups, such as IDUs.

Structurally, the HDV contains is a single strand of RNA, which is shorter than in any other animal-infecting virus. The particle is not capable of causing an infection. It becomes infectious when an external envelope of HB$_s$Ag, whose formation is controlled by the genome of HBV, covers the HDV protein core (the delta antigen) (see Figure 25.16).

HEPATITIS E

Hepatitis E is spread by fecal–oral transmission, much like hepatitis A, which it clinically resembles. The pathogen, known as *hepatitis E virus (HEV)*, is endemic in areas of the world with poor sanitation, especially India and southeast

APPLICATIONS OF MICROBIOLOGY

A SAFE BLOOD SUPPLY

Prior to blood banking, a physician typed the blood of a patient's friends until the proper blood type was found. In the 1940s, the discovery of many new blood antigens led to the development of new cross-matching techniques. As a result, blood banking became the role of specialists, not the primary care physician. Now in the United States alone, 25 million people donate blood or plasma, and 5 million patients receive whole blood or plasma each year. The safety of blood products is important for all people, especially those with hemophilia because they regularly receive transfusions of clotting factors.

The risk of transmitting hepatitis in blood products was recognized early. An important advance in protecting the blood supply from infectious agents, including hepatitis viruses, was the change to an all-volunteer donor system, which occurred in 1979. (Volunteer donors have a lower infection rate than paid donors.)

However, the large number of people with hemophilia who became infected with human immunodeficiency virus (HIV) in the early 1980s raised new questions regarding the safety of the blood supply. More sensitive donor screening was rapidly introduced to avoid future contamination. Serological

tests are now routinely performed on donated blood to detect the presence of six viruses (T-cell leukemia viruses, HTLV-1 and HTLV-2; AIDS viruses, HIV-1 and HIV-2; hepatitis viruses, HBV and HCV); and *Treponema pallidum* bacteria (the cause of syphilis).

Unfortunately, contamination in the blood of newly infected donors may not be detected by serological tests because there is a "window" of delay between the time of infection and the appearance of antibodies. Now, virtually all whole blood and plasma donations are screened for HCV, HIV, and West Nile virus by nucleic acid testing (NAT), which detects the virus nucleic acids directly, rather than antibodies. NAT has reduced the window of delay during which a newly acquired infection cannot be detected to approximately 25 days for HCV and 12 days for HIV. However, at present NAT takes several days to complete, so platelets, which become outdated in 5 days, are being released before NAT has been completed.

There is also concern over potential contamination of blood by newly discovered viruses. One response to the 2003 SARS outbreak was to defer anyone who has traveled in a SARS-affected area from making a donation for a period of 14 days. Technologies are therefore

being introduced to clean blood by removing 99.9% of white blood cells, which harbor many viruses, such as HIV and cytomegalovirus. Other new techniques are aimed at inactivating any bacteria or viruses in the blood. All of these treatments have certain limitations; for example, some are effective against selected organisms only, and others can be used only for platelets or plasma and not red blood cells. The American Red Cross already requires virus-inactivating treatment of blood plasma.

Unfortunately, there are no effective laboratory tests to screen blood for bloodborne protozoan diseases (i.e., Chagas' disease, malaria, babesiosis) and intracellular *Rickettsia* and *Ehrlichia*. Prevention of transmission relies on donors who have not traveled into endemic areas. To prevent trasmission of prions, blood donors with a family history of Creutzfeldt-Jakob disease, nerve cell transplant, or travel to Europe are required to wait at least six months or longer before giving blood.

A zero-risk blood supply is probably unattainable, but the goal is to make the blood supply as safe as possible. Synthetic blood substitutes are being developed and may one day replace the need for donor blood.

Asia. It resembles HAV in being a nonenveloped virus with a single strand of RNA but is not related serologically to it. Like hepatitis A virus, HEV does not cause chronic liver disease, but for some unexplained reason it is responsible for a mortality rate in excess of 20% in pregnant women.

OTHER TYPES OF HEPATITIS

New techniques in molecular biology and serology have provided evidence of blood-transmitted viruses known as *hepatitis F* (HFV) and *hepatitis G* (HGV). The HGV is found worldwide and in the United States is more prevalent than HCV. HGV is closely related to HCV and is sometimes called GB virus C (GBV-C). Apparently, however, it is so well adapted to its human hosts that it causes no

significant disease condition. About 5% of cases of chronic liver disease cannot be attributed to any known hepatitis in the series A through E. Whether these will eventually be attributed to HFV, HGV, or to some other member added to this alphabetic explosion is unknown.

VIRAL GASTROENTERITIS

LEARNING OBJECTIVE

• List the causative agents, mode of transmission, and symptoms of viral gastroenteritis.

Acute gastroenteritis is one of the most common diseases of humans, and about 90% of cases of acute viral gastroenteritis

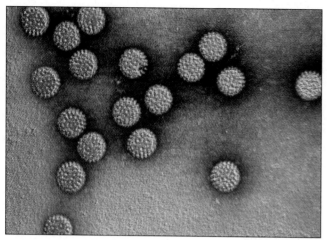

FIGURE 25.17 Rotavirus. This negatively stained electron micrograph shows the morphology of the rotavirus (*rota* = wheel), which gives the virus its name.

Q **What disease does rotavirus cause?**

are caused by either the rotavirus or the human caliciviruses, better known as the Norwalk family of viruses; collectively, the noroviruses.

Rotavirus (Figure 25.17) is probably the most common cause of viral gastroenteritis. It is estimated to cause about 3 million cases, but fewer than 100 deaths, every year in the United States. Mortality is much higher in less developed countries because rehydration therapy is not as available. More than 90% of children in the United States have been infected by the age of 3. In some cases parents also become infected. Immunity acquired then makes rotavirus infections, except for certain strains, much less common in adults. In most cases, following an incubation period of 2 to 3 days, the patient suffers from low-grade fever, diarrhea, and vomiting, which persists for about a week.

A vaccine was introduced to prevent this disease but was withdrawn a few years later when a relationship between the vaccine and a few cases of bowel obstruction in infants was observed. Some international health workers think that the vaccine would still be well worth the risk in parts of the world where the mortality rate from rotavirus infections is quite high. In fact, two newly developed rotavirus vaccines are in current clinical trials.

Noroviruses were first identified following an outbreak of gastroenteritis in Norwalk, Ohio, in 1968. The responsible agent was identified in 1972 and called the *Norwalk virus*. Several similar viruses were later identified, and this group was termed *Norwalk-like viruses*. All were determined to be members of the caliciviruses (named for the Latin *calyx*, meaning cup—cup-shaped depressions are visible on the viruses) and are now termed noroviruses.

They cannot be cultured nor do they infect the usual laboratory animals. Humans become infected by fecal–oral transmission from food and water and even aerosols from vomiting. The infective does may be as low as 10 viruses. The viruses continue to be shed for several days after the patient is asymptomatic. More than 20 million cases of norovirus gastroenteritis occur annually in the United States but only about 300 deaths. About half of adult Americans show serological evidence that they have been infected. Control of outbreaks (those occurring on cruise ships are particularly well known) is difficult because the viruses are unusually persistent on environmental surfaces, surviving for as long as three weeks. They will also survive exposure to 10 ppm chlorine solutions.

Following an incubation period of 18 to 48 hours, the patient suffers from vomiting and/or diarrhea for two or three days. The initial 1968 outbreak was originally called "winter vomiting disease." Vomiting is the most prevalent symptom in children; most adults experience diarrhea, although many adult patients experience only vomiting. The severity of symptoms often depends upon the size of the infective dose. Immunity to these viruses is poor, usually lasting only a few years. The recommended diagnostic method is currently PCR testing of stool or vomitus samples.

The only treatment for viral gastroenteritis is oral rehydration or, in exceptional cases, intravenous rehydration.

Viral diseases of the GI tract are summarized in Table 25.3.

FUNGAL DISEASES OF THE DIGESTIVE SYSTEM

LEARNING OBJECTIVE
- Identify the causes of ergot poisoning and aflatoxin poisoning.

Some fungi produce toxins called *mycotoxins*. When ingested, these toxins cause blood diseases, nervous system disorders, kidney damage, liver damage, and even cancer. Mycotoxin intoxication is considered when multiple patients have similar clinical signs and symptoms. Diagnosis is usually based on finding the fungi or mycotoxins in the suspected food (Table 25.4).

ERGOT POISONING

Some mycotoxins are produced by *Claviceps purpurea* (kla'vi-seps pŭr-pū-rēä), a fungus causing smut infections on grain crops. The mycotoxins produced by *C. purpurea* cause **ergot poisoning,** or *ergotism*, which results from the ingestion of rye or other cereal grains contaminated with the fungus.

TABLE 25.3	**Viral Diseases of the Lower Digestive System**				
	Pathogen	Symptoms	Incubation Period	Diagnostic Test	Treatment
Mumps	Mumps virus; Paramyxoviridae	Painful swelling of parotid glands	16–18 days	Symptoms; virus culture	Preventive vaccine
Viral gastroenteritis	Rotavirus	Vomiting, diarrhea for 1 week	1–3 days	Enzyme immunoassay for viral antigens in feces	Oral rehydration
	Norovirus	Vomiting, diarrhea for 2–3 days	18–48 hr.	PCR	Oral rehydration
Hepatitis (See Diseases in Focus Box on page 766.)					

Ergot poisoning was very widespread during the Middle Ages. The toxin can restrict blood flow in the limbs, with resulting gangrene. It may also cause hallucinogenic symptoms, producing bizarre behavior similar to that caused by LSD.

AFLATOXIN POISONING

Aflatoxin is a mycotoxin produced by the fungus *Aspergillus flavus*, a common mold. Aflatoxin has been found in many foods but is particularly likely to be found on peanuts. **Aflatoxin poisoning** can cause serious damage to livestock when their feed is contaminated with *A. flavus*. Although the risk to humans is unknown, there is strong evidence that aflatoxin contributes to cirrhosis of the liver and cancer of the liver in parts of the world, such as India and Africa, where food is subject to aflatoxin contamination.

PROTOZOAN DISEASES OF THE DIGESTIVE SYSTEM

LEARNING OBJECTIVE

- List the causative agents, modes of transmission, symptoms, and treatments for giardiasis, cryptosporidiosis, *Cyclospora* diarrheal infection, and amoebic dysentery.

Several pathogenic protozoa complete their life cycles in the human digestive system (Table 25.5). Usually they are ingested as resistant, infective cysts and are shed in greatly increased numbers as newly produced cysts.

GIARDIASIS

Giardia lamblia (frequently also known as *G. intestinalis*, which is the CDC usage, and occasionally as *G. duodenalis*) is a flagellated protozoan that is able to attach firmly to a human's intestinal wall (Figure 25.18). In 1681, van Leeuwenhoek described them as having "bodies . . . somewhat longer than broad and their belly, which was flatlike, furnished with sundry little paws."

G. lamblia is the cause of **giardiasis,** a prolonged diarrheal disease. Sometimes persisting for weeks, giardiasis is characterized by malaise, nausea, flatulence (intestinal gas), weakness, weight loss, and abdominal cramps. The distinctive odor of hydrogen sulfide can often be detected in the breath or stools. The protozoa, reproducing by binary fission, sometimes occupy so much of the intestinal wall that they interfere with food absorption.

Outbreaks of giardiasis in the United States occur often, especially during camping and swimming seasons. About 7% of the population are healthy carriers and shed

TABLE 25.4	**Fungal Diseases of the Lower Digestive System**				
	Pathogen	Symptoms	Intoxication/Infection	Diagnostic Test	Treatment
Ergot poisoning	*Claviceps purpurea*	Restricted blood flow to limbs; hallucinogenic	Mycotoxin produced by fungus growing on grains	Finding fungal sclerotia in food	None
Aflatoxin poisoning	*Aspergillus flavus*	Liver cirrhosis; liver cancer	Mycotoxin produced by fungus growing on food	Immunoassay for toxin in food	None

TABLE 25.5	Protozoan Diseases of the Lower Digestive System				
	Pathogen	Symptoms	Reservoir	Diagnostic Test	Treatment
Giardiasis	*Giardia lamblia*	Protozoan adheres to intestinal wall, may inhibit nutritional absorption; causes diarrhea	Water or mammals	FA	Metronidazole; quinacrine
Cryptosporidiosis	*Cryptosporidium hominis*	Shed in animal feces, protozoan enters water supply; causes self-limiting diarrhea but may be life-threatening if immunosuppressed	Cattle; water	Acid-fast stain; FA; ELISA	Oral rehydration
Cyclospora diarrheal infection	*Cyclospora cayetanensis*	Usually ingested with fruits and vegetables; causes watery diarrhea	Humans or birds	Microscopy	Trimethoprim and sulfamethoxazole
Amoebic dysentery (amoebiasis)	*Entamoeba histolytica*	Amoeba lyses epithelial cells of intestine, causes abscesses; significant mortality rate	Humans	Microscopy; serology	Metronidazole

the cysts in their feces. The pathogen is also shed by a number of wild mammals, especially beavers, and the disease occurs in backpackers who drink from untreated wilderness waters.

Most outbreaks are transmitted by contaminated water supplies. In a recent national survey of surface waters serving as sources for U.S. municipalities, the protozoan was detected in 18% of the samples. Because the cyst stage is relatively insensitive to chlorine, filtration or boiling of water supplies is usually necessary to eliminate the cysts from water.

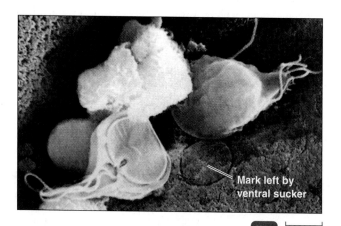

Mark left by ventral sucker

SEM ⊢—————⊣ 5 μm

FIGURE 25.18 The trophozoite form of *Giardia lamblia*, the flagellated protozoan that causes giardiasis. Notice the circular mark left on the intestinal wall by the ventral sucker disk the parasite uses to attach itself. The dorsal side is smoothly streamlined, and the intestinal contents move easily around the attached microorganism.

Q **What is the string test for giardiasis?**

Because *G. lamblia* is not reliably found in stools by microscopic examination, the *string test* is sometimes used for diagnosis. In this test, a gelatin capsule packed with about 140 cm of fine string is swallowed by the patient. One end of the string is taped to the cheek. The gelatin capsule dissolves in the stomach, and an enclosed weighted rubber bag attached to the other end of the string enters the upper bowel. After a few hours, the string is drawn up through the mouth and examined for trophozoite forms of *G. lamblia*. Several commercially available ELISA tests detect both ova and the parasite in stool specimens. The CDC currently recommends a direct fluorescent-antibody test (see Figure 18.11a, page 542) for detection of cysts. These tests are especially useful for epidemiological screening. Testing of drinking water for *Giardia* is difficult but often necessary to prevent or trace disease outbreaks. These tests are frequently combined with tests for *Cryptosporidium* protozoa, discussed in the next section.

Treatment with metronidazole or quinacrine hydrochloride is usually effective within a week. The FDA has recently approved a new oral drug, nitazoxanide (Alinia) for both crytosporidiosis (see below) and giardiasis. Like metronidazole, it affects anaerobic metabolic pathways, but it requires a shorter treatment regimen.

CRYPTOSPORIDIOSIS

Cryptosporidiosis is caused by the protozoan *Crypposporidium*. There have been recent changes in the taxonomy; the primary species affecting humans is now termed *C. hominis* (previously this was *C. parvum* genotype 1). It rarely infects animals. *C. parvum* (previously called *C. parvum* genotype 2) infects both humans and livestock. The term *cryptosporidiosis* describes infections by either organism. Infection occurs when humans ingest the cryptosporidian oocysts (Figure 25.19). The oocysts eventually release sporozoites into the small intestine. The motile sporozoites invade the epithelial cells of the intestine and undergo a cycle that eventually releases oocysts to be excreted in the feces. (Compare with the similar life cycle of *Toxoplasma gondii* in Figure 23.23, page 695). The disease is a cholera-like diarrhea of 10 to 14 days' duration. In immunodeficient individuals, including AIDS patients, the diarrhea becomes progressively worse and is life-threatening. The recommended drug for treatment is the newly introduced nitazoxanide, which is also effective in treating giardiasis.

The infection is transmitted to humans largely through recreational and drinking water systems contaminated with oocysts of *Cryptosporidium*, mostly from animal wastes, especially cattle. Studies in the United States show that many, if not most, lakes, streams, and even wells are contaminated. The oocysts, like the cysts of *G. lamblia,* are resistant to chlorination and must be removed from water by filtration. Even filtration sometimes fails. This is especially true of swimming pools, where both chlorination and filtration systems are ineffective in removing oocysts. An infective dose may be as low as ten oocysts. Fecal–oral transmission resulting from poor sanitation also occurs; many outbreaks have occurred in day-care settings.

Because of the low density of the parasite in most samples, concentration by some method is required. The parasite is then identified by a modified acid-fast staining procedure or some serological method such as fluorescent-antibody (FA) or ELISA assays. These diagnostic tests have a low sensitivity.

Testing of water is important, but the currently available methods have been described as being cumbersome, time-consuming, and inefficient. Most widely used is an FA test that can simultaneously detect both G. *lamblia* cysts and *Cryptosporidium* oocysts. Regular water testing will probably become mandatory, and research for simpler and more reliable methods has a high priority in public health science.

CYCLOSPORA DIARRHEAL INFECTION

A protozoan discovered in 1993 is responsible for a series of diarrheal disease outbreaks in recent years. The pathogen has since been named *Cyclospora cayetanensis.*

The symptoms of **Cyclospora diarrheal infection** are a few days of watery diarrhea, but in some cases it may persist for weeks. The disease is especially debilitating for immunosuppressed people, such as AIDS patients. Whether humans are the only host for the protozoan is uncertain. Most outbreaks have been associated with the ingestion of oocysts in water,

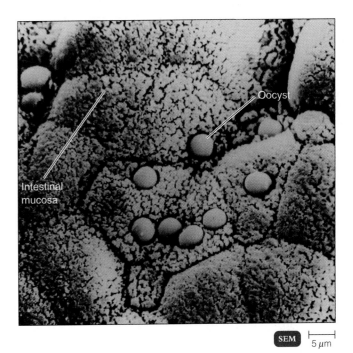

FIGURE 25.19 Cryptosporidiosis. Oocysts of *Cryptosporidium hominis* are shown here embedded in the intestinal mucosa.

Q **How is cryptosporidiosis transmitted?**

on contaminated berries, or similar uncooked foods. The foods are presumed to have been contaminated by oocysts shed in human feces or possibly from birds in the field.

Microscopic examination can identify the oocysts, which are about twice the diameter of those of *Cryptosporidium*. There is really no satisfactory test for contamination of foods. The antibiotic combination of trimethoprim and sulfamethoxazole is used for treatment.

AMOEBIC DYSENTERY (AMOEBIASIS)

Amoebic dysentery, or **amoebiasis,** is spread mostly by food or water contaminated by cysts of the protozoan amoeba *Entamoeba histolytica* (see Figure 12.18b, page 364). Although stomach acid can destroy trophozoites, it does not affect the cysts. In the intestinal tract the cyst wall is digested away, and the trophozoites are released. They then multiply in the epithelial cells of the wall of the large intestine. A severe dysentery results, and the feces characteristically contain blood and mucus. The trophozoites feed on tissue in the gastrointestinal tract (Figure 25.20).

Severe bacterial infections result if the intestinal wall is perforated. Abscesses might have to be treated surgically, and the invasion of other organs, particularly the liver, is not uncommon. Perhaps 5% of the U.S. population are asymptomatic carriers of *E. histolytica*. Worldwide, one

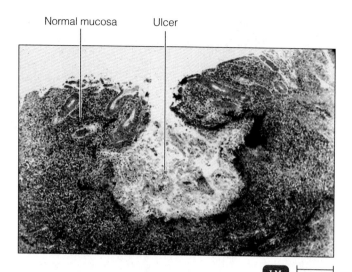

FIGURE 25.20 Section of intestinal wall showing a typical flask-shaped ulcer caused by *Entamoeba histolytica.*

Q If this lesion progressed far enough, could it be life-threatening?

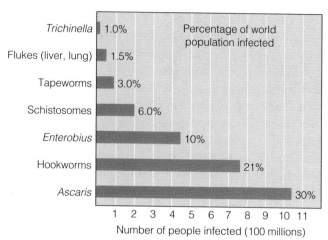

FIGURE 25.21 The worldwide prevalence of human infections with selected intestinal helminths.
SOURCE: World Health Organization.

Q How is each of these diseases transmitted?

person in ten is estimated to be infected, mostly asymptomatically, and about 10% of these infections progress to the more serious stages.

Diagnosis largely depends on recovering and identifying the pathogens in feces. (Red blood cells, ingested as the parasite feeds on intestinal tissue and observed within the trophozoite stage of an amoeba, help identify *E. histolytica*.) Several serological tests can also be used for diagnosis, including latex agglutination and fluorescent-antibody tests. Such tests are especially useful when the affected areas are outside the intestinal tract and the patient is not passing amoebae.

Metronidazole plus iodoquinol are the drugs of choice in treatment.

HELMINTHIC DISEASES OF THE DIGESTIVE SYSTEM

LEARNING OBJECTIVE

- List the causative agents, modes of transmission, symptoms, and treatments for tapeworms, hydatid disease, pinworms, hookworms, ascariasis, and trichinellosis.

Helminthic parasites are very common in the human intestinal tract, especially under conditions of poor sanitation. Figure 25.21 shows the worldwide incidence of infection with some intestinal helminths. In spite of their size and formidable appearance, they often produce few symptoms. They have become so well adapted to their

human hosts, and vice versa, that when their presence is revealed, it is often a surprise.

Table 25.6 summarizes the diseases of the digestive system caused by helminths.

TAPEWORMS

The life cycle of a typical **tapeworm** extends through three stages. The adult worm lives in the intestine of a human host, where it produces eggs that are excreted in the feces (see Figure 12.27, page 374). The eggs are ingested by animals such as grazing cattle, where the egg hatches into a larval form called a *cysticercus* (plural: *cysticerci*) that lodges in the animal's muscles. Human infections by tapeworms begin with the consumption of undercooked beef, pork, or fish containing cysticerci. The cysticerci develop into adult tapeworms that attach to the intestinal wall by suckers on the scolex (see the photo in Figure 12.27, page 374).

The adult beef tapeworm, *Taenia saginata* (te'-nē-ä sa-ji-nä'tä), can live in the human intestine for 25 years and reaches a length of 6 meters (18 feet) or longer. Even a worm of this size seldom causes significant symptoms beyond a vague abdominal discomfort. There is, however, psychological distress when a meter or more of detached segments (proglottids) break loose and unexpectedly slip out of the anus, which happens occasionally.

Taenia solium (sō'lē-um), the pork tapeworm, has a life cycle similar to that of the beef tapeworm. An important difference is that *T. solium* may produce the larval stage in the human host. **Taeniasis** develops when the adult tapeworm infects the human intestine. This is a generally benign, asymptomatic condition, but the host continuously

TABLE 25.6	Helminthic Diseases of the Lower Digestive System				
	Pathogen	Symptoms	Intermediate Host Definitive Host	Diagnostic Test	Treatment
Tapeworms	*Taenia saginata* (beef tapeworm); *T. solium* (pork tapeworm); *Diphyllobothrium latum* (fish tapeworm)	Helminth lives off undigested intestinal contents with few symptoms; pork tapeworm may cause larvae to form in many organs (neurocysticercosis) and cause damage; in this case, eggs are infectious; usually transmitted by ingesting larvae in meats	Intermediate host: cattle, pigs, fish Definitive host: humans	Mcroscopic exam of feces	Praziquantel; albendazole
Hydatid disease	*Echinococcus granulosus*	Larvae form in body; may be very large and cause damage; transmitted by ingesting tapeworm eggs	Intermediate host: humans Definitive host: dogs	Serology; X ray	Surgical removal; albendazole
Pinworms	*Enterobius vermicularis*	Itching around anus	Intermediate host: humans Definitive host: humans	Microscopic exam	Pyrantel pamoate
Hookworms	*Necator americanus, Ancyclostoma duodenale*	Larvae enter through skin; large infections may result in anemia	Larvae enter skin from soil Definitive host: human	Microscopic exam	Mebendazole
Ascariasis	*Ascaris lumbricoides*	Helminths live off undigested intestinal contents; transmitted by ingesting eggs from feces; usually few symptoms	Intermediate host: humans Definitive host: human	Microscopic exam	Mebendazole
Trichinellosis	*Trichinella spiralis*	Larvae encyst in striated muscle; transmitted by ingestion of larvae in meats; usually few symptoms, but large infections may be fatal	Intermediate host: mammals (including humans) Definitive host: mammals (including human)	Biopsy; ELISA	Mebendazole; corticosteroids

expels eggs of *T. solium*, which contaminate hands and food under poor sanitary conditions. **Cysticercosis,** infection with the larval stage, can develop when humans or swine ingest *T. solium* eggs. These eggs can leave the digestive tract and develop into larvae that lodge in tissue (usually brain or muscles). Cyticerci in muscle tissue are relatively benign and cause few serious symptoms, but the larvae occasionally lodge in an eye, causing **ophthalmic cysticercosis** and affecting vision (Figure 25.22). The most serious, and much more common disease is **neurocysticercosis,** which arises when the larvae develop in the central nervous system such as the brain. Neurocysticercosis, which is endemic in Mexico and Central America, has become a fairly common condition in parts of the United States with large Mexican and Central American immigrant populations.

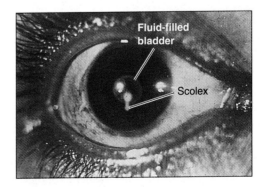

FIGURE 25.22 Ophthalmic cysticercosis. Some cases of cysticercosis affect the eye.

Q **What organ is most likely to be affected by neurocysticercosis?**

The symptoms often mimic those of epilepsy or a brain tumor. The number of cases reported reflects, in part, the use of CT (computed tomography) scanning in diagnosis. These expensive machines X ray the body in continuous "slices." In endemic areas, neurological patients can be screened with serological tests for antibodies to *T. solium*.

The fish tapeworm *Diphyllobothrium latum* (dī-fil-lō-bo'thrē-um lā'tum) is found in pike, trout, perch, and salmon. The CDC has issued warnings about the risks of fish tapeworm infection from sashimi and sushi (Japanese dishes prepared from raw fish), foods that have become increasingly popular. To relate a vivid example, about 10 days after eating, one person developed symptoms of abdominal distention, flatulence, belching, intermittent abdominal cramping, and diarrhea. Eight days later, the patient passed a tapeworm 1.2 m (4 ft) long, identified as a species of *Diphyllobothrium*.

Laboratory diagnosis of tapeworms consists of identifying the tapeworm eggs or segments in feces. Adult tapeworms in the intestinal stage can be eliminated with antiparasitic drugs such as praziquantel and albendazole. Cases of neurocysticercosis can sometimes be treated with drugs, but these often worsen the situation and surgery may be required for removal of cysticerci.

HYDATID DISEASE

Not all tapeworms are large. One of the most dangerous is *Echinococcus granulosus* (ē-kīn-ō-kok'kus gra-nū-lō'sus), which is only a few millimeters in length (see Figure 12.28a, page 375). Humans are not the definitive hosts. The adult form lives in the intestinal tract of carnivorous animals, such as dogs and wolves. Typically, humans become infected from the feces of a dog that has become infected by eating the flesh of a sheep or deer containing the cyst form of the tapeworm. Unfortunately, humans can be an intermediate host, and cysts can develop in the body. The disease occurs most frequently in people who raise sheep or hunt or trap wild animals.

Once ingested by a human, the eggs of *E. granulosus* may migrate to various tissues of the body. The liver and lungs are the most common sites, but the brain and numerous other sites also may be infected. Once in place, the egg develops into a **hydatid cyst** that can grow to a diameter of 1 cm in a few months (Figure 25.23). In some locations, cysts may not be apparent for many years. Some, where they are free to expand, become enormous, containing up to 15 liters (4 gallons) of fluid.

Damage may arise from the size of the cyst in such areas as the brain or the interior of bones. If the cyst ruptures in the host, it can lead to the development of a great many daughter cysts. Another factor in the pathogenicity of such cysts is that the fluid contains proteinaceous material to which the host becomes sensitized. If the cyst suddenly ruptures, the result can be life-threatening anaphylactic shock.

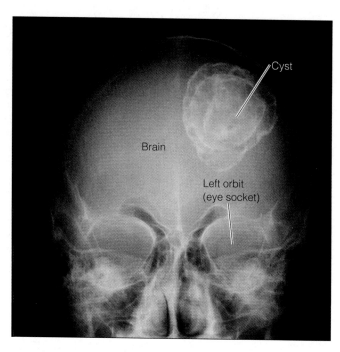

FIGURE 25.23 A hydatid cyst formed by *Echinococcus granulosus*. A large cyst can be seen in this X ray of the brain of an infected individual.

Q How do hydatid cysts affect the body?

For diagnosis, several serological tests that detect circulating antibodies are useful in screening. If available, physical imaging methods such as X rays, computed tomography, and magnetic resonance imaging are best.

Treatment is usually surgical removal, but care must be taken to avoid release of the fluid and the potential spread of infection or anaphylactic shock. If removal is not feasible, the drug albendazole can kill the cysts.

NEMATODES

PINWORMS

Most of us are familiar with the **pinworm**, *Enterobius vermicularis* (see Figure 12.29, page 376). This tiny worm (females are 8–13 mm in length, males 2–5 mm) migrates out of the anus of the human host to lay its eggs, causing local itching. Whole households may become infected. Diagnosis is usually based on finding eggs around the anus. These can be viewed by pressing transparent cellulose tape, sticky side down, against the skin, transferring the tape to a microscope slide and viewing the slide under a microscope. Such drugs as pyrantel pamoate (often available without a prescription) and mebendazole are usually effective in treatment.

HOOKWORMS

Hookworm infections were once a very common parasitic disease in the southeastern states. In the United States, the species most often seen is *Necator americanus* (see Figure 12.30,

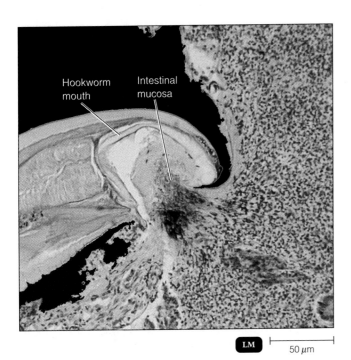

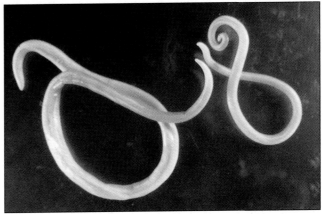

FIGURE 25.25 *Ascaris lumbricoides*, **the cause of ascariasis.** This photograph shows a male (the smaller worm with
the coiled end) and a female. These worms are up to 30 cm
(more than 1 foot) in length.

Q **What are the principal features of the life cycle of
A. lumbricoides?**

FIGURE 25.24 An *Ancylostoma* hookworm attached to
intestinal mucosa. Notice how the mouth of the worm is
adapted to feeding on the tissue.

Q **Why can a hookworm infection lead to anemia?**

page 377). Another species, *Ancyclostoma duodenale*, is
widely distributed around the world.

The hookworm attaches to the intestinal wall and feeds
on blood and tissue rather than on partially digested food
(Figure 25.24), so the presence of large numbers of worms can
lead to anemia and lethargic behavior. Heavy infections can
also lead to a bizarre symptom known as *pica*, a craving for
peculiar foods, such as laundry starch or soil containing a certain type of clay. Pica is a symptom of iron deficiency anemia.

Because the life cycle of the hookworm requires human
feces to enter the soil and bare skin to contact contaminated
soil, the incidence of the disease has declined greatly with
improved sanitation and the practice of wearing shoes.
Hookworm infections are diagnosed by finding parasite eggs
in feces and can be treated effectively with mebendazole.

ASCARIASIS

One of the most widespread helminthic infections is **ascariasis,** caused by *Ascaris lumbricoides*. This condition is
familiar to many American physicians. As described in
Chapter 12 (page 376), diagnosis is often made when an
adult worm emerges from the anus, mouth, or nose. These
worms can be quite large, up to 30 cm (about 1 ft) in
length (Figure 25.25). In the intestinal tract, they live on
partially digested food and cause few symptoms.

The worm's life cycle begins when eggs are shed in a
person's feces and, under poor sanitary conditions, are in-

gested by another person. In the upper intestine, the eggs
hatch into small wormlike larvae that pass into the bloodstream and then into the lungs. There they migrate into
the throat and are swallowed. The larvae develop into egg-
laying adults in the intestines. (All this migration just to
return to the place where they started!)

In the lungs, the tiny larvae may cause some pulmonary
symptoms. Extremely large numbers may block the intestine,
bile duct, or pancreatic duct. The worms do not usually cause
severe symptoms, but their presence can be manifested in
distressing ways. The most dramatic consequences of infection with *A. lumbricoides* are from the migrations of adult
worms. Worms have been known to leave the body of small
children through the umbilicus (navel) and to escape
through the nostrils of a sleeping person. Microscopic examination of feces for eggs is used for diagnosis. Once ascariasis
is diagnosed, it can be effectively treated with mebendazole.

TRICHINELLOSIS

Most infections by the small roundworm *Trichinella spiralis*,
called **trichinellosis** (formerly called trichinosis), are in-
significant. The larvae, in encysted form, are located in
muscles of the host. In 1970, routine autopsies of human
diaphragm muscles showed that about 4% of cadavers
tested carried this parasite.

The severity of the disease is generally proportional to the
number of larvae ingested. Ingesting undercooked pork is
probably the most common mode of infection (Figure 25.26),
but eating the flesh of animals that feed on garbage (bears,
for example) is an increasing cause of outbreaks. Quite a
few human cases of trichinellosis have occurred in France
from horsemeat infected in the United States and exported

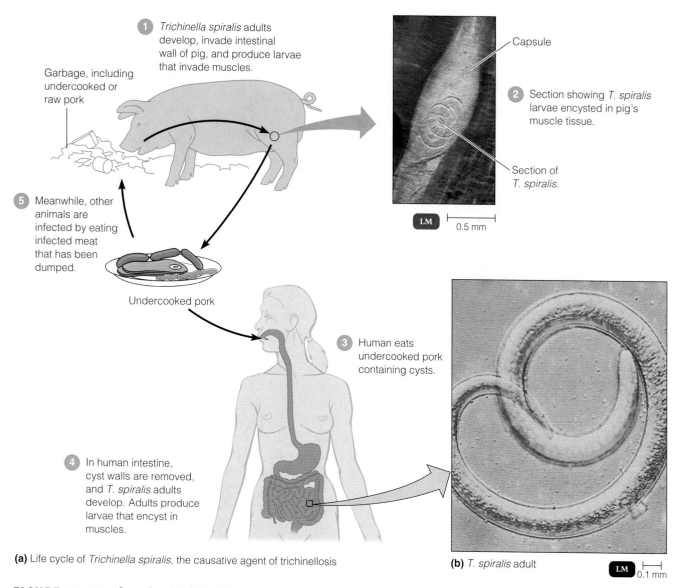

① *Trichinella spiralis* adults develop, invade intestinal wall of pig, and produce larvae that invade muscles.

Garbage, including undercooked or raw pork

Capsule

② Section showing *T. spiralis* larvae encysted in pig's muscle tissue.

Section of *T. spiralis*.

LM ├─────┤ 0.5 mm

⑤ Meanwhile, other animals are infected by eating infected meat that has been dumped.

Undercooked pork

③ Human eats undercooked pork containing cysts.

④ In human intestine, cyst walls are removed, and *T. spiralis* adults develop. Adults produce larvae that encyst in muscles.

(a) Life cycle of *Trichinella spiralis*, the causative agent of trichinellosis

(b) *T. spiralis* adult

LM ├─┤ 0.1 mm

FIGURE 25.26 Life cycle of *Trichinella spiralis*.

Q **What is the most common vehicle of infection of *T. spiralis*?**

to restaurants. Severe cases can be fatal—sometimes in only a few days.

Any ground meat can be contaminated from machinery previously used to grind contaminated meats. Eating raw sausage or hamburger is a risky habit. One person acquired trichinellosis by chewing the fingernails after handling infected pork. Freezing pork for prolonged periods [for example, $-23°C$ ($-10°F$) for 10 days] kills *T. spiralis*. However, some species found in wild game, such as *Trichinella nativa*, are not killed by freezing.

In the muscles of intermediate hosts such as pigs, the *T. spiralis* larvae are encysted in the form of short worms about 1 mm in length. When the flesh of an infected animal is ingested by humans, the cyst wall is removed by digestive action in the intestine. The organism then matures into the adult form. The adult worms spend only

about a week in the intestinal mucosa and produce larvae that invade tissue. Eventually, the encysted larvae localize in muscle (common sites include the diaphragm and eye muscles), where they are barely visible in biopsied specimens.

Symptoms of trichinellosis include fever, swelling around the eyes, and gastrointestinal upset. Small hemorrhages under the fingernails are often observed. Biopsy specimens, as well as a number of serological tests, can be used in diagnosis. Recently, a serological ELISA test that detects the parasite in meats has been developed. Treatment consists of administering mebendazole to kill intestinal worms and corticosteroids to reduce inflammation.

In the past 10 years, the number of cases reported annually in the United States has varied from 16 to 129. Deaths are rare, and in most years there are none.

STUDY OUTLINE

INTRODUCTION (p. 745)

1. Diseases of the digestive system are the second most common illnesses in the United States.
2. Diseases of the digestive system usually result from the ingestion of microorganisms or their toxins in food and water.
3. The fecal–oral cycle of transmission can be broken by the proper disposal of sewage, the disinfection of drinking water, and proper food preparation and storage.

STRUCTURE AND FUNCTION OF THE DIGESTIVE SYSTEM (p. 746)

1. The gastrointestinal (GI) tract, or alimentary canal, consists of the mouth, pharynx, esophagus, stomach, small intestine, and large intestine.
2. In the GI tract, with mechanical and chemical help from the accessory structures, large food molecules are broken down into smaller molecules that can be transported by blood or lymph to cells.
3. Feces, the solids resulting from digestion, are eliminated through the anus.

NORMAL MICROBIOTA OF THE DIGESTIVE SYSTEM (pp. 746–747)

1. Large numbers of bacteria colonize the mouth.
2. The stomach and small intestine have few resident microorganisms.
3. Bacteria in the large intestine assist in degrading food and synthesizing vitamins.
4. Up to 40% of fecal mass is microbial cells.

BACTERIAL DISEASES OF THE MOUTH (pp. 747–750)

DENTAL CARIES (TOOTH DECAY) (pp. 747–749)

1. Dental caries begin when tooth enamel and dentin are eroded and the pulp is exposed to bacterial infection.
2. *Streptococcus mutans*, found in the mouth, uses sucrose to form dextran from glucose and lactic acid from fructose.
3. Bacteria adhere to teeth and produce sticky dextran, forming dental plaque.
4. Acid produced during carbohydrate fermentation destroys tooth enamel at the site of the plaque.
5. Gram-positive rods and filamentous bacteria can penetrate into dentin and pulp.

6. Carbohydrates such as starch, mannitol, sorbitol, and xylitol are not used by cariogenic bacteria to produce dextran and do not promote tooth decay.
7. Caries are prevented by restricting the ingestion of sucrose and by the physical removal of plaque.

PERIODONTAL DISEASE (pp. 749–750)

8. Caries of the cementum and gingivitis are caused by streptococci, actinomycetes, and anaerobic gram-negative bacteria.
9. Chronic gum disease (periodontitis) can cause bone destruction and tooth loss; periodontitis is due to an inflammatory response to a variety of bacteria growing on the gums.
10. Acute necrotizing ulcerative gingivitis is often caused by *Prevotella intermedia*.

BACTERIAL DISEASES OF THE LOWER DIGESTIVE SYSTEM (pp. 750–762)

1. A gastrointestinal infection is caused by the growth of a pathogen in the intestines.
2. Incubation times range from 12 hours to 2 weeks. Symptoms of infection generally include a fever.
3. A bacterial intoxication results from the ingestion of preformed bacterial toxins.
4. Symptoms appear 1–48 hours after ingestion of the toxin. Fever is not usually a symptom of intoxication.
5. Infections and intoxications cause diarrhea, dysentery, or gastroenteritis.
6. These conditions are usually treated with fluid and electrolyte replacement.

STAPHYLOCOCCAL FOOD POISONING (STAPHYLOCOCCAL ENTEROTOXICOSIS) (pp. 751–752)

7. Staphylococcal food poisoning is caused by the ingestion of an enterotoxin produced in improperly stored foods.
8. *S. aureus* is inoculated into foods during preparation. The bacteria grow and produce enterotoxin in food stored at room temperature.
9. Boiling for 30 minutes is not sufficient to denature the exotoxin.
10. Foods with high osmotic pressure and those not cooked immediately before consumption are most often the source of staphylococcal enterotoxicosis.
11. Diagnosis is based on symptoms. Nausea, vomiting, and diarrhea begin 1–6 hours after eating and last about 24 hours.

12. Laboratory identification of *S. aureus* isolated from foods is used to trace the source of contamination.

13. Serological tests are available to detect toxins in foods.

SHIGELLOSIS (BACILLARY DYSENTERY) (pp. 752–753)

14. Shigellosis is caused by any of four species of *Shigella*.

15. Symptoms include blood and mucus in stools, abdominal cramps, and fever. Infections by *S. dysenteriae* result in ulceration of the intestinal mucosa.

16. Shigellosis is diagnosed by isolating and identifying the bacteria from rectal swabs.

SALMONELLOSIS (*SALMONELLA* GASTROENTERITIS) (pp. 753–754)

17. Salmonellosis, or *Salmonella* gastroenteritis, is caused by many *Salmonella enterica* serovars.

18. Symptoms include nausea, abdominal pain, and diarrhea and begin 12–36 hours after eating large numbers of *Salmonella*. Septic shock can occur in infants and in the elderly.

19. Fever might be caused by endotoxin.

20. Mortality is lower than 1%, and recovery can result in a carrier state.

21. Cooking food will usually kill *Salmonella*.

22. Laboratory diagnosis is based on isolating and identifying *Salmonella* from feces and foods.

TYPHOID FEVER (pp. 754–755)

23. *Salmonella typhi* causes typhoid fever; the bacteria are transmitted by contact with human feces.

24. Fever and malaise occur after a 2-week incubation period. Symptoms last 2–3 weeks.

25. *S. typhi* is harbored in the gallbladder of carriers.

26. Typhoid fever is treated with quinolones and cephalosporins; vaccines are available for high-risk people.

CHOLERA (pp. 755–758)

27. *Vibrio cholerae* O:1 and O:139 produce an exotoxin that alters the membrane permeability of the intestinal mucosa; the resulting vomiting and diarrhea cause a loss of body fluids.

28. The symptoms last for a few days. Untreated cholera has a 50% mortality rate.

29. Fluid and electrolyte replacement provide effective treatment.

NONCHOLERA VIBRIOS (p. 758)

30. Ingestion of other *V. cholerae* serotypes can result in mild diarrhea.

31. *Vibrio* gastroenteritis can be caused by *V. parahaemolyticus* and *V. vulnificus*.

32. These diseases are contracted by eating contaminated crustaceans or contaminated mollusks.

ESCHERICHIA COLI GASTROENTERITIS (pp. 758–759)

33. Traveler's diarrhea may be caused by enterotoxigenic or enteroinvasive strains of *E. coli*.

34. The disease is usually self-limiting and does not require chemotherapy.

35. Enterohemorrhagic *E. coli*, such as *E. coli* O157:H7, produces Shiga toxins that cause inflammation and bleeding of the colon, including hemorrhagic colitis and hemolytic uremic syndrome.

36. Shiga toxins can affect the kidneys to cause hemolytic uremic syndrome.

CAMPYLOBACTER GASTROENTERITIS (pp. 759–760)

37. *Campylobacter* is the second most common cause of diarrhea in the United States.

38. *Campylobacter* is transmitted in cow's milk.

HELICOBACTER PEPTIC ULCER DISEASE (p. 760)

39. *Helicobacter pylori* produces ammonia, which neutralizes stomach acid; the bacteria colonize the stomach mucosa and cause peptic ulcer disease.

40. Bismuth and several antibiotics may be useful in treating peptic ulcer disease.

YERSINIA GASTROENTERITIS (p. 760)

41. *Y. enterocolitica* and *Y. pseudotuberculosis* are transmitted in meat and milk.

42. *Yersinia* can grow at refrigeration temperatures.

CLOSTRIDIUM PERFRINGENS GASTROENTERITIS (pp. 760–761)

43. *C. perfringens* causes a self-limiting gastroenteritis.

44. Endospores survive heating and germinate when foods (usually meats) are stored at room temperature.

45. Exotoxin produced when the bacteria grow in the intestines is responsible for the symptoms.

46. Diagnosis is based on isolation and identification of the bacteria in stool samples.

CLOSTRIDIUM DIFFICILE–ASSOCIATED DIARRHEA (pp. 761–762)

47. Growth of *C. difficile* following antibiotic therapy can result in mild diarrhea or colitis.

48. The condition is usually associated with hospitalized patients and nursing home residents.

BACILLUS CEREUS GASTROENTERITIS (p. 762)

49. Ingesting food contaminated with the soil saprophyte *Bacillus cereus* can result in diarrhea, nausea, and vomiting.

VIRAL DISEASES OF THE DIGESTIVE SYSTEM (pp. 763–770)

MUMPS (pp. 763–764)

1. Mumps virus enters and exits the body through the respiratory tract.

2. About 16–18 days after exposure, the virus causes inflammation of the parotid glands, fever, and pain during swallowing. About 4–7 days later, orchitis may occur.

3. After onset of the symptoms, the virus is found in the blood, saliva, and urine.

4. A measles, mumps, rubella (MMR) vaccine is available.

5. Diagnosis is based on symptoms or an ELISA test is performed on viruses cultured in embryonated eggs or cell culture.

HEPATITIS (pp. 764–770)

6. Inflammation of the liver is called hepatitis. Symptoms include loss of appetite, malaise, fever, and jaundice.

7. Viral causes of hepatitis include hepatitis viruses, Epstein-Barr virus (EBV), and cytomegalovirus (CMV).

Hepatitis A (p. 764)

8. Hepatitis A virus (HAV) causes hepatitis A; at least 50% of all cases are subclinical.

9. HAV is ingested in contaminated food or water, grows in the cells of the intestinal mucosa, and spreads to the liver, kidneys, and spleen in the blood.

10. The virus is eliminated with feces.

11. The incubation period is 2–6 weeks; the period of disease is 2–21 days, and recovery is complete in 4–6 weeks.

12. Diagnosis is based on tests for IgM antibodies.

13. A vaccine is available; passive immunization can provide temporary protection.

Hepatitis B (pp. 764–768)

14. Hepatitis B virus (HBV) causes hepatitis B, which is frequently serious.

15. HBV is transmitted by blood transfusions, contaminated syringes, saliva, sweat, breast milk, and semen.

16. Blood is tested for HB_sAg before being used in transfusions.

17. The average incubation period is 3 months; recovery is usually complete, but some patients develop a chronic infection or become carriers.

18. A vaccine against HB_sAg is available.

Hepatitis C (p. 768)

19. Hepatitis C virus (HCV) is transmitted via blood.

20. The incubation period is 2–22 weeks; the disease is usually mild, but some patients develop chronic hepatitis.

21. Blood is tested for HCV antibodies before being used in transfusions.

Hepatitis D (Delta Hepatitis) (p. 768)

22. Hepatitis D virus (HDV) has a circular strand of RNA and uses HB_sAg as a coat.

Hepatitis E (pp. 768–769)

23. Hepatitis E virus (HEV) is spread by the fecal–oral route.

Other Types of Hepatitis (p. 769)

24. There is evidence of the existence of hepatitis types F and G.

VIRAL GASTROENTERITIS (pp. 769–770)

25. Viral gastroenteritis is most often caused by a rotavirus or norovirus.

26. The incubation period is 2–3 days; diarrhea lasts up to 1 week.

FUNGAL DISEASES OF THE DIGESTIVE SYSTEM (pp. 770–771)

1. Mycotoxins are toxins produced by some fungi.

2. Mycotoxins affect the blood, nervous system, kidneys, or liver.

ERGOT POISONING (pp. 770–771)

3. Ergot poisoning, or ergotism, is caused by the mycotoxin produced by *Claviceps purpurea*.

4. Cereal grains are the crop most often contaminated with the *Claviceps* mycotoxin.

AFLATOXIN POISONING (p. 771)

5. Aflatoxin is a mycotoxin produced by *Aspergillus flavus*.

6. Peanuts are the crop most often contaminated with aflatoxin.

PROTOZOAN DISEASES OF THE DIGESTIVE SYSTEM (pp. 771–774)

GIARDIASIS (pp. 771–772)

1. *Giardia lamblia* grows in the intestines of humans and wild animals and is transmitted in contaminated water.

2. Symptoms of giardiasis are malaise, nausea, flatulence, weakness, and abdominal cramps that persist for weeks.

3. Diagnosis is based on identification of the protozoa in the small intestine.

CRYPTOSPORIDIOSIS (pp. 772–773)

4. *Crytosporidium hominis* causes diarrhea; in immunosuppressed patients, the disease is prolonged for months.

5. The pathogen is transmitted in contaminated water.

6. Diagnosis is based on the identification of oocysts in feces.

CYCLOSPORA DIARRHEAL INFECTION (p. 773)

7. *C. cayetanensis* causes diarrhea; the protozoan was first identified in 1993.

8. It is transmitted in contaminated produce.

9. Diagnosis is based on the identification of oocysts in feces.

AMOEBIC DYSENTERY (AMOEBIASIS) (pp. 773–774)

10. Amoebic dysentery is caused by *Entamoeba histolytica* growing in the large intestine.

11. The amoeba feeds on red blood cells and GI tract tissues. Severe infections result in abscesses.

12. Diagnosis is confirmed by observing trophozoites in feces and by several serological tests.

HELMINTHIC DISEASES OF THE DIGESTIVE SYSTEM (pp. 774–778)

TAPEWORMS (pp. 774–776)

1. Tapeworms are contracted by the consumption of undercooked beef, pork, or fish containing encysted larvae (cysticerci).

2. The scolex attaches to the intestinal mucosa of humans (the definitive host) and matures into an adult tapeworm.

3. Eggs are shed in the feces and must be ingested by an intermediate host.

4. Adult tapeworms can be undiagnosed in a human.

5. Diagnosis is based on the observation of proglottids and eggs in feces.

6. Neurocysticercosis in humans occurs when the pork tapeworm larvae encyst in humans.

HYDATID DISEASE (p. 776)

7. Humans infected with the tapeworm *Echinococcus granulosus* might have hydatid cysts in their lungs or other organs.

8. Dogs and wolves are usually the definitive hosts, and sheep or deer are the intermediate hosts for *E. granulosus*.

NEMATODES (pp. 776–778)

Pinworms (p. 776)

9. Humans are the definitive host for pinworms, *Enterobius vermicularis*.

10. The disease is acquired by ingesting *Enterobius* eggs.

Hookworms (pp. 776–777)

11. Hookworm larvae bore through skin and migrate to the intestine to mature into adults.

12. In the soil, hookworm larvae hatch from eggs shed in feces.

Ascariasis (p. 777)

13. *Ascaris lumbricoides* adults live in human intestines.

14. Ascariasis is acquired by ingesting *Ascaris* eggs.

Trichinellosis (pp. 777–778)

15. *Trichinella spiralis* larvae encyst in muscles of humans and other mammals to cause trichinellosis.

16. The roundworm is contracted by ingesting undercooked meat containing larvae.

17. Adult females mature in the intestine and lay eggs; the new larvae migrate to invade muscles.

18. Symptoms include fever, swelling around the eyes, and gastrointestinal upset.

19. Biopsy specimens and serological tests are used for diagnosis.

STUDY QUESTIONS

Access more review material either online at **The Microbiology Place** (www.microbiologyplace.com) or with **The Microbiology Place CD-ROM** packaged with your new book. There you'll find activities, practice tests, quizzes, flashcards, case studies, and more to help you succeed.

Answers to the Study Questions can be found in Appendix G.

REVIEW

1. What properties of *S. mutans* implicate this bacterium in the formation of dental caries? Why is sucrose, more than any other carbohydrate, responsible for the formation of dental caries?

2. Complete the following table.

Disease	Causative Agent	Suspect Foods	Symptoms	Treatment
Staphylococcal food poisoning				
Shigellosis				
Salmonellosis				
Cholera				
Traveler's diarrhea				

3. List the general symptoms of gastroenteritis. Because there are many etiologies, what is the laboratory diagnosis usually based on?

4. Complete the following table:

Causative Agent	Suspect Foods	Treatment	Prevention
Vibrio parahaemolyticus			
V. vulnificus			
Enterotoxigenic E. coli			
Enteroinvasive E. coli			
Enterohemorrhagic E. coli			
Campylobacter jejuni			
Yersinia enterocolitica			
Clostridium perfringens			
Bacillus cereus			

5. Differentiate salmonellosis from typhoid fever.

6. E. coli is part of the normal microbiota of the intestines and can cause gastroenteritis. Explain why this one bacterial species is both beneficial and harmful.

7. What preventive treatments are currently used for hepatitis A? For hepatitis B? For hepatitis C?

8. How is blood that is to be used for transfusions tested for HBV? For HCV?

9. Define *mycotoxin*. Give an example of a mycotoxin.

10. Explain how the following diseases differ and how they are similar: giardiasis, amoebic dysentery, *Cyclospora* diarrheal infection, and cryptosporidiosis.

11. Differentiate between amoebic dysentery and bacillary dysentery (shigellosis).

12. Differentiate among the following factors of bacterial intoxication and bacterial infection: prerequisite conditions, causative agents, onset, duration of symptoms, and treatment.

13. Complete the following table.

Disease	Causative Agent	Mode of Transmission	Site of Infection	Symptoms	Prevention
Mumps					
Hepatitis A					
Hepatitis B					
Viral gastroenteritis					

14. Diagram the life cycle of a human tapeworm.

15. Diagram the life cycle of *Trichinella*, and include humans in the cycle.

16. How can bacterial and protozoan infections of the GI tract be prevented?

17. Look at your diagrams for questions 14 and 15. Indicate sequences in the life cycles that could be easily broken to prevent these diseases.

MULTIPLE CHOICE

1. All of the following can be transmitted by recreational (i.e., swimming) water sources *except*
   a. amoebic dysentery.
   b. cholera.
   c. giardiasis.
   d. hepatitis B.
   e. salmonellosis.

2. A patient with nausea, vomiting, and diarrhea within 5 hours after eating most likely has
   a. shigellosis.
   b. cholera.
   c. E. coli gastroenteritis.
   d. salmonellosis.
   e. staphylococcal food poisoning.

3. Isolation of E. coli from a stool sample is diagnostic proof that the patient has
   a. cholera.
   b. E. coli gastroenteritis.
   c. salmonellosis.
   d. typhoid fever.
   e. none of the above

4. Gastric ulcers are caused by
   a. stomach acid.
   b. Helicobacter pylori.
   c. spicy food.
   d. acidic food.
   e. stress.

5. Microscopic examination of a patient's fecal culture shows comma-shaped bacteria. These bacteria require 2–4% NaCl to grow. The bacteria probably belong to the genus
   a. Campylobacter.
   b. Escherichia.
   c. Salmonella.
   d. Shigella.
   e. Vibrio.

6. A recent cholera epidemic in Peru had all of the following characteristics. Which one *led* to the others?
   a. eating raw fish
   b. sewage contamination of water
   c. catching fish in contaminated water
   d. Vibrio in fish intestine
   e. including fish intestines with edibles

Use the following choices to answer questions 7–10:
   a. Campylobacter
   b. Cryptosporidium
   c. Escherichia
   d. Salmonella
   e. Trichinella

b 7. Identification is based on the observation of oocysts in feces.

e 8. A characteristic disease symptom caused by this microorganism is swelling around the eyes.

a 9. Microscopic observation of a stool sample reveals gram-negative helical cells.

d 10. This microbe is frequently transmitted to humans via raw eggs.

CRITICAL THINKING

1. Why is a human infection of trichinellosis considered a dead-end for the parasite?

2. Complete the following table:

Disease	Conditions Necessary for Microbial Growth	Basis for Diagnosis	Prevention
Staphylococcal food poisoning			
Salmonellosis			
C. difficile diarrhea			

3. Match the following foods with the genus of microorganism most likely to contaminate each:

Beef _____	**a.** *Vibrio*
Delicatessen meats _____	**b.** *Campylobacter*
Chicken _____	**c.** *E. coli* O157:H7
Milk _____	**d.** *Listeria*
Oysters _____	**e.** *Salmonella*
Pork _____	**f.** *Trichinella*

What disease does each microbe cause? How can these diseases be prevented?

4. Which diseases of the gastrointestinal tract can be acquired by swimming in a pool or lake? Why are these diseases not likely to be acquired while swimming in the ocean?

CLINICAL APPLICATIONS

1. In New York on April 26, patient A was hospitalized with a 2-day history of diarrhea. An investigation revealed that patient B had onset of watery diarrhea on April 22. On April 24, three other people (patients C, D, and E) had onset of diarrhea. All three had vibriocidal antibody titers ≥ 640. In Ecuador on April 20, B bought crabs that were boiled and shelled. He shared crabmeat with two people (F and G), then froze the remaining crab in a bag. Patient A returned to New York on April 21 with the bag of crabmeat in his suitcase. The bag was placed in a freezer overnight and thawed on April 22 in a double-boiler for 20 minutes. The crab was served 2 hours later in a crab salad. The crab was consumed during a 6-hour period by A, C, D, and E. Individuals F and G did not become ill. What is the etiology of this disease? How was it transmitted, and how could it have been prevented?

2. The 2130 students and employees of a public school system developed diarrheal illness on April 2. The cafeteria served chicken that day. On April 1, part of the chicken was placed in water-filled pans and cooked in an oven for 2 hours at a dial setting of 177°C. The oven was turned off, and the chicken was left overnight in the warm oven. The remainder of the chicken was cooked for 2 hours in a steam cooker and then left in the device overnight at the lowest possible setting (43°C). Two serotypes of a gram-negative, cytochrome oxidase-negative, lactose-negative rods were isolated from

32 patients. What is the pathogen? How could this outbreak have been prevented?

3. Staff members of one hospital ward noted an increase in the number of cases of HBV. Fifty cases occurred during a 6-month period compared with four cases during the previous 6 months. Between January 1 and 15, all 50 patients had multiple invasive procedures as shown below:
   a. Transfusion, fingerstick, IV catheter, heparin injection: 78%
   b. Transfusion, insulin injection, surgery, fingerstick: 64%
   c. Fingerstick, IV catheter, insulin injection, heparin injection: 80%
   d. Transfusion, heparin injection, surgery, IV catheter: 2%
   e. Heparin injection, IV catheter, insulin injection, surgery: 0%

   How did the patients acquire HBV? Provide an explanation for the 2% and 0%.

4. A 31-year-old male became feverish 4 days after arriving at a vacation resort in Idaho. During his stay, he ate at two restaurants that were not associated with the resort. At the resort, he drank soft drinks with ice, used the hot tub, and went fishing. The resort is supplied by a well that was dug 3 years ago. He went to the hospital when he developed vomiting and bloody diarrhea. Gram-negative, lactose-negative bacteria were cultured from his stool. The patient recovered after receiving intravenous fluids. What microorganism most likely caused his symptoms? How is this disease transmitted? What is the most likely source of his infection, and how would you verify the source?

5. Three to 5 days after eating Thanksgiving dinner at a restaurant, 112 people developed fever and gastroenteritis. All the food had been consumed except for five "doggie" bags. Bacterial analysis of the mixed contents of the bags (containing roast turkey, giblet gravy, and mashed potatoes) showed the same bacterium that was isolated from the patients. The gravy had been prepared from giblets of 43 turkeys that had been refrigerated for three days prior to preparation. The uncooked giblets were ground in a blender and added to a thickened hot stock mixture. The gravy was not reboiled and was stored at room temperature throughout Thanksgiving Day. What was the source of the illness? What was the most likely etiologic agent? Was this an infection or an intoxication?

6. An outbreak of typhoid fever occurred following a family gathering of 293 people. Cultures from 17 of these people yielded *Salmonella typhi*. Nine foods prepared by three food handlers were available at the event.

Foods Consumed	Percent Ill
Green salad and roast beef	60
Noodles and baked beans	42
Noodles and egg salad	12
Egg salad and roast beef	0
Baked beans and fruit	0

What is the most likely source of the *Salmonella*? Which tests would verify this?

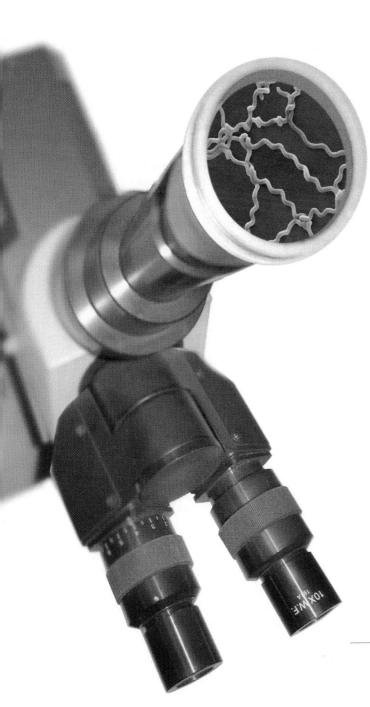

26

Microbial Diseases of the Urinary and Reproductive Systems

The **urinary system** is composed of organs that regulate the chemical composition and volume of the blood and as a result excrete mostly nitrogenous waste products and water. Because it provides an opening to the outside environment, the urinary system is prone to infections from external contacts. The mucosal membranes that line the urinary system are moist and, compared to skin, more supportive of bacterial growth.

The **reproductive system** shares several of the organs of the urinary system. Its function is to produce gametes to propagate the species and, in the female, support and nourish the developing embryo and fetus. In the same fashion as the urinary system, it provides openings to the external environment and is therefore prone to infections. This is especially true because there is often intimate sexual contact that promotes exchange of microbial pathogens between individuals. It is not surprising, therefore, that certain pathogens have adapted to this environment and a sexual mode of transmission. Often they have done this at the cost of an inability to survive in more rigorous environments.

UNDER THE MICROSCOPE

Leptospira interrogans. These tightly coiled bacterial pathogens cause the disease leptospirosis, which is usually transmitted to humans by contact with water contaminated by the urine of infected animals.

STRUCTURE AND FUNCTION OF THE URINARY SYSTEM

LEARNING OBJECTIVE
- List the antimicrobial features of the urinary system.

The urinary system consists of two *kidneys,* two *ureters,* a single *urinary bladder,* and a single *urethra* (Figure 26.1). Certain wastes, collectively called *urine,* are removed from the blood as it circulates through the kidneys. The urine passes through the ureters into the urinary bladder, where it is stored prior to elimination from the body through the urethra. In the female, the urethra conveys only urine to the exterior. In the male, the urethra is a common tube for both urine and seminal fluid.

Where the ureters enter the urinary bladder, physiological valves prevent the backflow of urine to the kidneys. This mechanism helps shield the kidneys from lower urinary tract infections. In addition, the acidity of normal urine has some antimicrobial properties. The flushing action of urine during urination also tends to remove potentially infectious microbes.

STRUCTURE AND FUNCTION OF THE REPRODUCTIVE SYSTEMS

LEARNING OBJECTIVE
- Identify the portals of entry for microbes into the female and male reproductive systems.

The **female reproductive system** consists of two *ovaries,* two *uterine (fallopian) tubes,* the *uterus,* including the *cervix,* the *vagina,* and *external genitals* (Figure 26.2). The ovaries produce female sex hormones and ova (eggs). When an ovum is released during the process of ovulation, it enters a uterine tube, where fertilization may occur if viable sperm are present. The fertilized ovum (zygote) descends the tube and enters the uterus. It implants in the inner wall of the uterus and remains there while it develops into an embryo and, later, a fetus. The external genitals (*vulva*) include the clitoris, labia, and glands that produce a lubricating secretion during copulation.

The **male reproductive system** consists of two *testes,* a system of *ducts, accessory glands,* and the *penis* (Figure 26.3). The testes produce male sex hormones and sperm. To exit the body, sperm cells pass through a series of ducts: the epididymis, ductus (vas) deferens, ejaculatory duct, and urethra.

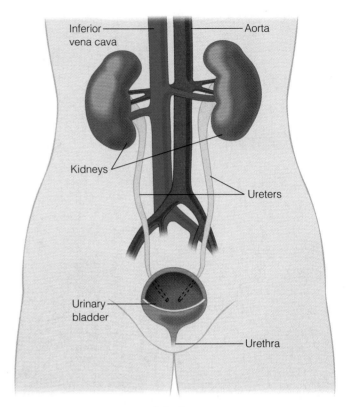

FIGURE 26.1 Organs of the human urinary system, shown here in the female.

 What anatomical features of the urinary system help prevent colonization by microbes?

NORMAL MICROBIOTA OF THE URINARY AND REPRODUCTIVE SYSTEMS

LEARNING OBJECTIVE
- Describe the normal microbiota of the upper urinary tract, the male urethra, and the female urethra and vagina.

Normal urine is sterile, but it may become contaminated with microbiota of the skin near the end of its passage through the urethra. Therefore, urine collected directly from the urinary bladder has fewer microbial contaminants than voided urine.

In the female genital system, the normal microbiota of the vagina are greatly influenced by sex hormones. For example, within a few weeks after birth, the female infant's vagina is populated by lactobacilli. This population grows because estrogens are transferred from maternal to fetal blood and cause glycogen to accumulate in the cells lining the vagina. Lactobacilli convert the glycogen to lactic

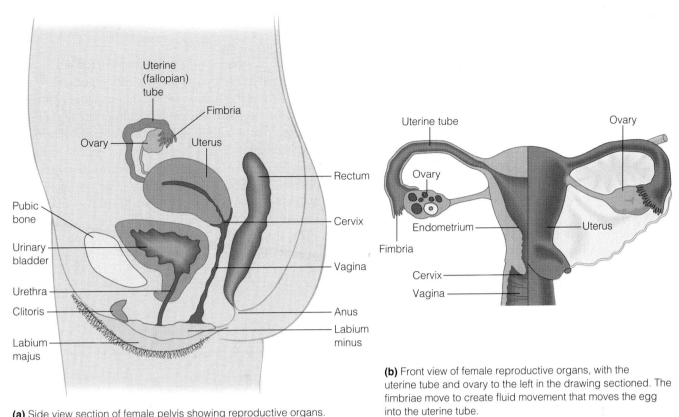

Uterine (fallopian) tube

Fimbria

Ovary

Uterus

Rectum

Pubic bone

Cervix

Urinary bladder

Vagina

Urethra

Clitoris

Anus

Labium majus

Labium minus

Uterine tube

Ovary

Ovary

Endometrium

Uterus

Fimbria

Cervix

Vagina

(a) Side view section of female pelvis showing reproductive organs.

(b) Front view of female reproductive organs, with the uterine tube and ovary to the left in the drawing sectioned. The fimbriae move to create fluid movement that moves the egg into the uterine tube.

FIGURE 26.2 Female reproductive organs.

Q Where are normal microbiota found in the female reproductive system?

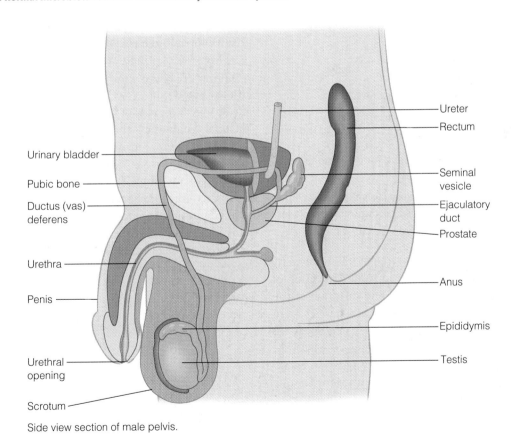

Urinary bladder

Ureter

Rectum

Pubic bone

Ductus (vas) deferens

Seminal vesicle

Ejaculatory duct

Prostate

Urethra

Penis

Anus

Urethral opening

Epididymis

Testis

Scrotum

Side view section of male pelvis.

FIGURE 26.3 Male reproductive and urinary organs. A side view section of a male pelvis.

Q What factors protect the male urinary and reproductive systems from infection?

acid, and the pH of the vagina becomes acidic. This glycogen–lactic acid sequence provides the conditions under which acid-tolerant normal microbiota grow in the vagina.

The physiological effects of estrogens diminish several weeks after birth, and other bacteria, including corynebacteria and a variety of cocci and bacilli, become established and dominate the microbiota. As a result, the pH of the vagina becomes more neutral until puberty. At puberty, estrogen levels increase, lactobacilli again dominate, and the vagina again becomes acidic. In the adult, a disturbance of this ecosystem by an increase in glycogen (caused by oral contraceptives or pregnancy, for ex-

ample) or elimination of the normal microbiota by antibiotics can lead to *vaginitis*, an inflammation of the vagina. When the female reaches menopause, estrogen levels again decrease, the composition of the microbiota returns to that of childhood, and the pH again becomes neutral. Pregnancy and menopause are factors that increase the risk of urinary tract infections, which are probably related to lowered acidity. The use of spermicides, which may inhibit lactobacilli, is also associated with urinary tract infections.

The male urethra is usually sterile, except for a few contaminating microbes near the external opening.

DISEASES OF THE URINARY SYSTEM

The urinary system normally contains few microbes, but it is subject to opportunistic infections that can be quite troublesome. Almost all such infections are bacterial, although occasional infections by pathogens such as schistosome parasites, protozoa, and fungi occur. In addition, as we will see in this chapter, sexually transmitted diseases often affect the urinary system as well as the reproductive system.

BACTERIAL DISEASES OF THE URINARY SYSTEM

> **LEARNING OBJECTIVES**
> - Describe the modes of transmission for urinary and reproductive system infections.
> - List the microorganisms that cause cystitis, pyelonephritis, and leptospirosis, and name the predisposing factors for these diseases.

Urinary system infections are most frequently initiated by an inflammation of the urethra, or *urethritis*. Infection of the urinary bladder is called *cystitis*, and infection of the ureters is *ureteritis*. The most significant danger from lower urinary tract infections is that they may move up the ureters and affect the kidneys, causing *pyelonephritis*. Occasionally the kidneys are affected by systemic bacterial diseases, such as *leptospirosis*. The pathogens causing these diseases are found in excreted urine.

Bacterial infections of the urinary system are usually caused by microbes that enter the system from external sources. In the United States there are about 7 million urinary tract infections each year. About 900,000 cases are of

nosocomial origin, and probably 90% of these are associated with urinary catheters. Because of the proximity of the anus to the urinary opening, intestinal bacteria predominate in urinary tract infections. Most infections of the urinary tract are caused by *E. coli*, mostly strains that have become adapted to colonizing these organs. Infections by *Pseudomonas*, because of their natural resistance to antibiotics, are especially troublesome.

Diagnosis of urinary tract infections is usually based on symptoms such as painful urination or a sensation that the urinary bladder does not feel empty even after urination. Urine may be cloudy or have a light bloody tinge. The traditional guideline—that urine containing more than 100,000 bacteria per milliliter is an indication of urinary tract infection—has been modified. Counts as low as 1000/ml of any single bacterial type or as few as 100/ml of coliforms (intestinal bacteria such as *E. coli*) are now considered an indication of significant infection, especially if leukocytes appear in the urine (*pyuria*). Before therapy is initiated, urine bacteria are usually cultured to determine antibiotic sensitivity.

CYSTITIS

Cystitis is a common inflammation of the urinary bladder in females. Symptoms often include *dysuria* (difficult, painful, urgent urination) and pyuria.

The female urethra is less than 2 inches long, and microorganisms traverse it readily. It is also closer than the male urethra to the anal opening and its contaminating intestinal bacteria. These considerations are reflected in the fact that the rate of urinary tract infections in women is eight times that of men. In either gender, most cases are due to infection by *E. coli*. The second most common bacterial cause is the coagulase-negative *Staphylococcus saprophyticus* (sap-rō-fit'i-kus).

Trimethoprim-sulfamethoxazole usually clears cases of cystitis quickly. Fluoroquinolone antibiotics or ampicillin are often successful if drug resistance is encountered.

PYELONEPHRITIS

In 25% of untreated cases, cystitis may progress to **pyelonephritis,** an inflammation of one or both kidneys. Symptoms are fever and flank or back pain. In females, it is often a complication of lower urinary tract infections. The causative agent in about 75% of the cases is *E. coli.* Pyelonephritis generally results in bacteremia; blood cultures and a Gram stain of the urine for bacteria are useful for diagnosis. If pyelonephritis becomes chronic, scar tissue forms in the kidneys and severely impairs their function. Because pyelonephritis is a potentially life-threatening condition, treatment usually begins with intravenous, extended-term administration of a broad-spectrum antibiotic, such as a second- or third-generation cephalosporin.

LEPTOSPIROSIS

Leptospirosis is primarily a disease of domestic or wild animals, but it can be passed to humans and sometimes causes severe kidney or liver disease. The causative agent is the spirochete *Leptospira interrogans* (in-tėr'rä-ganz), shown in Figure 26.4. *Leptospira* has a characteristic shape: an exceedingly fine spiral, only about 0.1 μm in diameter, wound so tightly that it is barely discernible under a darkfield microscope. Like other spirochetes, *L. interrogans* (so named because the hooked ends suggest a question mark) stains poorly and is difficult to see under a normal light microscope. It is an obligate aerobe that can be grown in a variety of artificial media supplemented with rabbit serum.

Animals infected with the spirochete shed the bacteria in their urine for extended periods. Humans become infected by contact with urine-contaminated water or soil or sometimes with animal tissue. People whose occupations expose them to animals or animal products are most at risk. Usually the pathogen enters through minor abrasions in the skin or mucous membranes. When ingested, it enters through the mucosa of the upper digestive system. In the United States, dogs and rats are the most common sources. Domestic dogs have a sizable rate of infection; even when immunized, they may continue to shed leptospira.

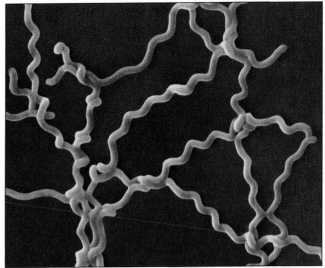

SEM ⊢—⊣ 0.2 μm

FIGURE 26.4 *Leptospira interrogans,* **the cause of leptospirosis.** This photo shows several of these tightly coiled spirochetes.

Q **On what basis is** *L. interrogans* **named?**

After an incubation period of 1 to 2 weeks, headaches, muscular aches, chills, and fever abruptly appear. Several days later, the acute symptoms disappear, and the temperature returns to normal. A few days later, however, a second episode of fever may occur. In a small number of cases the kidneys and liver become seriously infected (*Weil's disease*); kidney failure is the most common cause of death. Recovery results in a solid immunity, but only to the particular serovar involved. There are usually about 50 cases reported each year in the United States, but because the clinical symptoms are not distinctive, many cases are probably never diagnosed. A recent study in a clinic serving the urban poor in a large eastern U.S. city found that up to 16% of the patients tested positive for infection.

Most cases of leptospirosis are diagnosed by a serological test that is complicated and usually done by central reference laboratories. However, a number of rapid serological tests are available for a preliminary diagnosis. Also, a diagnosis can be made by sampling blood, urine, or other fluids for the organism or its DNA. Doxycyline (a tetracycline) is the recommended antibiotic for treatment; however, administration of antibiotics in later stages is often unsatisfactory. That immune reactions are responsible for pathogenesis in this stage may be an explanation.

DISEASES OF THE REPRODUCTIVE SYSTEMS

Microbes causing infections of the reproductive systems are usually very sensitive to environmental stresses and require intimate contact for transmission.

BACTERIAL DISEASES OF THE REPRODUCTIVE SYSTEMS

LEARNING OBJECTIVES

- List the causative agents, symptoms, methods of diagnosis, and treatments for gonorrhea, nongonococcal urethritis (NGU), pelvic inflammatory disease (PID), syphilis, lymphogranuloma venereum (LGV), chancroid, and bacterial vaginosis.

- List reproductive system diseases that can cause congenital and neonatal infections, and explain how these infections can be prevented.

Most diseases of the reproductive systems are transmitted by sexual activity and are called **sexually transmitted diseases (STDs).** More than 30 bacterial, viral, or parasitic diseases have been identified as sexually transmitted. In the United States, it is estimated that over 15 million new cases of STDs occur annually. Many of these diseases can be successfully treated with antibiotics and can be largely prevented by the use of condoms. However, over 60 million Americans have STDs, mostly viral, for which there is no effective cure.

GONORRHEA

One of the most common reportable, or notifiable, communicable diseases in the United States is **gonorrhea,** an STD caused by the gram-negative diplococcus *Neisseria gonorrhoeae.* An ancient disease, gonorrhea was described and given its present name by the Greek physician Galen in AD 150 (*gon* = semen + *rhea* = flow; a flow of semen—apparently, he confused pus with semen). The incidence of gonorrhea has tended to decrease in recent years, but more than 300,000 cases are still reported in the United States each year (Figure 26.5). The true number of cases is probably much larger, probably two or three times those reported (Figure 26.5b). More than 60% of patients with gonorrhea are aged 15 to 24.

In order to infect, the gonococcus must attach to the mucosal cells of the epithelial wall by means of fimbriae. The pathogen invades the spaces separating columnar epithelial cells, which are found in the oral-pharyngeal area,

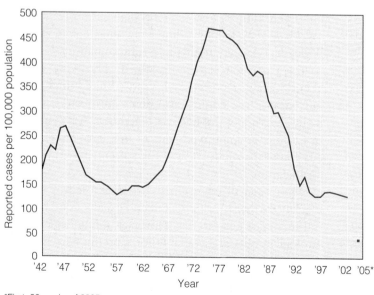

*First 26 weeks of 2005

(a) Incidence of gonorrhea in the United States from 1942 through the first 26 weeks of 2005

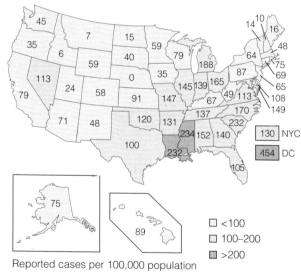

Reported cases per 100,000 population

(b) Geographical distribution of cases in 2004

FIGURE 26.5 The U.S. incidence and distribution of gonorrhea.
SOURCE: CDC, MMWR 52(54) (4/22/05); MMWR 53(52) (1/7/05); MMWR 54(26) (7/8/05).

Q **How do gonococci attach to mucosal epithelial cells?**

the eyes, rectum, urethra, opening of the cervix, and the external genitals of prepubertal females. The invasion sets up an inflammation and, when leukocytes move into the inflamed area, the characteristic pus formation results. In men, a single unprotected exposure results in infection with gonorrhea 20 to 35% of the time. Women become infected 60 to 90% of the time from a single exposure.

Males become aware of a gonorrheal infection by painful urination and a discharge of pus-containing material from the urethra (Figure 26.6). About 80% of infected males show these obvious symptoms after an incubation period of only a few days; most others show symptoms in less than a week. In the days before antibiotic therapy, symptoms persisted for weeks. A common complication is urethritis, although this is more likely to be the result of coinfection with *Chlamydia,* which will be discussed shortly. An uncommon complication is *epididymitis,* an infection of the epididymis. Usually only unilateral, this is a painful condition resulting from the infection ascending along the urethra and the vas deferens (see Figure 26.3).

In females, the disease is more insidious. Only the cervix, which contains columnar epithelial cells, is infected. The vaginal walls are composed of stratified squamous epithelial cells, which are not colonized. Very few women are aware of the infection. Later in the course of the disease, there might be abdominal pain from complications such as pelvic inflammatory disease (discussed on page 794).

In both males and females, untreated gonorrhea can disseminate and become a serious, systemic infection. Complications of gonorrhea can involve the joints, heart (*gonorrheal endocarditis*), meninges (*gonorrheal meningitis*), eyes, pharynx, or other parts of the body. *Gonorrheal arthritis,* which is caused by the growth of the gonococcus in fluids in joints, occurs in about 1% of gonorrhea cases. Joints commonly affected include the wrist, knee, and ankle.

If the mother is infected with gonorrhea, the eyes of the infant can become infected as it passes through the birth canal. This condition, **ophthalmia neonatorum,** can result in blindness. Because of the seriousness of this condition and the difficulty of being sure the mother is free of gonorrhea, antibiotics are placed in the eyes of all newborn infants. If the mother is known to be infected, an intramuscular injection of antibiotic is also administered to the infant. Some sort of prophylaxis is required by law in most states. Gonorrheal infections can also be transferred by hand contact from infected sites to the eyes of adults.

Gonorrheal infections can be acquired at any point of sexual contact; pharyngeal and anal gonorrhea are not uncommon. The symptoms of **pharyngeal gonorrhea** often resemble those of the usual septic sore throat. **Anal gonorrhea** can be painful and accompanied by discharges of

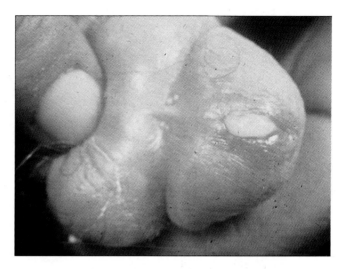

FIGURE 26.6 Pus-containing discharge from the urethra of a male with an acute case of gonorrhea.

Q   *What causes pus formation in gonorrhea?*

pus. In some cases, however, the symptoms are limited to itching.

Increased sexual activity with a series of partners, and the fact that the disease in the female may go unrecognized, contributed considerably to the increased incidence of gonorrhea and other STDs during the 1960s and 1970s. The widespread use of oral contraceptives also contributed to the increase. Oral contraceptives often replaced condoms and spermicides, which help prevent disease transmission.

There is no effective adaptive immunity to gonorrhea. The conventional explanation is that the gonococcus exhibits extraordinary antigenic variability—which is true. Lately, though, an alternative theory has appeared that provides an additional mechanism. The gonococcus is capable of producing several different opacity (Opa) proteins (see Chapter 15, page 456), which are required for the bacteria to adhere to and infect the cells lining the urinary and reproductive systems. These Opa proteins bind to a family of receptors on host cells; CD4 lymphocytes (a group including helper T cells and long-lived memory cells) express only one of this family of receptors. When this receptor on the lymphocyte binds to a particular Opa protein on the gonococcus, it prevents activation of the lymphocyte and turns off proliferation. This blocks development of an immunological memory against *N. gonorrhoeae.* (Experiments show that CD4 lymphocytes lacking this particular Opa receptor are stimulated to a strong immunological response). This mechanism that inhibits the adaptive immune response may also explain why infection with gonorrhea carries an increased risk of acquiring other STDs, including HIV.

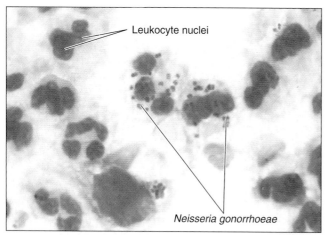

Leukocyte nuclei

Neisseria gonorrhoeae

LM ├──┤
10 μm

FIGURE 26.7 A smear of pus from a patient with gonor-rhea. The *Neisseria gonorrhoeae* bacteria are contained within phagocytic leukocytes. These gram-negative bacteria are visible here as pairs of cocci. The large stained bodies are the nuclei of the leukocytes.

Q How is gonorrhea diagnosed?

DIAGNOSIS

Gonorrhea in men is diagnosed by finding gonococci in a stained smear of pus from the urethra. The typical gram-negative diplococci within the phagocytic leukocytes are readily identified (Figure 26.7). It is uncertain whether these intracellular bacteria are in the process of being killed or whether they survive indefinitely. Probably at least a fraction of the bacterial population remains viable. Gram staining of exudates is not as reliable with women. Usually, a culture is taken from within the cervix and grown on special media. Cultivation of the nutritionally fastidious bacterium requires an atmosphere enriched in carbon dioxide. The gonococcus is very sensitive to adverse environmental influences (desiccation and temperature) and survives poorly outside the body. It even requires special transporting media to keep it viable for short intervals before the cultivation is under way. Cultivation has the advantage of allowing determination of antibiotic sensitivity.

Diagnosis of gonorrhea has been aided by the development of an ELISA that detects *N. gonorrhoeae* in urethral pus or on cervical swabs within about 3 hours with high accuracy. Other rapid tests now available use monoclonal antibodies against antigens on the surface of the gonococcus. Nucleic acid amplification tests are very accurate for identifying clinical isolates from suspected cases.

TREATMENT

Although penicillin has been an effective treatment for gonorrhea for many years, the dosages have had to be increased substantially because of penicillin-resistant bacteria. The current CDC recommendation for treatment of gonorrhea, which is periodically subject to change, is to use fluoroquinolone antibiotics such as ciprofloxacin; they are inexpensive and require only a single oral dose (see the box on the facing page). Resistance to fluoroquinolones is, however, becoming widespread, especially in Asia, the Pacific Islands, and some other areas. Third generation cephalosporins such as injected ceftriaxone are the recommended substitutes. Because of frequent coinfection with chlamydia, antibiotic treatment should also include an effective antichlamydial antibiotic, such as doxycycline or azithromycin (a broad-spectrum macrolide).

NONGONOCOCCAL URETHRITIS (NGU)

Nongonococcal urethritis (NGU), also known as **nonspecific urethritis (NSU),** refers to any inflammation of the urethra not caused by *Neisseria gonorrhoeae*. Symptoms include painful urination and a watery discharge.

The most common pathogen associated with NGU is *Chlamydia trachomatis*. Many people suffering from gonorrhea are coinfected with *C. trachomatis*, which infects the same columnar epithelial cells as the gonococcus. *C. trachomatis* is also responsible for the STD lymphogranuloma venereum (discussed on page 797) and trachoma (see page 636). This is the most common sexually transmitted pathogen in the United States, responsible for an estimated 3 to 4 million cases of NGU annually. Of special importance is the fact that five times as many cases are reported in females than males. In women, it is responsible for many cases of pelvic inflammatory disease (discussed on page 794), plus eye infections and pneumonia in infants born to infected mothers. Genital chlamydial infections are also associated with an increased risk of cervical cancer. It is uncertain whether chlamydial infection is an independent factor in this risk or whether it is associated with coinfections with human papillomavirus (page 801).

Because the symptoms are often mild in males and because females are usually asymptomatic, many cases of NGU go untreated. Although complications are not common, they can be serious. Males may develop inflammation of the epididymis. In females, inflammation of the uterine tubes may cause sterility by scarring the tubes. As many as 60% of such cases may be from chlamydial rather than gonococcal infection. It is estimated that about 50% of men and 70% of women are unaware of their chlamydial infection.

For diagnosis, culturing is the most reliable method, but this requires specialized cultivation methods, takes 24 to 72 hours, and is not always conveniently available. There are a number of new non–culture-based tests available. Several

CLINICAL PROBLEM SOLVING

AN ANTIBIOTIC-RESISTANT STD

You will see questions as you read through this problem. The questions are those that health care providers ask themselves and each other as they solve a clinical problem. Try to answer each question as you read through the problem.

1. On May 24, a 35-year-old man was seen at the San Francisco City and County Clinic with a history of painful urination and urethral discharge of approximately 1 month's duration. *What other information do you need about the patient's history?*

2. On March 11, he had returned from a "dating tour" in Thailand, during which he had sexual contact with seven or eight female prostitutes; he denied having had any sexual contact since returning to the United States. *What sample should be taken, and how should it be tested?*

3. *Neisseria gonorrhoeae* was identified by PCR of urethral discharge. He was treated with a single 500-mg dose of ciprofloxacin orally. *What is the advantage of PCR or enzyme immunoassay (EIA) over cultures for diagnosis?*

4. PCR and EIA provide results within a few hours, eliminating the need for the patient to return for treatment. In this case the patient returned to the clinic on June 7 with continuing symptoms. *N. gonorrhoeae* was again detected in urethral discharge. He denied having had any sexual contact since the previous visit. The attending physician contacted the staff at the microbiology laboratory to determine whether they could perform antimicrobial susceptibility testing of *N. gonorrhoeae* isolates. *Why would the physician be interested in culture and antimicrobial susceptibility test results on this patient's specimen?*

5. The patient failed therapy with ciprofloxacin. One reason for the patient's failure to respond to ciprofloxacin may be due to infection with a fluoroquinolone-resistant *Neisseria gonorrhoeae*. Susceptibility testing would be helpful to explore this possibility.

   The treatment and control of gonorrhea has been complicated by the ability of *Neisseria gonorrhoeae* to develop resistance to antimicrobial agents. Trends in antibiotics resistance are shown in the figure. The appearance of penicillinase-producing *N. gonorrhoeae* and chromosomally mediated penicillin- and tetracycline-resistant *N. gonorrhoeae* in the 1970s eventually led to the abandonment of these drugs as therapies for gonorrhea.

   The current CDC-recommended primary therapies for gonorrhea are two broad-spectrum cephalosporins (ceftriaxone and cefixime), and three fluoroquinolones (ciprofloxacin, ofloxacin, and levofloxacin). However, since the 1990s, fluoroquinolone-resistant *N. gonorrhoeae* (QRNG) have been reported from many parts of the world, including the United States. The increased prevalence of QRNG in Asia (where prevalence in several countries exceeds 40%), the Pacific Islands, Hawaii, and California, prompted CDC to recommend that fluoroquinolones not be used to treat patients with gonorrhea acquired in these areas. *How does antibiotic resistance emerge?*

6. In an environment filled with antibiotics, bacteria that have mutations for antibiotic resistance will have a selective advantage. The indiscriminate use of antibiotics has led to selection of antibiotic-resistant strains. *How is antibiotic susceptibility determined?*

7. *N. gonorrhoeae* must be grown in culture for disk-diffusion or broth dilution tests for antimicrobial susceptibility. The increasing use of nonculture methods for gonorrhea diagnosis such as PCR and EIA is a major challenge to monitoring antimicrobial resistance in *N. gonorrhoeae*.

   Although susceptibility testing is not routinely performed, CDC conducts surveillance nationwide to determine the incidence of resistance among *N. gonorrhoeae* causing infections. As part of the Gonococcal Isolate Surveillance Project (GISP), *N. gonorrhoeae* isolates are collected from the first 25 men with urethral gonorrhea attending STD clinics each month in approximately 29 cities in the United States.

   SOURCE: Data from CDC. *Sexually Transmitted Disease Surveillance 2003 Supplement: Gonococcal Isolate Surveillance Project (GISP) Annual Report–2003.*

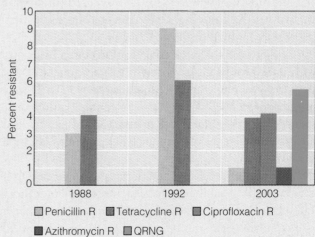

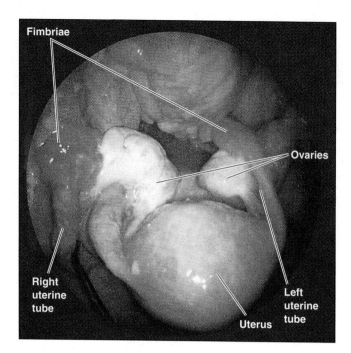

FIGURE 26.8 Salpingitis. This photograph, taken through a laparoscope (a specialized endoscope), shows an acutely inflamed right uterine tube and inflamed, swollen fimbriae and ovary, caused by salpingitis. The left tube is only mildly inflamed. (See Figure 26.2.) The use of a laparoscope is the most reliable diagnostic method for PID.

Q *What is PID?*

of them amplify and detect DNA or RNA sequences of *C. trachomatis*. These amplification tests can be done quickly and are very sensitive, in the range of 80 to 91%; their specificity is close to 100%. They are, however, relatively expensive and require a laboratory with specialized equipment. Urine samples can be used, but the sensitivity is lower than with swabs. The most recent development in amplification testing is to use swab specimens (urethral or vaginal, as the case might be) collected by the patient themselves—which they tend to prefer.

In view of the serious complications often associated with infections by *C. trachomatis*, it is recommended that physicians routinely screen sexually active women 25 years of age and younger for infection. Screening is also recommended for other higher-risk groups, such as persons who are unmarried, had a prior risk of STDs, and have multiple sexual partners.

Bacteria other than *C. trachomatis* can also be implicated in NGU. Another cause of urethritis and infertility is *Ureaplasma urealyticum* (ū-rē-ä-lit′i-kum). This pathogen is a member of the mycoplasma (bacteria without a cell wall). Another mycoplasma, *Mycoplasma hominis* (ho′mi-nis), commonly inhabits the normal vagina but can opportunistically cause uterine tube infection.

Both chlamydia and mycoplasma are sensitive to tetracycline-type antibiotics such as doxycycline, or macrolide-type antibiotics such as azithromycin.

PELVIC INFLAMMATORY DISEASE (PID)

Pelvic inflammatory disease (PID) is a collective term for any extensive bacterial infection of the female pelvic organs, particularly the uterus, cervix, uterine tubes, or ovaries. During their reproductive years, one in ten women suffers from PID, and one in four of these will have serious complications such as infertility or chronic pain.

Pelvic inflammatory disease is considered to be a polymicrobial infection—that is, a number of different pathogens might be the cause, including coinfections. The two most common microbes are *N. gonorrhoeae* and *C. trachomatis*. The onset of chlamydial PID is relatively more insidious, with fewer initial inflammatory symptoms than when caused by *N. gonorrhoeae*. However, the damage to the uterine tube may be greater with chlamydia, especially with repeated infections.

The bacteria may attach to sperm cells and be transported by them from the cervical region to the uterine tubes. Women who use barrier contraceptives, especially with spermicides, have a significantly lower rate of PID.

Infection of the uterine tubes, or **salpingitis,** is the most serious form of PID (Figure 26.8). Salpingitis can result in scarring that blocks the passage of ova from the ovary to the uterus, possibly causing sterility. One episode of salpingitis causes infertility in 10 to 15% of women; 50 to 75% become infertile after three or more such infections.

A blocked uterine tube may cause a fertilized ovum to be implanted in the tube rather than the uterus. This is called an *ectopic* (or *tubal*) *pregnancy*, and it is life-threatening because of the possibility of rupture of the tube and resulting hemorrhage. The reported cases of ectopic pregnancies have been increasing steadily, corresponding to the increasing occurrence of PID.

A diagnosis of PID depends strongly on signs and symptoms, in combination with laboratory indications of a gonorrheal or chlamydial infection of the cervix. The recommended treatment for PID is the simultaneous administration of doxycycline and cefoxitin (a cephalosporin). This combination is active against both the gonococcus and chlamydia. Such recommendations are constantly being reviewed.

SYPHILIS

The earliest reports of **syphilis** date back to the end of the fifteenth century in Europe, when the return of Columbus from the New World gave rise to a hypothesis that syphilis was introduced to Europe by his men. Another hypothesis

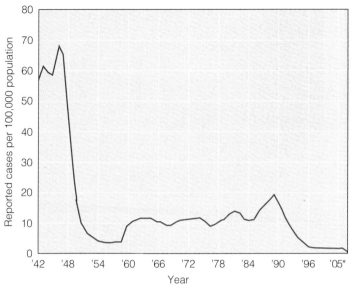

*First 26 weeks of 2005
(a) Incidence of syphilis in the United States from 1942 through the first 26 weeks of 2005

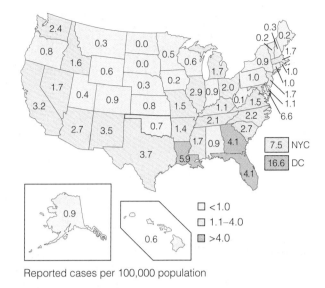

Reported cases per 100,000 population

☐ <1.0
☐ 1.1–4.0
▨ >4.0

(b) Geographical distribution of cases in 2004

FIGURE 26.9   The U.S. incidence and distribution of primary and secondary syphilis.
SOURCES: CDC, *MMWR* 52(54) (4/22/05); *MMWR* 53(52) (1/7/05); *MMWR* 54(26) (7/8/05).

Q **How is syphilis diagnosed?**

is that syphilis existed in Europe and Asia before the fifteenth century but became widespread only as urban living became more common. One English description of the "Morbus Gallicus" (French disease) seems clearly to describe syphilis as early as 1547 and ascribes its transmission in these terms: ". . . It is taken when one pocky person doth synne in lechery one with another."

The number of new syphilis cases in the United States has remained fairly stable (Figure 26.9) compared with gonorrhea (see Figure 26.5). The relative stability in the incidence of syphilis is remarkable because the epidemiology of the two diseases is quite similar, and concurrent infections are not uncommon. A factor in this is that syphilis results in a significant, if imperfect, immunity—compared to no conferred immunity from gonorrhea.

Many states discontinued the requirements for premarital syphilis tests because so few cases were detected. At present, the population most at risk is economically disadvantaged inner city residents, especially drug-using prostitutes of both sexes. It is relatively rare in affluent societies.

The causative agent of syphilis is a gram-negative spirochete, *Treponema pallidum* (Figure 26.10). Thin and tightly coiled, *T. pallidum* stains poorly with the usual bacterial stains. (The bacterial name is derived from the Greek words for twisted thread and pale.) *T. pallidum* lacks the enzymes necessary to build many complex molecules. There-

fore, it relies on the host for many of the compounds necessary for life. The virulent strains have been successfully cultured only in cell cultures, which is not very useful for routine clinical diagnosis. Separate strains of *T. pallidum* are responsible for certain tropically endemic skin diseases such as **yaws.** These diseases are not sexually transmitted.

Syphilis is transmitted by sexual contact of all kinds via syphilitic infections of the genitals or other body parts.

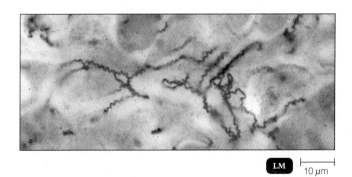

LM |—————| 10 μm

FIGURE 26.10   *Treponema pallidum*, the cause of syphilis. The microbes are made more visible in this brightfield micrograph by the use of a special silver stain.

Q **A diagnostic method for syphilis is the darkfield microscope. Why not use a brightfield microscope?**

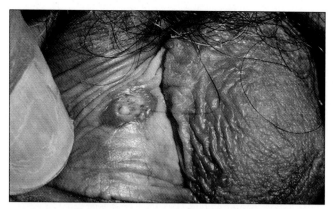

(a) Chancre of primary stage on a male in genital area.

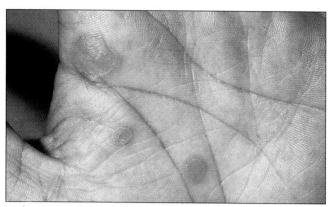

(b) Lesions of secondary syphilis rash on a palm; any surface area of the body may be afflicted with such lesions.

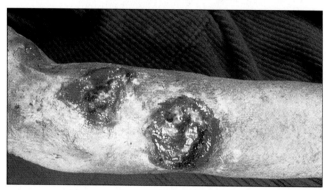

(c) Gummas of tertiary stage on the back of an arm; gummas such as these are rarely seen today in the era of antibiotics.

FIGURE 26.11 Characteristic lesions associated with various stages of syphilis.

 How are the primary, secondary, and tertiary stages of syphilis distinguished?

The incubation period averages 3 weeks but can range from 2 weeks to several months. The disease progresses through several recognized stages.

PRIMARY STAGE SYPHILIS

In the *primary stage* of the disease, the initial sign is a small, hard-based **chancre,** or sore, which appears at the site of infection 10 to 90 days following exposure—on average, about 3 weeks (Figure 26.11a). The chancre is painless, and an exudate of serum forms in the center. This fluid is highly infectious, and examination with a darkfield microscope shows many spirochetes. In a couple of weeks, this lesion disappears. None of these symptoms causes any distress. In fact, many women are entirely unaware of the chancre, which is often on the cervix. In males, the chancre sometimes forms in the urethra and is not visible. During this stage, bacteria enter the bloodstream and lymphatic system, which distribute them widely in the body.

SECONDARY STAGE SYPHILIS

Several weeks after the primary stage (the exact length of time varies), the disease enters the *secondary stage*, characterized mainly by skin rashes of varying appearance (Figure 26.11b). The damage done to tissues at this stage and the later tertiary stage is caused by an inflammatory response to circulating immune complexes that lodge at various body sites. Other symptoms often observed are the loss of patches of hair, malaise, and mild fever. The rash is widely distributed on the skin and is also found in the mucous membranes of the mouth, throat, and cervix.

At this stage, the lesions of the rash contain many spirochetes and are very infectious. Dentists and other health care workers coming into contact with fluid from these lesions can become infected by the spirochete entering through minute breaks in the skin. Such nonsexual transmission is possible, but the microbes do not survive long on environmental surfaces and are very unlikely to be transmitted via such objects as toilet seats. Secondary syphilis is a subtle disease; at least half of the patients diagnosed with this stage can recall no lesions at all.

LATENT PERIOD

The symptoms of secondary syphilis usually subside after a few weeks, and the disease enters a *latent period*. During this period, there are no symptoms. After 2 to 4 years of latency, the disease is not normally infectious, except for transmission from mother to fetus. The majority of cases do not progress beyond the latent stage, even without treatment.

TERTIARY STAGE SYPHILIS

Because the symptoms of primary and secondary syphilis are not disabling, people may enter the latent period without having received medical attention. In less than half of untreated cases, the disease reappears in a *tertiary stage.* This stage occurs only after an interval of many years from the onset of the latent phase.

T. pallidum has an outer layer of lipids that stimulates little effective immune response, especially from cell-destroying complement reactions. It has been described as a "Teflon pathogen." Nonetheless, most of the symptoms of tertiary syphilis are probably due to the body's immune reactions (of a cell-mediated nature) to surviving spirochetes.

Tertiary, or late-stage, syphilis can be classified generally by affected tissues or type of lesion. *Gummatous syphilis* is characterized by **gummas,** which are rubbery masses of tissue (Figure 26.11c) that appear in various organs (most commonly the skin, mucous membranes, and bones) after about 15 years. There they cause local destruction of these tissues but usually not incapacitation or death.

Cardiovascular syphilis results most seriously in a weakening of the aorta. In preantibiotic days, it was one of the more common symptoms of syphilis; it is now rare. *Neurosyphilis* occurs in up to 10% of patients if the disease is untreated. As parts of the central nervous system are affected, the result can be widely varying signs and symptoms. The patient can suffer from personality changes and other signs of dementia (*paresis*), seizures, loss of coordination of voluntary movement (*tabes dorsalis*), partial paralysis, loss of ability to use or comprehend speech, loss of sight or hearing, or loss of bowel and bladder control. Few, if any, pathogens are found in the lesions of the tertiary stage, and they are not considered very infectious. Today, it is rare for cases of syphilis to be allowed to progress to this stage.

CONGENITAL SYPHILIS

One of the most distressing and dangerous forms of syphilis, called **congenital syphilis,** is transmitted across the placenta to the unborn fetus. Damage to mental development and other neurological symptoms are among the more serious consequences. This type of infection is most common when pregnancy occurs during the latent period of the disease. A pregnancy during the primary or secondary stage is likely to produce a stillbirth.

DIAGNOSIS

Diagnosis of syphilis is complex because each stage of the disease has unique requirements. Tests fall into three general groups: visual microscopic inspection, nontreponemal serological tests, and treponemal serological tests. For preliminary screening, laboratories use either nontreponemal serological tests or microscopic examination of exudates from lesions when these are present. If a screening test is positive, the results are confirmed by treponemal serological tests.

Microscopic tests are important for screening for primary syphilis because serological tests for this stage are not reli-

able; antibodies take 1 to 4 weeks to form. The spirochetes can be detected in exudates of lesions by microscopic examination with a darkfield microscope (see Figure 3.4b, page 60). A darkfield microscope is necessary because the bacteria stain poorly and are only about 0.2 μm in diameter, near the lower limit of resolution for a brightfield microscope. Similarly, a direct fluorescent-antibody test (DFA-TP) using monoclonal antibodies (see Figure 18.11a, page 542) will both show and identify the spirochete. Figure 26.10 shows *T. pallidum* under brightfield illumination made possible by a special silver-impregnated stain.

At the secondary stage, when the spirochete has invaded almost all body organs, serological tests are reactive. *Nontreponemal serological tests* are so called because they are nonspecific; they do not detect antibodies produced against the spirochete itself but detect *reagin-type antibodies.* Generally, they are used for screening. Reagin-type antibodies are apparently a response to lipid materials the body forms as an indirect reaction to infection by the spirochete. The antigen used in such tests is thus not the syphilis spirochete but an extract of beef heart (cardiolipin) that seems to contain lipids similar to those that stimulated the reagin-type antibody production. These tests will detect only about 70 to 80% of primary syphilis cases, but they will detect 99% of secondary syphilis cases. An example of nontreponemal tests is the slide agglutination **VDRL test** (for Venereal Disease Research Laboratory). Also used are modifications of the **rapid plasma reagin (RPR) test,** which is similar. The newest nontreponemal test is an ELISA test that uses the VDRL antigen.

Treponemal-type serological tests that react directly with the spirochete are used to confirm the screening diagnosis (to check for false-positive nontreponemal tests) and to diagnose late-stage syphilis. Almost 30% of patients fail at this stage to respond to nontreponemal tests. Current treponemal tests detect antibodies in the patient with the use of antigens from *T. pallidum.* An example is the fluorescent treponemal antibody absorption test; **FTA-ABS** (an indirect fluorescent-antibody test (see Figure 18.11b, page 542). Treponemal tests are not used for screening because about 1% of the results will be false-positives, but a positive test with both treponemal and nontreponemal types is highly specific.

TREATMENT

Benzathine penicillin, a long-acting formulation that remains effective in the body for about 2 weeks, is the usual antibiotic treatment of syphilis. The serum concentrations achieved by this formulation are low, but the spirochete has remained very sensitive to this antibiotic.

For penicillin-sensitive people, several other antibiotics, such as azithromycin, doxycycline, and tetracycline, have also proven effective. Antibiotic therapy to treat gonorrhea and other infections will not likely eliminate

DISEASES IN FOCUS

CHARACTERISTICS OF THE MOST COMMON TYPES OF VAGINITIS AND VAGINOSIS

Vaginitis, or inflammation of the vagina, often accompanies vaginal infections. Vaginitis may be caused by microbial infections. Some infections are associated with sexual activity while others, such as vaginal candidiasis, are not. The cause of vaginitis cannot be determined on the basis of symptoms or physical examination alone. Usually, diagnosis involves examining a specimen of vaginal fluid under a microscope.

	Pathogen	Symptoms				Diagnosis	Treatment
		Odor **Color of** **discharge** **Consistency** **of discharge**	**Amount of** **discharge**	**Appearance** **of vaginal** **mucosa**	**pH** **(normal** **pH is** **3.8–4.2)**		
Candidiasis	Fungus, *Candida albicans*	Yeasty or none White Curdy	Varies	Dry, red	Below 4	Microscopic exam	Clotrimazole; fluconazole
Bacterial vaginosis	Bacterium, *Gardnerella vaginalis*	Fishy Gray-white Thin, frothy	Copious	Pink	Above 4.5	Presence of clue cells	Metronidazole
Trichomoniasis	Protozoan, *Trichomonas vaginalis*	Foul Greenish-yellow Frothy	Copious	Tender, red	5–6	Microscopic exam; DNA probes; monoclonal antibody	Metronidazole

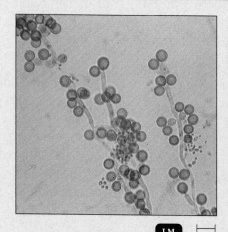

LM | 20 mm

Candida albicans

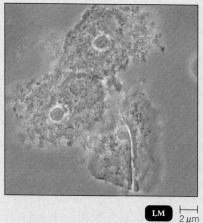

LM | 2 μm

Clue cells characteristic of vaginosis caused by *Gardnerella vaginalis*

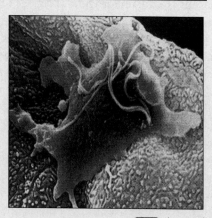

SEM | 1 μm

Trichomonas vaginalis

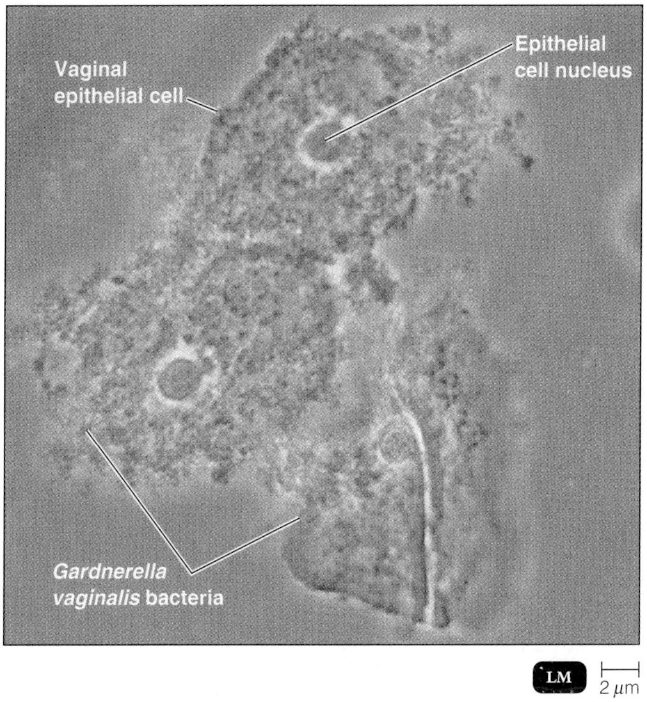

FIGURE 26.12 Clue cells. The outlines of the vaginal epithelial cells are blurred by a heavy coating of bacteria, mostly *Gardnerella vaginalis.*

Q **What symptoms would cause you to look for clue cells?**

syphilis as well because such therapy is usually administered for too short a period to affect the slow-growing spirochete.

LYMPHOGRANULOMA VENEREUM (LGV)

Several STDs that are uncommon in the United States occur frequently in the tropical areas of the world. For example, *Chlamydia trachomatis,* the cause of the eye infection trachoma and a major cause of NGU, is also responsible for **lymphogranuloma venereum (LGV),** a disease found in tropical and near-tropical regions. It is apparently caused by serovars of *C. trachomatis* that are invasive and tend to infect lymphoid tissue. In the United States, there are usually 200 to 400 cases each year, mostly in the southeast.

The microorganisms invade the lymphatic system, and the regional lymph nodes become enlarged and tender. Suppuration (a discharge of pus) may also occur. Inflammation of the lymph nodes results in scarring, which occasionally obstructs the lymph vessels. This blockage sometimes leads to massive enlargement of the external genitals in males. In females, rectal narrowing results from involvement of the lymph nodes in the rectal region. These conditions may eventually require surgery.

For diagnosis, pus can be aspirated from infected lymph nodes. When infected cells are properly stained with an iodine preparation, the clumped, intracellular chlamydias

can be seen as inclusions. The isolated organisms can also be grown in cell culture or in embryonated eggs. The drug of choice for treatment is doxycycline.

CHANCROID (SOFT CHANCRE)

The STD known as **chancroid (soft chancre)** occurs most frequently in tropical areas, where it is seen more often than syphilis. The number of reported cases in the United States has been declining from a peak of 5000 cases in 1988. Almost all occur in New York, Texas, California, Florida, and Georgia. Like syphilis, its incidence is strongly associated with drug use. Because chancroid is so seldom seen by some physicians and is difficult to diagnose, it is probably underreported. It is very common in Africa, Asia, and Latin America.

In chancroid, a swollen, painful ulcer that forms on the genitals involves an infection of the adjacent lymph nodes. Infected lymph nodes in the groin area sometimes even break through and discharge pus to the surface. Such lesions are an important factor in the sexual transmission of HIV, especially in Africa. Lesions might also occur on such diverse areas as the tongue and lips. The causative agent is *Haemophilus ducreyi* (dü-krā′ē), a small gram-negative rod that can be isolated from exudates of lesions. Symptoms and the culture of these bacteria are the primary means of diagnosis. The recommended antibiotics include erythromycin and ceftriaxone.

BACTERIAL VAGINOSIS

Inflammation of the vagina due to infection, or **vaginitis,** is most commonly caused by one of three organisms: the fungus *Candida albicans* (kan′did-ä al′bi-kans), the protozoan *Trichomonas vaginalis* (trik-ō-mōn′as va-jin-al′is), or the bacterium *Gardnerella vaginalis,* a small, pleomorphic gram-variable rod (see the box on the facing page). Most of these cases are attributed to the presence of G. *vaginalis* and are termed **bacterial vaginosis.** (Because there is no sign of inflammation, the term *vaginosis* is preferred to *vaginitis*).

The condition is something of an ecological mystery. It is believed that bacterial vaginosis is precipitated by some event that decreases the number of *Lactobacillus* vaginal bacteria that normally produce hydrogen peroxide and acid, which inhibit competition. This competitive change allows bacteria such as G. *vaginalis* to proliferate, producing amines that contribute to a further rise in pH. These microbes are a common inhabitant of the vaginas of asymptomatic women. In the United States, bacterial vaginosis accounts for 40 to 50% of cases of vaginitis. There is no corresponding disease condition in males, but G. *vaginalis* is often present in their urethras. Therefore, the condition may be sexually transmitted, but it also occurs occasionally in women who have never been sexually active.

Bacterial vaginosis is characterized by a vaginal pH above 4.5 and a copious, frothy vaginal discharge. When tested

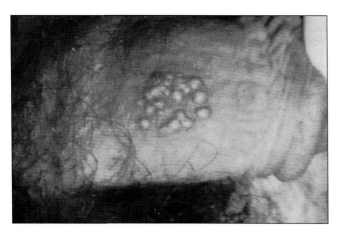

FIGURE 26.13   Vesicles of genital herpes on a penis.

Q *What microbe causes genital herpes?*

with a potassium hydroxide solution, these vaginal secretions emit a fishy odor from presence of the amines produced by *G. vaginalis.* Diagnosis is based on the vaginal pH, fishy odor (the *whiff test*), and microscopic observation of *clue cells* in the discharge. These clue cells are sloughed-off vaginal epithelial cells covered with bacteria, mostly *G. vaginalis* (Figure 26.12). The disease has been considered more of a nuisance than a serious infection, but it is now seen as a factor in many premature births and low birth-weight infants.

Treatment is primarily by metronidazole, a drug that eradicates the anaerobes essential to continuation of the disease but allows the normal lactobacilli to repopulate the vagina. Treatments designed to restore the normal population of lactobacilli, such as application of acetic acid gels and even yogurt (!), have not been shown conclusively to be effective.

VIRAL DISEASES OF THE REPRODUCTIVE SYSTEMS

LEARNING OBJECTIVE
- Discuss the epidemiology of genital herpes and genital warts.

Viral diseases of the reproductive system are difficult to treat, and so they represent an increasing health problem.

GENITAL HERPES

A much publicized STD is **genital herpes,** usually caused by *herpes simplex virus type 2 (HSV-2).* (The herpes simplex virus occurs as either type 1 or type 2.) Herpes simplex virus type 1 (HSV-1) is primarily responsible for cold sores or fever blisters (see page 626), but it can also cause genital herpes. The official names are human herpesvirus 1 and 2.

In the United States one in four persons over the age of 30 is infected with HSV-2—most are unaware they are in-

fected. There has been a marked increase in genital HSV-1 infections, which is usually acquired by oral–genital contact, and this now constitutes about 20% of cases of genital herpes in this country.

Genital herpes lesions appear after an incubation period of up to 1 week and cause a burning sensation. After this, vesicles appear (Figure 26.13). In both males and females, urination can be painful, and walking is quite uncomfortable; the patient is even irritated by clothing. Usually, the vesicles heal in a couple of weeks.

The vesicles contain infectious fluid, but many times the disease is transmitted when no lesions or symptoms are apparent. Semen may contain the virus. Condoms may not provide protection because in women the vesicles are usually on the external genitals (seldom on the cervix or within the vagina) and in men the vesicles may be on the base of the penis.

One of the most distressing characteristics of genital herpes is the possibility of recurrences. There is an element of truth in the medical adage that, unlike love, herpes is forever. As in other herpes infections, such as cold sores or chickenpox-shingles, the virus enters a lifelong latent state in nerve cells. Some people have several recurrences a year; for others, recurrence is rare. Men are more likely to experience recurrences than women. Reactivation appears to be triggered by several factors, including menstruation, emotional stress or illness (especially if accompanied by fever, a factor that is also involved in the appearance of cold sores), and perhaps just scratching the affected area. About 90% of patients with HSV-2 and about 50% of those with HSV-1 will have recurrences. Recurrence rates decrease over time, regardless of treatment.

Neonatal herpes is a serious consideration for women of childbearing age. Currently about 1500 cases a year are reported in this country. The virus can cross the placental barrier and affect the fetus. The result can be spontaneous abortion or serious fetal damage, such as mental retardation and defective vision and hearing. Herpes infection of the newborn is most likely to have serious consequences when the mother acquires the initial herpes infection during the pregnancy. Therefore, any pregnant woman without a history of genital herpes should avoid sexual contact with anyone who might be carrying the virus. Damage to the fetus or newborn is much less likely, by a factor of about ten, from exposure to recurrent or asymptomatic herpes. The reason is that maternal antibodies are apparently somewhat protective.

For practical purposes, the fetus is considered infected if the virus can be grown from amniotic fluid. This procedure is widely available and requires about 5 days. If the fetus is free of the virus, it still needs to be protected from infection during passage through the birth canal. Clinical signs of infection cannot always be seen, and tests to detect asymptomatic shedding of the viruses are not very reliable. Therefore, screening for them is of very limited use. If there are obvious viral lesions at delivery time, a cesarean section is probably wise. The operation should be done before the fetal membrane ruptures and the viruses spread to the uterus.

There is no cure for genital herpes, although research on its prevention and treatment is intensive. Discussions of chemotherapy use terms such as *suppression* or *management* rather than *cure*. Currently, the antiviral drugs acyclovir, famciclovir, and valacyclovir are recommended for treatment. They are fairly effective in alleviating the symptoms of a primary outbreak; there is some relief of pain and slightly faster healing. Taken over several months they lower the chances of recurrence during that time. Recent studies have indicated that valacyclovir taken daily can significantly cut sexual transmission of genital herpes.

A vaccine is undergoing clinical trials and has been "reasonably effective" in preventing both HSV-1 and HSV-2 disease. However, it was not effective in men, nor was it protective in women already carrying HSV-1 antibodies.

GENITAL WARTS

Warts are an infectious disease; since 1907 it has been known that they are caused by viruses known as papillomaviruses. It is probably less well known that warts can be transmitted sexually and that this is an increasing problem. Nearly a million new cases of **genital warts** are estimated to occur in the United States each year.

There are more than 60 serotypes of human papillomaviruses (HPV) and certain serotypes tend to be linked with certain forms of genital warts (technically, *condylomata acuminata*). Morphologically, some warts are extremely large and "warty" in appearance, with multiple fingerlike projections resembling cauliflower; others are relatively smooth or flat (Figure 26.14). The incubation period is usually a matter of a few weeks or months.

The greatest danger from genital warts is their connection to cervical cancer, which kills at least 4000 women annually in the United States. A few of the many serotypes, especially human papillomavirus (HPV) 16, are associated with a progression to cancer. In women this is usually cervical cancer, and in men it is usually cancer of the penis. Genital warts in women are much more likely to be precancerous than those in men. It is hoped that serological typing of warts will become a routine aid in determining which are the most dangerous, but at present this procedure is expensive and limited to relatively few laboratories. However, a test is currently available, if requested of the physician, that combines the common Pap smear test (the Papanicolaou test is intended to detect cervical cancer) with a test that detects the nucleic acids of some of the more likely cancer-causing strains of HPV.

As a general rule, warts can be treated but not cured (see the discussion on page 623). The available methods used for warts (such as surgery or cryotherapy) are not as

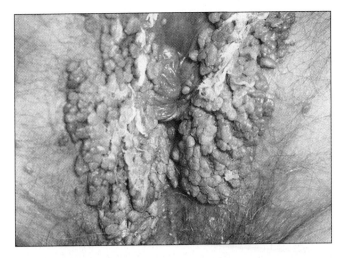

FIGURE 26.14 Genital warts on a vulva.

Q What is the relationship between genital warts and cervical cancer?

effective against genital warts. Two patient-applied gels, podofilox and imiquimod, are often useful treatments. Imiquimod (Aldara) stimulates the body to produce interferon (page 494), which appears to account for its antiviral activity. In 2005, an experimental vaccine aimed at four types of HPV was found to be very effective in preventing precancerous lesions. If approved in the United States, it will probably become a standard vaccine for young women. A similar vaccine, but targeting more viruses, is being submitted for approval in Europe and elsewhere.

AIDS

AIDS, or HIV infection, is a viral disease that is frequently transmitted by sexual contact. However, its pathogenicity is based upon damage to the immune system, so it was discussed on pages 566–576. It is important to remember that the lesions resulting from many of the diseases of bacterial and viral origin facilitate the transmission of HIV.

FUNGAL DISEASE OF THE REPRODUCTIVE SYSTEMS

LEARNING OBJECTIVE
• Discuss the epidemiology of candidiasis.

The fungal disease described here is the well-known *yeast infection* for which nonprescription treatments are advertised.

CANDIDIASIS

Vaginal infections by yeastlike fungi of the genus *Candida* are responsible for millions of physician office visits every

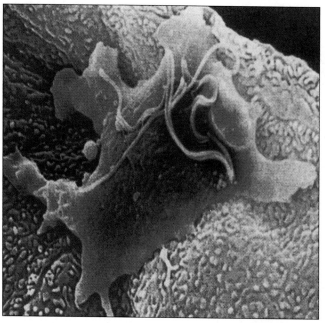

SEM $\vdash\!\!\!\!\dashv$ 1 μm

FIGURE 26.15 *Trichomonas vaginalis* **adhering to the surface of an epithelial cell in a cell culture preparation.** The flagella are clearly visible.
SOURCE: D. Petrin et al., "Clinical and Microbiological Aspects of *Trichomonas vaginalis*." *ASM Clinical Microbiology Reviews* 11 (1998):300–317.

Q Are there any harmful effects from infection by this protozoan?

year. By the time they reach the age of 25, an estimated half of college women will have had at least one physician-diagnosed episode. Nonprescription antifungal therapies to treat these infections are among the best-selling over-the-counter products in the United States. *Candida albicans* is the most common species, causing 85 to 90% of cases. Infections by other species, such as *C. glabrata*, are more likely to be resistant to antifungals and to be chronic or recurrent.

C. albicans often grows on mucous membranes of the mouth, intestinal tract, and genitourinary tract (Diseases in Focus box on page 799; see Figure 21.17, page 631). Infections are usually a result of opportunistic overgrowth when the competing microbiota are suppressed by antibiotics or other factors. As discussed in Chapter 21, *C. albicans* is the cause of **oral candidiasis,** or thrush. It is also responsible for occasional cases of NGU in males and for **vulvovaginal candidiasis,** which is the most common cause of vaginitis. About 75% of all women experience at least one episode.

The lesions of vulvovaginal candidiasis resemble those of thrush but produce more irritation; severe itching; a thick, yellow, cheesy discharge; and yeasty or no odor. *C. albicans,* the *Candida* species responsible for most cases, is an opportunistic pathogen. Predisposing conditions include the use

of oral contraceptives and pregnancy, which cause an increase of glycogen in the vagina (see the discussion of the normal vaginal microbiota earlier in this chapter). It is probable that hormones are a factor; candidiasis is much less common in girls before puberty or in women after menopause. Yeast infections are a frequent symptom in women suffering from uncontrolled diabetes; also, the use of broad-spectrum antibiotics suppresses the normal, competing bacterial microbiota, which leads to opportunistic fungal infections. Thus, diabetes and antibiotic therapy are predisposing factors to *C. albicans* vaginitis.

A yeast infection is diagnosed by microscopic identification of the fungus in scrapings of lesions and by isolation of the fungus in culture. Treatment usually consists of topical application of nonprescription antifungal drugs such as clotrimazole and miconazole. An alternative treatment is a single dose of oral fluconazole or other azole-type antifungal.

PROTOZOAN DISEASE OF THE REPRODUCTIVE SYSTEMS

LEARNING OBJECTIVE

- Discuss the epidemiology of trichomoniasis.

The only STD caused by a protozoan almost entirely affects only women. Although common, it is not widely known.

TRICHOMONIASIS

The anaerobic protozoan *Trichomonas vaginalis* is frequently a normal inhabitant of the vagina in females and of the urethra in many males (Figure 26.15). It is usually sexually transmitted. If the normal acidity of the vagina is disturbed, the protozoan may overgrow the normal microbial population of the genital mucosa and cause **trichomoniasis.** (Males rarely have any symptoms as a result of the presence of the protozoan.) It is often accompanied by a coinfection with gonorrhea. Its prevalence in certain STD clinics is 25% or higher. In response to the protozoan infection, the body accumulates leukocytes at the infection site. The resulting discharge is profuse, greenish yellow, and characterized by a foul odor. This discharge is accompanied by irritation and itching. Up to half the cases, however, are asymptomatic.

The incidence of trichomoniasis is higher than that of gonorrhea or chlamydia, but it is considered relatively benign and is not a reportable disease. It is known, however, to cause preterm delivery and problems associated with this, such as low birth weight.

Diagnosis is usually made by microscopic examination and identification of the organisms in the discharge. They can also be isolated and grown on laboratory media. The pathogen can be found in semen or urine of male carriers. New rapid tests making use of DNA probes and monoclonal antibodies are now available. Treatment is by oral metronidazole, administered to both sex partners, which readily clears the infection.

* * *

The major microbial diseases of the urinary and reproductive systems are summarized in Table 26.1.

TABLE 26.1 Microbial Diseases of the Urinary and Reproductive Systems

Disease	Pathogen	Symptoms	Diagnosis	Treatment
Bacterial Diseases of the Urinary System				
Cystitis (urinary bladder infection)	Escherichia coli, Staphylococcus saprophyticus	Difficulty or pain in urination	Evidence of bacteria in urine	Trimethoprim-sulfamethoxazole
Pyelonephritis (kidney infection)	Primarily E. coli	Fever; back or flank pain	Evidence of bacteria in blood and urine	Cephalosporin
Leptospirosis (kidney infection)	Leptospira interrogans	Headaches, muscular aches, fever; kidney failure a possible complication	Serological test, for most cases	Doxycyline
Bacterial Diseases of the Reproductive System				
Gonorrhea	Neisseria gonorrhoeae	Males: painful urination and discharge of pus Females: few symptoms but possible complications such as PID	PCR, ELISA	Fluoroquinolones
Nongonococcal urethritis (NGU)	Chlamydia trachomatis or other bacteria, including Mycoplasma hominis and Ureaplasma urealyticum	Painful urination and watery discharge. In females, possible complications such as PID.	PCR	Doxycyline; azithromycin
Pelvic inflammatory disease (PID)	N. gonorrhoeae, Chlamydia trachomatis	Chronic abdominal pain; possible infertility	Signs and symptoms in association with gonorrheal or chlamydial infections	Doxycyline and cefoxitin
Syphilis	Treponema pallidum	Initial sore at site of infection, later skin rashes and mild fever; final stages may be severe lesions, damage to cardiovascular and nervous systems; today, few progress to tertiary stage	FTA-ABS	Benzathine penicillin
Lymphogranuloma venereum (LGV)	Chlamydia trachomatis	Swelling in lymph nodes in groin	Observing intracellular bacteria; culture	Doxycyline
Chancroid (soft chancre)	Haemophilus ducreyi	Painful ulcers of genitals; swollen lymph nodes in groin	Symptoms and bacterial culture	Erythromycin; cetriaxone
Bacterial vaginosis	Gardnerella vaginalis	Fishy odor, frothy vaginal discharge	Presence of clue cells	Metronidazole

TABLE 26.1	**Microbial Diseases of the Urinary and Reproductive Systems** *(continued)*			
Disease	Pathogen	Symptoms	Diagnosis	Treatment
Viral Diseases of the Reproductive System				
Genital herpes	Herpes simples virus type 2; HSV type 1	Painful vesicles in genital area	Enzyme immunoassays; immunofluorescence; PCR	Acyclovir; preventive vaccine in clinical trials
Genital warts	Human papillomaviruses	Warts in genital area	Microscopic appearance of infected cells; PCR to identify viral strains	Podofilox; imiquimod. Preventive vaccine, in clinical trials
AIDS See Chapter 19, pp. 566–576				
Fungal Diseases of the Reproductive System				
Candidiasis	*Candida albicans*	Severe vaginal itching, yeasty odor, yellow discharge	Microscopic exam	Clotrimazole; fluconazole
Protozoan Diseases of the Reproductive System				
Trichomoniasis	*Trichomonas vaginalis*	Vaginal itching, greenish yellow discharge	Microscopic exam; DNA probes; monoclonal antibodies	Metronidazole

STUDY OUTLINE

INTRODUCTION (p. 785)

1. The urinary system regulates the chemical composition and volume of the blood and excretes nitrogenous waste and water.
2. The reproductive system produces gametes for reproduction and, in the female, supports the growing embryo.
3. Microbial diseases of these systems can result from infection from an outside source or from opportunistic infection by members of the normal microbiota.

STRUCTURE AND FUNCTION OF THE URINARY SYSTEM (p. 786)

1. Urine is transported from the kidneys through ureters to the urinary bladder and is eliminated through the urethra.
2. Valves prevent urine from flowing back to the urinary bladder and kidneys.
3. The flushing action of urine and the acidity of normal urine have some antimicrobial value.

STRUCTURE AND FUNCTION OF THE REPRODUCTIVE SYSTEMS (p. 786)

1. The female reproductive system consists of two ovaries, two uterine tubes, the uterus, the cervix, the vagina, and the external genitals.
2. The male reproductive system consists of two testes, ducts, accessory glands, and the penis; seminal fluid leaves the male body through the urethra.

NORMAL MICROBIOTA OF THE URINARY AND REPRODUCTIVE SYSTEMS (pp. 786–788)

1. The urinary bladder and upper urinary tract are sterile under normal conditions.
2. Lactobacilli dominate the vaginal microbiota during the reproductive years.
3. The male urethra is normally sterile.

DISEASES OF THE URINARY SYSTEM (pp. 788–789)

BACTERIAL DISEASES OF THE URINARY SYSTEM (pp. 788–789)

1. Urethritis, cystitis, and ureteritis are terms describing inflammations of tissues of the lower urinary tract.

2. Pyelonephritis can result from lower urinary tract infections or from systemic bacterial infections.

3. Opportunistic gram-negative bacteria from the intestines often cause urinary tract infections.

4. Nosocomial infections following catheterization occur in the urinary system. *E. coli* causes more than half of these infections.

5. More than 1000 bacteria of one species per milliliter of urine, or 100 coliforms per milliliter of urine, indicates an infection.

6. Treatment of urinary tract infections depends on the isolation and antibiotic sensitivity testing of the causative agents.

CYSTITIS (pp. 788–789)

7. Inflammation of the urinary bladder, or cystitis, is common in females.

8. Microorganisms at the opening of the urethra and along the length of the urethra, careless personal hygiene, and sexual intercourse contribute to the high incidence of cystitis in females.

9. The most common etiologies are *E. coli* and *Staphylococcus saprophyticus*.

PYELONEPHRITIS (p. 789)

10. Inflammation of the kidneys, or pyelonephritis, is usually a complication of lower urinary tract infections.

11. About 75% of pyelonephritis cases are caused by *E. coli*.

LEPTOSPIROSIS (p. 789)

12. The spirochete *Leptospira interrogans* is the cause of leptospirosis.

13. The disease is transmitted to humans by urine-contaminated water.

14. Leptospirosis is characterized by chills, fever, headache, and muscle aches.

15. Diagnosis is based on isolation of the bacteria and serological identification.

DISEASES OF THE REPRODUCTIVE SYSTEMS (pp. 790–804)

BACTERIAL DISEASES OF THE REPRODUCTIVE SYSTEMS (pp. 790–800)

1. Most diseases of the reproductive system are sexually transmitted diseases (STDs).

2. Most STDs can be prevented by the use of condoms and are treated with antibiotics.

GONORRHEA (pp. 790–792)

3. *Neisseria gonorrhoeae* causes gonorrhea.

4. Gonorrhea is a common reportable communicable disease in the United States.

5. *N. gonorrhoeae* attaches to mucosal cells of the oral-pharyngeal area, genitals, eyes, and rectum by means of fimbriae.

6. Symptoms in males are painful urination and pus discharge. Blockage of the urethra and sterility are complications of untreated cases.

7. Females might be asymptomatic unless the infection spreads to the uterus and uterine tubes (see pelvic inflammatory disease).

8. Gonorrheal endocarditis, gonorrheal meningitis, and gonorrheal arthritis are complications that can affect both sexes if gonorrheal infections are untreated.

9. Ophthalmia neonatorum is an eye infection acquired by infants during passage through the birth canal of an infected mother.

10. Gonorrhea is diagnosed by ELISA or nucleic acid amplification.

NONGONOCOCCAL URETHRITIS (NGU) (pp. 792–794)

11. Nongonococcal urethritis (NGU), or nonspecific urethritis (NSU), is any inflammation of the urethra not caused by *N. gonorrhoeae*.

12. Most cases of NGU are caused by *Chlamydia trachomatis*.

13. *C. trachomatis* infection is the most common STD.

14. Symptoms of NGU are often mild or lacking, although uterine tube inflammation and sterility may occur.

15. *C. trachomatis* can be transmitted to infants' eyes at birth.

16. Diagnosis is based on the detection of chlamydial DNA in urine.

17. *Ureaplasma urealyticum* and *Mycoplasma hominis* also cause NGU.

PELVIC INFLAMMATORY DISEASE (PID) (p. 794)

18. Extensive bacterial infection of the female pelvic organs, especially of the reproductive system, is called pelvic inflammatory disease (PID).

19. PID is caused by *N. gonorrhoeae*, *C. trachomatis*, and other bacteria that gain access to the uterine tubes. Infection of the uterine tubes is called salpingitis.

20. PID can result in blockage of the uterine tubes and sterility.

SYPHILIS (pp. 794–797)

21. Syphilis is caused by *Treponema pallidum*, a spirochete that has not been cultured in vitro. Laboratory cultures are grown in cell cultures.

22. *T. pallidum* is transmitted by direct contact and can invade intact mucous membranes or penetrate through breaks in the skin.

23. The primary lesion is a small, hard-based chancre at the site of infection. The bacteria then invade the blood and lymphatic system, and the chancre spontaneously heals.

24. The appearance of a widely disseminated rash on the skin and mucous membranes marks the secondary stage. Spirochetes are present in the lesions of the rash.

25. The patient enters a latent period after the secondary lesions spontaneously heal.

26. At least 10 years after the secondary lesion, tertiary lesions called gummas can appear on many organs.

27. Congenital syphilis, resulting from *T. pallidum* crossing the placenta during the latent period, can cause neurological damage in the newborn.

28. *T. pallidum* is identifiable through darkfield microscopy of fluid from primary and secondary lesions.

29. Many serological tests, such as VDRL, RPR, and FTA-ABS, can be used to detect the presence of antibodies against *T. pallidum* during any stage of the disease.

LYMPHOGRANULOMA VENEREUM (LGV) (pp. 797–798)

30. *C. trachomatis* causes lymphogranuloma venereum (LGV), which is primarily a disease of tropical and subtropical regions.

31. The initial lesion appears on the genitals and heals without scarring.

32. The bacteria are spread in the lymph system and cause enlargement of the lymph nodes, obstruction of lymph vessels, and swelling of the external genitals.

33. The bacteria are isolated and identified from pus taken from infected lymph nodes.

CHANCROID (SOFT CHANCRE) (p. 798)

34. Chancroid, a swollen, painful ulcer on the mucous membranes of the genitals or mouth, is caused by *Haemophilus ducreyi*.

BACTERIAL VAGINOSIS (pp. 798–800)

35. Bacterial vaginosis is an infection without inflammation caused by *Gardnerella vaginalis*.

36. Diagnosis of *G. vaginalis* is based on increased vaginal pH, fishy odor, and the presence of clue cells.

VIRAL DISEASES OF THE REPRODUCTIVE SYSTEMS (pp. 800–801)

GENITAL HERPES (pp. 800–801)

1. Herpes simplex viruses (HSV-1 and HSV-2) cause genital herpes.

2. Symptoms of the infection are painful urination, genital irritation, and fluid-filled vesicles.

3. Neonatal herpes is contracted during fetal development or birth. It can result in neurological damage or infant fatalities.

4. The virus might enter a latent stage in nerve cells. Vesicles reappear following trauma and hormonal changes.

GENITAL WARTS (p. 801)

5. Human papillomaviruses cause warts.

6. Some human papillomaviruses that cause genital warts have been associated with cancer of the cervix or penis.

AIDS (p. 801)

7. AIDS is a sexually transmitted disease of the immune system (see Chapter 19, pages 566–576).

FUNGAL DISEASE OF THE REPRODUCTIVE SYSTEMS (pp. 801–802)

CANDIDIASIS (pp. 801–802)

1. *Candida albicans* causes NGU in males and vulvovaginal candidiasis, or yeast infection, in females.

2. Vulvovaginal candidiasis is characterized by lesions that produce itching and irritation.

3. Predisposing factors are pregnancy, diabetes, tumors, and broad-spectrum antibacterial chemotherapy.

4. Diagnosis is based on observation of the fungus and its isolation from lesions.

PROTOZOAN DISEASE OF THE REPRODUCTIVE SYSTEM (pp. 802–803)

TRICHOMONIASIS (pp. 802–803)

1. *Trichomonas vaginalis* causes trichomoniasis when the pH of the vagina increases.

2. Diagnosis is based on observation of the protozoa in purulent discharges from the site of infection.

STUDY QUESTIONS

Access more review material either online at **The Microbiology Place** (www.microbiologyplace.com) or with **The Microbiology Place CD-ROM** packaged with your new book. There you'll find activities, practice tests, quizzes, flashcards, case studies, and more to help you succeed.

Answers to the Study Questions can be found in Appendix G.

REVIEW

1. List the members of the normal microbiota of the urinary system, and show their habitats in Figure 26.1.

2. List the normal microbiota of the reproductive system, and show their habitats in Figures 26.2 and 26.3.

3. How are urinary tract infections transmitted?

4. Explain why *E. coli* is frequently implicated in cystitis in females. List some predisposing factors for cystitis.

5. Name one organism that causes pyelonephritis. What are the portals of entry for microbes that cause pyelonephritis?

6. Complete the following table:

Disease	Causative Agent	Symptoms	Method of Diagnosis	Treatment
Bacterial vaginosis				
Gonorrhea				
Syphilis				
PID				
NGU				
LGV				
Chancroid				

7. Leptospirosis is a kidney infection of humans and other animals. How is this disease transmitted? What types of activities would increase one's exposure to this disease? What is the etiology?

8. Describe the symptoms of genital herpes. What is the causative agent? When is this infection least likely to be transmitted?

9. Name one fungus and one protozoan that can cause reproductive system infections. What symptoms would lead you to suspect these infections?

10. List the genital infections that cause congenital and neonatal infections. How can transmission to a fetus or newborn be prevented?

MULTIPLE CHOICE

1. Which of the following is usually transmitted by contaminated water?
   a. *Chlamydia*
   b. leptospirosis
   c. syphilis
   d. trichomoniasis
   e. none of the above

Use the following choices to answer questions 2–5:
   a. *Candida*
   b. *Chlamydia*
   c. *Gardnerella*
   d. *Neisseria*
   e. *Trichomonas*

e 2. Microscopic examination of vaginal smear shows flagellated eukaryotes.

a 3. Microscopic examination of vaginal smear shows ovoid eukaryotic cell.

c 4. Microscopic examination of vaginal smear shows epithelial cells covered with bacteria.

d 5. Microscopic examination of vaginal smear shows gram-negative cocci in phagocytes.

Use the following choices to answer questions 6–8:
   a. candidiasis
   b. bacterial vaginosis
   c. genital herpes
   d. lymphogranuloma venereum
   e. trichomoniasis

c 6. Difficult to treat with chemotherapy

c 7. Fluid-filled vesicles

e 8. Frothy, fishy discharge

Use the following choices to answer questions 9 and 10:
   a. *Chlamydia trachomatis*
   b. *Escherichia coli*
   c. *Mycobacterium hominis*
   d. *Staphylococcus saprophyticus*

b 9. The most common cause of cystitis.

a 10. In cases of NGU, diagnosis is made using PCR to detect microbial DNA.

CRITICAL THINKING

1. The tropical skin disease called yaws is transmitted by direct contact. Its causative agent, *Treponema pallidum pertenue*, is indistinguishable from *T. pallidum*. Syphilis epidemics in Europe coincided with the return of Columbus from the New World. How might *T. pallidum pertenue* have evolved into *T. pallidum* in the temperate climate of Europe?

2. Why can frequent douching be a predisposing factor to bacterial vaginosis, vulvovaginal candidiasis, or trichomoniasis?

3. *Neisseria* is cultured on Thayer-Martin media, consisting of chocolate agar and nystatin, incubated in a 5% CO_2 environment. How is this selective for *Neisseria*?

4. The list below is a key to selected microorganisms that cause genitourinary infections. Complete this key by listing genera discussed in this chapter in the blanks that correspond to their respective characteristics.

Gram-negative bacteria
  Spirochete
    Aerobic   _____
    Anaerobic   _____
  Coccus
    Oxidase-positive   _____
  Bacillus, nonmotile
    Requires X factor   _____
    Gram-positive wall   _____
  Obligate intracellular parasite _____
  Lacking cell wall
    Urease-positive   _____
    Urease-negative   _____
Fungus
  Pseudohyphae   _____
Protozoa
  Flagella   _____
No organism observed/cultured
  from patient   _____

CLINICAL APPLICATIONS

1. A previously healthy 19-year-old female was admitted to a hospital after 2 days of nausea, vomiting, headache, and neck stiffness. Cerebrospinal fluid and cervical cultures showed gram-negative diplococci in leukocytes; a blood culture was negative. What disease did she have? How was it probably acquired?

2. A 28-year-old woman was admitted to a Wisconsin hospital with a 1-week history of arthritis of the left knee. Four days later, a 32-year-old man was examined for a 2-week history of urethritis and a swollen, painful left wrist. A 20-year-old woman seen in a Philadelphia hospital had pain in the right knee, left ankle, and left wrist for 3 days. Pathogens cul-

tured from synovial fluid or urethral culture were gram-negative diplococci that required proline to grow. Antibiotic sensitivity tests gave the following results:

Antibiotic	MIC Tested (μg/ml)	Susceptible MIC (μg/ml)
Cefoxitin	0.5	≤2
Penicillin	8	≤0.06
Spectinomycin	64	≤32
Tetracycline	4	≤0.25

What is the pathogen, and how is this disease transmitted? Which of the antibiotics should be used for treatment? What is the evidence that these cases are related?

3. Using the following information, determine what the disease is and how the infant's illness might have been prevented:

May 11: A 23-year-old woman has her first prenatal examination. She is 4½ months pregnant. Her VDRL results are negative.

June 6: The woman returns to her physician complaining of a labial lesion of a few days' duration. A biopsy is negative for malignancy, and herpes test results are negative.

July 1: The woman returns to her physician because the labial lesion continues to cause some discomfort.

Sept. 15: The baby's father has multiple penile lesions and a generalized body rash.

Sept. 25: The woman delivers her baby. Her RPR is 1:32 and the infant's is 1:128.

Oct. 1: The woman takes her infant to a pediatrician because the baby is lethargic. She is told the infant is healthy and not to worry.

Oct. 2: The baby's father has a persistent body rash and plantar and palmar rashes.

Nov. 8: The infant becomes acutely ill with pneumonia and is hospitalized. The admitting physician finds signs of osteochondritis.

27

Environmental Microbiology

In previous chapters, we focused primarily on the disease-causing capabilities of microorganisms. In this chapter, you will learn about many of the positive functions microbes perform in the environment. Bacteria and other microorganisms are, in fact, essential to the maintenance of life on Earth.

Microbes, especially those that belong to the Domains Bacteria and Archaea, live in the most widely varied habitats on Earth. They are found in boiling hot springs, and as many as 5000 bacteria have been isolated from each milliliter of snow at the South Pole. Microbes have been recovered from minute opening in rocks a kilometer (0.62 mile) or more below the planet surface. There is even good evidence that bacteria are reproducing in water droplets of clouds sampled over mountain peaks in the Alps. Explorations of the deepest ocean have revealed large numbers of microbes living there, in eternal darkness and subject to incredible pressures (see the box in Chapter 6, page 164). Microbes are also found in clear mountain streams flowing from a melting glacier and in waters nearly saturated with salts, such as those of the Dead Sea (see the box in Chapter 5, page 147).

UNDER THE MICROSCOPE

A nitrogen-fixing symbiosis between a freshwater plant, *Azolla*, and a bacterium, *Anabaena azollae* (cyanobacteria that grow as chains of cells within a cavity in the plant leaf). This is an important factor in the cultivation of rice in Asia.

MICROBIAL DIVERSITY AND HABITATS

The diversity of microbial populations indicates that they take advantage of any niches found in their environment. Different amounts of oxygen, light, or nutrients may exist within a few millimeters in the soil. As a population of aerobic organisms uses up the available oxygen, anaerobes are able to grow. If the soil is disturbed by plowing, earthworms, or other activity the aerobes will again be able to grow to repeat this succession.

Microbes that live in extreme conditions of temperature, acidity, alkalinity, or salinity are called **extremophiles.** Most are members of the Archaea. The enzymes (**extremozymes**) that make growth possible under these conditions have been of great interest to industries because they can tolerate extremes of temperature, salinity, and pH that would inactivate other enzymes. The heat-resistant *Taq polymerase* enzyme from the source organism *Thermus aquaticus* (discussed in the context of PCR on page 258) is an example. An even more dramatic example of extremophiles capable of tolerating an extraordinary environment was the recent discovery of bacterial growth in nuclear-waste tanks in South Carolina. These tanks are a "witches' brew" of radioactive, chemically toxic wastes in which the radiation doses were many times that which would shatter the genetic structure of a human being.

Microorganisms live in an intensely competitive environment and must exploit any advantage they can. They may metabolize common nutrients more rapidly or use nutrients that competing organisms cannot metabolize. Some, such as the lactic acid bacteria that are so useful in making dairy products, are able to make an environmental niche inhospitable to competing organisms. The lactic acid bacteria are unable to use oxygen as an electron acceptor and are able to ferment sugars only to lactic acid, leaving most of the energy unused. However, the acidity inhibits the growth of more efficient, competing microbes.

SYMBIOSIS

Recall from Chapter 14 that **symbiosis** is the interaction between coexisting organisms or populations. Economically, the most important example of an animal-microbe symbiosis is that of the ruminants, animals that have a tanklike digestive organ called a *rumen*. Ruminants, such as cattle and sheep, graze on cellulose-rich plants. Bacteria in the rumen ferment the cellulose into compounds that are absorbed into the animal's blood and subsequently used for carbon and energy. Rumen protozoa keep the bacterial population under control by eating bacteria. Many of the microbes in the rumen also are digested for protein. (Examples of three animal-microbe symbioses are described in the boxes in Chapter 4, page 108; Chapter 6, page 164; and Chapter 11, page 319.)

A very important contribution to plant growth is made by **mycorrhizae**, or mycorrhizal symbionts (*myco* = fungus; *rhiza* = root). There are two primary types of these fungi: *endomycorrhizae*, also known as *vesicular-arbuscular mycorrhizae*; and *ectomycorrhizae*. Both types function as do root hairs on plants; that is, they extend the surface area through which the plant can absorb nutrients, especially phosphorus, which is not very mobile in soil.

Vesicular-arbuscular mycorrhizae form large spores that can be isolated easily from soil by sieving. The hyphae from these germinating spores penetrate into the plant root and form two types of structures: vesicles and arbuscules. **Vesicles** are smooth oval bodies that probably function as storage structures. **Arbuscules,** tiny bushlike structures, are formed inside plant cells (Figure 27.1a). Nutrients travel from the soil through fungal hyphae to these arbuscules, which gradually break down and release the nutrients to the plants. Most grasses and other plants are surprisingly dependent on these fungi for proper growth, and their presence is nearly universal in the plant kingdom.

Ectomycorrhizae mainly infect trees such as pine and oak. The fungus forms a mycelial *mantle* over the smaller roots of the tree (Figure 27.1b). Ectomycorrhizae do not form vesicles or arbuscules. Managers of commercial pine tree farms must ensure that seedlings are inoculated with soil containing effective mycorrhizae (Figure 27.2a).

Truffles, known as a food delicacy, are ectomycorrhizae, usually of oak trees (Figure 27.2b). In Europe, pigs or trained dogs are used to find them by smell and root them up. The surmise has been that the odor of the truffle is sexually attractive to the female pig because it is the odor of a potential mate. However, analysis indicates that although the saliva of a male pig contains such a hormone (which incidentally is also found in the armpit secretions of male humans), it is not the attractant. To a pig, male or female, the most important component of a truffle's odor is dimethyl sulfide, which is also responsible for the odor of cabbage. In nature, proliferation of the fungus depends on ingestion by an animal, which distributes the undigested spores into new locations. Increasingly, the cultivation of truffles is becoming a farming operation. Oak trees are planted in groves and artificially inoculated with fungal spores that are grown in the laboratory or extracted from ripe truffles.

SOIL MICROBIOLOGY AND BIOGEOCHEMICAL CYCLES

LEARNING OBJECTIVE

• Define *biogeochemical cycle.*

Billions of organisms, including those that are microscopic as well as comparatively huge insects and earthworms, form a vibrant living community in the soil. Typical soil has millions of bacteria in each gram. As Table 27.1 shows, the population is largest in the top few centimeters of soil and declines rapidly with depth. The most numerous organisms in soil are bacteria. Although actinomycetes are bacteria, they are usually considered separately. Many important antibiotics, such as streptomycin and tetracycline, were discovered by microbiologists investigating actinomycetes in soil.

Bacterial soil populations are usually estimated using plate counts on nutrient media, and the actual numbers are probably greatly underestimated by this method. No single nutrient medium or growth condition can possibly meet all the nutritional and other requirements of soil microorganisms.

We can think of soil as a "biological fire." A leaf falling from a tree is consumed by this fire as its organic matter is metabolized by microbes in the soil. Elements in the leaf enter the **biogeochemical cycles** for carbon, nitrogen, and sulfur that we will discuss in this chapter. In biogeochemical cycles, elements are oxidized and reduced by microorganisms to meet their metabolic needs. (See the discussion of oxidation-reduction in Chapter 5, page 123.) Without biogeochemical cycles, life on Earth would cease to exist.

THE CARBON CYCLE

LEARNING OBJECTIVE

• Outline the carbon cycle, and explain the roles of microorganisms in this cycle.

The primary biogeochemical cycle is the **carbon cycle** (Figure 27.3). All organisms, including plants, microbes, and animals, contain large amounts of carbon in the form of organic compounds such as cellulose, starches, fats, and proteins. Let's take a closer look at how these organic compounds are formed.

Recall from Chapter 5 that autotrophs perform an essential role for all life on Earth by reducing carbon dioxide

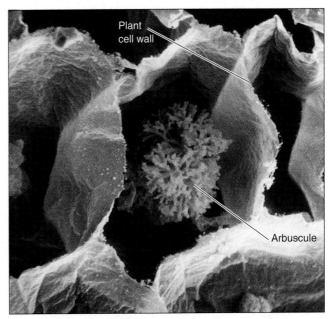

(a) Endomycorrhiza (vesicular-arbuscular mycorrhiza) A fully developed arbuscule of an endomycorrhiza in a plant cell. (The term *arbuscule* means little bush.) As the arbuscule decomposes, it releases nutrients for the plant.

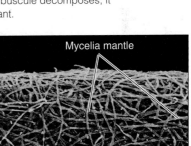

(b) Ectomycorrhiza The mycelial mantle of a typical ectomycorrhizal fungus surrounding a *Eucalyptus* tree root.

FIGURE 27.1 Mycorrhizae.

Q *Of what value is a mycorrhiza to a plant?*

to form organic matter. When you look at a tree, you might think that its mass is from the soil where it grows. In fact, its great mass of cellulose is derived from the 0.03% of carbon dioxide in the atmosphere. This occurs as a result of photosynthesis, the first step of the carbon cycle in which photoautotrophs such as cyanobacteria, green plants, algae, and green and purple sulfur bacteria *fix* (that is incorporate) carbon dioxide into organic matter using energy from sunlight. Chemoautotrophs such as *Thiobacillus* and *Beggiatoa* also fix carbon dioxide into organic matter, while metabolizing compounds such as hydrogen sulfide for energy.

(a) Growth of many plants is strongly influenced by infection by mycorrhizae. The relative growth of two pine seedlings is shown: the seedling on the left was inoculated with mycorrhizae; the seedling on the right was not.

(b) Truffles. Of the three truffles shown, one has been sliced to show the interior.

FIGURE 27.2 Mycorrhizae have considerable commercial value.

Q Why are mycorrhizae valuable for the uptake of phosphorus?

In the next step of the cycle, chemoheterotrophs such as animals and protozoa eat autotrophs and may in turn be eaten by other animals. Thus, as the organic compounds of the autotrophs are digested and resynthesized, the carbon atoms of carbon dioxide are transferred from organism to organism up the food chain.

Some of the organic molecules are used by chemoheterotrophs, including animals, to satisfy their energy requirements. When this energy is released through respiration, carbon dioxide immediately becomes available to start the cycle over again. Much of the carbon remains within the organisms until they excrete it as wastes or die. When plants and animals die, these organic compounds are decomposed by bacteria and fungi. During decomposition, the organic compounds are oxidized, and CO_2 is returned to the cycle.

Carbon is stored in rocks, such as limestone ($CaCO_3$), and is dissolved as carbonate ions (CO_3^{2-}) in oceans. Vast deposits of fossil organic matter exist in the form of fossil fuels, such as coal and petroleum. Burning these fossil fuels releases CO_2, resulting in an increased amount of CO_2 in the atmosphere. Many scientists believe the increased atmospheric carbon dioxide may be causing a **global warming** of the Earth.

An interesting aspect of the carbon cycle is methane (CH_4) gas. Sediments on the ocean floor contain an

TABLE 27.1	Microorganisms per Gram of Typical Garden Soil at Various Depths			
Depth (cm)	Bacteria	Actinomycetes*	Fungi	Algae†
3–8	9,750,000	2,080,000	119,000	25,000
20–25	2,179,000	245,000	50,000	5000
35–40	570,000	49,000	14,000	500
65–75	11,000	5000	6000	100
135–145	1400	—	3000	—

*Filamentous bacteria
†Algae found at lower depths are not metabolically active but were probably introduced by drainage or mechanical movement.

SOURCE: Adapted from M. Alexander, *Introduction to Soil Microbiology*, 2nd ed. New York: Wiley, 1991.

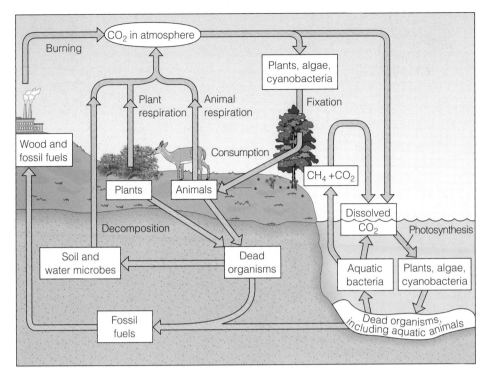

FIGURE 27.3 The carbon cycle. On a global scale, the return of CO_2 to the atmosphere by respiration closely balances its removal by fixation. However, the burning of wood and fossil fuels adds more CO_2 to the atmosphere; as a result, the amount of atmospheric CO_2 is steadily increasing.

Q **What effect does the accumulation of carbon dioxide in the atmosphere have on the Earth's climate?**

estimated 10 trillion tons of methane, about twice as much as the Earth's deposits of fossil fuels such as coal and petroleum. Furthermore, methanogenic bacteria in the ocean's depths are constantly producing more (see Seawater Microbiota on page 822). Methane is much more potent as a greenhouse gas than is carbon dioxide, and the Earth's environment would be dangerously altered if all this gas escaped to the atmosphere. Fortunately, hordes of sea-dwelling bacteria use escaping methane gas as an energy source, and it disappears before reaching the surface of the water and then the atmosphere. Because the metabolism of methane normally requires free oxygen, which is scarce in the ocean depths, it is uncertain how these bacteria do this.

THE NITROGEN CYCLE

LEARNING OBJECTIVES

- Outline the nitrogen cycle, and explain the roles of microorganisms in this cycle.
- Define *ammonification, nitrification, denitrification,* and *nitrogen fixation.*

The **nitrogen cycle** is shown in Figure 27.4. Nitrogen is needed by all organisms for the synthesis of protein, nucleic acids, and other nitrogen-containing compounds. Molecular nitrogen (N_2) makes up almost 80% of the Earth's atmosphere. For assimilation and use by plants, nitrogen must be fixed, that is, taken up and combined into organic compounds. The activities of specific microorganisms are important to the conversion of nitrogen to usable forms.

AMMONIFICATION

Almost all the nitrogen in the soil exists in organic molecules, primarily in proteins. When an organism dies, the process of microbial decomposition results in the hydrolytic breakdown of proteins into amino acids. In a process called **deamination,** the amino groups of amino acids are removed and converted into ammonia (NH_3). This release of ammonia is called **ammonification** (see Figure 27.4). Ammonification, brought about by numerous bacteria and fungi, can be represented as follows:

$$\text{Proteins from dead cells and waste products} \xrightarrow{\text{Microbial decomposition}} \text{Amino acids}$$

$$\text{Amino acids} \xrightarrow{\text{Microbial ammonification}} \text{Ammonia (NH}_3\text{)}$$

Microbial growth releases extracellular proteolytic enzymes that decompose proteins. The resulting amino acids are transported into the microbial cells, where ammonification occurs. The fate of the ammonia produced by ammonification depends on soil conditions (see the discussion of denitrification, which follows). Because ammonia is a gas, it rapidly disappears from dry soil, but in moist soil it becomes solubilized in water, and ammonium ions (NH_4^+) are formed:

$$NH_3 + H_2O \longrightarrow NH_4{}^+ OH \longrightarrow NH_4{}^+ + OH^-$$

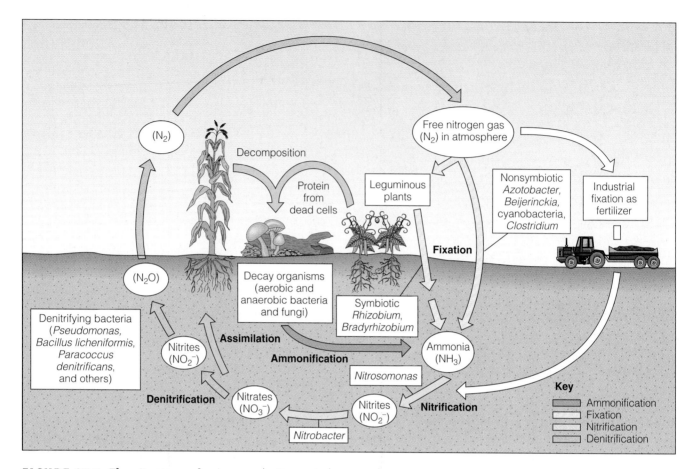

FIGURE 27.4   The nitrogen cycle. In general, nitrogen in the atmosphere goes through fixation, nitrification, and denitrification. Nitrates assimilated into plants and animals after nitrification go through decomposition, ammonification, and then nitrification again.

Q   **Which processes are performed exclusively by bacteria?**

Ammonium ions from this sequence of reactions are used by bacteria and plants for amino acid synthesis.

NITRIFICATION

The next sequence of reactions in the nitrogen cycle involves the oxidation of the nitrogen in the ammonium ion to produce nitrate, a process called **nitrification.** Living in the soil are autotrophic nitrifying bacteria, such as those of the genera *Nitrosomonas* and *Nitrobacter*. These microbes obtain energy by oxidizing ammonia or nitrite. In the first stage, *Nitrosomonas* oxidizes ammonium to nitrites:

$$\underset{\text{Ammonium ion}}{NH_4^+} \xrightarrow{\textit{Nitrosomonas}} \underset{\text{Nitrite ion}}{NO_2^-}$$

In the second stage, such organisms as *Nitrobacter* oxidize nitrites to nitrates:

$$\underset{\text{Nitrite ion}}{NO_2^-} \xrightarrow{\textit{Nitrobacter}} \underset{\text{Nitrate ion}}{NO_3^-}$$

Plants tend to use nitrate as their source of nitrogen for protein synthesis because nitrate is highly mobile in soil and is more likely to encounter a plant root than ammonium. Ammonium ions would actually make a more efficient source of nitrogen because they require less energy to incorporate into protein, but these positively charged ions are usually bound to negatively charged clays in the soil, whereas the negatively charged nitrate ions are not bound.

DENITRIFICATION

The form of nitrogen resulting from nitrification is fully oxidized and no longer contains any biologically usable energy. However, it can be used as an electron acceptor by microbes metabolizing other organic energy sources in the absence of atmospheric oxygen (see the discussion of anaerobic respiration in Chapter 5). This process, called **denitrification,** can lead to a loss of nitrogen to the atmosphere, especially as nitrogen gas. Denitrification can be represented as follows:

$$NO_3^- \longrightarrow NO_2^- \longrightarrow N_2O \longrightarrow N_2$$

Nitrate ion   Nitrite ion   Nitrous oxide   Nitrogen gas

Pseudomonas species appear to be the most important group of bacteria in denitrification in soils. Denitrification occurs in waterlogged soils where little oxygen is available. In the absence of oxygen as an electron acceptor, denitrifying bacteria substitute the nitrates of agricultural fertilizer. This converts much of the valuable nitrate into gaseous nitrogen that enters the atmosphere and represents a considerable economic loss.

NITROGEN FIXATION

We live at the bottom of an ocean of nitrogen gas. The air we breathe is about 79% nitrogen, and above every acre of soil (the area of an American football field from the goal line to the opposite 10-yard line, or 50.6 × 80 meters) stands a column of nitrogen weighing about 32,000 tons. But the only creatures on Earth that can use it directly as a nitrogen source are a few species of bacteria, including cyanobacteria. The process by which they convert nitrogen gas to ammonia is known as **nitrogen fixation.**

Bacteria that are responsible for nitrogen fixation all rely on the same enzyme, *nitrogenase.* It is estimated the Earth's entire supply of this essential enzyme could fit into a single large bucket. A characteristic of nitrogenase is that it is inactivated by oxygen. Therefore, it probably evolved early in the history of the planet, before the atmosphere contained much molecular oxygen and before nitrogen-containing compounds were available from decaying organic matter. Nitrogen fixation is brought about by two types of microorganisms: free-living and symbiotic. (Agricultural fertilizers are made up of nitrogen that has been fixed by industrial physical–chemical processes.)

Free-Living Nitrogen-Fixing Bacteria

Free-living nitrogen-fixing bacteria are found in particularly high concentrations in the *rhizosphere,* a region roughly 2 millimeters from the plant root. The rhizosphere represents something of a nutritional oasis in the soil, especially in grasslands. Among the free-living bacteria that can fix nitrogen are aerobic species such as *Azotobacter.* These aerobic organisms apparently shield the anaerobic nitrogenase enzyme from oxygen by, among other things, having a very high rate of oxygen use that minimizes the diffusion of oxygen into the interior of the cell, where the enzyme is located.

Another free-living obligate aerobe that fixes nitrogen is *Beijerinckia* (bī-yė-rink′ē-ä). Some anaerobic bacteria, such as certain species of *Clostridium,* also fix nitrogen. The bacterium C. *pasteurianum* (pas-tyėr-ē-ā′num), an obligately anaerobic, nitrogen-fixing microorganism, is a prominent example.

There are many species of aerobic, photosynthesizing cyanobacteria that fix nitrogen. Because their energy supply is independent of carbohydrates in soil or water, they are especially useful suppliers of nitrogen to the environment. Cyanobacteria usually carry their nitrogenase enzymes in specialized structures called **heterocysts** that provide anaerobic conditions for fixation (see Figure 11.13a, page 328).

Most of the free-living nitrogen-fixing bacteria are capable of fixing large amounts of nitrogen under laboratory conditions. However, in the soil there is usually a shortage of usable carbohydrates to supply the energy needed for the reduction of nitrogen to ammonia, which is then incorporated into protein. Nevertheless, these nitrogen-fixing bacteria make important contributions to the nitrogen economy of such areas as grasslands, forests, and the arctic tundra.

Symbiotic Nitrogen-Fixing Bacteria

Symbiotic nitrogen-fixing bacteria play an even more important role in plant growth for crop production. Members of the genera *Rhizobium, Bradyrhizobium,* and others infect the roots of leguminous plants, such as soybeans, beans, peas, peanuts, alfalfa, and clover. (These agriculturally important plants are only a few of the thousands of known leguminous species, many of which are bushy plants or small trees found in poor soils in many parts of the world.) Rhizobia, as these bacteria are commonly known, are specially adapted to particular leguminous plant species, on which they form **root nodules** (Figure 27.5). Nitrogen is then fixed by a symbiotic process of the plant and the bacteria. The plant furnishes anaerobic conditions and growth nutrients for the bacteria, and the bacteria fix nitrogen that can be incorporated into plant protein.

There are similar examples of symbiotic nitrogen fixation in nonleguminous plants, such as alder trees. These trees are among the first to appear in forests after fires or glaciation. The alder tree is symbiotically infected with an actinomycete (*Frankia*) and forms nitrogen-fixing root nodules. About 50 kg of nitrogen can be fixed each year by the growth of 1 acre of alder trees; the trees thus make a valuable addition to the forest economy.

Another important contribution to the nitrogen economy of forests is made by **lichens,** which are a combination fungus and an alga or a cyanobacterium in a mutualistic relationship (see Figure 12.10, page 356). When one

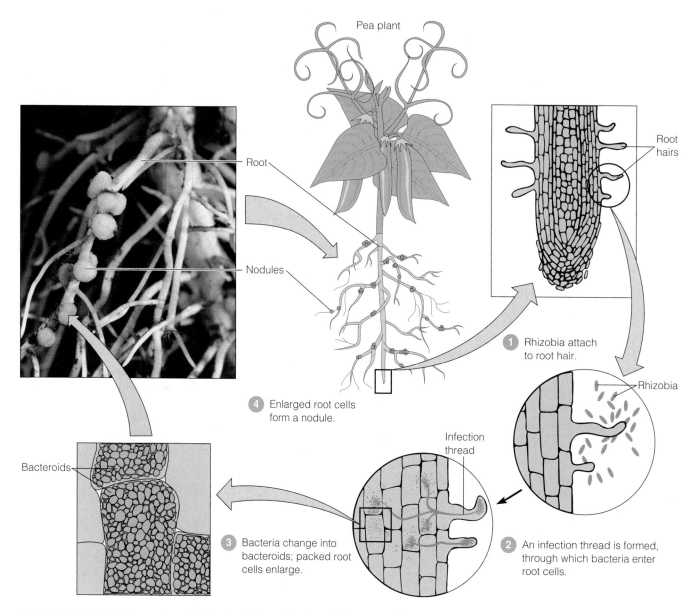

Pea plant

Root

Nodules

Root
hairs

1 Rhizobia attach
to root hair.

Rhizobia

4 Enlarged root cells
form a nodule.

Infection
thread

Bacteroids

3 Bacteria change into
bacteroids; packed root
cells enlarge.

2 An infection thread is formed,
through which bacteria enter
root cells.

FIGURE 27.5 The formation of a root nodule. Members of the nitrogen-fixing genera *Rhizobium* and *Bradyrhizobium* form these nodules on legumes. This mutualistic association is beneficial to both the plant and the bacteria.

Q **In nature, are leguminous plants most likely to be valuable in rich agricultural soils or poor desert soils?**

symbiont is a nitrogen-fixing cyanobacterium, the product is fixed nitrogen that eventually enriches the forest soil. Free-living cyanobacteria can fix significant amounts of nitrogen in desert soils after rains and on the surface of arctic tundra soils. Rice paddies can accumulate heavy growths of such nitrogen-fixing organisms. The cyanobacteria also form a symbiosis with a small floating fern, *Azolla,* which grows thickly in rice paddy waters (Figure 27.6). So much nitrogen is fixed by these microbes that other nitrogenous fertilizers are often unnecessary for rice cultivation.

THE SULFUR CYCLE

LEARNING OBJECTIVE

- Outline the sulfur cycle, and explain the roles of microorganisms in this cycle.

The **sulfur cycle** (Figure 27.7) and nitrogen cycle resemble each other in the sense that they represent numerous oxidation states of these elements. The most reduced forms of sulfur are the sulfides, such as the odorous gas hydrogen sulfide (H_2S). Like the ammonium ion of the nitrogen cycle, this is a reduced compound that generally forms under anaerobic

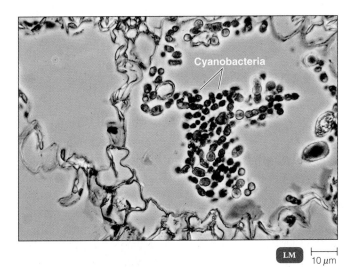

FIGURE 27.6 The *Azolla*–cyanobacteria symbiosis.
A section through the leaf of an *Azolla* freshwater fern. The cyanobacteria *Anabaena azollae* is visible as chains of cells within the leaf cavity.

Q **What is the major contribution of cyanobacteria as symbionts?**

conditions. In turn it represents a source of energy for autotrophic bacteria. These bacteria convert the reduced sulfur in H_2S into elemental sulfur granules and fully oxidized sulfates (SO_4^{2-}).

An interesting aspect of the sulfur cycle is the work of Sergei Winogradsky in first discovering chemoautotrophy, by which microbes obtain energy from inorganic chemicals. In Paris in the early 1900s, he was studying an

exceptionally large, filamentous bacterium called *Beggiatoa* (see Chapter 11, page 321), which grew on the surface of stagnant ponds and in the slime of sulfurous springs. Winogradsky allowed water containing H_2S to flow past the microbes, which formed numerous granules of sulfur within the cell. This was an oxidative change that may have indicated a reaction with the oxygen in the air. When water without H_2S was then allowed to flow past the cultures, the sulfur granules slowly disappeared but the bacteria continued to thrive. *Beggiatoa* was obviously obtaining energy from the reaction that converted H_2S into elemental sulfur. The elemental sulfur, he later determined, disappeared because it was being converted into fully oxidized sulfate ions that were rinsed away in the passing water. Therefore, both H_2S and elemental sulfur were being used as inorganic energy sources by *Beggiatoa* (see Figure 27.7). This discovery of the autotrophic physiology of bacteria is central to an understanding of biogeochemical cycles.

Frequently, elemental sulfur is released from decaying microbes. Elemental sulfur is essentially insoluble in temperate waters, and microbes have difficulty absorbing it. This is probably the origin of huge, prehistoric underground accumulations of sulfur.

Several phototrophic bacteria, such as the green and purple sulfur bacteria, also oxidize H_2S, forming colorful internal sulfur granules (see Figure 11.14, page 329). Like *Beggiatoa*, they can further oxidize the sulfur to sulfate ions. It is important to recognize that these organisms are using light for energy; the hydrogen sulfide is used to reduce CO_2 (see Chapter 5, page 141).

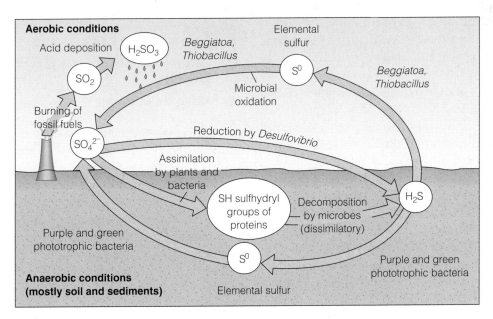

FIGURE 27.7 The sulfur cycle. Note the importance of aerobic and anaerobic conditions.

Q **Why is a source of sulfur necessary for all organisms?**

Hydrogen sulfide can be used as an energy source by *Thiobacillus* to produce sulfate ions and sulfuric acid. *Thiobacillus* can grow well at a pH as low as 2 and have practical uses in mining (see Figure 28.14, page 856). Sulfates are incorporated by plants and bacteria to become part of sulfur-containing amino acids for humans and other animals. There, they form disulfide links that give structure to proteins. As proteins are decomposed, in a process called **dissimilation,** the sulfur is released as hydrogen sulfide to reenter the cycle.

LIFE WITHOUT SUNSHINE

LEARNING OBJECTIVE

• Describe how an ecological community can exist without light energy.

Interestingly, it is possible for entire biological communities to exist without photosynthesis by exploiting the energy in H_2S. In Chapter 11 (page 329), equations are presented to show that photosynthesis and chemoautotrophic use of H_2S are similar in certain respects. An example of such a community around deep-sea vents is described in the box in Chapter 6 (page 164). Deep caves, totally isolated from sunlight, have been discovered that also support entire biological communities in a similar manner. The **primary producers** in these systems are chemoautotrophic bacteria rather than photoautotrophic plants or microbes.

Recently another microbial ecosystem operating far from sunlight has been discovered over 1 km deep within rocks, including shales, granites, and basalts. Such bacteria are called **endoliths** (inside rocks), which must grow in the near absence of oxygen and with minimal nutrient supplies. Sedimentary shales often contain trapped organic nutrients or sulfates that can support limited life. Rocks such as granite or basalt are minutely porous or have fractures that contain some water. Sulfate reduction is often a possible energy source. Also, in these rocks, chemical reactions and radioactivity produce hydrogen that can be used for energy by autotrophic endolithic bacteria. Carbon dioxide dissolved in the water serves as a carbon source, and cellular organic matter is produced. Some is excreted, or is released upon the death and lysis of the microbe, and becomes available for the growth of other microbes. Nutrient inputs, especially of nitrogen, are very small in this environment, and generation times may be measured in many years. Various survival strategies have developed for life with minimal nutrition. For example, suspended in a state between life and death, certain of these organisms become dramatically smaller. Ecologists speculating upon forms of life that might be found in the harsh environment of Mars are very interested in endoliths.

THE PHOSPHORUS CYCLE

LEARNING OBJECTIVE

• Compare and contrast the carbon cycle and the phosphorus cycle.

Another important nutritional element that is part of a biogeochemical cycle is phosphorus. The availability of phosphorus may determine whether plants and other organisms can grow in an area. The problems associated with excess phosphorus (eutrophication) are described later in the chapter.

Phosphorus exists primarily as phosphate ions (PO_4^{3-}) and undergoes very little change in its oxidation state. The **phosphorus cycle** instead involves changes from soluble to insoluble forms and from organic to inorganic phosphate, often in relation to pH. For example, phosphate in rocks can be solubilized by the acid produced by bacteria such as *Thiobacillus*. Unlike the other cycles, there is no volatile phosphorus-containing product to return phosphorus to the atmosphere in the way carbon dioxide, nitrogen gas, and sulfur dioxide are returned. Therefore, phosphorus tends to accumulate in the seas. It can be retrieved by mining the above-ground sediments of ancient seas, mostly as deposits of calcium phosphate. Seabirds also mine phosphorus from the sea by eating phosphorus-containing fish and depositing it as guano (bird droppings). Certain small islands inhabited by such birds have long been mined for these deposits as a source of phosphorus for fertilizers.

THE DEGRADATION OF SYNTHETIC CHEMICALS IN SOIL AND WATER

LEARNING OBJECTIVES

• Give two examples of the use of bacteria to remove pollutants.

• Define *bioremediation.*

We seem to take for granted that soil microorganisms will degrade materials entering the soil. Natural organic matter, such as falling leaves or animal residues, are in fact readily degraded. However, in this industrial age many chemicals that do not occur in nature (**xenobiotics**), such as plastics, enter the soil in large amounts. A new development is the production of plastic from natural chemicals of plants. The starchy sugars from plants is fed to lactic acid bacteria. The lactic acid is then reacted with a catalyst to make a biodegradable plastic. Many synthetic chemicals, such as pesticides, are highly resistant to degradation by microbial attack. A well-known example is the insecticide DDT, which proved so resistant that it accumulated to damaging levels in the environment. Polychlorinated biphenyls (PCBs), at one

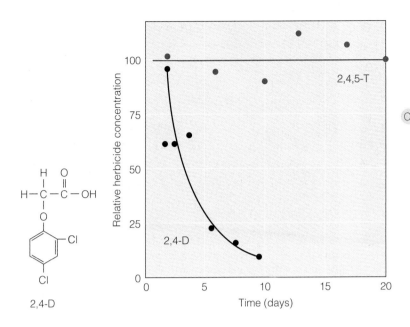

Enrichment (see page 172) with a bacterium that degrades pollutants is described in the box on page 33.

FIGURE 27.8 2,4-D (black) and 2,4,5-T (magenta). This graph shows the structures and rates of microbial decomposition of the herbicides 2,4-D (black) and 2,4,5-T (magenta).

Q **Which of these two herbicides is more easily degraded?**

time much used as a heat-transfer fluid in electrical transformers and a component in some plastics, resemble DDT in many ways. Both are resistant to biological degradation, and both tend to become concentrated in the final predators of the food chain—for example, fish-eating birds. Microbial ecologists are currently investigating the use of microbes to eliminate PCBs from contaminated soils and waters.

Some synthetic chemicals are made up of bonds and subunits that are subject to attack by bacterial enzymes. Small differences in chemical structure can make large differences in biodegradability. The classic example is that of two herbicides: 2,4-D (the common chemical used to kill lawn weeds) and 2,4,5-T (used to kill shrubs); both were components of Agent Orange, which was used to defoliate jungles during the Vietnam war. The addition of a single chlorine atom to the structure of 2,4-D extends its life in soil from a few days to an indefinite period (Figure 27.8).

A growing problem is the leaching into groundwaters of toxic materials that are not biodegradable or that degrade very slowly. The sources of these materials may include landfills, illegal industrial dumps, or pesticides applied to agricultural crops. Once groundwater becomes contaminated, the environmental and economic damage can be devastating. Researchers are developing processes and isolating bacteria that promote degradation and detoxification. Enrichment (see page 172) with a bacterium that degrades pollutants is described in the box on page 33.

BIOREMEDIATION

The use of microbes to detoxify or degrade pollutants is called **bioremediaton.** Oil spills from wrecked tankers represent some of the most dramatic examples of chemical pollution. The economic losses from contaminated fisheries and beaches can be enormous. To some degree, bioremediation occurs naturally, as microbes attack the petroleum if conditions are aerobic. However, microbes usually obtain their nutrients in aqueous solution, and oil-based products are relatively nonsoluble. Also, petroleum hydrocarbons are deficient in essential elements, such as nitrogen and phosphorus. Bioremediation of oil spills is greatly enhanced if the resident bacteria are provided with "fertilizer" containing nitrogen and phosphorus (Figure 27.9). Bioremediaton

FIGURE 27.9 Bioremediation at an oil spill in Alaska. The portion of beach on the left is uncleaned, the beach on the right has been treated with applications of carbon-free nutrients (fertilizer). However, below the surface layers, where conditions are anaerobic, oil often remains for much longer periods.

Q **Does the chemical formula of most petroleum products contain nitrogen, or phosphorus?** (*Hint:* See the box on page 33 in Chapter 2.)

(a) Solid municipal wastes being turned by a specially designed machine.

(b) Compost made from municipal wastes awaiting trucks to spread it on agricultural fields.

FIGURE 27.10 Composting municipal wastes.

Q A compost pile of grass and leaves is very high in carbon; does it have much nitrogen?

may also make use of microbes that have been selected for growth on a certain pollutant or of genetically modified bacteria that are specially adapted to metabolize petroleum products. The addition of such specialized microbes is called **bioaugmentation.**

Subsurface petroleum spills from leaking gasoline storage tanks, for example, can be removed from ground water by pumping the water into aerating tanks, providing fertilizer nutrients and returning the water after the gasoline has been decomposed. Similar principles are used to clean up pesticide and other chemical spills. Radiation-resistant bacteria have been genetically altered to make them more useful in cleaning up sites contaminated with radioactive solvents.

SOLID MUNICIPAL WASTE

Solid municipal waste (garbage) is most frequently placed into large compacted landfills. Conditions are largely anaerobic, and even presumably biodegradable materials such as paper are not very effectively attacked by microorganisms. In fact, recovering a 20-year-old newspaper in readable condition is not at all unusual. But such anaerobic conditions do promote the activity of the same methanogens that will be discussed with the operation of anaerobic sludge digesters in sewage treatment. The methane they produce can be tapped with drill holes and burned to generate electricity or purified and introduced into natural gas pipeline systems (see Figure 28.15, page 857). Such systems are part of the design of many large landfills in the United States, some of which provide energy for industrial plants and homes.

The amount of organic matter entering landfills can be considerably reduced if it is first separated from material that is not biodegradable and composted. **Composting** is a process used by gardeners to convert plant remains into the equivalent of natural humus (Figure 27.10). A pile of leaves or grass clippings will undergo microbial degradation. Under favorable conditions, thermophilic bacteria will raise the temperature of the compost to 55° to 60°C in a couple of days. After the temperature declines, the pile can be turned to renew the oxygen supply, and a second temperature rise will occur. Over time, the thermophilic microbial populations are replaced by mesophilic populations that slowly continue the conversion to a stable material similar to humus. Where space is available, municipal wastes are composted in windrows (long, low piles) that are distributed and periodically turned over by specialized machinery. Municipal waste disposal now also makes increasing use of composting methods.

AQUATIC MICROBIOLOGY AND SEWAGE TREATMENT

Aquatic microbiology refers to the study of microorganisms and their activities in natural waters, such as lakes, ponds, streams, rivers, estuaries, and oceans. Domestic and industrial wastewater enters lakes and streams, and its degradation and effects on the microbial life is an important factor in aquatic microbiology. We will also see that the method of treating wastewater by municipalities mimics a natural filtering process.

BIOFILMS

LEARNING OBJECTIVE
• Describe the importance of biofilms.

In nature, microorganisms seldom live in the isolated, single-species colonies that we see on laboratory plates.

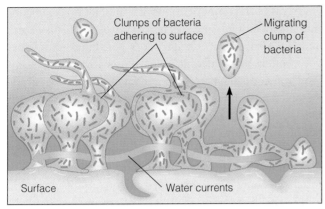

(a) Water currents move, as shown by the blue arrow, among pillars of slime formed by the growth of bacteria attached to solid surfaces. This allows efficient access to nutrients and removal of bacterial waste products. Individual slime-forming bacteria or bacteria in clumps of slime detach and move to new locations.

100 μm

(b) A bacterial biofilm that formed on a steel surface over two months in an industrial water system. The large oval bodies are diatoms that became entrapped in the sticky biofilm.

SEM   1 μm

FIGURE 27.11 Biofilms.

Q **What is the biofilm that forms on teeth called?**

They more typically live in slime communities, called **biofilms.** This fact was not well appreciated until the development of confocal microscopy (see page 64) made the three-dimensional structure of biofilms more visible. Cell-to-cell chemical communication, or *quorum sensing*, allows bacteria to coordinate their activity and group together into communities that provide benefits not unlike those of multicellular organisms (see the box in Chapter 3, page 57.). Therefore, biofilms are not just bacterial slime layers but biological systems; the bacteria are organized into coordinated, functional communities. Biofilms are usually attached to a surface, which might be a rock in a pond, a human tooth (plaque), or mucous membrane. However, some biofilm communities are assembled into floating bacterial particulate groupings. The floc that forms in certain types of sewage treatment (see Figure 27.22, page 832) is an example. Within a biofilm community, the bacteria are able to share nutrients and are somewhat sheltered from harmful factors in the environment, such as desiccation, antibiotics, and the body's immune system. The close proximity of microorganisms within a biofilm might also have the advantage of facilitating the transfer of genetic information by, for example, conjugation.

A biofilm usually begins to form when a free-swimming (planktonic) bacterium attaches to a surface. If these bacteria grew in a uniformly thick monolayer, they would become overcrowded, nutrients would not be available in lower depths, and toxic wastes could accumulate. Bacteria in biofilm communities avoid these problems by forming pillar-like structures (Figure 27.11a) with channels between them through which water can carry incoming nutrients and outgoing wastes. This constitutes a primitive circulatory system. Individual bacteria and clumps of slime occasionally leave the established biofilm and move to a new location where the biofilm becomes extended. A biofilm is generally composed of a surface layer about 10 μm thick, with pillars that extend up to 200 μm above it. In fast-flowing water, the biofilms may take the form of filamentous streamers attached at the upstream end.

The bacteria in biofilms can work cooperatively to carry out complex tasks. For example, the digestive systems of ruminant animals require at least five different microbial species to break down cellulose. The microbes in ruminants' digestive systems are located mostly within biofilm communities. Biofilms are also essential elements in the proper functioning of sewage systems, which we will be discussing in this chapter. They can also, however, be a problem in pipes and tubing (Figure 27.11b).

Biofilms are an important factor in human health. For example, microbes in biofilms are probably 1000 times more resistant to microbicides. Experts at the Centers for Disease Control and Prevention (CDC) estimate that 65% of human bacterial infections involve biofilms. Most nosocomial infections are probably related to biofilms on medical catheters (see Figure 21.3, page 619). In fact, biofilms form on almost all indwelling medical devices, which can include mechanical heart valves. Biofilms, which can include those formed by fungi such as *Candida*, are encountered in many disease conditions, such as infections related to the use of contact lenses, dental caries (see page 747), and infections by pseudomonad bacteria (see page 321).

One approach to preventing biofilm formation is to incorporate antimicrobials into surfaces on which biofilms might form (see page 57). Because the chemical signals that allow quorum sensing are essential to biofilm formation, research is underway to determine the makeup of these chemical signals and perhaps block them. Another approach is the discovery that biofilm formation can be inhibited by lactoferrin, which is abundant in many human secretions. Lactoferrin chelates iron, especially among the pseudomonads responsible for cystic fibrosis biofilms. The lack of iron inhibits the surface motility essential for the aggregation of the bacteria into biofilms.

AQUATIC MICROORGANISMS

LEARNING OBJECTIVE

- Describe the freshwater and seawater habitats of microorganisms.

Large numbers of microorganisms in a body of water generally indicate high nutrient levels in the water. Water contaminated by inflows from sewage systems or from biodegradable industrial organic wastes is relatively high in bacterial counts. Similarly, ocean estuaries (fed by rivers) have higher nutrient levels and therefore larger microbial populations than other shoreline waters.

In water, particularly water with low nutrient concentrations, microorganisms tend to grow on stationary surfaces and on particulate matter. In this way, a microorganism has contact with more nutrients than if it were randomly suspended and floating freely with the current. Many bacteria whose main habitat is water often have appendages and holdfasts that attach to various surfaces. One example is *Caulobacter* (see Figure 11.2, page 317).

FRESHWATER MICROBIOTA

Figure 27.12 shows a typical lake or pond that serves as an example to represent the various zones and the kinds of microbiota found in a body of fresh water. The **littoral zone** along the shore has considerable rooted vegetation, and light penetrates throughout it. The **limnetic zone** consists of the surface of the open water area away from the shore. The **profundal zone** is the deeper water under the limnetic zone. The **benthic zone** contains the sediment at the bottom.

Microbial populations of freshwater bodies tend to be affected mainly by the availability of oxygen and light. In many ways, light is the more important resource because photosynthetic algae are the main source of organic matter, and hence of energy, for the lake. These organisms are the primary producers of a lake that supports a population of bacteria, protozoa, fish, and other aquatic life. Photosynthetic algae are located in the limnetic zone.

Areas of the limnetic zone with sufficient oxygen contain pseudomonads and species of *Cytophaga, Caulobacter,* and *Hyphomicrobium.* Oxygen does not diffuse into water very well, as any aquarium owner knows. Microorganisms growing on nutrients in stagnant water quickly use up the dissolved oxygen in the water. In the oxygenless water, fish die, and odors are produced from anaerobic activity. Wave action in shallow layers, or water movement in rivers, tends to increase the amount of oxygen throughout the water and aid in the growth of aerobic populations of bacteria. Movement thus improves the quality of water and aids in the degradation of polluting nutrients.

Deeper waters of the profundal and benthic zones have low oxygen concentrations and less light. Algal growth near the surface often filters the light, and it is not unusual for photosynthetic microbes in deeper zones to use different wavelengths of light from those used by surface-layer photosynthesizers (see Figure 12.11a, page 358).

Purple and green sulfur bacteria are found in the profundal zone. These bacteria are anaerobic photosynthetic organisms that metabolize H_2S to sulfur and sulfate in the bottom sediments of the benthic zone.

The sediment in the benthic zone includes bacteria such as *Desulfovibrio* that use sulfate (SO_4^{2-}) as an electron acceptor and reduce it to H_2S. Methane-producing bacteria are also part of these anaerobic benthic populations. In swamps, marshes, or bottom sediments, they produce methane gas. *Clostridium* species are common in bottom sediments and may include botulism organisms, particularly those causing outbreaks of botulism in waterfowl.

SEAWATER MICROBIOTA

As knowledge of the microbial life of the oceans expands, largely identified by ribosomal RNA methods (see the discussion of FISH on page 304 in Chapter 10), biologists are becoming more conscious of the importance of oceanic microbes. One conclusion, so far, has been that nearly a third of all life on the planet consists of microbes that live, not in ocean waters, but under the seafloor. These microbes make immense amounts of methane gas that could be environmentally damaging if it were to be released into the atmosphere.

In the upper, sunlit waters of the ocean, cyanobacteria, especially the unicellular genus *Synechococcus* (sin'ē-kō-kok-kus), and the more recently discovered genus *Prochlorococcus* (prō-kilôr 'ō-kok-kus) comprise a large part of the population. The latter organism is tiny, not more than 0.7μm in diameter, but a drop of seawater sometimes contains 20,000 of their cells. An unseen population of such microscopic organisms fills the upper 200 meters of ocean and exerts a profound influence on life on Earth. The support of oceanic life is largely dependent upon such

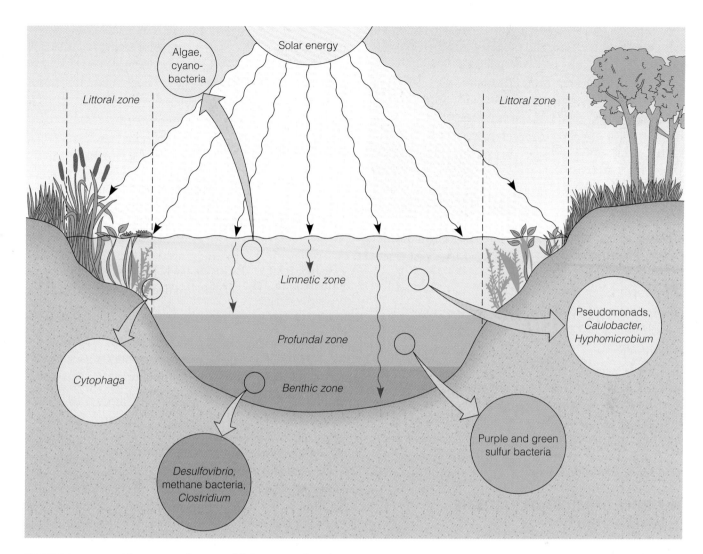

FIGURE 27.12   The zones of a typical lake or pond and some representative microorganisms of each zone. The microbes fill niches that vary in light, nutrients, and oxygen availability.

Q **Which of the microorganisms listed are most likely aerobes?**

photosynthetic microscopic life, the marine **phytoplankton** (a term derived from the Greek for wandering plants).

Photosynthetic bacteria such as these form the basis of the oceanic food chain. They fix carbon dioxide to form organic matter that is eventually released as dissolved organic matter and is used by the ocean's heterotrophic bacteria. Immense populations of another bacterium, *Pelagibacter ubique,* metabolize the waste products of these photosynthetic populations (see the discussion under Microbial Diversity on page 340). Bacteria of many kinds then serve as a particulate food source for a series of increasingly larger consumers. These are first the protozoa, which are in turn prey for multicellular zooplankton (plankton animal life such as shrimplike krill). These zooplankton are eventually prey for fish. Much of the carbon dioxide and mineral nutrients released by the metabolic activity of bacteria, protozoa, and

zooplankton is recycled into the photosynthetic phytoplankton.

In waters below about 100 meters, members of the Archaea begin to dominate microbial life. Planktonic members of this group of the genus *Crenarchaeota* (kren-ärk-e′ō-tä) account for much of the microbial biomass of the oceans. These organisms are well adapted to the cool temperatures and low oxygen levels of oceanic depths. Their carbon is primarily derived from dissolved CO_2.

Microbial **bioluminescence,** or light emission, is an interesting aspect of deep-sea life. Many bacteria are luminescent, and some have established symbiotic relationships with benthic-dwelling fish. These fish sometimes use the glow of their resident bacteria as an aid in attracting and capturing prey in the complete darkness of the ocean depths (Figure 27.13). These bioluminescent organisms have an enzyme called *luciferase* that picks up electrons

FIGURE 27.13 Bioluminescent bacteria as light organs in fish. This is a deep-sea flashlight fish (*Photoblepharon palpebratus*). The luminous organ under the eye can be covered by a tissue lid.

Q What enzyme is responsible for bioluminescence?

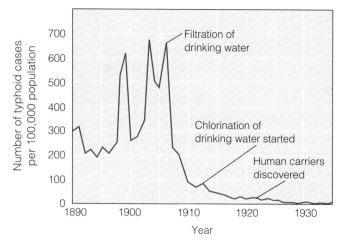

FIGURE 27.14 The incidence of typhoid fever in Philadelphia, 1890–1935. This graph clearly shows the effect of water treatments on incidence of typhoid.
SOURCE: E. Steel, *Water Supply and Sewerage*, New York: McGraw-Hill, 1953.

Q Why did the incidence of typhoid fever decrease?

from flavoproteins in the electron transport chain and then emits some of the electron's energy as a photon of light (see the box on page 826). An interesting question is why the bacteria should spend so much of their energy emitting light. A possible answer is that ultraviolet (UV) radiation reaches deep into the ocean and causes damage to DNA that can be repaired only by exposure to light—which at ocean depths the bacteria must supply themselves.

THE ROLE OF MICROORGANISMS IN WATER QUALITY

LEARNING OBJECTIVES

- Explain how wastewater pollution is a public health problem and an ecological problem.
- Discuss the causes and effects of eutrophication.
- Explain how water is tested for bacteriological quality.

Water in nature is seldom totally pure. Even rainfall is contaminated as it falls to Earth.

WATER POLLUTION

The form of water pollution that is our primary interest is microbial pollution, especially by pathogenic organisms.

The Transmission of Infectious Diseases Water that moves below the ground's surface undergoes a filtering that removes most microorganisms. For this reason, water from springs and deep wells is generally of good quality. The most dangerous form of water pollution occurs when feces enter the water supply. Many diseases are perpetuated by the fecal–oral route of transmission, in which a pathogen is shed in human or animal feces, contaminates water, and is ingested (see Chapter 25). The CDC estimates that in the

United States 900,000 people become ill each year from waterborne infections. Globally, it is estimated that waterborne diseases are responsible for over 2 million deaths each year, mostly among children under the age of 5. This is the equivalent of 20 jumbo jets crashing every day and represents about 15% of all child deaths in this age group.

Examples of such diseases are typhoid fever and cholera, caused by bacteria that are shed only in human feces. About 100 years ago, the *Journal of the American Medical Association* reported that the typhoid fever mortality rate in Chicago had declined from 159.7 per 100,000 people in 1891 to 31.4 per 100,000 in 1894. This advance in public health had been accomplished by extending the city water supply intake pipes in Lake Michigan to a distance of 4 miles from shore. The medical journal commented that this diluted the sewage contaminating the water supply, which at that time was not treated further. This same article speculated on the need to remove microorganisms that caused specific diseases. They suggested the use of sand filter beds, already widely used in Europe at the time. Sand filtration mimics the natural purification of spring water. Figure 27.14 illustrates the effect of the introduction of such filtration of water supplies on the incidence of typhoid fever in Philadelphia.

Table 27.2 summarizes recent disease outbreaks by microbial and chemical contamination of drinking water and recreational water. Infections associated with flood waters following Hurricane Katrina are discussed in the box on page 835.

Chemical Pollution Preventing chemical contamination of water is a difficult problem. Industrial and agricultural chemicals leached from the land enter water in great amounts and in forms that are resistant to biodegradation. Rural waters often have excessive amounts of nitrate from agricultural fertilizers. When ingested, the nitrate is converted to nitrite by bacteria in the gastrointestinal tract. Nitrite competes for oxygen in the blood and is especially likely to harm infants. Pesticides frequently contaminate water, and even fluorides, intended to prevent dental caries, may be accidentally added in excessive amounts. (See the box on page 826 for methods of detecting chemical pollutants.)

A striking example of industrial water pollution involved mercury in wastewater from paper manufacturing. The metallic mercury was allowed to flow into waterways as waste. It was assumed that the mercury was inert and would remain segregated in the sediments. However, bacteria in the sediments converted the mercury into a soluble chemical compound, methyl mercury, which was then taken up by fish and invertebrates in the waters. When such seafood is a substantial part of the human diet, the mercury concentrations can accumulate with devastating effects on the nervous system. At present, 33 American states have released warnings about consuming fish from mercury-contaminated waters. The FDA advises that pregnant or nursing women not eat certain fish, including swordfish and shark, that are likely to contain high levels of mercury. Bioremediation efforts using bacteria to detoxify mercury in one wildlife refuge are discussed in the box in Chapter 2 (page 33).

Another example of chemical pollution is the synthetic detergents developed immediately after World War II. These rapidly replaced many of the soaps then in use. Because these new detergents were not biodegradable, they rapidly accumulated in the waterways. In some rivers, large rafts of detergent suds could be seen traveling downstream. These detergents were replaced in 1964 by new biodegradable synthetic formulations.

Biogradable detergents, however, still present a major environmental problem because they often contain phosphates. Unfortunately, phosphates pass almost unchanged through sewage systems and can cause **eutrophication,** which is caused by an overabundance of nutrients in lakes and streams.

To understand the concept of eutrophication, recall that algae and cyanobacteria get their energy from sunlight and their carbon from carbon dioxide dissolved in water. In most waters only nitrogen and phosphorus supplies, therefore, remain inadequate for algal growth. Both of these nutrients can enter water from domestic, farm, and industrial wastes when waste treatment is absent or inefficient. These additional nutrients cause dense aquatic growths called **algal blooms.** Because many cyanobacteria can fix nitrogen from the atmosphere, these photosynthesizing organisms require only traces of phosphorus to initiate blooms. Once eutrophication results in blooms of algae or cyanobacteria, the eventual effect is the same as the addition of biodegradable organic matter. In the short run, these algae and

TABLE 27.2	**U.S. Outbreaks of Waterborne Diseases in Recent Years**

Waterborne Disease Outbreaks Associated with Drinking Water

Cause of Outbreak	Percentage
Unidentified	22.6
Legionella spp.	19.4
Parasitic (*Cryptosporidium, Giardia, Naegleria*)	16.1
Viral (Norovirus)	16.1
Chemical (copper, benzene, ethylene glycol)	16.1
Other bacteria (*Campylobacter, E. coli* O157:H7)	9.7

Waterborne Disease Outbreaks Associated with Recreational Water

Cause of Outbreak	Percentage
Cryptosporidium	36.7
Unknown	23.3
Norovirus	16.7
E. coli O157:H7	13.3
Shigella sonnei	6.7
Giardia	3.3

SOURCE: *MMWR* 53:SS-8 (10/22/2004)

cyanobacteria produce oxygen. However, they eventually die and are degraded by bacteria. During the degradation process, the oxygen in the water is used up, which may kill the fish. Undegraded remnants of organic matter settle to the bottom and hasten the filling of the lake.

Red tides of toxin-producing phytoplankton (Figure 27.15), which were mentioned in Chapter 12, are probably caused by excessive nutrients from oceanic upwellings or terrestrial wastes. In addition to eutrophication effects, this type of biological bloom can affect human health. Seafood, especially clams or similar mollusks, that ingest these plankton become toxic to humans.

Municipal waste containing detergents is likely to be the main source of phosphates in lakes and streams. As a result, phosphate-containing detergents and lawn fertilizers are banned in many places.

Coal-mining wastes, particularly in the eastern United States, are very high in sulfur content, mostly from pyrite (FeS_2). In the process of obtaining energy from the oxidation of the ferrous ion (Fe^{2+}), bacteria such as *Thiobacillus ferrooxidans* convert the (FeS_2) into

MICROBIOLOGY IN THE NEWS

BIOSENSORS: BACTERIA THAT DETECT POLLUTANTS AND PATHOGENS

Each year in the United States, industrial plants generate 265 million metric tons of hazardous waste, 80% of which make their way into landfills. Burying these chemicals does not remove them from the ecosystem, however; it just moves them to other places, where they may still find their way into bodies of water. Traditional chemical analyses to locate these chemicals are expensive and cannot distinguish between chemicals that affect biological systems from those that lie inert in the environment.

In response to this problem, scientists are developing biosensors, bacteria that can locate biologically active pollutants. Biosensors do not require costly chemicals or equipment, and they work quickly—within minutes.

In order to work, bacterial biosensors require both a receptor that is activated in the presence of pollutants and a reporter that will make such a change apparent. Biosensors use the *lux* operon from *Vibrio* or *Photobacterium* as a reporter. This operon contains inducer and structural genes for the enzyme luciferase. In the presence of a coenzyme called $FMNH_2$, luciferase reacts with the molecule in such a way that the enzyme–substrate complex emits blue-green light, which then oxidizes the $FMNH_2$ to produce FMN. Therefore, a bacterium containing the *lux* operon will emit visible light when the receptor is activated (see the photographs).

The *lux* operon is readily transferred to many bacteria. Scientists in several countries are investigating the use of *E. coli* containing the *lux* operon to detect hazardous chemicals in soil and water. The soil or water sample is placed in a tube containing the genetically modified *E. coli* bacteria. The bacteria will emit light as long as they are healthy but will stop emitting light if they have been killed by toxic pollutants.

In another application, *Lactococcus* bacteria containing the *lux* operon are being used to detect the presence of antibiotics in milk that is to be used for cheese production. (If milk contains antibiotics, the cheese starter cultures will not grow.) Because the emission of light requires a living cell, the presence of antibiotics is measured as a decrease in light output by the recombinant *Lactococcus* bacteria.

The Microtox System developed in Great Britain uses the marine bacterium *Photobacterium* directly to detect toxic pollutants. These bacteria cannot emit light if they are killed by pollutants.

Other biosensors use recombinant bacteria carrying a jellyfish gene for green fluorescent protein (gfp) and genes that are induced by pollutants or antibiotics. For example, *Pseudomonas* bacteria containing genes encoding toluene or benzene degradation and gfp will fluoresce in the presence of these pollutants.

Detecting harmful pollutants and pathogens in soil and water is necessary to protect humans and animals. However, after detection, bioremediation processes are still required to remove the pollutants.

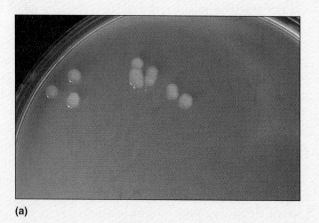

(a)

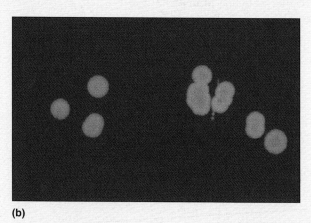

(b)

Vibrio fischeri emits light when energy is released by the transport of electrons to luciferase. Shown here are colonies of *V. fischeri* photographed (**a**) in daylight and (**b**) in the dark, illuminated by their own light.

sulfate. The sulfate enters streams as sulfuric acid, which lowers the pH of the water and damages aquatic life. The low pH also promotes the formation of insoluble iron hydroxides, which form the yellow precipitates often seen clouding such polluted waters.

WATER PURITY TESTS

Historically, most of our concern about water purity has been related to the transmission of disease. Therefore, tests have been developed to determine the safety of water; many of these tests are also applicable to foods.

FIGURE 27.15 A red tide. These blooms of aquatic growth are caused by excess nutrients in water. The color is from the pigmentation of the dinoflagellates.

Q What is the primary energy source of the dinoflagellates that cause such aquatic blooms?

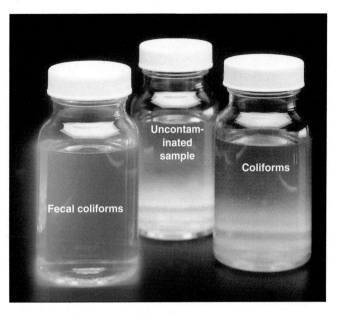

FIGURE 27.16 The ONPG and MUG coliform test. A yellow color (positive ONPG) indicates the presence of coliforms. Blue fluorescence (positive MUG) indicates the presence of the fecal coliform *E. coli*. The clear medium indicates an uncontaminated sample.

Q What causes the formation of the fluorescent compound in a positive MUG test?

It is not practical, however, to look only for pathogens in water supplies. For one thing, if we were to find the pathogens causing typhoid or cholera in the water system, the discovery would already be too late to prevent an outbreak of the disease. Moreover, such pathogens would probably be present only in small numbers and might not be included in tested samples.

The tests for water purity in use today are aimed instead at detecting particular **indicator organisms.** There are several criteria for an indicator organism. The most important criterion is that the microbe be consistently present in human feces in substantial numbers so that its detection is a good indication that human wastes are entering the water. The indicator organisms should also survive in the water at least as well as the pathogens would. The indicator organisms must be detectable by simple tests that can be carried out by people with relatively little training in microbiology.

In the United States, the usual indicator organisms are the *coliform bacteria.* **Coliforms** are defined as aerobic or facultatively anaerobic, gram-negative, non–endospore-forming, rod-shaped bacteria that ferment lactose to form gas within 48 hours of being placed in lactose broth at 35°C. Because some coliforms are not solely enteric bacteria but are more commonly found in plant and soil samples, many standards for food and water specify the identification of *fecal coliforms.* The predominant fecal coliform is *E. coli,* which constitutes a large proportion of the human intestinal population. There are specialized tests to distinguish between fecal coliforms and nonfecal coliforms. Note that coliforms are not themselves pathogenic under normal conditions, although certain strains can cause diarrhea (see Chapter 25, page 758) and opportunistic urinary tract infections (see Chapter 26, page 788).

The methods for determining the presence of coliforms in water are largely based on the lactose-fermenting ability of coliform bacteria. The multiple-tube method can be used to estimate coliform numbers by the most probable number (MPN) method (see Figure 6.18, page 181). The membrane filtration method is a more direct method of determining the presence and numbers of coliforms. This is possibly the most widely used method in North America and Europe. It makes use of a filtration apparatus similar to that shown in Figure 7.4 (page 194). In this application, though, the bacteria collected on the surface of a removable membrane filter are placed on an appropriate medium and incubated. Coliform colonies have a distinctive appearance (see Figure 6.9b and 6.9c, page 173), and are counted. This method is suitable for low turbidity waters that do not clog the filter and have relatively few noncoliform bacteria that would mask the results.

A newer and more convenient method of detecting coliforms, specifically the fecal coliform *E. coli,* makes use of media containing the two substrates *o*-nitrophenyl-β-D-galactopyranoside (ONPG) and 4-methylumbelliferyl-β-D-glucuronide (MUG). Coliforms produce the enzyme β-galactosidase, which acts on ONPG and forms a yellow color, indicating their presence in the sample. *E. coli* is unique among coliforms in almost always producing the enzyme β-glucuronidase, which acts on MUG to form a fluorescent compound that glows blue when illuminated by long-wave UV light (Figure 27.16). These simple tests, or variants of them, can detect the presence or absence of

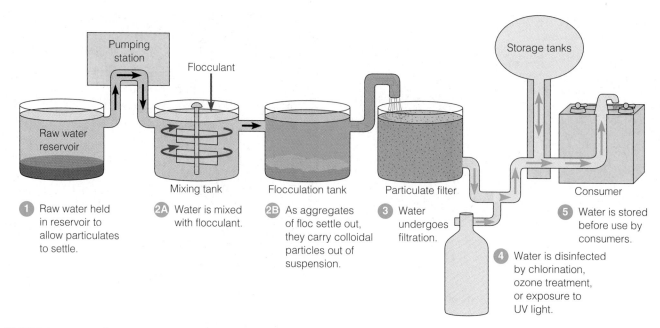

FIGURE 27.17 The steps involved in water treatment in a typical municipal water purification plant.

Q Does removal of "colloidal particles" by flocculation involve living organisms?

coliforms or *E. coli* and can be combined with the multiple-tube method to enumerate them. It can also be applied to solid media, such as in the membrane filtration method. The colonies fluoresce under UV light.

Coliforms have been very useful as indicator organisms in water sanitation, but they have limitations. One problem is the growth of coliform bacteria embedded in biofilms on the inner surfaces of water pipes. These coliforms do not, then, represent external fecal contamination of the water, and they are not considered a threat to public health. Standards governing the presence of coliforms in drinking water require that any positive water sample be reported, and occasionally these indigenous coliforms have been detected. This has led to unnecessary community orders to boil water.

A more serious problem is that some pathogens, especially viruses and protozoan cysts and oocysts, are more resistant than coliforms to chemical disinfection. Through the use of sophisticated methods of detecting viruses, it has been found that chemically disinfected water samples that are free of coliforms are often still contaminated with enteric viruses. The cysts of *Giardia lamblia* and oocysts of *Cryptosporidium* are so resistant to chlorination that completely eliminating them by this method is probably impractical; mechanical methods such as filtration are necessary. A general rule for chlorination is that viruses are more resistant to treatment than is *E. coli* and that the cysts of *Cryptosporidium* and *Giardia* are 100 times more resistant than viruses.

WATER TREATMENT

LEARNING OBJECTIVE

• Describe how pathogens are removed from drinking water.

When water is obtained from uncontaminated reservoirs fed by clear mountain streams or from deep wells, it requires minimal treatment to make it safe to drink. Many cities, however, obtain their water from badly polluted sources, such as rivers that have received municipal and industrial wastes upstream. The steps used to purify this water are shown in Figure 27.17. Water treatment is not intended to produce sterile water, but rather water that is free of disease-causing microbes.

COAGULATION AND FILTRATION

Very turbid (cloudy) water is allowed to stand in a holding reservoir for a time to allow as much particulate suspended matter as possible to settle out. The water then undergoes **flocculation,** the removal of colloidal materials such as clay, which is so small (smaller than 10 μm) that it would otherwise remain in suspension indefinitely. A flocculant chemical, such as aluminum potassium sulfate (alum), forms aggregations of fine suspended particles called *floc*. As these aggregations slowly settle out, they entrap colloidal material and carry it to the bottom. Large numbers of viruses and bacteria are also removed this way. Alum was used to clear muddy river water during the first half of the nineteenth century in the military forts of the

FIGURE 27.18 Ozone generation. Water treatment plants produce ozone by passing dry air between high-voltage electrodes in tanks called ozonators such as the two shown here.

Q **What is a major disadvantage of ozonation of water?**

American West, long before the germ theory of disease was developed.

After flocculation, the water is treated by **filtration**—that is, passing it through beds of 2 to 4 feet of fine sand or crushed anthracite coal. As mentioned previously, some protozoan cysts and oocysts are removed from water only by such filtration treatment. The microorganisms are trapped mostly by surface adsorption onto the sand particles. They do not penetrate the tortuous routing between the particles, even though the openings might be larger than the microbes that are filtered out. These filters are periodically backflushed to clear them of accumulations. Water systems of cities that have an exceptional concern for toxic chemicals supplement sand filtration with filters of activated charcoal (carbon). Charcoal removes not only particulate matter but also most dissolved organic chemical pollutants. A properly operated water treatment plant will remove viruses (which are harder to remove than bacteria and protozoa) with an efficiency of about 99.5%. Low-pressure *membrane filtration systems* are now coming into use. These systems have pore openings as small as 0.2 μm and are more reliable for removal of *Giardia* and *Cryptosporidium*.

DISINFECTION

Before entering the municipal distribution system, the filtered water is chlorinated. Because organic matter neutralizes chlorine, the plant operators must pay constant attention to maintaining effective levels of chlorine. There has been some concern that chlorine itself might be a health hazard because it could react with organic contaminants of the water to form carcinogenic compounds. At present, this possibility is considered an acceptable risk when compared with the proven usefulness of chlorinating of water.

As noted in Chapter 7 (page 205), another disinfectant for water is ozone treatment. Ozone (O_3) is a highly reactive form of oxygen that is formed by electrical spark discharges and UV light. (The fresh odor of air following an electrical storm or around a UV light bulb is from ozone.) Ozone for water treatment is generated electrically at the site of treatment (Figure 27.18). Ozone treatment is also valued because it leaves no taste or odor. Because it has little residual effect, ozone is usually used as a primary disinfectant treatment and is followed by chlorination. The use of UV light is also a supplement or alternative to chemical disinfection. Ultraviolet tube lamps are arranged so that water flows close to them. This is necessary because of the low penetrating power of UV radiation.

SEWAGE (WASTEWATER) TREATMENT

> **LEARNING OBJECTIVES**
>
> - Compare primary, secondary, and tertiary sewage treatment.
> - List some of the biochemical activities that take place in an anaerobic sludge digester.
> - Define *biochemical oxygen demand* (BOD), *activated sludge system, trickling filter, septic tank,* and *oxidation pond.*

Sewage, or wastewater, includes all the water from a household that is used for washing and toilet wastes. Rainwater flowing into street drains and some industrial wastes enter the sewage system in many cities. Sewage is mostly water and contains little particulate matter, perhaps only 0.03%. Even so, in large cities the solid portion of sewage can total more than 1000 tons of solid material per day.

Until environmental awareness intensified, a surprising number of large American cities had only a rudimentary sewage treatment system or no system at all. Raw sewage, untreated or nearly so, was simply discharged into rivers or oceans. A flowing, well-aerated stream is capable of considerable self-purification. Therefore, until expanding populations and their wastes exceeded this capability, this casual treatment of municipal wastes did not cause problems. In the United States, most cases of simple discharge have been improved. But this is not true in much of the world. Some 70% of the communities bordering the Mediterranean dump their unprocessed sewage into the sea. At one Asiatic tourist resort a hotel posted instructions that toilet paper was not to be flushed in the toilets—presumably because floating paper would make it clear that the sewage outlets were near the beach. In areas of Europe and South Africa where tourism is essential to the economy, local administrations are attempting to reassure visitors about bathing water quality with the Blue Flag campaign. The presence of

FIGURE 27.19 A beach displaying a blue flag.

Q **What sort of bacterial populations would need to be quantified to set standards for beach-area waters?**

the flag (Figure 27.19) shows that the coastal waters meet certain minimal standards of sanitation.

PRIMARY SEWAGE TREATMENT

The usual first step in sewage treatment is called **primary sewage treatment** (Figure 27.20a). In this process, large floating materials in incoming wastewater are screened out, the sewage is allowed to flow through settling chambers to remove sand and similar gritty material, skimmers remove floating oil and grease, and floating debris is shredded and ground. After this step, the sewage passes through sedimentation tanks, where more solid matter settles out. Sewage solids collecting on the bottom are called **sludge**—at this stage, *primary sludge.* About 40 to 60% of suspended solids are removed from sewage by this settling treatment, and flocculating chemicals that increase the removal of solids are sometimes added at this stage. Biological activity is not particularly important in primary treatment, although some digestion of sludge and dissolved organic matter can occur during long holding times. The sludge is removed on either a continuous or an intermittent basis, and the effluent (the liquid flowing out) then undergoes secondary treatment.

BIOCHEMICAL OXYGEN DEMAND

An important concept in sewage treatment and in the general ecology of waste management, **biochemical oxy-**

gen demand (BOD) is a measure of the biologically degradable organic matter in water. Primary treatment removes about 25 to 35% of the BOD of sewage.

BOD is determined by the amount of oxygen required by bacteria to metabolize the organic matter. The classic method of measurement is the use of special bottles with airtight stoppers. Each bottle is first filled with test water or dilutions. The water is initially aerated to provide a relatively high level of dissolved oxygen and is seeded with bacteria if necessary. The filled bottles are incubated in the dark for 5 days at 20°C, and the decrease in dissolved oxygen is determined by a chemical or electronic testing method. The more oxygen that is used up as the bacteria degrade the organic matter in the sample, the greater the BOD, which is usually expressed in milligrams of oxygen per liter of water. The amount of oxygen that normally can be dissolved in water is only about 10 mg/liter; typical BOD values of wastewater may be twenty times this amount. If this wastewater enters a lake, for example, bacteria in the lake begin to consume the organic matter responsible for the high BOD, rapidly depleting the oxygen in the lake water. (See the discussion of eutrophication earlier in the chapter, page 825.)

SECONDARY SEWAGE TREATMENT

After primary treatment, the greater part of the BOD remaining in the sewage is in the form of dissolved organic matter. **Secondary sewage treatment,** which is predominantly biological, is designed to remove most of this organic matter and reduce the BOD (see Figure 27.20b). In this process, the sewage undergoes strong aeration to encourage the growth of aerobic bacteria and other microorganisms that oxidize the dissolved organic matter to carbon dioxide and water. Two commonly used methods of secondary treatment are activated sludge systems and trickling filters.

In the aeration tanks of an **activated sludge system,** air or pure oxygen is passed through the effluent from primary treatment (Figure 27.21). The name is derived from the practice of adding some of the sludge from a previous batch to the incoming sewage. This inoculum is termed *activated sludge* because it contains large numbers of sewage-metabolizing microbes. The activity of these aerobic microorganisms oxidizes much of the sewage organic matter into carbon dioxide and water. Especially important members of this microbial community are species of *Zoogloea* bacteria, which form bacteria-containing masses in the aeration tanks called floc, or *sludge granules* (Figure 27.22). (See the discussion of floc earlier in the chapter.) Soluble organic matter in the sewage is incorporated into the floc and its microorganisms. Aeration is discontinued after 4 to 8 hours, and the contents of the tank are transferred to a settling tank, where the floc settles out, removing much of the organic matter. These solids are subsequently treated in an anaerobic sludge digester, which

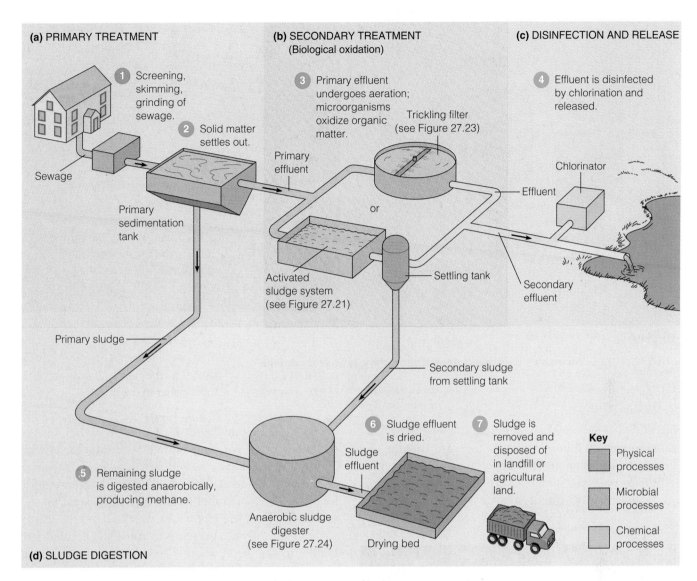

(a) PRIMARY TREATMENT

1 Screening, skimming, grinding of sewage.

2 Solid matter settles out.

Sewage

Primary sedimentation tank

Primary sludge

5 Remaining sludge is digested anaerobically, producing methane.

Anaerobic sludge digester (see Figure 27.24)

(d) SLUDGE DIGESTION

(b) SECONDARY TREATMENT
(Biological oxidation)

3 Primary effluent undergoes aeration; microorganisms oxidize organic matter.

Trickling filter (see Figure 27.23)

Primary effluent

or

Activated sludge system (see Figure 27.21)

Settling tank

Secondary sludge from settling tank

6 Sludge effluent is dried.

Sludge effluent

Drying bed

(c) DISINFECTION AND RELEASE

4 Effluent is disinfected by chlorination and released.

Chlorinator

Effluent

Secondary effluent

7 Sludge is removed and disposed of in landfill or agricultural land.

Key

Physical processes

Microbial processes

Chemical processes

FIGURE 27.20 The stages in typical sewage treatment. Microbial activity occurs aerobically in trickling filters or activated sludge aeration tanks and anaerobically in the anaerobic sludge digester. A particular system would use either activated sludge aeration tanks or trickling filters, not both, as shown in this figure. Methane produced by sludge digestion is burned off or used to power heaters or pump motors.

Q **Which processes require oxygen?**

will be described shortly. Probably more organic matter is removed by this settling-out process than by the relatively short-term aerobic oxidation by microbes. The clear effluent is disinfected and discharged.

Occasionally, the sludge will float rather than settle out; this phenomenon is called **bulking.** When this happens, the organic matter in the floc flows out with the discharge effluent, resulting in local pollution. Bulking is caused by the growth of filamentous bacteria of various types; *Sphaerotilus natans* and *Nocardia* species are frequent offenders. Activated sludge systems are quite efficient: they remove 75 to 95% of the BOD from sewage.

Trickling filters are the other commonly used method of secondary treatment. In this method, the sewage is sprayed over a bed of rocks or molded plastic (Figure 27.23a). The components of the bed must be large enough so that air penetrates to the bottom but small enough to maximize the surface area available for microbial activity. A biofilm of aerobic microbes grows on the rock or plastic surfaces (Figure 27.23b). Because air circulates throughout the rock bed, these aerobic microorganisms in the slime layer can oxidize much of the organic matter trickling over the surfaces into carbon dioxide and water. Trickling filters remove 80 to 85% of

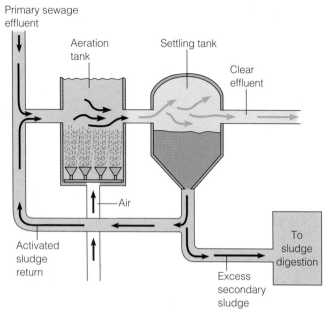

(a) Diagram of an activated sludge system

(b) An aeration tank, showing surface that is frothing from aeration

FIGURE 27.21   An activated sludge system of secondary sewage treatment.

Q   **What are the similarities between winemaking and activated sludge sewage treatment?**

the BOD, so they are generally less efficient than activated sludge systems. However, they are usually less troublesome to operate and have fewer problems from overloads or toxic sewage. Note that sludge is also a product of trickling filter systems.

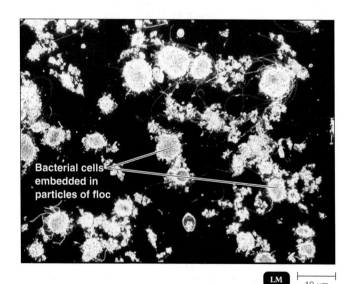

FIGURE 27.22   Floc formed by an activated sludge system. Gelatinous masses of floc are formed by a species of *Zoogloea* bacteria. If the filamentous bacteria visible in the photo predominate, the floc floats, called bulking—which is undesirable.

Q   **What happens to the suspended floc when aeration is ended in an activated sludge tank?**

Another biofilm-based design for secondary sewage treatment is the **rotating biological contactor** system. This is a series of disks several feet in diameter, mounted on a shaft. The disks rotate slowly, with their lower 40% submerged in wastewater. Rotation provides aeration and contact between the biofilm on the disks and the wastewater. The rotation also tends to cause the accumulated biofilm to slough off when it becomes too thick. This is about the equivalent of floc accumulation in activated sludge systems.

DISINFECTION AND RELEASE

Treated sewage is disinfected, usually by chlorination, before being discharged (see Figure 27.20c). The discharge is usually into an ocean or into flowing streams, although spray-irrigation fields are sometimes used to avoid phosphorus and heavy metal contamination of waterways.

Sewage can be treated to a level of purity that allows its use as drinking water. This is the practice now in some arid-area cities in the United States and will probably be expanded. In a typical system the treated sewage is filtered to remove microscopic suspended particles, then passed through a reverse osmosis purification system to remove microorganisms. Any remaining microorganisms are killed by exposure to UV light or other disinfectants. Scientists are very interested in sewage recycling during long-term stays in planned space stations. It costs thousands of dollars to send a gallon of drinking water into orbit.

(a) Rotating spray arm of a trickling filter system

(a) An anaerobic sludge digester at a California sewage-treatment plant. Much or all of a typical digester is below ground level, especially in cold climates. Methane from such a digester is often used to run pumps or heaters in the treatment plant. Excess methane is being burned off in the flame shown at the top of the digester.

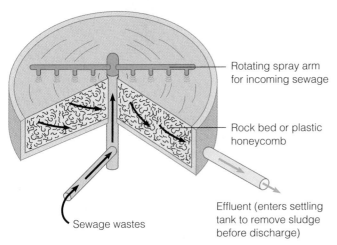

(b) A cutaway view of a trickling filter system

FIGURE 27.23   A trickling filter of secondary sewage treatment. The sewage is sprayed from the system of rotating pipes onto a bed of rocks or plastic honeycomb designed to have a maximum surface area and to allow oxygen to penetrate deeply into the bed.

Q **Which would make the most efficient bed for a trickling filter system, fine sand or golf balls?**

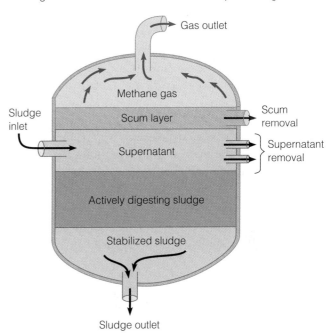

(b) Section of a sludge digester. The scum and supernatant layers are low in solids and are recirculated through secondary treatment.

FIGURE 27.24   Sludge digestion.

Q **What might be some uses for the stabilized sludge?**

SLUDGE DIGESTION

Primary sludge accumulates in primary sedimentation tanks; sludge also accumulates in activated sludge and in trickling filter secondary treatments. For further treatment, these sludges are often pumped to **anaerobic sludge digesters** (Figures 27.20d and 27.24). The process of sludge digestion is carried out in large tanks from which oxygen is almost completely excluded.

In secondary treatment, emphasis is placed on the maintenance of aerobic conditions so that organic matter is converted to carbon dioxide, water, and solids that can settle out. An anaerobic sludge digester, however, is designed to encourage the growth of anaerobic bacteria, especially methane-producing bacteria that decrease these organic solids by degrading them to soluble substances and gases,

mostly methane (60–70%) and carbon dioxide (20–30%). Methane and carbon dioxide are relatively innocuous end-products, comparable to the carbon dioxide and water from aerobic treatment. The methane is routinely used as a fuel for heating the digester and is also frequently used to run power equipment in the plant.

There are essentially three stages in the activity of an anaerobic sludge digester. The first stage is the production of carbon dioxide and organic acids from anaerobic

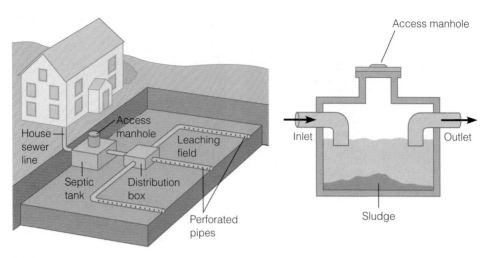

(a) Overall plan. Most soluble organic matter is disposed of by percolation into the soil.

(b) A section of a septic tank.

FIGURE 27.25 A septic tank system.

Q **Which type of soil would require the larger drainage area, clay or sandy?**

fermentation of the sludge by various anaerobic and facultatively anaerobic microorganisms. In the second stage, the organic acids are metabolized to form hydrogen and carbon dioxide, as well as such organic acids such as acetic acid. These products are the raw materials for a third stage, in which the methane-producing bacteria produce methane (CH_4). Most of the methane is derived from the energy-yielding reduction of carbon dioxide by hydrogen gas:

$$CO_2 + 4H_2 \longrightarrow CH_4 + 2H_2O$$

Other methane-producing microbes split acetic acid (CH_3COOH) to yield methane and carbon dioxide:

$$CH_3COOH \longrightarrow CH_4 + CO_2$$

After anaerobic digestion is completed, large amounts of undigested sludge still remain, although it is relatively stable and inert. To reduce its volume, this sludge is pumped to shallow drying beds or water-extracting filters. Following this step, the sludge can be used for landfill or as a soil conditioner. Sludge is assigned to two classes: class A sludge contains no detectable pathogens, class B sludge is treated only to reduce numbers of pathogens below certain levels. Most sludge is class B, and public access to application sites is limited. Sludge has about one-fifth the growth-enhancing value of normal commercial lawn fertilizers but has desirable soil-conditioning qualities, much as do humus and mulch. A potential problem is contamination with heavy metals that are toxic to plants.

SEPTIC TANKS

Homes and businesses in areas of low population density that are not connected to municipal sewage systems often use a **septic tank,** a device whose operation is similar in principle to primary treatment (Figure 27.25). Sewage enters a holding tank, and suspended solids settle out. The sludge in the tank must be pumped out periodically and disposed of. The effluent flows through a system of perforated piping into a leaching (soil drainage) field. The effluent entering the soil is decomposed by soil microorganisms. The microbial action necessary for proper functioning of a septic tank can be impaired by excessive amounts of products such as antibacterial soaps, drain cleaners, medications, "every flush" toilet bowl cleaners, and bleach.

These systems work well when not overloaded and when the drainage system is properly sized to the load and soil type. Heavy clay soils require extensive drainage systems because of the soil's poor permeability. The high porosity of sandy soils can result in chemical or bacterial pollution of nearby water supplies.

OXIDATION PONDS

Many industries and small communities use **oxidation ponds,** also called *lagoons* or *stabilization ponds,* for water treatment. These are inexpensive to build and operate but require large areas of land. Designs vary, but most incorporate two stages. The first stage is analogous to primary treatment; the sewage pond is deep enough that conditions are almost entirely anaerobic. Sludge settles out in this stage. In the second stage, which roughly corresponds to secondary treatment, effluent is pumped into an adjoining pond or system of ponds shallow enough to be aerated by wave action. Because it is difficult to maintain aerobic conditions for bacterial growth in ponds with so much organic matter, the growth of algae is encouraged to produce oxygen. Bacterial

MICROBIOLOGY IN THE NEWS

ILLNESSES AFTER HURRICANE KATRINA

After natural disasters that destroy sewage disposal and water purification systems, the risk for illness related to infectious diseases is a public health concern. When the number of infectious diseases increases after a natural disaster, they usually are caused by infectious agents normally present in the community or local environment.

By the end of 2005, 18 wound-associated *Vibrio* cases had been reported as part of the human aftermath of Hurricane Katrina; five of these patients died. Three deaths were associated with *V. vulnificus* and two with *V. parahaemolyticus*; this is an increase over the normal reported incidence of *Vibrio* wound infections in the Gulf Coast states. Four people were reported with gastroenteritis caused by nontoxigenic *V. cholerae* in Mississippi and Louisiana. One of the infections occurred in a 2-month-old boy with diarrhea, whose stool culture yielded both *Salmonella* group C2 and *V. cholerae* non-O1, non-

O139. He was hospitalized for two days in Mississippi. No deaths were associated with these nonwound cases.

Among hurricane evacuees from the New Orleans area, a cluster of infections with methicillin-resistant *Staphylococcus aureus* (MRSA) was reported in approximately 30 pediatric and adult patients at an evacuee facility in Dallas, Texas.

Over 1100 Hurricane evacuees residing in three facilities reported symptoms of acute gastroenteritis. Three-fourths of the patients with acute gastroenteritis symptoms were adults (over age 18). Noroviruses are the most common cause of outbreaks of acute gastroenteritis in the United States, and their presence was confirmed in patient stool samples by PCR. Conditions that might have facilitated virus transmission included population crowding, insufficient sanitation in lavatories, lack of an adequate number of hand-washing facilities, and delays in cleaning and decontaminating soiled

areas and bedding. Initial isolation procedures also were difficult to maintain over time because family members already traumatized by displacement and personal loss were separated from each other because of illness.

Excess moisture and standing water contribute to the growth of mold in homes and other buildings. Residents returning to a home that has been flooded need to be aware that mold may be present and may be a health risk. People who are sensitive to mold may experience stuffy nose, irritated eyes, wheezing, or skin irritation. People allergic to mold may have difficulty in breathing and shortness of breath. People with weakened immune systems and with chronic lung diseases may develop mold infections in their lungs. As building managers in areas affected by hurricanes attempt to re-start clean water systems, care should be taken to reduce the risk of legionellosis due to *Legionella* colonization in stagnant plumbing systems.

---

action in decomposing the organic matter in the wastes generates carbon dioxide. Algae, which use carbon dioxide in their photosynthetic metabolism, grow and produce oxygen, which in turn encourages the activity of aerobic microbes in the sewage. Large amounts of organic matter in the form of algae accumulate, but this is not a problem because the oxidation pond, unlike a lake, already has a large nutrient load.

Some small sewage-producing operations, such as isolated campgrounds and highway rest stop areas, use an *oxidation ditch* for sewage treatment. In this method, a small oval channel in the shape of a racetrack is filled with sewage water. A paddle wheel similar to that on an old-time Mississippi steamboat, but in a fixed location, propels the water in a self-contained flowing stream aerated enough to oxidize the wastes.

TERTIARY SEWAGE TREATMENT

As we have seen, primary and secondary treatments of sewage do not remove all the biologically degradable organic matter. Amounts of organic matter that are not excessive can be released into a flowing stream without causing a serious problem. Eventually, however, the pressures of increased

population might increase wastes beyond a body of water's carrying capacity, and additional treatments might be required. Even now, primary and secondary treatments are inadequate in certain situations, such as when the effluent is discharged into small streams or recreational lakes. Some communities have therefore developed **tertiary sewage treatment** plants. Lake Tahoe in the Sierra Nevada Mountains, surrounded by extensive development, is the site of one of the best-known tertiary sewage treatment systems.

The effluent from secondary treatment plants contains some residual BOD. It also contains about 50% of the original nitrogen and 70% of the original phosphorus, which can greatly affect a lake's ecosystem. Tertiary treatment is designed to remove essentially all the BOD, nitrogen, and phosphorus. Tertiary treatment depends less on biological treatment than on physical and chemical treatments. Phosphorus is precipitated out by combining with such chemicals as lime, alum, and ferric chloride. Filters of fine sands and activated charcoal remove small particulate matter and dissolved chemicals. Nitrogen is converted to ammonia and discharged into the air in stripping towers. Some systems encourage denitrifying bacteria to form volatile nitrogen gas. Finally, the purified water is chlorinated.

Tertiary treatment provides water that is suitable for drinking, but the process is extremely costly. Secondary treatment is less costly, but water that has undergone only secondary treatment still contains many water pollutants. Much work is being done to design secondary treatment plants in which the effluent can be used for irrigation. This design would eliminate a source of water pollution, provide nutrients for plant growth, and reduce the demand on already scarce water supplies. The soil to which this water is applied would act as a trickling filter to remove chemicals and microorganisms before the water reaches groundwater and surface water supplies.

* * *

We hope this chapter on environmental microbiology, as well as previous chapters in the book, has left you with a greater appreciation of the microbial influences around us. Without the natural and human-directed applications of microbes, life would be very different—and perhaps could not sustain itself at all.

STUDY OUTLINE

MICROBIAL DIVERSITY AND HABITATS

(pp. 810–811)

1. Microorganisms live in a wide variety of habitats because of their metabolic diversity and their ability to use a variety of carbon and energy sources and grow under different physical conditions.

2. Extremophiles live in extreme conditions of temperature, acidity, alkalinity, or salinity.

SYMBIOSIS (pp. 810–811)

3. Symbiosis is a relationship between two different organisms or populations.

4. Symbiotic fungi called mycorrhizae live in and on plant roots; they increase the surface area and nutrient absorption of the plant.

SOIL MICROBIOLOGY AND BIOGEOCHEMICAL CYCLES (pp. 811–820)

1. In biogeochemical cycles, certain chemical elements are recycled.

2. Microorganisms in the soil decompose organic matter and transform carbon-, nitrogen-, and sulfur-containing compounds into usable forms.

3. Microbes are essential to the continuation of biogeochemical cycles.

4. Elements are oxidized and reduced by microorganisms during these cycles.

THE CARBON CYCLE (pp. 811–813)

5. Carbon dioxide is incorporated, or fixed, into organic compounds by photoautotrophs and chemoautotrophs.

6. These organic compounds provide nutrients for chemoheterotrophs.

7. Chemoheterotrophs release CO_2 that is then used by photoautotrophs.

8. Carbon is removed from the cycle when it is in $CaCO_3$ and fossil fuels.

THE NITROGEN CYCLE (pp. 813–816)

9. Microorganisms decompose proteins from dead cells and release amino acids.

10. Ammonia is liberated by microbial ammonification of the amino acids.

11. The nitrogen in ammonia is oxidized to produce nitrates for energy by nitrifying bacteria.

12. Denitrifying bacteria reduce the nitrogen in nitrates to molecular nitrogen (N_2).

13. N_2 is converted into ammonia by nitrogen-fixing bacteria.

14. Nitrogen-fixing bacteria include free-living genera such as *Azotobacter*, cyanobacteria, and the symbiotic bacteria *Rhizobium* and *Frankia*.

15. Ammonium and nitrate are used by bacteria and plants to synthesize amino acids that are assembled into proteins.

THE SULFUR CYCLE (pp. 816–818)

16. Hydrogen sulfide (H_2S) is used by autotrophic bacteria; the sulfur is oxidized to form S^0 or SO_4^{2-}.

17. Winogradsky discovered that *Beggiatoa* bacteria oxidize sulfur (H_2S and S^0) for energy.

18. Plants and other microorganisms can reduce SO_4^{2-} to make certain amino acids. These amino acids are in turn used by animals.

19. H_2S is released by decay or dissimilation of these amino acids.

LIFE WITHOUT SUNSHINE (p. 818)

20. Chemoautotrophs are the primary producers in deep-sea vents and within deep rocks.

THE PHOSPHORUS CYCLE (p. 818)

21. Phosphorus (as PO_4^{3-}) is found in rocks and bird guano.

22. When solubilized by microbial acids, the PO_4^{3-} is available for plants and microorganisms.

23. Endolithic bacteria live in solid rock; these autotrophic bacteria use hydrogen as an energy source.

THE DEGRADATION OF SYNTHETIC CHEMICALS IN SOIL AND WATER (pp. 818–820)

24. Many synthetic chemicals, such as pesticides, are resistant to degradation by microbes.

25. Ecologists are trying to use bacteria to degrade PCBs.

26. The use of microorganisms to remove pollutants is called bioremediation.

27. The growth of oil-degrading bacteria can be enhanced by the addition of nitrogen and phosphorus fertilizer.

28. Municipal landfills prevent decomposition of solid wastes because they are dry and anaerobic.

29. In some landfills, methane produced by methanogens can be recovered for an energy source.

30. Composting can be used to promote biodegradation of organic matter.

AQUATIC MICROBIOLOGY AND SEWAGE TREATMENT (pp. 820–835)

BIOFILMS (pp. 820–822)

1. Microbes adhere to surfaces and accumulate as biofilms on solid surfaces in contact with water.

2. Biofilms form on teeth, contact lenses, and catheters.

3. Microbes in biofilms are more resistant to microbicides than are free swimming microbes.

AQUATIC MICROORGANISMS (pp. 822–824)

4. The study of microorganisms and their activities in natural waters is called aquatic microbiology.

5. Natural waters include lakes, ponds, streams, rivers, estuaries, and the oceans.

6. The concentration of bacteria in water is proportional to the amount of organic material in the water.

7. Most aquatic bacteria tend to grow on surfaces rather than in a free-floating state.

8. The number and location of freshwater microbiota depend on the availability of oxygen and light.

9. Photosynthetic algae are the primary producers of a lake; they are found in the limnetic zone.

10. Pseudomonads, *Cytophaga*, *Caulobacter*, and *Hyphomicrobium* are found in the limnetic zone, where oxygen is abundant.

11. Microbes in stagnant water use available oxygen and can cause odors and the death of fish.

12. The amount of dissolved oxygen is increased by wave action.

13. Purple and green sulfur bacteria are found in the profundal zone, which contains light and H_2S but no oxygen.

14. *Desulfovibrio* reduces SO_4^{2-} to H_2S in benthic mud.

15. Methane-producing bacteria are also found in the benthic zone.

16. Phytoplankton are the primary producers of the open ocean.

17. *Pelagibacter ubique* is a decomposer in ocean waters.

18. Archaea predominate below 100 m.

19. Some algae and bacteria are bioluminescent. They possess the enzyme luciferase, which can emit light.

THE ROLE OF MICROORGANISMS IN WATER QUALITY (pp. 824–828)

20. Microorganisms are filtered from water that percolates into groundwater supplies.

21. Some pathogenic microorganisms are transmitted to humans in drinking and recreational waters.

22. Resistant chemical pollutants may be concentrated in animals in an aquatic food chain.

23. Mercury is metabolized by certain bacteria into a soluble compound that is concentrated in animals.

24. Nutrients such as phosphates cause algal blooms, which can lead to eutrophication of aquatic ecosystems.

25. Eutrophication is the result of the addition of pollutants or natural nutrients.

26. *Thiobacillus ferrooxidans* produces sulfuric acid at coal-mining sites.

27. Tests for the bacteriological quality of water are based on the presence of indicator organisms, the most common of which are coliforms.

28. Coliforms are aerobic or facultatively anaerobic, gram-negative, non–endospore-forming rods that ferment lactose with the production of acid and gas within 48 hours of being placed in a medium at 35°C.

29. Fecal coliforms, predominantly *E. coli*, are used to indicate the presence of human feces.

WATER TREATMENT (pp. 828–829)

30. Drinking water is held in a holding reservoir long enough that suspended matter settles.

31. Flocculation treatment uses a chemical such as alum to coalesce and then settle colloidal material.

32. Filtration removes protozoan cysts and other microorganisms.

33. Drinking water is disinfected with chlorine to kill remaining pathogenic bacteria.

SEWAGE (WASTEWATER) TREATMENT (pp. 829–835)

34. Domestic wastewater is called sewage; it includes household water, toilet wastes, industrial wastes, and rainwater.

35. Primary sewage treatment is the removal of solid matter called sludge.

36. Biological activity is not very important in primary treatment.

37. Biochemical oxygen demand (BOD) is a measure of the biologically degradable organic matter in water.

38. Primary treatment removes about 25–35% of the BOD of sewage.

39. BOD is determined by measuring the amount of oxygen bacteria require to degrade the organic matter.

40. Secondary sewage treatment is the biological degradation of organic matter after primary treatment.

41. Activated sludge systems, trickling filters, and rotating biological contactors are methods of secondary treatment.

42. Microorganisms degrade the organic matter aerobically.

43. Secondary treatment removes up to 95% of the BOD.

44. Treated sewage is disinfected, usually by chlorination, before discharge onto land or into water.

45. Sludge is placed in an anaerobic sludge digester; bacteria degrade organic matter and produce simpler organic compounds, methane, and CO_2.

46. The methane produced in the digester is used to heat the digester and operate other equipment.

47. Excess sludge is periodically removed from the digester, dried, and disposed of (as landfill or soil conditioner) or incinerated.

48. Septic tanks can be used in rural areas to provide primary treatment of sewage.

49. Small communities can use oxidation ponds for secondary treatment.

50. These require a large area in which to build an artificial lake.

51. Tertiary sewage treatment uses physical filtration and chemical precipitation to remove all the BOD, nitrogen, and phosphorus from water.

52. Tertiary treatment provides drinkable water, whereas secondary treatment provides water usable only for irrigation.

STUDY QUESTIONS

Access more review material either online at **The Microbiology Place** (www.microbiologyplace.com) or with **The Microbiology Place CD-ROM** packaged with your new book. There you'll find activities, practice tests, quizzes, flashcards, case studies, and more to help you succeed. In addition, you'll find the following Interactive Tutorials: Nitrogen Cycle and Bioremediation.

Answers to the Study Questions can be found in Appendix G.

REVIEW

1. Give two examples of extremophiles.

2. The koala is a leaf-eating animal. What prediction can you make about its digestive system?

3. Give one possible explanation of why *Penicillium* would make penicillin, since the fungus does not get bacterial infections.

4. Diagram the carbon cycle in the presence and absence of oxygen. Name at least one microorganism that is involved at each step.

5. In the sulfur cycle, microbes degrade organic sulfur compounds, such as _Amino Acid_, to release H_2S, which can be oxidized by *Thiobacillus* to _SO_4^{2-}_. This ion can be assimilated into amino acids by _plant/bacter_ or reduced by *Desulfovibrio* to _H_2S_. H_2S is used by photoautotrophic bacteria as an electron donor to synthesize _carbohydrate_. The sulfur-containing by-product of this metabolism is _S^0_.

6. Why is the phosphorus cycle important?

7. Fill in the following table:

Process	Chemical Reactions	Microorganisms
Ammonification		
Nitrification		
Denitrification		
Nitrogen fixation		

8. The following organisms have important roles as symbionts with plants and fungi; describe the symbiotic relationship of each organism with its host: cyanobacteria, mycorrhizae, *Rhizobium*, *Frankia*.

9. Outline the treatment process for drinking water.

10. What is the purpose of a coliform count on water?

11. The following processes are used in wastewater treatment. Match the stage of treatment with the processes. Each choice can be used once, more than once, or not at all.

Processes	Treatment Stage
_b_ Leaching field	a. Primary
_a_ Removal of solids	b. Secondary
_b_ Biological degradation	c. Tertiary
_b_ Activated sludge system	
_c_ Chemical precipitation of phosphorus	
_b_ Trickling filter	
_c_ Results in drinking water	

12. Why is activated sludge a more efficient means of removing BOD than a sludge digester?

13. Why are septic tanks and oxidation ponds not feasible for large municipalities?

14. Explain the effect of dumping untreated sewage into a pond on the eutrophication of the pond. The effect of sewage that has primary treatment? The effect of sewage that has secondary treatment? Contrast your previous answers with the effect of each type of sewage on a fast-moving river.

15. Bioremediation refers to the use of living organisms to remove pollutants. Describe three examples of bioremediation.

MULTIPLE CHOICE

For questions 1–4, answer whether
    a. the process takes place under aerobic conditions.
    b. the process takes place under anaerobic conditions.
    c. the amount of oxygen doesn't make any difference.

  a **1.** Activated sludge system
  b **2.** Denitrification
  b **3.** Nitrogen fixation
  b **4.** Methane production

  5. The water used to prepare intravenous solutions in a hospital contained endotoxins. Infection control personnel performed plate counts to find the source of the bacteria. Their results:

	Bacteria/100 ml
Municipal water pipes	0
Boiler	0
Hot water line	300

All of the following conclusions about the bacteria can be drawn *except* which one?
    a. It was present as a biofilm in the pipes.
    b. It is gram-negative.
    c. It comes from fecal contamination.
    d. It comes from the city water supply.
    e. none of the above

Use the following choices to answer questions 6–8:
    a. aerobic respiration
    b. anaerobic respiration
    c. anoxygenic photoautotroph
    d. oxygenic photoautotroph

  c **6.** $CO_2 + H_2S \xrightarrow{\text{Light}} C_6H_{12}O_6 + S^0$
  b **7.** $SO_4^{2-} + 10H^+ + 10e^- \rightarrow H_2S + 4H_2O$
  b **8.** $CO_2 + 8H^+ + 8e^- \rightarrow CH_4 + 2H_2O$

  9. All of the following are effects of water pollution *except*
    a. the spread of infectious diseases.
    b. increased eutrophication.
    c. increased BOD.
    d. increased growth of algae.
    e. none of the above

  10. Coliforms are used as indicator organisms of sewage pollution because
    a. they are pathogens.
    b. they ferment lactose.
    c. they are abundant in human intestines.
    d. they grow within 48 hours.
    e. all of the above

CRITICAL THINKING

1. Here are the formulas of two detergents that have been manufactured:

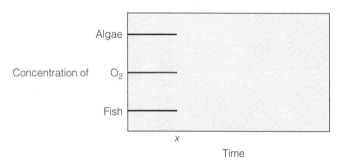

Which of these would be resistant, and which would be readily degraded by microorganisms? (*Hint:* Refer to the degradation of fatty acids in Chapter 5.)

2. Complete the following graph to show the effect of dumping phosphates into a body of water at time x.

Concentration of   Algae O_2 Fish x Time

CLINICAL APPLICATIONS

1. A patient with a heart pacemaker received antibiotic therapy for streptococcal bacteremia. One month later, he was treated for recurrence of the bacteremia. When he returned 6 weeks later, again with bacteremia, the physician recommended replacing the pacemaker. Why did this cure his condition?

2. The bioremediation process shown in the photograph is used to remove benzene and other hydrocarbons from soil contaminated by petroleum. The pipes are used to add nitrates, phosphates, oxygen, or water. Why are each of these added? Why is it not always necessary to add bacteria?

28

Applied and Industrial Microbiology

In the previous chapter on environmental microbiology, we saw that microbes are an essential factor in many natural phenomena that make life possible on Earth. In this chapter we will look at how microorganisms are harnessed in such useful applications as the making of food and industrial products. Many of these processes—especially baking, winemaking, brewing, and cheesemaking—have origins long lost in history (see the box in Chapter 1, page 3).

Modern civilization, with its large urban populations, could not be supported without methods of preserving food. In fact, civilization arose only after agriculture produced a year-round stable food supply so that people were able to give up a nomadic hunting-and-gathering way of life. It is also a fact that advances in microbiology, with its insight into spoilage processes and the possibility of disseminating diseases in preserved food, later became an essential element of this.

In Chapter 9, we discussed industrial applications of genetically modified microorganisms that are at the cutting edge of our knowledge of molecular biology. Many of these applications are now essential to modern industry.

UNDER THE MICROSCOPE

Bacteria that have been immobilized on silk fibers can promote a continuous conversion of substrate to product.

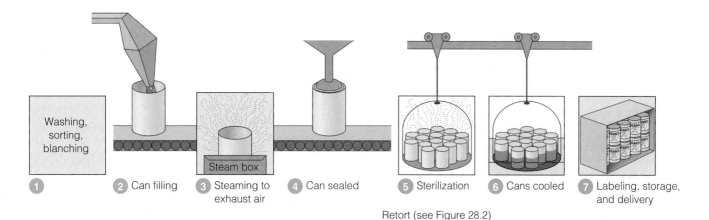

FIGURE 28.1 The commercial sterilization process in industrial canning. ❶ Blanching is a treatment with hot water or steam intended to soften the product so the can will fill better. It also destroys enzymes that might alter the color, flavor, or texture of the product and lower the microbial population. ❷ Cans are filled to capacity, leaving as little dead space as possible. ❸ To exhaust (drive out) most dissolved air, cans are heated in a steam box. ❹ The cans are sealed. ❺ Cans are sterilized by steam under pressure. ❻ Cans are cooled by submerging them or spraying them with water. ❼ Cans are labeled for sale.

Q **How does commercial sterilization differ from complete sterilization?**

FOOD MICROBIOLOGY

Many of the methods of food preservation used today were probably discovered by chance in centuries past. People in early cultures observed that dried meat and salted fish resisted decay. Nomads must have noticed that soured animal milk resisted further decomposition and was still palatable. Moreover, if the curd of the soured milk was pressed to remove moisture and allowed to ripen (in effect, cheesemaking), it was even more effectively preserved and tasted better. Farmers soon learned that if grains were kept dry, they did not become moldy.

FOODS AND DISEASE

As more food products are being prepared at central facilities and widely distributed, it is becoming more likely that food, like municipal water supplies, might be a source of widespread disease outbreaks. To minimize the potential for disease outbreaks, communities have established local agencies whose role is to inspect dairies and restaurants. The United States Food and Drug Administration (FDA) and Department of Agriculture (USDA) also maintain a system of inspectors at ports and central processing locations. A recent development in this field has been the introduction of the **Hazard Analysis and Critical Control Point (HACCP)** system, which is intended to safeguard food "from farm to fork." Before the introduction of the HACCP system, the primary role of governmental agencies was to conduct sampling to identify contaminated foods. Such sampling to identify contamination will still have its place, but the HACCP system is designed to prevent contamination by identifying points at which foods are most likely to be contaminated with harmful microbes. Monitoring of these control points can prevent the introduction of such microbes or, if they are present, arrest their proliferation. For example, the HACCP system can identify steps during processing at which meats are likely to become contaminated by the animal's intestinal contents. The HACCP system also requires monitoring of adequate temperatures to kill pathogens during processing and adequate storage temperatures to prevent their reproduction.

INDUSTRIAL FOOD CANNING

> **LEARNING OBJECTIVE**
> • Describe thermophilic anaerobic spoilage and flat sour spoilage by mesophilic bacteria.

In Chapter 7, you learned that preserving foods by heating a properly sealed container, as in home canning, is not difficult. The challenge in commercial canning is to use the right amount of heat necessary to kill spoilage organisms and dangerous microbes, such as the endospore-forming *Clostridium botulinum*, without degrading the appearance and palatability of food. Thus, much research is applied to determining the exact minimum heat treatment that will accomplish both these goals.

Industrial food canning is much more technically sophisticated than home canning (Figure 28.1). Industrially canned goods undergo what is called **commercial sterilization** by steam under pressure in a large **retort** (Figure 28.2) which operates on the same principle as an autoclave

FIGURE 28.2 Three commercial canning retorts. Note the worker at the extreme left.

Q **Is there any difference in principle between a canning retort and a hospital autoclave?**

(see Figure 7.2, page 191). Commercial sterilization is intended to destroy *C. botulinum* endospores and is not as rigorous as complete sterilization. The reasoning is that if *C. botulinum* endospores are destroyed, then any other significant spoilage or pathogenic bacteria will also be destroyed.

To ensure commercial sterilization, enough heat is applied for the **12D treatment** (12-decimal reductions, or *botulinal cook*), by which a theoretical population of *C. botulinum* endospores would be decreased by 12 logarithmic cycles. (See Figure 7.1 and Table 7.2, page 189.) What this means is that if there were 10^{12} (1,000,000,000,000) endospores in a can, after treatment there would be only one survivor. Because 10^{12} is an improbably large population, this treatment is considered quite safe. Certain thermophilic endospore-forming bacteria have endospores that are more resistant to heat treatment than those of *C. botulinum*. However, these bacteria are obligate thermophiles and generally remain dormant at temperatures lower than about 45°C (113°F). Therefore, they are not a spoilage problem at normal storage temperatures.

SPOILAGE OF CANNED FOOD

If canned foods are incubated at high temperatures, such as in a truck in the hot sun or next to a steam radiator, the thermophilic bacteria that often survive commercial sterilization can germinate and grow. **Thermophilic anaerobic spoilage** is therefore a fairly common cause of spoilage in low-acid canned foods. The can usually swells from gas, and the contents have a lowered pH and a sour odor. A number of thermophilic species of *Clostridium* can cause this type of spoilage. When thermophilic spoilage occurs but the can is not swollen by gas production, the spoilage

is termed **flat sour spoilage.** This type of spoilage is caused by thermophilic organisms such as *Geobacillus* (formerly *Bacillus*) *stearothermophilus* (ste-rō-thėr-mä′fil-us), which is found in the starch and sugars used in food preparation. Many industries have standards for the numbers of such thermophilic bacteria permitted in raw materials. Both types of spoilage occur only when the cans are stored at higher than normal temperatures, which permits the growth of bacteria whose endospores are not destroyed by normal processing.

Mesophilic bacteria can spoil canned foods if the food is underprocessed or if the can leaks. Underprocessing is more likely to result in spoilage by endospore formers; the presence of non–endospore-forming bacteria strongly suggests that the can leaks. Leaking cans are often contaminated during the cooling of cans after processing by heat. The hot cans are sprayed with cooling water or passed through a trough filled with water. As the can cools, a vacuum is formed inside, and external water can be sucked through a leak past the heat-softened sealant in the crimped lid (Figure 28.3). Contaminating bacteria in the cooling water are drawn into the can with the water. Spoilage from underprocessing or can leakage is likely to produce odors of putrefaction, at least in high-protein foods, and occurs at normal storage temperatures. In such types of spoilage, there is always the potential that botulinal bacteria will be present.

Some acidic foods, such as tomatoes or preserved fruits, are preserved by processing temperatures of 100°C or lower. The reasoning is that the only spoilage organisms that will grow in such acidic foods are easily killed by even 100°C temperatures. Primarily, these would be molds, yeasts, and certain vegetative bacteria.

Occasional problems in acidic foods develop from a few microorganisms that are both heat-resistant and acid-tolerant. Examples of heat-resistant fungi are the mold *Byssochlamys fulva* (bis-sō-klam′is fül′vä), which produces a *heat-resistant ascospore*, and a few molds, especially species of *Aspergillus*, that sometimes produce specialized resistant bodies called *sclerotia*. A spore-forming bacterium, *Bacillus coagulans* (kō-ag′ū-lanz), is unusual in that it is capable of growth at a pH of almost 4.0.

Table 28.1 lists types of spoilage in low- and medium-acid foods.

ASEPTIC PACKAGING

LEARNING OBJECTIVE

- Compare and contrast food preservation by industrial food canning, aseptic packaging, radiation, and high pressure.

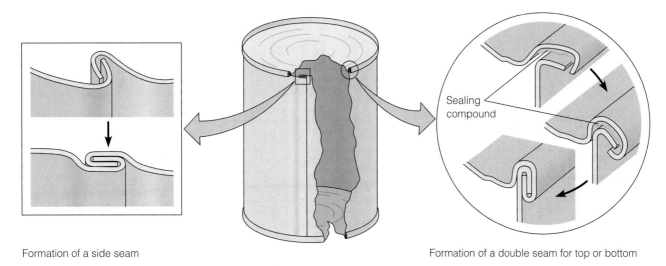

Formation of a side seam

Formation of a double seam for top or bottom

FIGURE 28.3 The construction of a metal can. Notice the seam construction, which was introduced about 1904. During cooling after sterilization (see Figure 28.1, step 6), the vacuum formed in the can may actually force contaminating organisms into the can along with water.

Q Why isn't the can sealed before it is placed in the steam box?

The use of **aseptic packaging** to preserve food has been increasing recently. Packages are usually made of some material that cannot tolerate conventional heat treatment, such as laminated paper or plastic. The packaging materials come in continuous rolls that are fed into a machine that sterilizes the material with a hot hydrogen peroxide solution, sometimes aided by ultraviolet (UV) light (Figure 28.4). Metal containers can be sterilized with superheated steam or other high-temperature methods. High-energy electron beams can also be used to sterilize the packaging materials. While still in the sterile environment, the material is formed into packages, which are then filled with liquid foods that have been conventionally sterilized by heat. The filled package is not sterilized after it is sealed.

RADIATION AND INDUSTRIAL FOOD PRESERVATION

It has long been recognized that irradiation is lethal to microorganisms; in fact, a patent was issued in Great Britain in 1905 for the use of ionizing radiation to improve the condition of foodstuffs. X rays were specifically suggested in 1921 as a way to inactivate the larvae in pork that are the cause of trichinellosis. Ionizing irradiation inhibits DNA synthesis and effectively prevents microorganisms, insects, and plants from reproducing. The ionizing irradiation is usually X rays or the gamma rays produced by radioactive cobalt-60. Up to certain energy levels, high energy electrons produced by electron accelerators are also used. The main practical difference

TABLE 28.1	Common Types of Spoilage in Low-Acid and Medium-Acid Canned Foods (pH above 4.5)		
	Indications of Spoilage		
Type of Spoilage	Appearance of Can	Contents of Can	
Flat sour (*Geobacillus stearothermophilus*)	Can not swollen	Appearance not usually altered; pH markedly lowered; sour; may have slightly abnormal odor; sometimes cloudy liquid	
Thermophilic anaerobic (*Thermoanaerobacterium thermosaccharolyticum*)	Swollen	Fermented, sour, cheesy, or butyric acid odor	
Putrefactive anaerobic (*Clostridium sporogenes*; possibly *C. botulinum*)	Swollen	May be partially digested; pH slightly above normal; typical putrid odor	

FIGURE 28.4 Aseptic packaging. Rolls of packaging material in foreground, filled packages at right center.

Q *Why is the use of this procedure increasing in recent years?*

TABLE 28.2	Approximate Doses of Radiation Needed to Kill Various Organisms (Prions Are Not Affected.)
Organisms	Dose (kGy)*
Higher animals (whole body)	0.005–0.1
Insects	0.01–1
Non–endospore-forming bacteria	0.5–10
Bacterial spores	10–50
Viruses	10–200

*Gray is a measure of ionizing irradiation; kGy is 1000 Grays.

SOURCE: J. Farkas, "Physical Methods of Food Preservation," in *Food Microbiology: Fundamentals and Frontiers*, 2d ed., M.P. Doyle et al. (eds) (Washington, DC: ASM Press, 2001).

is in penetration capabilities. These sources inactivate the target organisms and do *not* induce radioactivity in the food or packaging material. The relative doses of radiation needed to kill various organisms are presented in Table 28.2. Radiation is measured in Grays, named for an early radiologist—often in terms of thousands of Grays, abbreviated as kGy.

- *Low doses of irradiation (less than 1 kGy)* are used for killing insects (disinfestations) and inhibiting sprouting, as in stored potatoes. Similarly, it can delay ripening of fruits during storage.

- *Pasteurizing doses (1 to 10 kGy)* can be used on meats and poultry to eliminate or critically reduce the numbers of specific bacterial pathogens. At least 18 countries, including France and the Netherlands have such foods commercially available.

- *High doses (more than 10 kGy)* are used to sterilize, or to at least greatly lower the bacterial populations in many spices. Spices are often contaminated with 1 million or more bacteria per gram, although these are not considered to be normally hazardous to health.

A specialized use of irradiation has been to sterilize meats eaten by American astronauts, and a few health facilities have selectively used irradiation to sterilize foods ingested by immunocompromised patients. Table 28.3 lists foods that have been approved by the FDA to receive ionizing radiation. Irradiated food is marked in the United States with a radura symbol (Figure 28.5), and a printed notice. Unfortunately, this symbol has often been interpreted as a warning rather than the description of an

TABLE 28.3	Applications of Ionizing Radiation Accepted in the United States by the U.S. Food and Drug Administration
Product	Purpose
Wheat, wheat flour	Insect disinfestation
White potatoes	Sprout inhibition
Pork	*Trichinella spiralis* control
Enzymes (dehydrated)	Microbial control
Fruit	Insect disinfestation, ripening delay
Vegetables, fresh	Insect disinfestation
Herbs	Microbial control
Spices	Microbial control
Vegetable seasonings	Microbial control
Poultry, fresh or frozen	Microbial control
Meat, frozen, packaged*	Sterilization
Animal feed and pet food	*Salmonella* control
Meat, uncooked, chilled	Microbial control
Meat, uncooked, frozen	Microbial control

*For meats used solely in the National Aeronautics and Space Administration space flight programs.

approved processing treatment or preservative. In fact, irradiated foods are not radioactive; consider that the X-ray table in a hospital does not become radioactive from repeated daily exposure to ionizing radiation. Recently, the FDA has allowed, upon special approval, substitution of language such as "electronic pasteurization" rather than "irradiation."

When deep penetration is a requirement, the preferred method for irradiation is gamma rays produced by cobalt-60. However, this type of treatment requires several hours of exposure in isolation behind protective walls (Figure 28.6).

High-energy electron accelerators (Figure 28.7) are much faster and sterilize in a few seconds, but this treatment has low penetrating power and is suitable only for sliced meats, bacon, or similar thin products. Also, plasticware used in microbiology is usually sterilized in this way. Another recent application is to irradiate mail to kill possible bioterrorism agents that it might contain, such as anthrax endospores.

HIGH-PRESSURE FOOD PRESERVATION

A recent development in food preservation has been the use of a high-pressure processing technique. Prewrapped foods such as fruits, deli meats, and precooked chicken strips are submerged into tanks of pressurized water. The pressure can

FIGURE 28.5 Irradiation logo. This logo, the international radura symbol, indicates that a food has received irradiation treatment.

Q **Is irradiation the same as a chemical additive?**

reach 87,000 pounds per square inch (psi)—which has been compared to the equivalent of about three elephants standing on a dime. A laboratory autoclave operates at 15 psi. This process kills many bacteria, such as *Salmonella*, *Listeria*, and pathogenic strains of *E. coli*, by disrupting many cellular functions. It also kills nonpathogenic microorganisms that tend to shorten the shelf life of such products.

Because the process does not require additives, it does not require regulatory approval. It has the advantage of preserving colors and tastes of foods better than many other methods and does not provoke the concerns of irradiation.

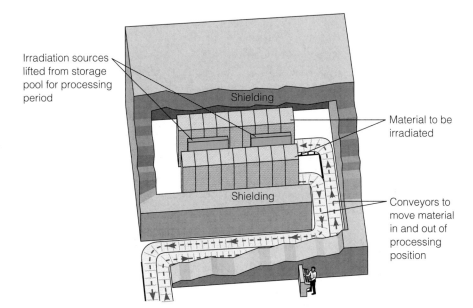

(a) An irradiation facility, showing the path of the material to be irradiated.

Irradiation sources lifted from storage pool for processing period

Shielding

Shielding

Material to be irradiated

Conveyors to move material in and out of processing position

(b) The irradiation source is in the lowered position in the storage pool. The blue glow is Cerenkov radiation caused by charged particles exceeding the speed of light in water.

FIGURE 28.6 A gamma-ray irradiation facility.

 Q **Can microwaves be used to sterilize foods?**

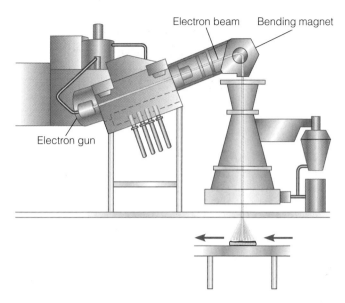

FIGURE 28.7 Electron-beam accelerator. These machines generate an electron stream that is accelerated down a long tube by electromagnets of the opposite charge. In the drawing, the electron beam is bent by a "bending magnet." This serves to filter out electrons of unwanted energy levels, providing a beam of uniform energy. The vertical beam is swept back and forth over the target as it is moved past the beam. The penetrating power of the beam is limited: if the target substance is expressed as an equivalent thickness of water, the maximum is about 3.9 cm (1.5 in.). In contrast, X rays will penetrate about 23 cm (9 in.).

Q **Are high energy electrons ionizing radiation?**

THE ROLE OF MICROORGANISMS IN FOOD PRODUCTION

LEARNING OBJECTIVE

• Name four beneficial activities of microorganisms in food production.

In the latter part of the nineteenth century, microbes used in food production were grown in pure culture for the first time. This development quickly led to an improved understanding of the relationships between specific microbes and their products and activities. This period can be considered the beginning of industrial food microbiology. For example, once it was understood that a certain yeast grown under certain conditions produced beer and that certain bacteria could spoil the beer, brewers were better able to control the quality of their products. Specific industries became active in microbiological research and selected certain microbes for their special qualities. For example, the brewing industry extensively investigated the isolation and identification of yeasts and selected those that could produce

more alcohol. In this section, we will discuss the role of microorganisms in the production of several common foods.

CHEESE

The United States leads the world in the making of cheese, producing millions of tons every year. Although there are many types of cheeses, all require the formation of a **curd,** which can be separated from the main liquid fraction, or **whey** (Figure 28.8). The curd is made up of a protein, **casein,** and is usually formed by the action of an enzyme, **rennin** (or chymosin), which is aided by acidic conditions provided by certain lactic acid–producing bacteria. These inoculated lactic acid bacteria also provide the characteristic flavors and aromas of fermented dairy products during the ripening process. The curd undergoes a microbial ripening process, except for a few unripened cheeses, such as ricotta and cottage cheese.

Cheeses are generally classified by their hardness, which is produced in the ripening process. The more moisture lost from the curd and the more the curd is compressed, the harder the cheese. Romano and Parmesan cheeses, for example, are classified as very hard cheeses; cheddar and Swiss are hard cheeses. Limburger, blue, and Roquefort cheeses are classified as semisoft; Camembert is an example of a soft cheese.

The hard cheddar and Swiss cheeses are ripened by lactic acid bacteria growing anaerobically in the interior. Such hard, interior-ripened cheeses can be quite large. The longer the incubation time, the higher the acidity and the sharper the taste of the cheese. A *Propionibacterium* (prō-pē-on-ē-bak-ti′rē-um) species produces carbon dioxide, which forms the holes in Swiss cheese. Semisoft cheeses, such as Limburger, are ripened by bacteria and other contaminating organisms growing on the surface. Blue and Roquefort cheeses are ripened by *Penicillium* molds inoculated into the cheese. The texture of the cheese is loose enough that adequate oxygen can reach the aerobic molds. The growth of the *Penicillium* molds is visible as blue-green clumps in the cheese. Camembert cheese is ripened in small packets so that the enzymes of *Penicillium* mold growing aerobically on the surface will diffuse into the cheese for ripening. The box on page 848 describes one use of the whey produced as a by-product by the dairy industry.

OTHER DAIRY PRODUCTS

Butter is made by churning cream until the fatty globules of butter separate from the liquid *buttermilk.* The typical flavor and aroma of butter and buttermilk are from *diacetyls,* a combination of two acetic acid molecules that is a metabolic end-product of fermentation by some

(a) The milk has been coagulated by the action of rennin (forming curd) and is inoculated with ripening bacteria for flavor and acidity. Here the workers are cutting the curd into slabs.

FIGURE 28.8 Making cheddar cheese.

Q Are there living bacteria in the final cheese product?

(b) The curd is chopped into small cubes to facilitate efficient draining of whey.

(c) The curd is milled to allow even more drainage of whey and is compressed into blocks for extended ripening. The longer the ripening period, the more acidic (sharper) the cheese.

lactic acid bacteria. Today, buttermilk is usually not a by-product of buttermaking but is made by inoculating skim milk with bacteria that form lactic acid and the di-acetyls. *Cultured sour cream* is made from cream inoculated with microorganisms similar to those used to make buttermilk.

A wide variety of slightly acidic dairy products— probably a heritage of a nomadic past—is found around the world. Many of them are part of the daily diet in the Balkans, eastern Europe, and Russia. One such product is *yogurt,* which is also popular in the United States. Commercial yogurt is made from milk, from which at least one-fourth of the water has been evaporated in a vacuum pan. The resulting thickened milk is inoculated with a mixed culture of *Streptococcus thermophilus,* primarily for acid production, and *Lactobacillus delbrueckii bulgaricus* (bul-gā′ri-kus), to contribute flavor and aroma. The temperature of the fermentation is about 45°C for several hours, during which time *S. thermophilus* outgrows *L. d. bulgaricus.* Maintaining the proper balance between the flavor-producing and the acid-producing microbes is the secret of making yogurt.

Kefir and *kumiss* are fermented milk beverages that are popular in eastern Europe. The usual lactic acid–producing bacteria are supplemented with a lactose-fermenting yeast to give these drinks an alcohol content of 1 to 2%.

NONDAIRY FERMENTATIONS

Historically, milk fermentation allowed dairy products to be stored and then consumed much later. Other microbial fermentations were used to make certain plants edible. For example, pre-Columbian people in Central and South America learned to ferment chocolate seeds before consumption. It is the microbial products released during fermentation that produce the chocolate flavor.

Microorganisms are also used in baking, especially for *bread.* The sugars in bread dough are fermented by yeasts. The species of yeast used in baking is *Saccharomyces cerevisiae.* This same species of yeast is also used in the brewing of beer from grains and the fermentation of wines from grapes. (At one time *S. cerevisiae* was classified as multiple species, such as *S. carlsbergensis, S. uvarum,* and *S. ellipsoideus;* these and a few other species names are often encountered in older literature). *S. cerevisiae* will grow readily under both aerobic and anaerobic conditions, although, unlike facultatively anaerobic bacteria such as *E. coli,* it cannot grow anaerobically indefinitely. Various strains of *S. cerevisiae* have been developed over the centuries and are highly adapted to certain fermentation uses.

Anaerobic conditions for the production of ethanol by the yeasts are mandatory for producing alcoholic beverages. In baking, carbon dioxide forms the typical bubbles of leavened bread. Aerobic conditions favor carbon dioxide

APPLICATIONS OF MICROBIOLOGY

FROM PLANT DISEASE TO SHAMPOO AND SALAD DRESSING

Xanthomonas campestris is a gram-negative rod that causes a disease called black rot in plants. After gaining access to a plant's vascular tissues, the bacteria use the glucose transported in those tissues to produce a sticky, gumlike substance. This substance builds up to form gumlike masses, which eventually block the plant's transport of nutrients. The gum that makes up these masses, xanthan, is composed of a high-molecular-weight polymer of mannose (see the photograph).

In contrast to its effects in plants, xanthan has no adverse effects when ingested by humans. Consequently, xanthan can be used as a thickener in foods such as dairy products and salad dressings, and in cosmetics such as cold creams and shampoos.

So when researchers at the U.S. Department of Agriculture (USDA) wanted to find some useful product that could be made out of whey, a liquid waste produced in abundance by the dairy industry, they thought of turning it into xanthan. However, because whey is mostly water and lactose, researchers had to figure out how to get X. campestris to produce xanthan using lactose rather than glucose.

One research team decided to engineer X. campestris to hydrolyze lactose more efficiently. They used an F$^+$ strain of E. coli that contains a plasmid with a lac

operon. Organisms of this strain were incubated with X. campestris, and, as proven by later restriction enzyme digests, the plasmids containing the lac gene were transferred by conjugation. However, deletion mutations subsequently occurred in the X. campestris DNA that reduced their β-galactosidase (lactose-utilizing) ability.

Another research team working with the USDA at Stauffer Chemical Company used a simpler approach—an enrichment based on satisfying only two requirements: that the bacteria grow on whey and make xanthan. First, they inoculated a whey medium with X. campestris and incubated it for 24 hours. Then they transferred an inoculum of this culture to a flask of lactose broth, to select a lactose-utilizing cell. The strain did not have to make xanthan from this broth; it only had to grow and use lactose.

A lactose-utilizing strain was isolated through serial transfers, selecting for the strain with the best ability to grow. After incubation for 10 days, an inoculum was transferred to another flask of lactose broth, and the procedure was repeated two more

times. When transferred to a flask of whey medium, the final lactose-utilizing bacteria grew in the whey, and the culture became extremely viscous—xanthan was being produced.

The process then had to be fine-tuned. The researchers addressed and solved problems such as how to sterilize the whey without destroying necessary ingredients and how to handle the extremely viscous fermentations that resulted from the procedure. The final result was a process in which 40 g/L of whey powder is converted into 30 g/L of xanthan gum. A quick survey of labels in your neighborhood supermarket will demonstrate just how successful this project was.

Xanthomonas campestris producing gooey xanthan.

---

production and are encouraged as much as possible. This is the reason the bread dough is kneaded repeatedly. Whatever ethanol is produced evaporates during baking. In some breads, such as rye or sourdough, the growth of lactic acid bacteria produce the typical tart flavor (see the box in Chapter 1, page 3).

Fermentation is also used in the production of such foods as *sauerkraut, pickles,* and *olives.*

Table 28.4 lists many fermented foods.

ALCOHOLIC BEVERAGES AND VINEGAR

Microorganisms are involved in the production of almost all alcoholic beverages. Beer and ale are products of grain starches fermented by yeast (Table 28.5). **Beer** is

fermented slowly with yeast strains that remain on the bottom (*bottom yeasts*). **Ale** is fermented relatively rapidly, at a higher temperature, with yeast strains that usually form clumps that are buoyed to the top by CO_2 (*top yeasts*). Because yeasts are unable to use starch directly, the starch from grain must be converted to glucose and maltose, which the yeasts can ferment into ethanol and carbon dioxide. In this conversion, called **malting,** starch-containing grains, such as malting barley, are allowed to sprout and then are dried and ground. This product, called **malt,** contains starch-degrading enzymes (amylases) that convert cereal starches into carbohydrates that can be fermented by yeasts. Light beers use amylases or selected strains of yeast to convert more of

TABLE 28.4	**Fermented Foods and Related Products**		
Foods and Products	Raw Ingredients	Fermenting Microorganism(s)	Location Produced
Dairy Products			
Cheeses (ripened)	Milk curd	*Streptococcus* spp., *Leuconostoc* spp.	Worldwide
Kefir	Milk	*Lactococcus lactis, Lactobacillus delbruekii bulgaricus, Candida* spp.	Primarily southwestern Asia
Kumiss	Mare's milk	*Lactobacillus d. bulgaricus, L. leichmannii, Candida* spp.	Russia
Yogurt	Milk, milk solids	*Streptococcus thermophilus, L. d. bulgaricus*	Worldwide
Meat and Fish Products			
Country-cured hams	Pork hams	*Aspergillus, Penicillium* spp.	Southern United States
Dry sausages	Pork, beef	*Pediococcus acidilactici*	Europe, United States
Fish sauces	Small fish	Halophilic *Bacillus* spp.	Southeast Asia
Nonbeverage Plant Products			
Cocoa beans (chocolate)	Cacao fruits (pods)	*Candida krusei, Geotrichum* spp.	Africa, South America
Coffee beans	Coffee cherries	*Erwinia dissolvens, Saccharomyces* spp.	Brazil, Congo, Hawaii, India
Kimchi (kim-chee)	Cabbage and other vegetables	Lactic acid bacteria	Korea
Miso	Soybeans	*Aspergillus oryzae, Zygosaccharomyces rouxii*	Primarily Japan
Olives	Green olives	*Leuconostoc mesenteroides, Lactobacillus plantarum*	Worldwide
Poi	Taro roots	Lactic acid bacteria	Hawaii
Sauerkraut	Cabbage	*Leuconostoc mesenteroides, Lactobacillus plantarum*	Worldwide
Soy sauce	Soybeans	*A. oryzae* or *A. sojae, Z. rouxii, Lactobacillus d. bulgaricus*	Japan, China, United States
Breads			
Rolls, cakes, breads, and so on	Wheat flours	*Saccharomyces cerevisiae*	Worldwide
San Francisco sourdough bread	Wheat flour	*Saccharomyces exiguus, Lactobacillus sanfranciscensis*	Northern California

the starch to fermentable glucose and maltose, resulting in fewer carbohydrates and more alcohol. The beer is then diluted to arrive at an alcohol percentage in the usual range. **Sake,** the Japanese rice wine, is made from rice without malting because the mold *Aspergillus* is first used to convert the rice's starch to sugars that can be fermented. (See the discussion of koji, page 855.) For

distilled spirits, such as *whiskey, vodka,* and *rum,* carbohydrates from cereal grains, potatoes, and molasses are fermented to alcohol. The alcohol is then distilled to make a concentrated alcoholic beverage.

Wines are made from fruits, typically grapes, that contain sugars that can be used directly by yeasts for fermentation; malting is unnecessary in winemaking. Grapes usually

TABLE 28.5	The Production of Alcoholic Beverages by Yeasts		
Beverage	Yeast	Method of Preparation	Function of Yeast
Beer and Wine			
Beer, lager	*Saccharomyces cerevisae* (bottom yeast)	Germinated barley releases starches and amylase enzymes (malting). Enzymes in malt hydrolyze starch to fermentable sugars (mashing). Liquid (wort) sterilized. Hops added for flavor. Yeast added, incubated at 3–10°C.	Converts sugar into alcohol and carbon dioxide; >6% alcohol. Yeast grows on bottom of fermenting vessel.
Beer, ale	*S. cerevisiae* (top yeast)	As in lager; incubated at 10–21°C.	Converts sugar into alcohol; and CO_2; <4% alcohol. Yeast grows at top of fermentation vessel.
Sake	*S. cerevisiae*	*Aspergillus oryzae* converts starch in steamed rice into sugar; yeast added; incubated at 20°C.	Converts sugar into alcohol; 14–16% alcohol.
Wine, natural	*S. cerevisiae*	Strain of grape provides various flavors and sugar concentrations. Grapes crushed into must; sulfur dioxide added to inhibit wild yeast; yeast added. Red wines: incubated at 25°C. Aged in oak 3–5 years and in bottle 5–15 years. White wines: incubated at 10–15°C. Aged 2–3 years in bottle.	Converts grape sugar into alcohol; 14% or less alcohol.
Wine, sparkling	*S. cerevisiae*	As natural wine, with secondary fermentation in bottle. 2.5% sugar and yeast added to bottled wine; incubated at 15°C; bottle inverted to collect yeast in neck.	In secondary fermentation, produces carbon dioxide; yeast settles quickly.
Distilled Beverages			
Rum	*Wild yeast*	Cane molasses inoculated from previous fermentation. Oak aging adds color. Distilled to concentrate.	Converts sugar to alcohol; 50–95% alcohol.
Brandy	*S. cerevisiae*	Fruits pressed; yeast added. Distilled to concentrate alcohol, blended with other brandies.	Converts sugar into alcohol; 40–43% alcohol.
Whiskey	*S. cerevisiae*	Wort (see beer) is fermented by yeast. Distilled to concentrate alcohol; aged in charred oak barrels.	Converts sugar to alcohol; 50–95% alcohol.

need no additional sugars, but other fruits might be supplemented with sugars to ensure enough alcohol production. The steps of winemaking are shown in Figure 28.9. Lactic acid bacteria are important when wine is made from grapes that are especially acidic from high concentrations of malic acid. These bacteria convert the malic acid to the weaker lactic acid in a process called **malolactic fermentation.** The result is a less acidic, better-tasting wine than would otherwise be produced.

Wine producers who allowed wine to be exposed to air found that it soured from the growth of aerobic bacteria that converted the ethanol in the wine to acetic acid. The result was *vinegar* (*vin* = wine; *aigre* = sour). The process is now used deliberately to make vinegar. Ethanol is first produced by anaerobic fermentation of carbohydrates by yeasts. The ethanol is then aerobically oxidized to acetic acid by acetic acid–producing bacteria of the genera *Acetobacter* and *Gluconobacter*.

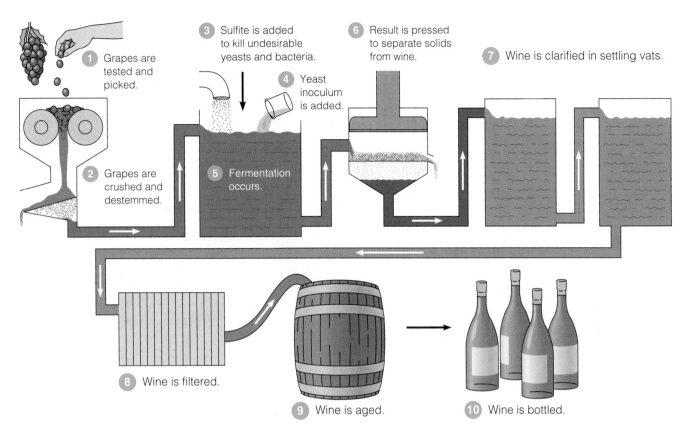

1 Grapes are tested and picked.

2 Grapes are crushed and destemmed.

3 Sulfite is added to kill undesirable yeasts and bacteria.

4 Yeast inoculum is added.

5 Fermentation occurs.

6 Result is pressed to separate solids from wine.

7 Wine is clarified in settling vats.

8 Wine is filtered.

9 Wine is aged.

10 Wine is bottled.

FIGURE 28.9 The basic steps in making red wine. For white wines, the pressing precedes fermentation so that the color is not extracted from the solid matter.

Q **What is the purpose of adding yeast in step 4?**

INDUSTRIAL MICROBIOLOGY

The industrial uses of microbiology had their beginnings in large-scale food fermentations that produced lactic acid from dairy products and ethanol from brewing. These two chemicals also proved to have many industrial uses unrelated to foods. During World Wars I and II, microbial fermentation and similar technologies were used in the production of armament-related chemical compounds such as glycerol and acetone. Present industrial microbiology largely dates from the technology developed to produce antibiotics following World War II. There is now renewed interest in some of these classic microbial fermentations, especially if they can be used as feedstocks, products that are renewable, or, ideally, products that would otherwise be wasted.

In recent years, industrial microbiology has been revolutionized by the application of genetically modified organisms. An example of a genetically engineered *biosensor* to detect pollution is explained in the box on page 826. In Chapter 9, we discussed the methods for making these modified organisms using recombinant DNA technology and described some of the products derived from them; this technology is now known as **biotechnology.**

FERMENTATION TECHNOLOGY

LEARNING OBJECTIVES
- Define *industrial fermentation* and *bioreactor.*
- Differentiate primary from secondary metabolites.

The industrial production of microbial products usually involves fermentation. *Industrial fermentation* is the large-scale cultivation of microbes or other single cells to produce a commercially valuable substance. (See the box in Chapter 5, page 137, for other definitions of fermentation). We have just discussed the most familiar examples: the anaerobic food fermentations used in the dairy, brewing, and winemaking industries. Much of the same technology, with the frequent addition of aeration, has been adapted to make other industrial products, such as insulin and human growth hormone, from genetically modified microorganisms. Industrial fermentation is also used in biotechnology to obtain useful products from genetically modified plant and animal cells (see Chapter 9). For example, animal cells are used to make monoclonal antibodies (see Chapter 18, page 534).

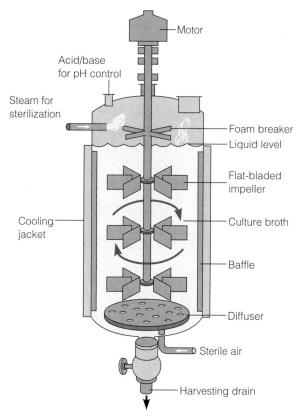

(a) Section of a continuously stirred bioreactor

(b) Bioreactor tank, at left

FIGURE 28.10   Bioreactors for industrial fermentations.

Q **Identify one essential difference between the bioreactor illustrated and a vat for brewing beer.**

Vessels for industrial fermentation are called **bioreactors;** they are designed with close attention to aeration, pH control, and temperature control. There are many different designs, but the most widely used bioreactors are of the continuously stirred type (Figure 28.10). The air is introduced through a diffuser at the bottom (which breaks up the incoming airstream to maximize aeration), and a series of impeller paddles and stationary wall baffles keep the microbial suspension agitated. Oxygen is not very soluble in water, and keeping the heavy microbial suspension well aerated is difficult. Highly sophisticated designs have been developed to achieve maximum efficiency in aeration and other growth requirements, including medium formulation. The high value of the products of genetically modified microorganisms and eukaryotic cells has stimulated the development of newer types of bioreactors and computerized controls for them.

Bioreactors are sometimes very large, holding as much as 500,000 liters. When the product is harvested at the completion of the fermentation, this is known as *batch production*. There are other designs of fermentors. For *continuous flow production*, in which the substrates (usually a carbon source) are fed continuously past immobilized enzymes or into a culture of growing cells, spent medium and desired product are continuously removed.

Generally speaking, the microbes in industrial fermentation produce either primary metabolites, such as ethanol, or secondary metabolites, such as penicillin. A **primary metabolite** is formed essentially at the same time as the new cells, and the production curve follows the cell population curve almost in parallel, with only minimal lag (Figure 28.11a). **Secondary metabolites** are not produced until the microbe has largely completed its logarithmic growth phase, known as the **trophophase,** and has entered the stationary phase of the growth cycle (Figure 28.11b). The following period, during which most of the secondary metabolite is produced, is known as the **idiophase.** The secondary metabolite may be a microbial conversion of a primary metabolite. Alternatively, it may be a metabolic product of the original growth medium that the microbe makes only after considerable numbers of cells and a primary metabolite have accumulated.

Strain improvement is also an ongoing activity in industrial microbiology. (A microbial **strain** differs physiologically in some significant way. For example, it has an enzyme to carry out some additional activity or lacks such an ability, but this is not enough difference to change its species identity). A well-known example is that of the mold used for penicillin production. The original culture of *Penicillium* did not produce penicillin in large enough quantities for commercial use. A more efficient culture was isolated from a moldy cantaloupe from a Peoria, Illinois, supermarket. This strain was treated variously with UV

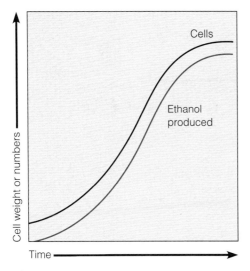

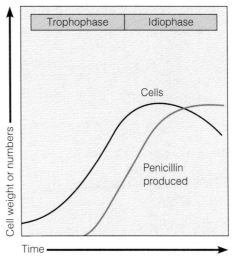

(a) A primary metabolite, such as ethanol from yeast, has a production curve that lags only slightly behind the line showing cell growth.

(b) A secondary metabolite, such as penicillin from mold, begins to be produced only after the logarithmic growth phase of the cell (trophophase) is completed. The main production of the secondary metabolite occurs during the stationary phase of cell growth (idiophase).

FIGURE 28.11 Primary and secondary fermentation.

 What is the origin of a secondary metabolite?

light, X rays, and nitrogen mustard (a chemical mutagen). Selections of mutants, including some that arose spontaneously, quickly increased the production rates by a factor of more than 100. Today, the original penicillin-producing molds produce, not the original 5 mg/L, but 60,000 mg/L. Improvements in fermentation techniques have nearly tripled even this yield. An example of a strain that was developed by enrichment and selection is described in the box on page 848.

IMMOBILIZED ENZYMES AND MICROORGANISMS

In many ways, microbes are packages of enzymes. Industries are increasing their use of free enzymes isolated from microbes to manufacture many products such as high-fructose syrups, paper, and textiles. The demand for such enzymes is high because they are specific and do not produce costly or toxic waste products. And, unlike traditional chemical processes that require heat or acids, enzymes work under moderate conditions and are safe and biodegradable. For most industrial purposes, the enzyme must be immobilized on the surface of some solid support or otherwise manipulated so that it can convert a continuous flow of substrate to product without being lost.

Continuous flow techniques have also been adapted to live whole cells, and sometimes even to dead cells (Figure 28.12). Whole-cell systems are difficult to aerate, and they lack the single-enzyme specificity of immobilized

enzymes. However, whole cells are advantageous if the process requires a series of steps that can be carried out by one microbe's enzymes. They also have the advantage of allowing continuous flow processes with large cell

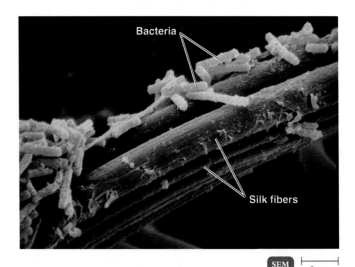

FIGURE 28.12 Immobilized cells. In some industrial processes, the cells are immobilized on surfaces such as the silk fibers shown here. The substrate flows past the immobilized cells.

How does this process resemble the action of a trickling filter in sewage treatment?

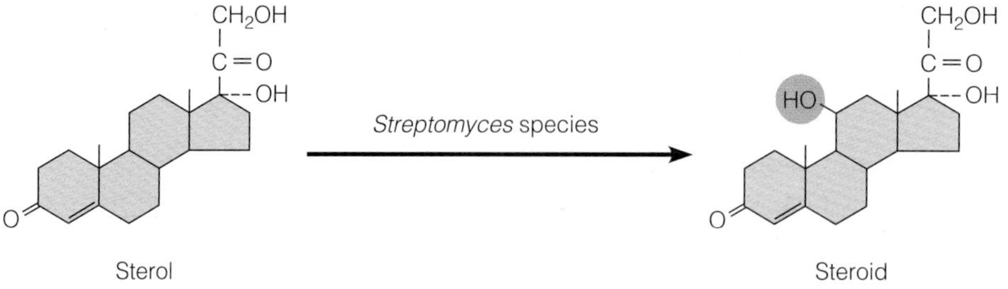

FIGURE 28.13 The production of steroids. Shown here is the conversion of a precursor compound such as a sterol into a steroid by *Streptomyces*. The addition of a hydroxyl group to carbon number 11 (highlighted in purple on the steroid) might require more than 30 steps by chemical means, but the microorganism can add it in only one step.

Q **Name a commercial product that is a steroid.**

populations operating at high reaction rates. Immobilized cells, which are usually anchored to microscopically small spheres or fibers, are currently used to make high-fructose syrup, aspartic acid, and numerous other products of biotechnology.

INDUSTRIAL PRODUCTS

> **LEARNING OBJECTIVE**
> • Describe the role of microorganisms in the production of industrial chemicals and pharmaceuticals.

As mentioned earlier, cheesemaking produces an organic waste called whey. The whey must be disposed of as sewage or dried and burned as solid waste. Both of these processes are costly and ecologically problematic. However, microbiologists have discovered an alternative use for whey, as discussed in the box on page 848. In this way, microbiologists are devising uses for old products and creating new ones. In this section, we will discuss some of the more important commercial microbial products and the growing alternative energy industry.

AMINO ACIDS

Amino acids have become a major industrial product from microorganisms. For example, over 600,000 tons of *glutamic acid* (L-glutamate), used to make the flavor enhancer monosodium glutamate, are produced every year. Certain amino acids, such as *lysine* and *methionine*, cannot be synthesized by animals and are present only at low levels in the normal diet. Therefore, the commercial synthesis of lysine and some of the other essential amino acids as cereal food supplements is an important industry. More than 70,000 tons each of lysine and methionine are produced every year.

Two microbially synthesized amino acids, *phenylalanine* and *aspartic acid* (L-aspartate), have become important

as ingredients in the sugar-free sweetener aspartame (NutraSweet). Some 3000 to 4000 tons of each of these amino acids are produced annually in the United States.

In nature, microbes rarely produce amino acids in excess of their own needs because feedback inhibition prevents wasteful production of primary metabolites (see Chapter 5, page 122). Commercial microbial production of amino acids depends on specially selected mutants and sometimes on ingenious manipulations of metabolic pathways. For example, in applications in which only the L-isomer of an amino acid is desired, microbial production, which forms only the L-isomer, has an advantage over chemical production, which forms both the **D-isomer** and the **L-isomer** (see Figure 2.13, page 45).

CITRIC ACID

Citric acid is a constituent of citrus fruits, such as oranges and lemons, and at one time these were its only industrial source. However, over 100 years ago, citric acid was identified as a product of mold metabolism. This discovery was first used as an industrial process when World War I interfered with the picking of the Italian lemon crop. Citric acid has an extraordinary range of uses beyond the obvious ones of giving tartness and flavor to foods. It is an antioxidant and pH adjuster in many foods, and in dairy products it often serves as an emulsifier. Well over 550,000 tons of citric acid are produced every year in the United States. Much of it is produced by a mold, *Aspergillus niger* (nī′jèr), using molasses as a substrate.

ENZYMES

Enzymes are widely used in different industries. For example, *amylases* are used in the production of syrups from corn starch, in the production of paper sizing (a coating for smoothness, as on this page), and in the production of glucose from starch. The microbiological production

TABLE 28.6	Microbial Enzymes Produced Commercially	
Enzyme	Microorganism	Use of Enzyme
α-Amylase	*Aspergillus* spp. (mold)	Laundry detergent
β-Amylase	*Bacillus subtilis*	Brewing
Cellulase	*Trichoderma viride* (yeast)	Fruit juices, coffee, paper
Invertase	*Saccharomyces cerevisiae* (yeast)	Candy manufacture
Lactase	*S. fragilis* (yeast)	Digestive aid, candy manufacture
Lipase	*Aspergillus niger*	Laundry detergent, tanning leather, cheese production
Oxidases	*A. niger*	Paper bleaching and fabrics, glucose test papers
Pectinase	*A. niger*	Fruit juice
Proteases	*A. oryzae*	Meat tenderizer, digestive aid, tanning leather
Rennin (chymosin)	*Mucor* (mold), *Escherichia coli*	Cheese production
Streptokinase	Group C beta-hemolytic *Streptococcus*	Lysis of blood clots

of amylase is considered to be the first biotechnology patent issued in the United States, which was to the Japanese scientist Jokichi Takamine. The basic process by which molds were used to make an enzyme preparation known as **koji** had been used for centuries in Japan to make fermented soy products. Koji is an abbreviation of a Japanese word meaning bloom of mold, reflecting the infiltration of a cereal substrate, either rice or a wheat-soybean mixture, with a filamentous fungus (*Aspergillus*). Primarily, the amylases in koji change starch into sugars, but koji preparations also contain proteolytic enzymes that convert the protein in soybeans into a more digestible and flavorful form. It is the basis of soybean fermentations that are staples of the Japanese diet, such as *soy sauce* and *miso* (a fermented paste of soybeans with a meaty flavor). *Sake*, the well-known Japanese rice wine, makes use of amylases of koji to change the carbohydrates of rice into a form that yeasts can use to produce alcohol. This is roughly the equivalent of the barley malt (page 848) used in beer brewing.

Glucose isomerase is an important enzyme; it converts the glucose that amylases form from starches into fructose, which is used in place of sucrose as a sweetener in many foods. Probably half of the bread baked in this country is made with the aid of *proteases*, which adjust the amount of glutens (protein) in wheat so that baked goods are improved or made uniform. Other proteolytic enzymes are used as meat tenderizers or in detergents as an additive to remove proteinaceous stains. About a third of all industrial enzyme production is for this purpose. *Rennin*, an enzyme used to form curds in milk, is usually produced commer-

cially by fungi but more recently by genetically modified bacteria. Several enzymes produced commercially by microorganisms are listed in Table 28.6.

VITAMINS

Vitamins are sold in large quantities combined in tablet form and are used as individual food supplements. Microbes can provide an inexpensive source of some vitamins. *Vitamin B$_{12}$* is produced by *Pseudomonas* and *Propionibacterium* species. *Riboflavin* is another vitamin produced by fermentation, mostly by fungi such as *Ashbya gossypii* (ash′bē-ä gos-sip′ē-ē). *Vitamin C* (ascorbic acid) is produced by a complicated modification of glucose by *Acetobacter* species.

PHARMACEUTICALS

Modern pharmaceutical microbiology developed after World War II, with the introduction of the production of antibiotics.

All antibiotics were originally the products of microbial metabolism. Many are still produced by microbial fermentations, and work continues on the selection of more productive mutants by nutritional and genetic manipulations. At least 6000 antibiotics have been described. One organism, *Streptomyces hygroscopius*, has different strains that make almost 200 different antibiotics. Antibiotics are typically made industrially by inoculating a solution of growth medium with spores of the appropriate mold or streptomycete and vigorously aerating it. After the antibiotic reaches a satisfactory concentration, it is extracted by solution, precipitation, and other standard industrial procedures.

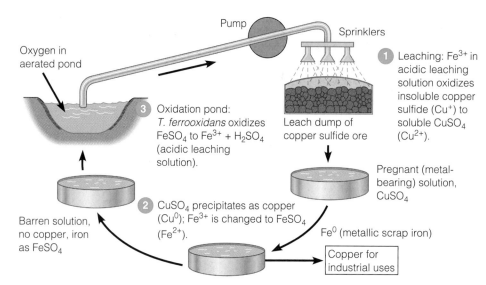

(b) Leaching solution being sprayed at top of ore dump

(a) Simplified copper ore leaching process

FIGURE 28.14 Biological leaching of copper ores. The chemistry of the process is much more complicated than shown here. Essentially, *Thiobacillus ferrooxidans* bacteria are used in a biological/chemical process that changes insoluble copper in the ore into soluble copper that leaches out and is precipitated as metallic copper. The solutions are continuously recirculated.

Q **Name another metal that is recovered by a similar process.**

Vaccines are a product of industrial microbiology. Many antiviral vaccines are mass-produced in chicken eggs or cell cultures. The production of vaccines against bacterial diseases usually requires the growth of large amounts of the bacteria. Recombinant DNA technology is increasingly important in the development and production of subunit vaccines (see Chapter 18, page 530).

Steroids are a very important group of chemicals that include *cortisone*, which is used as an anti-inflammatory drug, and *estrogens* and *progesterone*, which are used in oral contraceptives. Recovering steroids from animal sources or chemically synthesizing them is difficult, but microorganisms can synthesize steroids from sterols or from related, easily obtained compounds. For example, Figure 28.13 illustrates the conversion of a sterol into a valuable steroid.

COPPER EXTRACTION BY LEACHING

Thiobacillus ferrooxidans is used in the recovery of otherwise unprofitable grades of copper ore, which sometimes contain as little as 0.1% copper. At least 25% of the world's copper is produced this way. *Thiobacillus* bacteria get their energy from the oxidation of a reduced form of iron (Fe^{2+}) in ferrous sulfide to an oxidized form (Fe^{3+}) in ferric sulfate. Sulfuric acid (H_2SO_4) is also a product of the reaction. The acidic solution of Fe^{3+}-containing water is applied by sprinklers and allowed to percolate

downslope through the ore body (Figure 28.14). The ferrous iron, Fe^{2+}, and *T. ferrooxidans* are normally present in the ore and continue to contribute to the reactions. The Fe^{3+} in the sprinkling water reacts with insoluble copper (Cu^+) in *copper sulfides* in the ore to form soluble copper (Cu^{2+}), which takes the form of *copper sulfates*. In order to maintain a low enough pH, more sulfuric acid can be added. The soluble copper sulfate moves downslope to collection tanks where it contacts metallic scrap iron. The copper sulfates react chemically with the iron and precipitate out as metallic copper (Cu^0). In this reaction the metallic iron (Fe^0) is converted into ferrous iron (Fe^{2+}) that is recycled to an aerated oxidation pond where *Thiobacillus* bacteria use it for energy to renew the cycle. This process, although very time consuming, is economical and can recover as much as 70% of the copper in the ore. Uranium, gold, and cobalt ores are processed in a similar manner. The entire arrangement resembles a continuous flow bioreactor.

MICROORGANISMS AS INDUSTRIAL PRODUCTS

Microorganisms themselves sometimes constitute an industrial product. *Baker's yeast* (*S. cerevisiae*) is produced in large aerated fermentation tanks. At the end of the fermentation, the contents of the tank are about 4% yeast solids. The cells are harvested by continuous centrifuges and are pressed into the familiar yeast cakes or

Gas flaring stacks

Microturbines produce electricity from methane

FIGURE 28.15 Methane production from solid wastes in landfills. Methane accumulates in landfills and can be used for energy. This installation near Los Angeles has 50 microturbines that produce electricity from methane produced by the landfill. Immediately behind the microturbines are five gas flaring stacks that mask the flames from excess flared methane—a requirement so that aircraft will not confuse it with airport lighting.

Q **How is methane produced in a landfill?**

packets sold for home baking. Wholesale bakers purchase yeast in 50-lb boxes.

Other important microbes that are sold industrially are the symbiotic nitrogen-fixing bacteria *Rhizobium* and *Bradyrhizobium*. These organisms are usually mixed with peat moss to preserve moisture; the farmer mixes the peat moss and bacterial inoculum with the seeds of legumes to ensure infection of the plants with efficient nitrogen-fixing strains (see Chapter 27). For many years, gardeners have used the insect pathogen *Bacillus thuringiensis* to control leaf-eating insect larvae. This bacterium produces a toxin (Bt-toxin) that kills certain moths, beetles, and flies when ingested by their larvae. *B. thuringiensis* subspecies *israelensis* produces Bt-toxin that is especially active against mosquito larvae and is widely used in municipal control programs. Commercial preparations containing Bt-toxin and endospores of *B. thuringiensis* are available at almost any gardening supply store. For an example of microbes being developed to detect chemicals, see the box on page 848.

ALTERNATIVE ENERGY SOURCES USING MICROORGANISMS

LEARNING OBJECTIVE
• Define *bioconversion,* and list its advantages.

As our supplies of fossil fuels diminish or become more expensive, interest in the use of renewable energy resources will increase. Prominent among these is **biomass,** the collective organic matter produced by living organisms, including crops, trees, and municipal wastes. Microbes can be used for **bioconversion,** the process of converting biomass into alternative energy sources. Bioconversion can also decrease the amount of waste materials requiring disposal.

Methane is one of the most convenient energy sources produced from bioconversion. Many communities produce useful amounts of methane from wastes in landfill sites (Figure 28.15). Large cattle-feeding lots must dispose of immense amounts of animal manure, and much effort has been devoted to devising practical methods for producing methane from these wastes. A major problem with any scheme for large-scale methane production is the need to economically concentrate the widespread biomass material. If it could be economically concentrated, the animal and human wastes in the United States could provide much of our energy now supplied by fossil fuels and natural gas.

The agricultural industry has encouraged the production of **ethanol** from agricultural products. Gasoline containing ethanol (90% gasoline + 10% ethanol) is available in many parts of the United States and is used in automobiles worldwide. Some automobiles are able to use fuel that is 85% ethanol and this is available in some locations. Corn is currently the most frequently used substrate, but eventually agricultural waste products may be used.

INDUSTRIAL MICROBIOLOGY AND THE FUTURE

Microbes have always been exceedingly useful to humankind, even when their existence was unknown. They will remain an essential part of many basic food-processing technologies. The development of recombinant DNA technology has further intensified interest in industrial microbiology by expanding the potential for new products and applications (see the box in Chapter 9, page 267). As the supplies of fossil energy become more scarce, interest in renewable energy sources such as hydrogen and ethanol will increase. The use of specialized microbes to produce such products on an industrial scale will probably become more important. As new biotechnology applications and products enter the marketplace, they will affect our lives and well-being in ways that we can only speculate about today.

STUDY OUTLINE

FOOD MICROBIOLOGY (pp. 841–850)

1. The earliest methods of preserving foods were drying, the addition of salt or sugar, and fermentation.

2. Food safety is monitored by the FDA and USDA and also by use of the HACCP system.

INDUSTRIAL FOOD CANNING (pp. 841–842)

3. Commercial sterilization of food is accomplished by steam under pressure in a retort.

4. Commercial sterilization heats canned foods to the minimum temperature necessary to destroy *Clostridium botulinum* endospores while minimizing alteration of the food.

5. The commercial sterilization process uses sufficient heat to reduce a population of *C. botulinum* by 12 logarithmic cycles (12D treatment).

6. Endospores of thermophiles can survive commercial sterilization.

7. Canned foods stored above 45°C can be spoiled by thermophilic anaerobes.

8. Thermophilic anaerobic spoilage is sometimes accompanied by gas production; if no gas is formed, the spoilage is called flat sour spoilage.

9. Spoilage by mesophilic bacteria is usually from improper heating procedures or leakage.

10. Acidic foods can be preserved by heat of 100°C because microorganisms that survive are not capable of growth in a low pH.

11. *Byssochlamys*, *Aspergillus*, and *Bacillus coagulans* are acid-tolerant and heat-resistant microbes that can spoil acidic foods.

ASEPTIC PACKAGING (pp. 842–843)

12. Presterilized materials are assembled into packages and aseptically filled with heat-sterilized liquid foods.

RADIATION AND INDUSTRIAL FOOD PRESERVATION (pp. 843–845)

13. Gamma and X ray radiation can be used to sterilize food, kill insects and parasitic worms, and prevent the sprouting of fruits and vegetables.

HIGH-PRESSURE FOOD PRESERVATION (p. 845)

14. Pressurized water is used to kill bacteria in fruit and meat.

THE ROLE OF MICROORGANISMS IN FOOD PRODUCTION (pp. 846–850)

Cheese (p. 846)

15. The milk protein casein curdles because of the action by lactic acid bacteria or the enzyme rennin.

16. Cheese is the curd separated from the liquid portion of milk, called whey.

17. Hard cheeses are produced by lactic acid bacteria growing in the interior of the curd.

18. The growth of microbes in cheese is called ripening.

19. Semisoft cheeses are ripened by bacteria growing on the surface; soft cheeses are ripened by *Penicillium* growing on the surface.

Other Dairy Products (pp. 846–847)

20. Old-fashioned buttermilk was produced by lactic acid bacteria growing during the butter-making process.

21. Commercial buttermilk is made by letting lactic acid bacteria grow in skim milk for 12 hours.

22. Sour cream, yogurt, kefir, and kumiss are produced by lactobacilli, streptococci, or yeasts growing in low-fat milk.

Nondairy Fermentations (pp. 847–848)

23. Sugars in bread dough are fermented by yeast to ethanol and CO_2; the CO_2 causes the bread to rise.

24. Sauerkraut, pickles, olives, and soy sauce are products of microbial fermentations.

Alcoholic Beverages and Vinegar (pp. 848–850)

25. Carbohydrates obtained from grains, potatoes, or molasses are fermented by yeasts to produce ethanol in the production of beer, ale, sake, and distilled spirits.

26. The sugars in fruits such as grapes are fermented by yeasts to produce wines.

27. In winemaking, lactic acid bacteria convert malic acid into lactic acid in malolactic fermentation.

28. *Acetobacter* and *Gluconobacter* oxidize ethanol in wine to acetic acid (vinegar).

INDUSTRIAL MICROBIOLOGY (pp. 851–857)

1. Microorganisms produce alcohols and acetone that are used in industrial processes.

2. Industrial microbiology has been revolutionized by the ability of genetically modified cells to make many new products.

3. Biotechnology is a way of making commercial products by using living organisms.

FERMENTATION TECHNOLOGY (pp. 851–854)

4. The growth of cells on a large scale is called industrial fermentation.

5. Industrial fermentation is carried on in bioreactors, which control aeration, pH, and temperature.

6. Primary metabolites such as ethanol are formed as the cells grow (during the trophophase).

7. Secondary metabolites such as penicillin are produced during the stationary phase (idiophase).

8. Mutant strains that produce a desired product can be selected.

Immobilized Enzymes and Microorganisms (pp. 853–854)

9. Enzymes or whole cells can be bound to solid spheres or fibers. When substrate passes over the surface, enzymatic reactions change the substrate to the desired product.

10. They are used to make paper, textiles, and leather and are environmentally safe.

INDUSTRIAL PRODUCTS (pp. 854–857)

11. Most amino acids used in foods and medicine are produced by bacteria.

12. Microbial production of amino acids can be used to produce L-isomers; chemical production results in both D- and L-isomers.

13. Lysine and glutamic acid are produced by *Corynebacterium glutamicum*.

14. Citric acid, used in foods, is produced by *Aspergillus niger*.

15. Enzymes used in manufacturing foods, medicines, and other goods are produced by microbes.

16. Some vitamins used as food supplements are made by microorganisms.

17. Vaccines, antibiotics, and steroids are products of microbial growth.

18. The metabolic activities of *Thiobacillus ferrooxidans* can be used to recover uranium and copper ores.

19. Yeasts are grown for wine- and breadmaking; other microbes (*Rhizobium*, *Bradyrhizobium*, and *Bacillus thuringiensis*) are grown for agricultural use.

ALTERNATIVE ENERGY SOURCES USING MICROORGANISMS (p. 857)

20. Organic waste, called biomass, can be converted by microorganisms into alternative fuels, a process called bioconversion.

21. Fuels produced by microbial fermentation are methane, ethanol, and hydrogen.

INDUSTRIAL MICROBIOLOGY AND THE FUTURE (p. 857)

22. Recombinant DNA technology will continue to enhance the ability of industrial microbiology to produce medicines and other useful products.

STUDY QUESTIONS

Access more review material either online at **The Microbiology Place** (www.microbiologyplace.com) or with **The Microbiology Place CD-ROM** packaged with your new book. There you'll find activities, practice tests, quizzes, flashcards, case studies, and more to help you succeed. In addition, you'll find the following Activity: Bioreactor Simulation.

Answers to the Study Questions can be found in Appendix G.

REVIEW

1. What is industrial microbiology? Why is it important?

2. How does commercial sterilization differ from sterilization procedures used in a hospital or laboratory?

3. Why is a can of blackberries preserved by commercial sterilization typically heated to 100°C instead of at least 116°C?

4. Describe aseptic packaging.

5. Outline the steps in the production of cheese, and compare the production of hard and soft cheeses.

6. Outline the steps in the process of wine production.

7. Beer is made with water, malt, and yeast; hops are added for flavor. What is the purpose of the water, malt, and yeast? What is malt?

8. Why is a bioreactor better than a large flask for industrial production of an antibiotic?

9. The manufacture of paper includes the use of bleach and formaldehyde-based glue. The microbial enzyme xylanase whitens paper by digesting dark lignins. Oxidase causes the fibers to stick together, and cellulase will remove ink. List three advantages of using these microbial enzymes over traditional chemical methods for making paper.

10. Label the trophophase and idophase in this graph. Indicate when primary and secondary metabolites are formed.

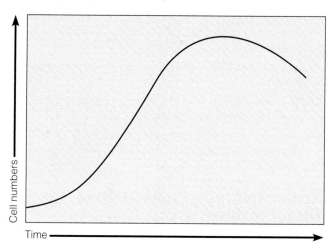

11. Describe an example of bioconversion. What metabolic processes can result in fuels?

MULTIPLE CHOICE

1. Foods packed in plastic for microwaving are
  a. dehydrated.
  b. freeze-dried.
  c. packaged aseptically.
  d. commercially sterilized.
  e. autoclaved.

2. *Acetobacter* is necessary for only one of the steps of vitamin C manufacture. The easiest way to accomplish this step would be to
  a. add substrate and *Acetobacter* to a test tube.
  b. affix *Acetobacter* to a surface and run substrate over it.
  c. add substrate and *Acetobacter* to a bioreactor.
  d. find an alternative to this step.
  e. none of the above

Use the following choices to answer questions 3–5:
  a. *Bacillus coagulans*
  b. *Byssochlamys*
  c. flat sour spoilage
  d. *Lactobacillus*
  e. thermophilic anaerobic spoilage

3. The spoilage of canned foods due to inadequate processing, accompanied by gas production.

4. The spoilage of canned foods caused by *Geobacillus stearothermophilus*.

5. A heat-resistant fungus that causes spoilage in acidic foods.

6. The term 12D treatment refers to
  a. heat treatment sufficient to kill 12 bacteria.
  b. the use of 12 different treatments to preserve food.
  c. a 10^{12} reduction in *C. botulinum* endospores.
  d. any process that destroys thermophilic bacteria.

7. Microorganisms themselves are industrial products. Which of the following pairs is mismatched?
  a. *Penicillium*—treatment of disease
  b. *S. cerevisiae*—for fermentation
  c. *Rhizobium*—increases nitrogen in the soil
  d. *B. thuringiensis*—insecticide

8. Which type of radiation is used to preserve foods?
  a. ionizing
  b. nonionizing
  c. radiowaves
  d. microwaves
  e. all of the above

9. Which of the following reactions is undesirable in wine-making?
  a. Sucrose → ethanol
  b. Ethanol → acetic acid
  c. Malic acid → lactic acid
  d. Glucose → pyruvic acid

10. Which of the following reactions is an oxidation carried out by *Thiobacillus ferrooxidans*?
  a. $Fe^{2+} \rightarrow Fe^{3+}$
  b. $Fe^{3+} \rightarrow Fe^{2+}$
  c. $CuS \rightarrow CuSO_4$
  d. $Fe^0 \rightarrow Cu^0$
  e. none of the above

CRITICAL THINKING

1. Which bacteria seem to be most frequently used in the production of food? Propose an explanation for this.

2. *Methylophilus methylotrophus* can convert methane (CH_4) into proteins. Amino acids are represented by this structure:

$$H_2N - \overset{\overset{\displaystyle H}{|}}{\underset{\underset{\displaystyle R}{|}}{C}} - C \overset{\displaystyle O}{\underset{\displaystyle OH}{}}$$

Diagram a pathway illustrating the production of at least one amino acid.

3. "Stone-washed" denim is produced with cellulase. How does cellulase accomplish the look and feel of stone-washing? What is the source of the cellulase?

CLINICAL APPLICATIONS

1. Suppose you are culturing a microorganism that produces enough lactic acid to kill itself in a few days.
  a. How can the use of a bioreactor help you maintain the culture for weeks or months? The graph on the next page shows conditions in the bioreactor:

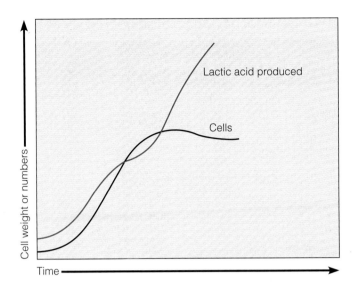

b. If your desired product is a secondary metabolite, when can you begin collecting it?

c. If your desired product is the cells themselves and you want to maintain a continuous culture, when can you begin harvesting?

2. Researchers at the CDC inoculated apple cider with 10^5 *E. coli* O157:H7 cells/ml to determine the fate of the bacteria in apple cider (pH 3.7). They obtained the following results:

	Number of *E. coli* O157:H7 cells/ml after 25 days
Apple cider at 25°C	10^4 (mold growth evident by 10 days)
Apple cider with potassium sorbate at 25°C	10^3
Apple cider at 8°C	10^2

What conclusions can you reach from these data? What disease is caused by *E. coli* O157:H7? (*Hint:* See Chapter 25.)

3. The antibiotic efrotomycin is produced by *Streptomyces lactamdurans*. *S. lactamdurans* was grown in 40,000 liters of medium. The medium consisted of glucose, maltose, soybean oil, $(NH_4)_2SO_4$, NaCl, KH_2PO_4, and Na_2HPO_4. The culture was aerated and maintained at 28°C. The following results were obtained from analyses of the culture medium during cell growth:

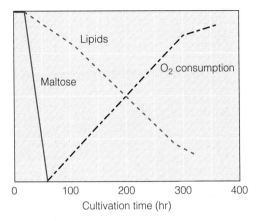

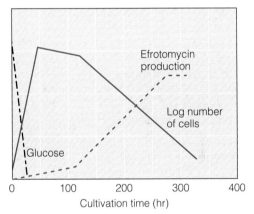

a. Under what conditions is the most efrotomycin produced? Is it a primary or secondary metabolite?

b. Which is used first, maltose or glucose? Suggest a reason for this.

c. What is the purpose of each ingredient in the growth medium? (*Hint:* See Chapter 6.)

d. What is *Streptomyces*? (*Hint:* See Chapter 11.)

APPENDIX A

CLASSIFICATION OF BACTERIA ACCORDING TO *BERGEY'S MANUAL**

Domain: Bacteria
  Phylum I: Aquificae
    Class I: Aquificae
      Order I: Aquificales
        Family I: Aquificaceae
          Aquifex
          Calderobacterium
          Hydrogenobacter
          Hydrogenobaculum
          Hydrogenothermus
          Persephonella
          Sulfurihydrogenibium
          Thermocrinis
        Genera incertae sedis
          Balnearium
          Desulfurobacterium
          Thermovibrio
  Phylum II: Thermotogae
    Class I: Thermotogae
      Order I: Thermotogales
        Family I: Thermotogaceae
          Fervidobacterium
          Geotoga
          Marinitoga
          Petrotoga
          Thermosipho
          Thermotoga
  Phylum III: Thermodesulfobacteria
    Order I: Thermodesulfobacteriales
      Family I: Thermodesulfobacteriaceae
        Thermodesulfobacterium
  Phylum IV: Deinococcus-Thermus
    Class I: Deinococci
      Order I: Deinococcales
        Family I: Deinococcaceae
          Deinococcus
      Order II: Thermales
        Family I: Thermaceae
          Marinithermus
          Meiothermus
          Oceanithermus
          Thermus
          Vulcanithermus
  Phylum V: Chrysiogenetes
    Class I: Chrysiogenetes
      Order I: Chrysiogenales
        Family I: Chrysiogenaceae
          Chrysiogenes
  Phylum VI: Chloroflexi
    Class I: Chloroflexi
      Order I: Chloroflexales
        Family I: Chloroflexaceae
          Chloroflexus
          Chloronema
          Heliothrix
          Roseiflexus
        Family II: Oscillochloridaceae
          Oscillochloris
      Order II: Herpetosiphonales
        Family I: Herpetosiphonaceae
          Herpetosiphon
    Class II: Anaerolineae
      Order I: Anaerolinaeles
        Family I: Anaerolinaceae
          Anaerolineae

Phylum VII: Thermomicrobia
  Class I: Thermomicrobia
    Order I: Thermomicrobiales
      Family I: Thermomicrobiaceae
        Thermomicrobium
Phylum VIII: Nitrospira
  Class I: Nitrospira
    Order I: Nitrospirales
      Family I: Nitrospiraceae
        Leptospirillum
        Magnetobacterium
        Nitrospira
        Thermodesulfovibrio
Phylum IX: Deferribacteres
  Class I: Deferribacteres
    Order I: Deferribacterales
      Family I: Deferribacteraceae
        Deferribacter
        Denitrovibrio
        Flexistipes
        Geovibrio
      Genera incertae sedis
        Caldithrix
        Synergistes
Phylum X: Cyanobacteria
  Class I: Cyanobacteria
    Subsection I
      Chamaesiphon
      Chroococcus
      Cyanobacterium
      Cyanobium
      Cyanothece
      Dactylococcopsis
      Gloeobacter
      Gloeocapsa
      Gloeothece
      Microcystis
      Prochlorococcus
      Prochloron
      Synechococcus
      Synechocystis
    Subsection II
      Chroococcidiopsis
      Cyanocystis
      Dermocarpella
      Myxosarcina
      Pleurocapsa
      Stanieria
      Xenococcus
    Subsection III
      Arthrospira
      Borzia
      Crinalium
      Geitlerinema
      Halospirulina
      Leptolyngbya
      Limnothrix
      Lyngbya
      Microcoleus
      Oscillatoria
      Planktothrix
      Prochlorothrix
      Pseudoanabaena
      Spirulina
      Starria

      Symploca
      Trichodesmium
      Tychonema
    Subsection IV
      Anabaena
      Anabaenopsis
      Aphanizomenon
      Calothrix
      Cyanospira
      Cylindrospermopsis
      Cylindrospermum
      Nodularia
      Nostoc
      Rivularia
      Scytonema
      Tolypothrix
    Subsection V
      Chlorogloeopsis
      Fischerella
      Geitleria
      Iyengariella
      Nostochopsis
      Stigonema
Phylum XI: Chlorobi
  Class I: Chlorobia
    Order I: Chlorobiales
      Family I: Chlorobiaceae
        Ancalochloris
        Chlorobaculum
        Chlorobium
        Chloroherpeton
        Pelodictyon
        Prosthecochloris
Phylum XII: Proteobacteria
  Class I: Alphaproteobacteria
    Order I: Rhodospirillales
      Family I: Rhodospirillaceae
        Azospirillum
        Inquilinus
        Magnetospirillum
        Phaeospirillum
        Rhodocista
        Rhodospira
        Rhodospirillum
        Rhodovibrio
        Roseospira
        Skermanella
        Thallassospira
        Tistrella
      Family II: Acetobacteraceae
        Acetobacter
        Acidiphilium
        Acidisphaera
        Acidocella
        Acidomonas
        Asaia
        Craurococcus
        Gluconacetobacter
        Gluconobacter
        Kozakia
        Muricoccus
        Paracraurococcus
        Rhodopila
        Roseococcus
        Rubritepida

---

*Bergey's Manual of Systematic Bacteriology, 2nd ed., 5 vols. (2004), is the reference for classification. Bergey's Manual of Determinative Bacteriology, 9th ed. (1994), should be used for identification of culturable bacteria and archaea.

Stella
Teichococcus
Zavarzinia
Order II: Rickettsiales
  Family I: Rickettsiaceae
    Orientia
    Rickettsia
  Family II: *Anaplasmataceae*
    Aegyptianella
    Anaplasma
    Cowdria
    Ehrlichia
    Neorickettsia
    Wolbachia
    Xenohaliotis
  Family III: Holosporaceae
    Holospora
  Genera incertae sedis
    Caedibacter
    Lyticum
    Odyssella
    Pseudocaedibacter
    Symbiotes
    Tectibacter
Order III: Rhodobacterales
  Family I: Rhodobacteraceae
    Ahrensia
    Albidovulum
    Amaricoccus
    Antarctobacter
    Gemmobacter
    Hirschia
    Hyphomonas
    Jannaschia
    Ketogulonicigenium
    Leisingera
    Maricaulis
    Methylarcula
    Oceanicaulis
    Octadecabacter
    Pannonibacter
    Paracoccus
    Pseudorhodobacter
    Rhodobaca
    Rhodobacter
    Rhodothalassium
    Rhodovulum
    Roseibium
    Roseinatronobacter
    Roseivivax
    Roseobacter
    Roseovarius
    Roseovivax
    Rubrimonas
    Ruegeria
    Sagittula
    Silicibacter
    Staleya
    Stappia
    Sulfitobacter
Order IV: Sphingomonadales
  Family I: Sphingomonodaceae
    Blastomonas
    Erythrobacter
    Erythromicrobium
    Erythromonas
    Novosphingobium
    Porphyrobacter
    Rhizomonas
    Sandaracinobacter
    Sphingobium
    Sphingomonas
    Sphingopyxis
    Zymomonas
Order V: Caulobacterales
  Family I: Caulobacteraceae
    Asticcacaulis
    Brevundimonas

Caulobacter
Phenylobacterium
Order VI: Rhizobiales
  Family I: Rhizobiaceae
    Agrobacterium
    Allorhizobium
    Carbophilus
    Chelatobacter
    Ensifer
    Rhizobium
    Sinorhizobium
  Family II: Aurantimonadaceae
    Aurantimonas
    Fulvimarina
  Family III: Bartonellaceae
    Bartonella
  Family IV: Brucellaceae
    Brucella
    Mycoplana
    Ochrobactrum
  Family V: Phyllobacteriaceae
    Aminobacter
    Aquamicrobium
    Defluvibacter
    Mesorhizobium
    Nitratireductor
    Phyllobacterium
    Paseuaminobacter
  Family VI: Methylocystaceae
    Albibacter
    Methylocystis
    Methylopila
    Methylosinus
    Terasakiella
  Family VII: Beijerinckiaceae
    Beijerinckia
    Chelatococcus
    Methylocapsa
    Methylocella
  Family VIII: Bradyrhizobiaceae
    Afipia
    Agromonas
    Blastobacter
    Bosea
    Bradyrhizobium
    Nitrobacter
    Oligotropha
    Rhodoblastus
    Rhodopseudomonas
  Family IX: Hyphomicrobiaceae
    Ancalomicrobium
    Ancylobacter
    Angulomicrobium
    Aquabacter
    Azorhizobium
    Blastochloris
    Devosia
    Dichotomicrobium
    Filomicrobium
    Gemmiger
    Hyphomicrobium
    Labrys
    Methylorhabdus
    Pedomicrobium
    Prosthecomicrobium
    Rhodomicrobium
    Rhodoplanes
    Seliberia
    Starkeya
    Xanthobacter
  Family X: Methylobacteriaceae
    Methylobacterium
    Microvirga
    Protomonas
    Roseomonas
  Family XI: Rhodobiaceae
    Rhodobium
    Roseospirillum

Order VII: Parvularculales
  Family I: Parvularculaceae
    Parvularcula
Class II: Betaproteobacteria
  Order I: Burkholderiales
    Family I: Burkholderiaceae
      Burkholderia
      Cupriavidus
      Lautropia
      Limnobacter
      Pandoraea
      Paucimonas
      Polynucleobacter
      Ralstonia
      Thermothrix
      Wautersia
    Family II: Oxalobacteraceae
      Duganella
      Herbaspirillum
      Janthinobacterium
      Massilia
      Oxalicibacterium
      Oxalobacter
      Telluria
    Family III: Alcaligenaceae
      Achromobacter
      Alcaligenes
      Bordetella
      Brackiella
      Oligella
      Pelistega
      Pigmentiphaga
      Sutterella
      Taylorella
    Family IV: Comamonadaceae
      Acidovorax
      Alicycliphilus
      Brachymonas
      Caldimonas
      Comamonas
      Delftia
      Diaphorobacter
      Hydrogenophaga
      Hylemonella
      Lampropedia
      Macromonas
      Ottowia
      Polaromonas
      Ramlibacter
      Rhodoferax
      Variovorax
      Xenophilus
    Genera incertae sedis
      Aquabacterium
      Ideonella
      Leptothrix
      Roseateles
      Rubrivivax
      Schlegelella
      Sphaerotilus
      Tepidimonas
      Thiomonas
      Xylophilus
  Order II: Hydrogenophilales
    Family I: Hydrogenophilaceae
      Hydrogenophilus
      Thiobacillus
  Order III: Methylophilales
    Family I: Methylophilaceae
      Methylobacillus
      Methylophilus
      Methylovorus
  Order IV: Neisseriales
    Family I: Neisseriaceae
      Alysiella
      Aquaspirillum
      Chromobacterium
      Eikenella

Formivibrio
Iodobacter
Kingella
Laribacter
Microvirgula
Morococcus
Neisseria
Prolinoborus
Simonsiella
Vitreoscilla
Vogesella
Order V: Nitrosomonadales
  Family I: Nitrosomonadaceae
    Nitrosolobus
    Nitrosomonas
    Nitrosospira
  Family II: Spirillaceae
    Spirillum
  Family III: Gallionellaceae
    Gallionella
Order VI: Rhodocyclales
  Family I: Rhodocyclaceae
    Azoarcus
    Azonexus
    Azospira
    Azovibrio
    Dechloromonas
    Dechlorosoma
    Ferribacterium
    Propionibacter
    Propionivibrio
    Quadricoccus
    Rhodocyclus
    Sterolibacterium
    Thauera
    Zoogloea
Order VII: Procabacteriales
  Family I: Procabacteriaceae
    Procabacter
Class III: Gammaproteobacteria
Order I: Chromatiales
  Family I: Chromatiaceae
    Allochromatium
    Amoebobacter
    Chromatium
    Halochromatium
    Isochromatium
    Lamprobacter
    Lamprocystis
    Marichromatium
    Nitrosococcus
    Pfennigia
    Rhabdochromatium
    Rheinheimera
    Thermochromatium
    Thioalkalicoccus
    Thiobaca
    Thiocapsa
    Thiococcus
    Thiocystis
    Thiodictyon
    Thioflavicoccus
    Thiohalocapsa
    Thiolamprovum
    Thiopedia
    Thiorhodococcus
    Thiorhodovibrio
    Thiospirillum
  Family II: Ectothiorhodospiraceae
    Alcalilimnicola
    Alkalispirillum
    Arhodomonas
    Ectothiorhodospira
    Halorhodospira
    Nitrococcus
    Thioalkalispira
    Thioalkalivibrio
    Thiorhodospira

Order II: Acidithiobacillales
  Family 1: Acidithiobacillaceae
    Acidithiobacillus
  Family II: Thermithiobacillaceae
    Thermithiobacillus
Order III: Xanthomonadales
  Family I: Xanthomonadaceae
    Frateuria
    Fulvimonas
    Luteimonas
    Lysobacter
    Nevskia
    Pseudoxanthomonas
    Rhodanobacer
    Schineria
    Stenotrophomonas
    Thermomonas
    Xanthomonas
    Xylella
Order IV: Cardiobacteriales
  Family I: Cardiobacteriaceae
    Cardiobacterium
    Dichelobacter
    Suttonella
Order V: Thiotrichales
  Family I: Thiotrichaceae
    Achromatium
    Beggiatoa
    Leucothrix
    Thiobacterium
    Thiomargarita
    Thioploca
    Thiospira
    Thiothrix
  Family II: Francisellaceae
    Francisella
  Family III: Piscirickettsiaceae
    Cycloclasticus
    Hydrogenovibrio
    Methylophaga
    Piscirickettsia
    Thioalkalimicrobium
    Thiomicrospira
Order VI: Legionellales
  Family I: Legionellaceae
    Legionella
  Family II: Coxiellaceae
    Aquicella
    Coxiella
    Rickettsiella
Order VII: Methylococcales
  Family I: Methylococcaceae
    Methylobacter
    Methylocaldum
    Methylococcus
    Methylomicrobium
    Methylomonas
    Methylosarcina
    Methylosphaera
Order VIII: Oceanospirillales
  Family I: Oceanospirillaceae
    Balneatrix
    Marinomonas
    Marinospirillum
    Neptunomonas
    Oceanobacter
    Oceanospirillum
    Oleispira
    Pseudospirillum
    Thalassolituus
  Family II: Alcanivoraceae
    Alcanivorax
    Fundibacter
  Family III: Hahellaceae
    Hahella
    Zooshikella
  Family IV: Halomonadaceae
    Halomonas

    Carnimonas
    Chromohalobacter
    Cobetia
    Deleya
    Zymobacter
  Family V: Oleiphilaceae
    Oleiphilus
  Family VI: Saccharospirillaceae
    Saccharospirillum
Order IX: Pseudomonadales
  Family I: Pseudomonadaceae
    Azomonas
    Azotobacter
    Cellvibrio
    Chryseomonas
    Flavimonas
    Mesophilobacter
    Pseudomonas
    Rhizobacter
    Rugamonas
    Serpens
  Family II: Moraxellaceae
    Acinetobacter
    Moraxella
    Psychrobacter
  Family III: Incertae sedis
    Enhydrobacter
Order X: Alteromonadales
  Family I: Alteromonadaceae
    Aestuariibacter
    Alishewanella
    Alteromonas
    Colwellia
    Ferrimonas
    Glaciecola
    Idiomarina
    Marinobacter
    Marinobacterium
    Microbulbifer
    Moritella
    Pseudoalteromonas
    Psychromonas
    Shewanella
    Thalassomonas
  Family II: Incerta sedis
    Teredinibacter
Order XI: Vibrionales
  Family I: Vibrionaceae
    Allomonas
    Catenococcus
    Enterovibrio
    Grimontia
    Listonella
    Photobacterium
    Salinivibrio
    Vibrio
Order XII: Aeromonadales
  Family I: Aeromonadaceae
    Aeromonas
    Oceanimonas
    Oceanisphaera
    Tolumonas
  Family II: Succinivibrionaceae
    Anaerobiospirillum
    Ruminobacter
    Succinomonas
    Succinivibrio
Order XIII: Enterobacteriales
  Family I: Enterobacteriaceae
    Alterococcus
    Arsenophonus
    Brenneria
    Buchnera
    Budvicia
    Buttiauxella
    Calymmatobacterium
    Cedecea
    Citrobacter

Edwardsiella
Enterobacter
Erwinia
Escherichia
Ewingella
Hafnia
Klebsiella
Kluyvera
Leclercia
Leminorella
Moellerella
Morganella
Obesumbacterium
Pantoea
Pectobacterium
Phlomobacter
Photorhabdus
Plesiomonas
Pragia
Proteus
Providencia
Rahnella
Raoultella
Saccharobacter
Salmonella
Samsonia
Serratia
Shigella
Sodalis
Tatumella
Trabulsiella
Wigglesworthia
Xenorhabdus
Yersinia
Yokenella
Order XIV: Pasteurellales
  Family I: Pasteurellaceae
    Actinobacillus
    Gallibacterium
    Haemophilus
    Lonepinella
    Pasteurella
    Mannheimia
    Phocoenobacter
Class IV: Deltaproteobacteria
  Order I: Desulfurellales
    Family I: Desulfurellaceae
      Desulfurella
      Hippea
  Order II: Desulfovibrionales
    Family I: Desulfovibrionaceae
      Bilophila
      Desulfovibrio
      Lawsonia
    Family II: Desulfomicrobiaceae
      Desulfomicrobium
    Family III: Desulfohalobiaceae
      Desulfohalobium
      Desulfomonaas
      Desulfonatronovibrio
      Desulfothermus
    Family IV: Desulfonatronumaceae
      Desulfonatronum
  Order III: Desulfobacterales
    Family I: Desulfobacteraceae
      Desulfatibacillum
      Desulfobacter
      Desulfobacterium
      Desulfobacula
      Desulfobotulus
      Desulfocella
      Desulfococcus
      Desulfofaba
      Desulfofrigus
      Desulfomusa
      Desulfonema
      Desulforegula
      Desulfosarcina

Desulfospira
Desulfotignum
    Family II: Desulfobulbaceae
      Desulfobulbus
      Desulfocapsa
      Desulfofustis
      Desulforhopalus
      Desulfotalea
    Family III: Nitrospinaceae
      Nitrospina
  Order IV: Desulfarcales
    Family I. Desulfarculaceae
      Desulfarculus
  Order V: Desulfuromonales
    Family I: Desulfuromonaceae
      Desulfuromonas
      Desulfuromusa
      Malonomonas
      Pelobacter
    Family II: Geobacteraceae
      Geobacter
      Trichlorobacter
  Order VI: Syntrophobacterales
    Family I: Syntrophobacteraceae
      Desulfacinum
      Syntrophobacter
      Desulforhabdus
      Desulfovirga
      Thermodesulforhabdus
    Family II: Syntrophaceae
      Desulfobacca
      Smithella
      Syntrophus
  Order VII: Bdellovibrionales
    Family I: Bdellovibrionaceae
      Bacteriovorax
      Bdellovibrio
      Micavibrio
      Vampirovibrio
  Order VIII: Myxococcales
    Family I: Cystobacteraceae
      Anaeromyxobacter
      Archangium
      Cystobacter
      Hyalangium
      Melittangium
      Stigmatella
    Family II: Myxococcaceae
      Corallococcus
      Myxococcus
      Pyxicoccus
    Family III: Polyangiaceae
      Byssophaga
      Chondromyces
      Haploangium
      Jahnia
      Polyangium
      Sorangium
    Family IV: Nannocystaceae
      Nannocystis
      Plesiocystis
    Family V: Haliangiaceae
      Haliangium
    Family VI: Kofleriaceae
      Kofleria
Class V: Epsilonproteobacteria
  Order I: Campylobacterales
    Family I: Campylobacteraceae
      Arcobacter
      Campylobacter
      Dehalospirillum
      Sulfurospirillum
    Family II: Helicobacteraceae
      Helicobacter
      Sulfurimonas
      Thiovulum
      Wolinella

Phylum XIII: Firmicutes
Class I: Clostridia
  Order I: Clostridiales
    Family I: Clostridiaceae
      Acetivibrio
      Acidaminobacter
      Alkaliphilus
      Anaerobacter
      Anaerotruncus
      Bryantella
      Caloramator
      Caloranaerobacter
      Caminicella
      Clostridium
      Coprobacillus
      Dorea
      Faecalibacterium
      Hespellia
      Natronincola
      Oxobacter
      Parasporobacterium
      Sarcina
      Soehngenia
      Tepidibacter
      Thermobrachium
      Thermohalobacter
      Tindallia
    Family II: Lachnospiraceae
      Acetitomaculum
      Anaerofilum
      Anaerostipes
      Butyrivibrio
      Catenibacterium
      Catonella
      Coprococcus
      Johnsonella
      Lachnobacterium
      Lachnospira
      Pseudobutyrivibrio
      Roseburia
      Ruminococcus
      Shuttleworthia
      Sporobacterium
    Family III: Peptostreptococcaceae
      Anaerococcus
      Filifactor
      Finegoldia
      Gallicola
      Fusibacter
      Helcococcus
      Micromonas
      Peptoniphilus
      Peptostreptococcus
      Sedimentibacter
      Sporanaerobacter
      Tissierella
    Family IV: Eubacteriaceae
      Acetobacterium
      Anaerovorax
      Eubacterium
      Mogibacterium
      Pseudoramibacter
    Family V: Peptococcaceae
      Carboxydothermus
      Dehalobacter
      Desulfitobacterium
      Desulfonispora
      Desulfosporosinus
      Desulfotomaculum
      Pelotomaculum
      Peptococcus
      Syntrophobotulus
      Thermoterrabacterium
    Family VI: Heliobacteriaceae
      Heliobacterium
      Heliobacillus
      Heliophilum
      Heliorestis

Family VII: Acidaminococcaceae
  Acetonema
  Acidaminococcus
  Allisonella
  Anaeroarcus
  Anaeroglobus
  Anaeromusa
  Anaerosinus
  Anaerovibrio
  Centipeda
  Dendrosporobacter
  Dialister
  Megasphaera
  Mitsuokella
  Papillibacter
  Pectinatus
  Phascolarctobacterium
  Propionispira
  Propionispora
  Quinella
  Schwartzia
  Selenomonas
  Sporomusa
  Succiniclasticum
  Succinispira
  Veillonella
  Zymophilus
Family VIII: Syntrophomonadaceae
  Acetogenium
  Aminobacterium
  Aminomonas
  Anaerobaculum
  Anaerobranca
  Caldicellulosiruptor
  Carboxydocella
  Dethiosulfovibrio
  Pelospora
  Syntrophomonas
  Syntrophospora
  Syntrophothermus
  Thermoaerobacter
  Thermanaerovibrio
  Thermohydrogenium
  Thermosyntropha
Order II: Thermoanaerobacteriales
Family I: Thermoanaerobacteriaceae
  Ammonifex
  Caldanaerobacter
  Carboxydibrachium
  Coprothermobacter
  Gelria
  Moorella
  Sporotomaculum
  Thermacetogenium
  Thermanaeromonas
  Thermoanaerobacter
  Thermoanaerobacterium
  Thermoanaerobium
  Thermovenabulum
Order III: Halanaerobiales
Family I: Halanaerobiaceae
  Halanaerobium
  Halocella
  Halothermothrix
Family II: Halobacteroidaceae
  Acetohalobium
  Halanaerobacter
  Halobacteroides
  Halonatronum
  Natroniella
  Orenia
  Selenihalanaerobacter
  Sporohalobacter
Class II: Mollicutes
Order I: Mycoplasmatales
Family I: Mycoplasmataceae
  Eperythrozoon
  Haemobartonella

  Mycoplasma
  Ureaplasma
Order II: Entomoplasmatales
Family I: Entomoplasmataceae
  Entomoplasma
  Mesoplasma
Family II: Spiroplasmataceae
  Spiroplasma
Order III: Acholeplasmatales
Family I: Acholeplasmataceae
  Acholeplasma
  Phytoplasma
Order IV: Anaeroplasmatales
Family I: Anaeroplasmataceae
  Anaeroplasma
  Asteroleplasma
Order V: Incertae sedis
Family I: Erysipelotrichaceae
  Bulleidia
  Erysipelothrix
  Holdemania
  Solobacterium
Class III: Bacilli
Order I: Bacillales
Family I: Bacillaceae
  Amphibacillus
  Anoxybacillus
  Bacillus
  Exiguobacterium
  Filobacillus
  Geobacillus
  Gracilibacillus
  Halobacillus
  Jeotgalibacillus
  Lentibacillus
  Marinibacillus
  Oceanobacillus
  Paraliobacillus
  Saccharococcus
  Salibacillus
  Ureibacillus
  Virgibacillus
Family II: Alicyclobacillaceae
  Alicyclobacillus
  Pasteuria
  Sulfobacillus
Family III: Caryophanaceae
  Caryophanon
Family IV: Listeriaceae
  Brochothrix
  Listeria
Family V: Paenibacillaceae
  Ammoniphilus
  Aneurinibacillus
  Brevibacillus
  Oxalophagus
  Paenibacillus
  Thermicanus
  Thermobacillus
Family VI: Planococcaceae
  Filibacter
  Kurthia
  Planococcus
  Planomicrobium
  Sporosarcina
Family VII: Sporolactobacillaceae
  Marinococcus
  Sporolactobacillus
Family VIII: Staphylococcaceae
  Gemella
  Jeotgalicoccus
  Macrococcus
  Salinicoccus
  Staphylococcus
Family IX: Thermoactinomycetaceae
  Thermoactinomyces
Family X: Turicibacteraceae
  Turicibacter

Order II: Lactobacillales
Family I: Lactobacillaceae
  Lactobacillus
  Paralactobacillus
  Pediococcus
Family II: Aerococcaceae
  Abiotrophia
  Aerococcus
  Dolosicoccus
  Eremococcus
  Facklamia
  Globicatella
  Ignavigranum
Family III: Carnobacteriaceae
  Agitococcus
  Alkalibacterium
  Allofustis
  Alloiococcus
  Carnobacterium
  Desemzia
  Dolosigranulum
  Granulicatella
  Isobaculum
  Lactosphaera
  Marinilactibacillus
  Trichococcus
Family IV: Enterococcaceae
  Atopobacter
  Enterococcus
  Melissococcus
  Tetragenococcus
  Vagococcus
Family V: Leuconostocaceae
  Leuconostoc
  Oenococcus
  Weissella
Family VI: Streptococcaceae
  Lactococcus
  Streptococcus
Family VII: Incertae sedis
  Acetoanaerobium
  Oscillospira
  Syntrophococcus
Phylum XIV: Actinobacteria
Class I: Actinobacteria
Order I: Acidimicrobiales
Family I: Acidimicrobiaceae
  Acidimicrobium
Order II: Rubrobacterales
  Conexibacter
  Rubrobacter
  Solirubrobacter
  Thermoleophilum
Order III: Coriobacteriales
Family I: Coriobacteriaceae
  Atopobium
  Collinsella
  Coriobacterium
  Cryptobacterium
  Denitrobacterium
  Eggerthella
  Olsenella
  Slackia
Order IV: Sphaerobacterales
Family I: Sphaerobacteraceae
  Sphaerobacter
Order V: Actinomycetales
Suborder: Actinomycineae
Family I: Actinomycetaceae
  Actinobaculum
  Actinomyces
  Arcanobacterium
  Mobiluncus
  Varibaculum
Suborder: Micrococcineae
Family I: Micrococcaceae
  Arthrobacter
  Citricoccus

Kocuria
Micrococcus
Nesterenkonia
Renibacterium
Rothia
Stomatococcus
Yania
Family II: Bogoriellaceae
Bogoriella
Family III: Rarobacteraceae
Rarobacter
Family IV: Sanguibacteraceae
Sanguibacter
Family V: Brevibacteriaceae
Brevibacterium
Family VI: Cellulomonadaceae
Cellulomonas
Oerskovia
Tropheryma
Family VII: Dermabacteraceae
Brachybacterium
Dermabacter
Family VIII: Dermatophilaceae
Dermatophilus
Kineosphaera
Family IX: Dermacoccaceae
Dermacoccus
Demetria
Kytococcus
Family X: Intrasporangiaceae
Arsenicicoccus
Intrasporangium
Janibacter
Nostocoidia
Ornithinicoccus
Ornithinimicrobium
Terrabacter
Terracoccus
Tetrasphaera
Family XI: Jonesiaceae
Jonesia
Family XII: Microbacteriaceae
Agrococcus
Agromyces
Aureobacterium
Clavibacter
Cryobacterium
Curtobacterium
Frigoribacterium
Leifsonia
Leucobacter
Microbacterium
Rathayibacter
Subtercola
Family XIII: Beutenbergiaceae
Beutenbergia
Georgenia
Salana
Family XIV: Promicromonosporaceae
Cellulosimicrobium
Promicromonospora
Xylanibacterium
Xylanimonas
Suborder: Corynebacterineae
Family I: Corynebacteriaceae
Corynebacterium
Family II: Dietziaceae
Dietzia
Family III: Gordoniaceae
Gordonia
Skermania
Family IV: Mycobacteriaceae
Mycobacterium
Family V: Nocardiaceae
Nocardia
Rhodococcus
Family VI: Tsukamurellaceae
Tsukamurella

Family VII: Williamsiaceae
Williamsia
Suborder: Micromonosporineae
Family I: Micromonosporaceae
Actinoplanes
Asanoa
Catellatospora
Catenuloplanes
Couchioplanes
Dactylosporangium
Micromonospora
Pilimelia
Spirilliplanes
Verrucosispora
Virgisporangium
Suborder: Propionibacterineae
Family I: Propionibacteriaceae
Luteococcus
Microlunatus
Propionibacterium
Propioniferax
Propionimicrobium
Tessaracoccus
Family II: Nocardioidaceae
Aeromicrobium
Actinopolymorpha
Friedmanniella
Hongia
Kribbella
Micropruina
Marmoricola
Nocardioides
Propionicimonas
Suborder: Pseudonocardineae
Family I: Pseudonocardiaceae
Actinoalloteichus
Actinopolyspora
Amycolatopsis
Crossiella
Kibdelosporangium
Kutzneria
Prauserella
Pseudonocardia
Saccharomonospora
Saccharopolyspora
Streptoalloteichus
Thermobispora
Thermocrispum
Family II: Actinosynnemataceae
Actinokineospora
Actinosynnema
Lechevalieria
Lentzea
Saccharothrix
Suborder: Streptomycineae
Family I: Streptomycetaceae
Kitasatospora
Streptomyces
Streptoverticillium
Suborder: Streptosporangineae
Family I: Streptosporangiaceae
Acrocarpospora
Herbidospora
Microbispora
Microtetraspora
Nonomuraea
Planobispora
Planomonospora
Planopolyspora
Planotetraspora
Streptosporangium
Family II: Nocardiopsaceae
Nocardiopsis
Streptomonospora
Thermobifida
Family III: Thermomonosporaceae
Actinomadura
Spirillospora

Thermomonospora
Suborder: Frankineae
Family I: Frankiaceae
Frankia
Family II: Geodermatophilaceae
Blastococcus
Geodermatophilus
Modestobacter
Family III: Microsphaeraceae
Microsphaera
Family IV: Sporichthyaceae
Sporichthya
Family V: Acidothermaceae
Acidothermus
Family VI: Kineosporiaceae
Cryptosporangium
Kineococcus
Kineosporia
Suborder: Glycomycineae
Family I: Glycomycetaceae
Glycomyces
Order VI: Bifidobacteriales
Family I: Bifidobacteriaceae
Aeriscardovia
Bifidobacterium
Falcivibrio
Gardnerella
Parascardovia
Scardovia
Family II: Unknown Affiliation
Actinobispora
Actinocorallia
Excellospora
Pelczaria
Turicella
Phylum XV: Planctomycetes
Order I: Planctomycetales
Family I: Planctomycetaceae
Gemmata
Isosphaera
Pirellula
Planctomyces
Phylum XVI: Chlamydiae
Order I: Chlamydiales
Family I: Chlamydiaceae
Chlamydia
Chlamydophila
Family II: Parachlamydiaceae
Neochlamydia
Parachlamydia
Family III: Simkaniaceae
Rhabdochlamydia
Simkania
Family IV: Waddliaceae
Waddlia
Phylum XVII: Spirochaetes
Class I: Spirochaetes
Order I: Spirochaetales
Family I: Spirochaetaceae
Borrelia
Brevinema
Clevelandina
Cristispira
Diplocalyx
Hollandina
Pillotina
Spirochaeta
Treponema
Family II: Serpulinaceae
Brachyspira
Serpulina
Family III: Leptospiraceae
Leptonema
Leptospira
Phylum XVIII: Fibrobacteres
Class I: Fibrobacteres
Family I: Fibrobacteraceae
Fibrobacter

Phylum XIX: Acidobacteria
    Family I: Acidobacteriaceae
      Acidobacterium
      Geothrix
      Holophaga
Phylum XX: Bacteroidetes
  Class I: Bacteroidetes
    Order I: Bacteroidales
      Family I: Bacteroidaceae
        Acetofilamentum
        Acetomicrobium
        Acetothermus
        Anaerophaga
        Anaerorhabdus
        Bacteroides
        Megamonas
      Family II: Rikenellaceae
        Alistipes
        Marinilabilia
        Rikenella
      Family III: Porphyromonadaceae
        Dysgonomonas
        Porphyromonas
        Tannerella
      Family IV: Prevotellaceae
        Prevotella
  Class II: Flavobacteria
    Order I: Flavobacteriales
      Family I: Flavobacteriaceae
        Aequorivita
        Arenibacter
        Bergeyella
        Capnocytophaga
        Cellulophaga
        Chryseobacterium
        Coenonia
        Croceibacter
        Empedobacter
        Flavobacterium
        Gelidibacter
        Gillisia

        Mesonia
        Muricauda
        Myroides
        Ornithobacterium
        Polaribacter
        Psychroflexus
        Psychroserpens
        Riemerella
        Saligentibacter
        Tenacibaculum
        Weeksella
        Zobellia
      Family II: Blattabacteriaceae
        Blattabacterium
  Class III: Sphingobacteria
    Order I: Sphingobacteriales
      Family I: Sphingobacteriaceae
        Pedobacter
        Sphingobacterium
      Family II: Saprospiraceae
        Haliscomenobacter
        Lewinella
        Saprospira
      Family III: Flexibacteraceae
        Belliella
        Cyclobacterium
        Cytophaga
        Dyadobacter
        Flectobacillus
        Flexibacter
        Hongiella
        Hymenobacter
        Meniscus
        Microscilla
        Reichenbachia
        Runella
        Spirosoma
        Sporocytophaga
      Family IV: Flammeovirgaceae
        Flammeovirga
        Flexithrix

        Persicobacter
        Thermonema
      Family V: Crenotrichaceae
        Chitinophaga
        Crenothrix
        Rhodothermus
        Salinibacter
        Toxothrix
Phylum XXI: Fusobacteria
  Class I: Fusobacteria
    Order I: Fusobacteriales
      Family I: Fusobacteriaceae
        Fusobacterium
        Ilyobacter
        Leptotrichia
        Propionigenium
        Sebaldella
        Sneathia
        Streptobacillus
      Family II: Incertae sedis
        Cetobacterium
Phylum XXII: Verrucomicrobia
  Class I: Verrucomicrobiae
      Opitutus
      Prosthecobacter
      Verrucomicrobium
      Victivallis
      Xiphinematobacter
Phylum XXIII: Dictyoglomi
  Class I: Dictyoglomi
    Order I: Dictyoglomales
      Family I: Dictyoglomaceae
        Dictyoglomus
Phylum XXIV: Gemmatimonadetes
  Class I: Gemmatimonadetes
    Order I: Gemmatimonadales
      Gemmatimonas

APPENDIX B

To diagnose a disease, it is often necessary to obtain a sample of material that may contain the pathogenic microorganism. Samples must be taken aseptically. The sample container should be labeled with the patient's name, room number (if hospitalized), date, time, and medications being taken. Samples must be transported to the laboratory immediately for culture. Delay in transport may result in the growth of some organisms, and their toxic products may kill other organisms. Pathogens tend to be fastidious and die if not kept in optimum environmental conditions.

In the laboratory, samples from infected tissues are cultured on differential and selective media in an attempt to isolate and identify any pathogens or organisms that are not normally found in association with that tissue.

UNIVERSAL PRECAUTIONS*

The following procedures should be used by all health care workers, including students, whose activities involve contact with patients or with blood or other body fluids. These procedures were developed to minimize the risk of transmitting HIV or AIDS in a health care environment, but adherence to these guidelines will minimize the transmission of *all* nosocomial infections.

1. Gloves should be worn when touching blood and body fluids, mucous membranes, and nonintact skin and when handling items or surfaces soiled with blood or body fluids. Gloves should be changed after contact with each patient.
2. Hands and other skin surfaces should be washed immediately and thoroughly if contaminated with blood or other body fluids. Hands should be washed immediately after gloves are removed.
3. Masks and protective eyewear or face shields should be worn during procedures that are likely to generate droplets of blood or other body fluids.
4. Gowns or aprons should be worn during procedures that are likely to generate splashes of blood or other body fluids.
5. To prevent needlestick injuries, needles should not be recapped, purposely bent or broken, or otherwise manipulated by hand. After disposable syringes and needles, scalpel blades, and other sharp items are used, they should be placed in puncture-resistant containers for disposal.
6. Although saliva has not been implicated in HIV transmission, mouthpieces, resuscitation bags, and other ventilation devices should be available for use in areas in which the need for resuscitation is predictable. Emergency mouth-to-mouth resuscitation should be minimized.
7. Health care workers who have exudative lesions or weeping dermatitis should refrain from all direct patient care and from handling patient-care equipment.
8. Pregnant health care workers are not known to have a greater risk of contracting HIV infection than health care workers who are not pregnant; however, if a health care worker develops HIV infection during pregnancy, the infant is at risk of infection. Because of this risk, pregnant health care workers should be especially familiar with, and strictly adhere to, precautions to minimize the risk of HIV transmission.

INSTRUCTIONS FOR SPECIFIC SAMPLING PROCEDURES

WOUND OR ABSCESS CULTURE

1. Cleanse the area with a sterile swab moistened in sterile saline.
2. Disinfect the area with 70% ethanol or iodine solution.
3. If the abscess has not ruptured spontaneously, a physician will open it with a sterile scalpel.
4. Wipe the first pus away.
5. Touch a sterile swab to the pus, taking care not to contaminate the surrounding tissue.
6. Replace the swab in its container, and properly label the container.

EAR CULTURE

1. Clean the skin and auditory canal with 1% tincture of iodine.
2. Touch the infected area with a sterile cotton swab.
3. Replace the swab in its container.

EYE CULTURE

This procedure is often performed by an ophthalmologist.

1. Anesthetize the eye with topical application of a sterile anesthetic solution.
2. Wash the eye with sterile saline solution.
3. Collect material from the infected area with a sterile cotton swab. Return the swab to its container.

BLOOD CULTURE

1. Close the room's windows to avoid contamination.
2. Clean the skin around the selected vein with 2% tincture of iodine on a cotton swab.
3. Remove dried iodine with gauze moistened with 80% isopropyl alcohol.
4. Draw a few milliliters of venous blood.
5. Aseptically bandage the puncture.

URINE CULTURE

1. Provide the patient with a sterile container.
2. Instruct the patient to first void a small volume from the urinary bladder before collection (to wash away extraneous bacteria of the skin microbiota) then to collect a midstream sample.
3. A urine sample may be stored under refrigeration (4°–6°C) for up to 24 hours.

FECAL CULTURE

For bacteriological examination, only a small sample is needed. This may be obtained by inserting a sterile swab into the rectum or feces. The swab is then placed in a tube of sterile enrichment broth for transport to the laboratory. For examination for parasites, a small sample may be taken from a morning stool. The sample is placed in a preservative (polyvinyl alcohol, buffered glycerol, saline, or formalin) for microscopic examination for eggs and adult parasites.

SPUTUM CULTURE

1. A morning sample is best because microorganisms will have accumulated while the patient is sleeping.
2. The patient should rinse his or her mouth thoroughly to remove food and normal microbiota.
3. The patient should cough deeply from the lungs and expectorate into a sterile glass wide-mouth jar.
4. Care should be taken to avoid contaminating health care workers.
5. In cases such as tuberculosis in which there is little sputum, stomach aspiration may be necessary.
6. Infants and children tend to swallow sputum. A fecal sample may be of some value in these cases.

---

*Source: Centers for Disease Control and Prevention and National Institutes of Health. *Biosafety in Microbiological and Biomedical Laboratories*. Available from www.cdc.gov.

APPENDIX C

METABOLIC PATHWAYS

FIGURE C.1 **The Calvin-Benson cycle for photosynthetic carbon metabolism.**

1–**3** The initial fixation and reduction of carbon occurs, generating the three-carbon compounds glyceraldehyde 3-phosphate and dihydroxyacetone phosphate, **4** which are interconvertible. **A**–**D** On average, 2 of every 12 three-carbon molecules are used in the synthesis of glucose. **5** Ten of every 12 three-carbon molecules are used to generate ribulose 5-phosphate by a complex series of reactions. **6** The ribulose 5-phosphate is then phosphorylated at the expense of ATP, forming ribulose 1,5-diphosphate, the acceptor molecule with which the sequence began. (See Figure 5.23, p. 143, for a simplified version of the Calvin-Benson cycle.)

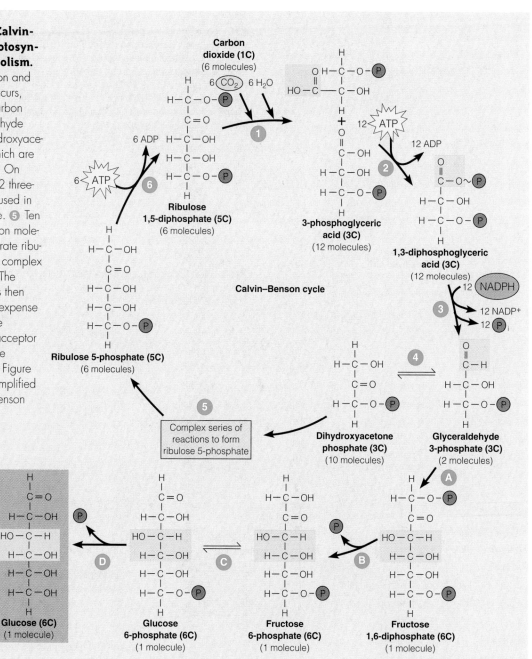

Carbon dioxide (1C) (6 molecules)

$6\ CO_2$ $6\ H_2O$

Ribulose 1,5-diphosphate (5C) (6 molecules)

6 ADP

$6\ ATP$ **6**

3-phosphoglyceric acid (3C) (12 molecules)

$12\ ATP$

12 ADP

Calvin–Benson cycle

1,3-diphosphoglyceric acid (3C) (12 molecules)

2

$12\ NADPH$

3

12 NADP+

12 P_i

Ribulose 5-phosphate (5C) (6 molecules)

5

Complex series of reactions to form ribulose 5-phosphate

Dihydroxyacetone phosphate (3C) (10 molecules)

4

Glyceraldehyde 3-phosphate (3C) (2 molecules)

A

B

Fructose 1,6-diphosphate (6C) (1 molecule)

C

Fructose 6-phosphate (6C) (1 molecule)

D

Glucose 6-phosphate (6C) (1 molecule)

Glucose (6C) (1 molecule)

871

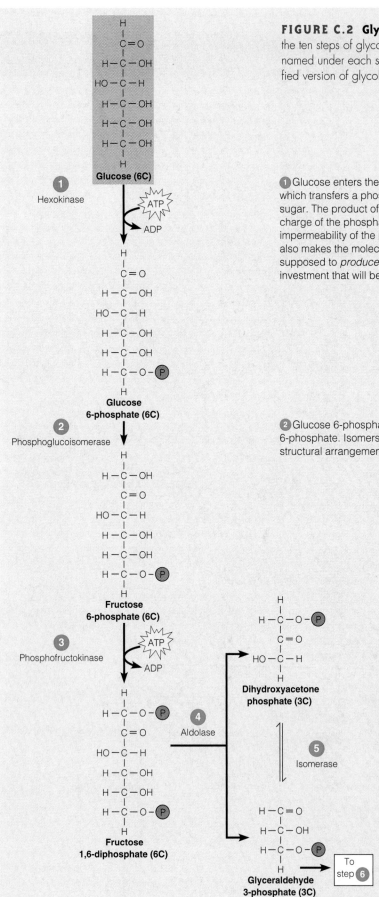

FIGURE C.2 Glycolysis (Embden-Meyerhof pathway). Each of the ten steps of glycolysis is catalyzed by a specific enzyme, which is named under each step number. (See Figure 5.12, p. 128, for a simplified version of glycolysis.)

1 Glucose enters the cell and is phosphorylated by the enzyme hexokinase, which transfers a phosphate group from ATP to the number 6 carbon of the sugar. The product of the reaction is glucose 6-phosphate. The electrical charge of the phosphate group traps the sugar in the cell because of the impermeability of the plasma membrane to ions. Phosphorylation of glucose also makes the molecule more chemically reactive. Although glycolysis is supposed to *produce* ATP, in step ①, ATP is actually consumed—an energy investment that will be repaid with dividends later in glycolysis.

2 Glucose 6-phosphate is rearranged to convert it to its isomer, fructose 6-phosphate. Isomers have the same number and types of atoms but in different structural arrangements.

3 In this step, still another molecule of ATP is invested in glycolysis. An enzyme transfers a phosphate group from ATP to the sugar, producing fructose 1,6-diphosphate.

4 This is the reaction from which glycolysis gets its name ("sugar splitting"). An enzyme cleaves fructose 1,6-diphosphate into two different three-carbon sugars: glyceraldehyde 3-phosphate and dihydroxyacetone phosphate. These two sugars are isomers.

5 The enzyme isomerase interconverts the three-carbon sugars. The next enzyme in glycolysis uses only glyceraldehyde 3-phosphate as its substrate. This pulls the equilibrium between the two three-carbon sugars in the direction of glyceraldehyde 3-phosphate, which is removed as fast as it forms.

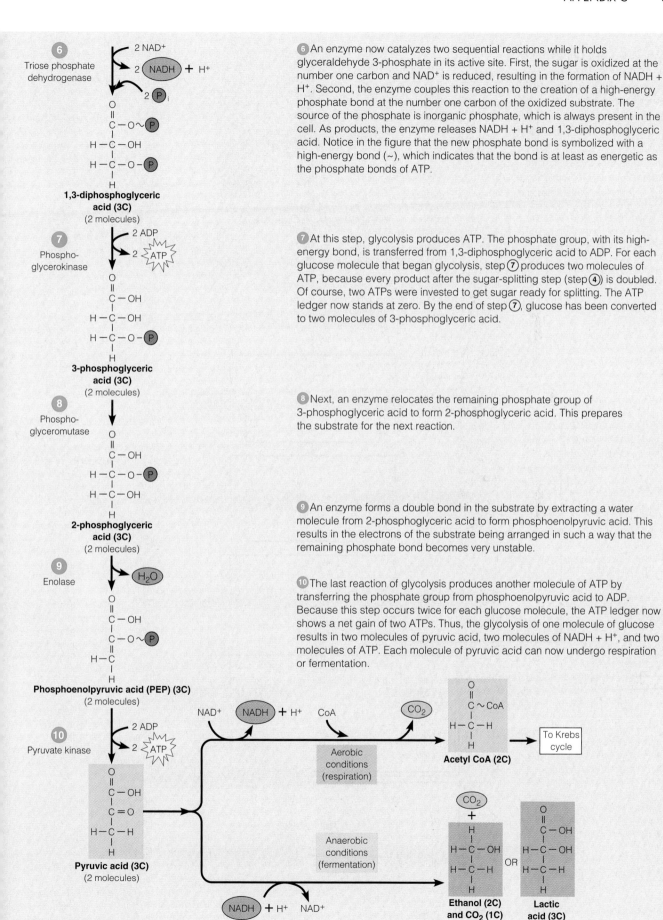

6
Triose phosphate dehydrogenase

2 NAD⁺
2 NADH + H⁺
2 Pᵢ

O
‖
C—O~P
|
H—C—OH
|
H—C—O—P
|
H

1,3-diphosphoglyceric acid (3C)
(2 molecules)

7
Phospho-glycerokinase

2 ADP
2 ATP

O
‖
C—OH
|
H—C—OH
|
H—C—O—P
|
H

3-phosphoglyceric acid (3C)
(2 molecules)

8
Phospho-glyceromutase

O
‖
C—OH
|
H—C—O—P
|
H—C—OH
|
H

2-phosphoglyceric acid (3C)
(2 molecules)

9
Enolase

H₂O

O
‖
C—OH
|
C—O~P
‖
H—C
|
H

Phosphoenolpyruvic acid (PEP) (3C)
(2 molecules)

10
Pyruvate kinase

2 ADP
2 ATP

O
‖
C—OH
|
C=O
|
H—C—H
|
H

Pyruvic acid (3C)
(2 molecules)

NAD⁺   NADH + H⁺   CoA   CO₂

Aerobic conditions (respiration)

O
‖
C~CoA
|
H—C—H
|
H

Acetyl CoA (2C)   → To Krebs cycle

Anaerobic conditions (fermentation)

NADH + H⁺   NAD⁺

CO₂
+
H
|
H—C—OH
|
H—C—H
|
H

Ethanol (2C) and CO₂ (1C)

OR

O
‖
C—OH
|
H—C—OH
|
H—C—H
|
H

Lactic acid (3C)

6 An enzyme now catalyzes two sequential reactions while it holds glyceraldehyde 3-phosphate in its active site. First, the sugar is oxidized at the number one carbon and NAD⁺ is reduced, resulting in the formation of NADH + H⁺. Second, the enzyme couples this reaction to the creation of a high-energy phosphate bond at the number one carbon of the oxidized substrate. The source of the phosphate is inorganic phosphate, which is always present in the cell. As products, the enzyme releases NADH + H⁺ and 1,3-diphosphoglyceric acid. Notice in the figure that the new phosphate bond is symbolized with a high-energy bond (~), which indicates that the bond is at least as energetic as the phosphate bonds of ATP.

7 At this step, glycolysis produces ATP. The phosphate group, with its high-energy bond, is transferred from 1,3-diphosphoglyceric acid to ADP. For each glucose molecule that began glycolysis, step ⑦ produces two molecules of ATP, because every product after the sugar-splitting step (step ④) is doubled. Of course, two ATPs were invested to get sugar ready for splitting. The ATP ledger now stands at zero. By the end of step ⑦, glucose has been converted to two molecules of 3-phosphoglyceric acid.

8 Next, an enzyme relocates the remaining phosphate group of 3-phosphoglyceric acid to form 2-phosphoglyceric acid. This prepares the substrate for the next reaction.

9 An enzyme forms a double bond in the substrate by extracting a water molecule from 2-phosphoglyceric acid to form phosphoenolpyruvic acid. This results in the electrons of the substrate being arranged in such a way that the remaining phosphate bond becomes very unstable.

10 The last reaction of glycolysis produces another molecule of ATP by transferring the phosphate group from phosphoenolpyruvic acid to ADP. Because this step occurs twice for each glucose molecule, the ATP ledger now shows a net gain of two ATPs. Thus, the glycolysis of one molecule of glucose results in two molecules of pyruvic acid, two molecules of NADH + H⁺, and two molecules of ATP. Each molecule of pyruvic acid can now undergo respiration or fermentation.

FIGURE C.3 The pentose phosphate pathway. This pathway, which operates simultaneously with glycolysis, provides an alternate route for the oxidation of glucose and plays a role in the synthesis of biological molecules, depending on the needs of the cell. Possible fates of the various intermediates are shown in blue. (See Chapter 5, p. 127.)

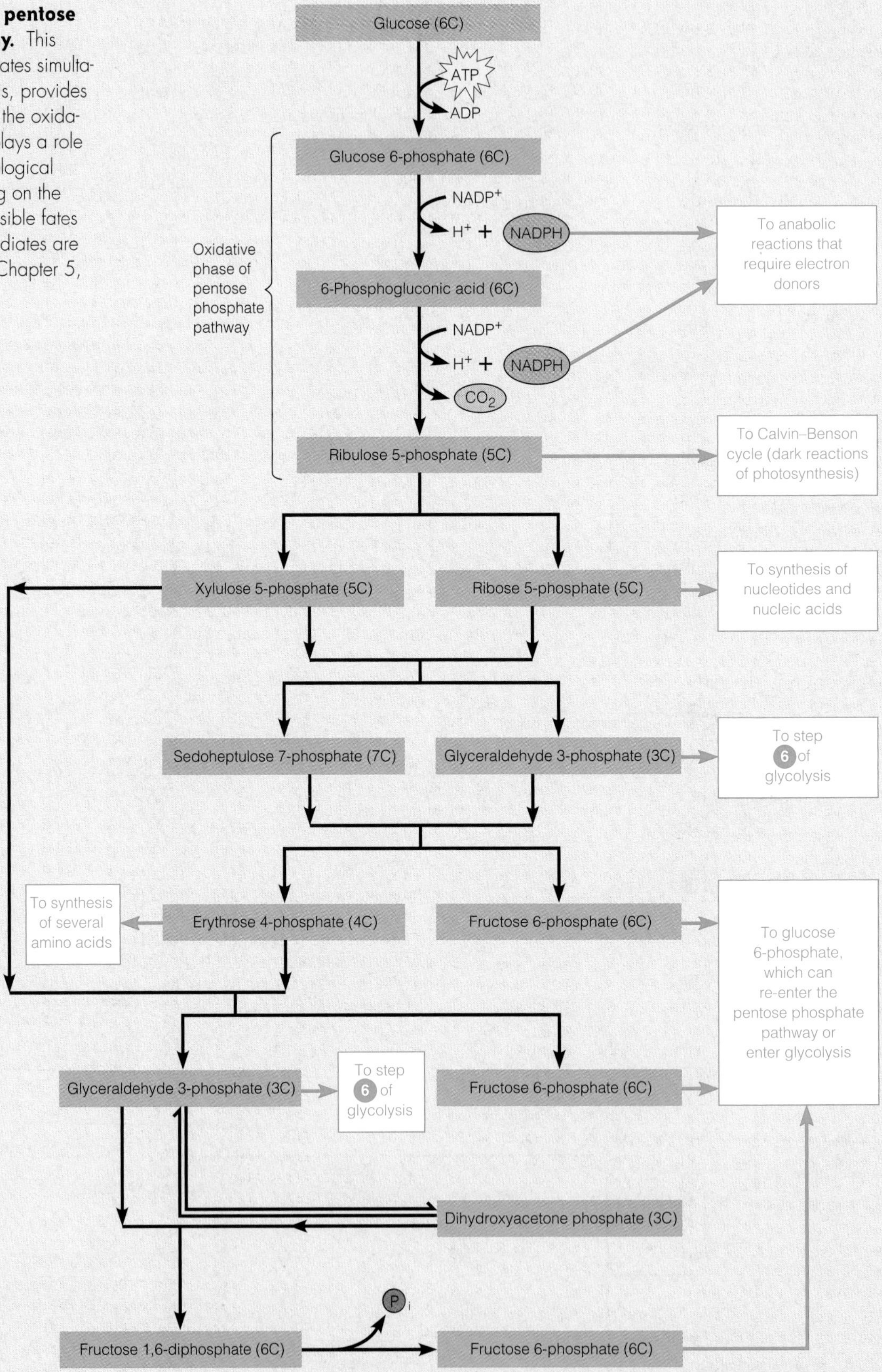

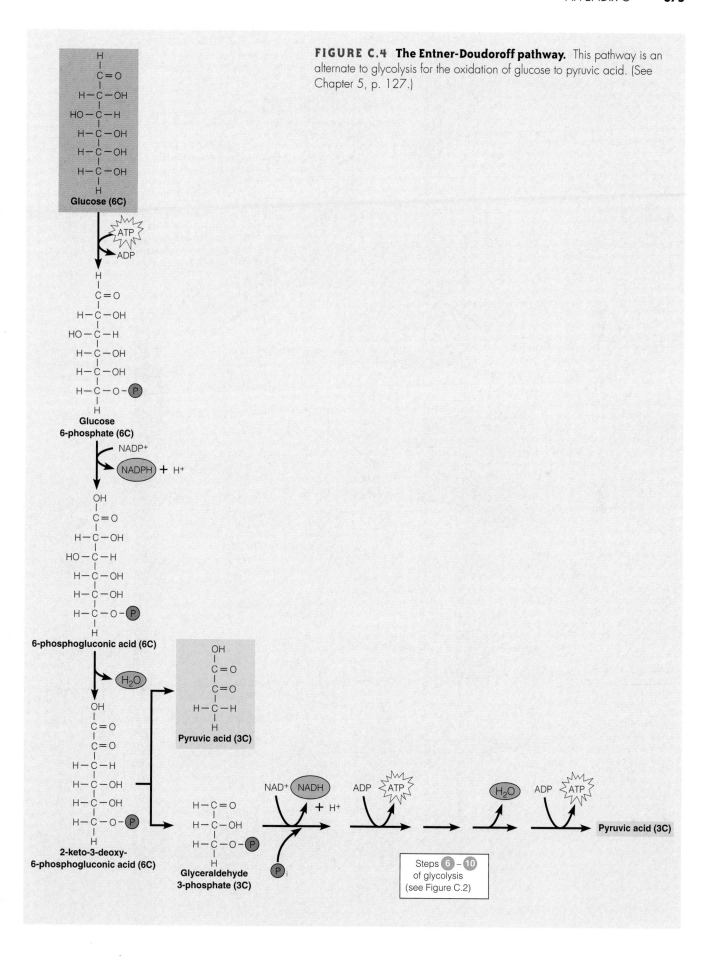

FIGURE C.4 The Entner-Doudoroff pathway. This pathway is an alternate to glycolysis for the oxidation of glucose to pyruvic acid. (See Chapter 5, p. 127.)

FIGURE C.5 The Krebs cycle. See Figure 5.13, p. 130, for a simplified version.

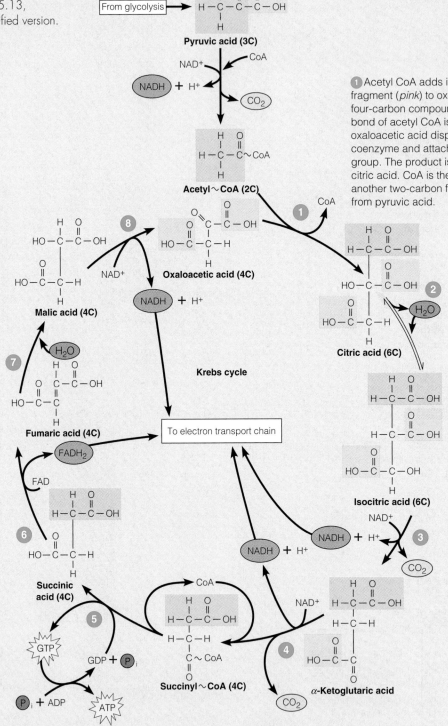

From glycolysis

Pyruvic acid (3C)

NAD⁺ → NADH + H⁺ + CO₂, CoA

Acetyl∼CoA (2C)

1 Acetyl CoA adds its two-carbon acetyl fragment (*pink*) to oxaloacetic acid, a four-carbon compound. The unstable bond of acetyl CoA is broken as oxaloacetic acid displaces the coenzyme and attaches to the acetyl group. The product is the six-carbon citric acid. CoA is then free to prime another two-carbon fragment derived from pyruvic acid.

8 The last oxidative step produces another molecule of NADH + H⁺ and regenerates oxaloacetic acid, which accepts a two-carbon fragment from acetyl CoA for another turn of the cycle.

Malic acid (4C)

Oxaloacetic acid (4C)

NAD⁺, NADH + H⁺

Citric acid (6C)

2 A molecule of water is removed, and another is added back. The net result is the conversion of citric acid to its isomer, isocitric acid.

7 Bonds in the substrate are rearranged in this step by the addition of a water molecule.

H₂O

Fumaric acid (4C)

Krebs cycle

To electron transport chain

Isocitric acid (6C)

NAD⁺

3 The substrate loses a CO₂ molecule (*pale violet*), and the remaining five-carbon compound is oxidized, reducing NAD⁺ to NADH + H⁺.

FADH₂

FAD

6 In another oxidative step, two hydrogens are transferred to FAD to form FADH₂. The function of this coenzyme is similar to that of NADH + H⁺, but FADH₂ stores less energy.

Succinic acid (4C)

GTP

GDP + Pᵢ

Pᵢ + ADP

ATP

Succinyl∼CoA (4C)

CoA

NADH + H⁺

NADH + H⁺, CO₂

α-Ketoglutaric acid

NAD⁺

CO₂

4 CO₂ (*pale violet*) is lost; the remaining four-carbon compound is oxidized by the transfer of electrons to NAD⁺ to form NADH + H⁺ and is then attached to CoA by an unstable bond.

5 Substrate-level phosphorylation occurs in this step. CoA is displaced by a phosphate group, which is then transferred to GDP to form guanosine triphosphate (GTP). GTP is similar to ATP, which is formed when GTP donates a phosphate group to ADP.

APPENDIX D

EXPONENTS, EXPONENTIAL LOGARITHMS, AND GENERATION TIME

EXPONENTS AND EXPONENTIAL NOTATION

Very large and very small numbers, such as 4,650,000,000 and 0.00000032, are cumbersome to work with. It is more convenient to express such numbers in exponential notation—that is, as a power of 10. For example, 4.65×10^9 is in standard exponential notation, or **scientific notation:** 4.65 is the *coefficient,* and 9 is the power or *exponent.* In standard exponential notation, the coefficient is always a number between 1 and 10, and the exponent can be positive or negative.

To change a number into exponential notation, follow two steps. First, determine the coefficient by moving the decimal point so there is only one nonzero digit to the left of it. For example,

The coefficient is 3.2. Second, determine the exponent by counting the number of places you moved the decimal point. If you moved it to the left, the exponent is positive. If you moved it to the right, the exponent is negative. In the example, you moved the decimal point seven places to the right, so the exponent is -7. Thus

$$0.00000032 = 3.2 \times 10^{-7}$$

Now suppose you are working with a larger number instead of a very small number. The same rules apply, but the exponential value will be positive rather than negative. For example,

$$4,650,000,000 = 4.65 \times 10^9$$

To multiply numbers written in exponential notation, multiply the coefficients and *add* the exponents. For example,

$$(3 \times 10^4) \times (2 \times 10^3) =$$
$$(3 \times 2) \times (10^{4+3}) = 6 \times 10^7$$

To divide, divide the coefficient and *subtract* the exponents. For example,

$$\frac{3 \times 10^4}{2 \times 10^3} = \frac{3}{2} \times 10^{4-3} = 1.5 \times 10^1$$

Microbiologists use exponential notation in many situations. For instance, exponential notation is used to describe the number of microorganisms in a population. Such numbers are often very large (see Chapter 6). Another application of exponential notation is to express concentrations of chemicals in a solution—chemicals such as media components (Chapter 6), disinfectants (Chapter 7), or antibiotics (Chapter 20). Such numbers are often very small. Converting from one unit of measurement to another in the metric system requires multiplying or dividing by a power of 10, which is easiest to carry out in exponential notation.

LOGARITHMS

A **logarithm (log)** is the power to which a base number is raised to produce a given number. Usually we work with logarithms to the base 10, abbreviated $\log_{10}$. The first step in finding the $\log_{10}$ of a number is to write the number in standard exponential notation. If the coefficient is exactly 1, the $\log_{10}$ is simply equal to the exponent. For example

$$\log_{10} 0.00001 = \log_{10} (1 \times 10^{-5})$$
$$= -5$$

If the coefficient is not 1, as is often the case, the logarithm function on a calculator must be used to determine the logarithm.

Microbiologists use logs for calculating pH levels and for graphing the growth of microbial populations in culture (see Chapter 6).

CALCULATING GENERATION TIME

As a cell divides, the population increases exponentially. Numerically this is equal to 2 (because one cell divides into two) raised to the number of times the cell divided (generations);

$$2^{\text{number of generations}}$$

To calculate the final concentration of cells:

$$\frac{\text{Initial number}}{\text{of cells}} \times 2^{\text{number of generations}} = \frac{\text{Number}}{\text{of cells}}$$

For example, if 5 cells were allowed to divide 9 times, this would result in

$$5 \times 2^9 = 2560 \text{ cells}$$

To calculate the number of generations a culture has undergone, cell numbers must be converted to logarithms. Standard logarithm values are based on 10. The log of 2 (0.301) is used because one cell divides into two.

$$\text{Number of generations} = \frac{\text{log number} - \text{log number of}}{0.301}$$

To calculate the generation of time for a population:

$$\frac{60 \text{ min/hr} \times \text{hours}}{\text{number of generations}} = \text{minutes/generation}$$

As an example, we will calculate the generation time if 100 bacterial cells growing for 5 hours produced 1,720,320 cells:

$$\frac{\log 1,720,320 - \log 100}{0.301} = 14 \text{ generations}$$

$$\frac{60 \text{ min/hr} \times 5 \text{ hours}}{14 \text{ generations}} = 21 \text{ minutes/generation}$$

A practical application of the calculation is determining the effect of a newly developed food preservative on the culture. Suppose 900 of the same species were grown under the same conditions as the previous example, except that the preservative was added to the culture medium. After 15 hours, there were 3,276,800 cells. Calculate the generation time, and decide whether the preservative inhibited growth.

Answer: 75 minutes/generation. The preservative did inhibit growth.

APPENDIX E

RULES OF PRONUNCIATION

The easiest way to learn new material is to talk about it, and that requires saying scientific names. Scientific names may look difficult at first glance, but keep in mind that generally every *syllable* is pronounced. The primary requirement in saying a scientific name is to communicate it.

The rules for the pronunciation of scientific names depend, in part, on the derivation of the root word and its vowel sounds. We have provided some general guidelines here. Pronunciations frequently do not follow the rules because a common usage has become "accepted" or the derivation of the name cannot be determined. For many scientific names there are alternative correct pronunciations.

VOWELS

Pronounce all the vowels in scientific names. Two vowels written together and pronounced as one sound are called a *diphthong* (for example, the *ou* in *sound*). A special comment is needed about the pronunciation of the vowel endings *-i* and *-ae*: There are two alternative ways to pronounce each of these. In this book, we usually give the pronunciation of a long *e* (ē) to the *-i* ending, and a long *i* (ī) to the *-ae* ending. However, the reverse pronunciations are also correct and in some cases are preferred. For example, *coli* is usually pronounced kō′lī.

CONSONANTS

When *c* or *g* is followed by *ae, e, oe, i,* or *y,* it has a soft sound. When *c* or *g* is followed by *a, o, oi,* or *u,* it has a hard sound. When a double *c* is followed by *e, i,* or *y,* it is pronounced as *ks* (e.g., cocci).

ACCENT

The accented syllable is usually the next-to-last or third-to-last syllable.

1. The accent is on the next-to-last syllable:
   a. When the name contains only two syllables. Example: pes′tis.
   b. When the next-to-last syllable is a diphthong. Example: a-kan-thä-mē′bä.
   c. When the vowel of the next-to-last syllable is long. Example: tre-pō-nē′mä. The vowel in the next-to-last syllable is long in words ending in the following suffixes:

Suffix	Example
-ales	Orders such as Eubacteriales
-ina	Sarcina
-anus, -anum	pasteurianum
-uta	diminuta

   d. When the word ends in one of the following suffixes:

Suffix	Example
-atus, -atum	caudatum
-ella	Salmonella

2. The accent is on the third-to-last syllable in family names. Families end in *-aceae,* which is always pronounced -ā′sē-ē.

PRONUNCIATION OF MICROORGANISMS IN THIS TEXT

Pronunciation key:

a	hat	ē	see	o	hot	th	thin
ā	age	ė	term	ō	go	u	cup
ã	care	g	go	ô	order	ů	put
ä	father	i	sit	oi	oil	ü	rule
ch	child	ī	ice	ou	out	ū	use
e	let	ng	long	sh	she	zh	seizure

Acanthamoeba polyphaga   a-kan-thä-mē′bä pol′if-ä-gä
Acetobacter   a-sē′tō-bak-tėr
Acinetobacter   a-si-ne′ tō-bak-tėr
Actinomyces israelii   ak-tin-ō-mī′ sēs is-rā′lē-ē
Aedes aegypti   ā′ē-dēz ē-jip′tē
A. albopictus   al-bō-pik′tus
Aeromonas hydrophilia   ār′ō-mō-nas hī′dro-fil-ē-ä
Afipia felis   ä-fi′pē-ä fē′lis
Agrobacterium tumefaciens   ag′rō-bak-ti′rē-um tü′me-fāsh-enz
Ajellomyces   ä-jel-lō-mī′sēs
Alcaligenes   al′kä-li-gen-ēs
Alexandrium   äl-eg-zan′drē-um
Amanita phalloides   am-an-ī′ta fal-loi′dēz
Anabaena   an-ä-bē′nä
Ancylostoma duodenale   an-sil-os′tō-mä dü′o-den-al-ē
Anopheles   an-of′e-l-ēz
Aquaspirillum bengal   ä-kwä-spī-ril′lum ben′gal
A. graniferum   gra-ni′fėr-um
A. serpens   sėr′pens
Arthroderma   är-thrō-dėr′mä
Ascaris lumbricoides   as′kar-is lum-bri-koi′dēz
Ashbya gossypii   ash′bē-ä gos-sip′ē-ē
Aspergillus flavus   a-spėr-jil′lus flā′vus
A. fumigatus   fü-mi-gä′tus
A. niger   nī′jėr
A. oryzae   ô′ri-zī
A. sojae   sō′gī
Azolla   ä-zō′lä
Azomonas   ā-zō-mō′nas
Azospirillum   ā-zō-spī′ril-lum
Azotobacter   ä-zo′tō-bak-tėr
Babesia microti   ba-bē′sē-ä mī-krō′tē
Bacillus anthracis   bä-sil′lus an-thrā′sis
B. cereus   se′rē-us
B. circulans   sėr′ku-lans
B. coagulans   kō-ag′ū-lanz
B. licheniformis   lī-ken-i-fôr′mis
B. megaterium   meg-ä-tėr′ē-um
B. sphaericus   sfe′ri-kus
B. subtilis   su′til-us
B. thuringiensis   thùr-in-jē-en′sis
Bacteroides hypermegas   bak-tė-roi′dēz hī-pėr′meg-äs
Balamuthia   bal′am-üth-ē-ä
Balantidium coli   bal-an-tid′ē-um kō′!ī (or kō′lē)
Bartonella henselae   bär′tō-nel-lä hen′sel-ī
Bdellovibrio bacteriovorus   del-lō-vib′rē-ō bak-tė-rē-o′vô-rus
Beauveria   bō-vär′ē-a
Beggiatoa   bej′jē-ä-tō-ä

Beijerinckia   bī-yė-rink′ē-ä
Bifidobacterium globosum   bī-fi-dō-bak-ti′rē-um glob-ō′sum
B. pseudolongum   sū-dō-lông′um
Blastomyces dermatitidis   blas-tō-mī′sēz dėr-mä-tit′i-dis
Blattabacterium   blat-tä-bak-ti′rē-um
Boletus edulis   bō-lē′tus e′dū-lis
Bordetella pertussis   bòr′de-tel-lä pėr-tus′sis
Borrelia burgdorferi   bôr′-rel-ē-ä burg-dôr′fėr-ē
Bradyrhizobium   brad-ē-rī-zō′bē-um
Brucella abortus   brü′sel-lä ä-bôr′tus
B. melitensis   me-li-ten′sis
B. suis   sü′is
Burkholderia   bėrk′hōld-ėr-ē-ä
B. andropogonis   ān′dro-po-gō-nis
B. cepacia   se-pā′sē-ä
B. pseudomallei   sū-dō-mal′le-ē
Byssochlamys fulva   bis-sō-klam′is fül′vä
Campylobacter fetus   kam′pi-lō-bak-tėr fē′tus
C. jejuni   jē-jū′nē
Candida albicans   kan′did-ä al′bi-kanz
C. glabrata   gla′brä-tä
C. krusei   krūs′ä-ē
C. oleophila   ō-lē-of′il-ä
C. utilis   ū′til-is
Canis familiaris   kānis fa-mil′ē-âr-is
Carpenteles   kär-pen′tel-ēz
Caulobacter   kô-lō-bak′tėr
Cephalosporium   sef-ä-lō-spô′rē-um
Ceratocystis ulmi   sē-rä-tō-sis′tis ul′mē
Chilomastix   kē′lō-ma-sticks
Chlamydia trachomatis   kla-mi′dē-ä trä-kō′mä-tis
Chlamydomonas   klam-i-dō-mō′näs
Chlamydophila pneumoniae   kla-mi-do′fil-ä nü-mō′nē-ī
C. psittaci   sit′tä-sē
Chlorobium   klô-rō′bē-um
Chloroflexus   klô-rō-flex′us
Chromatium vinosum   krō-mā′tē-um vi-nō′sum
Chroococcus turgidus   krō-ō-kok′kus tėr′ji-dus
Chrysops   krī′sops
Citrobacter   sit′rō-bak-tėr
Claviceps purpurea   kla′vi-seps pùr-pù-rē′ä
Clonorchis sinensis   klo-nôr′kis si-nen′sis
Clostridium acetobutylicum   klôs-tri′dē-um a-sē-tō-bū-til′li-kum
C. botulinum   bo-tū-lī′num
C. butyricum   bü-ti′ri-kum
C. difficile   dif′fi-sil
C. pasteurianum   pas-tyėr-ē-ā′num
C. perfringens   pėr-frin′jens
C. sporogenes   spô-rä′jen-ēz
C. tetani   te′tan-ē
Coccidioides immitis   kok-sid-ē-oi′dēz im′mi-tis
Corynebacterium diphtheriae   kôr′ī-nē-bak-ti-rē-um dif-thi′rē-ī
C. glutamicum   glü-tam′i-kum
C. xerosis   ze-rō′sis
Coxiella burnetii   käks′ē-el-lä bėr-ne′tē-ē
Crenarchaeota   kren-ärk-e′ ō-tä
Cristispira pectinis   kris-tē-spī′rä pek′tin-is
Crucibulum vulgare   krü-si-bū′lum vul′gär-ē
Cryphonectria parasitica   kri-fō-nek′trē-ä par-ä-si′ti-kä
Cryptococcus neoformans   krip′tō-kok-kus nē-ō-fôr′manz
Cryptosporidium hominis   krip′tō-spô-ri-dē-um ho′min-is
C. parvum   pär′vum
Culex   kū′leks
Culiseta   kū-li′se-tä
Cyclospora cayetanensis   sī′klō-spô-rä kī′ē-tan-en-sis
Cytophaga   sī-täf′äg-ä
Daptobacter   dap′to-bak-tėr
Dermacentor andersoni   dėr-mä-sen′tôr an-dėr-sōn′ē
D. variabilis   vär-ē-a′bil-is
Desulfotomaculum nigrificans   dē′sul-fō-to-ma-kū-lum nī′gri-fi kans
Desulfovibrio desulfuricans   dē′sul-fō-vib-rē-ō dē-sul-fėr′i-kans

Didinium nasutum   dī-di′nē-um nä-süt′um
Diphyllobothrium latum   dī-fil-lō-bo′thrē-um lā′tum
Dracunculus medininsis   dra-kun′ku-lus med-in′in-sis
Echinococcus granulosus   ē-kīn-ō-kok′kus gra-nū-lō′sus
Ectothiorhodospira mobilis   ek′tō-thī-ō-rō-dō-spī-rä mō′bil-is
Ehrlichia   ėr′lik-ē-ä
Emericella   em′ėr-ē-sel-lä
Emmonsiella   em′mon-sē-el-lä
Entamoeba dispar   en-tä-mē′bä dis′par
E. histolytica   his-tō-li′ti-kä
Enterobacter aerogenes   en-te-rō-bak′tėr ā-rä′jen-ēz
E. cloacae   klō-ā′kē
Enterobius vermicularis   en-te-rō′bē-us ver-mi-kū-lar′is
Enterococcus faecalis   en-te-rō-kok′kus fē-kā′lis
E. faecium   fē′sē-um
Entomophaga   en′tō-mo-fäg-ä
Epidermophyton   ep-i-dėr-mō-fī′ton
Epulopiscium   ep′ū-lō-pis-sē-um
Erwinia dissolvens   ėr-wi′nē-ä dis-solv′ens
Erysipelothrix rhusiopathiae   ār-i-si-pel′ō-thrix rus-ē-ō-path′ē-ī
Erysiphe grammis   âr′i-sīf gram′mis
Escherichia coli   esh-ė-rik′ē-ä kō′lī (or kō′lē)
Euglena   ū-glē′nä
Eupenicillium   ū-pen-i-sil′lē-um
Eurotium   yėr-ō′tē-um
Filobasidiella   fi-lō-ba-si-dē-el′lä
Fonsecaea pedrosoi   fon-se′kē-ä pe-drō′sō-ē
Francisella tularensis   fran′sis-el-lä tü′lä-ren-sis
Frankia   frank′ē-ä
Fusarium   fu′ sār-ē-um
Fusobacterium   fü-sō-bak-ti′rē-um
Gambierdiscus toxicus   gam′bē-ėr-dis-kus toks′i-kus
Gardnerella vaginalis   gärd-nė-rel′lä va-jin-al′is
Gelidium   jel-id′ē-um
Geobacillus stearothermophilus   gē′ō-bä-sil-lus ste-rō-thėr-mä′fil-us
Geotrichum   jē-ō-trik′um
Giardia lamblia   jē-är′dē-ä lam′lē-ä
Glaucocystis nostochinearum   glou′ko-sis-tis no′stok-in-ē-âr-um
Gloeocapsa   glē-ō-kap′sä
Glossina   gläs-sē′nä
Gluconacetobacter xylinus   glü′kon-a-sē-tō-bak-tėr zy′lin-us
Gluconobacter   glü′kon-ō-bak-tėr
Gracilaria   gra′sil-âr-ē-ä
Gymnoascus   jim-nō-as′kus
Gymnodynium breve   jim-nō-din′ē-um brev′ē
Haemophilus ducreyi   hē-mä′fil-us dü-krä′ē
Haloarcula   hä′lō-är-kū-lä
Halobacterium halobium   ha-lō-bak-ti′rē-um hal-ō′bē-um
Halococcus   hä′lō kok-kus
Helicobacter pylori   hē′lik-ō-bak-tėr pī′lô-rē
H. influenzae   in-flü-en′zī
H. vaginalis   va-jin-al′is
Histoplasma capsulatum   his-tō-plaz′mä kap-su-lä′tum
Holospora   hō-lo′spô-rä
Homo sapiens   hō′mō sā′pē-ens
Hydrogenomonas   hī-drō-je-nō-mō′näs
Hyphomicrobium   hī-fō-mī-krō′bē-um
Isospora   ī-so′spô-rä
Ixodes scapularis   iks-ō′dēs skap-ū-lār′is
I. pacificus   pas-i′fi-kus
Klebsiella pneumoniae   kleb-sē-el′lä nü-mō′nē-ī
Lactobacillus delbrueckii   lak-tō-bä-sil′lus del-brük′ē-ē
L. leichmannii   līk-man′nē-ē
L. plantarum   plan-tä′rum
L. sanfranciscensis   san-fran-si′sken-sis
Lactococcus lactis   lak-tō-kok′kus lak′tis
Laminaria japonica   lam′i-när-e-ä ja-pon′i-kä
Legionella pneumophila   lē-jä-nel′lä nü-mō′fi-lä
Leishmania braziliensis   lish′mä-nē-ä brä-sil′ē-en-sis
L. donovani   don′ō-van-ē
L. tropica   trop′i-kä

Leptospira interrogans   lep-tō-spī′rä in-tèr′rä-ganz
Leuconostoc mesenteroides   lü-kō-nos′tok mes-en-ter-oi′dēz
Listeria monocytogenes   lis-te′rē-ä mo-nō-sī-tô′je-nēz
Magnetospirillum magnetotacticum   mag-nē-tō-spī′ril-lum mag-ne-tō-tak′ti-kum
Malassezia furfur   mal′as-sēz-ē-a fur′fur
Mannheimia haemolytica   man-hī′me-ä hē′mō-li-ti-kä
Meniscus   mē-nis′kus
Methanobacterium   meth-a-nō-bak-tèr′ē-um
Methanosarcina   meth-a-nō-sär′sī-nä
Methylophilus methylotrophus   meth-i-lo′fi-lus meth-i-lō-trōf′us
Microcladia   mī-krō-klād′ē-ä
Micrococcus luteus   mī-krō-kok′kus lü′tē-us
Micromonospora purpurea   mī-krō-mo-nä′spô-rä pûr-pù-rē′ä
Microsporum gypseum   mī-krō-spô′rum jip′sē-um
Mixotricha   mix-ō-trik′ä
Moraxella catarrhalis   mô-raks-el′lä ka-tär′al-is
M. lacunata   la-kü-nä′tä
M. osloensis   os′lō-en-sis
Mucor indicus   mū′kôr in′di-kus
Mycobacterium avium-intracellulare   mī-kō-bak-ti′rē-um ā′vē-um-in′trä-cel-ū-lä-rē
M. bovis   bō′vis
M. leprae   lep′rī
M. smegmatis   smeg-ma′tis
M. tuberculosis   tü-bèr-kü-lō′sis
Mycoplasma hominis   mī-kō-plaz′mä ho′mi-nis
M. pneumoniae   nu-mō′nē-ī
Myxococcus fulvus   micks-ō-kok′kus ful′vus
Naegleria fowleri   nī-gle′rē-ä fou′lèr-ē
Nannizia   nan′nē-zē-ä
Necator americanus   ne-kā′tôr ä-me-ri-ka′nus
Neisseria gonorrhoeae   nī-se′rē-ä go-nôr-rē′ī
N. meningitidis   me-nin-ji′ti-dis
Nereocystis   nē-rē-ō-sis′tis
Nitrobacter   nī-trō-bak′tèr
Nitrosomonas   nī-trō-sō-mō′näs
Nocardia asteroides   nō-kär′dē-ä as-tèr-oi′dēz
Nosema locustae   nō′sē-mä lō′kust-ī
Oocystis   ō-ō-sis′tis
Opisthorchis sinensis   [opisthorchis] si-nen′sis
Ornithodorus   ôr-nith-ō′dô-rus
Paenibacillus polymixa   pi′nē-bä-sil-lus pò-lē-miks′ä
Pantoea agglomerans   pan′tō-ē-ä äg′glom-ér-anz
Paracoccus denitrificans   pär-ä-kok′kus dē-nī-tri′fi-kanz
Paragonimus westermani   pär-ä-gōn′e-mus we-stèr-ma′nē
Paramecium multimicronucleatum   pär-ä-mē′sē-um mul-tē-mī-krō-nü-klē-ä′tum
Pasteurella multocida   pas-tyèr-el′lä mul-tō′si-dä
Pediculus humanus corporis   ped-ik′ū-lus hü′ma-nus kôr′pô-ris
Pediococcus acidilactici   pe-dē-ō-kok′kus a-sid-i-lak′ti-sē
Pelagibacter ubique   pel-aj′ē-bak-tèr ū′bēk
Penicillium chrysogenum   pen-i-sil′lē-um krī-so′jen-um
P. griseofulvum   gri-sē-ō-fúl′vum
Peridinium   per-i-din′ē-um
Petriellidium   pet-rē-el-li′dē-um
Pfiesteria   fē′ster-ē-ä
Phormidium luridum   fôr-mi′dē-um ler′i-dum
Photobacterium   fō′tō-bak-ti-rē-um
Phytophthora cinnamoni   fī-tof′thô-rä cin′nä-mō-nē
P. infestans   in-fes′tans
P. ramorum   ra′môr-um
Phytoplasma asteris   fī′tō-plaz-mä as′tèr-is
Pityrosporum ovale   pit-i-ros′pô-rum ovale
Plasmodium falciparum   plaz-mō′dē-um fal-sip′är-um
P. malariae   mä-lā′rī-ī
P. ovale   ō-vä′lē
P. vivax   vī′vaks
Plesiomonas shigelloides   ple-sē-ō-mō′nas shi-gel-loi′des
Pneumocystis jiroveci   nü-mō-sis′tis ye-rō′vet-zē
Porphyromonas   pôr′fī-rō-mō-nas

Prevotella intermedia   prev′ō-tel-la in′tèr-mē-dē-ä
Prochlorococcus   prō-klôr′ō-kok-kus
Propionibacterium acnes   prō-pē-on′ē-bak-ti-rē-um ak′nēz
P. freudenreichii   froi-den-rīk′ē-ē
Proteus mirabilis   prō′tē-us mi-ra′bi-lis
P. vulgaris   vul-ga′ris
Prototheca   prō-tō-thā′kä
Pseudomonas aeruginosa   sü-dō-mō′nas ā-rü-ji-nō′sä
P. fluorescens   flôr-es′ens
P. syringae   sèr-in′jī
Pyridictum abyssi   pir-i′dik-tum a-bis′sē
Quercus   kwer′kus
Rhizobium japonicum   rī-zō′bē-um jap-on′i-kum
R. meliloti   mel-i-lot′ē
Rhizopus stolonifer   rī′zō-pùs stō′lon-i-fèr
R. oryzae   ō′rī-zī
Rhodococcus bronchialis   rō-dō-kok′kus bron-kē′al-is
R. erythropolis   er-i-throp′ō-lis
Rhodopseudomonas   rō-dō-su-dō-mō′nas
Rhodospirillum rubrum   rō-dō-spī-ril′um rūb′rum
Rickettsia prowazekii   ri-ket′sē-ä prou-wä-ze′kē-ē
R. rickettsii   ri-ket′sē-ē
R. typhi   tī′fē
Riftia   rift′ē-ä
Rosa multiflora   rō-sä mul-ti-flô′rä
Saccharomyces beticus   sak-ä-rō-mī′sēs bet′i-kus
S. cerevisiae   se-ri-vis′ē-ī
S. exiguus   egz-ij′ū-us
S. fragilis   fra′jil-is
Salmonella bongori   sal′mön-el-lä bon′gôr-ē
S. choleraesuis   kol-ér-ä-sü′is
S. enterica   en-ter′i-kä
S. enteritidis   en-tèr-ī′ti-dis
S. typhi   tī′fē
S. typhimurium   tī-fi-mùr′ē-um
Sarcoptes   sär-kop′tēs
Sargassum   sär-gas′sum
Sartorya   sär-tô′rē-ä
Schistosoma haemotobium   shis-tō-sō′mä or skis-tō-sō′mä hē′mō-tō-bē-um
Schizosaccharomyces   skiz-ō-sak-ä-rō-mī′sēs
Serratia marcescens   ser-rä′tē-ä mär-ses′sens
Shigella boydii   shi-gel′lä boi′dē-ē
S. dysenteriae   dis-en-te′rē-ī
S. flexneri   fleks′nèr-ē
S. sonnei   sōn′ne-ē
Sphaerotilus natans   sfe-rä′ti-lus nä′tans
Spirillum minor   spī′ril-lum mī′nôr
S. volutans   vō′lū-tans
Spiroplasma   spī-rō-plaz′mä
Spirosoma   spī-rō-sō′mä
Spirulina   spī-rü-lī′nä
Spongomorphora   spon′jō-môr-fô-rä
Sporothrix schenkii   spô-rō′thriks shen′kē-ē
Stachybotrys   stak′ē-bo-tris
Staphylococcus aureus   staf-i-lō-kok′kus ô′rē-us
S. epidermidis   e-pi-der′mi-dis
S. saprophyticus   sap-rō-fit′i-kus
Stella   stel′lä
Stigmatella aurantiaca   stig-mä′tel-lä ô-rän-tē′ä-kä
Streptobacillus moniliformis   strep-tō-bä-sil′lus mon′-il-i-fôr-mis
Streptococcus agalactiae   strep-tō-kok′kus ā′gal-act-ē-ī
S. mutans   mū′tans
S. pneumoniae   nü-mō′nē-ī
S. pyogenes   pī-äj′en-ēz
S. thermophilus   thèr-mo′fil-us
Streptomyces aureofaciens   strep-tō-mī′sēs ô-rē-ō-fa′si-ens
S. erythraeus   ā-rith′rē-us
S. fradiae   frā′dē-ī
S. griseus   gri′sē-us
S. hygroscopius   hī′grō-skō-pē-us
S. lactamdurans   lak′tam-dùr-anz

S. nodosus   nō-dō′sus
S. noursei   nôr′sē-ī
S. olivoreticuli   ō-liv-ō-re-tik′ū-lē
S. venezuelae   ve-ne-zü-e′lī
Sulfolobus   sul′fō-lō-bus
Synechococcus   sin′ē-kō-kok-kus
Taenia saginata   te′nē-ä sa-ji-nä′tä
T. solium   sō′lē-um
Talaromyces   ta-lä-rō-mī′sēs
Taxomyces   tacks′ō-mī-sēs
Tetrahymena   tet-rä-hī′me-nä
Thermoactinomyces vulgaris   thėr-mō-ak-tin-ō-mī′sēs vul-ga′ris
Thermoanaerobium thermosaccharolyticum   thėr′mō-an-e-rō-bē-um thėr-mō-sak-kär-ō-li′ti-kum
Thermoplasma   thėr-mō-plaz′mä
Thermotoga maritima   thėr′mō-tō-gä mar-it′ē-mä
Thiobacillus ferrooxidans   thī-ō-bä-sil′lus fer-rō-oks′i-danz
T. thiooxidans   thī-ō-oks′i-danz
Thiocapsa floridana   thī-ō-kap′sä flôr′i-dä-nä
Thiomargarita namibiensis   thī′ō-mär-gär-ē-tä na′mi-be-en-sis
Thunnus   tün′nus
Toxoplasma gondii   toks-ō-plaz′mä gon′dē-ē
Trachelomonas   trä-kel-ō-mōn′as
Treponema pallidum   tre-pō-nē′mä pal′li-dum
Triatoma   trī-ä-tō′ma
Tribonema vulgare   trī′bō-nē-mä vul′gär-ē
Trichinella nativa   trik-in-el′lä na′tē-vä

T. spiralis   spī-ra′lis
Trichoderma   trik′ō-dėr-mä
Trichomonas vaginalis   trik-ō-mōn′as va-jin-al′is
Trichonympha sphaerica   trik-ō-nimf′ä sfe′ri-kä
Trichophyton   trik-ō-fī′ton
Trichosporon   trik-ō-spôr′on
Trichuris trichiura   trik′ėr-is trik-ē-yėr′a
Tridacna   trī-dak′nä
Tropheryma   trō-fer-ē′mä
Trypanosoma brucei gambiense   tri-pa′nō-sō-mä brüs′ē gam-bē-ens′
T. brucei rhodesiense   rō-dē-sē-ens′
T. cruzi   kruz′ē
Ureaplasma urealyticum   ū-rē-ä-plaz′mä ū-rē-ä-lit′i-kum
Usnea   üs′nē-ä
Veillonella   vī-lo-nel′lä
Vibrio cholerae   vib′rē-ō kol′ėr-ī
V. parahaemolyticus   pa-rä-hē-mō-li′ti-kus
V. vulnificus   vul′ni-fi-kus
Wolbachia   wol-ba′kē-ä
Wuchereria bancrofti   vū-kėr-ār′ē-ä ban-krof′tē
Xanthomonas campestris   zan′thō-mō-nas kam′pe-stris
Xenopsylla cheopis   ze-nop′sil-lä kē-ō′pis
Yersinia enterocolitica   yėr-sin′ē-ä en′tėr-ō-kōl-it-ik-ä
Y. pestis   pes′tis
Y. pseudotuberculosis   sū-dō-tü-bėr-kū-lō′sis
Zoogloea   zō′ō-glē-ä
Zygosaccharomyces rouxii   zī′gō-sak-ä-rō-mī-ses rō′ē-ē

APPENDIX F

BACTERIA AND THE DISEASES THEY CAUSE
Proteobacteria
ALPHAPROTEOBACTERIA

Cat-scratch disease	*Bartonella henselae*	p. 683
Ehrlichiosis	*Ehrlichia* spp.	p. 687
Endemic murine typhus	*Rickettsia typhi*	p. 688
Epidemic typhus	*R. prowazekii*	pp. 687–688
Rocky Mountain spotted fever	*R. rickettsii*	p. 688
Brucellosis	*Brucella* spp.	p. 678

BETAPROTEOBACTERIA

Gonorrhea	*Neisseria gonorrhoeae*	pp. 790–792
Meningitis	*N. meningitidis*	pp. 645–646
Neonatal gonorrheal ophthalmia	*N. gonorrhoeae*	p. 636
Pelvic inflammatory disease	*N. gonorrhoeae*	p. 794
Melioidosis	*Burkholderia pseudomallei*	p. 730
Nosocomial infections	*Burkholderia* spp.	p. 320
Whooping cough	*Bordetella pertussis*	pp. 718–719
Rat-bite fever	*Spirillum minor*	p. 683

GAMMAPROTEOBACTERIA

Animal bites	*Pasteurella multocida*	p. 682
Bacillary dysentery	*Shigella* spp.	pp. 752–753
Epiglottitis	*Haemophilus influenzae*	p. 714
Meningitis	*H. influenzae*	p. 645
Otitis media	*H. influenzae*	p. 716
Pneumonia	*H. influenzae*	p. 726
Conjunctivitis	*H. influenzae*	pp. 634–635
Chancroid	*H. ducreyi*	p. 798
Cholera	*Vibrio cholerae*	pp. 755–758
Gastroenteritis	*V. parahaemolyticus*	p. 758
Gastroenteritis	*V. vulnificus*	p. 758
Cystitis	*Escherichia coli*	p. 788–789
Gastroenteritis	*E. coli*	pp. 758–759
Pyelonephritis	*E. coli*	p. 789
Dermatitis	*Pseudomonas aeruginosa*	p. 622
Otitis externa	*P. aeruginosa*	p. 622
Legionellosis	*Legionella pneumophila*	pp. 726–728
Plague	*Yersinia pestis*	pp. 683–684
Gastroenteritis	*Y. enterocolitica*	p. 760
Otitis media	*Moraxella catarrhalis*	p. 716
Q fever	*Coxiella burnetti*	pp. 729–730
Salmonellosis	*Salmonella enterica*	pp. 753–754, 756–757
Typhoid fever	*S. typhi*	pp. 754–755
Tularemia	*Francisella tularensis*	pp. 676–678

EPSILONPROTEOBACTERIA

Gastroenteritis	*Campylobacter jejuni*	pp. 759–760
Peptic ulcers	*Helicobacter pylori*	p. 760

Clostridia

Tetanus	*Clostridium tetani*	pp. 647–649
Gangrene	*C. perfringens*	pp. 681–682
Gastroenteritis	*C. perfringens*	pp. 760–761
Botulism	*C. botulinum*	pp. 649–651
Gastroenteritis	*C. difficile*	pp. 761–762

Mollicutes

Pneumonia	*Mycoplasma pneumoniae*	p. 726
Urethritis	*Mycoplasma, Ureaplasma*	p. 794

Bacilli

Anthrax	*Bacillus anthracis*	pp. 679–681
Gastroenteritis	*B. cereus*	p. 762
Listeriosis	*Listeria monocytogenes*	p. 647

Bacterial endocarditis	*Staphylococcus aureus*	p. 675
Folliculitis	*S. aureus*	pp. 615–619
Food poisoning	*S. aureus*	pp. 751–752
Impetigo	*S. aureus*	pp. 619–620
Scalded skin syndrome	*S. aureus*	p. 620
Toxic shock syndrome	*S. aureus*	p. 620
Cystitis	*S. saprophyticus*	pp. 788–789
Erysipelas	*Streptococcus pyogenes*	p. 621
Impetigo	*S. pyogenes*	p. 621
Necrotizing fasciitis	*S. pyogenes*	p. 621
Puerperal sepsis	*S. pyogenes*	pp. 674–675
Rheumatic fever	*S. pyogenes*	p. 676
Scarlet fever	*S. pyogenes*	pp. 714–715
Toxic shock syndrome	*S. pyogenes*	p. 622
Sepsis	*S. agalactiae*	p. 674
Sepsis	*Enterococcus* spp.	p. 674
Strep throat	*S. pyogenes*	p. 714
Meningitis	*S. pneumoniae*	p. 646
Otitis media	*S. pneumoniae*	pp. 715–716
Pneumonia	*S. pneumoniae*	pp. 724–726
Dental caries	*S. mutans*	pp. 747–749
Endocarditis	Alpha-hemolytic streptococci	p. 675

Actinobacteria

Acne	*Propionibacterium acnes*	pp. 622–623
Diphtheria	*Corynebacterium diphtheriae*	pp. 715–716
Leprosy	*Myobacterium leprae*	pp. 651–652
Tuberculosis	*M. tuberculosis*	pp. 719–723
Rapidly growing mycobacteria	*Mycobacterium* spp.	p. 203
Mycetoma	*Nocardia asteroides*	p. 335
Vaginosis	*Gardnerella vaginalis*	pp. 798–800

Chlamydiae

Inclusion conjunctivitis	*Chlamydia trachomatis*	p. 636
Lymphogranuloma venereum	*C. trachomatis*	pp. 797–798
Pelvic inflammatory disease	*C. trachomatis*	p. 794
Trachoma	*C. trachomatis*	p. 636
Urethritis	*C. trachomatis*	pp. 792–794
Pneumonia	*Chlamydophila pneumoniae*	p. 729
Psittacosis	*C. psittaci*	p. 727

Spirochetes

Leptospirosis	*Leptospira interrogans*	p. 789
Relapsing fever	*Borrelia* spp.	p. 684
Lyme disease	*B. burgdorferi*	pp. 684–687
Syphilis	*Treponema pallidum*	pp. 794–797

Bacteroidetes

Periodontal disease	*Porphyromonas* spp.	pp. 749–750
Acute necrotizing gingivitis	*Prevotella intermedia*	p. 750

Fusobacteria

Rat-bite fever	*Streptobacillus moniliformis*	p. 683

FUNGI AND THE DISEASES THEY CAUSE

Ascomycetes

Aspergillosis	*Aspergillus fumigatus*	p. 736
Blastomycosis	*Blastomyces dermatitidis*	p. 736
Histoplasmosis	*Histoplasma capsulatum*	pp. 734–735
Ringworm	*Microsporum, Trichophyton*	pp. 629–630

Anamorphs

Candidiasis	*Candida albicans*	pp. 630–631, 801–802
Coccidioidomycosis	*Coccidioides immitis*	p. 735
Pneumonia	*Pneumocystis jiroveci*	p. 736
Sporotrichosis	*Sporothrix schenckii*	p. 630

Basidiomycetes

Meningitis	*Cryptococcus neoformans*	p. 660
Mycotoxins		pp. 467, 770–771
Dandruff	*Malassezia furfur*	p. 615

PROTOZOA AND THE DISEASES THEY CAUSE

Archaezoa

Giardiasis	*Giardia lamblia*	pp. 771–772
Trichomoniasis	*Trichomonas vaginalis*	pp. 802–803

Apicomplexa

Babesiosis	*Babesia microti*	p. 699
Cryptosporidiosis	*Cryptosporidium* spp.	pp. 772–773
Cyclospora infection	*Cyclospora cayetanensis*	p. 773
Malaria	*Plasmodium* spp.	pp. 696–698
Toxoplasmosis	*Toxoplasma gondii*	pp. 695–696

Amoebozoa

Amoebic dysentery	*Entamoeba histolytica*	pp. 773–774
Keratitis	*Acanthamoeba* spp.	p. 637
	Alexandrium Pfiesteria	p. 360

Euglenzoa

African trypanosomiasis	*Trypanosoma brucei*	pp. 660–661
American trypanosomiasis	*T. cruzi*	pp. 693–695
Leishmaniasis	*Leishmania* spp.	pp. 367, 698–699
Meningoencephalitis	*Naegleria fowleri*	pp. 661–662

HELMINTHS AND THE DISEASES THEY CAUSE
Platyhelminths

Tapeworm infections	*Taenia* spp.	pp. 774–776
Hydatid disease	*Echinococcus granulosus*	p. 776
Schistosomiasis	*Schistosoma* spp.	p. 701
Swimmer's itch	Schistosomes	p. 701

Nematodes

Ascariasis	*Ascaris lumbricoides*	p. 777
Hookworm disease	*Necator americanus, Ancyclostoma*	pp. 776–777
Pinworm disease	*Enterobius vermicularis*	p. 776
Trichinellosis	*Trichinella spiralis*	pp. 777–778

ALGAE AND THE DISEASES THEY CAUSE
Red Algae, Diatoms, and Dinoflagellates   pp. 357–360
Oomycota   pp. 360–361

ARTHROPODS AND THE DISEASES THEY CAUSE

Pediculosis	*Pediculus humanus*	pp. 631–632
Scabies	*Sarcoptes scabiei*	p. 631

VIRUSES AND THE DISEASES THEY CAUSE
DNA Viruses

Fifth disease	Parvovirus	p. 628
Genital warts	Papovavirus	p. 801
Warts	Papovavirus	p. 623
Smallpox	Poxvirus	pp. 623–624
Burkitt's lymphoma	Herpesvirus	pp. 688–689
Chickenpox	Herpesvirus	pp. 624–626
Cold sores	Herpesvirus	pp. 626–627
Cytomegalic inclusion disease	Herpesvirus	p. 690
Genital herpes	Herpesvirus	pp. 800–801
Infectious mononucleosis	Herpesvirus	p. 690
Keratitis	Herpesvirus	p. 637
Roseola	Herpesvirus	p. 629
Shingles	Herpesvirus	pp. 624–626
Hepatitis B	Hepadnavirus	pp. 764–768

RNA Viruses

Encephalitis	Bunyavirus	pp. 658–660
Hantavirus pulmonary syndrome	Bunyavirus	p. 693
Gastroenteritis	Calcivirus	p. 770
Hepatitis E	Calcivirus	p. 768
Common cold	Coronavirus	p. 617
Hepatitis D	Deltavirus	p. 768
Encephalitis	Flavivirus	pp. 658–660
Hepatitis C	Flavivirus	p. 768
Yellow fever	Flavivirus	pp. 691–692
Hemorrhagic fever	Filovirus, Arenavirus	pp. 691–693
Influenza	Orthomyxovirus	pp. 406, 731–733
Common cold	Picornavirus	p. 617
Hepatitis A	Picornavirus	p. 764
Poliomyelitis	Picornavirus	pp. 652–654
Measles	Paramyxovirus	p. 627
Mumps	Paramyxovirus	pp. 763–764

PRIONS AND THE DISEASES THEY CAUSE

APPENDIX G

CHAPTER 1
Review

1. The observations of flies coming out of manure and maggots coming out of dead animals, and the appearance of microorganisms in liquids after a day or two, led people to believe that living organisms arose from nonliving matter.

2. Pasteur's S-neck flasks allowed air to get into the beef broth, but the curves of the S trapped bacteria before they could enter the broth.

3. **a.** Certain microorganisms cause diseases in insects. Microorganisms that kill insects can be effective biological control agents because they are specific for the pest and do not persist in the environment.

   b. Carbon, oxygen, nitrogen, sulfur, and phosphorus are required for all living organisms. Microorganisms convert these elements into forms that are useful for other organisms. Many bacteria decompose material and release carbon dioxide into the atmosphere for plants to use. Some bacteria can take nitrogen from the atmosphere and convert it into a form that can be used by plants and other microorganisms.

   c. Normal microbiota are microorganisms that are found in and on the human body. They do not usually cause disease, and can be beneficial.

   d. Organic matter in sewage is decomposed by bacteria into carbon dioxide, nitrates, phosphates, sulfate, and other inorganic compounds in a wastewater treatment plant.

   e. Recombinant DNA techniques have resulted in insertion of the gene for insulin production into bacteria. These bacteria can produce human insulin inexpensively.

   f. Microorganisms can be used as vaccines. Some microbes can be genetically engineered to produce components of vaccines.

4. **Matching**

a, c	Studies biodegradation of toxic wastes.
h	Studies the causative agent of *Hantavirus* pulmonary syndrome.
a, d, f	Studies the production of human proteins by bacteria.
b	Studies the symptoms of AIDS.
e	Studies the production of toxin by *E. coli*.
c	Studies the life cycle of *Cryptosporidium*.
b, d	Develops gene therapy for a disease.
g	Studies the fungus *Candida albicans*.

5. **Matching**

g	Archaea
d	Algae
c	Bacteria
b	Fungi
f	Helminths
e	Protozoa
a	Viruses

6. **Matching**

k	Avery, MacLeod, and McCarty
n	Beadle and Tatum
o	Berg
q	Ehrlich
c	Fleming
i	Hooke
j	Iwanowski
b	Jacob and Monod
a	Jenner
l	Koch
r	Lancefield
d	Lederberg and Tatum
g	Lister
e	Pasteur
f	Stanley
h	van Leeuwenhoek
m	Virchow
p	Weizmann

7. *Erwinia amylovora* is the correct way to write this scientific name. Scientific names can be derived from the names of scientists. In this case, *Erwinia* is derived from Erwin F. Smith, an American plant pathologist. Scientific names also can describe the organism, its habitat, or its niche. *E. amylovora* is a pathogen of plants (*amylo-* = starch; *vora* = eat).

8. **a.** *B. thuringiensis* is sold as a biological insecticide.

   b. *Saccharomyces* is the yeast sold for making bread, wine, and beer.

Multiple Choice

1. a	**6.** e
2. c	**7.** c
3. d	**8.** a
4. c	**9.** c
5. b	**10.** a

CHAPTER 2
Review

1. Atoms with the same atomic number and chemical behavior are classified as chemical elements.

2. Refer to Figure 2.1.

3. ^{14}C and ^{12}C are isotopes of carbon. ^{12}C has 6 neutrons in its nucleus and ^{14}C has 8 neutrons.

4. Hydrogen bonds.

5. **a.** Ionic
   b. Single covalent bond
   c. Double covalent bonds
   d. Hydrogen bond

6. 10^4 or 10,000 times.

7.

Element	Atomic Weight	×	Number of Atoms	=	Total Weight of That Element
C	12	×	6	=	72
H	1	×	12	=	12
O	16	×	6	=	96

The molecular weight of $C_6H_{12}O_6$ is 180 grams.

8. **a.** Synthesis reaction, condensation, or dehydration
   b. Decomposition reaction, digestion, or hydrolysis
   c. Exchange reaction
   d. Reversible reaction

9. The enzyme lowers the activation energy required for the reaction, and therefore speeds up this decomposition reaction.

10. **a.** Lipid
    b. Protein
    c. Carbohydrate
    d. Nucleic acid

11.

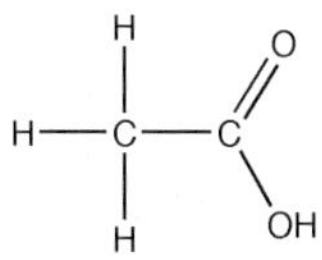

a. Acetic acid 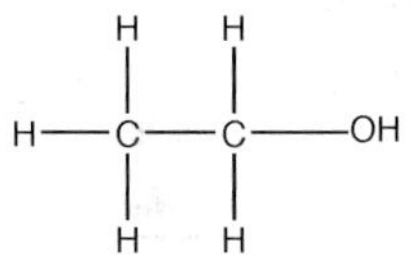   b. Ethyl alcohol 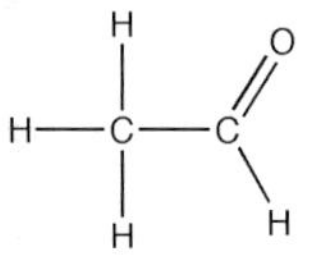   c. Acetaldehyde

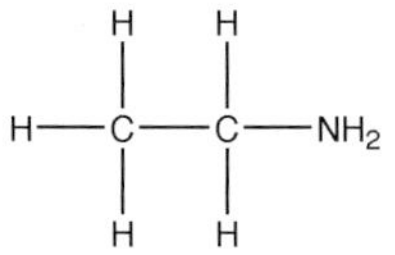

d. Ethanolamine 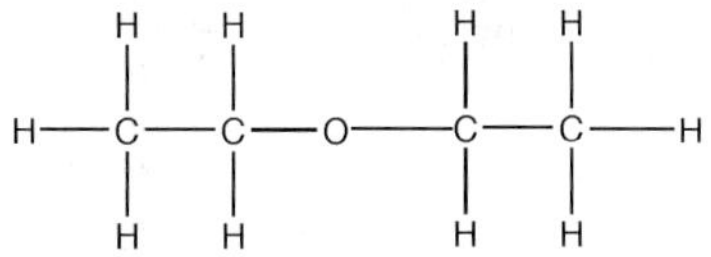   e. Diethylether

12. **a.** Amino acids
    b. Right to left
    c. Left to right

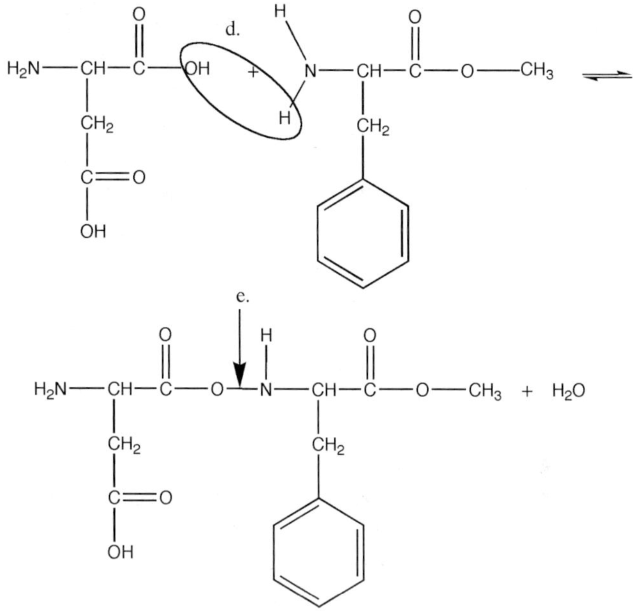

13. Breaking of bonds between <u>phosphorus and oxygen</u>. These are <u>covalent</u> bonds.

14. The entire protein shows tertiary structure. No quaternary structure.

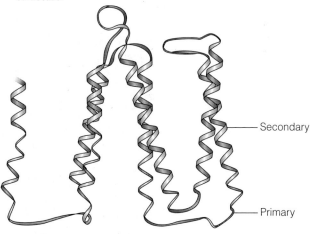

Secondary

Primary

15.

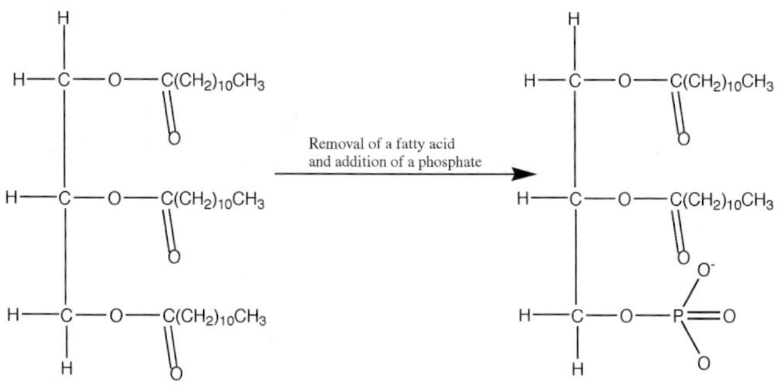

Removal of a fatty acid and addition of a phosphate

Multiple Choice

1. c	6. c
2. b	7. a
3. b	8. a
4. e	9. b
5. b	10. c

CHAPTER 3
Review

1. $1\ \mu m = 10^{-6}\ m$
   $1\ nm = 10^{-9}\ m$
   $1\ \mu m = 10^3\ nm$

2. **a.** Compound light microscope
   b. Darkfield microscope
   c. Phase-contrast microscope
   d. Fluorescence microscope
   e. Electron microscope
   f. Differential interference contrast microscope

3. **a.** Ocular lens
   b. Objective lens
   c. Diaphragm
   d. Condensor
   e. Illuminator

4.

Ocular lens magnification	×	Oil immersion lens magnification	=	Total magnification of speciman
10×		100×		1000×

5. . . . that a <u>beam of electrons</u> focused by <u>magnets</u> . . . on a <u>television-like screen or photographic plate</u>.

6.

Type of Microscope	Maximum Magnification	Resolution
Compound light	2,000×	0.2 μm
Electron	100,000×	0.0025 μm

7. Bacterial cells have a slightly negative charge, and the colored positive ion of a basic dye is attracted to the negative charge of the cell. Acid dyes do not stain bacterial cells because the negatively charged colored ion is repelled by the like charge of the cell.

8. **a.** A simple stain is used to determine cell shape and arrangement.
   b. A differential stain is used to distinguish kinds of bacteria based on their reaction to the differential stain.
   c. A negative stain does not distort the cell and is used to determine cell shape, size, and the presence of a capsule.
   d. A flagella stain is used to determine the number and arrangement of flagella.

9. In a Gram stain, the mordant combines with the basic dye to form a complex that will not wash out of gram-positive cells. In a flagella stain, the mordant accumulates on the flagella so that they can be seen with a light microscope.

10. A counterstain stains the colorless non–acid-fast cells so that they are easily seen through a microscope.

11. In the Gram stain, the decolorizer removes the color from gram-negative cells. In the acid-fast stain, the decolorizer removes the color from non–acid-fast cells.

12. Endospore: safranin is the *counterstain*.
    Gram: safranin is the *counterstain*.

13.

Steps	Appearance after this step of	
	Gram-positive cells	Gram-negative cells
Crysytal violet	Purple	Purple
Iodine	Purple	Purple
Alcohol-acetone	Purple	Colorless
Safranin	Purple	Red

Multiple Choice

1. e	6. e
2. c	7. d
3. b	8. b
4. a	9. a
5. a	10. c

CHAPTER 4
Review

1.

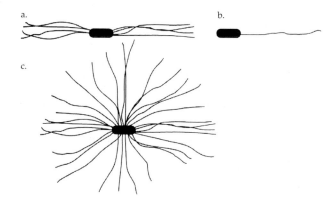

2. Endospore formation is called <u>sporogenesis</u>. It is initiated by <u>certain adverse environmental conditions</u>. Formation of a new cell from an endospore is called <u>germination</u>. This process is triggered by <u>favorable growth conditions</u>.

3.

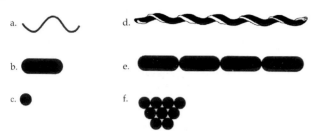

a. ∿   d.

b.   e.

c.   f.

4. Matching

__d__	Cell wall
__f__	Endospore
__a__	Fimbriae
__c__	Flagella
__a, e__	Glycocalyx
__i__	Pili
__b, h__	Plasma membrane
__g__	Ribosomes

5. An endospore is called a resting structure because it is a method of one cell "resting," or surviving, as opposed to growing and reproducing. The protective endospore wall allows a bacterium to withstand adverse conditions in the environment.

6. **a.** Both allow materials to cross the plasma membrane from a high concentration to a low concentration without expending energy. Facilitated diffusion requires carrier proteins.
   b. Both require enzymes to move materials across the plasma membrane. In active transport, energy is expended.
   c. Both move materials across the plasma membrane with an expenditure of energy. In group translocation, the substrate is changed after it crosses the membrane.

7. Mycoplasmas do not have cell walls.

8. **a.** Both lack cell walls. A spheroplast is a gram-negative cell whose wall has beeen destroyed by lysozyme. An L form is a cell that is not synthesizing a complete wall.
   b. Both lack cell walls. Mycoplasmas do not normally (genetically) make walls. L forms do not make walls because of environmental reasons, e.g., penicillin.

9. **a.** Diagram (a) refers to a gram-positive bacterium because the lipopolysaccharide–phospholipids–lipoprotein layer is absent.
   b. The gram-negative bacterium initially retains the violet stain, but it is released when the outer membrane is dissolved by the decolorizing agent. After the dye–iodine complex enters, it becomes trapped by the peptidoglycan of gram-positive cells.
   c. The outer layer of the gram-negative cells prevents penicillin from entering the cells.
   d. Essential molecules diffuse through the gram-positive wall. Porins and specific channel proteins in the gram-negative outer membrane allow passage of small water-soluble molecules.
   e. Gram-negative.

10. An extracellular enzyme (amylase) hydrolyzes starch into disaccharides (maltose) and monosaccharides (glucose). A carrier enzyme (maltase) hydrolyzes maltose and moves one glucose into the cell. Glucose can be transported by group translocation as glucose-6-phosphate.

11. Matching

__c__	Centriole
__d__	Chloroplasts
__g__	Golgi complex
__a__	Lysosomes
__f__	Mitochondria
__b__	Peroxisomes
__e__	Rough ER

12. A mitochondrion is an example of an organelle that resembles a prokaryotic cell. The inner membrane of a mitochondrion is arranged in folds similar to mesosomes. ATP is generated on this membrane just as it is in prokaryotic plasma membranes. Mitochondria can reproduce by binary fission, and they contain circular DNA and 70S ribosomes.

13. Phagocytosis. Pinocytosis.

14. Erythromycin inhibits protein synthesis in a prokaryotic cell; it will inhibit protein synthesis in mitochondria and chloroplasts.

Multiple Choice

1. e	6. e
2. d	7. b
3. b	8. e
4. a	9. a
5. d	10. b

CHAPTER 5
Review

1. Metabolism is the sum of all chemical reactions that occur within a living organism.

2. Catabolic reactions break down organic compounds and release energy, while anabolic reactions use the products of catabolism and energy to build cell material.

3.

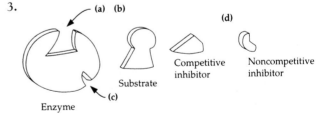

(a) (b)   (d)

Competitive inhibitor   Noncompetitive inhibitor

Substrate

Enzyme   (c)

4. **a.** When the enzyme and substrate combine, the substrate molecule will be transformed.
   b. When the competitive inhibitor binds to the enzyme, the enzyme will not be able to bind with the substrate.
   c. When the noncompetitive inhibitor binds to the enzyme, the active site of the enzyme will be changed so the enzyme cannot bind with the substrate.
   d. The noncompetitive inhibitor.

5. The optimum temperature for an enzyme is one that favors movement of molecules so the enzyme can "find" its substrate. Lower temperatures will decrease the rate of collisions and the rate of reactions. Increased temperatures will denature the enzyme.

6. Ethyl alcohol, lactic acid, butyl alcohol, acetone, and glycerol are some of the possible products. Refer to Table 5.4 and Figure 5.18b.

7.

Organism	Carbon Source	Energy Source
Photoautotroph	CO_2	Light
Photoheterotroph	Organic molecules	Light
Chemoautotroph	CO_2	Inorganic molecules
Chemoheterotroph	Organic molecules	Organic molecules

8. **(a)** is the Calvin–Benson cycle, **(b)** is glycolysis, and **(c)** is the Krebs cycle.

9. Glycerol is catabolized by pathway (b) as dihydroxyacetone phosphate. Fatty acids by pathway (c) as acetyl groups.

10. In pathway (c) at α-ketoglutaric acid.

11. Glyceraldehyde-3-phosphate from the Calvin–Benson cycle enters glycolysis. Pyruvic acid from glycolysis is decarboxylated to produce acetyl for the Krebs cycle.

12. In (a), between glucose and glyceraldehyde-3-phosphate.

13. The conversion of pyruvic acid to acetyl, isocitric acid to α-ketoglutaric acid, and α-ketoglutaric acid to succinyl~CoA.

14. By pathway (c) as acetyl groups.

15.

	Uses	Produces
Calvin–Benson cycle	6 NADPH	
Glycolysis		2 NADH
Pyruvic acid → acetyl		1 NADH
Isocitric acid → α-ketoglutaric acid		1 NADH
α-ketoglutaric acid → Succinyl~CoA		1 NADH
Succinic acid → Fumaric acid		1 FADH2
Malic acid → Oxaloacetic acid		1 NADH

16. Dihydroxyacetone phosphate; acetyl; oxaloacetic acid; α-ketoglutaric acid.

17.

ATP generated by	Reaction
Photophosphorylation	An electron, liberated from chlorophyll by light, is passed down an electron transport chain.
Oxidative phosphorylation	Cytochrome c passes two electrons to cytochrome a.
Substrate-level phosphorylation	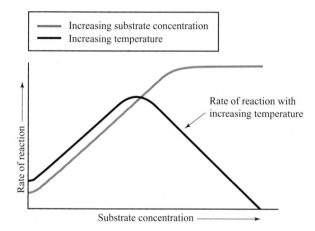

18. **a.** Oxidation–reduction: A coupled reaction in which one substance is oxidized and one is reduced.
    b. The final electron acceptor in aerobic respiration is molecular oxygen; in anaerobic respiration, it is another inorganic molecule.
    c. In cyclic photophosphorylation, electrons are returned to chlorophyll. In noncyclic photophosphorylation, chlorophyll receives electrons from hydrogen atoms.

19. The pentose phosphate pathway produces pentoses for the synthesis of nucleic acids, precursors for the synthesis of glucose by photosynthesizing organisms, precursors in the synthesis of certain amino acids, and NADPH.

20. Oxidation

21. Reactions requiring ATP are coupled with reactions that produce ATP.

22.

The reaction rate will increase until the enzymes are saturated.

Multiple Choice
1. a	**3.** b	**5.** c	**7.** b	**9.** c
2. d	**4.** c	**6.** b	**8.** a	**10.** b

CHAPTER 6
Review

1. In binary fission, the cell elongates and the chromosome replicates. Next, the nuclear material is evenly divided. The plasma membrane invaginates toward the center of the cell. The cell wall thickens and grows inward between the membrane invaginations; two new cells result.

2. Refer to Figure 6.14.

3. Carbon (C) is required for synthesis of molecules that make up a living cell. Carbon-containing compounds also are required as an energy source for heterotrophs.

4. Most bacteria grow best between pH 6.5 and 7.5.

5. The addition of salt or sugar to foods increases the osmotic pressure for microorganisms on the food. The resulting hypertonic environment causes plasmolysis of the microbial cells.

6. a. Catalyzes the breakdown of H_2O_2 to O_2 and H_2O.

  b. H_2O_2; peroxide ion is O_2^{2-}.

  c. Catalyzes the breakdown of H_2O_2;
  $$NADH + H^+ + H_2O_2 \xrightarrow{Peroxidase} NAD^+ + 2H_2O$$

  d. O_2^-; this diatom has one unpaired electron.

  e. Converts superoxide to O_2 and H_2O_2;
  $$2O_2^- + 2H^+ \xrightarrow{Superoxide\ dismutase} O_2 + H_2O_2$$

  The enzymes are important in protecting the cell from the strong oxidizing agents, peroxide and superoxide, that form during respiration.

7. Both environments prevent molecular oxygen from reaching the bacterial cells. In reducing media, thioglycolate combines with dissolved oxygen, thereby removing it from the medium. In an anaerobic incubator, air is replaced with an atmosphere of CO_2 (and N_2). *Clostridium* is an obligate anaerobe that lacks superoxide dismutase and catalase. Consequently, the accumulation of superoxides and peroxides will kill the cell in an aerobic environment.

8. Direct methods are those in which the microorganisms are seen and counted. Direct methods are direct microscopic count, plate count, filtration, and most probable number.

9. The growth rate of bacteria slows down with decreasing temperatures. Mesophilic bacteria will grow slowly at refrigeration temperatures and will remain dormant in a freezer. Bacteria will not spoil food quickly in a refrigerator.

10. Number of cells $\times\ 2^n$ generations = Total number of cells
$$6 \quad \times \quad 2^7 \quad = \quad 768$$

11. Petroleum can meet the carbon and energy requirements for an oil-degrading bacterium; however, nitrogen and phosphate are usually not available in large quantities. Nitrogen and phosphate are essential for making proteins, phospholipids, nucleic acids, and ATP.

12. A chemically defined medium is one in which the exact chemical composition is known. A complex medium is one in which the exact chemical composition is not known.

13.

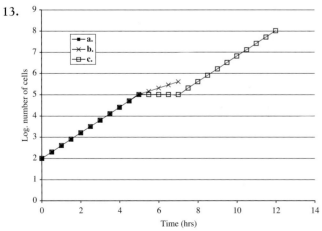

Multiple Choice
1. c	**6.** d
2. a	**7.** e
3. c	**8.** a
4. a	**9.** b
5. c	**10.** b

CHAPTER 7
Review

1. a. Lysis.

  b. Altered permeability and leakage of cell contents.

  c. Destruction of enzymes and structural proteins such as those in the plasma membrane.

  d. Interference with protein synthesis and cell division.

2. Autoclave. Due to the high specific heat of water, moist heat is readily transferred to cells.

3. Most organisms that cause disease or rapid spoilage of food are destroyed by pasteurization.

4. Variables that affect determination of the thermal death point are

  a. The innate heat resistance of the strain of bacteria.

  b. The past history of the culture, whether it was freeze-dried, wetted, etc.

  c. The clumping of the cells during the test.

  d. The amount of water present.

  e. The organic matter present.

  f. Media and incubation temperature used to determine viability of the culture after heating.

5. a. Ionizing radiation can break DNA directly. However, due to the high water content of cells, the formation of free radicals (H· and OH·) that break DNA strands is likely to occur.

  b. Ultraviolet radiation damages DNA by the formation of thymine dimers.

6. Microorganisms tend to die at a constant rate over a period of time. The constant rate is indicated by the solid line after exposure to the bactericidal compound.

7.

Sterilization Method	Temp.	Time	Type	Preferred Use	Mechanism of Action
Autoclaving	121°C	15 min	Moist	Media, equipment	Protein denaturation
Hot air	170°C	2 hr	Dry	Glassware	Oxidation
Pasteurization	72°C	15 sec	Moist	Milk, alcoholic drinks	Protein denaturation

8. All three processes kill microorganisms; however, as moisture and/or temperatures are increased, less time is required to achieve the same result.

9. Salts and sugars create a hypertonic environment. Salts and sugars (as preservatives) do not directly affect cell structures or metabolism; instead, they alter the osmotic pressure. Jams and jellies are preserved with sugar; meats are usually preserved with salt. Molds are more capable of growth in high osmotic pressure than bacteria.

10.
1. Acts rapidly.
2. Attacks all, or a wide range of, microbes.
3. Is able to penetrate.
4. Readily mixes with water.
5. Is not hampered by organic matter.
6. Stable.
7. Does not stain or corrode.
8. Nontoxic.
9. Pleasant odor.
10. Economical.
11. Safe to transport.

11.

Method of Action	Standard Use
a. Disrupts plasma membrane	Skin surfaces
b. Inhibits protein function	Antiseptic
c. Oxidation	Disinfect water
d. Denatures proteins, destroys lipids	Skin surfaces
e. Oligodynamic	AgNO₃ to prevent gonococcal eye infections
f. Inactivation of proteins	Chemical sterilizer
g. Denatures proteins	Chemical sterilizer
h. Oxidation	Antiseptic

12. Disinfectant B is preferable because it can be diluted more and still be effective.

13. Quaternary ammonium compounds are most effective against gram-positive bacteria. Gram-negative bacteria that were stuck in cracks or around the drain of the tub would not have been washed away when the tub was cleaned. These gram-negative bacteria could survive the washing procedure. Some pseudomonads can grow on quats that have accumulated.

Multiple Choice

1. d 6. b
2. b 7. b
3. d 8. a
4. d 9. a
5. b 10. b

CHAPTER 8
Review

1. DNA consists of a strand of alternating sugars (deoxyribose) and phosphate groups with a nitrogenous base attached to each sugar. The bases are adenine, thymine, cytosine, and guanine. DNA exists in a cell as two strands twisted together to form a double helix. The two strands are held together by hydrogen bonds between their nitrogenous bases. The bases are paired in a specific, complementary way: A-T and C-G. The information held in the sequence of nucleotides in DNA is the basis for synthesis of RNA and proteins in a cell.

2. a.

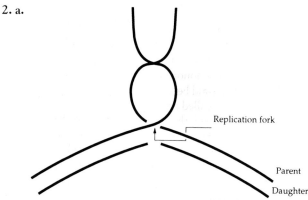

b. DNA polymerases synthesize a complementary strand of DNA from a DNA template. RNA polymerase starts each fragment of the lagging strand with an RNA primer.
c. Each new double-stranded DNA molecule contains one original strand and one new strand.

3. a. ATAT<u>TACTTTGCATGGACT</u>.
b. met-lys-arg-thr-(end).
c. TATAATGAAACGTTCCTGA.
d. No change.
e. Cysteine substituted for arginine.
f. Proline substituted for threonine (missense mutation).
g. Frameshift mutation.
h. Adjacent thymines might polymerize.
i. ACT.

4. One end of the mRNA molecule becomes associated with a ribosome. Ribosomes are composed of rRNA and protein. The anticodon of a tRNA with its activated amino acid pairs with the mRNA codon at the ribosome.

5. A mutant is isolated by direct selection because it grows on a particular medium. The colonies on an antibiotic-containing medium can be identified as resistant to that antibiotic.

A mutant is isolated by indirect selection because it does not grow on a particular medium. Replica plating could be employed to inoculate an antibiotic-containing medium. Colonies that did not grow on this medium can be isolated from the original plate and are antibiotic sensitive.

6. **Matching**

    b   A mutagen that is incorporated into DNA in place of a normal base.

    d   A mutagen that causes the formation of highly reactive ions.

    c   A mutagen that alters adenine so that it base-pairs with cytosine.

    a   A mutagen that causes insertions.

    e   A mutagen that causes the formation of pyrimidine dimers.

7. The basis for the Ames test is that a mutated cell can revert to a cell resembling the original, nonmutant cell by undergoing another mutation. The reversion rate of histidine auxotrophs of *Salmonella* in the presence of a mutagen will be higher than the spontaneous rate (in the absence of a mutagen).

8. Plasmids are small, self-replicating circles of DNA that are not associated with the chromosome. The F plasmid can be integrated into the chromosome. The F plasmid can be transferred from a donor to a recipient cell in conjugation. When the F plasmid becomes integrated into the chromosome, the cell is called an Hfr. During conjugation between an Hfr and an F cell, the chromosome of the Hfr cell, with its integrated F factor, replicates, and the new copy of the chromosome is transferred to the recipient cell.

9. **a.** . . . a <u>repressor</u> protein must be bound tightly to the <u>operator</u> site . . . it will bind to the <u>repressor</u> so that <u>transcription</u> can occur.

   b. . . . called a <u>corepressor</u>, causes the <u>repressor</u> to bind to the <u>operator</u>. Derepression is by removal of the <u>corepressor</u>, C in this case, when the <u>corepressor</u> is needed in the cell.

   c. None; constitutive enzymes are produced at certain necessary levels regardless of the amount of substrate or endproduct.

10. Light repair; dark repair; proofreading by DNA polymerase.

11. **a.** The genetic makeup of an organism.

   b. The external manifestations of the genotype.

   c. Rearrangement of genes to form new combinations; in nature, this usually occurs between members of the same species; in vitro, recombinant DNA is made from genes of different species.

12. CTTTGA. Endospores and pigments offer protection against UV radiation. Additionally, repair mechanisms can remove and replace thymine polymers.

13. **a.** Culture 1 will remain the same. Culture 2 will convert to F^+ but will have its original genotype.

   b. The donor and recipient cells' DNA can recombine to form combinations of $A^+B^+C^+$ and $A^-B^-C^-$. If the F plasmid also is transferred, the recipient cell may become F^+.

14. Semiconservative replication ensures the offspring cell will have one correct strand of DNA. Any mutations that may have occurred during DNA replication have a greater chance of being correctly repaired.

15. Mutation and recombination provide genetic diversity. Environmental factors select for the survival of organisms through natural selection. Genetic diversity is necessary for the survival of some organisms through the processes of natural selection. Organisms that survive may undergo further genetic change, resulting in the evolution of the species.

Multiple Choice

1. c   6. b
2. d   7. a
3. c   8. c
4. d   9. d
5. c   10. a

CHAPTER 9
Review

1. Recombinant DNA (rDNA) is DNA that is combined from different sources. In nature, rDNA results from conjugation, transduction, and tranformation. Genetic engineering is the artifical making of rDNA.

2. **a.** Both are DNA. cDNA is a segment of DNA made by RNA-dependent DNA polymerase. It is not necessarily a gene; a gene is a transcribable unit of DNA that codes for protein or RNA.

   b. Both are DNA. A restriction fragment is a segment of DNA produced when a restriction endonuclease hydrolyzes DNA. It is not usually a gene; a gene is a transcribable unit of DNA that codes for protein or RNA.

   c. Both are DNA. A DNA probe is a short, single-stranded piece of DNA. It is not a gene; a gene is a transcribable unit of DNA that codes for protein or RNA.

   d. Both are enzymes. DNA polymerase synthesizes DNA one nucleotide at a time using a DNA template; DNA ligase joins pieces (strands of nucleotides) together.

   e. Both are DNA. Recombinant DNA results from joining DNA from two different sources; cDNA results from copying a strand of RNA.

   f. The proteome is the expression of the genome. An organism's genome is one complete copy of its genetic information. The proteins encoded by this genetic material comprise the proteome.

3. **a.** A desired gene can be spliced into a plasmid and inserted into a cell by transformation.

   b. A desired gene can be spliced into a viral genome and inserted into a cell by transduction.

   c. Antibiotic resistance genes are used as markers or labels on plasmids so that the cell containing the plasmid can be found by direct selection on an antibiotic-containing medium.

   d. A genetically engineered bacterium should be producing a new protein product. Radioactively labeled antibodies against a specific protein can be used to locate the bacterial colony producing the protein.

4. Restriction fragments from one source can be cloned in microbial cells to make a gene library. Synthetic DNA is made in a lab.

5. In protoplast fusion, two wall-less cells fuse together to combine their DNA. A variety of genotypes can result from this process. In b, c, and d, specific genes are inserted directly into the cell.

6. *Bam*HI, *Eco*RI, and *Hind*III make sticky ends. Fragments of DNA produced with the same restriction enzyme will spontaneously anneal to each other at their sticky ends.

7. The gene can be spliced into a plasmid and inserted into a bacterial cell. As the cell grows, the number of plasmids will increase. The polymerase chain reaction can make copies of a gene using DNA polymerase in vitro.

8. In a eukaryotic cell, RNA polymerase copies DNA; RNA processing removes the introns, leaving the exons in the mRNA. cDNA can be made from the mRNA by reverse transcriptase.

9. See Tables 9.1 and 9.2.

10. You probably used a few plant cells in a Petri plate for your experiment. How will you select the plant cells that actually have the new Ti plasmid? You can grow these cells on plant-cell culture media with tetracycline. Only the cells with the new plasmid will grow.

11. In RNAi, siRNA binds mRNA creating double-stranded RNA, which is enzymatically destroyed.

Multiple Choice

1. b	6. d
2. b	7. c
3. b	8. b
4. b	9. e
5. c	10. a

CHAPTER 10
Review

1. Taxonomy is the science of classifying organisms to establish the relatedness between groups of organisms.

2. The three distinct chemical types of cells (see Table 10.1).

3. Living organisms cannot be grouped into two groups. For example, plant and animal is not acceptable because if fungi are grouped with plants, the definition of plants can't include *cellulose* and *photosynthesis*. If fungi are grouped with animals, the definition of animals can't include *no cell wall* and *ingestive*. The goal is to look for a "natural" scheme; that is, what criteria can be used to characterize all organisms.

4. Fungi: Unicellular or multicellular organisms that absorb organic nutrients; noncellulose cell walls; lack flagella.
Plantae: Multicellular eukaryotes with tissue formation; cellulose cell walls; generally photosynthetic.
Animalia: Multicellular eukaryotes with tissue formation; develop from an embryo (gastrula); lacking cell walls; ingest organic nutrients through a mouth of some kind.

5. a. Both are prokaryotic. They differ in composition of their cell walls, plasma membranes, and rRNAs.
   b. Both are bound by ester-linked plasma membranes. Eukarya have membrane-bound organelles.
   c. Both use methionine as the start signal. Eukarya have membrane-bound organelles and ester-linked membranes.

6. Binomial nomenclature is the system of assigning a genus and specific epithet to each organism.

7. Common names are not specific and can be misleading. According to the rules of scientific nomenclature, each organism has only one binomial.

8. The genus name must be written out so the reader knows what organism is being discussed, since the abbreviation for both of these species is *E. coli*.

9. Domain, kingdom, phylum, class, order, family, genus, species.

10. Domain: Bacteria
Phylum: Firmicutes
Class: Bacilli
Order: Bacillales
It is more related to *Gemella*. Family: Staphylococcaceae

11. A eukaryotic species is a group of closely related organisms having limited geographical distribution that interbreeds but does not breed with other species. Species can be distinguished morphologically. Because of the distinct differences between eukaryotic organisms and bacteria, a bacterial species is defined as a population of cells with similar characteristics. A viral species is a population of viruses with similar characteristics that occupies a particular ecological niche.

12. (See Table 10.5)
Used primarily for identification:
   morphological characteristics
   differential staining
   biochemical tests
   serology
   phage typing
   fatty acid profiles
Used primarily for taxonomic classification:
   flow cytometry
   DNA base composition
   DNA fingerprinting
   rRNA sequencing
   PCR
   nucleic acid hybridization
Data obtained from laboratory tests employing any (or all) of these twelve techniques can be assimilated using numerical taxonomy to provide information on classification.

13. Most microorganisms do not contain structures that are readily fossilized, making it difficult to obtain information on the evolution of microorganisms. Recent developments in molecular biology have provided techniques for determining evolutionary relationships amongst bacteria.

14. A and D appear to be most closely related because they have similar G-C moles %. No two are the same species.

Multiple Choice

1. b	**6.** a
2. e	**7.** a
3. d	**8.** e
4. b	**9.** a
5. e	**10.** b

CHAPTER 11
Review

1. I.   Gram-positive
      A. Endospore-forming rod
        1. *Clostridium*
        2. *Bacillus*
      B. Nonendospore-forming
        1. Cells are rods
          a. *Streptomyces*
          b. *Mycobacterium*
        2. Cells are cocci
          a. *Streptococcus*
          b. *Staphylococcus*

   II. Gram-negative
      A. Cells are helical or curved
        1. *Treponema*
        2. *Spirillum*
      B. Cells are rods
        1. *Pseudomonas*
        2. *Escherichia*

   III. Lack of cell walls
      A. *Mycoplasma*
      B. *Chlamydia*

   IV. Obligate intracellular parasites
      A. *Rickettsia*
      B. *Coxiella*

2. **a.** Both are oxygenic photoautotrophs. Cyanobacteria are prokaryotes; algae are eukaryotes.
   b. Both are chemoheterotrophs capable of forming mycelia; some form conidia. Actinomycetes are prokaryotes; fungi are eukaryotes.
   c. Both are large rod-shaped bacteria. *Bacillus* forms endospores, *Lactobacillus* is a fermentative non-endospore-forming rod.
   d. Both are small rod-shaped bacteria. *Pseudomonas* has an oxidative metabolism; *Escherichia* is fermentative. *Pseudomonas* has polar flagella; *Escherichia* has peritrichous flagella.
   e. Both are helical bacteria. *Leptospira* (a spirochete) has an axial filament. *Spirillum* has flagella.
   f. Both are gram-negative, rod-shaped bacteria. *Escherichia* are facultative anaerobes, and *Bacteroides* are anaerobes.
   g. Both are obligatory intracellular parasites. *Rickettsia* are transmitted by ticks; *Chlamydia* have a unique developmental cycle.
   h. Both lack peptidoglycan cell walls. *Ureaplasma* are archaea; *Mycoplasma* are bacteria (see Table 10.2).

3. Matching

Nitrogen-fixing	d	*Frankia*
Anoxygenic	i	Purple bacteria
Oxygenic	a	Cyanobacteria
Oxidize NO_2^-	h	*Nitrobacter*
Reduce CO_2	f	Methanogens
Sheath	b	*Cytophaga*
Myxospores	g	Myxobacteria
Anaerobic	c	*Desulfovibrio*
Thermophilic	k	*Sulfolobus*
Filaments	j	*Sphaerotilus*
Projections	e	*Hyphomicrobium*

Multiple Choice

1. d	**6.** c
2. b	**7.** e
3. e	**8.** b
4. a	**9.** b
5. b	**10.** a

CHAPTER 12
Review

1. Conidiospores are asexual spores formed by the aerial mycelia of one organism. Ascospores are sexual spores resulting from the fusion of the nuclei of two opposite mating strains of the same species of fungus.

2.

	Spore Type(s)	
Phylum	**Asexual**	**Sexual**
Zygomycota	Zygospore	Sporangiospore
Ascomycota	Blastoconidia, Arthroconidia	Ascospore
Basidiomycota	Blastoconidia	Basidiospore

3.

Genus	**Mycosis**
Blastomyces	Systemic
Sporothrix	Subcutaneous
Microsporum	Cutaneous
Trichosporon	Superficial
Aspergillus	Systemic

4. **a.** *E. coli*
   b. *P. chrysogenum*

5. The alga produces carbohydrates for the lichen, and the fungus provides both the holdfast and protection from desiccation.

6. As the first colonizers on newly exposed rock or soil, lichens are responsible for the chemical weathering of large inorganic particles and the consequent accumulation of soil.

7.

Phylum	Cell wall composition	Special features
Oomycotes	Chitin	Heterotrophic; flagellated spores.
Dinoflagellates*	Cellulose and silica	Some produce neurotoxins
Diatoms*	Pectin and silica	Produce much O_2; fossilized inclusions form petroleum
Red algae	Cellulose	Grow in deep water; source of agar
Brown algae	Cellulose and alginic acid	Source of algin
Green algae	Cellulose	Plantlike

———
*Unicellular

The green algae (Chlorophyta) could be placed in the plant kingdom. They have chlorophyll b, as do land plants, and have colonial forms. In the most advanced colonial form (*Volvox*), groups of *Chlamydomonas*-like cells live together; some are specialized for reproductive functions, which suggests a possible evolutionary route for the formation of plant tissue. The other algae are most often classified as protists.

8. Cellular slime molds exist as individual amoeboid cells. Plasmodial slime molds are multinucleate masses of protoplasm. Both survive adverse environmental conditions by forming spores.

9. Complete the following table.

Phylum	Method of Motility	One Human Parasite
Archaezoa	Flagella	*Giardia*
Microsporidia	None	*Nosema*
Amoebozoa	Pseudopods	*Entamoeba*
Apicomplexa	None	*Plasmodium*
Ciliophora	Cilia	*Balantidium*
Euglenozoa	Flagella	*Trypanosoma*

10. *Trichomonas* cannot survive for long outside of a host because it does not form a protective cyst. *Trichomonas* must be transferred from host to host quickly.

11. Asexual reproduction occurs in the human host and sexual reproduction takes place in the mosquito. The definitive host and the vector are the mosquito.

12. Ingestion.

13. This is a cestode. The encysted larva is called a cysticercus. Tapeworms are dorsoventrally flattened and have an incomplete digestive system.

scolex neck   proglottids

14. The male reproductive organs are in one individual, and the female reproductive organs in another. Nematodes belong to the Phylum Aschelminthes.

15.

Vector Type	Example	Disease
Mechanical	Housefly	Salmonellosis
Suitable for reproduction of parasite	*Ixodes*	Lyme disease
As a host	*Anopheles*	Malaria

Multiple Choice

1. c	6. b
2. b	7. a
3. b	8. c
4. a	9. a
5. d	10. d

CHAPTER 13
Review

1. The term filterable describes the property of passing through filters that retain bacteria. Viruses are too small to be seen with a light microscope, but their presence is known because material passed through a filter is still capable of causing a disease.

2. Viruses absolutely require living host cells to multiply.

3. A virus:
   a. Contains DNA or RNA;
   b. Has a protein coat surrounding the nucleic acid;
   c. Multiplies inside a living cell using the synthetic machinery of the cell; and
   d. Causes the synthesis of virions. A virion is a fully developed virus particle that transfers the viral nucleic acid to other cells and initiates multiplication.

4. Polyhedral (Fig. 13.2); helical (Figure. 13.4); enveloped (Figure 13.3); complex (Figure 13.4).

5. A sample of bacteriophage is mixed with host bacteria and melted nutrient agar. The mixture is then poured over a layer of nutrient agar in a Petri plate. Each phage infects a bacterium, multiplies, and releases new phages. These newly produced phages infect other bacteria, and more new viruses are produced. Following multiplication, the bacteria are destroyed. This produces a number of clearings or plaques in the layer of bacteria. The number of phages in the original sample can be estimated by counting the number of plaques.

6. Primary cell lines tend to die after a few generations. Continuous cell lines can be maintained through an indefinite number of generations. Continuous cell lines then allow long-term observations of viruses. Continuous cell lines are transformed cells.

7. A prophage gene codes for the cholera toxin. When phage DNA is incorporated into the cell's DNA, toxin can be produced.

8. **Adsorption**: The virus attaches to the cell membrane by means of spikes located on its envelope.
Penetration: The virus gains entrance by piriocytosis, or its envelope may fuse with the plasma membrane of the host cell, allowing the virus to enter the cell.
Uncoating: Uncoating refers to the separation of the capsid from the viral DNA.
Biosynthesis: Viral DNA is released into the cell's nucleus, and transcription and translation from viral DNA occur. Viral DNA is synthesized.
Maturation: Capsids form around strands of viral DNA.
Release: The assembled capsid-containing nucleic acid pushes through the plasma membrane; a portion of the plasma membrane adheres to the capsid, thus forming the envelope.

9. a. Viruses cannot easily be observed in host tissues. Viruses cannot easily be cultured in order to be inoculated into a new host. Additionally, viruses are specific for their hosts and cells, making it difficult to substitute a laboratory animal for the third step of Koch's postulates.
b. Some viruses can infect cells without inducing cancer. Cancer may not develop until long after infection. Cancers do not seem to be contagious.

10. Subacute sclerosing panencephalitis . . . common viruses . . . Students will have to suggest a mechanism to fill in the last blank; some examples are latent, in an abnormal tissue.

11. Provirus
TSTA appear on the host cell surface, or T antigens appear in the nucleus. Transformed cells do not exhibit contact inhibition.
RNA-containing oncogenic viruses produce a double-stranded DNA molecule using reverse transcriptase. The DNA is integrated into the host cell's DNA as a provirus. The provirus may transform the host cell into a tumor cell.

12. Prions are infectious proteins that appear to lack any nucleic acid. Viroids are infectious RNAs that do not have a protein coat. A prion causes CJD. A viroid causes potato spindle tuber viroid disease.

13. Of the rigid cell walls . . . vectors such as sap-sucking insects . . . plant protoplasts and insect cell cultures.

Multiple Choice

1. e	6. e
2. c	7. c
3. b	8. d
4. c	9. d
5. b	10. c

CHAPTER 14
Review

1. a. Etiology is the study of the cause of a disease, whereas pathogenesis is the manner in which the disease develops.
b. Infection refers to the colonization of the body by a microorganism. Disease is any change from a state of health. A disease may, but does not always, result from infection.
c. A communicable disease is a disease that is spread from one host to another, whereas a noncommunicable disease is not transmitted from one host to another.

2. Microorganisms that reside more or less permanently on the body are called normal microbiota. Microorganisms that are present for a few days or weeks are transient microbiota.

3. Symbiosis refers to different organisms living together. Commensalism—one of the organisms is benefited and the other is unaffected; e.g., corynebacteria living on the surface of the eye. Mutualism—both organisms are benefited; e.g., *E. coli* receives nutrients and a constant temperature in the large intestine and produces vitamin K and certain B vitamins that are useful for the human host. Parasitism—one organism benefits while the other is harmed; e.g., *Salmonella enterica* receives nutrients and warmth in the large intestine, and the human host experiences gastroenteritis or typhoid fever.

4. A reservoir of infection is a source of continual infection.

Matching

   b   Influenza   c   Rabies   a   Botulism

5. Koch's postulates shows that the microorganism caused the disease—not contact with a sick individual or environmental conditions.
Some organisms are not easily seen in a host. Some microorganisms cannot be cultured on laboratory media. And some microorganisms are specific for one host.

6. a. Some kind of body contact between an infected individual and a susceptible host is required.
b. Pathogens are transmitted from one host to another by fomites via indirect contact.
c. Arthropod vectors can transmit pathogens mechanically where the pathogen is carried on external body parts. When an arthropod ingests a pathogen and the pathogen reproduces in the vector, it is called biological transmission.

d. Also a method of contact transmission. Pathogens are transmitted by droplets of saliva or mucus.

e. Pathogens can be transmitted to a large number of individuals by food or water.

f. Spread of pathogens by droplet nuclei or dust.

7. Nutrition, fatigue, age, habits, lifestyle, occupation, preexisting illness, chemotherapy.

8. **a.** Acute
   b. Chronic
   c. Subacute

9. Hospital patients may be in a weakened condition and therefore predisposed to infection. Pathogenic microorganisms are generally transmitted to patients by contact and airborne transmission. The reservoirs of infection are the hospital staff, visitors, and other patients.

10. A disease constantly present in a population is an endemic disease. When many people acquire the disease in a relatively short time, it is an epidemic disease.

11. Epidemiology is the science dealing with when and where diseases occur and how they are transmitted in the human population. The Centers for Disease Control and Prevention (CDC) is a central source of epidemiological information.

12. Changes in body function felt by the patient are called symptoms. Symptoms such as weakness or pain are not measurable by a physician. Objective changes that the physician can observe and measure are called signs.

13. When microorganisms causing a local infection enter a blood or lymph vessel and are spread throughout the body, a systemic infection can result.

14. Mutualistic microorganisms are providing a chemical or environment that is essential for the host. Commensal organisms are not essential; another microorganism might serve the function as well.

15. Incubation period, prodromal period, period of illness, period of decline (may be crisis), period of convalescence.

Multiple Choice

1. a	6. c
2. b	7. d
3. a	8. a
4. d	9. c
5. b	10. b

CHAPTER 15
Review

1. Mucous membranes: Microorganisms can adhere to and then penetrate mucous membranes. Skin: Microorganisms can penetrate unbroken skin through hair follicles and sweat ducts. Parenteral route: Pathogens can be introduced into tissues beneath the skin and mucous membranes by punctures, injections, bites, and cuts.

2. The ability of a microorganism to produce a disease is called pathogenicity. The degree of pathogenicity is virulence.

3. **a.** Would prevent adherence by making the mannose attachment site unavailable.
   b. Would prevent adherence of *N. gonorrhoeae*.
   c. *S. pyogenes* would not be able to attach to host cells and would be more susceptible to phagocytosis.

4. Cytopathic effects are observable changes produced in cells infected by viruses. Five examples are:
   a. Cessation of mitosis.
   b. Autolysis.
   c. The presence of inclusion bodies.
   d. Cell fusion producing syncytia.
   e. Transformation.

5.

	Exotoxin	Endotoxin
Bacterial source	Gram +	Gram −
Chemistry	Proteins	Lipid A
Toxigenicity	High	Low
Pharmacology	Destroy certain cell parts or physiological functions	Systemic, fever, weakness, aches, and shock
Example	Botulinum toxin	Salmonellosis

6. Encapsulated bacteria can resist phagocytosis and continue growing. *Streptococcus pneumoniae* and *Klebsiella pneumoniae* produce capsules that are related to their virulence. M protein found in the cell walls of *Streptococcus pyogenes* and A protein in the cell walls of *Staphylococcus aureus* help these bacteria resist phagocytosis.

7. Hemolysins lyse red blood cells; hemolysis might supply nutrients for bacterial growth. Leukocidins destroy neutrophils and macrophages that are active in phagocytosis; this decreases host resistance to infection. Coagulase causes fibrinogen in blood to clot; the clot may protect the bacterium from phagocytosis and other host defenses. Bacterial kinases break down fibrin; kinases can destroy a clot that was made to isolate the bacteria, thus allowing the bacteria to spread. Hyaluronidase hydrolyzes the hyaluronic acid that binds cells together; this could allow the bacteria to spread through tissues. Siderophores take iron from host iron-transport proteins, thus allowing bacteria to get iron for growth. IgA proteases destroy IgA antibodies; IgA antibodies protect mucosal surfaces.

8. Pathogenic fungi do not have specific virulence factors; capsules, metabolic products, toxins, and allergic responses contribute to the virulence of pathogenic fungi. Some fungi produce toxins that, when ingested, produce disease. Protozoa and helminths elicit symptoms by destroying host tissues and producing toxic metabolic wastes.

9. *Legionella.*

10. Botulinum toxin is more potent than *Salmonella* toxin. A much smaller amount of botulinum toxin will kill 50 percent of the inoculated hosts.

11. Food infection refers to a disease that results from pathogens growing in the G-I tract. Food intoxication results from ingestion of a toxin formed in food. Pathogens grow in the food and excrete an exotoxin. The pathogens do not infect the host; symptoms are due to the toxin.

12. Viruses avoid the host's immune response by growing inside host cells; some can remain latent in a host cell for prolonged periods. Some protozoa avoid the immune response by mutations that change their antigens.

Multiple Choice

1. e	3. d	5. c	7. b	9. d
2. c	4. d	6. a	8. a	10. c

CHAPTER 16
Review

1. a. The ability of the human body to ward off diseases.
   b. The lack of resistance to an infectious disease.
   c. Antibody and T cell response to a specific infection.

2.

	Mechanical	Chemical
Skin	Dry, packed cells	Sebum
Eyes	Tears	Lysozyme
Digestive tract	Movement out	HCl
Respiratory tract	Ciliary escalator	
Urinary tract	Movement out	
Genital tract	Movement out	Acidic in female

3. See Table 16.1.

4. Phagocytosis is the ingestion of a microorganism or any foreign particulate matter by a cell.

5. Granulocytes have granules in the cytoplasm. Among the granulocytes, neutrophils have the most prominent phagocytic activity. Monocytes are agranulocytes (without granules) that develop into macrophages.
   When an infection occurs, granulocytes migrate to the infected area. Monocytes follow the granulocytes to the infected tissue. During migration, monocytes enlarge and develop into actively phagocytic cells called macrophages. Macrophages phagocytize dead or dying bacteria.

6. Phagocytic cells that migrate to the infected area are called wandering macrophages. Fixed macrophages remain in certain tissues and organs.

7. Refer to Figures 16.7 and 16.8.

8. Inflammation is the body's response to tissue damage. The characteristic symptoms of inflammation are redness, pain, heat, and swelling.

9. The functions of inflammation are:
   1. To destroy the injurious agent, if possible, and to remove it and its by-products from the body;
   2. If destruction is not possible, to confine or wall off the injurious agent and its by-products by forming an abscess;
   3. To repair or replace tissues damaged by the injurious agent or its by-products.

10. Leukocytic pyrogen, released from phagocytic granulocytes, has the ability to raise body temperature. The higher temperature is believed to inhibit the growth of some microorganisms. The higher temperature speeds up body reactions and may help body tissues to repair themselves more quickly.

11. The chill is an indication that body temperature is rising. Shivering and cold skin are mechanisms for increasing internal temperature. Crisis indicates body temperature is falling. The skin becomes warm as circulation is returned to it when the body attempts to dissipate extra heat.

12. Complement is a group of proteins found in normal blood serum. The classical pathway is activated by an antigen-antibody complex and C1 (see Figure 16.12). The alternative pathway is activated by microbial lipid-carbohydrate complex and factor B, factor D, and factor P (see Figure 16.13). The lectin pathway is initiated by carbohydrate-binding proteins (lectins) binding the mannose on microbes (see Figure 16.14).

13. Activation of complement can result in immune adherence and phagocytosis, local inflammation, and cell lysis.

14. Endotoxin binds C3b, which activates C5–C9 to cause cell lysis. This can result in free cell wall fragments, which bind more C3b, resulting in C5–C9 damage to host cell membranes.

15. Interferons are antiviral proteins produced by infected cells in response to viral infections. Alpha-IFN and beta-IFN induce uninfected cells to produce antiviral proteins. Gamma-IFN is produced by lymphocytes and activates neutrophils to kill bacteria.

Multiple Choice

1. a	3. c	5. b	7. c	9. d
2. d	4. d	6. a	8. b	10. e

CHAPTER 17
Review

1. a. Adaptive immunity is the resistance to infection obtained during the life of the individual; results from the production of antibodies and T cells. Innate immunity refers to the resistance of species or individuals to certain diseases that is not dependent on antigen-specific immunity.
   b. Humoral immunity is due to antibodies (and B cells). Cellular immunity is due to T cells.

c. Active immunity refers to antibodies produced by the individual who carries them. Passive immunity refers to antibodies produced by another source and then transferred to the individual who needs the antibodies.

d. T_H1 cells produce cytokines that activate T cells. Cytokines produced by T_H2 cells activate B cells.

e. Natural immunity is acquired naturally, i.e., from mother to newborn, or following an infection. Artificial immunity is acquired from medical treatment, i.e., by injection of antibodies or by vaccination.

f. T-dependent antigens: Certain antigens must combine with self-antigens to be recognized by T_H cells and then by B cells. T-independent antigens can elicit an antibody response without T cells.

g. T cells can be classified by their surface antigens: T_H cells possess the CD4 antigen; T_C cells have the CD8 antigen.

h. Immunoglobins = antibodies; TCRs = antigen-receptors on T cells.

2. Cellular immunity due to recognition of intracellular parasites.

3. The major histocompatability complex (MHC) are self-antigens. T_H cells react with MHC II; T_C cells react with MHC I.

4. An antigen is a chemical substance that causes the body to produce specific antibodies and can combine with these antibodies. A hapten is a low-molecular-weight substance that is not antigenic unless it is attached to a carrier molecule. Once an antibody has been formed against the hapten, the hapten alone will react with the antibodies independently of its carrier.

5. An antibody is a protein produced by the body in response to the presence of an antigen; it is capable of combining specifically with that antigen. Antibodies are proteins and usually consist of four polypeptide chains. Two of the chains are identical and are called heavy (H) chains. The other two chains are identical to each other but are of lower molecular weight and are called light (L) chains. The variable portions of the H and L chains are where antigen binding occurs. The variable portion is different for each kind of antibody. The remaining constant portions of each chain are identical for all of the antibodies in one class of antibody. Refer to Figure 17.3 for the structure of IgG antibodies.

6. Each person has a population of B cells with receptors for different antigens. When the appropriate antigen contacts the antigen receptor on a B cell, the cell proliferates to produce a clone of cells. Plasma cells in this clone produce antibodies specific to the antigen that caused their formation.

7. See Figure 17.18.

8. Activated T_C cells (CTLs) destroy target cells upon contact. T_H cells interact with an antigen to "present" it to a B cell for antibody formation. T_R cells suppress the immune response. Cytokines are chemicals released by cells that initiate a response by other cells. See Figure 17.3. Lymphokines cause an inflammatory response. An example of a cytokine is macrophage chemotactic factor, which attracts macrophages to the infection site. See Table 17.2 for functions of other cytokines.

9. **a.** Area *a* shows the primary response to the antigen. Area *b* shows the anamnestic response, in which the antibody titer is greater and remains high longer than in the primary response. The booster dose stimulated the memory cells to respond to the antigen.

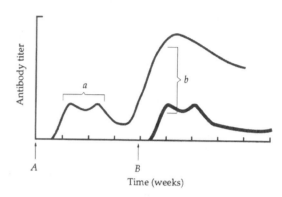

10. Neutralize toxins, inactivate viruses, fix complement to initiate cytolysis.

11. Surface recognition sites for antigen peptides and MHC proteins.

12. NK cells lyse target cells (usually tumor cells and virus-infected cells) on contact.

13. Both would prevent attachment of the pathogen; (a) interferes with the attachment site on the pathogen and (b) interferes with the pathogen's receptor site.

14. Rearrangement of the V region genes during embryonic development produces B cells with different antibody genes.

15. The person recovered because s/he produced antibodies against the pathogen. The memory response will continue to protect the person against that pathogen.

16. Human gamma globulin is the fraction of human serum in which antibodies are found. If antibodies against hepatitis are in the gamma globulin, this would be artificially acquired passive immunity.

17. Both create pores in the target cell's plasma membrane. CTLs release proteases that induce apoptosis; complement causes cell lysis.

Multiple Choice

1. d	3. b	5. d	7. c	9. c
2. e	4. c	6. e	8. d	10. d

CHAPTER 18
Review

1. **a.** Whole-agent. Live, avirulent virus that can cause the disease if it mutates back to its virulent state.

b. Whole-agent; (heat-) killed bacteria.

c. Subunit; (heat- or formalin-) inactivated toxin.

d. Subunit

e. Subunit

f. Conjugated

g. Nucleic acid

2. If excess antibody is present, an antigen will combine with several antibody molecules. Excess antigen will result in an antibody combining with several antigens. Refer to Figure 18.3.

3. Particulate antigens react in agglutination reactions. The antigens can be cells or soluble antigens bound to synthetic particles. Soluble antigens take part in precipitation reactions.

4. **a.** Some viruses are able to agglutinate red blood cells. This is used to detect the presence of large numbers of virions capable of causing hemagglutination (e.g., *Influenzavirus*).
   b. Antibodies produced against viruses that are capable of agglutinating red blood cells will inhibit the agglutination. Hemagglutination inhibition can be used to detect the presence of antibodies against these viruses.
   c. This is a procedure to detect antibodies that react with soluble antigens by first attaching the antigens to insoluble latex spheres. This procedure may be used to detect the presence of antibodies that develop during certain mycotic or helminthic infections.

5. See Figure 18.11.

6. **a.** Direct test (see Figure 18.11a)
   b. Indirect test (see Figure 18.11b)

7. An indirect ELISA test is used to detect the presence of antibodies. A direct ELISA test is used to detect the presence of an antigen (see Figure 18.14).

8. **a.** Direct test
   b. Indirect test
   The direct test provides definitive proof.

9. See Figure 18.2.

10. **Matching**
e	Precipitation
d, f	Immunoelectrophoresis
a	Agglutination
c	Complement fixation
f	Neutralization
b, d	ELISA

11.
e	Agglutination
c	Complement fixation
a	ELISA
f	FA
b	Neutralization
d	Precipitation

Multiple Choice
1. c   3. b   5. a   7. c   9. b
2. d   4. a   6. b   8. a   10. c

CHAPTER 19
Review

1. The immune state that results in altered immunologic reactions leading to pathogenic changes in tissue.

2.
Mediator	Function
Histamine	Increases blood capillary permeability, mucus secretion, and smooth muscle contraction.
Leukotrienes	Increase blood capillary permeability and smooth muscle contraction.
Prostaglandins	Increase smooth muscle contraction and mucus secretion.

3. Recipient's antibodies will react with donor's tissues.

4. The recipient will experience symptoms due to lysis of the donor RBCs. Hemolysis occurs because the antigen (donor RBCs)–antibody reaction fixes complement.

5. This condition develops when an Rh− mother becomes sensitized to the Rh+ antigens of her fetus. The mother's anti–Rh antibodies (IgG) can cross the placenta and react with fetal RBCs, causing their destruction. This condition can be prevented by passive immunization of the Rh− mother with anti-Rh antibodies shortly after birth. These anti-Rh antibodies combine with fetal Rh− RBCs, which may have entered maternal circulation, and enhance their clearance, thereby reducing the sensitization of the mother's immune system to this antigen.

6. Refer to Figure 19.7.
   a. The observed symptoms are due to lymphokines.
   b. When a person contacts poison oak initially, the antigen (catechols on the leaves) binds to tissue cells, is phagocytized by macrophages, and is presented to receptors on the surface of T cells. Contact between the antigen and the appropriate T cell stimulates the T cell to proliferate and become sensitized. Subsequent exposure to the antigen results in sensitized T cells releasing lymphokines, and a delayed hypersensitivity occurs.
   c. Small repeated doses of the antigen are believed to cause the production of IgG (blocking) antibodies.

7. Autografts and isografts are the most compatible. Xenotransplants are the least compatible.

8. **a.** Compatible. There are no Rh antigens on the donor's RBCs.
   b. Incompatible. The recipient will produce anti-Rh antibodies. If the recipient receives Rh+ RBCs in a subsequent transfusion, a hemolytic reaction will develop.
   c. Incompatible. The recipient has anti-A antibodies that will result in lysis of the donor's RBCs.

9. Immune responses against a person's own tissue antigen. During development, T cells that recognize self may not be eliminated. During adulthood, inactive T cells may become active or antibodies could cross-react with host cell antigens. New or altered antigens may be formed on the surface of host cells. These antigens may result from the use of certain drugs, or from infections by certain viruses.

10. Cytotoxic   Antibodies react with cell-surface antigens.

Immune complex Antibody–complement complexes deposit in tissues.

Cell-mediated   T cells destroy self cells.
See Table 19.3.

11. Natural
 Inherited
  Viral infections, most notably HIV
 Artificial
  Induced by immunosuppression drugs
 Result: Increased susceptibility to various infections depending on the type of immune deficiency.

12. Tumor cells have tumor-specific antigens such as TSTA and T antigen. Sensitized T_C cells may react with tumor-specific antigens, initiating lysis of the tumor cells.

13. Some malignant cells can escape the immune system by antigen modulation or immunological enhancement. Immunotherapy might trigger immunological enhancement. The body's defense against cancer is cell-mediated and not humoral. Transfer of lymphocytes could cause graft-versus-host disease.

14. AIDS is the last stage of an HIV infection. HIV is transmitted by sexual contact, by intravenous drug use, across the placenta, and in mother's milk. HIV is prevented by using condoms for hetero- and homosexual intercourse and oral and anal copulation, and by not re-using needles.

Multiple Choice

1. b	3. b	5. d	7. a	9. c
2. b	4. b	6. e	8. d	10. b

CHAPTER 20
Review

1.

Antimicrobial Agents	Synthetic or Antibiotic	Method of Action	Principal Use
Isoniazid	Synthetic	Inhibits mycolic acid synthesis	Tuberculosis
Sulfonamides	Synthetic	Inhibit folic acid synthesis	Gram-negative bacteria
Ethambutol	Synthetic	Competitive inhibitor	Tuberculosis
Trimethoprim	Synthetic	Inhibits folic acid synthesis	Broad spectrum
Fluoroquinolones	Synthetic	Inhibit DNA synthesis	Urinary tract infections
Penicillin, natural	Antibiotic	Inhibits cell wall synthesis	Gram-positive bacteria
Penicillin, semisynthetic	Antibiotic	Inhibits cell wall synthesis	Broad spectrum; penicillin-resistant bacteria
Cephalosporins	Antibiotic	Inhibit cell wall synthesis	Penicillin-resistant bacteria
Carbapenems	Antibiotic	Inhibit cell wall synthesis	Broad spectrum
Aminoglycosides	Antibiotic	Inhibit protein synthesis	Gram-negative bacteria
Tetracyclines	Antibiotic	Inhibit protein synthesis	Broad spectrum
Macrolides	Antibiotic	Inhibit protein synthesis	Gram-negative bacteria
Polypeptides	Antibiotic	Inhibit cell wall synthesis; injure plasma membrane	Gram-positive bacteria; gram-negative bacteria
Vancomycin	Antibiotic	Inhibits cell wall synthesis	Penicillin-resistant *Staphylococcus*
Rifamycins	Antibiotic	Inhibit mRNA synthesis	Tuberculosis
Polyenes	Antibiotic	Injure plasma membrane	Fungicide
Griseofulvin	Antibiotic	Inhibits mitosis	Antifungal
Amantadine	Synthetic	Blocks viral entry or uncoating	Influenza A
Zidovudine	Synthetic	Inhibits DNA synthesis	AIDS
Niclosamide	Synthetic	Inhibits oxidative phosphorylation	Tapeworms
Ketaconazole	Synthetic	Inhibits plasma membrane synthesis	Antifungal

2. A chemotherapeutic agent is a substance taken into the body to combat disease. A synthetic chemotherapeutic agent is prepared in a laboratory, whereas antibiotics are produced naturally by bacteria and some fungi.

3. **a.** Ehrlich discovered the first chemotherapeutic agent (salvarsan, which was used to treat syphilis).
   b. Fleming discovered the antibiotic penicillin.

4. Selective toxicity and broad spectrum.
   The drug should not produce hypersensitivity in the host, and drug resistance, should not harm normal microbiota.

5. Because a virus uses the host cell's metabolic machinery, it is difficult to damage the virus without damaging the host. Fungi, protozoa, and helminths possess eukaryotic cells. Therefore, antiviral, antifungal, antiprotozoan, and antihelminthic drugs must also affect eukaryotic cells.

6. Pyrimidine (idoxuridine) and purine (acyclovir) analogs. Prevent release of nucleic acid from viruses into the host cell (amantadine).
   Inhibition of infection of cells (interferon).
   Enzyme inhibitors (indinavir).

7. Both MIC and MBC can be determined with the broth dilution test. The disk-diffusion test only requires one 24-hour incubation.

8. In the disk-diffusion test, filter paper disks impregnated with chemotherapeutic agents are overlaid on an inoculated agar medium. During incubation, the agents diffuse from the disk and a zone of inhibition is observed in the area immediately around the disks. The zone of inhibition indicates susceptibility of the test organism to the agent tested; bacteriostatic or bactericidal cannot be determined.

9. Drug resistance is the lack of susceptibility of a microorganism to a chemotherapeutic agent. Drug resistance may develop when microorganisms are constantly exposed to an antimicrobial agent. The development of drug-resistant microorganisms can be minimized by judicious use of antimicrobial agents; following directions on the prescription; or by administering two or more drugs simultaneously.

10. **a.** Prevention of resistant strains of microorganisms;
    b. Take advantage of the synergistic effect;
    c. Provide therapy until a diagnosis is made; and
    d. Lessen the toxicity of individual drugs by reducing the dosage of each in combination.

11. **a.** Like polymyxin B, causes leaks in the plasma membrane.
    b. Interferes with translation.

12. **a.** Inhibits formation of peptide bond.
    b. Prevents translocation of ribosome along mRNA.
    c. Interferes with attachment of tRNA to mRNA-ribosome complex.
    d. Changes shape of 30S portion of ribosome, resulting in misreading mRNA.
    e. Prevents 70S ribosomal subunits from forming.
    f. Prevents release of peptide from ribosome.

13. DNA polymerase adds bases to the 3′–OH.

14. **a.** Penicillin inhibits bacterial cell wall synthesis. Echinocandin inhibits fungal cell wall synthesis.
    b. Imidazole interferes with fungal plasma membrane synthesis. Polymyxin B disrupts any plasma membrane.

Multiple Choice

1. b	6. d
2. a	7. e
3. a	8. b
4. b	9. c
5. a	10. d

CHAPTER 21
Review

1. Bacteria usually enter through inapparent openings in the skin. Fungal pathogens (except subcutaneous) often grow on the skin itself. Viral infections of the skin (except warts and herpes simplex) most often gain access to the body through the respiratory tract.

2. *Staphylococcus aureus; Streptococcus pyogenes.*

3.

Disease	Etiology	Symptoms	Treatment	Notes
Impetigo	Staphylococcus aureus	Vesicles that rupture and crust over	Hexachlorophene	May be epidemic
Erysipelas	Streptococcus pyogenes	Thickened red patches	Penicillin	May be endogenous

4.

Disease	Etiological Agent	Clinical Symptoms	Method of Transmission
Acne	*P. acnes*	Infected oil glands	Direct contact
Pimples	*S. aureus*	Infected hair follicles	Direct contact
Warts	Papovavirus	Benign tumor	Direct contact
Chicken pox	Herpesvirus	Vesicular rash	Respiratory route
Fever blisters	Herpesvirus	Recurrent "blisters"	Direct contact
Measles	Paramyxovirus	Papular rash, Koplik's spots	Respiratory route
Rubella	Togavirus	Macular rash	Respiratory route

5. Both are fungal infections. Sporotrichosis is a subcutaneous mycosis; athlete's foot is a cutaneous mycosis.

6. a. Conjunctivitis is an inflammation of the conjunctiva, and keratitis is an inflammation of the cornea.
b. Table 21.2.

7. Candidiasis is caused by *Candida albicans*. The yeast is able to grow when the normal microbiota are suppressed or when the immune system is suppressed. The yeast can be transferred from another person or be transient microbiota. White patches in the mouth or bright red areas of the skin and mucous membranes are signs of infection. Antifungal agents such as miconazole are used to treat candidiasis. Systemic infections are treated with oral ketoconazole.

8. The test determines the woman's susceptibility to rubella. If the test is negative, she is susceptible to the disease. If she acquires the disease during pregnancy the fetus could become infected. A susceptible woman should be vaccinated.

9.

Symptoms	Disease
Koplik's spots	Measles
Macular rash	Measles
Vesicular rash	Chickenpox
Small, spotted rash	German measles
"Blisters"	Cold sore
Corneal ulcer	Keratoconjunctivitis

10. The central nervous system can be invaded following keratoconjunctivitis; this results in encephalitis.

11. Attenuated measles, mumps, and rubella viruses.

12. Varicella-zoster virus appears to remain latent in nerve cells following recovery from a childhood infection of chickenpox. Later, the virus may be activated and cause a vesicular rash (shingles) in the area of the nerve.

13. To prevent neonatal gonorrheal ophthalmia. This is caused by *N. gonorrhoeae* contracted by the newborn during passage through the birth canal.

14. Trachoma.

15. Scabies is an infestation of mites in the skin. It is treated with permethrin insecticide or gamma benzene hexachloride. The presence of a six-legged arthropod (insect) indicates pediculosis (lice).

Multiple Choice

1. c	**6.** d
2. d	**7.** e
3. b	**8.** d
4. c	**9.** a
5. d	**10.** d

CHAPTER 22
Review

1. Meningitis is an infection of the meninges; encephalitis is an infection of the brain itself.

2.

Causative Agent	Susceptible Population	Mode of Transmission	Treatment
N. meningitidis	Children; military recruits	Respiratory	Penicillin
H. influenzae	Children	Respiratory	Rifampin
S. pneumoniae	Children; elderly	Respiratory	Penicillin
L. monocytogenes	Anyone	Foodborne	Penicillin
C. neoformans	Immunosuppressed individuals	Respiratory	Amphotericin B

3. "*Haemophilus*" refers to the requirement of this genus for growth factors found in blood (X and V factors) (Chapter 11). "*Influenzae*" because it was thought to be the causative agent of influenza.

4. The symptoms of tetanus are not due to bacterial growth (infection and inflammation) but to neurotoxin.

5.

	Salk	Sabin
Composition	Formalin-inactivated viruses	Live, attenuated viruses
Advantages	No reversion to virulence	Oral administration
Disadvantages	Booster dose needed; injected	Reversion to virulence

6. a. Vaccination with tetanus toxoid.
b. Immunization with antitetanus toxin antibodies.

7. "Cleaned" because C. *tetani* is found in soil that might contaminate a wound. "Deep puncture" because it is likely to be anaerobic. "No bleeding" because a flow of blood ensures an aerobic environment and some cleansing.

8. *Clostridium botulinum.* Canned foods. Paralysis. Supportive respiratory care; antitoxin. Anaerobic, nonacidic environment. Diagnosis is made by detecting toxin in foods or patient by inoculating mice with suspect samples. Prevention: use of adequate heat in canning; boiling food before consumption to inactivate toxin.

9. Etiology—*Mycobacterium leprae.*
Transmission—Direct contact.
Symptoms—Nodules on the skin; loss of sensation.
Treatment—Dapsone and rifampin.
Prevention—BCG vaccine.
Susceptible—People living in the tropics; genetic predisposition.

10. Etiology—Picornavirus (poliovirus).
Transmission—Ingestion of contaminated water.
Symptoms—Headache, sore throat, fever, nausea; rarely paralysis.
Prevention—Sewage treatment.
These vaccinations provide artificially acquired active immunity because they cause the production of antibodies, but they do not prevent or reverse damage to nerves.

11. Etiology—Rhabdovirus.
Transmission—Bite of infected animal; inhalation.
Reservoirs—Skunks, bats, foxes, raccoons.
Symptoms—Muscle spasms, hydrophobia, CNS damage.

12. Postexposure treatment—Passive immunization with antibodies followed by active immunization with HDCV. Preexposure treatment—Active immunization with HDCV. Following exposure to rabies, antibodies are needed immediately to inactivate the virus. Passive immunization provides these antibodies. Active immunization will provide antibodies over a longer period of time, but they are not formed immediately.

13.

Disease	Etiology	Vector	Symptoms	Treatment
Arboviral encephalitis	Togaviruses, Arboviruses	Mosquitoes (*Culex*)	Headache, fever, coma	Immune serum
African trypanosomiasis	*T. b. gambiense, T. b. rhodesiense*	Tsetse fly	Decreased physical activity and mental acuity	Suramm; melarsoprol

14. Most antibiotics cannot cross the blood-brain barrier.

15. The causative agent of Creutzfeldt–Jakob disease (CJD) is transmissible. Although there is some evidence for an inherited form of the disease, it has been transmitted by transplants. Similarities with viruses are (1) the prion cannot be cultured by conventional bacteriological techniques and (2) the prion is not readily seen in patients with CJD.

Multiple Choice

1. a	**3.** a	**5.** a	**7.** b	**9.** c
2. c	**4.** b	**6.** c	**8.** a	**10.** a

CHAPTER 23
Review

1. Fever, decrease in blood pressure, and lymphangitis. Septic shock occurs when low blood pressure cannot be controlled.

2. Bacteria can spread from an abscess with enzymes such as kinases and invade blood vessels.

3.

Disease	Causative Agent	Predisposing Conditions
p.s.	*S. pyogenes*	Abortion or childbirth
s.b.e.	alpha-hemolytic strep.	Preexisting lesions
a.b.e.	*S. aureus*	Abnormal heart valves

4. Rheumatic fever is an autoimmune disease that is precipitated by streptococcal sore throat. It is treated with anti-inflammatory drugs to relieve the symptoms. It is prevented by early diagnosis and treatment of streptococcal sore throat.

5. All are vectorborne rickettsial diseases. They differ from each other in (1) etiologic agent, (2) vector, (3) severity and mortality, and (4) incidence (e.g., epidemic, sporadic).

6.
Causative Agent	Vector	Symptoms	Treatment
Plasmodium	*Anopheles*	Recurrent fever, chills	Quinine derivative
Flavivirus	*Aedes aegypti*	Fever, nausea, jaundice	None
Flavivirus	*Aedes aegypti*	Muscle and joint pain	None
Borrelia	Soft ticks	Recurrent fever	Tetracycline
Leishmania	Sandflies	Fever, chills	Antimony

7.
Francisella tularensis	Animal reservoir, skin abrasions, ingestion, inhalation, bites	Rabbits	Small ulcer	Careful handling of animals
Brucella spp.	Animal reservoir, ingestion of milk, direct contact	Cattle	Undulant fever	Pasteurization of milk
Bacillus anthracis	Skin abrasions, inhalation, ingestion	Soil, cattle	Malignant pustule	Surveillance and vaccination of cattle
Borrelia burgdorferi	Tick bites	Deer, mice	Rash, neurologic; arthritis	Protection from ticks
Ehrlichia	Tick bites	Deer	Flulike	Protection from ticks
HHV5	Saliva, blood	Humans	Cytomegalic inclusion disease of the newborn	Ganciclovir

8. Plague
Causative agent—*Yersinia pestis*.
Vector—Rat flea.
Reservoir—Rodents.
Prognosis—Poor if untreated; good with antibiotic treatment.
Treatment—Tetracycline, streptomycin.
Control—Sanitation and ratproofing buildings.

9.
Causative Agent	Transmission	Reservoir	Endemic Area
Schistosoma spp.	Penetrate skin	Aquatic snail	Asia, South America
Toxoplasma gondii	Ingestion, inhalation	Cats	United States
Trypanosoma cruzi	"Kissing bug"	Rodents	Central America

10.
	Reservoir	Etiology	Transmission	Symptoms
Cat-scratch disease	Cats	*Bartonella henselae*	Scratch; touching eyes, fleas	Swollen lymph nodes, fever, malaise
Toxoplasmosis	Cats	*Toxoplasma gondii*	Ingestion	None, congenital infections, neurologic damage

11. Gangrenous tissue is anaerobic and has suitable nutrients for *C. perfringens*.

12. Infectious mononucleosis is caused by EB virus and transmitted in oral secretions.

13. Bubonic plague—Proliferation of bacteria in lymph vessels and lymph nodes. Enlarged lymph nodes (buboes). Transmission—Flea bites.
Pneumonic plague—Growth of bacteria in the lungs. Transmission—Respiratory route.

Multiple Choice

1. e	6. e
2. b	7. a
3. d	8. c
4. c	9. c
5. a	10. c

CHAPTER 24
Review

1. <u>Droplet infection</u>. Inhalation of cells and spores; ingestion of contaminated food.

2. Coarse hairs in the nose filter dust particles from inspired air. Mucus traps dust and microorganisms, and cilia move the trapped particles toward the throat for elimination. The ciliary escalator of the lower respiratory system moves particles toward the throat. Alveolar macrophages can phagocytize microorganisms that enter the lungs. IgA antibodies are found in mucus, saliva, and tears.

3. Beta-hemolytic streptococci inhibit growth of pneumococci; faster-growing organisms can compete with pathogens.

4. Mycoplasmal pneumonia is caused by *Mycoplasma pneumoniae* bacteria. Viral pneumonia can be caused by several different viruses. Mycoplasmal pneumonia can be treated with tetracyclines, whereas viral pneumonia cannot.

5. Bacteria infecting the nose and throat can move through the eustachian tube to the inner ear. Microorganisms can enter the ear directly via swimming pool water or injury to the eardrum or skull. The bacteria that most commonly cause otitis media are *S. aureus*, *Streptococcus pneumoniae*, beta-hemolytic streptococci, and *H. influenzae*. The middle ear is connected to the nose and throat.

6.

Upper Respiratory System

Common cold	Coronaviruses	Sneezing, excessive nasal secretions, congestion

Lower Respiratory System

Viral pneumonia	Several viruses	Fever, shortness of breath, chest pains
Influenza	Influenzavirus	Chills, fever, headache, muscular pains
RSV	Respiratory syncytial virus	Coughing, wheezing

Amantadine is used to treat influenza. Palivizumab, for life-threatening RSV.

7.

Disease	Symptoms
Streptococcal pharyngitis	Pharyngitis and tonsillitis
Scarlet fever	Rash and fever
Diphtheria	Membrane across throat
Whooping cough	Paroxysmal coughing
Tuberculosis	Tubercles, weight loss, and coughing
Pneumococcal pneumonia	Reddish lungs, fever
H. influenzae pneumonia	Similar to pneumococcal pneumonia
Chamydial pneumonia	Low fever, cough, and headache
Legionellosis	Fever and cough
Psittacosis	Fever and headache
Q fever	Chills and chest pain
Epiglottitis	Inflamed, abscessed epiglottis
Melioidosis	Pneumonia

8. Inhalation of large numbers of spores from *Aspergillus* or *Rhizopus* can cause infections in individuals with impaired immune systems, cancer, and diabetes.

9. No. Many different organisms (gram-positive bacteria, gram-negative bacteria, and viruses) can cause pneumonia. Each of these organisms is susceptible to different antimicrobial agents.

10.

Disease	Endemic Areas in the United States
Histoplasmosis	States adjoining the Mississippi and Ohio Rivers
Coccidioidomycosis	American Southwest
Blastomycosis	Mississippi
Pneumocystis	Ubiquitous

Refer to Table 24.2.

11. In the tuberculin test, purified protein derivative (PPD) from *M. tuberculosis* is injected into the skin. Induration and reddening of the area around the injection site indicates an active infection or immunity to tuberculosis.

12. Hypothesis 1: Close indoor contact in winter promotes epidemic transmission.
Hypothesis 2: The viruses grow best at slightly cooler temperatures.
Hypothesis 3: A physiological change in humans during winter allows viral growth.

13. Gram-positive cocci

Catalase-positive	*Staphylococcus aureus*
beta-hemolytic	*Streptococcus pyogenes*
alpha-hemolytic	*S. pneumoniae*

Gram-positive rods

Not acid-fast	*C. diphtheriae*
Acid-fast	*Mycobacterium tuberculosis*
Gram-negative cocci	*Moraxella catarrhalis*

Gram-negative rods

Aerobes

Coccobacilli	*Bordetella pertussis*

Rods

Grow on nutrient agar	*Burkholderia pseudomallei*
Require special media	*Legionella pneumophila*

Facultative anaerobes

Coccobacilli	*Haemophilus influenzae*

Intracellular

Elementary bodies	*Chlamydophila psittaci*
No elementary bodies	*Coxiella burnetii*
Wall-less	*Mycoplasma pneumoniae*

Multiple Choice

1. a	3. e	5. c	7. a	9. b
2. c	4. a	6. b	8. e	10. d

CHAPTER 25
Review

1. *S. mutans* becomes established in the mouth when the teeth erupt from the gums. A sticky capsule enables the bacteria to adhere to teeth. The dextran capsule is produced when the bacteria grow on sucrose. These bacteria and others that become trapped in the dextran produce lactic acid, which erodes tooth enamel.

2.
Disease	Suspect Foods	Symptoms
Staph	Not cooked prior to eating	Vomiting and diarrhea
Shigellosis	Contaminated water	Mucus and blood in stools
Salmonellosis	Poultry; contaminated water	Fever, vomiting, and diarrhea
Cholera	Contaminated water	Rice water stools
Traveler's	Contaminated water	Vomiting and diarrhea

Refer to Table 25.2.

3. Fever, nausea, abdominal pain, cramps, diarrhea. The diagnosis is based on isolation of the etiologic agent from leftover food or the patient's stools.

4. Complete the following table:

Causative Agent	Suspect Foods	Prevention
V. parahaemolyticus	Oysters, shrimp	Cooking
V. vulnificus	Contact with coastal waters	
Enterotoxigenic *E. coli*	Water, vegetables	Cooking
Enteroinvasive *E. coli*	Water, vegetables	Cooking
Enterohemorrhagic *E. coli*	Alfalfa sprouts, tomatoes	Cooking
C. jejuni	Chicken	Cooking
Y. enterocolitica	Meat, milk	Cooking
C. perfringens	Meat	Refrigeration after cooking
B. cereus	Rice dishes	Refrigeration after cooking

Refer to Table 25.2 to complete the table.

5. Both are caused by *Salmonella enterica*. However, typhoid fever is caused by a few strains of *S. enterica* that are invasive. The bacteria can cross the intestinal wall and can be disseminated throughout the body. Typhoid fever is characterized by fever and malaise without diarrhea.

6. Certain strains of *E. coli* may produce an enterotoxin or invade the epithelium of the large intestine.

7. At present there are no treatments for hepatitis. Exposed individuals can be given pooled immune globulin for hepatitis A or HBIG for passive immunity to hepatitis B. Vaccines can prevent hepatitis A and B.

8. Antibodies specific for HBSAg are used to screen blood for HBV. A viral protein is used to test blood for antibodies against HCV.

9. Toxins produced by fungi; see pp. 770–771.

10. All four are caused by protozoa. The infections are acquired by ingesting protozoa in contaminated water. Giardiasis is a prolonged diarrhea. Amoebic dysentery is the most severe dysentery, with blood and mucus in the stools. *Cryptosporidium* and *Cyclospora* produce severe diseases in persons with immune deficiencies.

11.
Disease	Etiologic Agent	Symptoms
Amoebic	*Entamoeba histolytica*	Blood and mucus in stools, perforation of the intestinal wall, abscesses
Bacillary	*Shigella* spp.	Leukocytes in feces in addition to the symptoms listed for amoebic dysentery

12. **Food intoxication**: Microorganisms must be allowed to grow in food from the time of preparation to the time of ingestion. This usually occurs when foods are stored unrefrigerated or improperly canned. The etiologic agents (*Staphylococcus aureus* or *Clostridium botulinum*) must produce an exotoxin. Onset: 1 to 48 hours. Duration: A few days. Treatment: Antimicrobial agents are ineffective. The patient's symptoms may be treated.

Food infection: Viable microorganisms must be ingested with food or water. The organisms could be present during preparation and survive cooking or be inoculated during later handling. The etiologic agents are usually gram-negative organisms (*Salmonella, Shigella, Vibrio,* and *Escherichia*) that produce endotoxins. *Clostridium perfringens* is a gram-positive bacterium that causes food infection. Onset: 12 hours to 2 weeks. Duration: Longer than intoxication because the microorganisms are growing in the patient. Treatment: Rehydration.

13.

Disease	Site	Symptoms
Mumps	Parotid glands	Inflammation of the parotid glands and fever
Infectious hepatitis	Liver	Anorexia, fever, diarrhea
Serum hepatitis	Liver	Anorexia, fever, joint pains, jaundice
Viral gastroenteritis	Lower GI tract	Nausea, diarrhea, vomiting

Refer to Table 25.3 to complete this question.

14. **Life cycle of the beef tapeworm, *Taenia saginata*.** The adult tapeworm lives in the intestine of the human, the definitive host. Tapeworm segments and eggs are eliminated with feces and are ingested by intermediate hosts, such as grazing cattle. The tapeworm eggs hatch, and cysticerci form in the animal's muscles, which are later consumed by humans. The pork tapeworm, *Taenia solium*, has a similar life cycle, except that cysticerci may also form in human tissue.

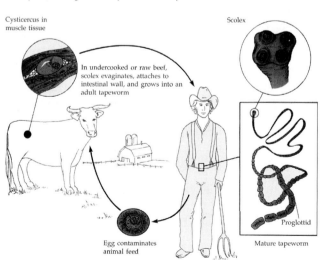

Cysticercus in muscle tissue

Scolex

In undercooked or raw beef, scolex evaginates, attaches to intestinal wall, and grows into an adult tapeworm

Egg contaminates animal feed

Mature tapeworm

Proglottid

15. Refer to Figure 25.26.

16. Adequate sewage treatment and sanitary living conditions.

17. Cook meat thoroughly. Eliminate the source of contamination to cattle and pigs.

Multiple Choice

1. d	3. e	5. e	7. b	9. a
2. e	4. c	6. b	8. e	10. d

CHAPTER 26
Review

1. Organs of the upper urinary tract are sterile. The resident microbiota of the urethra are *Streptococcus*, *Bacteroides*, *Mycobacterium*, *Neisseria*, and some enterobacteria.

2. Normal microbiota of the male genital system is the same as that of the urinary tract. During reproductive years, lactobacilli predominate in the vagina.

3. Urinary tract infections may be transmitted by improper personal hygiene and contamination during medical procedures. They are often caused by opportunistic pathogens.

4. The proximity of the anus to the urethra and the relatively short length of the urethra can allow contamination of the urinary bladder in females. Predisposing factors for cystitis in females are gastrointestinal infections and vaginal and urinary tract infections.

5. *Escherichia coli* causes about 75% of the cases. From lower urinary tract or systemic infections.

6.
Disease	Symptoms	Diagnosis
Gardnerella	Fishy odor	Odor, pH, clue cells
Gonorrhea	Painful urination	Isolation of *Neisseria*
Syphilis	Chancre	FTA–ABS
PID	Abdominal pain	Culture of pathogen
NGU	Urethritis	Absence of *Neisseria*
LGV	Lesion, lymph node enlargement	Observation of *Chlamydia* in cells
Chancroid	Swollen ulcer	Isolation of *Haemophilus*

7. Transmission—Water; enters via wounds.
   Activities—Water contact; contact with animals or rodent-infested places.
   Etiology—*Leptospira interrogans*.

8. Symptoms: Burning sensation, vesicles, painful urination.
   Etiology: Herpes simplex type 2 (sometimes type 1).
   When the lesions are not present, the virus is latent and noncommunicable.

9. *Candida albicans*: Severe itching; thick, yellow, cheesy discharge.
Trichomonas vaginalis: Profuse yellow discharge with disagreeable odor.

Disease	Prevention of Congenital Disease
Gonorrhea	Treatment of newborn's eyes
Syphilis	Prevention and treatment of mother's disease
NGU	Treatment of newborn's eyes
Genital herpes	Cesarean delivery during active infection

Multiple Choice

1. b	6. c
2. e	7. c
3. a	8. e
4. c	9. b
5. d	10. a

CHAPTER 27
Review

1. Extremophiles include thermophiles such as *Thermus aquaticus*, acidophiles such as *Thiobacillus*, halophiles such as *Halobacterium*, and endoliths.

2. The koala should have an organ housing a large population of cellulose-degrading microorganisms.

3. *Penicillium* might make penicillin to reduce competition from faster-growing bacteria.

4.

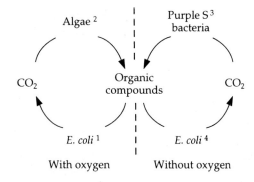

1—Any chemoheterotroph using aerobic respiration

2—Any aerobic autotroph

3—Any anaerobic autotroph

4—Any chemoheterotroph producing CO_2 via fermentation

5. Amino acids; SO_4^{2-}; plants and bacteria; H_2S; carbohydrates; S^0.

6. Phosphorus must be available for all organisms.

7.
Process	Reactions	Microorganisms
Ammonification	$-NH_2 \rightarrow NH_3$	Proteolytic bacteria
Nitrification	$NH_3 \rightarrow NO_2^-$	*Nitrosomonas*
	$NO_2 \rightarrow NO_3^-$	*Nitrobacter*
Denitrification	$NO_3 \rightarrow N_2$	*Bacillus*
N fixation	$N_2 \rightarrow NH_3$	*Rhizobium*

8. Cyanobacteria: With fungi, cyanobacteria act as the photoautotrophic partner in a lichen; they may also fix nitrogen in the lichen. With *Azolla*, they fix nitrogen.
Mycorrhizae: Fungi that grow in and on the roots of higher plants; increase absorption of nutrients.
Rhizobium: In root nodules of legumes; fix nitrogen.
Frankia: In root nodules of alders, roses, and other plants; fix nitrogen.

9. Settling
Flocculation treatment
Sand filtration (or activated charcoal filtration)
Chlorination
The amount of treatment prior to chlorination depends on the amount of inorganic and organic matter in the water.

10. A coliform count is used to determine the bacteriologic quality of water; that is, the presence of human pathogens or evidence of fecal contamination.

11.
b	Leaching field
a	Removal of solids
b	Biological degradation
b	Activated sludge
c	Chemical precipitation of phosphorus
b	Trickling filter
c	Results in drinking water

12. Activated sludge is an aerobic process that can result in complete oxidation of organic matter.

13. Both require large areas of land and can result in the pollution of surface or groundwater if they are overloaded.

14.

	BOD	Rate of Eutrophication	Dissolved Oxygen
Untreated	3+	3+	+
Primary	2+	2+	2+
Secondary	+	+	3+

Accumulation of BOD and loss of dissolved oxygen would be much less in a fast-moving river. Continual aeration caused by the river's movement would result in rapid oxidation of organic matter.

15. Biodegradation of sewage, herbicides, oil, or PCBs.

Multiple Choice

1. a	6. c
2. b	7. b
3. b	8. b
4. b	9. e
5. c	10. c

CHAPTER 28
Review

1. Industrial microbiology is the science of using microorganisms to produce products or accomplish a process. Industrial microbiology provides (1) chemicals such as antibiotics that would not otherwise be available, (2) processes to remove or detoxify pollutants, (3) fermented foods that have desirable flavors or enhanced shelf life, and (4) enzymes for manufacturing a variety of goods.

2. The goal of commercial sterilization is to eliminate spoilage and disease-causing organisms. The goal of hospital sterilization is complete sterilization.

3. The acid in the berries will prevent the growth of some microbes.

4. A presterilized package is aseptically filled with presterilized food.

5. Milk $\xrightarrow{\text{Lactic Acid Bacteria}}$ Curd + Whey
      $\downarrow$ $\downarrow$
    Cheese   Waste

   Hard cheese is ripened by lactic acid bacteria growing anaerobically in the interior of the curd. Soft cheese is ripened by molds growing aerobically on the outside of the curd.

6. Fruit juice $\xrightarrow{\text{Yeast}}$ Ethyl alcohol + CO_2

7. Nutrients must be dissolved in water; water is also needed for hydrolysis. Malt is the carbon and energy source that the yeast will ferment to make alcohol. Malt contains glucose and maltose from the action of amylases on starch in seeds (barley).

8. A bioreactor provides the following advantages over simple flask containers:
   - Larger culture volumes can be grown.
   - Process instrumentation for monitoring and controlling critical environmental conditions such as pH, temperature, dissolved oxygen, and aeration can be used.
   - Sterilization and cleaning systems are designed in place.
   - Aseptic sampling and harvest systems for in-process sampling exist.
   - Improved aeration and mixing characteristics result in improved cell growth and high final cell densities.
   - A high degree of automation is possible.
   - Process reproducibility is improved.

9. (1) Enzymes don't produce hazardous wastes; (2) Enzymes work under reasonable conditions, e.g., they don't require high temperatures or acidity; (3) Eliminates the need to use petroleum in chemical syntheses of solvents such as alcohol and acetone; (4) Enzymes are biodegradable; (5) Enzymes are not toxic.

10. A primary metabolite is produced during trophophase; a secondary metabolite, during idiophase.

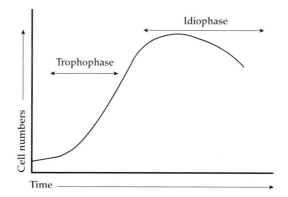

11. The production of ethyl alcohol from corn; or methane from sewage. Alcohols and hydrogen are produced by fermentation; methane is produced by anaerobic respiration.

Multiple Choice

1. c	3. e	5. b	7. a	9. b
2. b	4. c	6. c	8. a	10. a

12D treatment A sterilization process that would result in a decrease of the number of *Clostridium botulinum* endospores by 12 logarithmic cycles.

ABO blood group system The classification of red blood cells based on the presence or absence of A and B carbohydrate antigens.

abscess A localized accumulation of pus.

A-B toxin Bacterial exotoxins consisting of two polypeptides.

acellular vaccine A vaccine consisting of antigenic parts of cells.

acetyl group

$$H_3C - \overset{\overset{\displaystyle O}{\|}}{C} -$$

acid A substance that dissociates into one or more hydrogen ions (H^+) and one or more negative ions.

acid-fast stain A differential stain used to identify bacteria that are not decolorized by acid-alcohol.

acidic dye A salt in which the color is in the negative ion; used for negative staining.

acidophile A bacterium that grows below pH 4.

acquired immune deficiency The inability, obtained during the life of an individual, to produce specific antibodies or T cells, due to drugs or disease.

activated macrophage A macrophage that has increased phagocytic ability and other functions after exposure to mediators released by T cells after stimulation by antigens.

activated sludge system A process used in secondary sewage treatment in which batches of sewage are held in highly aerated tanks; to ensure the presence of microbes efficient in degrading sewage, each batch is inoculated with portions of sludge from a precious batch.

activation energy The minimum collision energy required for a chemical reaction to occur.

active site A region on an enzyme that interacts with the substrate.

active transport Net movement of a substance across a membrane against a concentration gradient; requires the cell to expend energy.

acute disease A disease in which symptoms develop rapidly but last for only a short time.

acute-phase proteins Serum proteins whose concentration changes by at least 25% during inflammation.

adaptive immunity The ability, obtained during the life of the individual, to produce specific antibodies and T cells.

adenosarcoma Cancer of glandular epithelial tissue.

adenosine diphosphate (ADP) The substance formed when ATP is hydrolyzed and energy is released.

adenosine triphosphate (ATP) An important intracellular energy source.

adherence Attachment of a microbe or phagocyte to another's plasma membrane or other surface.

adhesin A carbohydrate-specific binding protein that projects from prokaryotic cells; used for adherence, also called a ligand.

adjuvant A substance added to a vaccine to increase its effectiveness.

aerobe An organism requiring molecular oxygen (O_2) for growth.

aerobic respiration Respiration in which the final electron acceptor in the electron transport chain is molecular oxygen (O_2).

aerotolerant anaerobe An organism that does not use molecular oxygen (O_2) but is not affected by its presence.

aflatoxin A carcinogenic toxin produced by *Aspergillus flavus*.

agar A complex polysaccharide derived from a marine alga and used as a solidifying agent in culture media.

agglutination A joining together or clumping of cells.

agranulocyte A leukocyte without visible granules in the cytoplasm; includes monocytes and lymphocytes.

agranulocytosis Destruction of granulocytic white blood cells.

alcohol An organic molecule with the functional group—OH.

alcohol fermentation A catabolic process, beginning with glycolysis, that produces ethyl alcohol to reoxidize NADH.

aldehyde An organic molecule with the functional group

alga (plural: **algae**) A photosynthetic eukaryote; may be unicellular, filamentous, or multicellular but lack the tissues found in plants.

algal bloom An abundant growth of microscopic algae producing visible colonies in nature.

algin A sodium salt of mannuronic acid ($C_6H_8O_6$); found in brown algae.

allergen An antigen that evokes a hypersensitivity response.

allergy *See* hypersensitivity.

allograft A tissue graft that is not from a genetically identical donor (i.e., not from self or an identical twin).

allosteric inhibition The process in which an enzyme's activity is changed because of binding to the allosteric site.

allosteric site The site on an enzyme at which a noncompetitive inhibitor binds.

allylamines Antifungal agents that interfere with sterol synthesis.

amanitin A polypeptide toxin produced by *Amanita* spp., inhibits RNA polymerase.

Ames test A procedure using bacteria to identify potential carcinogens.

amination The addition of an amino group.

amino acid An organic acid containing an amino group and a carboxyl group.

aminoglycoside An antibiotic consisting of amino sugars and an aminocyclitol ring; for example, streptomycin.

amino group —NH_2.

ammonification The release of ammonia from nitrogen-containing organic matter by the action of microorganisms.

amphibolic pathway A pathway that is both anabolic and catabolic.

amphitrichous Having tufts of flagella at both ends of a cell.

anabolism All synthesis reactions in a living organism; the building of complex organic molecules from simpler ones.

anaerobe An organism that does not require molecular oxygen (O_2) for growth.

anaerobic respiration Respiration in which the final electron acceptor in the electron transport chain is an inorganic molecule other than molecular oxygen (O_2); for example, a nitrate ion or CO_2.

anaerobic sludge digester Anaerobic digestion used in secondary sewage treatment.

anal pore A site in certain protozoa for elimination of waste.

analytical epidemiology Comparison of a diseased group and a healthy group to determine the cause of the disease.

anamnestic response *See* memory response.

anamorph Ascomycete fungi that have lost the ability to reproduce sexually; the asexual stage of a fungus.

anaphylaxis A hypersensitivity reaction involving IgE antibodies, mast cells, and basophils.

animalia The kingdom composed of multicellular eukaryotes lacking cell walls.

anion An ion with a negative charge.

anoxygenic Not producing molecular oxygen; typical of cyclic photophosphorylation.

antagonism Active opposition; (1) When two drugs are less effective than either one alone. (2) Competition among microbes.

antibiotic   An antimicrobial agent, usually produced naturally by a bacterium or fungus.

antibody   A protein produced by the body in response to an antigen, and capable of combining specifically with that antigen.

antibody-dependent cell-mediated cytotoxicity (ADCC)   The killing of antibody-coated cells by natural killer cells and leukocytes.

antibody titer   The amount of antibody in serum.

anticodon   The three nucleotides by which a tRNA recognizes an mRNA codon.

antigen   Any substance that causes antibody formation; also called immunogen.

antigen-antibody complex   The combination of an antigen with the antibody that is specific for it; the basis of immune protection and many diagnostic tests.

antigen-binding sites   A site on an antibody that binds to an antigenic determinant.

antigenic determinant   A specific region on the surface of an antigen against which antibodies are formed; also called epitope.

antigenic drift   A minor variation in the antigenic makeup of influenza viruses that occurs with time.

antigenic shift   A major genetic change in influenza viruses causing changes in H and N antigens.

antigenic variation   Changes in surface antigens that occur in a microbial population.

antigen-presenting cell (APC)   A macrophage, dendritic cell, or B cell that engulfs an antigen and presents fragments to T cells.

antihuman immune serum globulin (anti-HISG)   An antibody that reacts specifically with human antibodies.

antimetabolite   A competitive inhibitor.

antimicrobial drug   A chemical that destroys pathogens without damaging body tissues.

antimicrobial peptide   An antibiotic that is bactericidal and has a broad spectrum of activity; see bacteriocin.

antisense DNA   DNA that is complementary to the DNA encoding a protein; the antisense RNA transcript will hybridize with the mRNA encoding the protein and inhibit synthesis of the protein.

antisense strand (– strand)   Viral RNA that cannot act as mRNA.

antisepsis   A chemical method for disinfection of the skin or mucous membranes; the chemical is called an antiseptic.

antiserum   A blood-derived fluid containing antibodies.

antitoxin   A specific antibody produced by the body in response to a bacterial exotoxin or its toxoid.

antiviral protein (AVP)   A protein made in response to interferon that blocks viral multiplication.

apoenzyme   The protein portion of an enzyme, which requires activation by a coenzyme.

apoptosis   The natural programmed death of a cell; the residual fragments are disposed of by phagocytosis.

aquatic microbiology   The study of microorganisms and their activities in natural waters.

arbuscule   Fungal mycelia in plant root cells.

archaea   Domain of prokaryotic cells lacking peptidoglycan; one of the three domains.

arthroconidia   An asexual fungal spore formed by fragmentation of a septate hypha.

Arthus reaction   Inflammation and necrosis at the site of injection of foreign serum, due to immune complex formation.

artificially acquired active immunity   The production of antibodies by the body in response to a vaccination.

artificially acquired passive immunity   The transfer of humoral antibodies formed by one individual to a susceptible individual, accomplished by the injection of antiserum.

artificial selection   Choosing one organism from a population to grow because of its desirable traits.

ascospore   A sexual fungal spore produced in an ascus, formed by the ascomycetes.

ascus   A saclike structure containing ascospores; found in the ascomycetes.

asepsis   The absence of contamination by unwanted organisms.

aseptic packaging   Commercial food preservation by filling sterile containers with sterile food.

aseptic surgery   Techniques used in surgery to prevent microbial contamination of the patient.

aseptic techniques   Laboratory techniques used to minimize contamination.

asexual spore   A reproductive cell produced by mitosis and cell division (eukaryotes) or binary fission (actinomycetes).

atom   The smallest unit of matter that can enter into a chemical reaction.

atomic force microscopy   See scanned-probe microscopy.

atomic number   The number of protons in the nucleus of an atom.

atomic weight   The total number of protons and neutrons in the nucleus of an atom.

attenuated whole-agent vaccine   A vaccine containing live, attenuated (weakened) microorganisms.

autoclave   Equipment for sterilization by steam under pressure, usually operated at 15 psi and 121°C.

autograph   A tissue graft from one's self.

autoimmune disease   Damage to one's own organs due to action of the immune system.

autotroph   An organism that uses carbon dioxide (CO_2) as its principal carbon source. See also chemoautotroph, photoautotroph.

auxotroph   A mutant microorganism with a nutritional requirement that is absent in the parent.

axial filament   The structure for motility found in spirochetes; also called endoflagellum.

azoles   Antifungal agents that interfere with sterol synthesis.

bacillus (plural: bacilli)   (1) Any rod-shaped bacterium. (2) When written as a genus (*Bacillus*) refers to rod-shaped, endospore-forming, facultatively anaerobic, gram-positive bacteria.

bacteremia   A condition in which there are bacteria in the blood.

bacteria   Domain of prokaryotic organisms, characterized by peptidoglycan cell walls; **bacterium** (singular) when referring to a single organism.

bacterial growth curve   A graph indicating the growth of a bacterial population over time.

bactericide   A substance capable of killing bacteria.

bacteriocin   An antimicrobial peptide produced by bacteria that kills other bacteria.

bacteriology   The scientific study of prokaryotes, including bacteria and archaea.

bacteriophage (phage)   A virus that infects bacterial cells.

bacteriostasis   A treatment capable of inhibiting bacterial growth.

base   A substance that dissociates into one or more hydroxide ions (OH^-) and one or more positive ions.

base pairs   The arrangement of nitrogenous bases in nucleic acids based on hydrogen bonding; in DNA, base pairs are A-T and G-C; in RNA, base pairs are A-U and G-C.

base substitution   The replacement of a single base in DNA by another base, causing a mutation; also called point mutation.

basic dye   A salt in which the color is in the positive ion; used for bacterial stains.

basidiospore   A sexual fungal spore produced in a basidium, characteristic of the basidiomycetes.

basidium   A pedestal that produces basidiospores; found in the basidiomycetes.

basophil   A granulocyte (leukocyte) that readily takes up basic dye and is not phagocytic; has receptors for IgE Fc regions.

batch production   An industrial process in which cells are grown for a period of time after which the product is collected.

B cell   A type of lymphocyte; differentiates into antibody-secreting plasma cells and memory cells.

BCG vaccine   A live, attenuated strain of *Mycobacterium bovis* used to provide immunity to tuberculosis.

beer Alcoholic beverage produced by fermentation of starch.

benthic zone The sediment at the bottom of a body of water.

Bergey's Manual *Bergey's Manual of Systematic Bacteriology*, the standard taxonomic reference on bacteria; also refers to *Bergey's Manual of Determinative Bacteriology*, the standard laboratory identification reference on bacteria.

β-lactam Core structure of penicillins.

beta oxidation The removal of two carbon units from a fatty acid to form acetyl CoA.

biguanide A group of antimicrobial chemicals, including chlorhexidine, especially useful on skin and mucous membranes.

binary fission Prokaryotic cell reproduction by division into two daughter cells.

binomial nomenclature The system of having two names (genus and specific epithet) for each organism; also called scientific nomenclature.

bioaugmentation The use of pollutant-acclimated microbes or genetically engineered microbes for bioremediation.

biochemical oxygen demand (BOD) A measure of the biologically degradable organic matter in water.

biocide A substance capable of killing microorganisms.

bioconversion Changes in organic matter brought about by the growth of microorganisms.

biofilm A microbial community that usually forms as a slimy layer on a surface.

biogenesis The theory that living cells arise only from preexisting cells.

biogeochemical cycle The recycling of chemical elements by microorganisms for use by other organisms.

bioinformatics The science of determining the function of genes through computer-assisted analysis.

biological transmission The transmission of a pathogen from one host to another when the pathogen reproduces in the vector.

bioluminescence The emission of light from the electron transport chain; requires the enzyme luciferase.

biomass Organic matter produced by living organisms and measured by weight.

bioreactor A fermentation vessel with controls for environmental conditions, e.g., temperature and pH.

bioremediation The use of microbes to remove an environmental pollutant.

biotechnology The industrial application of microorganisms, cells, or cell components to make a useful product.

biotype *See* biovar.

biovar A subgroup of a serovar based on biochemical or physiological properties; also called biotype.

bisphenol A group of antimicrobial chemicals composed of two phenolic groups; includes triclosan.

blade A flat leaflike structure of multicellular algae.

blastoconidium An asexual fungal spore produced by budding from the parent cell.

blood-brain barrier Cell membranes that allow some substances to pass from the blood to the brain but restrict others.

brightfield microscope A microscope that uses visible light for illumination; the specimens are viewed against a white background.

broad-spectrum antibiotic An antibiotic that is effective against a wide range of both gram-positive and gram-negative bacteria.

broth dilution test A method of determining the minimal inhibitory concentration by using serial dilutions of an antimicrobial drug.

bubo An enlarged lymph node caused by inflammation.

budding (1) Asexual reproduction beginning as a protuberance from the parent cell that grows to become a daughter cell. (2) Release of an enveloped virus through the plasma membrane of an animal cell.

budding yeast Following mitosis, a yeast cell that divides unevenly to produce a small cell (bud) from the parent cell.

buffer A substance that tends to stabilize the pH of a solution.

bulking A condition arising when sludge floats rather than settles in secondary sewage treatment.

bullae (singular: bulla) Large serum-filled vesicles in the skin.

burst size The number of newly synthesized bacteriophage particles released from a single cell.

burst time The time required from bacteriophage attachment to release.

Calvin-Benson cycle The fixation of CO_2 into reduced organic compounds; used by autotrophs.

capnophile A microorganism that grows best at relatively high CO_2 concentrations.

capsid The protein coat of a virus that surrounds the nucleic acid.

capsomere A protein subunit of a viral capsid.

capsule An outer, viscous covering on some bacteria composed of a polysaccharide or polypeptide.

carbapenems Antibiotics that contain a β-lactam antibiotic and cilastatin.

carbohydrate An organic compound composed of carbon, hydrogen, and oxygen, with the hydrogen and oxygen present in a 2:1 ratio; carbohydrates include starches, sugars, and cellulose.

carbon cycle The series of processes that converts CO_2 to organic substances and back to CO_2 in nature.

carbon fixation The synthesis of sugars by using carbons from CO_2. *See also* Calvin-Benson cycle.

carbon skeleton The basic chain or ring of carbon atoms in a molecule; for example,

$$-\overset{|}{\underset{|}{C}}-\overset{|}{\underset{|}{C}}-\overset{|}{\underset{|}{C}}-$$

carboxyl group

$$-C\overset{O}{\underset{OH}{\diagup}}$$

carboxysome A prokaryotic inclusion containing ribulose 1,5-diphosphate carboxylase.

carcinogen Any cancer-causing substance.

carrier Organisms (usually refers to humans) that harbor pathogens and transmit them to others.

casein Milk protein.

catabolism All decomposition reactions in a living organism; the breakdown of complex organic compounds into simpler ones.

catabolite repression Inhibition of the metabolism of alternate carbon sources by glucose.

catalase An enzyme that catalyzes the breakdown of hydrogen peroxide to water and oxygen.

catalyst A substance that increases the rate of a chemical reaction but is not altered itself.

cation A positively charged ion.

CD (cluster of determination) Number assigned to an epitope on a single antigen, for example, CD4 protein, which is found on helper T cells.

cDNA (complementary DNA) DNA made in vitro from an mRNA template.

cell culture Eukaryotic cells grown in culture media; also called tissue culture.

cell-mediated immunity An immune response that involves T cells binding to antigens presented on antigen-presenting cells; T cells then differentiate into several types of effector T cells.

cell theory All living organisms are composed of cells and arise from preexisting cells.

cellular respiration *See* respiration.

cell wall The outer covering of most bacterial, fungal, algal, and plant cells; in bacteria, it consists of peptidoglycan.

Centers for Disease Control and Prevention (CDC) A branch of the U.S. Public Health Service that serves a central source of epidemiological information.

central nervous system (CNS) The brain and the spinal cord. *See also* peripheral nervous system.

centriole A structure consisting of nine microtubule triplets, found in eukaryotic cells.

centrosome Region in a eukaryotic cell consisting of a pericentriolar area (protein fibers) and a pair of centrioles; involved in formation of the mitotic spindle.

cephalosporin An antibiotic produced by the fungus *Cephalosporium* that inhibits the synthesis of gram-positive bacterial cell walls.

cercaria A free-swimming larva of trematodes.

chancre A hard sore, the center of which ulcerates.

chemical bond An attractive force between atoms forming a molecule.

chemical element A fundamental substance composed of atoms that have the same atomic number and behave the same way chemically.

chemical energy The energy of a chemical reaction.

chemically defined medium A culture medium in which the exact chemical composition is known.

chemical reaction The process of making or breaking bonds between atoms.

chemiosmosis A mechanism that uses a proton gradient across a cytoplasmic membrane to generate ATP.

chemistry The science of the interactions between atoms and molecules.

chemoautotroph An organism that uses an inorganic chemical as an energy source and CO_2 as a carbon source.

chemoheterotroph An organism that uses organic molecules as a source of carbon and energy.

chemokine A cytokine that induces, by chemotaxis, the migration of leukocytes into infected areas.

chemotaxis Movement in response to the presence of a chemical.

chemotherapy Treatment of disease with chemical substances.

chemotroph An organism that uses oxidation-reduction reactions as its primary energy source.

chimeric monoclonal antibody A genetically engineered antibody made of human constant regions and mouse variable regions.

chlamydoconidium An asexual fungal spore formed within a hypha.

chloramphenicol A broad-spectrum bacteriostatic chemical.

chloroplast The organelle that performs photosynthesis in photoautotrophic eukaryotes.

chlorosome Plasma membrane folds in green sulfur bacteria containing bacteriochlorophylls.

chromatin Threadlike, uncondensed DNA in an interphase eukaryotic cell.

chromatophore An infolding in the plasma membrane where bacteriochlorophyll is located in photoautotrophic bacteria; also known as thylakoids.

chromosome The structure that carries hereditary information, chromosomes contain genes.

chronic disease An illness that develops slowly and is likely to continue or recur for long periods.

ciliary escalator Ciliated mucosal cells of the lower respiratory tract that move inhaled particulates away from the lungs.

cilium (plural: cilia) A relatively short cellular projection from some eukaryotic cells, composed of nine pairs plus two microtubules. *See* flagellum.

cistern A flattened membranous sac in endoplasmic reticulum and the Golgi complex.

clade A group of organisms that share a particular common ancestor; a branch on a cladogram.

cladogram A dichotomous phylogenetic tree that branches repeatedly, suggesting the classification of organisms based on the time sequence in which evolutionary branches arose.

class A taxonomic group between phylum and order.

clonal deletion The elimination of B and T cells that react with self.

clonal selection The development of clones of B and T cells against a specific antigen.

clone A population of cells arising from a single parent cell.

clue cells Sloughed-off vaginal cells covered with *Gardnerella vaginalis*.

coagulase A bacterial enzyme that causes blood plasma to clot.

coccobacillus (plural: coccobacilli) A bacterium that is an oval rod.

coccus (plural: cocci) A spherical or ovoid bacterium.

codon A sequence of three nucleotides in mRNA that specifies the insertion of an amino acid into a polypeptide.

coenocytic hypha A fungal filament that is not divided into uninucleate cell-like units because it lacks septa.

coenzyme A nonprotein substance that is associated with and that activates an enzyme.

coenzyme A (CoA) A coenzyme that functions in decarboxylation.

coenzyme Q *See* ubiquinone.

cofactor (1) The nonprotein component of an enzyme. (2) A microorganism or molecule that acts with others to synergistically enhance or cause disease.

coliforms Aerobic or facultatively anaerobic, gram-negative, nonendospore-forming, rod-shaped bacteria that ferment lactose with acid and gas formation within 48 hours at 35°C.

collagenase An enzyme that hydrolyzes collagen.

collision theory The principle that chemical reactions occur because energy is gained as particles collide.

colony A visible mass of microbial cells arising from one cell or from a group of the same microbes.

colony hybridization The identification of a colony containing a desired gene by using a DNA probe that is complementary to that gene.

colony-stimulating factor (CSF) A substance that induces certain cells to proliferate or differentiate.

commensalism A symbiotic relationship in which two organisms live in association and one is benefited while the other is neither benefited nor harmed.

commercial sterilization A process of treating canned goods aimed at destroying the endospores of *Clostridium botulinum*.

communicable disease Any disease that can be spread from one host to another.

competence The physiological state in which a recipient cell can take and incorporate a large piece of donor DNA.

competitive inhibitor A chemical that competes with the normal substrate for the active site of an enzyme. *See also* noncompetitive inhibitor.

complement A group of serum proteins involved in phagocytosis and lysis of bacteria.

complementary DNA (cDNA) DNA made in vitro from an mRNA template.

complement fixation The process in which complement combines with an antigen–antibody complex.

complex medium A culture medium in which the exact chemical composition is not known.

complex virus A virus with a complicated structure, such as a bacteriophage.

composting A method of solid waste disposal, usually plant material, by encouraging its decomposition by microbes.

compound A substance composed of two or more different chemical elements.

compound light microscope (LM) An instrument with two sets of lenses that uses visible light as the source of illumination.

compromised host A host whose resistance to infection is impaired.

condensation reaction A chemical reaction in which a molecule of water is released; also called dehydration synthesis.

condenser A lens system located below the microscope stage that directs light rays through the specimen.

confocal microscopy A light microscope that uses fluorescent stains and laser to make two- and three-dimensional images.

congenital Refers to a condition existing at birth; may be inherited or acquired in utero.

congenital immune deficiency The inability, due to an individual's genotype, to produce specific antibodies or T cells.

conidiophore An aerial hypha bearing conidiospores.

conidiospore *See* conidium.

conidium An asexual spore produced in a chain from a conidiophore.

conjugated monoclonal antibody *See* immunotoxin.

conjugated vaccine   A vaccine consisting of the desired antigen and other proteins.

conjugation   The transfer of genetic material from one cell to another involving cell-to-cell contact.

conjugative plasmid   A prokaryotic plasmid that carries genes for sex pili and for transfer of the plasmid to another cell.

constitutive enzyme   An enzyme that is produced continuously.

contact inhibition   The cessation of animal cell movement and division as a result of contact with other cells.

contact transmission   The spread of disease by direct or indirect contact or via droplets.

contagious disease   A disease that is easily spread from one person to another.

continuous cell line   Animal cells that can be maintained through an indefinite number of generations in vitro.

continuous flow   An industrial fermentation in which cells are grown indefinitely with continual addition of nutrients and removal of waste and products.

corepressor   A molecule that binds to a repressor protein, enabling the repressor to bind to an operator.

cortex   The protective fungal covering of a lichen.

counterstain   A second stain applied to a smear, provides contrast to the primary stain.

covalent bond   A chemical bond in which the electrons of one atom are shared with another atom.

crisis   The phase of a fever characterized by vasodilation and sweating.

crista (plural: cristae)   Folding of the inner membrane of a mitochondrion.

crossing over   The process by which a portion of one chromosome is exchanged with a portion of another chromosome.

culture   Microorganisms that grow and multiply in a container of culture medium.

culture medium   The nutrient material prepared for growth of microorganisms in a laboratory.

curd   The solid part of milk that separates from the liquid (whey) in the making of cheese, for example.

cutaneous mycosis   A fungal infection of the epidermis, nails, or hair.

cuticle   The outer covering of helminths.

cyanobacteria   Oxygen-producing photoautotrophic prokaryotes.

cyclic AMP (cAMP)   A molecule derived from ATP, in which the phosphate group has a cyclic structure; acts as a cellular messenger.

cyclic photophosphorylation   The movement of an electron from chlorophyll through a series of electron acceptors and back to chlorophyll; anoxygenic; purple and green bacterial photophosphorylation.

cyst   A sac with a distinct wall containing fluid or other material; also, a protective capsule of some protozoa.

cysticercus   An encysted tapeworm larva.

cytochrome   A protein that functions as an electron carrier in cellular respiration and photosynthesis.

cytochrome oxidase   An enzyme that oxidizes cytochrome c.

cytokine   A small protein released from human cells that regulates the immune response; directly or indirectly may induce fever, pain, or T-cell proliferation.

cytolysis   The destruction of cells, resulting from damage to their cell membrane, that causes cellular contents to leak out.

cytopathic effect (CPE)   A visible effect on a host cell, caused by a virus, that may result in host cell damage or death.

cytoplasm   In a prokaryotic cell, everything inside the plasma membrane; in a eukaryotic cell, everything inside the plasma membrane and external to the nucleus.

cytoplasmic streaming   The movement of cytoplasm in a eukaryotic cell.

cytoskeleton   Microfilaments, intermediate filaments, and microtubules that provide support and movement for eukaryotic cytoplasm.

cytosol   The fluid portion of cytoplasm.

cytostome   The mouthlike opening in some protozoa.

cytotoxic T (T_C) cells   A specialized T cell that destroys infected cells presenting antigens.

cytotoxic T lymphocytes (CTL)   An activated T_C cell; kills cells presenting endogenous antigens.

cytotoxin   A bacterial toxin that kills host cells or alters their functions.

darkfield microscope   A microscope that has a device to scatter light from the illuminator so that the specimen appears white against a black background.

deamination   The removal of an amino group from an amino acid to form ammonia. *See also* ammonification.

death phase   The period of logarithmic decrease in a bacterial population; also called logarithmic decline phase.

debridement   Surgical removal of necrotic tissue.

decarboxylation   The removal of CO_2 from an amino acid.

decimal reduction time (DRT)   The time (in minutes) required to kill 90% of a bacterial population at a given temperature; also called D value.

decolorizing agent   A solution used in the process of removing a stain.

decomposition reaction   A chemical reaction in which bonds are broken to produce smaller parts from a large molecule.

deep-freezing   Preservation of bacterial cultures at $-50°C$ to $-95°C$.

defensins   Small peptide antibiotics made by human cells.

definitive host   An organism that harbors the adult, sexually mature form of a parasite.

degeneracy   Redundancy of the genetic code; that is, most amino acids are encoded by several codons.

degerming   The removal of microorganisms in an area; also called degermation.

degranulation   The release of contents of secretory granules from mast cells or basophils during anaphylaxis.

dehydration synthesis   *See* condensation reaction.

dehydrogenation   The loss of hydrogen atoms from a substrate.

delayed-type hypersensitivity   Cell-mediated hypersensitivity.

denaturation   A change in the molecular structure of a protein, usually making it nonfunctional.

dendritic cell   A type of antigen-presenting cell characterized by long fingerlike extensions; found in lymphatic tissue and skin.

denitrification   The reduction of nitrogen in nitrate to nitrite or nitrogen gas.

dental plaque   A combination of bacterial cells, dextran, and debris adhering to the teeth.

deoxyribonucleic acid (DNA)   The nucleic acid of genetic material in all cells and some viruses.

deoxyribose   A five-carbon sugar contained in DNA nucleotides.

dermatomycosis   A fungal infection of the skin; also known as tinea or ringworm.

dermatophyte   A fungus that causes a cutaneous mycosis.

dermis   The inner portion of the skin.

descriptive epidemiology   The collection and analysis of all data regarding the occurrence of a disease to determine its cause.

desensitization   The prevention of allergic inflammatory responses.

desiccation   The removal of water.

diapedesis   *See* emigration.

dichotomous key   An identification scheme based on successive paired questions; answering one question leads to another pair of questions, until an organism is identified.

differential interference contrast (DIC) microscope   An instrument that provides a three-dimensional, magnified image.

differential medium   A solid culture medium that makes it easier to distinguish colonies of the desired organism.

differential stain   A stain that distinguishes objects on the basis of reactions to the staining procedure.

differential white blood cell count   The number of each kind of leukocyte in a sample of 100 leukocytes.

diffusion   The net movement of molecules or ions from an area of higher concentration to an area of lower concentration.

dimorphism   The property of having two forms of growth.

dioecious   Referring to organisms in which organs of different sexes are located in different individuals.

diplobacilli (singular: diplobacillus)   Rods that divide and remain attached in pairs.

diplococci (singular: diplococcus)   Cocci that divide and remain attached in pairs.

diploid cell   A cell having two sets of chromosomes; diploid is the normal state of a eukaryotic cell.

diploid cell line   Eukaryotic cells grown in vitro.

direct agglutination test   The use of known antibodies to identify an unknown cell-bound antigen.

direct contact transmission   A method of spreading infection from one host to another through some kind of close association between the hosts.

direct FA test   A fluorescent-antibody test to detect the presence of an antigen.

direct microscopic count   Enumeration of cells by observation through a microscope.

disaccharide   A sugar consisting of two simple sugars, or monosaccharides.

disease   An abnormal state in which part or all of the body is not properly adjusted or is incapable of performing normal functions; any change from a state of health.

disinfection   Any treatment used on inanimate objects to kill or inhibit the growth of microorganisms; a chemical used is called a disinfectant.

disk-diffusion method   An agar-diffusion test to determine microbial susceptibility to chemotherapeutic agents; also called Kirby-Bauer test.

D-isomer   A stereoisomer.

dissimilation   A metabolic process in which nutrients are not assimilated but are excreted as ammonia, hydrogen sulfide, and so on.

dissimilation plasmid   A plasmid containing genes encoding production of enzymes that trigger the catabolism of certain unusual sugars and hydrocarbons.

dissociation   The separation of a compound into positive and negative ions in solution. *See also* ionization.

disulfide bond   A covalent bond that holds together two atoms of sulfur.

DNA base composition   The moles-percentage of guanine plus cytosine in an organism's DNA.

DNA chip   A silica wafer that holds DNA probes; used to recognize DNA in samples being tested.

DNA fingerprinting   Analysis of DNA by electrophoresis of restriction enzyme fragments of the DNA.

DNA gyrase   *See* topoisomerase.

DNA ligase   An enzyme that covalently bonds a carbon atom of one nucleotide with the phosphate of another nucleotide.

DNA polymerase   Enzyme that synthesizes DNA by copying a DNA template.

DNA probe   A short, labeled, single strand of DNA or RNA used to locate its complementary strand in a quantity of DNA.

DNA sequencing   A process by which the nucleotide sequence of DNA is determined.

domain   A taxonomic classification based on rRNA sequences; above the kingdom level.

donor cell   A cell that gives DNA to a recipient cell during genetic recombination.

droplet transmission   The transmission of infection by small liquid droplets carrying microorganisms.

DTaP vaccine   A combined vaccine used to provide active immunity, containing diphtheria and tetanus toxoids and *Bordetella pertussis* cell fragments.

D value   *See* decimal reduction time.

dysentery   A disease characterized by frequent, watery stools containing blood and mucus.

echinocandins   Antifungal agents that interfere with cell wall synthesis.

eclipse period   The time during viral multiplication when complete, infective virions are not present.

ecology   The study of the interrelationships between organisms and their environment.

edema   An abnormal accumulation of interstitial fluid in body parts or tissues, causing swelling.

electron   A negatively charged particle in motion around the nucleus of an atom.

electron acceptor   An ion that picks up an electron that has been lost from another atom.

electron donor   An ion that gives up an electron to another atom.

electronic configuration   The arrangement of electrons in shells or energy levels in an atom.

electron microscope   A microscope that uses electrons instead of light to produce an image.

electron shell   A region of an atom where electrons orbit the nucleus, corresponding to an energy level.

electron transport chain, electron transport system   A series of compounds that transfer electrons from one compound to another, generating ATP by oxidative phosphorylation.

electroporation   A technique by which DNA is inserted into a cell using an electrical current.

elementary body   The infectious form of chlamydiae.

ELISA (enzyme-linked immunosorbent assay)   A group of serological tests that use enzyme reactions as indicators.

embryonic stem cell   A cell from an embryo that has the potential to become a wide variety of specialized cell types.

emerging infectious disease (EID)   A new or changing disease that is increasing or has the potential to increase in incidence in the near future.

emigration   The process by which phagocytes move out of blood vessels; also called diapedesis.

Embden-Meyerhof pathway   *See* glycolysis.

enanthem   Rash on mucous membranes. *See also* exanthem.

encephalitis   Infection of the brain.

encystment   Formation of a cyst.

endemic disease   A disease that is constantly present in a certain population.

endergonic reaction   A chemical reaction that requires energy.

endocarditis   Infection of the lining of the heart (endocardium).

endocytosis   The process by which material is moved into a eukaryotic cell.

endoflagellum   *See* axial filament.

endogenous   (1) Infection caused by an opportunistic pathogen from an individual's own normal microbiota. (2) Antigens, usually of viral origin and degraded into fragments, generated within a cell.

endolith   An organism that lives inside rock.

endoplasmic reticulum (ER)   A membranous network in eukaryotic cells connecting the plasma membrane with the nuclear membrane.

endospore   A resting structure formed inside some bacteria.

endosymbiotic theory   A model for the evolution of eukaryotes which states that organelles arose from prokaryotic cells living inside a host prokaryote.

endotoxic shock   *See* gram-negative sepsis.

endotoxin   Part of the outer portion of the cell wall (lipid A) of most gram-negative bacteria; released on destruction of the cell.

end-product inhibition   *See* feedback inhibition.

energy level   Potential energy of an electron in an atom. *See also* electron shell.

enrichment culture   A culture medium used for preliminary isolation that favors the growth of a particular microorganism.

enteric   The common name for a bacterium in the family Enterobacteriaceae.

enterotoxin   An exotoxin that causes gastroenteritis, such as those produced by *Staphylococcus*, *Vibrio*, and *Escherichia*.

Entner-Doudoroff pathway   An alternate pathway for the oxidation of glucose to pyruvic acid.

envelope   An outer covering surrounding the capsid of some viruses.

enzyme   A molecule that catalyzes biochemical reactions in a living organism, usually a protein. *See also* ribozyme.

enzyme-linked immunosorbent assay *See* ELISA.

enzyme–substrate complex A temporary union of an enzyme and its substrate.

eosinophil A granulocyte whose granules take up the stain eosin.

epidemic disease A disease acquired by many hosts in a given area in a short time.

epidemiology The science that studies when and where diseases occur and how they are transmitted.

epidermis The outer portion of the skin.

epitope *See* antigenic determinant.

equilibrium The point of even distribution.

equivalent treatments Different methods that have the same effect on controlling microbial growth.

ergot A toxin produced in sclerotia by the fungus *Claviceps purpurea* that causes ergotism.

E test An agar diffusion test to determine antibiotic sensitivity using a plastic strip impregnated with varying concentrations of an antibiotic.

ethambutol A synthetic antimicrobial agent that interferes with the synthesis of RNA.

ethanol

$$\begin{array}{c} \text{H} \quad \text{H} \\ | \quad | \\ \text{H}-\text{C}-\text{C}-\text{OH} \\ | \quad | \\ \text{H} \quad \text{H} \end{array}$$

etiology The study of the cause of a disease.

eukarya All eukaryotes (animals, plants, fungi, and protists); members of the Domain Eukarya.

eukaryote A cell having DNA inside a distinct membrane-enclosed nucleus.

eukaryotic species A group of closely related organisms that can interbreed.

eutrophication The addition of organic matter and subsequent removal of oxygen from a body of water.

exanthem Skin rash. *See also* enanthem.

exchange reaction A chemical reaction that has both synthesis and decomposition components.

exergonic reaction A chemical reaction that releases energy.

exon A region of a eukaryotic chromosome that encodes a protein.

exotoxin A protein toxin released from living, mostly gram-positive bacterial cells.

experimental epidemiology The study of a disease using controlled experiments.

exponential growth phase *See* log phase.

extracellular polysaccharide (EPS) A glycocalyx, composed of sugars, that permits bacteria to attach to various surfaces.

extreme halophile An organism that requires a high salt concentration for growth.

extreme thermophile *See* hyperthermophile.

extremophile A microorganism that lives in environmental extremes of temperature, acidity, alkalinity, salinity, or pressure.

extremozymes Enzymes produced by extremophiles.

facilitated diffusion The movement of a substance across a plasma membrane from an area of higher concentration to an area of lower concentration, mediated by transporter proteins.

facultative anaerobe An organism that can grow with or without molecular oxygen (O_2).

facultative halophile An organism capable of growth in, but not requiring, 1–2% salt.

FAD Flavin adenine dinucleotide; a coenzyme that functions in the removal and transfer of hydrogen ions (H^+) and electrons from substrate molecules.

FAME Fatty acid methyl ester; identification of microbes by the presence of specific fatty acids.

family A taxonomic group between order and genus.

feedback inhibition Inhibition of an enzyme in a particular pathway by the accumulation of the end-product of the pathway; also called endproduct inhibition.

fermentation The enzymatic degradation of carbohydrates in which the final electron acceptor is an organic molecule, ATP is synthesized by substrate-level phosphorylation, and O_2 is not required.

fermentation test Method used to determine whether a bacterium or yeast ferments a specific carbohydrate; usually performed in a peptone broth containing the carbohydrate, a pH indicator, and an inverted tube to trap gas.

fever An abnormally high body temperature.

F factor (fertility factor) A plasmid found in the donor cell in bacterial conjugation.

fibrinolysin A kinase produced by streptococci.

filtration The passage of a liquid or gas through a screenlike material; a $0.45-\mu m$ filter removes most bacteria.

fimbria (plural: fimbriae) An appendage on a bacterial cell used for attachment.

FISH Fluorescent in situ hybridization; use of rRNA probes to identify microbes without cutturing.

fission yeast Following mitosis, a yeast cell that divides evenly to produce two new cells.

fixed macrophage A macrophage that is located in a certain organ or tissue (e.g., liver, lungs, spleen, or lymph nodes); also called a histiocyte.

fixing (1) In slide preparation, the process of attaching a specimen to a slide. (2) Regarding chemical elements, combining elements so that a critical element can enter the food chain. *See also* Calvin-Benson cycle; nitrogen fixation.

flaccid paralysis Loss of muscle movement, loss of muscle tone.

flagellum (plural: flagella) A thin appendage from the surface of a cell; used for cellular locomotion; composed of flagellin in prokaryotic cells, composed of nine pairs plus two microtubules in eukaryotic cells.

flaming The process of sterilizing an inoculating loop by holding it in an open flame.

flat sour spoilage Thermophilic spoilage of canned goods not accompanied by gas production.

flatworm An animal belonging to the phylum Platyhelminthes.

flavin adenine dinucleotide *See* FAD.

flavin mononucleotide *See* FMN.

flavoprotein A protein containing the coenzyme flavin; functions as an electron carrier in electron transport chains.

flocculation The removal of colloidal material during water purification by adding a chemical that causes colloidal particles to coalesce.

flow cytometry A method of counting cells using a flow cytometer, which detects cells by the presence of a fluorescent tag on the cell surface.

fluid mosaic model A way of describing the dynamic arrangement of phospholipids and proteins comprising the plasma membrane.

fluke A flatworm belonging to the class Trematoda.

fluorescence The ability of a substance to give off light of one color when exposed to light of another color.

fluorescence-activated cell sorter (FACS) A modification of a flow cytometer that counts and sorts cells labeled with fluorescent antibodies.

fluorescence microscope A microscope that uses an ultraviolet light source to illuminate specimens that will fluoresce.

fluorescent-antibody (FA) technique A diagnostic tool using antibodies labeled with fluorochromes and viewed through a fluorescence microscope; also called immunofluorescence.

fluoroquinolone A synthetic antibacterial agent that inhibits DNA synthesis.

FMN Flavin mononucleotide; a coenzyme that functions in the transfer of electrons in the electron transport chain.

focal infection A systemic infection that began as an infection in one place.

folliculitis An infection of hair follicles, often occurring as pimples.

fomite A nonliving object that can spread infection.

forespore A structure consisting of chromosome, cytoplasm, and endospore membrane inside a bacterial cell.

frameshift mutation A mutation caused by the addition or deletion of one or more bases in DNA.

freeze-drying *See* lyophilization.

FTA-ABS test An indirect fluorescent-antibody test used to detect syphilis.

fulminating A condition that develops quickly and rapidly increases in severity.

functional group An arrangement of atoms in an organic molecule that is responsible for most of the chemical properties of that molecule.

fungus (plural: fungi) An organism that belongs to the Kingdom Fungi; a eukaryotic absorptive chemoheterotroph.

furuncle An infection of a hair follicle.

fusion The merging of plasma membranes of two different cells, resulting in one cell containing cytoplasm from both original cells.

gamete A male or female reproductive cell.

gametocyte A male or female protozoan cell.

gamma globulin *See* immune serum globulin.

gastroenteritis Inflammation of the stomach and intestine.

gas vacuole A prokaryotic inclusion for buoyancy compensation.

gel electrophoresis The separation of substances (such as serum proteins or DNA) by their rate of movement through an electrical field.

gene A segment of DNA (a sequence of nucleotides in DNA) encoding a functional product.

gene library A collection of cloned DNA fragments created by inserting restriction enzyme fragments in a bacterium, yeast, or phage.

gene therapy Treating a disease by replacing abnormal genes.

generalized transduction The transfer of bacterial chromosome fragments from one cell to another by a bacteriophage.

generation time The time required for a cell or population to double in number.

genetic code The mRNA codons and the amino acids they encode.

genetic engineering *See* recombinant DNA technology.

genetic recombination The process of joining pieces of DNA from different sources.

genetics The science of heredity and gene function.

genetic screening Techniques for determining which genes are in a cell's genome.

genome One complete copy of the genetic information in a cell.

genomics The study of genes and their function.

genotype The genetic makeup of an organism.

genus (plural: genera) The first name of the scientific name (binomial); the taxon between family and species.

germicide *See* biocide.

germination The process of starting to grow from a spore or endospore.

germ theory of disease The principle that microorganisms cause disease.

global warming Retention of solar heat by gases in the atmosphere.

globulin The class of proteins that includes antibodies. *See also* immunoglobulin.

glycocalyx A gelatinous polymer surrounding a cell.

glycolysis The main pathway for the oxidation of glucose to pyruvic acid; also called Embden-Meyerhof pathway.

Golgi complex An organelle involved in the secretion of certain proteins.

graft-versus-host (GVH) disease A condition that occurs when a transplanted tissue has an immune response to the tissue recipient.

gram-negative bacteria Bacteria that lose the crystal violet color after decolorizing by alcohol; they stain red after treatment with safranin.

gram-negative sepsis Septic shock caused by gram-negative endotoxins.

gram-positive bacteria Bacteria that retain the crystal violet color after decolorizing by alcohol; they stain dark purple.

gram-positive sepsis Septic shock caused by gram-positive bacteria.

Gram stain A differential stain that classifies bacteria into two groups, gram-positive and gram-negative.

granulocyte A leukocyte with visible granules in the cytoplasm; includes neutrophils, basophils, and eosinophils.

granzymes Proteases that induce apoptosis.

green nonsulfur bacteria Gram-negative, nonproteobacteria; anaerobic and phototrophic; use reduced organic compounds as electron donors for CO_2 fixation.

green sulfur bacteria Gram-negative, nonproteobacteria; strictly anaerobic and phototrophic; no growth in dark; use reduced sulfur compounds as electron donors for CO_2 fixation.

griseofulvin A fungistatic antibiotic; produced by *Penicillium griseofulvum*.

group translocation In prokaryotes, active transport in which a substance is chemically altered during transport across the plasma membrane.

gumma A rubbery mass of tissue characteristic of tertiary syphilis.

HAART (highly active antiretroviral therapy) A combination of drugs used to treat HIV infection.

halogen One of the following elements: fluorine, chlorine, bromine, iodine, or astatine.

haploid cell A eukaryotic cell or organism with one of each type of chromosome.

hapten A substance of low molecular weight that does not cause the formation of antibodies by itself but does so when combined with a carrier molecule.

Hazard Analysis and Critical Control Point (HACCP) System of prevention of hazards, for food safety.

helminth A parasitic roundworm or flatworm.

helper T (T_H) cell A specialized T cell that often interacts with an antigen before B cells interact with the antigen.

hemagglutination The clumping of red blood cells.

hemoflagellate A parasitic flagellate found in the circulatory system of its host.

hemolysin An enzyme that lyses red blood cells.

herd immunity The presence of immunity in most of a population.

hermaphroditic Having both male and female reproductive capacities.

heterocyst A large cell in certain cyanobacteria; the site of nitrogen fixation.

heterolactic Describing an organism that produces lactic acid and other acids or alcohols as end-products of fermentation; e.g., *Escherichia*.

heterotroph An organism that requires an organic carbon source; also called organotroph.

Hfr cell A bacterial cell in which the F factor has become integrated into the chromosome; Hfr stands for high frequency of recombination.

high-efficiency particulate air (HEPA) filter A screenlike material that removes particles larger than 0.3 μm from air.

high-temperature short-time (HTST) pasteurization Pasteurizing at 72°C for 15 seconds.

histamine A substance released by tissue cells that causes vasodilation, capillary permeability, and smooth muscle contraction.

histocompatibility antigen An antigen on the surface of human cells.

histone A protein associated with DNA in eukaryotic chromosomes.

holdfast The branched base of an algal stipe.

holoenzyme An enzyme consisting of an apoenzyme and a cofactor.

homolactic Describing an organism that produces only lactic acid from fermentation; e.g., *Streptococcus*.

horizontal gene transfer Transfer of genes between two organisms in the same generation. *See also* vertical gene transfer.

host An organism infected by a pathogen. *See also* definitive host; intermediate host.

host range The spectrum of species, strains, or cell types that a pathogen can infect.

hot-air sterilization Sterilization by the use of an oven at 170°C for approximately 2 hours.

H (hemagglutinin) spike Antigenic projections from the outer lipid bilayer of *Influenzavirus*.

human leukocyte antigen (HLA) complex Human cell surface antigens. *See also* major histocompatibility complex.

humanized antibody Monoclonal antibodies that are partly or fully human proteins.

humoral immunity Immunity produced by antibodies dissolved in body fluids, mediated by B cells; also called antibody-mediated immunity.

hyaluronidase An enzyme secreted by certain bacteria that hydrolyzes hyaluronic acid and helps spread microorganisms from their initial site of infection.

hybridoma A cell made by fusing an antibody-producing B cell with a cancer cell.

hydrogen bond A bond between a hydrogen atom covalently bonded to oxygen or nitrogen and another covalently bonded oxygen or nitrogen atom.

hydrolysis A decomposition reaction in which chemicals react with the H^+ and OH^- of a water molecule.

hydroxyl radical A toxic form of oxygen (OH•) formed in cytoplasm by ionizing radiation and aerobic respiration.

hyperacute rejection Very rapid rejection of transplanted tissue, usually in the case of tissue from nonhuman sources.

hyperbaric chamber An apparatus to hold materials at pressures greater than 1 atmosphere.

hypersensitivity An altered, enhanced immune reaction leading to pathological changes; also called allergy.

hyperthermophile An organism whose optimum growth temperature is at least 80°C; also called extreme thermophile.

hypertonic (hyperosmotic) solution A solution that has a higher concentration of solutes than an isotonic solution.

hypha A long filament of cells in fungi or actinomycetes.

hypotonic (hypoosmotic) solution A solution that has a lower concentration of solutes than an isotonic solution.

ID$_{50}$ The number of microorganisms required to produce a demonstrable infection in 50% of the test host population.

idiophase The period in the production curve of an industrial cell population in which secondary metabolites are produced; a period of stationary growth following the phase of rapid growth. *See also* trophophase.

IgA The class of antibodies found in secretions.

IgD The class of antibodies found on B cells.

IgE The class of antibodies involved in hypersensitivities.

IgG The most abundant class of antibodies in serum.

IgM The first class of antibodies to appear after exposure to an antigen.

imiquimod A substance that improves the body's natural response to infection and disease.

immune complex A circulating antigen-antibody aggregate capable of fixing complement.

immune deficiency The absence of an adequate immune response; may be congenital or acquired.

immune serum globulin The serum fraction containing immunoglobulins (antibodies); also called gamma globulin.

immune surveillance The body's immune response to cancer.

immunity The body's defense against particular pathogenic microorganisms.

immunization *See* vaccination.

immunodiffusion test A test consisting of precipitation reactions carried out in an agar gel medium.

immunoelectrophoresis The identification of proteins by electrophoretic separation followed by serological testing.

immunofluorescence *See* fluorescent-antibody technique.

immunogen *See* antigen.

immunoglobulin (Ig) A protein (antibody) formed in response to an antigen and can react with that antigen. *See also* globulin.

immunology The study of a host's defenses to a pathogen.

immunosuppression Inhibition of the immune response.

immunotherapy Making use of the immune system to attack tumor cells, either by enhancing the normal immune response or by using toxin-bearing specific antibodies. *See also* immunotoxin.

immunotoxin An immunotherapeutic agent consisting of a poison bound to a monoclonal antibody.

inapparent infection *See* subclinical infection.

incidence The fraction of the population that contracts a disease during a particular period of time.

inclusion Material held inside a cell, often consisting of reserve deposits.

inclusion body A granule or viral particle in the cytoplasm or nucleus of some infected cells; important in the identification of viruses that cause infection.

incubation period The time interval between the actual infection and first appearance of any signs or symptoms of disease.

indicator organism A microorganism, such as a coliform, whose presence indicates conditions such as fecal contamination of food or water.

indirect (passive) agglutination test An agglutination test using soluble antigens attached to latex or other small particles.

indirect contact transmission The spread of pathogens by fomites (nonliving objects).

indirect FA test A fluorescent-antibody test to detect the presence of specific antibodies.

inducer A chemical or environmental stimulus that causes transcription of specific genes.

induction The process that turns on the transcription of a gene.

infection The growth of microorganisms in the body.

infectious disease A disease in which pathogens invade a susceptible host and carry out at least part of their life cycle in the host.

inflammation A host response to tissue damage characterized by redness, pain, heat, and swelling; and sometimes loss of function.

innate immunity Host defenses that afford protection against any kind of pathogen. *See also* adaptive immunity.

innate resistance The resistance of an individual to diseases that affect other species and other individuals of the same species.

inoculum A culture medium in which microorganisms are implanted.

inorganic compound A small molecule that does not contain carbon and hydrogen.

insertion sequence (IS) The simplest kind of transposon.

integrase An enzyme produced by HIV that allows the integration of HIV DNA into the host cell's DNA.

interferon (IFN) A specific group of cytokines. Alpha- and beta-IFNs are antiviral proteins produced by certain animal cells in response to a viral infection. Gamma-IFN stimulates macrophage activity.

interleukin (IL) A chemical that causes T-cell proliferation. *See also* cytokine.

intermediate host An organism that harbors the larval or asexual stage of a helminth or protozoan.

intoxication A condition resulting from the ingestion of a microbially produced toxin.

intron A region in a eukaryotic gene that does not code for a protein or mRNA.

invasin A surface protein produced by *Salmonella typhimurium* and *Escherichia coli* that rearranges nearby actin filaments in the cytoskeleton of a host cell.

iodophor A complex of iodine and a detergent.

ion A negatively or positively charged atom or group of atoms.

ionic bond A chemical bond formed when atoms gain or lose electrons in the outer energy levels.

ionization The separation (dissociation) of a molecule into ions.

ionizing radiation High-energy radiation with a wavelength less than 1nm; causes ionization. X rays and gamma rays are examples.

ischemia Localized decreased blood flow.

isograft A tissue graft from a genetically identical source (i.e., from an identical twin).

isomer One or two molecules with the same chemical formula but different structures.

isoniazid (INH) A bacteriostatic agent used to treat tuberculosis.

isotonic (isosmotic) solution A solution in which, after immersion of a cell, osmotic pressure is equal across the cell's membrane.

isotope A form of a chemical element in which the number of neutrons in the nucleus is different from the other forms of that element.

karyogamy Fusion of the nuclei of two cells; occurs in the sexual stage of a fungal life cycle.

kelp A multicellular brown alga.

keratin A protein found in epidermis, hair, and nails.

ketolide Semi-synthetic macrolide antibiodies; effective against macrolide-resistant bacteria.

kinase (1) An enzyme that removes a $\textcircled{P}$ from ATP and attaches it to another molecule. (2) A bacterial enzyme that breaks down fibrin (blood clots).

kingdom A taxonomic classification between domain and phylum.

kinin A substance released from tissue cells that causes vasodilation.

Kirby-Bauer test *See* disk-diffusion method.

Koch's postulates Criteria used to determine the causative agent of infectious diseases.

koji A microbial fermentation on rice; usually *Aspergillus oryzae*; used to produce amylase.

Krebs cycle A pathway that converts two-carbon compounds to CO_2, transferring electrons to NAD^+ and other carriers; also called tricarboxylic acid (TCA) cycle or critic acid cycle.

lactic acid fermentation A catabolic process, beginning with glycolysis, that produces lactic acid to reoxidize NADH.

lagging strand During DNA replication, the daughter strand that is synthesized discontinuously.

lag phase The time interval in a bacterial growth curve during which there is no growth.

larva The sexually immature stage of a helminth or arthropod.

latent disease A disease characterized by a period of no symptoms when the pathogen is inactive.

latent infection A condition in which a pathogen remains in the host for long periods without producing disease.

LD$_{50}$ The lethal dose for 50% of the inoculated hosts within a given period.

leading strand During DNA replication, the daughter strand that is synthesized continuously.

lectin Carbohydrate-binding proteins on a cell.

lepromin test A skin test to determine the presence of antibodies to *Mycobacterium leprae*, the cause of leprosy.

leukocidins Substances produced by some bacteria that can destroy neutrophils and macrophages.

leukocyte A white blood cell.

leukotriene A substance produced by mast cells and basophils that causes increased permeability of blood vessels and helps phagocytes attach to pathogens.

L-form Prokaryotic cells that lack a cell wall; can return to walled state.

lichen A mutualistic relationship between a fungus and an alga or a cyanobacterium.

ligand *See* adhesin.

light-dependent reaction The process by which light energy is used to convert ADP and phosphate to ATP. *See also* photophosphorylation.

light-independent reactions The process by which electrons and energy from ATP are used to reduce CO_2 to sugar. *See also* Calvin-Benson cycle.

light-repair enzyme *See* photolyase.

limnetic zone The surface zone of an inland body of water away from the shore.

***Limulus* amoebocyte lysate (LAL) assay** A test to detect the presence of bacterial endotoxins.

lipase An enzyme that breaks down triglycerides into their component glycerol and fatty acids.

lipid A non–water soluble organic molecule, including triglycerides, phospholipids, and sterols.

lipid A A component of the gram-negative outer membrane; endotoxin.

lipid inclusion *See* inclusion.

lipopolysaccharide (LPS) A molecule consisting of a lipid and a polysaccharide, forming the outer membrane of gram-negative cell walls.

L-isomer A stereoisomer.

lithotroph *See* autotroph.

littoral zone The region along the shore of the ocean or a large lake where there is considerable vegetation and where light penetrates to the bottom.

local infection An infection in which pathogens are limited to a small area of the body.

localized anaphylaxis An immediate hypersensitivity reaction that is restricted to a limited area of skin or mucous membrane; for example, hayfever, a skin rash, or asthma. *See also* systemic anaphylaxis.

logarithmic decline phase *See* death phase.

log phase The period of bacterial growth or logarithmic increase in cell numbers; also called exponential growth phase.

lophotrichous Having two or more flagella at one or both ends of a cell.

luciferase An enzyme that accepts electrons from flavoproteins and emits a photon of light in bioluminescence.

lymphangitis Inflammation of lymph vessels.

lymphocyte A leukocyte involved in specific immune responses.

lyophilization Freezing a substance and sublimating the ice in a vacuum; also called freeze-drying.

lysis (1) Destruction of a cell by the rupture of the plasma membrane, resulting in a loss of cytoplasm. (2) In disease, a gradual period of decline.

lysogenic conversion The acquisition of new properties by a host cell infected by a lysogenic phage.

lysogenic cycle Stages in viral development that result in the incorporation of viral DNA into host DNA.

lysogeny A state in which phage DNA is incorporated into the host cell without lysis.

lysosome An organelle containing digestive enzymes.

lysozyme An enzyme capable of hydrolyzing bacterial cell walls.

lytic cycle A mechanism of phage multiplication that results in host cell lysis.

macrolide An antibiotic that inhibits protein synthesis; for example, erythromycin.

macromolecule A large organic molecule.

macrophage A phagocytic cell; a mature monocyte.

macule A flat, reddened skin lesion.

magnetosome An iron oxide inclusion, produced by some gram-negative bacteria, that acts like a magnet.

major histocompatibility complex (MHC) The genes that code for histocompatibility antigens; also known as human leukocyte antigen (HLA) complex.

malolactic fermentation The conversion of malic acid to lactic acid by lactic acid bacteria.

malt Germinated barley grains containing maltose, glucose, and amylase.

malting The germination of starchy grains resulting in glucose and maltose production.

margination The process by which phagocytes stick to the lining of blood vessels.

mast cell A type of cell found throughout the body that contains histamine and other substances that stimulate vasodilation.

matrix Fluid in mitochondria.

maximum growth temperature The highest temperature at which a species can grow.

M (microfold) cell Intestinal cells that take up and transfer antigens to lymphocytes.

mechanical transmission The process by which arthropods transmit infections by carrying pathogens on their feet and other body parts.

medulla A lichen body consisting of algae (or cyanobacteria) and fungi.

meiosis A eukaryotic cell replication process that results in cells with half the chromosome number of the original cell.

membrane attack complex (MAC) Complement proteins C5–C9, which together make lesions in cell membranes that lead to cell death.

membrane filter A screenlike material with pores small enough to retain microorganisms; a $0.45-\mu m$ filter retains most bacteria.

memory cells   A long-lived B or T cell responsible for the memory, or secondary, response.

memory response   A rapid rise in antibody titer following exposure to an antigen after the primary response to that antigen; also called anamnestic response or secondary response.

meningitis   Inflammation of the meninges, the three membranes covering the brain and spinal cord.

merozoite   A trophozoite of *Plasmodium* found in red blood cells or liver cells.

mesophile   An organism that grows between about 10°C and 50°C; a moderate-temperature–loving microbe.

mesosome   An irregular fold in the plasma membrane of a prokaryotic cell that is an artifact of preparation for microscopy.

messenger RNA (mRNA)   The type of RNA molecule that directs the incorporation of amino acids into proteins.

metabolic pathway   A sequence of enzymatically catalyzed reactions occurring in a cell.

metabolism   The sum of all the chemical reactions that occur in a living cell.

metacercaria   The encysted stage of a fluke in its final intermediate host.

metachromatic granule   A granule that stores inorganic phosphate and stains red with certain blue dyes; characteristic of *Corynebacterium diphtheriae*. Collectively known as volutin.

methane   The hydrocarbon CH_4, a flammable gas formed by the microbial decomposition of organic matter; natural gas.

methylate   Addition of a methyl group ($—CH_3$) to a molecule; methylated cytosine is protected from digestion by restriction enzymes.

microaerophile   An organism that grows best in an environment with less molecular oxygen (O_2) than is normally found in air.

micrometer (μm)   A unit of measurement equal to 10^{-6} m.

microorganism   A living organism too small to be seen with the naked eye; includes bacteria, fungi, protozoa, and microscopic algae; also includes viruses.

microtubule   A hollow tube made of the protein tubulin; the structural unit of eukaryotic flagella and centrioles.

microwave   Electromagnetic radiation with wavelength between 10^{-1} and 10^{-3} m.

minimal bactericidal concentration (MBC)   The lowest concentration of chemotherapeutic agent that will kill test microorganisms.

minimal inhibitory concentration (MIC)   The lowest concentration of a chemotherapeutic agent that will prevent growth of the test microorganisms.

minimum growth temperature   The lowest temperature at which a species will grow.

miracidium   The free-swimming, ciliated larva of a fluke that hatches from the egg.

missense mutation   A mutation that results in the substitution of an amino acid in a protein.

mitochondrion (plural: mitochondria)   An organelle containing Krebs cycle enzymes and the electron transport chain.

mitosis   A eukaryotic cell replication process in which the chromosomes are duplicated; usually followed by division of the cytoplasm of the cell.

mitosome   Eukaryotic organelle derived form degenrate mitochondria, found in *Trichomonas* and *Giardia*.

MMWR   *Morbidity and Mortality Weekly Report;* a CDC publication containing data on notifiable diseases and topics of special interest.

mole   An amount of a chemical equal to the atomic weights of all the atoms in a molecule of the chemical.

molecular biology   The science dealing with DNA and protein synthesis of living organisms.

molecular weight   The sum of the atomic weights of all atoms making up a molecule.

molecule   A combination of atoms forming a specific chemical compound.

monobactam   A synthetic antibiotic with a β-lactam ring that is monocyclic in structure, in contrast to the bicyclic β-lactam structure of the penicillins and cephalosporins.

monoclonal antibody (Mab)   A specific antibody produced in vitro by a clone of B cells hybridized with cancerous cells.

monocyte   A leukocyte that is the precursor of a macrophage.

monoecious   Having both male and female reproductive capacities.

monomer   A small molecule that collectively combines to form polymers.

mononuclear phagocytic system   A system of fixed macrophages located in the spleen, liver, lymph nodes, and red bone marrow.

monosaccharide   A simple sugar consisting of 3–7 carbon atoms.

monotrichous   Having a single flagellum.

morbidity   (1) The incidence of a specific disease. (2) The condition of being diseased.

morbidity rate   The number of people affected by a disease in a given period of time in relation to the total population.

mordant   A substance added to a staining solution to make it stain more intensely.

mortality   The number of deaths from a specific notifiable disease.

mortality rate   The number of deaths resulting from a disease in a given period of time in relation to the total population.

most probable number (MPN) method   A statistical determination of the number of coliforms per 100 ml of water or 100 g of food.

motility   The ability of an organism to move by itself.

M protein   A heat- and acid-resistant protein of streptococcal cell walls and fibrils.

mucous membranes   Membranes that line body openings, including the intestinal tract, open to the exterior; also called mucosa.

mutagen   An agent in the environment that brings about mutations.

mutation   Any change in the nitrogenous base sequence of DNA.

mutation rate   The probability that a gene will mutate each time a cell divides.

mutualism   A type of symbiosis in which both organisms or populations are benefited.

mycelium   A mass of long filaments of cells that branch and intertwine, typically found in molds.

mycolic acid   Long-chained, branched fatty acids characteristic of members of the genus *Mycobacterium*.

mycology   The scientific study of fungi.

mycorrhiza   A fungus growing in symbiosis with plant roots.

mycosis   A fungal infection.

mycotoxin   A toxin produced by a fungus.

NAD^+   A coenzyme that functions in the removal and transfer of hydrogen ion (H^+) and electrons from substrate molecules.

$NADP^+$ A coenzyme similar to NAD^+.

nanobacteria   Bacteria well below the generally accepted lower limit diameter (about 200 nanometres) for bacteria.

nanometer (nm)   A unit of measurement equal to 10^{-9} m, 10^{-3} μm.

natural killer (NK) cell   A lymphoid cell that destroys tumor cells and virus-infected cells.

naturally acquired active immunity   Antibody production in response to an infectious disease.

naturally acquired passive immunity   The natural transfer of humoral antibodies, for example, transplacental transfer.

natural penicillins   Penicillin molecules made by *Penicillium* spp.; penicillins G and V are examples. *See also* semisynthetic penicillins.

necrosis   Tissue death.

negative (indirect) selection   The process of identifying mutations by selecting cells that do not grow using replica plating.

negative staining   A procedure that results in colorless bacteria against a stained background.

neurotoxin   An exotoxin that interferes with normal nerve impulse conduction.

neutralization   An antigen–antibody reaction that inactivates a bacterial exotoxin or virus.

neutron   An uncharged particle in the nucleus of an atom.

neutrophil   A highly phagocytic granulocyte; also called polymorphonuclear leukocyte (PMN) or polymorph.

nicotinamide adenine dinucleotide *See* NAD$^+$.

nicotinamide adenine dinucleotide phosphate *See* NADP$^+$.

nitrification The oxidation of nitrogen in ammonia to produce nitrate.

nitrogen cycle The series of processes that converts nitrogen (N$_2$) to organic substances and back to nitrogen in nature.

nitrogen fixation The conversion of nitrogen (N$_2$) into ammonia.

nitrosamine A carcinogen formed by the combination of nitrite and amino acids.

nomenclature The system of naming things.

noncommunicable disease A disease that is not transmitted from one person to another.

noncompetitive inhibitor An inhibitory chemical that does not compete with the substrate for an enzyme's active site. *See also* allosteric inhibition; competitive inhibitor.

noncyclic photophosphorylation The movement of an electron from chlorophyll to NAD$^+$; plant and cyanobacterial photophosphorylation.

nonionizing radiation Short-wavelength radiation that does not cause ionization; ultraviolet (UV) radiation is an example.

non-nucleoside reverse transcriptase inhibitor A drug that binds with and inhibits the action of the HIV reverse transcriptase enzyme.

nonsense codon A codon that does not encode any amino acid.

nonsense mutation A base substitution in DNA that results in a nonsense codon.

normal microbiota The microorganisms that colonize a host without causing disease; also called normal flora.

nosocomial infection An infection that develops during the course of a hospital stay and was not present at the time the patient was admitted.

notifiable disease A disease that physicians must report to the U.S. Public Health Service; also called reportable disease.

N (neuraminidase) spikes Antigenic projections from the outer lipid bilayer of *Influenzavirus*.

nuclear envelope The double membrane that separates the nucleus from the cytoplasm in a eukaryotic cell.

nuclear pore An opening in the nuclear envelope through which materials enter and exit the nucleus.

nucleic acid A macromolecule consisting of nucleotides; DNA and RNA are nucleic acids.

nucleic acid hybridization The process of combining single complementary strands of DNA.

nucleic acid vaccine A vaccine made up of DNA, usually in the form of a plasmid; also called DNA vaccine.

nucleoid The region in a bacterial cell containing the chromosome.

nucleolus (plural: nucleoli) An area in a eukaryotic nucleus where rRNA is synthesized.

nucleoside A compound consisting of a purine or pyrimidine base and a pentose sugar.

nucleoside analog A chemical that is structurally similar to the normal nucleosides in nucleic acids but with altered base-pairing properties.

nucleoside reverse transcriptase inhibitor A nucleoside analog antiretroviral drug.

nucleotide A compound consisting of a purine or pyrimidine base, a five-carbon sugar, and a phosphate.

nucleotide excision repair (NER) The repair of DNA involving removal of defective nucleotides and replacement with functional ones.

nucleus (1) The part of an atom consisting of the protons and neutrons. (2) The part of a eukaryotic cell that contains the genetic material.

numerical identification Bacterial identification schemes in which test values are assigned a number.

nutrient agar Nutrient broth containing agar.

nutrient broth A complex medium made of beef extract and peptone.

objective lenses In a compound light microscope, the lenses closest to the specimen.

obligate aerobe An organism that requires molecular oxygen (O$_2$) to live.

obligate anaerobe An organism that does not use molecular oxygen (O$_2$) and is killed in the presence of O$_2$.

obligate halophile An organism that requires high osmotic pressures such as high concentrations of NaCl.

ocular lens In a compound light microscope, the lens closest to the viewer; also called the eyepiece.

oligodynamic action The ability of small amounts of a heavy metal compound to exert antimicrobial activity.

oligosaccharide A carbohydrate consisting of 2 to approximately 20 monosaccharides.

oncogene A gene that can bring about malignant transformation.

oncogenic virus A virus that is capable of producing tumors; also called oncovirus.

oocyst An encysted apicomplexan zygote in which cell division occurs to form the next infectious stage.

Opa A bacterial outer membrane protein; cells with Opa form opaque colonies.

operator The region of DNA adjacent to structural genes that controls their transcription.

operon The operator and promoter sites and structural genes they control.

opportunistic pathogen A microorganism that does not ordinarily cause a disease but can become pathogenic under certain circumstances.

opsonization The enhancement of phagocytosis by coating microorganisms with certain serum proteins (opsonins); also called immune adherence.

optimum growth temperature The temperature at which a species grows best.

order A taxonomic classification between class and family.

organelle A membrane-enclosed structure within eukaryotic cells.

organic compound A molecule that contains carbon and hydrogen.

organic growth factor An essential organic compound that an organism is unable to synthesize.

osmosis The net movement of solvent molecules across a selectively permeable membrane from an area of lower solute concentration to an area of higher solute concentration.

osmotic lysis Rupture of the plasma membrane resulting from movement of water into the cell.

osmotic pressure The force with which a solvent moves from a solution of lower solute concentration to a solution of higher solute concentration.

oxidation The removal of electrons from a molecule.

oxidation pond A method of secondary sewage treatment by microbial activity in a shallow standing pond of water.

oxidation-reduction A coupled reaction in which one substance is oxidized and one is reduced; also called redox reaction.

oxidative phosphorylation The synthesis of ATP coupled with electron transport.

oxygenic Producing oxygen, as in plant and cyanobacterial photosynthesis.

ozone O$_3$.

PABA Para-aminobenzoic acid; a precursor for folic acid synthesis.

pandemic disease An epidemic that occurs worldwide.

papule Small, solid elevation of the skin.

parasite An organism that derives nutrients from a living host.

parasitism A symbiotic relationship in which one organism (the parasite) exploits another (the host) without providing any benefit in return.

parasitology The scientific study of parasitic protozoa and worms.

parenteral route A portal of entry for pathogens by deposition directly into tissues beneath the skin and mucous membranes.

pasteurization The process of mild heating to kill particular spoilage microorganisms or pathogens.

pathogen A disease-causing organism.

pathogenesis The manner in which a disease develops.

pathogenicity The ability of a microorganism to cause disease by overcoming the defenses of a host.

pathology The scientific study of disease.

pellicle (1) The flexible covering of some protozoa. (2) Scum on the surface of a liquid medium.

penicillins A group of antibiotics produced either by *Penicillium* (natural penicillins) or by adding side chains to the β-lactam ring (semisynthetic penicillins).

pentose phosphate pathway A metabolic pathway that can occur simultaneously with glycolysis to produce pentoses and NADH without ATP production; also called hexose monophosphate shunt.

peptide bond A bond joining the amino group of one amino acid to the carboxyl group of a second amino acid with the loss of a water molecule.

peptidoglycan The structural molecule of bacterial cell walls consisting of the molecules N-acetylglucosamine, N-acetylmuramic acid, tetrapeptide side chain, and peptide side chain.

perforin Protein that makes a pore in a target cell membrane, released by T_C cells.

pericarditis Inflammation of the pericardium, the sac around the heart.

period of convalescence The recovery period, when the body returns to its predisease state.

peripheral nervous system (PNS) The nerves that connect the outlying parts of the body with the central nervous system.

periplasm The region of a gram-negative cell wall between the outer membrane and the cytoplasmic membrane.

peritrichous Having flagella distributed over the entire cell.

peroxidase An enzyme that breaks down hydrogen peroxide.

peroxide anion An oxygen anion consisting of two atoms of oxygen (O_2^{2-}).

peroxisome Organelle that oxidizes amino acids, fatty acids, and alcohol.

peroxygen A class of oxidizing-type sterilizing disinfectants.

persistent viral infection A disease process that occurs gradually over a long period.

pH The symbol for hydrogen ion (H^+) concentration; a measure of the relative acidity or alkalinity of a solution.

phage *See* bacteriophage.

phage conversion Genetic change in the host cell resulting from infection by a bacteriophage.

phage typing A method of identifying bacteria using specific strains of bacteriophages.

phagocyte A cell capable of engulfing and digesting particles that are harmful to the body.

phagocytosis The ingestion of solids by eukaryotic cells.

phagolysosome A digestive vacuole.

phagosome A food vacuole of a phagocyte; also called a phagocytic vesicle.

phalloidin A peptide toxin produced by *Amanita phalloides*, affects plasma membrane function.

phase-contrast microscope A compound light microscope that allows examination of structures inside cells through the use of a special condenser.

phenol ⬡—OH Also called carbolic acid.

phenolic A synthetic derivative of phenol used as a disinfectant.

phenotype The external manifestations of an organism's genotype, or genetic makeup.

phosphate group A portion of a phosphoric acid molecule attached to some other molecule, Ⓟ,

$$PO_4^{3-}, \quad {}^-O-\overset{\overset{\textstyle O}{\|}}{\underset{\underset{\textstyle O^-}{|}}{P}}-O^-$$

phospholipid A complex lipid composed of glycerol, two fatty acids, and a phosphate group.

phosphorous cycle The various solubility stages of phosphorus in the environment.

phosphorylation The addition of a phosphate group to an organic molecule.

photoautotroph An organism that uses light as its energy source and carbon dioxide (CO_2) as its carbon source.

photoheterotroph An organism that uses light as its energy source and an organic carbon source.

photolyase An enzyme that splits thymine dimers in the presence of visible light.

photophosphorylation The production of ATP in a series of redox reactions; electrons from chlorophyll initiate the reactions.

photosynthesis The conversion of light energy from the sun into chemical energy; the light-fueled synthesis of carbohydrate from carbon dioxide (CO_2).

phototaxis Movement in response to the presence of light.

phototroph An organism that uses light at its primary energy source.

phylogeny The evolutionary history of a group of organisms; phylogenetic relationships are evolutionary relationships.

phylum A taxonomic classification between kingdom and class.

phytoplankton Free-floating photoautotrophs.

pilus (plural: **pili**) An appendage on a bacterial cell used for the transfer of genetic material during conjugation.

pinocytosis The engulfing of fluid by infolding of the plasma membrane, in eukaryotes.

plankton Free-floating aquatic organisms.

plantae The kingdom composed of multicellular eukaryotes with cellulose cell walls.

plaque A clearing in a bacterial lawn resulting from lysis by phages. *See also* dental plaque.

plaque-forming units (pfu) Visible viral plaques counted.

plasma The liquid portion of blood in which the formed elements are suspended.

plasma cell A cell that an activated B cell differentiates into; plasma cells manufacture specific antibodies.

plasma (cytoplasmic) membrane The selectively permeable membrane enclosing the cytoplasm of a cell; the outer layer in animal cells, internal to the cell wall in other organisms.

plasmid A small circular DNA molecule that replicates independently of the chromosome.

plasmodium (1) A multinucleated mass of protoplasm, as in plasmodial slime molds. (2) When written as a genus, refers to the causative agent of malaria.

plasmogamy Fusion of the cytoplasm of two cells; occurs in the sexual stage of a fungal life cycle.

plasmolysis Loss of water from a cell in a hypertonic environment.

plate count A method of determining the number of bacteria in a sample by counting the number of colony-forming units on a solid culture medium.

pleomorphic Having many shapes, characteristic of certain bacteria.

pluripotent A cell that can differentiate into a many different types of tissue cells.

pneumonia Inflammation of the lungs.

point mutation *See* base substitution.

polar molecule A molecule with an unequal distribution of charges.

polyene antibiotic An antimicrobial agent that alters sterols in eukaryotic plasma membranes and contains more than four carbon atoms and at least two double bonds.

polymer A molecule consisting of a sequence of similar molecules, or monomers.

polymerase chain reaction (PCR) A technique using DNA polymerase to make multiple copies of a DNA template in vitro. *See also* cDNA.

polymorphonuclear leukocyte (PMN) *See* neutrophil.

polypeptide (1) A chain of amino acids. (2) A group of antibiotics.

polysaccharide A carbohydrate consisting of 8 or more monosaccharides joined through dehydration synthesis.

porins A type of protein in the outer membrane of gram-negative cell walls that permits the passage of small molecules.

portal of entry The avenue by which a pathogen gains access to the body.

portal of exit The route by which a pathogen leaves the body.

positive (direct) selection A procedure for picking out mutant cells by growing them.

pour plate method   A method of inoculating a solid nutrient medium by mixing bacteria in the melted medium and pouring the medium into a Petri dish to solidify.

precipitation reaction   A reaction between soluble antigens and multivalent antibodies to form visible aggregates.

precipitin ring test   A precipitation test performed in a capillary tube.

predisposing factor   Anything that makes the body more susceptible to a disease or alters the course of a disease.

prevalence   The fraction of a population having a specific disease at a given time.

primary cell line   Human tissue cells that grow for only a few generations in vitro.

primary infection   An acute infection that causes the initial illness.

primary metabolite   A product of an industrial cell population produced during the time of rapid logarithmic growth. *See also* secondary metabolite.

primary producer   An autotrophic organism, either chemotroph or phototroph, that converts carbon dioxide into organic compounds.

primary response   Antibody production in response to the first contact with an antigen. *See also* memory response.

primary sewage treatment   The removal of solids from sewage by allowing them to settle out and be held temporarily in tanks or ponds.

prion   An infectious agent consisting of a self-replicating protein, with no detectable nucleic acids.

privileged site (tissue)   An area of the body (or a tissue) that does not elicit an immune response.

probiotics   Microbes inoculated into a host to occupy a niche and prevent growth of pathogens.

prodromal period   The time following the incubation period when the first symptoms of illness appear.

profundal zone   The deeper water under the limnetic zone in an inland body of water.

proglottid   A body segment of a tapeworm containing both male and female organs.

prokaryote   A cell whose genetic material is not enclosed in a nuclear envelope.

prokaryotic species   A population of cells that share certain rRNA sequences; in conventional biochemical testing, it is a population of cells with similar characteristics.

promoter   The starting site on a DNA strand for transcription of RNA by RNA polymerase.

prophage   Phage DNA inserted into the host cell's DNA.

prophylactic   Anything used to prevent disease.

prostaglandin   A hormonelike substance that is released by damaged cells, intensifies inflammation.

prostheca   A stalk or bud protruding from a prokaryotic cell.

protease   An enzyme that digests protein (proteolytic enzymes).

protein   A large molecule containing carbon, hydrogen, oxygen, and nitrogen (and sulfur); some proteins have a helical structure and others are pleated sheets.

protein kinase   An enzyme that activates another protein by adding a Ⓟ from ATP.

proteobacteria   Gram-negative, chemoheterotrophic bacteria that possess a signature rRNA sequence.

proteomics   The science of determining all of the proteins expressed in a cell.

protist   Term used for unicellular and simple multicellular eukaryotes; usually protozoa and algae.

proton   A positively charged particle in the nucleus of an atom.

protoplast   A gram-positive bacterium or plant cell treated to remove the cell wall.

protoplast fusion   A method of joining two cells by first removing their cell walls; used in genetic engineering.

protozoan (plural: protozoa)   Unicellular eukaryotic organisms; usually chemoheterotrophic.

provirus   Viral DNA that is integrated into the host cell's DNA.

pseudohypha   A short chain of fungal cells that results from the lack of separation of daughter cells after budding.

pseudopod   An extension of a eukaryotic cell that aids in locomotion and feeding.

psychrophile   An organism that grows best at about 15°C and does not grow above 20°C; a cold-loving microbe.

pscyhrotroph   An organism that is capable of growth between about 0°C and 30°C.

purines   The class of nucleic acid bases that includes adenine and guanine.

purple nonsulfur bacteria   Alphaproteobacteria; strictly anaerobic and phototrophic; grow on yeast extract in dark; use reduced organic compounds as electron donors for CO_2 fixation.

purple sulfur bacteria   Gammaproteobacteria; strictly anaerobic and phototrophic; use reduced sulfur compounds as electron donors for CO_2 fixation.

pus   An accumulation of dead phagocytes, dead bacterial cells, and fluid.

pustule   A small pus-filled elevation of skin.

pyocyanin   A blue-green pigment produced by *Pseudomonas aeruginosa*.

pyrimidines   The class of nucleic acid bases that includes uracil, thymine, and cytosine.

quaternary ammonium compound (quat)   A cationic detergent with four organic groups attached to a central nitrogen atom; used as a disinfectant.

quinolone   An antibiotic whose mode of action is to inhibit DNA replication by interfering with the enzyme DNA gyrase.

quorum sensing   The ability of bacteria to communicate and coordinate behavior via signaling molecules.

R   Used to represent nonfunctional groups of a molecule. *See also* resistance factor.

random shotgun sequencing   A technique for determining the nucleotide sequence in an organism's genome.

rapid plasma reagin (RPR) test   A serological test for syphilis.

r-determinant   A group of genes for antibiotic resistance carried on R factors.

receptor   An attachment for a pathogen on a host cell.

recipient cell   A cell that receives DNA from a donor cell during genetic recombination.

recombinant DNA (rDNA)   A DNA molecule produced by combining DNA from two different sources.

recombinant DNA (rDNA) technology   Manufacturing and manipulating genetic material in vitro; also called genetic engineering.

recombinant vaccine   A vaccine made by recombinant DNA techniques.

redia   A trematode larval stage that reproduces asexually to produce cercariae.

redox reaction   *See* oxidation-reduction.

red tide   A bloom of planktonic dinoflagellates.

reducing medium   A culture medium containing ingredients that will remove dissolved oxygen from the medium to allow the growth of anaerobes.

reduction   The addition of electrons to a molecule.

refractive index   The relative velocity with which light passes through a substance.

regulatory T (T_R) cells   Lymphocytes that appear to suppress other T cells.

relative risk   A comparison of the risk of disease in two groups.

rennin   An enzyme that forms curds as part of any dairy fermentation product; originally from calves' stomachs, now produced by molds and bacteria.

replica plating   A method of inoculating a number of solid minimal culture media from an original plate to produce the same pattern of colonies on each plate.

replication fork   The point where DNA strands separate and new strands will be synthesized.

repression   The process by which a repressor protein can stop the synthesis of a protein.

repressor   A protein that binds to the operator site to prevent transcription.

reservoir of infection   A continual source of infection.

resistance   The ability to ward off diseases through innate and adaptive immunity.

resistance (R) factor   A bacterial plasmid carrying genes that determine resistance to antibiotics.

resistance transfer factor (RTF)   A group of genes for replication and conjugation on the R factor.

resolution   The ability to distinguish fine detail with a magnifying instrument; also called resolving power.

respiration   A series of redox reactions in a membrane that generates ATP; the final electron acceptor is usually an inorganic molecule.

restriction enzyme   An enzyme that cuts double-stranded DNA at specific sites between nucleotides.

reticulate body   The intracellular growing stage of chlamydiae.

retort   A device for commercially sterilizing canned food by using steam under pressure; operates on the same principle as an autoclave but is much larger.

reverse genetics   Genetic analysis that begins with a piece of DNA and proceeds to find out what it does.

reverse transcriptase   An RNA-dependent DNA polymerase; an enzyme that synthesizes a complementary DNA from an RNA template.

reversible reaction   A chemical reaction in which the end-products can readily revert to the original molecules.

RFLP   Restriction fragment length polymorphism; a fragment resulting from restriction-enzyme digestion of DNA.

Rh factor   An antigen on red blood cells of rhesus monkeys and most humans; possession makes the cells Rh^+.

rhizine   A rootlike hypha that anchors a fungus to a surface.

ribonucleic acid (RNA)   The class of nucleic acids that comprises messenger RNA, ribosomal RNA, and transfer RNA.

ribose   A five-carbon sugar that is part of ribonucleotide molecules and RNA.

ribosomal RNA (rRNA)   The type of RNA molecule that forms ribosomes.

ribosomal RNA (rRNA) sequencing   Determination of the order of nucleotide bases in rRNA.

ribosome   The site of protein synthesis in a cell, composed of RNA and protein.

ribozyme   An enzyme consisting of RNA that specifically acts on strands of RNA to remove introns and splice together the remaining exons.

rifamycin   An antibiotic that inhibits bacterial RNA synthesis.

ring stage   A young *Plasmodium* trophozoite that looks like a ring in a red blood cell.

RNAi   RNA interference; cellular degradation of double-stranded RNA, along with single-stranded RNA having the same sequence.

RNA primer   A short strand of RNA used to start synthesis of the lagging strand of DNA, and to start the polymerase chain reaction.

root nodule   A tumorlike growth on the roots of certain plants containing symbiotic nitrogen-fixing bacteria.

rotating biological contactor   A method of secondary sewage treatment in which large disks are rotated while partially submerged in a sewage tank exposing sewage to microorganisms and aerobic conditions.

rough ER   Endoplasmic reticulum with ribosomes on its surface.

roundworm   An animal belonging to the phylum Nematoda.

S (Svedberg unit)   Notes the relative rate of sedimentation during ultra-high-speed centrifugation.

salt   A substance that dissolves in water to cations and anions, neither of which is H^+ or OH^-.

sanitization   The removal of microbes from eating utensils and food preparation areas.

saprophyte   An organism that obtains its nutrients from dead organic matter.

sarcina (plural: sarcinae)   (1) A group of eight bacteria that remain in a packet after dividing. (2) When written as a genus, refers to gram-positive, anaerobic cocci.

saturation   The condition in which the active site on an enzyme is occupied by the substrate or product at all times.

saxitoxin   A neurotoxin produced by some dinoflagellates.

scanned probe microscopy   Microscopic technique used to obtain images of molecular shapes, to characterize chemical properties, and to determine temperature variations within a specimen.

scanning acoustic microscope (SAM)   A microscope that uses high-frequency ultrasound waves to penetrate surfaces.

scanning electron microscope (SEM)   An electron microscope that provides three-dimensional views of the specimen magnified 1000–10,000×.

scanning tunneling microscopy   *See* scanned-probe microscopy.

schizogony   The process of multiple fission, in which one organism divides to produce many daughter cells.

scientific nomenclature   *See* binomial nomenclature.

sclerotia   The compact mass of hardened mycelia of the fungus *Claviceps purpurea* that fills infected rye flowers; produces the toxin ergot.

scolex   The head of a tapeworm, containing suckers and possibly hooks.

secondary infection   An infection caused by an opportunistic microbe after a primary infection has weakened the host's defenses.

secondary metabolite   A product of an industrial cell population produced after the microorganism has largely completed its period of rapid growth and is in a stationary phase of the growth cycle. *See also* primary metabolite.

secondary response   *See* memory response.

secondary sewage treatment   Biological degradation of the organic matter in wastewater following primary treatment.

secretory vesicle   A membrane-enclosed sac produced by the ER; transports synthesized material into cytoplasm.

selective medium   A culture medium designed to suppress the growth of unwanted microorganisms and encourage the growth of desired ones.

selective permeability   The property of a plasma membrane to allow certain molecules and ions to move through the membrane while restricting others.

selective toxicity   The property of some antimicrobial agents to be toxic for a microorganism and nontoxic for the host.

self-tolerance   The ability of an organism to recognize and not make antibodies against self.

semiconservative replication   The process of DNA replication in which each double-stranded DNA molecule contains one original strand and one new strand.

semisynthetic penicillins   Modifications of natural penicillins by introducing different side chains that extend the spectrum of antimicrobial activity and avoid microbial resistance.

sense codon   A codon that codes for an amino acid.

sense strand (+ strand)   Viral RNA that can act as mRNA.

sentinel animal   An organism in which changes can be measured to assess the extent of environmental contamination and its implication for human health.

sepsis   The presence of a toxin or pathogenic organism in blood and tissue.

septa   A cross-wall in a fungal hypha.

septate hypha   A hypha consisting of uninucleate cell-like units.

septicemia   The proliferation of pathogens in the blood, accompanied by fever; sometimes causes organ damage.

septic shock   A sudden drop in blood pressure induced by bacterial toxins.

serial dilution   The process of diluting a sample several times.

seroconversion   A change in a person's response to an antigen in a serological test.

serological testing   Techniques for identifying a microorganism based on its reaction with antibodies.

serology   The branch of immunology that studies blood serum and antigen–antibody reactions in vitro.

serotype   *See* serovar.

serovar   A variation within a species; also called serotype.

serum   The liquid remaining after blood plasma is clotted; contains antibodies (immunoglobulins).

severe sepsis   Decreased blood pressure and dysfunction of at least one organ.

sexual dimorphism   The distinctly different appearance of adult male and female organisms.

sexual spore   A spore formed by sexual reproduction.

Shiga toxin   An exotoxin produced by *Shigella dysenteriae* and entero-hemorrhagic *E. coli*.

shock   Any life-threatening loss of blood pressure. *See also* septic shock.

shuttle vector   A plasmid that can exist in several different species; used in genetic engineering.

siderophore   Bacterial iron-binding proteins.

sign   A change due to a disease that a person can observe and measure.

simple stain   A method of staining microorganisms with a single basic dye.

singlet oxygen   Highly reactive molecular oxygen ($O_2{}^-$).

siRNA   Short interfering RNA; An intermediate in the RNAi process in which the long double-stranded RNA has been cut up into short (~21 nucleotides) double-stranded RNA.

site-directed mutagenesis   Techniques used to modify a gene in a specific location to produce the desired polypeptide.

slide agglutination test   A method of identifying an antigen by combining it with a specific antibody on a slide.

slime layer   A glycocalyx that is unorganized and loosely attached to the cell wall.

sludge   Solid matter obtained from sewage.

smear   A thin film of material containing microorganisms, spread over the surface of a slide.

smooth ER   Endoplasmic reticulum without ribosomes.

solute   A substance dissolved in another substance.

solvent   A dissolving medium.

Southern blotting   A technique that uses DNA probes to detect the presence of specific DNA in restriction fragments separated by electrophoresis.

specialized transduction   The process of transferring a piece of cell DNA adjacent to a prophage to another cell.

species   The most specific level in the taxonomic hierarchy. *See also* bacterial species; eukaryotic species; viral species.

specific epithet   The second or species name in a scientific binomial. *See also* species.

spectrum of microbial activity   The range of distinctly different types of microorganisms affected by an antimicrobial drug; a wide range is referred to as a broad spectrum of activity.

spheroplast   A gram-negative bacterium treated to damage the cell wall, resulting in a spherical cell.

spicule   One of two external structures on the male roundworm used to guide sperm.

spike   A carbohydrate-protein complex that projects from the surface of certain viruses.

spiral   *See* spirillum and spirochete.

spirillum (plural: spirilla)   (1) A helical or corkscrew-shaped bacterium. (2) When written as a genus, refers to aerobic, helical bacteria with clumps of polar flagella.

spirochete   A corkscrew-shaped bacterium with axial filaments.

splicing   A process by which an RNA molecule cuts out introns and rejoins exons to produce a molecule of mRNA.

spontaneous generation   The idea that life could arise spontaneously from nonliving matter.

spontaneous mutation   A mutation that occurs without a mutagen.

sporadic disease   A disease that occurs occasionally in a population.

sporangiophore   An aerial hypha supporting a sporangium.

sporangiospore   An asexual fungal spore formed within a sporangium.

sporangium   A sac containing one or more spores.

spore   A reproductive structure formed by fungi and actinomycetes.

sporogenesis   *See* sporulation.

sporozoite   A trophozoite of *Plasmodium* found in mosquitoes, infective for humans.

sporulation   The process of spore and endospore formation; also called sporogenesis.

spread plate method   A plate count method in which inoculum is spread over the surface of a solid culture medium.

staining   Colorizing a sample with a dye to view through a microscope or to visualize specific structures.

staphylococci (singular: staphylococcus)   Cocci in a grapelike cluster or broad sheet.

stationary phase   The period in a bacterial growth curve when the number of cells dividing equals the number dying.

stem cell   An undifferentiated cell that gives rise to a variety of specialized cells.

stereoisomers   Two molecules consisting of the same atoms, arranged in the same manner but differing in their relative positions; mirror images; also called D-isomer and L-isomer.

sterile   Free of microorganisms.

sterilization   The removal of all microorganisms, including endospores.

steroid   A specific group of lipids, including cholesterol and hormones.

stipe   A stemlike supporting structure of multicellular algae and basidiomycetes.

storage vesicle   Organelles that form from the Golgi complex; contain proteins made in the rough ER and processed in the Golgi complex.

strain   Genetically different cells within a clone. *See* serovar.

streak plate method   A method of isolating a culture by spreading microorganisms over the surface of a solid culture medium.

streptobacilli (singular: streptobacillus)   Rods that remain attached in chains after cell division.

streptococci (singular: streptococcus)   (1) Cocci that remain attached in chains after cell division. (2) When written as a genus, refers to gram-positive, catalase-negative bacteria.

streptogramin   Antibiotics that block protein synthesis at 70S ribosomes.

streptokinase   A blood-clot dissolving enzyme, produced by beta-hemolytic streptococci.

streptolysin   A hemolytic enzyme, produced by streptococci.

structural gene   A gene that determines the amino acid sequence of a protein.

subacute disease   A disease with symptoms that are intermediate between acute and chronic.

subclinical infection   An infection that does not cause a noticeable illness; also called inapparent infection.

subcutaneous mycosis   A fungal infection of tissue beneath the skin.

substrate   Any compound with which an enzyme reacts.

substrate-level phosphorylation   The synthesis of ATP by direct transfer of a high-energy phosphate group from an intermediate metabolic compound to ADP.

subunit vaccine   A vaccine consisting of an antigenetic fragment.

sulfa drug   A bacteriostatic compound that interferes with folic acid synthesis by competitive inhibition.

sulfhydryl group   —SH.

sulfur cycle   The various oxidation and reduction stages of sulfur in the environment, mostly due to the action of microorganisms.

sulfur granule   *See* inclusion.

superantigen   An antigen that activates many different T cells, thereby eliciting a large immune response.

superficial mycosis   A fungal infection localized in surface epidermal cells and along hair shafts.

superinfection   The growth of a pathogen that has developed resistance to an antimicrobial drug being used; the growth of an opportunistic pathogen.

superoxide dismutase (SOD)   An enzyme that destroys superoxide free radicals.

superoxide free radical   A toxic form of oxygen ($O_2{}^-$) formed during aerobic respiration.

surface-active agent   Any compound that decreases the tension between molecules lying on the surface of a liquid; also called surfactant.

susceptibility   The lack of resistance to a disease.

symbiosis   The living together of two different organisms or populations.

symptom   A change in body function that is felt by a patient as a result of a disease.

syncytium   A multinucleated giant cell resulting from certain viral infections.

syndrome   A specific group of signs or symptoms that accompany a disease.

synergism   (1) The effect of two microbes working together that is greater than the effect of either acting alone. (2) The principle whereby the effectiveness of two drugs used simultaneously is greater than that of either drug used alone.

synthesis reaction   A chemical reaction in which two or more atoms combine to form a new, larger molecule.

synthetic drug   A chemotherapeutic agent that is prepared from chemicals in a laboratory.

systematics   The science organizing groups of organisms into a hierarchy.

systemic anaphylaxis   A hypersensitivity reaction causing vasodilation and resulting in shock; also called anaphylactic shock.

systemic (generalized) infection   An infection throughout the body.

systemic mycosis   A fungal infection in deep tissues.

tachyzoite   A rapidly growing trophozoite form of a protozoan.

T antigen   An antigen in the nucleus of a tumor cell.

tapeworm   A flatworm belonging to the class Cestoda.

target cell   An infected body cell to which defensive cells of the immune system bind.

taxa   Subdivisions used to classify organisms, e.g., domain, kingdom, phylum.

taxis   Movement in response to an environmental stimulus.

taxonomy   The science of the classification of organisms.

T cell   A type of lymphocyte, which develops from a stem cell processed in the thymus gland, that is responsible for cell-mediated immunity.

T-dependent antigen   An antigen that will stimulate the formation of antibodies only with the assistance of helper T cells. *See also* T-independent antigen.

teichoic acid   A polysaccharide found in gram-positive cell walls.

teleomorph   The sexual stage in the life cycle of a fungus; also refers to a fungus that produces both sexual and asexual spores.

temperature abuse   Improper food storage at a temperature that allows bacteria to grow.

terminator   The site on a DNA strand at which transcription ends.

tertiary sewage treatment   A method of waste treatment that follows conventional secondary sewage treatment; nonbiodegradable pollutants and mineral nutrients are removed, usually by chemical or physical means.

tetracycline   Broad-spectrum antibiotics that interfere with protein synthesis, produced by *Streptomyces* spp.

tetrad   A group of four cocci.

thallus   The entire vegetative structure or body of a fungus, lichen, or alga.

thermal death point (TDP)   The temperature required to kill all the bacteria in a liquid culture in 10 minutes.

thermal death time (TDT)   The length of time required to kill all bacteria in a liquid culture at a given temperature.

thermoduric   Heat resistant.

thermophile   An organism whose optimum growth temperature is between 50°C and 60°C; a heat loving microbe.

thermophilic anaerobic spoilage   Spoilage of canned foods due to the growth of thermophilic bacteria.

thylakoid   A chlorophyll-containing membrane in a chloroplast. A bacterial thylakoid is also known as a chromatophore.

tincture   A solution in aqueous alcohol.

T-independent antigen   An antigen that will stimulate the formation of antibodies without the assistance of helper T cells. *See also* T-dependent antigen.

tinea   Fungal infection of hair, skin, or nails.

Ti plasmid   An *Agrobacterium* plasmid carrying genes for tumor induction in plants.

titer   An estimate of the amount of antibodies or viruses in a solution; determined by serial dilution and expressed as the reciprocal of the dilution.

toll-like rceptors (TLRs)   Transmembrane proteins of immune cells that recognize pathogens and activate an immune responses directed against those pathogens.

topoisomerase   Enzyme that relaxes supercoiling of DNA ahead of replication form; separates DNA circles at the end of DNA replication.

total magnification   The magnification of a microscopic specimen, determined by multiplying the ocular lens magnification by the objective lens magnification.

toxemia   The presence of toxins in the blood.

toxigenicity   The capacity of a microorganism to produce a toxin.

toxin   Any poisonous substance produced by a microorganism.

toxoid   An inactivated toxin.

trace element   A chemical element required in small amounts for growth.

transamination   The transfer of an amino group from an amino acid to another organic acid.

transcription   The process of synthesizing RNA from a DNA template.

transduction   The transfer of DNA from one cell to another by a bacteriophage. *See also* generalized transduction; specialized transduction.

transferrin   A human iron-binding protein that reduces iron available to a pathogen.

transfer RNA (tRNA)   The type of RNA molecule that brings amino acids to the ribosomal site where they are incorporated into proteins.

transfer vesicle   Membrane-bound sacs that move proteins from the Golgi complex to specific areas in the cell.

transformation   (1) the process in which genes are transferred from one bacterium to another as "naked" DNA in solution. (2) The changing of a normal cell into a cancerous cell.

transient microbiota   The microorganisms that are present in an animal for a short time without causing a disease.

translation   The use of mRNA as a template in the synthesis of protein.

transmission electron microscope (TEM)   An electron microscope that provides high magnifications (10,000–100,000×) of thin sections of a specimen.

transport vesicle   Membrane-bound sacs that move proteins from the rough ER to the Golgi complex.

transporter protein   A carrier protein in the plasma membrane.

transposon   A small piece of DNA that can move from one DNA molecule to another.

trickling filter   A method of secondary sewage treatment in which sewage is sprayed out of rotating arms onto a bed of rocks or similar materials, exposing the sewage to highly aerobic conditions and microorganisms.

triglyceride   A simple lipid consisting of glycerol and three fatty acids.

triplex agent   A short segment of DNA that binds to a target area on a double strand of DNA blocking transcription.

trophophase   The period in the production curve of an industrial cell population in which the primary metabolites are formed; a period of rapid, logarithmic growth. *See also* idiophase.

trophozoite   The vegetative form of a protozoan.

tuberculin skin test   A skin test used to detect the presence of antibodies to *Mycobacterium tuberculosis*.

tumor necrosis factor (TNF)   A polypeptide released by phagocytes in response to bacterial endotoxins; induces shock; also called cachectin.

tumor-specific transplantation antigen (TSTA)   A viral antigen on the surface of a transformed cell.

turbidity   The cloudiness of a suspension.

turnover number   The number of substrate molecules acted on per enzyme molecule per second.

ubiquinone   A low–molecular weight, nonprotein carrier in an electron transport chain; also called coenzyme Q.

ultra-high-temperature (UHT) treatment   A method of treating food with high temperatures (140–150°C) for very short times to make the food sterile so that it can be stored at room temperature.

uncoating   The separation of viral nucleic acid from its protein coat.

undulating membrane A highly modified flagellum on some protozoa.

use-dilution test A method of determining the effectiveness of a disinfectant using serial dilutions.

vaccination The process of conferring immunity by administering a vaccine; also called immunization.

vaccine A preparation of killed, inactivated, or attenuated microorganisms or toxoids to induce artificially acquired active immunity.

vacuole An intracellular inclusion, in eukaryotic cells, surrounded by a plasma membrane; in prokaryotic cells, surrounded by a proteinaceous membrane.

valence The combining capacity of an atom or a molecule.

vancomycin An antibiotic that inhibits cell wall synthesis.

variolation An early method of vaccination using infected material from a patient.

vasodilation Dilation or enlargement of blood vessels.

VDRL test A rapid screening test to detect the presence of antibodies against Treponema pallidum. (VDRL stands for Venereal Disease Research Laboratory.)

vector (1) A plasmid or virus used in genetic engineering to insert genes into a cell. (2) An arthropod that carries disease-causing organisms from one host to another.

vegetative Referring to cells involved with obtaining nutrients, as opposed to reproduction.

vehicle transmission The transmission of a pathogen by an inanimate reservoir.

vertical gene transfer Transfer of genes from an organism or cell to its offspring.

vesicle (1) A small serum-filled elevation of the skin. (2) Smooth oval bodies formed in plant roots by mycorrhizae.

V factor NAD^+ or $NADP^+$.

vibrio (1) A curved or comma-shaped bacterium. (2) When written as a genus (*Vibrio*), a gram-negative, motile, facultatively anaerobic curved rod.

viral hemagglutination The ability of certain viruses to cause the clumping of red blood cells in vitro.

viral hemagglutination inhibition test A neutralization test in which antibodies against particular viruses prevent the viruses from clumping red blood cells in vitro.

viral species A group of viruses sharing the same genetic information and ecological niche.

viremia The presence of viruses in the blood.

virion A complete, fully developed viral particle.

viroid Infectious RNA.

virology The scientific study of viruses.

virulence The degree of pathogenicity of a microorganism.

virus A submicroscopic, parasitic, filterable agent consisting of a nucleic acid surrounded by a protein coat.

volutin Stored inorganic phosphate in a prokaryotic cell. See also metachromatic granule.

wandering macrophage A macrophage that leaves the blood and migrates to infected tissue.

Western blotting A technique that uses antibodies to detect the presence of specific proteins separated by electrophoresis.

whey The fluid portion of milk that separates from curd.

xenobiotics Synthetic chemicals that are not readily degraded by microorganisms.

xenodiagnosis A method of diagnosis based on exposing a parasite-free normal host to the parasite and then examining the host for parasites.

xenotransplantation product A tissue graft from another species; also called xenotransplant.

X factor Substances from the heme fraction of blood hemoglobin.

yeast Nonfilamentous, unicellular fungi.

yeast infection Disease caused by growth of certain yeasts in a susceptible host.

zone of inhibition The area of no bacterial growth around an antimicrobial agent in the disk-diffusion method.

zoonosis A disease that occurs primarily in wild and domestic animals but can be transmitted to humans.

zoospore An asexual algal spore; has two flagella.

zygospore A sexual fungal spore characteristic of the zygomycetes.

zygote A diploid cell produced by the fusion of two haploid gametes.

CREDITS

Illustration Credits

All illustrations by Precision Graphics unless otherwise noted.

2.1, 2.3, 2.9, 2.12, 2.14, 2.15 G. J. Tortora and S. R. Grabowski, *Principles of Anatomy and Physiology*, 8th ed. 2.1, 2.4, 2.6, 2.9, 2.12, 2.13, 2.14. © Biological Sciences Textbooks, 1996. Reproduced by permission of Addison Wesley Longman. Illustrations, Jared Schneidman Design.

4.22, 4.24, 4.25, 4.26 Adapted from G. J. Tortora and S. R. Grabowski, *Principles of Anatomy and Physiology*, 9th ed. 3.1, 3.25, 3.20, 3.21. © Biological Sciences Textbooks, 2000. Reproduced by permission of Wiley.

5.1 G. J. Tortora and S. R. Grabowski, *Principles of Anatomy and Physiology*, 8th ed. 25.1. © Biological Sciences Textbooks, 1996. Reproduced by permission of Addison Wesley Longman. Illustration, Page Two Associates.

8.3 Adapted from E. N. Marieb, *Human Anatomy and Physiology*, 3rd ed. © Benjamin Cummings, 1995, 3.27; 8.5, 8.6 Adapted from N. Campbell, *Biology*, 5th ed. © Benjamin Cummings, 1999, F16.11; 8.7 P. Berg and M. Singer, *Dealing With Genes: The Language of Heredity*. © University Science Books, 1992, 13.1; 8.16 Adapted from N. Campbell, *Biology*, 4th ed. © Benjamin Cummings, 1996, 16.23.

9.4 M. Bloom, G. Frever, and D. Micklos, *Laboratory DNA Science*. © Benjamin Cummings, 1996, p. 283; 9.12 Adapted from N. Campbell, *Biology*, 4th ed. © Benjamin Cummings, 1996, 19.1, 19.6; 9.16 From J. D. Watson et al., *Recombinant DNA*, 2nd ed. © W. H. Freeman, 1992.

13.T2 From R. I. B. Francki et al., eds., "Classification and Nomenclature of Viruses, Fifth Report of the Intl. Comm. on Taxonomy of Viruses," *Archives of Virology: Supplementum 2*. © Springer-Verlag, 1991.

14.4 CDC

16.1 Adapted from N. Campbell, *Biology*, 4th ed. © Benjamin Cummings, 1996, 39.1; 16.5, 16.7 G. J. Tortora and S. R. Grabowski, *Principles of Anatomy and Physiology*, 7th ed., 22.1, 22.10. © Biological Sciences Textbooks, 1993. Reproduced by permission of Addison Wesley Longman.

19.12 Adapted from F. J. Hoth Jr., M. W. Meyers, and D. S. Stein, "Current Status of HIV Therapy," *Hospital Practice* 27:145. Illustration, Alan D. Iselin. Reproduced with permission; 19.13, 19.14 P. D. Greenberg, "Immunopathogenesis of HIV Infection," *Hospital Practice* 27:109. Illustrations, Ilil Arbel. Reproduced with permission; 19.16 Adapted from UNAIDS data by the Map Design Unit of the World Bank. Reproduced by permission of Confronting AIDS: Public Priorities in a Global Epidemic and the U. of California San Francisco data from HIVInSite, 2005; 19.T02 Adapted from E. N. Marieb, *Human Anatomy and Physiology*, 3rd ed. © Benjamin Cummings, 1995, T18.4.

20.20 T. D. Brock et al., *Biology of Microorganisms*, 7th ed. © Prentice-Hall, 1993, p. 410. Reproduced with permission.

21.12, 22.12 Adapted from P. R. Murphy et al., *Medical Microbiology*, 3rd ed. © Williams & Wilkins, 1993. Reproduced by permission.

23.1 Adapted from N. Campbell, *Biology*, 3rd ed. © Benjamin Cummings, 1993, 38.5a; 23.13 A. C. Steere, "Current Understanding of Lyme Disease," *Hospital Practice* 28:37. Illustration, Nancy Lou Makris Riccio. Reproduced with permission.

24.2 Adapted from N. Campbell, *Biology*, 3rd ed. © Benjamin Cummings, 1993, 38.22; 24.10 A. M. Dannenberg, Jr., "Pulmonary Tuberculosis," *Hospital Practice* 28:51. Illustrations, Seward Hung and Laura Pardi Duprey. Reproduced with permission; 24.20 CDC; 24.21 Adapted from J. W. Smith and M. S. Bartlett, *Laboratory Medicine* © American Society of Clinical Pathologists, 1979, 10:430–35, 1979. Illustration, Gwen Gloege. Reproduced with permission.

25.8, 25.9 M. Schaechter et al., eds., *Mechanisms of Microbial Disease*, 2nd ed. © Williams & Wilkins, 1993, 18.1; 25.14 From "Helicobacter Pylori Infection," *Hospital Practice* 26:1. Illustration, Laura Pardi Duprey. Reproduced with permission.

Photo Credits

Ch. Opener 1, 1.7 Steve Gschmeissner/Photo Researchers; 1.1a CNRI/PSL/Photo Researchers; 1.1b Biophoto Associates/Photo Researchers; 1.1c K. W. Jean/Visuals Unlimited; 1.1d Stephen Durr; 1.1e Lee D. Simon/Photo Researchers; 1.2a Registered trademark of Pfizer, Inc. Reproduced with permission; 1.2b, 1.5 Christine Case; 1.3 Charles O'Rear/Bettmann/Corbis; 1.4 (top & middle) Bettmann/ Corbis; 1.4 (bottom) Rockefeller Archive Center; 1.6 Michael M. Kliks/Donald Heyneman; AM box Ken Karp

Ch. Opener 2 Ron Dengler/Visuals Unlimited; AM box Ken Graham/Ken Graham Agency

Ch. Opener 3, 3.9b, Table 3.2.9 Karl Aufderheide/Visuals Unlimited; 3.1 Leica Microsystems; 3.4 & 3.5 David M. Phillips/Visuals Unlimited; 3.6b, Table 3.2.5 CDC; 3.7, Table 3.2.6 Dennis Kunkel Microscopy; 3.8, Table 3.2.7 U. of Missouri, Kansas City, Center for Research on Interfacial Structure and Properties (UMKC-CRISP), Paulette Spencer, Director; 3.9a Phototake Electra/Phototake NYC; 3.10a, Table 3.2.10 M. Amrein et al., "Scanning Tunneling Microscopy of recA-DNA Complexes Coated with a Conducting Film," *Science*, 1988 Apr 22; 240(4851):514–6. Reprinted with permission. © AAAS, 1988; 3.10b, Table 3.2.11 D. M. Czajkowsky, et al., "Vertical Collapse of a Cytolysin Prepore Moves its Transmembrane Beta-hairpins to the Membrane," *The EMBO Journal*, 2004 Aug 18; 23(16):3206–15. Epub 2004 Aug 5. Image provided by Zhifeng Shao, U. of Virginia; 3.11, 3.13a Jack Bostrack/Visuals Unlimited;

3.12 Carmen Espinoza, Louisiana State U. Medical Center; 3.13b Joseph W. Duris and Silvia Rossbach, Western Michigan U.; 3.13c Eric Graves/Photo Researchers; AM box (left) Eshel Ben-Jacob, School of Physics and Astronomy, Tel Aviv U., Israel; AM box (middle) Patricia L. Grilione/Phototake; AM box (right) E. C. Cole et al., "Pseudomonas Pellicle in Disinfectant Testing: Electron Microscopy, Pellicle Removal, and Effect on Test Results," *Applied Environmental Microbiology*, 1989 Feb; 55(2):511–3, F1A; Table 3.2.1–4 David M. Phillips/Visuals Unlimited; Table 3.2.8 Phototake Electra/Phototake NYC; End of Ch. Biophoto Associates/Science Source/Photo Researchers

Ch. Opener 4 Dennis Kunkel/Phototake; 4.1.a (top) David M. Phillips/Visuals Unlimited; 4.1.a (bottom) Oliver Meckes and Nicole Ottawa/Photo Researchers; 4.1b&c G. Shih and R. Kessel/Visuals Unlimited; 4.1d David Scharf/Peter Arnold; 4.2a&b Manfred Kage/Peter Arnold; 4.2c Dennis Kunkel/Phototake NYC; 4.2d Microworks Color/Phototake NYC; 4.3 Jeffrey C. Burnham, Medical College of Ohio/ASM News; 4.4a London School of Hygiene/Photo Researchers; 4.4b Stanley Flegler/Visuals Unlimited; 4.4c Charles Stratton/Visuals Unlimited; 4.5a Horst Volker and Heinz Schlesner, Institut fur Allgemeine Mikrobiologie, Kiel/Michael Thomm; 4.5b H. W. Jannasch, Woods Hole Oceanographic Institution; 4.6b Ralph A. Slepecky/Visuals Unlimited; 4.7a Cabisco/Visuals Unlimited; 4.7b Ed Reschke/Peter Arnold; 4.7c Michael Abbey/Visuals Unlimited ; 4.7d Science Source/Photo Researchers; 4.9 Lee D. Simon/Science Source/Photo Researchers; 4.10 Custom Medical Stock Photo; 4.11 Kwangshin Kim/Photo Researchers; 4.14 T. J. Beveridge/Biological Photo Service; 4.15 H. S. Pankratz and R. L. Uffen, Michigan State U./Biological Photo Service; 4.16 Christine Case; 4.20 D. Balkwill and D. Maratea; 4.21 Visuals Unlimited; 4.22b (left) Biophoto Associates/Photo Researchers; 4.22b (right) D.W. Fawcett/Photo Researchers; 4.23 David M. Phillips/ Visuals Unlimited; 4.24 CNRI/SPL/ Photo Researchers; 4.25 R. Bolender and D. Fawcett/Photo Researchers; 4.26 M. Powell/Visuals Unlimited; 4.27 Keith Porter/Photo Researchers; 4.28 E. H. Newcomb and W. P. Wergin/Biological Photo Service; Table 4.1 Manfred Kage/Peter Arnold

Ch. Opener 5 The Culture Collection of Algae (UTEX); 5.4 Science/Visuals Unlimited; 5.22 & 5.23 Christine Case; AM box Helen E. Carr/Biological Photo Service

Ch. Opener 6, 6.8, 6.9 & 6.10 Christine Case; 6.6 Lester Lefkowitz/Corbis; 6.11 Lee D. Simon/ Photo Researchers; 6.17a Pall/ Visuals Unlimited; 6.17b K. Taiaro/ Visuals Unlimited; AM box Peter Batson/imagequestmarine.com

Ch. Opener 7 Dennis Kunkel/ Phototake NYC; 7.3 & 7.6 Christine Case; 7.8 CDC

Ch. Opener 8, 8.1 Gopal Murti/ Photo Researchers; 8.6 Visuals Unlimited; 8.7 Martin Guthold, Dept. of Physics, Wake Forest U.; 8.10 Visuals Unlimited; 8.25 Dennis Kunkel/Phototake NYC; 8.28 Gopal Murti/Phototake NYC

Ch. Opener 9 Secchi-Lecaque/ Roussel–UCLA/SPL/Photo Researchers; 9.5 Brad Metz; 9.6, 9.10 Matt Meadows/Peter Arnold; 9.7 Institute Pasteur/ Phototake NYC; 9.13 Secchi-Lecaque/Roussel–UCLA/SPL/ Photo Researchers; 9.17 CDC ; 9.18 R. S. Oremland et al., "Structural and Spectral Features of Selenium Nanospheres Produced by Se-respiring Bacteria," *Applied Environmental Microbiology*, 2004 Jan; 70(1):52–60, F1A. © American Society for Microbiology, 2004; 9.19 Holt Studios International Ltd./ Alamy; End of Ch. 1 Christine Case; End of Ch. 2 Mike Zeller, Office of Biotechnology, Iowa State U.

Ch. Opener 10 A. B. Dowsett/ Photo Researchers; 10.3 Peter Siver/Visuals Unlimited; 10.4a M. D. Maser/Visuals Unlimited; 10.4b S. M. Awramik, U. of California/Biological Photo Service; 10.10 Christine Case; 10.11a Sinclair Stammers/Photo Researchers; 10.11b Colin

Cuthbert/Photo Researchers; 10.12 CDC ; 10.13 Chris Jones, U. of Leeds; 10.14 Pascal Goetgheluck/SPL/Photo Researchers; 10.17a Volker Steger/SPL/Photo Researchers; 10.17d James Liao's Lab, UCLA DNA Microarray Core Facility; 10.18a&b V. A. Kempf, K. Trebesius, I. B. Autenrieth, "Fluorescent In Situ Hybridization Allows Rapid Identification of Microorganisms in Blood Cultures," *Journal of Clinical Microbiology*, 2000 Feb; 38(2):830–8, F1. © American Society for Microbiology, 2000; MN box Darryl W. Bush, Marine World/Africa USA, Vallejo, CA; Table 10.1 Ralph Robinson/ Visuals Unlimited; Table 10.2 A. B. Dowsett/Science Source/ Photo Researchers; Table 10.3 Eugene McArdle/Custom Medical Stock Photo

Ch. Opener 11, 11.8 Dennis Kunkel/Visuals Unlimited; 11.1 USDA/APHIS/Animal and Plant Health Inspection Service; 11.2 Yves V. Brun; 11.3 Biological Photo Service; 11.4 John D. Cunningham/Visuals Unlimited; 11.5 Science VU/Visuals Unlimited; 11.6, 11.19, 11.22 David M. Phillips/ Visuals Unlimited; 11.7 Linda Stannard, U. of Cape Town/Photo Researchers; 11.9a Institute Pasteur/CNRI/Phototake NYC; 11.9b Pablo Zunino, Laboratorio de Microbiología, Instituto de Investigaciones Biológicas Clemente Estable; 11.10 S. Rendulic, J. Berger, and S. Schuster, Max-Planck-Institute; 11.11 Jonathan Eisenback/Phototake NYC; 11.12 B. Dowsett, CAMR/Photo Researchers; 11.13a Robert Calentine/Visuals Unlimited; 11.13b Ron Dengler/Visuals Unlimited; 11.13c Susan M. Barnes, Los Alamos National Laboratory; 11.14 Paul Johnson/ Biological Photo Service; 11.15 Cabisco/Visuals Unlimited; 11.16 Esther R. Angert ; 11.17a From Hannay and Fitz James, *Canadian Journal of Microbiology* 1, 1955/National Researcher Council of Canada; 11.17b From R. E. Strange and J. R. Hunter, in G. W. Gould and A. Hurst, eds., *The Bacterial Spore*, 1969, p.461, F4. © Academic Press, 1969; 11.18 Tony Brain/SPL/ Custom Medical Stock Photo; 11.20a From D. C. Krause and D. Taylor-Robinson, "Mycoplasmas which Infect Humans," F1. in J. Maniloff et al., eds., *Mycoplasmas: Molecular Biology and Pathogenesis*,

1992. Reproduced by permission of the American Society for Microbiology; 11.20b Michael Gabridge/Visuals Unlimited; 11.21 Frederick P. Mertz/Visuals Unlimited; 11.23 Kurt Reed, Marshfield Medical Research Foundation; 11.24 J. A. Breznak and H. S. Pankratz/Biological Photo Service; 11.25 Sydney Finegold; 11.26 Karl O. Stetter; 11.27 Heide Schulz/Max-Planck-Institute

Ch. Opener 12 M. F. Brown/Visuals Unlimited; 12.2, 12.4, 12.9, 12.10, 12.11, 12.23 Christine Case; 12.3 David Scharf/Peter Arnold; 12.5a, c&d, 12.17b David M. Phillips/Visuals Unlimited; 12.5b M. F. Brown/Visuals Unlimited; 12.5e & 12.6 (left) G. Shih and R. Kessel/Visuals Unlimited; 12.6 (right), 12.7 (right) M. F. Brown/ Visuals Unlimited; 12.7 (left) Manfred Kage/Peter Arnold; 12.8 Biophoto Associates/Photo Researchers; 12.12 Jacqui Hurst/Corbis; 12.13, 12.18 Manfred Kage/Peter Arnold; 12.15 Davy Reynolds; 12.16 M. Abbey/Visuals Unlimited; 12.17c Mary Anne Harrington, TML/MSH Shared Microbiology Service; 12.17d E. Koneman/Visuals Unlimited; 12.20 E. R. Degginger/Photo Researchers; 12.21 Michael Abbey/Visuals Unlimited; 12.22 Cabisco/Visuals Unlimited; 12.24 D. O. Rosenberry, "Malformed Frogs in Minnesota: An Update: USGS Water Fact Sheet" FS-043-01, 2001. Photo by David Hoppe; 12.25 Steve J. Upton, Parasitology Research, Div. of Biology, Kansas State U.; 12.27 Stanley Flegler/Visuals Unlimited; 12.28a M. B. Hildreth, M. D. Johnson, K. R. Kazacos, "Echinococcus Multilocularis: A Zoonosis of Increasing Concern in the United States, *Compendium on Continuing Education for the Practicing Veterinarian*, 1991 May; 13(5):727–41; 12.28b & 12.29 Robert Calentine/Visuals Unlimited; 12.30 David Scharf/Peter Arnold; 12.31 Hans Pfletschinger/Peter Arnold; 12.32 Tom Murray/BugGuide.Net; CP box DPDx: CDC's Web site for parasitology identification, www.dpd.cdc.gov/dpdx

Ch. Opener 13 C. Garon and J. Rose, CDC; 13.2b R. C. Valentine and H. G. Pereira, *Journal of*

Molecular Biology/Biological Photo Service; 13.3, 13.18b K. G. Murti/ Visuals Unlimited; 13.4 Science Source/Photo Researchers; 13.5a Eye of Science/Photo Researchers; 13.5b Hans Gelderblom/ Visuals Unlimited; 13.6 Christine Case; 13.9 G. Steven Martin/ Visuals Unlimited; 13.14a Chris Bjornberg/Photo Researchers; 13.14b D. O. White and F. J. Fenner, *Medical Virology*, 4th eds. © Academic Press, 1994; 13.16a C. Garon and J. Rose, CDC; 13.16b Linda Stannard, U. of Cape Town/Photo Researchers; 13.18a Alfred Pasieka/Science Photo/Photo Researchers; 13.18c Linda Stannard, U. of Cape Town/Photo Researchers; 13.20 Visuals Unlimited; 13.23 T. O. Diener, USDA/Agricultural Research Service Honey Breeding

Ch. Opener 14 David M. Phillips/ Photo Researchers; 14.1a Dennis Kunkel/Visuals Unlimited; 14.1b P. Motta, Dept. of Anatomy, U. "La Sapienza," Rome/ Photo Researchers; 14.1c Steve Gschmeissner/Photo Researchers; 14.6a Stone/GettyImages; 14.6b Helen King/Corbis; 14.6c Stockbyte Platinum/ Alamy; 14.6d Andrew Davidhazy, Photo Arts and Sciences at Rochester Institute of Technology; 14.7a&b The Image Bank/ GettyImages; 14.7c David Hoffman Photo Library/Alamy; 14.8 Imagebroker/Alamy

Ch. Opener 15 C. C. Ginocchio et al., "Contact with Epithelial Cells Induces the Formation of Surface Appendages on Salmonella Typhimurium," *Cell*, 1994 Feb 25; 76(4):717–24; 15.1b SPL/Photo Researchers; 15.1c David M. Phillips/ Visuals Unlimited ; 15.2 C. C. Ginocchio et al., "Contact with Epithelial Cells Induces the Formation of Surface Appendages on Salmonella Typhimurium," *Cell*, 1994 Feb 25; 76(4):717–24; 15.7a Frederick A. Murphy, School of Veterinary Medicine, U. of California Davis; 15.7b Diana Hardie, U. of Cape Town Medical School, South Africa; 15.8 John P. Bader/Biological Photo Service

Ch. Opener 16, 16.6 Eye of Science/Photo Researchers; 16.2 Ed Reschke/Peter Arnold; 16.4 R. G. Kessel and R. H. Kardon, *Tissues and Organs*. © W.H. Freeman, 1979/Visuals Unlimited; 16.10 *Journal of Experimental Education*; Table 16.1.1–4, 6 (left),

INDEX

Note: A *t* following a page number indicates tabular material, a *f* following a page number indicates a figure, a *b* following a page number indicates a boxed feature, and a page number in **boldface** indicates a definition.

abacavir, 598*t*
ABO blood group system, **555**, 555*t*
abscess, **487**, 488*f*, **619**
absorbance (optical density/OD), 182
A-B toxins, **460**, 460*f*, 462*t*
Acanthamoeba, 364, 367*t*, 661, 662, 666*t*
acanthamoeba keratitis, **637**, 637*t*
acceptors, proton, 36
acellular vaccines, **530**
acetate kinase, 117*t*
acetic acid, 136*f*, 139*t*
Acetobacter, 139*t*, 140, 314*t*, 316
acetoin, 136*f*
acetone, 2, 136*f*, 139*t*
acetyl CoA (coenzyme A), 119, 125–126, 129, 130*f*, 135*f*
acetyl CoA synthetase, 117*t*
acetyl group, 117*t*, 129, 130*f*
acid-anionic sanitizers, 202, 208*t*
acid-base balance, 36–37, 37*f*
acid-fast bacteria, 89
acid-fast stains, **70**–71, 71*f*, 72*t*
acid fuchsin dye, 69
acidic solutions, 162
  pH and, 36–37, 37*f*
acidophiles, **162**
acids, **35**–36, 36*f*
Acinetobacter, 436*t*
acne, 37, 615, 617*t*, **622**–623
acquired immunity. *See* adaptive immunity
acquired immunodeficiencies, **566**, 567*t*
acquired immunodeficiency syndrome. *See* AIDS
acridine dyes, 234
actinobacteria (high G + C gram-positive bacteria), 315*t*
Actinomyces, adherence and, 455
Actinomyces genus, 315*t*, 334, 335, 336
  plate counts of, 183
  reproduction in, 174–175
Actinomyces israelii, 335
Actinomycetales, 315*t*
Actinomycetes, antibiotics produced by, 582, 582*t*
actinomycosis, 335
Activase, 268*t*
activated macrophages, **516**–517, 517*f*, 517*t*
activated sludge system, 830–831, 831*f*, 832*f*
activation energy, **116**, 117*f*
active site of enzymes, 116, **119**–121, 119*f*, 121*f*
active transport, 92, **94**, 148
  vs. group translocation, 94
acute disease, **428**
acute necrotizing ulcerative gingivitis, **750**, 750*t*
acute-phase proteins, in inflammatory response, **487**, 488*f*
acute *vs.* chronic inflammation, 486–487

acyclovir, 589*t*, 599, 599*f*
  spectrum of activity and, 583–584, 583*t*
ADA (adenosine deaminase) deficiency, 18
adaptive immunity, **475**, 475*f*, **502**–522
  antigens, **504**–505, 505*f*
  cellular component, 503–504
  dual nature of, 503
  humoral (antibody-mediated), 503, 509–511
  types of, 520, 520*f*, 522*f*
ADCC (antibody-dependent cell-mediated cytotoxicity), **517**, 518*f*
Addison's disease, 561*t*
adefovir dipivoxil, 589*t*, 600
adenine (A), 47–49, 48*f*
adenine nucleotide, of ATP, 49
adenocarcinomas, **410**
adenosine deaminase (ADA) deficiency, 18
adenosine diphosphate (ADP), **49**
adenosine diphosphoglucose (ADPG), 148, 148*f*
adenosine monophosphate (AMP), 49
adenosine triphosphatase (ATP synthase), 132, 133*f*
adenosine triphosphate, **49**, 49*f*.
  See also ATP
  structure, 49, 49*f*
Adenoviridae, 392*t*, 404, 404*f*
adenoviruses, 388*f*, 389*f*, 391, 425
  cytopathic effects, 467*t*
adherence
  in pathogenicity, **455**, 455*f*, 469*f*
  in phagocytosis, **484**, 485*f*
adhesins (ligands), **455**, 455*f*
adjuvants, **533**
ADP (adenosine diphosphate), **49**, 124, 125, 128*f*
  anabolic reactions and, 115, 115*f*
  Krebs cycle and, 129, 130*f*
ADPG (adenosine diphosphoglucose), 148, 148*f*
Aedes aegypti (mosquito), 435*t*
Aedes (mosquito), 378*t*, 435*t*
aerial hypha, 346, 347*f*
aerobes, **129**, 166, 166*t*
  obligate, 166, 166*t*
aerobic respiration, **129**, 135*f*
  anaerobic respiration *vs.*, 139*t*
  ATP yield during, 134*t*
  electron transport train (system), 130–134, 131*f*, 133*f*
  fermentation *vs.*, 139*t*
  glycolysis, 125, 126*f*, **127**, 128*f*, 134*t*
  hydrogen peroxide produced during, neutralized by catalase, 167
  Krebs cycle, **129**, 130*f*
  summary, 133–134, 135*f*
aerotolerant anaerobes, 166*t*, **167**
aflatoxin, 234, **467**
aflatoxin poisoning, **771**, 771*f*
African trypanosomiasis (sleeping sickness), 228, 367*t*, 368, 378*t*, 435*t*, 457, 468, **660**–661, 666*t*
agar, 10*f*, 163, **168**
  blood, 172, 172*f*, 333
  nutrient (recipe), 169*t*
  properties, 168
  red algae and, 357–358
agglutination, **511**, 512*f*

agglutination tests, **537**–538, 537*f*, 538*f*
agranulocytes, **479**, 481*f*, 484*t*
agranulocytosis, **557**
Agre, Peter, 15*t*
agricultural applications of biotechnology, 18, 273–275
  important products of genetic engineering, 276*t*
agricultural wastes, fermented, 139*t*
Agrobacterium genus, 314*t*, 318
  Entner-Doudoroff pathway and, 127
Agrobacterium tumefaciens, crown gall and, 273–274, 274*f*, 318
AIDS, **21**, 408–409, **566**–576, 567*t*.
  See also HIV; HIV infection
  as an STD, 801
  diseases associated with, 572*t*
  epidemic incidence of (reported cases in U.S.), 428, 428*f*
  health care workers and, 573*b*
  interleukin-12 and, 519*b*
  origin of, 566–567
  Pneumocystis as opportunistic pathogen, 353
  portals of entry, 454*t*
  prevention, 575
  scientific research and, 576
airborne microbes, as demonstrated by Pasteur, 8–9
airborne transmission of disease, 433
air conditioning systems, legionellales and, 322
air filters, 194
Ajellomyces, 354*t*
alanine (Ala), 44*t*, 45*f*
alanine deaminase, 117*t*
alanine racemase, 117*t*
alarmone, 231
albendazole, 590*t*, 601
alcohol
  early beliefs about how it was produced, 9
  peroxisomes oxidization of, 106
alcohol fermentation, 136*f*, **137**, 138*f*, 139*t*
alcohols, **201**, 201*t*
  mechanism of action, 201, 201*t*, 207*t*
alcohols functional group, 38
  chemical structure/compounds found in, 38*t*
alcohol washes
  in fixing specimens, 68
  in Gram staining, 69–70, 70*f*, 86
aldehyde functional group, chemical structure/compounds found in, 38*t*
aldehydes, **204**, 208*t*
Alexandrium, 360, 367*t*
algae/alga, **4**, 5*f*, **357**–361
  as eukaryotic cell, 77
  as members of Eukarya domain, 6
  as photoautotrophs, 144–146, 145*f*
  as photosynthesizers, 4
  as symbionts, 361
  cellular arrangement, compared to other eukaryotes, 345*t*
  cell wall composition, 4, 100
  characteristics of selected Phyla, 359*t*
  chloroplast organelle of, 105–106, 107*f*
  classified by energy and carbon sources, 145*f*
  embryo formation and, 345*t*

food acquisition method, 345*t*
  metachromatic granules in, 96
  multicellularity, compared to other eukaryotes, 345*t*
  naming rules, 287
  neurotoxins produced by, 468
  nutritional needs, 4
  photosynthesis and, 17, 114*f*, 141–143, 146*t*
  pond, 5*f*
  protoplast fusion in, 261*f*
  reproductive methods, 4
  role in nature, 361
  shapes of, 4, 5*f*
  typical eukaryotic cell structure, 4, 99*f*
  where found, 4, 5*f*
algal blooms, 361, **825**
algal plant cell, typical structures, 99*f*
algicides, copper sulfate as, 208*t*
Algin, **357**
alkaline (basic) solutions, 36–37, 37*f*
alkaline habitats, cyanobacteria and, 37
alkaline solutions, microbial growth and, 162
alkylation, 205
allergen, **551**
allergic contact dermatitis, **558**, 560*f*
allergy (hypersensitivity), **551**–558
  anaphylactic, 551–554, 551*t*, 552*f*
  types of, 551*t*
allografts, **563**
allosteric inhibition, 121*f*, **122**
  in feedback (end-product) inhibition, 122–123
allosteric site, 121*f*, **122**
allylamines, 597
alpha-amino acids, 43, 43*f*
alpha-carbon atom, 43, 43*f*
alpha-hemolytic streptococci, 333
alpha-interferon
  genetically engineered, 268*t*
  to treat viral hepatitis, 589*t*, 600
alpha-ketoglutaric acid, 129, 130*f*
Alphaproteobacteria, **313**, 314*t*, 316–318
alpha-tumor necrosis factor (alpha-TNF), 489, 518–519, 518*t*
Alphavirus, 392*t*, 435*t*
Alveolata, **367**
Amanita muscaria, 353*f*
amanitin, **467**
amantadine, 589*t*
amebic meninogencephalitis, **661**–662, 666*t*
American Academy of Microbiology, 273
American Official Analytical Chemist's use-dilution test, 198
American trypanosomiasis (Chagas' disease), 10*f*, 367*t*, 368, 378*t*, 435*t*, **693**–694, 694*f*, 703*t*
Ames test, **237**–238, 239*f*
amination, **149**, 150*f*
amino acids, **43**–47
  as an industrial product, 854
  as D-isomers, 43, 45, 45*f*, 85
  as L-isomers, 43, 45, 45*f*, 85
  as organic growth factors, 167
  biosynthesis of, 149, 150*f*
    intermediates in Krebs cycle and, 129, 149
  catabolism of, 138, 140*f*
    biochemical tests to detect, 139, 141*f*

Dear Student:

As the authors of *Microbiology: An Introduction,* **Ninth Edition**, we set out to write a book that explains core microbiology topics in a way that is easy for students to understand and remember. We would appreciate hearing about your experience with this textbook and its multimedia support, and we invite your suggestions for improvements!

Many thanks,

Gerard J. Tortora   Bert Funke   Christine Case

1. What did you like most about *Microbiology: An Introduction,* **Ninth Edition**? Please provide three examples, in order of priority.

_____

_____

_____

_____

2. Do you have suggestions to improve this book? Please write your specific suggestions, citing page numbers if appropriate.

_____

_____

_____

_____

3. Did you use the Microbiology Place Website and CD-ROM?   ☐ Yes ☐ No
   If yes, which topics or activities were most useful to you? Are there topics that should be added? How can we improve the content of the website and CD-ROM?

_____

_____

_____

4. How helpful were the following features on the Microbiology Place Website and CD-ROM?

Used Regularly

Animations   ☐ _____
Interactive Tutorials ☐ _____
Case Studies   ☐ _____
Activities   ☐ _____
Chapter Quizzes   ☐ _____
Practice Tests   ☐ _____
Microbe Review   ☐ _____

Suggestions for improvements:

_____

_____

_____

School: ______________________________

Optional:

Your name: ___________________________

Email: ______________________________

Date: _______________________________

May Benjamin Cummings have permission to quote your comments in promotions for ***Microbiology: An Introduction***? ☐ Yes ☐ No

BUSINESS REPLY MAIL

FIRST-CLASS MAIL PERMIT NO. 275 SAN FRANCISCO CA

POSTAGE WILL BE PAID BY ADDRESSEE

NO POSTAGE NECESSARY IF MAILED IN THE UNITED STATES

BENJAMIN CUMMINGS
PEARSON EDUCATION
1301 SANSOME STREET
SAN FRANCISCO CA 94111-9328

eo- dawn, early. Example: *Eobacterium*, a 3.4-billion-year-old fossilized bacterium.

epi- upon, over. Example: epidemic, number of cases of a disease over the normally expected number.

erythro- red. Example: erythema, redness of the skin.

eu- well, proper. Example: eukaryote, a proper cell.

exo- outside, outer layer. Example: exogenous, from outside the body.

extra- outside, beyond. Example: extracellular, outside the cells of an organism.

firmi- strong. Example: *Bacillus firmus* forms resistant endospores.

flagell- a whip. Example: flagellum, a projection from a cell; in eukaryotic cells, it pulls cells in a whiplike fashion.

flav- yellow. Example: *Flavobacterium* cells produce yellow pigment.

fruct- fruit. Example: fructose, fruit sugar.

-fy to make. Example: magnify, to make larger.

galacto- milk. Example: galactose, monosaccharide from milk sugar.

gamet- to marry. Example: gamete, a reproductive cell.

gastr- stomach. Example: gastritis, inflammation of the stomach.

gel- to stiffen. Example: gel, a solidified colloid.

-gen an agent that initiates. Example: pathogen, any agent that produces disease.

-genesis formation. Example: pathogenesis, production of disease.

germ, germin- bud. Example: germ, part of an organisms capable of developing.

-gony reproduction. Example: schizogony, multiple fission producing many new cells.

gracili- thin. Example: *Aquaspirillum gracile*, a thin cell.

halo- salt. Example: halophile, an organism that can live in high salt concentrations.

haplo- one, single. Example: haploid, half the number of chromosomes or one set.

hema-, hemato-, hemo- blood. Example: *Haemophilus*, a bacterium that requires nutrients from red blood cells.

hepat- liver. Example: hepatitis, inflammation of the liver.

herpes creeping. Example: herpes, or shingles, lesions appear to creep along the skin.

hetero- different, other. Example: heterotroph, obtains organic nutrients from other organisms; other feeder.

hist- tissue. Example: histology, the study of tissues.

hom-, homo- same. Example: homofermenter, an organism that produces only lactic acid from fermentation of a carbohydrate.

hydr-, hydro- water. Example: dehydration, loss of body water.

hyper- excess. Example: hypertonic, having a greater osmotic pressure in comparison with another.

hypo- below, deficient. Example: hypotonic, having a lesser osmotic pressure in comparison with another.

im- not, in. Example: impermeable, not permitting passage.

inter- between. Example: intercellular, between the cells.

intra- within, inside. Example: intracellular, inside the cell.

io- violet. Example: iodine, a chemical element that produces a violet vapor.

iso- equal, same. Example: isotonic, having the same osmotic pressure when compared with another.

-itis inflammation of. Example: colitis, inflammation of the large intestine.

-karyo, -caryo a nut. Example: eukaryote, a cell with a membrane-enclosed nucleus.

kin- movement. Example: streptokinase, an enzyme that lyses or moves fibrin.

lacti- milk. Example: lactose, the sugar in milk.

lepis- scaly. Example: leprosy, disease characterized by skin lesions.

lepto- thin. Example: *Leptospira*, thin spirochete.

leuko- whiteness. Example: leukocyte, a white blood cell.

lip-, lipo- fat, lipid. Example: lipase, an enzyme that breaks down fats.

-logy the study of. Example: pathology, the study of changes in structure and function brought on by disease.

lopho- tuft. Example: lophotrichous, having a group of flagella on one side of a cell.

luc-, luci- light. Example: luciferin, a substance in certain organisms that emits light when acted upon by the enzyme luciferase.

lute-, luteo- yellow. Example: *Micrococcus luteus*, yellow colonies.

-lysis loosening, to break down. Example: hydrolysis, chemical decomposition of a compound into other compounds as a result of taking up water.

macro- largeness. Example: macromolecules, large molecules.

mendosi- faculty. Example: mendosicutes, archaeobacteria lacking peptidoglycan.

meningo- membrane. Example: meningitis, inflammation of the membranes of the brain.

meso- middle. Example: mesophile, an organism whose optimum temperature is in the middle range.

meta- beyond, between, transition. Example: metabolism, chemical changes occurring within a living organism.

micro- smallness. Example: microscope, an instrument used to make small objects appear larger.

-mnesia memory. Examples: amnesia, loss of memory; anamnesia, return of memory.

molli- soft. Example: Mollicutes, a class of wall-less eubacteria.

-monas a unit. Example: *Methylomonas*, a unit (bacterium) that utilizes methane as its carbon source.

mono- singleness. Example: monotrichous, having one flagellum.

morpho- form. Example: morphology, the study of the form and structure of organisms.

multi- many. Example: multinuclear, having several nuclei.

mur- wall. Example: murein, a component of bacterial cell walls.

mus-, muri- mouse. Example: murine typhus, a form of typhus endemic in mice.

mut- to change. Example: mutation, a sudden change in characteristics.

myco-, -mycetoma, -myces a fungus. Example: *Saccharomyces*, sugar fungus, a genus of yeast.

myxo- slime, mucus. Example: Myxobacteriales, an order of slime-producing bacteria.

necro- a corpse. Example: necrosis, cell death or death of a portion of tissue.

-nema a thread. Example: *Treponema* has long, threadlike cells.

nigr- black. Example: *Aspergillus niger*, a fungus that produces black conidia.

ob- before, against. Example: obstruction, impeding or blocking up.

oculo- eye. Example: monocular, pertaining to one eye.

-oecium, -ecium a house. Examples: perithecium, an ascus with an opening that encloses spores; ecology, the study of the relationships among organisms and between an organism and its environment (household).

-oid like, resembling. Example: coccoid, resembling a coccus.

oligo- small, few. Example: oligiosaccharide, a carbohydrate composed of a few (7–10) monosaccharides.

-oma tumor. Example: lymphoma, a tumor of the lymphatic tissues.

-ont being, existing. Example: schizont, a cell existing as a result of schizogony.

ortho- straight, direct. Example: orthomyxovirus, a virus with a straight, tubular capsid.

-osis, -sis condition of. Examples: lysis, the condition of loosening; symbiosis, the condition of living together.

pan- all, universal. Example: pandemic, an epidemic affecting a large region.

para- beside, near. Example: parasite, an organism that "feeds beside" another.

peri- around. Example: peritrichous, projections from all sides.

phaeo- brown. Example: Phaeophyta, brown algae.

phago- eat. Example: phagocyte, a cell that engulfs and digests particles or cells.

philo-, -phil liking, preferring. Example: thermophile, an organism that prefers high temperatures.

-phore bears, carries. Example: conidiophore, a hypha that bears conidia.

-phyll leaf. Example: chlorophyll, the green pigment in leaves.

-phyte plant. Example: saprophyte, a plant that obtains nutrients from decomposing organic matter.

pil- a hair. Example: pilus, a hairlike projection from a cell.

plankto- wandering, roaming. Example: plankton, organisms drifting or wandering in water.

plast- formed. Example: plastid, a formed body within a cell.

-pnoea, -pnea breathing. Example: dyspnea, difficulty in breathing.

pod- foot. Example: pseudopod, a footlike structure.

poly- many. Example: polymorphism, many forms.

post- after, behind. Example: posterior, a place behind a (specific) part.

pre-, pro- before, ahead of. Examples: prokaryote, a cell with the first nucleus; pregnant, before birth.

pseudo- false. Example: pseudopod, false foot.

psychro- cold. Example: psychrophile, an organism that grows best at low temperatures.

-ptera wing. Example: Diptera, the order of true flies, insects with two wings.

pyo- pus. Example: pyogenic, pus-forming.

rhabdo- stick, rod. Example: rhabdovirus, an elongated, bullet-shaped virus.

rhin- nose. Example: rhinitis, inflammation of mucous membranes in the nose.

rhizo- root. Examples: *Rhizobium*, a bacterium that grows in plant roots; mycorrhiza, a fungus that grows in or on plant roots.

rhodo- red. Example: *Rhodospirillum*, a red-pigmented, spiral-shaped bacterium.

rod- gnaws. Example: rodents, the class of mammals with gnawing teeth.

rubri- red. Example: *Clostridiium rubrum*, red-pigmented colonies.

rumin- throat. Example: *Ruminococcus*, a bacterium associated with a rumen (modified esophagus).

saccharo- sugar. Example: disaccharide, a sugar consisting of two simple sugars.